Plant Biodiversity

Monitoring, Assessment and Conservation

Plant Biodiversity

Monitoring, Assessment and Conservation

Edited by

Abid A. Ansari
Department of Biology, Faculty of Science, University of Tabuk, Saudi Arabia

Sarvajeet S. Gill
Stress Physiology and Molecular Biology Lab, Centre for Biotechnology, MD University, Haryana, India

Zahid Khorshid Abbas
Department of Biology, Faculty of Science, University of Tabuk, Saudi Arabia

and

M. Naeem
Plant Physiology Section, Department of Botany, Aligarh Muslim University, India

CABI is a trading name of CAB International

CABI
Nosworthy Way
Wallingford
Oxfordshire OX10 8DE
UK

Tel: +44 (0)1491 832111
Fax: +44 (0)1491 833508
E-mail: info@cabi.org
Website: www.cabi.org

CABI
745 Atlantic Avenue
8th Floor
Boston, MA 02111
USA

Tel: +1 (617) 682-9015
E-mail: cabi-nao@cabi.org

A catalogue record for this book is available from the British Library, London, UK.

Library of Congress Cataloging-in-Publication Data

Names: Ansari, Abid A., editor. | Gill, Sarvajeet Singh, editor. | Abbas, Zahid Khorshid, editor. | Naeem, M., 1980- editor.
Title: Plant biodiversity : monitoring, assessment, and conservation / edited by Abid A. Ansari, Sarvajeet S. Gill, Zahid Khorshid Abbas, and M. Naeem.
Description: Wallingford, Oxfordshire ; Boston, MA : CABI International, [2016] | Includes bibliographical references and index.
Identifiers: LCCN 2016041120 (print) | LCCN 2016042121 (ebook) | ISBN 9781780646947 (hbk : alk. paper) | ISBN 9781780646961 (ePDF) | ISBN 9781780646954 (ePub)
Subjects: LCSH: Plant diversity. | Plant diversity conservation. | MESH: Plant Physiological Phenomena | Biodiversity | Ecosystem
Classification: LCC QK46.5.D58 P526 2016 (print) | LCC QK46.5.D58 (ebook) | NLM QK 717 | DDC 581.7--dc23
LC record available at https://lccn.loc.gov/2016041120

ISBN-13: 978 1 78064 694 7

Commissioning editor: Ward Cooper
Editorial assistant: Emma McCann
Production editor: Tim Kapp

Typeset by SPi, Pondicherry, India
Printed and bound by Gutenberg Press Limited, Tarxien, Malta

Contents

Contributors

Zahid Khorshid Abbas, Department of Biology, Faculty of Science, University of Tabuk, Tabuk-71491, Saudi Arabia

Sulaiman Mohammad Al-Ghanim, Department of Biology, Faculty of Science, University of Tabuk, Tabuk-71491, Saudi Arabia

Volkan Altay, Biology Department, Mustafa Kemal University, İskenderun, Hatay, Turkey

Ernaz Altundag, Biology Department, Duzce University, Duzce, Turkey

Fahad Mohammed Alzeibar, Department of Biology, Faculty of Science, University of Tabuk, Tabuk-71491, Saudi Arabia

Francisco Amich, Grupo de Investigación de Recursos Etnobiológicos del Duero-Douro (GRIRED), Department of Botany, Faculty of Biology, University of Salamanca, Campus Unamuno, Salamanca, 37007, Spain

Abid Ali Ansari, Department of Biology, Faculty of Science, University of Tabuk, Tabuk-71491, Saudi Arabia. E-mail: aansari@ut.edu.sa, aaansari40@gmail.com

Lia Ascensão, CESAM – Centre for Environmental and Marine Studies, Faculdade de Ciências, University of Lisbon, 1749-016 Lisbon, Portugal. E-mail: lia.ascensao@fc.ul.pt

Tarique Hassan Askary, Division of Entomology, Sher-e-Kashmir University of Agricultural Sciences and Technology, Main Campus, Shalimar, Srinagar-190025, Jammu and Kashmir, India. E-mail: tariq_askary@rediffmail.com

P. Azmath, Herbal Drug Technological Laboratory, Department of Studies in Microbiology, University of Mysore, Manasagangotri, Mysore-570006, Karnataka, India

Syed Baker, Herbal Drug Technological Laboratory, Department of Studies in Microbiology, University of Mysore, Manasagangotri, Mysore-570006, Karnataka, India

Shafat Ahmad Banday, Division of Fruit Science, Sher-e-Kashmir University of Agricultural Sciences and Technology, Main Campus, Shalimar, Srinagar-190025, Jammu and Kashmir, India

Kathryn E. Barry, Department of Biological Sciences, Marquette University, Milwaukee, Wisconsin 53201, USA. E-mail: kebarry@uwm.edu

Ana Maria Carvalho, Mountain Research Centre (CIMO), School of Agriculture, Polytechnic Institute of Bragança, Campus de Santa Apolónia, 5300-253 Bragança Portugal

Sumit Chakravarty, Department of Forestry, Uttar Banga Krishi Viswavidyalaya, Pundibari-736165 (Cooch Behar) West Bengal, India. E-mail: c_drsumit@yahoo.com

Otília Correia, Centre for Ecology, Evolution and Environmental Changes (cE3c), Faculdade de Ciências, University of Lisbon, 1749-016 Lisbon, Portugal. E-mail: odgato@fc.ul.pt

Mudasir Irfan Dar, Department of Botany, Aligarh Muslim University, Aligarh-202002, Uttar Pradesh, India

Martin T. Dokulil, EX Research Institute Mondsee, Mondseestrasse 9, A-5310 Mondsee, Austria. E-mail: martin.dokulil@univie.ac.at

Gisela Gaio-Oliveira, Centre for Ecology, Evolution and Environmental Changes (cE3c), Faculty of Sciences, University of Lisbon, 58, 1250-102 Lisbon, Portugal

Petros Ganatsas, Laboratory of Silviculture, Aristotle University of Thessaloniki, 54124, Greece. E-mail: pgana@for.auth.gr

Saikat Gantait, AICRP on Groundnut, Directorate of Research, Bidhan Chandra Krishi Viswavidyalaya, Kalyani, Nadia, West Bengal 741235, India. E-mail: saikatgantait@yahoo.com

Sankar K. Ghosh, Department of Biotechnology, Assam University, Silchar 788011, Assam, India

Sarvajeet Singh Gill, Stress Physiology and Molecular Biology Lab, Centre for Biotechnology, MD University, Rohtak 124001, Haryana, India

José Antonio González, Grupo de Investigación de Recursos Etnobiológicos del Duero-Douro (GRIRED), Department of Botany, Faculty of Biology, University of Salamanca, Campus Unamuno, Salamanca, 37007, Spain. E-mail: ja.gonzalez@usal.es

Salih Gucel, Institute of Environmental Sciences, Near East University, Lefkoşa, Northern Cyprus

B.P. Harini, Department of Zoology, Bangalore University, Jnanabharathi Campus, Bangalore-560056, Karnataka, India

Umar Iqbal, Division of Fruit Science, Sher-e-Kashmir University of Agricultural Sciences and Technology, Main Campus, Shalimar, Srinagar-190025, Jammu and Kashmir, India

Saiful Islam, Forest Department Headquarters, Ban Bhaban, Agargaon, Dhaka-1207, Bangladesh. E-mail: sislam47@hotmail.com

Disha Jaggi, Directorate of Weed Research, Jabalpur-482004, Madhya Pradesh, India

K.S. Kavitha, Herbal Drug Technological Laboratory, Department of Studies in Microbiology, University of Mysore, Manasagangotri, Mysore-570006, Karnataka, India

Akeel A. Khan, GF College affiliated to MJP Rohilkhand University, Shahjahanpur, U.P., India

Fareed Ahmad Khan, Department of Botany, Aligarh Muslim University, Aligarh-202002, Uttar Pradesh, India

Bhumesh Kumar, Directorate of Weed Research, Jabalpur-482004, Madhya Pradesh, India. E-mail: kumarbhumesh@yahoo.com

Sanjeev Kumar, Chief Conservator of Forests, Working Plans Jamshedpur-831001, Jharkhand, India. E-mail: sanjeevkumar201@gmail.com

Maria Amélia Martins-Loução, Centre for Ecology, Evolution and Environmental Changes (cE3c), Faculty of Sciences, University of Lisbon, 58, 1250-102 Lisbon, Portugal. E-mail: maloucao@fc.ul.pt

Sheikh Mehraj, Division of Fruit Science, Sher-e-Kashmir University of Agricultural Sciences and Technology, Main Campus, Shalimar, Srinagar-190025, Jammu and Kashmir, India

Mohammad Maqbool Mir, Division of Fruit Science, Sher-e-Kashmir University of Agricultural Sciences and Technology, Main Campus, Shalimar, Srinagar-190025, Jammu and Kashmir, India. E-mail: mmaqboolmir@rediffmail.com

Shabir Ahmad Mir, Department of Food Technology, Islamic University of Sciences & Technology, Awantipura, Pulwama-191122, Jammu and Kashmir, India

Sunit Mitra, Department of Botany, Ranaghat College, Nadia – 741201, West Bengal, India. E-mail: sunit_mitra2003@yahoo.co.in

Mohammad Mobin, Department of Biology, Faculty of Science, University of Tabuk, Tabuk-71491, Saudi Arabia. E-mail: mhasa@ut.edu.sa

Sobhan Kumar Mukherjee, Department of Botany, Ranaghat College, Nadia-741201, West Bengal, India. E-mail: sobhankr@gmail.com

Mohd Irfan Naikoo, Department of Botany, Aligarh Muslim University, Aligarh-202002, Uttar Pradesh, India. E-mail: mdirfanmsbo@gmail.com

K.S. Negi, Regional Station, Bhowali, Niglat, District Nainital-263132, Uttarakhand, India

Fouzia Nousheen, Department of Botany, Aligarh Muslim University, Aligarh-202002, Uttar Pradesh, India

Robin Owings, 409 Pineview Drive, West Blocton, Alabama 35184, USA

Munir Ozturk, Botany Department, Ege University, Bornova, 35040 Bornova/İzmir, Turkey. E-mail: munirozturk@gmail.com

Saurabh Pagare, Directorate of Weed Research, Jabalpur-482004, Madhya Pradesh, India

Nazir A. Pala, Department of Forestry, Uttar Banga Krishi Viswavidyalaya, Pundibari-736165 (Cooch Behar) West Bengal, India

Anjula Pandey, Plant Exploration and Germplasm Collection Division, National Bureau of Plant Genetic Resources, New Delhi 110012, India. E-mail: anjuravinder@yahoo.com

K. Pradheep, Plant Exploration and Germplasm Collection Division, National Bureau of Plant Genetic Resources, New Delhi 110012, India

Nishanta Rajakaruna, College of the Atlantic, 105 Eden Street, Bar Harbor, Maine 04609, USA and Unit for Environmental Sciences and Management, North-West University, Private Bag X6001, Potchefstroom 2520, South Africa

D. Rakshith, Herbal Drug Technological Laboratory, Department of Studies in Microbiology, University of Mysore, Manasagangotri, Mysore-570006, Karnataka, India
H.C. Yashavantha Rao, Herbal Drug Technological Laboratory, Department of Studies in Microbiology, University of Mysore, Manasagangotri, Mysore-570006, Karnataka, India
Ghulam Hasan Rather, Division of Fruit Science, Sher-e-Kashmir University of Agricultural Sciences and Technology, Main Campus, Shalimar, Srinagar-190025, Jammu and Kashmir, India
Meenal Rathore, Directorate of Weed Research, Jabalpur-482004, Madhya Pradesh, India
Farha Rehman, Department of Botany, Aligarh Muslim University, Aligarh-202002, Utter Pradesh, India
Hasibur Rehman, Department of Biology, Faculty of Science, University of Tabuk, Tabuk-71491, Saudi Arabia
Munib ur Rehman, Division of Fruit Science, Sher-e-Kashmir University of Agricultural Sciences and Technology, Main Campus, Shalimar, Srinagar-190025, Jammu and Kashmir, India
Shalini Saggu, Department of Biology, Faculty of Science, University of Tabuk, Tabuk-71491, Saudi Arabia
Narayan Chandra Sahu, Sasya Shyamala Krishi Vigyan Kendra, Ramakrishna Mission Vivekananda University, Arapanch, Sonarpur, Kolkata 700150, West Bengal, India
Abir Sarraj-Laabidi, Institut Supérieur des Sciences Biologiques Appliquées de Tunis (ISSBAT), University of Tunis El Manar 1 (UTM), Tunis, Tunisia
S. Satish, Herbal Drug Technological Laboratory, Department of Studies in Microbiology, University of Mysore, Manasagangotri, Mysore-570006, Karnataka, India, and Department of Plant Pathology, University of Georgia, Athens, Georgia 30602, USA. E-mail: satish.micro@gmail.com
Stefan A. Schnitzer, Department of Biological Sciences, Marquette University, Milwaukee, Wisconsin 53201, USA. E-mail: S1@uwm.edu
Asma Hammami-Semmar, Institut National des Sciences Appliquées et de Technologie (INSAT), Tunis, Tunisia
Nabil Semmar, Institut Supérieur des Sciences Biologiques Appliquées de Tunis (ISSBAT), Université de Tunis El Manar 1 (UTM), Tunis, Tunisia. E-mail: nabilsemmar@yahoo.fr
Jyoti K. Sharma, Environmental Sciences & Natural Resource Management, School of Natural Sciences, Shiv Nadar University, Village Chithera, Dadri, Gautam Budh Nagar-201314, India. E-mail: jyoti.sharma@snu.edu.in, jyotikumarsharma@gmail.com
Gopal Shukla, Department of Forestry Uttar Banga Krishi Viswavidyalaya, Pundibari-736165 (Cooch Behar) West Bengal, India. E-mail: gopalshukla12@gmail.com
Raghwendra Singh, Directorate of Weed Research, Jabalpur-482004, Madhya Pradesh, India
Uma Rani Sinniah, Department of Crop Science, Faculty of Agriculture, Universiti Putra Malaysia, 43400 Serdang, Selangor, Malaysia
K. Subashree, Department of Ecology and Environmental Sciences, Pondicherry University, Puducherry-605014, India
S.M. Sundarapandian, Department of Ecology and Environmental Sciences, Pondicherry University, Puducherry-605014, India. E-mail: smspandian65@gmail.com
Pundarikakshudu Tetali, Temple Rose Livestock Farming Pvt. Ltd 201 – Amelia, Lakaki Road, Shivaji Nagar, Model Colony, Pune-411016, Maharashtra, India
Sujata Tetali, Agharkar Research Institute, G. G. Agarkar Road Pune-411 004, Maharashtra, India. E-mail: sujatatetali@gmail.com
Meg Trau, 19 Plain Street, Natick, Massachusetts 01760, USA. E-mail: mtrau@coa.edu
Niraj Tripathi, Directorate of Weed Research, Jabalpur-482004, Madhya Pradesh, India
Subrata Trivedi, Department of Biology, Faculty of Science, University of Tabuk, Tabuk-71491, Saudi Arabia. E-mail: strivedi@ut.edu.sa
Mayank Varun, Directorate of Weed Research, Jabalpur-482004, Madhya Pradesh, India

Abbreviations

~	approximately
$\Delta^{13}C$	carbon isotope discrimination
Ψ_{Min}	minimum water potential recorded at solar noon
Ψ_{PD}	predawn water potential measured at sunrise
α	alpha
Al	aluminium
Au	gold
β	beta
BGCI	Botanic Gardens Conservation International
C	C-atom
CBD	Convention on Biological Diversity
Chl	chlorophyll content
Co	coenzyme A
CPD	Continuous Professional Development
CWR	Crop Wild Relatives
EPCS	European Plant Conservation Strategy
FTIR	Fourier Transform Infrared Radiation
Fv/Fm	maximal photochemical efficiency measured before sunrise
γ	gamma
g	gravity
GCE	glassy carbon electrode
G_{max}	maximum stomatal conductance
GNP	gold nanoparticle
GSPC	Global Strategy for Plant Conservation
H^+	hydrogen ion
HR-SEM	High-Resolution Scanning Electron Microscope
HR-TEM	High-Resolution Transmission Electron Microscope
IBSE	inquiry-based science education
IC	inhibitory concentration
ICT	Information and Communication Technologies
ITPGRFA	International Treaty on Plant Genetic Resources for Food and Agriculture
IUCN	International Union for Conservation of Nature
λ	lambda
L	Linnaeus
LAI	leaf area index
LBG	Lisbon Botanic Garden
LD	lethal dose
LMA	leaf mass per area
LOtC	Learning Outside of the Classroom
mg/kg^{-1}	milligram per kilogram inverse
µg	microgram
µg/ml	microgram per millilitre
MIC	minimum inhibitory concentration

μl	microlitre
ml	millilitre
mM	millimolar
μm	micrometre
MTCC	microbial type
n	normal
N	normality
NAD	nicotinamide adenine dinucleotide
NADP	nicotinamide adenine dinucleotide phosphate
Ni	nickel
nm	nanometre
NMR	nuclear magnetic resonance
NP	nanoparticle
NP_{max}	maximum photosynthetic rate
PAR	photosynthetic active radiation
Pi	internal CO_2 partial pressure
ppm	parts per million
pv	pathovar
RPW	Red Palm Weevil
SEM	Scanning Electron Microscopy
SNP	silver nanoparticle
sp.	species (singular)
spp.	species (plural)
ssp.	subspecies
TEM	Transmission Electron Microscopy
TLC	thin-layer chromatography
Tr_{max}	maximum transpiration rate
TTC	trichlorotriphenyl tetrazolium chloride
UV	ultraviolet
var.	variety
V/V	volume by volume
WHP	wild harvested plants
WHO	World Health Organization
WUE	water-use efficiency expressed as the rate of assimilated CO_2 per water loss
w/v	weight per volume
XRD	X-ray diffraction
ZnO	zinc oxide

1 New Challenges to Promote Botany's Practice Using Botanic Gardens: The Case Study of the Lisbon Botanic Garden

MARIA AMÉLIA MARTINS-LOUÇÃO* AND GISELA GAIO-OLIVEIRA

Centre for Ecology, Evolution and Environmental Changes, Faculty of Sciences, University of Lisbon, Lisbon, Portugal

Abstract

Botanic Gardens are living plant museums, where plants have been arranged according to previous research and discovery in order to be used as a display for public appreciation and learning. Thus, more than ever, they can serve to improve the practice of botany, being able: (i) to drive the continuous exchange of ideas using the 'old and new' botany approach; (ii) to show the importance of the plant world in alleviating regional and global changes; (iii) to expand forward-thinking in science promotion. The Global Strategy of Plant Conservation (GSPC) has promoted global action in environment conservation and study. From the beginning, it has been adopted by Botanic Gardens Conservation International (BGCI) to function as a *leitmotiv* for botanic gardens best practice. In the present century, a new endeavour is required to facilitate the appeal of botany to younger generations. Links between taxonomic knowledge, molecular studies, information and communication technologies, and social sciences are urgently needed. Lisbon Botanic Garden has been working on these challenges for the past six years.

1.1 Introduction

Plants have a key role in ecosystem balance and, thus, are vital for environmental sustainability. Besides that ecological value, plants represent an economic and cultural value for humans, since they give us food, feed for animals, fibres and pharmaceutical products. It is not surprising that the World Health Organization (WHO) has estimated that up to 80% of the world's human population depend on plants (or plant products) for their primary health care (Bodeker *et al.*, 2003), while playing an expanding role in its development. Although the value of plants is fully recognized, no effective action is taken to prevent the continuous decline in plant diversity and, consequently, in ecosystem services provided by them. In addition, climate change is altering the habitat of plants, with unknown consequences (Thuiller *et al.*, 2005; Blackmore *et al.*, 2011; Merritt and Dixon, 2011).

In spite of this impending crisis, public concern with biodiversity remains limited. Although many people feel that biodiversity is a heritage that should be preserved as fully as possible, the awareness tends to be centred on a few 'charismatic' species and ecosystems. But biodiversity also includes thousands of less conspicuous species and less spectacular ecosystems. Thus, we need to better inform society about plants' significance and the importance of biodiversity for humanity.

Within the core missions of botanic gardens, educating people about the role of biodiversity and the need for plant conservation represent important issues with which to engage broader audiences (www.bgci.org). Botanic gardens are the modern-day arks; they

*E-mail: maloucao@fc.ul.pt

 Plant Biodiversity: Monitoring, Assessment and Conservation (eds A.A. Ansari, S.S. Gill, Z.K. Abbas and M. Naeem)

have received economically significant plants from the colonies and developing scientific programmes for the planting and cultivation of medicinal herbs. In the 16th century, the first botanic gardens provided plants for medical research (Rinker, 2002). From the 17th to the 19th centuries, botanic gardens functioned as displays of colonial diversity, as well as taxonomic inventories contributing to the knowledge of tropical plant diversity (BGCI, 2014). During the 20th century, botanic gardens were transformed into specialized research centres (BGCI, 2014), particularly focused on the global loss of biodiversity and, thus, included scientific research on plant conservation and public education (BGCI, 2012).

Today, there are few nations in the world without botanic gardens. Besides conserving *ex situ* biodiversity, botanic gardens develop research on plant systematics, using ecological, evolutionary and phylogenetic approaches. Before the implementation of conservation measures, either *in situ* or *ex situ,* we need to identify, recognize and study plant species' ecology. Plant conservation is an important issue for botanic gardens, particularly after the adoption of the Global Strategy for Plant Conservation (GSPC) (CBD, 2002). This strategy was launched in 2002 by the Convention on Biological Diversity (CBD) that 'marked an important advance in raising awareness of the threats faced by plants worldwide, as well as providing, for the first time, a coherent framework for policy and action needed to halt the loss of plant diversity' (CBD, 2002). In spite of the successes brought by this convention, it was not enough to mobilize effective actions for diminishing plant biodiversity loss. New achievements should be reached by 2020, according to the CBD revision in 2011 (Sharrock, 2011).

For the success of GSPC, botanic gardens need to strength their scientific contribution, as well as their influence in government policies. This requires a better development and coordination of tasks within and among botanic gardens to underpin conservation actions according to GSPC's targets. Only with this renewed challenge botanic gardens can underline their social role. This raises four critical questions, which we review in this chapter:

- Can botanic gardens pave the way for plant learning?
- What kind of strategic mechanisms do botanic gardens need to develop?
- How important is the role of science promoters and communication services to further raise botanic gardens' social relevance?
- How important are the activities offered by botanic gardens' education services to increase public awareness of plant diversity loss and the need for conservation?

Firstly, we are focusing on the unique opportunity botanic gardens have to play a key-role in plant education through complementing academic programmes at either graduate or postgraduate level. We then consider the strategic approaches that botanic gardens have developed and the role they need to play, through systematic research and conservation assessments. Thirdly, we examine the role of science promotion and communication as a fundamental area to raise awareness for plant diversity and conservation. Finally, we consider new methods offered by education services to contribute to capacity building and training in plant science and conservation. We provide examples of each of these, based on the experiences of the Lisbon Botanic Garden (LBG).

The Lisbon Botanic Garden (LBG) is a 136-year-old institution, belonging now to the National Museum of Natural History and Science, as part of the University of Lisbon. It is a scientific garden, classified as a national heritage monument in 2010 and a member of the World Monuments Watch since 2011. It was developed during the 19th century (1873–1878), when scientific and cultural progress was emerging in Portugal. The Lisbon Botanic Garden (LBG) has always been recognized as an idyllic spot within the centre of Lisbon, particularly by foreign tourists, probably following the beautiful and romantic words expressed by Thomas Mann, in his novel *Confessions of Felix Krull* (Mann, 1955). Over four hectares, this botanic garden features an important and diverse living collection, where the Portuguese colonial memory is imprinted. Being part of the University of Lisbon, taxonomic expertise has been preserved to support adequate conservation policies as well as to maintain scientific collections. Plant conservation and sustainable use of resources are at the core of all scientific activity. The environmental consequences of the loss of biodiversity and the role of humankind on this global problem have been presented to the LBG public, benefiting from researchers' scientific knowledge and optimized through the launch of an education department and the development of a strategic plan of communication.

The above four key questions are behind the mission of Botanic Gardens Conservation International (BGCI), a global network derived from the International Union for Conservation of Nature (IUCN) in support of botanic gardens, and particularly focused on connecting people with nature and finding solutions for sustainable livelihoods. A primary concern of BGCI has been to provide a means for botanic gardens in all parts of the globe to share information and new advances to benefit conservation and education (Wyse Jackson and Sutherland, 2000). After the Earth Summit in Rio de Janeiro 1992, the Plant Conservation Report in 2009 and the adoption of the Global Strategy for Plant Conservation (GSPC), BGCI has explored different ways to engage botanic gardens with the various objectives and targets of the GSPC. Therefore, plant diversity learning, conservation policies and education practices are interrelated keywords, encompassing the mission of botanic gardens all over the world.

1.2 Botanic Gardens and Plant Learning

For the public visiting botanic gardens, there is a natural attraction rather than a need to understand the plant collections. This is because the majority of visitors do not have motivation to learn *per se* while visiting these institutions; rather, to enjoy the garden as a place of peace and relaxation (Darwin Edwards, 2000). Motivation for visiting may also be affected by other factors: namely, the existence of entrance fees for the majority of botanic gardens, the availability of free-access green urban areas and the association of the outdoors more with physical activity than enjoyment of nature. Moreover, many people living in urban areas are experiencing a growing disconnection from nature. Richard Louv, in his 2005 book, *Last Child in the Woods*, calls attention to this problem, referring to the 'nature-deficit disorder', or our lack of relationship to the environment, which is increasingly noticeable in children. Besides this, most people find animals much more interesting than plants. This is what James Wandersee, of Louisiana State University, and Elizabeth Schussler, of the Ruth Patrick Science Education Center, called 'plant blindness' (Wandersee and Schussler, 1999; Wandersee and Schussler, 2001): the inability of people to recognize the importance of plants in the biosphere and to appreciate the aesthetic and unique biological features of these life forms (Wandersee and Schussler, 1999).

Botanic gardens can be important as a 'cure' for such blindness and nature-deficit disorder, through their role in plant learning and informal education to increase the knowledge of plant biology (Wandersee and Schussler, 2001; Powledge, 2011). They can serve this mission through informal education aimed at schools, hands-on activities, open days and guided tours for the general public. Moreover, botanic gardens offer research programmes to attract university students to plant biology, systematics and taxonomy, since they house botanical reference collections such as herbaria.

However, while taxonomy flourished in the 18th century, the era of exploration, expedition and identification, at the end of the 19th century it went out of fashion (Granjou *et al.*, 2014). In the 20th century, with the increase of laboratory work and technological advances that allowed the rise of new biochemical and molecular approaches, taxonomists have been regarded as consultants or mere collectors of living things rather than as researchers (Sluys, 2013). Today, however, with apprehensions about biodiversity and ecosystem services, taxonomy may be viewed as a crucially relevant scientific field with new opportunities (Mallet and Willmott, 2003; Tautz *et al.*, 2003).

In the 21st century, botanic gardens are able to complement the work developed at universities and colleges, paving the way for plant learning, including the traditional taxonomic approaches but also encompassing systematics, with phylogenetic reconstruction.

Besides displaying plant collections, LBG was created to promote botany learning within the University of Lisbon, a connection that still exists today. Science research in LBG is mainly focused on studying the ecology and taxonomy of vascular and non-vascular flora towards *in situ* and *ex situ* conservation. Plant collections in the garden (> 1500 species) as well as in the herbarium (> 250,000 exsiccates) serve various fields of botanic research. In recent years, two different learning approaches were developed in LBG in relation to students from the university: plant learning for biology students or a source of inspiration for fine art students (Martins-Loução *et al.*, 2014b).

Botanic gardens, particularly those with a strong research activity, have always collaborated with colleges, universities or other science centres. Some of them offer specialized courses or postgraduate programmes, particularly in the UK or USA. Similar to many others (Smith, 2008), LBG promotes botanic garden and academy networks at national and international level through project collaborations, joint fieldwork and training programmes. Training ranges from informal instruction in the field to laboratory practices on plant identification in a straight collaboration with the University of Lisbon's Faculty of Sciences. From 2002 to 2008 more than 200 people received education in plant taxonomy and Mediterranean vegetation. Former graduate students were the drivers of such training courses, giving rise, recently, to the establishment of an online database of Portuguese flora (www.flora-on.pt), which fulfils the first target of GSPC (CBD, 2002; CBD, 2012).

Following the GSPC and the mission statement of BGCI (www.bgci.org), LBG re-stated its mission: contributing to understanding and documenting plant diversity (objective I), conserving plant diversity (objective II) and promoting education and awareness about plant diversity (objective IV) (Martins-Loução *et al.*, 2010; Martins-Loução *et al.*, 2014b). The taxonomists and curators who worked on plant collections, either in the garden or the herbaria, were encouraged to strengthen research interactions to attract more postgraduate students. It has been a challenge to move students from the sophisticated and high technology laboratory at the university campus to the old collections-based facilities of the garden. A positive relationship between well-known research teams and students' interest was observed. This was particularly relevant for the non-vascular flora group that was, and still is, very well established and involved in a number of inventories and biomonitoring studies based on bryophytes (Sérgio *et al.*, 2007). The Portuguese bryoflora comprises about 40% of European species and almost 65% of all Iberian bryophytes. These organisms are identified as one of the oldest groups of plants, with great ecological importance, fundamental for ecosystem biodiversity maintenance (Sérgio *et al.*, 2000; Luís *et al.*, 2012) and serving as bioindicators of environmental quality and climate change (Figueira *et al.*, 2009; Sérgio *et al.*, 2014). *The Checklist of Bryophytes of Portugal* includes all the bryoflora of mainland Portugal and was published in the *Atlas and Red Data Book of Threatened Bryophytes of Portugal* (Sérgio *et al.*, 2013). These publications were possible due to the scientific knowledge of Portuguese bryophytes as well as the maintenance of herbaria collections, from the first ones in the 19th century to the most recent studies, adopting criteria and categories proposed by the IUCN. To increase the scientific impact, and responding to student demand, this non-vascular plant group at LBG has enlarged its focus of plant systematics to include phylogenetic studies and modelling approaches (Draper *et al.*, 2003; Sérgio *et al.*, 2007; Stech *et al.*, 2011; Martins *et al.*, 2012; Sérgio *et al.*, 2014). Both fieldwork and herbaria collections have been the bases of all these works showcasing plant taxonomists and LBG at international level. Besides bryophytes, lichen and fungi collections have been enlarged and studied either from a taxonomic or systematic perspective, contributing to the knowledge of the biology of those organisms and the development of conservation strategies (Melo *et al.*, 2006; Carvalho, 2012; Ryvarden and Melo, 2014). Traditional aspects of botanic gardens research in plant taxonomy and biosystematics continue to underpin much work in biodiversity, and they remain major botanic garden priorities (Wyse Jackson and Sutherland, 2000). However, the continuation of these studies depends on funding and, unfortunately, taxonomic studies are normally underfunded in Portugal as well as in Europe and USA (Mallet and Willmott 2003; Granjou *et al.*, 2014).

The Mediterranean region is important for the diversity of crops and their wild relatives (Kell *et al.*, 2008), as well as an area of plant diversity (Myers *et al.*, 2000). The development of species checklists and inventories is, thus, seen as a first step to effective conservation (CBD, 2002; CBD, 2012), as well as habitat characterization (Rosselló-Graell, 2003; Vieira *et al.*, 2004; Stofer *et al.*, 2006). A checklist allows the characterization of a country's rich resources and, thus, needs data to be organized in a logical and retrievable way (Prendergast, 1995). Non-organized taxonomic data, as well as dispersed biological literature, are major obstructions to the production of complete national inventories. Additional information on native and cultivated status, ethnobotanical uses, national and global distribution, *in situ* and *ex situ* conservation status, threat assessment, and legislation is sparse and difficult to access (Magos Brehm *et al.*, 2008b). However, in Portugal, the development of the crop wild relatives (CWR) and wild harvested plants (WHP) inventory (Magos Brehm *et al.*, 2008a) together with the evaluation of inter- and intra-population genetic diversity of some endemic threatened plant species (Magos Brehm *et al.*, 2012) was possible, due to a collaborative PhD programme between LBG and Birmingham University (Magos Brehm, 2009). The connection to herbaria collections present at LBG, good plant identification skills of the student and the facilities of molecular laboratories and CWR knowledge background at Birmingham University were crucial for the development of such a programme, launching a collaboration that outlined the PhD thesis. The inventory database includes a detailed evaluation of the current utilization of CWR (Magos Brehm *et al.*, 2008b). But the most important implication of this kind of work is that Portugal have produced baseline data about the taxa present and their actual and potential value, to be further used as a conservation tool to complement the setting of conservation priorities (Magos Brehm *et al.*, 2008a; Brehm *et al.*, 2010; Magos Brehm *et al.*, 2012). The establishment of conservation priorities among species is also an important tool in the implementation of any conservation strategy, since financial resources are generally limited (Brehm *et al.*, 2010).

Similarly to what has been seen in many other botanic gardens during the past 30 years (Maunder, 2008), LBG has become key to documenting plant diversity and promoting plant conservation, becoming an important Portuguese player in the United Nations' Global Strategy for Plant Conservation (Martins-Loução *et al.*, 2010; Martins-Loução *et al.*, 2014a). The institutional dependence of LBG on the University of Lisbon is, thus, found to be strategically important for the ongoing role of the botanic garden on plant learning. LBG with its herbaria and seed bank collections are the repositories of vast amounts of taxonomic and botanical diversity information. Apart from the collections, the research developed in LBG with its partner institutions, within and outside the country, as well as the complementary formation classes within the garden give an important and complete taxonomic knowledge to graduate and postgraduate students.

1.3 Strategies to Implement Conservation Policies

Because of biodiversity loss and the rapid disappearance of natural ecosystems and their species, botanic gardens, arboretums and other facilities bear a heavier responsibility for preserving plant genetic resources. Botanic gardens may use this conservation mission as a flag to attract new researchers and students, but also to appeal for new sources of financial support. This means that while the basis of conservation can be taught in universities, botanic gardens must draw attention to their practical role and publicize their work through *in situ* and *ex situ* conservation programmes.

Consistent with the claim for the development of a global programme on conservation of useful plants and traditional knowledge (Barve *et al.*, 2013), the scientific community needs to create a concerted effort focused on the loss of basic knowledge about plants and their uses, especially at the local and regional level. The second phase of GSPC implementation for the period 2011 to 2020 (CBD, 2012) is therefore necessary to minimize the loss of diversity and threatened plant-based ecosystems worldwide, as well as to safeguard tens of thousands of plant species close to extinction. For botanic gardens, GSPC is a catalyst for working together at all levels – local, national, regional and global – in order to understand, conserve and promote the sustainable use of plant diversity while promoting awareness (CBD, 2012). At the Iberian level, Associação Ibero Macaronesica de Jardins Botánicos (the Ibero-Macaronesian Association, that includes Portugal, Spain and Cape Verde, www.jbotanicos.org), has held several meetings to explore the ways in which botanic gardens can engage with the various objectives and targets of the Strategy. This Association has played an important role in the dissemination of research projects and educational activities through an annual online journal, *El/O Botanico* (www.elbotanico.org).

When we think in terms of conservation, it is important to face the fact that a great number of European plant species are vulnerable and at risk of extinction due to climate change (Thuiller *et al.*, 2005). Moreover, changes observed in the use of land, as well as other environmental threats, such as nitrogen deposition, may alter competitive interactions in plant communities, yielding novel patterns of dominance and ecosystem function (Cruz *et al.*, 2010; Dias *et al.*, 2013; Branquinho *et al.*, 2014). For plant species with orthodox seeds (seeds that tolerate a decrease in moisture content), seed banks provide the most practical method for preserving large amounts of genetic material in a small space. The ultimate goal is seed preservation for several centuries, or perhaps even millennia (Gómez-Campo, 2006). Botanic gardens with seed banks and/or seed conservation expertise have a significant contribution to GSPC (objective II). The LBG seed bank (> 3600 seeds) represents an important and effective asset (Draper *et al.*, 2007; Clemente *et al.*, 2011; Clemente *et al.*, 2012; Clemente and Martins-Loução, 2013) with already more than 55% of threatened Portuguese flora represented. In a decade, the LBG seed bank increased the number of taxa and threatened endemic species banked in *ex situ* conservation (Clemente and Martins-Loução, 2013), contributing to target 8 of GSPC: the *ex situ* conservation of 75% of threatened flora in each country by 2020. Besides this conservation effort, the expertise of curators has also served in habitat restoration (Clemente *et al.*, 2004; Draper *et al.*, 2004; Draper *et al.*, 2007; Marques *et al.*, 2007a; Oliveira *et al.*, 2012; Oliveira *et al.*, 2013; Pinto *et al.*, 2013), extension of plant collections outside the LBG (Magos Brehm and Martins-Loução, 2011) and *in situ* conservation (Draper *et al.*, 2003; Rosselló-Graell, 2003; Pinto *et al.*, 2012; Pinto *et al.*, 2013). Every year, seeds of threatened plant species are collected, as well as seeds from plants growing in threatened habitats. Help is also given to Portuguese-speaking countries with which some project collaborations have been established (e.g. collaboration with Mozambique: Niassa Reserve). This is also a way of engaging local people while increasing the capacity for implementing the strategy (CBD, 2012).

With the increasing demands posed by needing to feed the world, it is vital that we address agricultural adaptation more comprehensively, facing the present global changes, climate and land use changes. Currently, agronomists and plant physiologists are investing in better use of crops. However, there is a need to look for new crop varieties that can be productive in the climate of the future (Howden *et al.*, 2007). Evidence has shown that more than half of the wild relatives of 29 globally important crops are not adequately conserved in seed banks (Gewin, 2013). Because of this, a programme was launched, headed by Global Crop Diversity Trust, based in Bonn, Germany, in partnership with Millennium Seed Bank at the Royal Botanic Gardens in Kew, UK, financed by the Norwegian Government, to conserve crop wild relatives (CWR). It is the largest international initiative so far established for this purpose, with a large worldwide consortium, where LBG was invited to represent Portugal.

Ex situ conservation may also serve minimization actions of impacted construction projects, since seed banking can be considered 'warehouses' of biological information or a 'backup' of genetic resources (Draper *et al.*, 2007; Gonzalez *et al.*, 2009). The LBG seed bank is able to function both as a gene bank for restoration programmes (Clemente *et al.*, 2004; Draper *et al.*, 2007; Oliveira *et al.*, 2013) and as a source of scientific expertise to support the best seed samples and germination procedures (Marques *et al.*, 2007b; Oliveira *et al.*, 2011; Oliveira *et al.*, 2012). This is possible due to its participation in a EU-funded coordination action, ENSCONET, in the 6th EU framework programme (2004 to 2009). Until very recently, seed banks in Europe acted in an uncoordinated way at local or national levels, adopting different working standards. As a result of this EU-funded action, which for the first time unified all key facilities for European native seeds, much has changed over the last five years. Today, native seed research facilities in Europe have defined high-level, common working standards, for collecting, cleaning, checking quality and viability (ENSCONET, 2009a, b).

Apart from seed conservation, associated knowledge on plant biology and propagation found in seed banks and botanic gardens are invaluable for those who need to implement conservation practices (Smith, 2008). As suggested by Maxted (Maxted *et al.*, 1997) there is a need for a multiple-tier approach to conservation and, ultimately, to prioritization measures. This means that for a correct characterization of a plant species conservation status, a good knowledge about the ecology and genetic variability of species is required. Also, the Convention on Biological Diversity (CBD) (CBD, 1992), the GSPC (CBD, 2002; CBD, 2012), the European Plant Conservation Strategy (EPCS) (2002) and the International Treaty on Plant Genetic Resources for Food and Agriculture (ITPGRFA) (FAO, 2001) address the need for conserving the genetic diversity of plants, either wild species, crops or the wild relatives of crops. This means that a species-targeted conservation strategy should, whenever possible, include information on the genetic diversity of the target taxa to maximize the intra-specific diversity and potentially useful genes conserved (Magos Brehm *et al.*, 2012). This kind of approach can be used to avoid founder effects, determine effective population size and avoid outbreeding or inbreeding depression, anyone of which can influence a reintroduction programme (Smith, 2008).

Thus, apart from plant systematics, molecular biology has found application in botanic gardens research directed towards conservation. Molecular biology lab facilities were developed at LBG in 2008. Before that, international collaborations and networks both with other museum departments and international partners were the best strategy found to update the skills present at LBG seed bank. Besides seed banking, it was also possible to start an endemic Portuguese plant DNA collection that has been growing slowly (Clemente *et al.*, 2012). DNA barcoding is presently an important tool in conservation strategies and, interestingly, many of the plant group researchers are based in botanic gardens (Cbol *et al.*, 2009). DNA barcoding is currently an important tool in biodiversity assessment for both animal and plant communities. This is because it either provides insights into the process of species definition or helps in the identification of unknown specimens (Cbol *et al.*, 2009). However, in plants, DNA barcoding has been challenging and its success varies among plant lineages. Though sequencing nrITS from environmental or floristic-pooled samples is now almost routine, the great difficulty is to obtain sequence data from numerous unlinked single-copy nuclear markers. Only this can allow high-resolution species discrimination to cope with closely related species assemblages (Cbol *et al.*, 2009). The collaboration with Royal Botanic Gardens, in Kew, UK, and Canary Botanic Gardens, Viera y Clavijo, in Las Palmas, Spain, has been crucial for the development of this expertise. The work done in a group of closely related species of *Caryophyllaceae* and sympatric hybridizing orchids enabled LBG with experience of sequencing several different types of markers to discriminate species (Cotrim *et al.*, 2010).

At present, the research community must pay attention to the development of locally adapted varieties, either wild or wild relatives of crops (CWR) as well as its evolutionary mechanisms. The biodiversity stored in gene banks may fuel advances in plant breeding, as the genetic sequence data provides a genomic 'parts list' that can help to decipher mechanisms that enable plants to adapt to myriad environments (McCouch *et al.*, 2013). Banking of seeds is a good asset but seed banking is not the major focus of conservation strategies. According to the GSPC's objective II (CBD, 2012), plant diversity conservation includes ecological regions or vegetation types (targets 4 and 5) as well as productive lands (target 6). These must be secured through an effective management and/or restoration for conserving plants and their genetic diversity, besides *in situ* (target 7) and *ex situ* (target 8) conservation. Particular attention is also paid to the genetic diversity of crops, including their wild relatives (target 9), and management plans to prevent new biological invasions (target 10). This means that conservation entails safeguarding of species diversity, both *in situ* and *ex situ*, ecological systems and evolutionary processes in nature. It is, thus, a complex 'mission' that should be an integral *raison d'être* for botanic gardens.

Before the development of any conservation strategy, it is important to understand the ecology, adaptation and speciation mechanisms of plants (Marques, 2010; Serrano *et al.*, 2014b). Four factors can be pointed out to influence plant evolutionary interplays (Marques, 2010) as well as endemism or rarity (H.C. Serrano, Ecology of the rare endemic Plantago: understanding the limiting factors towards its conservation. Lisbon, Portugal, 2015, ULisboa PhD thesis): demographic (e.g. gene flow, dispersal, competition), plant/animal (e.g. pollination, predation), edaphic (e.g. nutrient availability, presence of toxic elements) and bioclimatic factors (e.g. drought, cold stress). Following the indications of GSPC, LBG is focused on studying threatened Portuguese or Iberian endemic species through best conservation strategies. Multidisciplinary works have been developed to further understand the threats of Portuguese endemic species, namely, *Narcissus cavanillesii, Plantago almogravensis* and *Ononis hackelii* (Marques, 2010; Pinto *et al.*, 2012; Rosselló-Graell *et al.*, 2003; H.C. Serrano, Ecology of the rare endemic Plantago: understanding the limiting factors towards its conservation. Lisbon, Portugal, 2015, ULisboa PhD thesis). The work on *Narcissus* was developed in collaboration with Real Jardín Botánico, in Madrid, Spain, and intended to understand in what way breeding systems could be a threat to the endemic *N. cavanillesii* (Marques, 2010; Marques *et al.*, 2010), particularly after the recognition of *N. x perezlarae*, a hybrid of *N. cavanillesii x N. serotinus*, as a new taxon for Portugal (Marques *et al.*, 2005). Results based on reproductive biology, molecular studies and demographical monitoring suggest that hybridization does not represent a threat to the survival of parental species (Marques, 2010; Marques *et al.*, 2011; Marques *et al.*, 2012a, b) and its outcomes are mostly dependent on pre- rather than early post-zygotic isolation mechanisms (Marques *et al.*, 2007a; Marques *et al.*,

2012a). On the contrary, *N. cavanillesii* seems to benefit from occurring in sympatric populations, probably due to an increase in pollination success (Marques *et al.*, 2012a). From a conservation perspective and facing the threat status of *N. cavanillesii,* the results of these studies suggest that conservation efforts should be focused on preserving habitats and their interactions or restoring communities in the case of any disturbance (Draper *et al.*, 2004; Rosselló-Graell *et al.*, 2007; Marques, 2010). All these results attribute an important role to hybridization as an evolutionary mechanism, especially because it can lead to an increase in genetic diversity (Rieseberg, 1997). Nevertheless, the possible outcomes may be different according to the provenance of their parents (Tauleigne Gomes and Lefèbvre, 2005; Tauleigne Gomes and Lefèbvre, 2008).

This *in situ* conservation perspective, focused on preserving habitats rather than being species-specific, may not be generalized. For each threatened species it is important to understand its ecologic limiting factors that may have been driving the species to a rarity status near extinction, before the development of conservation guidelines. In the case of *Plantago almogravensis*, an endemic and rare Portuguese species, the plant lost part of its genetic variability (H.C. Serrano, Ecology of the rare endemic Plantago: understanding the limiting factors towards its conservation. Lisbon, Portugal, 2015, unpublished thesis) in favour of the site-adapted traits. This represents increased tolerance to aluminium (Al) (Serrano *et al.*, 2011), associated to Al hyper accumulation (Branquinho *et al.*, 2007) and development of nanism habit (H.C. Serrano, Ecology of the rare endemic Plantago: understanding the limiting factors towards its conservation. Lisbon, Portugal, 2015, ULisboa PhD thesis). This rare plant species found its ecological niche, outside its optimal performance site, due to trade-offs with environmental constrains (Serrano *et al.*, 2014a). The expansion of this species outside its present habitat has been proved to be a good strategy in the enlargement of its distribution area (Pinto *et al.*, 2013).

Although there is a lot of conservation work going on in botanic gardens, conservation is rarely seen as the main priority. This is because most botanic gardens are primarily visitor attractions, from where they get important revenues (Powledge, 2011; Martins-Loução *et al.*, 2014b). However, it is the use of the conservation work that botanic gardens develop that can charm new students and attract new projects, connecting and cooperating, at the same time, with conservation practitioners and the private sector. It is by broadening their contacts that botanic gardens can find real social relevance in this demanding contemporary world. Although GSPC is focused on conservation efforts in the country of origin, it also advises the support and development of regional efforts. This is what the Millennium Seed Bank (Royal Botanic Gardens, Kew, UK) has been doing, offering its knowledge, innovation and practices to indigenous communities, through technology transfer and information sharing. So, one of the strategies of broadening the visibility of botanic gardens is to spread their capacity building and toolkits of the best conservation practices within and outside their country of origin.

1.4 Communication in Botanic Gardens

At present, botanic gardens create new proactive ways of communication more because they need visitors to raise their revenues than because they need to be considered as scientific centres. All over the world, botanic gardens are becoming sophisticated business entities that increasingly depend upon the financial patronage of the public (Maunder, 2008). This produced a profound change both in the way curators and horticulturists see visitors and in governmental authorities viewing botanic gardens as tourist attractions. Botanic gardens very often recognize that their future depends on the way they are able to transmit the importance of their work and on the links they establish with their local community (Maunder, 2008; Powledge, 2011). This new social role was completely embraced by LBG, exploring more a subtle art to attract visitors that is based on staff skills and creativity rather than on improving garden infrastructures, which is completely dependent on financial investments (Martins-Loução *et al.*, 2014b).

LBG is located in the centre of Lisbon, and is enclosed within the walls of the National Museum of Natural History and Science and private buildings, along a deep slope facing upstream of the Tagus river. Some collections, such as palms, ficus and cycads, are the garden's hallmarks. The outstanding diversity of species from all continents, growing outside greenhouses, lends an unexpectedly tropical atmosphere to several spots in the garden. In addition, its Lisbon location allowed the development of several microclimatic niches where flora from different bioclimatic regions live side by side in a wild environment that is very much appreciated by tourists. But there was also a need to attract Portuguese visitors to the garden. For that, different strategies have been utilized, namely, the development of new interpretation panels, new temporary exhibitions, art performances in the garden and guided visits (Gaio-Oliveira *et al.*, 2013; Martins-Loução *et al.*, 2014b). Collaborations between botanic garden curators and artists from the Beaux Art Faculty were established. Also, other projects such as the European Youth Volunteer project and the Leonardus programme brought to the garden young students with different backgrounds, who offered their expertise and creativity to establish more activities either focused on plant diversity awareness or artistic expression (Martins-Loução *et al.*, 2008; Martins-Loução *et al.*, 2014b).

The strategies used by botanic gardens to attract and communicate to the public are very important and have enlarged the type of personal resources at the garden. Besides scientists, curators or horticulturists, botanic gardens need professionals dedicated to science communication and education. These professionals are responsible for the interaction between science and society; they know how to meet the expectations of the public, how to change public behaviour towards biodiversity loss awareness and to mobilize action for a sustainable future. Communication and science promotion are important issues to be used by botanic gardens to approach the public and attract visitors (Gaio-Oliveira, personal unpublished data). There is a need to improve scientific literacy in order to increase public awareness for plant diversity and conservation. However, different findings put botanic gardens in alignment with urban gardens for public attraction: the main factors that motivate people to visit these gardens are the appreciation of the aesthetic and rare qualities of plants; interest in garden design and landscaping; admiration of the scenery; and the pleasure of being outdoors (Neves, 2009; Villagra-Islas, 2011). Still, some people have the mistaken idea that botanic gardens are passive places, where plants are identified with strange and difficult names, as was common in 19th-century museums (Rinker, 2002). But today, botanic gardens should be looked upon as global treasures in a century of ecological crisis.

In the absence of focused governmental policies, botanic gardens can play important roles, both increasing consciousness about the need for better plant knowledge and supporting policy regulations on ecosystems conservation. A solid commitment to education and ethics should be established for botanic gardens, enabling them to enlarge their ecological leadership (Rinker, 2002; Powledge, 2011). Social interaction and scientific dissemination of knowledge for the general public ought to be key aspects of gardens visits. Besides visits, other different approaches can be used to communicate with a new public. An interesting example is the use of art. Art *per se* (e.g. art exhibitions, concerts, performances) may be considered an effective tool in attracting visitors to increase revenue, associating the arts with plant awareness (Martins-Loução *et al.*, 2014b). Since botanic garden displays are also a form of exhibition, creative and well thought out landscape design can be another powerful tool for connecting people with nature (Villagra-Islas, 2011). This approach often brings about debate over whether these activities deviate from the organizational mission related to plant conservation and engaging the public with plants. It is important to bear in mind that, taking a holistic perspective, binding science and art with creativity can promote a better public understanding for ecological values.

However, botanic gardens improve their public relevance by proving that they can answer everyday problems. They need to show they are accredited centres on plant knowledge and conservation practices, as well as the refuge of some threatened species (Rinker, 2002). As well as this, they also need to admit that they may be indirectly responsible for the spread of invasive species (BGCI, 2012; Heywood and Sharrock, 2013). Because of this, it was crucial to launch a code of conduct (Heywood and Sharrock, 2013) and disseminate it as widely as possible. Nowadays, all botanic gardens may share information about the impacts of invasive alien species and advice on their control as it is stated in GSPC (CBD, 2012). One example was the invasion of red palm weevil (RPW), which threatened the LBG palms collection. The pest, commonly called the red palm weevil (RPW), the *Curculionidae* (Coleoptera) *Rhynchophorus ferrugineus* Olivier, was accidentally introduced into several European countries of the Mediterranean Basin, and is affecting palm trees all over the Mediterranean region (Hallet *et al.*, 1999). As a consequence of commercial exchanges of palm trees from contaminated areas of North Africa, this pest is putting at risk large collections of palms within botanic gardens.

Before pest arrival, LBG established careful monitoring in its palm collection (more than 30 species grown outside) and mapping of many others surrounding the garden. An intensive outreach programme was established to share information about the risks of current management practices (objective IV, GSPC), to raise awareness among the owners of private gardens about the danger and consequences of pest arrival, and to alert local authorities and companies to help prevent the arrival of red palm weevil in gardens (target 10, GSPC) (Pinto *et al.*, 2011). Although it did not prevent the pest arriving, this science communication programme was very useful for community engagement and alerting people about the risks of invasive species in general, either plants or animals.

Botanic gardens can therefore make social changes if they become leader institutions for research and education about the plant kingdom. But, as in many other places, science communication needs to be promoted, engaging new professionals in one of the most challenging societal issues: awareness of the need for biodiversity conservation and sustainable use of resources. Moreover, they can help botanic gardens in taking action by maintaining continuous communication with local communities. Unfortunately, in many botanic gardens, as for LBG, this type of work is often achieved through short-term funded projects. Botanic gardens need to be much more proactive and explore new ways of funding if they want to become more socially responsible. This requires a continuous understanding of community needs and very clear expression of the values, social role and responsibilities of the organization (Dodd and Jones, 2011).

1.5 Education in Botanic Gardens

Defining the level of success of public education among botanical gardens is difficult, given the diversified features of the gardens (Kneebone, 2006; Gaio-Oliveira, personal unpublished data). Dependent on their resources, either human, financial or both, all botanic gardens aim at accomplishing GSPC target 14 – promoting education on plants and awareness on human impacts in plant diversity loss.

Botanic gardens offer learning opportunities for practical and multi-sensory activities with plants. However, all these educative offers vary from garden to garden and there is, still, limited evidence of research into learning experiences. In 2003 LBG launched an education department and since 2008 more than 8000 students, every year, have visited the botanic garden to attend educational programmes and learning activities. For schools, it has been described as an ideal place for students to take part in scientific activities that have been specifically planned to fit the national curriculum.

This success brought an increasing concern about the assessment of educational activities, discussion about the best practices for reaching young students and promoting the role of botanic gardens in plant education. This is one of the concerns about GSPC indicators. More than numbers, it is crucial to assess the quality of these activities, the scientific content as well as the results (CBD, 2012). Different projects have been launched in LBG, focused on involving young children in research activities (Barata and Martins-Loução, 2009), and on understanding changing attitudes and behaviours after environmental education projects (Barata, 2014). This last study has tested the effect of psychosocial factors that may promote environmental action. Results demonstrate the potential of using botanic gardens in environmental education and suggest that associating these educational practices with psychosocial processes, such as public commitments, will promote a change in pro-environmental behaviours among teenagers. This is an important issue bearing in mind that botanic gardens are to become more proactive in social contexts.

The education provided by botanic gardens should be beyond environmental education. Thus, to increase concern about plant diversity they need to establish thoughtful relations with educational agents, either schools or universities, in order to encourage students' motivation and interest in science. While at universities the teaching of science often focuses on the accumulation of facts at botanic gardens, informal education can present the problems faced by society daily. The European Union has recommended that teachers should promote high-quality education (European Commission, 2012). Thus, improving teachers' academic and professional training, through the creation of continuous professional development (CPD) programmes and making the profession more attractive to young people, were considered vital in European recommendations (Council of the European Union, 2014).

Scientific studies from many countries suggest that both students and teachers find the engagement with the epistemology and practices of scientific inquiry very challenging (Dillon *et al.*, 2006; Barata *et al.*, 2013). Policy orientations in Europe as well as in North America encourage teachers to use inquiry-based learning as an efficient method in trying to get science students doing the very things scientists do (Bybee, 2010; European Commission, 2013). LBG, in Portugal, was one of the partners of the INQUIRE project (SCIENCE-IN-SOCIETY-2010 no 266616) designed to reinvigorate Inquiry-Based Science Education (IBSE) in formal and 'Learning Outside the Classroom' (LOtC) educational systems throughout Europe. In this project, LBG functioned both as a LOtC and a science research institution. The project aimed to design a CPD programme to help teachers move from more traditional pedagogies to inquiry-based ones. It was particularly focused on teaching biodiversity and climate change and directed at teachers of 5th to the 9th grade students of basic and secondary education (Gaio-Oliveira *et al.*, 2012; Martins-Loução *et al.*, 2012). LBG results asserted that the CPD course was able to change teaching conceptions through the establishment of teachers' learning communities of practice outside schools (Martins-Loução *et al.*, 2013). Teachers were able to evaluate and reflect on their own practice so that, at the end, they developed new approaches to apply scientific inquiry while teaching biodiversity and climate change in the context of their own classrooms (Barata *et al.*, 2013).

Educational departments of botanic gardens can, therefore, encourage the implementation of the IBSE method, both in garden activities and in teachers' professional development programmes (Martins-Loução *et al.*, 2012). Another strategy is to help schools to use playgrounds as LOtC spaces. Botanic gardens should promote this approach in several ways; for example, by developing courses for teachers interested in improving (or creating from scratch) playground gardens in order to integrate them in the curriculum; or by developing good practice manuals that allow school communities to use the best playgrounds as LOtC areas (Gaio-Oliveira and Garcia, 2014).

Botanic gardens may also use novel technologies to introduce young people to the wonders of the plant kingdom. However, while digital technology is changing how we live and communicate, botanic gardens are still experiencing this fast paced digital age (see Roots, 2013, vol. 10, http://www.bgci.org/education/roots/). Nevertheless, science education can be more effective when information and communication technologies (ICT) are used, an important asset for reaching a broader audience, according to the indicators of target 14 (GSPC). Within the EU programme, 'Natural Europe: Natural History & Environmental Cultural Heritage in European Digital Libraries for Education' project (CIP-ICT-PSP- 250579, www.natural-europe.eu) aimed to develop digital libraries and to produce online educational pathways based on natural history museum exhibitions, including inquiry-based activities and their assessment. The results of this study have shown that online supported guided tours have proven to offer new positive ways of exploring information and preparing visiting tours (Barata *et al.*, 2012). The results reported from this EU project emphasize the crucial role that botanic gardens may have in the preparation of inquiry-based activities, tools and materials, profiting from online resources as a driver for visits to real contexts in the garden.

Informal education in botanic gardens may, thus, complement formal education in schools, helping teachers to motivate students for science, engage them for improving their understanding and to inspire their personal involvement for creative experiences (Gano and Kinzler, 2011). In the current technological world, the use of ICT tools may benefit student's engagement to contribute positively for science learning. The importance of plant diversity and the need for its conservation incorporated into communication, education and public awareness programmes, as it is stated in target 14 of GSPC (CBD, 2012) must be the driver of different creative initiatives within botanic gardens.

1.6 Conclusions

Even during the problems caused by a global financial crisis, botanic gardens are more than live botanical collections. Through scientific research, these are true institutions for plant education, projecting their knowledge as a continuous dialogue between taxonomy and systematics to explain evolution and phylogeography (Cbol *et al.*, 2009; Chase and Reveal, 2009; Nieto Feliner, 2014). This scientific knowledge is pivotal for a better understanding of plants, their evolution and consequent development of conservation tools. This demonstrates that the maintenance of plant science research at botanic gardens paves the way for plant education. Also, by promoting systematic research on plants, botanic gardens contribute to fundamental knowledge about how plants maintain the planet's environmental balance and ecosystem stability.

Thus, by maintaining their status as research centres, botanic gardens may better promote plant diversity and conservation practices, support postgraduate students, or implement capacity building of professionals' as well as teachers' training workshops. Strategically, botanic gardens need to be engaged in real societal problems, such as the consequences of climate change and the established 'planet boundaries' (Rockström *et al.*, 2009) to help construct a better sustainable future. Addressing these problems will enable intervention at science centres and universities and help restrain the loss of species and ecosystems. Because of this, botanic gardens must play an important role in underpinning conservation action. However, their successes need to be better understood and better disseminated, independently of any failure to reduce the loss of biodiversity (Blackmore *et al.*, 2011). Also, strategically, botanic gardens need to demonstrate to their public how they are involved in science, conservation and educational activities (Powledge, 2011). Many botanic gardens are already doing this (e.g. new installations of Millennium Seed Bank, Royal Botanic Gardens, Kew, UK, New York Botanic Garden, USA), but most of them safeguard the laboratories and maintain hidden historical collections. Of course, herbaria and laboratories need to be preserved and clean, for the safety of materials and research experiences, but opening large windows allowing visitors to observe researchers at work, could be organized sometimes, with small investments. Another possibility is to promote open-day laboratories to make people aware of the scientific work developed behind closed doors. However, it is advisable that new exhibition spaces must be designed as showcases of the vital scientific work (Maunder, 2008).

Changes include the presence of science promoters and the implementation of a communication department or group. Botanic gardens are true plant diversity arks that, to be fully understood, need a twofold process of biological learning and social engagement. Visits to the garden bring different audiences, most of them drawn by the aesthetic value (Ballantyne *et al.*, 2008), but the act of communicating, the concept, the rationale, influences how people perceive the message (Hall *et al.*, 2014). To align the language to a broader audience, it requires understanding of perspectives and simplified contexts within the social setting to effectively contribute to a change in attitude (Hall *et al.*, 2012). This is a new and important perception for the community. However, if botanic gardens really want to promote their social relevance, raise awareness about plants and effectively disseminate GSPC, a good collaboration with social sciences, humanities and communication professionals is needed. Besides, the strategy of plant conservation also emphasises the need to look for "indicators of quality rather than showing only numbers concerning public engagement actions" (CBD, 2012). Creative approaches, binding science with art, will also benefit message dissemination while attracting a new public to the garden.

Last, but not least, the educational services of botanic gardens need to take care with messages and the way they promote botany education. More hands-on educators should be involved in new pedagogic approaches to help children and young students learn about how plants support and improve our livelihoods. Inquiry-based approaches should be widespread as well as the recognition of botanic gardens as places of learning outside the classroom (Dillon *et al.*, 2006). In these places the students should have the opportunity to use the language of science and to increase their interest in learning, according to EU reflections (Osborne and Dillon, 2008). Establishing partnerships with university research centres and strengthening the collaboration with schools will contribute to better contents that fit scholarly curricula; diverse pedagogic offerings drive youths' enquiring minds and, at the same time, increase teachers' motivation for plant education.

Acknowledgements

Gisela Gaio-Oliveira acknowledges support from Fundação para a Ciência e Tecnologia (SFRH/BPD/65886/2009). Part of the education research was developed within the EU INQUIRE project (FP7-Science-In-Society-2010, under contract no. 266616).

The authors also acknowledge all the work developed by the following research team: Adelaide Clemente, Alexandra Escudeiro, Ana Isabel Dias Correia, Ana Julia Pinheiro, Ana Margarida Francisco, Ana Raquel Barata, Antónia Rosseló-Graell, Cecilia Sérgio, César Garcia, Cristina Branquinho, Cristina Ramalho, Cristina Tauleigne-Gomes, David Draper, Helena Cotrim, Helena Serrano, Isabel Marques, Ireneia Melo, Joana Magos Brehm, Joana Camejo, Leena Luís, Manuel João Pinto, Manuela Sim-Sim, Marco Jacinto, Maria Teresa Antunes, Miguel Porto, Pedro Pinho, Nuno Carvalho, Palmira Carvalho, Rui Figueira.

References

Ballantyne, R., Packer, J. and Hughes, K. (2008) Environmental awareness, interests and motives of botanic gardens visitors: implications for interpretive practice. *Tourism Management* 29(3), 439–444.

Barata, A.R. (2014) A educaçãoambiental no contexto da sociedade: comopromovercomportamentospró-ambientais? PhD thesis. Instituto Universitário de Lisboa, Lisbon, Portugal, p. 248.

Barata, A.R. and Martins-Loução, M.A. (2009) Ao ritmo das plantas no Jardim Botânico. *El/O Botanico* 3, 52–53.

Barata, R., Paulino, I., Ribeiro, B., Serralheiro, F., Lopes, L.F. and Alves, M.J. (2012) Digital natural history repositories and tools for inquiry- based education. *International Congress on ICT and Education towards Education 20*. Instituto de Educação, Universidade de Lisboa, Lisbon, Portugal, pp. 1468–1483.

Barata, R., Carvalho, N., Paulino, I., Gaio-Oliveira, G., Alves, M. and Martins-Loução, M. (2013) The use of IBSE for improving science literacy and education at MUHNAC. *Inquire Conference 2013 Raising Standards through Inquiry: Professional Development in the Natural Environment*. BGCI, Royal Botanic Gardens, Kew, London, UK, pp. 122–128.

Barve, V., Bhatti, R., Bussmann, R., Bye, R., Chen, J., Dullo, E., Giovannini, P., Linares, E., Magill, R., Roguet, D., Salick, J., Van On, T., Vandebroek, I., Wightman, G. and Jackson, P.W. (2013) *Call for a Global Program on Conservation of Useful Plants and Traditional Knowledge*. Available at: www.plants2020.net/news/1037 (accessed 20 May 2016).

BGCI (2012) *International Agenda for Botanic Gardens in Conservation*, 2nd edn. Botanic Gardens Conservation International, Richmond, UK.

BGCI (2014) *Resource Centre: The History of Botanic Gardens, 2014*. Available at: www.bgci.org/resources/history (accessed 20 May 2016).

Blackmore, S., Gibby, M. and Rae, D. (2011) Strengthening the scientific contribution of botanic gardens to the second phase of the global strategy for plant conservation. *Botanical Journal of the Linnean Society* 166(3), 267–281.

Bodeker, G., Bhat, K., Burley, J. and Vantomme, P. (2003) *Medicinal Plants for Forest Conservation and Health Care*. Food and Agriculture Organization, Rome, Italy, p. 158.

Branquinho, C., Serrano, H.C., Pinto, M.J. and Martins-Loucao, M.A. (2007) Revisiting the plant hyperaccumulation criteria to rare plants and earth abundant elements. *Environmental Pollution* 146(2), 437–443.

Branquinho, C., Gonzalez, C., Clemente, A., Pinho, P. and Correia, O. (2014) The impact of the rural land-use on the ecological integrity of the intermittent streams of the Mediterranean 2000 Natura network. In: Sutton, M., Mason, K., Sheppard, L., Sverdrup, H., Haeuber, R. and Hicks, W. (eds), *Nitrogen Deposition, Critical Loads and Biodiversity*. Springer, Heidelberg, Germany.

Brehm, J.M., Maxted, N., Martins-Loucao, M.A. and Ford-Lloyd, B.V. (2010) New approaches for establishing conservation priorities for socio-economically important plant species. *Biodiversity and Conservation* 19(9), 2715–2740.

Bybee, R. (2010) *The Teaching of Science: 21st Century Perspectives*. NSTA Press, Arlington, Virginia, USA.

Carvalho, P. (2012) Collema. Liquenologia, SEd, editor. SEL, Pontevedra, Spain, p. 52.

CBD (1992) *Convention on biological diversity: text and annexes*. Secretariat of the Convention on Biological Diversity, United Nations, Montreal, Canada. Available at: www.cbd.int/doc/legal/cbd-en.pdf (accessed 20 May 2016).

CBD (2002) *Global strategy for plant conservation*. Secretariat of the Convention on Biological Diversity, United Nations, Montreal, Canada. Available at: www.cbd.int/decisions/cop/?m=cop-06 (accessed 25 August 2016).

CBD (2012) *Global strategy for plant conservation: 2011–2020*. Secretariat of the Convention on Biological Diversity, United Nations, Montreal. Canada.

Cbol, Plant Working Group C, Hollingsworth, P.M., Forrest, L.L., Spouge, J.L., Hajibabaei, M., Ratnasingham, S. *et al.* (2009) A DNA barcode for land plants. *Proceedings of the National Academy of Sciences USA* 106(31), 12794–12797.

Chase, M.W. and Reveal, J.L. (2009) A phylogenetic classification of the land plants to accompany APG III. *Botanical Journal of the Linnean Society* 161(2), 122–127.

Clemente, A. and Martins-Loução, M.A. (2013) Banco de Sementes do Jardim Botânico MNHNC: o balanço de uma década. *El/O Botanico* 7, 17–19.

Clemente, A.S., Werner, C., Máguas, C., Cabral, M.S., Martins-Loução, M.A. and Correia, O. (2004) Restoration of a limestone quarry: effect of soil amendments on the establishment of native Mediterranean sclerophyllous shrubs. *Restoration Ecology* 12(1), 20–28.

Clemente, A., Magos Brehm, J. and Martins-Loução, M.A. (2011) Conservação ex situ salva espécies ameaçadas da Flora Portuguesa. *El/O Botanico* 5, 48–49.

Clemente, A., Cotrim, H., Magos Brehm, J., Dias, S., Costa, C. and Martins-Loução, M.A. (2012) As Colecções da Flora Portuguesa ameaçada no Banco de Germoplasma do Jardim Botânicodo Museu Nacional de História Natural e da Ciência da Universidade de Lisboa. *El/O Botanico* 6, 18–19.

Cotrim, H., Santos, P. and Martins-Loução, M.A. (2010) *Testing plant DNA barcode regions for species discrimination in Silene sect SiphonomorphaOtth*. The European Consortium for the Barcode of Life (ECBOL 2), Braga, Portugal.

Council of the European Union (2014) *Conclusions on Efficient and Innovative Education and Training to*

Invest in Skills - supporting the 2014 European Semester. Available at: http://www.consilium.europa.eu/uedocs/cms_data/docs/pressdata/en/educ/141138.pdf (accessed 26 August 2016)

Cruz, C., Dias, T., Pinho, P., Branquinho, C., Máguas, C. *et al.* (2010) Policies for plant diversity conservation on a global scale: a Nitrogen driver analysis. *Kew Bulletin* 65(4), 525–528.

Darwin Edwards, I. (2000) Education by stealth: the subtle art of educating people who didn't come to learn. *Roots* 20(7), 37–40.

Dias, T., Oakley, S., Alarcón-Gutiérrez, E., Ziarelli, F., Trindade, H. *et al.* (2013) N-driven changes in a plant community affect leaf-litter traits and may delay organic matter decomposition in a Mediterranean maquis. *Soil Biology and Biochemistry* 58, 163–171.

Dillon, J., Rickinson, M., Teamey, K., Morris, M., Choi, M. *et al.* (2006) The value of outdoor learning: evidence from research in the UK and elsewhere. *School Science Review* 87(320), 107–111.

Dodd, J. and Jones, C. (2011) Towards a new social purpose: the role of botanic gardens in the 21st century. *Roots* 8(1), 1–5.

Draper, D., Rosselló-Graell, A., Garcia, C., Tauleigne Gomes, C. and Sérgio, C. (2003) Application of GIS in plant conservation programmes in Portugal. *Biological Conservation* 113(3), 337–349.

Draper, D.M., Rosselló-Graell, A., Marques, I. and Iriondo, J.M. (2004) Translocation action of *Narcissus cavanilesii*: selecting and evaluating the receptor site. In: *4th Conference on the Conservation of Wild Plants, 2004*. Available at: www.nerium.net/Plantaeuropa/Download/Procedings/Draper_Rossello_Et_Al.pdf.

Draper, D., Marques, I., Rosselló-Graell, A. and Martins-Loução, M.A. (2007) Role of a Gene Bank as a mitigation tool: the case of Alqueva Dam (Portugal). *Bocconea* 21, 385–390.

European Commission (2012) *Rethinking Education: Investing in Skills for Better Socio-economic Outcomes*. Available at: www.cedefop.europa.eu/files/com669_en.pdf (accessed 26 August 2016).

European Commission (2013) *Reducing early school leaving: key messages and policy support. Final Report of the Thematic Working Group on Early School Leaving*. Available at: https://ec.europa.eu/education/policy/strategic-framework/doc/esl-group-report_en.pdf (accessed 26 August 2016).

ENSCONET (2009a) *Curation Protocols & Recommendations.* EU, 6th Framework Program.

ENSCONET (2009b) *Seed Collecting Manual for Wild Species*. EU, 6th Framework Program.

FAO (2001) International treaty on plant genetic resources for food and agriculture. Available at: http://www.planttreaty.org/ (accessed 25 August 2016).

Figueira, R., Tavares, P.C., Palma, L., Beja, P. and Sérgio, C. (2009) Application of indicator kriging to the complementary use of bioindicators at three trophic levels. *Environmental Pollution* 157, 2689–2696.

Gaio-Oliveira, G. and Garcia, C. (2014) Science arrives to schoolyards. *El/O Botanico* 8, 65–67. Available at: www.elbotanico.org/revista8_articulos/science-arrives-to-schoolyards.pdf.

Gaio-Oliveira, G., Barata, A.R., Carvalho, N. and Martins-Loução, M.A. (2012) Science teachers' continuing professional development in inquiry based education on plant diversity and conservation. *El/O Botanico* 6, 38–39.

Gaio-Oliveira, G., Martins-Loução, M.A. and Melo, I. (2013) A communication strategy developed by the botanic garden from the National Museum of Natural History and Science (Lisbon University, Portugal) for the promotion of plant diversity. *El/O Botanico* 7, 44–45.

Gano, S. and Kinzler, R. (2011) Bringing the museum into the classroom. *Science* 331, 1028–1029.

Gewin, V. (2013) Weeds warrant urgent conservation. *Nature* 22 July. Available at: www.nature.com/news/weeds-warrant-urgent-conservation-1.13422 (accessed 20 May 2016).

Gómez-Campo, C. (2006) Erosion of genetic resources within seed genebanks: the role of seed containers. *Seed Science Research* 16, 291–294.

Gonzalez, C., Clemente, A., Nielsen, K., Branquinho, C. and Santos, R.F. (2009) Human–nature relationship in Mediterranean streams: integrating different types of knowledge to improve water management. *Ecology and Society* 14(2), 35.

Granjou, C., Mauz, I., Barbier, M. and Breucker, P. (2014) Making taxonomy environmentally relevant. Insights from an all taxa biodiversitory inventory. *Environmental Science and Policy* 38, 254–262. Available at: dx.doi.org/10.1016/j.envsci.2014.01.004 (accessed 20 May 2016).

Hall, D.M., Gilbertz, S., Horton, C. and Peterson, T. (2012) Culture as a means to contextualize policy. *Journal of Environmental Studies and Science* 2(3), 222–233.

Hall, D., Lazarus, E.D. and Swannack, T.M. (2014) Strategies for communicating system models. *Environmental Modelling & Software* 55, 70–76.

Hallet, R., Oehlschlager, A. and Borden, J. (1999) Pheromone trapping protocols for the Asian palm weevil, *Rhynchophorus ferrugineus* (Coleoptera: Curculionidae). *International Journal of Pest Management* 45(3), 231–237.

Heywood, V. and Sharrock, S. (2013) *European Code of Conduct for Botanic Gardens on Invasive Alien Species*. Botanic Gardens Conservation International, Richmond, Strasbourg, France.

Howden, S.M., Soussana, J.F., Tubiello, F.N., Chhetri, N., Dunlop, M. and Meinke, H. (2007) Adapting agriculture to climate change. *Proceedings of the National Academy of Sciences USA* 104(50), 19691–19696.

Kell, S., Knüpffer, H., Jury, S., Ford-Lloyd, B.V. and Maxted, N. (2008) Crops and wild relatives of the Euro-Mediterranean region: making and using a conservation catalogue. In: Maxted, N., Ford-Lloyd, B.V., Kell, S., Iriondo, J.M., Dullo, M. and Turok, J. (eds) *Crop Wild Relative Conservation and Use*. CAB International, Wallingford, Oxfordshire, UK, pp. 69–109.

Kneebone, S. (2006) *Education Centre: A Global Snapshot of Botanic Garden Education Provision*. Available at: www.bgci.org/education/global_snapshot_edu_provis.

Louv, R. (2005) *Last Child in the Woods*. Workman Publishing Company, New York.

Luís, L., Hughes, S.J. and Sim-Sim, M. (2012) Bryofloristic evaluation of the ecological status of Madeiran streams: towards the implementation of the European Water Framework Directive in Macaronesia. *Nova Hedwigia* 96(1–2), 181–204.

Magos Brehm, J. (2009) Conservation of wild plant genetic resources in Portugal. PhD thesis. University of Birmingham, Birmingham, UK.

Magos Brehm, J. and Martins-Loução, M.A. (2011) Aextensão da coleção do Jardim Botânico do Museu Nacional de História Natural da Universidade de Lisboa: o Jardim de Famões. *El/O Botanico* 5, 10–11.

Magos Brehm, J., Maxted, N., Ford-Lloyd, B.V. and Martins-Loução, M.A. (2008a) National inventories of crop wild relatives and wild harvested plants: case-study for Portugal. *Genetic Resources and Crop Evolution* 55(6), 779–796.

Magos Brehm, J., Mitchell, M., Maxted, N., Ford-Lloyd, B.V. and Martins-Loução, M.A. (2008b) IUCN Red listing of crop wild relatives: is a national approach as difficult as some think? In: Maxted, N., Ford-Lloyd, B.V., Kell, S., Iriondo, J.M., Dullo, M. and Turok, J. (eds), *Crop Wild Relative Conservation and Use*. CAB International, Wallingford, Oxfordshire, UK, pp. 211–242.

Magos Brehm, J., Ford-Lloyd, B.V., Maxted, N. and Martins-Loução, M.A. (2012) Using neutral genetic diversity to prioritize crop wild relative populations: a Portuguese endemic case study for *Dianthus cintranus* Boiss. & Reut. Subsp. Barbatus R. Fern. & Franco. In: Maxted, N. and Ford-Lloyd, B.V. (eds) *Agrobiodiversity Conservation: Securing the Diversity of Crop Wild Relatives and Landraces*. CAB International, Wallingford, Oxfordshire, UK, pp. 193–210.

Mallet, J. and Willmott, K. (2003) Taxonomy: renaissance or Tower of Babel? *Trends in Ecology and Evolution* 18 (2), 57–59.

Mann, T. (1955) *Confessions of Felix Krull, Confidence Man: the Early Years*. Knopf, New York.

Marques, I. (2010) Evolutionary outcomes of natural hybridization in Narcissus (Amaryllidaceae): the case of *N. xperezlaraes*. I. European PhD, University of Lisbon, Lisbon, Portugal, p. 179.

Marques, I., Rosselló-Graell, A. and Draper Munt, D. (2005) *Narcissus xperezlarae* Font Quer (Amaryllidaceae). A new taxon for the Portuguese flora. *Flora Mediterranea* 28, 196–197.

Marques, I., Rosselló-Graell, A., Draper Munt, D. and Iriondo, J.M. (2007a) Pollination patterns limit hybridisation between two sympatric species of Narcissus (Amayllidaceae). *American Journal of Botany* 94, 1352–1359.

Marques, I., Draper, D., Rosselló-Graell, A. and Martins-Loução, M.A. (2007b) Germination behaviour of seven mediterranean grassland species. *Bocconea* 21, 367–372.

Marques, I., Feliner, G.N., Draper Munt, D., Martins-Loução, M.A. and Aguilar, J.F. (2010) Unraveling cryptic reticulate relationships and the origin of orphan hybrid disjunct populations in narcissus. *Evolution* 64(8), 2353–2368.

Marques, I., Nieto Feliner, G., Martins-Loução, M.A. and Fuertes Aguilar, J. (2011) Fitness in Narcissus hybrids: Low fertility is overcome by early hybrid vigour, absence of exogenous selection and high bulb propagation. *Journal of Ecology* 99(6), 1508–1519.

Marques, I., Aguilar, J., Martins-Loução, M. and Feliner, G. (2012a) Spatial–temporal patterns of flowering asynchrony and pollinator fidelity in hybridizing species of Narcissus. *Evolutionary Ecology* 26(1), 1–18.

Marques, I., Nieto Feliner, G.,Martins-Loução, M.A. and Fuertes Aguilar, J. (2012b) Genome size and base composition variation in natural and experimental Narcissus (Amaryllidaceae) hybrids. *Annals of Botany* 109(1), 257–264.

Martins, A., Figueira, R., Sousa, A.J. and Sérgio, C. (2012) Spatio-temporal patterns of Cu contamination in mosses using geostatistical estimation. *Environmental Pollution* 170(1), 276–284.

Martins-Loução, M.A., Escudeiro, A. and Barata, A.R. (2008) Voluntários europeus no Jardim Botânico: uma experiência enriquecedora. *El/O Botanico* 2, 26.

Martins-Loução, M., Sérgio, C., Melo, I., Correia, A.I., Escudeiro, A. *et al*. (2010) O contributo do Jardim Botânico de Lisboa para a Estratégia Global para a Conservação de Plantas (2003–2009). *El/O Botanico* 4, 10–11.

Martins-Loução, M., Gaio-Oliveira, G., Barata, R. and Carvalho, N. (2012) The use of IBSE as a tool for the development of teachers' curriculum: challenges and opportunities offered by LOtC institutions. In: Kapelari, S., Jeffreys, D., Willison, J., Vergou, A., Regan, E., Dillon, J., Bromley, G.and Bonomi, C. (eds) *International Congress on ICT and Education. Towards Education 2.0, 2012, Lisbon*. Instituto de Educação, University of Lisbon, Lisbon, Portugal, pp. 2803–2811.

Martins-Loução, M., Gaio-Oliveira, G., Barata, R., Carvalho, N. and Zoccoli, M. (2013) How can LOtC institutions provide a change in teaching methodology to promote students' engagement in natural sciences? The Lisbon Botanic Garden as a case study. *Inquire Conference 2013 Raising Standards Through Inquiry: Professional Development in the Natural Environment, 2013*. BGCI, Royal Botanic Gardens, Kew, London, UK, pp. 95–100.

Martins-Loução, M.A., Clemente, A., Escudeiro, A., Correia, A.I., Sérgio, C. *et al.* (2014a) O Jardim Botânico da Universidade de Lisboa e a Estratégia Global para a Conservação de Plantas (2011–2020). *El/O Botanico* 8, 7–8.

Martins-Loução, M.A., Gaio-Oliveira, G., Melo, I. and Antunes, M.T. (2014b) The subtle art of attracting people to Lisbon Botanic Garden. *Roots* 11(2), 21–24.

Maunder, M. (2008) Beyond the greenhouse. *Nature* 455, 596–597.

Maxted, N., Hawkes, J., Guarino, L. and Sawkins, M. (1997) The selection of criteria of taxa for plant genetic conservation. *Genetic Resources and Crop Evolution* 44, 337–348.

McCouch, S., Baute, G.J., Bradeen, J., Bramel, P., Bretting, P.K. *et al.* (2013) Agriculture: feeding the future. *Nature* 499(7456), 23–24.

Melo, I., Salcedo, I. and Tellería, M.T. (2006) Contribution to the knowledge of Tomentelloid fungi in the Iberian Peninsula. *Nova Hedwigia* 82, 167–187.

Merritt, D. J. and Dixon, K. W. (2011) Restoration seed banks – a matter of scale. *Science* 332(6028), 424–425.

Myers, N., Mittermeier, C., da Fonseca, G.A.B. and Kent, J. (2000) Biodiversity hotspots for conservation priorities. *Nature* 403, 853–858.

Neves, K. (2009) Urban Botanical Gardens and the aesthetics of ecological learning: a theoretical discussion and preliminary insights from Montreal's Botanical Garden. *Anthropologica* 51(1), 145–157.

Nieto Feliner, G. (2014) Patterns and processes in plant phylogeography in the Mediterranean Basin. A review. *Perspectives in Plant Ecology. Evolution and Systematics* 16(5), 265–278.

Oliveira, G., Clemente, A., Nunes, A. and Correia, O. (2011) Effect of substrate treatments on survival and growth of Mediterranean shrubs in a revegetated quarry: an eight-year study. *Ecological Engineering* 37(2), 255–259.

Oliveira, G., Clemente, A., Nunes, A. and Correia, O. (2012) Testing germination of species for hydroseeding degraded Mediterranean areas. *Restoration Ecology* 20, 623–630.

Oliveira, G., Clemente, A., Nunes, A. and Correia, O. (2013) Limitations to recruitment of native species in hydroseeding mixtures. *Ecological Engineering* 57, 18–26.

Osborne, J. and Dillon, J. (2008) *Science Education in Europe: Critical Reflections. A Report to Nuffield Foudation.* Nuffield Foundation, London, UK, p. 32.

Pinto, M.J., Antunes, M.T. and Martins-Loução, M.A. (2011) Estratégia de prevenção e controlo da praga das palmeiras provocada por *Rhynchophorus ferrugineus* (Olivier, 1790) no Jardim Botânico (MNHN-UL). *El/O Botanico* 5, 38–39.

Pinto, M.J., Serrano, H.C., Antunes, C., Branquinho, C. and Martins-Loução, M.A. (2012) Monitorização de species raras e isolados populacionais do sudoeste de Portugal. *El/O Botanico* 6, 16–17.

Pinto, M.J., Serrano, H.C., Branquinho, C. and Martins-Loução, M.A. (2013) Éxito en la translocación de una-planta con restricciones dispersivas. *Conservación Vegetal* 17, 8–9.

Powledge, F. (2011) The evolving role of Botanical Gardens. Hedges against extinction, showcases for botany? *Biological Science* 61(10), 743–749.

Prendergast, H. (1995) Published sources of information on wild plant species. In: Guarino, L., RamanathaRao, V. and Reid, R. (eds), *Collecting Plant Genetic Diversity: Technical Guidelines*. CAB International, Wallingford, Oxfordshire, UK, pp. 153–179.

Rieseberg, L. (1997) Hybrid origins of plant species. *Annual Review of Ecology, Evolution, and Systematics* 28, 359–389.

Rinker, H.B. (2002) *The Weight of a Petal: The Value of Botanical Gardens*. Available at: www.actionbioscience.org/biodiversity/rinker2.html (accessed 25 August 2016).

Rockström, J., Steffen, W., Noone, K., Persson, Å., Chapin, F.S.I. *et al.* (2009) Planetary boundaries: exploring the safe operating space for humanity. *Ecology and Society* 14(2), 32–63.

Rosselló-Graell, A. (2003) Caracterização fito-ecológica das lagoas temporárias do Campo Militar de Santa Margarida (Ribatejo, Portugal). *Portugaliae Acta Biologica* 21, 245–278.

Rosselló-Graell, A., Draper Munt, D., Salvado, F., Albano, S., Ballester, S. and Correia, A.I. (2003) Conservation programme for *Narcissus cavanillesii* A. Barra & G. López (Amaryllidaceae) in Portugal. *Bocconea* 16(2), 853–856.

Rosselló-Graell, A., Marques, I., Draper Munt, D. and Iriondo, J.M. (2007) The role of breeding system in the reproductive success of *Narcissus cavanillesii* A. Barra & G. López (Amaryllidaceae). *Bocconea* 21, 359–365.

Ryvarden, L. and Melo, I. (2014) *Poroid Fungi of Europe. Synopsis Fungorum*, vol. 31. Fungiflora, Oslo, Norway.

Sérgio, C., Araújo, M. and Draper, D. (2000) Portuguese bryophytes diversity and priority areas for conservation. *Lindbergia* 25, 116–123.

Sérgio, C., Figueira, R., Draper, D., Menezes, R. and Sousa, A. (2007) The use of herbarium data for the assessment of red list categories: modelling bryophyte distribution based on ecological information. *Biological Conservation* 135, 341–351.

Sérgio, C., Garcia, C.A., Sim-Sim, M., Vieira, C., Hespanhol, H. and Stow, S. (2013) *Atlas e Livro Vermelho dos Briófitos ameaçados de Portugal (Atlas and Red Data Book of Endangered Bryophytes of Portugal).* Edições Documenta, MUHNAC/CBA, Lisbon, Portugal.

Sérgio, C., Garcia, C.A., Vieira, C., Hespanhol, H., Sim-Sim, M. *et al.* (2014) Conservation of Portuguese red listed bryophytes species in Portugal. Promoting a shift in perspective on climate changes. *Plant Biosystems* 148(3–4), 837–850.

Serrano, H.C., Pinto, M.J., Martins-Loução, M.A. and Branquinho, C. (2011) How does an Al-hyperaccumulator plant respond to a natural field gradient of soil phytoavailable Al? *Science of the Total Environment* 409(19), 3749–3756.

Serrano, H.C., Antunes, C., Pinto, M.J., Máguas, C., Martins-Loução, M.A. and Branquinho, C. (2014a) The ecological performance of metallophyte plants thriving in geochemical islands is explained by the Inclusive Niche Hypothesis. *Journal of Plant Ecology* 1, 41–45. DOI: 10.1093/jpe/rtu007.

Serrano, H.C., Pinto, M.J., Cotrim, H., Branquinho, C. and Martins-Loução, M.A. (2014b) A investigação ecológica como base de estratégias de conservação. *El/O Botanico* 8, 37–39.

Sharrock, S. (ed.) (2011) *Global Strategy for Plant Conservation: A Guide to the GSPC. All the Targets, Objectives and Facts*. Botanic Gardens Conservation International, Kew, Surrey, UK. Available at: cncflora.jbrj.gov.br/portal/static/pdf/documentos/GSPC.pdf (accessed 26 August 2016).

Sluys, R. (2013) The unappreciated, fundamentally analytical nature of taxonomy and the implications for the inventory of biodiversity. *Biodiversity Conservation* 22, 1095–2105.

Smith, P.P. (2008) Ex situ conservation of wild species: services provided by Botanic Gardens. In: Maxted, N., Ford-Lloyd, B.V., Kell, S., Iriondo, J.M., Dullo, M. and Turok, J. (eds), *Crop Wild Relative Conservation and Use*. CAB International, Wallingford, Oxfordshire, UK, pp. 407–412.

Stech, M., Werner, O., González-Mancebo, J.M., Patiño, J., Sim-Sim, M., Fontinha, S., Hildebrandt, I. and Ros, R.M. (2011) Phylogenetic inference in Leucodon (Leucodontaceae, Bryophyta) in the North Atlantic region. *Taxon* 60(1), 79–88.

Stofer, S., Bergamini, A., Aragon, G., Carvalho, P., Coppins, B. *et al.* (2006) Species richness of lichen functional groups in relation to land use intensity. *Lichenologist* 38(4), 331–353.

Tauleigne Gomes, C. and Lefèbvre, C. (2005) Natural hybridisation between two coastal endemic species of America (Plumbaginaceae) from Portugal. 1. Populational in situ investigations. *Plant Systematics and Evolution* 250, 215–230.

Tauleigne Gomes, C. and Lefèbvre, C. (2008) Natural hybridisation between two coastal endemic species of America (Plumbaginaceae) from Portugal. 2. Ecological investigations on a hybrid zone. *Plant Systematics and Evolution* 273, 225–236.

Tautz, D., Arctander, P., Minelli, A., Thomas, R.H. and Vogler, A.P. (2003) A plea for DNA taxonomy. *Trends in Ecology and Evolution* 18(2), 70–74.

Thuiller, W., Lavorel, S., Araujo, M.B., Sykes, M.T. and Prentice, I.C. (2005) Climate change threats to plant diversity in Europe. *PNAS* 102(23), 8245–8250.

Vieira, C., Séneca, A. and Sérgio, C. (2004) The bryoflora of Valongo. The refuge of common and rare species. *Boletin de la Sociedad de Biologia de Concepcion* 25, 1–15.

Villagra-Islas, P. (2011) Newer plant displays in botanical gardens: the role of design in environmental interpretation. *Landscape Research* 36(5), 573–597.

Wandersee, J.H. and Schussler, E.E. (1999) Preventing plant blindness. *The American Biology Teacher* 61, 82–86.

Wandersee, J.H. and Schussler, E.E. (2001) Toward a theory of plant blindness. *Plant Science Bulletin* 47, 2–9.

Wyse Jackson, P. and Sutherland, L.A. (2000) *International Agenda for Botanic Gardens in Conservation*. BGCI, Kew, UK.

2 New Horizons in Diversification of Temperate Fruit Crops

Mohammad Maqbool Mir[1*], Umar Iqbal[1], Sheikh Mehraj,[1] Shabir Ahmad Mir[2], Munib ur Rehman[1], Shafat Ahmad Banday[1] and Ghulam Hasan Rather[1]

[1]*Division of Fruit Science, Sher-e-Kashmir University of Agricultural Sciences and Technology, Srinagar, Jammu and Kashmir, India;* [2]*Department of Food Technology, Islamic University of Sciences & Technology, Pulwama, Jammu and Kashmir, India*

Abstract

Diversification of fruits refers to expansion in different kinds and varieties grown in a specific geographical region, area or zone with the aim of increasing productivity and marketability and showing a direct impact on the overall economy of the grower. Due to the increase in population and shrinkage in land holdings, it has become necessary to find a solution to this alarming situation. There is yet another important factor of paramount importance, i.e. the gradual change in the environment, which has markedly affected the potential productivity of temperate fruit crops. Cultivation of one or more varieties of a specific kind of fruit on commercial lines over a huge area has posed great challenges in combating pest and diseases. To improve fruit cultivation and meet international standards in terms of quality and production, major restructuring and planning is urgently required. Introduction of a new technology or concept into the existing system must be sustainable. In the present context, diversification, which is a necessary change in the conventional cropping system with a well-defined objective or purpose, is one of the possible solutions to the challenges enumerated above. The adverse impacts of mono-cropping, the balanced supply of diversified fruit crops to local as well as distant markets for fresh consumption, and ways to enhance the availability spectrum of temperate fruits into the market needs to be highlighted. The total area of major fruits in the world is around about 55 million ha with a production of 600 million t. Furthermore, India is the second largest producer of fruits after China with an area of 6.4 million ha and a production of 75 million t. These fruits are easy to grow and hardy in nature, producing a crop even under adverse soil and climatic conditions. Most of them are very rich sources of vitamins, minerals and other nutrients, such as carbohydrates, proteins and fats. This increase in production in India is accounted for by enhanced productivity owing to the adoption of different high-yielding varieties and improved cultural practices. Major temperate fruits include apple, pear, quince, peach, plum, apricot, cherry, nuts and some underutilized fruit crops.

2.1 Introduction

Generally agriculture as a whole and horticulture in particular are the backbone of the Indian economy and play a major role in the country's socio-economic development. 51% of the total geographical area is made up of arable land and about 100 million ha is potential irrigated area, the highest for any country in the world. Unfortunately, agriculture contribution towards the gross domestic product is below 22%.

India is endowed with a wide spectrum of agro-climatic conditions, and a large variety of fruit crops, ranging from tropical, subtropical, temperate and arid, are grown. In India, the area used for fruit crops is 6.98 million ha with a production of 81.82 million t, accounting for 12.56% of

*E-mail: mmaqboolmir@rediffmail.com

 Plant Biodiversity: Monitoring, Assessment and Conservation (eds A.A. Ansari, S.S. Gill, Z.K. Abbas and M. Naeem)

global production and ranking second-highest in the world (Anonymous, 2014b). The country has become self-sufficient in food grain production and is the second-largest producer of fruit and vegetables, but the per capita consumption per day is around 70 g for fruits and 140 g for vegetables, which is far below the national dietary requirement of 120 g and 280 g respectively (National Institute of Nutrition, Hyderabad, India). Currently, diversification in fruit crops, along with the introduction of up-to-date practices in the existing system, with a well-defined objective or purpose, are possible solutions to the major challenges (Gill, 2002). Diversification refers to expansion in the number of crops grown in a region with the objective of increasing overall productivity and marketability to bridge the gap between increasing demand and the supply of food (Small, 1999). In the state of Jammu and Kashmir, apple growing occupies (45.30%) of the total area, with production of (79.44%) out of a total area of 161,000 ha and production of 1.65 million t. Compare this with other fruit crops: pears are 4.16% of total area and 3.54% of production, apricots are 1.81% area and 0.82% production, cherries are 1.11% area and 0.64% production; other fruits (both fresh and dried) are viewed as of minor importance because of the dominance, even mono-cropping, of apples (Anonymous, 2014a). Moreover, the Delicious family of apples constitute nearly 90% of total apple production. Monoculture of a particular fruit crop or cultivar is more vulnerable to spread of diseases, insect pest and other natural calamities, thus risk-bearing capacity is minimal in a single fruit or a single variety of a particular fruit crop, e.g. pests and diseases such as apple scab, sanjo scale, woolly aphid, anar butterfly, etc. have destroyed crops in the valley. Monoculture leads to genetic degradation in the different cultivars, and poses a threat due to their susceptibility to various biotic stresses. Crop production is vital to any country, contributing to the domestic product. There are strong links with other sectors, i.e. livestock, industry, trade, etc., whose output is strongly influenced by the capacity and capability of the crop sector. In a country such as India, there is a need for improvement in food security, income growth, employment generation, water and land resources, sustainable agricultural development and other environment cum ecological factors. So, in this case, diversification of crops could be an important tool, and this has been envisaged as a movement towards improving stable productivity for making Indian agriculture more competitive in international markets.

However, even though diversification of crops is a crucial tool for economic development, the capability of any country to diversify for attaining various goals will largely depend upon the potential for diversification and the reactions of farmers to these activities. Future opportunities that would benefit from diversification are changes in government policy, patterns of demand, improvements in irrigation, and other infrastructure changes due to technological breakthroughs. In order to avoid a glut of fruit in the market, extending the harvesting period and efficient use of the crop could be another strategy for crop production planning. This staggering of fruit production would be achieved by utilizing fruit crops/cultivars with different maturity periods on the same site. About 50% of the area under fruit cultivation would thus be devoted to the production of early and late maturing cultivars, so that the fruit marketing season is extended from the present two months to about five or six months. The time to reach fruit maturity and ripening for different cultivars can be best utilized for extending the apple and pear season by varietal diversification and extending the fruit season by fruit diversification to regulate the market. The present monoculture of fruits in an orchard has made the horticulture/fruit industry more subject to the hazards of nature, pest and diseases. With this in mind, it is always better to aim for diversification in fruit production for sustained horticulture development and market stability by growing disease and pest resistance with high-quality crops/cultivars. Therefore, by growing different varieties resistant to epidemics, we will be able to stabilize our fruit market as well as sustain fruit productivity.

A sizeable area in our country is either lying barren or does not produce satisfactory yields, for one reason or another. Such types of soil are usually referred to as problematic soils or waste lands. Since India's productive land is shrinking day by day due to increasing population pressure, this increases the demand for fruit. Thus, fruit cultivation on waste or barren land is one way to increase fruit production in countries like India; this can be achieved by growing different fruits suitable to these areas by diversification.

The new orientation for future planning of scientific production requires a detailed survey of the area for assessing the potential of the particular crop on the basis of micro-climatic condition and delineation of the area for the production of different fruit varieties for different uses. Similarly, it should be possible to produce delineation of the area for growing fruits for fresh, domestic and export purposes and also meet the needs of the processing industry. So from the above factors, it is clear that in Jammu and Kashmir, diversification would be an effective tool for improving the economy of the rural people in particular and urban in general.

The total area of Jammu and Kashmir state is 22.2236 million ha, including land occupied by Pakistan and China. The net geographical area of the state occupied by India is 10.1387 million ha, 19.95% of which is forest area ranging from 909 to 6908 m above mean sea level. More than 70% of the population of the state gets their livelihood through agriculture and its allied sectors. There are 700,000 families, comprising 3.5 million people, who are directly or indirectly associated with horticulture, particularly fruit crops. So there is a great demand for diversification of minor/underutilized fruit crops to boost the economy for those living on the edge of poverty, especially in rural areas (Anonymous, 2014b).

2.2 Need for Diversification

Indian agriculture in general and horticulture in particular are fraught with risk and uncertainty, since more than two-thirds of cultivable land is dependent on the monsoon rains. For farmers the outcome of their efforts is uncertain due to weather and market-induced risk. Hence, it is very important that farmers should try to change their crops or cropping system using the available resources, by changing and modifying the sequence, trend and time options of crop/cropping operations and at the same time taking care to maintain an ecological balance and conserving and enhancing natural resources. This trend to shift the system from less profitable and sustainable crops or cropping systems to more profitable and sustainable ones is the main theme of crop diversification. For temperate fruits, the introduction of non-traditional fruit crops, new varieties with varying chilling hour requirements, or lesser-known fruits, has become an essential strategy. It can increase the farmer's income, minimize risk due to crop failures associated with mono-cropping, extend the availability of different fruits for a longer period of time, thus sustaining our fruit market, and above all, appeal to an international market and bring new money into the country. In temperate fruits, there are a number of varieties within a particular fruit crop that could be best utilized in the diversification of temperate horticulture, particularly fruits.

2.3 Diversification and Prolonged Fruit Season

One of the most important advantages of diversification is the possibility of extending the availability of different types of fruits over a longer span of time, thus indirectly benefiting the farming economy and other related sectors (Table 2.1).

2.4 Underutilized/Minor Fruits

Researchers and commercial companies are working on minor fruit crops and often present them as 'new crops' (Vietmeyer, 1990). The decreasing trend of these crops of new generations due to loss of local knowledge and other factors leads to a misleading image. A crop can be completely new to an area because it has been introduced there recently from a distant place or country, as in the case of the kiwi fruit (Ferguson, 1999). Many

Table 2.1. Availability of fruits in different seasons.

Kind of fruit	Time of availability
Strawberry	April
Cherry, Apricot, Nectarine	May–June
Apricot, Plum, Nectarine, Peach, Pear	June–July
Plum, Pear, Apple, Fig, Loquat	July–August
Apple, Pear	August–October
Persimmon, Hazelnut, Apple	September
Pomegranate, Quince, Apple, Walnut, Hazelnut	September–October
Kiwi fruit, Pecan nut, Chestnut	October–November

underutilized fruit crops that were once grown judiciously today fall into disuse, due to cultural, economic and other factors. Farmers are cultivating these types of crops less frequently because they are getting more profitable income from other crops in the same environment with less input. The declining popularity of these crops will have an adverse impact on genetic conservation, utilization and may prevent the use of important elements in the improvement of the crop programme (Padulosi *et al.*, 2002). These underutilized crop species contribute a lot in respect of food security and household food to small or poor farmers and help to increase their income, because in such areas the cultivation of major crops is economically marginal.

A global plan of action for conservation and sustainable utilization of plant genetic resources has found that one of the priorities is promotion and commercialization of underutilized crops (FAO, 1996b). The major goals identified by different governments were to provide more food security, to highlight and promote the conservation and management of these crop species for diversification, and to improve the market demand.

2.5 Underutilized Fruits – What are they?

Within the fruit crop industry perhaps no term has caused more controversy than the word 'underutilized'. The word is generally used for crops whose ability has not been fully understood due to geographical, social and economical factors. It is not surprising, when such things are being debated in different national or international forums, that there is a call for clarification over the exact meaning of terminology. With respect to geographical distribution, a crop could be underutilized in one place but fully utilized elsewhere.

Regarding socio-economic implications, many species represent an important component of the daily diet of millions of people, e.g. leafy vegetables in sub-Saharan Africa, but the poor return makes them underutilized economically (Guarino, 1997). The technological marketing system in advanced countries makes such species more profitable than in developing countries, where the same crop is poorly managed and marketed. Such is the case for crops like chestnut, fig, persimmon, loquat, quince and pomegranate in the Kashmir Valley, while these crops have more economical value in other parts of the world. In the state of Jammu and Kashmir, these fruits are underutilized because of poor management and market interventions.

A similar case is that of hulled wheat (*Triticum monococcum*) and *T. spelta*, an important crop in European countries, particularly Italy where *in situ* and *ex situ* conservation are being supported with considerable effort (Padulosi *et al.*, 1996). Due to confusion among different workers between the terms 'underutilized' and 'neglected', definitions are being reported by the International Plant Genetic Resources Institute (IPGRI) for these two categories of crops (Eyzaguirre *et al.*, 1999). There are some neglected crops, grown and distributed by traditional farmers, that are an important crop to the local community. Some species of crops may be distributed throughout the world but tend to occupy the niche of centre of excellence in a particular region. However, these crops continue to be maintained by different practices and remain neglected by researchers and policy makers due to the lack of consumer demand.

2.6 Challenges in the Promotion of Underutilized Species

Though it is clear that research funding for these minor species around the world is inadequate, their importance should not be underestimated, nor the fact that these crops feed a large proportion of people. During a participatory conference in 1998 in Syria on priority setting for underutilized and neglected plant species of the Mediterranean region (Padulosi, 1999a), ten constraints for the publicity of these underutilized species were identified, debated and discussed by eminent policy makers and scholars (Lazaroff, 1989; von Maydell, 1989; FAO, 1996b; Maxted *et al.*, 1997; Monti, 1997; Heywood, 1999; Padulosi, 1999a).

With special reference to agriculture, there must be a debate over the issue of globalization and whether or not we should promote underutilized species. A number of world-class policy makers and researchers see no future for these crops. There should be a definitive argument for securing a resource base of these crops, particularly in developing countries, in order to maintain the safety net of diversified food (Eyzaguirre *et al.*, 1999). Generally, diversification in agriculture is important for those fragile groups who are not able to afford large commodities and to whom the more diversified is the portfolio, the greater will be the self-sustainability.

In simple language, minor crop species play a very important role for the rural community by reducing poverty and allowing the pursuit of resource-based rather than commodity-based improvements (Burgess, 1994; Blench, 1997). In addition, it is not only poor people who benefit from underutilized species. All people benefit in terms of diet, income, better maintenance of agro-ecosystems and use of marginal lands for cultural identity (Padulosi, 1999b).

2.7 Securing the Genetic Base of Minor Fruits

Securing the resource base of underutilized species is a vital component of the promotion line and is central to the concerns of the IPGRI. The provision of diversification in the crop improvement programme was a main element of the green revolution (CGIAR, 1994). Key to this is full recognition of these crops and how to safeguard this genetic germplasm. Establishment of germplasm conservation for underutilized species has been advocated in the Global Plan of Action of FAO (FAO, 1996a) and by several international organizations. Another crucial aspect for promotion of these minor species is obtaining more return on the existing crop. For instance, cape gooseberry (physalis), strawberry and some medicinal and aromatic plants can be valuable in intercropping systems, particularly during the nonbearing stage. There is also a need to gather information on distribution and traditional knowledge of these crops, so that data can be generated for future access by policy makers and others.

2.8 The Importance of Collaboration

Without doubt, the work on underutilized fruit species is the most challenging task of plant genetic resources since the early 1970s, a period that witnessed a race to rescue major crops (Pistorius, 1997). Underutilized species will never command such attention as major crops did, and thus need a specific approach for their promotion. This strategy is known as 'filiere', a French word meaning to make a bridge between initiators, leading to good policy decisions for improvement of the product.

Thus filiere would bring greater participation of local players for addressing needs efficiently. There should be greater participation of stakeholders to make right decisions at the right time, so that the marketing and processing sectors will be enhanced (Padulosi, 1999c). In the context of Jammu and Kashmir, the significant variations in the agro-climatic conditions in different regions do not reveal much about the regional pattern of crop diversification. For example, there are 22 districts in Jammu and Kashmir state, of which ten districts of Kashmir province are typically temperate, while ten districts of Jammu province are subtropical to intermediate, and the remaining two districts of Ladakh region are purely cold arid land. The temperate region of the state has the potential to grow temperate fruits and is actually experiencing changes in the cropping pattern. As far as the subtropical part is concerned, there are fruits which are grown commercially, i.e. strawberry, peaches, olives, etc. The various types of minor/underutilized fruits suitable for growing under changing climatic condition of the Kashmir zone include pomegranate, kiwi, quince, grapes, hazelnut, chestnut, pistachio nut, pecan nut, persimmon, fig, loquat, etc. The area of Jammu and Kashmir state used for fruit growing is 325,133 ha, of which 43% is used for apples. Table 2.2 and Table 2.3 show the area and production of different fruits grown in Jammu and Kashmir state – there is an urgent need

Table 2.2. Area (ha) under minor fruit crops in Jammu and Kashmir.

No.	Kind of fruit	Total Kashmir	Total Jammu	Total J&K
1	Grapes	275	152	427
2	Olive	101	456	557
3	Kiwi	01	19	20
4	Pomegranate	111	0	111
5	Strawberry	51	0	51
6	Quince	180	0	180
7	Other fresh	1316	6568	7884
8	Pecan nut	0	619	619
9	Other dry	172	24	196

Table 2.3. Production (MT) of minor fruits in Jammu and Kashmir (source: Anonymous, 2014a).

No.	Kind of fruit	Total Kashmir	Total Jammu	Total J&K
1	Grapes	491	251	742
2	Olive	2	43	45
3	Kiwi	9	2	11
4	Pomegranate	341	0	341
5	Strawberry	341	0	341
6	Quince	951	0	951
7	Other fresh	6789	10486	17275
8	Pecan nut	0	12	12
9	Other dry	315	16	331

to improve the area that grows these crops so the economy of poor people will be improved.

2.9 Achievements

In an effort to diversify the crops and cropping pattern, a number of innovations have been made among and within the temperate fruits.

2.10 Pome Fruits

2.10.1 Apple

Apple (*Malus domestica*) is the main fruit crop of the temperate regions and in India the leading apple-producing states are Jammu and Kashmir and Himachal Pradesh, though apples are also grown in the hills of Uttar Pradesh and parts of North East India. For years, Red Delicious and Starking Delicious cultivars of apple, which are fairly sensitive to weather fluctuations, particularly during flowering time, have been the main commercial cultivars. Monoculture, as well as other factors, have led to a slow decline in the apple productivity. However, with the introduction of many other varieties whose chilling requirements vary from low to very high, apple cultivation has extended to even non-traditional areas. The availability period has also extended tremendously to the benefit of the growers, by way of earning better economic returns. Some of the cultivars that are gaining popularity among the growers are Royal Delicious, American Apirouge, Vance Delicious, Oregon Spur, Scarlet Spur, Red Chief, Gala Must, Early Red One, Top Red, Super Chief, Scarlet Spur, Ginger Gold, Gala Group, Fuji Group, Golden Spur, Red Gold, Lal Ambri, Sunhari, Shalimar-1, Shalimar-2, Granny Smith, Red Fuji, Starkrimson, Red Velox, etc. Among the low chilling cultivars, Anna and Dorset Golden appear to have a good potential for cultivation in the warmer areas. Under the mid-hill zone of Himachal Pradesh, particularly some areas of Solan district, Gale Gala, Early Red One, Oregon Spur and Scarlet Spur have been observed to have great potential as remunerative crops and a diversification option for the area.

2.10.2 Pear

Pear (*Pyrus communis*) is the second most important fruit crop among the pome fruits. It has a wide adaptability and can be a very good option for diversification of fruit crops, as it can be easily and profitably grown in areas with varying climatic conditions. Among the early season cultivars, China Pear, Beurre de Amanlis, Celmar de Ete, Laxton's Superb and Flemish Beauty are popular. Among the mid-season cultivars, Clapp's Favourite, Max Red Bartlett, Red Bartlett, Starkrimson, William Bartlett and, among the late season, Vicar of Winkfield, Fertility, Leconte, Anjou, Conference, Doyenne du Comice, Beurre Hardy, Winter Nellis and Chinese Sandy Pear are commonly grown. In low hill country, Punjab Nakh, Pathar Nakh, Punjab Nectar, Punjab Beauty, Baggugosha, Nijisseki and Punjab Soft are promising cultivars.

2.10.3 Quince

The genus consists of a single species, *Cydonia oblonga Mill*, native to warmer regions of southeastern Europe and Asia Minor. It is a small deciduous tree that bears a pome fruit and grows as a single or multi-stemmed shrub. The fruits are either apple- or pear-shaped. The immature fruit is green with dense grey-white pubescence, which normally changes when the fruit reach maturity from green to yellow colour. Quince is a deciduous tree known to the countries around the Mediterranean since antiquity. Nowadays, quince is used as an ornamental plant, and as a root stock for pear trees, with its fruit being used mainly for production of jam and sweets rather than for raw consumption. In the Kashmir Valley, there is a scattered plantation of the trees, plus some established orchards belonging to the government and progressive farmers where the land is wet. Some indigenous superior selections (round and rectangular) are spread throughout the valley.

2.10.4 Loquat

The loquat (*Eriobotrya japonica L.*) is one of the pome fruits and is closely related to apple, pear and quince. It can be successfully grown from subtropical to temperate regions. In India, it is commercially cultivated in Uttar Pradesh, Uttaranchal, Punjab and Himachal Pradesh and not much work has been done on its cultivation and standardization of varieties. In the state of Jammu and Kashmir, plants are found scattered in the valley, particularly in the famous Mughal gardens and in the backyards of kitchen gardens.

2.11 Health Benefits of Pome Fruits

Pome fruits are very popular to eat, because of their taste, crispness, high vitamin C content, fibre and ability to reduce cholesterol levels. Many dentists advise patients to eat an apple after every meal if they are unable to brush their teeth. There is a large demand for apple products in European countries in the form of slices, pie fillings, dried apples, sauces, juice and cider. Consuming pears may help in weight loss and reduce the rate of hypertension, cancer, diabetes and heart disease in accordance with an overall healthy diet. The cooling effect of pear is excellent in relieving fever. The possible health benefits from quince fruit are due to various antioxidants and other important constituents. Vitamin C is an important antioxidant and scavenges the free radicals produced in the body. It improves healing of wounds, and also aids gum and tooth health, the immune system, lung health, reduces the adverse effects of ageing, lowers the risk of some cancers, protects against colds and certain other infections. It is an important source of dietary fibre, which has many advantages as far health is concerned, normalizing blood sugar levels and cholesterol levels, helping to prevent constipation and haemorrhoids, and supporting weight loss. Some recent investigation shows that quince leaf has anticancer properties (Carvalho *et al.*, 2010; Osman *et al.*, 2010). The quince seed mucilage in 10–20% concentration is good for wound healing. It is also used against sore throat, cough, pneumonia and other intestinal and lung diseases. It is also a good source of copper and helps to reduce tissue damage; it supports bone health, thyroid gland function and nerve health, increases good cholesterol (HDL), decreases bad cholesterol (LDL), and reduces fatigue and weakness (Sajid *et al.*, 2015). The leaves contain compounds with antioxidant, antimicrobial and anticancerous attributes that have been a subject of recent research for pharmaceutical and medical uses as well as for food preservatives. Loquat fruit is a delicious and refreshing fruit and has more than 60% pulp. The fruit contains good amounts of carbohydrates, proteins and is generally used for dessert purposes as well as for jelly, jam, juice, canning and chutneys. It is considered to be a sedative and is used in allaying vomiting and thirst; an infusion of leaves is used against diarrhoea.

2.12 Stone Fruits

Among the stone fruits, peach, plum, apricot and cherry are commonly grown in the temperate region. Apricot, peach and plum were once very remunerative crops of the mid-to-high hills of Jammu and Kashmir and they heralded a revolution in the stone fruit industry. But again due to the monoculture of Elberta, Quetta and some indigenous selections of peach, Santa Rosa, Burbank and Wickson varieties of plum, and some varieties of apricot, there was a glut in the market and subsequently poor returns. The area under these crops started to decline, reaching a stage where growers preferred to cut down their trees to open up more area for the expansion of vegetable crop cultivation. Lack of expertise in crop management threatened the fate of the stone fruit industry in the area. However, after the introduction of many new cultivars and improvements in management, the growers are slowly converting their land back into fruit growing.

2.12.1 Peach

Peach (*Prunus persica*) requires the warmest climate among the temperate fruits and is generally grown in the low and mid hills (1000–2000 m above mean sea level), except for cultivars belonging to the low chilling group, which can be grown under the subtropical conditions. Apart from July Elberta, which was introduced earlier, some of the cultivars introduced later on and found to be highly suitable are as follows: World's Earliest, Red June, Andross, Pratap, Early Grande, Quetta, Shan-e-Punjab, Redhaven, Sun Crest, Glo Haven, Early Elberta, Miss Italia, Lucia, etc.

2.12.2 Nectarine

Nectarine (*Prunus persica* var. nectarine), which was earlier known as fuzzless peach, is a new introduction in the state and has been found to be very remunerative and a good option for crop diversification. The variety May Fire, which ripens in June, can fetch a good price in the market, being an early crop and having a very attractive deep red colour. The only problem, however, is that it tends to overcrop, leading to small-sized fruits. But by following a thinning schedule, this problem can be easily solved and a good attractive size can be obtained. Silver King and Snow Queen, which ripen in mid-to-late June, are among the best cultivars of nectarine, with an attractive colour, large-sized fruits and early cropping compared to the traditional one, July Elberta. Independence and Red Gold are the other cultivars that have been tested and found suitable as varietal diversification options.

2.12.3 Plum

Plum (*Prunus salicina*) grown in the temperate region of India mostly belong to the Japanese group, requiring more chilling hours than peach but fewer than the plum belonging to the European group. In the higher ranges, however, European plum, particularly prunes, are being grown successfully, since they have more sugar content and are mostly sold as dried plums. A new and good introduction, Black Amber, appears to be more suitable for high hills. Some of the cultivars found suitable for the mid hills are Red Beaut, Beauty, New Plum, Formosa, Sharp's Early and Sutlej Purple as early cultivars; Shiro, Duarte, Santa Rosa, Satsuma, Burbank and Frontier as mid-season cultivars and Mariposa, Stanley Prune, Grande Duke as late cultivars. Sutlej Purple, Alucha Purple, Titron, Kala Amritsari can also be successfully grown in the low hills and valley areas.

2.12.4 Apricot

Apricot (*Prunus armeniaca*) resembles Japanese plums in its climatic requirement except for the white-fleshed, sweet-kernel apricots, which require a colder climate and are more suitable for the high hills, where they are mostly grown for their dried form. Apart from Kaisha, Shakarpara, Nugget, etc., two new introductions from Armenia – Sateni and Yeravani – can be successfully grown for their dried forms in the high hills. In the mid hills, apart from New Castle, some of the other cultivars for diversification include Haricot, Turkey, Early Shipley, Charmagz, Gilgati Sweet, Amba, Kaisha, Australian Sweet and Quetta.

2.12.5 Almond

Almond (*Prunus dulcis*) can be grown successfully in the dry temperate zone of Jammu and Kashmir and also in some parts of Himachal Pradesh. However, it can also be grown in the mid-hill sub-humid zone of Himachal Pradesh, but profitably only as green almond. Some of the cultivars suitable for the dry temperate zone are Ne-Plus-Ultra, Texas, Thin Shelled; for High and mid hills: Merced, Non Pariel and for the low hills and valley areas: Drake, Merced, Pranyaj, Primorskij, Jordanolo, Makhdoom, Waris, Shalimar, Parbat, etc.

2.12.6 Cherry

In India, Jammu and Kashmir state has the monopoly on cherry cultivation and is the second fruit crop onto the market after strawberry. The major fruit-growing belts of the state are higher areas of Srinagar, Ganderbal and Baramulla (Mir *et al.*, 2009). Cherry (*Prunus avium*) requires a climatic condition quite different from the other stone fruits and that is why its successful cultivation is restricted to a few areas only. Some of the promising cultivars are Bing, Napolean, Tartarian, Sam, Sue, Stella, Van, Lambert, Black Republican. Early season: Guigne Pourpera Precece, Guigne Noir Gross Lucenta, Black Heart; mid season: Guigne Noir Hative, Black Heart; and late season: Bigarreau Napoleon, Bigarreau Noir Grossa, Stella. The sour and duke cherries are less delicate than the sweet types. Cherry is a very attractive and delicious fruit, being used in the ripe form and for culinary purposes. In Europe, a wine called Kirschwasser is distilled from the fermented fruit pulp of cherries. Nowadays the demand for canned cherries has also increased enormously because of limited availability of fruit in the market.

2.13 Health Benefits of Stone Fruits

The stone fruits are juicy and nutritious and constitute an important source of minerals, sugars, vitamins and organic acids, in addition to fat and carbohydrates. They are popular in the market as fresh as well as canned products, which makes them in demand in the international market. Sweet cherry raisins are a popular product in North America. They are helpful in reducing cardiovascular strokes, the adverse effects of ageing, muscle pain, gout, strokes and colon cancer. In western countries, a good-quality brandy and wine are prepared from plums, having the high sugar content with low sorbitol necessary for good fermentation. A common therapeutic use of peach is for the ailment of urinary blood accumulation and indigestion. The flowers of the peach have also sedative, antispasmodic, anthelmintic and laxative properties (Joshi and Bhutani, 1995).

2.14 Nut Crops

2.14.1 Walnut

Walnut (*Juglans regia*) is the most commonly grown nut crop in the hilly region of Jammu and Kashmir, Himachal Pradesh and Uttarakhand. It is grown as a boundary tree or as a scattered plantation because of its huge size. Being rich in omega 3-fatty acids, the nut is popular among the health-conscious. Promising cultivars are Gobind, Hamdan, Suliman, Eureka, Pratap, Wilson Wonder, Franquette, Lake English, and some SKAUW selections of the agricultural university.

2.14.2 Pecan nut

Pecan nut (*Carya illinoensis*) is another nut crop which is rich in omega 3-fatty acids and it has a good scope for being grown in the lower elevations (500–2000 m above mean sea level) and areas considered too hot for walnut. In the USA, pecan is considered the 'queen of nuts' because of its value both as wild and as a cultivated nut of acceptable quality (Woodroof, 1979). Besides the USA, the worldwide distribution of pecan is confined to Australia, Canada, Egypt, India, Israel, Mexico, Morocco, Peru, Turkey and South Africa. Pecan nut is one of the most important nuts grown in India. In India it is grown in Jammu and Kashmir, Himachal Pradesh, Uttaranchal, and Nilgiri Hills. In Himachal Pradesh, pecan plantations remain confined to Kangra, Mandi, Kullu and Solan districts which constitute approximately 700 ha (Mehta and Tahkur, 2004). Recently the area under pecan nut has been increased due to its high economic returns and adaptation to the intermediate zone of Jammu (Rajouri and Poonch). Jammu and Kashmir and Himachal Pradesh are the principal pecan-producing states in India. Despite many constraints, the agro-climatic conditions of Jammu and Kashmir are ideally suited to pecan growth and production. It can also be easily grown in drought-prone areas due to its long tap-root system. Some of the promising cultivars are Mahan, Nellis, Burkett, Colby, Western Schley, Kiowa, Kanza, Chocktaw, Cheyenne and Pawnee.

2.14.3 Hazelnut

The hazelnuts belong to the family Corylaceae. The species are found growing in wild forms in temperate zones extending from Japan, China, Turkey, India, and Europe to North America. The nuts are small-sized and have a hard shell. Turkey is the world's leading producer of filberts/hazelnuts and other growers are Italy, Spain and the USA (Mitra and Bose, 1991). In India the hazelnut exists in the Western Himalayas, particularly Jammu and Kashmir, in its wild form. In the Kashmir Valley, hazelnut diversity is available in Dachigam National Park and other higher elevations of the state in the wild form. Trees are cultivated for nuts, timber and for ornamental purposes (K. Ahmad, Documentation and evaluation of hazelnut germplasm in Dachigam area of Kashmir. Division of Pomology, Sher-e-Kashmir University of Agricultural Sciences and Technology of Kashmir, Kashmir, India, 2009, unpublished thesis). Its wild type is found growing in the Pangi area of Chamba, in the high hills of Kullu and Shimla District of Himachal Pradesh and some parts of Meghalaya. Some of the improved cultivars introduced from Italy are Tonda, Gentile Della Langhe, Tonda Romana Tonda Giffoni and some superior indigenous selections.

2.14.4 Chestnut

Chestnuts (*Castanea* sp.) belonging to the family Fagaceae have been an important item of the human diet in the northern hemisphere since time immemorial. The remains of carbonized plants recovered from the sites destroyed in AD 79 furnish evidence of many staple food plants used then and chestnut is one of them. It is native to Western Asia, Europe, and Northern America. It is mainly grown in China, USA, USSR, Turkey, Thailand, Japan and Portugal. According to FAO statistics, world chestnut production is estimated to be 2.01 million tons distributed over an area of 3.8 million ha (Anonymous, 2013). China is the world's largest producer of chestnut with an annual production of around 1.7 million tons, which represents approximately 70% of world production. Europe is responsible for 17% of worldwide chestnut production, especially in Turkey (5%), Italy (4%) and Portugal (3%). In India, it is grown throughout the Himalayas up to Assam and Meghalaya at an altitude of 2000–3000 m above mean sea level (amsl) (K. Amardeep, Survey and selection of promising chestnuts in district Srinagar. Sher-e-Kashmir University of Agriculture Science and Technology of Kashmir, Shalimar Srinagar, India, 2008, unpublished thesis). The important chestnut-producing states in India are Jammu and Kashmir, Himachal Pradesh and some pockets of the north-western Uttar Pradesh. In Jammu and Kashmir the profile of the fruit industry has always been dominated by apple. However, a few chestnut orchards were established by the State Horticulture production department in a few areas of Srinagar (Harwan), Anantnag and Budgam for public interest (S. Hussain, Standardization of propagation techniques in chestnut (*Castanea sativa* M.). Sher-e-Kashmir University of Agriculture, Science and Technology of Kashmir, Shalimar, Srinagar, Kashmir, 2014, unpublished thesis). In the past there were plantations, almost forests, of chestnuts stretching from the Bay of Biscay in France to Switzerland, and poor people consumed these nuts as their staple food during autumn and winter. Chestnuts have high nutritive value and many uses. Carbohydrates comprise about 85% of the dry matter of fresh chestnuts, predominantly starch.

2.15 Health Benefits of Nut Crops

The consumption of nuts is increasing day by day due to being labelled as healthy foods by modern science. Some nuts are highly rich in oil and this could be considered a drawback but for the fact more than 70% of these fatty acids are polyunsaturated, out of which 57% are linolenic and 13% are linoleic, which gives the nuts valuable properties. These characteristics decrease cholesterol levels and lower the risk of cardiovascular diseases. Pecan flour is reported to have high oil content. Unlike other nuts, chestnuts are rich in carbohydrates and moderate in protein content. Furthermore, when nuts are consumed with carbohydrate-rich foods, they blunt the postprandial glycaemic response of the carbohydrate and are useful against type 2 diabetes (Kendall *et al.*, 2010). The positive health benefits of tree nuts have been reasonably attributed to their composition of vitamins, minerals, etc. Polyphenols and physterols have been found to be effective in reducing the number of chronic diseases, including anti-inflammatory activity, serum low-density lipoprotein (LDL) and colon, breast and prostate cancers (Amaral *et al.*, 2003; Chen and Blumberg, 2008). Diets high in monounsaturated fatty acids have a positive effect on the ratio of total cholesterol to high-density lipoprotein (HDL) cholesterol (Maguire *et al.*, 2004). Sugars such as raffinose, stachyose and sucrose are present in chestnuts. Nuts also contain squalene and tocopherols. Low amounts of tocopherol obtained from nuts are beneficial for heart disease (Ryan *et al.*, 2006; Kornsteiner *et al.*, 2006). Nuts are also used in manufacturing of different confectionery items, syrup, soup and chutneys.

2.16 Other Fruits

2.16.1 Kiwi fruit

The kiwi fruit (*Actinida deliciosa* L.) belongs to family Actinidiacae and is native to China. It is known as 'China's miracle fruit' and 'The horticultural wonder of New Zealand' and has gained enormous popularity in the past two decades in many countries of the world. In fact, no other fruit has attracted so much attention in such a short period in the history of commercial fruit production (Mitra and Bose, 1991). Up to 1960, the fruit was known as Chinese gooseberry in the majority of countries including New Zealand. However, in order to promote its sale, it was named kiwi fruit or kiwi berry by an importer in the USA because of its brownish colour and hairy appearance, which was thought to resemble the kiwi, the national symbol of New Zealand where the fruit was grown

(Ferguson, 1984). It is a recent introduction in India from New Zealand. It is being grown commercially in New Zealand, Italy, USA, Japan, Australia, France, Chile and Spain. In India, the kiwi is commercially cultivated in Jammu and Kashmir, Himachal Pradesh and some mid hills of Uttar Pradesh, Sikkim, Meghalaya, Arunachal Pradesh and Nilgiri areas. In the Kashmir Valley, fruit orchards have been established at Parihaspora and Zangam nursery (Pattan) in the district of Baramulla and some areas of Budgam, Srinagar and Ramban districts. Kiwi is mainly grown for its fruits, which are highly nutritious. Kiwi fruit is one of the most recent introductions among the temperate fruit plants. It is being grown profitably in the mid hills, low hills and valley area of Jammu and Kashmir, Himachal Pradesh, hills of Uttar Pradesh and parts of the North Eastern States of India. The plant resembles the grape vine. It is a deciduous climber, having male and female flowers borne on separate plants. For effective pollination and fruit set, it is very important to have a male:female ratio of 1:9. It is astringent when unripe, hence it can be consumed only when softened. It is highly nutritious, rich in antioxidants and having anti-cancerous properties. Today, it is a much sought-after fruit, especially by the health-conscious. Some of the cultivars recommended for cultivation are Allison, Abbott, Bruno, Hayward, Monty, Allison (male), Tomuri (male).

2.16.2 Prospects of kiwi fruit production

There are several factors that make the cultivation of kiwi fruit in hilly areas of the North Eastern states particularly productive and profitable. The Jammu and Kashmir state has diverse climatic conditions in the foothill, mid-hill and higher hills of this subtropical region. Kiwi fruit can grow well wherever, the climate is warm and humid. Thus, kiwi fruit cultivation can successfully be adapted in all the North Eastern states including Assam. Kiwi fruit requires 700–800 chilling hours below 7°C and the summer temperature should not go above 35°C (Singh *et al.*, 2008). More than 90% of the fruit is edible, as in fact, except for the skin, is the whole fruit, along with the seed. It contains a good amount of fibre and is a rich source of sugars and several minerals such as phosphorus, potassium and calcium. It is a rich source of vitamins C and E, and low in calories. Fresh kiwi is a very rich source of potassium, an important component of cell and body fluids, that helps control heart rate and blood pressure by countering the effects of sodium (Berkow and Barnard, 2005). The ripe fruit has an excellent aroma, a delicate flavour and a good refreshing quality.

2.16.3 Persimmon

Persimmon (*Diospyros kaki*) is a deciduous tree that is grown successfully in kitchen gardens in some parts of Jammu and Kashmir, particularly in temperate areas at an elevation ranging from 1500 to 1800 m above mean sea level (amsl). There are the astringent types of persimmon, as well as types that can be eaten raw. Some of the astringent cultivars commonly grown and that can be eaten only when ripened and softened are Hachiya, Hyakume and Jiro, while Fuyu is a non-astringent, raw-eating cultivar, which can be eaten when hard. All these cultivars are usually ready for harvest by September. Most of the indigenous selections from the areas of Ganderbal and Srinagar in the Kashmir valley have been found to be good pollinators and have been recommended for effective pollination, fruit set and fruit retention in persimmon.

2.16.4 Health benefits of persimmon

The fruit is a rich source of vitamin A and C and is eaten fresh, stewed or cooked as jam. It can be used in baked goods, puddings and other desserts. The fruit has a high glucose and low protein content. An Israeli-bred cultivar, Sharon, has been found to reduce heart attacks if used regularly. The fruit has a high tannin content, which makes the fruit astringent, but levels decrease after ripening. A decoction of the fruit stem and calyx can be taken to relieve hiccups, coughs and breathing difficulties.

2.16.5 Pomegranate

Pomegranate (*Punica granatum* L.) is one of the options for diversification and is becoming popular in Jammu and Kashmir state because of its hardy nature. It is a well-known fruit of tropical and subtropical zones of the world. Some of the cultivars performing better are Dholka, Bedana, Kandhari, G-137 and some promising selections made by a variety of universities and institutes. Pomegranate behaves as a deciduous species in a temperate climate but in subtropical and tropical climates, it behaves as an evergreen or partially deciduous. The central area of Jammu is rich in wild pomegranate plantations and the fruit is processed for its seeds (anardana) (Shant and Sapru, 1993). In India, it is considered a crop of the arid and semi-arid regions, because it withstands different soil and climatic stresses (Kaulgud, 2001). In the Kashmir Valley, the fruit was grown on a large scale in the early 20th century in the district of Srinagar and adjoining areas of the district of Ganderbal, but the plantation was a local one which succumbed to pest attack, while a second one was developed for residential and commercial establishments (Mir *et al.*, 2007).

2.16.6 Health benefits of pomegranate

Pomegranate has been used for centuries to confer health benefits in a number of inflammatory diseases. Fruits are widely consumed either fresh or as a beverage. Pomegranate is an important source of anthocyanins, the hydrolysable tannins punicalagin and punicalin, ellagic and gallic acids (Lansky and Newman, 2007; Turk *et al.*, 2008). The main polyphenols in pomegranate that have proven to have antioxidant and anti-inflammatory bioactivities include the tannins and anthocyanins, which are concentrated in the peel and pith of the fruit (Basu and Penugonda, 2009). Pomegranate rind is tannin-rich and its microbial activity has been observed against pathogenic bacteria (Yehia *et al.*, 2011). This bacterium infects the intestinal tract and causes infections of various organs (Marchi *et al.*, 2015). The juice contains more than 15% sugar, with citric and malic acids (Ulrich, 1970; Whiting, 1970). The fruit juice also lowers cholesterol level in diabetic patients and reduces cancers due to the presence of flavonoids, anthocyanins, acids, etc. (Esmaillzadeh *et al.*, 2006).

2.16.7 Strawberry

Among the small fruits, strawberry (*Fragaria ananasa*) is most commonly grown in Jammu and Kashmir, Himachal Pradesh and some areas of Uttrakhand and Uttar Pradesh. It is a low-volume and high-value cash crop and can also be grown in kitchen gardens for home consumption. It is harvested quite early in April and May and is among the first fruits to come to the market. But it is highly perishable and needs to be transported to the market as early as possible for better returns. Hence, another good option is to produce planting material (runners), as very healthy runners can easily be produced in Jammu and Kashmir because of the climatic conditions prevailing there. Moreover, there is an ever-increasing demand for planting material from the neighbouring states, where it is almost impossible to raise healthy and economical planting material, so it can be a very viable venture. Some of the recommended and promising cultivars are Chandler, Senga Sengana, Tioga, Torrey, Selva, Pajaro, Belrubi, Camarosa, Sweet Charlie and Redcoat.

2.16.8 Health benefits of strawberry

The strawberry is a delicious fruit and is a rich source of minerals and vitamins. The crop is in great demand in the processing industry, and has excellent aroma properties. Due to its high content of antioxidants, polyphenols and ellagic acid, it slows down the adverse effects of ageing, prevents urinary tract infections and helps to reduce blood pressure.

2.16.9 Grape

The grape (*Vitis vinfera* L.) belongs to family Vitaceae and originated in Asia Minor. The history of the grape is as old as that of man. Grapes were introduced into India by Muslim invaders from Afghanistan and Persia around 1300. In India grapes are grown in Maharashtra, Karnataka, Andhra Pradesh and Tamil Nadu. Commercial cultivation of grapes under North Indian conditions began only a few decades ago (Bose and Mitra, 2001). Grapes are grown in Jammu and Kashmir and, in the Kashmir Valley, the Ripura area of Ganderbal district is known for its grape cultivation. The Kashmir Valley has an advantage for growing grapes: in other parts of the country, grapes are mainly harvested during February to June, but in the Kashmir Valley grapes are harvested during August to September, which is an off-season for the rest of the country. Hence grapes have a huge market outside Kashmir and command much higher prices. Varieties like Thomson Seedless, Sahibi, Perlett, Himrod and Kali Sahibi have great potential in the temperate region of the state. Grapes are mainly utilized for wine production. In addition to wine, a considerable quantity of grapes are used in other ways: for the table, as raisins, for juice and for canning.

2.16.10 Health benefits of grapes

Grapes are a very important fruit crop and beneficial for reducing different diseases. The skins contain resveratrol, which confers a small benefit to those who already have heart disease. In addition, low doses can respond to low doses of cytotoxic drugs and play a part in an innovative plan to increase the efficiency of anti-cancer therapy. Resveratrol seems to be a good tool in chemopreventive or chemotherapeutic strategies (Delmas *et al.*, 2006). Because of the natural antioxidant properties of resveratrol, grapes have a positive effect against cancers, arthritis, asthma allergies, joint pain, obesity, heart diseases and have a significant effect on ageing. The antioxidants have a key role on genes that impact the process of ageing (Zern *et al.*, 2005).

2.16.11 Olive

The olive (*Olea europaea* L.) is a symbol of prosperity and peace. It is a subtropical, evergreen tree and is also grown in some temperate regions on a small scale, but requires chilling for fruiting like other deciduous fruits. The tree is relatively slow-growing, long-lived and attains a height of three to 15 metres, or even more. The wood is resistant to decay and if the top dies, a new trunk will often grow from the roots. The fruit is mostly used for oil extraction purposes and a small quantity is used for the table, as well as being medicinally important. The olive most probably originated in the Mediterranean region (Zeven and Zhukovsky, 1975), and the majority of today's olive-producing countries lie in the Mediterranean basin and contribute 98% of the world's total production of olive and olive products (Singh *et al.*, 1986). These countries include Spain, Italy, Greece, Turkey, Portugal, Tunisia, France, Morocco, USA, Jordan and Cyprus. In Asia, olive cultivation is confined to Palestine, Iraq, Iran and China. In India there is great scope for olive cultivation in the Himalayan region encompassing the northern states, particularly Jammu and Kashmir, lying between 30° and 35° north latitudes and between altitudes ranging from 1000 to 1300 m above mean sea level (Mitra and Bose, 1991). As far as Kashmir is concerned, with a climate changing from temperate to warm temperate, olive is a good option for diversification. Many oil-bearing cultivars like Coratina, Frantio, Moralina, Zaituna and Pendolina have performed well in parts of the Kashmir Valley.

2.16.12 Health benefits of olive

Olive and its oil are very beneficial to the health of humans. Owen *et al.* (2004) reported that olive and extra virgin olive oil contain high concentrations of phenolic antioxidants and squalene. Squalene is transferred to the skin of the body and protects it against skin cancer. Therefore, in an area such as the Mediterranean basin, where olives are liberally consumed and oil is the major cooking and garnishing fat of choice, intake of phenolic antioxidants and squalene in the diet is likely to be considerably higher than in other areas. Probably this is one of the major factors which determine the lower incidence of cancer in the regions where the oil is mainly consumed (Newmark, 1997). Olive oil is the main source of fat and could be an important tool against cardiovascular disease. The benefits of olive oil consumption are beyond mere reduction of LDL cholesterol. A recent study conducted by researchers found that essential fatty acids found in olive oil were similar to those present in human milk and this helps in the assimilation of vitamins and minerals. One of the studies conducted in southern Greece reported that people who take good amounts of olive oil with vegetables may have a lower risk of rheumatoid arthritis. An olive-rich diet will also help to maintain blood-sugar levels and reduce diabetes by preventing insulin resistance (Waterman and Lockwood, 2007). The wide range of anti-atherogenic effects related to this oil may contribute to the low rate of cardiovascular mortality found in Southern Europe (Covas, 2007).

2.16.13 Fig

Fig (*Ficus carica* L.) is a crop of the warm temperate zone (Westwood, 1993; Murli and Balachandaran, 1997) and is one of the earliest cultivated fruit trees in the world (Solomon *et al.*, 2006). It is one of the most salt- and drought-tolerant of the fruit trees. In India the crop is commercially grown in Pune (Maharashtra) and a few sporadic plantations are found in Bangalore, Sreeranghpattanum, Bellary and some isolated patches in Andhra Pradesh and Tamil Nadu, some parts of Uttar Pradesh, Punjab and Himachal Pradesh (Sharma and Badiyala,

2006). Fig names are based on ground colour, internal flesh colour, maturity date and country of origin (Aljane *et al.*, 2008). In Jammu and Kashmir, some rare plants are found in kitchen gardens or scattered at the periphery of orchards, and can be grown in waste land. There is a crucial need for conservation and improvement of the crop (Sadder and Ateyyeh, 2006; Rout and Mohapatra, 2008). Varietal discrimination and identification can be achieved either by morphological or molecular markers (Saddoud *et al.*, 2008).

Fig is a rich source of amino acids, minerals, carbohydrates and fibre, and is consumed fresh, dried, preserved or canned, all of these being easily digested and palatable. It also possesses some medicinal properties, being of benefit in healing boils and some infections (Wang *et al.*, 2003). Higher quantities of fibre, potassium, calcium and iron make it more nutritious than bananas, grapes and oranges (Chessa, 1997; Michailides, 2003). The fruit is an important source of antioxidants, such as vitamin C, tocopherols, carotenoids and phenolics that modify the process of tumour cells and avoid neurochemical changes related to ageing (Kader, 2001; Shukitt Hale *et al.*, 2007). The varieties with a dark skin colour contain a higher amount of anthocyanins, polyphenols and flavonoids along with better antioxidant properties than those with lighter-coloured skins (Solomon *et al.*, 2006).

2.17 Emerging Challenges

All is not well with the present process of crop diversification in the state of Jammu and Kashmir. Faulty practices of cropping sequences over the years have caused great losses to crop diversity as well as soil fertility and other related factors. The indiscriminate application of different chemicals on fruit crops has also had a negative impact on human health and the environment, and there are other impending challenges.

The first of these is the cultivation of high-value cash crops, which leads to a level of unsustainability through decreasing soil fertility, erratic weather conditions, emergence of new pests and disease through various sources. Monocropping has led to the deterioration of soil health and a negative effect on the farming economy. This alarming problem has arisen due to the application of spurious agrochemicals in the post-liberalization period. There has been an exponential growth of the sectors supplying these products. The high incidence of disease and pest infestation has led to an excessive use of chemicals, which, in turn, has given rise to falling productivity: a vicious circle. This tragedy has not only increased the production cost but also affects the ecology.

Second, cheaper imports of many commodities have caused a significant reduction in their prices and this has become a serious issue for the future.

Third, the shift of fruit growing from niche locations with a comparative natural advantage towards less naturally suitable areas, but which exploit the latest technology, has posed new problems. Some circumstances arise in hilly areas that mean they have lost their comparative advantage. The process is likely to be accentuated with the intensification of globalization that brings new improvements (Jodha, 2000).

Fourth, drastic changes in climate and weather now pose a major threat to the cultivation of some of the high-value cash crops. For example, during the last five to ten years, the cultivation of cherry, peach and plum has shifted along the foothills and other specific areas, primarily because of excess or low precipitation in peak time, meaning that they are not fulfilling their potential. This has obliged orchardists in some areas of the Kashmir Valley to switch over to the cultivation of high-value crops.

Fifth, the market yards do not have adequate space for sale of the product. There is a lack of infrastructure such as high-tech packhouses and other related facilities needed to cope with the pressures of modern technology.

Sixth, low and stagnant productivity of crops with input cost is another worrying factor. These need to be replaced by high-yield, drought-resistant and dwarf varieties.

2.18 Summing Up

The crop diversification in India that started with the introduction of apples in the 1960s and vegetable crops in the 1980s, has made rapid progress in the areas enjoying favourable temperate climatic conditions. The evidence from these areas shows that all types of landowners have adopted the cultivation of high-value vegetables as intercrops. These intercrops have made a tremendous impact on the income of farmers. Factors that triggered the process of crop diversification include fluctuation of prices in the market, availability of a huge market outside the crop-growing states, and creation of self-help groups that increase the economy of the area in general and rural people in particular. The crop diversification of the temperate states such as Jammu and Kashmir, Himachal Pradesh and Uttarakhand throws up some important points. First, the process cannot be initiated without a bundle of support facilities, which requires political and social intervention. Second, the provision of basic infrastructural facilities, such as roads, markets, cold stores and sound R&D for harnessing of niche crops from far-flung areas. Third, the farmer should be supported by self-help groups to withstand risk by providing some type of incentives. Finally, sociopolitical dispensation, continuous technological backup and open trade regimes need to ensure economic viability for the sound diversification of fruits.

2.19 Constraints in Crop Diversification

Diversification in fruit crops is very important for the sustainability development of the temperate horticulture sector. However, there are some major constraints that may hamper this industry. Most of the land in the temperate zone is either rain-fed or occupied by hilly tracts, which makes it difficult for farmers to grow sensitive fruit crops. There is also a lack of quality plant material of superior varieties for the farmers. Poor basic facilities like roads from orchards to the fruit markets, lack of refrigerated vans, and small holdings of lands also discourages modernization and mechanization of horticultural crops. Due to inadequate post-management practices and facilities, most of the perishable fruit crop deteriorates in a short time, causing a collapse in the fruit-growing economy. Furthermore, weak linkages between researchers and extension workers, less investment in the horticulture sector and overuse of vital resources are further constraints.

2.20 Future Thrust

There is a need for inclusion of high-value crops through both horizontal and vertical diversification approaches. There is also a need to evaluate and propagate the high-value nut crops in foothill areas where other fruit crops are difficult to grow. The fruit crops with varying harvesting times and high potential yield need to be identified for the future.

2.21 Conclusion

Diversification of temperate fruit crops is the new blueprint for sustainable horticulture. It is not a change of traditional and less remunerative to more remunerative fruits, but should be demand-driven, needs-based, situation-specific and meet the standards of international markets. In the present global scenario, addition of more fruit crops, particularly underutilized ones, will go a long way in improving the socio-economic status of the farmers. In addition to major fruit crops, temperate regions of India benefit from a suitable environment for growing different minor fruits to increase nutritional food security and save money on imports. In the present context of climate change, growing minor temperate fruits will be helpful in utilizing the huge wastelands available in the foothills of the Himalayas. This improvement will not only reduce the risk of the main fruit crop but increase the livelihood of farmers.

References

Aljane, F., Ferchichi, A. and Boukhris, M. (2008) Pomological characteristics of local fig (*Ficus carica*) cultivars in Southern Tunisia. *Acta Horticulturae* 798, 123–128.

Amaral, S., Casal, S., Pereira, J.A., Seabra, R.M. and Oliveira, B.P.P. (2003) Determination of sterol and fatty acid compositions, oxidative stability and nutritional value of six walnut (*Juglans regia* L.) cultivars grown in Portugal, *Journal of Agriculture and Food Chemistry* 51(26), 7698–7702.

Anonymous (2013) *FAO Statistical Databases*. Food and Agriculture Organization of the United Nations. Available at: faostat.fao.org/site/339/default.aspx (accessed 20 May 2016).

Anonymous (2014a) Department of Horticulture, Government of Jammu and Kashmir. Available at: www.hortikashmir.gov.in (accessed 20 May 2016).

Anonymous (2014b) Economic Survey 2014–15, Volume I. Directorate of Economics and Statistics, Government of Jammu and Kashmir. Available at: www.ecostatjk.nic.in/ecosurvey/EcoSurvey201415vol1.pdf (accessed 20 May 2016).

Basu, A. and Penugonda, K. (2009) Pomegranate juice: a heart healthy fruit juice. *Nutrition Review* 67, 49–56.

Berkow, S.E. and Barnard, N.D. (2005) Blood pressure regulation and vegetarian diets. *Nutrition Review* 63, 1–8.

Blench, R.M. (1997) Neglected Species, Livelihood and Biodiversity in Difficult Areas: How Should the Public Sector Respond. *Natural Resources Perspective No. 23*. Overseas Development Institute, London, UK.

Bose, T.K. and Mitra, S.K. (2001) *Fruits: Tropical and Subtropical*. Partha Sankar Basu NAYA UDYOG 206, Bidhan Sarani Calcutta, India, pp. 331–340.

Burgess, M.A. (1994) Cultural responsibility in the preservation of local economic plant resources. *Biodiversity and Conservation* 3, 126–136.

Carvalho, M., Silva, B.M., Silva, R., Valentão, P., Andrade, P.B. and Bastos, M.L. (2010) First report on *Cydonia oblonga* Miller anticancer potential: differential antiproliferative effect against human kidney and colon cancer cells. *Journal of Agricultural and Food Chemistry* 58, 3366–3370.

CGIAR (1994) *Challenging Hunger – The Role of the CGIAR*. Consultative Group on International Agricultural Research, Washington DC, USA.

Chen, C.Y.O and Blumberg, J.B. (2008) Phytochemical composition of nuts. *Asian Pacific Journal of Clinical Nutrition* 17, 329–332.

Chessa, I. (1997). Fig. In: Mitra, S. (ed.) *Postharvest Physiology and Storage of Tropical and Subtropical Fruits*. CAB International, Wallingford, Oxfordshire, UK, pp. 245–268.

Covas, M. (2007) Olive oil and the cardiovascular system. *Elsevier Pharmacological Research* 55, 175–186.

Delmas, D., Lancon, A., Cohn, D., Jannin, B. and Latruffe, N. (2006) Resveratrol as a chemo preventive agent: a promising molecule for fighting cancer. *Current Drugs Targets* 4, 423–442.

Esmaillzadeh, A., Tahbaz, F., Gaieni, I., Alavi-Majd, H. and Azadbakht, L. (2006) Cholesterol lowering effect of concentrated pomegranate juice consumption in type II diabetic patients with hyperlipidemia. *International Journal for Vitamin and Nutrition Research* 76(3), 147–151.

Eyzaguirre, P., Padulosi, S. and Hodgkin, T. (1999) IPGRI's strategy for neglected and underutilized species and the human dimension of agro biodiversity. In: Padulosi, S. (ed.) *Priority Setting for Underutilized and Neglected Plant Species of the Mediterranean Region. Report of the IPGRI Conference, 9–11 February 1998, ICARDA, Aleppo, Syria*. International Plant Genetic Resources Institute, Rome, Italy, pp. 1–20.

FAO (1996a) *Report on the State of the World's Plant Genetic Resources for Food and Agriculture, prepared for the International Technical Conference on Plant Genetic Resources, Leipzig, Germany, 17–23 June 1996*. Food and Agriculture Organization of the United Nations, Rome, Italy.

FAO (1996b) *Global Plan of Action for the Conservation and Sustainable Utilization of Plant Genetic Resources for Food and Agriculture and Leipzig Declaration, adopted by the International Technical Conference on Plant Genetic Resources, Leipzig, Germany, 17–23 June 1996*. Food and Agriculture Organization of the United Nations, Rome, Italy.

Ferguson, A.R. (1984) Kiwifruit: a botanical review. *Horticulture Review* 6, 1–64.

Ferguson, A.R. (1999) New temperate fruits: *Actinidia chinensis* and *Actinidia deliciosa*. In: Janick, J. (ed.) *Perspective on New Crops and New Uses. Proceedings of the Fourth National Symposium on New Crops and New Uses Biodiversity and Agricultural Sustainability*, 8–11 November 1998, Phoenix, Arizona, pp. 342–347.

Gill, A.S. (2002) Diversified farming systems for sustainable agriculture. *Agriculture Today* 5, 53–55.

Guarino, L. (ed.) (1997) Traditional African vegetables. Promoting the conservation and use of underutilized and neglected crops 16. In: *Proceedings of the IPGRI International Workshop on Genetic Resources of Traditional Vegetables in Africa: Conservation and Use*. 29–31 August 1995, ICRAF-HQ, Nairobi, Kenya. Institute of Plant Genetics and Crop Plant Research, Gatersleben/International Plant Genetic Resources Institute, Rome, Italy.

Heywood, V. (1999) Use and potential of wild plants in farm households. FAO Farm System Management Series No. 15. FAO, Rome, Italy. *FAO/IBPGR Plant Genetic Resources Newsletter* 71, 37.

Jodha, N.S. (2000) Globalisation and fragile mountain environment: policy challenges and choices. *Mountain Research and Development* 20(4), 296–299.

Joshi, V.K. and Bhutani, V.P. (1995) Peaches. In: Salunkhe, D.K. and Kadam, S.S. (eds) *Handbook of Fruit Science and Technology: Cultivation, Storage and Processing*. Marcel Dekker, New York, USA, pp. 243–296.

Kader, A. (2001) Importance of fruits, nuts, and vegetables in human nutrition and health. *Perishables Handling Quality* 106, 4–6.

Kaulgud, S.N. (2001) Pomegranate. In: Chadha, K.L. (ed.) *Handbook of Horticulture*. ICAR, New Delhi, India, pp. 297–304.

Kendall, W.C., Esfahani, A., Traun, J., Srichankul, K. and Jenkins, D.A. (2010) Health benefits of nuts in prevention and management of diabetes. *Asian Pacific Journal of Clinical Nutrition* 19(1), 110–116.

Kornsteiner, M., Wagner, K.H. and Elmadfa, I. (2006) Tocopherols and total phenolics in 10 different nut types. *Food Chemistry* 98, 381–387.

Lansky, E.P. and Newman, R.A. (2007) Pomegranate (*Punica granatum* L.) and its potential for prevention and treatment of inflammation and cancer. *Journal of Ethnobotany and Pharmacology* 109, 177–206.

Lazaroff, L. (1989) Strategies for development of a new crop. In: Wickens, G.E., Haq, N. and Day, P. (eds) *New Crops for Food and Industry*. Chapman & Hall, London, UK, pp. 108–119.

Maguire, L.S., O'Sullivan, S.M., Galvin, K., O'Connor, T.P. and O'Brien, N.M. (2004) Fatty acid profile, squalene and phytosterol content of walnuts, almonds, peanuts, hazelnuts and the macadamia nut. *International Journal of Food Science and Nutrition* 55(3), 171–178.

Marchi, L.B., Monteiro, A.R.G., Mikcha, J.M.G., Santos, A.R., Chinellato, M.M. *et al.* (2015) Evaluation of antioxidant and antimicrobial capacity of pomegranate peel extract (*Punica granatum* L.) under different drying temperatures. *Chemical Engineering Transactions* 44, 121–126.

Maxted, N., Hawkes, J.G., Guarino, L. and Sawkins, M. (1997) Towards the selection of taxa for plant genetic conservation. *Genetic Resources and Crop Evolution* 44, 337–348.

Mehta, K.A. and Tahkur, B.S. (2004) Pecan (*Carya illinoensis* (Wang) K. Koch) cultivation in India – problems, prospects and future strategies to enhance productivity. *Proceedings Compendium of Summer School on Enhancement of Temperate Fruit Production in Changing Climate, 10 September–9 October 2001*, pp. 231–238.

Michailides, T.J. (2003) Diseases of fig. In: Ploetz, R.C. (ed.) *Diseases of Tropical Fruit Crops*. CAB International, Wallingford, Oxfordshire, UK, pp. 253–273.

Mir, M.M., Sofi, A.A, Singh, D.B. and Khan, F.U. (2007) Evaluation of pomegranate cultivars under temperate conditions of Kashmir Valley. *Indian Journal of Horticulture* 64(2), 150–154.

Mir, M.M., Askary, T.H. and Hassan, W. (2009) Cultivation of cherry in Kashmir Valley. *Indian Farmer's Digest* 42(6), 34–36.

Mitra, S.K. and Bose, T.K. (1991) *Temperate Fruits*. Horticulture and Allied Publishers, 27/3, Chakraberia Lane, Calcutta, India.

Monti, L. (ed.) (1997) *Proceedings of the CNR International Workshop on Neglected Plant Genetic Resources with a Landscape and Cultural Importance for the Mediterranean Region. 7–9 November 1996, Naples, Italy*.

Murli, T.P. and Balachandaran, M. (1997) Fig. In: Chattopadhyay, T.K.A. (ed.) *Textbook of Pomology*, Vol. III. Kalyani Publishers, Ludhiana, India, pp. 189–204.

Newmark, H.L. (1997) Squalene, olive oil and cancer risk: a review and hypothesis. *Cancer Epidemiological Biomarkers and Prevention* 6, 1101–1103.

Osman, A.G., Koutb, M. and Sayed, A.E. (2010) Use of hematological parameters to assess the efficiency of quince (*Cydonia oblonga* Miller) leaf extract in alleviation of the effect of ultraviolet-A radiation on African catfish *Clarias gariepinus* (Burchell, 1822). *Journal of Photochemistry and Photobiology B* 99, 1–8.

Owen, R.W., Haubner, R., Wurtele, G., Hull, W.E., Speigeihalder, B. and Bartsch, H. (2004) Olives and olive oil in cancer prevention. *European Journal of Cancer Prevention* 13, 319–326.

Padulosi, S. (1999a) *Priority Setting for Underutilized and Neglected Plant Species of the Mediterranean Region. Report of the IPGRI Conference, 9–11 February 1998, ICARDA, Aleppo, Syria*. International Plant Genetic Resources Institute, Rome, Italy.

Padulosi, S. (1999b) Criteria for priority setting in initiatives dealing with underutilized crops in Europe. *Proceedings of the European Symposium, 30 June–3 July 1998, Braunschweig, Germany*. International Plant Genetic Resources Institute, Rome, Italy.

Padulosi, S. (1999c) Partners and partnership. In: Swaminathan, M.S. (ed.) *Enlarging the Basis of Food Security: The Role of Underutilized Species. International Workshop held at the M.S. Swaminathan Research Foundation, 17–19 February, Chennai, India*.

Padulosi, S, Hammer, K. and Heller, J. (1996) Hulled wheats: promotion of conservation and use of valuable underutilized species. *Proceedings of the First International Workshop on Hulled Wheats, 21–22 July 1995, Castelvecchio Pascoli, Tuscany, Italy.* International Plant Genetic Resources Institute, Rome, Italy.

Padulosi, S., Hodgkin, T., Williams, J.T. and Haq, N. (2002) Underutilised crops: trends, challenges and opportunities in the 21st century. In: Engels, J.M.M., Ramanatha, R.V., Brown, A.H.D. and Jackson, M.T. (eds.) *Managing Plant Genetic Diversity*, Vol. 30. International Plant Genetic Resources Institute, Rome, Italy, pp. 323–338.

Pistorius, R. (1997) *Scientists, Plants and Politics – A History of the Plant Genetic Resources Movement*. International Plant Genetic Resources Institute, Rome, Italy.

Rout, G.R. and Mohapatra, A. (2008) Use of molecular markers in ornamental plants: a critical reappraisal. *European Journal of Horticultural Science* 71(2), 53–68.

Ryan, E., Galvin, K., O'Connor, T.P., Maguire, A.R. and O'Brien, N.M. (2006) Fatty acid profile, squalene and phytosterol content of Brazil, pecan, pine, pistachio and cashew nuts. *International Journal of Food Science and Nutrition* 57(1/4), 219–228.

Sadder, M.T. and Ateyyeh, A.F. (2006) Molecular assessment of polymorphism among local Jordanian genotypes of the common fig (*Ficus carica* L.). *Science of Horticulture* 107, 347–351.

Saddoud, O. Baraket, G. Chatti, K. Trifi, M. Marrakchi, M. Salhi-Hannachi, A. and Mars, M. (2008) Morphological variability of fig (*Ficus carica* L.) cultivars. *International Journal of Fruit Science* 8, 35–51.

Sajid, M.S., Zubair, M., Waqas, M., Nawaz, M. and Ahmad, Z. (2015) A review on quince (*Cydonia oblonga*). A useful medicinal plant. *Global Veterinaria* 14(4), 517–524.

Shant, P.S. and Sapru, B.L. (1993) Grafting technique in pomegranate. In: Singh S.P. (ed.) *Scientific Horticulture*. Scientific Publishers, Jodhpur, India, pp. 45–46.

Sharma, S.K. and Badiyala, S.D. (2006) Variability studies in common fig in Hamirpur district of Himachal Pradesh. *Indian Journal of Horticulture* 63(2), 159–161.

Shukitt Hale, B., Carey, A.N., Jenkins, D., Rabin, B.M and Joseph, J.A. (2007) Beneficial effects of fruit extracts on neuronal function and behaviour in a rodent model of accelerated aging. *Neurobiology Aging* 28, 1187–1194.

Singh, R.P., Chadha, T.R. and Singh, J.M. (1986) Performance of some olive (*Olea europaea* L.) cultivars. In: Chadha, T.R., Bhutani, V.P. and Kaul J.L. (eds) *Advances in Research on Temperate Fruits*. Dr Y.S. Parmar University of Horticulture and Forestry, Solan, Himachal Pradesh, pp. 55–59.

Singh, A., Patel, R.K. and Verma, M.R. (2008) Popularizing kiwifruit cultivation in North East India. *Environmental Bulletin* 16, 1.

Small, E. (1999) New crops for Canadian Agriculture. In: Janick, J. (ed.) *Perspectives on New Crops and New Uses*. ASHS Press, Alexandria, Virginia, USA, pp. 15–52.

Solomon, A., Golubowicz, S., Yablowicz, Z., Grossman, S., Bergma, M., Gottlieb, H.E., Altman, A., Kerem, Z. and Flaishman, M.A. (2006) Antioxidant activities and anthocyanin content of fresh fruits of common fig (*Ficus carica* L.). *Journal of Agriculture and Food Chemistry* 54, 7717–7723.

Turk, G., Sonmez, M., Aydin, M., Yuce, A., Gur, S., Yuksel, M., Aksu, E.H. and Aksoy, H. (2008) Effects of pomegranate juice consumption on sperm quality, spermatogenic cell density, antioxidant activity and testosterone level in male rats. *Clinical Nutrition* 27, 289–296.

Ulrich, R. (1970) Organic acids. In: Hulme, A.C. (ed.) *The Biochemistry of Fruits and Their Products*. Academic Press, New York, USA and London, UK, pp. 89–118.

Vietmeyer, N. (1990) The new crops era. In: Janick, J. and Simon, J. (eds) *Advances in New Crops. Proceedings of the First National Symposium on New Crops: Research, Development, Economics.* Indianapolis, Indiana, 23–26 October 1988. Timber Press, Portland, Oregon, USA, pp. xviii–xxii.

Von Maydell, H.J. (1989) Criteria for the selection of food producing trees and shrubs in semi arid regions. In: Wickens, N, Haq, N. and Day, P. (eds) *New Crops for Food and Industry*. Chapman and Hall, London, UK, pp. 66–75.

Wang, L., Jiang, W., Ma, K., Ling, Z. and Wang, Y. (2003) The production and research of fig (*Ficus carica* L.) in China. *Acta Horticulturae* 605, 191–196.

Waterman, E. and Lockwood, B. (2007) Active components and clinical applications of olive oil. *Alternative Medicine Review* 12, 331–342.

Westwood, M.N. (1993) *Temperate Zone Pomology, Physiology and Culture*, 3rd edn. Timber Press, Portland, Oregon, USA.

Whiting, G.C. (1970) Sugars. In: Hulme A.C. (ed.) *The Biochemistry of Fruits and Their Products*. Academic Press, New York, USA, pp. 1–31.

Woodroof, J.G. (1979) *Tree Nuts*. AVI Publishing Company, Westport, Connecticut, USA.

Yehia, H.M., Elkhadragy, M.F. and Moneim, A.E.A. (2011) Antimicrobial activity of pomegranate rind peel extracts. *African Journal of Microbiological Research* 4(22), 3664–3668.

Zern, T.L., Wood, R.J., Greene, C. and West, K.L. (2005) Grape polyphenols exert a cardioprotective effect in pre- and postmenopausal women by lowering plasma lipids and reducing oxidative stress. *Journal of Nutrition* 135(8), 1911–1917.

Zeven, A.C. and Zhukovsky, P. (1975) *Dictionary of Cultivated Plants and their Centres of Diversity*. Centre for Agricultural Publication and Documentation, Wageningen, Netherlands.

3 Asteraceae of India: its Diversity and Phytogeographical Affinity

SUNIT MITRA AND SOBHAN KUMAR MUKHERJEE*

Department of Botany, Ranaghat College, Nadia, West Bengal, India

Abstract

Asteraceae (*nom. alt.* Compositae), with its approximately 1600–1700 genera and more than 24,000 species, is the largest family of flowering plants (Funk *et al.*, 2009). The members of this group are found to occur in all the regions of the globe except Antarctica (Anderberg *et al.*, 2007). India, with an area of 3,287,263 km^2, is the seventh-largest country of the world and occupies about 2.46% of the land area. Biogeographically, the Indian subcontinent can be divided into several regions or provinces primarily based on climatic conditions, soil types and the floristic composition. Present studies reveal that Asteraceae, with its 1314 taxa under 204 genera distributed in to 20 tribes, is the most diversified Angiospemic plant family in Indian flora, followed by the Poaceae (1291), Orchidaceae (1229), and Fabaceae (1192). Asteraceae is a predominant temperate family, and most of the members of this family are distributed in the temperate regions of the globe. In India, most of the taxa (955) of Asteraceae, which is about 72.67% of the total Asteraceae in India, are found to be located in the temperate regions of Himalaya and the north-east part of India. The chief centre of diversity of the Indian Asteraceae is the Himalayan biogeographic zone. This is due to its variable climatic condition and altitudes, which in turn have resulted in diverse habitats, from the cold deserts of Ladakh in the Trans-Himalayan biogeographic zone to the evergreen tropical and temperate forests of north-east India. There are 202 taxa of Asteraceae which are endemic to the Indian region. All the exotic taxa of Asteraceae have been broadly divided into 18 major categories. It is also very interesting to note that there are 30 taxa which are considered as rare, endangered and threatened. Based on the distribution of these 1314 taxa of Asteraceae, the Indian region can be divided in to 12 phytogeographical zones, in comparison to the classification of the Indian region into 11 phytogeographical zones as suggested by Balakrishnan in 1996.

3.1 Introduction

To begin with, we would like to share a few lines of Li *et al.* (1975), where he stated:

> Rather than waiting for a definitive work in the distant future, we here propose to summarize our current knowledge in a form that can be of some help in meeting immediate needs. This work can also be used as a basis for further improvement.

The sunflower family or the Asteraceae (*nom. alt.* Compositae), with its approximately 1600–1700 genera and more than 24,000 species, is the largest family of flowering plants (Funk *et al.*, 2009). The members of this group are found to occur in all the regions of the globe except Antarctica (Anderberg *et al.*, 2007).

The family is especially diverse in the tropical and subtropical regions of North America, the Andes, eastern Brazil, southern Africa, the Mediterranean region, central Asia, and south-western China. The majority of Asteraceae species are herbaceous, yet an important component of the family comprises shrubs or even trees occurring primarily in the tropical regions of North and South America, Africa and Madagascar and on isolated islands in the Atlantic and Pacific Oceans. Many species of sunflowers are ruderal and especially abundant in disturbed areas, but a significant number of them, especially in mountainous tropical regions, are narrow endemics. Because of the relentless habitat transformation precipitated by human expansion in montane tropical regions, a number of these species are consequently in danger of extinction.

*E-mail: sobhankr@gmail.com

India, with an area of 32,87,263 km^2 is the seventh-largest country in the world and occupies about 2.46% of the land area on the globe. It lies between 6° 45' N to 37° 6' N latitude and 68° 51' E to 97° 25' E longitude. It is situated entirely to the north of the Equator with a greater area in the subtropical zone.

With regard to the natural boundary, the great wall of the Himalaya lies at the north. The Thar Desert of Rajasthan stretches across the north–west. In the south, it narrows to form the great Indian Peninsula, which ends up in the Indian Ocean with Cape Comorin at its southernmost tip. The Bay of Bengal lies at the eastern side of the Peninsula, wherein lie the Andaman and Nicobar Islands. In the west, is the Arabian Sea with another group of islands known as the Lakshadweep. In the east, the rest of the country is linked by the long mountain stretches of the Khasi and Jaintia Hills from Myanmar. The countries bordering India from east to west are Myanmar, Bangladesh, China, Tibet, Bhutan, Nepal, Afghanistan and Pakistan (Fig. 3.1).

Physiographically, India can be divided in to four major zones.

3.1.1 The Northern Mountains

To the north and arching south-east tower the young Himalayan peaks, formed in the Upper Tertiary when geosyncline deposits were compressed, folded and upraised by the northward movement of the Deccan block from the ancient continent of Gondwana.

The eastern Himalayas rise from the Brahmaputra valley through a zone of swamp forest that gives way from about 200–900 m to broad-leaved tropical evergreen rainforests, topped by subtropical grasslands and subtropical forests up to about 1800 m and succeeded by temperate forests to about 3500 m.

In the humid Eastern Himalayas the timber line reaches 4570 m, but is only 3600 m in the drier western ranges.

The Eastern Himalayan region has provided a gateway for the migration of plants and is considered a meeting ground of the Indo-Malaysian and Sino-Japanese floras. The Eastern Hills, running in a north-east-to-south-west direction, are a series of ranges, averaging 1500–1800 m, which divide India from Burma, and support tropical evergreen forests and bamboo thickets in previously cleared sections.

The Brahmaputra, or Assam Valley is the corridor that links Arunachal Pradesh to the rest of India.

The Assam Plateau situated between Bangladesh and the Burma is a fragment of the Deccan block and it is broken by the Khasi, Garo, and Jaintia Hills with an average height of 1200–1800 m.

The Jhelum river flows through the Kashmir Valley, which is a heavily farmed intermontane basin situated below the icy peak of the Western part of the Himalaya. Himalayan conifers of tropical and subtropical types grow on the slopes of Kashmir.

Many parts of this region bordering Jammu and Kashmir, as well as the dry deciduous forests of the Pir Panjal range to the west, along the border of Pakistan, are floristically underexplored.

3.1.2 The North Indian Plain

Parallel to the mountain wall of the north, to the south are the lowlands of the alluvial Indo-Gangetic Plain, that sweeps more than 3200 km across the subcontinent, from the Arabian Sea on the west through the deserts of Rajasthan, across the floodplain bluffs and gullies around Delhi to the rice-bowl muds and clays of the Ganges-Brahmaputra delta and ends in the tidal mangrove forests of the Sundarbans.

The Indo-Gangetic region is divided into several climatic and topographic zones. From the Pakistan border eastward to Delhi is a submontane strip of dry deciduous forest dropping off from the Siwalik Hills. The region becomes increasingly dry towards the south, but as the divide is a great crossroads of civilizations, the area has been long and densely populated. The divide gives way in a south-eastward arc to the Upper Ganges plain, which receives between 64 cm and 100 cm of rainfall a year and is traversed by three major rivers, the Jamuna, Ganges and Gagra, flowing south-eastward in parallel.

The plain is largely featureless except for the relief provided by the bluffs of older alluvial deposits, the bhangar, and the newer lowland deposits, the khadar – a relief largely absent from the floodplains of the Middle Ganges region, where rainfall increases to as much as 180 cm a year.

3.1.3 The peninsular plateau

Just to the north of the Tropic of Cancer, the alluvial plains begin to rise following a complex topography of hill country, interrupted by the Vindhya Range escarpment, to the east–west Satpura Range, still forested and with heights of up to 1350 m, which marks the meeting point of northern and southern India, at the northern extent of peninsular India proper.

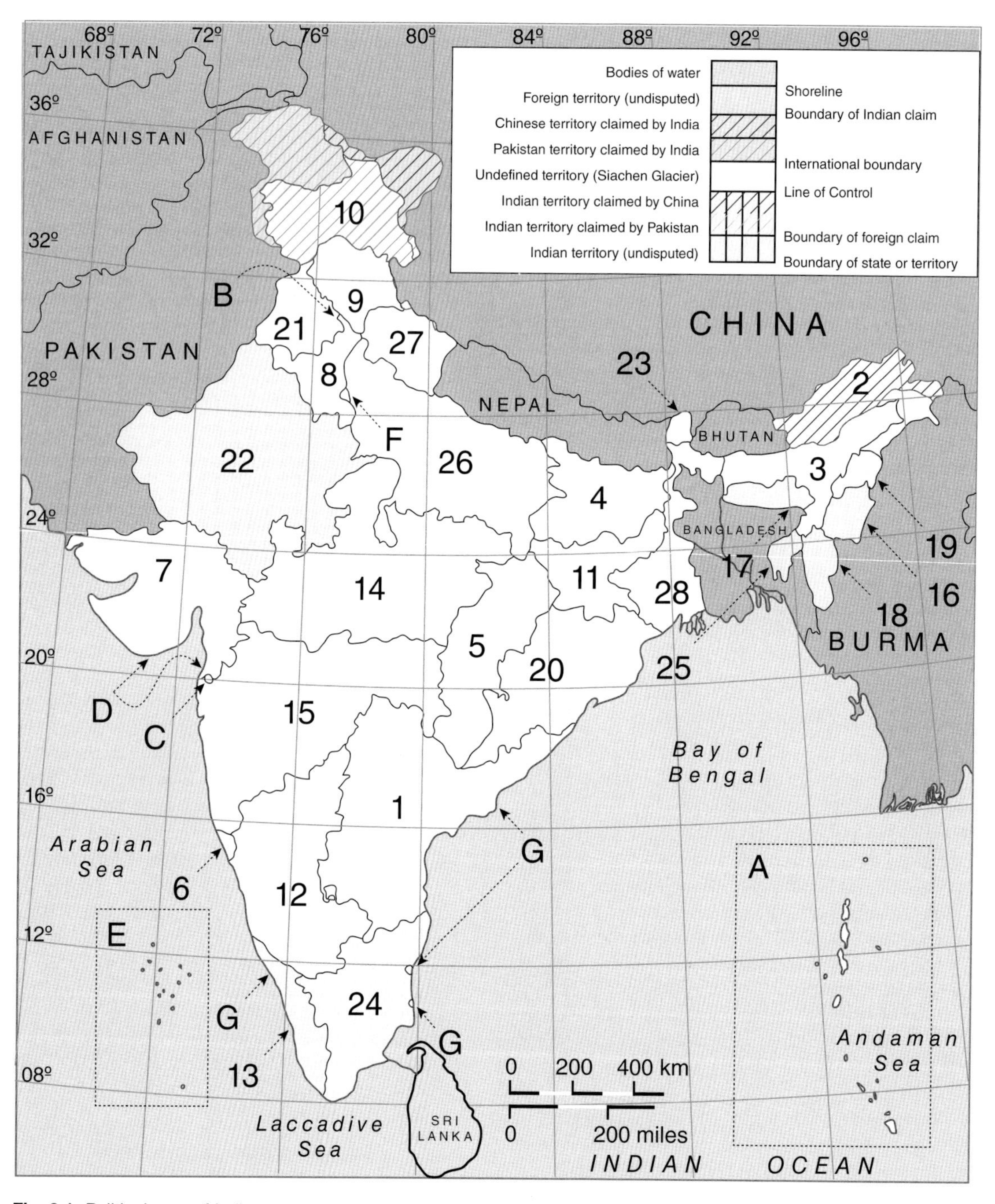

Fig. 3.1. Political map of India.

Rainfall in the hill country ranges from more than 100 cm a year in the east, where dry deciduous forest dominates, to not more than 25 cm westward in the Aravalli Hills, where thorn forest and scrub take over.

Lying across the Tropic of Cancer is the Malwa Plateau, which is drained by the Chambal river, and is fairly fertile in the south, where the soil is derived from the Deccan lavas.

Peninsular India (Fig. 3.2) is dominated by the great triangle of the eastward-tilting Deccan Plateau, an Archaean shield of some 39,000 km^2, extending from about 22° N to 8° N and with an average elevation of 460–600 m. Although built fundamentally on the ancient crystalline rocks of Gondwana, the north-west Deccan in the State of Maharashtra is characterized by thick flows of basaltic lavas from the Cretaceous era that have formed clay 'black cotton' soil or regur, that retains the little moisture, the region receives (about 100 cm a year, as it lies in the rain shadow of the Western Ghats). Just east of the Ghats is an area that gets no more than 50–80 cm of rain a year on average.

To the north-east, the topography is more varied with hills, plateaus, basins and valleys, and the rainfall is more certain and abundant. The Bastar-Orissa highland, at the great eastern bend of the peninsula, is still under monsoon forest cover, while to the south the silty Mahanadi basin and the rice-growing Godavari basin are under cultivation. The Eastern Ghats, fold ridges that peak in several points at more than 1500 m, are also still forested, although the tribal groups inhabiting the uplands practice shifting agriculture. Separating the Mahanadi and Godavari basins on either side of 200° N is the Bastar Plateau.

In the south the plains of the plateau, sliced by outcropping of granite or gneissic rock, rise through dry deciduous woodland, thorn forests and scrub to more than 900 m where (in the state of Karnataka) the Eastern and Western Ghats converge. The Cauvery River valley separates the Deccan from a small plateau broken by the Nilgiri and Cardamom hills, which themselves are separated by the Palghat Gap. The higher elevations of the Western Ghats and Nilgiri hills, which reach up to 2500 m, catch the relief rains, recording up to 280 cm of rain during the monsoon months from June to October, and are wooded to about 1500 m with semi-evergreen and subtropical broadleaved hill forests, especially in Sholas.

The eastern Coromandel Coast on the Bay of Bengal is divided into two regions roughly demarcated by Madras at 13° N. North of these are low delta lands where rice is grown without irrigation, while to the south are plains that receive only 100 cm or less of rainfall a year and most of that from the retreating monsoon of October to December.

The western coastland, from the Gulf of Cambay to Cape Comorin some 1600 km south, forms a strip up to 50–65 km wide at the bottom of the steep Deccan escarpment. Below the transitional zone of Goa and Kanara, the Ghats retreat inland; giving way to the Malabar Coast that stretches from the slopes through low lateritic terraces and hill country to alluvial flats forming shallow lagoons and mud banks for long stretches. The region above 16° N is narrow and broken by ravines, lying at the foot of the Ghat scarps. Between the Gulf of Cambay and the border of Pakistan is the peninsula of Saurashtra, built of Deccan lava. It is an area of sandy valleys and rocky ridges that supports scanty woodlands. Lying along the border of Pakistan is the Thar desert with about 25 cm of rainfall a year.

3.1.4 The coastal regions and islands

The Andaman and Nicobar Islands lie in a chain in the north-east Indian Ocean on the southward trend line of the Himalayan range.

The Andamans comprise 204 islands and are thought to contain India's main surviving primary forests – tropical evergreen rainforest dominated by Dipterocarpus and rich in Pterocarpus and *Mesua ferrea*.

The Nicobars are composed of 22 volcanic islands south of the Andamans, and the major upper canopy genera are Hopea, Alstonia, Calophyllum, and Terminalia. The low float islands of the northern Nicobars support scrub forest. Coastal areas in both island groups are bordered with mangrove forests and littoral vegetation.

The Laccadives are a group of 16 coral atolls totalling only 28 km^2 about 300 km off southern India's Malabar Coast. Ten of the islands are inhabited, while three are small keys with no vascular plants. Only about 40 of the 130 or so recorded species are indigenous to the islands and none are endemics.

Biogeographically, the Indian subcontinent can be divided into several regions or provinces primarily based on the climatic condition, soil types and floristic composition (Hooker (1854, 1906) Clarke (1898), Chatterjee (1940, 1962), Razi (1955), Puri (1960), Rao (1974), Rodgers and Panwar (1988), Balakrishnan (1996)). In the present work, for discussion of the distribution and diversity of the species of Asteraceae, we follow the phytogeographical classification of Balakrishnan (1996), with some modifications as needed.

India's rich floristic diversity is undoubtedly due to climatic and altitudinal variation, coupled with varied ecological habitats. There are almost rainless

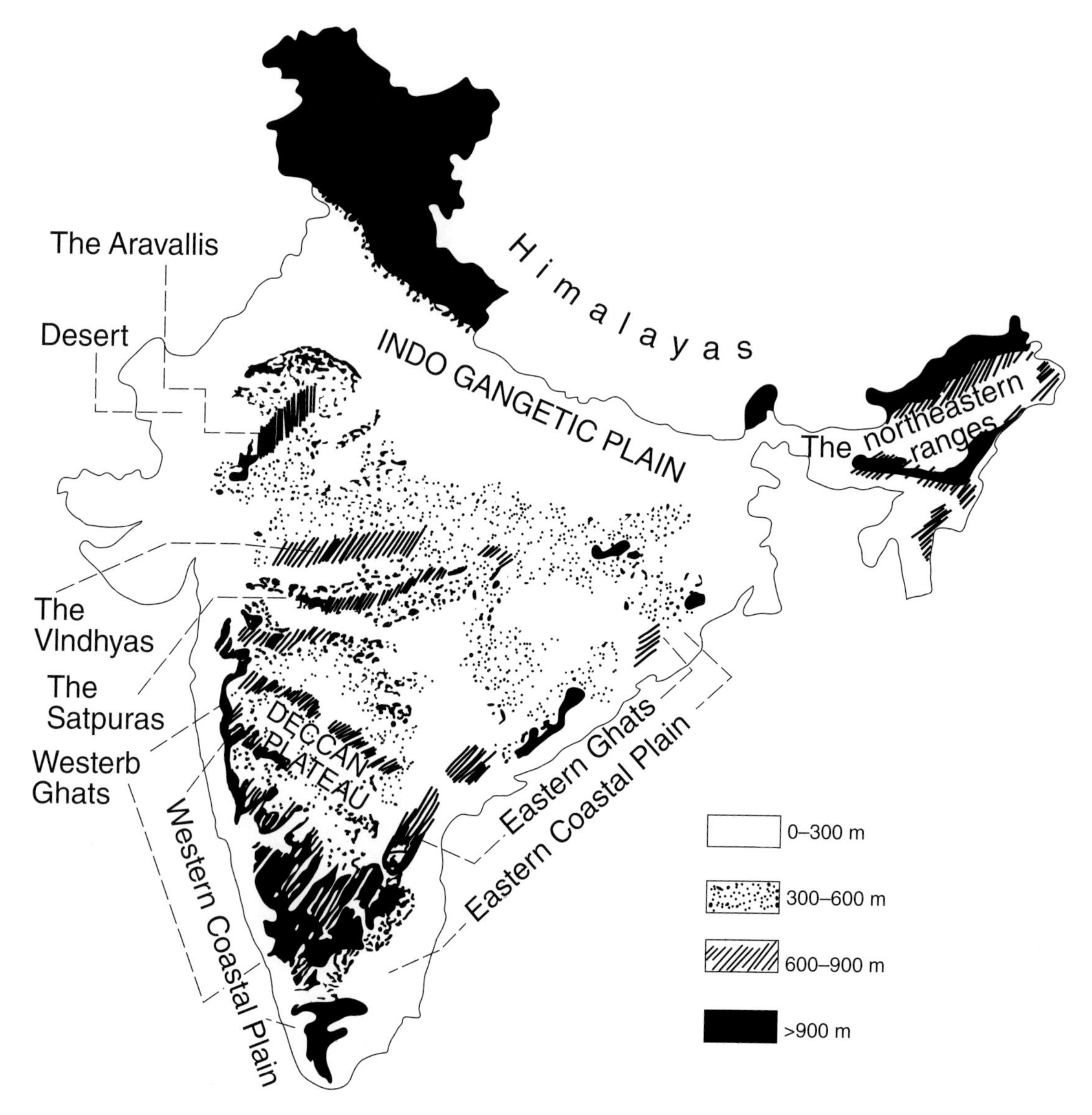

Fig. 3.2. Physical region of India.

zones in the west of the country and the highest rainfall zone in the world at its eastern corner. The altitude varies from sea level to some of the highest mountains ranges of the world. The habitation also varies from the humid tropical Western Ghats to the Cold Desert of Ladakh and the snow-clad mountains of the Himalayas, as well as the long warm coastal stretches of peninsular India. All these factors greatly favour the distribution, diversification and spread of the family Asteraceae (*nom. alt.* Compositeae).

3.2 Diversity and Distribution of Asteraceae in India

Van Rheede (1690), in the tenth volume of the *Hortus Malabaricus*, describes about 14 species of Asteraceae from the peninsular region of the Indian subcontinent. Linneaeus (1753) introduced 72 species from this region in his *Species Plantarum*, while Burman (1768) described 56 Asteraceae from India. However, the first important contribution on this family was made by Don (1825) and de Candolle (1838). In 1831

Nathaniel Wallich catalogued about 400 species of Asteraceae, a majority of which were new, although a small proportion were treated as synonyms by subsequent workers. William Roxburgh has been named as the Father of Indian Botany for the third edition of his *Flora Indica* (1832), which identified 89 species of Asteraceae from eastern India.

Before Independence, the major contribution to this family was made by Clarke (1876, 1881), who enumerated about 566 species under 122 genera, belonging to 12 tribes. In *Flora of British India*, the monumental seven-volume work of Sir J.D. Hooker (Hooker, 1875–1897), Clarke (1881) treated 608 species of Asteraceae under 128 genera, belonging to 12 tribes. After the publication of *Flora of British India*, different authors, such as Collett (1902), Prain (1903), Duthie (1902), Cooke (1906), Fyson (1914), Gamble (1921), Haines (1922), Fischer (1938), Calder *et al.* (1926), Kanjilal *et al.* (1939), Razi (1959), Nayer and Ramamurthy (1973), Ghosh and Datta (1976), Ghosh (1977,1979), Karthikeyan (2000) and Sharma and Singh (2001), contributed a great deal on the Indian Asteraceae, adding about 450 species of Asteraceae to the existing list.

Rao *et al.* (1988) published an account of this group, enumerating 1052 taxa of Asteraceae belonging to 177 genera. Hajra *et al.* (1995a and 1995b) subsequently revised the family in detail, and mentioned 892 species, 37 sub-species, 123 varieties, and 13 forma under 167 genera, belonging to 12 tribes and 17 subtribes. In their treatment Hajra *et al.* (1995a, 1995b) excluded 41 species belonging to 10 genera, which are mainly cultivated in different parts of the country as ornamental plants. Economic and medicinal properties of the family are enumerated by Chopra *et al.* (1956).

Karthikeyan *et al.* (2009) enumerated 1149 taxa under 193 genera of this family from within the political boundary of India. On the basis of this enumeration, this family stood fourth in the list of the most diversified family of India after Poaceae (1291), Orchidaceae (1229), and Fabaceae (1192). The chief centre of the diversity of the Indian Asteraceae is the Himalayan biogeographic zone – its variable climatic condition and altitudes have resulted in the diverse habitats described above. The present observations also help us to predict the same – Western Himalaya includes about 580 species of Asteraceae, Eastern Himalaya and north-east India have over 375 species of Asteraceae, and the Deccan peninsula (including the Western Ghats hot spots zone) comprises about 258 species of Asteraceae. It is also very interesting to note that in the Andaman and Nicobar Islands there are only 25 species of Asteraceae.

From the pattern of distribution of the Asteraceae genera, it is also very interesting to note that the most diversified genera of the Asteraceae like *Taraxacum* (95), *Saussurea* (77) and *Artemisia* (72) are distributed in the high mountains of the Himalayas.

From the distributional pattern it is predicted that most of the genera are adventive, consisting of weeds of adventive nature.

3.2.1 Taxonomic treatment

In this present work for the delimitation of the tribe, subtribes and genera, the taxonomic treatment of Kadreit and Jeffrey (2007) has been followed.

According to this treatment, ten genera which have been previously treated as valid genera are merged with other genera: see Table 3.1.

Regarding the tribal level, Kadreit and Jeffrey (2007) treated the tribe *Lactuceae* as tribe *Cichoreae* which is followed here.

Recently Maity and Maiti (2010) segregated the genus *Youngia* Cassini (see Cassini, 1826–1834),

Table 3.1. Name of the genera of Asteraceae and their present correct entities.

Sl. No.	Name of the previous valid Genera	Currently merged with the genus (after Kadreit and Jeffrey, 2007)
1	*Brachyactis* Ledeb.	*Symphytotrichum* Nees
2	*Hemistepta* Bunge	*Saussurea* DC.
3	*Homognaphalium* Kirpiczn.	*Gnaphalium* L.
4	*Jurinella* Jaub. & Spach	*Jurinea* Cass.
6	*Lipschitziella* Kamelin	*Saussurea* DC.
7	*Mulgedium* Cass.	*Lactuca* L.
8	*Omalotheca* Cass.	*Gnaphalium* L.
9	*Piloselloides* (Less.) C. Jeffrey ex Cufod.	*Gerbera* L.
10	*Struchium* P. Browne	*Sparganophoros* Vaill.

into six distinct genera, these are *Askelia* W. A. Weber, *Crepedifolium* Sennikov, *Faberia* Hemsley *ex* Forb. and Hemsley, *Pseudoyoungia* D. Maity & Maiti, *Tibetoseris* Sennikov *amend* D. Maity & Maiti and *Youngia* Cassini (*s.s*), but here, for easier understanding, the generic delimitation of the genus *Youngia* Cassini, as considered by Bremer *et al.* (1992), Bremer (1987, 1994, 1996), and Kadreit and Jeffrey (2007), has been followed.

Besides these, in the present treatment the genus *Calyptocarpus* has been included in the list of the Indian Asteraceae. So, in the present study a total number of 1314 taxa of Asteraceae have been considered in 204 genera which are distributed in 20 tribes, after Kadreit and Jeffrey (2007).

Statistical analysis of the 20 tribes of Asteraceae present in India and the numbers of genera and species under each tribe is given in Table 3.2.

A detailed conspectus of the generic distribution of this group in India and their distributional pattern in India and the world is given in Table 3.3.

By analysing the data obtained from Table 3.3, it is clear to us that the position of the dominant genera under the present study differ completely from the dominant genera of the previous study. A comparative account of the ten dominant genera of the present study and that observed by Singh *et al.* (1999) is given in Table 3.4 below.

3.3 Phytogeography of Asteraceae of India

Several workers have attempted to divide the world's flora into different phytogeographic zones or divisions based on the comparative study. Takhtajan (1986) based on the climatic parameter, geographic position, evolutionary history and taxonomic affinities, divided the world's flora into six kingdoms which are as follows.

Holarctic Kingdoms (Holarctics)
Palaeotropical Kingdom (Paleotropics)
Neotropical Kingdom (Neotropics)
Cape Kingdom (Capensis)
Australian Kingdom (Australia)
Holantartic Kingdom (Holantics).

C.B. Clarke (1876), divided the then Indian subcontinent of the British throne in to nine phytogeographical divisions, which are as follows.

1. Deserta
2. North-west Himalaya
3. Central Himalaya
4. East Himalaya
5. Khasia
6. Gangetic Plain
7. Coromandeliana
8. Malabaria
9. Ceylon.

After independence, the political boundary of British-ruled India changed enormously. Balakrishnan (1996) classified present-day India into 11 phytogeographic regions (see Fig. 3.3), based on the phytogeographical zones of the world as mentioned by Takhtajan (1986):

1. North West Himalaya
2. Indo-Gangetic Plain
3. Eastern Himalaya (Arunachal Pradesh, Sikkim)
4. Assam (excluding Arunachal Pradesh and North Bengal of the West Bengal State)
5. Central India
6. Arid Zone
7. North Western Ghats and Northern West Coast
8. South Western Ghats and West Coast and Lakshadweep
9. Deccan
10. Eastern Ghats and Coromandel Coast
11. Andaman and Nicobar Islands.

By giving a critical comparison of the phytogeographic zonation of Takhtajan (1986) and Balakrishnan (1996), it has been decided that present-day India can be divided into 12 phytogeographic zones, in which the Andaman and Nicobar groups of Islands are divided into two separate zones: Andaman Province, in the Indo-Chinese Region, and the Nicobar Islands in Sumatran Province of the Malaysian region. So in the present study the Indian subcontinent is divided in to 12 phytogeographic zones, as follows.

1. North West Himalaya
2. Indo-Gangetic Plain
3. Eastern Himalaya (Arunachal Pradesh, Sikkim)
4. Assam (excluding Arunachal Pradesh and North Bengal of the West Bengal State)
5. Central India
6. Arid Zone
7. North Western Ghats and Northern West Coast
8. South Western Ghats and West Coast and Lakshadweep
9. Deccan
10. Eastern Ghats and Coromandel Coast
11. Andaman Islands
12. Nicobar Islands.

Table 3.2. Conspectus of the tribes of Asteraceae in India.

Sl. No.	Name of the tribe	Genera in India	Genera in the world	% of Indian genera in relation to world	Species in India	Species in the world	% of the Indian species in relation to world species	Distribution in India	Distribution in the world
1	*Anthemideae* Cass.	15	111	13.5	140	1800	7.8	Himalaya, NE India, Gangetic Plain, Himalayan Plateau	Extra Tropical, with main concentration at the Central Asia, Mediterranean region, South Africa
2	*Arctotideae* Cass.	1	17	5.9	2	215	1	Cultivated	South Africa, Nambia, Angola
3	*Astereae* Cass.	24	205	11.7	123	3080	3.9	Himalaya, Tropical parts of the India, South India, North East India, Nilgiri Hills	Cosmopolitan in distribution, often in cultivation, in Africa, America, Asia, Europe, and one species in Arctic region also
4	*Athroismeae* Panero	3	6	50	4	59	6.8	Throughout India, including Himalayan region, South Western Ghats	Eastern Africa mostly, a few in Western Africa, Madagascar, SE Asia
5	*Calenduleae* Cass.	2	12	17	3	120	2.5	Mostly cultivated	Macronesia, N Africa, Mediterranean region, South and Central Europe, Yemen, Iraq to Iran
6	*Cardueae* Cass.	26	73	35.6	164	2360	6.9	Throughout India	Mediterranean region, Central Asia
7	*Cichorieae* Lam. & DC.	27	86	31.4	242	1500	16.1	Mostly concentrated in the Himalayan region, South India, Nilgiri Hills, one species throughout India	Mostly distributed in the Northern Hemisphere and a few species are distributed in the tropical region
8	*Coreopsideae* Lindl.	7	30	23.3	29	550	5.27	Throughout India and often cultivated	Worldwide in distribution, mostly concentrated in America
9	*Eupatorieae* Cass.	8	182	4.39	36	2200	1.7	Throughout India	Western Hemisphere, Pantropical or Subtropical in distribution
10	*Gnaphalieae* (Cass.) Lecoq. & Juillet.	12	185	7	86	1240	7	Throughout India, in NE India and S India	Cosmopolitan but mostly distributed in Australia and in S Africa
11	*Helenieae* Lindl.	1	13	7.6	3	120	2.5	Mostly cultivated	SE Canada, N Mexico, SW America, Temperate E South America
12	*Heliantheae* Cass.	19	113	16.9	50	1500	3.33	Mostly throughout India, including Andaman and Nicobar Islands	Pantropical with most species in Mexico

Continued

Table 3.2. Continued.

Sl. No.	Name of the tribe	Genera in India	Genera in the world	% of Indian genera in relation to world	Species in India	Species in the world	% of the Indian species in relation to world species	Distribution in India	Distribution in the world
13	*Inuleae* Cass.	16	66	24.2	112	687	16.3	Throughout India, mostly concentrated in the Himalayan region	Eurasia, East and South Africa
14	*Madieae* Jepson	1	36	2.7	1	200	0.05	NW Himalaya	W America, SW Canada, NW Mexico
15	*Millerieae* Lindl.	6	34	17.6	8	400	2	Throughout India, often cultivated	Central Mexico, N Andes, Africa, Asia
16	*Mutisieae* Cass.	7	82	8.5	19	950	2	Himalaya, NE India, one species cultivated throughout India	Mostly in South America, but often found in Africa, Madagascar, Asia
17	*Neurolaeneae* Rydb.	1	5	20	1	150	0.6	Throughout India	Tropical Mexico and South America
18	*Senecioneae* Cass.	16	150	10.7	176	3500	5	Mostly in the Himalayan region and often throughout India as cultivated	SW America, Mexico
19	*Tageteae* Cass.	3	32	9.3	8	270	3	Rajasthan, cultivated in most of India	SW America, Mexico
20	*Vernonieae* Cass.	9	118	7.6	90	1000	9	Throughout India	Tropical parts of the world

Note: Based on this table, it is clear that the genera are predominant in five (5) dominant tribes: *Cichorieae* (27), *Cardueae* (26), *Astereae* (24), *Heliantheae* (19); *Inuleae* and *Senecioneae* occupy the fifth position jointly, with 16 genera each.

Table 3.3. Conspectus of the genera of Asteraceae in India and their comparative distribution in India and throughout the world.

Sl. No.	Name of the genus	Number of taxa: Species	Number of taxa: Sub Species	Number of taxa: Variety	Number of taxa: Forma	Total no. of taxa in India	Total no. of taxa in the world (Mabberley, 2008)	Distribution in India	Distribution in the world (Mabberley, 2008)
1	*Acanthospermum* Schrank	1	1	0	0	2	6	Throughout India	Neotropical
2	*Achilla* L.	4	0	0	0	4	115	Eastern Himalaya, NE India, W Himalaya	Europe, Asia, Africa, Some species are naturalized in America, Australia, New Zealand and S Africa
3	*Acroptilon* Cass.	1	0	0	0	1	1	Alpine W Himalaya	Pakistan, Iran, North America
4	*Acmella* Rich ex Pers.	3	0	0	0	3	30	Throughout India, Andaman and Nicobar Islands, E, N, S India, Himalaya	New World
5	*Adenocaulon* Hook.	2	0	0	0	2	5	Temperate Himalaya, C to E Himalaya	Argentina, Chile, Eastern Asia, Nepal, Canada
6	*Adenoon* Dalzell	1	0	0	0	1	0	Endemic in Peninsular India	
7	*Adenostemma* J. R.& G Forst	2	8	0	0	10	26	Throughout India	Pantropical
8	*Ageratina* Spach.	4	0	0	0	4	265	Throughout tropical India to NE India	Tropics and Subtropics of the New World
9	*Ainsliaea* DC.	3	0	0	0	3	50	NE India	China, India, Japan, Nepal, SE Asia
10	*Allardia* Decne	7	0	0	0	7	8	W Himalaya, E Himalaya	Afghanistan, Central Asia, Mongolia, China
11	*Amberboa* (Pers.) Less.	1	0	0	0	1	6	Punjab	E Europe, Caucasus, SW Asia
12	*Ambrosia* L.	2	0	0	0	2	4	NE India	N Mexico, SW USA
13	*Anaphalis* DC.	37	6	0	0	43	110	Throughout India	Asia, N and S America
14	*Anthemis* L.	1	0	0	0	1	175	NW Himalaya	Europe, SW Asia, NE Africa
15	*Arctium* L.	1	0	0	0	1	27	C to W Himalaya	Temperate Eurasia, Introduced elsewhere
16	*Arctotis* L.	2	0	0	0	2	60	Cultivated	South Africa, Namibia, Angola
17	*Arnica* L.	1	0	0	0	1	30	NW Himalaya	In Northern Hemisphere
18	*Artemisia* L.	46	20	6	0	72	522	Himalaya and Himalayan Plateau	Northern Hemisphere, S America, S Africa, Pacific Islands

Continued

Table 3.3. Continued.

Sl. No.	Name of the genus	Number of taxa				Total no. of taxa in India	Total no. of taxa in the world (Mabberley, 2008)	Distribution in India	Distribution in the world (Mabberley, 2008)
		Species	Sub Species	Variety	Forma				
19	*Artemisiella* Ghafoor	1	0	0	0	1	1	W Himalaya	Ladakh, Tibet, Nepal, Bhutan, S China
20	*Aster* L.	28	5	4	3	40	180	Himalaya	Mostly in Eurasia, Africa, one species in Arctic region
21	*Athroisma* DC.	1	0	0	0	1	12	Himalaya to Sub Himalaya	Africa, India, Indonesia, Madagascar, SE Asia
22	*Ayapana* Spach.	1	0	0	0	1	16	Introduced	Neotropics
23	*Bellis* L.	1	0	0	0	1	8	Cultivated	Europe, widely introduced
24	*Bidens* L.	12	2	0	0	14	280	Cosmopolitan	Cosmopolitan, most Species in America
25	*Blainvillea* Cass.	1	0	0	0	1	10	Throughout India	Pantropical
26	*Blepharispermum* DC.	2	0	0	0	2	15	SW Ghats, Peninsular India	Africa, Arabian Peninsula, India, Madagascar, Sri Lanka
27	*Blumea* DC.	32	6	0	0	38	100	Throughout India and Himalaya	Tropical and Subtropical Asia
28	*Blumeopsis* Gagnep.	1	0	0	0	1	1	Throughout India	E Asia
29	*Brachycome* Cass.	2	0	0	0	2	70	NE India	Australia, New Guinea, New Zealand, New Caledonia
30	*Caesulia* Roxb.	1	0	0	0	1	1	Throughout India	Endemic in India
31	*Calendula* L.	2	0	0	0	2	15	Cultivated	Micronesia, N Africa, Mediterranean regions, S and C Europe, Yemen, Iraq and Iran
32	*Callistephus* Cass.	1	0	0	0	1	1	Cultivated	Northern China, Cultivated
33	*Calotis* R. Br.	1	0	0	0	1	30	Himalaya	Australia mostly
34	*Calyptocarpus* Less.	2	0	0	0	2	3	Himalaya	Texas to Guatemala
35	*Carduus* L.	5	0	0	0	5	90	Himalaya	Eurasia and N and C Africa
36	*Carpesium* L.	6	4	0	0	10	25	Himalaya	Mainly Asian, extending up to Europe and Australia
37	*Carthamus*	3	0	0	0	3	20	Himalaya and Sub Himalaya	Iran, Caucasus, E Mediterranean region to C Asia
38	*Catamixis* Thom.	1	0	0	0	1	1	Himalaya	India Himalaya; Endemic
39	*Cavea* W. W. Smith & Small	1	0	0	0	1	1	E Himalaya	NE India to SW China
40	*Centaurea* L.	7	0	0	0	7	250	Throughout India	Eurasia, especially in Irano-Turanian and Mediterranean Regions
41	*Centipeda* Lour.	1	0	0	0	1	10	Throughout India	Southern Hemisphere, and SE Asia
42	*Centrantherum* Cass.	1	0	0	0	1	3	Introduced	Tropical America, Philippines, Australia

43	*Cephalorrhynchus* Boiss	1	0	0	0	1	14	Himalaya	SE Europe to China
44	*Chaetoseris* C. Shih	2	2	0	0	4	18	Himalaya and S India	E Asia
45	*Chondrilla* L.	3	2	0	0	5	25	Himalaya	Eurasia, Mediterranean region
46	*Chromolaena* DC.	1	0	0	2	3	165	Throughout India	One species, Pantropical Weed, others tropical to subtropical
47	*Chrysanthellum* Rich.	1	0	0	0	1	11	Throughout India	Pantropical
48	*Chrysanthemum* Rich	14	0	0	0	14	41	Himalaya	Asia, Europe
49	*Cicerbita* Wallr.	4	0	0	0	4	5	Himalaya	Eurasia
50	*Cichorium* L.	2	0	0	0	2	6	Cultivated	Europe, Mediterranean Region
51	*Cirsium* Mill.	14	7	0	0	21	250	Himalaya	Eurasia, N America, N and E Africa
52	*Cissampelos* (DC) Miq.	8	0	0	0	8	10	Himalaya and S India	E and SE Asia
53	*Conyza* Less.	9	1	0	0	10	100	Tropical India	Tropical and Subtropical Region, Probably native to Africa
54	*Coreopsis* L.	3	1	0	0	4	70	Cultivated	America
55	*Cosmos* Cav.	3	0	0	0	3	28	Cultivated	Mexico, Central America, S America
56	*Cotula* L.	6	0	0	0	6	55	Cultivated	S and E Africa, Australia, S America, Mexico
57	*Cousinia* Cass.	9	0	0	0	9	700	Himalaya	C and W Asia
58	*Crassocephalum* Moench.	1	0	0	0	1	24	Throughout India	Tropical Africa, Madagascar
59	*Cremanthodium* Benth.	16	2	0	0	18	75	Himalaya	China, Tibet, Himalaya
60	*Crepis* L.	12	6	0	0	18	200	Himalaya	Cosmopolitan, except in Australia
61	*Crupina* (Pers.) DC.	1	0	0	0	1	3	W Himalaya	Mediterranean and Irano-Turanian Regions up to Central Asia
62	*Cyanthillium* Bl.	3	0	4	0	7	7	Throughout India	Indian Ocean Island, Indonesia, SE Asia, Tropical Africa
63	*Cyathocline* Cass.	2	0	3	0	5	3	South India	Tropical Asia
64	*Cyanara* L.	2	0	0	0	2	8	Cultivated	Mediterranean region
65	*Dahlia* Cav.	3	0	0	0	3	40	Cultivated	Mexico, Central America, Columbia
66	*Dichrocephala* Hier ex DC.	4	2	0	0	6	10	Himalaya	Africa, Madagascar, Indonesia, Tropical and SW Asia
67	*Dicoma* Cass.	2	0	0	0	2	65	Throughout India	S Africa, Tropical Africa, Madagascar, NE Africa, Asia
68	*Dimorphotheca Vail.*	1	0	0	0	1	20	NW Himalaya	S Africa, Zimbabwe, Angola
69	*Dolomiaea* DC.	2	0	0	0	2	12	Himalaya	C Asia
70	*Doronicum* L.	4	0	1	0	5	40	Himalaya	Eurasia, N Africa
71	*Dubyaea* DC.	2	2	0	0	4	14	Himalaya	Himalaya – China

Continued

Table 3.3. Continued.

Sl. No.	Name of the genus	Number of taxa				Total no. of taxa in India	Total no. of taxa in the world (Mabberley, 2008)	Distribution in India	Distribution in the world (Mabberley, 2008)
		Species	Sub Species	Variety	Forma				
72	*Duhaldea* DC.	9	0	1	0	10	14	Himalaya	Asia, Probably one species in Africa
73	*Echinopsis* L.	4	0	0	0	4	120	Himalaya	Eurasia, N and C Africa
74	*Eclipta* L.	1	0	0	0	1	5	Throughout India	America, Australia
75	*Elephantopus* L.	1	0	0	0	1	12	Throughout India	Pantropical
76	*Eleuthanthera* Poil ex Bose	1	0	0	0	1	1	Throughout India	Neotropical, Adventive at the tropical regions of the Old World
77	*Emilia* Cass.	8	0	2	0	10	100	Peninsular India	Tropical regions mostly, and Africa
78	*Enydra* Lour.	1	0	0	0	1	10	Throughout India	Pantropical
79	*Epaltes* Cass.	3	0	0	0	3	14	South India	Pantropical
80	*Epilasia* (Bunge.) Benth.	1	0	0	0	1	3	Punjab	Caucasus, Near East Central Asia
81	*Erechtites* Raf.	2	0	0	0	2	5	Throughout India	N and S America, West Indies
82	*Erigeron* L.	21	0	3	0	24	400	Himalayan region, Mostly cultivated	N and S America, West Indies, Galapagos Islands, Eurasia
83	*Ethulia* L. f.	2	0	0	0	2	19	Himalaya, E and NE	Africa, S Asia, Indonesia
84	*Eupatorium* L.	11	0	2	0	13	45	Himalayan	Eurasia, N America
85	*Felicia* Cass.	1	0	0	0	1	85	Cultivated	Africa and Arabica
86	*Filago* L.	4	0	0	0	4	46	NW and W Himalaya	Asia, Europe, N Africa, N America
87	*Flaveria* Juss.	2	0	0	0	2	22	Rajasthan	America, one species in Australia
88	*Gaillardia* Foug.	3	0	0	0	3	20	Cultivated	SE Canada, N Mexico, SW America, temperate E South America
89	*Galinsuga* Ruiz.	2	0	0	0	2	15	Throughout India	Neotropical, species is adventive
90	*Gamochaeta* Wedd.	1	0	0	0	1	80	Throughout India	S America
91	*Garhadilous* Jaub.	1	0	0	0	1	2	NW Himalaya	Caucasus, E Asia
92	*Gerbera* L.	5	0	1	0	6	30	Cultivated mostly in Himalaya	China, Kenya, Malawi, Madagascar, Mozambique, S Africa
93	*Glossocardia* Cass.	2	0	0	0	2	11	Throughout India	E Africa, SE Asia, Australia, Pacific Island
94	*Gnaphalium* L.	5	0	0	0	5	80	Throughout India	Cosmopolitan
95	*Gochnatia* Kunth.	2	0	0	0	2	60	NE India, CW Himalaya	Mostly from Cuba, Caribbean Islands
96	*Goniocaulon Cass.*	1	0	0	0	1	1	Throughout India	Tropical Africa , India
97	*Grangea* L.	1	0	0	0	1	10	Throughout India	Africa, Madagascar, Tropical Asia
98	*Guizotia* Cass.	1	0	0	0	1	6	Cultivated	Mostly Eastern Africa, Africa and India

99	*Gynura* Cass.	10	0	2	0	12	40	Andaman and Nicobar Islands and cultivated	Palaeotropical
100	*Helianthus* L.	10	0	0	0	10	50	Cultivated	North America
101	*Helichrysum* Mill.	6	0	2	0	12	600	SW India	Africa, Madagascar, Asia, Europe
102	*Helipterum* DC.	2	0	0	0	2	25	Cultivated	Southern Africa
	1.**Hemistepta Bunge* ex Fisch.	1	0	0	0	1	1	E and NE India	India, Australia
103	*Heteropappus* Less.	3	0	0	0	3	20	Himalaya	C and E Asia
104	*Hieracium* L.	6	0	2	0	8	90	Himalaya	Eurasia, N and S America
105	*Hippolytia* Polij.	1	0	0	0	1	19	C to E Himalaya	Central Asia, Mongolia, China–Himalaya
	2**Homognaphalium* Kirp.	1	0	0	0	1	1	Throughout India	North Africa, Asia
106	*Hypochaeris* L.	2	0	0	0	2	50	N and NE India	Eurasia, N Africa, Columbia, W Venezuela, Chile, Argentina
107	*Ifloga* Cass.	1	0	0	0	1	6	NE India	Asia, Europe, Middle East, Africa
108	*Inula* L.	17	0	3	0	20	100	Throughout India	Old World
109	*Ixeridium* (A. Gray) Tzvelaga	3	0	1	0	4	13	C and E to NE Himalaya	Asia, Malaysia, New Guinea
110	*Ixeris* L.	1	0	0	0	1	10	Himalaya and NE India	Asia, Malaysia, New Guinea
111	*Jurinea* Cass.	5	0	4	0	9	200	W to C Himalaya	Western and Central Asia, Europe, N Africa
112	*Kalimeris* (Cass.) Cass.	1	0	0	0	1	8	NE India	E Asia
113	*Kleinia* Mill.	5	0	0	2	7	50	Cultivated in S India mostly	Africa, Madagascar, Arabia East, India, Sri Lanka
114	*Loeipinia* Pall.	1	0	0	0	1	5	Alpine India	Mediterranean Region, Near to East Central Asia
115	*Lactuca* L.	18	0	4	0	22	60	Himalaya	N Hemisphere, Asia–New Guinea, N America, Mexico
116	*Lagascea* Cav.	1	0	0	0	5	9	Weed	Pantropical
117	*Lagenophora* Cass.	1	0	0	0	1	14	Western India	SE Asia, Australia, Pacific Island, Central America, S America
118	*Laggera* Sch. Bip.	2	0	0	0	2	17	Throughout India	Tropical Africa, Arabia
	3**Lamprachaenium* Benth.	1	0	0	0	1	1	W Ghats	
119	*Lapsana* L.	1	0	0	0	1	1	NW Himalaya	Europe
120	*Lasiopogon* Cass.	1	0	0	0	1	8	E and N India	Africa and Middle East

Continued

Table 3.3. Continued.

Sl. No.	Name of the genus	Number of taxa				Total no. of taxa in India	Total no. of taxa in the world (Mabberley, 2008)	Distribution in India	Distribution in the world (Mabberley, 2008)
		Species	Sub Species	Variety	Forma				
121	*Launea* Cass.	9	0	0	0	9	54	Throughout India	Old World, one species in Australia
122	*Leibnitzia* Cass.	3	0	0	0	3	6	Himalaya	China, Guatemala, India, Japan, Nepal, Siberia, Taiwan, Mexico
123	*Leontopodium* (Pers.) R. Br. Ex Cass.	11	0	2	0	13	58	Himalaya	Asia and Europe
124	*Leucanthemum* Mill.	1	0	0	0	1	1	Cultivated	N Africa, E Mediterranean
125	*Ligularia* Cass.	18	2	4	1	25	125	Himalaya	Eurasia
126	*Logfia* Cass.	1	0	0	0	1	9	Himalaya to NW India	Europe, Middle East and N Africa
127	*Matricaria* L.	4	0	0	0	4	6	Himalaya to N India	Europe, N Africa, Micronesia, N America
128	*Melampodium* L.	1	0	0	0	1	40	Cultivated	Neotropical
129	*Microglossa* DC.	1	0	0	0	1	19	E Himalaya–NE India	Africa, Madagascar, Tropical Asia
130	*Mikania* Willd.	2	0	0	0	2	100	Throughout India, including Andaman and Nicobar Islands	Pantropical weed with few inhabited at Old World
131	*Montonoa* La Llave & Lex.	2	0	0	0	2	25	Throughout India	Mexico, C America, N and S America, N Peru
132	*Moonia* Arn.	2	0	0	0	2	1	S India and W Ghats	SE India, Sri Lanka
	4**Mulgedium* Cass.	3	0	2	0	5	60	Himalaya C E and W India	N Hemisphere, Asia, New Guinea, N America, Mexico, Tropical, S Africa
133	*Myriactis* Less.	5	0	1	0	6	27	Himalaya Nilgiri Hills	Tropical America, E Asia Indonesia, Philippines
134	*Nannoglotis* Maxim.	2	0	0	0	2	9	E–C Himalaya	South Central China
135	*Nanothamus* Thomson	1	0	0	0	1	1	Endemic in W India	India only
136	*Ochrocephala* Dittrich	1	0	0	0	1	1	Himalaya	C and E Africa and India
137	*Olgaea* Iljin	1	0	0	0	1	16	W Himalaya	Afghanistan, Central and E Asia
138	*Oligochaeta* (DC) K. Koch	1	0	0	0	1	4	Throughout India	Irano-Turanian Region, Caucasus, Afghanistan, India
139	*Onopordum* L.	1	0	0	0	1	60	NW Himalaya	WC Asia, Europe, N Africa

	5**Paramicrorhynchus* Krip.	1	0	0	0	1	8	Throughout India	Micronesia, Mediterranean area, E–S Tropical Africa
140	*Parasenecio* W. W. Sm. & Small	4	0	0	0	4	70	Himalaya	Eurasia, Russia, E Asia
141	*Parthenium* L.	1	0	0	0	1	16	Throughout India	America
142	*Pegolettia* Cass.	1	0	0	0	1	9	Rajasthan	S Africa
143	*Pentanema* Cass.	3	0	1	0	4	20	Throughout India	Turkey, C Asia, India, Sri Lanka–Africa
144	*Pericallis* Webb.	1	0	0	0	1	15	Rajasthan	Canary Islands, Madeira
145	*Petasites* Mill.	2	0	0	0	2	20	Himalaya–NE India	Eurasia, N America
146	*Phagnalon* Cass.	3	0	0	0	3	43	W Himalaya	W Asia, Europe N Africa
147	*Phyllocephalum* Bl.	9	0	1	0	10	10	Mainly Peninsular India and Western part of India	India–Java
148	*Picris* L.	1	0	1	0	2	50	Himalaya–Nilgiri Hills	Eurasia, S Tropical Africa, E Asia to Australia and New Zealand
149	*Pluchea* Cass.	7	0	5	0	12	80	Throughout India	Pantropical
150	*Prenanthes* L.	5	0	0	0	5	8	Himalaya–NE India	Eurasia to tropical Africa
151	*Pseudoelephantopus* Rohr.	1	0	0	0	1	2	E and N India	Tropical America
152	*Pseudoconyza* Cuatrec	1	0	0	0	1	1	Indian Plains	Central America, Africa, Asia
153	*Pseudognaphalium* Kirp.	3	0	1	0	4	90	Throughout India	Africa, Asia, Central N and S America, New Zealand
154	*Psiadia* Jacq.	1	0	1	0	2	60	SW Ghats	Tropical Africa, Madagascar
155	*Psychrogeton* Boiss	1	0	2	0	3	20	Alpine W Himalaya	Central Asia, W China
156	*Pterocaulon* Elliot.	1	0	0	0	1	18	Orissa	N and S America, Australia
157	*Pterocypsela* C. Shih	1	0	0	0	1	7	E Himalaya–NE India	E Asia, China, Malaysia–New Guinea
158	*Pulicaria* Gaertn.	12	0	0	0	12	85	NW Himalaya Rajasthan, often in other places	Europe, Africa, Arabia, Asia
159	*Rhaponticum* Vaill	1	0	0	0	1	26	Alpine W Himalaya	Europe, probably native of Australia
160	*Reichardia* Roth.	1	0	1	0	2	8	E to NW India	Micronesia, Mediterranean, E–S Tropical Africa
161	*Rhyncospermum* Reinw.	1	0	0	0	1	1	NW India to temperate Himalaya	E and SE Asia
162	*Rudbeckia* L.	1	0	0	0	1	17	Cultivated	N America
163	*Saussurea* DC.	70	0	7	0	77	300	Himalaya	Temperate Eurasia, N America, probably a native of Australia

Continued

Table 3.3. Continued.

Sl. No.	Name of the genus	Number of taxa				Total no. of taxa in India	Total no. of taxa in the world (Mabberley, 2008)	Distribution in India	Distribution in the world (Mabberley, 2008)
		Species	Sub Species	Variety	Forma				
164	*Sclerocarpus* Jacq.	1	0	0	0	1	8	Throughout India	SW America, C America, Africa
165	*Scorzonera* L.	8	0	0	0	8	180	Himalaya	Eurasia, N Africa
166	*Senecio* L.	49	0	6	0	55	250	In Himalaya, Peninsular India, NE India, Rajasthan	Cosmopolitan, S America, S Africa
167	*Serphidium* Poljakov.	4	0	0	0	4	522	C–NW Himalaya	N Hemisphere, S America, S Africa, Pacific Islands
168	*Serratula* L.	1	0	0	0	1	4	C–W Himalaya	Eurasia,
169	*Sigesbeckia* L.	1	0	0	0	1	8	Throughout India	Pantropical; most of the species in Mexico
170	*Silybum* Adnas.	1	0	0	0	1	3	NW Himalaya	Mediterranean
171	*Solidago* L.	5	0	1	0	6	100	Himalaya, NE India, Cultivated	N America,
172	*Solvia* Ruiz. & Pav.	1	0	0	0	1	1	C to W Himalaya and N India	Widespread as weed
173	*Sonchus* L.	5	1	1	0	7	80	Throughout India	Old World
174	*Soroseris* Stebbins	7	0	1	0	8	9	Throughout India	China–Himalaya Mountain
175	*Sphaeranthus* L.	4	0	0	0	4	41	Throughout India and Andaman and Nicobar Islands	Old World tropics
	6**Sparganophoros* Vail.	1	0	0	0	1	1	Nicobar Islands and S India	Pantropical
176	*Striodiscus* Less.	1	0	0	0	1	5	Cultivated	Southern Africa
177	*Stenoseris* C. Shih	1	0	0	0	1	6	E Himalaya	E Asia
178	*Stevia* Cav.	2	0	0	0	2	230	W Himalaya	Mexico–S America
179	*Spillanthus* Jacq.	5	0	0	0	5	6	Throughout India	Pantropical
180	*Symphyotrichus* Ness.	3	0	0	0	3	92	Himalayan	North American, Mexico S America
181	*Synedrella* Gaertn.	2	0	0	0	1	2	Himalaya and Indian Plains	Neotropical
182	*Synotis* (Clarke) Jeff. & Chen.	23	0	1	0	24	55	Himalaya and NE India	E Asia, especially in Sino-Himalayan Region
183	*Tagetes* L.	5	0	0	0	5	50	Cultivated	Tropical and Subtropical America
184	*Tanacetum* L.	19	0	0	0	19	160	W Himalaya, N India, Upper Gangetic Plain	Europe, N Africa, N America

185	*Taraxacum* Webber & Wigg.	93	0	2	0	95	Species not designated	Himalaya, mostly in W and NW Himalaya	Mainly in N Hemisphere
186	*Thespis* DC.	1	0	0	0	1	3	E, N and NE India	SE Asia
187	*Thymophylla* Lag.	1	0	0	0	1	18	Cultivated	SW USA, Mexico, NW Argentina
188	*Tithonia* Desf. ex Juss.	2	0	0	0	2	11	Throughout India	Pantropical
189	*Tragopogon* L.	5	0	0	0	5	110	N–W Himalaya	Eurasia
190	*Tricholepis* DC	13	0	0	0	13	20	C–NW Himalaya, C India, Peninsular India, W Ghats, Rajasthan	Central Asia, Afghanistan, Nepal, Pakistan, India
191	*Tridax* L.	1	0	0	0	1	30	Throughout India	Neotropical
192	*Tussilago* L.	1	0	0	0	1	1	W Himalaya	Eurasia
193	*Uechtritzia* Feryn	2	0	0	0	2	3	NW Himalaya	NW India
194	*Vaethemia* DC.	1	0	0	0	1	1	NW Himalaya	Iran, Pakistan, Afghanistan
195	*Verbesina* L.	1	0	0	0	1	300	C, E, N and S India	America, Mexico and Andes
196	*Vernonia* Schreb.	55	1	7	0	63	22	Himalaya, N and S India, Andaman and Nicobar Islands, W Ghats, NE India	SE USA, Central Mexico
197	*Vittadinia* Rich.	1	0	0	0	1	20	Cultivated	Australia
198	*Wedellia* Jacq.	6	0	2	0	8	110	E Himalaya SW Ghats, Peninsular, India, Andaman and Nicobar Islands	Pantropical
199	*Wollastonia* DC ex Decne	1	0	0	0	1	1	A and N Island, E Himalaya	Indospecific, Coastal Region
200	*Xanthium* L.	4	0	0	0	4	6	Throughout India, Andaman and Nicobar Islands	Warm temperate region of the world
201	*Xanthopthalmum* Sch.	1	0	0	0	1	2	Cultivated	Europe, Northern Africa, Caucasus, S and N Asia
202	*Youngia* Cass.	12	0	2	0	14	30	Himalaya, NE India, S India	Asia
203	*Zinnia* L.	3	0	0	0	3	25	Cultivated	SW USA, Neotropical Region
204	*Zoegea* L.	1	0	0	0	1	3	NW Himalaya	Turkey, Middle East, Irano-Turanian Region and Central Asia
	Total Genera = 204	Total taxa of Asteraceae in India				1314	+ 6 genera are found in India which are marked with * but have not been accepted by Kadreit and Jeffrey (2007)		

Table 3.4. Dominant genera of Asteraceae based on the number of species.

Present study			Previous study (Singh *et al.*, 2001)		
Sl. No.	Name of the genera	Number of species	Sl. No.	Name of the genera	Number of species
1	*Taraxacum* Webber & Wigg.	95	1	*Taraxacum* Webber & Wigg.	81
2	*Saussurea* DC.	77	2	*Saussurea* DC.	61
3	*Artemisia* L.	72	3	*Vernonia* Schreb.	53
4	*Vernonia* Schreb.	63	4	*Senecio* L.	43
5	*Senecio* L.	55	5	*Artemisia* L.	34
6	*Anaphalis* DC	43	6	*Anaphalis* DC	31
7	*Aster* L.	40	7	*Blumea* DC.	29
8	*Blumea* DC.	38	8	*Lactuca* L.	24
9a	*Chondrilla* L.	25	9	*Aster* L.	23
9b	*Ligularia* Cass.	25	10	*Inula* L.	20
10a	*Erigeron* L.	24			
10b	*Synotis* (Clarke) Jeff. & Chen.	24			

Note: From Table 3.3, again it is clear that there are 81 genera, which are represented by only one species in India, of which 12 genera are monotypic. These 12 monotypic genera are given in Table 3.5. Distribution of genera has been verified with the help of Mabberley's *Plant-Book* (2008).

Table 3.5. Monotypic genera of Asteraceae present in India and their distribution in India and throughout the world.

Sl. No.	Name of the genera	Distribution in India	Distribution in the world
1	*Acroptilon* Cass.	Alpine W Himalaya (Jammu and Kashmir)	Pakistan, Iran, N America
2	*Adenoon* Dalzell	Tamilnadu, Maharastra, Karnataka, Goa: endemic	
3	*Blumeopsis* Gagnep.	Throughout India (West and Eastern Himalaya, Bihar, Karnataka, Andaman and Nicobar Islands)	Bangladesh, Malaya, Sumatra
4	*Caesulia* Roxb.	Uttar Pradesh, Madhya Pardesh, Rajasthan, Punjab, West Bengal, Bihar, Orissa, Andhra Pradesh, Karnataka, Maharastra, Goa	Bangladesh, Nepal, Myanmar
5	*Callistephus* Cass.	Throughout India (Cultivated)	Northern China, cultivated
6	*Catamixis* Thom.	Uttaranchal, Uttar Pradesh, Haryana	Nepal Himalayas
7	*Cavea* W. W. Smith & Small	Eastern Himalaya	China
8	*Eleuthanthera* Poil ex Bose	Throughout India (mainly West Bengal, Bihar, Orissa)	America
9	*Goniocaulon Cass.*	Uttar Pradesh, Madhya Pradesh, Bihar, West Bengal	Pakistan, Africa
10	*Nanothamus* Thomson	Karnataka, Maharastra	Endemic in India
11	*Rhyncospermum* Reinw.	Jammu and Kashmir, Himachal Pradesh, Uttaranchal, Sikkim, Meghalaya, Nagaland	Myanmar, Japan, Malaya, Java
12	*Tussilago* L.	Jammu and Kashmir, Himachal Pradesh, Uttaranchal	NW Asia, N Africa, Europe

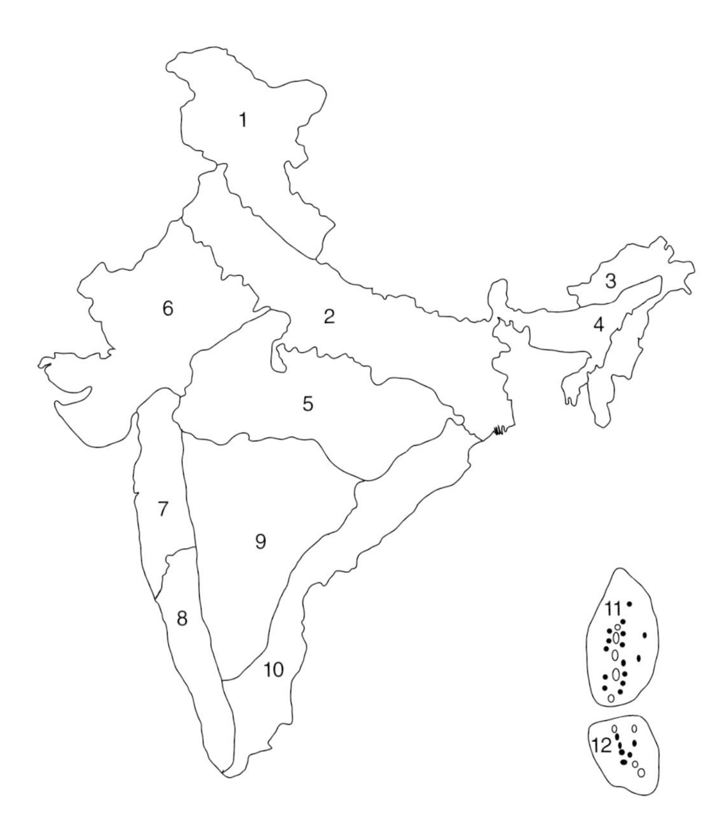

Fig. 3.3. Phytogeographic zones of India (after Balakrishnan, 1996).

India is situated in the confluence of the three major global biogeographic realms: Afrotropical, Eurasian and Indo-Malayan, and the flora of India shows an admixture of the floristic components of these three zones of the globe. Due to territorial contiguity of the country with the Middle East, Central Asia, China, and the Eastern Asiatic region, there is a close affinity between the Asteraceae of India and the Asteraceae of those regions. Based on the migration routes of the floristic elements from the adjoining area to the Indian subcontinent, as shown by Chatterjee (1940 and 1962; see Fig. 3.4), it has been observed that the members of the Asteraceae present in the Western Himalayas have a similarity to those of Central Asia, the Mediterranean region and Europe; in the same way, Asteraceae of the Eastern Himalayas and of the north-east region show a similarity with those of the Indo-Malayan and Indo-Chinese regions. Besides these, Indian Asteraceae is also characterized by the presence of floristic elements of distant places, which have been found in the country for many years. These are represented by the Australian elements, African elements, American elements, etc. A list of such 38 species of Asteraceae with disjunct distribution is given in Table 3.6.

According to Steere and Inoue (1972), such discontinuous distribution of a particular species, or vicariads, in countries now widely separated by vast oceanic barriers has considerable phytogeographical significance, as it provides some suggestive insight into different episodes of the climatic and geographical history of the earth that might have influenced the process of evolution, migration, and even survival of the floras.

Rao *et al.* (1996) show that the Indian Asteraceae consists of 16 different floristic groups based on phytogeographical affinity. In the present study, it has been observed that Indian Asteraceae possesses 18 different groups. Two new groups which are considered here are marked with an asterisk (*). All the genera are arranged alphabetically under each group for better understanding.

Widespread:

Ambrosia L., *Bidens* L., *Chrysanthellum* Rich., *Conyza* Less., *Cotula* L., *Eclipta* L., *Enydra* Lour., *Erigeron* L., *Gnaphalium* L., *Hypochaeris* L., *Pluchea* Cass., *Wedellia* Jacq., *Xanthium* L., etc.

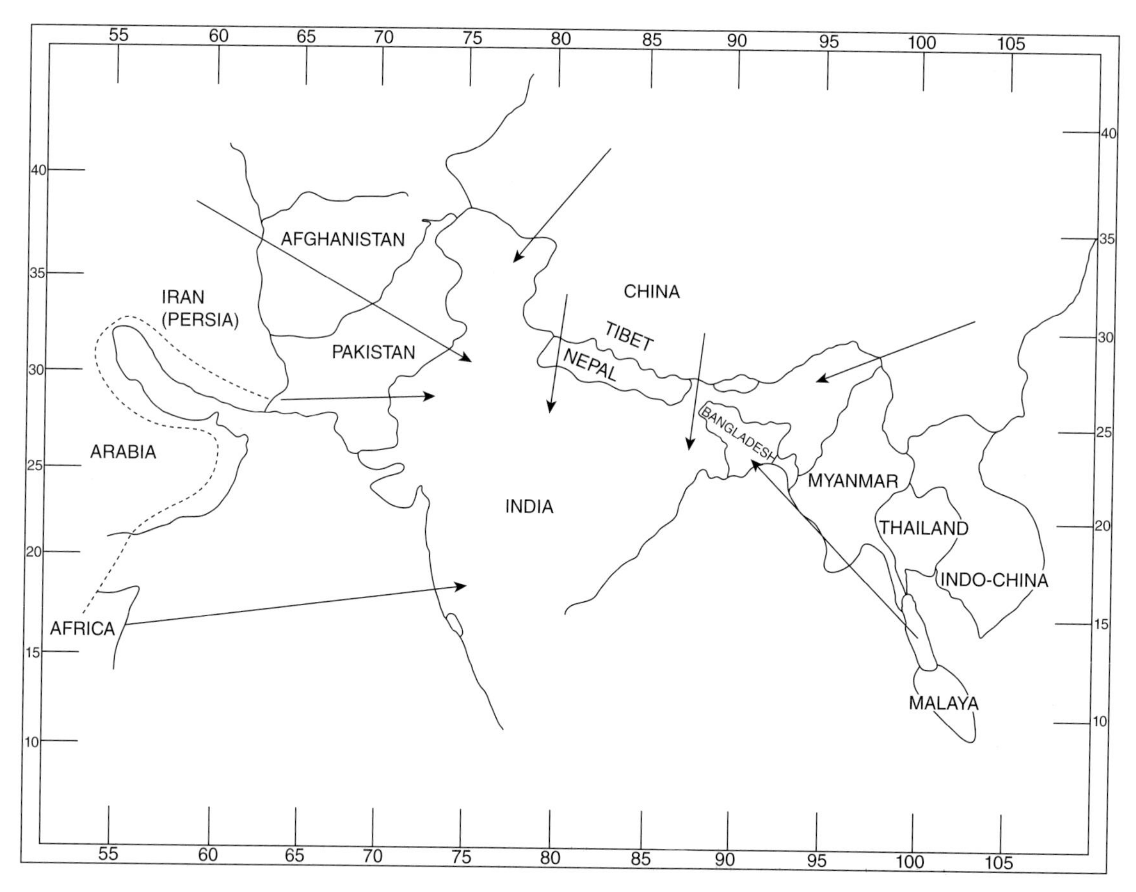

Fig. 3.4. The routes of migration of the floristic elements from adjacent regions to the Indian subcontinent (after Chatterjee, 1940 and 1962).

Pantropical:

Adenostemma J. R.& G Forst, *Blainvillea* Cass., *Coreopsis* L., *Eupatorium* L., *Sigesbeckia* L., *Spillanthus* Jacq., *Verninia* Schreb., etc.

Mediterranean:

Amberboa (Pers.) Less., *Carduus* L., *Ifloga* Cass., *Matricaria* L., *Pentanema* Cass., *Reichardia* Roth., *Scorzonera* L., etc.

Palaeotropical:

Crepis L., *Crassocephalum* Moench., *Dicoma* Cass., *Dimorphotheca* Vail., *Gerbera* L., *Guizotia* Cass., *Helichrysum* Mill., *Kleinia* Mill., *Laggera* Sch. Bip., *Pegolettia* Cass., *Psiadia* Jacq., etc.

Eurasian:

Chondrilla L., *Doronicum* L. , *Lactuca* L., *Lapsana* L., *Ligularia* Cass., *Picris* L., *Pulicaria* Gaertn., *Sonchus* L., *Tragopogon* L., *Tussilago* L. etc.

North temperate:

Achilla L., *Adenocaulon* Hook., *Arnica* L., *Aster* L., *Artemisia* L., *Cicerbita* Wallr., *Chrysanthemum* Rich, *Cirsium* Mill., *Tanacetum* L., *Taraxacum* Webber & Wigg. etc.

Neotropical:

Acanthospermum Schrank., *Dahlia* Cav., *Galinsuga* Ruiz, *Lagascea* Cav., *Montonoa* La Llave & Lex., *Sclerocarpus* Jacq., *Tithonia* Desf. ex Juss., *Tridax* L., *Zinnia* L., etc.

Australian:

Brachycome Cass., *Calotis* R. Br., *Erechtites* Raf., etc.

***Asiatic:**

Caesulia Roxb., *Cyathocline* Cass., *Heteropappus* Less., *Ixeris* L., etc.

North America:

Flaveria Juss., *Gaillardia* Foug., *Prenanthes* L., *Solidago* L., etc.

Table 3.6. Disjunct distribution of some species of Asteraceae in India and the world.

Sl. No.	Name of the species	Distribution in India	Distribution in the world
1	*Acanthospermum hispidum* DC.	Western Himalaya, Eastern Himalaya, Rajasthan, Bihar, West Bengal, Orissa	Africa, South America, Nepal
2	*Achillea alpina* L.	Meghalaya	Europe
3	*Achiellea millefolium* L.	Western Himalaya, Tamil Nadu	Central Asia, Europe, America
4	*Ainsliaea aptera* DC.	Western Himalaya, Eastern Himalaya	Pakistan, Bhutan, China
5	*Ambrosia artemisiifolia* L.	Meghalaya	America, China
6	*Anaphilis margaritacea* (L.) Benth.	Western Himalaya, Eastern Himalaya, North East India	South East Asia, North America
7	*Arctium lappa* L.	Western Himalaya	China, Tibet, West Asia, Europe
8	*Artemisia dracunculus* L.	Western Himalaya	China, Tibet, West Asia
9	*Artemisia nilagirica* (C.B. Clarke) Pamp.	Western Himalaya, Eastern Himalaya, Western Ghats	Central Asia, Africa, Nepal, Sri Lanka, Java
10	*Bidens biternata* (Lour.) Merr. & Sherff.	India	Africa, Sri Lanka, Australia
11	*Bidens sulphurea* (Cav.) Sch. – Bip.	India	Africa, Central and South America
12	*Blainvillea acmella* (L.) Philipson	Gangetic plain, Western Himalaya	Africa, America, Australia, Sri Lanka
13	*Crupina vulgaris* Cass.	Jammu and Kashmir, Gujarat	Europe
14	*Elephantopus scaber* L.	India	Africa, Australia
15	*Eleutheranthera ruderalis* (Sw.) Sch. – Bip.	West Bengal, Bihar, Orissa	America
16	*Emilia sonchifolia* (L.) DC.	India	China, Africa, Central Asia, Australia
17	*Erechtites valerianaefolia* (Wolf) DC.	Uttar Pradesh, Tamil Nadu, Madhya Pradesh	Sri Lanka, Central and South America, China, Japan, Australia
18	*Erigeron karvinskianus* DC.	Western Himalaya, North East India	Europe, America, Australia, Nepal, Bhutan, Myanmar, Japan, Mexico
19	*Ethulia megacephala* Sch.- Bip. ex Miq.	Western Himalaya, Eastern Himalaya, North East India	Europe, China
20	*Eupatorium nodiflorum* Wall. ex DC.	Jammu and Kashmir, North East India	Central Asia, Europe
21	*Filago hurdwarica* (Wall. ex DC.) Wagenitz	Western Himalaya	West Asia, Tibet, Europe
22	*Gnaphalium polycaulon* Pers.	India	Australia, Myanmar, China, Japan, Senegal
23	*Hemistipta lyrata* Bunge ex Fisch.	Western Himalaya, Eastern Himalaya, North East India	Nepal, Myanmar, China, Australia
24	*Lentopodium brachyactis* Gaud.	Western Himalaya, Eastern Himalaya	Central Asia
25	*Picris hieracioides* L.	Western Himalaya	Europe, Australia, New Zealand
26	*Senecio vulgaris* L.	Jammu and Kashmir, Gujarat, Tamil Nad	North Africa, Europe
27	*Siegesbeckia orientalis* L.	In different hill stations	North Africa, Europe
28	*Silybum marianum* (L.) Gaertn.	Western Himalaya, Tamil Nadu	Europe, Africa
29	*Solvia anthemifolia* (Juss.) R. Br.	Uttar Pradesh	South America, Australia
30	*Struchium sparganophorum* (L.) Kuntze	Western Ghats, Andaman and Nicobar Islands	Africa, Central and South America
31	*Syndrella vallis* (Less.) A. Gray	Maharashtra	America
32	*Taraxacum baltistanicum* van Soest	Western Himalaya	Central Asia, Australia
33	*Taraxacum ceratophorum* (Ledeb.) DC.	Jammu and Kashmir	Europe, America

Continued

Table 3.6. Continued.

Sl. No.	Name of the species	Distribution in India	Distribution in the world
34	*Taraxacum duplidens* Lindberg f.	Himachal Pradesh	France, Germany
35	*Tussilago farfara* L.	Western Himalaya	Europe, Africa, America
36	*Vittadinia australis* A. Rich.	Eastern Himalaya	Africa, Australia, New Zealand, Sri Lanka
37	*Wedelia biflora* (L.) DC.	West Bengal, Maharashtra	Africa, Indo-China, Japan, Australia
38	*Zoegia purpurea* Fresen.	Jammu and Kashmir	Central Asia

Western and Central Asiatic:
Acroptilon Cass., *Allardia* Decne, *Cousinia* Cass., *Garhadilous* Jaub., *Olgaea* Iljin, *Phagnalon* Cass., *Psychrogeton* Boiss., *Zoegea* L., etc.

Afro – Asiatic – Australian:
Blumea DC., *Blepharispermum* DC., *Echinopsis* L., *Glossocardia* Cass., *Grangea* L., *Gynura* Cass., *Myriactis* Less., *Sphaeranthus* L., etc.

Indo-Malaysian:
Adenoon Dalzell, *Centipeda* Lour., *Goniocaulon* Cass., *Moonia* Arn., etc.

***Micronesian:**
Microglossa DC., *Phyllocephalum* Bl., etc.

Indian:
Adenoon Dalzell, *Caesulia* Roxb., *Catamixis* Thom., *Cyathocline* Cass., *Nanothamus* Thomson, etc.

India-wide:
Blumeopsis Gagnep., *Dubyaea* DC., *Cremanthodium* Benth., *Rhyncospermum* Reinw., *Tricholepis* DC., *Youngia* Cass., etc.

Euro-Siberian:
Anaphalis DC., *Anthemis* L., *Centaurea* L., *Cichorium* L., *Inula* L., *Leontopodium* (Pers.) R. Br. *ex* Cass., *Serratula* L., etc.

Sino-Japanese:
Soroseris Stebbins, *Thespis* DC. etc.

3.4 Endemism in the Indian Asteraceae

When a species becomes confined to a particular geographical area, it is known as the endemic species. Phytogeographically, endemism plays a significant role as it provides an insightful view of the centre of diversity, vicariance, and the adaptive evolution of the floristic elements of a particular area, and highlights the indigenous nature of the plant diversity of that country or region.

The strategic position of the Indian subcontinent at the northern hemisphere, and its wide latitudinal distribution, helps to share a large number of plant taxa in its adjoining areas. These numbers of foreign elements are so high that it was believed Indian flora was nothing but an admixture of floras of adjoining countries (Hooker, 1906, Champion and Trevor, 1938). But Chatterjee (1940) by his analysis of the Indian flora was able to change this ideology and prove that India has a flora of its own.

According to Karthikeyan (2000), out of an estimated 20,074 Indian Angiosperms, 5752 taxa (28.65%) are endemic elements of the Indian flora and are distributed in different phytogeographical regions of the Indian subcontinent. A critical perusal of the literature (Nayar and Karthikeyan, 1981; Jain and Sastry, 1983; Jain and Sastry, 1984; Ahmedullah and Nayar, 1987; Nayar, 1996; Nayar and Sastry, 1987–1990; Rao and Datt, 1996; Lakshminarasimhan and Rao, 1996; Ahmedullah, 2000; Sharma and Singh, 2001) reveals that India contains about 242 taxa of endemic Asteraceae within the political boundary of the Indian subcontinent. This present study reveals that there are 228 endemic taxa of Asteraceae present within the Indian subcontinent, of which 108 taxa are distributed in the Himalayan region (Tables 3.7 and 3.8), 89 taxa present in the Western Ghats and Deccan Peninsula (Table 3.9), 16 species distributed in the north-east region (Table 3.10) and the residual 15 taxa (Table 3.11) distributed in other parts of the Indian subcontinent. There are 14 such species which lost their endemic status when they were found to be distributed in other localities, or their taxonomic delimitation has been changed. More critical and detailed studies will change the present status.

It is very interesting to note that genera such as *Caesulia* Roxb., *Catamixis* Thomson, *Cavea* W. W. Smith & Small, which are previously known as the endemic members of Asteraceae in the Indian subcontinent, lost their endemic status when their new distributional area was recorded. So at present, out of the 204 genera of the Asteraceae recorded from India, four genera are endemic to

Table 3.7. Endemic species of Asteraceae distributed in Western and North Western Himalayas.

Sl. No.	Name of the taxa	Distribution
1	*Aster incise* Pamp. var. *kunawarensis* Pamp.	WH (Himachal Pradesh at 3000 m)
1a	*Aster laka* C.B. Clarke	NW Himalaya (Jammu and Kashmir)
2	*Aster molliiusculus* (Lindl. ex DC.) C.B. Clarke var. *minor* Grierson.	NW Himalaya (Himachal Pradesh)
3	*Anaphalis himachalensis* Goel	NW Himalaya (Lahul–Spiti)
4	*Artemisia elegantissima* Pamp. var. *kumaonensis* Pamp.	WH (Uttaranchal; Kumaon: Askote)
5	*Artemisia eriocephala* Pamp.	WH (Uttaranchal; Garhwal: Niti)
6	*Artemisia filiformilobulata* Y.R. Ling & Puri	WH (Himachal Pradesh, Uttaranchal)
7	*Artemisia incisa* Pamp. var. *kunawarensis* Pamp.	NW Himalaya (Himachal Pradesh)
8	*Artemisia strongylocephala* Pamp.	NW Himalaya
9	*Bidens minuta* Miré & H. Gillet	NW Himalaya (Jammu and Kashmir)
10	*Cacalia levingii* (C.B. Clarke) R. Mathur	WH (Jammu and Kashmir)
11	*Chondrilla setulosa* C.B. Clarke & Hook. *f.*	WH (Jammu and Kashmir)
12	*Cicerbita filicina* (Duthie ex Stebbins) Aswal & Goel	NW Himalaya
13	*Cremanthodium arnicoides* R.D. Good	NW Himalaya
14	*Crepis dachigamensis* Singh	WH (Jammu and Kashmir)
15	*Doronicum roylei* DC. var. *epapposa* Hook. *f.*	WH (Jammu and Kashmir)
16	*Doronicum pardalianches* L.	WH (Jammu and Kashmir)
17	*Lactuca benthamii* C.B. Clarke	WH (Jammu and Kashmir)
18	*Lactuca decipiens* C.B. Clarke var. *multifida* Hook. *f.*	WH (Jammu and Kashmir)
19	Ligularia jacquemontiana (Decne.) M.A. Rau	NW Himalaya
20	*Olgaea thomsonii* (Hook. f.) Iljin	WH (Jammu and Kashmir)
21	*Saussurea atkinsonii* C.B. Clarke	WH (in cold desert)
22	*Saussurea sudhanshui* Hajra	WH (Uttaranchal)
23	*Taraxcum ambylepidocarpum* van Soset	WH (Jammu and Kashmir)
24	*Taraxacum aurorum* van Soset	WH (Jammu and Kashmir, Himachal Pradesh)
25	*Taraxacum banhyhalensis* van Soset	WH (Jammu and Kashmir)
26	*Taraxacum coronatum* Hand. Mazz.	WH (Jammu and Kashmir)
27	*Taraxacum eriocarpum* H. Hartm.	WH (Jammu and Kashmir)
28	*Taraxacum flavum* van Soset	WH (Jammu and Kashmir)
29	*Taraxacum forestii* van Soset	WH (Jammu and Kashmir)
30	*Taraxacum fulvescens* van Soset	WH (Jammu and Kashmir)
31	*Taraxacum falvobrunneum* van Soset	WH (Jammu and Kashmir)
32	*Taraxacum harbhajan-singhii* van Soset	WH (Jammu and Kashmir)
33	*Taraxacum harbhajan-singhii* van Soset var. *pahalgamense* van Soset	WH (Jammu and Kashmir)
34	*Taraxacum helianthum* van Soset	WH (Jammu and Kashmir)
35	*Taraxacum heteroloma* Hand. Mazz.	WH (Jammu and Kashmir)
36	*Taraxacum heybroeckii* van Soset	WH (Jammu and Kashmir)
37	*Taraxacum laevigatum* (Willd.) DC.	WH (Jammu and Kashmir)
38	*Taraxacum lahulense* van Soset	WH (Jammu and Kashmir)
39	*Taraxacum latibasis* van Soset	WH (Jammu and Kashmir)
40	*Taraxacum lobbichleri* van Soset	WH (Jammu and Kashmir)
41	*Taraxacum longicarpum* van Soset	WH (Jammu and Kashmir, Uttaranchal)
42	*Taraxacum melleum* van Soset	WH (Jammu and Kashmir)
43	*Taraxacum nagaricum* van Soset	WH (Jammu and Kashmir)
44	*Taraxacum nivale* J. Lange	WH (Jammu and Kashmir)
45	*Taraxacum padulosum* (Scop.) Schlech. var. *tenuifolium* (Hoppe) Hand. Mazz.	WH (Jammu and Kashmir)
46	*Taraxacum parvuliforme* van Soset	WH (Jammu and Kashmir)
47	*Taraxacum phoenicolepis* van Soset	WH (Jammu and Kashmir)
48	*Taraxacum polyodon* Dahlstedt	WH (Jammu and Kashmir)
49	*Taraxacum primogenium* Hand. Mazz.	WH (Jammu and Kashmir)

Continued

Table 3.7. Continued.

Sl. No.	Name of the taxa	Distribution
50	*Taraxacum pseudobicorne* van Soset	WH (Jammu and Kashmir)
51	*Taraxacum pseudoeriopodum* van Soset	WH (Jammu and Kashmir)
52	*Taraxacum pseudostenolepium* van Soset	WH (Jammu and Kashmir, Uttaranchal)
53	*Taraxacum pseudostevenii* van Soset	WH (Jammu and Kashmir)
54	*Taraxacum sherriffi* van Soset	WH (Himachal Pradesh)
55	*Taraxacum spiticum* van Soset	WH (Jammu and Kashmir; Himachal Pradesh)
56	*Taraxacum staticifolium* van Soset	WH (Jammu and Kashmir)
57	*Taraxacum stereodiforme* van Soset	WH (Jammu and Kashmir)
58	*Taraxacum stevenii* (Spreng.) DC.	WH (Jammu and Kashmir)
59	*Taraxacum stewartii* van Soset	WH (Jammu and Kashmir)
60	*Taraxacum tenebrystylum* van Soset	WH (Jammu and Kashmir)
61	*Taraxacum tricuspidatum* van Soset	WH (Jammu and Kashmir)
62	*Taraxacum violaceo-maculatum* van Soset	WH (Jammu and Kashmir)
63	*Taraxacum vulpinum* van Soset	WH (Jammu and Kashmir)
64	*Taraxacum xanthophyllum* Haglund	WH (Jammu and Kashmir)
65	*Tricholepis elongata* DC.	WH (Jammu and Kashmir)
66	*Uechtritzia lacei* (G. Watt) C. Jeffrey	NW Himalaya

Table 3.8. Endemic species Asteraceae distributed in Eastern Himalayas.

Sl.No.	Name of the taxa	Distribution
1	*Aster platylepis* Y.-L.Chen	Eastern Himalaya (Sikkim)
2	*Ainsliaea angustifolia* Hook.f. & Thomson ex Clarke	Eastern Himalaya
3	*Artemisia thellungiana* Pamp.	Eastern Himalaya (Sikkim)
4	*Blumea sikkimensis* Hook.f.	Eastern Himalaya (Sikkim)
5	*Cacalia chola* (W.W.Sm.) R. Mathur	Eastern Himalaya (Sikkim)
6	*Inula macrosperma* Hook.f.	Eastern Himalaya (Alpine region of Sikkim)
7	*Lactuca cooperi* J. Anthony	Eastern Himalaya (Sikkim)
8	*Ligularia dux* (C.B.Clarke) R.Mathur	Eastern Himalaya (Sikkim)
9	*Ligularia hookeri* (C.B. Clarke) Hand.-Mazz.	Eastern Himalaya (Sikkim)
10	*Ligularia kingiana* (W.W. Sm.) R. Mathur	Eastern Himalaya (Sikkim)
11	*Ligularia pachycarpa* (C.B. Clarke ex Hook.f.) Kitam.	Eastern Himalaya (Sikkim)
12	*Saussurea andersonii* C.B. Clarke	Eastern Himalaya (Sikkim and Darjeeling of West Bengal)
13	*Saussurea gossypiphora* D. Don. var. *lilliputa* Lipschitz	Eastern Himalaya (Sikkim)
14	*Saussurea laneana* W.W. Sm.	Eastern Himalaya (Sikkim)
15	*Saussurea nimborum* W.W. Sm.	Eastern Himalaya (Sikkim)
16	*Saussurea obscura* Lipsch.	Eastern Himalaya (Sikkim)
17	*Saussurea pantlingiana* W.W.Sm.	Eastern Himalaya (Sikkim)
18	*Senecio tetrandrus* Buch.-Ham. ex Wall.	Eastern Himalaya (Sikkim and Darjeeling of West Bengal)
19	*Taraxacum insigne* Dahlst. ex G.E. Haglund	Eastern Himalaya (Darjeeling of West Bengal)

this country: *Adenoon* Dalzell, *Lamprachenium* Benth. *Leucoblepharis* Arn. and *Nannothamus* Thomson.

Another feature revealed from the list of the endemic Asteraceae taxa of India is that, based on the species number, *Phyllocephalum* Blume contains the highest number of endemic species. The five most dominant genera on the basis of numbers of endemic species are listed in Table 3.12.

It is also interesting to note that there is not a single endemic species present on the Nicobar Islands.

3.5 Exotic Asteraceae

Diversity in the soil types, climatic condition, altitudinal variation and other variation of other ecological factors have favoured the establishment of

Table 3.9. Endemic plants of peninsular India, including Western Ghats (WG).

Sl. No.	Name of the taxa	Distribution
1	*Adenoon indicum* Dalzell	Tamil Nadu and Kerala
2	*Adenostemma lavenia* (L.) Kuntze var. *rugosum* (DC.) Ram Lal	Maharastra, Tamil Nadu and Karnataka
3	*Anaphalis aristata* DC.	WG and C. India (Rajasthan, Tamil Nadu and Kerala)
4	*Anaphalis barnesii* Fischer	WG (Kerala)
5	*Anaphalis barnesii* C.E.C. Fisch.	WG (Idduki district, Travancore)
6	*Anaphalis beddomei* Hook. *f.*	WG (Tamil Nadu and Kerala; rare)
7	*Anaphalis elliptica* DC.	WG (Tamil Nadu)
8	*Anaphalis leptophylla* (DC.) DC.	WG (Kerala, Tamil Nadu)
9	*Anaphalis meeboldii* W.W. Smith	WG (Kerala, Tamil Nadu)
10	*Anaphalis neelgerryana* (Sch.-Bip. ex DC.) DC.	WG (Kerala, Tamil Nadu)
11	*Anaphalis notoniana* (DC.) DC.	WG (Kerala, Tamil Nadu)
12	*Anaphalis travancorica* W.W. Smith	WG (Kerala, Tamil Nadu and Karnataka)
13	*Anaphalis wightiana* (DC.) DC.	WG (Kerala, Tamil Nadu and Karnataka)
14	*Blepharispermum subsessile* DC.	Madhya Pradesh, Maharashtra and Karnataka
15	*Blumea belangeriana* DC.	WG (Maharastra), Nagaland
16	*Blumea venkataramanii* R.S. Rao & Hemadri	WG (Maharastra)
17	*Carpesium cernuum* L. var. *ciliatum* Hook. *f.*	WG (Kerala)
18	*Cyathocline lutea* Law ex Wight	WG (Deccan Plataeu)
19	*Cyathocline perpurea* (Buch.-Ham.) var. *alba* Santapau	WG (Maharastra)
20	*Cyathocline perpurea* (Buch.-Ham.) var. *bicolor* Santapau	WG (Maharastra)
21	*Cissampelopsis ansteadi* (Tad. & Jacob) C. Jeffrey & Y.L. Chen	WG (Tamil Nadu)
22	*Cissampelopsis calcadensis* (Ramaswami) C. Jeffrey & Y.L. Chen	WG (Tamil Nadu)
23	*Emilia ramulosa* Gamble	WG (Madras, Travancore)
24	*Emilia sonchifolia* (L.) DC. var. *mucronata* C.B. Clarke	WG (Tamil Nadu)
25	*Emilia zeylanica* C.B. Clarke var. *paludosa* Gamble	WG (Tamil Nadu)
26	*Erigeron wightii* DC.	WG (Tamil Nadu)
27	*Epaltes pygmaea* DC.	WG (Tamil Nadu)
28	*Gynura nitida* DC.	WG (Tamil Nadu)
29	*Gynura travancorica* W.W. Sm.	WG (Travancore, Tamil Nadu)
30	*Helichrysum perlanigerum* Gamble	WG (Madras, Travancore)
31	*Helichrysum wightii* C.B. Clarke ex Hook.f.	WG (Tamil Nadu)
32	*Kleinia balsamica* (Dalzell & A. Gibson) M.R. Almeida	WG (Maharastra)
33	*Kleinia shevaroyensis* (Fyson) Uniyal	WG Shevaroy Hills
34	*Myriactis wightii* DC. var. *bellidioides* Hook.f.	WG (Tamil Nadu, Nilgiri Hills, at 2600 m)
35	*Nanothamnus sericeus* Thomson	WG (Tamil Nadu)
36	*Phyllocephalum hookeri* (C.B. Clarke) Uniyal	WG (Maharastra)
37	*Phyllocephalum mayurii* (C. Fischer) Narayana	WG (Karnataka)
38	*Phyllocephalum rangacharii* (Gamble) Narayana	WG (Tamil Nadu, Kerala)
39	*Phyllocephalum ritchiei* (Hook.f.) Narayana	WG (Maharastra, Karnataka)
40	*Phyllocephalum sengaltherianum* (Narayana) Narayana	WG (Tamil Nadu)
41	*Phyllocephalum tenue* (C.B. Clarke) Narayana	Maharastra, Goa and Karnataka
42	*Pseudojacoboea lavandulaefolius* (DC.) R. Mathur	C and S India (Rajasthan, Tamil Nadu, Kerala)
43	*Psidia ceylanica* (Arn.) Grierson var. *beddomei* (Gamble) Chandrasekaran	WG and Deccan region (Tamil Nadu, Kerala)
44	*Senecio belgaumensis* C.B. Clarke	Maharastra, Goa and Karnataka
45	*Senecio candicans* DC.	WG (Tamil Nadu)
46	*Senecio gibsoni* Hook.f.	WG (Maharastra)
47	*Senecio hewrensis* Hook.f.	WG (Maharastra)
48	*Senecio hohenackeri* Hook.f.	WG (Andhra Pradesh, Tamil Nadu)
49	*Senecio intermedicus* Wight	WG (Tamil Nadu)
50	*Senecio kundaicus* C. Fischer	WG (Tamil Nadu)

Table 3.9. Continued.

Sl. No.	Name of the taxa	Distribution
51	*Senecio lawsoni* Gamble	WG (Tamil Nadu)
52	*Senecio lessingianus* C.B. Clarke	WG (Tamil Nadu)
53	*Senecio mayurii* C. Fischer	WG (Tamil Nadu)
54	*Senecio multiceps* Balakr.	WG (Tamil Nadu)
56	*Senecio neelgherryanus* DC.	WG (Tamil Nadu)
57	*Sonchus jaini* Chandrabose et. al.	WG (Tamil Nadu, on grassy slopes up to an altitude of 2000 m)
58	*Tricholepis amplexicaulis* C.B. Clarke	Maharashtra, Karnataka and Kerala
59	*Tricholepis angustifolia* DC.	Karnataka and Kerala
60	*Vernonia anamallica* Beddome ex Gamble	WG (Kerala)
61	*Vernonia anaimudica* Balakr. & Nair	WG (Kerala)
62	*Vernonia beddomei* Hook.f.	WG (Kerala)
63	*Vernonia bourdillonii* Gamble	WG (Kerala)
64	*Vernonia bourneana* W.W. Smith	WG (Kerala)
65	*Vernonia comorinensis* W.W. Smith	WG (Kerala)
66	*Vernonia fysonii* Calder	WG (Tamil Nadu)
67	*Vernonia gossypina* Gamble	WG (Tamil Nadu)
68	*Vernonia heynei* Beddome ex Gamble	WG (Tamil Nadu)
69	*Vernonia indica* Wallich ex C.B. Clarke	WG (Tamil Nadu)
70	*Vernonia malabarica* Hook.f.	WG (Tamil Nadu)
71	*Vernonia meeboldii* W.W. Smith	WG (Tamil Nadu)
72	*Vernonia multibracteata* Gamble	WG (Tamil Nadu)
73	*Vernonia peninsularis* (C.B. Clarke) C.B. Clarke ex Hook.f.	WG (Tamil Nadu)
74	*Vernonia pulneyensis* Gamble	WG (Tamil Nadu)
75	*Vernonia ramaswamii* Hutch.	WG (Tamil Nadu)
76	*Vernonia rauii* Uniyal	WG (Tamil Nadu)
77	*Vernonia saligna* DC. var. *nilghirensis* Hook.f.	WG (Tamil Nadu)
78	*Vernonia salvifolia* Wight	WG (Tamil Nadu)
79	*Vernonia shevaroyensis* Gamble	WG (Tamil Nadu)
80	*Vernonia travancorica* Hook.f.	WG (Tamil Nadu)
81	*Youngia nilgiriensis* Babcock	WG (Tamil Nadu)

Table 3.10. List of endemic Asteraceae of North East India, including Assam.

Sl. No.	Name of the taxa	Distribution
1	*Aster trinervius* Roxburgh ex D. Don var. wattii (C.B. Clarke) Griers.	NE India (Nagaland, Manipur)
2	*Artemisia indica* var. *dissecta*	NE India (Assam)
3	*Dichrocephala hamiltonii* Hook.f.	NE India (Assam)
4	*Inula kalapani* C.B. Clarke	NE India (Khasi Hills; Meghalaya)
5	*Ligularia japonica* Less.	NE India (Meghalaya)
6	*Myriactis assamensis* C.E.C. Fisch.	NE India (Meghalaya, Nagaland)
7	*Petasites kamengicus* Deb	NE India (Arunachal Pradesh)
8	*Prenanthes khasiana* C.B. Clarke	NE India (Meghalaya)
9	*Saussurea nagensis* C.E.C. Fisch.	NE India (Assam, Nagaland)
10	*Senecio mishmi* C.B. Clarke	NE India (Arunachal Pradesh)
11	*Senecio rhabdos* C.B. Clarke	NE India (Nagaland)
12	*Synotis borii* (Raizada) R. Mathur	NE India
13	*Synotis jowaiensis* (Balakr.) R. Mathur	NE India (Jowai, Meghalaya)
14	*Synotis lushaensis* (C. E. C. Fisch.) C. Jeffrey & Y.L. Chen	NE India (Mizoram)
15	*Synotis simonsii* (C. B. Clarke) C. Jeffrey & Y.L. Chen	NE India (Assam)
16	*Vernonia parryae* C.E.C. Fisch.	NE India (Assam, Mizoram)

Table 3.11. Endemic plants of Asteraceae distributed in other parts of India, including the Andaman Islands.

Sl. No.	Name of the taxa	Distribution
1	*Anaphalis aristata* DC.	Western Ghats, Central India, Rajasthan, (common near waterfalls)
2	*Anaphalis lawii* (Hook.f.) Gamble	Western Ghats, Orissa, West Bengal
3	*Blumea eriantha* DC.	Uttar Pradesh, Bihar, Orissa, Madhya Pradesh, Rajasthan, Maharastra, Goa
4	*Blumea hookeri* C.B. Clarke ex Hook.f.	Sikkim, Meghalaya, the Andaman and Nicobar Islands
5	*Blumea malcolmii* (C.B. Clarke) Hook.f.	Madhya Pradesh, Western Ghats
6	*Blumea membranacea* DC. var. jacquemontii (Hook.f.) Randeria	Uttar Pradesh, Bihar, West Bengal, Orissa, Madhya Pradesh, Rajasthan, Maharastra, Tamil Nadu
7	*Helichrysum cutchicum* (C.B. Clarke) R. Rao ex Desh	Gujrat
8	*Phyllocephalum phyllolaenum* (DC.) Narayana	Western Ghats, Rajasthan
9	*Phyllocephalum scabridum* (DC.) Kirkman	Western Ghats, Gujrat, Madhya Pradesh
10	*Pulicaria rajputanae* Blatt. & Hallb.	Rajasthan
11	*Senecio bombayensis* Balakr.	Western Ghats, Gujrat, Rajasthan, Madhya Pradesh
12	*Tricholepis glaberrima* DC.	Western Ghats, Madhya Pradesh, Rajasthan
13	*Vernonia andamanica* Balakr. & Nair	The Andaman Islands
14	*Vernonia patula* (Dryand.) Merr.	The Andaman and Nicobar Islands
15	*Glossocardia setosa* Blatt. & Hallb.	Rajasthan

Table 3.12. Dominant endemic genera based on number of endemic species.

Sl. No.	Name of the genus	Number of taxa present in India	Number of endemic taxa in India	Percentage (%)
1	*Phyllocephalum* Bl.	9	8	88.88
2	*Taraxacum* Webber & Wigg.	95	44	46.31
3	*Vernonia* Schreb.	63	27	42.85
4	*Senecio* L.	55	21	38.18
5	*Anaphalis* DC.	43	11	25.58

the Asteraceae family from different corners of the globe, including some which are invasive weeds. According to Rao (1994, 1996), Sharma and Singh (2001),Vasudeva Rao (1986), Rao and Datt (1996), the majority of the Asteraceae weeds migrated during the early part of 15th century with the Portuguese settlement in India. At present, most of these species are invaders in agricultural fields, waste grounds, roadsides, fallow land, gardens, parks and railway tracks. Not only have they migrated, but they have taken over. Masking the native floristic elements with their highly alleopathic effects (Rao and Rao, 1977), they are able to replace the native floristic composition in many areas.

Many years ago, Darwin (1872) and Wallace (1902) recognized the role of such exotic species in the extinction of the native species. In the Convention on Biological Diversity (1992) it has also been recognized that alien species are the most important threat in species extinction after habitat destruction.

Maheshwari (1962) indicated that different human activities, such as shifting cultivation, faulty pasturage, etc. were the most important causes of the spread and naturalization of the exotic species to different parts of the country. But the rise of long-distance transport facilities and the increase in global trade has seen plant migration also increase in proportion.

Among the most widely introduced Asteraceae members, *Parthenium hysterophorous* is the most recognizable, and this American species is now observed in every part of the country, from sea level to 2300 m high in the Himalayas. Other invasive weeds are *Mikania cordata, Eupatorium odoratum, Ageratum conyzoides, Bidens pilosa* and *Xanthium strumarium*. Besides the destruction of the natural vegetation, these exotic weeds often create health hazards to humans and cattle; for example, *Amborsia artemisiifolia*, a North American species introduced into India is reported to be a causal factor for hay fever (Heywood, 1993).

A list of some very common exotic weeds of the Asteraceae reported in Indian territory is given in Table 3.13 below. It is interesting to note that most of the exotic weeds are migrants from tropical America, Mexico and Africa.

3.6 Rare and Endangered Asteraceae

Members of the Asteraceae are very adaptive in nature and also show a high level of ecological amplitude, which enables them to adapt to a new place very easily. But there are some species of this family which are unable to survive tough competition with the local native vegetation because of their low level of genetic diversity. Such species are becoming rare and endangered; another cause is that they are comparatively new, so have not had time to spread to a wide area. For example, *Anaphalis barnesii* Fischer and *Lactuca benthamii* C.B. Clarke.

At present, habitat destruction is one of the most significant causes of the rarity of the species and many species have become extinct due to destruction of their habitat; for example, *Vernonia recurva*, an endemic species from the Anaimalai Hills, was last collected in 1857. It is considered that this is probably extinct (Vivekanathan, 1987).

Over-exploitation of the species may cause rarity of the species: *Saussurea costus* is the best example of this.

A list of Rare, Endangered and Threatened (RET) plants arranged alphabetically is given in Table 3.14.

3.7 Concluding Remarks

Recent study has revealed that Asteraceae, with its 1314 taxa under 204 genera, distributed into 20 tribes is the most diversified Angiospemic plant family of the Indian flora, followed by the Poaceae (1291), Orchidaceae (1229), and Fabaceae (1192). In addition to 204 genera, another six genera (1.**Hemistepta* Bunge ex Fisch.; 2.**Homognaphalium* Kirp.; 3.**Lamprachaenium* Benth.; 4.**Mulgedium* Cass.; 5.**Paramicrorhynchus* Krip.; 6.**Sparganophores* Vail.) have also been found in India, but have not been considered by Kadreit and Jeffrey (2007). These genera have been indicated in the list of Indian genera with asterisks (*).

Asteraceae is a predominantly temperate family and most of the members are distributed in the temperate regions of the globe. In India most of the taxa (955) of Asteraceae, or about (72.67%) of the total Asteraceae in India, are found in the temperate regions of Himalaya and the north-east part of India.

India, by virtue of its unique phytogeographical location at the confluence of three major biogeographic realms: Afrotropical, Eurasian and Indo-Malayan, enables the floristic admixture of these three biogeographic zones.

The flora of any country are best understood when the indigenous floristic elements are compared with exotic floristic elements. Observing the affinities of Asteraceae in the Indian subcontinent to Asteraceae in other parts of the world, we can detect climatological as well as floral similarities.

The phytogeographical affinity of India and adjacent countries belonging to different phytogeographical and biogeographic realms has been governed entirely by the lofty Himalayan Mountains in the north.

The separation of the Indian plate from the massive Gondwanaland and its movement northwards (according to the continental drift theory of Wegener, 1924); the shrinking of the Tethys; the formation of the Basin (Lakhanpal, 1988) extending from west to east; the complete obliteration of the Tethys; the uplift of the Himalayas; the formation of a land bridge and land connection and migration of the flora of the neighbouring countries – all these factors have changed the floristic composition of the Indian subcontinent entirely.

According to Ohba (1988), the flora of the Arctic region, Europe, parts of Africa, the Mediterranean, Iran, the Indo-Turanian area, East Tibet and South China migrated into the Himalayan region through the high-altitude Central Asian Corridor.

Regarding the phytogeographical affinity of the central and eastern Himalayas, Kitamura (1955) commented that the Sino-Japanese and Sino-Himalayan elements are present in this region. Besides these, the Himalayas also play an important role in preventing the migration of Indian flora, especially the flora of peninsular India, to Central Asia.

Critical perusal of the distributional pattern of the taxa of Asteraceae in India reveals that the Iranian, West Asian, Afghanistani, Pakistani, Eurasian, Siberian, Mediterranean and Indo-Tibetan genera, such as *Allardia* Decne, *Amberboa* (Pers.) Less, *Arctium* L., *Artemisiella* Ghafoor, *Carduus* L., *Chondrilla* L., *Parasenecio* W. W. Sm. & Small,

Table 3.13. Exotic species of Asteraceae of Indian flora.

Sl. No.	Name of the taxa	Distribution in India	Native country or region
1	*Acanthospermum hispidum* DC.	Himachal Pradesh, U. P. Arunachal Pradesh, Rajasthan, Tamil Nadu	Brazil
2	*Adenostemma laevenia* (L.) Kuntz.	Throughout India	South America
3	*Ageratina adenophora* (Spreng.) R.M. King & H. Rob.	Throughout Indian Himalayan region	Mexico
4	*Ageratum conyzoides* L.	Throughout India	South America
5	*Ambrosia artemisiifolia* L.	North East India	South America
6	*Bidens biternata* (Lour.) Merr. & Sherff.	Uttar Pradesh, Punjab	South America
7	*Bidens pilosa* L.	More or less throughout India	South America
8	*Chrysanthemum cinerarifolium* Vis.	Cultivated	Europe (Yugoslavia, Italy)
9	*Conyza canadensis* (L.) Cronq.	Various states of India	South America
10	*Conyza bonariensis* (L.) Cronq.	Mainly Temperate Himalayas, (in different states)	South America
11	*Cotula australis* (Spreng) Hook.f.	In many states of India	Australia (Southern temperate region)
12	*Crassocephalum crepidioides* S. Moore	In many states of India	Tropical America
13	*Elephantopus scaber* L.	In many states of India	Australia, America
14	*Emilia sonchifolia* (L.) DC.	In many states of India	Afro-Asian
15	*Erechtites valerianaefolia* (Wolf) DC.	In many states of India	Tropical America
16	*Erigeron karvinskianus* DC.	In many states of India	Mexico
17	*Eupatorium adenophorum* Hort. Berol. ex Kunth.	In many states of India	Mexico
18	*Eupatorium odoratum* L.	In many states of India	Tropical America
19	*Eupatorium riparium* Regel.	In many states of India	Tropical America
20	*Flaveria trinervia* (Spreng.) C. Mohr.	In many states of India	America
21	*Galinsoga quadriradiata* Ruiz. & Pav.	In many states of India	Central America
22	*Galinsoga parviflora* Cav.	In many states of India	South America
23	*Hypochaeris radicata* L.	In many states of India	Europe
24	*Laggera aurita* (L.f.) Sch.-Bip. ex C.B. Clarke	In many states of India	Afro-Asian
25	*Lagascea mollis* Cav.	In many states of India	Mexico
26	*Mikania micrantha* Kunth.	In many states of India	Tropical America
27	*Parthenium hysterophorus* L.	In many states of India	Tropical America
28	*Sclerocarpus africanus* Jack.	In many states of India	Tropical Africa
29	*Sonchus oleraceous* L.	In many states of India	Europe
30	*Sonchus wightianus* DC.	In many states of India	Africa
31	*Sphaeranthus indicus* L.	In many states of India	Africa
32	*Synedrella nodiflora* (L.) Gaertn.	In many states of India	Tropical Africa
33	*Tagetes minuta* L.	In many states of India	Tropical Africa
34	*Taraxacum officinale* Weber	In many states of India	Europe
35	*Tithonia diversifolia* (Hemsl.) A. Gray	In many states of India	Mexico
36	*Tridax procumbens* L.	Throughout India	Mexico
37	*Wedelia chinensis* (Osbeck.) Merr.	In many states of India	Afro-Asian
38	*Xanthium strumarium* L.	In many states of India	South America

Pentanema Cass., *Serratula* L., *Serphidium* Poljakov., and *Vaethemia* DC. have migrated through the Kargil–Karakoram molassic region, and also migrated to the Western and Central Himalayan region, as far as Sikkim in the Eastern Himalayas.

The African, Turkish, Russian, Caucassusian, Temperate European, and Irano-Turanian regions, represented by genera such as *Anthemis* L., *Centaurea* L., *Chondrilla* L., *Dichrocephala* Hier ex DC., *Dimorphotheca Vail.*, *Epilasia* (Bunge) Benth., *Filago* L., *Flaveria* Juss., *Lapsana* L., *Lasiopogon* Cass., *Pegolettia* Cass., *Serphidium* Poljakov., *Silybum* Adnas., *Stevia* Cav., *Tragopogon* L., *Tussilago* L., *Uechtritzia* Feryn and *Zoegea* L.,

Table 3.14. List of the RET plants of Asteraceae.

Sl. No.	Name of the taxa	Distribution in India	Status	Remarks
1	*Anaphalis barnesii* C.F.C. Fisch	Kerala (Iduki)	Endangered	Endemic, known from the type only
2	*Anaphalis brevifolia* DC.	Tamil Nadu (Anamallai Hills); Western Ghats	Rare; endangered	
3	*Aster heliopsis* Grierson	Sikkim	Very rare	Not reported in any Indian herbarium
4	*Catamixis baccharoides* Thomson	Uttaranchal	Vulnerable	
5	*Chondrilla setulosa* Clarke ex Hook.f.	Western Himalaya (Jammu and Kashmir)	Rare	Endemic, known by the type only
6	*Cicerbita filicina* (Duthie ex Stebbins) Aswal & Goel	Uttaranchal	Endangered	Endemic, known by the type only
7	*Cremanthodium plantagineum* Maxim. forma *ellistii* (Hook.f.) R. Good	Jammu and Kashmir, Himachal Pradesh	Rare	
8	*Cyathocline lutea* Law ex Wight	Western Ghat (Maharastra)	Rare	Endemic
9	*Helichrysum cutchicum* (C.B. Clarke) R.S. Rao & Deshp.	Gujrat	Rare	Endemic
10	*Helichrysum perlanigerum* Gamble	Southern Western Ghats	Rare	Endemic
11	*Inula racemosa* Hook.f.	Jammu and Kashmir	Vulnerable	Endemic
12	*Inula kalapani* C.B. Clarke	Meghalaya (Khasi Hills)	Rare	Endemic
	Inula macrosperma Hook. f.	Alpine region of Sikkim	Rare	Endemic
13	*Lactuca benthamii* C.B. Clarke	Kashmir	Endangered	Endemic, known by the type only
14	*Lactuca cooperi* J. Anthony	Sikkim	Endangered	Endemic, known by the type only
15	*Lactuca undulata* Ledeb.	Kashmir	Endangered	
16	*Nanothamnus sericeus* Thomson	Western Ghats (Maharastra and Karnataka)	Endangered	Endemic
17	*Saussurea bracteata* Decne.	Western Himalaya (Jammu and Kashmir, Himachal Pradesh, Uttaranchal)	Rare	Endemic
18	*Saussurea clarkei* Hook.f.	Kashmir	Rare	Endemic
19	*Saussurea costus* (Falc.) Lipsch.	Western Himalaya (Jammu and Kasjmir, Himachal Pradesh, Uttaranchal)	Critically Endangered	
20	*Senecio kundaicus* C.E.C. Fisch.	Nilgiri	Endangered	Endemic, known by the type only
21	*Senecio mayurii* C.E.C. Fisch.	Karnataka	Rare	Endemic, known by the type only
22	*Senecio mishmi* C.B. Clarke	Arunachal Pradesh (Mishmi Hills)	Vulnerable	Endemic, known by the type only
23	*Senecio rhabdos* C.B. Clarke	Manipur, Nagaland	Rare	Endemic, known by the type only
24	*Synotis simonsii* (C.B. Clarke) C. Jeffrey & Y.L. Chen	Assam	Rare	Endemic, known by the type only
25	*Vernonia andamanica* N.P. Balakr. & N.G. Nair	North Andaman (Saddle Park)	Rare	Endemic, known by the type only

Continued

Table 3.14. Continued.

Sl. No.	Name of the taxa	Distribution in India	Status	Remarks
26	*Vernonia multibracteata* Gamble	Western Ghats	Endangered	Endemic, known by the type only
27	*Vernonia pulneyensis* Gamble	Western Ghats (Pulney Hills)	Endangered	Endemic
28	*Vernonia recurva* Gleason	Western Ghats (Annamalai Hills)	Endangered	Endemic, it was collected once in the year 1857 and no other collection (possibly extinct)
29	*Vernonia shevaroyensis* Gamble	Maharastra (Shevaroy Hills)	Rare	Endemic
30	*Youngia nilgiriensis* Babc.	Western Ghats (Sispara)	Endangered	Endemic, known by the type only

migrated through the Western Corridor and the Siwalik molassic of the high-altitude Central Asian Corridor, and distributed in North West Himalaya and in the Rajasthan, Punjab and Upper Gangetic Plain region. But those genera which entered through the Western Corridor are very poorly distributed in the eastern part of the country, as their spread is restricted by the arid climatic conditions of the Thar Desert of Rajasthan. Similarly, the spread of these genera in peninsular India is restricted by the 3500-km-wide Indo-Gangetic Plain and the Vindhya–Satpura Plateau.

Migration of the Chinese, Australian, Indonesian, Tibetan and Sino-Japanese species, such as *Carpesium* L., *Cavea* W. W. Smith & Small, *Cousinia* Cass., *Leibnitzia* Cass., *Microglossa* DC., *Nannoglotis* Maxim., *Prenanthes* L. and *Pseudoelephantopus* Rohr. takes place through the Tibetean-Creask, a natural creek in the Himalayas, and thus, these species are distributed in the Eastern Himalayas and in North East India.

Geologically, the Chinese mountains are older than the Himalayas, so there is a direct land connection between the Chinese mountains and the Khasi and Jaintia Hills of North East India, which has been considered as the Chinese–Myanmar migration route. Beside this, the uplift of the Himalayas from Tethys Bay formed a direct land connection between Malaya and North East India, through the Andaman Islands. The Australian, Indo-Pacific, Malayan, Indonesian, New Zealand, Macronesian and Micronesian region elements represented by genera such as *Ainsliaea* DC., *Glossocardia* Cass., *Ixeridium* (A. Gray) Tzvelaga, *Ixeris* L., *Kalimeris* (Cass.) Cass. *Reichardia* Roth., *Sphaeranthus* L., *Thespis* DC., *Vittadinia* A. Rich. and *Wollastonia* DC ex Decne, migrate via this route and then to the Indian Plain.

Due to the Gondwanaland connection between peninsular India and Australia, the Malay Peninsula, Ceylon and Africa, several members of the Asteraceae family are found in peninsular India.

According to Meusel (1952), the woody habit in an endemic group is a relict character, while Carlquist (1966), suggested that the woody habit is a derived characteristic resulting from insular isolation. Most species of the genus *Vernonia* Schreb. are herbaceous in nature. But the species existing in the shola forest of peninsular India, which is a natural abode of the relict vegetation, include some woody species of *Vernonia* Schreb., such as *Vernonia shevaroyensis* Gamble and *Vernonia travancorica* Hook. f. This portion of peninsular India shows phytogeographical affinity with North East India, indicated by the presence of the same woody habit of *Vernonia parryae* C. Fisher (Mizoram). Similarly, the presence of *Vernonia andamanica* Balak., a woody species of *Vernonia* in the Andaman Islands, shows the affinity between the Andaman Islands and peninsular or North East India.

In the present study, it has been considered that India phytogeographically can be divided into 12 phytogeographical regions in comparison to the previous 11 phytogeographical divisions of India (Balakrishnan, 1996).

The Nicobar Islands, being volcanic, are climatologically as well as geomorphologically different from the rest of the Andaman group of islands, but show a similarity with the southern part of India (especially the Deccan): the floristic composition of this region differs greatly from the Andaman group of islands.

So, the Nicobar Islands are considered as the 12th phytogeographical province of India (Fig. 3.3).

Acknowledgements

The authors are grateful to the Director and the curators of the various offices of the Botanical Survey of India for their encouragement and for providing assistance in the libraries and Herbaria.

References and bibliography

Ahmedullah, M. (2000) Endemism in Indian flora. In: Singh, N.P., Singh, D.K., Hajra P.K. and Sharma, B.D. (eds) *Flora of India. Introductory Volume Part II.* Botanical Survey of India, Calcutta, India, pp. 246–265.

Ahmedullah, M. and Nayar, M.P. (1987) *Endemic Plants of Indian Region.* BSI, Calcutta, India.

Anderberg, A.A., Baldwin, B.G., Bayer, R.J., Breitwieser, I., Jeffrey, C. *et al.* (2007) *Compositae.* In: Kadreit, J.W. and Jeffrey, C. (eds) *The Families and Genera of Vascular Plants* (Kubitzki, K., ed.) Vol. VIII. *Flowering Plants. Eudicots, Asterales.* Springer, Berlin, Germany, pp. 61– 588.

Balakrishnan, N.P. (1996) Phytogeographic divisions: general considerations. In: Hajra, P.K., Sharma, B.D., Sanjappa, M. and Sastry, A.R.K. (eds) *Flora of India. Introductory Volume* (Part 1). Botanical Survey of India, Calcutta, India.

Bremer, K. (1987) Tribal interrelationships of the Asteraceae. *Cladistics* 3 (3), 210–253.

Bremer, K. (1994) *Asteraceae. Cladistics & Classification.* Timber Press, Portland, Oregon, USA.

Bremer, K. (1996) Major clades and grades of the Asteraceae. In: Hind, D.J.N. (ed.) *Compositae: Systematics. Proceedings of the International Compositae Conference* 1, 1–7. Royal Botanic Gardens, Kew, Surrey, UK.

Bremer, K., Jansen, R.K., Karis, P.O., Kallersjo, M., Keeley S.C., Kim, K.J., Michaels, H.J. Palmer, J.D. and Wallace, R.S. (1992) A review of the phylogeny and classification of the *Asteraceae. Nordic Journal of Botany* 12, 141–148.

Burman, N.L. (1768; repr. 1984) *Flora Indica.* Bishen Singh and Mahendra Pal Singh, Dehradun, Uttarakhand, India.

Calder, C.C., Narayanaswami, V. and Ramaswami, M.S. (1926) List of species and genera of Indian Phanerogams not included in J.D. Hooker's Flora of British India. *Records of Botanical Survey of India* 2(1), 1–157.

Carlquist, S. (1966) Wood anatomy of *Compositae.* A summary, with comments on factors controlling wood evolution. *Aliso* 6, 25–44.

Cassini, H. (1826–1834) *Opuscules Phytologiques.* Vols I–II, 1826. Vol. III, 1834. Paris, France.

Champion, H.G. and Trevor, C. (1938) *Manual of Indian Silviculture.* Oxford University Press, London, UK.

Chatterjee, D. (1940) Studies on the endemic flora of India and Burma. *Journal of Royal Asiatic Society and Bengal* 5, 19–67.

Chatterjee, D. (1962) Floristic patterns of Indian vegetation. In: Maheshwari, P., Johri, B.M. and Vasil, I.K. (eds) *Proceedings of Summer School in Botany, Darjeeling.* Ministry of Scientific Research and Cultural Affairs, Government of India, New Delhi, India, pp. 32–42.

Chopra, R.N., Nayar, S.L. and Chopra, I.C. (1956) *Glossary of Indian Medicinal Plants.* CSIR, New Delhi, India.

Clarke, C.B. (1876) *Compositae Indica Descriptae et Secus Genera Benthamii Ordinatae.* Thacker, Spink and Co., Calcutta, India.

Clarke, C.B. (1881) *Compositae.* In: Hooker, J.D. *Flora of British India. Vol. III.* L. Reeve & Co., London, UK, pp. 219–419.

Clarke, C.B. (1898) On the sub-areas of British India, illustrated by the detailed distribution of the *Cyperaceae* in that Empire. *Journal of the Linnean Society (Botany)* 34, 1–146.

Collett, C.S.H. (1902) *Flora Simalensis.* Simla, India.

Cooke, T. (1906) *Flora of the Presidency of Bombay.* Taylor & Francis, London, UK.

Darwin, C.R. (1872) *The Expression of the Emotions in Man and Animals.* John Murray, London, UK.

de Candolle, A.P. (1838) *Prodromus Systematis Naturalis* 6. Treuttel and Würtz, Paris, France.

Don, D. (1825) *Prodromus Florae Nepalensis.* J. Gale & Company, London, UK.

Duthie, J.F. (1902) *The Flora of Upper Gangetic Plain and the Adjacent Siwalik and sub-Himalayan Tracts.* Superintendent of Government Printing, Calcutta, India.

Fischer, C.E.C. (1938) Flora of the Lushai Hills. *Records of Botanical Survey of India* 12, 75–16l.

Funk, V.A., Susanna, A., Stuessy, T.F. and Bayer, R.J. (2009) *Systematics, Evolution, and Biogeography of Compositae.* International Association for Plant Taxonomy, Vienna, Austria.

Fyson, P.F. (1914) *The Flora of the Nilgiri and Pulney Hill Tops.* Superintendent of Government Printing, Madras, India.

Gamble, J.S. (1921) *Flora of the Presidency of Madras.* 2. West, Newman and Adlard, London, UK.

Ghosh, R.B. (1977) Sixth list of genera and species of Angiosperms not included in the Flora of British India. *Bulletin of Botanical Society of Bengal* 31, 84–89.

Ghosh, R.B. (1979) Seventh list of Angiosperms not included in the Flora of British India. *Bulletin of Botanical Society of Bengal* 33, 87–93.

Ghosh, R.B. and Datta, S.C. (1976) Angiosperms not included in the Flora of British India: List V. *Bulletin of Botanical Society of Bengal* 30, 1291–1333.

Haines, H.H. (1922) *The Botany of Bihar and Orissa.* Adlard & Son & West Newman, London, UK.

Hajra, P.K., Rao, R.R., Singh, D.K. and Uniyal, B.P. (1995a) *Flora of India*. 12. Asteraceae (Anthemidae – Heliantheae). Botanical Survey of India, Calcutta, India.

Hajra, P.K., Rao, R.R., Singh, D.K. and Uniyal, B.P. (eds) (1995b) *Flora of India*. 13. Asteraceae (Inulae – Vernonieae). Botanical Survey of India, Calcutta, India.

Heywood, V.H. (1993) *Flowering Plants of the World*. Oxford University Press, New York, pp. 73–77.

Hooker, J.D. (1875–1897) *Flora of British India*. Vols I–VII. London, UK.

Hooker, J.D. (1854) *Himalayan Journal*. John Murray, London, UK.

Hooker, J.D. (1881) Compositae. *The Flora of British India*. Vol. 3. Lovell Reeve and Co., London, UK.

Hooker, J.D. (1906) *A Sketch of the Flora of British India*. Clarendon Press, Oxford, UK.

Jain, S.K. and Sastry, A.R.K. (1983) *Materials for a Catalogue of Threatened Plants of India*. Botanical Survey of India, Calcutta, India.

Jain, S.K. and Sastry, A.R.K. (1984) *The Indian Plant Red Data Book – I*. Botanical Survey of India, Calcutta, India.

Kadreit, J.W. and Jeffrey, C. (2007) The families and genera of vascular plants (Kubitzki, K., ed.) Vol. VIII. *Flowering Plants. Eudicots, Asterales*. Springer, Berlin, Germany.

Kanjilal, U.N., Das, A., Kanjilal, P.C. and De, R.N. (1939) *Flora of Assam*, Vol. III. Government Press, Writers Building, Calcutta, India.

Karthikeyan, S. (2000) A statistical analysis of flowering plants of India. In: Singh, N.P., Singh, D.K., Hjara, P.K. and Sharma, B.D. (eds) *Flora of India Introductory Volume, Part II*. Botanical Survey of India, New, Delhi, India, pp. 201–217.

Karthikeyan, S., Sanjappa, M. and Moorthy, S. (2009) *Flowering Plants of India. Dicotyledons*, Vol. I. *Acanthaceae – Aviciniaceae*. Botanical Survey of India, Calcutta, India, pp. 184–299.

Kitamura, S. (1955) Flowering plants and ferns. In: Kihara, H. (ed.) *Fauna and Flora of Nepal*, Vol. 1: *Himalaya*. Kyoto University Press, Kyoto, Japan, pp. 73–77.

Lakhanpal, R.N. (1988) The advent of temperate elements in the Himalayan flora. In: Whyte, P. and Aigner, J.S. (eds) *The Palaeoenvironment of East Asia from the Mid-Tertiary: Geology, Sea Level Changes, Palaeoclimatology and Palaeobotany*. Occasional Papers and Monographs 77. Centre of Asian Studies, University of Hong Kong, Hong Kong, China, pp. 673–679.

Lakshminarasimhan, P. and Rao, P.S.N. (1996) A supplementary list of Angiosperms recorded (1983–1993) from Andaman and Nicobar Islands. *Journal of Economic and Taxonomic Botany* 20, 175–185.

Li, H.L. *et al.* (1975) *Flora of Taiwan 1*. Epoch Pub. Co. Ltd.

Linnaeus, C. (1753) *Species Plantarum.* I–II. Linnaean Society, London, UK.

Mabberley, D.J. (2008) *Mabberley's Plant-Book*, 3rd edn. Cambridge University Press, Cambridge, UK.

Maheshwari, J.K. (1962) Studies on the naturalised flora of India. In: Maheshwari, P. (ed.) *Proceedings of Summer School in Botany.* Darjeeling, India, pp. 154–170.

Maity, D. and Maiti, G.G. (2010) Taxonomic delimitation of the genus *Tibetoseris* Sennikov and the new genus *Pseudoyoungia* of the Compositae – Cichorieae from Eastern Himalaya. *Compositae Newsletters* 48, 22–42.

Meusel, A. (1952) *Thomas Müntzer und seine Zeit*. Aufbau-Verlag, Idstein, Hesse, Germany.

Nayar, M.P. (1996) *Hotspots of Endemic Plants of India, Nepal and Bhutan*. Tropical Botanic Garden and Research Institute, Thiruvananthpuram, India.

Nayar, M.P. and Karthikeyan, S. (1981) Fourth list of species and genera of Indian Phanerogams not included in J.D. Hooker's Flora of British India (excluding Bangladesh, Burma, Sri Lanka, Malayan Peninsula and Pakistan). *Records of Botanical Survey of India* 21(2), 129–152.

Nayar, M.P. and Ramamurthy, K. (1973) Third list of species and genera of Indian Phanerogams not included in J.D. Hooker's Flora of British India (excluding Bangladesh, Burma, Ceylon, Malayan Peninsula and Pakistan). *Bulletin of Botanical Survey of India* 15, 204–234.

Nayar, M.P. and Sastry, A.R.K. (1987–1990) *Red Data Book of Indian Plants*. Vols 1–3. Botanical Survey of India, Calcutta, India.

Ohba, H. (1988) The Alpine flora of the Nepal Himalayas: an introductory note. In: Ohba, H. and Malla S.B. (eds) *The Himalayan Plants*, Vol. 1. Bull. No. 31. University Museum, University of Tokyo, Japan. Available at: www.um.u-tokyo.ac.jp/publish_db/Bulletin/no31/no31005.html (accessed 26 May 2015).

Prain, D. (1903) *Bengal Plants*, Vols 1–2. Bishen Singh Mahendra Pal Singh, Dehradun, Uttarakhand, India.

Puri, G.S. (1960) *Indian Forest Ecology*, Vols 1–2. Oxford and IBH, New Delhi, India.

Rao, A.S. (1974) The vegetation and phytogeography of Assam-Bunna. In: Mani, M.S. (ed.) *Ecology and Biogeography in India*. Junk, The Hague, Netherlands, pp. 204–246.

Rao, R.R. (1994) *Biodiversity in India*. Bishen Singh Mahendra Pal Singh, Dehradun, Uttarakhand, India.

Rao, R.R. (1996) Compositae in the conservation of genetic diversity in wild plants in India. In: Caligari, P.D.S., Hind, D.I.N. and and Beentje, H.J. (eds) *Compositae: Biology and Utilization. Proceedings of the International Compositae Conference 1994*. Royal Botanic Gardens, Kew, Surrey, UK, pp. 269–275.

Rao, R.R. and Datt, B. (1996) Diversity and phytogeography of Indian compositae. In: Caligari, P.D.S., Hind, D.I.N. and Beentje, H.J. (eds) *Compositae: Biology and Utilization*. Proceedings of the International

Compositae Conference 1994. Royal Botanic Gardens, Kew, Surrey, UK, pp. 445–461.

Rao, A.S. and Rao, R.R. (1977) Changing pattern in the Indian flora. *Bulletin of Botanical Survey of India* 19, 156–166.

Rao, R.R., Choudhery, H.J., Hajra, P.K., Kumar, S., Pant, A.C., Naithani, B.D., Uniyal, B.P., Mathur, R. and Mamgain, S.K. (1988) *Florae Indicae Enumeratio*. Asteraceae. Botanical Survey of India, Calcutta, India.

Razi, B.A. (1955) Some observations on plants of south Indian hill tops and their distribution. *Proceedings of Nature Institute of Science India* 21B(2), 79–89.

Razi, B.A. (1959) A second list of species and genera of Indian Phanerogams not included in J.D. Hooker's Flora of British India. *Records of Botanical Survey of India* 18(1), 1–56.

Rodgers, WA. and Panwar, H.S. (1988) *Planning a Wildlife Protected Area Network in India (Vols 1–2)*. Wildlife Institute of India, Dehradun, Uttarakhand, India (mimeographed).

Roxburgh, W. (1832) *Flora Indica*, ed. Carey, W. and Wallich, N., 2nd edn. Mission Press, Serampore, West Bengal, India.

Sharma, J.R. and Singh, D.K. (2001) Status of plant diversity in India: an overview. In: Roy, P.S., Singh, S. and Toxopeus, A.G. (eds) *Biodiversity & Environment*. Indian Institute of Remote Sensing, Dehradun, Uttarakhand, India, pp. 69–105.

Singh, D.K., Uniyal, B.P. and Mathur, R. (1999) Jammu and Kashmir. In: Mudgal, V. and Hajra, P.K. (eds) *Floristic Diversity and Conservation Strategies in India*. II. Botanical Survey of India, Dehradun, Uttarakhand, India, pp. 905–974.

Steere, W.C. and Inoue, H. (1972) Distribution patterns and speciation of bryophytes in the circumspecific regions: Introduction 1. *Hattori Botanical Laboratory* 35, 1–2.

Takhtajan, A.L. (1986) *Floristic Regions of the World*. University of California Press, Berkeley, California, USA.

United Nations (1992) *Convention on Biological Diversity*. Available at: https://www.cbd.int/doc/legal/cbd-en.pdf. (accessed 28 September 2016).

Van Rheede, H.A. (1690) *Hortus Malabaricus*. 10. Amsterdam, Netherlands.

Vasudeva Rao, M.K. (1986) A preliminary report on the angiosperms of Andaman and Nicobar Islands. 1. *Economic Taxonomy Botany* 8(1), 107–184.

Vivekanathan, K. (1987) *Vernonia recurva* Bedd. ex S. Moore. In: Nayar, M.P. and Sastry, A.R.K. (eds) *Red Data Book of Indian Plants*, Vol. 1. Botanical Survey of India, Calcutta, India.

Wallace, A.R. (1902) *Man's Place in the Universe*. Chapman and Hall Ltd, London, UK.

Wegener, A. (1924) *The Origin of Continents and Oceans*. Methuen, London, UK.

4 Maintenance of Plant Species Diversity in Forest Ecosystems

Kathryn E. Barry and Stefan A. Schnitzer*

Department of Biological Sciences, Marquette University, Milwaukee, Wisconsin, USA

Abstract

One of the major questions in ecology is how plant species coexist and thus how diversity is maintained. While there are many theories to explain the maintenance of plant species diversity, compelling empirical support exists for very few of them. Here we summarize four major putative theories to explain the maintenance of forest plant species diversity, each of which has ample empirical support. These theories are: 1) niche differentiation; 2) negative density dependence; 3) disturbance; and 4) neutral dynamics with respect to competition. We also present a literature review comprising 51 studies that explicitly examined the maintenance of plant species diversity, published between 2000 and 2015. In the literature review, we include only studies that stated, either as part of their introduction or in clearly stated objectives, that the hypothesis was the maintenance of diversity mechanism. We found that there has been a huge amount of progress in the research on the maintenance of species diversity. An overwhelming majority (all but three) found significant evidence for the mechanism that they tested. The large majority (64%) of the studies were conducted in tropical regions (27% in Panama alone). Trees were the main focal group, comprising 75% of the studies, 14% focused on shrubs, and 7% on lianas. Only 4% of the studies focused on non-woody plants. Seedlings were also a main focus, and 61% of the studies focused explicitly on the seedling life stage.

Our findings lead us to conclude that the study of the maintenance of species diversity has progressed rapidly over the past 15 years. However, there is still a substantial amount of additional research to be done. Future studies may benefit from explicitly examining key theoretical requirements of the mechanism that they are investigating. The use of spatial and temporal heterogeneity may be critical and, in many cases, would improve the precision of the study. Most tests have focused on single plant groups and single mechanisms for the maintenance of species diversity, but moving into the future, a focus on multi-mechanism models for the maintenance of diversity using plant groups that differ in multiple axes of life-history strategy will be crucial to understanding how species coexist.

4.1 Introduction

Understanding the maintenance of diversity is one of the central goals of ecology. By understanding how species diversity is maintained, we gain valuable insight into the forces and mechanisms that determine the distribution of species, as well as how communities are assembled, regenerate and function. Over the past century, more than 100 hypotheses have been proposed as potential explanations for the maintenance of species diversity (Palmer, 1994), with the vast majority of these hypotheses aimed at understanding the coexistence of species within a single trophic level (Wright, 2002).

For forest plants, the maintenance of species diversity can be reduced to four major hypotheses for the maintenance of species diversity, which have been commonly tested empirically in temperate and tropical forests. The four hypotheses are: 1) niche differentiation; 2) negative density dependence; 3) disturbance; and 4) a neutral dynamics model in terms of species competition. While these hypotheses have been examined numerous times, the extent of support that they have received relative to each other is not well known. That is, there is little information about which hypotheses have received strong empirical support and which

*E-mail: s1@marquette.edu

have not. By determining the level of support for each hypothesis, we can begin to determine which hypotheses we should believe and which need more examination.

There are three other key issues that are also poorly understood, which can be summarized in the following questions.

1) Are the studies biased with regards to geographic region? Forests in different regions have very different dynamics. For example, currently some species are increasing in abundance in tropical forests that are experiencing more droughts while others are decreasing, which may change the species composition and richness differentially between geographic regions (Feeley *et al.*, 2011; Schnitzer and Bongers, 2011; Enquist and Enquist, 2011).
2) Are the studies biased with regards to plant life-history stage? Mechanistic hypotheses on the maintenance of species are community-level hypotheses that describe the ability of plant species to be maintained over long periods. Focusing on a single life-history stage, when a matrix of life-history stages occur throughout the community and all must be maintained, may lead only to proximate answers for the maintenance of diversity of a particular life-history stage (seedling, sapling or adult) while losing sight of the community.
3) Are the studies biased with regards to functional group? Plants demonstrate a wide variety of functional traits that dictate how they interact with their environment. Groups of plants that demonstrate similar traits are likely to demonstrate similar mechanisms, while other groups may not (McCarthy-Neumann and Kobe, 2008). Focusing on a single group is limiting for maintenance of diversity studies, as these groups may respond to different mechanisms.

In this chapter, we provide a description of the four key hypotheses for the maintenance of species diversity. We then review the literature published over the past 15 years, addressing these four diversity maintenance mechanisms. We used Google Scholar to search for studies using the following search terms: 1) 'niche', 'maintenance of diversity', forest; 2) 'negative density dependence', 'maintenance of diversity', forest; 3) 'negative feedback', 'maintenance of diversity', forest; 4) 'Janzen-Connell hypothesis', 'maintenance of diversity', forest; 5) 'disturbance' 'maintenance of diversity', forest. We restricted the publications to the years 2000 to 2014 inclusive (1 January 2000–31 December 2014) and we excluded review articles or studies based strictly on mathematical modelling. We also excluded studies that were not testing the hypothesis as a maintenance of diversity mechanism, and any study that was not focused on a forest ecosystem.

4.2 Four Major Hypotheses for the Maintenance of Forest Plant Species Diversity

4.2.1 Niche assembly

A fundamental theory for the maintenance of species diversity is niche differentiation or niche assembly (e.g. Hutchinson, 1957; Hutchinson, 1959; MacArthur and Levins, 1967; MacArthur, 1969; MacArthur, 1972). The theory states that no two species can coexist if they have the exact same niche requirements. Competition for shared resources results in the specialization of species on a unique combination of resources, thereby allowing species to avoid competitive exclusion (or succumbing to competitive exclusion if the species does not specialize sufficiently). Thus, according to this theory, ecologically similar species are able to stably occupy the same community by specializing on a different set of resources and requirements (niches) and thereby limiting direct competition with similar species (Hutchinson, 1957; Tilman, 1982). See Chase and Leibold (2003) for a detailed discussion on the definitions and importance of niches in ecology.

Niche differentiation has been found, to some degree, in nearly every ecosystem. For example, many tropical plants have known associations with specific microhabitat factors, suggesting that they are specializing on that microhabitat niche (Clark *et al.*, 1998; Clark *et al.*, 1999; Svenning, 1999; John *et al.*, 2007). In Borneo, up to 80% of tree species may be specialized on edaphic factors (Russo *et al.*, 2008). In a congener-rich oak-dominated subtropical forest, Cavender-Bares *et al.* (2004) found that oaks fell into three clear community types based on available soil moisture, nutrient availability and fire regime. Experimental evidence from grasslands suggests that plants strongly partition resources, particularly resources that are limiting (Tilman and Wedin, 1991). Indeed, even among ectomycorrhizal fungal communities, differentiation for vertical space on roots has been shown (Dickie *et al.*, 2002). While there is strong evidence for the existence of niche partitioning, it is still unknown whether the strength of niche differentiation as a central mechanism for the maintenance of species diversity has received more support than other diversity maintenance mechanisms.

4.2.2 Negative density dependence

Another leading hypothesis for plant diversity maintenance is negative density dependence (NDD) due to the presence of species-specific enemies (Janzen, 1970; Connell, 1971). Theoretically, negative density dependence can maintain stable, community-level diversity because seedlings growing near conspecific adults suffer reduced growth and survival due to the accumulation of species-specific enemies (e.g. pathogens, parasites and herbivores) around the adult. The reduction of conspecific seedlings increases the probability that adults are replaced by a different (heterospecific) species. If adults are consistently replaced by heterospecific species, then rare species are favoured because there is a higher number of suitable recruitment sites, and dominant species are limited because there are fewer suitable recruitment sites. Both of these conditions are critical for community-level diversity maintenance (Mangan *et al.*, 2010).

There is now compelling evidence that NDD is operating in both temperate and tropical plant forests (Harms *et al.*, 2000; Packer and Clay, 2000; Hille Ris Lambers *et al.*, 2002, Wright, 2002; Petermann *et al.*, 2008). A plethora of studies have confirmed the presence of negative density dependent patterns. For example, Connell *et al.* (1984) demonstrated that individuals near a member of their own species were more likely to suffer from increased mortality and decreased growth. Harms *et al.* (2000) found that all species in the Barro Colorado 50 ha plot demonstrated negative density dependence during recruitment. Indeed, both experimental and observational studies support the evidence of NDD (e.g. Comita *et al.*, 2010; Mangan *et al.*, 2010; Ledo and Schnitzer, 2014).

4.2.3 Disturbance

Disturbance may maintain plant species diversity in several different ways. For example, tree-fall gaps and similar smaller-scale disturbances may maintain diversity by providing an equilibrium state in which both late and early successional species coexist, and the system remains in a stable dynamic state (Schnitzer *et al.*, 2008). The creation of a tree-fall gap begins a process of succession. The persistence of tree-fall gaps creates a regeneration niche for shade-intolerant species to persist in the population rather than becoming competitively excluded (Denslow, 1987; Swaine and Whitmore, 1988; Whitmore, 1989; Dalling *et al.*, 1998). Disturbance may also alter the distribution of nutrients across both time and space, creating a broad spectrum of disturbance niches for plants to capitalize on (Chase and Leibold, 2003). This niche-driven disturbance approach provides an equilibrium-based hypothesis that requires disturbance to create a gradient of resource availability from mid-gap to non-gap (Ricklefs, 1977; Denslow, 1987).

Another hypothesis for how disturbance enables the maintenance of plant species was formalized by Connell in 1978 as the Intermediate Disturbance Hypothesis (IDH). This hypothesis states that species diversity will be highest when disturbance is intermediate in size, in time since disturbance, and in frequency. The IDH relies on an intermediate amount of disturbance to provide disturbed habitat for early successional species, but not so much disturbance that late successional species cannot establish. Disturbance may, in fact, maintain diversity in some plant groups, but not in others. In Panama, lianas (woody vines) are able to capitalize tree-fall gaps (Schnitzer and Carson, 2000; Schnitzer and Carson, 2001), but trees are not (Hubbell *et al.*, 1999; Brokaw and Busing, 2000). Obiri and Lawes (2004) examined the extent to which species richness in coastal scarp forests in South Africa was determined by specialization along a gap niche axis and found little evidence that competition for resources along a niche-forest gradient maintained the diversity of tree seedlings and saplings.

4.2.4 Neutral competition dynamics

The idea that competitively neutral species could coexist was popularized by several authors, most notably and comprehensively by Hubbell in 2001 (but see also Hubbell, 1979; Hubbell, 1997; Bell, 2001). The 'neutral model' provides both a hypothesis for the maintenance of species diversity and a potential null distribution for examining the validity of other mechanisms. The neutral model says that species diversity can be maintained simply through a combination of dispersal of propagules and mortality. In this model, space in a community is limited, and that space will be colonized by the species that arrives in the space first. The space is then vacated when the species dies. The neutral model allows for diversity to be maintained through simple dispersal and local extinction, with rare instances of speciation (Hubbell, 2001). This approach derives from models of island biogeography proposed by MacArthur and Wilson (1963, 1967), and much of the historical evidence for neutral processes in ecology comes from island communities. For example, Simberloff and Wilson (1970) found that in red mangrove islands that had been experimentally defaunated, species diversity was largely limited by area and distance from mainland populations, leading to the conclusion that these communities were largely dispersal assembled (see also Simberloff, 1974).

Many studies have reported that competitively neutral processes, such as recruitment limitation, rather than deterministic processes, such as niche differentiation, determine diversity (e.g. Hubbell *et al.*, 1999; Brokaw and Busing, 2000; Obiri and Lawes, 2004). Furthermore, for tropical seedlings, species richness may be determined more by competitively neutral factors, such as proximity of seed sources, efficient dispersal mechanisms and favourable establishment conditions, rather than deterministic processes (Denslow and Guzman, 2000; Paine *et al.*, 2008).

4.3 Empirical Support for the Four Diversity Maintenance Hypotheses

We reviewed the literature on the four diversity maintenance hypotheses and found 51 studies that met our criteria and were published between 1 January 2000 and 31 December 2014. We found that the most commonly invoked mechanism for the maintenance of species diversity was negative density dependence, which represented 73% of the studies included in this review (n=37). Niche differentiation comprised nearly 40% of tests (n=20), with neutral (n=11) and disturbance (n=4) representing a combined 30% of studies. Of the 51 studies, 29 (57%) examined the role of two mechanisms in maintaining diversity, and only two studies examined the role of three or more mechanisms.

4.3.1 Evidence for niche differentiation

Over the 15 years that we surveyed, several studies found compelling evidence for niche differentiation (Debski *et al.*, 2002; Newmaster *et al.*, 2003; Harms *et al.*, 2001; Gazol and Ibáñez, 2010; Paine *et al.*, 2008; Kraft *et al.*, 2008; Kursar *et al.*, 2009; Chuyong *et al.*, 2011; Metz, 2012). These studies fell into two general categories: 1) studies that utilized a trait-based approach to understand how plants were adapted to utilize resources or defend against herbivory; and 2) studies that examined general spatial patterns of habitat specialization among species. For example, Kraft *et al.* (2008) found compelling evidence for niche differentiation in a diverse tropical forest in Yasuni, Ecuador, using functional trait-based analyses. All traits measured, except for wood density, were more evenly distributed across species in a 25 ha plot than would be predicted by a neutral model, suggesting that co-occurring species diverge in their strategies for coping with limited resources, the cornerstone prediction of niche differentiation. Also in Parque Nacional Yasuni, Ecuador, Metz (2012) found that 90% of 136 species demonstrated habitat specialization during at least one time period over nine years, and that 60% of the species had significant habitat associations during at least half of the census periods. Likewise, in the 50 ha plot on Barro Colorado Island in Panama, around 50% of the trees demonstrated habitat specialization with respect to soil chemistry (John *et al.*, 2007). A study by Kursar *et al.* (2009) in both Panama and Peru found that coexisting adult plants within the genus *Inga* were more divergent in their defence strategies against herbivory than would be predicted by neutral dynamics. In Cameroon, 63% of tree species showed significant positive associations with at least one of five habitat types (Chuyong *et al.*, 2011). In adult understorey trees in Malaysia, spatial point pattern analysis elucidated a similar pattern of significant habitat specialization that contributed to coexistence (Debski *et al.*, 2002). However, this pattern appears to vary across some plant groups. Dalling *et al.*, (2012) compare the degree to which lianas and trees specialize on topographic and soil chemistry habitats, and found that significantly fewer lianas specialized on topographic (44%) or soil chemistry (21%) derived habitat values vs 66% and 52% respectively for trees (see also Ledo and Schnitzer, 2014).

Several studies clearly connected niche differentiation to the maintenance of species richness. Two studies of temperate zone forest sedges found that environmental heterogeneity and interspecific microhabitat preferences were important for the maintenance of local understorey sedge species diversity. Furthermore, species diversity was greatest when the habitat was most variable (i.e. when there are the highest potential number of niches: Bell *et al.*, 2000; Vellend *et al.*, 2000).

4.3.2 Evidence for negative density dependence

The literature examined provides compelling evidence for both NDD (the pattern) and putative mechanisms in tropical and temperate trees, shrubs and palms (Hubbell *et al.*, 2001; Blundell and Peart, 2004; Wyatt and Silman, 2004; Norghauer *et al.*, 2006; Comita *et al.*, 2007; Queenborough *et al.*, 2007; Yamazaki *et al.*, 2009; Chen *et al.*, 2010; Comita *et al.*, 2010; Mangan *et al.*, 2010; Bagchi *et al.*, 2010; Metz *et al.*, 2010; Bai *et al.*, 2012; Johnson *et al.*, 2012; Wang *et al.*, 2012; Jurinitz *et al.*, 2013; Xu *et al.*, 2014). Comita *et al.* (2010) found evidence for a pattern consistent with NDD in 180 tropical tree and shrub species at Barro Colorado Island, Panama. Furthermore, Comita *et al.* (2010) demonstrated that, contrary to the original predictions of Janzen (1970) and Connell (1971), relatively rare species experienced stronger NDD than more common species. John *et al.* (2002) found similar patterns in tropical dry forests in India. Mortality in small diameter trees (1–10 cm dbh) was negatively correlated with the abundance of conspecifics. In subtropical China, seedlings of legume species were less likely to survive in areas surrounding adults of their species (Liu *et al.*, 2012). In a temperate tree species (*Prunus serotina*), Packer and Clay (2000, 2004) demonstrated that seedlings were significantly less likely to survive near conspecific adults.

Soil microbes are a potential mechanism driving the presence of negative density dependent patterns. In Panama, Mangan *et al.* (2010) demonstrated that rare tree species experience stronger NDD than common ones, and that soil biota were responsible for the NDD pattern. Bagchi *et al.* (2014) removed insects and fungus using soil pesticide treatments and found that seedling diversity (in terms of the inverse Simpson index) was lower when soil fungi were eliminated. Xu *et al.* (2014) tested NDD across an elevational gradient in subtropical China and found that negative feedback from soil-borne enemies caused significantly decreased growth and survival of seedlings, and that the effect of soil-borne enemies decreased with increasing elevation. In the temperate zone, Packer and Clay (2000) established that seedlings of the temperate tree species *Prunus serotina* were less likely to survive in the presence of a single soil pathogen; further, that this was a likely causative agent of NDD (Packer and Clay, 2004).

Evidence from this review also suggests that life-history characteristics determine the strength of the NDD pattern. In a wet forest in Costa Rica, McCarthy-Neumann and Kobe (2008) found that shade tolerance interacted with the strength of negative density dependence. Specifically, seedlings of species with characteristics indicative of shade tolerance (high wood density, large seed size) tended to demonstrate weak or no NDD. Also in Costa Rica, Kobe and Vriesendorp (2011) reported that in seedlings of canopy, subcanopy and understorey tree species, differences in negative density dependence were more strongly related to physiologically based life-history traits than biotic feedback related to community abundance. Bai *et al.* (2012) reported that NDD was strongest in seedlings and saplings of trees and shrubs with gravity-dispersed seeds compared to wind- and animal-dispersed seeds, and also that shrubs were significantly less likely to demonstrate NDD when compared to trees. Other factors such as drought, logging activity and fire also seem to be negatively associated with the strength of NDD (John *et al.*, 2002; Bunker and Carson, 2005; Bagchi *et al.*, 2011).

NDD does not appear to be associated with all growth forms. For example, in a forest in Panama, Ledo and Schnitzer (2014) found that lianas have a positive density dependent pattern (and thus are not maintained by NDD), whereas trees were consistently negatively density dependent. Bai *et al.* (2012) found a similar result in temperate tree and shrub species in China. Tree seedlings and saplings were negatively affected by the presence of nearby conspecific adults, while shrub seedlings and saplings were far less likely to demonstrate this negative effect.

4.3.3 Evidence for disturbance hypotheses

Disturbance appears to maintain a portion of forest plant species diversity. For example, Schnitzer and Carson (2001) examined the extent to which disturbance maintains species diversity for lianas, and reported that liana species diversity was higher in tree-fall gaps than in intact forest, even when controlling for liana density. Ledo and Schnitzer

(2014) and Dalling *et al.* (2012) provided further evidence that lianas are maintained by disturbance events. Dalling *et al.* (2012) examined the degree to which lianas and trees specialize on specific habitat values and found that 63% of liana species had an affinity for areas of the BCI 50 ha plot in Panama, with low canopy height indicative of recent canopy gap. Ledo and Schnitzer (2014) confirmed this finding using point pattern analysis to determine that only disturbance explained the distribution of lianas in the BCI 50 ha plot. Additionally, ter Steege and Hammond (2001) found that community attributes that are sensitive to disturbance, such as seed mass and wood density, are strongly correlated with tree diversity.

There is little evidence to suggest that disturbance maintains diversity in the way that Connell (1978) suggested in his intermediate disturbance hypothesis. Obiri and Lawes (2004) measured gap size and estimated time since gap creation and examined diversity patterns in the colonizing saplings of coastal scarp forests in South Africa. This study reported only a slight (non-significant) increase in species diversity in gaps of intermediate size and time since creation. Furthermore, disturbance did not seem to maintain tree diversity. In the 50 ha plot on Barro Colorado Island (Panama), Brokaw and Busing (2000) found that tree seedlings appeared to colonize in gaps by chance dispersal and establishment rather than being particularly well adapted to the gap environment. Additionally, Schnitzer and Carson (2001) found no difference in the species richness and density of shade-tolerant non-pioneer tree species between gap and non-gap sites in the BCI 50 ha plot.

4.3.4 Evidence for neutral processes

Many studies during the time period of this survey provided evidence that neutral processes explain small parts of diversity maintenance patterns, but few provide compelling evidence that it is a major mechanism driving the maintenance of diversity. The findings of Comita *et al.* (2007) suggest that recruitment limitation determines which seedlings are able to establish in tropical tree and shrub communities in Panama. Obiri and Lawes (2004) also provide compelling evidence that neutral processes (i.e. recruitment limitation) determine the plants that colonize gaps in South Africa rather than specialization on a regeneration niche. Denslow and Guzman (2000) found that tropical tree seedling species richness was a function of not only seedling density, but the presence of proximate seed sources with efficient dispersal mechanisms and appropriate establishment conditions, suggesting that recruitment limitation plays a strong role in seedling community establishment. Seidler and Plotkin (2006) confirmed this finding in adults and saplings in Malaysia and Panama, documenting that both the extent and the scale of conspecific spatial aggregation is strongly correlated with the mode of seed dispersal. By contrast, Uriarte *et al.* (2005) found that in a hurricane-affected system in Puerto Rico, simple recruitment limitation was not sufficient to determine seedling species composition, and models that included NDD with recruitment limitation provided more explanatory power.

4.4 Bias in the maintenance of diversity literature

The studies included in our review were strongly biased towards tropical regions, with 64% of the studies being conducted in the tropics, 13% in the subtropics, and 22% in the temperate zone (see Fig. 4.1). Of the tropical countries, Panama was the largest contributor to this bias, accounting for 27% of the 51 studies. China contributed 16% of the studies, and Ecuador contributed 8% (see Fig. 4.2).

Maintenance of diversity studies were biased on their selection of growth form. Three-quarters (38) of the studies sampled were conducted on tree species alone (see Fig. 4.3).

The bias towards tree species may be a limiting factor for a general understanding of diversity maintenance. In tropical forests, tree species represent a small fraction of vascular plant species diversity (Gentry and Dodson, 1987). Tree diversity in temperate forests is less than 20% of vascular plant species diversity (Gilliam, 2007), which is even higher than in tropical forests. Recent evidence presented by Ledo and Schnitzer (2014) shows that tropical tree species in the 50 ha plot on Barro Colorado Island (Panama) are maintained in the community via a combination of niche differentiation and negative density dependence. By contrast, liana species diversity is maintained by a different set of mechanisms, and most lianas species tended to have a clumped rather than an overdispersed distribution that is consistent with negative density dependence. Shrubs in the 50 ha plot on Barro Colorado Island were also significantly clumped rather than overdispersed (Hubbell, 2001).

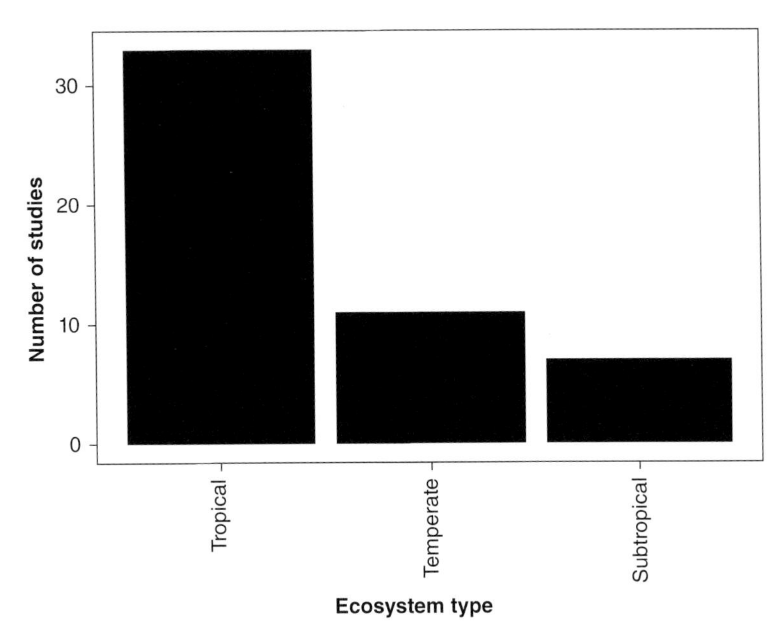

Fig. 4.1. Ecosystem type used to test four major hypotheses for the maintenance of species diversity. Some studies identified the ecosystem as a subregion of the broader ecosystem type: subtropical montane (one study), subtropical monsoon (one study), and tropical dry (one study).

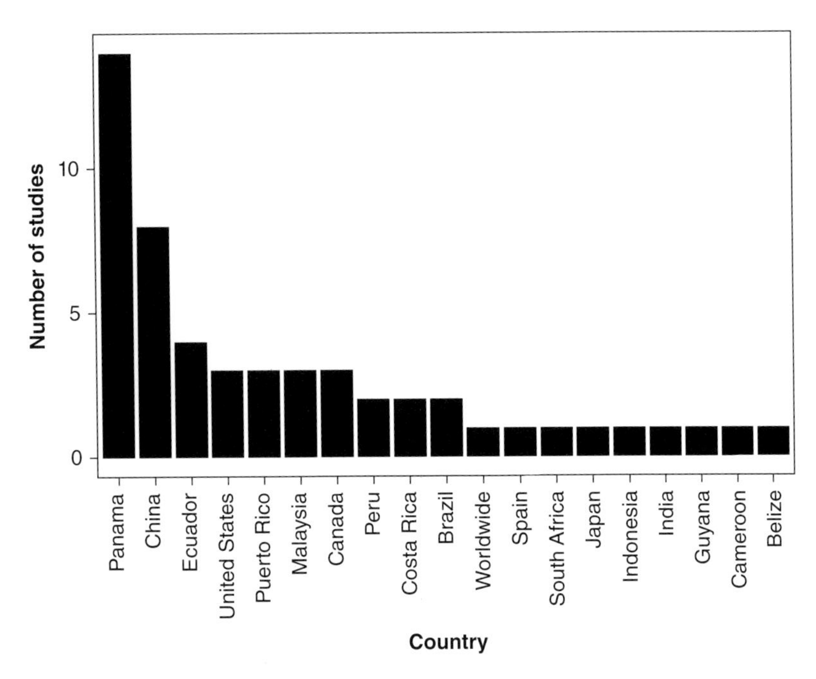

Fig. 4.2. The countries in which the four major hypotheses for forest plant diversity maintenance were tested. All studies that reported results from two separate countries were counted towards both countries separately, with the exception of the one study that reported results from major countries worldwide.

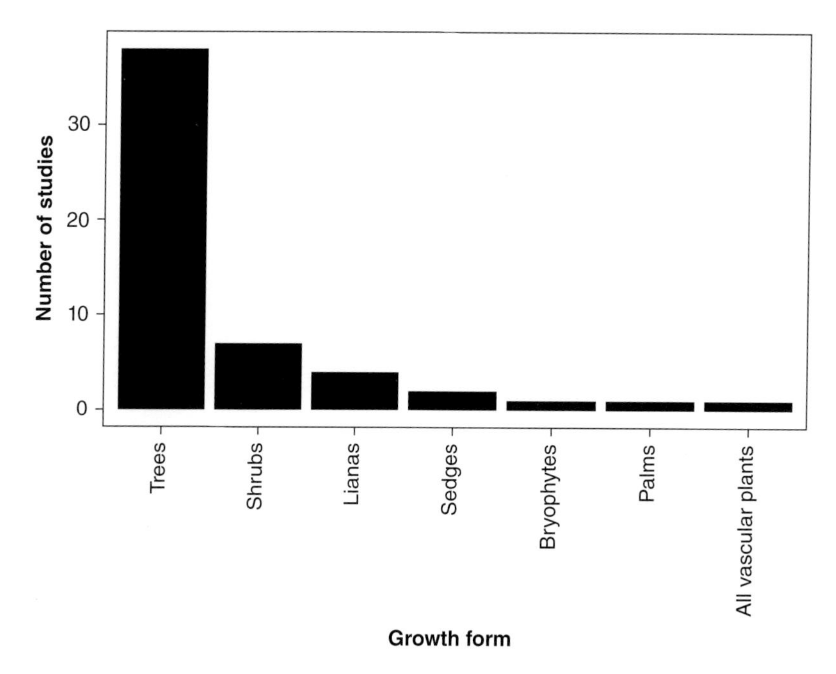

Fig. 4.3. Plant group used for testing the four major hypotheses for forest plant diversity maintenance. Studies that reported trends from multiple growth forms were counted in both categories.

Species' functional traits may determine their susceptibility to specific mechanisms (Denslow and Guzman, 2000; Seidler and Plotkin, 2006; McCarthy-Neumann and Kobe, 2008). Dispersal ability provides an ultimate limit to how far individuals may be from one another. Species with shorter dispersal distances would not persist in a community if they were not adapted to the heightened pest and pathogen pressure near their parent plants. Thus, dispersal ability may be a large determinant of whether a species demonstrates NDD (Denslow and Guzman, 2000; Seidler and Plotkin, 2006). Different growth forms tend to have different suites of traits, and thus they may respond differently to the mechanisms that maintain their diversity.

Studies on the maintenance of plant species diversity tended to focus on seedlings, and over 60% of these studies utilized only individuals that had been established for less than one year (Fig. 4.4). The bias towards seedlings may be particularly problematic.

The seed to seedling stage represents an ephemeral transition in the life history of a plant community. Many seedlings recruit into a community only to perish within the first several years after establishment. Seedlings have poorly developed root systems and little storage, and thus are especially vulnerable to the vagaries of temporal variation. The vulnerability to temporal variation in climate, as well as stochastic events, makes it especially important for studies to incorporate data from over many seasons. One-third of the studies examined seedling growth and mortality over multiple years. For example, Metz *et al.* (2010) reported that the strength of NDD and niche differentiation on seedling recruitment varied annually. Similarly, Zhu *et al.* (2013) found that all studied species were associated with at least one environmental variable during at least one of their life stages, but that the frequency of habitat association and negative density dependence decreased with plant age.

4.5 Conclusion

Ecologists have made great advances in determining the factors and mechanisms responsible for the maintenance of species diversity. The positive signal found for multiple mechanisms suggests that many mechanisms are working at different scales (Metz, 2012) and may combine to maintain

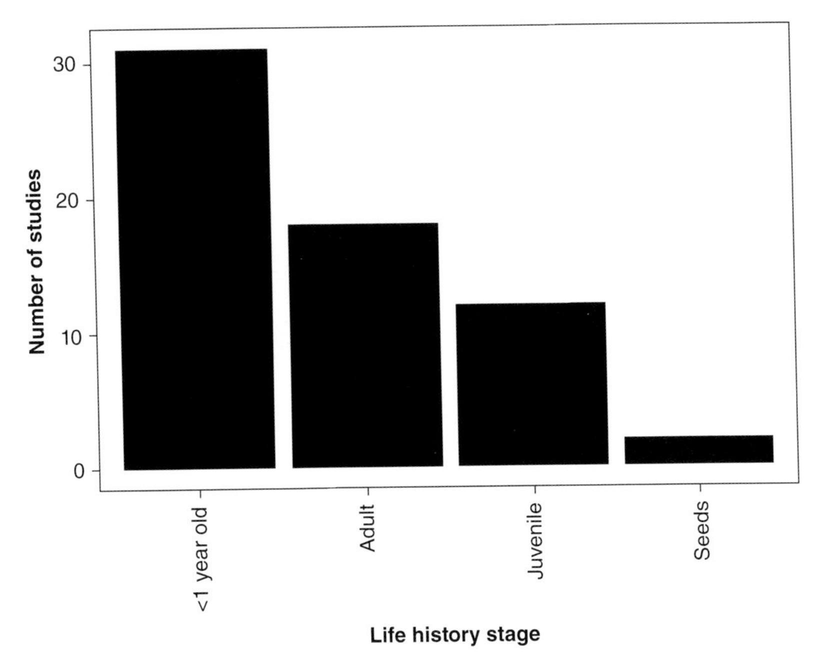

Fig. 4.4. Life-history stage of individuals included in each of the 51 studies reported on in a review of studies on the four major hypotheses for the maintenance of plant species diversity. Studies that included more than one life-history stage were counted in both categories.

diversity (Wills *et al.*, 2006). However, there are still a number of issues yet to be resolved about the strength of diversity maintenance mechanisms. Few studies have accounted for multiple life-history stages and multiple growth forms. Even fewer studies have compared the relative strengths of diversity maintenance mechanisms in a single study. Only two studies that we are aware of combined both of these approaches (Dalling *et al.*, 2012; Ledo and Schnitzer, 2014), and found evidence for multiple mechanisms that interact to maintain diversity differentially for different plant groups. Furthermore, only three studies focused on non-woody species, which represent the largest portion of diversity in all forests. Consequently, a more well-rounded approach to testing diversity mechanisms is necessary to understand how diversity is maintained.

References

Bagchi, R., Swinfield, T., Gallery, R.E., Lewis, O.T., Gripenberg, S., Narayan, L. and Freckleton, R.P. (2010) Testing the Janzen–Connell mechanism: pathogens cause overcompensating density dependence in a tropical tree. *Ecology Letters* 13(10), 1262–1269.

Bagchi, R., Philipson, C.D., Slade, E.M., Hector, A., Phillips, S., Villanueva, J.F. and Lewis, O.T. (2011) Impacts of logging on density-dependent predation of dipterocarp seeds in a South-East Asian rainforest. *Philosophical Transactions of the Royal Society B: Biological Sciences* 366(1582), 3246–3255.

Bagchi, R., Gallery, R.E., Gripenberg, S., Gurr, S.J, Narayan, L., Addis, C.E., Freckleton, R.P. and Lewis, O.T. (2014) Pathogens and insect herbivores drive rainforest plant diversity and composition. *Nature* 506 (7486), 85–88.

Bai, X., Queenborough, S.A., Wang, X., Zhang, J., Li, B., Yuan, Z, Xing, D., Lin, F., Ye, J. and Hao, Z. (2012) Effects of local biotic neighbors and habitat heterogeneity on tree and shrub seedling survival in an old-growth temperate forest. *Oecologia* 170(3), 755–765.

Bell, G. (2001) Neutral macroecology. *Science* 293(5539), 2413–2418.

Bell, G., Lechowicz, M.J. and Waterway, M.J. (2000) Environmental heterogeneity and species diversity of forest edges. *Journal of Ecology* 88(1), 67–87.

Blundell, A.G. and Peart, D.R. (2004) Density-dependent population dynamics of a dominant rain forest canopy tree. *Ecology* 85(3), 704–715.

Brokaw, N. and Busing, R.T. (2000) Niche versus chance and tree diversity in forest gaps. *Trends in Ecology and Evolution* 15(5), 183–188.

Bunker, D.E. and Carson, W.P. (2005) Drought stress and tropical forest woody seedlings: effect on community structure and composition. *Journal of Ecology* 93(4), 794–806.

Cavender-Bares, J., Kitajima, K. and Bazzaz, F.A. (2004) Multiple trait associations in relation to habitat differentiation among 17 Floridian oak species. *Ecological Monographs* 74(4), 635–662.

Chase, J.M. and Leibold, M.A. (2003) *Ecological Niches: Linking Classical and Contemporary Approaches*. University of Chicago Press, Chicago, USA and London, UK.

Chen, L., Mi, X., Comita, L.S., Zhang, L., Ren, H. and Ma, K. (2010) Community-level consequences of density dependence and habitat association in a subtropical broad-leaved forest. *Ecology Letters* 13(6), 695–704.

Chuyong, G.B., Kenfack, D., Harms, K.E., Thomas, D.W., Condit, R. and Comita, L.S. (2011) Habitat specificity and diversity of tree species in an African wet tropical forest. *Plant Ecology* 212(8), 1363–1374.

Clark, D.B., Clark, D.A. and Read, J.M. (1998) Edaphic variation and the mesoscale distribution of tree species in a neotropical rain forest. *Journal of Ecology* 86(1), 101–112.

Clark, D.B., Palmer, M.W. and Clark, D.A. (1999) Edaphic factors and the landscape-scale distributions of tropical rain forest trees. *Ecology* 80(8), 2662–2675.

Comita, L.S., Aguilar S., Perez R., Lao, S. and Hubbell, S.P. (2007) Patterns of woody plant species abundance and diversity in the seedling layer of a tropical forest. *Journal of Vegetation Science* 18(2), 163–174.

Comita, L.S., Muller-Landau, H.C., Aguilar, S. and Hubbell, S.P. (2010) Asymmetric density dependence shapes species abundances in a tropical tree community. *Science* 329(5989), 330–332.

Connell, J.H. (1971) On the role of natural enemies in preventing competitive exclusion in some marine animals and in rain forest trees. *Dynamics of Populations* 298–312.

Connell, J.H. (1978) Diversity in tropical rain forests and coral reefs. *Science* 199(4335), 1302–1310. DOI: 10.1126/science.199.4335.1302.

Connell, J.H., Tracey, J.G. and Webb, L.J. (1984) Compensatory recruitment, growth, and mortality as factors maintaining rain forest tree diversity. *Ecological Monographs* 54(2), 141–164. DOI: 10.2307/1942659.

Dalling, J.W., Hubbell, S.P. and Silvera, K. (1998) Seed dispersal, seedling establishment and gap partitioning among tropical pioneer trees. *Journal of Ecology* 86(4), 674–689. DOI: 10.1046/j.1365-2745.1998.00298.x.

Dalling, J.W., Schnitzer, S.A., Baldeck, C., Harms, K.E., John, R., Mangan, S.A., Lobo, E., Yavitt, J.B. and Hubbell, S.P. (2012) Resource-based habitat associations in a neotropical liana community: habitat associations of lianas. *Journal of Ecology* 100(5), 1174–1182. DOI: 10.1111/j.1365-2745.2012.01989.x.

Debski, I., Burslem, D.F.R.P., Palmiotto, P.A., LaFrankie, J.V., Lee, H.S. and Manokaran, N. (2002) Habitat preferences of Aporosa in two Malaysian forests: implications for abundance and coexistence. *Ecology* 83(7), 2005–2018.

Denslow, J.S. (1987) Tropical rainforest gaps and tree species diversity. *Annual Review of Ecology and Systematics*, 431–451.

Denslow, J.S. and Guzman, G. (2000) Variation in stand structure, light and seedling abundance across a tropical moist forest chronosequence, Panama. *Journal of Vegetation Science* 11(2), 201–212.

Dickie, I.A., Xu, B. and Koide, R.T. (2002) Vertical niche differentiation of ectomycorrhizal hyphae in soil as shown by T-RFLP analysis. *New Phytologist* 156(3), 527–535.

Enquist, B.J. and Enquist, C.A.F. (2011) Long-term change within a neotropical forest: assessing differential functional and floristic responses to disturbance and drought. *Global Change Biology* 17(3), 1408–1424.

Feeley, K.J., Davies, S.J., Perez, R., Hubbell, S.P. and Foster, R.B. (2011) Directional changes in the species composition of a tropical forest. *Ecology* 92(4), 871–882.

Gazol, A., and Ibáñez, R. (2010) Variation of plant diversity in a temperate unmanaged forest in northern Spain: behind the environmental and spatial explanation. *Plant Ecology* 207(1), 1–11.

Gentry, A.H. and Dodson, C.H. (1987) Diversity and biogeography of neotropical vascular epiphytes. *Annals of the Missouri Botanical Garden* 74, 205–233.

Gilliam, F.S. (2007) The ecological significance of the herbaceous layer in temperate forest ecosystems. *BioScience* 57(10), 845.

Harms, K.E., Wright, S.J., Calderón, O., Hernández, A. and Herre, E.A. (2000) Pervasive density-dependent recruitment enhances seedling diversity in a tropical forest. *Nature* 404(6777), 493–495.

Harms, K.E., Condit, R.S., Hubbell, P. and Foster, R.B. (2001) Habitat associations of trees and shrubs in a 50-ha neotropical forest plot. *Journal of Ecology* 89(6), 947–959.

Hille Ris Lambers, J., Clark, J.S. and Beckage, B. (2002) Density-dependent mortality and the latitudinal gradient in species diversity. *Nature* 417(6890), 732–735.

Hubbell, S.P. (1979) Tree dispersion, abundance, and diversity in a tropical dry forest. *Science* 203(4387), 1299–1309.

Hubbell, S.P. (1997) A unified theory of biogeography and relative species abundance and its application to tropical rain forests and coral reefs. *Coral Reefs* 16:S9.

Hubbell, S.P. (2001) *The Unified Neutral Theory of Biodiversity and Biogeography*. Princeton University Press, Princeton, New Jersey, USA.

Hubbell, S.P., Foster, R.B., O'Brien, S.T., Harms, K.E., Condit, R., Wechsler, B., Wright, S.J. and Loo de Lao, S.

(1999) Light-gap disturbances, recruitment limitation, and tree diversity in a neotropical forest. *Science* 283 (5401), 554–557. DOI: 10.1126/science.283.5401.554.

Hubbell, S.P., Ahumada, J.A., Condit, R. and Foster, R.B. (2001) Local neighborhood effects on long-term survival of individual trees in a neotropical forest. *Ecological Research* 16(5), 859–875.

Hutchinson, G.E. (1957) Concluding remarks. In: *Proceedings of the Cold Spring Harbor Symposium on Quantitative Biology* 22, 415–427.

Hutchinson, G.E. (1959) Homage to Santa Rosalia or Why are there so many kinds of animals? *American Naturalist* 93(870), 145–159.

Janzen, D.H. (1970) Herbivores and the number of tree species in tropical forests. *American Naturalist* 104 (940), 501–528.

John, R., Dattaraja, H.S., Suresh, H.S. and Sukumar, R. (2002) Density-dependence in common tree species in a tropical Dry Forest in Mudumalai, Southern India. *Journal of Vegetation Science* 13(1), 45–56.

John, R., Dalling, J.W., Harms, K.E., Yavitt, J.B., Stallard, R.F. *et al.* (2007) Soil nutrients influence spatial distributions of tropical tree species. *Proceedings of the National Academy of Sciences* 104(3), 864–869.

Johnson, D.J., Beaulieu, W.T., Bever, J.D. and Clay, K. (2012) Conspecific negative density dependence and forest diversity. *Science* 336(6083), 904–907.

Jurinitz, C.F., Oliveira, A.A. and Bruna, E.M. (2013) Abiotic and biotic influences on early-stage survival in two shade-tolerant tree species in Brazil's Atlantic forest. *Biotropica* 45(6), 728–736.

Kobe, R.K. and Vriesendorp, C.F. (2011) Conspecific density dependence in seedlings varies with species shade tolerance in a wet tropical forest. *Ecology Letters* 14(5), 503–510.

Kraft, N.J.B., Valencia, R. and Ackerly, D.D. (2008) Functional traits and niche-based tree community assembly in an Amazonian forest. *Science* 322(5901), 580–582.

Kursar, T.A., Dexter, K.G., Lokvam, J., Pennington, R.T., Richardson, J.E. *et al.* (2009) The evolution of antiherbivore defenses and their contribution to species coexistence in the tropical tree genus *Inga*. *Proceedings of the National Academy of Sciences* 106(43), 18073–18078.

Ledo, A. and Schnitzer, S.A. (2014) Disturbance and clonal reproduction determine liana distribution and maintain liana diversity in a tropical forest. *Ecology* 95(8), 2169–2178. DOI: 10.1890/13-1775.1.

Liu, Y., Shixiao, Y., Zhi-Ping, X. and Staehelin, C. (2012) Analysis of a negative plant–soil feedback in a subtropical monsoon forest. *Journal of Ecology* 100(4), 1019–1028.

MacArthur, R.H. and Wilson, E.O. (1963) An equilibrium theory of insular zoogeography. *Evolution* 17, 373–387.

MacArthur, R.H. and Levins, R. (1967) The limiting similarity, convergence, and divergence of coexisting species. *American Naturalist* 101(921), 377–385.

MacArthur, R.H. (1969) Patterns of communities in the tropics. *Biological Journal of the Linnean Society* 1 (1–2), 19–30.

MacArthur, R.H. (1972) *Geographical Ecology*. Harper and Row, New York, USA.

Mangan, S.A., Schnitzer, S.A., Herre, E.A., Mack, K.M.L., Valencia, M.C., Sanchez, E.I and Bever, J.D. (2010) Negative plant–soil feedback predicts tree-species relative abundance in a tropical forest. *Nature* 466(7307), 752–755.

McCarthy-Neumann, S. and Kobe, R.K. (2008) Tolerance of soil pathogens co-varies with shade tolerance across species of tropical tree seedlings. *Ecology* 89(7), 1883–1892.

Metz, M.R. (2012) Does habitat specialization by seedlings contribute to the high diversity of a lowland rain forest? *Journal of Ecology* 100(4), 969–679.

Metz, M.R., Sousa, W.P. and Valencia, R. (2010) Widespread density-dependent seedling mortality promotes species coexistence in a highly diverse Amazonian rain forest. *Ecology* 91(12), 3675–3685.

Newmaster, S.G., Belland, R.J., Arsenault, A. and Vitt, D.H. (2003) Patterns of bryophyte diversity in humid coastal and inland cedar Hemlock Forests of British Columbia. *Environmental Reviews* 11(S1), S159–S185.

Norghauer, J.M., Malcolm, J.R. and Zimmerman, B.L. (2006) Juvenile mortality and attacks by a specialist herbivore increase with conspecific adult basal area of Amazonian Swietenia macrophylla (Meliaceae). *Journal of Tropical Ecology* 22(4), 451–460.

Obiri, J.A.F. and Lawes, M.J. (2004) Chance versus determinism in canopy gap regeneration in coastal scarp forest in South Africa. *Journal of Vegetation Science* 15(4), 539–547.

Packer, A. and Clay, K. (2000) Soil pathogens and spatial patterns of seedling mortality in a temperate tree. *Nature* 404(6775), 278–281.

Packer, A. and Clay, K. (2004) Development of negative feedback during successive growth cycles of black cherry. *Proceedings of the Royal Society of London. Series B: Biological Sciences* 271(1536), 317–324.

Paine, C.E.T., Harms, K.E., Schnitzer, S.A. and Carson, W.P. (2008) Weak competition among tropical tree seedlings: implications for species coexistence. *Biotropica* 40(4), 432–440.

Palmer, M.W. (1994) Variation in species richness: towards a unification of hypotheses. *Folia Geobotanica et Phytotaxonomica* 29(4), 511–530.

Petermann, J.S., Fergus, A.J.F., Turnbull, L.A. and Schmid, B. (2008) Janzen–Connell effects are widespread and strong enough to maintain diversity in grasslands. *Ecology* 89(9), 2399–2406.

Queenborough, S.A., Burslem, D.F.R.P., Garwood, N.C. and Valencia, R. (2007) Neighborhood and community interactions determine the spatial pattern of tropical tree seedling survival. *Ecology* 88(9), 2248–2258.

Ricklefs, R.E. (1977) Environmental heterogeneity and plant species diversity: a hypothesis. *American Naturalist* 111(978), 376–381.

Russo, S.E., Brown, P., Tan, S. and Davies, S.J (2008) Interspecific demographic trade-offs and soil-related habitat associations of tree species along resource gradients. *Journal of Ecology* 96(1), 192–203.

Schnitzer, S.A, Mascaro, J. and Carson, W.P (2008) Treefall gaps and the maintenance of plant species diversity in tropical forests. In: Carson, W.P. and Schnitzer, S.A. (eds) *Tropical Forest Community Ecology*. Wiley-Blackwell, Chichester, West Sussex, UK and Malden, Massachusetts, USA, pp. 196–210.

Schnitzer, S.A. and Bongers, F. (2011) Increasing Liana abundance and biomass in tropical forests: emerging patterns and putative mechanisms. *Ecology Letters* 14(4), 397–406.

Schnitzer, S.A. and Carson, W.P. (2000) Have we missed the forest because of the trees? *Trends in Ecology and Evolution* 15, 376–377.

Schnitzer, S.A. and Carson, W.P. (2001) Treefall gaps and the maintenance of species diversity in a tropical forest. *Ecology* 82(4), 913–919.

Seidler, T.G. and Plotkin, J.B. (2006) Seed dispersal and spatial pattern in tropical trees. *PLoS Biology* 4(11), e344.

Simberloff, D.S. (1974) Equilibrium theory of island biogeography and ecology. *Annual Review of Ecology and Systematics* 5, 161–182.

Simberloff, D.S. and Wilson, E.O. (1970) Experimental zoogeography of islands. A two-year record of colonization. *Ecology* 51, 934–937.

Svenning, J.-C. (1999) Microhabitat specialization in a species-rich palm community in Amazonian Ecuador. *Journal of Ecology* 87(1), 55–65.

Swaine, M.D. and Whitmore, T.C. (1988) On the definition of ecological species groups in tropical rain forests. *Vegetatio* 75(1–2), 81–86.

ter Steege, H. and Hammond, D.S. (2001) Character convergence, diversity, and disturbance in tropical rain forest in Guyana. *Ecology* 82(11), 3197–3212.

Tilman, D. (1982) *Resource Competition and Community Structure*. Princeton University Press, Princeton, New Jersey, USA.

Tilman, D. and Wedin, D. (1991) Dynamics of nitrogen competition between successional grasses. *Ecology* 72(3), 1038–1049.

Uriarte, M., Canham, C.D., Thompson, J., Zimmerman, J.K. and Brokaw, N. (2005) Seedling recruitment in a hurricane-driven tropical forest: light limitation, density-dependence and the spatial distribution of parent trees. *Journal of Ecology* 93(2), 291–304.

Vellend, M., Lechowicz, M.J. and Waterway, M.J. (2000) Environmental distribution of four Carex species (Cyperaceae) in an Old-Growth Forest. *American Journal of Botany* 87(10), 1507–1516.

Wang, X., Comita, L.S., Hao, Z., Davies, S.J., Ye, J., Lin, F. and Yuan, Z. (2012) Local-scale drivers of tree survival in a temperate forest. *PLoS ONE* 7(2), e29469. DOI: 10.1371/journal.pone.0029469.

Whitmore, T.C. (1989) Canopy gaps and the two major groups of forest trees. *Ecology* 70(3), 536–538.

Wills, C., Harms, K.E., Condit, R., King, D., Thompson, J. *et al.* (2006) Nonrandom processes maintain diversity in tropical forests. *Science* 311(5760), 527–531. DOI: 10.1126/science.1117715.

Wright, J.S. (2002) Plant diversity in tropical forests: a review of mechanisms of species coexistence. *Oecologia* 130(1), 1–14.

Wyatt, J.L. and Silman, M.R. (2004) Distance-dependence in two Amazonian palms: effects of spatial and temporal variation in seed predator communities. *Oecologia* 140(1), 26–35.

Xu, M., Wang, Y., Liu, Y., Zhang, Z. and Yu, S. (2014) Soil-borne pathogens restrict the recruitment of a subtropical tree: a distance dependent effect. *Oecologia* 177(1), 723–732.

Yamazaki, M., Iwamoto, S. and Seiwa, K. (2009) Distance-and density-dependent seedling mortality caused by several diseases in eight tree species co-occurring in a temperate forest. *Plant Ecology* 201(1), 181–96.

Zhu, Y., Getzin, S., Wiegand, T., Ren, H. and Ma, K. (2013) The relative importance of Janzen–Connell effects in influencing the spatial patterns at the Gutianshan Subtropical Forest. *PloS One* 8(9), e74560.

5 Plant Diversity of the Drylands in Southeastern Anatolia-Turkey: Role in Human Health and Food Security

Munir Ozturk[1]*, Volkan Altay[2], Salih Gucel[3] and Ernaz Altundag[4]

[1]*Botany Department, Ege University, Izmir, Turkey;* [2]*Biology Department, Mustafa Kemal University, Hatay, Turkey;* [3]*Institute of Environmental Sciences, Near East University, Lefkoşa, Northern Cyprus;* [4]*Biology Department, Düzce University Düzce, Turkey*

Abstract

Two of the gene centres, the Mediterranean and the Near East, meet in Turkey, which comprises the Irano-Turanian, Mediterranean and Euro-Siberian phytogeographical divisions. The country is situated on the crossroads of important migratory routes and has been home to several civilizations, therefore increasing its significance for plant diversity. It is accepted as the centre of origin for several plants like pea, wheat, flax, lentil, chickpea, beet, tuberous species, herbaceous species like clover, medics, oats, together with woody species like pistachios, pear, vines, apple, plum and pomegranate. Wheat and barley are said to have been first cultivated in the fertile crescent. Very recent studies have revealed that wheat was cultivated for the first time at Karacadağ and its environs located in Southeastern Anatolia. In this study we have therefore included Diyarbakır, Gaziantep, Kahramanmaraş, Mardin, Şanlıurfa, Adıyaman, Siirt, Şırnak and Hakkari States from the Southeastern Anatolia Region. The plants distributed in the region were evaluated for their role in food security. The references available on this topic were fully surveyed and current use by the local inhabitants was recorded together with the way they use these species.The plant taxa distributed in the region were studied and their potential as animal feed evaluated. Generally these belong to the families of Fabaceae and Poaceae. Our investigations showed that the taxa such as *Allium scorodoprasum, Anethum graveolens, Capparis spinosa* var. *spinosa, Crataegus monogyna* ssp. *monogyna, Geranium tuberosum, Glycyrrhiza glabra, Gundelia tournefortii* var. *armata, Lepidium sativum* ssp. *sativum, Malva sylvestris, M. neglecta, Mentha pulegium, Morus nigra, Nasturtium officinale, Nigella sativa, Olea europaea, Orchis coriophora, Ornithogalum narbonense, Rheum ribes, Rhus coriaria, Pistacia khinjuk, P. vera, Portulaca oleracea, Rubus sanctus, Rumex acetosella, R. pulcher, Thymbra spicata* var. *spicata, Thymus* sp., *Trigonella foenum-graecum, Urtica dioica* and *U. urens* are used by the locals as food, in salad and spices, and also consumed as tea. In addition to these, taxa such as *Capparis ovata, C. spinosa, Cerasus mahaleb, Glycyrrhiza glabra, Pistacia khinjuk, P. terebinthus, Rhus coriaria* and *Thymbra spicata* are collected from the wild and sold in the country; also exported. Many taxa distributed in the region are used in traditional folk medicine. These are given alphabetically with their botanical name, part of the plant that is used, ailment treated and information on the preparations used.The taxa used as dye plants were also recorded.This investigation is expected to serve as a basis for future food security questions in the region.

5.1 Introduction

Turkey is a meeting place of three different plant geographical divisions with varying floral as well as climatic diversity. These phytogeographical regions are European-Siberian in the north, the Mediterranean in the west and Irano-Turanian in the central and east-south-east regions (Zohary, 1973). This has enabled the development of different

*E-mail: munirozturk@gmail.com

types of ecosystems with different soil and vegetational characteristics (Yiğit *et al.*, 2002; Avcı, 2005; Ozturk *et al.*, 2006a, b; Ozturk *et al.*, 2012c). The country also has two important gene centres of biodiversity with a transition between the Mediterranean, Near East and nine other continental countries (Vavilov, 1951).

The climatic and topographic features across the country have resulted in a remarkable biodiversity. However, the population explosion followed by heavy industrialization, intensive land use, deforestation, poor irrigation practices and overgrazing have created a great pressure on its biodiversity (Reynolds and Smith, 2002; Archer, 2004; Wang *et al.*, 2005; Ozturk *et al.*, 2006a, b; Zheng *et al.*, 2006; Xu *et al.*, 2008; Ozturk *et al.*, 2012c). These impacts have led to the loss and reduction of organic matter in the soil of areas with an arid or semi-arid climate (Perevolotsky and Seligman, 1998; Wallace, 2000; Yates *et al.*, 2000; Tellawi, 2001; Wang, 2003; Yang *et al.*, 2005; Wang *et al.*, 2006a, b; Xu *et al.*, 2008; Jeddi and Chaieb, 2010). Physical and chemical properties of the soil have been degraded as a result of anthropogenic influence, which is closely associated with the vegetation in the area (Ozturk *et al.*, 2006a, b).

Therefore, determination of the ecological structure of plant cover in natural and semi-natural areas and land-use practices carried out on such habitats is of paramount importance in determining the extent of the changes occurring from an ecological point of view. Further economical evaluation of these areas is essential for a sustainable future (Karabulut, 2006).

In this chapter, nine states from Southeastern Anatolia – Diyarbakır, Gaziantep, Kahramanmaraş, Mardin, Şanlıurfa, Adıyaman, Siirt, Şırnak and Hakkari – have been included. This area is also known as Upper Mesopotamia, and lately it has come under great pressure due to the construction of dams, highways and other activities (Ozturk *et al.*, 2006a, b; Ozturk *et al.*, 2012c).

The medicinal and economic potential of plants distributed on the dry lands of this area is discussed here in detail in the light of studies undertaken here during the last few decades. Moreover, the evaluation has been considered in the light of future food security and its importance for the region.

5.2 Study Area

The Southeastern Anatolian Region covers an area extending to the southern border of Syria and Iraq, formed by the chain of south-eastern Taurus mountains. A major part of this area consists of rough plains with very few mountainous areas (Demir, 2003). Nearly half of the total land in the region is favourable for agricultural production. There is a great difference in the summer and winter temperatures: summers are hot and dry, with an impact from the climatic conditions observed under the Mediterranean precipitation regime. The climate, soils and the vegetation differ from other arid and semi-arid parts of Turkey. In general, the summer temperatures are around 25–30°C. The average temperature in July is 31°C in Diyarbakir, 30.4°C in Siirt, and 27.3°C in Gaziantep. The highest temperatures are generally above 40°C (46.2°C in Diyarbakir, 44°C in Gaziantep, 46°C in Siirt). In winter the weather is very cold and frost is common due to the effects of the continental climate. The average January temperature on low plateaus descends to 5°C and towards the north it drops below 0°C. The lowest temperatures are around –20°C due to the continental climatic impacts (–24.2°C in Diyarbakir, –19.3°C in Siirt and –17.5 °C in Gaziantep). The temperature goes down to –30°C on the higher altitudes of Hakkari State. The Southeastern Anatolia Region is under the influence of Mediterranean precipitation regime, with rainy winters and fairly dry summers. The average annual rainfall varies between 400 mm and 1200 mm. The plateaus of Gaziantep and Sanliurfa, together with Diyarbakir basin, receive a rainfall ranging between 400 mm and 600 mm (549 mm in Gaziantep, 726 mm in Siirt, 491 mm in Diyarbakir), which increases towards the lower parts of Taurus Mountains to 1000 mm. The highest precipitation in the region is recorded in the State of Hakkari, going up to 1000 mm on the southern slopes.

The region enters the Mesopotamian basin in the south, dominated by arid and red-coloured steppe soils, and the amount of lime in the soils is very high due to low rainfall. The area is dominated by plant species of xeric character, except for the higher altitudes of the mountains like Karacadağ, Mazı and Midyat. The steppe vegetation includes species of Arabian origin. On the edge of the steppes, the forests start from 700–850 m, with a domination by shrub such as oak, due to land degradation. Major species found here are *Quercus infectoria, Q. brantice* and *Q. libani.* The famous Tigris river occupies the middle of Diyarbakir basin, among the steppes together with other small and large branches. *Salix triandra* and *Populus euphratica* are distributed all along the two sides of the river.

The presence of *Olea europaea* in the arid steppe parts of the Southeastern Anatolian Region among the general vegetation is highly noteworthy. The olive groves extend to Kilis, Viranşehir and Ceylanpınar. However, due to a very low relative humidity, the plants have difficulty establishing. In the western part of the steppe area, especially west of the Euphrates, the vegetation changes and at around 500–600 m, the limestone plateaus are covered by *Olea europaea* and *Pistacia vera* trees. *Pistacia vera* plantations are observed commonly on the plateaus around Gaziantep-Şanlıurfa. From Gaziantep onwards, dry steppes dominate (İnandık, 1965).

In spite of the presence of poor steppe species cover in the Southeastern Anatolia Region, it has a rich diversity of Poaceae and Fabaceae members. Of ten naturally occurring wheat (*Triticum*) cultivars in Turkey, half are found to grow around the Karacadağ area. Fabaceae from the wild species of lentils, vetch, peas, sainfoin and alfalfa from Fabaceae are also distributed in this area. Many species of *Cicer, Lens, Lathyrus, Vicia, Pisum, Onobrychis, Lotus, Medicago, Trigonella* and *Trifolium* are also commonly seen in Southeastern Anatolia, and some of these are endemic to the region.

Major plant species dominating the steppe vegetation of Southeastern Anatolia are *Acanthophyllum verticillatum, Achillea bantolina, Alhagi maurorum, Astragalus gummifer, Avena barbata, Bromus macrostaphyus, Cichorium glandulosum, Convolvulus reticulatus, Dianthus multipunctatus, Delphinium peregrinum, Eryngium campestre, Euphorbia aleppica, Gentiana olivieri, Hordeum leporinum, Onosma giganteum, Silene kotschyi, Thymus syriacus, Trifolium campestre, Centaurea hypericum, Trifolium campestre* and different species of *Salvia* and *Verbascum*. The major shrubs found in the region are *Amygdalus arabica, Cerasus microcarpa, Cercis siliquastrum, Ficus caria, Acer monspessulanum, Cerasus mahalep, Crataegus aronica, Pyrus syriaca, Celtis tournefortii, Pistacia khinjuk* and *P. vera*.

5.3 Plant Diversity

The study area has attracted the attention of large numbers of investigators due to its historical importance. The floristic studies have been undertaken by Davis (1965–1985), Davis *et al.* (1988), Güner *et al.* (2000), Aslan and Türkmen (2001), Aslan and Türkmen (2003), Yıldız (2001), Tatlı *et al.* (2002), Varol (2003), Akan *et al.* (2005a), Aydoğdu and Akan (2005), Aytaç and Duman (2005), Yapıcı and Saya (2007), Balos and Akan (2008), Eker *et al.* (2008), Atamov *et al.* (2009), Kaya and Ertekin (2009), Özuslu and Iskender (2009), Ozuslu and Tel (2010), and Ozturk *et al.* (2012c).

An evaluation of the floristic studies has revealed that a total of 3914 vascular plant taxa belonging to 121 families and 793 genera show distribution in the region, including 450 endemics. The families with the highest number of taxa are Asteraceae (499 taxa), Fabaceae (467 taxa), Lamiaceae (266 taxa), Brassicaceae (220 taxa) and Poaceae (201 taxa). The genera with the highest number of taxa are; *Astragalus* (152 taxa), *Allium* (65 taxa), *Centaurea* (58 taxa), *Trifolium* (55 taxa) and *Silene* (49 taxa) (see Table 5.1).

In addition to the floristic studies, many investigations on the forest, steppe, halophytic, ruderal, dry stream, meadow and water-marsh vegetation of this region have also been undertaken (Varol and Tatlı, 2001; Varol and Tatlı, 2002; M. Yavuz, Şanlıurfa'nın Akçakale İlçesi'ndeki halofitik alanların florası ve vejetasyonu. Department of Biology, Institute of Science and Technology, Harran University, 2005, unpublished thesis; Atamov *et al.*, 2006; Varol *et al.*, 2006; Kaya *et al.*, 2009; Kaya, 2010; Kaya *et al.*, 2010; Tel *et al.*, 2010; Yıldırım and Cansaran, 2010; Tel and Tak, 2012; Ozturk *et al.*, 2014). These vegetation studies have revealed that a total of 46 plant associations have been described by the authors cited here. Out of these, the forest vegetation is represented by 17,

Table 5.1. The families and genera in Southeastern Anatolia with the highest numbers of taxa.

	Family	Number of taxa	Genera	Number of taxa
1	Asteraceae	499	*Astragalus*	152
2	Fabaceae	467	*Allium*	65
3	Lamiaceae	266	*Centaurea*	58
4	Brassicaceae	220	*Trifolium*	55
5	Poaceae	201	*Silene*	49
6	Apiaceae	197	*Euphorbia*	48
7	Liliaceae	193	*Verbascum*	47
8	Caryophyllaceae	189	*Veronica*	45
9	Scrophulariaceae	153	*Onosma*	44
10	Boraginaceae	149	*Salvia*	43

steppe by 16, halophytic vegetation by 9, ruderal by 2, dry streams by 1, and meadows and water-marsh vegetation by 1 association. These associations are summarized below.

5.3.1 Forest vegetation

Pistacio khynjuki-Cotinetum coggyriae Tel *et al.*, 2010; *Astragalo lamarckii-Quercetum brantii* Tel *et al.*, 2010; *Lonicero ibericae-Aceretum cinerascentis* Tel *et al.*, 2010; *Centaureo lycopifoliae-Pinetum brutiae* Varol and Tatlı, 2001; *Dorcynio hirsuti-Populetum tremulae* Varol and Tatlı, 2001; *Galio tenuissimi-Quercetum cerridis* Varol and Tatlı, 2001; *Potentillo crantzii-Fagetum orientalis* Varol and Tatlı, 2001; *Galio ibicini-Quercetum pinnatilobae* Varol and Tatlı, 2001; *Lagoecio cuminoides-Sytracetum officinalii* Varol and Tatlı, 2001; *Thlaspio microstyli-Cedretum libani* Varol and Tatlı, 2001; *Gastridio ventricosi-Pinetum pineae* Varol and Tatlı, 2002; *Medicagini coronatae-Pinetum brutiae* Varol *et al.*, 2006; *Potentillo calycinae-Pinetum brutiae* Varol *et al.*, 2006; *Verbasco amani-Abietum ciliciae* Varol *et al.*, 2006; *Centaureo lycopifoliae-Pinetum pallasianae* Varol *et al.*, 2006; *Nepeto trachionatae-Quercetum brantii* Kaya *et al.*, 2009; *Teucrio multicauli-Crataegetum aroniae* Kaya *et al.*, 2009.

5.3.2 Steppe vegetation

Astragalo cuspistipulati-Acantholimetum acerosi Varol and Tatlı, 2001; *Achilleo grandifoliae-Micromerietum brachycalicii* Varol and Tatlı, 2001; *Phlomo lineari-Astragaletum kurdicii* Varol and Tatlı, 2001; *Achilleo pseudoaleppicae-Astragaletum diphtheritae* Kaya, 2010; *Sideritido microchlamydis-Convolvuletum oxysepali* Kaya *et al.*, 2010; *Phlomido capitati-Lagoeicetum cominoidis* Tel and Tak, 2012; *Cardo braviphyllaris-Phletum boissierii* Tel and Tak, 2012; *Onobrycho caput-galli-Picnometum acarnae* Tel and Tak, 2012; *Salvio palaestinae-Tragopogetum pterocarpi* Tel and Tak, 2012; *Ainsworthio trachycarpae-Elymetum erosiglumis* Tel and Tak, 2012; *Balloto brachyodontae-Stachietum cataonicae* Tel and Tak, 2012; *Phlomido capitati-Picnometum acarnae* Tel and Tak, 2012; *Astragalo compacti-Amygdaletum arabicae* Tel *et al.*, 2010; *Helichryso aucheri-Thymetum kotschyani* Tel *et al.*, 2010; *Verbasco diversifoliae-Astragaletum cephalotis* Tel *et al.*, 2010; *Phlomido capitatae-Thymetum migrici* Tel *et al.*, 2010.

5.3.3 Halophytic vegetation

Halothamno hierochunticae-Salsoletum incanescentis Kaya *et al.*, 2010; *Hymenolobo procumbentis-Aeluropetum lagopoidis* Kaya *et al.*, 2010; *Aeluropuseto lagopoidesae-Chenopodietum vulvariae* Atamov *et al.*, 2006; *Frankenieto pulverulentae-Salsoletum sodae* Atamov *et al.*, 2006; *Cressa creticae-Aeluropusetum lagopoidesae* Atamov *et al.*, 2006; *Prosopo farctae-Hordetum murinumae* Atamov *et al.*, 2006; *Cresso creticae-Hordetum murinumae* Atamov *et al.*, 2006; *Alhago manniferae-Hordetum murinumae* Atamov *et al.*, 2006; *Ammio visnagae-Hordetum murinumae* Atamov *et al.*, 2006.

5.3.4 Ruderal vegetation

Frankenio pulverulentae-Chenopodietum albi Kaya *et al.*, 2010; *Prosopo farctae-Alhagietum manniferae* Kaya *et al.*, 2010.

5.3.5 Dry stream vegetation

Acantho dioscoridi-Viticetum agni-casti Kaya *et al.*, 2009

5.3.6 Meadows and water-marsh vegetation

Phragmitetum australisae Koch, 1926

5.4 Economically Important Vascular Plants in Southeastern Anatolia

The plants in this region have been used as potential forage crops, food plants, for medicinal purposes, fuel, dye, basket making and other handicrafts, such as musical instrument making, as well as ornaments by different civilizations throughout history (Ozturk and Ozcelik, 1991; Plotkin, 2000; Ozturk *et al.*, 2011; Ozturk *et al.*, 2012a, b). The ethnobotanical studies published from the region indicate that 367 taxa have been evaluated as medicinals and aromatics, 225 taxa as food plants, 156 taxa as fodder, 19 taxa as herbal drinks, 19 taxa as spices, and 25 taxa used in cheese making. In addition to these, nearly 159 taxa have been used for other economic purposes (fuel, ornaments, dye, musical instruments, handicrafts, brooms, baskets, toys) (see Fig 5.1). These studies reveal that the widest use is in medicinal and aromatic applications (38%), followed by food plants

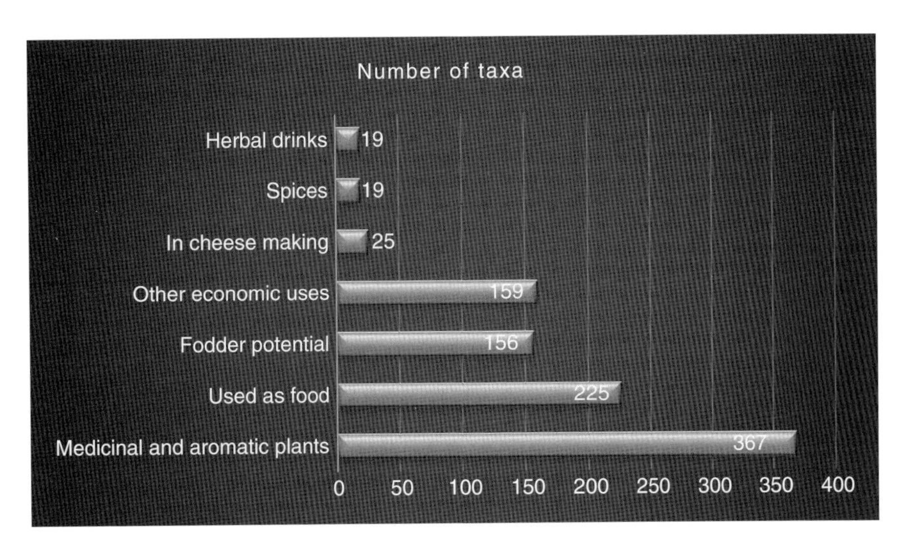

Fig. 5.1. The distribution pattern of ethnobotanical uses of plant taxa from Southeastern Anatolia.

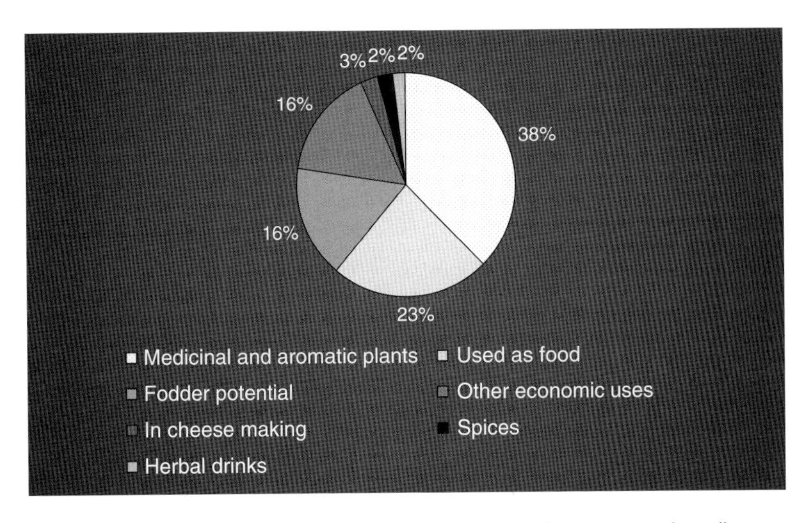

Fig. 5.2. The distribution pattern of ethnobotanical uses of plant taxa from Southeastern Anatolia on a percentage basis.

(23%), fodder species (16%) and other economic uses (16%) (see Fig 5.2).

5.5 Medicinal and Aromatic Plants in Southeastern Anatolia

An evaluation of the ethnobotanical uses highlights the fact that people in this region have been using the plants as a major source for the treatment of diseases, as in other parts of the world (Blumenthal, 1998; Plotkin, 2000; Kurt *et al.*, 2004; Algier *et al.*, 2005; Newman and Cragg, 2007; Kendir and Güvenç, 2010; Ozturk *et al.*, 2011; Pleskanovskaja *et al.*, 2011; Ozturk *et al.*, 2012a, b; Ozturk *et al.*, 2014). This type of use dates back thousands of years (De Silva *et al.*, 2009). In all, 367 medicinal and aromatic plant taxa have been reported from the study area. These are given alphabetically with their botanical name, part used, ailment treated and information on the preparations used (see Table 5.2).

An evaluation of these on the basis of disease shows that a large number of taxa are used for urinary system disorders (108 taxa), followed by stomach

Table 5.2. List of medicinal and aromatic plants distributed in Southeastern Anatolia.

	Family / Plant taxa	Treatment	Part used	Preparation	Source
	ADIANTHACEAE				
1	*Adianthum capillus-veneris*	Appetite, shortness of breath, kidney stone, expectorant, cough, urinary inflammations	AP	BO, IN, RW	8, 10
	AMARANTHACEAE				
2	*Amaranthus retroflexus*	Stomach, digestion	AP	CO	18
	ANACARDIACEAE				
3	*Cotinus coggyria*	Mouth sore	FR	GA	8
4	*Pistacia eurycarpa*	Scorpion bite, antiseptic soap	SE, ST	GU	14
5	*Pistacia khinjuk*	Stomach	ST	GU	12
6	*Pistacia terebinthus* ssp. *palaestina*	Wounds, mouth sore, anti-inflammatory, stomach, sore throat	BR, FR, LE, RO, SE, ST	DC, RW, GU, PM	6, 8, 10, 17
7	*Pistacia vera*	Wounds	LE	PM	17
8	*Rhus coriaria*	Wounds, antiseptic, diarrhoea, antipyretic, styptic, mouth sore	FR	DC, PW	1, 8, 10
	APIACEAE				
9	*Anethum graveolens*	Cholesterol, flatulence, halitosis	AP	BO, RW	1, 12
10	*Apium nodiflorum*	Appetite	SH	RW	10
11	*Bifora testiculata*	Stomach ache	AP	IN	4
12	*Diplotaenia cachrydifolia*	Rheumatism, diabetes	RO	BO	12
13	*Echinophora tenuifolia* ssp. *sibthorpiana*	Indigestion	SH	DU	17
14	*Eryngium billardieri*	Sinusitis, antifungal, bronchitis, stomach ache	RO	BO, RW	12, 14
15	*Erygnium campestre* var. *virens*	Peptic ulcer, cardiac ailments	AP, SE	AT, CO	17, 18
16	*Ferula hausknechtii*	Wounds	RO		12
17	*Ferula longipedunculata*	Aphrodisiac	RO	PW	5, 8
18	*Ferula meifolia*	Aphrodisiac	RO	PW	18
19	*Ferula orientalis*	Haemorrhoids, skin inflammation, snake bite, scorpion bite, bee stings, wormy wound (in animal)	RO	BO, PL	12, 14
20	*Ferulago cassia*	Eye infections, to encourage milk secretion in new mothers	SE	CR, IH	8
21	*Foeniculum vulgare*	Abdominal pain (in baby)	SE	BO, PW	10, 12
22	*Grammosciadium platycarpum*	Stomach ache	AP	BO	12
23	*Johrenia dichotoma* ssp. *sintenisii*	Wounds	AP	PW	12
24	*Lecokia cretica*	Hypertension	SE	CR	8
25	*Malabaila dasyanthal*	Haemorrhoids	LE	DR, FE	14
26	*Pimpinella peregrina*	Flatulence	FR	BO	14
27	*Prangos pabularia*	Wound	RO	PL	12
28	*Prangos peucedanifolia*	To prevent itching	FL		4

29	*Smyrnium connatum*	Dyspnoea (breathlessness)	ST	FE	8
30	*Torilis leptophylla*	Kidney disorders	FR, LE		4
31	*Turgenia latifolia*	Pain relief for babies	FR, LE		4
	APOCYNACEAE				
32	*Nerium oleander*	Cancer, itching	LE, RO	BO	9–11
	ARACEAE				
33	*Arum conophalloides*	Diabetes, sedative	TU	BO	14
34	*Arum detruncatum* var. *virescens*	Diarrhoea, tension, diabetes	LE	DC	13, 14
35	*Arum maculatum*	Stomach, digestive, intestines, haemorrhoids	LE, TU	CO, SW	8, 18
36	*Biarum carduchorum*	Haemorrhoids	TU	PW	2
	ARALIACEAE				
37	*Hedera helix*	Cardiac diseases	LE	IN	8
	ARISTOLOCHIACEAE				
38	*Aristolochia bottae*	Wound healing, snake bites, parasite, haemorrhoids, wound (in animal)	AP, RO	AT, BO, DR, PT, PW	6, 12, 14
	ASCLEPIDACEAE				
39	*Vincetoxicum canescens* ssp. *canescens*	Scabies	BU	PO	13
	ASPLENIACEAE				
40	*Ceterach officinarum*	Kidney stone, diuretic, urinary inflammations	AP, LE	IN	8, 18
	ASTERACEAE				
41	*Achillea biebersteinii*	Fungal ailments, menstrual pain, abdominal pain, haemorrhoids, urinary tract disorders	AP, FL, LE, SH	IN, PM	8, 11, 17
42	*Achillea goniocephala*	Asthma, rheumatism, cold, women's ailments	FL	AT	5
43	*Achillea kotschyi* ssp. *kotschyi*	Menstrual pain, abdominal pain	AP	IN	8
44	*Achillea millefolium* ssp. *millefolium*	Stomach, kidney, diabetes, tension, jaundice, typhoid	FL	AT	14, 18
45	*Achillea setacea*	Menstrual pain, abdominal pain	AP	IN	8
46	*Achillea vermicularis*	Kidney pains, diarrhoea	FL	AT, DR	14
47	*Anthemis austriaca*	Cold	FL	AT	12
48	*Anthemis cotula*	Stomach pain, cold, stomach, bronchitis, hair loss	AP, FL	BO	6, 12, 14
49	*Anthemis nobilis*	Diuretic, stomach ache	FL	IN	13
50	*Anthemis tinctoria* var. *pallida*	Stomach ache, cold and flu, peptic ulcer	AP, SH	IN	8, 17
51	*Anthemis wiedemanniana*	Pain	FL	IN	11
52	*Artemisia abrotanum*	Analgesic, parasitic, sedative, stomach	ST	BO	14
53	*Arctium minus* ssp. *pubens*	Haemorrhoids, eye ailments	FL, ST	DU, PO	12, 13
54	*Arctium tomentosum* var. *glabrum*	Abscesses	RO	PO	13
55	*Artemisia absinthium*	Shortness of breath, diabetes, cold	AP	BO, IN	8, 12

Continued

Table 5.2. Continued.

	Family / Plant taxa	Treatment	Part used	Preparation	Source
56	*Artemisia spicigera*	Rheumatism, stomach ache, abdominal pain, headache, cough	AP	BO	12
57	*Bellis perennis*	Cold and flu, diuretic, tonic, skin diseases, bronchitis, stomach ache, dyspnoea, urinary inflammations	AP, FL	BO, IN, PT	8-10, 18
58	*Carlina lanata*	Wounds	WP	PM	17
59	*Centaurea consanguinea*	Tuberculosis	SE, SH	PI	17
60	*Centaurea depressa*	Anxiety (children)	AP	IN	4
61	*Centaurea glastifolia*	Prostate treatment	AP	BO	12
62	*Centaurea iberica*	Goitre, nerves, diabetes, snake bite	LE	RW	10, 12, 14
63	*Centaurea kurdica*	Kidney ailment	FR	BO	6
64	*Centaurea lycopifolia*	Healing wounds	AP	DC, IN	8
65	*Centaurea pterocaula*	Diarrhoea	LE	BO	12
66	*Centaurea rigida*	Snake bite	LE	BO	6
67	*Centaurea saligna*	Coagulation, wounds	LE	DR	14
68	*Centaurea virgata*	Abdominal pain	WP	IN	11
69	*Chrysophthalmum montanum*	Sinusitis	FL, LE	PW	11
70	*Cichorium intybus*	Ulcer, asthma, sedative, analgesic, abdominal pain, prostate treatment, hypertension, diabetes, cardiac diseases, liver diseases, cough, bronchitis	AP, RO	AT, BO, CO, DC, DR, IN, RW	8, 10, 12, 14
71	*Cichorium pumilum*	Liver diseases	RO		19
72	*Cirsium pubigerum* var. *spinosum*	Swelling	RO	PT	12
73	*Crepis sancta*	Eye ailments	FL	RW	10
74	*Gundelia tournefortii*	Stomach, strengthening of gums, appetite	AP	CO	17, 18
75	*Helianthus tuberosus*	Diabetes	TU	RW	12
76	*Helichrysum arenarium* ssp. *aucheri*	Kidney stones	AP	DC	8
77	*Helichrysum armenium* ssp. *armenum*	Urinary tract disorders, kidney stone, earache	FL	DC, IN	8, 11
78	*Helichrysum plicatum* ssp. *plicatum*	Kidney ailments, anti-inflammatory, kidney stone, earache, cancer, tumour	AP, FL, SH	AT, BO, DC, DR, DU, IH, IN	8, 12-14, 17
79	*Helichrysum stoechas*	Stomach, kidney	FL	AT	18
80	*Inula helenium* ssp. *vanensis*	Haemorrhoids	AP	PW	12
81	*Inula montbretiana*	Cancers, cold	AP	DC	8
82	*Inula oculus-christi*	Wounds	FL	BO	14
83	*Lactuca saligna*	Hypertension	BU	RW	12
84	*Leontodon hispidus* var. *hispidus*	Eye ailments	FL	RW	10

85	*Matricaria aurea*	Asthma, cold, stomach ache, bronchitis, flatulence, cardiac ailments	AP, LE	BO, IN	6, 17
86	*Matricaria chamomilla*	Shortness of breath, cold, stomach, diuretic, appetite	FL	AT	17, 18
87	*Onopordum carduchorum*	Asthma, hepatitis, liver disorders, cancer	FL, LE, SE	BO, DC	14, 17
88	*Scorzonera cana* var. *cana*	Intestinal parasites	RO	FE	14
89	*Scorzonera latifolia*	Pain, sterility	RO	GU	12, 13
90	*Scorzonera mirabilis*	Headache	LE	FE	14
91	*Scorzonera tomentosa*	Fungal infections	LE	CR	8
92	*Senecio vernalis*	Eye ailments	FL	RW	10
93	*Tanacetum argenteum* ssp. *argenteum*	Pain	FL	IN	4
94	*Tanacetum argyrophyllum* var. *agrophyllum*	Diabetes	AP	BO	12
95	*Tanacetum balsamita* ssp. *balsamitoides*	Diuretic, kidney stones, parasites, wounds	FL, LE	BO, PT	12, 14
96	*Tanacetum chilliophyllum* var. *chilliophyllum*	Diabetes	AP	BO	12
97	*Taraxacum montanum*	Eye ailments, wounds	LA		11, 12
98	*Taraxacum sintenisii*	Increase milk yield (cow)	AP		4
99	*Tragopogon dubius*	Gastrointestinal disorders	AP	IN	8
100	*Tragopogon pratensis*	Stomach	LE	CO	18
101	*Tripleurospermum parviflorum*	Cold, antipyretic, pain	FL	IN	10, 11
102	*Tussilago farfara*	Cough, expectorant, bronchitis	AP	IN	8
103	*Xanthium spinosum*	Urinary tract disorders	AP	BO	6
104	*Xanthium strumarium* ssp. *cavanillesii*	Kidney ailments	WP	DC	17
	BERBERIDACEAE				
105	*Bongardia chrysogonum*	Haemorrhoids, urinary antiseptic, cancers	TU	DR	14, 18
	BORAGINACEAE				
106	*Alkanna orientalis*	Stomach ache	RO	BO	14
107	*Alkanna tinctoria* ssp. *anatolica*	Peptic ulcer, wounds, burn	RB	CO, PM	17
108	*Anchusa azurea* var. *azurea*	Wounds, cancer, stomach, rheumatism	AP, RO	BO, CR	1, 6
109	*Anchusa azurea* var. *kurdica*	Diuretic, rheumatism, snake bite, swelling	FL	BO	10, 14
110	*Anchusa strigosa*	Cancer, diuretic	AP	IN	6, 11
111	*Echium italicum*	Wounds, pruritus	AP, RO	DC	8, 13
112	*Heliotropium circinatum*	Kidney ailments	WP	DC	17
113	*Heliotropium europaeum*	Antipyretic, expectorant			19
114	*Nonea pulla*	Snake bite	LE	RW	12
	BRASSICACEAE				
115	*Alyssum pateri* ssp. *pateri*	Cardiac, kidney, stomach ailments, diarrhoea	AP	BO	12

Continued

Table 5.2. Continued.

	Family / Plant taxa	Treatment	Part used	Preparation	Source
116	*Capsella bursa-pastoris*	Rheumatism	AP	RW	10
117	*Cardaria draba*	Pains	AP	BO	6
118	*Crambe orientalis*	Sedative	FL	AT	14
119	*Fibigia clypeata*	Animal diseases	AP	IN	8
120	*Lepidium sativum* ssp. *sativum*	Urinary antiseptic, appetite, diuretic, wounds	AP, FL	FE	1, 14
121	*Nasturtium officinale*	Appetite, hypertension	AP	CO, RW	8-10, 18
122	*Sinapis arvensis*	Diabetes, headache, rheumatism	AP, LE	CO, RW	9, 10, 18
	CAMPANULACEAE				
123	*Campanula hakkiarica*	Increase milk production, kidney stone	LE, RO	BO	14
124	*Michauxia campanuloides*	Healing wounds	LE	CR	8
	CAPPARACEAE				
125	*Capparis ovata* var. *palaestina*	Bronchitis, sedative, aphrodisiac, abdominal pain, diabetes	AP, FL, SE	AT, DR, RW	6, 10, 14
126	*Capparis spinosa* var. *spinosa*	Diuretic, aphrodisiac, appetite, ulcer	BU, FR	BO, DC, RW	5, 11
	CAPRIFOLIACEAE				
127	*Lonicera nummulariifolia*	Sedative, antidepressant	FL	AT, DR	14
128	*Sambucus ebulus*	Cancers, immune system, haemorrhoids, vasodilator, rheumatism	FR, RO, SE	BO, CR	8, 14
	CARYOPHYLLACEAE				
129	*Agrostemma githago*	Diabetes, urinary tract disorders	RO, WP	AT	18
130	*Dianthus lactiflorus*	Expectorant	FL	AT, DR	14
131	*Gypsophila nabelekii*	Diuretic, antispasmotic	RH	BO	14
132	*Silene vulgaris*	Rheumatism, sedative	BR, LE	BO	14
133	*Telephium imperati* ssp. *orientale*	Wounds, acne, chilblains, haemorrhoids	LE, WP	BO, PM	13, 17
	CHENOPODIACEAE				
134	*Chenopodium botrys*	Headache, digestive system diseases, antihelmintic, laxative	ST	AT, CO, DR	5, 14
	CONVOLVULACEAE				
135	*Convolvulus galaticus*	Purgative	RO	BO	14
	CORNACEAE				
136	*Cornus mas*	Diabetes	FR	DC	8
	CRASSULACEAE				
137	*Sedum tenellum*	Cancer, diabetes	LE	RW	14
	CUCURBITACEAE				
138	*Bryonia multiflora*	Nevres, sedative, stomach ache, constipation, haemorrhoids	RO, SE	AT, BO, DR, FE, PI,	12, 14
139	*Cucurpita pepo*	Kidney stone	SE		14
140	*Ecballium elaterium*	Sinusitis	FR	DU, PO	6, 8, 11, 17
	CUPRESSACEAE				
141	*Juniperus drupacea*	Cough, bronchitis, asthma	CO	DC	8

142	*Juniperus excelsa*	Rheumatism, diuretic	CO, LE	BO	14
143	*Juniperus oxycedrus* ssp. *oxycedrus*	Cough, psoriasis	CO	DC, IN	6, 8, 11
	CUSCUTACEAE				
144	*Cuscuta* spp.	Liver diseases, knee pain	AP	IN	15, 16
	CYPERACEAE				
145	*Cyperus longus*	Abdominal pain (in baby), halitosis, hepatic steatosis, diabetes, stomach	TU	RW	9, 10
146	*Cyperus rotundus*	Abdominal pain (in baby), halitosis, hepatic steatosis, diabetes, stomach	TU	RW	9, 10
	DIPSACACEAE				
147	*Scabiosa argentea*	Diuretic, wound healing	RO		18
	ELAEAGNACEAE				
148	*Elaeagnus angustifolia*	Cold and flu, pulmonary analgesic	FR	AT, DR	14
	EQUISETACEAE				
149	*Equisetum arvense*	Diuretic, kidney stone	AP	BO, IN	5, 12, 13
150	*Equisetum fluviatile*	Anxiety, kidney ailments	AP	BO	12
151	*Equisetum ramosissimum*	Diuretic, kidney stone, stomach, urinary inflammations	AP, ST	DR, IN	8, 14
	EUPHORBIACEAE				
152	*Andrache telephioides*	Wounds	FL, LE	PT	11
153	*Euphorbia cheiradenia*	Constipation	LA		10
154	*Euphorbia denticulata*	Abdominal pain, diarrhoea	LA		12
155	*Euphorbia helioscopia*	Curing warts	LA		8
156	*Euphorbia heteradena*	Constipation	LA		13
157	*Euphorbia kotschyana*	Curing warts	LA		8
158	*Euphorbia macrocarpa*	Wound	LA		12
159	*Euphorbia macroclada*	Constipation, warts	LA		11, 14
160	*Euphorbia macrostegia*	Curing warts	LA		8
161	*Euphorbia peplus* var. *minima*	Curing warts	LA		8
162	*Euphorbia virgata*	Constipation	LA		12
	FABACEAE				
163	*Alhagi pseudalhagi*	Diuretic, shortness of breath, itching, asthma, diarrhoea	AP, FR	BO, DR, RW	3, 4, 10
164	*Argyrolobium crotalarioides*	Stomach ache	FL	RW	3
165	*Astragalus diptherites* var. *diptherites*	Wounds	AP	AS	3
166	*Astragalus eriocephalus*	Antibacterial, tuberculosis	RO	BO	14
167	*Astragalus gummifer*	Hair loss	ST	BO	11
168	*Astaragalus microcephalus*	Pulmonary, antibacterial	ST	GU	14
169	*Colutea cilicica*	Sedative, purgative	FL, LE	AT	14
170	*Glycyrrhiza glabra* var. *glabra*	Cold, expectorant, lung relief, bronchitis, kidney stone, diuretic, hypertension	RO	BO	1, 3, 4, 10, 11

Continued

Table 5.2. Continued.

	Family / Plant taxa	Treatment	Part used	Preparation	Source
171	*Glycyrrhiza glabra* var. *glandulifera*	Sedative, angina, pharyngitis, cold and flu, diabetes, heart ailments	RO	BO	9, 12, 14
172	*Medicago orbicularis*	Increase milk yield (cow)	AP		4
173	*Medicago sativa* ssp. *sativa*	Styptic	AP	PT	12
174	*Melilotus elegans*	Kidney ailments	WP	DC	17
175	*Melilotus officinalis*	Diuretic, aphrodisiac	FL	AT	14
176	*Onobrychis gracilis*	Cold and flu	AP	DC	8
177	*Onobrychis megataphros*	Heart and vascular diseases	AP	BO, DR	3
178	*Onobrychis sulphurea* var. *vanensis*	Diuretic, kidney stone	FL	BO	14
179	*Ononis spinosa* ssp. *leiosperma*	Diuretic, kidney stone	FL, LE, RO	BO, IN	8, 14
180	*Prosopis farcta*	Diarrhoea	FR	RW	3, 10
181	*Robinia pseudoacacia*	Diuretic	FL	IN	11
182	*Sophora alopecuroides*	Aphrodisiac, pulmonary	RO	AT, DR	14
183	*Trifolium repens* var. *giganteum*	Stomach ache	AP	BO	12
184	*Trigonella coelesyriaca*	Eye infections	AP		4
185	*Trigonella foenum-graecum*	Hypoglycaemic	SE	PO	13
186	*Trigonella mesopotamica*	Eye infections	AP		4
	FAGACEAE				
187	*Quercus brantii*	Diarrhoea, diabetes, anxiety	FR	RW	6, 7, 10, 11
188	*Quercus infectoria* ssp. *boissieri*	Diabetes,	FR	DC	7, 17
189	*Quercus ithaburensis* ssp. *macrolepis*	Diabetes, anxiety	FR	RW	9, 10
190	*Quercus petraea* ssp. *pinnotiloba*	Ulcer, haemorrhoids, coagulation	GD, LE		14
191	*Quercus robur* ssp. *robur*	Diabetes	FR	DC	17
	GENTIANACEAE				
192	*Gentiana olivieri*	Diabetes, aphrodisiac	FL, RO	BO, IN	14, 17
	GERANIACEAE				
193	*Geranium stepporum*	Diabetes, stomach, coagulative	ST	AT, DR	14
194	*Pelargonium endlicherianum*	Antihelmintic	WP	RW	11
195	*Pelargonium quercetorum*	Antihelmintic (in children)	RO	BO	12
	HYPERICACEAE				
196	*Hypericum capitatum* var. *capitatum*	Pain (animal)	AP		4
197	*Hypericum capitatum* var. *luteum*	Pain (animal)	AP		4

198	*Hypericum hyssopifolium* ssp. *elongatum*	Haemorrhoids, kidney stone, purgative	FL	AT	14
199	*Hypericum lydium*	Sedative, expectorant, rheumatism, ulcer	FL	AT	5
200	*Hypericum perforatum*	Stomach ache, burns	AP, FL	DC, IN	8, 13, 18
201	*Hypericum retusum*	Stomach, appetite, expectorant, antipyretic, pain (animal)	AP, FL, LE		3, 18
202	*Hypericum scabrum*	Appetite	FL	IN	11
203	*Hypericum triquetrifolium*	Diuretic, heart pain, diabetes, stomach, pain (animal)	FL, WP	BO, DC	4, 6, 9, 10, 17
	ILLECEBRACEAE				
204	*Paronychia kurdica*	Wounds healing, gall bladder disease	AP	BO, RW	6
	IRIDACEAE				
205	*Gladiolus kotschyanus*	Aphrodisiac	TU	BO	14
	JUGLANDACEAE				
206	*Juglans regia*	Rheumatism, hair loss, purgative, antiseptic, hair loss, haemorrhoids, appetite, tonic, diabetes, wounds	FB, FR, LE, SE	DR, PM, PO, RW	7, 9–14
	JUNCACEAE				
207	*Juncus inflexus*	Kidney stone, foot pruritus, scabies	FL, RO	BO, IN	8, 12
	LAMIACEAE				
208	*Ajuga chamaepitys* ssp. *chia* var. *ciliata*	Tonic, wounds, diaphoretic, haemorrhoids	FL	IN	8, 18
209	*Cyclotrichum niveum*	Shortness of breath, stomach, anxiety, cold	ST	DC, IN	11, 18
210	*Lavandula stoechas* ssp. *stoechas*	Cold	WP	IN	17
211	*Marrubium cuneatum*	Kidney stone	ST	BO	11
212	*Marrubium parviflorum* ssp. *parviflorum*	Cold, stomach ache	WP	IN	4, 17
213	*Melissa officinalis* ssp. *inodora*	Vasodilator, headache, cardiac diseases	AP	DC, IN	8
214	*Melissa officinalis* ssp. *officinalis*	Sedative, gastrointestinal	LE	BO	14
215	*Mentha aquatica*	Stomach ache	LE	AT	18
216	*Mentha longifolia*	Cancer, tuberculosis, sedative, stomach, kidney, cold and flu, stomach ache	AP, LE, RO	BO, RW	9, 14
217	*Mentha longifolia* ssp. *longifolia*	Cold, rheumatism, allergy, headache	AP, RO	BO, PW	10, 12
218	*Mentha longifolia* ssp. *typhoides* var. *typhoides*	Rheumatism, allergy, blood coagulant, headache	AP, LE, RO	BO, CR, PW	8, 10
219	*Mentha pulegium*	Spasm, dyspnoea, stomach ache, cold, rheumatism, cough, headache	AP, FL, LE	IN, PM	1, 17, 18
220	*Mentha spicata* ssp. *spicata*	Antipyretic, stomach	LE	IN	11
221	*Micromeria congesta*	Cough, respiratory disorders	RO	BO	2
222	*Micromeria dolichodonta*	Dyspnoea, eye ailments	SH	DU, PM	17

Continued

Table 5.2. Continued.

	Family / Plant taxa	Treatment	Part used	Preparation	Source
223	*Micromeria fruticosa* ssp. *brachycalyx*	Cold and flu	AP	IN	8
224	*Nepeta flavida*	Colds	AP	IN	8
225	*Origanum majorana*	Sedative, diaphoretic, stomach ache	ST	FE	13
226	*Origanum syriacum* var. *bevanii*	Colds	LE	IN	8
227	*Origanum vulgare* ssp. *gracile*	Cold and flu, stomach, intestinal	FL, LE	IN	11, 18
228	*Phlomis kurdica*	Asthma, shortness of breath	FL		19
229	*Phlomis russeliana*	Prostate disease	LE	IN	8
230	*Prunella vulgaris*	Abdominal pain	AP	BO	12
231	*Rosmarinus officinalis*	Stomach ache	AP	AT	18
232	*Salvia multicaulis*	Wounds, cough, cold, respiratory disorders, urinary tract disorders, stomach ache	AP	BO, DU	4, 6, 18
233	*Salvia sclarea*	Cold, stomach pain	LE, SH	AT, DC, DR	14, 17
234	*Salvia cryptantha*	Cough, bronchitis	AP	IN	8
235	*Salvia syriaca*	Wounds	LE	PT	10
236	*Salvia verbenaca*	Fungal infections	LE	IN	8
237	*Salvia virgata*	Facilitate the digestion	AP		4
238	*Satureja cilicica*	Menstrual pain, abdominal pains	LE	IN	8
239	*Sideritis libanotica* ssp. *linearis*	Cold and flu	AP	IN	8
240	*Sideritis libanotica* ssp. *microchlamys*	Stomach	FL, ST	IN	11
241	*Sideritis montana*	Stomach, anxiety, cold, heart ailments	AP	AT	18
242	*Sideritis perfoliata*	Cold	AP	IN	8
243	*Sideritis syriaca* ssp. *nusairiensis*	Cold, stomach ache	SH	IN	17
244	*Stachys brantii*	Wounds	SH	PM	17
245	*Stachys lavandulifolia*	Stomach, appetite	ST	IN	11
246	*Teucrium chamaedrys* ssp. *chamaedrys*	Halitosis, toothache, appetite	LE, SH	RW	9, 10
247	*Teucrium chamaedrys* ssp. *sinuatum*	Rheumatism, stomach ache, cancer, heart and vascular disorders	AP	BO, DC, RW	11–13
248	*Teucrium chamaedrys* ssp. *tauricolum*	Haemorrhoids	AP	IN	8
249	*Teucrium polium*	Diabetes, cold, spasm, diarrhoea, rheumatism, abdominal pain, stomach ache, wounds, gastrointestinal, kidney, cancer, flatulence, antihelmintic, constipation, respiratory, ulcer, fever lowering, lung inflammations	AP, FL, LE, WP	AT, BO, DC, DR, IN, PM	1, 4, 6, 8, 11, 12, 14, 17
250	*Thymbra sintenisii* ssp. *sintenisii*	Stomach ache, tension, appetite	AP	AT	6, 19

251	*Thymbra spiacata* var. *spicata*	Cold, stomach ache, diabetes, antiseptic, halitosis, sedative, cough, toothache	FL, LE, WP	DC, IN	1, 11, 13, 17
252	*Thymus fallax*	Sedative, haemorrhoids	LE	AT, FE	14
253	*Thymus kotschyanus* var. *kotschyanus*	Bronchitis, sedative, shortness of breath	AP, FL, LE	AT, BO, DR	12, 14
254	*Thymus sipyleus* ssp. *rosulans*	Cold	AP	IN	8
	LILIACEAE				
255	*Allium akaka*	Diabetes, cancer	BL	RW	14
256	*Allium cepa*	Wounds	BL	CO, RW	9, 10, 14
257	*Asparagus acutifolius*	Kidney stones, urinary inflammation, flu	AP	IN	8
258	*Colchicum szovitsii*	Sterility, rheumatism	BL	BO, DR, FE	14
259	*Danae racemosa*	Kidney stone	RO	IN	8
260	*Eremurus spectabilis*	Rheumatism, gastrointestinal, stomach	LE, RO	BO, CO	14, 18
261	*Muscari tenuiflorum*	Antibiotic, rheumatism, tumour	BL	DR, FE	8
262	*Polygonatum multiflorum*	Aphrodisiac, infertility	RH	PW	8
263	*Ruscus aculeatus* ssp. *angustifolius*	Kidney stone, urinary inflammation	RO	DC, IN	8
	LINACEAE				
264	*Linum mucronatum* ssp. *mucronatum*	For sterility (women)	FL	IN	4
265	*Linum pubescens*	Wounds, bronchitis, cough	SE		18
	LORANTHACEAE				
266	*Viscum album* ssp. *album*	Cancer, immune system, epilepsy, diabetes, cholesterol lowering, kidney stones	LE, ST	BO, IN	8, 14
	MALVACEAE				
267	*Alcea apterocarpa*	Asthma, coughing	FL		4
268	*Alcea calvertii*	Kidney stone	RO	IN	13
269	*Alcea digitata*	Emollient, asthma, diuretic, cold, wound, enteritis	AP	IN	5
270	*Alcea fasciculiflora*	Abscesses, itching of scabies	RO		13
271	*Alcea flavovirens*	Kidney, urinary tract diseases	LE	BO	12
272	*Alcea hohenackeri*	Kidney stone, urinary tract diseases, cold and flu, stomach ache	FL, LE, RO	BO, CO	10, 12
273	*Alcea kurdica*	Cancer, sedative	SE	AT, DR	14
274	*Alcea pallida*	Cough, bronchitis	FL	IN	8
275	*Alcea setosa*	Wounds	FL, FR, RO	CR	6
276	*Alcea striata*	Cough, flu, wounds, expectorant, bronchitis	FL, RO	BO, CR	6, 18
277	*Althaea armeniaca*	Asthma, coughing	FL		4
278	*Althea officinalis*	Dyspnoea, diuretic, antilithic	AP, FL	DC, IN	13, 17
279	*Malva neglecta*	Haemorrhoid, stomach ache, kidney stone, diuretic, wounds, anti-inflammatory, tension regulation, blood sugar lowering, cold, abscess	AP, LE, RO	BO, CO, DR, IN, PT, RW	1, 6, 8, 10, 11-14

Continued

Table 5.2. Continued.

	Family / Plant taxa	Treatment	Part used	Preparation	Source
280	*Malva sylvestris*	Diuretic, wounds, anti-inflammatory	AP		1
	MORACEAE				
281	*Morus alba*	Blood-forming, antipyretic, pulmonary system	FR	DR, FE	7, 10, 14
282	*Morus nigra*	Throat spasm, haematinic, antipyretic, aphrodisiac, wounds (in baby)	FR	FE, DR, DU	7, 9, 10, 14, 17
	MORINACEAE				
283	*Morina persica* var. *persica*	Cold and flu	FL	DC	8
	MYRTACEAE				
284	*Myrtus communis* ssp. *communis*	Blood sugar lowering	FR	DC	8
	OLEACEAE				
285	*Fraxinus excelsior*	Diuretic, antipyretic	LE		7
286	*Fraxinus ornus* ssp. *cilicica*	Influenza	SB	IN	8
	ORCHIDACEAE				
287	*Orchis palustris*	Aphrodisiac	TU	AT, DR	14
288	*Orchis simia*	Diabetes	TU		19
	PAPAVERACEAE				
289	*Fumaria asepala*	Eczema, itching	FL	IN	4, 18
290	*Fumaria microcarpa*	Heart, vein	FL, ST	BO	14
291	*Fumaria officinalis*	Diabetes	AP	CO	8
292	*Glaucium corniculatum* ssp. *corniculatum*	Pain	FL	IN	4
293	*Papaver arenarium*	Pain	FL		4
294	*Papaver rhoeas*	Sedative, depression, lung relief	FR, LE	BO, CO	14, 18
	PINACEAE				
295	*Abies cilicica*	Stomach ache, ulcer, colds and flu, menstrual pain	CO, RE		8
296	*Pinus brutia*	Haemorrhoids, burns, ulcer, stomach ache, tuberculosis diseases,wound healing	BU, BR, CO, RE	DC, DU, IN, PW	8, 11, 17
	PLANTAGINACEAE				
297	*Plantago atrata*	Wounds	LE	FE	13
298	*Plantago lanceolata*	Stomach ache, diabetes, wounds	LE, RO	AT, BO, FE, PT, RW	11–14, 18, 18
299	*Plantago major* ssp. *major*	Insect bites, wounds, asthma, blood coagulant, abscess	LE	CO, CR, DR, IN, RW	8, 12–14, 17
300	*Plantago maritima*	As wash for cancerous uterus	LE	DC	13
	PLATANACEAE				
301	*Platanus orientalis*	Hepatitis, angina, flu	SB	IN	14, 17
	POACEAE				
302	*Cynodon dactylon* var. *dactylon*	Diuretic, kidney stones	RH	BO, IN	8, 14

303	*Triticum sativum*	Cure abscess			14
304	*Zea mays*	Haemorrhoids, kidney, diuretic, antilithic	FR, LE, SE	BO, IN	10, 13, 14
	POLYGONACEAE				
305	*Polygonum bellardii*	Kidney stones	AP	IN	8
306	*Rheum ribes*	Diabetes, hypertension, stomach ailment, ulcer, diarrhoea, anthelmintic, digestion, osteoporosis	RH, RO, ST, SB	BO, DC, DR, IN	8, 9–14
307	*Rumex crispus*	Diuretic, kidney stones	LE	CO	18
308	*Rumex tuberosus* ssp. *horizontalis*	Wounds	LE	PT	12
	PORTULACACEAE				
309	*Portulaca oleracea*	Dysmenorrhoea (menstrual pain), diabetes, to strengthen bones (children), respiratory	AP	BO, CO, RW	9, 10, 18
	PUNICACEAE				
310	*Punica granatum*	Immune system, aphrodisiac, headache, diarrhoea	FR	FW	10, 14
	RANUNCULACEAE				
311	*Nigella sativa*	Intestinal disorders, flatulence	SE	RW	1
312	*Ranunculus constantinopolitanus*	Rheumatism, swollen feet	LE	CR	8
313	*Ranunculus kotschyi*	Rheumatism	LE	PT	12
	RESEDACEAE				
314	*Reseda lutea* var. *Lutea*	Stomach ache	RO	FE	13
	RHAMNACEAE				
315	*Paliurus spina christii*	Wounds, fungal ailments, cough (animal), antipyretic, stomach ulcer, urinary inflammation, haemorrhoids	BR, FR, RO, SE	BO, DC, IN, PM, OI	8, 11, 12, 17, 18
	ROSACEAE				
316	*Alchemilla hessii*	Wounds	LE	PT	12
317	*Amygdalus communis*	Diabetes, kidney	FR	RW	11, 18
318	*Cerasus mahalep* var. *mahaleb*	Diabetes, tonic, expectorant, cough	FR, LE, SE	BO, RW, SI	6, 8, 11
319	*Cerasus microcarpa*	Prostate ailments	FR	RW	12
320	*Crataegus aronia* var. *aronia*	Heart diseases, muscle spasm, hypertension, stomach ache	FR, LE	DC, RW	8, 10, 17
321	*Crataegus curvisepala*	Coughing	FR	AT	14
322	*Crataegus monogyna* ssp. *monogyna*	Heart ailments, sedative, antispasmodic, hypertension, diuretic, diarrhoea	FL, FR	DR, IN, RW	1, 11, 13
323	*Crataegus orientalis* var. *orientalis*	Hypertension	FR	RW	10
324	*Cydonia oblonga*	Sore throat	LE	BO	6
325	*Eriolobus trilobatus* var. *trilobatus*	Cardiac diseases, diabetes, dyspnoea	FR	DC	8
326	*Persica vulgaris*	Diabetes	SE	RW	10
327	*Potentilla erecta*	Kidney stone	AP	AT	18

Continued

Table 5.2. Continued.

	Family / Plant taxa	Treatment	Part used	Preparation	Source
328	*Potentilla speciosa* var. *speciosa*	Cancer	AP	DC	8
329	*Prunus armeniaca*	Constipation, antihelmintic	FR, SE	RW	9, 10
330	*Prunus divaricata* ssp. *divaricata*	Diabetes	FR	RW, SI	8, 11
331	*Pyrus elaeagnifolia*	Constipation	FR	RW	11
332	*Rosa canina*	Aphrodisiac, diabetes, cholesterol, cold, haemorrhoids, cough, stomach ache, antihelmintic, weight loss	FR	BO, DC, DR, IN	5, 8–14, 18
333	*Rosa damascena*	Cold	FR	BO, DR	10
334	*Rosa foetida*	Heart pain, cancer	FR	BO	6
335	*Rosa heckeliana* ssp. *vanheurckiana*	Cold, cough	FR	DR	12
336	*Rubus caesius*	Kidney stones	SE	RW	12
337	*Rubus canescens* var. *canescens*	Diabetes	RO	BO	10
338	*Rubus discolor*	Diabetes	RO	IN	11
339	*Rubus sanctus*	Infertility, kidney stones	RO	DC	17, 18
	RUBIACEAE				
340	*Galium consanguineum*	Haemorrhoids	AP	DU	12
	SALICACEAE				
341	*Salix aegyptiaca*	Toothache	LE	CR	12
342	*Salix alba*	Rheumatism, haemorrhoids, toothache	LE		7, 12
	SCROPHULARIACEAE				
343	*Linaria kurdica*	Heart, vein	LE, SE	AT	14
344	*Verbascum asperuloides*	Wounds	SH	PM	17
345	*Verbascum lasianthum*	Haemorrhoids, wounds	RO	DR	6
346	*Verbascum oreophilum*	Heart, vein	SE	BO	14
347	*Verbascum pinetorum*	Wound healing	LE	CR	8
348	*Verbascum speciosum*	Rheumatism	AP	BO	12
349	*Verbascum splendidum*	Haemorrhoids	FL	DU	17
350	*Veronica anagallis-aquatica* ssp. *oxycarpa*	Appetite	AP	RW	10
351	*Veronica orientalis*	Gum disorders	ST	BO	14
	SOLANACEAE				
352	*Hyoscyamus niger*	Sedative, depression, toothache, tooth cavity, lung and throat inflammation	LE, SE, SH	DR, FE, IH	8, 13, 14
353	*Hyoscyamus reticulatus*	For intoxication	RO, SH	FE	13

	TAMARICACEAE				
354	*Tamarix smyrnensis*	Diuretic, constipation	LE	BO	14
355	*Tamarix tetrandra*	Appetite, constipation	LE, SB		7
	THYMELACEAE				
356	*Daphne mucronata*	Toothache, rheumatism, cough	BR, SB	BO	12
	URTICACEAE				
357	*Parietaria judaica*	Cancer, sedative	SE, ST	DR	14
358	*Urtica dioica*	Cancer, rheumatism, haemorrhoids, diabetes, gynaecological diseases, anti-inflammatory, tonic, diuretic, kidney stones, hair loss, to encourage milk secretion in new mothers, sedative, cold, pain, aphrodisiac, purifies the blood, psoriasis, tuberculosis, embolism	AP, LE, SE, SH, ST	BO, CO, CR, DC, DR, DU, IN, RW	1, 5, 6, 8–14, 17, 18
359	*Urtica urens*	Diuretic, cancer, gynaecological diseases, anti-inflammatory, tonic, kidney stones, hair loss	LE, SE	AT	1, 13
	VALERIANACEAE				
360	*Valeriana dioscoridis*	Tension lowering	FL	IN	8
361	*Valeriana officinalis*	Sedative	RO		18
	VERBENACEAE				
362	*Verbena officinalis*	Analgesic, infertility, cardiac diseases, cancers, kidney stones	AP, WP	DC, IN	8, 17
363	*Vitex agnus-castus*	Itching	AP	AT	18
	VIOLACEAE				
364	*Viola kitaibeliana*	Wound healing, cough	FL, LE	CR, DC	8
365	*Viola odorata*	Vascular, stomach, kidney ailments	FL, WP	BO, RW	12
	ZYGOPHYLLACEAE				
366	*Peganum harmala*	Sedative, depression, epilepsy, haemorrhoids, prostatitis, urinary incontinence	RO, SE	BO, DC	13, 14
367	*Tribulus terrestris*	Heart ailment, atherosclerosis, cysts, kidney stone, diuretic, embolism, shortness of breath	AP, SH	BO, DC, IN	6, 8, 11, 12, 17, 18

Notes

Part used: AP: Aerial parts; BL: Bulb; BR: Branches; BU: Bud; CO: Cones; FB: Fruit bark; FL: Flower; FR: Fruit; GD: Gall dust; LA: Latex; LE: Leaves; RB: Root bark; RE: Resin; RH: Rhizome; RO: Root; SB: Stem bark; SE: Seed; SH: Shoot; ST: Stem; TU: Tuber; WP: Whole plant

Preparation: AS: Ash; AT: As tea; BO: Boiled; CO: Cooked; CR: Crushed; DC: Decoction; DR: Dried; DU: Direct use; FE: Fresh; FW: Fruit water; GA: Gargle; GU: Gum; IH: Inhalation; IN: Infusion; OI: Oil; PI: Pill; PL: Plant water; PM: Pomade; PO: Pounded; PT: Poultice; PW: Powdered; RW: Raw; SI: Syrup; SW: Swallowed

Sources: 1: Akan *et al.*, 2005b; 2: Akan *et al.*, 2008; 3: Akan *et al.*, 2013b; 4: Akan *et al.*, 2013c; 5: Akdoğan and Akgün, 2006; 6: Akgül, 2008; 7: Aslan *et al.*, 2011; 8: Demirci and Özhatay, 2012; 9: A. Gelse, Adıyaman çevresinin etnobotanik özellikleri [Ethnobotanical Properties of the Adiyaman Environment]. Department of Biology, Institute of Science and Technology, Yüzüncü Yıl University, 2012, unpublished thesis; 10: A. Gençay, Cizre (şırnak)'nin etnobotanik özellikleri. Department of Biology, Institute of Science and Technology, Yüzüncü Yıl University, 2007, unpublished thesis; 11: N. Güldaş, Adıyaman İlinde etnobotanik değeri olan bazı bitkilerin kullanım alanlarının tespiti. Department of Biology, Institute of Science and Technology, Fırat University, 2009, unpublished thesis; 12: İ. Kaval, Geçitli (Hakkari) ve çevresinin etnobotanik özellikleri. Department of Biology, Institute of Science and Technology, Yüzüncü Yıl University, 2011, unpublished thesis; 13: Özgökçe and Özçelik, 2004; 14: Öztürk and Ölçücü, 2011; 15: Şekeroğlu *et al.*, 2011; 16: Şığva and Seçmen, 2009; 17: Tursun, 2001; 18: Yapıcı *et al.*, 2009

(101 taxa) and respiratory disorders (84 taxa) (see Fig 5.3). On the basis of parts used, mainly aerial parts are used (100 taxa), leaves (84 taxa) and flowers (71 taxa) (see Fig 5.4). The most common preparation methods are boiled (20%), followed by infusion (18%), raw (11%) and decoction (10%). Other uses and their percentages are given in Fig 5.5.

5.6 Plants Consumed as a Source of Food

Lately, with an understanding of the use of synthetic foods as a cause of obesity, people have started going back to natural plant foods. These have become an indispensable part of daily human nutrition, because of their richness in minerals, fibre

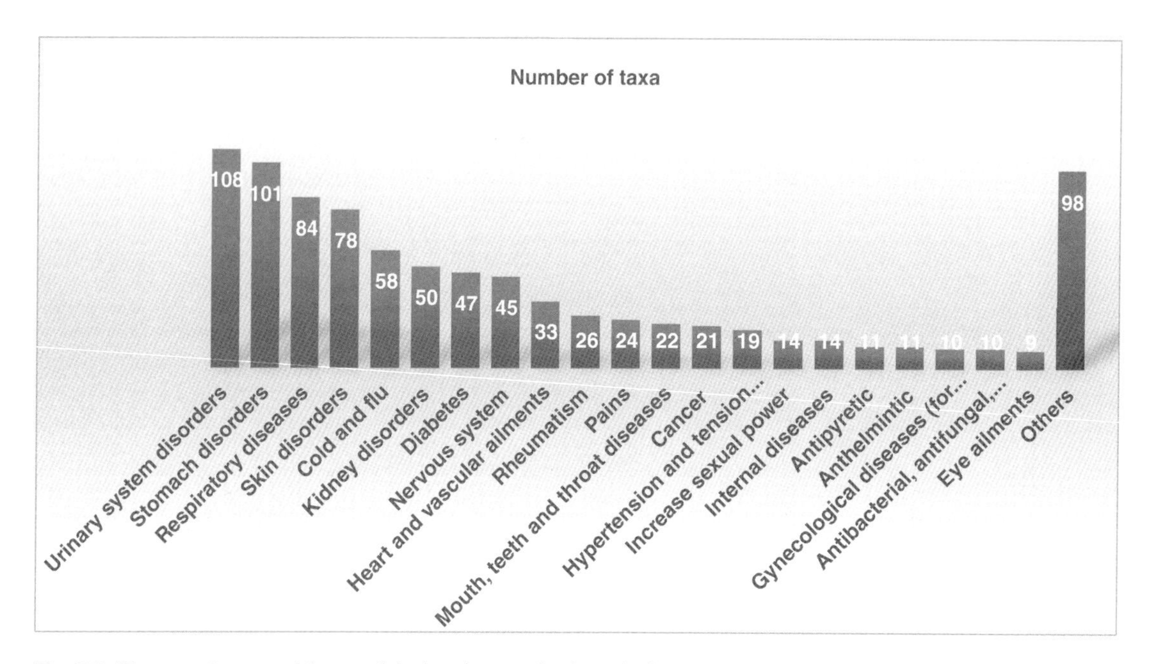

Fig. 5.3. Therapeutic uses of the medicinal and aromatic plants in Southeast Anatolia

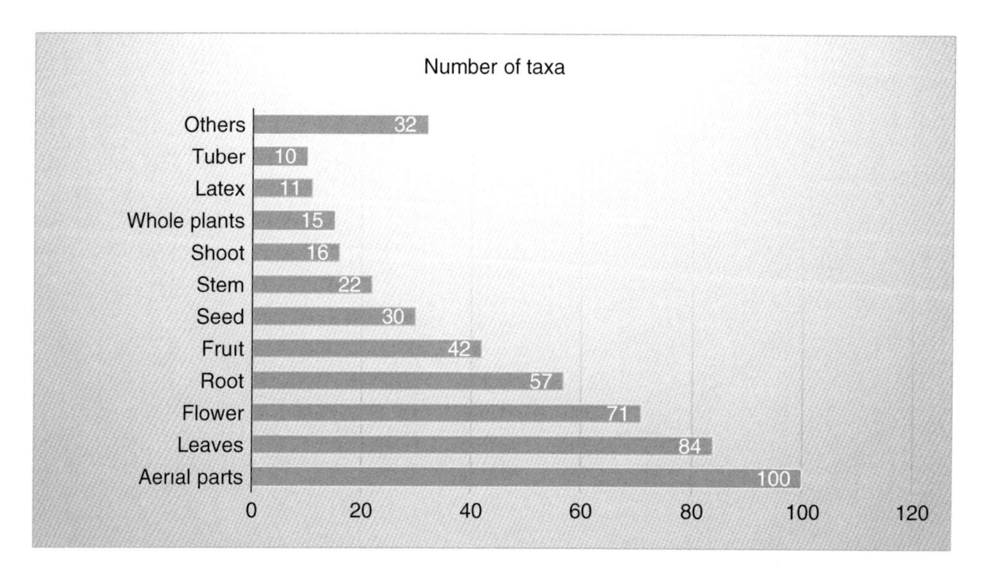

Fig. 5.4. The parts of the medicinal and aromatic plants used in Southeastern Anatolia.

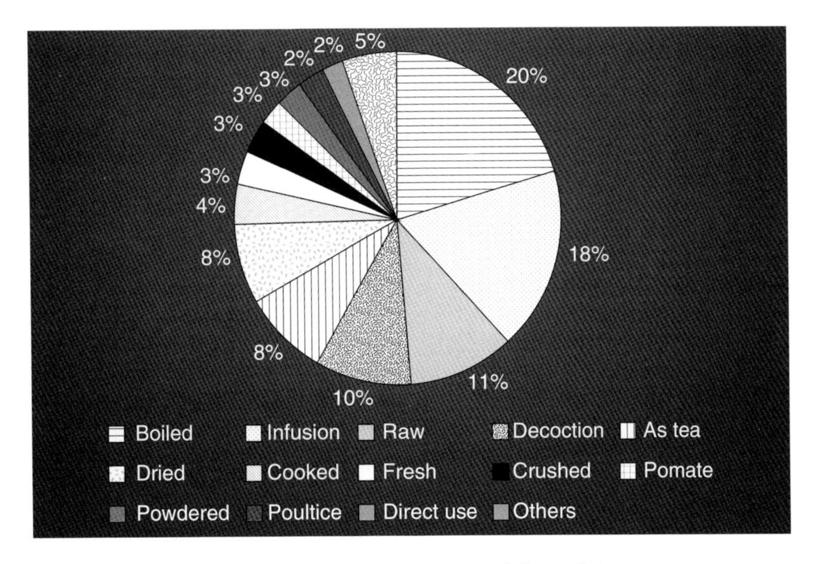

Fig. 5.5. The percentages of the preparations used in the traditional folk medicine.

and vitamins (Tukan *et al.*, 1998). The collection and use of plants as food in Turkey is a very old custom (Ozturk and Ozcelik, 1991; Ozturk *et al.*, 2012a). The population has tried to satisfy nutritional needs by collecting plants from their surrounding mountains and forests. This tradition still continues in rural areas (Ozturk *et al.*, 2011; Ozturk *et al.*, 2012a, b), and is especially common in Southeastern Anatolia (Ozturk and Ozcelik, 1991; Ozturk *et al.*, 2014). In the light of studies undertaken in this region, a total of 225 taxa have been recorded as being used as food plants (see Table 5.3). On the basis of consumption of plant parts, mainly fruits have been used (61 taxa), followed by aerial parts (49 taxa) and leaves (44 taxa). Other current uses are given in Fig. 5.6. The plants are consumed mainly raw (63.71%) or cooked (30.24%). Other types of consumption are boiled (2.02%), as gum (2.02%), as nectar (1.61%) and as latex (0.4%).

The plants consumed in daily use as foods are collected during the appropriate seasons and sold at the local markets (Surmeli *et al.*, 2001). These plants include taxa such as *Capparis ovata*, *C. spinosa*, *Cerasus mahaleb*, *Glycyrrhiza glabra*, *Pistacia khinjuk*, *P. terebinthus*, *Rhus coriaria* and *Thymbra spicata*. Some of these are also exported. Although the use as food plants varies from state to state, a large number of these are also used as spices and as flavouring agents in cheese making or as herbal drinks. The plants used as spices include 19 taxa (see Table 5.4); 25 taxa in cheese making (see Table 5.5), and 13 taxa as herbal teas (see Table 5.6).

In addition to the teas prepared from different plant taxa (Ozturk *et al.*, 2011), special coffees are prepared from the fruits of *Pistacia khinjuk* (Akan *et al.*, 2005b) and *P. terebinthus* (A. Gençay, Cizre (Şırnak)'nin etnobotanik özellikleri. Department of Biology, Institute of Science and Technology, Yüzüncü Yıl University, 2007, unpublished thesis; A. Gelse, Adıyaman çevresinin etnobotanik özellikleri [Ethnobotanical Properties of the Adiyaman Environment]. Department of Biology, Institute of Science and Technology, Yüzüncü Yıl University, 2012, unpublished thesis), the seeds of *Ricinus communis* (İ. Kaval, Geçitli (Hakkari) ve çevresinin etnobotanik özellikleri. Department of Biology, Institute of Science and Technology, Yüzüncü Yıl University, 2011, unpublished thesis). The fruit juices of *Morus nigra* and *Punica granatum* (A. Gelse, Adıyaman çevresinin etnobotanik özellikleri [Ethnobotanical Properties of the Adiyaman Environment]. Department of Biology, Institute of Science and Technology, Yüzüncü Yıl University, 2012, unpublished thesis) are sold commonly, together with the syrup prepared from the roots of *Glycyrrhiza glabra* (Akan *et al.*, 2005b; A. Gelse, Adıyaman çevresinin etnobotanik özellikleri [Ethnobotanical Properties of the Adiyaman Environment]. Department of Biology, Institute of Science and Technology, Yüzüncü Yıl University, 2012, unpublished thesis).

Table 5.3. The plants consumed as foods in Southeastern Anatolia.

	Plant taxa	Part used	Preparation	Source
1	*Alcea flavovirens*	LE	Cooked	1
2	*Alkanna froedinii*	AP	Raw	1
3	*Alkanna orientalis* var. *orientalis*	AP	Cooked	2
4	*Alkanna trichophylla* var. *mardinensis*	FL	Nectar	3
5	*Alliaria petiolata*	LE		4
6	*Allium akaka*	LE	Raw	1
7	*Allium ampeloprasum*	LE	Cooked	5, 6
8	*Allium cepa*	BL	Raw, Cooked	2, 5
9	*Allium giganteum*	LE	Raw	1
10	*Allium kharputense*	AP	Raw	2
11	*Allium scorodoprasum* ssp. *rotundum*	AP		7
12	*Allium stamineum*	LE	Raw, Cooked	6
13	*Amaranthus viridis*	LE, SH	Cooked	2, 5
14	*Amygdalus communis*	FR	Raw	1, 2, 5, 6
15	*Anchusa azurea* var. *azurea*	AP	Cooked	7, 8
16	*Anchusa azurea* var. *kurdica*	AP	Cooked	2
17	*Anchusa strigosa*	LE	Cooked	6
18	*Andrache telephioides*	SH	Raw	6
19	*Anethum graveolens*	AP, LE, ST	Raw	1, 8
20	*Anthemis hyalina*	ST		7
21	*Argyrolobium crotalarioides*	SE	Raw	9
22	*Arum conophalloides* var. *conophalloides*	AP	Boiled, Cooked	1
23	*Arum italicum*	AP	Cooked	5
24	*Arum maculatum*	LE	Cooked	6
25	*Astragalus hamosus*	FR	Raw	3
26	*Berberis crataegina*	SE	Raw	2
27	*Brassica nigra*	AP	Raw	8
28	*Campanula sclerotricha*	LE	Cooked	1
29	*Capparis ovata* var. *palaestina*	BU, FR	Raw	2, 5, 7
30	*Capparis spinosa* var. *spinosa*	BU		10
31	*Capsella bursa-pastoris*	AP, FR, SH	Cooked, Raw	2, 3, 5
32	*Cardaria draba*	AP	Boiled	3
33	*Cardemine uliginosa*	AP	Raw	1
34	*Carduus nutans* ssp. *leiophyllus*	ST	Eaten	11
35	*Carduus nutans* ssp. *nutans*	ST	Eaten	7
36	*Carduus pycnocephalus* ssp. *albidus*	ST	Eaten	7
37	*Celtis glabrata*	FR	Raw	1
38	*Celtis tournefortii*	FR	Raw	8
39	*Centaurea cynarocephala*	RO	Raw	3
40	*Centaurea hyalolepis*	FL		11
41	*Centaurea iberica*	SH	Cooked	2
42	*Centaurea polypodiifolia* var. *szovitsiana*	SH	Cooked	6
43	*Centaurea solstitialis* ssp. *solstitialis*	AP		7
44	*Centaurea triumfettii*	AG	Gum	7
45	*Centranthus longiflorus* ssp. *longiflorus*	LE	Raw	1
46	*Cerasus avium*	FR	Raw, Cooked	2
47	*Cerasus microcarpa*	FR	Raw	1
48	*Chaerophyllum macropodum*	ST	Raw	1
49	*Chaerophyllum macrospermum*	AP	Boiled, Cooked	1
50	*Chondrilla juncea* var. *juncea*	RO	Gum	3
51	*Cicer arietinum*	FR, SE	Raw, Cooked	2, 5
52	*Cichorium intybus*	AP, SH	Raw, Cooked	2, 3

Continued

Tabel 5.3. Continued.

	Plant taxa	Part used	Preparation	Source
53	*Cirsium arvense* ssp. *arvense*	AP, ST	Raw	2, 5
54	*Cirsium lappaceum*	LE	Raw	10
55	*Cirsium pubigerum* var. *spinosum*	ST	Raw	1
56	*Cirsium vulgare*	ST	Fresh	7
57	*Citrullus lanatus*	FR, SE	Raw, Dried	2, 5
58	*Convolvulus arvensis*	LE	Raw	1
59	*Convolvulus betonicifolius* ssp. *peduncularis*	LE	Raw	1
60	*Convolvulus stachydifolius*	LE	Cooked	2
61	*Coriandrum sativum*	LE, SE	Raw	1
62	*Coronilla scorpioides*	SE	Raw	9
63	*Crataegus aronia* var. *aronia*	FR	Raw	2, 3
64	*Crataegus monogyna* ssp. *monogyna*	FR	Raw	6, 8
65	*Crataegus orientalis* var. *orientalis*	FR	Raw	2, 5
66	*Crataegus pontica*	FR	Raw	1
67	*Crepis sancta*	FL	Raw	5
68	*Crocus cancellatus* ssp. *damacenus*	BL		11
69	*Crocus pallasii*	BL		11
70	*Cucumis melo*	FR, SE	Raw, Dried	2, 5
71	*Cucurbita pepo*	FR	Dried, Cooked	2, 5
72	*Cyperus longus*	TU	Raw	2
73	*Cyperus rotundus*	TU	Raw	2
74	*Dianthus strictus* var. *strictus*	FL		11
75	*Diospyros kaki*	FR	Raw	2, 5
76	*Diplotaenia cachrydifolia*	AP	Cooked	1
77	*Dranculus vulgaris*	AP	Cooked	2
78	*Echinophora tenuifolia* ssp. *sibthorpiana*	AG, LE		7, 10
79	*Echinops heterophyllus*	CA, ST	Raw	1
80	*Echinops orientalis*	CA	Raw	1
81	*Echinops pungens* var. *adenoclades*	CA	Raw	2, 5
82	*Echinops sphaerocephalus* ssp. *sphaerocephalus*	CA	Raw	2
83	*Elaeagnus angustifolia*	FR	Raw	2, 5
84	*Eremurus spectabilis*	SH	Cooked	1, 2
85	*Erodium cicutarium* ssp. *cicutarium*	FR	Raw	2, 3
86	*Erophila verna* ssp. *verna*	LE		7
87	*Eryngium billardieri*	RO, ST	Raw	1, 2, 5
88	*Eryngium campestre* var. *virens*	ST	Raw	2, 3, 5
89	*Erysimum repandum*	AP	Cooked	2
90	*Euphorbia cheriradenia*	LA		5
91	*Falcaria vulgaris*	AP	Raw, Cooked	1
92	*Ferula orientalis*	SH	Raw, Cooked	1
93	*Ficus carica* ssp. *carica*	FR	Raw, Cooked	2, 5
94	*Foeniculum vulgare*	AP	Raw	6
95	*Geranium tuberosum* ssp. *deserti-syriacum*	TU	Raw	11
96	*Gladiolus atroviolaceus*	FL	Raw	3
97	*Gundelia tournefortii* var. *armata*	RO, SH	Raw, Cooked	7, 8
98	*Gundelia tournefortii* var. *tournefortii*	RO, ST	Raw, Latex, Gum, Cooked	1–3, 5, 6
99	*Helianthus annuus*	SE	Raw	1, 2, 5
100	*Helianthus tuberosus*	TU	Raw	1
101	*Hibiscus esculentis*	FR	Cooked	2, 5
102	*Hordeum bulbosum*	BL	Raw	1, 2
103	*Hyacinthella nervosa*	AG		7
104	*Imperata cylindrica*	FL	Raw	2
105	*Iris masia*	FL	Raw	2

Continued

Tabel 5.3. Continued.

	Plant taxa	Part used	Preparation	Source
106	*Iris persica*	FL, RH	Raw	5, 6
107	*Iris reticulata* var. *reticulata*	FL	Boiled	3
108	*Ixiolirion tataricum* ssp. *montanum*	FL	Nectar	3, 7
109	*Juglans regia*	FR	Raw	5, 6
110	*Jurinea pulchella*	SH	Cooked	2
111	*Lactuca serriola*	AG, LE		7, 10
112	*Lathyrus annuus*	FR	Raw	2
113	*Lathyrus cicera*	FR, SE	Raw, Cooked	2, 9
114	*Lathyrus gorgoni* var. *gorgoni*	SE	Raw	2
115	*Lathyrus inconspicuus*	FR, SE	Raw	2, 3
116	*Lathyrus palustris* ssp. *palustris*	SE	Raw	2
117	*Lathyrus sativus*	SE	Raw	9
118	*Lens culinaris*	SE	Cooked	5
119	*Lens orientalis*	SE	Raw	9
120	*Lepidium sativum* ssp. *sativum*	AP	Cooked, Raw	2, 5, 8
121	*Malus sylvestris* ssp. *orientalis*	FR	Raw, Cooked	1, 2, 5
122	*Malva neglecta*	AP, FR, LE	Cooked, Raw	2–4, 6
123	*Malva sylvestris*	FR, LE	Cooked	10
124	*Malvella sherardiana*	AP	Raw	11
125	*Matricaria aurea*	FR, LE	Raw	3
126	*Mentha longifolia* ssp. *longifolia*	LE	Raw	2
127	*Mentha longifolia* ssp. *typhoides* var. *typhoides*	LE	Raw, Cooked	2, 5
128	*Mentha pulegium*	LE	Direct use	10
129	*Mentha spicata* ssp. *spicata*	AP	Raw	6
130	*Michauxia laevigata*	ST	Raw	1
131	*Morus alba*	FR	Eaten fresh, Cooked	1, 2, 5, 12
132	*Morus nigra*	FR	Raw	2, 5, 10, 12
133	*Myrtus communis*	FR	Raw	2
134	*Nasturtium officinale*	AP	Cooked, Raw	2, 5, 8
135	*Nigella sativa*	SE	Raw	8
136	*Nonea pulla*	LE	Raw	1
137	*Notobasis syriaca*	SH	Raw, Cooked	2, 5, 7
138	*Olea europaea*	FR	Raw	2, 5, 10
139	*Onobrychis crista-galli*	SE	Raw	9
140	*Ononis spinosa*	LE	Eaten, Dried	1
141	*Onopordum acanthium*	ST	Raw	2, 5
142	*Onopordum carduchorum*	ST	Raw	3
143	*Onosma molle*	FL	Nectar	11
144	*Onosma roussaei*	FL	Nectar	3
145	*Onosma sericeum*	AP	Cooked, Gum	2, 7
146	*Ornithogalum narbonense*	AP, LE, ST	Cooked, Raw	4, 8
147	*Oryza sativa*	SE	Cooked	2, 5
148	*Papaver rhoeas*	BU, FL, LE	Cooked, Raw	2, 3
149	*Pelargonium quercetorum*	LE	Raw, Cooked	1
150	*Persica vulgaris*	FR	Raw, Cooked	2, 5
151	*Pimpinella anthriscoides* var. *anthriscoides*	SH	Cooked	1
152	*Pimpinella eriocarpa*	AG		7
153	*Pistacia khinjuk*	FR	Raw	1, 8
154	*Pistacia terebinthus* ssp. *palaestina*	FR, SH, ST	Raw, Gum	2, 5, 6
155	*Pistacia vera*	FR	Raw	10
156	*Pisum sativum* ssp. *elatius* var. *pumilio*	SE	Raw	9
157	*Polygonum arenastrum*	AP	Raw, Cooked	2, 5
158	*Polygonum cognatum*	AP	Raw, Cooked	2

Continued

Tabel 5.3. Continued.

	Plant taxa	Part used	Preparation	Source
159	*Portulaca oleracea*	AP, SH	Raw, Cooked	1–3, 5, 6, 8, 10
160	*Prangos pabularia*	AP	Cooked	2
161	*Prosopis farcta*	FR		2
162	*Prunella vulgaris*	SH		1
163	*Prunus armeniaca*	FR	Raw, Cooked	2, 5
164	*Prunus divaricata*	FR	Raw	6
165	*Punica granatum*	FR	Raw	2, 5
166	*Pyracantha coccinea*	FR	Raw	2
167	*Pyrus commnis* ssp. *communis*	FR	Raw, Cooked	2, 3, 5
168	*Pyrus elaeagnifolia*	FR	Raw, Cooked	6
169	*Pyrus syriaca* var. *syriaca*	FR	Raw, Cooked	1
170	*Quercus brantii*	FR	Raw	1
171	*Quercus infectoria* ssp. *boissieri*	LE	Boiled	2
172	*Ranunculus kochii*	LE	Raw	1
173	*Rheum ribes*	SH, ST	Raw	1, 2, 5, 6
174	*Rhus coriaria*	FR	Raw	1, 2
175	*Rosa canina*	FL, FR	Cooked, Raw	1, 6
176	*Rubus caesius*	FR	Raw	1
177	*Rubus canescens* var. *canescens*	FR	Raw, Cooked	2, 5
178	*Rubus discolor*	FR	Raw, Cooked	6
179	*Rubus sanctus*	FR, RO	Raw	1, 3, 10
180	*Rumex acetosella*	LE	Raw	8
181	*Rumex alpinus*	LE	Raw	1
182	*Rumex conglomeratus*	LE	Raw, Cooked	2, 5
183	*Rumex crispus*	LE	Raw, Cooked	2
184	*Rumex scutatus*	LE	Raw	6
185	*Rumex tuberosus* ssp. *horizontalis*	LE	Cooked	1
186	*Salvia poculata*	LE	Cooked	1
187	*Scandix stellata*	AP	Cooked	3
188	*Scorpiurus muricatus* ssp. *villosus*	FR		2
189	*Scorzonera mollis* ssp. *mollis*	RO	Raw, Cooked	1
190	*Scorzonera pseudolanata*	TU	Raw	3
191	*Sesamum indicum*	SE	Raw	5
192	*Silene coniflora*	FL		7
193	*Silene dichotoma* ssp. *dichotoma*	AP	Cooked	3
194	*Silene vulgaris* var. *vulgaris*	WP	Raw	6
195	*Sinapis alba*	AP	Raw	8
196	*Sinapis arvensis*	LE, ST	Raw	5, 6, 7
197	*Sisymbrium altissimum*	AP	Raw, Cooked	3
198	*Smyrnium olusantrum*	ST	Raw	1
199	*Symphytum kurdicum*	LE	Cooked	1
200	*Thymbra sintenisii* ssp. *sintenisii*	AP	Raw	3
201	*Thymbra spicata* var. *spicata*	AP	Raw	8
202	*Torilis tenella*	AP	Raw, Cooked	3
203	*Tragopogon buphthalmoides* var. *buphthalmoides*	LE, RO	Raw	1, 6, 10
204	*Tragopogon buphthalmoides* var. *latifolius*	AP	Raw	1
205	*Tragopogon longirostris* var. *longirostris*	SH	Raw	2
206	*Tragopogon pusillus*	FR		11
207	*Triticum aestivum*	SE	Cooked	5
208	*Urtica dioica*	AP, WP	Raw, Cooked	2, 5, 10
209	*Vaccaria pyramidata* var. *grandiflora*	SE		2
210	*Veronica anagallis-aquatica* ssp. *oxycarpa*	AP	Raw	2
211	*Vicia aintabensis*	SE	Raw	9

Continued

Tabel 5.3. Continued.

	Plant taxa	Part used	Preparation	Source
212	*Vicia alpestris* **ssp.** *alpestris*	SE	Raw	1
213	*Vicia anatolica*	SE	Raw	9
214	*Vicia cracca* ssp. *tenuifolia*	SE	Raw	1
215	*Vicia ervilia*	SE	Raw	2
216	*Vicia faba*	SE	Raw	2
217	*Vicia hybrida*	SE	Raw	9
218	*Vicia narbonensis* var. *narbonesis*	FR, SE	Raw	3, 9
219	*Vicia pannonica* var. *purpurascens*	FR	Raw	3
220	*Vicia sativa ssp. nigra* var. *nigra*	FR, SE	Raw	2, 9
221	*Vicia sativa ssp. nigra* var. *segetalis*	FR	Raw	2
222	*Vicia sativa* ssp. *sativa*	FR	Raw	2
223	*Vitis vinifera*	FR, LE	Raw, Cooked	1–3, 5
224	*Zea mays*	FR	Cooked	1, 2, 5
225	*Ziziphora tenuior*	AG		7

Notes

Part used: AG: Above ground; AP: Aerial parts; BL: Bulb; BU: Bud; CA: Capitulum; FL: Flower; FR: Fruit; LA: Latex; LE: Leaves; RH: Rhizome; RO: Root; SE: Seed; SH: Shoot; ST: Stem; TU: Tuber; WP: Whole plant

Sources: 1: İ. Kaval, Geçitli (Hakkari) ve çevresinin etnobotanik özellikleri. Department of Biology, Institute of Science and Technology, Yüzüncü Yıl University, 2011, unpublished thesis; 2: A. Gençay, Cizre (Şırnak)'nin etnobotanik özellikleri. Department of Biology, Institute of Science and Technology, Yüzüncü Yıl University, 2007, unpublished thesis; 3: Akgül, 2008; 4: Yapıcı *et al.*, 2009; 5: A. Gelse, Adıyaman çevresinin etnobotanik özellikleri [Ethnobotanical Properties of the Adiyaman Environment]. Department of Biology, Institute of Science and Technology, Yüzüncü Yıl University, 2012, unpublished thesis; 6: N. Güldaş, Adıyaman İlinde etnobotanik değeri olan bazı bitkilerin kullanım alanlarının tespiti. Department of Biology, Institute of Science and Technology, Fırat University, 2009, unpublished thesis; 7: Akan *et al.*, 2013c; 8: Akan *et al.*, 2005b; 9: Akan *et al.*, 2013b; 10: Şığva and Seçmen, 2009; 11: Akan *et al.*, 2008; 12: Aslan *et al.*, 2011.

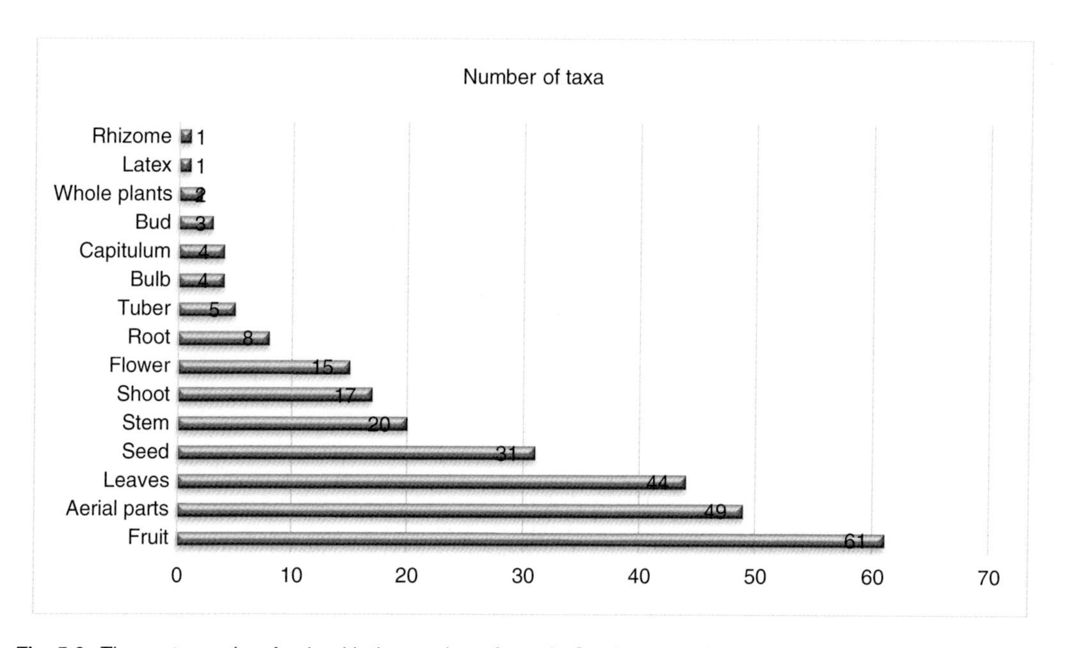

Fig. 5.6. The parts used as foods with the number of taxa in Southeastern Anatolia

Table 5.4. Plants used as spices in Southeast Anatolia.

	Plant taxa	Part used	Source
1	*Allium ampeloprasum*	Shoot	1, 2
2	*Allium longicuspis*	Aerial parts	1
3	*Filipendula ulmaria*	Aerial parts	1
4	*Foeniculum vulgare*	Aerial parts	1, 2
5	*Mentha aquatica*	Leaves	3
6	*Mentha longifolia* ssp. *longifolia*	Aerial parts	1, 2
7	*Mentha longifolia* ssp. *typhoides* var. *calliantha*	Leaves	1, 4
8	*Mentha pulegium*	Leaves	3
9	*Mentha spicata* ssp. *spicata*	Leaves	5
10	*Ocimum basilicum*	Aerial parts	1
11	*Origanum vulgare* ssp. *gracile*	Leaves, Flower	5
12	*Primula auriculata*	Aerial parts	1
13	*Ranunculus kotschyi*	Leaves	1
14	*Rhus coriaria*	Fruit	1, 2, 4–7
15	*Salvia syriaca*	Fruit	7
16	*Scilla persica*	Leaves	1
17	*Thymbra sintenisii* ssp. *sintenisii*	Aerial parts	7
18	*Thymbra spicata* var. *spicata*	Leaves, Flower	5, 8
19	*Thymus kotschyanus* var. *kotschyanus*	Aerial parts	1

Sources: 1: İ. Kaval, Geçitli (Hakkari) ve çevresinin etnobotanik özellikleri. Department of Biology, Institute of Science and Technology, Yüzüncü Yıl University, 2011, unpublished thesis; 2: A. Gençay, Cizre (Şırnak)'nin etnobotanik özellikleri. Department of Biology, Institute of Science and Technology, Yüzüncü Yıl University, 2007, unpublished thesis; 3: Tursun, 2001; 4: A. Gelse, Adıyaman çevresinin etnobotanik özellikleri [Ethnobotanical Properties of the Adiyaman Environment]. Department of Biology, Institute of Science and Technology, Yüzüncü Yıl University, 2012, unpublished thesis; 5: N. Güldaş, Adıyaman İlinde etnobotanik değeri olan bazı bitkilerin kullanım alanlarının tespiti. Department of Biology, Institute of Science and Technology, Fırat University, 2009, unpublished thesis; 6: Şığva and Seçmen, 2009; 7: Akgül, 2008; 8: Akan *et al.*, 2005b

Table 5.5. The plants added to cheese in Southeastern Anatolia.

	Plant taxa	Part used	Source
1	*Allium giganteum*	Aerial parts	1
2	*Allium kharputense*	Aerial parts	2
3	*Allium scorodoprasum* ssp. *rotundum*	Shoot	1
4	*Allium trachycoleum*	Shoot	1
5	*Allium vinele*	Shoot	2
6	*Chaerophyllum macrospermum*	Aerial parts	1
7	*Diplotaenia cachrydifolia*	Aerial parts	1
8	*Eremurus spectabilis*	Aerial parts	2
9	*Euphorbia cheiradenia*	Latex	2
10	*Ferula haussknechtii*	Shoot	3
11	*Ferula orientalis*	Shoot	1
12	*Ferulago angulata* ssp. *angulata*	Aerial parts	1
13	*Ferulago angulata* ssp. *carduchorum*	Aerial parts	1
14	*Ferulago stellata*	Aerial parts	1
15	*Gundelia tournefortii* var. *tournefortii*	Shoot	1
16	*Heracleum persicum*	Aerial parts	1
17	*Medicago sativa* ssp. *sativa*	Root	1
18	*Pimpinella kotschyana*	Aerial parts	1
19	*Prangos pabularia*	Aerial parts	2
20	*Primula auriculata*	Aerial parts	1

Continued

Table 5.5. Continued.

	Plant taxa	Part used	Source
21	*Prunella vulgaris*	Shoot	1
22	*Ranunculus fenzlii*	Aerial parts	2
23	*Sium sisarum* var. *lancifolium*	Aerial parts	1
24	*Thymus kotschyanus* var. *kotschyanus*	Aerial parts	1
25	*Trigonella foenum-graecum*	Aerial parts	4

Sources: 1: İ. Kaval, Geçitli (Hakkari) ve çevresinin etnobotanik özellikleri. Department of Biology, Institute of Science and Technology, Yüzüncü Yıl University, 2011, unpublished thesis; 2: A. Gençay, Cizre (Şırnak)'nin etnobotanik özellikleri. Department of Biology, Institute of Science and Technology, Yüzüncü Yıl University, 2007, unpublished thesis; 3: Özgökçe and Özçelik, 2004; 4: N. Güldaş, Adıyaman İlinde etnobotanik değeri olan bazı bitkilerin kullanım alanlarının tespiti. Department of Biology, Institute of Science and Technology, Fırat University, 2009, unpublished thesis

Table 5.6. The plants used as herbal teas in Southeastern Anatolia.

	Plant taxa	Part used	Source
1	*Anthemis hyalina*	Flower	1
2	*Cyclotrichum leucotrichum*	Aerial parts	1
3	*Lamium macrodon*	Flower	2
4	*Micromeria cristata*	Stem	3
5	*Phlomis armeniaca*	Flower, Leaves	2, 4
6	*Salvia multicaulis*	Flower, Leaves	2, 4
7	*Salvia syriaca*	Flower	2
8	*Scutellaria tomentosa*	Above ground	5
9	*Sideritis libanotica ssp. linearis*	Aerial parts	6
10	*Sideritis libanotica ssp. microchlamys*	Stem	3
11	*Stachys lavandulifolia*	Stem	3
12	*Teucrium polium*	Flower	7
13	*Thymbra spiacata* var. *spicata*	Leaves	8, 9

Sources: 1: Akan *et al.*, 2008; 2: A. Gençay, Cizre (Şırnak)'nin etnobotanik özellikleri. Department of Biology, Institute of Science and Technology, Yüzüncü Yıl University, 2007, unpublished thesis; 3: N. Güldaş, Adıyaman İlinde etnobotanik değeri olan bazı bitkilerin kullanım alanlarının tespiti. Department of Biology, Institute of Science and Technology, Fırat University, 2009, unpublished thesis; 4: A. Gelse, Adıyaman çevresinin etnobotanik özellikleri [Ethnobotanical Properties of the Adiyaman Environment]. Department of Biology, Institute of Science and Technology, Yüzüncü Yıl University, 2012, unpublished thesis; 5: Akan *et al.*, 2013c; 6: Akgül, 2008; 7: Akan *et al.*, 2013c; 8: Akgül, 2008; 9: Şığva and Seçmen, 2009.

5.7 Plants Used as Fodder in Southeastern Anatolia

Meadows and pastures are an important genetic source for cultivated plants, contribute much to biological diversity, serve as areas of shelter for animals, and shield land from erosion, in addition to serving as the most important natural sources of forage plants for animals (Aydın and Uzun, 2002; Ozturk *et al.*, 2012c). These habitats are also very important as a cheaper feed-source for animal nutrition and animal health in terms of the quality of animal products. Therefore, pasture and meadow areas and their efficiency is of paramount importance (Kaya *et al.*, 2001; Babalık and Sönmez, 2009).

Harlan (1983) has separated the fodder plants into four geographical regions. He has reported that Turkey includes three of these regions: Europe, the Mediterranean and the Middle East. He reports that the some of the species of *Lolium, Trifolium, Medicago, Dactylis, Festuca, Avena, Phleum, Lupinus* are spread out in the centre of Europe, whereas some species of *Dactylis, Festuca, Avena, Phleum, Lupinus* are from the Mediterranean and some species of *Trifolium, Medicago, Onobrychis, Agropyron, Festuca, Bromus* come from the Middle East.

Turkey is known as the first cultivation centre of Leguminosae members like *Vicia*, *Pisum*, *Lupinus*, and *Lens* (Harlan, 1971; Zohary and Hopf, 1994; Ozturk *et al.*, 2012d). *Hordeum spontaneum* is believed to be the ancestor of barley, *Avena strigosa* the ancestor of oats, and *Secale anatolicum*, *S. montanum* and *S. segatale* the ancestors of rye. *Lens orientalis* is the ancestor of lentil, *Vicia galilae* the ancestor of beans, *Pisum elatius* and *P. humile* are ancestors of pea. These have all spread across Anatolia. Şanlıurfa and its environs are known as a gene resource for plants, especially for wheat (*Triticum*) and lentil (*Lens*) (Ekim, 1994; Cevheri and Çetin, 2010).

In this context, and in the light of studies conducted in the region, Southeastern Anatolia shows a potential spread of fodder plants with a total of 156 taxa (see Table 5.7). Nearly 50% of these taxa (78) belong to the Fabaceae family including 15, 13, 10, 7, 7, 6 and 6 taxa respectively from *Trifolium*, *Vicia*, *Medicago*, *Lathyrus*, *Trigonella*, *Astragalus* and *Onobrychis*.

Table 5.7. The potential fodder plants from Southeastern Anatolia.

	Plant taxa	Part used	Source
1	*Acanthus syriacus*	Aerial parts	1
2	*Aegilops triuncialis* ssp. *triuncialis*	Aerial parts	2
3	*Alcea setose*	Aerial parts	3
4	*Alhagi pseudalhagi*	Fruits, Aerial parts	4, 5
5	*Alkanna orientalis*	Leaves	4
6	*Anchusa azurea* var. *azurea*	Leaves	2, 4
7	*Anchusa azurea* var. *kurdica*	Leaves	2
8	*Anchusa strigosa*	Aerial parts	2
9	*Astragalus chiristianus*	Aerial parts	3
10	*Astragalus gummifer*	Aerial parts	6
11	*Astragalus hamosus*	Aerial parts	7
12	*Astragalus lamarckii*	Root	7
13	*Astragalus russelii*	Root	1, 7
14	*Astragalus xylobasis* var. *angustus*	Aerial parts	1
15	*Avena sterilis* var. *sterilis*	Aerial parts	1, 5
16	*Bromus japonicus* ssp. *japonicus*	Aerial parts	2
17	*Bunium paucifolium* var. *paucifolium*	Aerial parts	1
18	*Centaurea iberica*	Aerial parts	5
19	*Centaurea stapfiana*	Aerial parts	3
20	*Cephalaria hakkarica*	Aerial parts	8
21	*Cephalaria procera*	Aerial parts	9
22	*Cephalaria setose*	Aerial parts	5
23	*Chrozophora tinctoria*	Aerial parts	2
24	*Chrysopogon gryllus* ssp. *gryllus*	Aerial parts	2
25	*Cicer arietinum*	Leaves, Aerial parts	4, 5
26	*Cicer echinospermum*	Aerial parts	7
27	*Cichorium intybus*	Aerial parts	5
28	*Convolvulus arvensis*	Aerial parts	1, 3
29	*Coronilla scorpioides*	Aerial parts	1, 7
30	*Crepis sancta*	Aerial parts	4, 5
31	*Cynodon dactylon* var. *villosus*	Aerial parts	2
32	*Cynosurus effuses*	Aerial parts	2
33	*Daucus broteri*	Above ground	1
34	*Echinops sphaerocephalus* ssp. *sphaerocephalus*	Aerial parts	5
35	*Echium italicum*	Aerial parts	3
36	*Eremopyrum bonaepartis* ssp. *bonaepartis*	Above ground	1
37	*Erysimum repandum*	Aerial parts	5
38	*Euphorbia microsphaera*	Above ground	1
39	*Ferula hausknechtii*	Aerial parts	8

Continued

Table 5.7. Continued.

	Plant taxa	Part used	Source
40	*Ferula orientalis*	Aerial parts	8
41	*Galium aparine*	Aerial parts	8
42	*Geranium tuberosum* ssp. *tuberosum*	Aerial parts, Tuber	1, 3
43	*Gundelia tournefortii*	Aerial parts	5
44	*Gypsophila viscose*	Above ground	1
45	*Hedysarum pannosum*	Above ground	1, 7
46	*Helianthus annuus*	Stem, Leaves	4, 8
47	*Heliotropium europaeum*	Aerial parts	4
48	*Hippocrepis unisiliquosa* ssp. *unisiliquosa*	Aerial parts	1, 7
49	*Hordeum bulbosum*	Aerial parts	4, 5
50	*Hordeum murinum* ssp. *leporinum* var. *leporinum*	Above ground	1
51	*Hordeum spontaneum*	Above ground	1
52	*Hymenocarpus circinnatus*	Aerial parts	7
53	*Jurinea pulchella*	Stem	5
54	*Lactuca undulata*	Aerial parts	2
55	*Lathyrus annuus*	Aerial parts	5
56	*Lathyrus aphaca* var. *modestus*	Above ground	1
57	*Lathyrus cicera*	Aerial parts	5, 7, 9
58	*Lathyrus gorgoni* var. *gorgoni*	Aerial parts	5
59	*Lathyrus inconspicuus*	Aerial parts, Fruits	3, 5
60	*Lathyrus palustris* ssp. *palustris*	Aerial parts	5
61	*Lathyrus sativus*	Above ground	1
62	*Lens culinaris*	Fruits	4, 5
63	*Lens orientalis*	Aerial parts	7
64	*Leontodon hispidus* var. *hispidus*	Aerial parts	5
65	*Lithospermum purpurocaeruleum*	Aerial parts	3
66	*Lotus aegaeus*	Aerial parts	7
67	*Medicago lupulina*	Aerial parts	7
68	*Medicago minima* var. *minima*	Above ground	1, 7
69	*Medicago noeana*	Aerial parts	7
70	*Medicago orbicularis*	Aerial parts	7, 9
71	*Medicago polymorpha* var. *vulgaris*	Above ground	1
72	*Medicago radiate*	Above ground	1, 3, 5, 7
73	*Medicago rigidula* var. *cinerascens*	Aerial parts	5
74	*Medicago rigidula* var. *rigidula*	Above ground	1, 4, 7
75	*Medicago rigidula* var. *submitis*	Aerial parts	3
76	*Medicago sativa* ssp. *sativa*	Aerial parts	7, 8
77	*Melica ciliata* ssp. *ciliata*	Above ground	1
78	*Melilotus officinalis*	Aerial parts	7
79	*Notobasis syriaca*	Shoot	5
80	*Onobrychis altissima*	Above ground	1
81	*Onobrychis armena*	Above ground	1, 4, 5
82	*Onobrychis cornuta*	Aerial parts	8
83	*Onobrychis crista-galli*	Aerial parts	7
84	*Onobrychis kotschyana*	Above ground	1, 7
85	*Onobrychis megataphros* var. *podperae*	Above ground	1
86	*Onopordum acanthium*	Aerial parts	4, 5
87	*Onosma sericeum*	Aerial parts	5
88	*Oryza sativa*	Aerial parts	4, 5
89	*Paliurus spina-christii*	Leaves	8
90	*Papaver rhoeas*	Aerial parts	3
91	*Phlomis bruguieri*	Above ground	1

Continued

Table 5.7. Continued.

	Plant taxa	Part used	Source
92	*Phlomis kurdica*	Above ground	1
93	*Pisum sativum* ssp. *elatius* var. *pumilio*	Above ground	1
94	*Platanus orientalis*	Leaves	4, 5
95	*Prangos pabularia*	Whole plants	8
96	*Prosopis farcta*	Aerial parts	5
97	*Pterocephalus pyrethrifolius*	Flower	3
98	*Quercus brantii*	Leaves, Fruit	5
99	*Quercus ilex*	Leaves, Fruit	5
100	*Quercus infectoria* ssp. *boissieri*	Leaves, Fruit	4, 5, 8
101	*Quercus ithaburensis* ssp. *macrolepis*	Leaves, Fruit	4, 5
102	*Salvia multicaulis*	Aerial parts	3
103	*Salvia syriaca*	Above ground	1
104	*Scandix stellata*	Above ground	1
105	*Scorpiurus muricatus* var. *subvillosus*	Above ground	1, 7
106	*Scorzonera kotschyi*	Above ground	1
107	*Scorzonera laciniata* ssp. *laciniata*	Aerial parts	2
108	*Senecio vernalis*	Aerial parts	4, 5
109	*Smyrniopsis aucheri*	Aerial parts	8
110	*Sorghum halepense* var. *muticum*	Aerial parts	4, 5
111	*Tanacetum argyrophyllum* var. *agrophyllum*	Aerial parts	8
112	*Taraxacum sintenisii*	Above ground	1
113	*Tragopogon longirostris* var. *longirostris*	Shoot	3, 5, 9
114	*Trifolium angustifolium* var. *angustifolium*	Aerial parts	4, 5, 7
115	*Trifolium boissieri*	Aerial parts	7
116	*Trifolium bullatum*	Aerial parts	7
117	*Trifolium campestre*	Above ground	1, 4, 5
118	*Trifolium dasyurum*	Above ground	1
119	*Trifolium echinatum* var. *echinatum*	Above ground	1, 5
120	*Trifolium hybridum var. hybridum*	Aerial parts	5
121	*Trifolium leucanthum*	Above ground	1, 7
122	*Trifolium nigrescens* ssp. *petrisavii*	Aerial parts	5, 9
123	*Trifolium physodes* var. *psilocalyx*	Aerial parts	5
124	*Trifolium pilulare*	Above ground	1, 7
125	*Trifolium resupinatum* var. *resupinatum*	Above ground	1
126	*Trifolium spumosum*	Above ground	1
127	*Trifolium stellatum* var. *stellatum*	Aerial parts	7
128	*Trifolium tomentosum* var. *tomentosum*	Above ground	1, 5, 7
129	*Trigonella capitata*	Aerial parts	5
130	*Trigonella coelesyriaca*	Above ground	1, 7
131	*Trigonella filipes*	Above ground	1, 7
132	*Trigonella mesopotamica*	Above ground	1
133	*Trigonella monantha* **ssp**. *monantha*	Above ground	1, 7
134	*Trigonella monspeliaca*	Above ground	1
135	*Trigonella spruneriana* var. *spruneriana*	Above ground	1, 5, 7
136	*Tripleurospermum parviflorum*	Flower, aerial parts	4, 5
137	*Triticum aestivum*	Aerial parts	2
138	*Triticum dicoccoides*	Above ground	1
139	*Turgenia latifolia*	Above ground	1
140	*Vaccaria pyramidata* var. *grandiflora*	Aerial parts	3, 5
141	*Vaccaria pyramidata* var. *oxyodonta*	Above ground	1
142	*Vicia anatolica*	Above ground	1, 7
143	*Vicia assyriaca*	Aerial parts	7

Continued

Table 5.7. Continued.

	Plant taxa	Part used	Source
144	*Vicia cracca* ssp. *tenuifolia*	Aerial parts	8
145	*Vicia ervilia*	Aerial parts	5, 7
146	*Vicia faba*	Aerial parts	5
147	*Vicia hybrid*	Aerial parts	3
148	*Vicia narbonensis* var. *narbonesis*	Fruits	3
149	*Vicia palaestina*	Aerial parts	7
150	*Vicia pannonica* var. *purpurascens*	Fruits	3
151	*Vicia sativa* ssp. *nigra* var. *nigra*	Above ground	1, 5
152	*Vicia sativa* ssp. *nigra* var. *segetalis*	Above ground	1, 5
153	*Vicia sativa* ssp. *sativa*	Above ground	1, 5
154	*Vicia villosa*	Aerial parts	7
155	*Vitex pseudo-negundo*	Leaves	5
156	*Zea mays*	Aerial parts	4, 5

Sources: 1: Akan *et al*., 2013c; 2: Akan *et al*., 2008; 3: Akgül, 2008; 4: A. Gelse, Adıyaman çevresinin etnobotanik özellikleri [Ethnobotanical Properties of the Adiyaman Environment]. Department of Biology, Institute of Science and Technology, Yüzüncü Yıl University, 2012, unpublished thesis; 5: A. Gençay, Cizre (Şırnak)'nin etnobotanik özellikleri. Department of Biology, Institute of Science and Technology, Yüzüncü Yıl University, 2007, unpublished thesis; 6: N. Güldaş, Adıyaman İlinde etnobotanik değeri olan bazı bitkilerin kullanım alanlarının tespiti. Department of Biology, Institute of Science and Technology, Fırat University, 2009, unpublished thesis; 7: Akan *et al*., 2013b; 8: İ. Kaval, Geçitli (Hakkari) ve çevresinin etnobotanik özellikleri. Department of Biology, Institute of Science and Technology, Yüzüncü Yıl University, 2011, unpublished thesis; 9: Yapıcı *et al*., 2009.

5.8 Other Economic Plants Used in Southeastern Anatolia

Apart from the uses given above, plants in the region are also used in basket making, toys, brooms, musical instruments, handicrafts, in house constructions, decorations, as natural dyes, and as fuel. A total of 159 taxa are used for this purpose (see Table 5.8).

The majority are used as fuel (49 taxa: 22.38%), or as ornaments (35 taxa: 15.98%), followed by handcrafts (35 taxa: 15.98%) (see Fig 5.7). The number of taxa and their percentages in the production of dyes is 27 taxa (12.33%), musical instruments 25 taxa (11.42%) and brooms 15 taxa (6.85%) (see Fig 5.7).

Trigonella monantha ssp. *noeana*, *Ballota saxalis* ssp. *saxalis*, *Cyclotrichum leucotrichum*, *Matricaria aurea* and *Parietaria judaica* are used as aromatic plants; *Ammi visnaga*, *Verbascum orientale* and *Verbascum kotschyi* as insecticides; *Pistacia terebinthus* ssp. *palaestina* and *Pistacia khinjuk* in soap making; *Ammi visnaga* (A. Gelse, Adıyaman çevresinin etnobotanik özellikleri [Ethnobotanical Properties of the Adiyaman Environment]. Department of Biology, Institute of Science and Technology, Yüzüncü Yıl University, 2012, unpublished thesis) and *Daucus littoralis* in making toothpicks; *Alcea hohenackeri* as detergent; *Cyperus longus* for rope and *Eremurus spectabilis* for gum production (Akan *et al*., 2005b; A. Gençay, Cizre (Şırnak)'nin etnobotanik özellikleri. Department of Biology, Institute of Science and Technology, Yüzüncü Yıl University, 2007, unpublished thesis; Akan *et al*., 2008; N. Güldaş, Adıyaman İlinde etnobotanik değeri olan bazı bitkilerin kullanım alanlarının tespiti. Department of Biology, Institute of Science and Technology, Fırat University, 2009, unpublished thesis; İ. Kaval, Geçitli (Hakkari) ve çevresinin etnobotanik özellikleri. Department of Biology, Institute of Science and Technology, Yüzüncü Yıl University, 2011, unpublished thesis; A. Akgül, Midyat (Mardin) civarında etnobotanik. Graduate School of Science, Ege University, Izmir, Turkey, 2008, unpublished thesis). The total percentage of these uses in general does not go beyond 6.37%. Photographs of some of the representative medicinal plants and other aspects can be found in Fig. 5.8.

5.9 Conclusions

A determination of the drought-tolerant plant species which will be suitable for dry conditions in the future needs to be considered for food security (Ozturk *et al*., 2011; Ozturk *et al*., 2012a, b, c). Ecological sustainability is another important factor,

Table 5.8. Other economic uses of plants in Southeastern Anatolia.

	Plant taxa	Fuel	Ornamental	Dye	Musical Instruments	Handcrafts	Brooms	Baskets	Toys	Others	Source
1	*Abies cilicica*				x						1
2	*Acer monspessulanum ssp. cinerascens*					x					2
3	*Acer pseudoplatanus*				x						1
4	*Adianthum capillaris*		x								3
5	*Alcea hohenackeri*		x							x	4, 5
6	*Alhagi pseudalhagi*						x				6, 7
7	*Alkanna hirsutissima*			x							7
8	*Alkanna megacarpa*			x							7
9	*Alkanna orientalis var. orientalis*			x							3, 4
10	*Alkanna tinctoria ssp. anatolica*			x							8
11	*Amaranthus patulus*		x								3, 4
12	*Amaranthus viridis*		x								3, 4
13	*Ammi visnaga*									x	3, 4, 9
14	*Amygdalus communis*	x			x	x					1, 2, 4
15	*Anagyris foetida*							x			10
16	*Anchusa azurea var. Azurea*					x					11
17	*Anemone coronaria*		x								4
18	*Anthemis tinctoria var. Pallida*			x							8
19	*Artemisia annua*		x				x				3, 4
20	*Arundo donax*				x	x					1, 2, 4
21	*Astragalus aduncus*	x									6
22	*Astragalus amblolepis*	x									4
23	*Astragalus gaziantepicus*	x									6
24	*Astragalus karabaghensis*	x									2
25	*Astragalus microcephalus*	x									12
26	*Astragalus pycnocephalus var. pycnocephalus*	x									2
27	*Astragalus russelii*	x							x		6, 7
28	*Ballota saxalis ssp. Saxalis*									x	11
29	*Berberis crataegina*	x				x					4
30	*Buxus sempervirens*				x						1
31	*Cardaria draba*	x									7
32	*Carpinus orientalis*	x			x	x					1, 13
33	*Carthamus tinctorius*			x							14
34	*Castanea sativa*				x						1

Continued

Table 5.8. Continued.

	Plant taxa	Fuel	Ornamental	Dye	Musical Instruments	Handcrafts	Brooms	Baskets	Toys	Others	Source
35	*Celtis australis*							x			10
36	*Centaurea iberica*	x									3
37	*Cerasus mahalep var. Mahalep*					x					2, 11
38	*Cerasus microcarpa*				x						1
39	*Convolvulus betonicifolius*		x								3
40	*Convolvulus holosericeus*						x				15
41	*Convolvulus stachydifolius*		x								3, 4
42	*Cornus mas*				x						1
43	*Cotinus coggyria*			x							11
44	*Crataegus aronia var. Aronia*	x									4, 8
45	*Crataegus orientalis var. orientalis*	x							x		3, 4
46	*Crupina crupinastrum*						x				11
47	*Cyclotrichum leucotrichum*									x	11
48	*Cyperus longus*									x	2
49	*Datura innoxia*		x								3, 4
50	*Daucus littoralis*									x	4
51	*Delphinium peregrinum*		x								15
52	*Diospros ebenum*				x						1
53	*Elaeagnus angustifolia*		x								3, 4
54	*Eminium rauwolfii var. rauwolfii*			x							7
55	*Eminium spiculatum var. spiculatum*			x							7
56	*Eremurus spectabilis*									x	2
57	*Erodium cicutarium ssp. cicutarium*								x		7
58	*Eucalyptus camaldulensis*	x				x					4, 13
59	*Fagus orientalis*				x						1
60	*Fraxinus excelsior*	x			x	x					1, 5, 13
61	*Fritallaria imperialis*		x								2, 3, 5
62	*Fritillaria persica*		x								3, 5
63	*Geranium dissectum*										15
64	*Gladiolus atroviolaceus*		x						x		3, 4
65	*Gladiolus micranthus*		x								7
66	*Gleditsia triancanthos*	x	x								4
67	*Gossypium herbaceum*	x									3, 4
68	*Helianthus annuus*	x									3, 4

69	*Iberis acutiloba*						x			7
70	*Imperata cylindrica*					x				4
71	*Ipomoea purpurea*		x							2
72	*Iris masia*		x			x				4
73	*Iris persica*		x							3, 4, 7
74	*Ixiolirion tataricum ssp. tataricum*		x							7
75	*Jasminum fruticans*			x						8
76	*Juglans regia*			x	x	x				1, 2, 4, 13
77	*Juncus inflexus*					x				2, 3, 4
78	*Juniperus excelsa*				x					1
79	*Kochia scoparia*						x			15
80	*Lagoecia cuminoides*		x							4
81	*Lantana camara*		x							4
82	*Malva neglecta*							x		11
83	*Matricaria aurea*								x	11
84	*Medicago noeana*							x		7
85	*Morus alba*	x			x	x		x		1, 2, 3, 13
86	*Morus nigra*	x			x	x				1, 2, 13
87	*Narcissus tazetta ssp. Tazetta*		x							4
88	*Nerium oleander*	x	x			x				3, 4, 5
89	*Olea europaea*	x								8
90	*Onobrychis cornuta*	x								2
91	*Paliurus spina christii*			x		x				8, 11
92	*Papaver rhoeas*		x					x		3
93	*Parietaria judaica*								x	11
94	*Peganum harmala*					x				4, 11
95	*Phlomis brugueri*	x								15
96	*Phragmites australis*	x				x				4, 8
97	*Picea orientalis*				x					1
98	*Pinus brutia*	x			x					1, 8
99	*Pinus nigra ssp. Pallasiana*	x			x	x				1, 13
100	*Pistacia khinjuk*								x	9
101	*Pistacia terebinthus ssp. palaestina*								x	4, 11
102	*Platanus orientalis*	x			x	x				1, 3, 4, 8, 13
103	*Polygala supina*						x			15
104	*Polygonum bellardii*						x			11
105	*Populus alba*	x			x	x				1, 13
106	*Populus euphratica*	x	x			x		x		3, 4, 11, 13

Continued

Table 5.8. Continued.

	Plant taxa	Fuel	Ornamental	Dye	Musical Instruments	Handcrafts	Brooms	Baskets	Toys	Others	Source
107	*Populus nigra ssp. Nigra*	x				x		x	x		2, 5, 10, 13
108	*Prosopis farcta*	x					x				4
109	*Prunus armeniaca*			x							4
110	*Prunus spinosa*				x						1
111	*Pterocarya fraxinifolia*			x							5
112	*Punica granatum*				x			x			1, 10
113	*Pyracantha coccinea*		x								4
114	*Pyrus syriaca var. Syriaca*	x									2
115	*Quercus brantii*	x		x		x					2, 4, 11, 13
116	*Quercus ilex*	x		x							4
117	*Quercus infectoria ssp. boissieri*	x		x		x					2, 3, 4, 8, 13
118	*Quercus ithaburensis ssp. macrolepis*	x		x							3, 4
119	*Quercus robur ssp. Robur*	x									8
120	*Ranunculus asiaticus*		x								4
121	*Rhus coriaria*	x		x							3, 8
122	*Rosa canina*		x	x							2, 4
123	*Rosa damascena*		x		x						1, 4
124	*Rubia tenuifolia ssp. Doniettii*								x		11
125	*Rubia tinctorium*			x							2
126	*Rubus canescens*		x								3
127	*Rubus sanctus*			x							2
128	*Rumex tuberosus ssp. horizontalis*			x							2
129	*Salix aegyptiaca*	x				x					2
130	*Salix alba*	x				x		x			2, 3, 4, 13
131	*Salix babylonica*	x				x					4
132	*Salix excelsa*	x				x					4
133	*Salix viminalis*							x			10
134	*Salsola tragus*	x									15
135	*Salvia verticillata ssp. amasiaca*			x							2
136	*Salvia verticillata ssp. verticillata*			x							2
137	*Salvia virgata*			x							2

138	*Scabiosa argentea*						x			7, 11
139	*Scrophularia striata*						x			11
140	*Senecio vernalis*			x						3
141	*Sideritis libanotica ssp. linearis*						x			11
142	*Suaeda altissima*						x			3, 4
143	*Tagetes erecta*		x							2
144	*Tamarix smyrnensis*		x							4
145	*Tamarix tetrandra*					x				13
146	*Tilia rubra*				x					1
147	*Trigonella monantha ssp. noeana*								x	11
148	*Triticum aestivum*					x		x		3, 4, 11
149	*Triticum vulgare*							x		3
150	*Tulipa julia*		x							3, 4
151	*Typha angustifolia*					x				2
152	*Ulmus glabra*	x				x				13
153	*Verbascum kotschyi*								x	15
154	*Verbascum orientale*								x	15
155	*Vitex pseudo-negundo*	x				x	x			4, 11
156	*Vitis vinifera*	x								11
157	*Washingtonia filifera*		x							3, 4
158	*Xeranthemum annuum*						x			2, 11, 12
159	*Zea mays*	x								2, 3

Sources: 1: Akan *et al.*, 2013a; 2: İ. Kaval, Geçitli (Hakkari) ve çevresinin etnobotanik özellikleri. Department of Biology, Institute of Science and Technology, Yüzüncü Yıl University, 2011, unpublished thesis; 3: A. Gelse, Adıyaman çevresinin etnobotanik özellikleri [Ethnobotanical Properties of the Adiyaman Environment]. Department of Biology, Institute of Science and Technology, Yüzüncü Yıl University, 2012, unpublished thesis; 4: A. Gençay, Cizre (Şırnak)'nin etnobotanik özellikleri. Department of Biology, Institute of Science and Technology, Yüzüncü Yıl University, 2007, unpublished thesis; 5: N. Güldaş, Adıyaman İlinde etnobotanik değeri olan bazı bitkilerin kullanım alanlarının tespiti. Department of Biology, Institute of Science and Technology, Fırat University, 2009, unpublished thesis; 6: Akan *et al.*, 2013b; 7: Akan *et al.*, 2013c; 8: Şığva ve Seçmen, 2009; 9: Akan *et al.*, 2005b; 10: Akan, 2013; 11: Akgül, 2008; 12: Yapıcı *et al.*, 2009; 13: Aslan *et al.*, 2011; 14: Özgökçe and Özçelik, 2004; 15: Akan *et al.*, 2008.

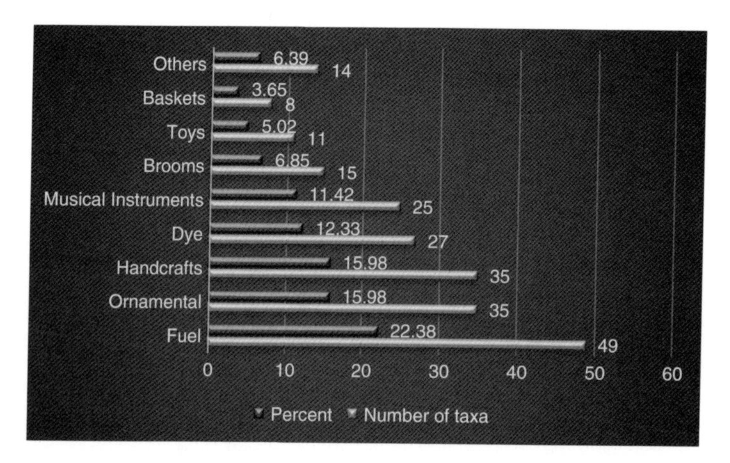

Fig. 5.7. Number and percentages of Southeastern Anatolian plant taxa with other economic uses.

Fig. 5.8. Some commonly used medicinal plants from the study area: 1. *Achillea biebersteinii*; 2. *Hypericum capitatum* var. *capitatum*; 3. *Salvia syriaca*; 4. *Anthemis tinctoria*; 5. *Cichorium intybus*; 6. *Bellis perennis*; 7. *Glycyrrhiza glabra* var. *glandulifera*; 8. *Hypericum perforatum*; 9. *Myrtus communis*; 10. *Pistacia terebinthus* ssp. *palaestina*; 11. *Viscum album*; 12. *Thymbra spicata*.

especially for medicinal, aromatic and other economically important plants. It is not possible to place a price on these, but they have great economic value (Costanza and Farber, 2002; Farber *et al.*, 2006). The use of such plants in agricultural activities carries great weight for humans as well as domesticated animals, in particular because both south-west Asia and Turkey are gene centres of these plants; therefore it is imperative to identify these taxa. Conservation of the economically important medicinal and other plants in this context is also high priority. It is necessary during conservation to look at how these plants are utilized by local people. The protection of genetic resources in Turkey is perhaps the highest priority (Pleskanovskaja *et al.*, 2011).

The laws for 'nature and species protection' cannot be applied effectively unless alternatives are developed for collecting plants from nature. The best option in this connection is to employ cultivation practices for such plants. Endangered species and those threatened with extinction should be considered in terms of their contribution to the natural environment and the economy of the country. A sustainable conservation of genetic resources from our natural wealth and for future research is also very important (Bayram *et al.*, 2010).

For medicinal and aromatic plants, as well as other plants of economic importance for relevant stakeholders and industry, a long-term plan is a pre-requisite. For economically important plants, it is essential that we study their behaviour under future climate-change scenarios, together with drought, flooding, erosion, other natural disasters, ecosystem viability and sustainable land management. We must work on the market preferences and demand trends of genetic resources and biodiversity, their varietal development, organic products, and the planning related to their production based on industry issues. There is also a need to create research inventories and collaboration platforms. In addition, each organization needs close cooperation and communication among themselves as well as with the local inhabitants (Bayram *et al.*, 2010).

The world today is facing the crisis of antibiotic-resistant strains of viruses. The major hindrance for herbal therapies is the lack of amalgamation of indigenous knowledge with modern medical practices, because little or no scientific data are available regarding the safety and efficacy of herbal drugs. There is an urgent need to document and authenticate the available indigenous knowledge with modern scientific principles.

During the next 50 years, the global population is expected to reach a level of 9 billion. This will lead to a decrease in renewable natural resources. Housing and farmland use will increase, leading towards to a decrease in the number of species. The area of fertile soils will dwindle and deforestation will add to the species loss. Climate change will add to this loss with a depletion of water resources. Undoubtedly all of these factors will pose great threats for future generations (Pleskanovskaja *et al.*, 2011). Out of our natural resources, in particular, availability of medicinal and aromatic plants together with other consumable herbals will suffer greater loss (Ozturk *et al.*, 2011; Ozturk *et al.*, 2012a, b, c).

References

Akan, H. (2013) An ethnobotanical investigation on the baskets of Mardin, South East Anatolia. *ADYÜTAYAM* 1(1), 21–30.

Akan, H., Aslan, M. and Balos, M.M. (2005a) Şanlıurfa kent merkezindeki semt pazarlarında satılan bazı bitkiler ve kullanım amaçları. *Ot Sistematik Botanik Dergisi* 12(2), 43–58.

Akan, H., Kaya, Ö.F., Eker, İ. and Cevheri, C. (2005b) The flora of Kaşmer Dağı (Şanlıurfa, Turkey). *Turkish Journal of Botany* 29, 291–310.

Akan, H., Korkut, M.M. and Balos, M.M. (2008) Arat Dağı ve çevresinde (Birecik, Şanlıurfa) etnobotanik bir araştırma. *Science and English Journal of Fırat University* 20(1), 67–81.

Akan, H., Balos, M.M. and Aslan, M. (2013a) An ethnobotanical research on handmade musical instruments in Şanlıurfa, South East Anatolia, Turkey. *Biological Diversity and Conservation* 6(1), 93–100.

Akan, H., Balos, M.M. and Tel, A.Z. (2013b) The ethnobotany of some Legume plants around Birecik (Şanlıurfa). *ADYÜTAYAM* 1(1), 31–39.

Akan, H., Aydoğdu, M., Korkut, M.M. and Balos, M.M. (2013c) An ethnobotanical research of the Kalecik mountain area (Şanlıurfa, South-East Anatolia). *Biological Diversity and Conservation* 6(2), 84–90.

Akdoğan, H. and Akgün, B. (2006) *Göksun (Kahramanmaraş) çevresinde halk ilacı olarak kullanılan bazı bitkisel gıdalar*. Türkiye 9. Gıda Kongresi (Food Congress), 24–26 May, Bolu, Turkey, pp. 183–186.

Algier, A.A., Hanoğlu, Z., Özden, G. and Kara, F. (2005) The use of complementary and alternative (non-conventional) medicine in cancer patients in Turkey. *European Journal of Oncology Nursing* 9, 138–146.

Archer, E.R.M. (2004) Beyond the climate versus grazing impasse: using remote sensing to investigate the effects of grazing system choice on vegetation cover

in the eastern Karoo. *Journal of Arid Environments* 57, 381–408.
Aslan, M. and Türkmen, N. (2001) New floristic records for C7 grid square. *Ot Sistematik Botanik Dergisi* 8(2), 69–73.
Aslan, M. and Türkmen, N. (2003) New floristic records for squares C6 and C7 from Turkey. *Ot Sistematik Botanik Dergisi* 10(2), 163–168.
Aslan, M., Akan, H. and Balos, M.M. (2011) Şanlıurfa'da bazı odunsu bitkilerin etnobotaniği üzerine bir araştırma. *Ot Sistematik Botanik Dergisi* 18(1), 117–136.
Atamov, V., Aktoklu, E., Çetin, E., Aslan, M. and Yavuz, M. (2006) Halophytication in Harran (Şanlıurfa) and Amik Plain (Hatay) in Turkey. *Phytologia Balcanica* 12(3), 401–412.
Atamov, V., Aydın, N. and Aslan, M. (2009) Direkli Tepeleri (Şanlıurfa) florası. *Ot Sistematik Botanik Dergisi* 16(1), 97–114.
Avcı, M. (2005) Diversity and endemism in Turkey's vegetation. *Coğrafya Dergisi* 13, 27–55.
Aydoğdu, M. and Akan, H. (2005) The flora of Kalecik Mountain (Şanlıurfa, Turkey). *Turkish Journal of Botany* 29, 155–174.
Aydın, İ. and Uzun, F. (2002) *Çayır-Mera Amenajmanı ve ıslahı, Textbook No. 9*. Faculty of Agriculture, Ondokuz Mayıs University, Samsun, Turkey.
Aytaç, Z. and Duman, H. (2005) The steppic flora of high Mounts Ahir, Öksüz and Binboğa (Kahramanmaraş-Kayseri, Turkey). *Flora Mediterranea* 15, 121–178.
Babalık, A.A. and Sönmez, K. (2009) Otlatılan ve korunan mera kesimlerinde bakı faktörünün topraküstü biomass miktarı üzerine etkileri. *Süleyman Demirel Üniversitesi Orman Fakültesi Dergisi, Seri A* 1, 52–58.
Balos, M.M and Akan, H. (2008) Flora of the region between Zeytinbahçe and Akarçay (Birecik, Şanlıurfa-Turkey). *Turkish Journal of Botany* 32, 201–226.
Bayram, E., Kırıcı, S., Tansı, S., Yılmaz, G., Arabacı, O., Kızıl, S. and Telci, İ. (2010) *Tıbbi ve Aromatik Bitkiler Üretiminin Arttırılması Olanakları*. Turkey Agricultural Engineering Technical Conference Proceedings, Ankara, Turkey, pp. 437–456.
Blumenthal, M. (1998) *The Complete German Commission E Monographs: Therapeutic Guide to Herbal Medicines*. Integrative Medicine Communications Boston-Massachusetts, American Botanical Council, Austin, Texas, USA.
Cevheri, C. and Çetin, E. (2010) Genetic resources of fodder plants in Şanlıurfa. *Ot Sistematik Botanik Dergisi* 17(1), 191–198.
Costanza, R. and Farber, S. (2002) Introduction to the special issue on the dynamics and value of ecosystem services: integrating economic and ecological perspectives. *Ecological Economics* 41, 367–373.
Davis, P.H. (1965–1985) *Flora of Turkey and the East Aegean Islands*. Vols 1–9. Edinburgh University Press, Edinburgh, UK.
Davis, P.H., Mill, R.R. and Tan, K. (1988) *Flora of Turkey and the East Aegean Islands (Supplement)*. Vol. 10. Edinburgh University Press, Edinburgh, UK.
Demir, E. (2003) The contribution of the Southeastern Anatolian project to the domestic economy and its effect on the settlements areas. Gazi University. *Gazi Eğitim Fakültesi Dergisi* 23(3), 189–205.
Demirci, S. and Özhatay, N. (2012) An ethnobotanical study in Kahramanmaraş (Turkey); wild plants used for medicinal purpose in Andırın, Kahramanmaraş. *Turkish Journal of Pharmacological Science* 9(1), 75–92.
De Silva, T., Bahorun, T., Sahu, M. and Huong, L.M. (2009) *Traditional and Alternative Medicine–Research and Policy Perspectives*. NAM S & T Centre, Daya Publishing House, Delhi, India, p. 594.
Eker, İ., Koyuncu, M. and Akan, H. (2008) The geophytic flora of Şanlıurfa Province, Turkey. *Turkish Journal of Botany* 32, 367–380.
Ekim, T. (1994) *GAP Bölgesi bitkileri, GAP Bölgesinde bitki örtüsü ve ormanlar*. Turkey Environment Foundation Publications, Ankara, Turkey, pp. 9–22.
Farber, S., Costanza, R. and Childers, D.L. (2006) Linking ecology and economics for ecosystem management. *BioScience* 56, 121–133.
Güner, A., Özhatay, N., Ekim, T. and Başer, K.H.C. (2000) *Flora of Turkey and the East Aegean Islands (Supplement)*. Vol. 11. Edinburgh University Press, Edinburgh, UK.
Harlan, J.R. (1971) Agriculture origins: centres and non-centres. *Science* 174, 468–474.
Harlan, J.R. (1983) The scope for collection and improvement of forage plants. In: McIvor, J.G. and Bray, R.A. (eds) *Genetic Resources of Forage Plants*. CSIRO, Vega Press Pty Ltd, Blackburn, Victoria, Australia, pp. 3–14.
İnandık, H. (1965) *Türkiye Bitki Coğrafyasına giriş* [Introduction to Plant Geography of Turkey]. Institute of Geography Publications, Istanbul University, Istanbul, Turkey.
Jeddi, K. and Chaieb, M. (2010) Changes in soil properties and vegetation following livestock grazing exclusion in degraded arid environments of South Tunisia. *Flora* 205, 184–189.
Karabulut, M. (2006) An examination and monitoring vegetation conditions in Turkey using NOAA AVHRR data. *Coğrafi Bilimler Dergisi* 4(1), 29–42.
Kaya, Ö.F. (2010) Kaşmer Dağı (Şanlıurfa)'nın step vejetasyonu üzerine sintaksonomik bir çalışma. *Kastamonu Üniversitesi Orman Fakültesi Dergisi* 10(1), 1–11.
Kaya, Ö.F. and Ertekin, A.S. (2009) Flora of the protected area at the Tektek dağları (Şanlıurfa-Turkey). *Ot Sistematik Botanik Dergisi* 16(2), 79–96.
Kaya, İ., Öncüler, A., Ünal, Y. and Yıldız, S. (2001) Kars yöresi çayır-mera otlarının botaniksel bileşimi ve farklı biçim besin madde düzeyleri [Botanical composition

and nutrient levels of the different forms of Kars herbaceous grassland taxa]. National Animal Nutrition Congress, Elazığ, Turkey, pp. 100–108.

Kaya, Ö.F., Ketenoğlu, O. and Bingöl, M.Ü. (2009) A phytosociological investigation on forest and dry stream vegetation of Karacadağ (Şanlıurfa-Diyarbakır). *Kastamonu Üniversitesi Orman Fakültesi Dergisi* 9(2), 157–170.

Kaya, Ö.F., Çetin, E., Aydoğdu, M., Ketenoğlu, O. and Atamov, V. (2010) Syntaxonomical analyses of the secondary vegetation of Harran Plain (Şanlıurfa-Turkey) ensuing excessive ırrigation by using GIS and remote sensing. *Ekoloji* 19(75), 1–14.

Kendir, G. and Güvenç, A. (2010) Etnobotanik ve Türkiye'de Yapılmış Etnobotanik Çalışmalara Genel Bir Bakış. *Hacettepe Üniversitesi Eczacılık Fakültesi Dergisi* 30(1), 49–80.

Kurt, E., Bavbek, S. and Pasaoglu, G. (2004) Use of alternative medicines by allergic patients in Turkey. *Allergol Immunopathology* 32, 289–294.

Newman, D.J. and Cragg, G. (2007) Natural products as sources of new drugs over the last 25 years. *Journal of Natural Products* 70(3), 461–477.

Özgökçe, F. and Özçelik, H. (2004) Ethnobotanical aspects of some taxa in East Anatolia, Turkey. *Economy Botany* 58(4), 697–704.

Öztürk, F. and Ölcücü, C. (2011) Ethnobotanical features of some plants in the district of Şemdinli (Hakkari-Turkey). *International Journal of Academy Research* 3(1), 120–125.

Ozturk, M. and Ozcelik, H. (1991) *Useful Plants of East Anatolia*. Siskav, Ankara, Turkey.

Ozturk, M., Guvensen, A., Aksoy, A. and Beyazgul, M. (2006a) An overview of the soils and sustainable land use in Turkiye. In *Proceedings of the Fifth International GAP Engineering Congress*, Şanlıurfa, Turkey, pp. 1548–1555.

Ozturk, M., Guvensen, A., Sakcalı, S. and Bahadir, H. (2006b) An overview of the land degradation problems in East Anatolia. In *Proceedings of the Fifth International GAP Engineering Congress*, Şanlıurfa, Turkey, pp. 1556–1561.

Ozturk, M., Gucel, S., Altundag, E. and Celik, S. (2011) Turkish Mediterranean medicinal plants in the face of climate change. In: Ahmad, A., Siddiqi, T.O. and Iqbal, M. (eds) *Medicinal Plants in Changing Environment*. Capital Publishing Company, New Delhi, India, pp. 50–71.

Ozturk, M., Gucel, S., Altundag, E., Mert, T., Gork, C., Gork, G. and Akcicek, E. (2012a) An overview of the medicinal plants of Turkey. In: Singh, R. (ed.) *Genetic Resources, Chromosome Engineering and Crop Improvement: Medicinal Plants*. CRC Press, LLC, Taylor & Francis, Boca Raton, Florida, USA, pp. 181–206.

Ozturk, M., Gucel, S., Celik, A., Altundag, E., Mert, T., Akcicek, E. and Celik, S. (2012b) *Myrtus communis* in phytotheraphy in the Mediterranean. In: Singh, R. (ed.) *Genetic Resources, Chromosome Engineering and Crop Improvement: Medicinal Plants*. CRC Press, LLC, Taylor & Francis, Boca Raton, Florida, USA, pp. 923–934.

Ozturk, M., Kebapçı, U., Gücel, S., Çetin, E. and Altundag̃, E. (2012c) Biodiversity and land degradation in the lower Euphrates subregion of Turkey. *Journal of Environmental Biology* 33, 311–323.

Ozturk, M., Altay, V., Gucel, S. and Aksoy, A. (2012d) Aegean grasslands as endangered ecosystems in Turkey. *Pakistan Journal of Botany* 44, 7–17.

Ozturk, M., Altay, V., Gucel, S. and Guvensen, A. (2014) Halophytes in the East Mediterranean – their medicinal and other economical values. In: Khan, M.A., Böer, B., Öztürk, M., Al Abdessalaam, T.Z., Clüsener-Godt, M. and Gul, B. (eds) *Sabkha Ecosystems, Vol. IV: Cash Crop Halophyte Biodiversity and Conservation*. Tasks for Vegetation Science Series. Springer, Dordrecht, Netherlands, pp. 247–272.

Özuslu, E. and Iskender, E. (2009) Geophytes of Sof Mountain (Gaziantep-Turkey). *Biological Diversity and Conservation* 2(2), 78–84.

Ozuslu, E. and Tel, A.Z. (2010) Karkamış Sulak alanının (Gaziantep-Türkiye) Biyolojik çeşitliliği. *Türk Bilimsel Derlemeler Dergisi* 3(2), 9–30.

Perevolotsky, A. and Seligman, N.G. (1998) Role of grazing in Mediterranean rangeland ecosystems. *BioScience* 48, 1007–1017.

Pleskanovskaja, S.A., Mamedova, G.A., Ozturk, M., Gucel, S. and Ashyraliyeva, M. (2011) An overview of ethnobotany of Turkmenistan and use of *Juniperus turcomanica* in phytotherapy. In: Singh, R.J. (ed.) *Genetic Resources, Chromosome Engineering and Crop Improvement: Medicinal Plants*. CRC Press, LLC, Taylor & Francis, Boca Raton, Florida, USA, pp. 207–220.

Plotkin, M.J. (2000) *Doğada İlaç Arayışı. Türkiye Çevre Vakfı Yayını, 164 (Çeviren-Belkıs Dişbudak, 2004).* Önder Press, Ankara, Turkey, p. 208.

Reynolds, J.F. and Smith, D.M.S. (2002) *Global Desertification: Do Humans Cause Deserts?* Vol. 88. Dahlem University Press, Berlin, Germany.

Şekeroğlu, N., Meraler, S.A. and Koca, U. (2011) Mardin'in Şifalı Bitkileri. Mardin Haber Günlük Yerel süreli yayın. *Yazı Dizisi* 8, 9–11.

Şığva, H.Ö. and Seçmen, Ö. (2009) Ethnobotanic survey of Işıklı (Çarpın), Dağdancık and Tokdemir in Gaziantep, Turkey. IUFS. *Journal of Biology* 68(1), 19–26.

Surmeli, B., Sakcali, S., Ozturk, M. and Serin, M. (2001) *Preliminary Studies on the Medicinal Plants of Kilis.* XIII. Plant Raw Materials Meeting. Marmara University, Istanbul, Turkey, pp. 211–220.

Tatlı, A., Akan, H., Tel, A.Z. and Kara, C. (2002) The flora of Upper Ceyhan Valley (Kahramanmaraş-Turkey). *Turkish Journal of Botany* 26, 259–275.

Tel, A.Z. and Tak, M. (2012) Perre (Pirin) Antik Şehri (Adıyaman) vejetasyonu. *Biyoloji Bilimleri Araştırma Dergisi* 5(2), 45–62.

Tel, A.Z., Tatlı, A. and Varol, Ö. (2010) Phytosociological structure of Nemrut Mountain (Adıyaman-Turkey). *Turkish Journal of Botany* 34, 417–434.

Tellawi, A.M. (2001) *Conservation and Sustainable Use of Biological Diversity in Jordan*. GCEP, Amman, Jordan.

Tukan, S.K., Takruri, H.R. and Al-Eisavi, D.M. (1998) The use of wild plant edible plants in the Jordanian diet. *International Journal of Food Sciences and Nutrition* 49, 225–235.

Tursun, N. (2001) Kahramanmaraş ilinde tıbbi amaçla kullanılan yabancı ot nitelikli bitkilerinin belirlenmesi. *Türkiye Herboloji Dergisi* 4(1), 30–38.

Varol, Ö. (2003) Flora of Başkonuş Mountain (Kahramanmaraş). *Turkish Journal of Botany* 27, 117–139.

Varol, Ö. and Tatlı, A. (2001) The vegetation of Çimen Mountain (Kahramanmaraş). *Turkish Journal of Botany* 25, 335–358.

Varol, Ö. and Tatlı, A. (2002) Phytosociological investigations of a *Pinus pinea* L. forest in the Eastern Mediterranean Region (K. Maraş-Turkey). *Plant Ecology* 158, 223–228.

Varol, Ö., Ketenoğlu, O., Bingöl, Ü., Geven, F. and Güney, K. (2006) A phytospciological study on the coniferous forests of Bas¸konuş MTS, Anti-Taurus, Turkey. *Acta Botanica Hungarica* 48(1–2), 195–211.

Vavilov, N.I. (1951) The Origin, Variation, Immunity and Breeding of Cultivated Plants. *The Chronica Botanica*, Cambridge, UK. pp. 293–350.

Wallace, J.S. (2000) Increasing agricultural water use efficiency to meet future food production. *Agriculture, Ecosystems & Environment* 82, 105–119.

Wang, T. (2003) Study on sandy desertification on China. 1. Definition of sandy desertification and its connotation. *Journal of Desert Research* 23, 477–482 (in Chinese).

Wang, X., Chen, H.F., Dong, Z. and Xia, D. (2005) Evolution of the southern Mu Us desert in north China over the past 50 years: an analysis using proxies of human activity and climate parameters. *Land Degradation & Development* 16, 351–366.

Wang, T., Chen, G.T., Zhao, H.L., Dong, Z.B., Zhang, X.Y., Zheng, X.J. and Wang, N.A. (2006a) Research progress on aeolian desertification process and controlling in North of China. *Journal of Desert Research* 26, 507–516 (in Chinese).

Wang, X.M., Chen, H.F. and Dong, Z.B. (2006b) The relative role of climatic and human factors in desertification in semiarid China. *Global Environmental Change* 16, 48–57.

Xu, D.Y., Kang, X.W., Zhuang, D.F. and Pan, J.J. (2008) Multi-scale quantitative assessment of the relative roles of climate change and human activities in desertification – A case study of the Ordos Plateau, China. *Journal of Arid Environments* 74, 498–507.

Yang, X., Zhang, K., Jia, B. and Ci, L. (2005) Desertification assessment in China: an overview. *Journal of Arid Environments* 63, 517–531.

Yapıcı, İ.Ü. and Saya, Ö. (2007) Kurtalan (Siirt) ilçesinin florasına katkılar. *Ot Sistematik Botanik Dergisi* 14(1), 47–60.

Yapıcı, İ.Ü., Hoşgören, H. and Saya, Ö. (2009) Kurtalan (Siirt) ilçesinin etnobotanik özellikleri. *Dicle Üniversitesi Ziya Gökalp Eğitim Fakültesi Dergisi* 12, 191–196.

Yates, C.J., Norton, D.A. and Hobbs, R.J. (2000) Grazing effects on plant cover, soil and microclimate in fragmented woodlands in south-western Australia: implications for restoration. *Austral Ecology* 25, 36–47.

Yiğit, N., Çolak, E., Ketenoğlu, O., Kurt, L., Sözen, M., Hamzaoğlu, E., Karataş, A. and Özkurt, Ş. (2002) *Çevresel Etki Değerlendirme 'ÇED'* [Environmental Impact Assessment 'EIA'], Kılavuz Paz. Tic. Ltd Şti., Ankara, Turkey (in Turkish).

Yıldırım, C. and Cansaran, A. (2010) A study on the floristical, phytosociological and phytoecological structure of Turkish *Astragalus angustifolius* ssp. *angustifolius* associations. *Kastamonu Üniversitesi Orman Fakültesi Dergisi* 10(2), 164–171.

Yıldız, B. (2001) Floristical characteristic of Berit Dağı (Kahramanmaraş). *Turkish Journal of Botany* 25, 63–102.

Zheng, Y.R., Xie, Z.X., Robert, C., Jiang, L.H. and Shimizu, H. (2006) Did climate drive ecosystem change and induce desertification in Otindag sandy land, China over the past 40 years? *Journal of Arid Environments* 64, 523–541.

Zohary, M. (1973) *Geobotanical Foundations of the Middle East*, Vols I–II. Gustav Fischer Verlag, Stuttgart, Germany.

Zohary, M. and Hopf, M. (1994) *Domestication of Plants in the Old World*. 2nd edn. Oxford Science Publication, Clarendon Press, Oxford, UK and New York, USA.

6 Observations on Some Ethnomedicinal Plants of Jharkhand

SANJEEV KUMAR*

Chief Conservator of Forests, Working Plans Jamshedpur, Jharkhand, India

Abstract

This chapter describes the indigenous knowledge associated with medicinal plants used by the tribal people of Jharkhand. During an ethnobotanical survey, 100 plants were recorded. Such information can be utilized to improve the economy of the tribes by organizing the systematic collection of medicinal plants and their parts, and establishing cottage industries based on them. Conservation of biodiversity is always linked with tradition, hence such a study helps in developing strategy in this direction.

6.1 Introduction

People living in and around forests have been dependent upon them for most of human history. In fact, the genesis of ethnobotany goes back to early humans, who started using plants for various purposes, including food, medicine, bark (as cloth), weapons to hunt animals, and other uses. Traditional knowledge evolved by trial and error. But ethnobotany in the modern era started only a century ago. In 1896, American botanist John Harshberger coined the term 'ethnobotany' to mean the study of plants used by primitive and aboriginal people. Since then, in broad terms, it has been defined as the relationship of humans with plants using their indigenous knowledge. Ethnobotany includes all types of relationships between people and plants, and plays an important role in the conservation of nature and traditional human cultures. Indigenous knowledge has become recognized worldwide, due to its intrinsic value and being a potential tool for conservation effort of biodiversity. Over the last century, ethnobotany has evolved into a scientific discipline that has resulted in the development of subdisciplines, such as ethnoforestry, ethnoagriculture, ethnosilviculture, ethnomedicine, etc. Ball, 1867; Archer, 1947; Bodding, 1925, Bodding, 1927; Gupta, 1964; Ghoshal, 1965; Bhomik and Chaudhuri, 1966; Gupta, 1969a; Gupta, 1969b; Jain and Tarafder, 1970; Chandra and Pandey, 1984; Chandra and Pandey, 1985; Ghosh and Sahu, 1986; Gupta, 1987; Goel and Mudgal, 1988; Jain, 1989; Hembrom, 1991; Hembrom, 1994; Chandra, 1995; Hembrom, 1996; Jain, 1996; Jha *et al.*, 1997; Varma *et al.*, 1999; Jain, 2003; Jaipuriar, 2003; Hembrom and Goel, 2005; Kumar, 2007; Kumar, 2014; Anonymous (1948–1972); Bangali *et al.*, 2009; Forest Working Plan, 1966–1967, 1985–1986; Haines, 1908; Haines, 1921–1925; Jain, 1991; Pal and Jain, 1998; Roy, 1915; Roychaudhary, 1957a, 1957b; Srinivasan, 1956 are examples of studies by ethnobotanists who have worked in Jharkhand.

6.2 Study Area

Jharkhand has an area of 79,714 km^2 or 30,778 $miles^2$. The name 'Jharkhand' means 'The Land of Forests'. The recorded forest area of the state is 23,605 km^2, which is 29.61% of the geographical area of the state. The state has five forest types: Moist Peninsular Low Level Sal; Dry Peninsular Sal; Northern Dry Mixed Deciduous Forest; Dry Deciduous Scrub; and Dry Bamboo Brakes. These belong to two major forest type groups: Tropical Moist Deciduous and Tropical

*E-mail: sanjeevkumar201@gmail.com

Dry Deciduous Forests. Examples of the important species which constitute the forests are Sal, Teak, Mahua, Asan, Dhaura, Gamhar, Kusum, Palas, Arjun and Chiraunji. The study was conducted in East Singhbhum (Jamshedpur), West Singhbhum (Chaibasa), and the Seraikela-Kharsawan districts of Singhbhum. Jharkhand has 32 tribal groups. These are the Asur, Baiga, Banjara, Bathudi, Bedia, Binjhia, Birhor, Birjia, Chero, Chick-Baraik, Gond, Gorait, Ho, Karmali, Kharia, Kharwar, Khond, Kisan, Kora, Korwa, Lohra, Mahli, Mal-Paharia, Munda, Oraon, Pahariya, Santal, Sauria-Paharia, Savar, Bhumij, Kol and Kanwar. They have very rich cultures. The people not only obtain all the resources necessary for their survival from their surrounding forests, but they also have a sacred cultural tie to these forests. This chapter outlines the variety of ways in which these people use plants for medicinal purposes.

6.3 Materials and Methods

The study was conducted in East Singhbhum (Jamshedpur), West Singhbhum (Chaibasa), and the Seraikela-Kharsawan districts of the Singhbhum area of Jharkhand. It was carried out during 2011–2012. The data and information presented in this chapter have been collected after discussion with local people, local traditional medicine practitioners and members of the Village Forest Protection Committees. This study consists of the enumeration of recorded medicinal plants, traditional knowledge about them, vernacular name, families, plant part used and mode of administration in the treatment of various ailments. These have also been verified through consultation with other literature and floras.

6.4 Results and Discussion

As a result of the study, various types of ethnobotanical relation have been recorded. Plants are used in various ways by local people. Accordingly they have been placed into one of 12 groups: 1. Edible products, 2. Grasses, 3. Mats, ropes and baskets, 4. Medicinal plants, 5. Oil seeds, 6. Tans and dyes 7. Fodder trees, 8. Gums and resins, 9. Fibres, 10. Animal products, 11. Cultural items, 12. Colours and minerals. This chapter focuses on important medicinal plants used by the rural and tribal population of Jharkhand.

6.4.1 Ethno medicinal plants

The use of folk medicines is attributed to local people's decades-long experience of and faith in herbal treatments. Local people depend mostly on medicinal plants because these plants are good and cheap sources of materials needed in primary health care. Locals use medicinal plants for the treatment of various ailments on the basis of indigenous knowledge passed from generation to generation. They also use medicinal plants on the advice of elders, such as herbalists and traditional practitioners. Medicinal plants come from all plant types: trees, shrubs, herbaceous flowering plants, ferns, mosses, lichens and fungi. Medicines are prepared and applied in various ways, depending upon the injury or ailment to be treated. An extract or infusion, where plant parts are immersed in hot water, is often made to treat internal ailments or as a general tonic. A poultice, where the plant is mashed or ground into a powder and applied to the body, is often used to treat external ailments. Colds and respiratory ailments are sometimes treated by inhaling steam from boiling water containing the medicinal plants.

The list below of 100 ethnomedicinal plants is a humble attempt: there are many more plants which are used in the treatment of various ailments. In Section 6.4.2 below, the scientific names of plants have been arranged in alphabetical order, along with the name of the plant family in parenthesis, followed by the local name (LN). Images of some important ethnomedicinal plants are shown in Fig. 6.1.

6.4.2 Enumeration of medicinal plants

1. *Abelmoschus crinitus* Wall. (Malvaceae) LN: Bankapas
Fruit and root are used in the treatment of cough and cold. Roots are also used in a tonic.
2. ***Abrus precatorius*** L. (Fabaceae) LN: Ghunchi, Ratti
Seeds and fruit syrup are used to treat stomach disorders.
3. ***Abutilon indicum*** (L) Sweet. (Malvaceae) LN: Kanghi
The plant is used as a febrifuge. Decoction of leaves is used to treat toothache. Bark is used as a diuretic.
4. ***Achyranthus aspera L.*** (Amaranthaceae) LN: Apamarg
All parts of the plant are used as medicine.
5. ***Acorus calamus*** *L.* (Araceae) LN: Bach

Fig. 6.1. Images of some important ethnomedicinal plants of Jharkhand. (A) *Costus specious*, (B) *Gloriosa superba*, (C) *Curculigo orchioides*, (D) *Rauvolfia serpentina*, (E) *Holarrhena antidysentrica.*

Rhizome and roots of the plant are used as medicine. Roots are used to treat cough and fever. Rhizome is used as a laxative and a diuretic.

6. ***Adina cordifolia*** **(Willd. ex Roxb.) Benth. & Hook. f. ex Brandis** (Rubiaceae) LN: Karam
Bark paste is used to kill worms.

7. ***Adhatoda zeylanica*** **Medic.** (Acanthaceae) LN: Basak
A decoction of leaves is used to treat coughs and colds.

8. ***Adiantum lunulatum*** **Burn.** (Adiantaceae) LN: Dodhari
A decoction of root is given to treat throat infections.

9. ***Aegle marmelos*** **(L.) Corr.** (Rutaceae) LN: Sinjo, Bel
Pulp of fruit is used as a laxative, or given as a cooling drink. Leaves are used to treat jaundice.

10. ***Aerva lanata*** **(L.)Juss.ex Schult.** (Amaranthaceae) LN: Lupora, Bahara
Powder of the plant is administered to treat dysentery. Also used to treat malarial fever.

11. ***Ageratum conyzoides*** L. (Asteraceae) LN: Puru
Roots are bitter but used as a digestive and appetizer. Paste of leaves mixed with Karanj (*Pongamia pinnata*) is applied to wounds and sores.

12. ***Ailanthus excelsa*** **Roxb.** (Simaroubaceae) LN: Ghor-karanj
The bark is powdered and eaten to cure rheumatism.

13. ***Alangium salvifolium*** **(L.f.) Wang** (Alangiaceae) LN: Dhela
The bark and roots are eaten to treat jaundice.

14. ***Albizzia lebbek*** **Benth.** (Mimosaceae) LN: Siris
Powdered root bark is used to strengthen gums and to cure itching. Leaves and fruits are boiled together, and the infusion is given in cases of anaemia.

15. ***Allemanda cathartica*** L. (Apocynaceae) LN: Pilajara
Flowers are used to treat jaundice.

*16. Alstonia scholaris (*L.) **R.Br.** (Apocynaceae) LN: Chhatni
Bark is used to treat gastrointestinal problems. It is also useful in treating malarial fever.
17. Amaranthus spinosus **L.** (Amaranthaceae) LN: Katali chaulai
The plant has laxative and cooling properties. Its paste is also applied to eczema.
18. Amaranthus tricolor L. (Amaranthaceae) LN: Lal sag
The root is used as an antidote to scorpion venom. It is also used to increase the appetite and treat indigestion.
19. Andrographis paniculata **(Burm. f.) Nees** (Acanthaceae) LN: Kalmegh
A decoction of the whole plant is given in cases of fever and malaria. It is also used to treat skin irritation.
20. Argemone mexicana L. (Papaveraceae) LN: Rengini kanta
Leaf juices are applied to cuts and wounds.
21. Aristolochia indica L. (Aristolochiaceae) LN: Godh
Used as an antidote to snakebite.
22. Artocarpus lakoocha **Roxb.** (Moraceae) LN: Barhar
The seed is ground and given to children as a purgative.
23. Asparagus racemosus **Willd.** (Asparagaceae) LN: Satavar
Tuberous root is taken for physical strength and to treat urinary diseases.
24. Bacopa monniera **(L.)** (Scrophulariaceae) LN: Brahmi
Used as a nerve tonic. Also used to treat anaemia, cough and fever.
25. Baliospermum montanum **(Willd.) Muell.** (Euphorbiaceae) LN: Danti
Seeds are used to treat gastric disorders.
26. Bauhinia purpurea L.(Caesalpiniaceae) LN: Koinar
Root, bark, flower are used to treat stomach disorders.
27. Bombax ceiba **L.** (Bombacaceae) LN: Semal
Young tap roots are used to treat dysentery. Paste of flowers and leaves are applied to skin problems.
28. Boswellia serrata **Roxb. ex Colebar** (Bursaraceae) LN: Salai
The juice of the bark is used for curing opthalmia. The gum is given in cases of lack of vigour and for acute bronchitis. Also used as an ointment for sores.
29. Bridelia retusa **Spreng.** (Euphorbiaceae) LN: Kaj
The gum is ground and mixed with water and given in cases of lack of vigour.
30. Bryonia palmata L. (Cucurbitaceae) LN: Shivlingi
Paste of leaves is used to treat headaches. A powder of seeds and root is given to aid conception in women.
31. Butea frondosa **Koen. ex Roxb.** (Fabaceae) LN: Palas or paras
Juice of leaves is mixed with milk curd and turmeric and is given to treat sunburn in children. Dried flowers (made into a paste with water and applied over the body) are used to prevent sunstroke.
32. Caesalpinia cristata L.(Caesalpiniaceae) LN: Kat karanj
Seeds are used to treat malaria.
33. Calotropis gigantia **R. Br.** (Asclepiadaceae) LN: Akwan, madar
Plant juice is used as a remedy for rheumatism. Flower is given as a remedy for coughs.
34. Canscora decussate **(Roxb.) Roem. & Schult.** (Gentianaceae) LN: Katchirata, Sankhahuli
Stem soaked in water is used as a tonic and blood purifier. Also used to treat fever.
35. Capparis zeylanica L. (Capparaceae) LN: Buru
A paste made of the plant is applied to blisters and boils.
36. Casearia elliptica **Willd.** (Flacourtiaceae) LN: Chorcho,
Bark is applied to treat dropsy and fever. Root powder is taken to treat stomach pain.
37. Cassia fistula L. (Caesalpiniaceae) LN: Amaltas, Bandarlati, Dhanras.
The fruit is used as a purgative.
38. Cassia tora L. (Caesalpinaceae) LN: Chakor
Seed paste is used to treat ringworm.
39. Catunaregam nutans **(DC) Tiruv** (Rubiaceae) LN: Boi-Bindi, Mauna
Roots are given to treat malarial fever.
40. Celastrus paniculata Willd. (Celastraceae) LN: Kujri
Seed oil used to treat skin disease and tuberculosis.
41. Centella asiatica **(L.) Urban.** (Apiaceae) LN: Ben sag
The leaves are used as a heart tonic and diuretic. Also used to treat jaundice.
42. Chlorophytum arundinaceum **Baker.** (Liliaceae) LN: Safed Musli
Tuber is used as a tonic and in the treatment of pain.
43. Clerodendrum serratum **(L.) Moon.** (Verbenaceae) LN: Bharangi
Roots are used to treat coughs and colds. Used to treat fever and eye diseases also.

*44. **Costus spesiosus*** **(Koen. ex Retz.) Sm.** (Zingiberaceae) LN: Kevuk kand (See Fig. 6.1A)
Rhizomes are eaten as a purgative and tonic.
*45. **Cordia myxa*** **Roxb. non L.** (Ehretiaceae) LN: Lasora
Seeds are used as an anthelmintic. Leaves are applied to treat ulcers and headaches.
*46. **Curculigo orchioides*** **Gaertn.** (Hypoxidaceae) LN: Kali musali, Turum (See Fig. 6.1C)
A paste of root is applied to wounds. It is also given to treat fever. Rhizome is also used in tonics.
*47. **Cryptolepis buchananii*** **Roem. & Schult.** (Periplocaceae) LN: Dudhilata
A preparation of stem is given to nursing mothers.
*48. **Cuscuta reflexa*** **Roxb.** (Cuscutaceae) LN: Akas Bel
Used to cure anaemia.
*49. **Cynodon dactylon*** **(L.) Pers.** (Poaceae) LN: Dub
Used to treat piles. Rhizomes are given to treat wounds and fever. For headache, it is ground with gram and made into a paste, which is applied to the head.
*50. **Dalbergia sissoo*** **Roxb.** (Fabaceae) LN: Sissu
The leaves are boiled, and the infusion is drunk in cases of lack of vigour.
*51. **Dendrophthoe falcate*** **(L.f.) Ettingh** (Loranthaceae) LN: Banda
Stem and leaf pastes are used to treat skin disease. Stem paste is also given to aid fertility in women.
*52. **Desmodium pulchellum*** **Backer** (Fabaceae) LN: Jeetedari
Roots are taken to treat stomach problems. Flowers are applied to teeth for dental problems.
*53. **Embelia ribes*** **Burm. f.** (Myrsinaceae) LN: Bhaherung
Fruit is used as a purgative.
*54. **Eclipta alba*** **Hassk.** (Bhringraj) LN: Asteraceae
Mainly used to treat jaundice.
*55. **Euphorbia hirta*** **L.** (Euphorbiaceae) LN: Dudhi
Juice of plant is given to treat dysentery. Root of plant is also used to treat bronchial infections and asthma.
*56. **Gloriosa superba*** **L.** (Liliaceae) LN: Kalihari (See Fig. 6.1B)
Root powder is given to treat rheumatic fever.
*57. **Hemidesmus indicus*** **R. Br.** (Perplocaceae) LN: Anantmula
Roots are used to treat skin diseases, fever, asthma and bronchitis. Also acts as blood purifier.
*58. **Holarrhena antidysentrica*** **Willd.** *(*Apocynaceae) LN: Kutaj (See Fig. 6.1E)
The bark is used to treat dysentery and diarrhoea.
*59. **Jatropha curcas*** **L.** (Euphorbiaceae) LN: Ratanjot
Fresh twig is used for brushing teeth.
*60. **Lagerstroemia parviflora*** **Roxb.** (Lythraceae) LN: Sidha
The bark is used as a cure for itching.
*61. **Lawsonia inermis*** **L.**(Lythraceae) LN: Mehandi
The seed is used as a medicine for lack of vigour. Bark is used as a sedative.
*62. **Limonia acidissima*** **L.** (Rutaceae) LN: Kat-bel
Fruit pulp is used to treat asthma. Also used to treat gum infection.
*63. **Litsea monopetala*** **(Roxb.) Pers.** (Lauraceae) LN: Meda
Bark and leaf are pounded and applied to sores.
*64. **Ludwigia octovalvis*** **(Jacq.) Raven** (Onagraceae). LN: Parsati
A decoction of root is taken to treat fever. A paste of the entire plant is used to treat skin diseases.
*65. **Mallotus philippensis*** **Muell.-Arg.** (Euphorbiaceae) LN: Rori
Fruits are used to treat intestinal worms.
*66. **Melia azadirachta*** **L.** *Azadirachta indica* A. Juss. (Meliaceae) LN: Neem
Leaves are used to treat ulcers and eczema. Dried flowers mixed with candy are taken improve appetite and promote digestion.
*67. **Melia azedarach*** **L.** (Meliaceae) LN: Bakain
The seed and bark are ground and given for lack of vigour and are also used as a poultice for curing pain.
*68. **Nyctanthes arbortristis*** **L.**(Oleaceae) LN: Harsingar
Decoction of leaves are taken to treat coughs and colds.
*69. **Ocimum sanctum*** **L.** (Lamiaceae) LN: Kali Tulsi
Used to treat fever, coughs and colds.
*70. **Oroxylon indicum*** **Vent.** (Bignoniaceae) LN: Sonachal
Root bark is given to treat dysentery and diarrhoea. Seeds are purgative.
*71. **Pholidota imbricata*** **(Roxb.) Lindl.** (Orchidaceae) LN: Patthar Kela
Used as a remedy for rheumatism.
*72. **Phyllantus emblica*** **L.**(Euphorbiaceae) LN: Aonla
The ripe fruit is eaten to whet the appetite. Is also used with other remedies to treat dysentery.
*73. **Pureria tuberose*** **DC.** (Leguminosae) LN: *Patal Kohara,*
The tuber and other parts are used to treat boils and other swellings.
*74. **Rauwolfia serpentina*** **Benth.ex Kurtz.** (Apocynaceae) LN: *Sarpgandha* (See Fig. 6.1D)

Root is mainly used as a sedative. It is also used as an antidote to snake venom.
75. Schleichera oleosa **(Lour) Oken** (Sapindaceae) LN: Kusum
The bark is astringent. As a rub with oil, it is used as a cure for itching.
76. Scoparia dulcis L. (Scrophulariaceae) LN: Mirchi
A decoction of the plant is used as a tonic. It is also used to test for pregnancy. The plant extract is taken orally to improve digestion.
77. Semecarpus anacardium **Linn. f.** (Anacardiaceae) LN: Bhelwa
Fried in Til oil (extracted from *Sesamum indicum* seeds), it is used as a remedy for rheumatism. The fruits are used to treat dyspepsia and piles.
78. Shorea robusta **Roxb. ex Gaertn. f.** (Dipterocarpaceae) LN: Sal
Used to cure dysentery. The leaves and bark are burnt and mixed with Til oil and used for healing burns.
79. Sida acuta **Burm.f.** (Malvaceae) LN: Bir muru baha
Fruits are used to treat boils.
80. Smilax ovalifolia **Roxb.** (Smilacaceae) LN: Atkir, Ramdatwan
Roots are used to treat ulcers and rheumatism.
81. Solanum surattense **Burm. f.** (Solanaceae) LN: Kantakari
Fruit is boiled in ghee and is given to treat coughs and colds.
82. Soymida febrifuga **(Roxb.) Juss.** (Meliaceae) LN: Ruhen, Rohan
The bark is used as an astringent. A decoction is given to treat rheumatic swellings.
83. Sphaeranthus indicus L. (Asteraceae) LN: Gorakhmundi
Leaves are used to treat malarial fever and headache.
84. Spondias pinnata **Kurz.** (Anacardiaceae) LN: Amra
The ripe fruit is used to improve the appetite.
85. Sterculia urens **Roxb.** (Sterculiaceae) LN: Keonji
The gum called 'katila' is given to treat cases of dysentery and lack of vigour.
86. Tamarindus indica L. (Caesalpiniaceae) LN: Tetar (Imli)
The fruit is used as a carminative and digestive. Poultice of leaves are applied to inflammatory swellings to relieve pain.
87. Tectona grandis **Linn. f.** (Verbenaceae) LN: Sagwan, Teak
Flowers are given to treat bronchitis and urinary discharges.
88. Terminalia balerica **Roxb.** (Combretaceae) LN: Bahera
The kernel of seed is ground and mixed with water, and given as a remedy for worms. The seed is taken as a purgative. Oil is used on the body to cool it.
89. Terminalia chebula **Retz.** (Combretaceae) LN: Harre
The fruit is ground and used as a purgative.
90. Terminalia tomentosa **Bedd.** (Combretaceae) LN: Asan
The bark is burnt and mixed with Til oil, and used as a cure for itching.
91. Thespesia lampas **(cav.) Dalz. & Gibs.** (Malvaceae) LN: Bankapasi
The roots and fruit are given to treat gonorrhoea. Also used to treat wounds and sores.
92. Thysanolaena maxima **(Roxb.) O. Ktze** (Poaceae) LN: Phul-jharu
Preparations from roots are given to treat fever and asthma.
93. Tinospora cordifolia **(Willd.) Miers ex Hook. f & Thoms.** (Menispermaceae) LN: Gurich
A paste of the plant is applied to treat rheumatism. Also used as a tonic.
94. Vernonia albicans DC. LN: Jhurjhuri, Barangom, Asteraceae
A decoction of the leaf is given to treat fever and dysentery.
95. Urginea indica **Kunth.** (Liliaceae) LN: Jangali pyaj
A mashed bulb is given to treat fever and pneumonia.
96. Viscum articulatum **Burm.f.** (Loranthaceae) LN: Kathkomjunga
Leaf paste is used to treat hemiplegia. Also used to treat fever.
97. Vitex negundo L. (Verbenaceae) LN: Sindwar
Paste of leaf is used as a germicide.
98. Woodfordia fruticosa **Kurtz.** (Lythraceae) LN: Dhawai
An infusion of the flower is given to treat diarrhoea. It is also dried and mixed with bhang (a preparation of cannabis) and is said to be cooling.
99. Zingiber officinale **Rose** (Zingiberaceae) LN: Ade, Adrak
The rhizome is given to treat coughs, colds and bronchitis.
100. Ziziphus xylopyra **(Retz.) Willd.** (Rhamnaceae) LN: Kat-ber, Karkat
Leaves are used to cure sore throat and dysentery.

6.4.3 Conclusion

This chapter reveals the close relationship between the tribes of Jharkhand and local plant diversity. Such ethnobotanical surveys among tribal populations can be utilized to improve the local economy by organizing the systematic collection and marketing of medicinal plants and establishing cottage industries to create value. Their indigenous knowledge can also be used in scientific research and in developing new concepts for the sustainable conservation of natural resources.

Acknowledgements

The author is grateful to members of the Village Forest Management and Protection Committees, local vaidyas (herbalists) and other villagers for sharing valuable facts.

References

Anonymous (1948–1972) *The Wealth of India: A Dictionary of Indian Raw Material and Natural Products*, Vols 1–9. CSIR, New Delhi, India.

Archer, W.G. (1947) The Santal treatment of Witchcraft. *Man in India* 27, 103–121.

Ball, V. (1867) On the jungle products used as articles of food by the inhabitants of the districts of Manbhoom and Hazaribagh. *Journal of the Asiatic Society* 2, 73–82.

Bangali B., Singh, J.P. and Jaiswal, A.K. (2009) *Lac in Jharkhand, A Status Report*. IINRG, Namkum, Ranchi, Jharkhand, India.

Bhomik, P.K. and Chaudhuri, B. (1966) Some aspects of magico-religious beliefs and the practices of the Mundas. *Folklore* 8(3), 100–107.

Bodding, P.O. (1925) Studies in Santal medicine and connected folklore – II, Santals and diseases. *Memoirs of Asiatic Society of Bengal* 10(1), 1–132.

Bodding, P.O. (1927) Studies in Santal medicine and connected folklore – II, Santals and diseases. *Memoirs of Asiatic Society of Bengal* 10(2), 134–426.

Chandra, K. (1995) An ethnobotanical study on some medicinal plants of district Palamau (Bihar). *Sachitra Ayurveda* 48, 311–314.

Chandra, K. and Pandey, B.N. (1984) Some folk medicine of Singhbhum. *Sachitra Ayurveda* 36(4), 253–257.

Chandra, K. and Pandey, B.N. (1985) Medicinal plants of Santal Pargana, district Dumka (Bihar) Part I. *Sachitra Ayurveda* 37(3), 307–314.

Forest Working Plan (1966–1967, 1985–1986). Department of Forests, Gumla, Jharkhand, India.

Ghosh, T.K. and Sahu, S.C. (1986) Plants used by Mundras of Chotanagpur for preparation of country liquor (Handia). *Mendel* 3(2), 79–82.

Ghosal, S. (1965) Tree in folk life. *Folklore* 6, 179–190.

Goel, A.K. and Mudgal, V. (1988) A survey of medicinal plants used by the tribals of Santal Pargana (Bihar). *Journal of Economic and Taxonomic Botany* 12(2), 329–335.

Gupta, S.P. (1964) An appraisal of the food habits and nutritional state among the Asur, the Korwa and the Sauria Pahariya of Chhotanagpur Plateau. *Bulletin of the Bihar Tribal Research Institute* 6(1), 127–179.

Gupta, S.P. (1969a) Adivasi mandia beverage. *Adivasi* 12(1–4), 123–126.

Gupta S.P. (1969b) *Tribes of Chotanagpur Plateau. An Ethnonutritional and Pharmacological Cross-section*. Bihar Tribal Welfare Research Institute, Ranchi, Jharkhand, India.

Gupta, S.P. (1987) Study of plants during ethnobiological research among tribals. In: Jain, S.K. (ed.) *A Manual of Ethnobotany*. Scientific Publishers, Jodhpur, India, pp. 12–22.

Haines, H.H. (1908) *A Forest Flora of Chotanagur including Gangpur and the Santal Parganas*. Bishen Singh Mahendrapal Singh, Dehradun, India.

Haines, H.H. (1921–1925) *The Botany of Bihar and Orissa*, Vols 1–3. Adlard & Son & West Newman Ltd, London, UK.

Hembrom, P.P. (1991) Tribal medicine in Chotanagpur and Santal Parganas of Bihar. *India Ethnobotany* 3, 97–99.

Hembrom, P.P. (1994) *Adivasi Oushadh* [Tribal Medicines] (in Hindi). Paharia sewa samiti, Satia. DST, Government of India, New Delhi, India.

Hembrom, P.P. (1996) Contact therapy practised by Mundas of Chotanagpur, Bihar. *Ethnobotany* 8(1–2), 36–39.

Hembrom, P.P. and Goel, A.K. (2005) Horopathy: ethnomedicine of Mundas. *Ethnobotany* 17, 89–95.

Jain, S.K. (1991) *Dictionary of Indian Folk Medicine and Ethnobotany*. Deep Publications, New Delhi, India.

Jain, S.K. and Tarafder, C.R. (1970) Medicinal plant-lore of the Santals (a revival of P.O. Bodding's work). *Economic Botany* 24(3), 241–278.

Jain S.P. (1989) Tribal remedies from Saranda Forest, Bihar, India – 1. *International Journal of Crude Drug Research* 27(1), 29–32.

Jain, S.P. (1996) Ethno-medico-botanical survey of Chaibasa, Singhbhum district, Bihar. *Journal of Economic and Taxonomic Botany* Add. Ser. 12, 403–407.

Jain, S.P. (2003) Comparative ethno-medico-botanical studies of tribes of Netarhat Plateau, Chhotanagpur and Singhbhum districts of Jharkhand, India. *Journal of Economic and Taxonomic Botany* 27(2), 295–299.

Jaipuriar, M.K. (2003) Threatened herbal heritage of tribal land Jharkhand. *Indian Forester* 129(1), 48–54.

Jha, P.K., Chaudhari, R.S. and Chaudhari, S.K. (1997) Studies of medicinal plants of Palamau (Bihar) – (IInd part). *Biojournal* 9(1–2), 21–38.

Kumar, S. (2007) NTFP and tribal life in Chotanagpur, Jharkhand, India. In: Das, A.P. and Pandey, A.K. (eds) *Advances in Ethnobotany*. Bishen Singh Mahendrapal Singh, Dehradun, India, pp. 253–265.

Kumar, S. (2014) *Ethnobotanical Studies in India*. Deep Publications, New Delhi, India.

Pal, D.C. and Jain, S.K. (1998) *Tribal Medicine*. Naya Prakashan, Calcutta, India.

Roy, S.C. (1915) *The Oraons of Chotanagpur*. 'Man in India Office', 18 Church Road, Ranchi, Jharkhand, India.

Roychaudhary, P.C. (1957a) Bihar District Gazetteers: Hazaribagh. Government of Bihar, Jharkhand, India.

Roychaudhary, P.C. (1957b) Bihar District Gazetteers: Singhbhum. Government of Bihar, Jharkhand, India.

Srinivasan, M.M. (1956) Host plants of Lac insects. *Indian Forester* 82(4), 180–189.

Varma, S.K., Sriwastawa, D.K. and Pandey, A.K. (1999) *Ethnobotany of Santhal Pargana*. Narendra, Delhi, India.

7 Plant Diversity: Envisioning Untold Nanofactories for Biogenic Synthesis of Nanoparticles and their Applications

SYED BAKER[1], K.S. KAVITHA[1], P. AZMATH[1], D. RAKSHITH[1], B.P. HARINI[3] AND S. SATISH[1&2*]

[1]Herbal Drug Technological Laboratory, Department of Studies in Microbiology, University of Mysore, Mysore, Karnataka, India; [2]Department of Plant Pathology, University of Georgia, Athens, Georgia, USA; [3]Department of Zoology, Bangalore University, Jnanabharathi Campus, Bangalore, Karnataka, India

Abstract

Plants, owing to their rich biodiversity, form almost unlimited natural resources on the planet. Exploitation of plants has been happening since ancient times, and has shaped the biosphere and its inhabitants. Plants have been serving mankind in various ways since life arose and man has been continuously using them for various requirements. A closer understanding of the association of humans with their surrounding flora is essential for better utilization of plants. The recent implementation of new technologies and improved scientific knowledge related to plant biology have been the focus of much attention, with the intention of bioprospecting and reformulating plants for diverse applications. One such area gaining importance is the evaluation of nanoparticle synthesis. The process of plant-mediated nanoparticles can be termed phytosynthesis of nanoparticles, wherein metal salts are efficiently reduced to materials at a nanoscale. At this size, the materials often exhibit significant and enhanced properties compared to its bulk material. In recent years there has been a significant interest in scientific communities towards plant mediated nanoparticles, especially noble metallic nanoparticles such as silver, gold, platinum and bimetallic ones. Owing to the fact that nanoparticles have been used in innumerable applications in various fields of sciences such as pharmaceuticals, agriculture, electronics, food packaging, biosensors, industrial spares components, textiles and anti-infective agents, nanoparticles have been envisioned as the particles of the century.

7.1 Introduction

Plants, because of their rich biodiversity, form one of the most abundant natural resources on the planet. Exploitation of plants can be traced back to ancient times, shaping the biosphere and its inhabitants since life arose, while humans have been continuously using them for various purposes. A closer understanding of the association of man with surrounding flora is essential for better utilization of plants (Satish *et al.*, 1999; Christaki *et al.*, 2012; Katiyar *et al.*, 2012). Contact with plants was perhaps the key to human development, and enabled us to surpass all other intelligent life on the planet. As scientific knowledge developed, interest focused on the importance of plants, which had so many applications in various disciplines: industry,

*E-mail: satish.micro@gmail.com and satish@uga.edu

medicine, agriculture, environment and many other beneficial uses (Fig. 7.1) (Somerville and Bonetta, 2001; Satish *et al.*, 2008; Kavitha and Satish, 2013).

Thus plants have directly or indirectly influenced human culture. Plants are an excellent source of various value-added products enhancing the welfare of humankind. Plants offer a repertoire of bioactive metabolites, which can provide opportunities in drug discovery (Pan *et al.*, 2010). Many drugs commonly used today for human health care are of plant origin. The World Health Organization (WHO) estimates that more than 80% of the world population currently use herbal medicine for some aspects of primary health care. Plant products also play an important role in the health care systems of the remaining 20% of the population (Cragg *et al.*, 1999).

The recent implementation of new technologies and improved scientific knowledge related to plant biology have drawn unprecedented attention to bioprospecting and reformulating plants for multiple applications. One such area gaining importance is the evaluation of plants for the synthesis of nanoparticles. The process of creating plant-mediated nanoparticles can be termed as phytosynthesis of nanoparticles, wherein the metal salts are efficiently reduced to materials at a nanoscale (Archana and Kasote, 2013; Arunachalam *et al.*, 2013; Baker *et al.*, 2013a). At this size, the material often exhibits significant and enhanced properties compared to its bulk material. In recent years, there has been a significant interest by scientific communities in plant-mediated nanoparticles, especially noble metallic ones such as silver, gold, platinum and bimetallic (Baishya *et al.*, 2012). Owing to the fact that these nanoparticles have been employed in innumerable applications in various fields of science such as the pharmaceutical industry, agriculture, electronics, food packaging, biosensors, industrial spare components, textiles and anti-infective agents in medicine (Fig. 7.2). These applications of nanoparticles have been envisioned as the particles of the century (Baker *et al.*, 2013a, b).

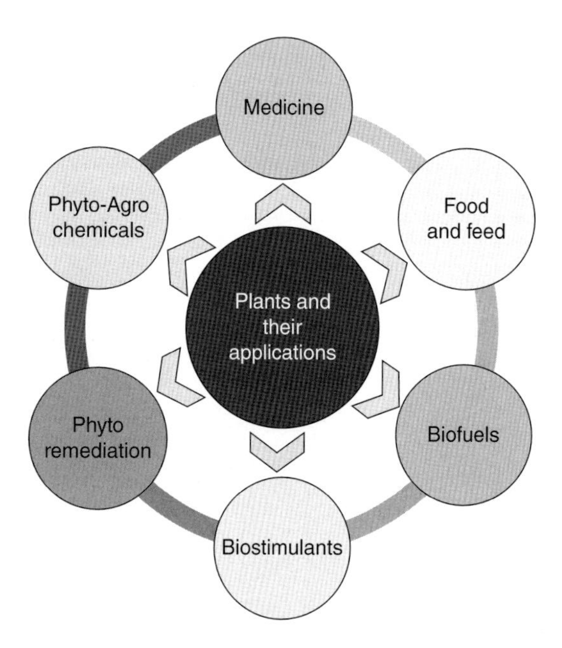

Fig. 7.1. Plants and their applications.

7.2 Nanoparticles and Their History

Nanoparticles are ultrafine particles which are said to have at least one dimension, i.e. depth, length or width, at a nanoscale. At this nano-size, a collection of atoms are bonded together with a structural radius bearing less than 100 nm (Baker *et al.*, 2013a). Nanoparticles have emerged as the most inspiring materials in the technical world (Fendler, 1996; Rotello, 2004; Kreuter, 2007; Edwards *et al.*, 2007). The concept of nanoparticles was first expounded by Paul Ehrlich to Karl Maria during Weber's opera *Der Freischütz* (Greiling, 1954; Kreuter, 2007). In this opera, so-called 'Freikugeln' (literally 'free bullets'), made by summoning the spirit of the devil, can hit the target even if marksmen fail to aim properly (Kreuter, 2007). This inspired Paul Ehrlich to develop the concept of targeted delivery and what he termed 'magic bullets', a breakthrough in drug therapy (Kreuter, 2007). In 1950, tremendous progress was made in the world of medicine with the development of biopharmaceutics and pharmacokinetics to control drug release, which became a major focus of attention. This was further developed by miniaturized delivery systems by a group of pioneers led by Paul Speiser at the Swiss Federal Institute of Technology in Zurich, to investigate and develop polyacrylic beads and develop microcapsulation for oral administration (Prathna *et al.*, 2010). In 1960, the first nanoparticles were designed for drug delivery of vaccines (Prathna *et al.*, 2010; Tyagi *et al.*, 2012). In recent years, due to their unique properties nanoparticles are being leveraged for an array of applications, which are being constantly explored

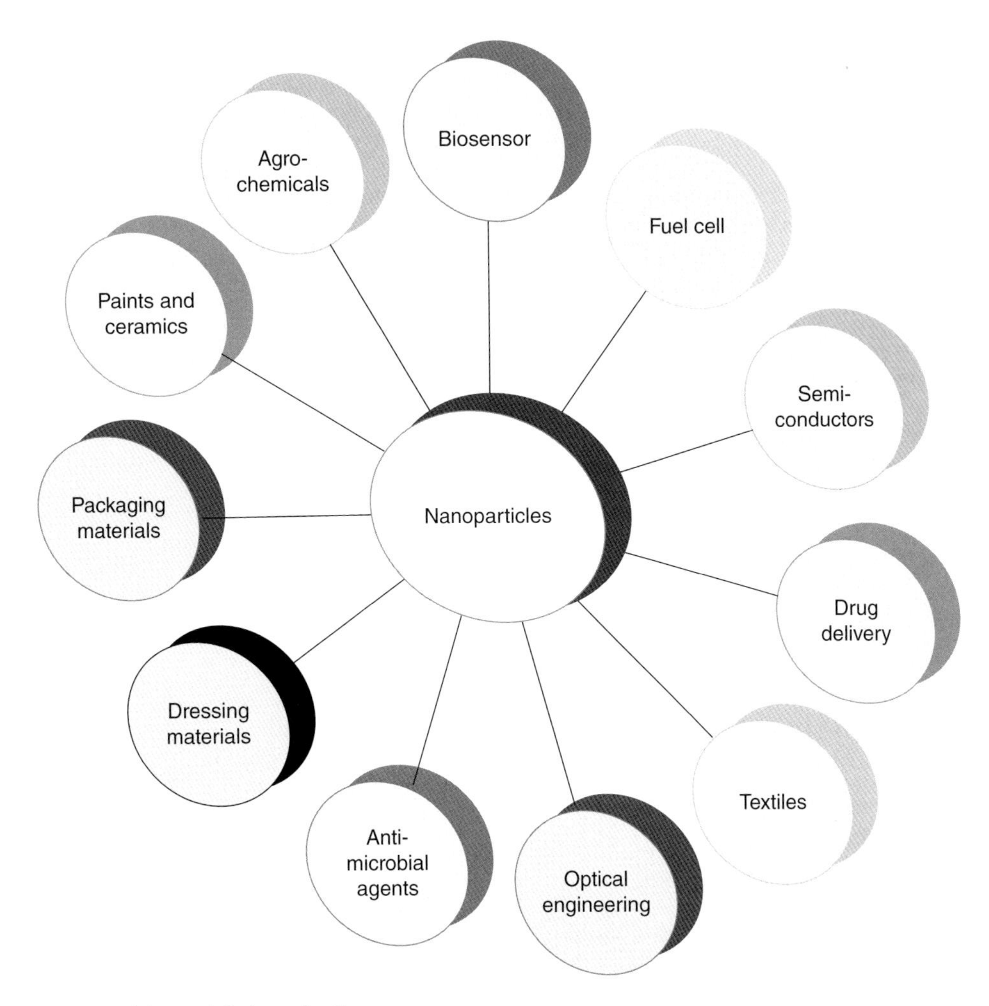

Fig. 7.2. Nanoparticles and their applications.

(Khanna and Speiser, 1969; Khanna *et al.*, 1970; Merkle and Speiser, 1973; Kreuter, 2007).

7.3 Nanoparticles: prime properties

Particles at nanoscale tend to differ from the bulk material; often they are structured in a hierarchical manner with intricately arranged atoms and molecules that ultimately makes up a particle at a nanoscale. This produces a myriad of different functional properties, such as an increase in the ratio of surface area to volume, which exhibits realm-predominate quantum effects, with dominance of the atoms acting on the surface of the particle compared to the atoms present inside the particles. These factors greatly influence the chemical reactivity, mechanical, optical, electric and magnetic properties of materials (Buzea *et al.*, 2007; Saman *et al.*, 2013). Because of this, the number of atoms present at the surface of a nanoparticle tend to be greater than for large particles, thus resulting in high surface area and particle number per unit mass (Buzea *et al.*, 2007; Saman *et al.*, 2013). As the surface area increases, the reactivity of the particles is also enhanced, as the atoms at this size have less chance of binding to their neighbour than atoms of larger particles. This reduces the binding energy and affects the melting point, quantum confinement and reflects the catalytic ability (Buzea *et al.*, 2007; Saman *et al.*, 2013). Another fascinating property of the nanoparticle is its dimensions below the critical wavelength of visible light, which renders it transparent (Roduner, 2006; Buzea *et al.*, 2007; Morris, 2008).

7.4 Synthesis of Nanoparticles

The synthesis of nanoparticles can be achieved by various conventional methods, with size-dependent properties, which determine their applications. Based on the type of protocols, these conventional methods are divided into 'top-down processes' and 'bottom-up processes' (Baker *et al.*, 2013b). Whereas the top-down processes break down the bulk materials into the desired size and shape, in the bottom-up approach there is a molecular organization followed by formation of nuclei and crystal growth, which aggregates to form nanoparticles of the desired size and shape (Fig. 7.3). Some of the most popular techniques of the bottom-up approach include the template-based process, sol-gel fabrication, inert gas condensation, solvothermal reaction, pyrolysis, flame spray synthesis and biological-based synthesis. Similarly, top-down processes include attrition or milling, polishing, lithographic techniques, laser-beam processing film deposition and growth. But the majority of these conventional techniques are limited by such factors as expense, toxicity risks to health and environmental contaminants. Hence there is a need for an eco-friendly, facile and biogenic-based process in nanomaterial synthesis. In this context, the evaluation of biological entities has gained attention across the scientific community to cope with the demand and alternatives for widely used conventional methods in synthesis of nanoparticles (Shankar *et al.*, 2004; Busnaina, 2006).

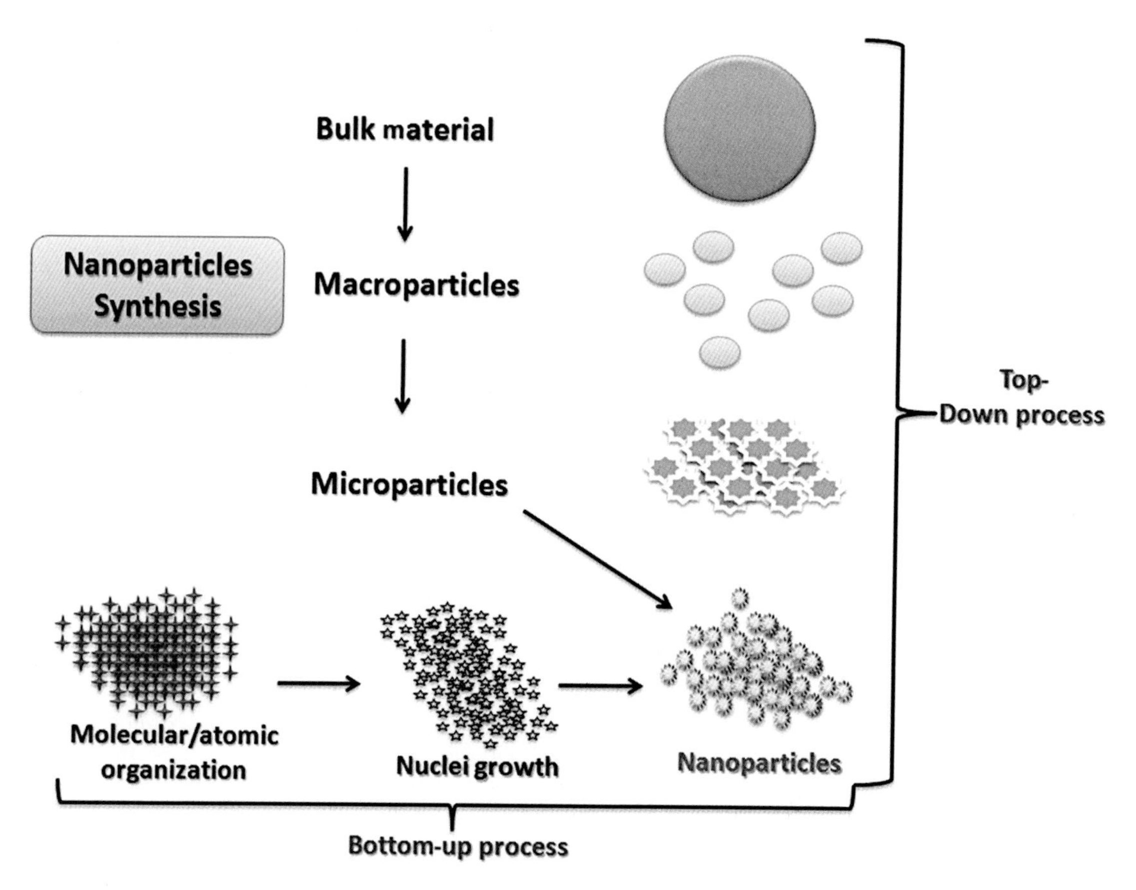

Fig 7.3. Top-down and bottom-up processes for nanoparticle synthesis.

7.5 Concept of Biological Entities in Nanoparticle Synthesis

The concept of biological entities in synthesis of nanoparticles was based on the biosorption of metals by Gram-negative and Gram-positive bacteria. The sorption of metals was described using Freundlich adsorption isotherm. Further, industrially important metals such as silver, cadmium and copper were evaluated to reveal the concept of biogenic principle synthesis. Silver was precipitated as discrete colloidal aggregates at the cell surface and occasionally in the cytoplasm (Singh *et al.*, 2011). With the help of electron microscopy, results indicated these aggregates, which suggested that bacterial cells had the capability of binding metal (Beveridge and Murray, 1976; Doyle *et al.*, 1980; Beveridge and Fyfe, 1985; Mullen *et al.*, 1989). Ten years later, these silver-based crystals were identified as nanoparticles which exhibited certain unique properties, by using *Pseudomonas stutzeri* AG259, a bacterial strain that was originally isolated from a silver mine (Nair and Pradeep, 2002; Zhang *et al.*, 2005). These ideas are being honed as scientists exploit better tools and develop new hypotheses, revealing even deeper ties between the biologist and nanotechnologist and inspiring interdisciplinary research outcomes resulting in envisioning biological entities in a facile route to nanoparticle synthesis. Recent development and the implementation of new technologies is leading to a 'nano-revolution', which seems to have drawn unprecedented attention. A biogenic route to synthesize nanoparticles using a biological system has created a paradigm shift towards greener principle protocols, which has generated interest among researchers in fabricating nanoparticles using eco-friendly and safe methods. Biological systems may vary from simple prokaryotic bacteria to multi-cellular eukaryotic organisms. These bio-systems act as nanofactories with the ability to produce the desired shape, and can influence the topography of inorganic nanomaterials in defining size. Thus the bio-mimetic approach has great potential for the successful mastering of technical advances, innovative approaches, and understanding the holistic mechanism underlying the biosynthesized nanoparticles (Pugazhenthiran *et al.*, 2009; Narayanan and Sakthivel, 2010; Baker and Satish, 2012; Baker *et al.*, 2013b).

7.6 Plant-mediated Synthesis of Nanoparticles

Plants and their products have exhibited tremendous potential in the reduction of metal salts to synthesize nanoparticles. This unique property of plants can be traced to their biological roles in phytoremediation and phytomining; as plants encounter various environmental pollutants during the course of their growth, they develop tolerant mechanisms against these pollutants. In most scenarios, these pollutants can be heavy metals, which leads to the conversion of these metals into nanoparticles (Ponarulselvam *et al.*, 2012; Patra and Baek, 2014). The concept of plant-mediated nanoparticles emerged with the work carried out by Gardea-Torresdey and his group (Gardea-Torresdey *et al.*, 1999) which demonstrated the synthesis of silver and gold nanoparticles within live plants. In *Medicago sativa*, the plant's growing medium was enriched with metal salts of silver and gold. The plant's uptake of silver and gold ions produced nanoparticles. The formation of nanoparticles was confirmed using hyphenated techniques such as X-ray absorption spectroscopy and transmission electron microscopy. Since then, huge interest in biogenic synthesis has led to an increase in the scientific study of nanoparticle synthesis, with different sizes and shapes seen to bear valuable applications (Ponarulselvam *et al.*, 2012). The use of plants in the synthesis of nanoparticles is more valuable than other biological entities, as this plant-based process is a single-step protocol. Using microbes is laborious and during the course of time they may mutate and lose their effectiveness. On the other hand, the plant-based process can overcome the limitations of conventional methods, and forms an eco-friendly, reliable source for the synthesis of nanoparticles. The plant-based process not only synthesizes nanoparticles, but phytomolecules also stabilize the nanoparticles by capping around them, as has been acknowledged in various scientific studies (Herlekar *et al.*, 2014). Interesting facets of plant-based nanoparticles include the one-step process of synthesis and their size-tuning properties by varying different parameters (Baker *et al.*, 2013b; Kavitha *et al.*, 2013).

7.7 Protocols and Different Parameters Influencing Plant-mediated Nanoparticles

Protocols for the synthesis of nanoparticles using plants are very simple and compatible with other biological systems. According to the scientific literature, in many cases, the whole plant is evaluated for production of nanoparticles. The plant is grown in the solid medium enriched with metal salts and there is an uptake of metal ions; nanoparticles are formed inside the plants, creating biofactories (Gardea-Torresdey *et al.*, 1999). In some cases, the plant materials are shade-dried and ground to a fine powder, which is then treated with metal salts at different concentrations, resulting in the reduction of metal salts to form nanoparticles. A known amount of plant materials are first processed to obtain a liquid extract, which is then treated with the metal salts at different concentrations to create the desired nanoparticles. For efficient nanoparticle synthesis, optimized variables play a vital role: for instance, pH, temperature, concentration of metal salts, reaction time, and ratio of metal salts to plant material. According to Makarov *et al.* (2014), variation of pH in the reaction mixture can disturb the binding of the phytochemicals to metal salts, which affects the shape and size of nanoparticles. In the study conducted by Vanaja *et al.* (2014), the production of silver nanoparticles was at its maximum at pH 7.8, compared to acidic pH 5.8 and alkaline pH 8.8. The study concluded that under an acidic pH level, reduction of the metal salt was suppressed due to the lower availability of functional groups. Similarly, Gardea-Torresdey *et al.* (1999) observed the pH-dependent property of alfalfa biomass. When combined with the metal salt gold chloroaurate there was an increased binding affinity of metal salts when pH was increased from 3.0 to 7.0. A similar study (Sathishkumar *et al.*, 2009) reports that nanocrystalline silver was synthesized using *Cinnamon zeylanicum*, and a pH variation between 1 and 11 resulted in maximum synthesis at the lower pH, with well-dispersed nanoparticles, compared to the higher pH. The study also concluded that at higher pH, large numbers of functional groups were available for binding the metal ions, resulting in a large number of spherical nanoparticles. Finally, the study highlights the important role of pH in nanoparticle synthesis (Sathishkumar *et al.*, 2009).

Similarly, temperature is one of the factors influencing the synthesis protocol. Most plant-mediated synthesis of nanoparticles is carried out at ambient and room temperature (Prasad, 2014). But in recent years, studies conducted on temperatures have revealed that synthesis of nanoparticles was at its maximum level at elevated temperatures, i.e. 50 °C and above. This is due to the fact that rates of nucleation and crystal formation increase as the temperature increases, preventing the secondary reduction process (Baker *et al.*, 2015). According to the study conducted by Vanaja *et al.* (2013a), the synthesis of silver nanoparticles using *Cissus quadrangularis* extract, during which the temperature was varied from 30 to 70°C, resulted in maximum synthesis at 70°C, thus confirming temperature as a critical factor in achieving synthesis. A similar result was obtained when using olive leaf extract, which was subjected to temperatures from 30 to 90°C (Khalil *et al.*, 2014). Gold nanoparticle synthesis was achieved with extract of *Momordica charantia* at 100°C to form stable nanoparticles. The study highlights the role of increased temperature versus incubation time for nanoparticle formation (Pandey *et al.*, 2012). Sporadic studies in the literature so far confirm that 1 mM metal salt concentration is optimal to achieve maximum synthesis of various nanoparticles. According to Banerjee *et al.* (2014) leaf extracts of available Indian plants showed reduction of metal salts at a concentration of 1 mM when treated with 5% aqueous extract. Similarly, Vanaja *et al.* (2013b) conducted a kinetic study on the formation of silver nanoparticles using leaf extract of *Coleus aromaticus*, which resulted in maximum nanoparticle synthesis at a concentration of 1 mM silver nitrate. Studies also confirm that maintaining the ratio of metal salt with different concentrations of plant extracts is highly essential to achieve maximum synthesis. The majority of these reports reveal that a 9:1 ratio (metal salts: plant extract) influences rapid and maximum synthesis of nanoparticles under optimized parameters. The optimization studies mainly depend on the type of plant species selected and their part evaluated for the synthesis of nanoparticles.

7.8 Different Plant Parts Evaluated for the Synthesis of Nanoparticles

Plants are known to possess an array of phytocomponents bearing high significant values (Bhatia *et al.*, 2011; Lu *et al.*, 2014). For nanoparticle synthesis, the selection of plants and their components influences the synthesis of nanoparticles (El-Shahaby *et al.*, 2013). Almost every plant part has been evaluated for the synthesis of nanoparticles, including roots, stem, leaves, bark, fruits, seeds, latex and vegetables. The phytocomponents in each plant part vary greatly from species to species. Thus, wise selection of plant species and their components is important in reducing the metal salts and synthesizing nanoparticles.

According to Tho *et al.* (2013), seed extract of *Nelumbo nucifera* was capable of synthesizing silver nanoparticles which were stable for two months. Fourier transform infrared (FTIR) analysis revealed the amines, alkynes and aromatic functional groups responsible for the synthesis of silver nanoparticles. The study conducted by Jagtap and Bapat (2013) reported synthesis of silver nanoparticles using *Artocarpus heterophyllus* seed powder extract. This contains Jacalin as a major component, and its functional group was identified using FTIR, which revealed the role of amino acids and amide group I in the reduction of metal salts and stabilization of nanoparticles. Similarly, synthesis of silver nanoparticles was achieved by treating aqueous seed extract of *Jatropha curcas* with silver nitrate at different concentrations. The FTIR analysis confirms the role of primary and secondary amide, along with fatty acids, in mediating the synthesis (Bar *et al.*, 2009). According to the study conducted by Dubey *et al.* (2010), both silver and gold nanoparticles were synthesized using a leaf extract of *Rosa rugosa* by treating the metal salts at different concentrations and under optimized parameters to achieve polydispersed nanoparticles. FTIR analysis revealed the presence of amine, hydroxyl and carbonyl groups, responsible for nanoparticle synthesis and their stabilization. Similarly, gold chloroaurate was reduced using leaf extract of *Cassia auriculata* to produce gold nanoparticles within ten minutes of reaction time. FTIR analysis revealed that amide and hydroxyl groups were responsible for the reduction to form nanoparticles (Kumar *et al.*, 2011). Aqueous root extract of *Morinda citrifolia*, when challenged with 1 mM of gold chloroaurate, resulted in the synthesis of polydispersed cubic gold nanoparticles. FTIR analysis revealed that protein was responsible for mediating and stabilizing gold nanoparticles (Suman *et al.*, 2014). Latex of *Jatropha gossypifolia*, when evaluated and challenged with metal salts, was capable of synthesizing gold nanoparticles with a size range of 30–50 nm (Borase *et al.*, 2014). When stem and leaf extract of *Piper nigrum* was treated with silver nitrate, this resulted in the synthesis of silver nanoparticles with a size range of 7–50 nm and 9–30 nm for leaf and stem extract respectively. FTIR revealed that amide groups present in the stem extract were responsible for mediating and stabilizing nanoparticles (Paulkumar *et al.*, 2014). The fruit extract of *Vitis vinifera* synthesized silver nanoparticles with a size range of 1–100 nm (Gnanajobitha *et al.*, 2013). Silver and gold nanoparticles were synthesized using *Averrhoa bilimbi* L. FTIR analysis revealed that the presence of primary alcohols, tertiary amides and phenols were responsible for the synthesis of nanoparticles (Isaac *et al.*, 2013). The whole plant *Cacumen Platycladi* was sun-dried and ground into powder and evaluated to synthesize platinum nanoparticles by combining sodium tetrachloroplatinate with the extract under optimized conditions. FTIR analysis revealed that the presence of reducing sugars and flavonoids mediated the synthesis and stabilized nanoparticles (Zheng *et al.*, 2013).

7.9 Mechanism for the Synthesis of Plant-mediated Nanoparticles

The mechanism behind the plant-mediated synthesis of nanoparticles is yet to be completely elucidated. But scientific studies postulate that phytochemicals, including reducing sugars, non-reducing sugars, alkaloids, polyphenols, terpenoids and proteins in plants are responsible for mediating the synthesis. The overall mechanism relies on activation and binding of biomolecules to the metal salts, followed by redox reaction to form nucleation and crystal growth, resulting in Ostwald ripening to form small crystals which transfer their mass and aggregates to form nanostructured particles as shown in Fig. 7.4 (Baker *et al.*, 2015). Most of the studies also confirm that these diverse classes of phytochemicals not only reduce the metal ions, but also stabilize the synthesized nanoparticles and prevent secondary aggregation to form microparticles, which is one of the important factors in maintaining the properties of nanoparticles.

For instance, the study conducted by Kesharwani *et al.* (2009) reported that phytochemicals such as alkaloids, amino acids, sugars and alcohols were responsible for synthesis and stabilization of nanoparticles. Whereas, according to Ahmad *et al.* (2010), reduction of silver ions to form nanoparticles was due to the presence of H+ ions, NAD+ and ascorbic acid in the extract of *Desmodium trifolium*. Similarly, ferulic acid and chlorogenic acid in *Ananas comosus* efficiently act as reducing agents, which are oxidized by silver nitrate to form silver nanoparticles (Ahmad and Sharma, 2012). According to Vilchis-Nestora *et al.* (2008), the

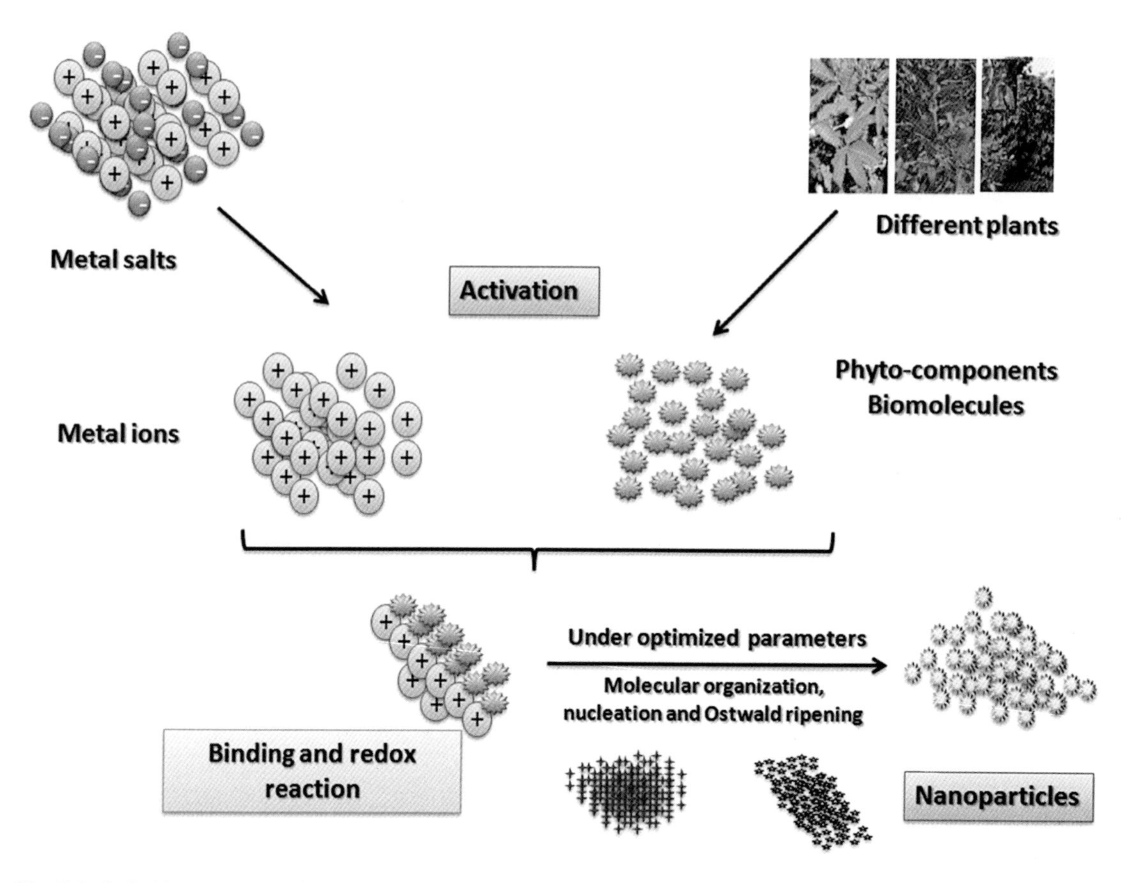

Fig. 7. 4. Probable mechanism for synthesis of nanoparticles.

biomolecules caffeine and theophylline present in the *Camellia sinensis* extract were responsible for production and stabilization of silver and gold nanoparticles.

7. 10 Characterization of Nanoparticles

Nanoparticle synthesis is one of the prime aspects of nanotechnology, due to their unique properties, which have been valuable in various applications in the last few years (Baker and Satish, 2012). It is worth noting that nanoparticle technology will remain in its infancy without an accurate and well-defined characterization of nanoparticles (Anumolu and Pease, 2012). The characterization of nanoparticles ensures and determines the application, dependent on the nanoparticle properties. There are various hyphenated techniques (Fig. 7.5) for biophysical characterization of nanoparticles, which elucidate size, shape, chemical composition, etc. For metallic nanoparticles, such as silver and gold, initial preliminary confirmation can be achieved by visual observation of a change in colour (Kavitha *et al.*, 2013).

Characterization of nanoparticles is essential to reveal their physico-chemical properties, which are important when considering their applications. There are various tools that help to reveal the properties of nanoparticles: for instance, preliminary confirmation of nanoparticles can be achieved using the UV-visible spectroscopic technique. This is based on the well-known phenomenon of light absorption, wherein nanoparticles have different optical properties compared to the bulk material, which are sensitive due to surface plasmon resonance, i.e. collective oscillations of conduction band electrons in response to electromagnetic waves (Lopez-Serrano *et al.*, 2014; Patra and Baek, 2014). Nanoparticles exhibit specific absorbance bands in their spectra when the light passes through the sample, and the results are based on

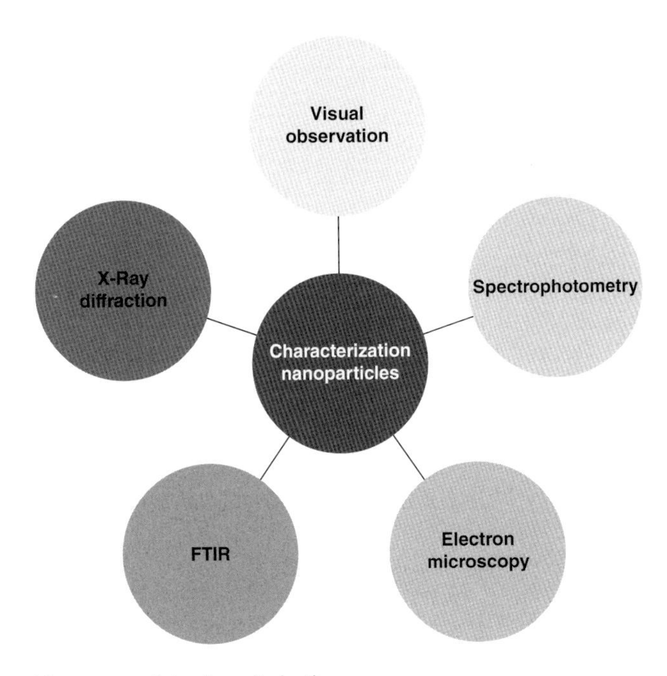

Fig. 7.5. Techniques used for nanoparticle characterization.

the band energy gap, which evaluates the dispersion and aggregation of the nanoparticles (Daniel and Astruc, 2004; Roy and Fendler, 2004; Norman *et al.*, 2005; Van Dijk *et al.*, 2006; Patra and Baek, 2014; Lopez-Serrano *et al.*, 2014). Microscopic techniques, such as scanning electron microscopy, transmission electron microscopy and atomic force microscopy, are employed to reveal the size and shape of nanoparticles (Baker *et al.*, 2013b; Kavitha *et al.*, 2013). Accurate measurement of particle size, morphology, surface texture and roughness is an important pre-requisite for application of nanoparticles in science and technology (Pletikapic *et al.*, 2012; Lopez-Serrano *et al.*, 2014). Particles smaller than 1.0 μm cannot be observed with light, due to diffraction effects restricting the resolution of optical microscopy. Hence, to observe an object at nanoscale, higher resolution is required: electromagnetic radiation of shorter wavelengths must be employed. The basic principle is that an electron beam interacts with a sample, many measurable signals are generated, and electrons can be transmitted, back-scattered and diffracted to give an image. Electron microscopic techniques have been routinely employed to achieve magnifications and disclose details with a resolution of up to about 0.1 nm (Herrera and Sakulchaicharoen, 2009). Further chemical composition and nature of the nanoparticles can be attenuated with FTIR spectroscopy, which measures infrared intensity versus wavelength of light, resulting in a prediction of those biomolecules associated with nanoparticles. In FTIR, interaction between matter and electromagnetic fields in the Infrared region takes place, during which electromagnetic waves couple with molecular vibrations. This results in the excitation of molecules to a higher vibrational state by absorbing infrared radiation, causing vibrational changes and rotational status of the molecules (Kalisz *et al.*, 2008; Davis *et al.*, 2010). X-ray diffraction (XRD) is generally a highly reliable analytical technique that is widely used to predict the crystalline nature of particles. Complete elucidation on surface properties and coatings, crystallographic structure or elemental composition of material can be achieved with the XRD principle (Lopez-Serrano *et al.*, 2014). XRD is also used to evaluate single crystal or polycrystalline materials. The X-ray diffraction technique uses monochromatic X-rays generated by a cathode ray tube, which passes through the samples. Deviations in these rays are collectively studied using Bragg's law, which in turn determines the characteristics of the sample by generating diffraction peaks (Lopez-Serrano *et al.*, 2014).

Table 7.1. Plant-mediated nanoparticles and their applications.

Sl no.	Plants	Nanoparticles	Applications	References
1	*Abelmoschus moschatus* L.	Silver	Antimicrobial	Rane *et al.* (2014)
2	*Abutilon indicum* (L.) Sweet.	Silver	Antibacterial	Prathap *et al.* (2014)
3	*Aegle marmelos* L.	Gold	Urinary catheter infection	Arunachalam *et al.* (2013)
4	*Aloe*	Zinc oxide	Antibacterial, antifungal	Gunalan *et al.* (2012)
5	*Aloe vera* (L.) Burm.f.	Indium oxide	Optical	Maensiri *et al.* (2008)
6	*Alternanthera sesslis* (L.) DC.	Silver	Antioxidant, antimicrobial	Niraimathi *et al.* (2012)
7	*Ananas comosus* (L.) Merr.	Gold	Antibacterial	Bindhu and Umadevi (2014)
8	*Andrographis paniculata* Nees.	Silver	Hepatocurative, anti-parasitic	Suriyakalaa *et al.* (2013) Panneerselvam *et al.* (2011)
9	*Arabidopsis*	Palladium	Catalytic	Parker *et al.* (2014)
10	*Artemisia annua* L.	Palladium	Catalytic	Edayadulla *et al.* (2015)
11	*Bacopa monnieri* L.	Platinum	Parkinson's disease, antioxidant	Nellore *et al.* (2013)
12	*Barbated Skullcup*	Gold	Electrochemistry	Wang *et al.* (2009)
13	*Boswellia serrate* Triana & Planch.	Silver	Antibacterial	Kora *et al.* (2012)
14	*Camellia sinensis* (L.) Kuntze.	Zinc oxide	Antimicrobial	Senthilkumar and Sivakumar (2014)
15	*Carica papaya* L.	Copper oxide	Catalytic	Sankar *et al.* (2014)
16	*Cassia auriculata* L.	Gold	Antidiabetic	Kumar *et al.* (2011)
17	*Cassia fistula* L.	Gold	Antidiabetic	Daisy and Saipriya (2012)
18	*Coriandrum Sativum* L.	Silver	Optical	Sathyavathi *et al.* (2010)
19	*Costus ingneus*	Silver	Antidiabetic	Sataraddi and Nandibewoor (2012)
20	*Citrullus colocynthis* (L.) Schrad.	Silver	Antioxidant, anticancer	Satyavani *et al.* (2011)
21	*Citrus limon* (L.) Burm.f.	Silver	Treatment of fabrics and antifungal	Vankar and Shukla (2012)
22	*Cyamopsis tetragonoloba* (L.) Taub.	Silver	Biosensor to determine ammonia	Pandey *et al.* (2012)
23	*Dioscorea batatas*	Silver	Antimicrobial	Nagajyothi and Lee (2011)
24	*Dioscorea bulbifera*	Silver	Antimicrobial	Ghosh *et al.* (2012a)
25	*Eucalyptus chapmaniana*	Silver	Antimicrobial, cytotoxic	Sulaiman *et al.* (2013)
26	*Euphorbia prostrata*	Silver and titanium dioxide	Antileishmanial	Zahir *et al.* (2014)
27	*Euphorbia prostrate*	Silver	Pesticidal activity	Zahir *et al.* (2012)
28	Flaxseed	Silver	Antimicrobial activity	Sharbidre and Kasote (2013)
29	*Gelcemium semervirens*	Silver	Cytotoxicity	Das *et al.* (2012)
30	*Geraniol*	Silver	Cytotoxicity	Safaepour *et al.* (2009)
31	*Gloriosa superba* L.	Copper oxide	Antibacterial	Naika *et al.* (2015)
32	*Gnidia glauca*	Gold	Chemocatalytic	Ghosh *et al.* (2012b)
33	*Grapes*	Gold	Antimicrobial, anticancer	Lokina and Narayanan (2013)
34	*Gum karaya*	Copper oxide	Antibacterial	Padil and Černík (2013)
35	*Hydrastis Canadensis* L.	Silver	Cytotoxicity	Das *et al.* (2012)
36	*Lemongrass*	Cerium oxide	Optical	Maensiri *et al.* (2007)
37	*Magnolia kobus*	Copper	Antibacterial	Lee *et al.* (2013)
38	*Mentha piperita*	Silver and gold	Antibacterial	Ali *et al.* (2011)
39	*Moringa oleifera*	Silver	Antimicrobial	Prasad and Elumalai (2011)

Continued

Table 7.1. Continued.

Sl no.	Plants	Nanoparticles	Applications	References
40	*Musa paradisiaca* L.	Silver	Acaricidal, larvicidal	Jayaseelan *et al.* (2012)
41	*Nerium oleander*	Copper	Antibacterial	Gopinath *et al.* (2014)
42	*Ocimum sanctum* (Holy basil)	Platinum	Water electrolysis	Soundarrajan *et al.* (2012)
43	Olive	Silver	Antibacterial	Khalil *et al.* (2014)
44	*Opuntia dilenii haw*	Nickel aluminate	Catalytic	Chinnadurai *et al.* (2014)
45	*Parthenium hysterophorus* L.	Silver	Larvicidal	Naba *et al.* (2014)
46	*Parthenium hysterophorus* L.	Zinc oxide	Antibacterial	Rajiv *et al.* (2014)
47	*Pulicaria glutinosa*	Palladium	Catalytic	Khan *et al.* (2014)
48	*Rumex dentatus*	Silver	Antimicrobial	El-Shahaby *et al.* (2013)
49	*Sesbania*	Gold	Catalytic	Sharma *et al.* (2007)
50	*Solanum trilobatum*	Titanium dioxide	Larvicidal	Rajakumar *et al.* (2014)
51	*Sterculia foetida* L.	Silver	Antimicrobial	Singh and Vidyasagar (2014)
52	*Tabernaemontana divaricate*	Copper oxide	Antibacterial	Sivaraj *et al.* (2014)
53	Tea leaf and coffee	Copper oxide	Antibacterial	Sutradhar *et al.* (2014)
54	*Tridax procumbens*	Copper oxide	Antibacterial	Gopalakrishnan *et al.* (2012)
55	*Trigonella foenum graecum*	Silver	Catalytic	Aromal and Philip (2012)
56	*Triticum aestivum*	Silver	Catalytic and toxicology	Waghmode *et al.* (2013)
57	Watermelon rind	Palladium	Catalytic	Lakshmipathy *et al.* (2014)
58	*Withania somnifera* (L.) Dunal.	Silver	Antimicrobial	Nagati *et al.* (2012)
59	*Zingiber officinale* L.	Copper	Antimicrobial	Subhankari and Nayak (2013)
60	*Ziziphus mauritiana* Lam.	Chitosan	Immune-restorative effect	Bhatia *et al.* (2011)

7.10.1 Some of the prime applications of phytosynthesized nanoparticles

According to the study conducted by Rajiv *et al.* (2014), highly stable zinc oxide nanoparticle synthesis was achieved using *Parthenium* leaf extract. Synthesized nanoparticles were spherical, with an approximate size of 27 nm. The synthesized nanoparticles exhibited antibacterial properties at 25 and 50 μg/ml concentration, with the highest activity observed for *Klebsiella pneumonia*, followed by *Bacillus subtilis*. The study concluded that zinc oxide nanoparticles act as a significant antibacterial agent. Similarly, synthesis of copper oxide nanoparticles was achieved with aqueous extract of *Gloriosa superba* L. The size of these nanoparticles was between 5 to 10 nm, with the majority being spherical. The synthesized nanoparticles exhibited antibacterial properties against selected pathogenic bacteria. The study highlighted that copper oxide nanoparticles exhibited significant antibacterial property (Naika *et al.*, 2015). Similarly, copper oxide nanoparticles were synthesized using gum extract of Karaya, a natural non-toxic hydrocolloid. The synthesized nanoparticles exhibited superior antibacterial qualities against important pathogenic bacteria. The study highlights that the size depends on the antibacterial property of the copper oxide nanoparticles and confirms that nanoparticles with a size of 4.8 nm showed the maximum zone of inhibition, compared to nanoparticles of 7.8 nm (Padil and Černík, 2013). Anti-*Helicobacter pylori* activity was confirmed with monodispersed silver nanoparticles synthesized using methanolic extract of *Solanum xanthocarpum* berry. Morphological characteristics revealed nanoparticles of 10 nm in size, which displayed activity against 24 clinical isolates. The inhibition effect of nanoparticles was validated with standard silver nitrate and four different antibiotics. The study concluded that synthesized nanoparticles showed significant activity when compared to silver nitrate and metronidazole (Amin *et al.*, 2012). Cytotoxic activity of biologically synthesized silver nanoparticles was achieved using plant extracts of *Pleargonium graveonlens* and *Azadirachta indica*. The synthesized silver nanoparticles were of uniform size and shape, ranging from 1 to 10 nm with average size of 6 nm. Silver nanoparticles conferred against cancer cell lines Fibrosarcoma-Wehi 164. The analysis revealed the direct dose–response relationship of the nanoparticles and cell lines. As the concentration of silver nanoparticles increased, the cytotoxicity also increased. It was observed that silver nanoparticles concentration at 1 µg/ml inhibited the cell line's growth up to 30%, at 5 µg/ml resulted in inhibition up to 60% and, finally, the concentration necessary to produce 50% cell death was 2.6 µg/ml, thus confirming the cytotoxic activity of the nanoparticles (Safaepour *et al.*, 2009). Silver nanoparticles were synthesized using the aqueous extracts of *Euphorbia prostrata*. The synthesized nanoparticles were evaluated for pesticidal activity against *Sitophilus oryzae*. The pesticidal activity was conducted for 14 days at different concentrations; LD50 values were 213.32, 24.90 and 44.69 mg/kg^{-1}. Similarly, D90 values were 1648.08, 2675.13 and 168.28 mg/kg^{-1}. The study concluded that the biologically synthesized silver nanoparticles had the potential to control *Sitophilus oryzae* (Zahir *et al.*, 2012). Further studies have also confirmed antiplasmodial activity of silver nanoparticles synthesized from the leaf extract of *Andrographis paniculata* against *Plasmodium falciparum*. The parasitic activity was analysed by IC50 values of silver nanoparticles. The results showed that silver nanoparticles at a concentration of 25 µg/ml exhibited 26±0.2% and at 100 µg/ml were 83±0.5% parasitic activity (Panneerselvam *et al.*, 2011). Similarly, antifungal activity of silver nanoparticles synthesized using a leaf extract of *Citrus limon* was observed against the important fungal pathogens *Fusarium oxysporum* and *Alternaria brassicicola*. The results indicated that synthesized nanoparticles suppressed the growth of fungal pathogens (Vankar and Shukla, 2012). Gold nanoparticles were synthesized using the bark and stem extract of *Cassia fistula*. The synthesized gold nanoparticles exhibited a hypoglycaemic effect, with a significant decrease in serum biochemistry parameters. Further increase in body weight, total protein levels and high-density lipoprotein in rats with streptozotocin-induced diabetes was observed with gold nanoparticles treatment. Thus, the study highlights the antidiabetic properties of biologically synthesized nanoparticles (Daisy and Saipriya, 2012). Similarly, various biological activities were achieved with silver nanoparticles synthesized using *Costus igneus* extract. The synthesized silver nanoparticles showed significant antidiabetic activity, and maximum amylase inhibition up to 87% at 100 µg/ml (Sataraddi and Nandibewoor, 2012). Larvicidal activity of biologically synthesized silver nanoparticles was achieved by using aqueous root extract of *Parthenium hysterophorus* (*P. hysterophorus*). Silver nanoparticles were evaluated against *Culex quinquefasciatus*. The results exhibited maximum efficacy up to 60.18% against the larvae by using silver nanoparticles and the study forms the first report on the mosquito larvicidal activity (Mondal *et al.*, 2014).

Triticum aestivum (khapali ghahu) extract was investigated for synthesis of silver nanoparticles within 15 minutes, which resulted in polydispersed silver nanoparticles. The synthesized nanoparticles displayed excellent catalytic activity at 200 µl by reducing hydrogen peroxide. The study also investigated the preliminary toxicological activity against earthworm with 1500 ppm concentration (Waghmode *et al.*, 2013). According to Sharma *et al.* (2007), growth of Sesbania seedlings in chloroaurate solution resulted in the accumulation of gold with the formation of stable gold nanoparticles in plant tissues. The study demonstrated a significant degree of efficiency for biotransforming Au(III) into Au(0) by plant tissues. The synthesized gold nanoparticles exhibited catalytic function with nanoparticle-rich biomass by reducing aqueous 4-nitrophenol. Gold nanoparticles were synthesized using Barbated Skullcup, a dried whole plant of *Scutellaria barbata* D. Don, as the reducing agent to mediate gold chloroaurate into nanoparticles. The synthesized nanoparticles were coated and modified onto the glassy carbon electrode (GCE), and exhibited enhanced electro transmission activity between electrode and the *p*-nitrophenol (Wang *et al.*, 2009). A plant extract of *Opuntia dilenii haw* was evaluated to synthesize nickel aluminate by a microwave combustion method. The synthesized nanoparticles were confirmed using hyphenated techniques and formation of pure nickel aluminate phase was confirmed by FTIR and formation of NiAl2O4 nanoparticles was confirmed by HR-SEM and HR-TEM. The study was compared with the conventional method, which resulted in an eco-friendly and facile route towards synthesis of nanoparticles, leading to improved performance in the selective oxidation of benzyl alcohol to benzaldehyde (Ragupathia *et al.*, 2014). Biosynthesis of palladium nanoparticles was achieved using *Artemisia annua* leaf extracts. The synthesized nanoparticles varied between 20 to 30 nm. Thermogravimetric analysis exhibited that bioactive molecules were capped onto the surface of synthesized nanoparticles. These nanoparticles displayed significant catalytic activity for the synthesis of several (indolyl) indolin-2-ones in high yield in aqueous medium. The study also revealed that the catalyst was recycled five times without any significant loss of catalytic activity (Edayadulla *et al.*, 2015). Similarly, use of *Ocimum sanctum* as a reducing agent resulted in the synthesis of platinum nanoparticles. FTIR analysis revealed ascorbic acid, gallic acid, terpenoids, certain proteins and amino acids as major components responsible for synthesis of nanoparticles. The study also displayed significant catalytic activity of platinum nanoparticles (Soundarrajan *et al.*, 2012). Phytosynthesis of copper oxide nanoparticles was achieved using *Centella asiatica* (L.) extract at room temperature. The synthesized nanoparticles exhibited photocatalytic degradation of methyl orange to its leuco form in the aqueous medium without any addition of reducing agents, thus forming an economic process compared to other conventional methods (Devi and Singh, 2014). Copper oxide (CuO) nanoparticles were synthesized using *Carica papaya* leaf extract. The FTIR analysis revealed the presence of bioactive compounds responsible for the synthesis of nanoparticles. The results also showed the effective Coomassie brilliant blue R-250 dye degrading activity beneath sunlight (Sankar *et al.*, 2014). Gold nanoparticles were synthesized using *Gnidia glauca* flower extract in a process based on green principles. The synthesized nanoparticles were spherical in shape and approximately 10 nm. The synthesized nanoparticles exhibited significant chemocatalytic properties in the reduction of 4-nitrophenol to 4-aminophenol by $NaBH4$ (sodium borohydride) in aqueous phase (Ghosh *et al.*, 2012a).

7.11 Future Prospects for Plant-mediated Nanoparticles

Phytosynthesized nanoparticles have contributed significantly towards the growing knowledge of plant biology in recent years. Even though increasing numbers of scientific papers are reporting diverse plant species in reduction of metal salts to synthesize nanoparticles with desired properties, many more studies should be carried out in this area to facilitate the commoditization of nanoparticle-based products. Different types of nanoparticles are being employed in innumerable applications, but few reports are available on plant-based nanoparticles in the agricultural sector, an area ripe for research, since these plant-mediated nanoparticles may be very good alternatives to pesticides. The unique properties of nanoparticles make them potential candidates for use in multi-applications, especially when compared to chemical ingredients, which are mostly synthetic and contain substances that restrict their use in biomedical applications. Plant-based nanoparticles come into their own here, owing to the fact that these phytosynthesized nanoparticles are capped with biomolecules of plant origin, which overcome the side effects of conventionally synthesized nanoparticles. Plant-based nanoparticles can also play a role in developing dressing materials for enhanced wound healing properties. Other possibilities are a disinfectant with plant-based nanoparticles for sanitation, and the development of cosmetics based on green principles, containing nanoparticles encapsulated with therapeutic agents. Further plant-mediated nanoparticles can also be used in developing packaging and storage materials, such as food containers for longer preservation. The plant-derived metallic nanoparticles have a projected impact on the diagnosis and treatment of various diseases with controlled side effects. At the same time, the selection of plant species becomes an important parameter, as harvesting of endangered species can pose a risk and imbalance in plant diversity. In future decades, plant-based nanoparticles can be tailored with other biomolecules, which will open a new avenue in enhancing the properties of nanoparticles. New multifunctional roles of these biologically synthesized nanoparticles could include the development of new potent antimicrobial agents against multidrug resistant microorganisms.

7.12 Conclusion

This chapter summarizes the significance of plant diversity and plants' emerging role in biogenic nanofactories, which forms one of the most facile processes of synthesizing nanoparticles as an alternative approach to the most commonly used conventional methods, which are bound with various limitations. The chapter highlights some of the reported scientific literature discussing synthesis of myriad nanoparticles with multiple applications. Overall, the chapter envisions the future for plant-mediated nanoparticles, with pointers towards formulating new strategies in the coming decades by citing the important parameters for rapid and efficient synthesis. The compilation of information in the present chapter can offer insight to a wider audience of scientific communities working on phytosynthesis of nanoparticles.

References

Ahmad, N. and Sharma, S. (2012) Green synthesis of silver nanoparticles using extracts of *Ananas comosus*. *Green Sustainable Chemistry* 2, 141–147.

Ahmad, N., Sharma, S., Singh, V., Shamsi, S., Fatma, A. and Mehta, B.R. (2010) Biosynthesis of silver nanoparticles from *Desmodium triflorum*: a novel approach towards weed utilization. *Biotechnology Research International* 2011, 1–8.

Ali, D.M., Thajuddin, N., Jeganathan, K. and Gunasekaran, M. (2011) Plant extract mediated synthesis of silver and gold nanoparticles and its antibacterial activity against clinically isolated pathogens. *Colloids and Surfaces B: Biointerfaces* 85(2), 360–365.

Amin, M., Anwar, F., Janjua, M.R., Iqbal, M.A. and Rashid, U. (2012) Green synthesis of silver nanoparticles through reduction with *Solanum xanthocarpum* L. berry extract: characterization, antimicrobial and urease inhibitory activities against *Helicobacter pylori*. *International Journal of Molecular Science* 13(8), 9923–9941.

Anumolu, A. and Pease, L.F. (2012) Rapid nanoparticle characterization. In: Hashim, A.A. (ed.) *The Delivery of Nanoparticles*. InTech, Rijeka, Croatia. DOI: 10.5772/35031. Available at: www.issp.ac.ru/ebooks/books/open/The_Delivery_of_Nanoparticles.pdf (accessed 26 May 2016).

Archana, A.S. and Kasote, D.M. (2013) Synthesis of silver nanoparticles using flaxseed hydroalcoholic extract and its antimicrobial activity. *Current Biotechnology* 2, 1–5.

Aromal, S.A. and Philip, D. (2012) Green synthesis of gold nanoparticles using *Trigonella foenum-graecum* and its size-dependent catalytic activity. *Spectrochimica Acta Part A Molecular and Biomolecular Spectroscopy* 97, 1–5.

Arunachalam, K., Annamalai, S.K., Aarrthy, M., Arunachalam Raghavendra, R. and Kennedy, S. (2013) One step green synthesis of phytochemicals mediated gold nanoparticles from *Aegle marmelos* for the prevention of urinary catheter infection. *International Journal of Pharmacology and Pharmaceuticals Science* 6(1), 700–706.

Baishya, D., Sharma, N. and Bora, R. (2012) Green synthesis of silver nanoparticle using *Bryophyllum pinnatum* (Lam.) and monitoring their antibacterial activities. *Archives of Applied Science and Research* 4(5), 2098–2104.

Baker, S. and Satish, S. (2012) Endophytes: towards a vision in synthesis of nanoparticle for future therapeutic agents. *International Journal of Bio-inorganic Hybrid Nanomaterials* 1, 67–77.

Baker, S., Harini, B.P., Rakshith, D. and Satish, S. (2013a) Marine microbes: invisible Nanofactories. *Journal of Pharmacy Research* 6, 383–388.

Baker, S., Rakshith, D., Kavitha, K.S., Santosh, P., Kavitha, H.U. and Yashavantha Rao, H.C. (2013b) Plants: emerging as nanofactories towards facile route in synthesis of nanoparticles. *BioImpacts* 3, 111–117.

Baker, S., Kumar, K.M., Santosh, P., Rakshith, D. and Satish, S. (2015) Extracellular synthesis of silver nanoparticles by novel *Pseudomonas veronii* AS 41G inhabiting *Annona squamosa* L. and their bactericidal activity. *Spectrochimica Acta Part A: Molecular and Biomolecular Spectroscopy* 136, 1434–1440.

Banerjee, P., Satapathy, M., Mukhopahayay, A. and Das, P. (2014) Leaf extract mediated green synthesis of silver nanoparticles from widely available Indian plants: synthesis, characterization, antimicrobial property and toxicity analysis. *Bioresources and Bioprocessing* 1(3), 1–10.

Bar, H., Bhui, D.K., Sahoo, G.P., Sarkar, P., Pyne, S. and Misra, A. (2009) Green synthesis of silver nanoparticles using seed extract of *Jatropha curcas*. *Colloids and Surfaces A: Physicochemical and Engineering Aspects* 348, 212–216.

Beveridge, T.J. and Fyfe, W.S. (1985) Metal fixation by bacterial cell walls. *Canadian Journal of Earth Science* 22, 1893–1898.

Beveridge, T.J. and Murray, R.G.E. (1976) Uptake and retention of metals by cell walls of *Bacillus subtilis*. *Journal of Bacteriology* 127, 1502–1518.

Bhatia, A., Shard, P., Chopra, D. and Mishra, T. (2011) Chitosan nanoparticles as carrier of Immunorestoratory plant extract: synthesis, characterization and immunorestoratory efficacy. *International Journal of Drug Delivery* 3, 381–385.

Bindhu, M.R. and Umadevi, M. (2014) Antibacterial activities of green synthesized gold nanoparticles. *Materials Letters* 120, 122–125.

Borase, H.P., Patil, C.D., Salunkhe, R.B., Suryawanshi, R.K., Salunke, B.K. and Patil, S.V. (2014) Phytolatex synthesized gold nanoparticles as novel agent to enhance sun protection factor of commercial sunscreens. *International Journal of Cosmetics Science* 36(6), 571–578.

Busnaina, A. (2006) *Nano-manufacturing Handbook*. CRC Press, Taylor & Francis Group, Boca Raton, Florida, USA.

Buzea, C., Pacheco, I. and Robbie, K. (2007) Nanomaterials and nanoparticles: sources and toxicity. *Biointerphases* 2(4), 17–71.

Chinnadurai, R., Vijaya, J.J., Kennedy, L.J. (2014) Preparation, characterization and catalytic properties of nickel aluminate nanoparticles: A comparison between conventional and microwave method. *Journal of Saudi Chemical Society* 58. DOI: 10.1016/j.jscs.2014.01.00635.

Christaki, E., Bonos, E., Giannenas, I. and Florou-Paneri, P. (2012) Aromatic plants as a source of bioactive compounds. *Agriculture* 2, 228–243.

Cragg, G.M., Boyd, M.R., Khanna, R., Kneller, R., Mays, T.D. and Mazan, K.D., Newman, D.J. and Sausville, E.A. (1999) International collaboration in drug discovery

and development: the NCI experience. *Pure and Applied Chemistry* 71(9), 1619–1633.
Daisy, P. and Saipriya, K. (2012) Biochemical analysis of *Cassia fistula* aqueous extract and phytochemically synthesized gold nanoparticles as hypoglycemic treatment for diabetes mellitus. *International Journal of Nanotechnology* 7, 1189–1202.
Daniel, M.-C. and Astruc, D. (2004) Gold nanoparticles: assembly, supramolecular chemistry, quantum-size related properties and applications toward biology, catalysis, and nanotechnology. *Chemical Reviews* 104, 293–346.
Das, R.K., Gogoi, N., Babu, P.J., Sharma, P., Mahanta, C. and Bora, U. (2012) The synthesis of gold nanoparticles using *Amaranthus spinosus* leaf extract and study of their optical properties. *Advances in Material Physics and Chemistry* 2, 275–281.
Davis, R., Irudayaraj, J., Reuhs, B.L. and Mauer, L.J. (2010) Detection of E. coli O157:H7 from ground beef using Fourier transform infrared (FT-IR) spectroscopy and chemometrics. *Journal of Food Science* 75, M340–M346.
Devi, H.S. and Singh, T.D. (2014) Synthesis of copper oxide nanoparticles by a novel method and its application in the degradation of methyl orange. *Advances in Electronic and Electric Engineering* 4(1), 83–88.
Doyle, R.J., Matthews, T.H. and Streips, U.N. (1980) Chemical basis for selectivity of metal ions by the *Bacillus subtilis* cell wall. *Journal of Bacteriology* 143(1), 471–480.
Dubey, S.P., Lahtinen, M. and Sillanpa, M. (2010) Green synthesis and characterizations of silver and gold nanoparticles using leaf extract of *Rosa rugosa. Colloids and Surfaces A: Physicochemical and Engineering Aspects* 364, 34–41.
Edayadulla, N., Basavegowda, N. and Lee, Y.R. (2015) Green synthesis and characterization of palladium nanoparticles and their catalytic performance for the efficient synthesis of biologically interesting di(indolyl) indolin-2-ones. *Indian Journal of Engineering and Chemistry* 21, 1365–1372.
Edwards, E.W., Wang, D.Y. and Mohwald, H. (2007) Hierarchical organization of colloidal particles: from colloidal crystallization to supraparticle chemistry. *Macromolecular Chemistry and Physics* 208(5), 439–445.
El-Shahaby, O., El-Zayat, M., Salih, E., El-Sherbiny, I.M. and Reicha, F.M. (2013) Evaluation of antimicrobial activity of water Infusion plant-mediated silver nanoparticles. *Journal of Nanomedicine and Nanotechology* 4(4), 1–7.
Fendler, J.H. (1996) Self-assembled nanostructured materials. *Chemistry of Materials* 8(8), 1616–1624.
Gardea-Torresdey, J.L., Tiemann, K.J., Gamez, G., Dokken, K. and Pingitore, N.E. (1999) Recovery of gold (III) by alfalfa biomass and binding characterization using X-ray microfluorescence. *Advance Environmental Research* 3, 83–93.
Ghosh, S., Patil, S., Ahire, M., Kitture, R., Gurav, D.D. and Jabgunde, A.M. (2012a) Gnidia glauca flower extract mediated synthesis of gold nanoparticles and evaluation of its chemocatalytic potential. *Journal of Nanobiotechnology* 10(17). DOI: 10.1186/1477-3155-10-17.
Ghosh, S., Patil, S., Ahire, M., Kitture, R., Kale, S. and Pardesi, K. (2012b) Synthesis of silver nanoparticles using *Dioscorea bulbifera* tuber extract and evaluation of its synergistic potential in combination with antimicrobial agents. *International Journal of Nanomedicine* 7, 483–496.
Gnanajobitha, G., Paulkumar, K., Vanaja, M., Rajeshkumar, S., Malarkodi, C., Annadurai, G. and Kannan, C. (2013) Fruit-mediated synthesis of silver nanoparticles using *Vitis vinifera* and evaluation of their antimicrobial efficacy. *Journal of Nanostructure Chemistry* 3(67), 1–6.
Gopalakrishnan, K., Ramesh, C., Ragunathan, V. and Thamilselvan, M. (2012) Antibacterial activity of Cu_2O nanoparticles on *E.coli* synthesized from *Tridax procumbens* leaf extract and surface coating with polyaniline. *Digital Journal of Nanomaterials and Biostructure* 7(2), 833–839.
Gopinath, M., Subbaiya, R., Selvam, M.M. and Suresh, D. (2014) Synthesis of copper nanoparticles from *Nerium oleander* leaf aqueous extract and its antibacterial activity. *International Journal of Current Microbiology and Applied Science* 3(9), 814–818.
Greiling, W. (1954) *Paul Ehrlich*. Econ Verlag, Dusseldorf, Germany, p. 48.
Gunalan, S., Sivaraj, R. and Rajendran, V. (2012) Green synthesized ZnO nanoparticles against bacterial and fungal pathogens. *Progress in Natural Science: Materials International* 22(6), 693–700.
Herlekar, M., Barve, S. and Kumar, R. (2014) Plant-mediated green synthesis of iron nanoparticles. *Journal of Nanoparticle Research* 2014, 1–9.
Herrera, J.E. and Sakulchaicharoen, N. (2009) Microscopic and spectroscopic characterization of nanoparticles. In: Pathak, Y. and Thassu, D. (eds) *Drug Delivery Nanoparticles Formulation and Characterization*. New York, USA, Informa Healthcare Inc., pp. 239–251.
Isaac, R.S.R., Sakthivel, G. and Murthy, C. (2013) Green synthesis of gold and silver nanoparticles using Averrhoa bilimbi fruit extract. *Journal of Nanotechnology* 2013, 1–6.
Jagtap, U. and Bapat, V.A. (2013) Green synthesis of silver nanoparticles using *Artocarpus heterophyllus* Lam. seed extract and its antibacterial activity. *Industrial Crops and Products* 46, 132–137.
Jayaseelan, C., Rahuman, A.A., Rajakumar, G., Santhoshkumar, T., Kirthi, A.V. and Marimuthu, S. (2012) Efficacy of plant-mediated synthesized silver

nanoparticles against hematophagous parasites. *Parasitological Research* 111(2), 921–933.

Kalisz, S., Svoboda, K., Robak, Z., Baxter, D. and Andersen L.K. (2008) Application of FTIR absorption spectroscopy to characterize waste and biofuels for pyrolysis and gasification. *Archives of Waste Management* 8, 51–62.

Katiyar, C., Gupta, A., Kanjilal, S. and Katiyar, S. (2012) Drug discovery from plant sources: an integrated approach. *Ayurveda* 33, 10–19.

Kavitha, K.S. and Satish, S. (2013) Evaluation of antimicrobial and antioxidant activities from *Toona ciliata* Roemer. *Journal of Analytical Science and Technology* 4(23), 1–7.

Kavitha, K.S., Baker Rakshith, D., Kavitha, H.U., Yashwantha Rao, H.C. and Harini, B.P. (2013) Plants as green source towards synthesis of nanoparticles. *International Research Journal of Biological Science* 2, 66–76.

Kesharwani, J., Yoon, K.Y., Hwang, J. and Rai, M. (2009) Phytofabrication of silver nanoparticles by leaf extract of Datura metel: hypothetical mechanism involved in synthesis. *Journal of Bionanoscience* 3(1), 39–44.

Khalil, M.M.H., Ismail, E.H., El-Baghdady, K.Z. and Mohamed, D. (2014) Green synthesis of silver nanoparticles using olive leaf extract and its antibacterial activity. *Arab Journal of Chemistry* 7(6), 1131–1139.

Khan, M., Khan, M., Kuniyil, M., Adil, S.F., Al-Warthan, A., Alkhathlan, H.Z., Tremel, W. and Tahir, M.N. (2014) Biogenic synthesis of palladium nanoparticles using *Pulicaria glutinosa* extract and their catalytic activity towards the suzuki coupling reaction. *Dalton Transactions* 43(24), 9026–9031.

Khanna, S.C. and Speiser, P. (1969) Epoxy resin beads as a pharmaceutical dosage form I: methods of preparation. *Journal of Pharmacological Sciences* 58, 1114–1117.

Khanna, S.C., Jecklin, T. and Speiser, P. (1970) Bead polymerisation technique for sustained release dosage form. *Journal of Pharmacological Science* 59, 614–618.

Kora, A.J., Sashidhar, R.B. and Arunachalama, J. (2012) Aqueous extract of gum olibanum (*Boswellia serrata*): a reductant and stabilizer for the biosynthesis of antibacterial silver nanoparticles. *Processes in Biochemistry* 47(10), 1516–1520.

Kreuter, J. (2007) Nanoparticles – a historical perspective. *International Journal of Pharmacology* 331, 1–10.

Kumar, V.G., Gokavarapu, S.D., Rajeswari, A., Dhas, T.S., Karthick, V., Kapadia, Z., Shrestha, T., Barathy, I.A. and Roy, A. (2011) Facile green synthesis of gold nanoparticles using leaf extract of antidiabetic potent *Cassia auriculata. Colloids and Surfaces B: Biointerfaces* 87(1), 159–163.

Lakshmipathy, R., Reddy, B.P., Saradha, N.C., Chidambaram, K. and Pasha, S.K. (2014) Watermelon rind-mediated green synthesis of noble palladium nanoparticles: catalytic application. *Applied Nanoscience* 5(2), 223–228.

Lee, H.J., Song, J.Y. and Kim, B.S. (2013) Biological synthesis of copper nanoparticles using Magnolia kobus leaf extract and their antibacterial activity. *Journal of Chemical Technology and Biotechnology* 88(11), 1971–1977.

Lokina, S. and Narayanan, V. (2013) Antimicrobial and anticancer activity of gold nanoparticles synthesized from grapes fruit extract. *Chemical Science Transactions* 2(1), 105–110.

Lopez-Serrano, A., Olivas, R.M., Landaluze, J.S. and Camara, C. (2014) Nanoparticles: a global vision. Characterization, separation, and quantification methods. Potential environmental and health impact. *Analytical Methods* 6, 38–56.

Lu, F., Sun, D., Huang, J., Du, M., Yang, F., Chen, H. and Hong, Y. (2014) Plant-mediated synthesis of Ag–Pd alloy nanoparticles and their application as catalyst toward selective hydrogenation. *ACS Sustainable Chemical Engineering* 2(5), 1212–1218.

Maensiri, S., Masingboon, C., Laokul, P., Jareonboon, W., Promarak, V., Anderson, P.L. and Seraphin, S. (2007) Egg white synthesis and photoluminescence of platelike clusters of CeO_2 nanoparticles. *Crystal Growth and Design* 7(5), 950–955.

Maensiri, S., Laokul, P., Klinkaewnarong, J., Phokha, S., Promarak, V. and Seraphin, S. (2008) Indium oxide (In_2O_3) nanoparticles using *Aloe vera* plant extract: synthesis and optical properties. *Optoelectronics and Advanced Materials* 2(3), 161–165.

Makarov, V.V., Love, A.J., Sinitsyna, O.V., Makarova, S.S., Yaminsky, I.V., Taliansky, M.E. and Kalinina, N.O. (2014) 'Green' nanotechnologies: synthesis of metal nanoparticles using plants. *Acta Naturae* 6(1), 35–44.

Merkle, H.P. and Speiser, P. (1973) Preparation and *in vitro* evaluation of cellulose acetate phthalate coacervate microcapsules. *Journal of Pharmacological Science* 62(9), 1444–1448.

Mondal, N.K., Chowdhury, A., Dey, U., Mukhopadhya, P., Chatterjee, S., Das, K. and Datta, J.K. (2014) Green synthesis of silver nanoparticles and its application for mosquito control. *Asian Pacific Journal of Tropical Disease* 4(Suppl 1), S204–S210.

Morris, J.E. (2008) *Nanopackaging: Nanotechnologies and Electronics Packaging*. Springer, New York, USA, pp. 93–107.

Mullen, M.D., Wolf, D.C., Ferris, F.G., Beveridge, T.J., Flemming, C.A. and Bailey, G.W. (1989) Bacterial sorption of heavy metals. *Applied Environmental Microbiology* 55(12), 3143–3149.

Nagajyothi, P.C. and Lee, K.D. (2011) Synthesis of plant-mediated silver nanoparticles using *Dioscorea batatas* rhizome extract and evaluation of their antimicrobial activities. *Journal of Nanomaterials* 2011, 1–7.

Nagati, V.B., Alwala, J., Koyyati, R., Donda, M.R., Banala, R. and Padigya, P.R.M. (2012) Green synthesis of plant-mediated silver nanoparticles using *Withania somnifera* leaf extract and evaluation of their antimicrobial activity. *Asian Pacific Journal of Tropical Biomedicine* 2, 1–5.

Naika, H.R., Lingaraju, K., Manjunath, K., Kumar, D., Nagaraju, G., Suresh, D. and Nagabhushana, H. (2015) Green synthesis of CuO nanoparticles using *Gloriosa superba* L. extract and their antibacterial activity. *Journal of Taibah University of Sciences* 9(1) 7–12.

Nair, B. and Pradeep, T. (2002) Coalescence of nanoclusters and formation of submicron crystallites assisted by *Lactobacillus* strains. *Crystal Growth and Design* 2(4), 293–298.

Narayanan, K.B. and Sakthivel, N. (2010) Biological synthesis of metal nanoparticles by microbes. *Advances in Colloid Interface Science* 156(1–2), 1–13.

Nellore, J., Pauline, C. and Amarnath, K. (2013) *Bacopa monnieri* phytochemicals mediated synthesis of platinum nanoparticles and its neurorescue effect on 1-methyl 4-phenyl 1,2,3,6 tetrahydropyridine-induced experimental parkinsonism in zebrafish. *Journal of Neurodegenerative Diseases* 2013, 1–8.

Niraimathi, K.L., Sudha, V., Lavanya, R. and Brindha, P. (2012) Biosynthesis of silver nanoparticles using *Alternanthera sessilis* (Linn.) extract and their antimicrobial, antioxidant activities. *Colloids and Surfaces B: Biointerfaces* 102, 288–291.

Norman, T.J., Grant, C.D., Schwartzberg, A.M. and Zhang, J.Z. (2005) Structural correlations with shift in the extended plasma resonance of gold nanoparticle aggregates. *Optical Materials* 27, 1197–1203.

Padil, V.V.T. and Černík, M. (2013) Green synthesis of copper oxide nanoparticles using gum karaya as a biotemplate and their antibacterial application. *International Journal of Nanomedicine* 8, 889–898.

Pan, L., Chai, H. and Kinghorn, A.D. (2010) The continuing search for antitumor agents from higher plants. *Phytochemistry Letters* 3(1), 1–8.

Pan, S.Y., Pan, S., Yu, Z.L., Ma, D.L., Chen, S.B., Fong, W.F., Han, Y.F. and Ko, K.M. (2010) New perspectives on innovative drug discovery: an overview. *Journal of Pharmacy and Pharmaceutical Science* 13, 450–471.

Pandey, S., Oza, G., Mewada, A. and Sharon, M. (2012) Green synthesis of highly stable gold nanoparticles using *Momordica charantia* as nano fabricator. *Archives of Applied Science and Research* 4(2), 1135–1141.

Panneerselvam, C., Ponarulselvam, S. and Murugan, K. (2011) Potential anti-plasmodial activity of synthesized silver nanoparticle using *Andrographis paniculata* Nees (Acanthaceae). *Archives of Applied Science and Research* 3(6), 208–217.

Parker, H.L., Rylott, E.L., Hunt, A.J., Dodson, J.R., Taylor, A.F. and Bruce, N.C. (2014) Supported palladium nanoparticles synthesized by living plants as a catalyst for suzuki-miyaura reactions. *PLoS One* 9(1), e87192.

Patra, J.K. and Baek. K-H. (2014) Green nanobiotechnology: factors affecting synthesis and characterization techniques. *Journal of Nanomaterials*, Article ID 417305. 12 pp. Available at: www.hindawi.com/journals/jnm/2014/417305 (accessed 26 May 2016).

Paulkumar, K., Gnanajobitha, G., Vanaja, M., Rajeshkumar, S., Malarkodi, C. and Pandian, K. (2014) *Piper nigrum* leaf and stem assisted green synthesis of silver nanoparticles and evaluation of its antibacterial activity against agricultural plant pathogens. *Science World Journal* 2014, 1–9.

Pletikapić, G., Žčutić, V., Vinković Vrček, I. and Svetličić, V. (2012) Atomic force microscopy characterization of silver nanoparticles interactions with marine diatom cells and extracellular polymeric substance. *Journal of Molecular Recognition* 25, 309–317.

Ponarulselvam, S., Panneerselvam, C., Murugan, K., Aarthi, N., Kalimuthu, K. and Thangamani, S. (2012) Synthesis of silver nanoparticles using leaves of *Catharanthus roseus* Linn. G. Don and their antiplasmodial activities. *Asian Pacific Journal of Tropical Biomedicine* 2(7), 574–580.

Prasad, R. (2014) Synthesis of silver nanoparticles in photosynthetic plants. *Journal of Nanoparticles* 2014, 1–8.

Prasad, T.N.V.K.V. and Elumalai, E.K. (2011) Biofabrication of Ag nanoparticles using *Moringa oleifera* leaf extract and their antimicrobial activity. *Asian Pacific Journal Tropical Biomedicine* 1(6), 439–442.

Prathap, M., Alagesan, A. and Kumari, B.D.R. (2014) Antibacterial activities of silver nanoparticles synthesized from plant leaf extract of *Abutilon indicum* (L.) sweet. *Journal of Nanostructure Chemistry* 4, 106.

Prathna, T.C., Mathew, L., Chandrasekaran, N., Raichur, A.M. and Mukherjee, A. (2010) Biomimetic synthesis of nanoparticles: science, technology and applicability. In: Mukherjee, A. (ed.) *Biomimetics, Learning from Nature*. InTech, Rijeka, Croatia, pp. 25–40.

Pugazhenthiran, N., Anandan, S., Kathiravan, G., Udayaprakash, N.K., Crawford, S. and Kumar A (2009) Microbial synthesis of silver nanoparticles by *Bacillus* sp. *Journal of Nanoparticles Research* 11, 1811–1815.

Ragupathia, C., Vijayaa, J.J. and Kennedy, L.J. (2014) Preparation, characterization and catalytic properties of nickel aluminate nanoparticles: a comparison between conventional and microwave method. *Journal of Saudi Chemical Society* (in press).

Rajakumar, G., Rahuman, A.A., Jayaseelan, C., Santhoshkumar, T., Marimuthu, S. and Kamaraj, C. (2014) *Solanum trilobatum* extract-mediated synthesis of titanium dioxide nanoparticles to control *Pediculus humanus capitis*, *Hyalomma anatolicum*

anatolicum and *Anopheles subpictus*. *Parasitological Research* 113(2), 469–479.
Rajiv, P., Sivaraj, R., Priya, R.S.V. and Vanathi, P. (2014) Synthesis and characterization Parthenium mediated zinc oxide nanoparticles and assessing its medicinal properties. *International Conference on Advances in Agricultural, Biological and Environmental Sciences (AABES-2014)* 2014, 80–82.
Rane, R.V., Meenakshi, K., Shah, M. and George, I.A. (2014) Biological synthesis of silver nanoparticles using *Abelmoschus moschatus*. *Indian Journal of Biotechnology* 13, 342–346.
Roduner, E. (2006) Size matters: why nanomaterials are different. *Chemical Society Reviews* 35, 583–592.
Rotello, V. (2004) *Nanoparticles: Building Blocks for Nanotechnology*. Springer, New York, USA, p. 284.
Roy, D. and Fendler, J. (2004) Reflection and absorption techniques for optical characterization of chemically assembled nanomaterials. *Advanced Materials* 16, 479–508.
Safaepour, M., Shahverdi, A.R., Shahverdi, H.R., Khorramizadeh, M.R. and Gohari, A.R. (2009) Green synthesis of small silver nanoparticles using geraniol and its cytotoxicity against Fibrosarcoma-Wehi 164. *Avicenna Journal of Medical Biotechnology* 1(2), 111–115.
Saman, S., Moradhaseli, S., Shokouhian, A., and Ghorbani, M. (2013) Histopathological effects of ZnO nanoparticles on liver and heart tissues in wistar rats. *Advances in Bioresearch* 4, 83-88.
Sankar, R., Manikandan, P., Malarvizhi, V., Fathima, T., Shivashangari, K.S. and Ravikumar, V. (2014) Green synthesis of colloidal copper oxide nanoparticles using Carica papaya and its application in photocatalytic dye degradation. *Spectrochimica Acta Part A: Molecular Biomolecular and Spectroscopy* 18(121), 746–750.
Sataraddi, S.R. and Nandibewoor, S.T. (2012) Biosynthesis, characterization and activity studies of Ag nanoparticles, by (*Costus ingneus*) insulin plant extract. *Der Pharmacia Lettre* 4(1), 152–158.
Sathishkumar, M., Sneha, K., Won, S.W., Cho, C.W., Kim, S. and Yun, Y.S. (2009) *Cinnamon zeylanicum* bark extract and powder mediated green synthesis of nano-crystalline silver particles and its bactericidal activity. *Colloids and Surfaces B: Biointerfaces* 73(2), 332–338.
Sathyavathi, R., Krishna, M.B., Rao, S.V., Saritha, R. and Rao, D.R. (2010) Biosynthesis of silver nanoparticles using *Coriandrum Sativum* leaf extract and their application in nonlinear optics. *Advanced Science Letters* 3(2), 138–143.
Satish, S., Raveesha, K.A. and Janardhana, G.R. (1999) Antibacterial activity of plant extracts on phytopathogenic *Xanthomonas campestris* pathovars. *Letters of Applied Microbiology* 28, 145–147.
Satish, S., Raghavendra, M.P. and Raveesha, K.A. (2008) Evaluation of the antibacterial potential of some plants against human pathogenic bacteria. *Advances in Biological Research* 2(3–4), 44–48.
Satyavani, K., Gurudeeban, S., Ramanathan, T. and Balasubramanian, T. (2011) Biomedical potential of silver nanoparticles synthesized from calli cells of *Citrullus colocynthis* (L.) Schrad. *Journal of Nanobiotechnology* 9, 43.
Senthilkumar, S.R. and Sivakumar, T. (2014) Green tea (*Camellia sinensis*) mediated synthesis of zinc oxide (ZnO) nanoparticles and studies on their antimicrobial activities. *International Journal of Pharmacology and Pharmaceutical Science* 6(6), 461–465.
Shankar, S.S., Rai, A., Ankamwar, B., Singh, A., Ahmad, A. and Sastry, M. (2004) Biological synthesis of triangular gold nanoprisms. *Nature Materials* 3, 482–488.
Sharbidre, A.A. and Kasote, D.M. (2013) Synthesis of silver nanoparticles using flaxseed hydroalcoholic extract and its antimicrobial activity. *Current Biotechnology* 2(2), 1–5.
Sharma, N.C., Sahi, S.V., Nath, S., Parsons, J.G., Gardea-Torresdey, J.L. and Pal, T. (2007) Synthesis of plant-mediated gold nanoparticles and catalytic role of biomatrix-embedded nanomaterials. *Environmental Science Technology* 41(14), 5137–5142.
Singh, P.S. and Vidyasagar, G.M. (2014) Green synthesis, characterization and antimicrobial activity of silver nanoparticles by using *Sterculia foetida* L. young leaves aqueous extract. *International Journal of Green Chemistry and Bioproducts* 4(1), 1–5.
Singh, C., Sharma, V., Naik, P.K., Khandelwal, V. and Singh, H. (2011) A green biogenic approach for synthesis of gold and silver nanoparticles using *Zingiber officinale*. *Digest Journal of Nanomaterials and Biostructures* 6, 535–542.
Sivaraj, R., Rahman, P.K., Rajiv, P., Salam, H.A. and Venckatesh, R. (2014) Biogenic copper oxide nanoparticles synthesis using *Tabernaemontana divaricate* leaf extract and its antibacterial activity against urinary tract pathogen. *Spectrochimica Acta, Part A, Molecular and Biomolecular Spectroscopy* 10(133), 178–181.
Somerville, C.R. and Bonetta, D. (2001) Plants as factories for technical materials. *Plant Physiology* 125, 168–171.
Soundarrajan, C., Sankari, A., Dhandapani, P., Maruthamuthu, S., Ravichandran, S., Sozhan, G. and Palaniswamy, N. (2012) Rapid biological synthesis of platinum nanoparticles using *Ocimum sanctum* for water electrolysis applications. *Bioprocesses Biosystems and Engineering* 35(5), 827–833.
Subhankari, I. and Nayak, P.L. (2013) Antimicrobial activity of copper nanoparticles synthesised by ginger (*Zingiber officinale*) extract. *World Journal of Nano Science and Technology* 2(1), 10–13.
Sulaiman, G.M., Mohammed, W.H., Marzoog, T.R., Al-Amiery, A.A., Kadhum, A.A., Mohamad, A.B. and

Bagnati, R. (2013) Green synthesis, antimicrobial and cytotoxic effects of silver nanoparticles using *Eucalyptus chapmaniana* leaves extract. *Asian Pacific Journal of Tropical Biomedicine* 3, 58–63.

Suman, T.Y., Rajasree, S.R.R., Ramkumar, R., Rajthilak, C. and Perumal, P. (2014) The Green synthesis of gold nanoparticles using an aqueous root extract of *Morinda citrifolia* L. *Spectrochimica Acta Part A* 118, 11–16.

Suriyakalaa, U., Antony, J.J., Suganya, S., Siva, D., Sukirtha, R. and Kamalakkannan, S. (2013) Hepatocurative activity of biosynthesized silver nanoparticles fabricated using *Andrographis paniculata*. *Colloids and Surfaces B: Biointerfaces* 102, 189–194.

Sutradhar, P., Saha, M. and Maiti, D. (2014) Microwave synthesis of copper oxide nanoparticles using tea leaf and coffee powder extracts and its antibacterial activity. *Journal of Nanostructure Chemistry* 4, 86–92.

Tho, N.T., An, T.N., Tri, M.D., Sreekanth, T.V., Lee, J.S. and Nagajyothi, P.C. (2013) Green synthesis of silver nanoparticles using *Nelumbo nucifera* seed extract and its antibacterial activity. *Acta Chimica Slovenica* 60(3), 673–678.

Tyagi, P.K., Shruti, S.S. and Ahuja, A. (2012) Synthesis of metal nanoparticles: a biological prospective for analysis. *International Journal of Pharmaceutical Innovations* 2, 48–60.

Van Dijk, M.A., Tchebotareva, A.L., Orrit, M., Lippitz, M., Berciaud, S., Lasne, D., Cognet, L., and Lounis, B. (2006) Absorption and scattering microscopy of single metal nanoparticles. *Physical Chemistry Chemical Physics* 8, 3486–3495.

Vanaja, M., Gnanajobitha, G., Paulkumar, K., Rajeshkumar, S., Malarkodi, C. and Annadurai, G. (2013a) Phytosynthesis of silver nanoparticles by Cissus quadrangularis: influence of physicochemical factors. *Journal of Nanostructure Chemistry* 3(17), 1–8.

Vanaja, M., Rajeshkumar, S., Paulkumar, K., Gnanajobitha, G., Malarkodi, C. and Annadurai, G. (2013b) Kinetic study on green synthesis of silver nanoparticles using *Coleus aromaticus* leaf extract. *Advance and Applied Science Research* 4(3), 50–55.

Vanaja, M., Paulkumar, K., Gnanajobitha, G., Rajeshkumar, S., Malarkodi, C. and Annadurai, G. (2014) Herbal plant synthesis of antibacterial silver nanoparticles by *Solanum trilobatum* and its characterization. *International Journal of Metals* 2014, 1–8.

Vankar, S.P. and Shukla, D. (2012) Biosynthesis of silver nanoparticles using lemon leaves extract and its application for antimicrobial finish on fabric. *Applied Nanoscience* 2, 163–168.

Vilchis-Nestora, A.R., Sanchez-Mendieta, V., Camacho-Lopez, M.A., Gomez-Espinosaa, R.M., Camacho-Lopez, M.A. and Arenas-Alatorre, J. (2008) Solventless synthesis and optical properties of Au and Ag nanoparticles using *Camellia sinensis* extract. *Materials Letters* 62, 3103–3105.

Waghmode, S., Chavan, P., Kalyankar, V. and Dagade, S. (2013) Synthesis of silver nanoparticles using *Triticum aestivum* and its effect on peroxide catalytic activity and toxicology. *Journal of Chemistry* 2013, 265864–265869.

Wang, Y., He, X., Wang, K., Zhang, X. and Tan, W. (2009) Barbated Skullcup herb extract-mediated biosynthesis of gold nanoparticles and its primary application in electrochemistry. *Colloids and Surfaces Part B: Biointerfaces* 73(1), 75–79.

Zahir, A.A., Bagavan, A., Kamaraj, C., Elango, G. and Rahuman, A.A. (2012) Efficacy of plant-mediated synthesized silver nanoparticles against *Sitophilus oryzae*. *Journal of Biopesticides* 5, 95–102.

Zahir, A.A., Chauhan, I.S., Bagavan, A., Kamaraj, C., Elango, G. and Shankar, J. (2014) Synthesis of nanoparticles using *Euphorbia prostrata* extract reveals a shift from apoptosis to G0/G1 arrest in *Leishmania donovani*. *Journal of Nanomedicine & Nanotechnology* 5, 1–4.

Zhang, H., Li, Q., Lu, Y., Sun, D., Lin, X. and Deng, X. (2005) Biosorption and bioreduction of diamine silver complex by *Corynebacterium*. *Journal of Chemical Technology and Biotechnology* 80(3), 285–290.

Zheng, B., Kong, T., Jing, X., Odoom-Wubah, T., Xianxue, L., Daohua, S. Fenfen, L., Zheng, Y. and Huang, J. (2013) Plant-mediated synthesis of platinum nanoparticles and its bioreductive mechanism. *Journal of Colloid and Interface Science* 396, 138–145.

8 Plant Diversity Repertoire of Bioactive Triterpenoids

K.S. Kavitha[1], Syed Baker[1], D. Rakshith[1], P. Azmath[1], B.P. Harini[3] and S. Satish[1&2*]

[1]*Herbal Drug Technological Laboratory, Department of Studies in Microbiology, University of Mysore, Mysore, Karnataka, India;* [2]*Department of Plant Pathology, University of Georgia, Athens 30602, Georgia, USA;* [3]*Department of Zoology, Bangalore University, Jnanabharathi Campus, Bangalore-560056, Karnataka, India*

Abstract

Pharmaceutical biology perceives medicinal plants as a rich source of bioactive compounds bearing biological activities which can be traced back to the dawn of life. Plants form one of the most abundant and diverse living systems in nature. Improved scientific knowledge in plant biology has led scientific communities to gain in-depth knowledge of plants and their metabolites by screening and characterization of novel phytochemicals with innumerable valuable roles, which has had a huge impact on all human life. Owing to this, research on plant-based natural products has generated tremendous interest with numerous studies highlighting new secondary metabolites of plant origin. Among these secondary metabolites in plants, triterpenoids form a prominent group of bioactive compounds, widely distributed among diverse plants species. Triterpenoids have myriad biological activities, including antioxidative, antimutagenic, anticancer, antidiabetic, anti-inflammatory, antibacterial, antifungal, antiviral, antimalarial, immunomodulatory and many more. Triterpenoids are highly multifunctional and have great chemical diversity; their potential anti-therapeutic agents could be exploited for various biological activities. This chapter explores the valuable scientific literature pertaining to triterpenoids.

8.1 Introduction

Pharmaceutical biology perceives medicinal plants as a rich source of bioactive compounds bearing biological activities which can be traced back to the dawn of life. Plants form one of the most abundant and diverse living systems in nature. Even before the scientific era, traditional knowledge on the role of plants in curing various illness was well documented. This led to advances in plant research, transforming these glimpses of traditional knowledge into phyto-derived products with different biological activities, resulting in commercial products in various sectors, such as pharmaceutical, agricultural and industrial. This scientific research gained momentum and transformed the global market for plant-derived therapeutic agents (Kavitha and Satish, 2013; Lin *et al.*, 2013). The excellence of medicinal plants as a new source of drugs is constantly being explored: a large number of drugs are of plant origin, since plants are richly diverse, with more than 500,000 different species (Dhanalakshmi *et al.*, 2013). The potential of plants to deliver important pharmaceutical products is based on the fact that plants lack an immune response but myriad phytochemicals are responsible for defence mechanisms and help in plant growth and development. These phytochemicals belong to structurally diverse classes and are produced during different growth stages, with each part of the plant capable of producing different phytocomponents. With interdisciplinary knowledge, such as combinatorial chemistry and molecular modelling, and other

*E-mail:satish.micro@gmail.com and satish@uga.edu

conventional methods, scientists have found it simple to isolate individual phytocomponents. Despite advances in synthetic chemistry related to drug discovery, even more interest has focused on plant-derived drugs (Wilson, 2013). Plants bear a vast repertoire of phytochemicals, including sugars, proteins, amino acids, enzymes, fatty acids, waxes, vitamins, alkaloids, tannins, flavonoids, sterols and triterpenes (Revathi *et al.*, 2014). These secondary metabolites are reported to have significant applications in various sectors. Terpenoids are found to be prominent phytocomponents in plants. It is estimated that more than 40,000 terpenoids are reported in different biological entities, with plants among diverse classes of terpenoids (Rohdich *et al.*, 2005). Terpenoids have been classified according to the number of carbon building-blocks into monoterpenes (C10), sesquiterpene (C15), diterpenes (C20), sesterterpenes (C25), triterpenes (C30), tetraterpenes (C40) and polyterpenes (C>40) (Finefield *et al.*, 2012).

Among the different classes of terpenoids, triterpenoids are widely distributed in different plant species: for instance, seaweeds, in the wax coating of fruits, and herbs (Rabi and Bishayee, 2009; Gadhvi *et al.*, 2013). Triterpenoids have been found to possess various biological and pharmacological activities, as represented in Fig. 8.1. Many more scientific studies are discovering the previously unknown roles of these secondary metabolites, making them much more valuable compounds. Triterpenoids are also synthesized chemically by modifying the structure of naturally occurring triterpenoids; during the synthesis some of the analogues become more potent and active, which results in bioactivity. But screening of plants, isolation and characterization of new classes of triterpenoids are essential to trace their multifunctional biological activities and thus establish potential strategies for developing new drug derivatives for product development. This chapter explores the identification and characterization of triterpenoids in plants, and their applications in the various fields of drug discovery.

Triterpenoids can be isolated and characterized using different hyphenated techniques, and during the isolation, selection of plant species is one of the prime aspects. Techniques best suited for the purification of triterpenoids involve combinational chromatographic techniques. Chemical profiling can be achieved using various spectroscopic techniques and structure can be elucidated (Fig. 8.2).

8.2 Biosynthesis of Triterpenoids

Triterpenoids are biosynthesized in cytosol by utilizing IIP and its isomer DMAPP derived from acetyl coA via the cytosolic mevalonic acid MVA pathway. One molecule of IIP (5C) is converted into monoterpene (10C) by the enzyme prenyl transferase. Then

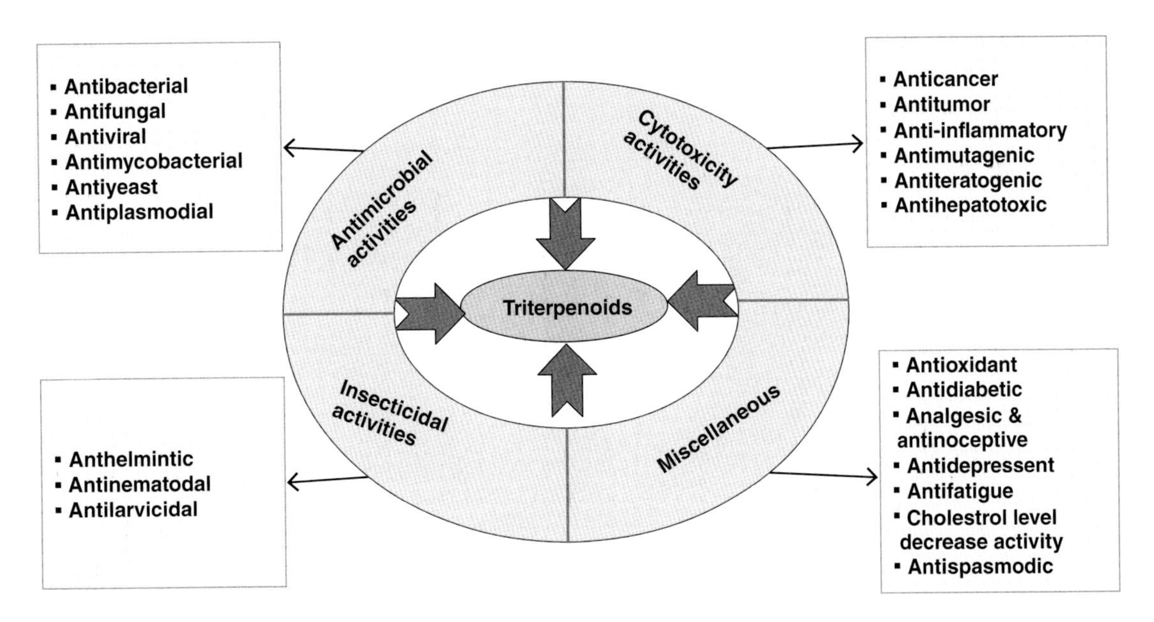

Fig. 8.1. Biological activities of triterpenoids.

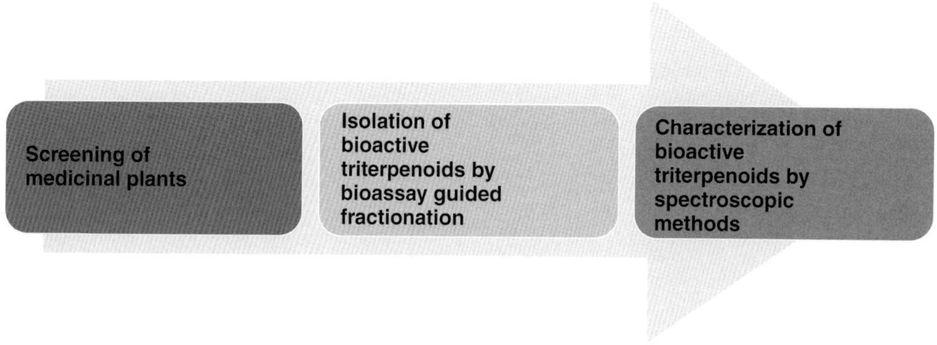

Screening of medicinal plants

Isolation of bioactive triterpenoids by bioassay guided fractionation

Characterization of bioactive triterpenoids by spectroscopic methods

Fig. 8.2. Schematic representation for the isolation of triterpenoids from plants.

this monoterpene is condensed with one more IIP to form sesquiterpene (15C) by the enzyme prenyl transferase. A further two molecules of sesquiterpene are condensed to form triterpene squalene (30C) by the enzyme squalene synthase. This condensation is a dimerisation reaction in which one molecule of NADPH is involved and it is eliminated as NADP. Squalene is then converted to 2,3-oxidosqualene by squalene epoxygenase, involves one molecule of NADPH and an O_2 molecule. It was reported that cyclization of 2,3-oxidosqualene by catalytic activity of oxidosqualene cyclase (OSC) resulted in formation of the skeleton for teriterpenoids synthesis. Later, the skeleton is modified to a hydrophobic aglycone, which is further oxidized by cytochrome P450 monooxygenase (P450), resulting in further modifications of the skeleton to form O-glycosylation, leading to the formation of type triterpenoids. In recent years, molecular detection of OSC's genes forms the base to determine terpenoid-producing plants. Higher plants have several OSCs, not only for sterol biosynthesis, such as cycloartenol synthase (CAS) and LAS, but also for triterpenoid biosynthesis (Zhao *et al.*, 2010a). Triterpenoids are thus further sub-classified, based on chemical diversity, i.e. lanostanes, cucurbitanes, cycloartanes, dammaranes, ergostanes, etc. (Table 8.1) (Cassels and Asencio, 2011).

8.3 Pharmacological Properties of Triterpenoids

8.3.1 Antibacterial and antifungal

Increase in the emergence of drug-resistant microorganisms has led to dangerous situations, especially in the developing countries, wherein nosocomal infections are growing at an alarming pace, due to their resistance to the available standard antibiotics. Hence, scientific communities are in search of new classes of antibiotics, which can combat the drug-resistant microorganisms. For much of history, plants have served as part of the antimicrobial repertoire, and most of the available antibiotics on the market today originated from plants, whose analogues are modified to improve their efficacy. Various scientific studies highlight the ways in which different triterpenoids have antimicrobial effects against a series of pathogenic microorganisms. According to the report of Prachayasittikul *et al.* (2010), five lupane-type triterpenoids, lupeol, lupenone, betulin, lupeol acetate, 28-O-acetylbetulin isolated from *Diospyros rubra* Lec., exhibited significant antibacterial activity against *Corynebacterium diphtheria* NCTC 10356, with minimal inhibition concentration in a range from 64 to 256 µg/ml, whereas betulin displayed activity against *Streptococcus pyogenes* at 85 µg/ml. Ghosh *et al.* (2010) isolated friedelin and cerin from *Quercus suber*. Friedelin exhibited significant antimicrobial activity against *Fusarium equiseti*, *Curvularia eragrostidies*, *Colletrichum gleosproide*, *Ralstonia solanacearum*, *Pseudomonas syringae*, *Erwinia caratovora*, *Xanthomonas* sp., compared to cerin.

De Leon *et al.* (2010) studied antibacterial activity against *Staphylococcus aureus* using *Maytenus blepharodes*, resulting in the isolation of zeylasterone and demethylzeylasterone, wherein zeylasterone showed more significant antibacterial activity than demethylzeylasterone at 30 µg/ml.

Duraipandiyan *et al.* (2010) determined the antifungal activity of friedelin isolated from *Azima tetracantha* against *Trichophyton rubrum*, and *Curvularia lunata* exhibited a minimum inhibitory

Table 8.1. List of different types of triterpenoids isolated from various plants.

Sl. No.	Types of triterpenoids on the basis of their different carbon frameworks	Plants	Part used	Triterpenoid profile	References
1	Apotirucallane	*Luvunga sarmentosa* (Bl.) Kurz.	Leaves	Luvungins A–G, and 1-α-acetoxyluvungin A	Lien *et al.* (2002)
2	Cucurbitane	*Momordica charantia* Linn.		Charantin	Lee *et al.* (2009)
3	Cycloartane	*Polygonum bistorta* (L.) Samp.	Rhizomes	24(E)-ethylidenecycloartanone, 24(E)-ethylidenecycloartan-3α-ol, cycloartane-3,24-dione, 24-methylenecycloartanone, γ-sitosterol, β-sitosterol, β-sitosterone, friedelin, 3β-friedelinol	Manoharan *et al.* (2005)
4	Cyclopropyl	*Ficus microcarpa* L.f.	Aerial roots	27-nor-3β-hydroxy-25-oxocycloartane, (22E)-25,26,27-trinor-3β-hydroxycycloart-22-en-24-al, 3βacetoxy-15α-hydroxy-13,27-cyclours-11-ene, 3β-acetoxy-12α-formyloxy-13,27-cycloursan-11α-ol	Chiang *et al.* (2001)
5	Dammarane	*Cordia multispicata* Cham.		Cordianols A–I and cordialin A	Kuroyanagi *et al.* (2003)
6	Fernane	*Euphorbia supina* Raf.		Supinenolones	Tanaka and Matsunaga (1987)
7	Friedelane	*Hibiscus tiliaceus* L.	Stem and bark	27-oic-3-oxo-28-friedelanoic acid	Li *et al.* (2006)
8	Hopane	*Tripterygium regelii* Sprag. & Tekeda.	Roots	Zeorin	Inayama *et al.* (1989)
9	Lanostane	*Klainedoxa gabonensis* Pierre.	Stem bark	2-hydroxy-24-methylenelanostan-1,8-dien-3-one (klainedoxalanostenone)	Nkanwen *et al.* (2013)
10	Lupane	*Liquidambar formosana* Hance.	Roots	3β,6β-dihydroxylup-20(29)-en-28-oic acid β-glucopyranosyl ester, 2α-acetoxyl-3β,6β-dihydroxylup-20(29)-en-28-oic acid β-glucopyranosyl ester	Yu *et al.* (2012)

Table 8.1. Continued.

Sl. No.	Types of triterpenoids on the basis of their different carbon frameworks	Plants	Part used	Triterpenoid profile	References
11	Multiflorane	*Sandoricum indicum* Cav.	Stem bark	Sandorinic acids A–C	Tanaka *et al.* (2001)
12	Oleanane	*Liquidambar formosana* Hance.	Fruits	Liquidambaric acid, Oleanolic acid, 3α-acetoxy-25-hydroxy-olean-12-en-28-oic acid and lantanolic acid	Dat *et al.* (2004)
13	Onoceranoid	*Lycopodium obscurum* L.		(3α,8β,14α,21β)-26,27-dinoronocerane-3,8,14,21-tetrol and 26-nor-8β-hydroxy-α-onocerin	Zhao *et al.* (2010b)
14	Taraxasteryl	*Pergularia tomentosa* L.		Pergularine A and pergularine B	Babaamer *et al.* (2012)
15	Tirucallane	*Castela texana* (T. & G.) Rose.	Aerial parts	23*R*,24*R*,25-trihydroxytirucall-7-en-3-one (24-*epi*-piscidinol A)	McChesney *et al.* (1997)
16	Ursane	*Salvia santolinifolia* Boiss.	Whole plant	2α,3β,9α-trihydroxyurs-12-en-28-oic acid (Santolinoic acid)	Ahmad *et al.* (2007)

concentration (MIC) at 62.5 μg/ml. Antibacterial activity of camaric acid and betulinic acid was extracted from *Lantana viburnoides* sp. by Innocent *et al.* (2011). Camaric acid showed significant activity against *Salmonella typhi*, *Streptococcus faecalis*, *Pseudomonas aeruginosa*, *Staphylococcus aureus* and *Bacillus subtilis*, with MIC values of 19.5, 19.5, 9.76, 4.88 and 19.5 μg/ml respectively. Moodley *et al.* (2011) reported β-amyrin, methyl oleanolate, oleanolic acid and 3β-hydroxyolean-11-en-28,13β-olide from the fruit and ursolic acid from the leaves of *Carissa macrocarpa*. The antibacterial activity of all isolated compounds exhibited the least inhibitory activity in the range from 0.12 to 1.0 mg/ml, except 3β-hydroxyolean-11-en-28,13β-olide, which showed MIC ranging from 0.06 to 0.12 mg/ml against *S. aureus*, *E. faecium*, *S. saprophyticus* *E. coli*, *K. pneumonia* and *P. aeruginosa*. Another study by Dong *et al.* (2011) reported that lamesticumin A, lamesticumin A, lamesticumins B-F, lansic acid 3-ethyl ester and ethyl lansiolate had been found in *Lansium domesticum* Corr. The compounds displayed moderate antibacterial activity against *S. aureus, S. epidermidis, M. luteus, B. subtilis, M. pyogenes* and *B. cereus*, with MIC ranging from >50 to 3.12 μg/ml, whereas *E. coli, S. flexneri, P. aeruginosa, S. marcescens* and *A. faecalis* showed activity at MIC of >50 μg/ml. Similarly, Kiplimo *et al.* (2011) investigated *Vernonia auriculifera* for antibacterial activity, resulting in the identification of lupenyl acetate, oleanolic acid, β-amyrin acetate, β-amyrin, friedelanone, friedelin acetate, α-amyrin and β-sitosterol. The isolated triterpenoids exhibited moderate antibacterial activity; α-and β-amyrin showed MIC of 0.25 μg/ml against *Staphylococcus aureus*, *Bacillus subtilis*, *Enterococcus faecium* and *Staphylococcus saprophyticus*, whereas lupenyl acetate and oleanolic acid exhibited MIC of 0.25 μg/ml against *Stenotrophomonas maltophilia*.

Yinusa *et al.* (2012) employed clinical pathogens to check the effectiveness of Betunilic acid isolated from the stem bark of *Sarcocephalus latifolius*. The compounds showed MIC of 12.5 μg/ml against *Staphylococcus aureus*, *Streptococcus pyrogenes*, *Bacillus subtilis*, *Escherichia coli*, *Proteus vulgaris*, *Proteus mirabilis*, *Pseudomonas aeruginosa*, MRSA, *Candida albicans*, *Candida tropicalis* and *Candida virusei*, while *Salmonella typhi* and *Shigellia dysenteric* exhibited MIC values of 6.25 μg/ml, and no inhibitory action was observed for *Corynebacterium ulcerans*.

Antifungal activity of lupeol and β-amyrin found in *Cariniana domestica* (Mart) Miers. showed MIC at 125 mg/ml against *Saccharomyces cerevisiae* (Janovik *et al.*, 2012). The aerial parts of *Eriope blanchetii* displayed the distribution of betulinic, ursolic and oleanolic acids. These compounds under esterification form the derivatives: 3β-3-chlorobenzoyl betulinic acid, 3β-3-chlorobenzoyl ursolic acid and 3β-3-chlorobenzoyl oleanolic acid, respectively. The isolated compounds were reported to exhibit a more significant inhibition against *E. coli* and *S. aureus* than its derivatives (de Silva *et al.*, 2012). A moderate zone of inhibition against test pathogens was reported by Zhao *et al.* (2012) using olean-9(11), 12-dien-3-O-palmitate found in *Anemone rivularis*, resulting in MIC 50 and 75 μg/ml against *B. subtilis* and *S. aureus*, respectively. Similarly, Choi *et al.* (2012) studied antibacterial activity of 22-dehydroclerosterol, and β-amyrin isolated from *Clerodendron trichotomum* showed moderate activity on *E. coli*, *S. aureus* and *H. pylori*. In another investigation, Manzano *et al.* (2013) reported that a pentacyclic group of triterpenoids: lupeol, epilupeol and acetyl lupeol, were derived from *Vernonanthura patens* (Kunth) H. Rob. Significant antifungal activity against *Fusarium oxysporum* and *Penicillium notatum* was observed with lupeol and epilupeol, whereas acetyl lupeol was found to be inactive. A new oleanane-type triterpenoid,3-O-[β-D-glucoronopyranosyl-(1→3)-α–L-rhamnopyranosyl-(1→2)β-D-glucopyranosyl]-4α,20α-hydroxy methyl olean-12-ene-28-oic acid was characterized by Acharya and Khan (2013) from *Ricinus communis*. The compound was found to be active against *E. coli*, *Klebsiella pneumonia* and *Staphylococcus aureus*, with MIC values of 260, 235 and 350 μg/ml, respectively.

8.3.2 Anti-tuberculosis

Tuberculosis is one of the major diseases caused by *Mycobacterium tuberculosis*. It is believed that one-third of the world's population is infected with this bacterium, causing serious concern, especially among immune-compromised patients. To date the drugs available to combat tuberculosis have a number of side effects, so there is a need to search for natural products without any adverse effects. Plant-based natural products offer a great source of bioactive compounds with an anti-tuberculosis effect. Most of the triterpenoids are known to be effective against this bacterium. Truong *et al.* (2011) reported that bonianic acid A, and bonianic acid B and 3-O-acetyluncaric acid, from the leaves and twigs of *Radermachera boniana*, resulted in significant activity against *Mycobacterium tuberculosis* H37Rv. Similarly, Mann *et al.* (2011) showed that friedelin from the root bark of *Terminalia avicennioides* Guill. and Perr. exhibited significant activity against Bacillus Calmette *Guerin* (BCG) with MIC value of 4.9 µg/ml. In 2012, Mann *et al.* discovered three more bioactive triterpenoids: arjunolic acid, α-amyrin and 2,3,23-trihydroxylolean-12-ene, in *T. avicennioides*. Arjunolic acid showed moderate activity, having MIC of 156 µg/ml against *Mycobacterium bovis* (BCG). Jiménez-Arellanes *et al.* (2013) obtained ursolic acid and oleanolic acid from the hexane extracts of *Chamaedora tepejilote* and *Lantana hispida*, respectively. The compounds, at low dosages, exhibited significant intracellular activity against *M. tuberculosis* H37Rv strain.

8.3.3 Anti-HIV

Human Immunodeficiency Virus (HIV) is the etiological agent of Acquired Immunodeficiency Syndrome that created great socio-economic difficulties to patients in obtaining treatment. There is a vital need for new, safe, and cheap anti-HIV agents. Traditional systems of medicine in India, including herbs or natural plant products, are used to minimize the consequences of the syndrome or prolong life. It is therefore not surprising that these traditional medicines, having anti-HIV properties, may offer alternatives to expensive medicines in the future. Over the past two decades, notable progress has been made in research on the secondary phytoconstituents of plants possessing anti-HIV activity. A variety of secondary metabolites obtained from natural origins showed moderate to good anti-HIV activity (Dahake *et al.*, 2013). Chen *et al.* (2009) reported 14 cucurbitane triterpenoids from *Momordica charantia*: kuguacins F–S, that includes two pentanorcucurbitacins, one octanorcucurbitacin, and two trinorcucurbitacins, in addition to six known analogues wherein the isolated compounds showed least anti-HIV-1 activity in *in vitro* studies. The plant *M. charantia* also revealed the presence of five cucurbitacins, kuguacins A–E, in addition to three analogues: 3b,7b,25-trihydroxycucurbita-5,(23E)-diene-19-al, 3b,25-dihydroxy-5b,19-epoxycucurbita-6,(23E)-diene and momordicine I from the root extracts. Kuguacins C and E showed moderate anti-HIV-1 activity with EC_{50} values of 8.45 and 25.62 µg/ml respectively. Three new highly oxygenated nortriterpenoids named wilsonianadilactones D–F were reported from the leaves and stems of *Schisandra wilsoniana* by Yang *et al.* (2011). All the three compounds showed weak anti-HIV-1 activity, with therapeutic index (TI) values (CC_{50}/EC_{50}) greater than 8.16, 14.7, and 17.5 µg/ml, respectively. Another report by Gao *et al.* (2013) disclosed that compounds present in *Schisandra wilsoniana* fruits, characterized as schilancidilactones V and W, also resulted in moderate anti-HIV-1 activity, with EC_{50} 3.05 and 2.87µg/ml, respectively.

8.3.4 Antimalarial activity

Malaria is the world's most significant tropical disease. In South East Asia 100 million malaria cases occur every year and 70% of these are reported from India. *Plasmodium falciparum*, the most widespread causative agent for human malaria, has become increasingly resistant to standard antimalarials such as chloroquine and antifolates. Currently, studies on designing new drug molecules or drug formulations from plants that are effective, safe, and can target different sites and exhibit different modes of action are in progress (Bagavan *et al.*, 2011). In this view, some of the research findings have proved that the plant-derived triterpenoids exhibit anti-malarial properties.

Chianese *et al.* (2010) reported eight known triterpenoids: azadirone, azadiradione, epoxyazadiradione, gedunin, deacetylgedunin, desmethyllimocin B (6), protoxylocarpin G, and spicatin, along with two new triterpenoids: neemfruitin A and neemfruitin B, from the fruit of *Azadirachta indica*. Azadirone, gedunin and neemfruitin A showed significant anti-plasmodial activity. Another study by Ramalhete *et al.* (2011) reported cucurbitane-type triterpenoids, characterized as balsaminol F and balsaminoside B, in addition to known glycosylated cucurbitacins, cucurbita-5, 24-diene-3b, 23(R)-diol-7-O-b-D-glucopyranoside and kuguaglycoside A, from *Momordica balsamina*. Compound balsaminol F, on acylation, resulted in the formation of triacetylbalsaminol F and tribenzoylbalsaminol F. The study showed that the triterpenoid, triacetylbalsaminol F, was found to exhibit significant anti-malarial activity against *Plasmodium falciparum* chloroquine-sensitive strain 3D7 and the chloroquine-resistant strain Dd2 with IC_{50} values of 0.4 and 0.2 μM, respectively. The compound also exhibited a significant effect on *P. berghei* liver stages without causing toxicity. Ruphin *et al.* (2013) reported the quinone methide pentacyclic triterpenoid derivative, in *Salacia leptoclada*. The compound exhibited significant activity against *Plasmodium falciparum* chloroquine-resistant strain FC29 with IC_{50} 0.052±0.030 μg/ml.

8.3.5 Anthelminthic activity

Helminthes infections are extremely pervasive, affecting a large proportion of the world's population. The helminthes parasites mainly subsist in the human intestinal tract, but they are also found in tissue, as their larvae migrate towards them. Diseases are of a chronic, incapacitate nature. Normally helminths are a class of eukaryotic parasites. They can be divided into three groups: cestodes, or tapeworms; nematodes, or roundworms; and trematodes, or flukes (Dwivedi *et al.*, 2010). According to WHO, only a few drugs are used in treatment, and research on anthelmintic drugs is sparse. Due to the emergence of anthelmintic resistance to available drugs, plant-derived compounds could play a major role in the treatment. Enwerem *et al.* (2001) isolated betulinic acid from *Berlina grandiflora* and examined it for anthelmintic activities against *Caenorhabditis elegans*. The compound showed significant activity at 100 and 500 ppm comparable to piperazine. Similarly, Lasisi and Kareem (2011) reported betuline and glucoside of betulinic acid from the stem bark of *Bridelia ferruginae*. The compounds exhibited moderate anthelmintic activity against *Fasciola gigantical* (liver fluke), *Taenia solium* (tapeworm) and *Pheritima posthuma* (earthworm, annelid) and also revealed that an increase in concentration of compounds causes death of the parasite. Only a few reports on the efficacy of triterpenoids for anthemintic activity have been documented as per our literature survey; much more research should focus on the discovery of new anthemintic drugs.

8.3.6 Insecticidal activity

In recent years, plant extracts and their derivatives have been of interest for use in insecticides and repellents. Among plant-derived compounds, triterpenoids have received great attention due to the variety of their biological activities. The few reports of triterpenoids as insecticidal agents provide a starting point for further exploitation of bioactive compounds. Mayanti *et al.* (2011) reported tetranortriterpenoids, named as, kokosanolide A and C, isolated from seeds, and onoceranoid-type triterpenoids such as kokosanolide B, 8,14-secogammacera-7,14-diene-3,21-dione and 8,14-secogammacera-7, 14(27)-diene-3,21-dione from the bark of *Lansium domesticum* (kokossan). Among the isolated compounds, kokosanolide B and C showed significant anti-feedant activity against the fourth instar larvae of *Epilachna vigintioctopunctata* at a concentration of 1%. Another study by Lingampally *et al.* (2012), reported betulinic acid from the bark extract of *Ziziphus jujuba* exhibited significant growth-regulating effect against the stored grain pest *Tribolium confusum* [Duval] [*Coleoptera*: Tenebrionidae]. The fifth instar, sixth instar larvae and pupae of test grain pest were treated with 1 μg/μl of betulinic acid and resulted in the abnormal pupae, pupal-adult intermediates and deformed adults, leading to suppression of the pest population. Amin *et al.* (2012) reported the nine compounds from the alcohol extract of *Acanthus montanus* (Nees) T. Anders. Of these, eight compounds resulted in variable degrees of insecticidal activity. β-sitosterol-3-O-β –D-glucoside exhibited potent mosquitocidal activity (100% mortality) against adult *Aedes aegypti* at 1.25 μg/mg, followed by palmitic acid (90%), linaroside (80%), and acetoside (70%) respectively.

8.3.7 Nematicidal activity

Nematodes attack a wide range of organisms; some parasitize and cause severe diseases in plants, making them a great challenge to humans and agriculture. An estimated 80 billion dollars of crop damage annually is caused by root-knot nematodes (Rodriguez-Kabana and Canullo, 1992). Control of disease depends on the fumigation of infected parts of the plant or applications of synthetic fertilizer and pesticides. Due to increased awareness of human health and environmental concerns, much scientific research focuses on the search for new, safer and eco-friendly alternatives in the area of nematode management. There are few scientific reports on triterpenoids as nematicidal agents. Camarinic acid from *Lantana camara* resulted in the death of *Meloidogyne incognita* juveniles at 1.0% concentration (Begum *et al.*, 2000). Similarly, in 2008, Begum *et al.* reported ten more triterpenoids from the same plant, with 1.0 mg/ml concentrations of pomolic acid, lantanolic acid and lantoic acid showing 100% mortality after 24 h; and camarin, lantacin, camarinin, and ursolic acid showing 100% mortality of the root-knot nematode *Meloidogyne incognita* after 48 h compared to furadan. The same group of researchers reported new compounds from different parts of *Cordia latifolia*, such as cordinoic acid from the stem; cordicilin, cordinol and cordicinol from the leaves; cordioic and cordifolic acid from the stem bark; and latifolicin A–D and rosmarinic acid from the fruit. The investigation revealed 100% nematicidal activity with cordinoic acid at 0.5% after 24 h, and cordicilin exhibited more significant nematicidal activity than the free acid against *M. incognita* (Begum *et al.*, 2011). Another investigation by Li *et al.* (2013) reported 23 oleanane-type and eight lupane-type triterpenoid saponins from the roots of *Pulsatilla koreana*, of which compounds such as hederacholchiside E, hederacoside B, raddeanoside R13, hederoside C and pulsatilla saponin D exhibited significant nematicidal activity, with LC_{50} values ranging from 70.1 to 94.7 μg/ml after 48 h against *M. incognita*.

8.3.8 Anticancerous activity

Cancer research is increasing as it becomes clear that a lack of early detection and limited therapeutic drugs are major obstacles to controlling cancer. Recently, drug resistance has become a significant health problem because of inappropriate use of drugs, many side effects of chemo- and radiotherapeutic agents, prolonged use of drugs, etc. To minimize these drug-associated problems, research in search of plant-derived anticancer drugs is gaining in importance. Moulisha *et al.* (2010) studied anticancer activities using the leaves of *Terminalia arjuna*. Their study revealed the presence of ursolic acid. The compound showed significant anticancer activity against cancer cell line, K562 with IC_{50} 7.40 mM. Wang *et al.* (2011) isolated new triterpenoids named as 2a, 3a, 19b, 23b-tetrahydroxyurs-12-en-28-oic acid (THA) from *Sinojackia sarcocarpa*, displaying significant cytotoxicity against cancer cell lines A2780 and HepG2. Anticancerous activity using Pipoly roots revealed the presence of two new triterpenoids: magnosides A and B, which exhibited an inhibitory effect on cancer cell lines with IC_{50} values of 4 and 33mM, respectively (Haddad *et al.*, 2013); similarly lupine-2,3-diol from *Salvia leriifolia* Benth. revealed significant antiproliferative activity against the prostate cancer cell lines (PC3) with IC_{50} 2.7± 0.1mM (Choudhary *et al.*, 2013).

8.3.9 Cytotoxic activity

Medicinal plants are reported to have anticancer, anti-inflammatory, antimicrobial, antiviral, antifungal and antitumor activities and cytotoxic properties in the Ayurvedic system of medicine (Nemati *et al.*, 2013). Various studies on bioactive triterpenoids isolated from plants for cytoxicity assays using different cell lines have been reported. The new tirucallanetype triterpene, 3-α-tigloylmelianol, isolated from the fruit of *Melia azedarach* by Ntalli *et al.* (2010) exhibited moderate antiproliferative activity against the human lung adenocarcinoma epithelial cell line A549. Innocent *et al.* (2011) reported camaric acid and Betulinic acid from the roots of *Lantana viburnoides* sp. that showed high toxicity at LC_{50} = 4.1 μg/ml against brine shrimp larvae. Zhang *et al.* (2012a) reported that the compounds Toonaciliatavarins D and E, isolated from *Toona ciliata* Roem. var. *henryi.* exhibited moderate cytotoxic activity against K562 (leukaemia), SMMC-7721 (hepatocellular carcinoma), MCF-7 (breast cancer) and KB (oral epithelial cancer) human cell lines as well as multi-drug-resistant cell lines MCF-7/ADM and KB/VCR. Similarly, compounds such as 2α,3α,19α-trihydroxy-urs-12-en-24-formyl-28-oic acid, 2α,3β, 21α-trihydroxy-urs-12-en-28-oic acid and 3β-acetyloxy-1-oxo-olean-12-en-28-oic acid, isolated from *Berberis koreana*, were reported to exhibit significant cytotoxicity activity against A549, SK-OV-3, SK-MEL-2, and HCT-15 human tumour cell lines with IC_{50} varying from 7.17 to 48.73 μM (Kim *et al.*, 2012). Another study on cytotoxicity activity by Lavoie *et al.* (2013) reported least activity for lung carcinoma (A549) with IC_{50} 22 μM, using abiesonic acid isolated from *Abies balsamea*. Ruphin *et al.* (2013) isolated new quinone methide pentacyclic triterpenoid derivative compounds from *Salacia leptoclada*, and revealed a cytotoxic effect on murine P388 leukemia cells with IC_{50} 0.041±0.020 μg/ml. Similarly, Pan *et al.* (2014) reported myriaboric acid from the roots of *Croton lachnocarpus* Benth., displaying cytotoxicity with an IC_{50} 42.2 μM against human hepato-cellular carcinoma SMMC-7721 cell line. Currently, researchers are much focused on designing new bioactive potent compounds and their mode of action in combating a number of serious diseases.

Triterpenoids could also be one of the major groups of bioactive compound that can serve as subjects for future research.

8.3.10 Anti-inflammatory activity

Inflammation is a complex biological response by vascular tissue to harmful stimuli caused by injury, cellular changes, malignant infections and environmental agents. It is a protective attempt by the body to remove injurious foreign particles as well as initiate the healing process in the tissue. The attention of researchers focuses on finding out safer and more potent anti-inflammatory drugs. Natural products seem to offer safe and environmentally friendly alternatives to synthetic drugs such as non-steroidal anti-inflammtory drugs (NAIDs), a class that includes ibuprofen and naproxen, which, although they are readily available for the treatment of inflammation, are associated with high-risk side effects (Kumar *et al.*, 2013).

Araruna and Carlos (2010) investigated the compounds 3,14-dihydroxy-11,13-dihydrocostunolide and 8-tigloyl-15-deoxyl-actucin, isolated from *Lactuca sativa*, for anti-inflammatory activity, finding that both compounds exhibited substantial lipoxygenase inhibitory activity and also revealed significant ($p<0.05$) *in vivo* assay, based on the carrageenan-induced paw edema model. Lucetti *et al.* (2010) isolated lupeol acetate for the first time from the latex of *Himatanthus drasticus*. At different concentrations of lupeol acetate, i.e. 10, 25 and 50 mg/kg, these displayed inhibition of first (neurogenic for 0–5 min) and second (inflammatory for 20–25 min) phases. Lupeol acetate also significantly inhibited carrageenan- and dextran-induced paw edemas, as well as the neutrophil migration to the peritoneal cavity, and found it to be a potent anti-inflammatory compound. Three plant species of *Ligustrum*, *L. lucidum* Ait. (LL), *L. pricei* Hayata. (LP) and *L. sinensis* Lour. (LS), were investigated for analgesic/anti-inflammatory activities by Wu *et al.* (2011). Among the plant species, LP revealed the presence of amyrin, betulinic acid and lupeol in higher amounts, and also potential analgesic and anti-inflammatory activity. Another study by Kouam *et al.* (2012) reported four acyclic triterpene derivatives, sapelenins G–J, along with sapelenins A–D, ekeberin D2, (+)-catechin and epicatechinand anderolide G, from the stem bark of *Entandrophragma cylindricum* Sprague. All the compounds showed moderate to significant anti-inflammatory activity, with sapelenin G exhibiting significant inhibitory activity by suppressing the interleukins-17 secretion by PBMCs, without causing any cytotoxic effects. Borges *et al.* (2013) reported that the compounds chiococcasaponin I, III and IV, isolated from the roots of *Chiococca alba* (L.) Hitchc., showed anti-inflammatory properties against *in vitro* lipopolysaccharide-induced inflammation. *Elaeagnus oldhamii* Maxim. leaves extract revealed that isoamericanol B, in addition to four triterpenoids: 1) cis-3-O-p-hydroxycinnamoyl oleanolic acid, 2) trans-3-O-p-hydroxycinnamoyl oleanolic acid, 3) cis-3-O-p-hydroxycinnamoyl ursolic acid and 4) trans-3-O-p-hydroxycinnamoyl ursolic acid, resulted in significant inhibition of the expression of nitric oxide produced in lipopolysaccharide (LPS) stimulated RAW 264.7 cells. Isoamericanol B showed IC_{50} of 10.3± 0.4 μg/ml for inhibition of nitric acid, and also did not show significant changes in cell viability (Liao *et al.*, 2013). Similarly, Wu *et al.* (2014) reported that heptursosides B, heptoleosides B and C, and asiaticoside D, from the stem bark of *Schefflera heptaphylla*, were found to have potent anti-inflammatory activities on lipopolysaccharide-induced nitric oxide production in RAW264.7 cells, without causing a cytotoxic effect.

8.3.11 Antidiabetic activity

Diabetes mellitus is a common cause of death: the number of diabetes cases is predicted to rise from 171 million in 2000 to 366 million by 2030. Long-term medications can contribute to the increase in mortality and morbidity. Two types of diabetes mellitus (DM) are categorized: type 1 is insulin dependent (IDDM), in which the body does not produce insulin; this is commonly found in children and young adults and constitutes about 5–10% of all cases. Type 2 is noninsulin-dependent (NIDDM), in which the body does not produce enough or there is inadequate utilization of secreted insulin; this accounts for 90–95% of diabetes, and is common in older adults due to the greater prevalence of obesity and sedentary lifestyles (Suganya *et al.*, 2014). Conventional treatment regimens such as insulin, oral antidiabetic agents and some enzyme (α-glucosidase) inhibitors such as acarbose etc. have proved to possess limited efficacy, tolerability and significant mechanism-based side effects. Thus, research was initiated to investigate bioactive anti-diabetic compounds of plant origin (Shyam and Ganapaty, 2013). Roots of *Sanguisorba tenuifolia* revealed the presence of two new triterpenoids characterized as 2-oxo-3β, 19α-dihydroxyolean-12-en-28-oic acid β-D-gluco-pyranosyl ester and 2α,19α-dihydroxy-3-oxo-12-ursen-28-oic acid β-D-glucopyranosyl ester, which exhibited significant α-glucosidase inhibitory activity with IC_{50} ranging from 0.62 to 3.62 mM (Kuang *et al.*, 2011). The bacosine, a triterpenoid isolated from *Bacopa monnieri* by Ghosh *et al.* (2011) resulted in the prevention of elevation of glycosylated haemoglobin *in vitro* with IC_{50} 7.44 μg/ml and showed the increase of glycogen content in the liver of diabetic rats and peripheral glucose utilization in the diaphragm of diabetic rats *in vitro* compared to insulin. Kumar *et al.* (2013) reported that antidiabetic activity using bioactive compounds such as betulinic acid, n-heptacosan-7-one, n-nonatriacontan-18-one, quercetin, β sitosterol, stigmasterol, and stigmasteryl palmitate isolated from *Dillenia indica* displayed 47.4, 55.2, 48.8, and 44.3% of α-amylase inhibition and 52.2, 78.2, 52.5, and 34.2% α-glucosidase inhibition activities with betulinic acid, quercetin, β-sitosterol and stigmasterol, respectively at 50μg/kg doses. Whereas, Koneri *et al.* (2014) reported oleanane-type triterpenoid saponins from *Momordica cymbalaria* roots as a potent antidiabetic agent with respect to insulin secretion.

8.3.12 Antioxidant activity

A free radical is defined as any chemical that contains unpaired electrons. The unpaired electron produces a highly reactive free radical, which reacts with inhaled oxygen in our biological system and produces ROS (reactive oxygen species), commonly termed as oxidative stress (Winyard *et al.*, 1988). The increased release of free radicals causes imbalances in the body and leads to deprivation of essential macromolecules such as lipids, protein and DNA. Synthetic antioxidants such as butylated hydroxytoluene and butylated hydroxyanisole have recently been reported to be dangerous for human health. This has led researchers to discover safe, potent and natural antioxidants. Many more studies have been carried out using many *in vitro* preliminary assays, including 2,2-diphenyl-1-picryl-hydrazyl (DPPH), superoxide anion scavenging, hydroxyl radical (OH), nitric oxide radical (NO), ferric-reducing antioxidant power (FRAP), metal chelating activities, etc. Grace-Lynn *et al.* (2012) reported lantadene A, pentacyclic triterpenoids from *Lantana camara*, as potent antioxidant agents. Similarly, Hashem *et al.* (2012) reported 100% free radical scavenging activity with isolated compounds such as sitosterol, erythrodiol, oleanolic acid, ursolic acid and 5',6' norigeumone from *Mayodendron igneum* Kurz. Latif *et al.* (2014) isolated a new triterpenoid, sorbanolic acid, and used *Sorbus lanata*, along with three known compounds, i.e. 3β,23-dihydroxy-lup-20(29)ene-28-oic acid-23-caffeate, 3β,23-dihydroxy-lup-20(29)ene-28-oic acid-3β-caffeate and lyoniside, resulting in significant free radical scavenging assay with IC_{50} values of 24.2, 9.2, 6.0, 21.2 μM, respectively. Liu *et al.* (2014) investigated *Agrimonia pilosa* Ledeb. for antioxidant activity and revealed that the presence of 1β, 2β, 3β, 19α tetrahydroxy-12-en-28-oic acid and corosolic acid exhibited weaker radical scavenging activity with EC_{50} values on DPPH radical, hydroxyl radical and ABTS radical of >100.0, 35.2 and >100.0 μg/mL, respectively.

8.3.13 Anti-platelet activity

Platelet aggregation beyond the purpose of haemostasis is the underlying cause of blood-clotting-related diseases. Platelets can adhere to the walls of the blood vessels, release bioreactive compounds and aggregate to each other. An increase in platelet aggregation leads to the conditions of arterial thrombosis and atherogenesis. Therefore, medicinal plants which are rich in bioactive compounds that inhibit platelet function are of great interest, and it has also been shown that some plants, such as garlic and tomato, may be beneficial in protecting against cardiovascular diseases as a result of platelet aggregation inhibition (Olas *et al.*, 2005). Lupenol isolated from *Elephantopus scaber* L. showed significantly inhibited thrombin-induced platelet aggregation at 100 μg/ml; at 25 μg/ml there was less activity. It was also reported that higher lupenol concentration displayed significant anti-platelet activity, with IC_{50} varied from 60 to 95% (Sankaranarayanan *et al.*, 2010). Similarly, Mosa *et al.* (2011) isolated 3-oxo-5α-lanosta-8, 24-dien-21-oic acid and 3β-hydroxylanosta-9,24-dien-24-oic acid from *Protorhus longifolia* and showed significant thrombin-induced platelet aggregation with IC_{50} of 0.99 mg/ml. Saputri and Jantan (2012) reported selective inhibition on ADP-induced platelet aggregation with garcihombronane D and F isolated from the twigs of *Garcinia hombroniana*.

8.3.14 Other enzyme assays

Triterpenoids also play an important role in various enzymatic inhibitory activities, in which enzymes are the biocatalysts that alter the rate of reaction involved in the different metabolic pathways. Plant-derived products could be potent alternatives to inhibit various enzymes responsible for pathological conditions. Da Silva *et al.* (2007) reported the antiproteolytic and antihaemorrhagic properties of macrolobin A and B isolated from *Pentaclethra macroloba* against *Bothrops* snake venom, acting as inhibitors to neutralize the haemorrhagic, fibrinolytic, and proteolytic activities of class P-I and P-III metalloproteases isolated from *B. neuwiedi* and *B. jararacussu* venoms. Thuong *et al.* (2008) isolated triterpenoids from *Diospyros kaki*, finding that triterpenoids with a 3β-hydroxy group exhibited significant protein tyrosine phosphatase 1B (PTP1B) inhibitory activity, with IC_{50} ranging from 3.1±0.2 to 18.8±1.3 μM, whereas those with a 3α-hydroxy moiety did not inhibit enzyme activity. Compounds such as arloside A methyl ester, 3-O-β-D-xylopyranosyl(1→2)-β-D-glucopyranosyl-28-O-β-D-glucopyranosyl oleanolic acid and chikusetsusaponin IVa isolated from *Panax stipuleanatus* showed significant nuclear factor (NF)-κB activity with IC_{50} values of 6.3, 3.1 to 16.7 μM, respectively, and also reduced TNF-α-induced NF-κB activation by 36.5%, 68.5% and 38.7% at 10 mM concentration in a dose-dependent manner (Liang *et al.*, 2010).

8.4 New Triterpenoid Compounds Isolated from Plants

In the past few decades, many triterpenoids and derivatives have been isolated from various parts of different plants. Many investigations on bioactive triterpenoids of plants have documented numerous new triterpenoids, yet exploitation of such compounds for various biological activities has to be carried out to check its potential. A new triterpenoid saponin, characterized as 3-O-β-D-glucopyranosyl-(1→4)- α-L-arabinopyranosyl-bayogenin-28-α-L-rhamnopyranosyl-(1→4)-β-D-glucopyranosyl-(1→6)-β-D- glucopyranosyl ester was isolated from *Pulsatilla cernua* by Xu *et al.* (2010). Three compounds, 3β-O-(E)-(4-methoxy) cinnamoyl-15 α-hydroxyl β-amyrin, adian-5-en 3-ol and lupeol were reported by Rohini and Das (2010) from *Rhizophora mucronata*. Li *et al.* (2010) reported platycoside N, oleanane-type triterpenoid saponins from *Platycodon grandiflorum* roots. *Euphorbia hirta* revealed the presence of ß-amyrin (Sharma *et al.*, 2010), oleanolic acid from the fruit of *Lagenaria siceraria* (Gangwal *et al.*, 2010), α-amyrin eicosanoate from *Saussurea lappa* roots (Robinson *et al.*, 2010).

Compounds such as clethric acid-28-O-β-d-glucopyranosyl ester, mussaendoside T, β-stigmasterol, hederagenin, ursolic acid, clethric acid, 3β,6β, 19α,24-tetrahydroxyurs-12-en-28-oic acid, mussaendoside I, and cadambine were isolated for the first time by Xu *et al.* (2011) using the leaves of *Anthocephalus chinensis*. Zhou *et al.* (2011) reported alianthusaltinin A, tirucallane-type triterpenoid, from *Ailanthus altissima*; similarly *Crataeva nurvala* revealed betulinic acid and lupeol (Parvin *et al.*, 2011), leaves of *Olax mannii* Oliv. bears glutinol and rhoiptelenol, as reported by Sule *et al.* (2011). Compounds such as α-amyrin caprylate, lupeol, nor-α-amyrin and 3β,28-di hydroxyl olean-12-enyl-palmitate were reported from *Bauhinia variegate* (Saha *et al.*, 2011)

Zhang *et al.* (2012b) reported two new apotirucallane triterpenoids, chisiamols G and H, from the twigs of *Chisocheton paniculatus.* Zuo *et al.* (2012) reported two new cycloartane triterpenoid glycosides, curculigosaponin N and curculigosaponin O, from the rhizomes of *Curculigo orchioides* Gaertn.

Domingues *et al.* (2012) reported ursolic, betulinic, oleanolic, betulonic, 3-acetylursolic, and 3-acetyloleanolic acids from the bark of *Eucalyptus globules.* Jamal *et al.* (2012) isolated glochidonol, Stigmast-22-en-3-ol and Stigmast-5-en-3-ol from *Phyllanthus reticulates.* Dammarane-type triterpenoids such as ginsenosides Rh(18), Rh(19) and Rh(20), along with two new triterpene sapogenins, 12β,23(R)-epoxydammara-24-ene-3β,6α,20(S)-triol and dammara-(20E)22,25-diene-3β,6α,12β, 24S-tetrol, were reported by Li *et al.* (2012) from the stems and leaves of *Panax ginseng* C. A. Mey. Chien *et al.* (2012) reported 2α,3α -dihydroxyolean-12-en-28-al, 3α-hydroxyolean-12-en-30-ol and 3α-hydroxyolean-2-oxo-12-en-28-al from the bark of *Liquidambar formosana* Hance.

Root barks of *Cassine xylocarpa* and *Celastrus vulcanicola* revealed the presence of xyloketal, -3α, 25-epoxy-olean-12-ene, and 3β,21α-dihydroxy-glut-5-ene (Nunez *et al.*, 2013). One new cycloartane-type triterpene, hirtinone, was reported by Vieira *et al.* (2013) from *Trichilia hirta.* A pentacyclic triterpenoid, prunol, isolated from *Prunus cerasoides* D.Don. (Ali and Shaheen, 2013), glaucescic acid from the root extract of *Terminalia glaucescens* (Aiyelaagbe *et al.*, 2014), four newly reported triterpenoids, 2-O-acetyl-3-O-(4′-O-acetyl)- α-arabinopyranosyl maslinic acid, 2-O-acetyl-3-O-(3′-O-acetyl)- α-l-arabinopyranosylmaslinic acid, 2-O-acetyl-3-O-(3′,4′-O-diacetyl)- α-l-arabinopyranosylmaslinic acid, and 3-O-(3′-O-acetyl)3-α-l-arabinopyranosyloleanolic acid, were isolated from the ethyl acetate extract of *Garcinia hanburyi* resin by Wang *et al.* (2014). These untested natural triterpenoids could be significant in the quest for new drugs in further research studies.

8.5 Conclusion and Future Perspectives

Medicinal plants are universally considered as natural reservoirs consisting of numbers of biologically active chemical substances with potential therapeutic effects. The triterpenoids are considered to be among the major bioactive constituents. Furthermore, an extensive characterization of different triterpenoids and their derivatives from plants may provide greater insight into the field of drug discovery. Although many of the experimental results validate that triterpenoids exhibit significant pharmacological effects, few reports about toxicity evaluations have been documented. The study of the biosynthetic pathway and defensive role of triterpenoids and their derivatives may encourage further study and make a valuable contribution to phytochemical analysis. Continuing study into the bioactive phyto-compounds is a fascinating field of research, because of their pharmacologically and uniquely complex structures. Thus, detailed investigations in this field would help to develop effective new drugs, which could satisfy the demand of sectors such as pharmaceuticals, medicine, agriculture and industry.

References

Acharya, C. and Khan, N.A. (2013) A triterpenoid saponin from the seeds of *Ricinus communis* and its antimicrobial activity. *Chemistry of Natural Compounds* 49, 54–57.

Ahmad, Z., Mehmood, S., Ifzal, R., Malik, A., Afza, N. and Ashraf, M. (2007) A new ursane-type triterpenoid from *Salvia santolinifolia*. *Turkish Journal of Chemistry* 31, 495–502.

Aiyelaagbe, O., Olaoluwa, O., Oladosuand, I. and Gibbons, S. (2014) A new triterpenoid from *Terminalia glaucescens* (Planch. ex Benth.). *Records of Natural Products* 8, 7–11.

Ali, L. and Shaheen, F. (2013) Isolation and structure elucidation of a new triterpenoid from *Prunus cerasoides* D. Don. *Records of Natural Products* 7, 80–85.

Amin, E., Radwan, M.M., El-Hawary, S.S., Fathy, M.M., Mohammed, R. and Becnel, J.J. (2012) Potent insecticidal secondary metabolites from the medicinal plant *Acanthus montanus*. *Records of Natural Products* 6, 301–305.

Araruna, K. and Carlos, B. (2010) Anti-inflammatory activities of triterpene lactones from *Lactuca sativa*. *Phytopharmacology* 1, 1–6.

Babaamer, Z.Y., Sakhri, L., Al-Jaber, H.I., Al-Qudah, M.A. and Abu Zarga, M.H. (2012) Two new taraxasterol-type triterpenes from *Pergularia tomentosa* growing wild in Algeria. *Journal of Asian Natural Products Research* 14, 1137–1143.

Bagavan, A., Rahuman, A.A., Kaushik, N.K. and Sahal, D. (2011) *In vitro* antimalarial activity of medicinal plant extracts against *Plasmodium falciparum*. *Parasitological Research* 108, 15–22.

Begum, S., Wahab, A., Siddiqui, B.S. and Qamar, F. (2000) Nematicidal constituents of the aerial parts of *Lantana camara*. *Journal of Natural Products* 63, 765–767.

Begum, S., Zehra, S.Q., Siddiqui, B.S., Fayyaz, S. and Ramzan, M. (2008) Pentacyclic triterpenoids from the aerial parts of *Lantana camara* and their nematicidal activity. *Chemical Biodiversity* 5, 1856–1866.

Begum, S., Perwaiz, S., Siddiqui, B.S., Khan, S., Fayyaz, S. and Ramzan, M. (2011) Chemical constituents of *Cordia latifolia* and their nematicidal activity. *Chemical Biodiversity* 8, 850–861.

Borges, R.M., Valença, S.S., Lopes, A.A., Barbi, N. and Silva, A.J. (2013) Saponins from the roots of *Chiococca alba* and their *in vitro* anti-inflammatory activity. *Phytochemistry Letters* 6, 96–100.

Cassels, B.K. and Asencio, M. (2011) Anti-HIV activity of natural triterpenoids and hemisynthetic derivatives 2004–2009. *Phytochemistry Reviews* 10, 545–564.

Chen, J.C., Liu, W.Q., Lu, L., Qiu, M.H., Zheng, Y.T. and Yang, L.M. (2009) Kuguacins F-S, cucurbitane triterpenoids from *Momordica charantia*. *Phytochemistry* 70, 133–140.

Chianese, G., Yerbanga, S.R., Lucantoni, L., Habluetzel, A., Basilico, N. and Taramelli, D. (2010) Antiplasmodial triterpenoids from the fruits of Neem, *Azadirachta indica*. *Journal of Natural Products* 73, 1448–1452.

Chiang, Y.M., Su, J.K., Liu, Y.H. and Kuo, Y.H. (2001) New cyclopropyl-triterpenoids from the aerial roots of *Ficus microcarpa*. *Chemical Pharmacy Bulletin* 49, 581–583.

Chien, S., Xiao, J., Tseng, Y., Kuo, Y. and Wang, S. (2012) Composition and antifungal activity of balsam from *Liquidambar formosana* Hance. *Holzforschung* 67, 345–351.

Choi, J.W., Cho, E.J., Lee, D.G., Choi, K., Ku, J. and Park, K. (2012) Antibacterial activity of triterpenoids from *Clerodendron trichotomum*. *Journal of Applied Biology and Chemistry* 55, 169–172.

Choudhary, M.I., Hussain, A., Adhikari, A., Marasini, B.P., Sattar, S.A. and Atiat-tul-Wahab (2013) Anticancer and α-chymotrypsin inhibiting diterpenes and triterpenes from *Salvia leriifolia*. *Phytochemistry Letters* 6, 139–143.

Dahake, R., Roy, S., Patil, D., Rajopadhye, S., Chowdhary, A. and Deshmukh, R.A. (2013) Potential anti-HIV activity of *Jatropha curcas* Linn. Leaf Extracts. *Journal of Antivirus and Antiretrovirus* 5(7), 160–165.

Dat, N.T., Lee, I.S., Cai, X.F., Shen, G. and Kim, Y.H. (2004) Oleanane triterpenoids with inhibitory activity against NFAT transcription factor from *Liquidambar formosana*. *Biological and Pharmaceutical Bulletin* 27, 426–428.

da Silva, J.O., Fernandes, R.S., Ticli, F.K., Oliveira, C.Z., Mazzi, M.V. and Franco, J.J. (2007) Triterpenoid saponins, new metalloprotease snake venom inhibitors isolated from *Pentaclethra macroloba*. *Toxicon* 50, 283–91.

De Leon, L., Lopez, M.R. and Moujir, L. (2010) Antibacterial properties of zeylasterone, a triterpenoid isolated from *Maytenus blepharodes*, against *Staphylococcus aureus*. *Microbiological Research* 165, 617–626.

de Silva, M.L., David, J.P., Silva, L.C., Santos, R.A., David, J.M. and Lima, L.S. (2012) Bioactive oleanane, lupane and ursane triterpene acid derivatives. *Molecules* 17, 12197–12205.

Dhanalakshmi, D., Dhivya, R. and Manimegalai, K. (2013) Antibacterial activity of selected medicinal plants from South India. *Hygeia Journal of Drugs and Medicine* 1, 63–68.

Domingues, R.M., Oliveira, E.L., Freire, C.S., Couto, R.M., Simões, P.C. and Neto, C.P. (2012) Supercritical fluid extraction of *Eucalyptus globulus* bark – a promising approach for triterpenoid production. *International Journal of Molecular Science* 13, 7648–7662.

Dong, S.H., Zhang, C.R., Dong, L., Wu, Y. and Yue, J.M. (2011) Onoceranoid-type triterpenoids from *Lansium domesticum*. *Journal of Natural Products* 74, 1042–1048.

Duraipandiyan, V., Gnanasekar, M. and Ignacimuthu, S. (2010) Antifungal activity of triterpenoid isolated from *Azima tetracantha* leaves. *Folia Histochemica et Cytobiologica* 48, 311–313.

Dwivedi, G., Rawal, D., Nagda, S. and Jain, T. (2010) Anthelmintic activity of Tea (*Camellia sinensis*) extract. *International Journal of Pharma Science and Research* 1(11), 451–453.

Enwerem, N.M., Okogun, J.I., Wambebe, C.O., Okorie, D.A. and Akah, P.A. (2001) Anthelmintic activity of the stem bark extracts of *Berlina grandiflora* and one of its active principles, Betulinic acid. *Phytomedicine* 8, 112–114.

Finefield, J.M., David, S.H., Kreitman, M. and Williams, R.M. (2012) Enantiomeric natural products: occurrence and biogenesis. *Angewandte Chemie International Edition in English* 51, 4802–4836.

Gadhvi, R., Reddy, M.N. and Mishra, G.J. (2013) Isolation of terpenoids constituents from *Lippia nodiflora* by preparative HPTLC method. *International Journal of Medicinal Plants and Alternative Medicine* 1, 104–109.

Gangwal, A., Parmar, S.K. and Sheth, N.R. (2010) Triterpenoid, flavonoids and sterols from *Lagenaria siceraria* fruits. *Der Pharmacia Lettre* 2, 307–317.

Gao, X., Li, Y., Shu, L., Shen, Y., Yang, L. and Yang, L. (2013) New triterpenoids from the fruits of *Schisandra wilsoniana* and their biological activities. *Bulletin of the Korean Chemical Society* 34, 827–830.

Ghosh, P., Mandal, A., Chakraborty, M. and Saha, A. (2010) Triterpenoids from *Quercus suber* and their antimicrobial and phytotoxic activities. *Journal of Chemical Pharmacy Research* 2, 714–721.

Ghosh, T., Maity, T.K. and Singh, J. (2011) Antihyperglycemic activity of bacosine, a triterpene from *Bacopa monnieri*, in alloxan-induced diabetic rats. *Planta Medica* 77, 804–808.

Grace-Lynn, C., Darah, I., Chen, Y., Latha, L.Y., Jothy, S.L. and Sasidharan, S. (2012) *In vitro* antioxidant activity potential of Lantadene A, a pentacyclic triterpenoid of *Lantana* plants. *Molecules* 17, 11185–11198.

Haddad, M., Lelamer, A., Banuls, L.M., Vasquez, P., Carraz, M. and Vaisberg, A. (2013) *In vitro* growth inhibitory effects of 13,28-epoxyoleanane triterpenes saponins in cancer cells. *Phytochemistry Letters* 6, 128–134.

Hashem, F.A., Sengab, A.E., Shabana, M.H. and Khaled, S. (2012) Antioxidant activity of *Mayodendron igneum* Kurz and the cytotoxicity of the isolated terpenoids. *Journal of Medicinally Active Plants* 1, 88–97.

Inayama, S., Hori, H., Pang, G.M., Nagasawa, H. and Ageta, H. (1989) Isolation of a hopane-type triterpenoid, Zeorin, from a higher plant, *Tripterygium regelii*. *Chemical and Pharmaceutical Bulletin* 37, 2836–2837.

Innocent, E., Shah, T., Nondo, R.S.O. and Moshi, M.J. (2011) Antibacterial and cytotoxic triterpenoids from *Lantana viburnoides* ssp. *viburnoides* Var. *kisi*. *Spatula DD* 1, 213–218.

Jamal, A., Nasser, W.A. and Yaacob, L.B.D. (2012) Triterpenoids from *Phyllanthus reticulatus*, *International Conference on Applied Chemical Pharma Science*, pp. 19–20.

Janovik, V., Boligon, A.A., Frohlich, J.K., Schwanz, T.G., Pozzebon, T.V. and Alves, S.H. (2012) Isolation and chromatographic analysis of bioactive triterpenoids from the bark extract of *Cariniana domestica* (Mart) Miers. *Natural Products Research* 26, 66–71.

Jiménez-Arellanes, A., Luna-Herrera, J., Cornejo-Garrido, J., López-García, S., Castro-Mussot, M.E. and Meckes-Fischer, M. (2013) Ursolic and oleanolic acids as antimicrobial and immunomodulatory compounds for tuberculosis treatment. *BMC Complementary and Alternative Medicine* 13, 258–269. DOI: 10.1186/1472-6882-13-258.

Kavitha, K.S. and Satish, S. (2013) Evaluation of antimicrobial and antioxidant activities from *Toona ciliate* Roemer. *Journal of Analytical Science and Technology* 4, 23. DOI: 10.1186/2093-3371-4-23.

Kim, K.H., Choi, S.U., Kim, C.S. and Lee, K.R. (2012) Cytotoxic steroids from the trunk of *Berberis koreana*. *Bioscience Biotechnology and Biochemistry* 76, 825–827.

Kiplimo, J.J., Koorbanally, N.A. and Chenia, H. (2011) Triterpenoids from *Vernonia auriculifera* Hiern exhibit antimicrobial activity. *African Journal of Pharmacy and Pharmacology* 5, 1150–1156.

Koneri, R.B., Suman, S. and Ramaiah, C.T. (2014) Antidiabetic activity of a triterpenoid saponin isolated from *Momordica cymbalaria* Fenzl. *Indian Journal of Experimental Biology* 52, 46–52.

Kouam, S.F., Kusari, S., Lamshöft, M., Tatuedom, O.K., Spiteller, M. and Sapelenins, G.J. (2012) Acyclic triterpenoids with strong anti-inflammatory activities from the bark of the Cameroonian medicinal plant *Entandrophragma cylindricum*. *Phytochemistry* 83, 79–86.

Kuang, H., Li, H., Wang, Q., Yang, B., Wang, Z. and Xia, Y. (2011) Triterpenoids from the roots of *Sanguisorba tenuifolia* var. Alba. *Molecules* 16, 4642–4651.

Kumar, S., Kumar, V. and Prakash, O. (2013) Enzymes inhibition and antidiabetic effect of isolated constituents from *Dillenia indica*. *BioMed Research International* 2013, 1–7.

Kuroyanagi, M., Kawahara, N., Sekita, S., Satake, M., Hayashi, T. and Takase, Y. (2003) Dammarane-type triterpenes from the Brazilian medicinal plant *Cordia multispicata*. *Journal of Natural Products* 66, 1307–1312.

Lasisi, A.A. and Kareem, S.O. (2011) Evaluation of anthelmintic activity of the stem bark extract and chemical constituents of *Bridelia ferruginae* (Benth) Euphorbiaceae. *African Journal of Plant Science* 5, 469–474.

Latif, A., Hussain, S.H., Ali, M., Arfan, M., Ahmed, M. and Cox, R.J. (2014) A new antioxidant triterpenoid from the stem wood of *Sorbus lanata*. *Records of Natural Products* 8, 19–24.

Lavoie, S., Gauthier, C., Legault, J., Mercier, S., Mshvildadze, V. and Pichette, A. (2013) Lanostane- and cycloartane-type triterpenoids from *Abies balsamea* oleoresin. *Beilstein Journal of Organic Chemistry* 9, 1333–1339.

Lee, S.Y., Eom, S.H., Kim, Y.K., Park, N.I. and Park, S.U. (2009) Cucurbitane-type triterpenoids in *Momordica charantia* Linn. *Journal of Medicinal Plants Research* 3, 1264–1269.

Li, L., Huang, X., Sattler, I., Fu, H., Grabley, S. and Lin, W. (2006) Structure elucidation of a new friedelane triterpene from the mangrove plant *Hibiscus tiliaceus*. *Magnetic Resonance Chemistry* 44, 624–628.

Li, W., Zhang, W., Xiang, L., Wang, Z., Zheng, Y. and Wang, Y. (2010) Platycoside N: a new oleanane-type triterpenoid saponin from the roots of *Platycodon grandiflorum*. *Molecules* 15, 8702–8708.

Li, K.K., Yang, X.B., Yang, X.W., Liu, J.X. and Gong, X.J. (2012) New triterpenoids from the stems and leaves of *Panax ginseng*. *Fitoterapia* 83, 1030–1035.

Li, W., Sun, Y.N., Yan, X.T., Yang, S.Y., Lee, S.J. and Byun, H.J. (2013) Isolation of nematicidal triterpenoid saponins from *Pulsatilla koreana* root and their activities against *Meloidogyne incognita*. *Molecules* 18, 5306–5316.

Liang, C., Ding, Y., Nguyen, H.T., Kim, J.A., Boo, H.J. and Kang, H.K. (2010) Oleanane-type triterpenoids from *Panax stipuleanatus* and their anticancer activities. *Bioorganic and Medicinal Chemistry Letters* 20(23), 7110–7115.

Liao, C.R., Ho, Y.L., Huang, G.J., Yang, C.S., Chao, C.Y. and Chang, Y.S. (2013) One lignanoid compound and four triterpenoid compounds with anti-inflammatory activity from the leaves of *Elaeagnus oldhamii* Maxim. *Molecules* 18, 13218–13227.

Lien, T.P., Kamperdick, C., Schmidt, J., Adam, G. and Van Sung, T. (2002) Apotirucallane triterpenoids from *Luvunga sarmentosa* (Rutaceae). *Phytochemistry* 60, 747–754.

Lin, Y., Wang, C., Chen, I., Jheng, J., Li, J. and Tung, C. (2013) TIPdb: a database of anticancer, antiplatelet, and antituberculosis phytochemicals from indigenous plants in Taiwan. *The Scientific World Journal* 2013, Article ID 736386, 4. DOI: 10.1155/2013/736386.

Lingampally, V., Solanki, V.R., Jayaram, V., Kaur, A. and Raja, S.S. (2012) Betulinic acid: a potent insect growth regulator from *Ziziphus jujuba* against *Tribolium confusum* [Duval]. *Asian Journal of Plant Science and Research* 2, 198–206.

Liu, X., Zhu, L., Tan, J., Zhou, X., Xiao, L., Yang, X. and Wang, B. (2014) Glucosidase inhibitory activity and antioxidant activity of flavonoid compound and triterpenoid compound from *Agrimonia pilosa* Ledeb. *BMC Complementary and Alternative Medicine* 14, 12.

Lucetti, D.L., Lucetti, E.C.P., Bandeira, M.A., Veras, H., Silva, A.H. and Leal, L.K. (2010) Anti-inflammatory effects and possible mechanism of action of lupeol acetate isolated from *Himatanthus drasticus* (Mart.) Plumel. *Journal of Inflammation* 7, 60–71.

Mann, A., Ibrahim, K., Oyewale, A.O., Amupitan, J.O., Fatope, M.O. and Okogun, J.I. (2011) Antimycobacterial friedelane-terpenoid from the root bark of *Terminalia avicennioides*. *American Journal of Chemistry* 1, 52–55.

Mann, A., Ibrahim, K., Oyewale, A.O., Amupitan, J.O., Fatope, M.O. and Okogun, J.I. (2012) Isolation and elucidation of three triterpenoids and its antimycobacterial activity of *Terminalia avicennioides*. *American Journal of Organic Chemistry* 2, 14–20.

Manoharan, K.P., Benny, T.K. and Yang, D. (2005) Cycloartane type triterpenoids from the rhizomes of *Polygonum bistorta*. *Phytochemistry* 66, 2304–2308.

Manzano, P.I., Miranda, M., Abreu-Payrol, J., Silva, M., Sterner, O. and Peralta, E.L. (2013) Pentacyclic triterpenoids with antimicrobial activity from the leaves of *Vernonanthura patens* (Asteraceae). *Emirites Journal of Food Agriculture* 25, 539–543.

Mayanti, T., Tjokronegoro, R., Supratman, U., Mukhtar, M.R., Awang, K. and Hadi, A.H. (2011) Antifeedant triterpenoids from the seeds and bark of *Lansium domesticum* cv *kokossan* (Meliaceae). *Molecules* 16, 2785–2795.

McChesney, J.D., Dou, J., Sindelar, R.D., Goins, D.K., Walker, L.A. and Rogers, R.D. (1997) Tirucallane-type triterpenoids: NMR and X-ray diffraction analyses of 24-epi-piscidinol A and piscidinol A. *Journal of Chemical Crystallography* 27, 283–290.

Moodley, R., Chenia, H., Jonnalagadda, S.B. and Koorbanally, N. (2011) Antibacterial and anti-adhesion activity of the pentacyclic triterpenoids isolated from the leaves and edible fruits of *Carissa macrocarpa*. *Journal of Medicinal Plants Research* 5, 4851–4858.

Mosa, R.A., Oyedeji, A.O., Shode, F.O., Singh, M. and Opoku, A.R. (2011) Triterpenes from the stem bark of Protorhus longifolia exhibit anti-platelet aggregation activity. *African Journal of Pharmaceuticals and Pharmacology* 5, 2698–2714.

Moulisha, B., Kumar, G.A. and Kanti, H.P. (2010) Anti-leishmanial and anti-cancer activities of a pentacyclic triterpenoid isolated from the leaves of *Terminalia arjuna* Combretaceae. *Tropical Journal of Pharmaceutical Research* 9, 135–140.

Nemati, F., Dehpouri, A.A., Eslami, B., Mahdavi, V. and Mirzanejad, S. (2013) Cytotoxic properties of some medicinal plant extracts from Mazandaran, Iran. *Iran Red Crescent Medical Journal* 15(11), e8871.

Nkanwen, E.R., Gojayev, A.S., Wabo, H.K., Bankeu, J.J., Iqbal, M.C. and Guliyev, A.A. (2013) Lanostane-type triterpenoid and steroid from the stem bark of *Klainedoxa gabonensis*. *Fitoterapia* 86, 108–114.

Ntalli, N.G., Cottiglia, F., Bueno, C.A., Alché, L.E., Leonti, M. and Vargiu, S. (2010) Cytotoxic tirucallane triterpenoids from *Melia azedarach* fruits. *Molecules* 15, 5866–5877.

Nunez, M.J., Ardiles, A.E., Martinez, M.L., Torres-Romero, D., Jimenez, I.A. and Bazzocchi, I.L. (2013) Triterpenoids from *Cassine xylocarpa* and *Celastrus vulcanicola* (Celastraceae). *Phytochemistry Letters* 6, 148–151.

Olas, B., Wachowicz, B., Stochmal, A. and Oleszek, W. (2005) Inhibition of blood platelet adhesion and secretion by different phenolics from *Yucca schidigera* Roezl. bark. *Nutrition* 45, 199–206.

Pan, Z.H., Ning, D.S., Liu, J.L., Pan, B. and Li, D.P. (2014) A new triterpenoid saponin from the root of *Croton lachnocarpus* Benth. *Natural Products Research* 28, 48–51.

Parvin, S., Kader, M.A., Muhit, M.A., Haque, M.E., Mosaddik, M.A. and Wahed, M.I. (2011) Triterpenoids and phytosteroids from stem bark of *Crataeva nurvala* Buch Ham. *Journal of Applied Pharmacy Science* 1, 47–50.

Prachayasittikul, S., Saraban, P., Cherdtrakulkiat, R., Ruchirawat, S. and Prachayasittikul, V. (2010) New bioactive triterpenoids and antimalarial activity of *Diospyros rubra* Lec. *EXCLI Journal* 9, 1–10.

Rabi, T. and Bishayee, A. (2009) Terpenoids and breast cancer prevention. *Breast Cancer Research and Treatment* 115, 223–239.

Ramalhete, C., Cruz, F.P., Lopes, D., Mulhovo, S., Rosário, V.E. and Prudêncio, M. (2011) Triterpenoids as inhibitors of erythrocytic and liver stages of *Plasmodium* infections. *Bioorganic and Medicinal Chemistry* 19, 7474–7481.

Revathi, P., Jeyaseelansenthinath, T. and Thirumalai-kolundhusubramaian, P. (2014) Preliminary phytochemical screening and GC-MS analysis of ethanolic extract of mangrove plant-*Bruguiera cylindrica* (Rhizho) L. *International Journal of Pharmacognosy and Phytochemical Research* 6(4), 729–740.

Robinson, A., Yashvanth, S., Babu, S.K., Roa, J.M. and Madhavendra, S.S. (2010) Isolation of α-amyrin eicosanoate, a triterpenoid from the roots of *Saussurea lappa* Clarke – differential solubility as an aid. *Journal of Pharmaceutical Science and Technology* 2, 207–212.

Rodriguez-Kabana, R. and Canullo, G.H. (1992) Cropping systems for the management of phytonematodes. *Phytoparasitica* 20(3), 211–224.

Rohdich, F., Bacher, A. and Eisenreich, W. (2005) Isoprenoid biosynthetic pathways as anti-infective drug targets. *Biochemical Society Transactions* 33, 785–791.

Rohini, R.M. and Das, A.K. (2010) Triterpenoids from the stem bark of *Rhizophora mucronata*. *Natural Products Research* 24, 197–202.

Ruphin, F.P., Baholy, R., Emmanue, A., Amelie, R., Martin, M.T. and Koto-te-Nyiwa, N. (2013) Antiplasmodial, cytotoxic activities and characterization of a new naturally occurring quinone methide pentacyclic triterpenoid derivative isolated from *Salacia leptoclada* Tul. (Celastraceae) originated from Madagascar. *Asian Pacific Journal of Tropical Biomedicine* 3, 780–784.

Saha, S., Subrahmanyam, E.V.S., Kodangalad, C. and Shastry, S. (2011) Isolation and characterization of triterpenoids and fatty acid ester of triterpenoid from leaves of *Bauhinia variegate*. *Der Pharma Chemica* 3, 28–37.

Sankaranarayanan, S., Bama, P., Ramachandra, J., Jayasimman, R., Kalaichelvan, P.T. and Deccaraman, M. (2010) *In vitro* platelet aggregation inhibitory effect of triterpenoid compound from the leaf of *Elephantopus scaber* Linn. *International Journal of Pharmacy and Pharmaceutical Sciences* 2, 49–51.

Saputri, F.C. and Jantan, I. (2012) Inhibitory activities of compounds from the twigs of *Garcinia hombroniana* Pierre on human low-density lipoprotein (LDL) oxidation and platelet aggregation. *Phytothermal Research* 26, 1845–1850.

Sharma, S., Vijayvergia, R. and Singh, T. (2010) Extraction and identification of pentacyclic compound ß-Amyrin (Terpenoid). *Archives of Applied Science Research* 2, 124–126.

Shyam, T. and Ganapaty, S. (2013) Evaluation of antidiabetic activity of methanolic extracts from the aerial parts of *Barleria montana* in Streptozotocin induced diabetic rats. *Journal of Pharmacognosy and Phytochemistry* 2(1), 187–192.

Suganya, G., Kumar, S.P., Dheeba, B. and Sivakumar, R. (2014) *In vitro* antidiabetic, antioxidant and anti-inflammatory activity *of Clitoria ternatea* L. *International Journal of Pharmacy and Pharmaceutical Sciences* 6, 342–347.

Sule, M.I., Hassan, H.S., Pateh, U.U. and Ambi, A.A. (2011) Triterpenoids from the leaves of *Olax mannii* Oliv. *Nigerian Journal of Basic and Applied Science* 19, 193–196.

Tanaka, R. and Matsunaga, S. (1987) Supinenolones A, B and C, fernane type triterpenoids from *Euphorbia supine*. *Phytochemistry* 28, 3149–3154.

Tanaka, T., Koyano, T., Kowithayakorn, T., Fujimoto, H., Okuyama, E. and Hayashi, M. (2001) New multiflorane-type triterpenoid acids from *Sandoricum indicum*. *Journal of Natural Products* 64, 1243–1245.

Thuong, P.T., Lee, C.H., Dao, T.T., Nguyen, P.H., Kim, W.G. and Lee, S.J. (2008) Triterpenoids from the leaves of *Diospyros kaki* (persimmon) and their inhibitory effects on protein tyrosine phosphatase 1B. *Journal of Natural Products* 71, 1775–1778.

Truong, N.B., Pham, C.V., Doan, H.T., Nguyen, H.V., Nguyen, C.M. and Nguyen, H.T. (2011) Antituberculosis cycloartane triterpenoids from *Radermachera boniana*. *Journal of Natural Products* 74, 1318–1322.

Vieira, I.J., Ode, A.A., de Souza, J.J., Braz-Filho, R., Mdos, G.S. and de Araújo, M.F. (2013) Hirtinone, a novel cycloartane-type triterpene and other compounds from *Trichilia hirta* L. (Meliaceae). *Molecules* 18, 2589–2597.

Wang, O., Liu, S., Zou, J., Lu, L., Chen, L. and Qiu, S. (2011) Anticancer activity of 2α, 3α, 19β, 23β-Tetrahydroxyurs-12-en-28-oic acid (THA), a novel triterpenoid isolated from *Sinojackia sarcocarpa*. *PLoS One* 6, e21130.

Wang, H.M., Liu, Q.F., Zhao, Y.W., Liu, S.Z., Chen, Z.H. and Zhang, R.J. (2014) Four new triterpenoids isolated from the resin of *Garcinia hanburyi*. *Journal of Asian Natural Products Research* 16, 20–28.

Wilson, A.D. (2013) Diverse applications of electronic-nose technologies in agriculture and forestry. *Sensors* 13, 2295–2348.

Winyard, P.G., Arundel, L.A. and Blake, D.R. (1988) Lipoprotein oxidation and induction of feroxidase activity in stored human extracellular fluids. *Free Radical Research* 5, 227–235.

Wu, C., Hseu, Y., Lien, J., Lin, L., Lin, Y. and Ching, H. (2011) Triterpenoid contents and anti-inflammatory properties of the methanol extracts of *Ligustrum* species leaves. *Molecules* 16, 1–15.

Wu, C., Duan, Y.H., Tang, W., Li, M.M., Wu, X. and Wang, G.C. (2014) New ursane-type triterpenoid saponins from the stem bark of *Schefflera heptaphylla*. *Fitoterapia* 92, 127–132.

Xu, Y., Bai, L., Liu, Y., Liu, Y., Xu, T. and Xie, S. (2010) A new triterpenoid saponin from *Pulsatilla cernua*. *Molecules* 15, 1891–1897.

Xu, X.Y., Yang, X.H., Li, S.Z. and Song, Q.S. (2011) Two new triterpenoid glycosides from the leaves of *Anthocephalus chinensis*. *Journal of Asian Natural Products Research* 13, 1008–1013.

Yang, G., Li, Y., Zhang, X., Li, X., Yang, L. and Shi, Y. (2011) Three new nortriterpenoids from *Schisandra wilsoniana* and their anti-HIV-1 activities. *Natural Products and Bioprospecting* 1, 33–36.

Yinusa, I., George, N.I. and Amupitan, J.O. (2012) Isolation and bioactivity of pentacyclic triterpenoid (Betunilic acid) from the bark of *Sarcocephalus latifolius* (Smith Bruce). *Journal of Natural Sciences Research* 2, 13–23.

Yu, J., Liu, S. and Xuan, L. (2012) Two new lupene-type triterpenoids from the roots of *Liquidambar formosana*. *Natural Products Research* 26, 630–636.

Zhang, F., Wang, J.S., Gu, Y.C. and Kong, L.Y. (2012a) Cytotoxic and anti-inflammatory triterpenoids from *Toona ciliata*. *Journal of Natural Products* 75, 538–546.

Zhang, F., Xiu-Feng, H.E., Wen-Bin, W.U., Chen, W. and Yue, J. (2012b) New apotirucallane-type triterpenoids from *Chisocheton paniculatus*. *Natural Products and Bioprospecting* 2, 235–239.

Zhao, C.L., Cui, X.M., Chen, Y.P. and Liang, Q. (2010a) Key enzymes of triterpenoid saponin biosynthesis and the induction of their activities and gene expressions in plants. *Natural Product Communications* 5, 1147–1158.

Zhao, Y.H., Deng, T.Z., Chen, Y., Liu, X.M. and Yang, G.Z. (2010b) Two new triterpenoids from *Lycopodium obscurum* L. *Journal of Asian Natural Products Research* 12, 666–671.

Zhao, C., Shao, J. and Fan, J. (2012) A new triterpenoid with antimicrobial activity from *Anemone rivularis*. *Chemistry of Natural Compounds* 48, 803–805.

Zhou, X., Xu, M., Li, X., Wang, Y., Gao, Y. and Cai, R. (2011) Triterpenoids and sterones from the stem bark of *Ailanthus altissima*. *Bulletin of the Korean Chemical Society* 32, 127–130.

Zuo, A.X., Shen, Y., Jiang, Z.Y., Zhang, X.M., Zhou, J. and Lü, J. (2012) Two new triterpenoid glycosides from *Curculigo orchioides*. *Journal of Asian Natural Products Research* 14, 407–412.

9 Roles of Secondary Metabolites in Protection and Distribution of Terrestrial Plants under Climatic Stresses

ASMA HAMMAMI-SEMMAR[1] AND NABIL SEMMAR[2]*

[1]*Institut National des Sciences Appliquées et de Technologie (INSAT), Tunis, Tunisia;* [2]*Institut Supérieur des Sciences Biologiques Appliquées de Tunis (ISSBAT), Université de Tunis El Manar (UTM), Tunis, Tunisia*

Abstract

Climatic stress includes several physical-chemical factors which permanently exert pressure on the plant world. These factors include air temperature, light, drought, CO_2 and ozone concentrations, and show either (i) rapid or regular fluctuations or (ii) long and cumulative change trends. These variations lead to alarming or threatening environmental conditions that plants face through curative, defensive or anticipative responses based on biosynthesis, ratio regulation and storage of secondary metabolites (SMs). SMs concern terpenes, phenolic and nitrogen-containing compounds which are highly active and qualitatively (structurally) diversified natural compounds. These quantitative and qualitative chemical variations are specific to plant species and closely dependent on type and level of environmental conditions. On this basis, specific chemical features could be attributed to different plant–environment interactions. These SM features provide reliable and flexible tools to dynamically survey spatio-temporal variations of plant diversity.

9.1 Introduction

Terrestrial plants are permanently exposed to multiple environmental stresses exerting selective pressure on biodiversity. Efficient protections against these pressures are provided by secondary metabolites (SMs), which are minor compounds highly diversified by their chemical structures, roles and biological activities. Permanent, systematic and sudden external variations are often experienced by plants from different climatic stresses. Climatic stresses include low and high temperatures, drought, light radiation, carbon dioxide and ozone excesses, etc. Protective roles of SMs against these threat factors include antioxidative, volatile fresh and osmoregulator compounds belonging to different chemical classes. There are three main classes of SMs: terpenes, phenols and nitrogen compounds (Semmar *et al.*, 2008a).

Production, storage (accumulation), emission and metabolic ratio regulation of SMs represent different protective and warning responses of plants to environmental stresses. At plant community scale, different protective roles can be attributed to different classes of secondary metabolites. This was highlighted by some general variation trends observed in a given class of SMs under a given stress type. Repeated observations confirming such metabolic responses help to clarify the roles of SMs in plant populations. At a lower scale, i.e. at plant population scale, the chemical structures, levels and variation rates of SMs provide a large range of biochemical features, characterizing each plant species within the plant kingdom. This provides interesting ways to correlate biodiversity states with environmental conditions.

*E-mail: nabilsemmar@yahoo.fr

This chapter is organized into two parts.

1) The first main part focuses on the general protective role of secondary metabolites in the plant kingdom against physical (9.2) and chemical (9.3) climatic stresses. It is illustrated by several specific responses characterizing different plant species under each stress type. This part gives synthetic idea on different phytochemical responses for plant protection against local stress conditions.
2) The second part (9.4) introduces correlations between phytochemical features and air compositions of wide geographical areas.

9.2 Protective Roles of Secondary Metabolites in Plants against Physical Climatic Stresses

9.2.1 Temperature stress

Temperature stress can originate from cold or warm air. Plants acclimate to cold or warm air by producing appropriate SMs at efficient levels (amounts).

9.2.1.1 Cold stress

Under low temperatures, plants show several biochemical and biophysical changes, including lower metabolic rates as well as variations of protein conformations and membrane fluidity. Plants can either acclimate to cold air or anticipate freezing before freezing temperatures occur (Levitt, 1980; Janská *et al.*, 2010).

To avoid internal water freezing, several plants decrease their water osmotic potential via accumulation of anthocyanins (Fig. 9.1) (Chalker-Scott, 2002). Many plants avoid damage of sensitive tissues through water super-cooling as low as −41°C before freezing (Burke and Stushnoff, 1979; Gilmour *et al.*, 1988). Such super-cooling is often induced by increasing solute levels in xylem ray parenchyma cells, dormant flower buds and leaf tissues (Ishikawa, 1984; Chalker-Scott, 1992; Chalker-Scott, 2002). Epidermic accumulations of anthocyanins help plants to reduce cell osmotic potential, leading to anticipated resistance against frost-originated freezing. This process is linked to the fact that anthocyanins (which are glycosides) are extremely soluble in water, favouring their accumulation in vacuoles (Robinson, 1991; Chalker-Scott, 2002). In this way, anthocyanin-rich vacuoles will acquire a lower water-freezing temperature, helping plants to endure cold conditions.

This anti-freezing protection was observed in *Arabidopsis thaliana* (Brassicaceae), *Cichorium intybus* (Asteraceae) and *Plantago lanceolata* (Plantaginaceae) (Fig. 9.1) (Gilmour *et al.*, 1988; McKown *et al.*, 1996). After exposure to low temperatures for 48 h, *A. thaliana* showed an important development of anthocyanin-rich cortical cells (Leyva *et al.*, 1995; Ruelland *et al.*, 2009). Apart from *A. thaliana*, anthocyanin levels increased in *Plantago lanceolata* grown under cool temperatures (15°C for 16 h day and 10°C for 8 h night) compared with high-temperature conditions (27°C for 16 h day and 22°C for 8 h night) (Jaakola and Hohtola, 2010). In *Cichorium intybus* (chicory), temperature significantly influenced anthocyanin biosynthesis, which was (i) maximal under 10°C (day) vs 15°C (night); (ii) lower as day and night temperatures increase until quasi-total inhibition at 30°C (day) vs 25°C (night) (Boo *et al.*, 2006). Chemical analyses of immature fruit peel in pear and apple revealed anthocyanin biosynthesis induction subsequently to cold fronts (Steyn *et al.*, 2009). Moreover, the beneficial role of anthocyanin accumulation against cold conditions has been shown in maple, pine, plantain and maize (Walter, 1967; Creasy, 1968; Saure, 1990; Krol *et al.*, 1995; Curry, 1997; Oren-Shamir and Levi-Nissim, 1997a; Oren-Shamir and Levi-Nissim, 1997b; William *et al.*, 2001; Chalker-Scott, 2002; Steyn *et al.* 2004; Stilesa *et al.*, 2007; Hao *et al.*, 2009).

In *Arabidopsis thaliana*, low temperatures induce phenylalanine ammonium lyase (PAL) and chalcone synthase (CHS) by a light-dependent method (Leyva *et al.*, 1995; Ruelland *et al.*, 2009). PAL and CHS are key enzymes of phenolic metabolism in plants (Rhodes and Wooltorton, 1977; Tanaka and Uritani, 1977; Graham and Patterson, 1982; Vogt, 2010), and are transcriptionally regulated (Dangl, 1992; Salinas, 2002; Yamaguchi-Shinozaki and Shinozaki, 2006; Van Buskirk and Thomashow, 2006). Under light conditions, PAL and CHS mRNAs of *A. thaliana* accumulated at 4°C and decreased at 20°C (Leyva *et al.*, 1995). Comparative experiments showed that *A. thaliana* living under low temperature accumulated significantly more PAL and CHS transcripts in continuous light than in dark (Leyva *et al.*, 1995; Ruelland *et al.*, 2009). Accumulation of PAL protein was also demonstrated in *Brassica napus* and maize living under low temperatures (Parra *et al.*, 1990; Solecka and Kacperska, 2003).

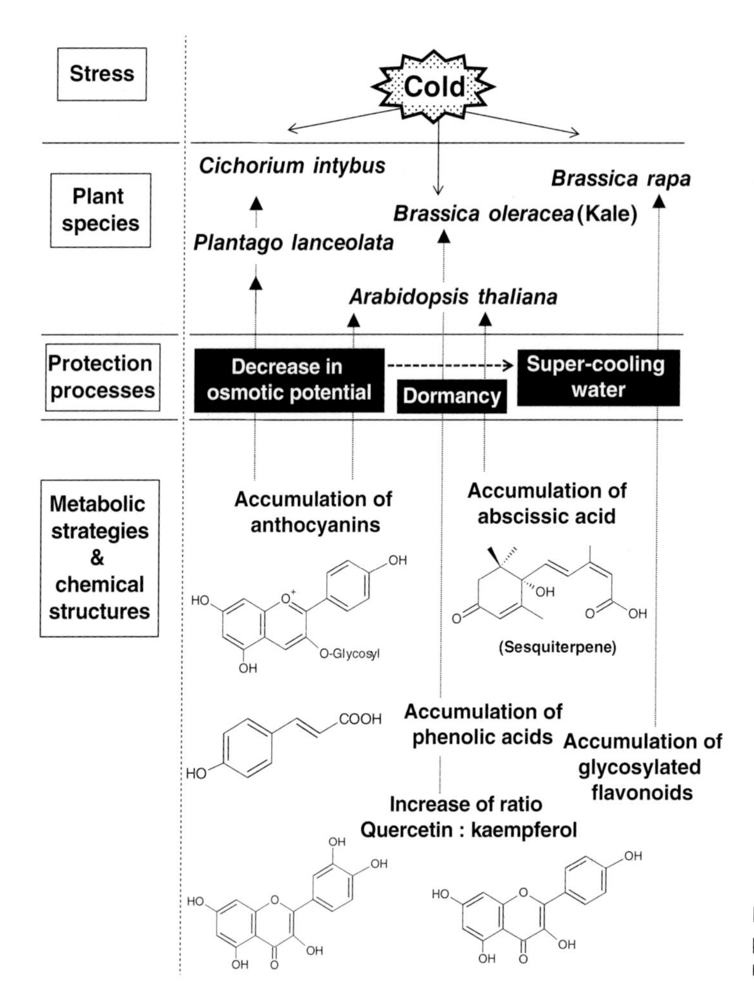

Fig. 9.1. Study cases illustrating protective roles played by secondary metabolites against cold stress.

Apart from the anthocyanins, other flavonoids showed significant variation of concentrations under low temperatures. Kaempferol glycosides of *Brassica rapa* (pak choi) accumulated more at 9°C compared to 22°C (Fig. 9.1). The ratio of quercetin to kaempferol showed an increasing trend with lower external temperatures (Jaakola and Hohtola, 2010): for example, in kale, this metabolic ratio significantly increased when temperature declined from 9.7 to 0.3°C (Fig. 9.1) (Schmidt *et al.*, 2010). Also, this species was found to accumulate more phenolic acids for seven days under 20–24 kJm^{-2} day^{-1} UV-B radiation at low temperatures, compared to the same light condition at higher temperatures. The significant increase in concentration concerned caffeoyl-malate, hydroxyferuloyl-malate, coumaroyl-malate, feruloyl-malate and sinapoyl-malate (Fig. 9.2) (Harbaum-Piayda *et al.*, 2010).

Apart from phenolic compounds, cold conditions increase the biosynthesis of abscissic acid (ABA) (Fig. 9.1). This metabolite is a sesquiterpenic hormone playing a role in plant dormancy through cell division inhibition and leaf fall. Several studies revealed the importance of abscissic acid in plant freezing tolerance: in *Arabidopsis thaliana*, exogenous treatment with ABA enhanced freezing tolerance (Lang *et al.*, 1989; Gusta *et al.*, 2005). ABA is a sesquiterpene playing a key hormonal role in the dormancy mechanism. ABA-defective *Arabidopsis* mutants such as los5 and frs1 (freezing sensitive 1) showed significantly lower cold acclimation than wild type (Heino *et al.*, 1990; Llorente *et al.*, 2000). In these mutants, ABA and transcripts of cold-inducible genes showed low levels, causing reduced freezing tolerance (Llorente *et al.*, 2000; Xiong *et al.*, 2001).

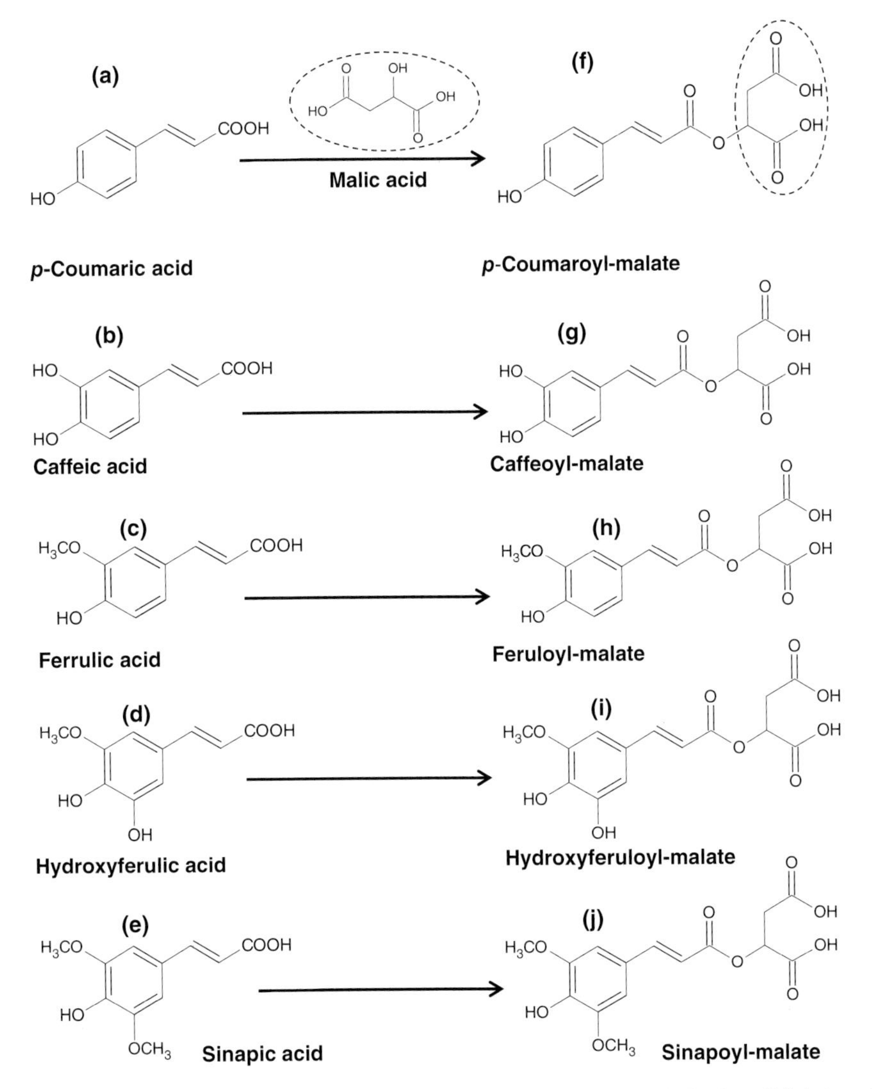

Fig. 9.2. Chemical structures of cinnamic acids (C6C3) (a–e) and their malonilated derivatives (f–j) induced in the kale *Brassica oleracea* living in cold environmental conditions.

9.2.1.2 Warm stress

Elevated temperatures induce in several plant species accumulation and/or emission of volatile terpenes, among which isoprene is the most dominant compound (Fig. 9.3). Isoprene or 2-methyl, 1,3-butadiene is a chloroplastic C5 hydrocarbon; it is widely found in the troposphere because of its emission by plants, among other volatile compounds. Biogenic volatile organic compounds (classically abbreviated as BVOCs) can be emitted by storing or non-storing plant species; the first emitters have accumulation compartments of synthesized BVOCs by opposition to non-storing emitters. These compartments are specialized tissues that can be ducts, glands, etc.

Studies of terpene-storing species such as *Pinus halepensis* (Llusià and Peñuelas, 1998; Blanch *et al.*, 2011), *Pinus elliottii* (Tingey *et al.*, 1980), *Salvia mellifera* and *Mentha* x *piperita* (Tingey *et al.*, 1991) have shown that temperature strongly affects emission. In *Pinus halepensis*, monoterpene emission rates increased exponentially with increasing temperature (Fig. 9.3) (Llusià and Peñuelas, 1998). In *Ponderosa* pine, the increase in needle temperature resulted in higher monoterpene emission (Lerdau *et al.*, 1994).

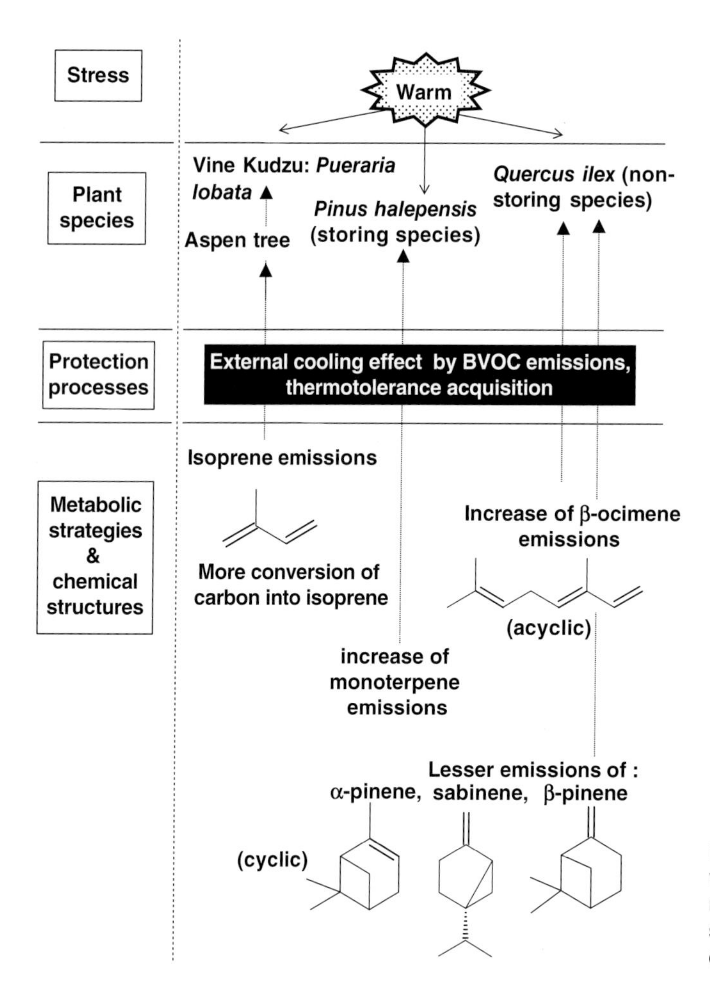

Fig. 9.3. Different study case highlighting warm-protective roles of secondary metabolites in different terrestrial plant species. Legend: BVOC, biogenic volatile organic compounds.

The Mediterranean non-storing oak species *Quercus ilex* is known to have light-dependent emission. In a laboratory study, it showed some temperature-dependent behaviour, emitting monoterpenes as follows: the two acyclic isomers cis-β-ocimene and trans-β-ocimene were hardly detected below 35°C, but their emission rates increased above this temperature (Fig. 9.3) (Staudt and Bertin, 1998). However, the cyclic monoterpenes, α-pinene, sabinene and β-pinene, had very reduced emission rates at 35°C, whereas they accounted for 85% of total emitted monoterpenes (4.1 ± 0.75 nmol.m^{-2}.s^{-1}) at 30°C and saturating light (Fig. 9.3). A recent study on *P. sylvestris* showed that the terpenoids emitted by heated needles are markedly higher than those emitted by needles in the natural environment (Zhao *et al.*, 2012). At a temperature of 200°C (simulating forest fire conditions) within 15 min, the emission was considerable: 16.314 μg.g^{-1} DW for total and 10.321 μg.g^{-1} DW for α-pinene (major emitted compound).

In the vine kudzu (*Pueraria lobata*), isoprene emissions decreased with cold conditions and increased with the temperature: the disappearance of isoprene emissions was observed in the leaves of plants growing at 19°C (Loreto and Sharkey, 1990). Transition from 20 to 30°C resulted in a small improvement of isoprene emission rate (0.5 nmol.$m^{-2}s^{-1}$) (Sharkey and Loreto, 1993). However, plants growing at 24°C increased their isoprene emission within 24 h after the external temperature was equal to 30°C. At 40°C, isoprene emission included 30% of taken up photosynthetic carbon under light flux of 2800 μmol photons $m^{-2}s^{-1}$; this was compatible with the decreasing and increasing effects of high temperatures on photosynthesis and isoprene emission, respectively (Fig. 9.3) (Sharkey and Loreto, 1993).

In other experiments, one leaf of kudzu was subjected to 30°C versus 20°C for the rest of the plant (Sharkey and Loreto, 1993). After the leaves were held at 30°C, isoprene production occurred at 4.5 h and continued for 9 h after the temperature fell to 20°C. The other leaves of the canopy did not produce isoprene. The same study reported that in the range of temperatures inducing isoprene emission, increasing by 10°C resulted in emission increase by about eight times in kudzu (Sharkey and Loreto, 1993). This increasing rate is higher than values reported in other plant species: it was reported that positive increment of 10°C results in two to four times higher isoprene emission (Tingey *et al.*, 1979, Tingey *et al.*, 1987; Monson and Fall, 1989; Loreto and Sharkey, 1990).

Isoprene emission subsequent to increasing temperatures was also shown in Aspen trees originating from high altitudes of the Front Range rocky mountain (3000 m) (Monson *et al.*, 1994). The plants do not initiate isoprene emission until six weeks after emergence. However, leaves suddenly exposed to warm temperatures increased their isoprene emission rates within minutes. Moreover, the basal emission rate continued to increase on each successive day during which warm temperatures were experienced, up to a maximum. This was explained by some thermotolerance effects acquired by trees through their isoprenoid emissions under high temperatures (Singsaas *et al.*, 1997; Delfine *et al.*, 2000): this adaptive mechanism could be linked to temperature lowering of emitting leaves as a consequence of the evaporation of isoprenoids. Shallcross and Monks (2000) suggested a surrounding cooling effect subsequent to isoprene emission Thus, the high amounts of emitted isoprenoids could make the photosynthetic apparatus more resistant to high temperatures.

9.2.2 Light stress

9.2.2.1 Direct UV-B stress

UV radiations can be classified into three types defined by three different wavelength ranges: UV-A (320–400 nm), UV-B (280–320 nm) and UV-C (≤ 280 nm). UV-C are highly energetic and extremely damaging to biological systems. They are strongly absorbed by stratospheric ozone and oxygen and subsequently removed from sunlight reaching the earth surface (Caldwell *et al.*, 1989). UV-A are substantially less damaging than UV-B and UV-C; their level reaching the earth surface is independent from ozone concentration because they are not attenuated by the ozone layer (Caldwell *et al.*, 1989; Madronich *et al.*, 1998). UV-B are harmful and they are primarily absorbed by stratospheric ozone, but they can reach the earth's surface with variable levels depending on several factors: altitude, solar zenith angle, clouds, aerosol, surface albedo, stratospheric ozone layer thickness, etc. (Webb, 1997). Release of man-made chlorine and bromide compounds (e.g. chlorofluorocarbons, CFCs) depleted stratospheric ozone causing a hole, leading to more UV-B reaching the earth. UV-B radiation also causes damage to nucleic acids, proteins and lipids, exposing the biological (plant) world to serious threats (Larson and Berenbaum, 1988; Huang *et al.*, 1993; Paul *et al.*, 1997). Moreover, UV-B radiation induces the formation of free radicals, which are potential chain-oxidative agents. In response to UV-B exposure, plants produce several types of protective secondary metabolites, which are UV-absorbing compounds and free radical scavengers (Searles *et al.*, 2001; Caldwell *et al.*, 2003). These metabolites are efficient UV-B-absorbing compounds with generally phenolic structures (phenolic acids or flavonoids) (Fig. 9.4). Subsequently, increased epidermal levels of these compounds should provide effective screening for the mesophyll (Mabry *et al.*, 1970; Ryan *et al.*, 1998; Fischbach *et al.*, 1999; Treutter, 2006). Absorption efficiency of UV-B has been found to depend on different factors, including the presence of substitutions (hydroxyls, sugars, acids) at some positions of absorbing molecules, combined with metabolic ratios between molecules. Apart from phenolic compounds (flavonoids, phenolic acids, coumarins, tannins), carotenoids (tetraterpenes) and alkaloids are other types of metabolites which vary significantly in different plant species exposed to UV-B (Fig. 9.4).

Concerning the flavonoids, up-regulation of ortho-dihydroxyflavonols represents a general acclimation process to UV-B exposure in different plant species such as *Arabidopsis*, *Petunia* and *Trifolium* (Hofmann *et al.*, 2000; Ryan *et al.*, 2002a; Ryan *et al.*, 2002b). Mutant plants in which flavonoids were poorly dihydroxylated showed lower growth rates than wild plants (Ryan *et al.*, 2002a; Ryan *et al.*, 2002b). Moreover, an increase of ortho-dihydroxylated flavonoids relative to monohydroxylated ones seemed to be a general phytochemical response to UV-B exposure in several plant species (Ryan *et al.*, 1998; Markham *et al.*, 1998a; Olsson *et al.*, 1999a;

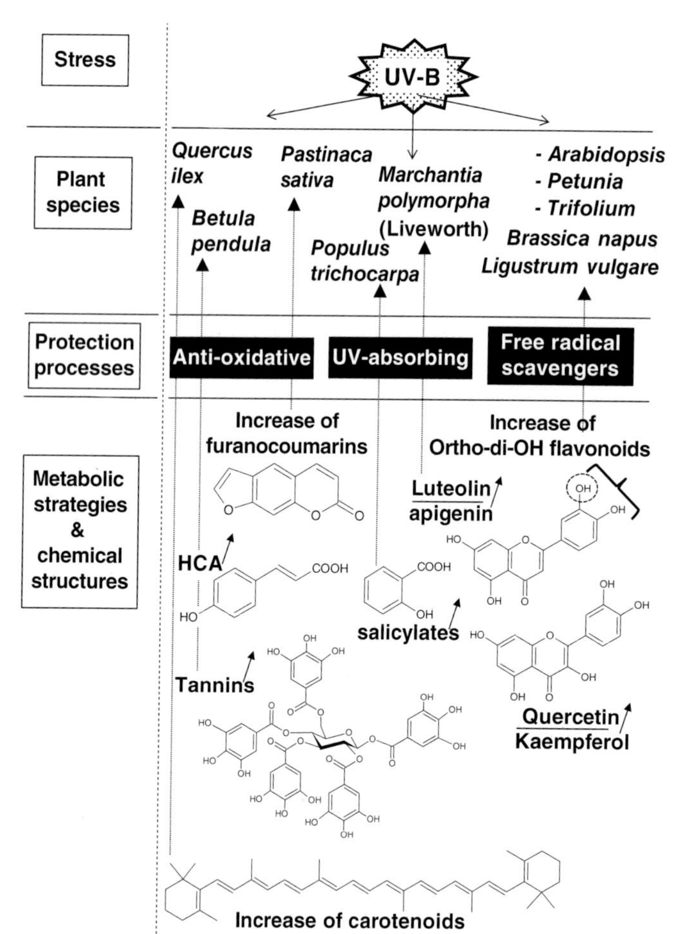

Fig. 9.4. Different study case highlighting UV-protective roles of secondary metabolites in different terrestrial plant species. Legend: HCA, hydroxycinnamic acid.

Hofmann *et al.*, 2000; Reifenrath and Müller, 2007; Winter and Rostàs, 2008; Agati and Tattini, 2010). For example, in *Marchantia polymorpha* (liverwort), UV-B exposure increases the ratio of luteolin to apigenin (Ryan *et al.*, 1998). Also, in responses of plants to UV-B stress, several plant species showed an elevated ratio of quercetin to kaempferol, highlighting the importance of ortho-dihydroxylation (Fig. 9.4) (Ryan *et al.*, 1998; Olsson *et al.*, 1999a; Hofmann *et al.*, 2000). In *Brassica napus* (Brassicaceae), UV-B exposure resulted in an increase of rutin (Fig. 9.5a) and quercetin-3-*O*-D-sophoroside-7-*O*-*D*-glucoside (Fig. 9.5c) among the flavonol pool (Gerhardt *et al.*, 2008). In *Ligustrum vulgare* (Oleaceae), the ortho-dihydroxylated flavonol and flavone glycosides _ quercetin-3-O-rutinoside (or rutine) (Fig. 9.5a) and luteolin-7-O-glucoside (Fig. 9.5b) showed accumulation after the plant was exposed to sunlight excess.

This may be due to the high dissipative ability of harmful UV-B radiation by dihydroxyflavones and flavonols compared to monohydroxyflavones (Smith and Markham, 1998; Tattini *et al.*, 2004). Phenolic compounds with ortho-dihydroxylated structures benefit from higher antioxidative activity than monohydroxylated cases (Montesinos *et al.*, 1995; Smith and Markham, 1998; Markham *et al.*, 1998b; Olsson *et al.*, 1999b; Hofmann *et al.*, 2000; Hofmann *et al.*, 2003; Keski-Saari *et al.*, 2005). Such antioxidative capacity was demonstrated by metal-chelating and free radical scavenging abilities in the ortho-dihydroxylated structures (Rice-Evans *et al.*, 1996). At cellular level, flavonoids are glycosylated to be stored in the vacuole; this avoids contact between unused or reserve flavonoids and the superoxide radical (O_2^-) (Olsson et al., 1999a; Neill and Gould, 2003). The O_2^- is a thermodynamically unstable radical in an aqueous solution, and

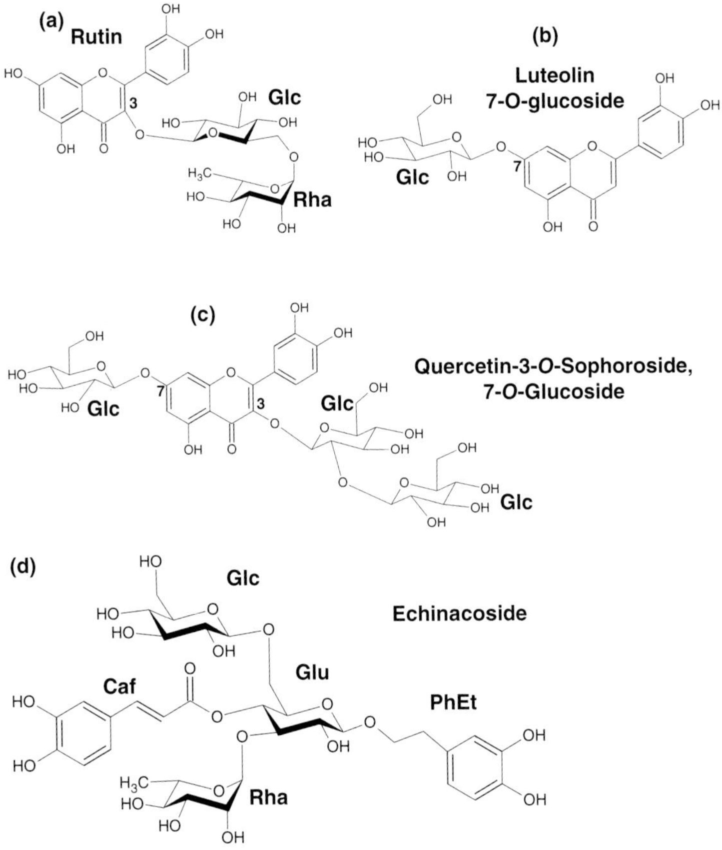

Fig. 9.5. Chemical structures of different phenolic compounds playing a protective role against UV-B stress in terrestrial plants. Legend: Caf, cafeic acid; Glc, glucose; Rha, rhamnose; PhEt, phenylethanoid.

its formation occurs widely in nature from the respiration process. Nevertheless, under severe light stress, H_2O_2 (formed from O_2^- by superoxide dismutase action) can enter the vacuole by tonoplastic diffusion; in these cases, quercetin-3-O-rutinoside (Fig. 9.5c) and luteolin-7-O-glucoside (Fig. 9.5b) can play efficient protective roles by their antioxidant properties (Yamasaki *et al.*, 1997; Olsson *et al.*, 1999a; Neill and Gould, 2003;).

The variation of phenolic metabolism in response to UV-B exposure was shown to be modulated by the acylation degree and type, as follows.

In *Ligustrum vulgare* (Oleaceae), excess of light induced a decrease in the ratio of hydroxycinnamates to total polyphenols. This was mainly due to a decrease in *p*-coumaric acid content (Fig. 9.2a). This decrease was compensated by a higher echinacoside biosynthesis, providing strong scavenging activity against the radical O_2^- (Tattini *et al.*, 2004).

Phytochemical responses to UV-B exposure also included accumulation of phenolic acids: under UV-B intensification, barley and birch (*Betula pendula* and *Betula resinifera*) up-regulated their hydroxycinnamic acid (HCA; also called coumaric acid) (Fig. 9.4) (Liu *et al.*, 1995; Lavola, 1998). In *Populus trichocarpa* (a tree species), salicylate level increased under UV-B exposure (Fig. 9.4).

Ligustrum vulgare (Oleaceae) exposed to excess of light showed well-structured quantitative and qualitative distribution of flavonoids and HCA derivatives in some tissues (Christensen *et al.*, 1998; Blount *et al.*, 2000; Ruegger and Chapple, 2001; Tattini *et al.*, 2004): with exposure to the sun, flavonoids were revealed to be present only in the abaxial epidermal layer of sharply sloped leaves, whereas hydroxycinnamates were found in flatly angled leaves. Among phenolic compounds, furanocoumarins show plant-species dependent variations under UV-B exposure (Fig. 9.4): the concentrations of bergapten, isopimpinellin and xanthotoxin were shown to be increased in wild parsnip (*Pastinaca sativa*) (Fig. 9.6b–d) (Zangerl and Berenbaum, 1987).

However, enhanced UV-B led to a decrease in furanocoumarins in *Citrus jambhiri* (Asthana *et al.*, 1993). On the other hand, such irradiation conditions decreased the ratio of bergapten to psoralen in this plant species (Fig. 9.6a, b) (McCloud *et al.*, 1992).

Among the plant responses to UV-B exposures, tannins, carotenoids and alkaloids have been reported to vary significantly in some species (Fig. 9.4). Tannins are phenolic polymers derived from two possible precursors: gallic acid and flavan-3-ol (Fig. 9.7a, b). Polymerizations of these two phenolic compounds give hydrolysable and condensed tannins, respectively (Fig. 9.7c, 9.7d). Tannins were found to increase in *Laurus nobilis* and *Betula pendula* under the effects of UV-B enhancement (Grammatikopoulos et al., 1998; Lavola, 1998) contrary to the species *Calluna vulgaris*, *Ceratonia siliqua*, *Phlomis fructicosa*, *Rubus chamaemorus* and *Vaccinium* spp. (Gehrke *et al.*, 1995; Levizou and Manetas, 2001).

(a) Psoralene

(b) OCH_3 Bergaptene

(c) OCH_3 Xanthotoxin

(d) OCH_3 OCH_3 Isopimpinellin

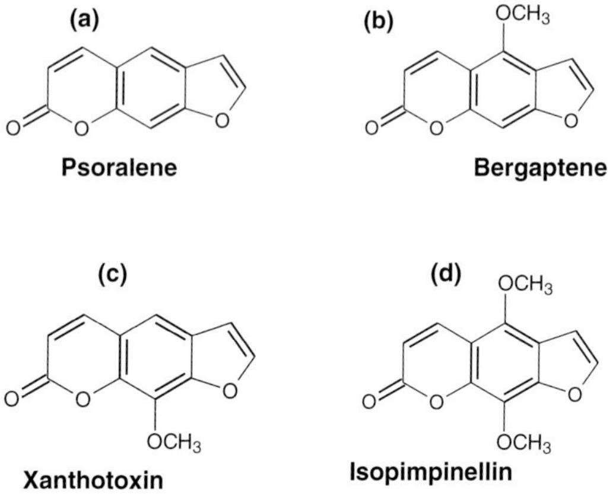

Fig. 9.6. Chemical structures of different furanocoumarins playing protective roles in terrestrial plants against UV-B stress.

Under enhanced UV-B, carotenoids increased in *Quercus ilex* (Fig. 9.5) (Filella and Penuelas, 1999) and decreased in soybean (*Glycine max*) (Vu *et al.*, 1981; Vu *et al.*, 1982; Singh, 1996) and rice (*Oryza sativa*) (Ambasht and Agrawal, 1997). In polar bryophytes and angiosperms, increasing doses of UV-B had a stimulatory effect on carotenoids (Newsham and Robinson, 2009).

Alkaloids represent another class of secondary metabolites characterized by the presence of intracyclic or lateral chain nitrogen (Fig. 9.8). These metabolites were found in the leaves of *Catharanthus roseus* exposed to UV-B enhancement (Hirata *et al.*, 1992; Hirata *et al.*, 1993; Ouwerkerk *et al.*, 1999).

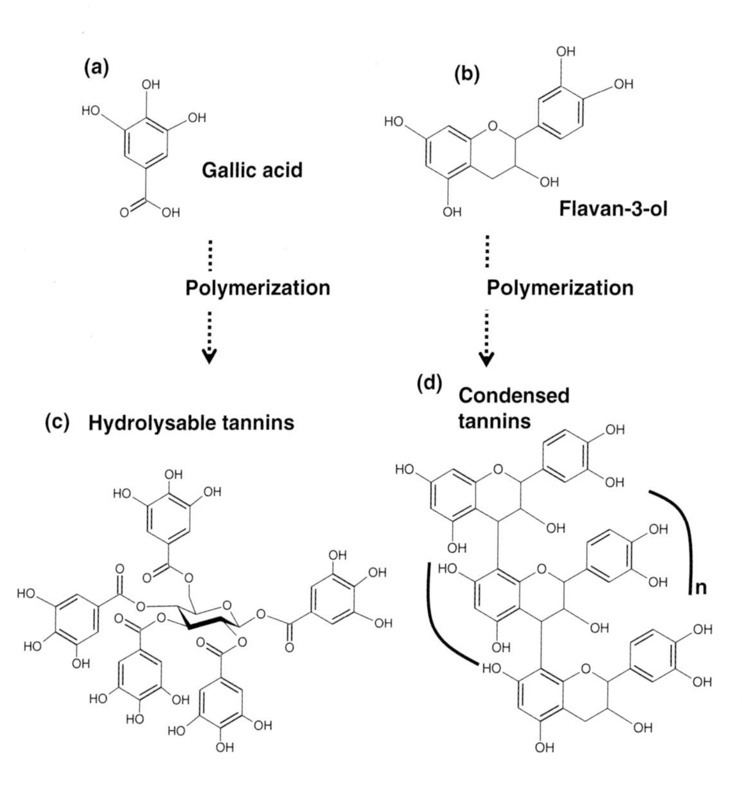

Fig. 9.7. Basic chemical structures of hydrolysable (c) and condensed (d) tannins deriving from gallic acid (a) and flavan-3-ol (b), respectively. Tannins have strong antioxidative activity because of their polyhydroxylation and play a protective role against UV-B in plants.

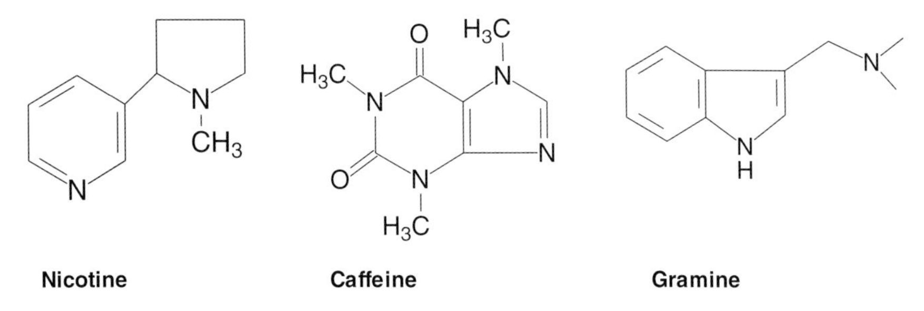

Fig. 9.8. Some alkaloid structures showing the occurrence of nitrogen in heterocycle or lateral chain.

This was linked to UV-B sensitive expression of the gene encoding TRYPD (Ouwerkerk *et al.*, 1999). However, high UV-B due to mountain elevation resulted in destruction of alkaloids in *Aquilegia caerulea* (Larson *et al.*, 1990).

9.2.2.2 *Light stresses resulting from interactions of UV with other radiations*

Interaction effects between UV-B and other radiation types on SMs have been subjected to different studies (Fig. 9.9): *Arabidopsis thaliana* produced more quercetin derivatives when it was strongly exposed to both UV-B and PPFD (photosynthetic photon flux density) than under high UV-B combined with low PPFD (Goetz *et al.*, 2010). In *Brassica napus*, exposure of plants to UV-A + PAR prior to UV-B + PAR (PAR: photosynthetically active radiations) resulted in down-regulation of some UV-B induced flavonoids (Wilson *et al.*, 2001).

These observations showed some 'memory' or historical effect, associating some plant metabolic responses to some temporal successions of light conditions. In another study on *Brassica napus*, far red (FR) radiation suppressed the UV-B dependent accumulation of most quercetin and kaempferol derivatives, possibly due to the phytochrome effect. The ratio of quercetin to kaempferol decreased by a factor of two under an increasing FR level, both in the presence and absence of UV-B (Gerhardt *et al.*, 2008). The low red : far red ratio is indicative of shade conditions where UV-B is low. Therefore, it was suggested that under high levels of FR (shade conditions), the phytochrome system might reduce the UV-B-protective flavonoids (Fiscus and Booker, 1995). Suppressing the biosynthesis of the UV-B protective flavonoids when they are not needed could represent an efficient resource/energy management method for shaded leaves or plants. This FR-dependent phytochrome system would minimize unnecessary utilization of metabolic precursors in the flavonoid biosynthetic pathways.

9.3 Protective Roles of Secondary Metabolites in Plants against Chemical Climatic Stresses

9.3.1 Water stress

Water stress is directly linked to water availability and can be characterized by three parameters: drought level, drought duration and past weather (or 'plant history'). These three factors significantly influence emission and storage levels of terpenes among other plant metabolic responses. These phytochemical responses showed high variability depending on plant species, co-occurring populations and living area. Several studies showed that drought conditions increase terpene concentrations in storing plant species (Hodges and Lorio, 1975; Gershenzon *et al.*, 1978; Kainulainen *et al.*, 1991): withholding watering conditions induced an increase in terpene concentrations in *Pinus halepensis*, *Pistacia lentiscus* and *Cistus monspeliensis* by 45–90% compared with control plants (Fig. 9.10) (Llusià and Peñuelas, 1998). In *Satureja douglassi*, a decrease in leaf water potential from −6 to −9 bars resulted in a significant increase in terpene concentrations from 17 to 21 mg.g^{-1} dry weight of leaf (Llusià and Peñuelas, 1998).

In addition to the variation of terpene production levels, water stress seems to have a significant effect on different types of stored terpenes: in *Cistus monspeliensis*, irrigation conditions seemed to favour storage only of α-pinene (monoterpene) (Fig. 9.10). However, drought treatment induced storage of both α-pinene and caryophyllene (a sesquiterpene) (Fig. 9.10) (Llusià and Peñuelas,

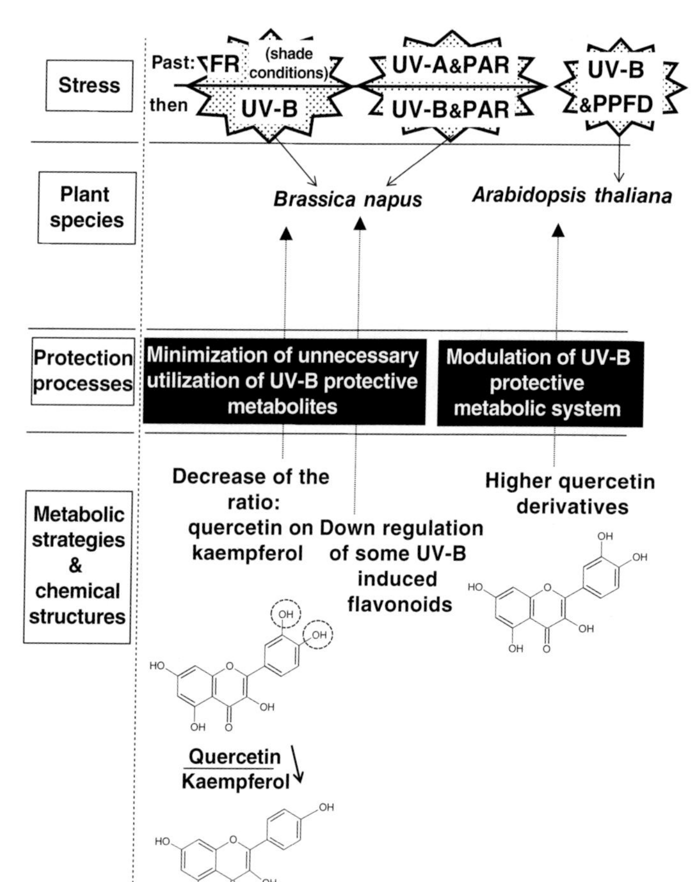

Fig. 9.9. Different study cases highlighting chemical responses of plants under interactive light conditions between UV and other radiation types. Legend: FR, far red; PAR, photosynthetically active radiations; PPFD, photosynthetic photon flux density.

1998). Different plant species react to water stress by storing and/or by varying their emissions of volatile terpenes. Under severe drought conditions, some Mediterranean plant species were shown to dramatically decrease their terpenic emissions. Such phytochemical responses are compatible with water stress-induced reductions of stomacal conductance and net photosynthetic rates. (Tenhunen *et al.*, 1990; Peñuelas *et al.*, 1998). In some plant species, such as *Quercus coccifera* and *Pistacia lentiscus*, terpenes were emitted despite negative net photosynthetic rates; this indicated that even under water stress, plants allocated a part of photosynthetically fixed carbon for terpenic production and emission (Llusià and Peñuelas, 1998). Leaf isoprene emission showing rapid and dramatic rate variations tends to be considered as a general and efficient response by which plants can cope with temperature and water stresses (Sharkey and Loreto, 1993).

Water stress had a positive effect for seven days on monoterpene emissions (MEs) in the leaves of *Pinus halepensis* (Aleppo pine) (Pinaceae) exposed to water withholding for 11 days (Ormeño *et al.*, 2007). From day 1 (t1) to day 11 (t11), MEs significantly increased in comparison with control plants. Moreover, these emissions were significantly correlated to plant water potential (Ψ) (measuring the plant water status). More precisely, MEs increased gradually from t1 until a peak at the seventh day (t7); in the second period (t7–t11), water stress persistence was followed by a slow decline in MEs. During the first phase (t1–t7), increase of MEs was due to specialized storing structures (in the leaves of *P. halepensis*) from which temperature-dependent volatilization of terpenes can occur (Llusià and Peñuelas, 2000). Moreover, water stress results in higher temperatures, positively influencing terpene emissions. In the slash pine (*Pinus elliottii*), monoterpenic emission proved to be uncoupled

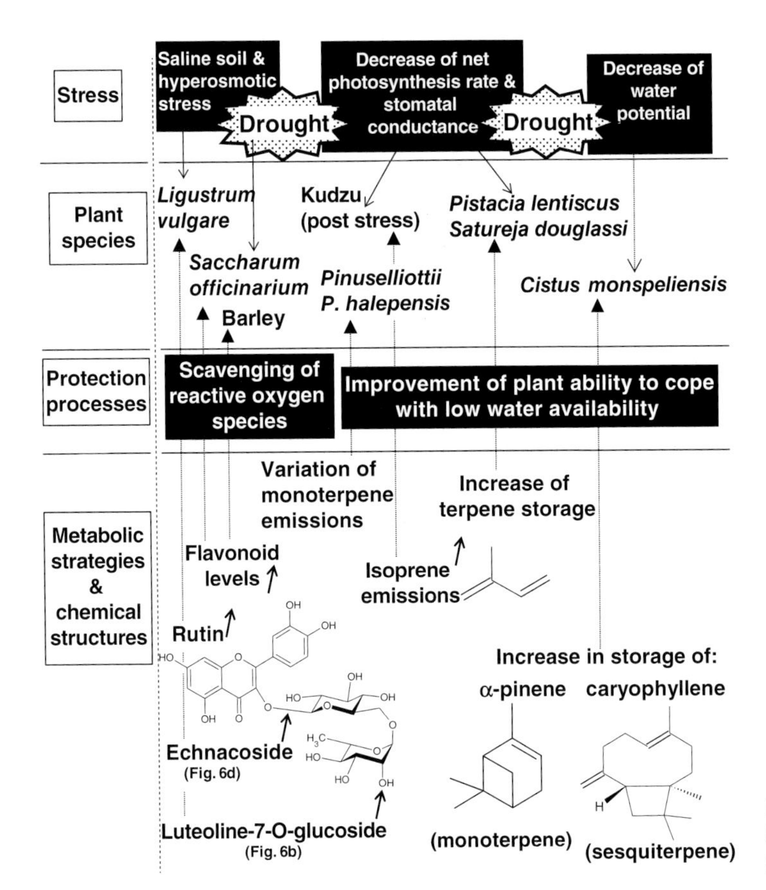

Fig. 9.10. Different study cases highlighting chemical responses of plants living in drought conditions.

from synthesis because of the presence of storage compartments (Llusià and Peñuelas, 1998; Staudt *et al.*, 2002; Tingey *et al.*, 1980). However, in the second phase (t7–t11), the slow decline of MEs was attributed to a lack of photosynthesis substrate due to a prolonged water deficit (Fig. 9.10) (Ormeño *et al.*, 2007).

A study on kudzu showed that concentrations and emissions of BVOCs are influenced by previously experienced water stress in the plant: several days of water withholding resulted in photosynthesis reduction to almost nil, but isoprene emission was less affected (Sharkey and Loreto, 1993). After the alleviation of water stress, photosynthesis returned to a normal level, whereas isoprene emission was amplified by a factor of five and was maintained at a high level for a more or less long period (depending on the stress background). In kudzu, periods preceding and following (for eight days) the water stress were characterized by lower and higher productions of isoprene evaluated to 24% and 67% of photosynthesis, respectively.

Water deficit is directly responsible for an increase in soil salinity, leading to ion imbalance and hyperosmotic stress (Kukreja *et al.*, 2005). Elevated salt concentrations were found to induce accumulation of reactive oxygen species, placing plants under oxidative stress. Oxidative stress associated with drought conditions is overcome by plants though up-regulation of antioxidant metabolites combined with photosynthesis down-regulation (Smirnoff, 1993; Mackerness *et al.*, 2001; Rizhsky *et al.*, 2002; Casati and Walbot, 2003): accumulation of antioxidant flavonoids was reported in *Ligustrum vulgare* L. and concerned both leaves directly exposed to solar radiation and those living in drought shade conditions (Fig. 9.10) (Tattini *et al.*, 2004). Drought conditions induced significantly higher quercetin-3-O-rutinoside (rutin) (Fig. 9.5a) and luteolin-7-O-glucoside (Fig. 9.5b) in the leaves of *L. vulgare* than watered conditions (Tattini *et al.*, 2004). Moreover, under drought (and light) stress, carbon was shown to be preferentially used by *L. vulgare* for the synthesis of more O_2^- scavengers (echinacoside, quercetin-3-*O*-rutinoside, luteolin-7-*O*-glucoside) (Fig. 9.10; Fig. 9.5a, b, d) than *p*-coumaric acid and monohydroxyflavones (showing less *in vitro* antioxidant activities) (Fig. 9.2a) (Tattini *et al.*, 2004). Flavonoid level in barley seeding (*Hordeum vulgare* L.) notably increased under sodium chloride stress (Fig. 9.10) (Ali and Abbas, 2003). Another saline conditions study on sugar cane (*Saccharum officinarum* L.) showed that the levels of soluble phenolics, anthocyanins and flavones levels were 2.5, 2.8, 3.0 times greater, respectively, in salt-tolerant clones compared to sensitive clones (Wahid and Ghazanfar, 2006).

9.3.2 Carbon dioxide stress

Carbon dioxide (CO_2) is chemically nonreactive with an atmospheric half-life approaching 100 years. Predictive simulation models estimated that CO_2 concentrations are expected to double in the next century (2100); at the same time, annual average temperatures of the northern hemisphere are predicted to significantly increase (by 2–6°C) particularly in high latitudes (IPCC, 2007). Current rising CO_2 levels and the global increase in temperature (due to the greenhouse effect) may affect plant composition (community changes) on the time scale of centuries (Monson *et al.*, 1995).

Several studies highlighted the variation of secondary metabolite levels as responses of plants to elevated tropospheric CO_2 concentrations. Such responses include emission and accumulation of BVOCs as well as the production of specific phenolic compounds.

Elevated air CO_2 concentrations induced an increase in monoterpene emissions in non-storing plant species as *Quercus ilex* and *Betula pendula* (Fig. 9.11) (Staudt *et al.*, 2001; Vuorinen *et al.*, 2005), whereas *Pinus ponderosa* and *Pseudotsuga menziesii* which are storing species, did not show such responses (Constable *et al.*, 1999). The Scots pine, *Pinus sylvestris*, grown for five years in elevated CO_2 concentration chambers, showed an increase in monoterpene emission (Fig. 9.11) (Räisänen *et al.*, 2008).

Grown for five years under both high temperature and high CO_2 levels, *P. sylvestris* showed monoterpene emissions which were 126% higher than control conditions (Räisänen *et al.*, 2008). Moreover, the emission rate of monoterpenes considerably increased (+23%) compared with the ambient conditions.

In *Acacia nigrescens* Oliv., an increase in air CO_2 concentrations induced a decrease of isoprene emission, probably due to a decrease in isoprene synthase activity (Fig. 9.11) (Sanadze, 1964; Monson and Fall, 1989; Loreto and Sharkey, 1990; Possell *et al.*, 2005; Wilkinson *et al.*, 2009; Scholefield *et al.*, 2004). Lower isoprene emissions could be due in this case to a variation in the distribution of the phosphoenol pyruvate between chloroplastic and cytosolic compartments; this leads to quantitative variation of pyruvate which affects the chloroplastic MEP pathway (biosynthetic pathway of isoprene) (Possell *et al.*, 2005).

The relationship between ambient CO_2 level and isoprene emission in plants seems to be more complex than a simple direct effect: elevated CO_2 tends to increase the foliar density, which is a favourable factor for isoprene emission (Fig. 9.11). Decrease of isoprene emission by leaf area unit was found to be compensated by a positive effect on leaf area under elevated CO_2 concentrations in air (Centritto *et al.*, 2004). This mechanism concerned more particularly leaves located at open canopies benefiting from high incident light for photosynthesis. Moreover, isoprene emission can be modulated by some interaction between water and CO_2 stresses: in *Populus deltoides*, water stress is overcome by a reduction of stomatal conductance, leading to lower concentration of internal CO_2 attenuating in this way the negative effect of elevated air CO_2 on isoprene emission (Pegoraro *et al.*, 2004).

An increase of CO_2 level induced the replacement of plants with C4 photosynthetic pathway (currently abundant in tropical grasslands) by plants with C3 metabolic pathway (Fig. 9.11) (Ehleringer and Monson, 1993). The C3 plants are more BVOC emitters than C4 plants, and their development could positively influence BVOC emissions (and air composition) in a given region (Monson *et al.*, 1995).

Scutellaria barbata (Lamiaceae) grown for 49 days in enhanced CO_2 chambers showed higher concentrations of six flavones in its vegetative tissue (Fig. 9.11): the total concentration grouping scutellarein, baicalin, apigenin, baicalein, wogonin and chrysin increased by 48% at 1200 and 81% at 3000 μmol.mol^{-1} CO_2 (Stutte *et al.*, 2008). By considering individual flavonoids, an increase in CO_2 from 400 to 1200 μmol.mol^{-1} resulted in 78%, 55% and 39% increase of scutellarein, baicalin and apigenin, respectively. Similar experiments based on variation of air CO_2 levels showed a more pronounced increase of flavonoids in *Scutellaria lateriflora* than *S. barbata*

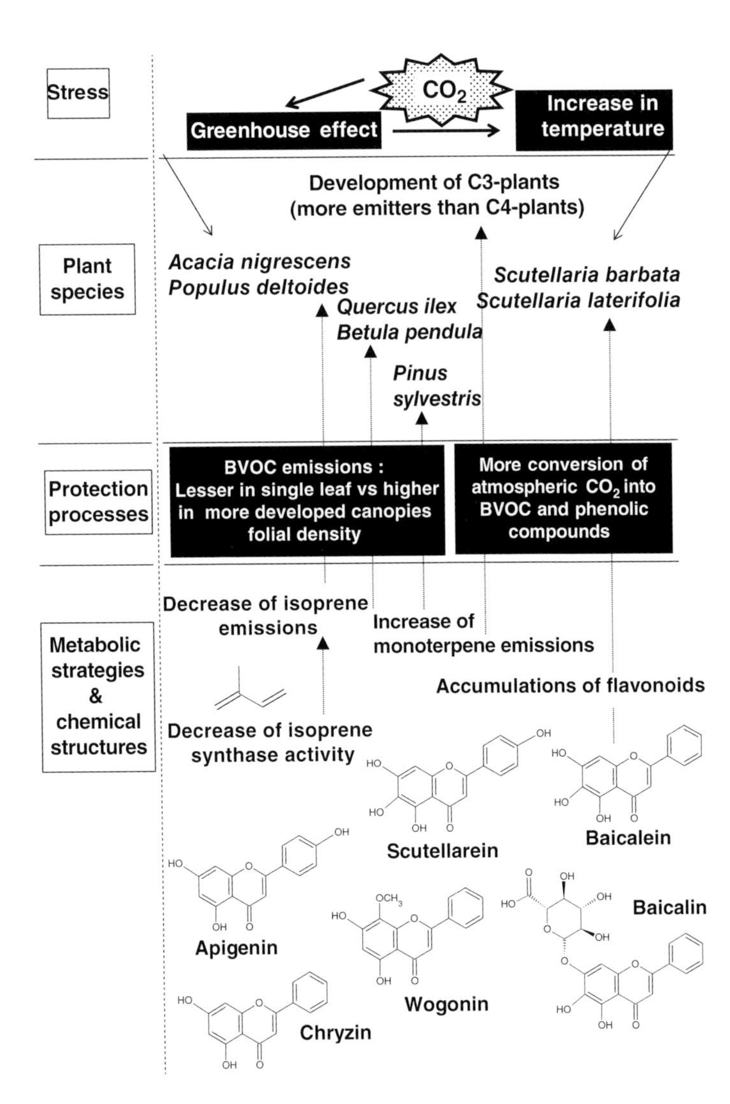

Fig. 9.11. Different study cases highlighting chemical responses of plants under elevated atmospheric CO_2 concentrations. Legend: BVOC, biogenic volatile organic compounds.

(Stutte *et al.*, 2008): for example, an increase in CO_2 from 400 to 1200 and 3000 µmol.mol^{-1} resulted in 4.2- and 13.7-fold increases in total flavonoid content, respectively (in *S. lateriflora*). Moreover at 1200 µmol.mol^{-1} CO_2, baicalin increased 2.8-fold in *S. lateriflora* vs an additive factor of 0.55 for *S. barbata*.

9.3.3 Ozone stress

Secondary metabolites and atmospheric ozone (O_3) are closely linked through different photochemical reactions of BVOCs with nitrogen oxides (NO_x) leading to O_3 formation (Trainer *et al.*, 2001; Roselle, 1994; Hodnebrog *et al.*, 2012). In addition to their role of O_3-precursors, BVOCs represent significant sources of reactive carbon into the atmosphere (Guenther *et al.*, 1995; Odum *et al.*, 1996; Wilson *et al.*, 1998). This means that BVOCs might be more or less involved in some air pollution. The presence of hydroxyl radicals (•OH) in air induces reactions with BVOCs, leading to a shift in the equilibrium between NO, NO_2 and O_3 in favour of higher concentration of tropospheric O_3 (Jenkin and Hayman, 1999). Ozone stress concerns either O_3-excess (i) or deficit (ii) (Fig. 9.12).

(i) Ozone represents a strong photo-oxidant responsible for air pollution (Roselle, 1994). It was found to be the most problematic source of air pollution in the Eastern United States (Sharkey

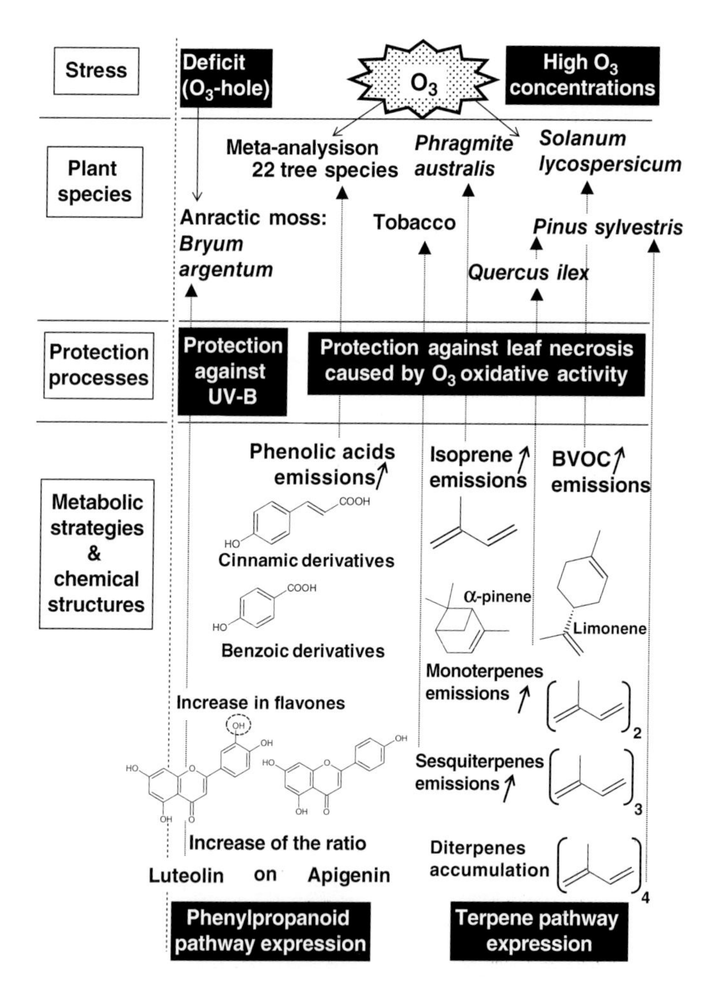

Fig. 9.12. Different study cases highlighting chemical responses of plants under high and low atmospheric ozone concentrations.

and Loreto, 1993). Its high reactivity means that it stays a short time (between hours and weeks) in the atmosphere. In plants, O_3 was shown to stimulate phenylpropanoid synthesis through up-regulation induction of associated genes and enzymes (Kangasjärvi *et al.*, 1994; Kangasjärvi *et al.*, 2005).

(ii) In the stratosphere, biologically threatening UV-B radiations are efficiently filtered by ozone (Weatherhead and Andersen, 2006). O_3 depletion over the last 40 years (starting in the 1970s) has resulted in more irradiation by UV-B of Antarctica (and the whole planet).

Plants showed diversified responses to tropospheric ozone stress including BVOC emissions and non-volatile compound concentrations (Fig. 9.12).

In tomatoes, *Solanum lycospersicum*, an increase in tropospheric O_3 concentrations induced an increase in BVOC emission magnitude (Peñuelas *et al.*, 1999). Leaves of the reed, *Phragmites australis*, exposed to short-term elevated ozone levels, showed increases in isoprene emissions (Fig. 9.12) (Velikova *et al.*, 2005). The isoprene emission was rapid and persisted for hours after reduction of ozone levels. However, *Picea rubens* exposed to elevated ozone for a longer period (three weeks) did not show significant variation in isoprene emission (Ennis *et al.*, 1990). This may indicate that exposure is a duration-dependent process where isoprene emissions could be more sensitive to short-term ozone fluctuations than to longer periods of exposure. High reactivity of isoprene is responsible for the scavenging ability of hydroxyl radicals and its short life span in air (minutes to hours) (Thompson, 1992). It seems to play an efficient protective role through the smothering of air radicals and the destruction of O_3 within the leaves. In this way, isoprene could efficiently contribute to necrosis reduction in leaves exposed to oxidative stress linked to high levels of surrounding O_3 (Vickers *et al.*, 2009; Jardine *et al.*, 2011). Recent experimental evidence suggests that stabilization of thylakoid membranes by isoprene reduces the formation of reactive oxygenated species (Velikova *et al.*, 2011; Velikova *et al.*, 2012).

Quercus ilex emitted relatively more monoterpene after exposure to ozone for a short duration (Fig. 9.12) (Loreto *et al.*, 2004). Another study, based on ozone fumigation, showed seasonal increases in limonene and α-pinene emissions (Rinnan *et al.*, 2005), i.e. different emission peaks were observed at different months during the study year. In the pine, *Pinus sylvestris*, long-term ozone treatment resulted in 40% higher emission of monoterpenes, without visible damage in the needles. In tobacco, high tropospheric O_3 concentrations induced emission of sesquiterpenes (Fig. 9.12) (Heiden *et al.*, 1999).

On the other hand, in *Pinus pinea* L., increase in O_3 concentrations in air was followed by decreased emissions of different monoterpenes such as linalool, β-myrcene, limonene and γ-terpinene by 30–65% (Fig. 9.13) (Hewitt *et al.*, 1994). Also in *Olea europea*, O_3-fumigation led to a decrease in α-pinene emissions during some months of the year (Llusià *et al.*, 2002).

Finally, a meta-analysis on 22 tree species showed that elevated ozone induced an increase in concentrations of phenolic acids, flavonoids and tannins in exposed plants (Valkama *et al.*, 2007). The analysis showed that angiosperms were more responsive than gymnosperms to elevated O_3. In gymnosperms, *Pinus sylvestris* showed a significant increase in diterpene concentrations (acid resins synthesis) (Valkama *et al.*, 2007).

Under an opposite stress, due to the Antarctica O_3 hole, *Bryum argenteum* (Antarctic moss) showed an increase in its flavone concentrations (Ryan *et al.*, 2009). Up-regulated flavonoids in *B. argenteum* seem to be primarily induced by an increase in UV-B exposure favoured by a decrease in stratospheric O_3 levels (Ryan *et al.*, 2009). From this herbarium specimens-based study, the flavones luteolin and apigenin were present in equal concentrations during the pre-ozone-hole period (1956–1965) (Ryan *et al.*, 2009). Then, the decrease in O_3 levels over the last few decades was correlated with the increase in the luteolin : apigenin ratio (Fig. 9.12). Another long-term moss study (on *Lycopodium* spore walls) showed that the O_3-induced change in UV-B fluxes from 1907 to 1993 resulted in a variation of concentrations of plant phenolics which may be subsequently used as an indicator of historical ozone trends (Lomax *et al.*, 2008).

9.4 Geographical Distributions of Plant Populations Characterized by Different Secondary Metabolite Patterns

Permanently emitted BVOCs by plant populations can significantly influence the chemical composition of air through different reaction types and reactivity levels depending on metabolite class and isomer type. The taxonomic compositions of such populations represent a key factor determining the type of BVOCs. This means that phytochemical patterns play a useful role in recognition of geographical origins or living conditions of studied plant species (Semmar *et al.*, 2008b).

Tropical forests represent a great source of BVOC emission, whereas the replacement farmlands are usually composed of corn and other grain crops, all low BVOC emitters (Monson *et al.*, 1995). A recent model estimated the contribution of tropics to ~80% of global annual isoprene emission (Kulmala *et al.*,

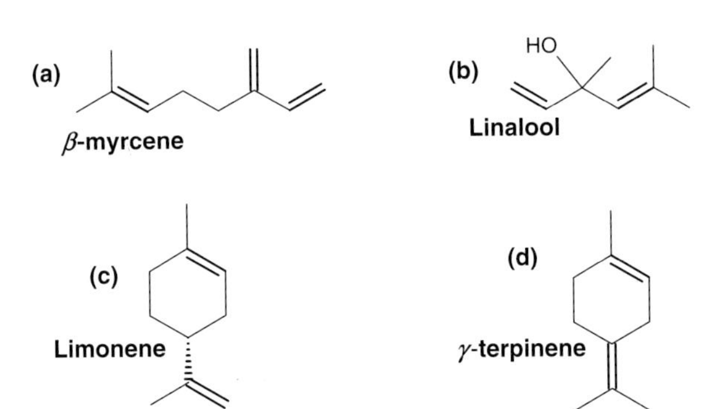

Fig. 9.13. Monoterpenes showing concentration decreasing in *Pinus pinea* (Pinaceae) under external oxidative ozone stress (Hewitt *et al.*, 1994).

2004). Peat lands are important producer ecosystems of isoprene at levels that could be compared to boreal forests (Janson and De Serves, 1998; Janson *et al.*, 1999). Broadleaf trees and shrubs have been estimated to be the highest emitters of isoprene (Guenther *et al.*, 2006; Millet *et al.*, 2008). However, coniferous trees (needle evergreen trees) are high emitters of monoterpenes and sesquiterpenes, with significantly higher emission rates during early summer (Sabillòn and Cremades, 2001; Kim *et al.*, 2005; Matsunaga *et al.*, 2013).

Mediterranean woody plants have been characterized by major emitted monoterpenes including α-pinene, β-pinene, β-phellandrene and limonene (Fig. 9.14a, b, d, e) (Llusià and Peñuelas, 1998). Moreover, these plants emitted relatively more caryophyllene compared to other sesquiterpenes (Fig. 9.14f). In this plant community, *Quercus ilex*, *Pinus lentiscus* and *Pinus halepensis* proved to be the highest emitters of α-pinene (Llusià and Peñuelas, 1998). *Cistus albidus* was characterized by high emission of caryophyllene (Fig. 9.14f), whereas *Quercus albidus* and *Phillygrea latifolia* emitted high amounts of limonene (Fig 9.14d).

Pinus sylvestris L. (Scots pine) specifically emits monoterpenes which mainly consist of α-pinene and $^{3}\Delta$-carene (Fig. 9.14a, c) (Kesselmeier and Staudt, 1999; Räisänen *et al.*, 2008). In eucalyptus, terpenic emissions are dominated by isoprene and monoterpenes (Guenther *et al.*, 1991). The development of large eucalyptus plantations in the Amazon Basin may increase BVOC emissions from tropical areas (Houghton, 1991).

Pueraria lobata (called kudzu) is a vine with rapid development, so that it exclusively occupies certain south-eastern American areas by forming large stands (Forseth and Teramura, 1986). Its high isoprene emissions transformed this region from a low to high BVOC source (Monson *et al.*, 1995). Moreover, isoprene emission in kudzu is characterized by high variability from day-to-day and leaf-to-leaf (Sharkey and Loreto, 1993). In Texas, BVOC emissions were dominated by isoprene in both the urban and rural land use categories, whereas more emissions of monoterpenes were apparent in croplands (Wiedinmyer *et al.*, 2000). Analysis of sesquiterpenes in two different Himalayan subpopulations of *Waldheimia glabra* (Decne) Regel (Asteraceae) highlighted higher levels in plants sampled at 5200 m than in those at 5000 m of altitude.

9.5 Conclusion

Terrestrial plants demonstrate a high flexibility in acclimation to different climatic stresses via strong qualitative and quantitative variations of secondary metabolites (SMs). These two variations concern synthesis and level regulations of chemical structures, respectively. Protective values of SMs are linked to their inductivity and reactivity in two ways.

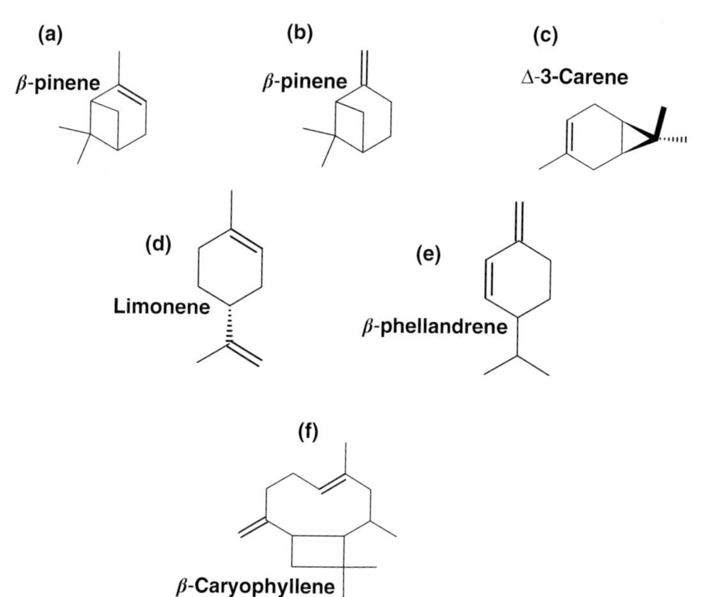

Fig. 9.14. Monoterpenes (a–e) and sesquiterpene (f) dominating emitted volatile organic compounds of different plant geographical populations.

1) SMs are sensitive toward quantitative and qualitative variations of environmental conditions: differential variations and changes in environmental conditions induce the synthesis and govern the regulation of SMs in plants.
2) SMs are active, reactive and support specific signals and, in this way, they play key roles in the life maintenance of plants and in their relationship with the environment: this is directly linked to the high structural diversity of SMs.

SMs' protective activities against climatic stresses relate to four types of damaging factors: oxidations, high temperatures, freezing and osmotic potential (Fig. 9.15).

Oxidations are directly originated from high light radiations (e.g. UV-B) or reactive chemicals (e.g. O_3, free radicals, etc.). Also, they may be indirectly linked to low tropospheric O_3 concentrations (favouring UV-B infiltration) or drought conditions (resulting in oxidative salt stresses). Plants stop harmful oxidative processes by increasing their antioxidant metabolites consisting of phenolic compounds in general. Metabolic regulation ratios generally vary in higher ortho-dihydroxylated structures at the expense of mono-hydroxyl or non-ortho-diOH compounds.

Elevated temperatures can originated from heat waves or they may be indirectly linked to water deficit (drought conditions) or the greenhouse effect (subsequent to high atmospheric CO_2 concentrations). Emissions of volatile compounds seemed to be an efficient method of cooling for leaves of plants suffering from high temperatures. Isoprene is the most abundant volatile compound and is widely emitted by tropical broadleaf trees. Other plant communities are characterized by relatively high emissions of monoterpenes and sesquiterpenes, including Mediterranean coniferous trees.

Freezing is a problem that is generally overcome by plants via accumulations of vacuolar anthocyanins. This results in lower osmotic potential, which reduces the water-freezing temperature. Moreover, abscisic acid (a sesquiterpene) is a plant hormone which is generally produced to initiate the dormancy mechanism, leading to freezing tolerance.

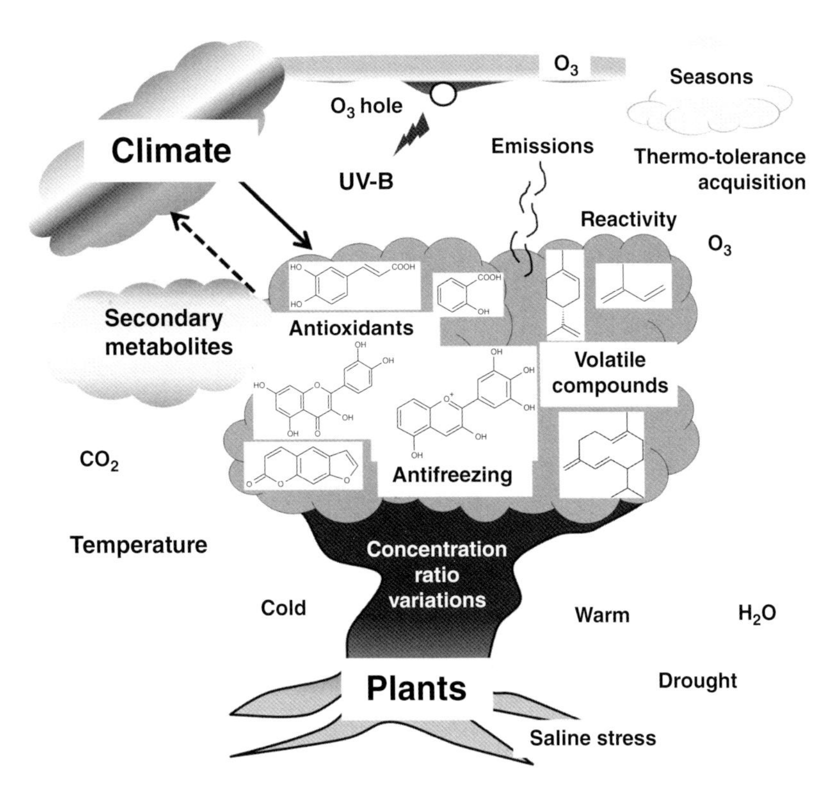

Fig. 9.15. Synthetic figure showing different climatic stress and different metabolic plant responses based on synthesis, reactivity and concentration variations of secondary metabolites.

Hyperosmotic stress and ion imbalance can be due to saline stress in the water deficit environment. Plants face this stress by up-regulating antioxidative compounds to protect against reactive oxygen species subsequent to high salt level.

The fact that SMs are environmentally inducible on the one hand and highly variable between and within plant species on the other hand means that they are useful for defining specific chemical features of different plant–environment interactions. Qualitative variations (biosynthesis, ratio regulations, accumulations) are defined by chemical structures, which are associated with different traits of metabolites (major, minor, inducible, reactive, etc.) to characterize different plant living conditions.

References

Agati, G. and Tattini, M. (2010) Multiple functional roles of flavonoids in photoprotection. *New Phytology* 186, 786–793.

Ali, R.M. and Abbas, H.M. (2003) Response of salt stressed barley seedlings to phenylurea. *Plant Soil and Environment* 49, 158–162.

Ambasht, N.K. and Agrawal, M. (1997) Influence of supplemental UV-B radiation on photosynthetic characteristics of rice plants. *Photosynthetica* 34, 401–408.

Asthana, A., McCloud, E.S., Berenbaum, M.R. and Tuveson, R.W. (1993) Phototoxicity of *Cirrus jambhiri* to fungi under enhanced UVB radiation: role of furanocoumarins. *Journal of Chemical Ecology* 19, 2813–2830.

Blanch, J.S., Llusia, J., Niinemets, U, Noe, S.M. and Penuelas, J. (2011) Instantaneous and historical temperature effects on α-pinene emissions in *Pinus halepensis* and *Quercus ilex*. *Journal of Environment Biology* 32, 1–6.

Blount, J.W., Korth, K.L., Masoud, S.A., Rasmussen, S., Lamb, C. and Dixon, R.A. (2000) Altering expression of cinnamic acid 4-hydroxylase in transgenic plants provides evidence for a feedback loop at the entry point into the phenylpropanoid pathway. *Plant Physiology* 122, 107–116.

Boo, H.O., Chon, S.U. and Lee, S.Y. (2006) Effects of temperature and plant growth regulators on anthocyanin synthesis and phenylalanine ammonia-lyase activity in chicory (*Cichorium intybus* L.). *Journal of Horticultural Science & Biotechnology* 81, 478–482.

Burke, M.J. and Stushnoff, C. (1979) Frost hardiness: a discussion of possible molecular causes of injury with particular reference to deep supercooling of water. In: Mussell, H. and Staples, R.C. (eds) *Stress Physiology in Crop Plants*. John Wiley, New York, USA, pp. 197–225.

Caldwell, M.M., Teramura, A.H. and Tevini, M. (1989) The changing solar ultraviolet climate and the ecological consequences for higher plants. *Trends in Ecology and Evolution* 4, 363–367.

Caldwell, M.M., Ballaré, C.L., Bornman, J.F., Flint, S.D., Björn, L.O., Teramura, A.H., Kulandaivelu, G. and Tevini, M. (2003) Terrestrial ecosystems, increased solar ultraviolet radiation and interactions with other climatic change factors. *Photochemical and Photobiological Science* 2, 29–38.

Casati, P. and Walbot, V. (2003) Gene expression profile in response to ultraviolet radiation in maize genotypes with varying flavonoid content. *Plant Physiology* 132, 1739–1754.

Centritto, M., Nascetti, P., Petrilli, L., Raschi, A. and Loreto, F. (2004) Profiles of isoprene emission and photosynthetic parameters in hybrid poplars exposed to free-air CO_2 enrichment. *Plant, Cell and Environment* 27(4), 403–412.

Chalker-Scott, L. (1992) Disruption of an ice-nucleation barrier in cold hardy *Azalea* buds by sublethal heat stress. *Annals of Botany* 70, 409–418.

Chalker-Scott, L. (2002) Do anthocyanins function as osmoregulators in leaf tissues? *Advances in Botanical Research* 37, 103–127.

Christensen, A.B., Gregersen, P.L., Schröder, J. and Collinge D.B. (1998) A chalcone synthase with an unusual substrate preference is expressed in barley leaves in response to UV light and pathogen attack. *Plant Molecular Biology* 37, 849–857.

Constable, J.V.H., Litvak, M.E., Greenberg, J.P. and Monson R.K. (1999) Monoterpene emission from coniferous trees in response to elevated CO_2 concentration and climate warming. *Global Change Biology* 5, 255–267.

Creasy, L.L. (1968) The role of low temperature in anthocyanin synthesis in McIntosh apples. *Proceedings of American Society of Horticultural Science* 93, 716–724.

Curry, E.A. (1997) Temperatures for optimal anthocyanin accumulation in apple tissue. *Journal of Horticultural Science* 72, 723–729.

Dangl, J.L. (1992) Regulatory elements controlling developmental and stress induced expression of phenylpropanoid genes. In: Boller, T. and Meins F (eds) *Plant Gene Research: Genes Involved in Plant Research,* Vol. 8. Springer-Verlag, Vienna, Austria, pp. 303–336.

Delfine, S., Csiky, O., Seufert, G. and Loreto, F. (2000) Fumigation with exogenous monoterpenes of a non-isoprenoid-emitting oak (*Quercus suber*): monoterpene acquisition, translocation, and effect on the photosynthetic properties at high temperatures. *New Phytology* 146, 27–36.

Ehleringer, J.R. and Monson, R.K. (1993) Evolutionary and ecological aspects of photosynthetic pathway variation. *Annual Reviews on Ecological Systems* 24, 411–440.

Ennis, C.A., Lazrus, A.L, Zimmerman, P.R. and Monson, R.K. (1990) Flux determination and physiological response in exposure of red spruce to gaseous hydrogen peroxide, ozone, and sulfur dioxide. *Tellus B: Chemical and Physical Meteorology* 42, 183–199.

Filella, I. and Penuelas, J. (1999) Altitudinal differences in UV absorbance, UV reflectance and related morphological traits of *Quercus ilex* and *Rhododendron ferrugineum* in the Mediterranean region. *Plant Ecology* 145, 157–165.

Fischbach, R.J., Kossmann, B., Panten, H., Steinbrecher, R., Heller, W., Seidlitz, H.K., Sandermann, H., Hertkorn, N. and Schnitzler, J.P. (1999) Seasonal accumulation of ultraviolet-B screening pigments in needles of Norway spruce (Piceaabies (L.) Karst.). *Plant, Cell and Environment* 22, 27–37.

Fiscus, E.L. and Booker, F.L. (1995) Is increased UV-B a threat to crop photosynthesis and productivity? *Photosynthesis Research* 43, 81–92.

Forseth, I.N. and Teramura, A.H. (1986) Kudzu leaf energy budget and calculated transpiration: the influence of leaflet orientation. *Ecology* 67, 564–571.

Gehrke, C., Johanson, U., Callaghan, T.V., Chadwick, D. and Robinson, D.H. (1995) The impact of enhanced UV-B radiation on litter quality and decomposition processes in *Vaccinium* leaves from the subarctic. *Oikos* 71, 213–222.

Gerhardt, K.E., Lampi, M.A. and Greenberg, B.M. (2008) The effects of far-red light on plant growth and flavonoid accumulation in *Brassica napus* in the presence of ultraviolet B radiation. *Photochemistry and Photobiology* 84, 1445–1454.

Gershenzon, J., Lincoln, D.E. and Langenheim, J.H. (1978) The effect of moisture stress on monoterpenoid yield and composition in *Satureja douglasii*. *Biochemical and Systematic Ecology* 6, 33–43.

Gilmour, S.J., Hajela, R.K. and Thomashow, M.F. (1988) Cold acclimation in *Arabidopsis thaliana*. *Plant Physiology* 87, 745–750.

Goetz, M., Abert, A., Stich, S., Heller, W., Scherb, H. *et al.* (2010) PAR modulation of the UV-dependent levels of flavonoid metabolites in *Arabidopsis thaliana* (L.) Heynh. Leaf rosettes: cumulative effects after a whole vegetative growth period. *Protoplasma* 243, 95–103.

Graham, D. and Patterson, B.D. (1982) Responses of plants to low, nonfreezing temperatures: proteins, metabolism and acclimation. *Annual Review of Plant Physiology* 33, 347–372.

Grammatikopoulos, G., Kyparissis, A., Drilias, P., Petropoulou, Y. and Manetas, Y. (1998) Effects of UV-B radiation on cuticle thickness and nutritional value of leaves in two Mediterranean evergreen sclerophylls. *Journal of Plant Physiology* 153, 506–512.

Guenther, A., Hewitt, C.N., Erickson, D., Fall, R., Geron, C. *et al.* (1995) Global model of natural volatile organic compound emissions. *Journal of Geophysical Research* 100, 8873–8892.

Guenther, A.B., Karl, T., Harley, P., Wiedinmyer, C., Palmer, P.I. and Geron, C. (2006) Estimates of global terrestrial isoprene emissions using MEGAN (Model of Emissions of Gases and Aerosols from Nature). *Atmospheric Chemistry and Physics* 6, 3181–3210.

Guenther, A.B., Monson, R.K. and Fall, R. (1991) Isoprene and monoterpene emission rate variability: observations with Eucalyptus and emission rate algorithm development. *Journal of Geophysical Research* 96, 10799–10808.

Gusta, L.V., Trischuk, R. and Weiser, C.J. (2005) Plant cold acclimation: the role of abscisic acid. *Journal of Plant Growth Regulation* 24, 308–318.

Hao, W., Arora, R., Yadav, A.K. and Joshee, N. (2009) Freezing tolerance and cold acclimation in guava (*Psidium guajava* L.). *HortScience* 44(5), 1258–1266.

Harbaum-Piayda, B., Walter, B., Bengtsson, G.B., Hubbermann, E.M., Bilger, W. and Schwarz, K. (2010) Influence of pre-harvest UV-B irradiation and normal or controlled atmosphere storage on flavonoid and hydroxycinnamic acid contents of pak choi (*Brassica campestris* L. ssp. *chinensis* var. *communis*). *Postharvest Biology and Technology* 56, 202–208.

Heiden, A.C., Hoffman, T., Kahl, J., Kley, D., Klockow, D. *et al.* (1999) Emission of volatile organic compounds from ozone-exposed plants. *Ecological Applications* 9, 1160–1167.

Heino, P., Sandman, G., Lang, V., Nordin, K. and Palva, E.T. (1990) Abscisic acid deficiency prevents development of freezing tolerance in *Arabidopsis thaliana* (L.) Heynh. *Theoretical and Applied Genetics* 79, 801–806.

Hewitt, C.N., Owen, S., Boissard, C. and Csisky, O. (1994) Biogenic emissions in the Mediterranean area: BEMA-Project. *Report on the 1st BEMA measuring campaign at Castelporziano, Rome (Italy), May 1994*. EUR 16293 EN. Environment Institute, European Commission, Luxembourg, pp. 137–150.

Hirata, K., Horiuchi, M., Asada, M., Ando, T., Miyamoto, K. and Miura, Y. (1992) Stimulation of dimeric alkaloid production by near ultraviolet light in multiple shoot cultures of *Carharanthus roseus*. *Journal of Fermentation and Bioengineering* 74, 222–225.

Hirata, K., Asada, M., Yatani, E., Miyamoto, K. and Miura, Y. (1993) Effects of near-ultraviolet light on alkaloid production in *Catharanthus roseus* plants. *Planta Medica* 59, 46–50.

Hodges, J. and Lorio, P. (1975) Moisture stress and xylem oleoresin in loblolly pine. *Forensic Science* 21, 283–290.

Hodnebrog, Ø., Solberg, S., Stordal, F., Svendby, T.M., Simpson, D. *et al.* (2012) A model study of the Eastern Mediterranean ozone levels during the hot summer of 2007. *Atmospheric Chemistry and Physics Discussions*, 12, 7617–7675.

Hofmann, R.W., Swinny, E.E., Bloor, S.J., Markham, K.R., Ryan, K.G. *et al.* (2000) Responses of nine *Trifolium repens* L. populations to ultraviolet-B radiation: differential flavonol glycoside accumulation and biomass production. *Annals of Botany* 86, 527–537.

Hofmann, R.W., Campbell, B.D., Bloor, S.J., Swinny, E.F., Markham, K.R., Ryan, K.G. and Fountain, D.W. (2003) Responses to UV-B radiation in *Trifolium repens* L. – physiological links to plant productivity and water availability. *Plant, Cell and Environment* 26, 603–612.

Houghton, R.A. (1991) Tropical deforestation and atmospheric carbon dioxide. *Climatic Change* 19, 99–118.

Huang, X.D., Dixon, D.G. and Greenberg, B.M. (1993) The impacts of UV-radiation and photomodification on the toxicity of PAHS to the higher plant *Lemna gibba* (Duckweed). *Environmental and Toxicological Chemistry* 12, 1067–1077.

IPCC (2007) Climate change: the physical science basis. In: Solomon, S., Qin, D., Manning, M., Chen, Z., Marquis, M., Averyt, K.B., Tignor, M. and Miller, H.L. (eds) *Contribution of Working Group I to the Fourth Assessment Report of the Intergovernmental Panel of Climate Change.* Cambridge University Press, Cambridge, UK and New York, USA.

Ishikawa, M. (1984) Deep supercooling in most tissues of wintering *Sasa senanensis* and its mechanism in leaf blade tissues. *Plant Physiology* 75, 196–202.

Jaakola, L. and Hohtola, A. (2010) Effect of latitude on flavonoid biosynthesis in plants. *Plant, Cell and Environment* 33, 1239–1247.

Janská, A., Marsík, P., Zelenková, S. and Ovesná, J. (2010) Cold stress and acclimation – what is important for metabolic adjustment? *Plant Biology* 12, 395–405.

Janson, R. and De Serves, C. (1998) Isoprene emissions from boreal wetlands in Scandinavia. *Journal of Geophysical Research* 103(D19), 25513–25517.

Janson, R., De Serves, C. and Romero, R. (1999) Emission of isoprene and carbonyl compounds from a boreal forest and wetland in Sweden. *Agricultural and Forest Meteorology* 98(9), 671–681.

Jardine, K.J., Monson, R.K., Abrell, L., Saleska, S.R., Arneth, A. *et al.* (2011) Within-plant isoprene oxidation confirmed by direct emissions of oxidation products methyl vinyl ketone and methacrole. *Global Change Biology* 18, 973–984.

Jenkin, M.E. and Hayman, G.D. (1999) Photochemical ozone creation potentials for oxygenated volatile organic compounds: sensitivity to variations in kinetic and mechanistic parameters. *Atmosphere and Environment* 33, 1275–1293.

Kainulainen, P., Okanen, J., Paloäki, V., Holopainen, J.K. and Holopainen T. (1991) Effect of drought and waterlogging stress on needle monoterpene of *Piceaabies*. *Canadian Journal of Botany* 70, 1613–1616.

Kangasjärvi, J., Talvinen, J., Utriainen, M. and Karjalainen, R. (1994) Plant defense systems induced by ozone. *Plant, Cell and Environment* 17, 783–794.

Kangasjärvi, J., Jaspers, P. and Kollist, H. (2005) Signalling and cell death in ozone-exposed plants. *Plant, Cell and Environment* 28, 1021–1036.

Keski-Saari, K., Pusenius, J. and Julkunen-Tiitto, R. (2005) Phenolic compounds in seedlings of *Betulapubescens* and *B. pendula* are affected by enhanced UVB radiation and different nitrogen regimens during early ontogeny. *Global Change Biology* 11, 1180–1194.

Kesselmeier, J. and Staudt, M. (1999) Biogenic volatile organic compounds (VOC): an overview on emission, physiology and ecology. *Journal of Atmospheric Chemistry* 33, 23–88.

Kim, J.C., Kim, K.J., Kim, D.S. and Han J.S. (2005) Seasonal variations of monoterpene emissions from coniferous trees of different ages in Korea. *Chemosphere* 59, 1685–1696.

Krol, M., Gray, G.R., Hurry, V.M., Oquist, G., Malek, L. and Huner, N.P.A. (1995) Low-temperature stress and photoperiod affect an increased tolerance to photoinhibition in *Pinus banksiana* seedlings. *Canadian Journal of Botany* 73, 1119–1127.

Kukreja, S., Nandwal, A.S., Kumar, N., Sharma, S.K., Sharma, S.K., Unvi, V. and Sharma P.K. (2005) Plant water status, H_2O_2 scavenging enzymes, ethylene evolution and membrane integrity of *Cicer arietinum* roots as affected by salinity. *Biology of Plants* 49(2), 305–308.

Kulmala, M., Suni, T., Lehtinen, K.E.J., Dal Maso, M., Boy, M. *et al.* (2004) A new feedback mechanism linking forests, aerosols, and climate. *Atmospheric Chemistry and Physics* 4(2), 557–562.

Lang, V., Heino, P. and Palva, E.T. (1989) Low temperature acclimation and treatment with exogenous abscisic acid induce common polypeptides in *Arabidopsis thaliana* (L.) Heinh. *Theoretical and Applied Genetics* 77, 729–734.

Larson, R.A. and Berenbaum, M.R. (1988) Environmental phototoxicity. Solar ultraviolet radiation affects the toxicity of natural and man-made chemical. *Environmental Science Technology* 22, 354–360.

Larson, R.A., Garrison, W.J. and Carlson, R.W. (1990) Differential responses of alpine and non-alpine *Aquilegia* species to increased ultraviolet-B radiation. *Plant, Cell and Environment* 13, 983–987.

Lavola, A. (1998) Accumulation of flavonoids and related compounds in birch induced by UV-B irradiance. *Tree Physiology* 18, 53–58.

Lerdau, M.T., Dilts, S.B., Westberg, H., Lamb, B.K. and Allwine, E.J. (1994) Monoterpene emission from *Ponderosa* pine. *Journal of Geophysical Research* 99, 16609–16615.

Levitt, J. (1980) Responses of plants to environmental stresses: chilling, freezing and high temperatures stresses. In: Kozlowsky, T.T. (ed.) *Physiological Ecology: A Series of Monographs, Texts and Treatises*, Vol. 1, 2nd edn. Academic Press, New York, USA, pp. 23–64.

Levizou, E. and Manetas, Y. (2001) Combined effects of enhanced UV-B radiation and additional nutrients on growth of two Mediterranean plant species. *Plant Ecology* 154, 181–186.

Leyva, A., Jarillo, J.A., Salinas, J. and Martinez-Zapater, J.M. (1995) Low temperature induces the accumulation of phenylalanine ammonia-lyase and chalcone synthase mRNAs of *Arabidopsis thaliana* in a light-dependent manner. *Plant Physiology* 108, 39–46.

Liu, L., Gitz III, D.C. and McClure, J.W. (1995) Effects of UV-B on flavonoids, ferulic acid, growth and photosynthesis in barley primary leaves. *Physiologia Plantarum* 93, 725–733.

Llorente, F., Oliveros, J.C., Martínez-Zapater, J.M. and Salinas, J. (2000) A freezing-sensitive mutant of *Arabidopsis*, frs1, is a new aba3 allele. *Planta* 211, 648–655.

Llusià, J. and Peñuelas, J. (1998) Changes in terpene content and emission in potted Mediterranean woody plants under severe drought. *Canadian Journal of Botany* 76: 1366–1373.

Llusià, J. and Peñuelas, J. (2000) Seasonal patterns of terpene content and emission from seven Mediterranean woody species in field conditions. *American Journal of Botany* 87, 133–140.

Llusià, J., Peñuelas, J. and Gimeno, B.S. (2002) Seasonal and species-specific response of VOC emissions by Mediterranean woody plant to elevated ozone concentrations. *Atmospheric Environment* 36, 3931–3938.

Lomax, B.H., Fraser, W.T. and Sephton, M.A. (2008) Plant spore walls as a record of long term changes in ultraviolet-B radiation. *Nature Geoscience* 1, 592–596.

Loreto, F. and Sharkey, T.D. (1990) A gas-exchange study of photosynthesis and isoprene emission in *Quercus rubra* L. *Planta* 182, 523–531.

Loreto, F., Pinelli, P., Manse, F. and Kollist, H. (2004) Impact of ozone on monoterpene emissions and evidence for an isoprene-like antioxidant action of monoterpenes emitted by *Quercus ilex* leaves. *Tree Physiology* 24(4), 361–367.

Mabry, T.J., Markham, K.R. and Thomas, M.B. (1970) *The Systematic Identification of Flavonoids.* Springer, New York, USA.

Mackerness, S.A.H., John, F.C., Jordan, B.R. and Thomas, B. (2001) Early signalling components in ultraviolet-B responses: distinct roles for different reactive oxygen species and nitric oxide. *FEBS Letters* 489, 237–242.

Madronich, S., McKenzie, R.L., Björn, L.O. and Caldwell, M.M. (1998) Changes in biologically active ultraviolet radiation reaching the Earth's surface. *Journal of Photochemistry and Photobiology B: Biology* 46(1–3), 5–19.

Markham, K.R., Ryan, K.G., Bloor, S.J. and Mitchell, K. (1998a) An increase in the luteolin : apigenin ratio in *Marchantia polymorpha* on UVB enhancement. *Phytochemistry* 48, 791–794.

Markham, K.R., Tanner, G.J., Caasi-Lit, M., Whitecross, M.F., Nayudu, M. and Mitchell, K.A. (1998b) Protective role for 3',4'-dihydroxyflavones induced by UV-B in a tolerant rice cultivar. *Phytochemistry* 49, 1913–1919.

Matsunaga, S.N., Niwa, S., Mochizuki, T., Tani, A., Kusumoto, D., Utsumi, Y., Enoki, T. and Hiura, T. (2013) Seasonal variation in basal emission rates and composition of mono- and sesquiterpenes emitted from dominant conifers in Japan. *Atmospheric Environment* 69, 124–130.

McCloud, E.S., Berenbaum, M.R. and Tuveson, R.W. (1992) Furanocourmarin content and phototoxicity of rough lemon (*Cirrus jambhiri*) foliage exposed to enhanced ultraviolet-B (UVB) irradiation. *Journal of Chemical Ecology* 18, 1125–1137.

McKown, R., Kuroki, G. and Warren, G. (1996) Cold responses of *Arabidopsis* mutants impaired in freezing tolerance. *Journal of Experimental Botany* 47(12), 1919–1925.

Millet, D.B., Jacob, D.J., Boersma, K.F., Fu, T.M, Kurosu, T.P., Chance, K., Heald, C.L and Guenther, A. (2008) Spatial distribution of isoprene emissions from North America derived from formaldehyde column measurements by the OMI satellite sensor. *Journal of Geophysical Research: Atmospheres* 113(2), D02307.

Monson, R.K. and Fall, R. (1989) Isoprene emission from Aspen leaves. The influence of environment and relation to photosynthesis and photorespiration. *Plant Physiology* 90, 267–274.

Monson, R.K., Harley, P.C., Litvak, M.E., Wildermuth, M., Guenther, A.B., Zimmerman, P.R. and Fall, R. (1994) Environmental and developmental controls over the seasonal pattern of isoprene emission from aspen leaves. *Oecologia* 99, 260–270.

Monson, R.K., Lerdau, M.T., Sharkey, T.D., Schimel, D.S. and Fall, R. (1995) Biological aspects of constructing volatile organic compound emission inventories. *Atmospheric Environment* 29(21), 2989–3002.

Montesinos, M.C., Ubeda, A., Terencio, M.C., Paya, M. and Alcaraz, M.J. (1995) Antioxidant profile of mono- and dihydroxylated flavone derivatives in free radical generating systems. *Zeitscrift für Naturforschung* 50c, 552–560.

Neill, S.O. and Gould, K.S. (2003) Anthocyanins in leaves: light attenuators or antioxidants. *Functional Plant Biology* 30, 865–873.

Newsham, K.K. and Robinson, S.A. (2009) Responses of plants in polar regions to UVB exposure: a meta-analysis. *Global Change Biology* 15(11), 2574–2589.

Odum, J.R., Hoffmann, T., Bowman, F., Collins, D., Flagan, R.C., Steinfeld, J.H. (1996) Gas/particle partitioning and secondary aerosol formation. *Environmental Science and Technology* 30, 2580–2585.

Olsson, L.C., Veit, M., Weissenböck, G. and Bornman, J.F. (1999a) Differential flavonoid response to enhanced UV-B radiation in *Brassica napus*. *Phytochemistry* 49, 1021–1028.

Olsson, L.C., Veit, M. and Bornman, J.F. (1999b) Epidermal transmittance and phenolic composition in leaves of atrazine-tolerant and atrazine sensitive cultivars of *Brassica napus* grown under enhanced UV-B radiation. *Physiologia Plantarum* 107, 259–266.

Oren-Shamir, M. and Levi-Nissim, A. (1997a) UV-light effect on the leaf pigmentation of *Cotinus coggygria* 'Royal Purple'. *Scientia Horticulturae* 71(1–2), 59–66.

Oren-Shamir, M. and Levi-Nissim, A. (1997b) Temperature effect on the leaf pigmentation of *Cotinus coggygria* 'Royal Purple'. *Scientia Horticulturae* 72(3), 425–432.

Ormeño, E., Mévy, J.P., Vila., B., Bousquet-Mélou, A., Greff, S., Bonin, G. and Fernandez, C. (2007) Water deficit stress induces different monoterpene and sesquiterpene emission changes in Mediterranean species. Relationship between terpene emissions and plant water potential. *Chemosphere* 67, 276–284.

Ouwerkerk, P.B.F., Hallard, D., Verpoorte, R. and Memelink, J. (1999) Identification of UV-B light-responsive regions in the promoter of the tryptophan decarboxylase gene from *Catharanthus roseus*. *Plant Molecular Biology* 41, 491–503.

Parra, C., Sdez, J., Pérez, H., Alberdi, M., Delseny, M., Hubert, E. and Meza-Basso, L. (1990) Cold resistance in rapeseed (*Brassica napus*) seedlings. Searching biochemical markers of cold-tolerance. *Archives of Biology and Experimental Medicine* 23, 187–194.

Paul, N.D., Rasanayagam, S., Moody, S.A., Hatcher, P.E. and Ayers, P.G. (1997) The role of interaction between trophic levels in determining the effects of UV-B on terrestrial ecosystems. *Plant Ecology* 128, 296–308.

Pegoraro, E., Rey, A., Bobich, E., Barron-Gafford, G., Grieve, A., Malhi, Y. and Murthy, R. (2004) Effect of elevated CO_2 concentration and vapour pressure deficit on isoterpene emission from leaves of *Populus deltoides* during drought. *Functional Plant Biology* 31(12), 1137–1147.

Peñuelas, J., Filella, I., Llusià, J., Siscart, D. and Piñol, J. (1998) Comparative field study of spring and summer leaf gas exchange and photobiology of the Mediterranean trees *Quercus ilex* and *Phillyrea latifolia*. *Journal of Experimental Botany* 49, 229–238.

Peñuelas, J., Llusià, J. and Gimeno, B.S. (1999) Effects of ozone concentrations on biogenic volatile organic compounds emission in the Mediterranean region. *Environmental Pollution* 105, 17–23.

Possell, M., Hewitt, C.N. and Beerling, D.J. (2005) The effects of glacial atmospheric CO_2 concentrations and climate on isoprene emissions by vascular plants. *Global Change Biology* 11, 60–69.

Räisänen, T., Ryyppö, A. and Kellomäki, S. (2008) Effects of elevated CO_2 and temperature on monoterpene emission of Scots pine (*Pinus sylvestris* L.). *Atmospheric Environment* 42, 4160–4171.

Reifenrath, K. and Müller, C. (2007) Species-specific and leaf-age dependent effects of ultraviolet radiation on two Brassicaceae. *Phytochemistry* 68, 875–885.

Rhodes, M.J.C. and Wooltorton, L.S.C. (1977) Changes in the activity of enzymes of phenylpropanoid metabolism in tomatoes stored at low temperatures. *Phytochemistry* 16, 655–659.

Rice-Evans, C.A., Miller, N.J. and Paganga, G. (1996) Structure–antioxidant activity relationships of flavonoids and phenolic acids. *Free Radical Biology and Medicine* 20(7), 933–956.

Rinnan, R., Rinnan, Å., Holopainen, T., Holopainen, J.K. and Pasanen, P. (2005) Emission of non-methane volatile organic compounds (VOCs) from boreal peatland microcosms – effects of ozone exposure. *Atmospheric Environment* 39, 921–930.

Rizhsky, L., Liang, H. and Mittler, R. (2002) The combined effect of drought stress and heat shock on gene expression in tobacco. *Plant Physiology* 130, 1143–1151.

Robinson, T. (1991) *The Organic Constituents of Higher Plants*. Cordus Press, North Amherst, Massachusetts, USA.

Roselle, S.J. (1994) Effects of biogenic emission uncertainties on regional photochemical modelling of control strategies. *Atmospheric Environment* 28, 1757–1772.

Ruegger, M. and Chapple, C. (2001) Mutations that reduce sinapoylmalate accumulation in *Arabidopsis thaliana* define loci with diverse roles in phenylpropanoid metabolism. *Genetics* 159, 1741–1749.

Ruelland, E., Vaultier, M.N., Zachowski, A. and Hurry, V. (2009) Cold signalling and cold acclimation in plants. *Advances in Botanical Research* 49, 135–150.

Ryan, K.G., Markham, K.R., Bloor, S.J., Bradley, J.M., Mitchell, K.A. and Jordan, B.R. (1998) UVB radiation-induced increase in quercetin : kaempferol ratio in wild-type and transgenic lines of *Petunia*. *Photochemistry and Photobiology* 68(3), 323–330.

Ryan, K.G., Swinny, E.E., Markham, K.R. and Winefield, C. (2002a) Flavonoid gene expression and UV photoprotection in transgenic and mutant *Petunia* leaves. *Phytochemistry* 59, 23–32.

Ryan, K.G., Swinny, E.E., Winefield, C., Markham, K.R. (2002b) Flavonoids and photoprotection in *Arabidopsis* mutants. *Zeitscrift für Naturforschung* 56c, 745–754.

Ryan, K.G., Burne, A. and Seppelt, R.D. (2009) Historical ozone concentrations and flavonoid levels in herbarium specimens of the Antarctic moss *Bryum argenteum*. *Global Change Biology* 15, 1694–1702.

Sabillòn, D. and Cremades, L.V. (2001) Diurnal and seasonal variation of monoterpene emission rates for two typical Mediterranean species (*Pinus pinea* and *Quercus ilex*) from field measurements-relationship with temperature and PAR. *Atmospheric Environment* 35, 4419–4431.

Salinas, J. (2002) Molecular mechanisms of signal transduction in cold acclimation. In: Hames, B.D. and Glover, D.M. (eds) *Frontiers in Molecular Biology*. Oxford University Press, Oxford, UK, pp. 116–139.

Sanadze, G.A. (1964) Light-dependent excretion of isoprene by plants. *Photosynthesis Research* 2, 701–707.

Saure, M.C. (1990) External control of anthocyanin formation in apple. *Scientia Horticulturae* 42, 181–218.

Schmidt, S., Zietz, M., Schreiner, M., Rohn, S., Kroh, L.W. and Krumbein, A. (2010) Genotypic and climatic influences on the concentration and composition of flavonoids in kale (*Brassica oleracea* var. sabellica). *Food Chemistry* 119, 1293–1299.

Scholefield, P.A., Doick, K.J., Herbert, B.M.J., Hewitt, C.N., Schnitzler, J.P., Pinelli, P. and Loreto, F. (2004) Impact of rising CO_2 on emissions of volatile organic compounds: isoprene emission from *Phragmites australis* growing at elevated CO_2 in a natural carbon dioxide spring. *Plant, Cell and Environment* 27(4), 393–401.

Searles, P.S., Flint, S.D. and Caldwell, M.M. (2001) A meta-analysis of plant field studies stimulating stratospheric ozone depletion. *Oecologia* 127, 1–10.

Semmar, N., Jay, M., Farman, M. and Roux, M. (2008a) A new approach to plant diversity assessment combining HPLC data, simplex mixture design and discriminant analysis. *Environmental Modeling and Assessment* 13(1), 17–33.

Semmar, N., Nouira, S. and Farman, M. (2008b) Variability and ecological significances of secondary metabolites in terrestrial biosystems. In: Aronoff, J.B. (ed.) *Handbook of Nature Conservation*. Nova Science Publishers, New York, USA, pp. 1–89.

Shallcross, D.E. and Monks, P.S. (2000) A role for isoprene in biosphere-climate-chemistry feedbacks? *Atmospheric Environment* 34, 1659–1660.

Sharkey, T.D. and Loreto, F. (1993) Water stress, temperature, and light effects on the capacity for isoprene emission and photosynthesis of kudzu leaves. *Oecologia* 95, 328–333.

Singh, A. (1996) Growth, physiological, and biochemical responses of three tropical legumes to enhanced UV-B radiation. *Canadian Journal of Botany* 74, 135–139.

Singsaas, E.L., Lerdau, M., Winter, K. and Sharkey, T.D. (1997) Isoprene increases thermotolerance of isoprene-emitting species. *Plant Physiology* 115, 1413–1420.

Smirnoff, N. (1993) The role of active oxygen in the response of plants to water deficit and desiccation. *New Phytology* 125, 27–58.

Smith, G.J. and Markham, K.R. (1998) Tautomerism of flavonol glucosides: relevance to plant UV protection and flower colour. *Journal of Photochemistry and Photobiology A: Chemistry* 118(2), 99–105.

Solecka, D. and Kacperska, A. (2003) Phenylpropanoid deficiency affects the course of plant acclimation to cold. *Physiologia Plantarum* 119(2), 253–262.

Staudt, M. and Bertin, N. (1998) Light and temperature dependence of the emission of cyclic and acyclic monoterpenes from holm oak (*Quercus ilex* L.) leaves. *Plant, Cell and Environment* 21, 385–395.

Staudt, M., Joffre, R., Rambal, S. and Kesselmeier, J. (2001) Effect of elevated CO_2 on monoterpene emission of young *Quercus ilex* trees and its relations to structural and ecophysiological parameters. *Tree Physiology* 21, 437–445.

Staudt, M., Rambal, S., Joffre, R. and Kesselmeier, J. (2002) Impact of drought on seasonal monoterpene emissions from *Quercus ilex* in southern France. *Journal of Geophysical Research* 107, 4602–4608.

Steyn, W.J., Holcroft, D.M., Wand, S.J.E. and Jacobs, G. (2004) Regulation of pear color development in relation to activity of flavonoid enzymes. *Journal of the American Society for Horticultural Science* 129, 1–6.

Steyn, W.J., Wand, S.J.E., Jacobs, G., Rosecrance, R.C. and Roberts, S.C. (2009) Evidence for a photoprotective function of low-temperature-induced anthocyanin accumulation in apple and pear peel. *Physiologia Plantarum* 136, 461–472.

Stilesa, E.A., Cecha, N.B., Deea, S.M. and Lace, E.P. (2007) Temperature-sensitive anthocyanin production in flowers of *Plantago lanceolata*. *Physiologia Plantarum* 129(4), 756–765.

Stutte, G.W., Eraso, I. and Rimando, A.M. (2008) Carbon dioxide enrichment enhances growth and flavonoid content of two *Scutellaria* species. *Journal of the American Society for Horticultural Science* 133(5), 631–638.

Tanaka, Y. and Uritani, I. (1977) Purification and properties of phenylalanine ammonia-lyase in cut-injured sweet potato. *Journal of Biochemistry* 81, 963–970.

Tattini, M., Galardi, C., Pinelli, P., Massai, R., Remorini, D. and Agati, G. (2004) Differential accumulation of flavonoids and hydroxycinnamates in leaves of *Ligustrum vulgare* under excess light and drought stress. *New Phytology* 163, 547–561.

Tenhunen, J.D., Reynolds, J.F., Lange, O.L., Dougherty, R.L., Harley, P.C., Kummerow, J. and Rambal, S. (1990) QUINTA: a physiologically based growth simulator for drought adapted woody plant species. In: Pereira, J.S. and Landsberg, J.J. (eds) *Biomass Production by Fast-Growing Trees*. Kluwer, Dordrecht, the Netherlands, pp. 135–168.

Thompson, A.M. (1992) The oxidizing capacity of the earth's atmosphere: probable past and future changes. *Science* 256, 1157–1165.

Tingey, D.T., Manning, M., Grothaus, L.C. and Burns, W.F. (1979) The influence of light and temperature on isoprene emission rates from live oak. *Plant Physiology* 47, 112–118.

Tingey, D., Manning, M., Grothaus, L. and Burns, W. (1980) Influence of light and temperature on monoterpene emission rates from slash pine. *Plant Physiology* 65, 797–801.

Tingey, D.T., Evans, R.C., Bates, E.H. and Gumpertz, M.L. (1987) Isoprene emissions and photosynthesis in three ferns – the influence of light and temperature. *Plant Physiology* 69, 609–616.

Tingey, D.T., Turner, D.P. and Weber J.A. (1991) Factors controlling the emissions of monoterpenes and other volatile organic compounds. In: Sharkey, T.D., Holland, E.A. and Mooney, H.A. (eds) *Trace Gas Emissions from Plants*. Academic Press, San Diego, USA, pp. 93–119.

Trainer, M., Parrish, D.D., Buhr, M.P., Norton, R.B., Fehsenfeld, F.C. *et al.* (2001) Correlation of ozone with NO_y in photochemically aged air. *Journal of Geographical Research* 98, 2917–2925.

Treutter, D. (2006) Significance of flavonoids in plant resistance: a review. *Environmental Chemistry Letters* 4, 147–157.

Valkama, E., Koricheva, J. and Oksanen, E. (2007) Effects of elevated O_3, alone and in combination with elevated CO_2, on tree leaf chemistry and insect herbivore performance: a meta-analysis. *Global Change Biology* 13, 184–201.

Van Buskirk, H.A. and Thomashow, M.F. (2006) *Arabidopsis* transcription factors regulating cold acclimation. *Physiologia Plantarum* 126, 72–80.

Velikova, V., Pinelli, P., Pasqualini, S., Reale, L., Ferranti, F. and Loreto, F. (2005) Isoprene decreases the concentration of nitric oxide in leaves exposed to elevated ozone. *New Phytology* 166(2), 419–426.

Velikova, V., Varkonyi, Z., Szabo, M., Maslenkova, L., Nogues, I. *et al.* (2011) Increased thermostability of thylakoid membranes in isoprene-emitting leaves probed with three biophysical techniques. *Plant Physiology* 157, 905–916.

Velikova, V., Sharkey, T.D. and Loreto, F. (2012) Stabilization of thylakoid membranes in isoprene-emitting plants reduces formation of reactive oxygen species. *Plant Signaling and Behavior* 7, 139–141.

Vickers, C.E., Gershenzon, J., Lerdau, M.T. and Loreto, F. (2009) A unified mechanism of action for volatile isoprenoids in plant abiotic stress. *Natural and Chemical Biology* 5, 283–291.

Vogt, T. (2010) Phenylpropanoid Biosynthesis. *Molecular Plant* 3(1), 2–20.

Vu, C.V., Men Jr, L.H. and Garrard, L.A. (1981) Effects of supplemental UV-B radiation on growth and leaf photosynthetic reactions of soybean (*Glycine max*). *Physiologia Plantarum* 52, 353–362.

Vu, C.V., Allen Jr, A.H. and Garrard, L.A. (1982) Effects of UV-B radiation (280–320 nm) on photosynthetic constituents and processes in expanding leaves of soybean [*Glycine max* (L.) Merr.] *Environmental and Experimental Botany* 22, 465–473.

Vuorinen, T., Nerg, A.M., Vapaavuori, E. and Holopainen, J.K. (2005) Emission of volatile organic compounds from two silver birch (*Betula pendula* Roth) clones grown under ambient and elevated CO_2 and different O_3 concentrations. *Atmospheric Environment* 39, 1185–1197.

Wahid, A. and Ghazanfar, A. (2006) Possible involvement of some secondary metabolites in salt tolerance of sugarcane. *Journal of Plant Physiology* 163, 723–730.

Walter, T.E. (1967) Factors affecting fruit colour in apples: a review of world literature. In: *Annual Report of East Malling Research Station for 1966*, East Malling Research Station, Maidstone, UK, pp. 70–82.

Weatherhead, E.C. and Andersen, S.B. (2006) The search for signs of recovery of the ozone layer. *Nature* 441, 39–45.

Webb, A.R. (1997) Monitoring changes in UV-B radiation. In: Limsden, P.J., (ed.) *Plants and UV-B: Responses to Environmental Change*. Cambridge University Press, Cambridge, UK, pp. 13–30.

Wiedinmyer, C., Strange, I.W., Estes, M., Yarwood, G. and Allen D.T. (2000) Biogenic hydrocarbon emission estimates for North Central Texas. *Atmospheric Environment* 34, 3419–3435.

Wilkinson, M.J., Monson, R.K. and Trahan, N. (2009) Leaf isoprene emission rate as a function of atmospheric CO_2 concentration. *Global Change Biology* 15, 1189–1200.

William, A., Hoch Zeldin, E.L. and Mccown, B.H. (2001) Physiological significance of anthocyanins during autumnal leaf senescence. *Tree Physiology* 21, 1–8.

Wilson, K.E.,Wilson, M.I. and Greenberg, B.M. (1998) Identification of the flavonoid glycosides that accumulate in *Brassica napus* L. cv. Topas specifically in response to ultraviolet B radiation. *Photochemistry and Photobiology* 67, 547–553.

Wilson, K.E., Thompson, J.E., Huner, N.P. and Greenberg, B.M. (2001) Effects of ultraviolet-A exposure on ultraviolet-B-induced accumulation of specific flavonoids in *Brassica napus*. *Photochemistry and Photobiology* 73, 15–21.

Winter, T.R. and Rostàs, M. (2008) Ambient ultraviolet radiation induces protective responses in soybean but does not attenuate indirect defense. *Environmental Pollution* 155, 290–297.

Xiong, L., Ishitani, M., Lee, H. and Zhu, J.K. (2001) The *Arabidopsis* LOS5/ABA3 locus encodes a molybdenum cofactor sulfurase and modulates cold stress- and osmotic stress-responsive gene expression. *Plant Cell* 13, 2063–2083.

Yamaguchi-Shinozaki, K. and Shinozaki, K. (2006) Transcriptional regulatory networks in cellular responses and tolerance to dehydration and cold stresses. *Annual Review of Plant Biology* 57, 781–803.

Yamasaki, H., Sakihama, Y. and Ikehara, N. (1997) Flavonoid-peroxidase reaction as a detoxification mechanism of plant cells against H_2O_2. *Plant Physiology* 115, 1405–1417.

Zangerl, A.R. and Berenbaum, M.R. (1987) Furanocoumarins in wild parsnip: effects of photosynthetically active radiation, ultraviolet light and nutrition. *Ecology* 68, 516–520.

Zhao, F.J., Shu, L.F. and Wang, Q.H. (2012) Terpenoid emissions from heated needles of *Pinus sylvestris* and their potential influences on forest fires. *Acta Ecologica Sinica* 32, 33–37.

10 Summer Semi-deciduous Species of the Mediterranean Landscape: A Winning Strategy of *Cistus* Species to Face the Predicted Changes of the Mediterranean Climate

OTÍLIA CORREIA[1] AND LIA ASCENSÃO[2]*

[1]*cE3c – Centre for Ecology, Evolution and Environmental Changes, Faculdade de Ciências, University of Lisbon, Lisbon, Portugal;* [2]*CESAM – Centre for Environmental and Marine Studies, Faculdade de Ciências, University of Lisbon, Lisbon, Portugal*

Abstract

Plants dominating the later stages of succession in Mediterranean shrublands, well adapted to summer drought, are frequently sclerophyllous and regenerate after disturbance through sprouting (obligate resprouters). In contrast, summer semi-deciduous species such as Cistaceae and Lamiaceae are obligate seeders, do not sprout, and fire stimulates the germination of their seeds. It is thought that these species evolved during the Pleistocene, when the Mediterranean climate was formed. This review focuses on four *Cistus* species: *C. albidus*, *C. ladanifer*, *C. monspeliensis* and *C. salviifolius*, which are frequent in the Mediterranean basin. The morpho-functional and physiological traits implicated in the adaptive strategies to face the predicted changes of the Mediterranean climate are analysed. Briefly, these species have similar morpho-functional traits, possibly indicative of the low soil fertility which is usually linked to early successional stages and xeric conditions. All the leaf traits seem to indicate that the summer semi-deciduous species are adapted to face the predicted changes of the Mediterranean climate, in the sense of increasing temperature and unpredictability of rain distribution. In fact, these species, with high autecological plasticity, seem to exhibit an opportunistic behaviour that allows them to respond to climate unpredictability, with photosynthetic organs ready to work whenever the climatic conditions become favourable.

10.1 Rockrose Shrublands and Thickets of the Iberian Peninsula

The high diversity that characterizes the Mediterranean region has been related to the presence of man since ancient times, and to the geomorphological characteristics, topography, soil and climate of these Mediterranean landscapes (Blondel and Aronson, 1999).

Rockroses (hereafter a general term to name any *Cistus* species) are important members of two typical Mediterranean ecosystems: the 'maquis' and the 'garrigue' from the landscape of the Mediterranean basin. From a phytosociological perspective, these shrublands belong to the *Cisto-Lavanduletea*, alliance *Cistion laurifolii* class (Rivas-Martínez, 1979).

Rockrose shrublands are a type of Mediterranean shrubland occupying vast areas of the Iberian Peninsula, as a consequence of deforestation or abandonment of agricultural or pasture lands. Moreover, due to their pioneer characteristics, they are able to colonize areas where the original vegetation was destroyed by fire or other kinds of disturbance.

*E-mail:lia.ascensao@fc.ul.pt

For this reason, they became dominant in many landscapes which would otherwise be exposed to erosion. According to Juhren (1966), rockroses colonize degraded areas and form one of the first successional shrub stages.

Cistaceae (Angiosperm, Malvales, APG II, 2003) is a small family distributed throughout the northern hemisphere (except for three South American species), with the genera *Cistus*, *Fumana*, *Halimium*, *Helianthemum*, and *Xolantha* (*Tuberaria*) almost restricted to the Mediterranean basin (Castroviejo *et al.*, 1993). Rockroses are relatively small, branched shrubs, although some specimens of *C. ladanifer* may reach 2.5 m in height. Leaves are opposite, generally small, simple and with an entire margin. Flowers are big and allogamic, very conspicuous, white, pink or purple-coloured, rarely with nectaries with numerous pollen producing stamens that attract numerous pollinating insects (Talavera *et al.*, 1988; López González, 1991). In these species, bees and several Dipterans are the main pollinators. Stamen sensitivity to insect touch is a facilitating cross-pollination process, keeping the flower's pollen away from the stigma. However, *C. albidus* do not have sensitive stamens, as its styles and stamens are almost the same size (Brandt and Gottsberger, 1988).

In Cistaceae, floral traits such as diameter of the corolla or the number of anthers per flower, are continuous and highly correlated variables, i.e. species with larger flowers also have more and larger anthers and more ovules (Herrera, 1992a). *Cistus albidus*, *C. crispus*, *C. creticus* and *C. heterophyllus* are the only rockroses with pink or purple flowers. Recently, some studies, using plant chemotypes and molecular approaches, confirmed this phenotype-based genus taxonomy, dividing *Cistus* into three subgenera (Guzmán and Vargas, 2005; Guzmán *et al.*, 2009; Barrajón-Catalán *et al.*, 2011).

In these species, flowering occurs in spring and the flowers are short-lived, lasting less than one day (12–24 h) – they open in the early morning, earlier in lighter sites and later in shady ones, and by the afternoon the petals begin to fall and the calyx closes towards the centre of the flower, pressing the anthers against the stigma (Martín Bolaños and Guinea, 1949; Brandt and Gottsberger, 1988; Blasco and Mateu, 1995).

The morphology of Cistaceae fruit is the clue for the family and genus names. The Greek *kisthos*, *Kistê* or Latin *Chistos*, means basket, box or capsule, a reference to the form of the fruits, a globose lignified and dehiscent capsule, opening through 6–12 valves, from the apex to the base, and containing minute seeds (López González, 1991; Castroviejo *et al.*, 1993).

The *Cistus* genus is distributed all over the Mediterranean area, from the Canary Islands and Portugal to the East, and from North Africa to the western French coast, and is particularly well represented in the Iberian Peninsula (Castroviejo *et al.*, 1993).

The most common species in Portugal are *Cistus albidus* L., *C. crispus* L., *C. ladanifer* L., *C. laurifolius* L., *C. libanotis* L., *C. monspeliensis* L., *C. populifolius* L., *C. psilosepalus* L. and *C. salviifolius* L.. On the rocky platforms of south-western Portugal there is also an endemic species designated as *Cistus palhinhae* (Franco, 1984) that is nowadays considered a subspecies of *C. ladanifer* (*C. ladanifer* ssp. *sulcatus*) (Demoly and Monserrat, 1993; Carlier *et al.*, 2008).

The apparent uselessness of these shrublands hides a world of culture and wisdom, offered by the people who live surrounded by extensive rockrose shrublands. Cistaceae are well known in Mediterranean cultures for medicinal uses. In *C. ladanifer*, nearly all plant organs are used: roots, stems, leaves, flowers and fruits. Its wood is considered as one of the hardest; with excellent heat power, it is frequently used in traditional bakery ovens. Leaves impart the unique Mediterranean fragrance which corresponds to rockrose oleoresin, the labdanum, also called ladanum or ladan. Almost all *Cistus* species produce, in different amounts, resinous exudates with strong amber and balsamic odour. However, the resin is mainly extracted in the western Mediterranean from *C. ladanifer* and in the eastern Mediterranean from *C. creticus*. Labdanum has been known since ancient times for its aromatic and pharmacological properties. Some authors state that labdanum may be the mysterious onycha, an ingredient in the holy incense, mentioned in the Old Testament (Abrahams, 1979), and others think that it may correspond to balm and myrrh, gum resins also referred to in the Bible (Juniper and Jeffree, 1983). The method of harvesting labdanum has greatly changed through history. In the past, one of the most frequent methods was to drive goats and sheep to graze in a rockrose shrubland; the resin stuck to the animals' hair and was later retrieved when the goats were sheared (López González, 1991). Nowadays, labdanum is extracted through hydrodistillation, by immersion of leaves in boiling water, or by vapour distillation with organic solvents. The resin and the essential oil are mainly used in the perfume and cosmetic industry, to fix fragrances and to make soaps and deodorants (López González, 1991).

C. ladanifer is still subject to substantial exploitation by multinational companies in Spain, particularly in Andalusia, and is becoming a rare raw material in those landscapes. Spain is the world's top producer of labdanum and labdanum absolute. The chemical composition of the essential oil of *C. ladanifer*, obtained from plants growing in different regions of the Mediterranean, is reported in several studies that highlight the interest in this oil for perfumery (Ramalho *et al.*, 1999; Gomes *et al.*, 2005; Teixeira *et al.*, 2007).

Several species of *Cistus* have been traditionally used in folk medicine for a wide range of ailments, especially for treatment of skin diseases, bacterial and fungal infections, digestive problems, diarrhoea, cold, flu and respiratory tract disorders (Attaguile *et al.*, 2000). In the last two decades, numerous phytochemical studies have been carried out on the genus *Cistus* to evaluate their biological and pharmacological properties. These studies have showed that the main constituents of the aqueous extracts of *Cistus* species are flavonoids, whereas organic extracts are rich in labdane-type diterpenes, compounds with antioxidant capacity, anti-inflammatory properties and antibacterial, antifungal, antiproliferative and cytotoxic activities (Papaefthimiou *et al.*, 2014 and references cited there).

Others *Cistus* species, such as *C. albidus* (in Algeria), *C. salviifolius* (in Greece), *C. crispus* (in Tunisia) and *C. incanus* were used in the past as flavours or substitutes for tea (Heywood, 1979). Besides these uses, several *Cistus* species and hybrids are also grown as ornamentals (Heywood, 1979). Since the 18th century, hybrids of *Cistus* have been developed to survive cold winters in northern Europe and to provide labdanum. Some of the hybrids produce attractive flowers, with crepe-textured petals of pure white, bright pink or purple colours which embellish many European gardens (see: www.cistuspage.org.uk).

10.2 Interactions in the Rockrose Shrublands

There is a surprisingly rich fungal diversity colonizing the rockrose shrublands, especially when we consider that humidity is not predominant in these environments. For that reason, the microbiota present in rockrose shrublands is quite specific and interesting from the scientific point of view, due to the environment's xericity. Endomycorrhiza as well as ectomycorrhiza have been reported from Cistaceae (Comandini *et al.*, 2006; de Vega *et al.*, 2010). More than 200 fungal species belonging to 40 genera establish mycorrhizae with *Cistus* (symbiotic links with the roots of these shrubs) (Boursnell, 1950; Comandini *et al.*, 2006) and many others associate with other plants which occur in the neighbourhood of rockroses. The mycobionts more common in *Cistus* species are from Cortinariaceae and Russulaceae, some of them being specific to these species. Therefore, the names of some of these mycorrhizal fungi include the Latin epithet *cistophyllum* or *cistophillus*, as in *Lactarius cistophillus*, *Entoloma cistophillum* or *Russula cistoadelta* (Comandini *et al.*, 2006). In some cases, spontaneous mycorrhization is decisive for the distribution of the different *Cistus* species in their natural habitat (Berliner *et al.*, 1986). Some of the mycorrhizal fungi associated with *Cistus* species present characteristics of pioneer species, often typical of disturbed habitats and associated with different hosts (Comandini *et al.*, 2006; Martín-Pinto *et al.*, 2006).

Mycorrhizal fungi are particularly important in these stressful habitats, providing plant hosts with water and nutrient uptake and protection against pathogens (Smith and Read, 1997).

Edible mycorrhizal mushrooms are not only a gourmet food, but also a source of income for those who cultivate or collect them from the wild. It is worth highlighting that truffles (genus *Tuber*) are able to mycorrhize many rockroses, so that these shrubs may be used to produce several species of edible truffles (*Tuber melanosporum, T. aestivum, T. brumale* and *T. albidum*) (Giovannetti and Fontana, 1982). *Cistus* species are used to form mycorrhizae and, therefore, truffle carpophores may be used to colonize poor and abandoned soils, degraded land where no other shrub species thrive and as an alternative to other shrub or larger tree species, to produce edible truffles (Giovannetti and Fontana, 1982). Similarly, the desert truffle (*Terfezia* species), an edible fungus, that includes species typically distributed in regions with arid and semi-arid climates of the western Mediterranean basin, establishes symbiotic mycorrhiza with species of the genera *Cistus* and *Helianthemum* (Díez *et al.*, 2002). Indeed, from an ecological point of view, the truffles grow in moderately calcareous sandy soils, slightly alkaline and poor in organic and mineral matter. In terms of climate, these fungi grow in a hot arid climate, as long as occasional rains occur in autumn–winter, then periods of drought follow. So, this mutualistic interaction between Cistaceae and fungi may play an important role in the maintenance and dynamics of Mediterranean shrublands, contributing to the prevention of erosion and desertification (Honrubia *et al.*, 1992).

One of the parasites specific to Cistaceae is *Cytinus hypocistis*, a holoparasitic species included in Cytinaceae family (formerly Raflesiaceae), that grows endophytically within the tissues of the host plant (de Vega *et al.*, 2007; Thorogood and Hiscock, 2007; de Vega *et al.*, 2009). This parasite-plant with scale-like leaves of yellow-orange colour, conspicuous flowers and a sweet taste is common in rockrose shrublands, being used in popular medicine as an astringent and anti-diarrhoeic. Flowering of *C. hypocistis* occurs in spring, coinciding with its host's flowering (López González, 1991). At this time their inflorescences burst through the host tissues at ground level and are mainly pollinated by ants, producing fruits with thousands of small seeds that are dispersed by a beetle species (López González, 1991; de Vega *et al.*, 2007; de Vega *et al.*, 2009). This plant–animal interaction may facilitate the seed dispersal and germination near the roots of the host plants. This kind of mutualistic interaction, a tripartite association among *Cytinus*, its host plant and the mycorrhizal fungi, can be very important in terms of maintaining biodiversity and ecosystem functioning (de Vega *et al.*, 2011).

10.3 Cistus Species and Their Role in the Ecosystem

The most dominant growth forms in the Mediterranean ecosystems are sclerophyllous evergreen and semi-deciduous species. Plants dominating the later stages of succession in Mediterranean shrublands are well adapted to summer drought. They are frequently sclerophyllous evergreens that regenerate after disturbance through resprouting from below-ground organs (resprouters) (Keeley, 1986; Trabaud, 1995; Clemente *et al.*, 1996; Clemente, 2002). In contrast, Cistaceae and Lamiaceae are 'seeders', are killed by fire, do not resprout, and fire stimulates the germination of their seeds from the soil seed bank (Trabaud, 1995; Clemente *et al.*, 1996; Clemente *et al.*, 2007). These species dominate the thickets and are found in earlier successional stages, being characteristic of a proclimax.

For a long time, Cistaceae were regarded as pyrophytic species, i.e. necessarily associated with fire. However, they are not exclusively dependent on fire and may be considered rather as heliophytes. After disturbance, they colonize open spaces where plant cover is generally scarce, and especially where competition with other species is reduced (Trabaud, 1995). In obligate seeders, fire-stimulation of germination may be triggered by high temperatures, which promote seed dispersal, break seed coat impermeability and/or destroy germination inhibitors in the soil (Corral *et al.*, 1990; Roy and Sonié, 1992; Thanos *et al.*, 1992; Trabaud, 1995; Ne'eman *et al.*, 1999; Paula and Pausas, 2008). On the other hand, fire modifies the sunlight spectrum at the soil level due to the destruction of the plant canopy, changing the light quality (red/far-red) which controls germination through phytochrome conversion (Smith, 1982; Roy and Arianoutsou-Faraggitaki, 1985).

Fire has been a major driving force in the selection of communities adapted to fire and the phylogenetic structure of communities reflects not only the filtering processes but also the diversification of seeder species (Verdú and Pausas, 2007). Also the two main fire response traits, resprouting and seeder regeneration strategies, showed a negative evolutionary correlation in woody plants of the Mediterranean basin. The seeder strategy is very rare in old lineages, appearing concomitantly with fire during the Quaternary (Pausas and Verdú, 2005; Saura-Mas and Lloret, 2007). Groups of species with different post-fire regenerative strategies have specific functional traits related to water use, which are linked to their life-history characteristics (Saura-Mas and Lloret, 2007; Vilagrosa *et al.*, 2014).

Obligate seeders accumulate a large number of dormant seeds in the soil (seed bank), and exhibit a pronounced increment in seedling density during the first years after a fire, resulting in high seedling mortality following the first summer stress after fire (Keeley, 1986; Clemente *et al.*, 1996; Clemente *et al.*, 2005; Clemente *et al.*, 2007). Because these species are often shallow-rooted (Keeley, 1986; Silva *et al.*, 2002), they have very low water potentials in summer, which is associated with the high hydraulic conductivity of *Cistus*. This is likely to lead to xylem embolism (Correia and Catarino, 1994), a mechanism often responsible for high seedling mortality in Mediterranean ecosystems

(Williams *et al.*, 1997; Clemente *et al.*, 2005). After disturbance, they colonize vast areas, often forming pure communities which persist for 10–15 years, the longevity of these species (Juhren, 1966; Roy and Sonié, 1992). The decline of these communities is due to the senescence of the individuals and/or to the absence of regeneration, leading to community auto-succession (Luis-Calabuig *et al.*, 2000). According to some authors, lack of germination at later successional stages is due to the absence of stimulation by fire and to germination-inhibiting conditions, deriving from light quality (red-far red) under *Cistus* canopies (Roy and Sonié, 1992), or even to allelopathic or autoallelopathic effects. Fire frequency is also an important shaper of Mediterranean communities, since the various species display different resiliencies to recurrent fires. In general, obligate seeders need 5–25 years to rebuild the soil seed bank, but as they are rather short-lived and seedling recruitment depends on fire occurrence, these communities predominate in zones where fire is relatively frequent.

10.4 Adaptations of Cistus to the Mediterranean Climate

Briefly, all *Cistus* species have similar morpho-functional characteristics, possibly indicative of the low soil fertility which is generally associated with early successional stages and xeric conditions. It is thought that the genus has evolved during the Quaternary (Pliocene), when the Mediterranean climate was formed, and therefore will have acquired specific characteristics well-adjusted to the particularities of this type of climate (di Castri, 1981; Herrera, 1984; Herrera, 1992b; Pausas and Verdú, 2005).

Many existing species in Mediterranean type-ecosystems originated in the Tertiary and now are mixed with more recent taxa that evolved in the Quaternary, during a period when the global climate became much more arid. Integrating paleobotanical, ecological and phylogenetic analysis, Valiente-Banuet *et al.* (2006) suggest that a large number of species from the Tertiary appear to have been preserved due to a facilitative interaction or 'nurse' effect of more modern species of the Quaternary.

Rockroses, semi-deciduous species, exhibit intermediate features between sclerophylls (regarded as stress tolerant) and summer deciduous species (stress avoiders) (Mooney and Kummerow, 1971). Semi-deciduous species and sclerophylls differ in several morphological and physiological adaptations to the water and nutrient constraints of the Mediterranean ecosystems, which affect their phenology and carbon gain (Kummerow, 1981; Correia *et al.*, 1987; Correia *et al.*, 1992; Diaz Barradas *et al.*, 1999; Werner *et al.*, 1999; Werner, 2000; Zunzunegui *et al.*, 2005).

Cistus species have developed two kinds of strategy. First, they avoid summer stress by shedding a large proportion of leaves and branches during summer. Second, the remaining leaves (summer leaves) have stress-tolerance mechanisms which enable them to endure very low water potentials, reducing stomatal conductance and water loss through transpiration (Correia *et al.*, 1987; O. Correia, Contribuição da fenologia e ecofisiologia em estudos da sucessão e dinâmica da vegetação mediterrânica [Contribution of phenology and physiological ecology in succession studies and dynamics of Mediterranean vegetation]. University of Lisbon, Lisbon, Portugal, 1988, unpublished thesis). Rockroses are viewed as summer semi-deciduous, or species with leaf dimorphism (Margaris and Vokou, 1983; Harley *et al.*, 1987), developing summer and winter leaves that exhibit not only morphological modifications, but also structural and ecophysiological adaptations throughout the year. Leaf phenology of rockroses is typical of summer semi-deciduous species: in summer, a considerable part of the leaves and branches are shed, and only small leaves remain attached to the standing branches; these summer leaves are formed by the end of spring or early summer (Correia *et al.*, 1992). These two leaf types – summer and winter leaves – exhibit not only morphological modifications, but also structural and ecophysiological adaptations throughout the year. Some morphological and physiological characteristics related with photosynthetic variables are presented on summer and winter leaves of the most frequent *Cistus* species.

10.5 Morphological and Structural Characteristics

All Cistus species have pubescent leaves, but while *C. albidus* bear a woolly-tomentose indumentum on both leaf surfaces (Fig. 10.1a, b), that gives the leaves a silver-grey appearance, *C. ladanifer*, *C. monspeliensis* and *C. salviifolius* only show a dense tomentose indumentum on the abaxial leaf surface (Figs 10.2d, 10.3d, 10.4b).

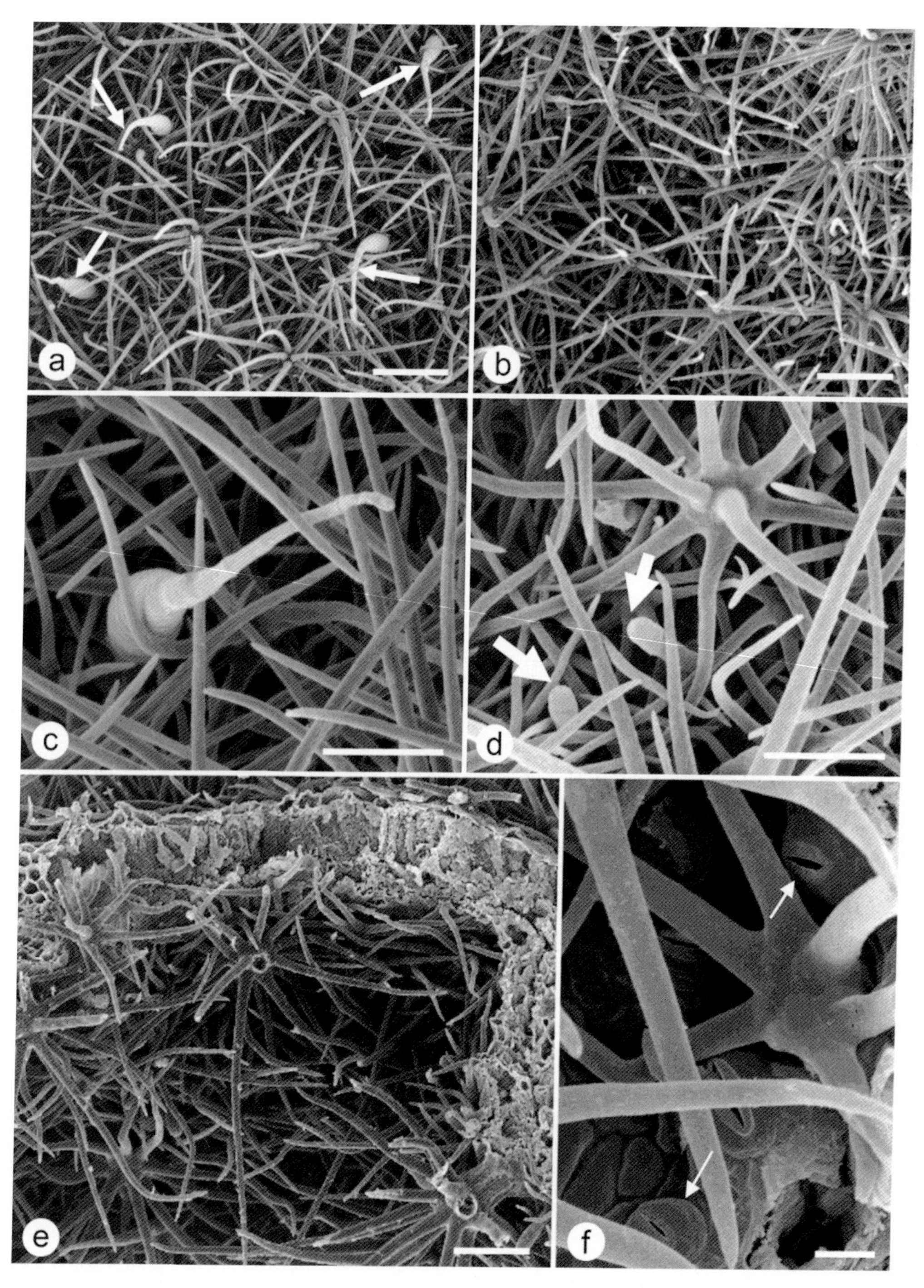

Fig. 10.1. SEM micrographs of mature leaves of *Cistus albidus*. a and b: leaf adaxial and abaxial surfaces showing the dense woolly-tomentose indumentum. Bottle-shaped glandular trichomes (arrows) are seen among non-glandular trichomes on the leaf adaxial surface. c: detail of a bottle-shaped glandular trichome. d: bulbous glandular trichomes (bold arrows) on the leaf abaxial surface. e: the dense tomentose indumentum on both leaf surfaces. f: stomata (thin arrows) are clearly seen on the leaf abaxial surface. Scale bars: (a, b) = 200 μm; (c, d, e) = 100 μm; (f) = 20 μm.

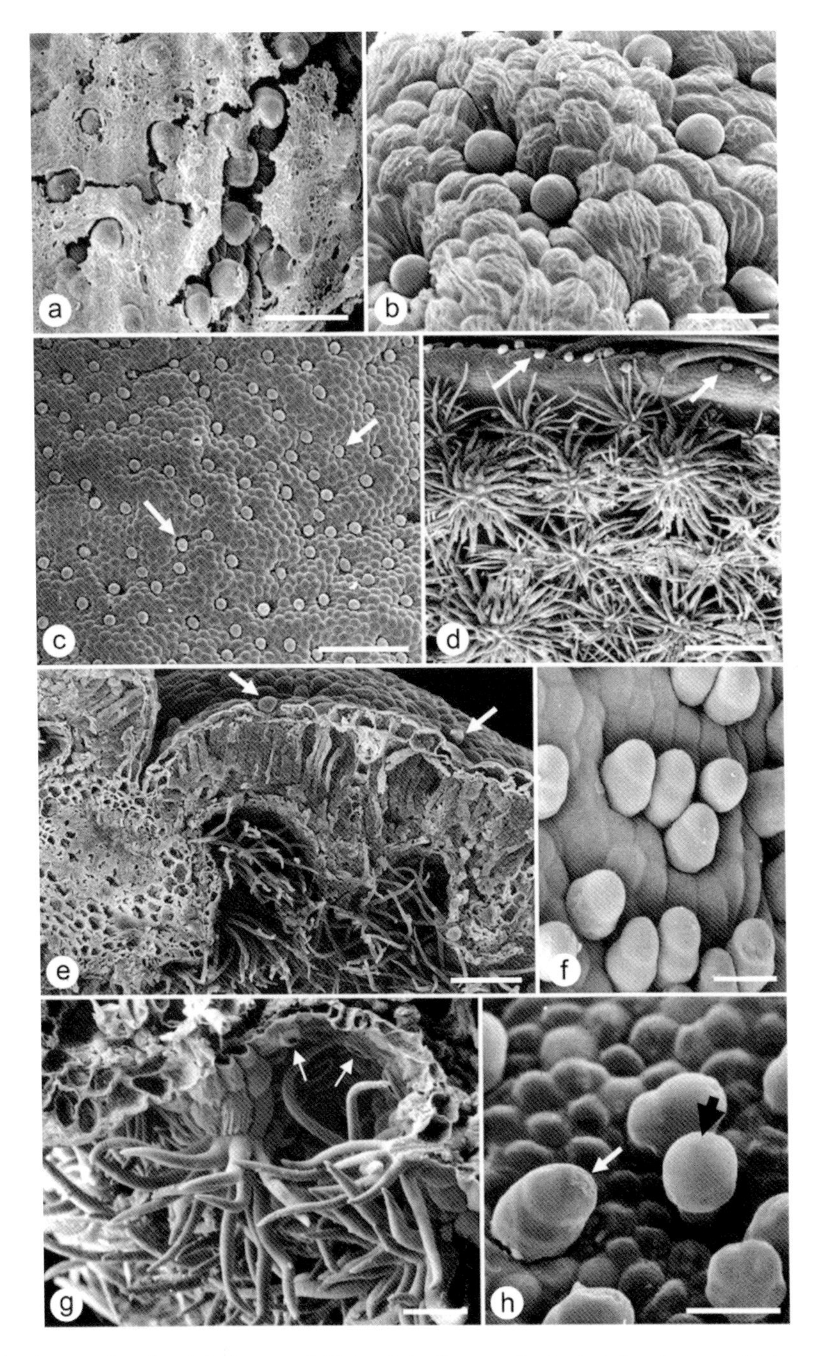

Fig. 10.2 SEM micrographs of mature leaves of *Cistus ladanifer*. a–c: leaf adaxial surfaces. a: thick layer of oleoresin almost completely covering the glandular trichomes. b and c: glandular trichomes (arrows) are observed after oleoresin withdrawal. d: stellate non-glandular trichomes on the leaf abaxial surface and oleoresin glandular trichomes (arrows) on the leaf margins. e: in a leaf cross-section, oleoresin trichomes (arrows) and deep leaf crypts nearly occluded by non-glandular trichomes are clearly seen. f–h: leaf abaxial surfaces. f: oleoresin trichomes showing the three-celled glandular head. g: stomata (thin arrows) concentrated in a leaf crypt. h: details of a bulbous trichome (bold black arrow) and oleoresin trichomes at the post-secretory stage. Cuticular cracks (white arrow) are observed at the apical glandular cell. Scale bars: (a, b) = 50 μm; (c, d) = 200 μm; (e) = 100 μm; (f, g, h) = 30 μm.

The aerial organs of *C. ladanifer* and *C. monspeliensis* are covered by a resinous secretion, labdanum, with a rather characteristic fragrance. To the naked eye, especially in summer due to the high temperature, this abundant and sticky exudate appears as a shiny film, a varnish that tends to harden and become brittle when dry. Observations by scanning electron microscopy (SEM) show a more or less homogenous oleoresin layer that nearly covers the epidermal leaf surface, hiding the glandular trichomes (Figs 10.2a, 10.3a). In fact, the quantity of secreted resin may reach 10–14% per dry weight unit in *Cistus* species such as *C. ladanifer*, *C. monspeliensis*, *C. laurifolius* and *C. palhinhae*, while other species have an intermediate percentage of resin of 5–6% or even lower, ranging from 0.5 to 3%, in species such as *C. albidus*, *C. crispus*, *C. incanus*, *C. psilosepalus* and *C. salviifolius* (Gülz *et al.*, 1996). In species producing great amounts of resin, the adaxial leaf surface is nearly covered by this secretion and, to observe the glandular trichomes, resin must be removed. Leaf washing with organic solvents prior to fixation leads to trichome collapse (Gülz *et al.*, 1996), so we have tried long glutaraldehyde fixations with several fixative substitutions, which seems to preserve the trichome structure (Figs 10.2b, 10.3b).

The leaf indumentum of *C. albidus*, *C. ladanifer*, *C. monspeliensis* and *C. salviifolius* consists of stellate non-glandular trichomes mixed with different types of glandular trichomes. In general, stellate trichomes with a variable number of single-celled arms (usually 12–16), mostly distributed on the abaxial surface, are mixed with dendritic trichomes (Figs 10.1d, 10.3f, 10.4b, d, e).

Besides these types of non-glandular trichomes, very long, simple, pointed trichomes, with their straight tips directed toward the leaf apex, are also present (Figs 10.2d, 10.3b–d). On *C salviifolius* leaves stellate trichomes of the porrect type (with an enlarged central arm perpendicular to the leaf surface) are also frequent on the adaxial surface (Fig. 10.1a, c).

In *C. Albidus*, both leaf surfaces are entirely covered by a thick layer of interwoven non-glandular trichomes, hiding completely the epidermis (Fig. 10.1a–e). Conversely, in *C. monspeliensis* and *C. ladanifer*, non-glandular trichomes are nearly confined to the leaf abaxial surface forming a dense mat (Figs 10.2d, 10.3d). On the adaxial surface only a few very elongated, simple tip-pointed trichomes occur, especially on the leaf margins and main vein (Figs 10.2d, 10.3b, c). The leaves of *C. salviifolius* are slightly pubescent on the adaxial surface (Fig. 10.4a, c), showing only a denser indumentum on the abaxial one (Fig. 10.4b, d).

The oleoresin glandular trichomes of these four *Cistus* species focused on in this chapter are of two different types. On *C. Monspeliensis*, due to its morphological appearance, they have been named bottle-shaped trichomes (Uphof, 1962). They are uniseriate trichomes with a two-or three-celled stalk and a multicellular glandular head which has a swollen basal region and ends in a tapering cylindrical appendage with three to five cells (Fig. 10.2b, c, f). A mature trichome has 8–12 glandular cells (not shown), and is about 80–100 μm in height and 40 μm in maximum diameter at the swollen basal region of the glandular head. At the secretory phase, the apical cell tip of the appendage shows a bulb-like structure and, at this level, a subcuticular space develops to store the secretion products (Fig. 10.2b). Similar glandular trichomes are also observed in *C. albidus* leaves, but they are bigger and a have a greater number of glandular cells (Fig. 10.1a, c). They are about 250–300 μm in height and 60 μm in maximum diameter in the glandular head. Bottle-shaped trichomes have been previously studied by Ascenção and Pais (1981), who named them capitate trichomes, described their morphology, characterized histochemically the oleoresin and related the ultrastructural features with the secretion process. Recently, bottle-shaped trichomes with variable dimensions have been observed in several *Cistus* species and named 'long ball-headed tubes' (Gülz *et al.*, 1996; Paolini *et al.*, 2009). However, this designation does not define properly that secretory structure, since the trichome is not a hollow gland and 'the ball' on the tip of the glandular head is a drop of oleoresin, only present at the secretory stage.

The oleoresin trichomes of *C. ladanifer* leaves are capitate, uniseriate trichomes with a short unicellular stalk often sunken in the epidermis and a three-celled glandular head (Fig. 10.2b, f, h). Mature trichomes are about 40 μm in height and 30 μm in maximum horizontal diameter of the glandular head. Oleoresin accumulates temporarily on a subcuticular space forming in the upper cell of the glandular head and is released by cuticle rupture (not shown). In Fig. 10.2h, the cracks in the cuticle after secretion (arrow) are visible. Oleoresin trichomes, although distributed on both leaf surfaces, are more abundant on the adaxial surface (Fig. 10.1a–e).

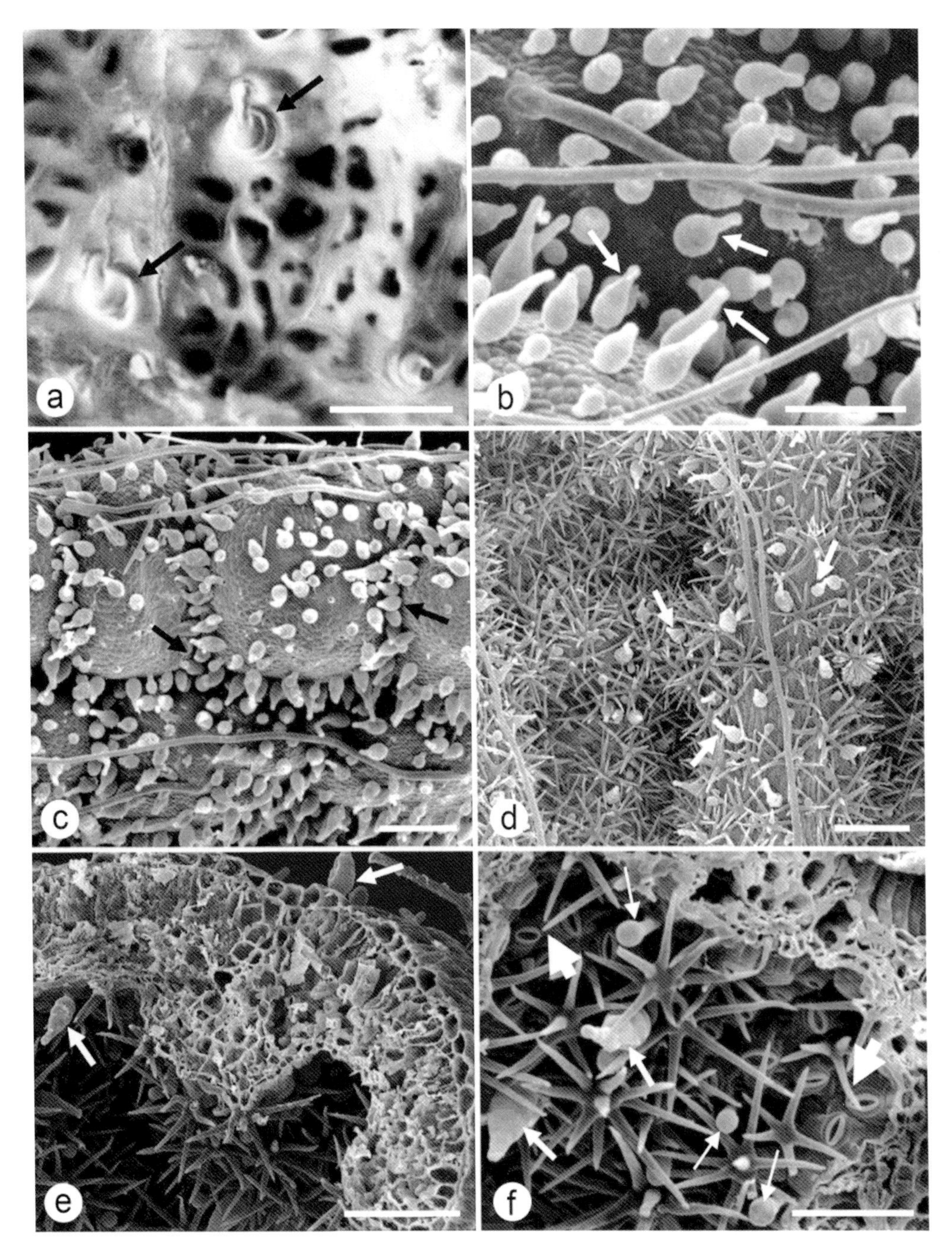

Fig. 10.3. SEM micrographs of mature leaves of *Cistus monspeliensis*. a–c, leaf adaxial surfaces. a: glandular trichomes (arrows) embedded in the oleoresin layer. b and c: bottle-shaped glandular trichomes (arrows) and long non-glandular trichomes are evident after oleoresin withdrawal. d–f: leaf abaxial surfaces. d: dense cover of glandular (arrows) and non-glandular trichomes. e: in a leaf cross-section, bottle-shaped trichomes (arrows) on both leaf surfaces and deep leaf crypts bordered by stellate non-glandular trichomes can be seen. f: stomata (thick arrows) partially covered by stellate non-glandular trichomes in a leaf crypt. Bottle-shaped and bulbous glandular trichomes (thin arrows) are scattered among the non-glandular indumentum. Scale bars: (a, b, g) = 50 μm; (c, d, e, f) = 150 μm.

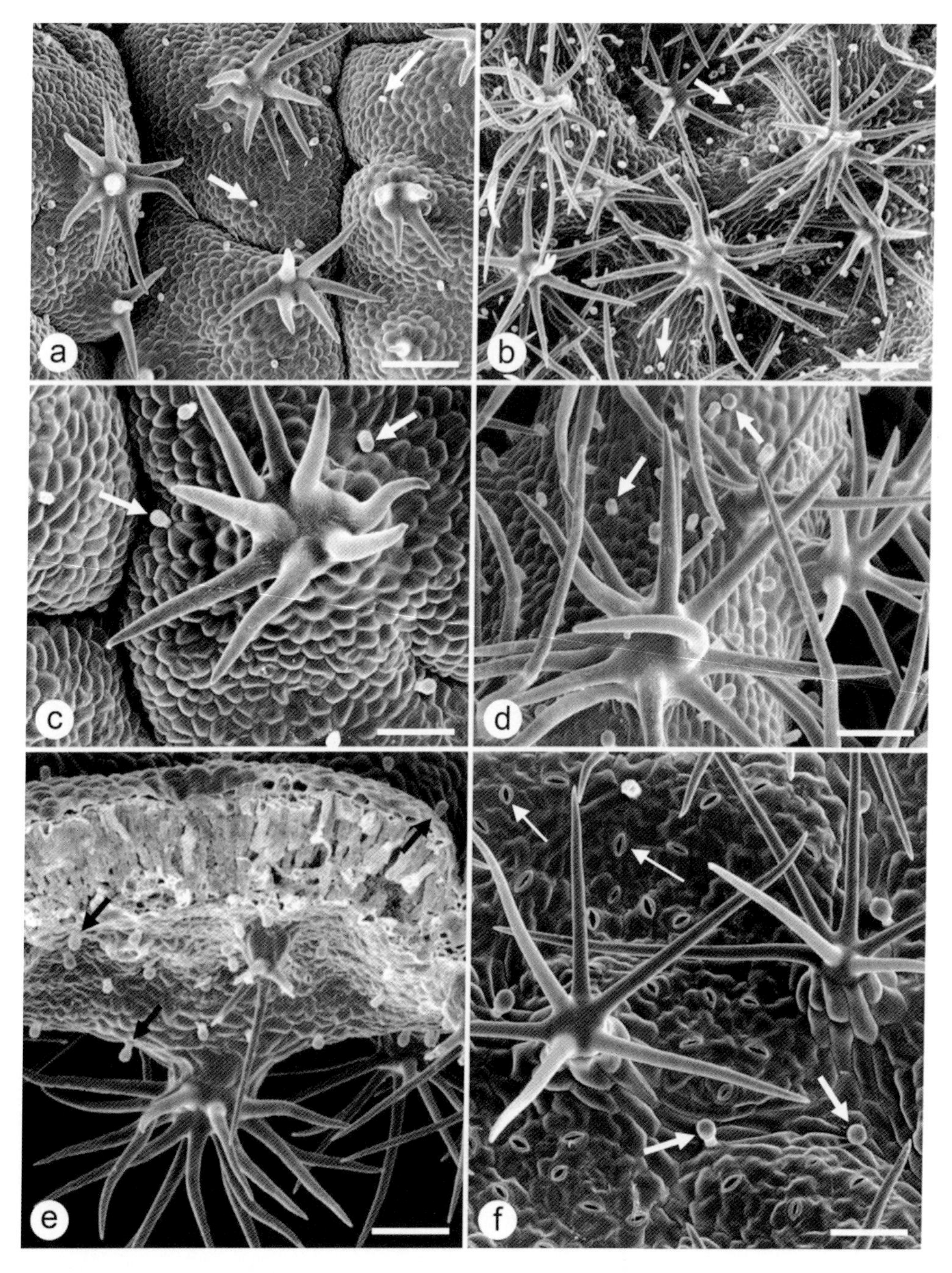

Fig. 10.4. SEM micrographs of mature leaves of *Cistus salviifolius*. a and c: small capitate glandular trichomes (arrows) and stellate non-glandular trichomes, some of the porrect type, on the adaxial surface of the crimped leaf lamina. b and d: capitate glandular trichomes (arrows) and stellate non-glandular trichomes mixed with dendritic on the veins of the leaf abaxial surface. e: leaf cross-section showing capitate glandular trichomes (arrows) on both leaf surfaces and large stellate trichomes of dendritic type. f: stomata (thin arrows) and capitate glandular trichomes (arrows) associated with veins can be seen on the abaxial epidermis. Scale bars: (a, b) = 200 μm; (c, d) = 100 μm; (e) = 300 μm; (f) = 150 μm.

We have also observed on the leaves of these four species another type of glandular trichome, a small capitate uniseriate trichome, with a long stalk cell and a two-celled bulbous glandular head about 20 μm in horizontal diameter (Figs 10.1d, 10.3b, f). These trichomes were difficult to observe on *C. albidus*, *C. ladanifer* and *C. monspeliensis* due to their small dimensions and the dense tomentose indumentum of these species, but are easily visible on both leaf surfaces of *C. salviifolius* clearly associated with the veins (Fig. 10.4a–d). Its occurrence on *C. monspeliensis* leaves had been noticed earlier by Ascenção and Pais (1981), who named them bulbous glandular trichomes. The ultrastructural study of these trichomes, carried out by the same authors, have shown that the glandular cells are typical transfer cells, with labyrintiform walls and a great enlargement of the plasma membrane following the cell wall along its whole length. Based on these ultrastructural features, Ascenção and Pais (1981) have assumed that bulbous glandular trichomes are involved in a rapid absorption or secretion of water solutions. However, further studies are needed to elucidate the putative role of these trichomes. Recently, Tattini *et al.* (2007) have described on *C. salviifolius* leaves a trichome type morphologically similar to those and have assumed that they were hydathodes.

The dense indumentum, found especially on the leaf abaxial surface of *C. albidus*, *C. ladanifer* and *C. Monspeliensis*, precluded the counting of stomata, although their presence is obvious in SEM observations. In these three species, as well as in *C. salviifolius*, the stomata concentrated in the leaf crypts on the abaxial surface are nearly hidden by the overlapping of stellate trichomes (Figs 10.1f, 10.2g, 10.3f, 10.4f). Although the leaves of all *Cistus* species are considered hypostomatic, stomata on the upper leaf surface have also been found in *C. salviifolius* (Tattini *et al.*, 2007; O. Correia, Contribuição da fenologia e ecofisiologia em estudos da sucessão e dinâmica da vegetação mediterrânica [Contribution of phenology and physiological ecology in succession studies and dynamics of Mediterranean vegetation]. University of Lisbon, Lisbon, Portugal, 1988, unpublished thesis).

According to Orshan's classification (1986), based on leaf area, *C. salviifolius*, *C. monspeliensis* and *C. albidus* are nanophyles (leaf area between 0.25 and 2.25 cm^2), while *C. ladanifer* is a nano-microphyle (leaf area between 2.25 and 12.25 cm^2) (Table 10.1). In these four species, the leaf length : width ratio varies between 1.45 for round-shaped leaves like in *C. salviifolius* and 6.69 for species with long leaves as in *C. ladanifer* (O. Correia, Contribuição da fenologia e ecofisiologia em estudos da sucessão e dinâmica da vegetação mediterrânica [Contribution of phenology and physiological ecology in succession studies and dynamics of Mediterranean vegetation]. University of Lisbon, Lisbon, Portugal, 1988, unpublished thesis). Leaf thickness and leaf mass area (LMA) have much lower values than those found in sclerophyllous species, but similar to those recorded for mesophytic species (Krause and Kummerow, 1977; Christodoulakis and Mitrakos, 1987; Gratani and Varone, 2006). *C. albidus*, for example, among the other three species with the thinnest leaves, lowest sclerophylly index and highest surface : volume ratio (Table 10.1), has characteristics tending to mesomorphism. However, due to the very dense leaf pubescence, the amount of reflected radiation

Table 10.1. Ecomorphological features of *Cistus albidus*, *C. ladanifer*, *C. monspeliensis* and *C. salviifolius* leaves. Mean values ±std.

	C. albidus	*C. ladanifer*	*C. monspeliensis*	*C. salviifolius*
Length (cm)	2.16±0.86	3.89±5.38	2.14±0.77	1.28±0.56
Width (cm)	0.99±0.43	0.81±0.38	0.32±0.12	0.87±0.36
Area (cm^2)	1.94±1.22	2.49±0.48	0.64±0.14	1.06±0.83
Thickness (μm)	140±24	239±78	165±24	253±39
LMA (g m^{-2})	126.3±45.2	217.4±39.6	212.4±34.9	168.5±48.8
Leaves/branch (n°)	9.75±3.30	6.88±1.76	10.70±1.99	6.50±1.91
Chl (a+b)* (mg dm^{-2})	3.60	6.58	5.18	3.83
Absorptance* (%)	85	71±2	84±3	89
Angle* (°)	>75	<25	>75	variable

LMA (leaf mass area or sclerophylly index). Values from Correia (1988). Mean values recorded for all the months of the year. *C. albidus*, n=39; *C. ladanifer*, n=117; *C. monspelienesis*, n=204; *C. salviifolius* n=109. * values recorded in September.

Table 10.2. Some morphological and physiological characteristics of *Cistus albidus* and *C. salviifolius* of winter leaves (WL, December–February) and summer leaves (SL, July–September). Mean values ±std.

	Cistus albidus			*Cistus salviifolius*		
	WL	SL	P	WL	SL	P
Leaf area (cm^2)	2.16±1.2	1.56±1.11	NS	1.43±1.18	0.67±0.37	**
LMA ($g\ m^{-2}$)	99.1±28.7	251.5±8.9	***	142.3±42.1	244.3±15.1	*
Total N (% dry weight)	1.76±0.26	1.05±0.07	*	1.84±0.15	0.95±0.07	***
Chl (a+b) ($mg\ dm^{-2}$)	3.91±0.43	2.87±0.64	*	5.01±0.84	2.59±1.32	*
Absorptance (%)	–	71±2	–	–	84±3	–

p – significance of the difference between WL and SL * $p \leq 0.05$, ** $p<0.01$, *** $p<0.001$, NS non-significant at $p>0.05$). Leaf area (n=15-30); leaf mass area (LMA, n=3-6); total nitrogen (N) and total chlorophyll (Chl a+b, n=3-6), and leaf absorptance (400–700 nm, n=6).

is higher, and consequently *C. albidus* is the species with the lowest leaf absorptance (71%), whereas the other *Cistus* species studied have values averaging 85% (84–89%), typical of most green plants (Ehleringer, 1981; Ehleringer and Comstock, 1987).

Cistus species are summer semi-deciduous species with foliar dimorphism; during the driest period of the year they shed larger winter mesomorphic leaves growing on dolichoblasts (long twigs), leaving only the smaller summer xeromorphic leaves on new brachyblasts (short twigs). The two leaf types display different morpho-functional characteristics. In general, summer leaves have smaller surface, higher leaf mass area (sclerophylly index), higher leaf pubescence and leaf angle and lower nitrogen and chlorophyll contents. In some species, leaves curl over the lower surface. In Table 10.2 some morphological and physiological characteristics of winter and summer leaves of *C. albidus* and *C. salviifolius* are shown.

Under the Mediterranean climate, when summer soil water stress is accompanied by high solar radiation, temperature and air evaporative demand, the morphological and structural adaptations of leaves play an important role to keep the energy balance. When plants are not able to efficiently use all the solar radiation at the leaf level, the combination of these stress factors may favour photoinhibition (Björkman and Powles, 1984; Valladares and Pearcy, 1997; Werner *et al.*, 1999; Werner *et al.*, 2002). To avoid photoinhibition, one of the plant best adaptations is to control leaf angle and pubescence, which allow the reduction of the amount of intercepted and absorbed radiation (Werner *et al.*, 1999; Werner *et al.*, 2002).

Cistus albidus and *C. monspeliensis* add to a high pubescence a pronounced leaf angle during summer (Fig. 10.5, Table 10.3).

The wide variation of leaf angle exhibited throughout the year by these two species may reduce the photoinhibition degree under water stress conditions associated with high light intensities. The more upward the arrangement of leaves during summer, the greater is the exposure to radiation when the sun angle is lower (early morning and late afternoon) and thus, a higher absorptance when the air evaporative demand is lower, which contributes to an increase in water-use efficiency. Also, the pronounced leaf shedding during summer, which may reach 80%, leads to a leaf area index (LAI) reduction (Correia *et al.*, 1992; Zunzunegui *et al.*, 2011; O. Correia, Contribuição da fenologia e ecofisiologia em estudos da sucessão e dinâmica da vegetação mediterrânica [Contribution of phenology and physiological ecology in succession studies and dynamics of Mediterranean vegetation]. University of Lisbon, Lisbon, Portugal, 1988, unpublished thesis). By the end of summer, *Cistus albidus* and *C. monspeliensis* present a LAI reduction of 50% and 40% respectively (Fig. 10.6). At this time, there is also a change in the structural pattern of the shrubs, since only small leaf clusters remain at the apex of branches. Besides, this structural change also involves some advantages for water economy, by reducing the transpiring surface during the period of maximum water stress (Werner, 2000). After the first autumn rains, the new leaves emerge and the leaf area index returns to its spring values.

Rockrose leaves are well protected against ultraviolet radiation, not only due to its dense pubescence, but also to the flavonoid compounds present on the oleoresin (Vogt *et al.*, 1987; Gülz *et al.*, 1996; Chaves *et al.*, 1997a, b; Robles and Garzino, 1998; Saracini *et al.*, 2005). Flavonoids are effective UV-B absorbers (UV-B; 280–320 nm) transmitting

Fig. 10.5. *Cistus albidus* (a, b) and *C. monspeliensis* (c, d) showing a general view of leaves and leaf angles in winter (a, c) and summertime (b, d).

Table 10.3. Variation of leaf angle in *Cistus albidus* and *C. monspeliensis* over the year (adapted from Werner *et al.*, 2002). Mean values ±std.

	C. albidus	*C. monspeliensis*
spring	34.9±18.8	31.6±13.6
summer	76.0±12.1	71.7±12.0
autumn	33.8±18.6	27.5±8.8

C. albidus, n = 247 – 580 and *C. monspeliensis*, n = 497 – 550)

visible light and photosynthetically active radiation (PAR) (Karabourniotis *et al.*, 1992; Day *et al.*, 1993; Day *et al.*, 1994; Grammatikopoulos and Manetas, 1994; Karabourniotis *et al.*, 1995). In fact, according to Day *et al.* (1993), the epidermis efficiency to reduce this radiation strongly depends both on the leaf thickness and on the concentration of such compounds, which absorb light at specific wavelengths. Phenolic compounds are primarily located in epidermal cell vacuoles (Caldwell *et al.*, 1983), cuticles (Wollenweber, 1985) and trichome cell walls (Karabourniotis *et al.*, 1992). The leaf content of flavonoids is seasonally variable, reaching a peak in summer as a result of UV-B radiation increase, which avoids or reduces changes in the photosynthetic apparatus (Chaves *et al.*, 1997a, b). In fact, in *C. ladanifer*, the flavonoid content in the oleoresin secretion present a three- or fourfold increase from spring to summer. This increase in flavonoids seems to be more related to enhanced UV-B radiation in this period than to other summer stress factors (Chaves *et al.*, 1997a). In rockroses the induction of flavonoid secretion by UV-B radiation during summer sustains the hypothesis of an ecological role of flavonoids as protecting filters against the harmful effects of this radiation on photosynthetic pigments or on DNA (Vogt *et al.*, 1987; Chaves *et al.*, 1997a, b).

On the other hand, there is evidence that plant morphogenesis may also be affected by flavonoids (Rozema *et al.*, 1997). Considering the differences among species, it is reasonable to expect

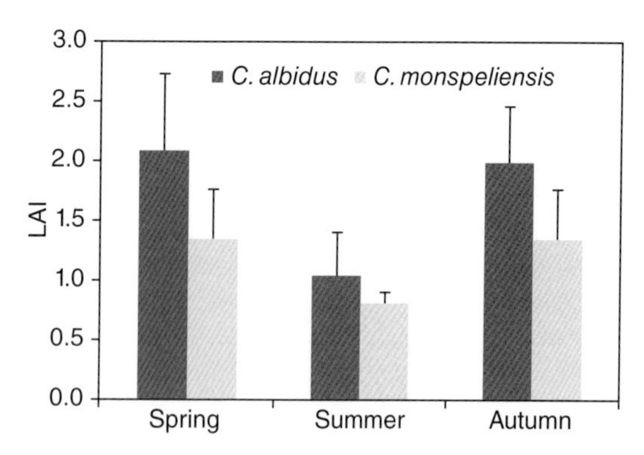

Fig. 10.6. Seasonal variation of leaf area index (LAI) in *Cistus albidus* and *C. monspeliensis* in spring (March–May), summer (August–September) and autumn (October–November). Mean values ± standard deviations of 5–8 plants (adapted from Werner, 2000).

a modification in the competitive capacities of species co-occurring in a given environment (Caldwell *et al.*, 1989). UV-B radiation also seems to affect the reproduction of some species through increased flowering and fruiting (Grammatikopoulos *et al.*, 1998; Stephanou and Manetas, 1998). In *Cistus* species, it was observed that UV-B radiation enhanced pollination success (higher number of seeds per fruit) and reproductive effort (total number of seeds per plant) (Stephanou and Manetas, 1998; Stephanou *et al.*, 2000).

10.6 Physiological Characteristics

One of the most widely established features of the Mediterranean climate is the occurrence of a more or less extended dry period in summer, with very high temperature and solar radiation, and lack of rainfall. Variations in physiological activity, namely in photosynthesis and water-use efficiency, together with morphological characteristics, play a key role in *Cistus* species' adaptation to this kind of climate, and allow them to dominate along an aridity gradient.

During the favourable season (with high water availability), semi-deciduous species exhibit higher photosynthetic rates together with rapid growth, flowering and fruiting (Correia *et al.*, 1987; O. Correia, Contribuição da fenologia e ecofisiologia em estudos da sucessão e dinâmica da vegetação mediterrânica [Contribution of phenology and physiological ecology in succession studies and dynamics of Mediterranean vegetation]. University of Lisbon, Lisbon, Portugal, 1988, unpublished thesis). A comparison between two *Cistus* species, in terms of annual patterns of maximum rates of net photosynthesis (NP_{max}), transpiration rates (Tr_{max}) and stomatal conductance (Gs_{max}) and predawn water potential (Ψ_{PD}) is shown in Fig. 10.7.

In both species, the patterns of these physiological parameters are similar, although with different absolute values. A sharp decrease is observed for all these variables in summer, corresponding to leaves with different morphological characteristics. The values obtained in March from winter leaves, when environmental conditions are optimal, are higher than those recorded for sclerophylls (Table 10.4) (Beyschlag *et al.*, 1987; Harley *et al.*, 1987; Tenhunen *et al.*, 1987; Gulias *et al.*, 2009; Galle *et al.*, 2011; Vilagrosa *et al.*, 2014). A similar pattern was observed by Werner and Máguas (2010) for carbon isotope discrimination ($\Delta^{13}C$) in winter and summer-grown leaves in *C. monspeliensis* and *C. albidus*. The leaf characteristics, such as lower LMA (leaves with slightly thick mesophyll) and higher nitrogen concentration (which is related to Rubisco content and photosynthetic capacity) suggest a higher internal CO_2 conductance and hence higher values in $\Delta^{13}C$ in winter leaves.

During summer, *C. monspeliensis* and *C. albidus* attain very low water potential with extremely negative values (Correia and Catarino, 1994; Werner *et al.*, 1999; Zunzunegui *et al.*, 2011). This reduction in water potential seems to be correlated with a greater concentration of proline, which induces an increase in osmotic potential or may be the expression of a protective mechanism (Zunzunegui *et al.*, 2005; Zunzunegui *et al.*, 2011). When water potentials and stomatal conductance to CO_2 are very low, the leaves of these two species show a strong photoinhibition (a reduction of the maximal photochemical efficiency of photosystem II, Fv/Fm) which are in agreement with the data reported for other studies (Werner *et al.*, 1999; Werner *et al.*, 2002; Grant and Incoll, 2005; Zunzunegui *et al.*, 2011; Grant *et al.*, 2014). A full recovery of this efficiency occurs after the first autumn rains (Fig. 10.8) (Werner *et al.*, 1999).

Despite their pronounced leaf angle, leaves remaining at the shoot apexes are still subject to very high radiation levels, which exceed their capacity for energy dissipation at the PSII reaction centres. In general, photoinhibition level depends on the incident radiation and other concurrent summer stresses. Therefore, although rockroses benefit from

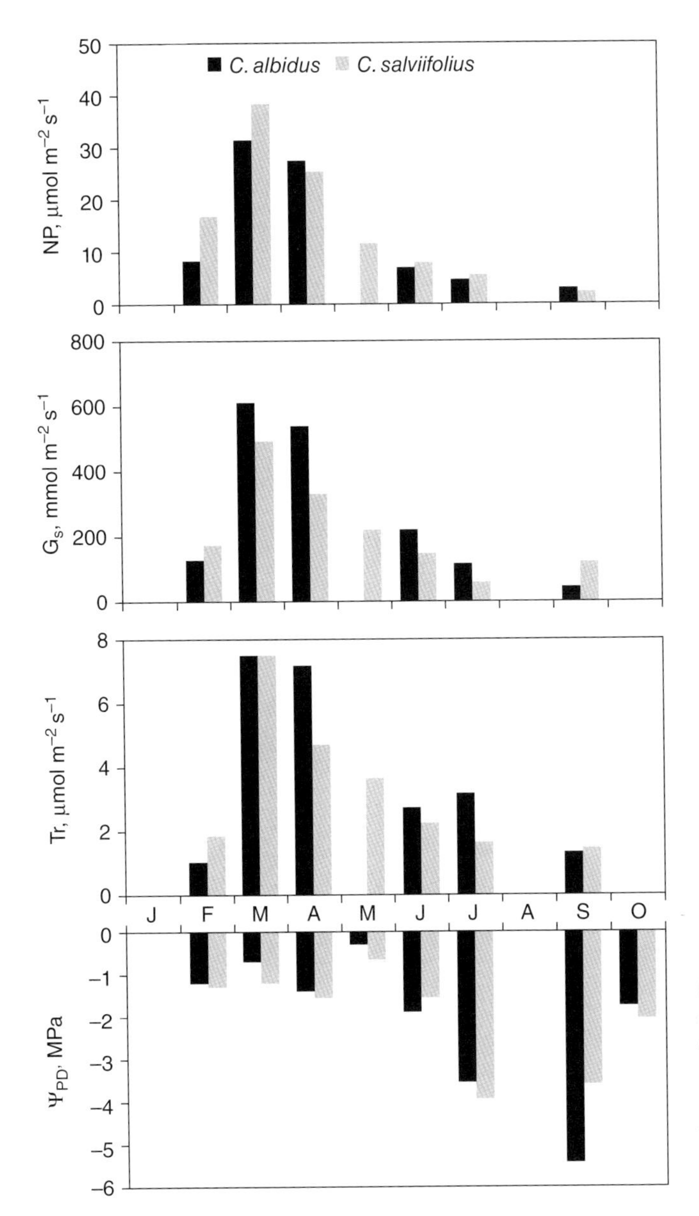

Fig. 10.7 Seasonal variation of maximum diurnal values of photosynthetic rates (NP_{max}), stomatal conductance (G_{max}), transpiration (Tr_{max}) and predawn water potential (Ψ_{PD}). Measurements in *Cistus albidus* and *C. salviifolius* under natural conditions in Serra da Arrábida, Portugal (adapted from Correia, 1988).

several morphological adaptations, such as foliar dimorphism, high leaf pubescence and presence of leaf resin layer, these characteristics are not enough to cope with the very high summer light intensities. In fact, leaf shedding in this season, especially of horizontal leaves produced in winter, is caused by chronic photoinhibition (Werner *et al.*, 1999; Werner *et al.*, 2002). This shedding of leaves or even apical twigs, caused by cavitation phenomena at the petiole level, allows the plants to reduce water loss through transpiration (Tyree *et al.*, 1993), a useful mechanism to eliminate leaves or distal branches while preserving other components. The decrease in light interception in the most upright summer leaves, especially during the hottest period of the day when the radiation peaks occur, reduces the overloading of photosynthetic reaction centres, avoiding a potentially lethal overheating of leaves (Werner *et al.*, 1999). The reduction of chlorophyll content may also add some protection by decreasing the amount of absorbed radiation (Kyparissis and Manetas, 1993a; Kyparissis *et al.*, 1995;

Table 10.4. Some physiological characteristics of *Cistus albidus* and *C. salviifolius* of winter leaves (WL, March) and summer leaves (SL, September). Mean values ±std.

	Cistus albidus			*Cistus salviifolius*		
	WL	SL	P	WL	SL	P
NP$_{max}$ (µmol m^{-2} s^{-1})	31.35±4.24	2.38±0.44	*	38.41±3.66	2.08±0.95	**
Pi (Pa)	20±2	28±2	***	17±3	26±5	**
WUE (µmol $mmol^{-1}$)	5.12±1.52	1.55±1.12	***	5.59±2.21	1.67±1.19	***
Ψ_{PD} (MPa)	–0.7	–5.4	–	–0.6	–3.9	–
Ψ_{MIN} (MPa)	–1.5	–9.9	–	–1.5	–5.7	–

NP_{max} – maximum photosynthetic rate; Pi – internal CO_2 partial pressure (values calculated during the light-saturation phase; PAR > 500 µmol m-2 s-1, from all the diurnal values); WUE – water-use efficiency (expressed as the rate of assimilated CO_2 µmoles per mmol of lost H_2O); (Ψ_{PD}) – predawn water potentials, measured at sunrise; (Ψ_{MIN}) – minimum water potentials recorded at solar noon.
p – significance of the difference between WL and SL *p≤0.05,** p<0.01, *** p>0.001. n – sample size (n=3)

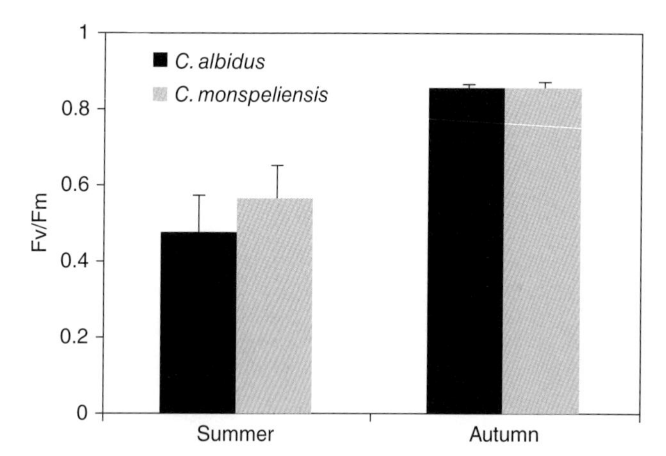

Fig. 10.8 Variation of maximal photochemical efficiency measured before sunrise (Fv/Fm) in *Cistus albidus* and *C. monspeliensis* in summer (August–September) and in autumn (October–November). Mean values ± standard deviations of 5–15 measurements per species (adapted from Werner *et al.*, 1999).

Kyparissis *et al.*, 2000). Although leaves undergo photoinhibition in summer, their recovery after the first September rains is very rapid, which emphasises their high plasticity (Werner *et al.*, 1999; Werner *et al.*, 2002; Grant *et al.*, 2014).

These phenological and structural characteristics exhibited by *Cistus* species and other Cistaceae also seem to play an important role during winter, when low temperatures linked to very high radiation levels may affect the photochemical efficiency of these plants, in a similar way to that occurring in summer (Zunzunegui *et al.*, 1999; Oliveira and Peñuelas, 2000; Oliveira and Peñuelas, 2002). Semi-deciduous species reduce their maximum photochemical efficiency in winter more dramatically than sclerophylls, and leaf angle also seems to play a very important photoprotective role during this season (Karavatas and Manetas, 1999; Oliveira and Peñuelas, 2000; Oliveira and Peñuelas, 2002).

The characteristics presented by species with leaf hairiness such as *C. albidus* may indicate that this species behaviour resembles that of poikilohydric (dehydration-tolerant) species, as reported in other Mediterranean semi-deciduous species with leaf dimorphism and high pubescence, like *Phlomis fruticosa* and *Lavandula stoechas*, that can absorb dew water directly through the numerous trichomes present on their leaves (Kyparissis and Manetas, 1993b; Grammatikopoulos and Manetas, 1994; Munné-Bosch *et al.*, 1999).

It is curious to notice that the leaves of the most pubescent *Cistus* species, *C. albidus*, seem to act like sponges, swelling during the wet season, probably due to the dew or rain water absorption to restore plant water status, similar to reports for other species (Grammatikopoulos *et al.*, 1994; Grammatikopoulos *et al.*, 1995). In fact, summer leaves of *C. albidus* and *C. salviifolius* withstand very low water potentials (Ψ_{MIN} = −9.9 MPa and −5.7 MPa respectively) as shown in Fig. 10.7 and Table 10.4. Apparently, these leaves reduce all their metabolic activity to a minimum level (like dormant plants), avoiding very sophisticated protective mechanisms against photoinhibition. These leaves seem to behave like lichens and other poikilohydric plants, which are able to reduce metabolic activity and remain relatively invulnerable to the environmental stress. It is remarkable that after the first autumn rains, the leaves of *C. albidus* swell, increasing in surface and volume and recovering their photosynthetic activity within a few days. These leaves' behaviour present an extraordinary resemblance with the poikilohydric capacities of angiosperms from semi-arid regions that are often named 'resurrection plants' (Gaff, 1971; Schwab *et al.*, 1989; Gaff, 2011; Grant *et al.*, 2014). The ability of *C. albidus* leaves to absorb dew, fog or rain water may be, at least in part, accomplished by the bulbous glandular trichomes observed on their leaves, which are morphologically similar to those previously described on *C. monspeliensis* leaves (Ascenção and Pais, 1981). However, the mechanisms of water absorption by plant surfaces are not yet well known and detailed studies are needed to clarify this process.

The maintenance of summer leaves during the drought and hot season seems to be a more advantageous strategy than the total leaf shedding to avoid stress. The rapid recovery of leaves in autumn makes this strategy more efficient than to produce new leaves, which would take more time and energy. On the other hand, it is the photosynthetic activity of these 'recovered' summer leaves that will support the development of winter leaves, which also starts with the first rains after summer. *Cistus* species are also characterized by a superficial and highly branched root system that ensures a prompt use of the first autumn rains, as well as of occasional rains of the unpredictable Mediterranean climate.

10.7 What Are the Implications? Future Perspectives

The Mediterranean region is marked by a high number of species, contrasting with temperate regions and deserts. This great biodiversity corresponds to a diversity of micro-sites, in topographic, edaphic and climatic terms.

Phylogenetic classifications of plants proposed by taxonomists do not reflect the ecological functions of species that determine their distribution and evolution. Therefore, classifications based on species functional characters rather than on their phylogenetic relations have been proposed. As stated by Whittaker (1975: 60): 'We can say something about the community by giving a list of its species composition, but a community is poorly described by such a list alone.' According to many ecologists, the characteristics of a community or ecosystem will be better understood and managed if the species are grouped according to complementary behavioural characteristics. Such ecological groups are named 'adaptive syndromes' or functional groups. Cistaceae were included in the plant group Type I, characterized by large flowers, non-edible fruits, small seeds and pioneer characters (Herrera, 1984; Herrera, 1992a). A commonly used classification is the one which differentiates two functional groups – the sclerophyllous species and the summer semi-deciduous species – these being the extremes of a continuum of adaptations according to the habitat characteristics. Díaz Barradas *et al.* (1999) consider six functional groups in Mediterranean vegetation, according not only to their morphological characteristics, but also the physiological ones. *Cistus* species and other Cistaceae are included in a group of species with pubescent leaves and very low water potentials in summer, adaptive characteristics to face the Mediterranean stresses. The regeneration strategy, seeders and resprouters can be considered two separate syndromes, with different functional characteristics at early stages (Vilagrosa *et al.*, 2014). As described by these authors, seeders show a range of traits that allow them to deal better with drought and unpredictable Mediterranean conditions, taking full advantage of periods with higher water availability.

Data obtained in the above-mentioned studies and from others suggest that summer semi-deciduous species like rockroses present a marked plasticity of response to environmental changes, through morphological, structural and physiological modifications. The short leaf longevity of summer semi-deciduous species, with low construction costs, may also be an advantage in habitats under intense stress.

The high biomass and nutrient turnover observed in these species, with a maximum leaf fall in May–June, just before the onset of the dry period, considerably reduces nutrient loss through rain-driven leaching, a pattern similar to that observed in many other summer semi-deciduous species (Núñez-Olivera and Escudero-Garcia, 1990; Valladares, 2000;

O. Correia, Contribuição da fenologia e ecofisiologia em estudos da sucessão e dinâmica da vegetação mediterrânica [Contribution of phenology and physiological ecology in succession studies and dynamics of Mediterranean vegetation]. University of Lisbon, Lisbon, Portugal, 1988, unpublished thesis). Nutrient translocation from senescing leaves to perennial organs also seems to be a quite efficient mechanism in this group of plants. Stored nutrients during summer are immediately used for autumn growth following the first rains, a similar pattern to that used by deciduous plants. However, summer semi-deciduous species have an advantage over the deciduous ones, since they can rapidly resume their activity during moderately favourable periods. On the other hand, the annual leaf fall and the development of smaller leaves in summer not only favour water economy through higher water retention in the soil and reduced transpiration, but also play an important role in the ecosystem. Litter fall and root decomposition represent the main input of organic matter and nutrients into the soil, which is important for recycling and ecosystem functioning. As a result, the summer semi-deciduous species contributing to soil formation show a high capacity to colonize degraded and disturbed areas.

Features like crown architecture, leaf area index, seasonal production cycle and leaf shedding also have a strong influence on the productivity of Mediterranean species (Valladares, 2000; Correia *et al*., 2014). What distinguishes the summer semi-deciduous species is their excellent structural and chemical photoprotection, such as leaf morphology and presence of oleoresins and flavonoids, which contribute to the avoidance of damage caused by excessive radiation loads and UV-B radiation.

Summer semi-deciduous species, with high autecological plasticity, exhibit opportunistic behaviour, with year-round characteristics that allow them to respond to climate unpredictability, and photosynthetic organs ready to work whenever the climatic conditions become favourable. The superficial root system and the capacity to absorb water through the leaves also seem to be processes that respond efficiently to the unpredictable conditions of the Mediterranean climate.

All these characteristics seem to indicate that *Cistus* species or the summer semi-deciduous species are fit to face the predicted changes of the Mediterranean climate, in the sense of increasing temperature and unpredictability of rain distribution. Nevertheless, it is not simply the adaptation of species to climate conditions that determine the distribution of communities in a future scenario of global change, but also the interactions that take place between them. In fact, summer semi-deciduous species (e.g. Cistaceae and Lamiaceae species) which developed with the advent of the Mediterranean climate, were determinants of composition and current diversity of the Mediterranean area. Their facilitation mechanisms with other species for millions of years allowed the maintenance of most ancient species. It is expected that the capability for mutualistic interaction of these species will play a major role in their survival during the predicted climate change.

References

Abrahams, H.J. (1979) Onycha, ingredient of the ancient Jewish incense: an attempt at identification. *Economic Botany* 33(2), 233–236.

Angiosperm Phylogeny Group II (2003) An update of the Angiosperm Phylogeny Group classification for the orders and families of flowering plants: APG II. *Botanical Journal of Linnaeus Society* 141, 399–436.

Ascenção, L. and Pais, M.S. (1981) Ultrastructural aspects of secretory trichomes in *Cistus monspeliensis*. In: Margaris, N.S. and Mooney, H.A. (eds) *Components of Productivity of Mediterranean-Climate Regions Basic and Applied Aspects*. Dr W. Junk Publishers, The Hague, Netherlands, pp. 27–31.

Attaguile, L.G., Russo, A., Campisi, A., Savoca, F., Acquaviva, R., Ragusa, N. and Vanella, A. (2000) Antioxidant activity and protective effect on DNA cleavage of extracts from *Cistus incanus* L. and *Cistus monspeliensis* L. *Cell Biology Toxicology* 16, 83–90.

Barrajón-Catalán, E., Fernández-Arroyo, S., Roldán, C., Guillén, E., Saura, D., Segura-Carretero, A. and Micola, V. (2011) A systematic study of the polyphenolic composition of aqueous extracts deriving from several *Cistus* genus species: evolutionary relationship. *Phytochemical Analysis* 22, 303–312.

Berliner, R., Jacoby, B. and Zamski, E. (1986) Absence of *Cistus incanus* from basaltic soils in Israel: effect of mycorrhizae. *Ecology* 67(5), 1283–1288.

Beyschlag, W., Lange, O.L. and Tenhunen, J.D. (1987) Diurnal patterns of leaf internal CO_2 partial pressure of the sclerophyll shrub *Arbutus unedo* growing in Portugal. In: Tenhunen, J.D., Catarino, F.M., Lange, O.L. and Oechel, W.C. (eds) *Plant Response to Stress Functional Analysis in Mediterranean Ecosystems*. Springer, Berlin, Germany, pp. 355–368.

Björkman, O. and Powles, S.B. (1984) Inhibition of photosynthetic reactions under water stress: interaction with light level. *Planta* 161, 490–504.

Blasco, S. and Mateu, I. (1995) Flowering and fruiting phenology and breeding system of *Cistus albidus* L. *Acta Botanica Gallica* 142(3), 245–251.
Blondel, J. and Aronson, J. (1999) *Biology and Wildlife of the Mediterranean Region*. Oxford University Press, New York, USA.
Boursnell, J.G. (1950) The symbiotic seed-born fungus in the *Cistaceae*. *Annals of Botany* 14, 217–243.
Brandt, U. and Gottsberger, G. (1988) Flower phenology, pollinating insects and breeding systems in *Cistus, Halimium* and *Tuberaria* species in Portugal. *Lagascalia* 15, 625–634.
Caldwell, M.M., Robberecht, R. and Flint, S.D. (1983) Internal filters: prospects for UV-acclimation in higher plants. *Physiologia Plantarum* 58(3), 445–450.
Caldwell, M.M., Teramura, A.H. and Tevini, M. (1989) The changing solar ultraviolet climate and the ecological consequences for higher plants. *Trends in Ecology and Evolution* 4, 363–367.
Carlier, J., Leitão, J. and Fonseca, F. (2008) Population genetic structure of *Cistus ladanifer* L. (Cistaceae) and genetic differentiation from co-occurring *Cistus* species. *Plant Species Biology* 23, 141–151.
Castroviejo, S., Aedo, C., Cirujano, S., Laínz, M., Montserrat, P., Morales, R., Muñoz Garmendia, F., Navarro, C., Paiva, J. and Soriano, C. (1993) *Flora Ibérica. Plantas Vasculares de la Península Ibérica e Islas Baleares. Vol. III. Plumbaginaceae – Capparaceae*. Real Jardín Botánico, CSIC, Madrid, Spain.
Chaves, N., Escudero, J.C. and Gutiérrez-Merino, C. (1997a) Quantitative variation of flavonoids among individuals of a *Cistus ladanifer* population. *Biochemical and Systematic Ecology* 25(5), 429–435.
Chaves, N., Escudero, J.C. and Gutiérrez-Merino, C. (1997b) Role of ecological variables in the seasonal variation of flavonoid content of *Cistus ladanifer* exudate. *Journal of Chemical Ecology* 23(3), 579–603.
Christodoulakis, N.S. and Mitrakos, K.A. (1987) Structural analysis of sclerophylly in eleven evergreen phanerophytes in Greece. In: Tenhunen, J.D., Catarino, F.M., Lange, O.L. and Oechel, W.C. (eds) *Plant Response to Stress Functional Analysis in Mediterranean Ecosystems*. Springer, Berlin, Germany, pp. 547–551.
Clemente, A.S. (2002) Vegetation dynamics after fire in Serra da Arrábida. PhD dissertation, University of Lisbon, Lisbon, Portugal.
Clemente, A.S., Rego, F.C. and Correia, O.A. (1996) Demographic patterns and productivity of post-fire regeneration in Portuguese Mediterranean maquis. *International Journal of Wildland Fire* 6(1), 5–12.
Clemente, A.S., Rego, F.C. and Correia, O. (2005) Growth, water relations and photosynthesis of seedlings and resprouts after fire. *Acta Oecologia* 27, 233–243.
Clemente, A., Rego, F.C. and Correia, O. (2007) Seed bank dynamics of two obligate seeders, *Cistus monspeliensis* and *Rosmarinus officinalis*, in relation to time since fire. *Plant Ecology* 190, 175–188.
Comandini, O., Contu, M. and Rinaldi, A.C. (2006) An overview of *Cistus* ectomycorrhizal fungi. *Mycorrhiza* 16, 381–395.
Corral, R., Pita, J.M. and Perez-Garcia, F. (1990) Some aspects of seed germination in four species of *Cistus* L. *Seed Science Technology* 18, 321–325.
Correia, O. and Catarino F. (1994) Seasonal changes in soil-to-leaf resistance in *Cistus* sp. and *Pistacia lentiscus*. *Acta Oecologia* 15(3), 289–300.
Correia, O.A., Catarino, F.M., Tenhunen, J.D. and Lange, O.L. (1987) Regulation of water use by four species of *Cistus* in the scrub vegetation of the Serra da Arrábida, Portugal. In: Tenhunen, J.D., Catarino, F.M., Lange, O.L. and Oechel, W.C. (eds) *Plant Response to Stress Functional Analysis in Mediterranean Ecosystems*. Springer, Berlin, Germany, pp. 247–258.
Correia, O.A., Martins, A.C. and Catarino, F.M. (1992) Comparative phenology and seasonal foliar nitrogen variation in Mediterranean species of Portugal. *Ecologia Mediterranea* 18, 7–18.
Correia, A.C., Costa e Silva, F., Correia, A.V., Hussain, M.Z., Rodrigues, A.D., David, J.S. and Pereira, J.S. (2014) Carbon sink strength of a Mediterranean cork oak understorey: how do semi-deciduous and evergreen shrubs face summer drought? *Journal of Vegetation Science* 25(2), 411–426.
Day, T.A., Martin, G. and Vogelmann, T.C. (1993) Penetration of UV-B radiation in foliage: evidence that the epidermis behaves as a non-uniform filter. *Plant Cell and Environment* 16, 735–741.
Day, T.A., Howells, B.W. and Rice, W.J. (1994) Ultraviolet absorption and epidermal-transmittance spectra in foliage. *Physiologia Plantarum* 92(2), 207–218.
Demoly, J.P. and Montserrat, P. (1993) *Cistaceae*. In: Castroviejo, S., Aedo, C., Cirujano, S., Laínz, M., Montserrat, P., Morales, R., Muñoz Garmendia, F., Navarro, C., Paiva, J. and Soriano, C. (eds) *Flora Ibérica, Plantas Vasculares de la Península Ibérica Y Islas Baleares, Vol. III – Plumbaginaceae (partim)-Capparaceae*. Madrid, Real Jardín Botánico, CSIC, Madrid, Spain, pp. 318–337.
de Vega, C., Ortiz, P.L., Arista, M. and Talavera, S. (2007) The endophytic system of Mediterranean *Cytinus* (Cytinaceae) developing on five host *Cistaceae* species. *Annals of Botany* 100, 1209–1217.
de Vega, C., Arista, M., Ortiz, P.L., Herrera, C.M. and Talavera, S. (2009) The ant-pollination system of *Cytinus hypocistis* (Cytinaceae), a Mediterranean root holoparasite. *Annals of Botany* 103, 1065–1075.
de Vega, C., Arista, M., Ortiz, P.L. and Talavera, S. (2010) Anatomical relations among endophytic holoparasitic angiosperms, autotrophic host plants and mycorrhizal fungi: a novel tripartite interaction. *American Journal of Botany* 97, 730–737.
de Vega, C., Arista, M., Ortiz, P.L., Herrera, C.M. and Talavera, S. (2011) Endozoochory by beetles: a novel

seed dispersal mechanism. *Annals of Botany* 107, 629–637.
Di Castri, F. (1981) Mediterranean-type shrublands of the world. In: Di Castri, F., Goodall, D.W. and Specht, R.L. (eds) *Mediterranean-Type Shrublands. Ecosystems of the World*. Vol. 11. Elsevier, Amsterdam, Netherlands, pp. 1–52.
Díaz Barradas, M.C., Zunzunegui, M., Tirado, R., Ain-Lhout, F. and García Novo, F. (1999) Plant functional types and ecosystem function in Mediterranean shrubland. *Journal of Vegetation Science* 10, 709–716.
Díez, J., Manjón, J.L. and Martin, F. (2002) Molecular phylogeny of the mycorrhizal desert truffles (*Terfezia* and *Tirmania*), host specificity and edaphic tolerance. *Mycologia* 94(2), 247–259.
Ehleringer, J.R. (1981) Leaf absorptances of Mohave and Sonoran Desert plants. *Oecologia* 49, 366–370.
Ehleringer, J.R. and Comstock, J.P. (1987) Leaf absorptance and leaf angle: mechanisms for stress avoidance. In: Tenhunen, J.D., Catarino, F.M., Lange, O.L. and Oechel, W.C. (eds) *Plant Response to Stress Functional Analysis in Mediterranean Ecosystems*. Springer, Berlin, Germany, pp. 55–76.
Franco, J.A. (1984) *Nova Flora de Portugal (Continente e Açores)*. Vol. I. Escolar Editora, Lisbon, Portugal.
Gaff, D.F. (1971) Desiccation-tolerant flowering plants in southern Africa. *Science* 174, 1033–1034.
Gaff, D.F. (2011) Mechanisms of desiccation tolerance in resurrection vascular plants. In: Barsa, A.S. and Barsa, R. (eds) *Mechanisms of Environmental Stress Resistance in Plants*. Harwood Academic Publishers, Amsterdam, Netherlands, pp. 43–58.
Galle, A., Florez-Sarasa, I., El Aououad, H. and Flexas, J. (2011) The Mediterranean evergreen *Quercus ilex* and the semideciduous *Cistus albidus* differ in their leaf gas exchange regulation and acclimation to repeated drought and re-watering cycles. *Journal of Experimental Botany* 62(14), 5207–5216.
Giovannetti, G. and Fontana, A. (1982) Mycorrhizal synthesis between Cistaceae and *Tuberaceae*. *New Phytology* 92(4), 533–537.
Gomes, P.B., Mata, V.G. and Rodrigues, A.E. (2005) Characterization of the Portuguese-grown *Cistus ladanifer* essential oil. *Journal of Essential Oil Research* 17, 160–165.
Grammatikopoulos, G. and Manetas, Y. (1994) Direct absorption of water by hairy leaves of *Phlomis fruticosa* and its contribution to drought avoidance. *Canadian Journal of Botany* 72, 1805–1811.
Grammatikopoulos, G., Karabourniotis, G., Kyparissis, A., Petropoulou, Y. and Manetas, Y. (1994) Leaf hairs of olive (*Olea europaea*) prevent stomatal closure by ultraviolet-B radiation. *Australian Journal of Plant Physiology* 21, 293–301.
Grammatikopoulos, G., Kyparissis, A. and Manetas, Y. (1995) Seasonal and diurnal gas exchange characteristics and water relations of drought semi-deciduous shrub *Phlomis fruticosa* L. under Mediterranean field conditions. *Flora* 190, 71–78.
Grammatikopoulos, G., Karousou, R., Kokkini, S. and Manetas, Y. (1998) Differential effects of enhanced UV-B radiation on reproductive effort in two chemotypes of *Mentha spicata* under field conditions. *Australian Journal of Plant Physiology* 25(3), 345–351.
Grant, O.M. and Incoll, L.D. (2005) Photochemical efficiency is an important component of ecophysiological variation of *Cistus albidus* between habitats in south-east Spain. *Functional Plant Biology* 32, 107–115.
Grant, O.M., Tronina, L., García-Plazaola, J.I., Esteban, R., Pereira, J.S. and Chaves, M.M. (2014) Resilience of a semi-deciduous shrub, *Cistus salvifolius*, to severe summer drought and heat stress. *Functional Plant Biology* 42(2), 219–228.
Gratani, L. and Varone, L. (2006) Long-time variations in leaf mass and area of Mediterranean evergreen broad-leaf and narrow-leaf maquis species. *Photosynthetica* 44(2), 161–168.
Gulías, J., Cifre, J., Jonasson, S., Medrano, H. and Flexas, J. (2009) Seasonal and inter-annual variations of gas exchange in thirteen woody species along a climatic gradient in the Mediterranean island of Mallorca. *Flora* 204, 169–181.
Gülz, P.G., Herrmann, T. and Hangst, K. (1996) Leaf trichomes in the genus *Cistus*. *Flora* 191, 85–104.
Guzmán, B. and Vargas, P. (2005) Systematics, character evolution, and biogeography of *Cistus* L. (Cistaceae) based on ITS, trnL-trnF, and matK sequences. *Molecular Phylogenetics and Evolution* 37, 644–660.
Guzmán, B., Lledó, M.D. and Vargas, P. (2009) Adaptive Radiation in Mediterranean *Cistus* (Cistaceae). *PLoS One* 4(7), e6362. DOI: 10.1371/journal.pone.0006362.
Harley, P.C., Tenhunen, J.D., Beyschlag, W. and Lange, O.L. (1987) Seasonal changes in net photosynthesis rates and photosynthetic capacity in leaves of *Cistus salviifolius*, a European Mediterranean semi-deciduous shrub. *Oecologia* 74, 380–388.
Herrera, C. (1984) Tipos morfológicos y funcionales en plantas del matorral mediterráneo en sur de España. *Studia Oecologia* 5, 7–34.
Herrera, C. (1992a) Historical effects and sorting processes as explanations for contemporary ecological patterns: character syndromes in Mediterranean woody plants. *American Journal of Nature* 140, 421–446.
Herrera, J. (1992b) Flower variation and breeding system in the Cistaceae. *Plant Systematic and Evolution* 179, 245–255.
Heywood, V.H. (1979) *Flowering Plants of the World*. Oxford University Press, Oxford and London, UK; Melbourne, Australia.
Honrubia, M., Cano, A. and Molina-Niñirola, C. (1992) Hypogeous fungi from Southern Spanish semiarid lands. *Persoonia* 14(4), 647–653.

Juhren, M.C. (1966) Ecological observations on *Cistus* in the Mediterranean vegetation. *Forest Science* 12(4), 415–426.

Juniper, B.E. and Jeffree, C.E. (1983) *Plant Surfaces*. Edward Arnold Publishers, London, UK.

Karabourniotis, G., Papadopoulos, K., Papamarkou, M. and Manetas, Y. (1992) Ultraviolet-B radiation absorbing capacity of leaf hairs. *Physiologia Plantarum* 86, 414–418.

Karabourniotis, G., Kotsabassidis, D. and Manetas, Y. (1995) Trichome density and its protective potential against ultraviolet-B radiation damage during leaf development. *Canadian Journal of Botany* 73, 376–383.

Karavatas, S. and Manetas, Y. (1999) Seasonal patterns of photosystem 2 photochemical efficiency in evergreen sclerophylls and drought semi-deciduous shrubs under Mediterranean field conditions. *Photosynthetica* 36(1–2), 41–49.

Keeley, J.E. (1986) Resilience of Mediterranean shrub communities to fires. In: Dell, B., Hopkins, A.J.M. and Lamont, B.B. (eds) *Resilience in Mediterranean-Type Ecosystems*. Dr. W. Junk Publishers, Dordrecht, Germany, pp. 95–112.

Krause, D. and Kummerow, J. (1977) Xeromorphic structure and soil moisture in the chaparral. *Oecologia Plantarum* 12, 133–148.

Kummerow, J. (1981) Structure of roots and root systems. In: Di Castri, F., Goodall, D.W. and Specht, R.L. (eds) *Mediterranean-Type Shrublands. Ecosystems of the World*. Vol.11. Elsevier, Amsterdam, Netherlands, pp. 269–288.

Kyparissis, A. and Manetas, Y. (1993a) Seasonal leaf dimorphism in a semi-deciduous Mediterranean shrub: ecophysiological comparisons between winter and summer leaves. *Acta Oecologia* 14, 23–32.

Kyparissis, A. and Manetas, Y. (1993b) Autumn revival of summer leaves in the seasonal dimorphic, drought semi-deciduous Mediterranean shrub *Phlomis fruticosa* L. *Acta Oecologia* 14, 725–737.

Kyparissis, A., Petropoulou, Y. and Manetas, Y. (1995) Summer survival of leaves in a soft-leaved shrub (*Phlomis fruticosa* L., Labiatae) under Mediterranean field conditions: avoidance of photoinhibitory damage through decreased chlorophyll contents. *Journal of Experimental Botany* 46(293), 1825–1831.

Kyparissis, A., Drilias, P. and Manetas, Y. (2000) Seasonal fluctuations in photoprotective (xanthophyll cycle) and photoselective (chlorophylls) capacity in eight Mediterranean plant species belonging to two different growth forms. *Australian Journal of Plant Physiology* 27, 265–272.

López González, G. (1991) *La guía Incafo de los árboles y arbustos de la Península Ibérica*. 3rd edn. INCAFO, SA, Madrid, Spain.

Luis-Calabuig, E., Tárrega, R., Calvo, L., Marcos, E. and Valbuena, L. (2000) History of landscape changes in northwest Spain according to land use and management. In: Trabaud, L. (ed.) *Life and Environment in the Mediterranean*. Advances in Ecological Sciences. WIT Press, Boston, Massachusetts, USA, pp. 43–86.

Margaris, N.S. and Vokou, D. (1983) Structural and physiological features of woody plants in a phryganic ecosystem related to the adaptive mechanism. *Ecologia Mediterranea* 8, 449–459.

Martín Bolaños, M. and Guinea, E. (1949) Jarales y jaras (Cistografia Hispánica). *Boletin – Instituto Forestal de Investigaciones y Experiencias* 49, 1–221.

Martín-Pinto, P., Vaquerizo, H., Peñalver, F., Olaizola, J. and Oria-de-Rueda, J.A. (2006) Early effects of a wildfire on the diversity and production of fungal communities in Mediterranean vegetation types dominated by *Cistus ladanifer* and *Pinus pinaster* in Spain. *Forest Ecological Management* 225, 296–305.

Mooney, H.A. and Kummerow, J. (1971) The comparative water economy of representative evergreen sclerophyll and drought deciduous shrubs of Chile. *Botany Gazette* 132, 245–252.

Munné-Bosch, S., Nogués, S. and Alegre, L. (1999) Diurnal variations of photosynthesis and dew absorption by leaves in two evergreen shrubs growing in Mediterranean field conditions. *New Phytology* 144, 109–119.

Ne'eman, G., Henig-Sever, N. and Eshel, A. (1999) Regulation of the germination of *Rhus coriaria*, a post-fire pioneer, by heat, ash, pH, water potential and ethylene. *Physiologia Plantarum* 106, 47–52.

Núñez-Olivera, E. and Escudero-Garcia, J.C. (1990) Índice de esclerofilia, área media foliar y concentración de clorofilas en hojas maduras de *Cistus ladanifer* L. variaciones estacionales. *Studia Oecologica* 7, 63–75.

Oliveira, G. and Peñuelas, J. (2000) Comparative photochemical and phenomorphological responses to winter stress of an evergreen (*Quercus ilex* L.) and a semi-deciduous (*Cistus albidus* L.) Mediterranean woody species. *Acta Oecologia* 21, 97–107.

Oliveira, G. and Peñuelas, J. (2002) Comparative protective strategies of *Cistus albidus* and *Quercus ilex* facing photoinhibitory winter conditions. *Environmental and Experimental Botany* 47, 281–289.

Orshan, G. (1986) Plant form as describing vegetation and expressing adaptation to environment. *Annals of Botany* 44, 1–38.

Paolini, J., Falchi, A., Quilichini, Y., Desjobert, J.M., De Cian, M.C., Varesi, L. and Costa, J. (2009) Morphological, chemical and genetic differentiation of two subspecies of *Cistus creticus* L. (*C. creticus* subsp. *eriocephalus* and *C. creticus* subsp. *corsicus*). *Phytochemistry* 70, 1146–1160.

Papaefthimiou, D., Papanikolaou, A., Falara, V., Givanoudi, S., Kostas, S. and Kanellis, A.K. (2014) Genus *Cistus*: a model for exploring labdane-type diterpenes' biosynthesis and a natural source of high value products with biological, aromatic, and pharmacological properties. *Frontiers in Chemistry* 2, 1–19.

Paula, S. and Pausas, J.G. (2008) Burning seeds: germination responses to heat treatments in relation to resprouting ability. *Journal of Ecology* 96, 543–552.
Pausas, J.G. and Verdu, M. (2005) Plant persistence traits in fire-prone ecosystems of the Mediterranean basin: a phylogenetic approach. *Oikos* 109, 196–202.
Ramalho, P.S., de Freitas, V.A.P., Macedo, A., Silva, G. and Silva, A.M.S. (1999) Volatile components of *Cistus ladanifer* leaves. *Flavour and Fragrance Journal* 14, 300–302.
Rivas-Martínez, S. (1979) Brezales y jarales de Europa occidental (Revisión Fitosociológica de las clases *Calluno-Ulicetea y Cisto-Lavanduletea*). *Lazaroa* 1, 5–127.
Robles, C. and Garzino, S. (1998) Essential oil composition of *Cistus albidus* leaves. *Phytochemistry* 48(8), 1341–1345.
Roy, J. and Arianoutsou-Faraggitaki, M. (1985) Light quality as the environmental trigger for the germination of the fire-promoted species *Sarcopterium spinosum* L. *Flora* 177, 345–349.
Roy, J. and Sonié, L. (1992) Germination and population dynamics of *Cistus* species in relation to fire. *Journal of Applied Ecology* 29, 647–655.
Rozema, J., Van de Staaij, J., Björn, L.O. and Caldwell, M. (1997) UV-B as an environmental factor in plant life: stress and regulation. *Trends in Ecology and Evolution* 12, 22–28.
Saracini, E., Tattini, M., Traversi, M.L., Vincieri, F.F. and Pinelli, P. (2005) Simultaneous LC-DAD and LC-MS determination of ellagitannins, flavonoid glycosides, and acyl-glycosyl flavonoids in *Cistus salvifolius* L. leaves. *Chromatographia* 62(5–6), 245–249.
Saura-Mas, S. and Lloret, F. (2007) Leaf and shoot water content and leaf dry matter content of Mediterranean woody species with different post-fire regenerative strategies. *Annals of Botany* 99, 545–554
Schwab, K.B., Schreiber, U. and Heber, U. (1989) Response of photosynthesis and respiration of resurrection plants to desiccation and rehydration. *Planta* 177, 217–227.
Silva, J.S., Rego, F.C. and Martins-Loução, M.A. (2002) Belowground traits of Mediterranean woody plants in a Portuguese shrubland. *Ecologia Mediterranea* 28(2), 5–13.
Smith, H. (1982) Light quality, photoperception and plant strategy. *Annual Reviews in Plant Physiology* 33, 481–518.
Smith, S.E. and Read, D.J. (1997) *Mycorrhizal Symbiosis*. 2nd edn. Academic Press, London, UK. 605 pp.
Stephanou, M. and Manetas, Y. (1998) Enhanced UV-B radiation increases the reproductive effort in the Mediterranean shrub *Cistus creticus* under field conditions. *Plant Ecology* 134, 91–96.
Stephanou, M., Petropoulou, Y., Georgiou, O. and Manetas, Y. (2000) Enhanced UV-B radiation, flower attributes and pollinator behaviour in *Cistus creticus*: a Mediterranean field study. *Plant Ecology* 147, 165–171.
Talavera, S.J., Herrera, J., Arroyo, J., Ortiz, P.L. and Devesa, J.A. (1988) Estudio de la flora apícola de Andalucia Occidental. *Lagascalia* 15, 567–592.
Tattini, M., Matteini, P., Saracini, E., Traversi, M.L., Giordano, C. and Agati, G. (2007) Morphology and biochemistry of non-glandular trichomes in *Cistus salvifolius* L. leaves growing in extreme habitats of the Mediterranean basin. *Plant Biology* 9, 411–419.
Teixeira, S., Mendes, A., Alves, A. and Santos, L. (2007) Simultaneous distillation–extraction of high-value volatile compounds from *Cistus ladanifer* L. *Analytica Chimica Acta* 584, 439–446.
Tenhunen, J.D., Beyschlag, W., Lange, O.L. and Harley, P.C. (1987) Changes during summer drought in leaf CO_2 uptake rates of macchia shrubs growing in Portugal: limitations due to photosynthetic capacity, carboxilation efficiency, and stomatal conductance. In: Tenhunen, J.D., Catarino, F.M., Lange, O.L. and Oechel, W.C. (eds) *Plant Response to Stress: Functional Analysis in Mediterranean Ecosystems*. Springer, Berlin, Germany, pp. 305–327.
Thanos, C.A., Georghiou, K., Kadis, C. and Pantazi, C. (1992) Cistaceae: a plant family with hard seeds. *Israel Journal of Botany* 41, 251–263.
Thorogood, C.J. and Hiscock, S.J. (2007) Host specificity in the parasitic plant *Cytinus hypocistis*. *Research Letters in Ecology* 2007, Article ID 84234, 4. DOI: 10.1155/2007/84234.
Trabaud, L. (1995) Modalités de germination des cistes et des pins méditerranéens et colonisation des sites perturbés. *Revue d'Ecologie* 50, 3–14.
Tyree, M.T., Cochard, M., Cruiziat, P., Sinclair, B. and Ameglio, T. (1993) Drought-induced leaf shedding in walnut: evidence for vulnerability segmentation. *Plant, Cell and Environment* 16, 879–882.
Uphof, J.C.T. (1962) Plant hairs. In: *Encyclopaedia of Plant Anatomy IV*. Vol. 5. Gebrüder Borntraeger, Berlin, Germany, p. 116.
Valiente-Banuet, A., Rumebe, A.V., Verdú, M. and Callaway, R.M. (2006) Modern quaternary plant lineages promote diversity through facilitation of ancient tertiary lineages. *Proceedings of the National Academy of Sciences* 103(45), 16812–16817.
Valladares, F. (2000) Características mediterráneas de la conversión fotosintética de la luz en biomasa: de órgano a organismo. In: Zamora, R.R. and Pugnaire, F.I. (eds) *Ecosistemas Mediterráneos. Análise Funcional*. Castillo y Edisart, Albolote, Granada, Spain, pp. 67–93.
Valladares, F. and Pearcy, R.W. (1997) Interactions between water stress, sun-shade acclimation, heat tolerance and photoinhibition in the sclerophyll *Heteromeles arbutifolia*. *Plant, Cell and Environment* 20, 25–36.

Verdú, M. and Pausas, J.G. (2007) Fire drives phylogenetic clustering in Mediterranean basin woody plant communities. *Journal of Ecology* 95, 1316–1323.

Vilagrosa, A., Hernandez, E.I., Luis, V.C., Cochard, H. and Pausas, J.G. (2014) Physiological differences explain the co-existence of different regeneration strategies in Mediterranean ecosystems. *New Phytology* 201, 1277–1288.

Vogt, T., Proksch, P. and Gülz, P.G. (1987) Epicuticular flavonoid aglycones in the genus *Cistus*, *Cistaceae*. *Journal of Plant Physiology* 131, 25–36.

Werner, C. (2000) Evaluation of structural and functional adaptations of mediterranean macchia species to drought stress with emphasis on the effects of photoinhibition on whole-plant carbon gain. A Modelling approach. PhD dissertation, University of Bielefeld, Bielefeld, Germany.

Werner, C. and Máguas, C. (2010) Carbon isotope discrimination as a tracer of functional traits in a Mediterranean macchia community. *Functional Plant Biology* 37, 467–477.

Werner, C., Correia, O. and Beyschlag, W. (1999) Two different strategies of Mediterranean macchia plants to avoid photoinhibitory damage by excessive radiation levels during summer drought. *Acta Oecologia* 20(1), 15–23.

Werner, C., Correia, O. and Beyschlag, W. (2002) Characteristic patterns of chronic and dynamic photoinhibition of different functional groups in a Mediterranean ecosystem. *Functional Plant Biology* 29, 999–1011.

Whittaker, R.H. (1975) *Communities and Ecosystems*. Macmillan, New York, USA.

Williams, J.E., Davis, S.D. and Portwood, K. (1997) Xylem embolism in seedlings and resprouts of *Adenostoma fasciculatum* after fire. *Australian Journal of Botany* 45, 291–300.

Wollenweber, E. (1985) Exudate flavonoids in higher plants of arid regions. *Plant Systematics and Evolution* 150, 83–88.

Zunzunegui, M., Fernandez Baco, L., Diaz Barradas, M.C. and Garcia Novo, F. (1999) Seasonal changes in photochemical efficiency in leaves of *Halimium halimifolium*, a Mediterranean semideciduous shrub. *Photosynthetica* 37(1), 17–31.

Zunzunegui, M., Díaz Barradas, M.C., Ain-Lhout, F., Clavijo, A. and Garcia Novo, F. (2005) To live or to survive in Doñana dunes: adaptive responses of woody species under a Mediterranean climate. *Plant and Soil* 273, 77–89.

Zunzunegui, M., Díaz Barradas, M.C., Ain-Lhout, F., Alvarez-Cansino, L., Esquivias, M.P. and Garcia Novo, F. (2011) Seasonal physiological plasticity and recovery capacity after summer stress in Mediterranean scrub communities. *Plant Ecology* 212, 127–142.

11 Aquatic Plant Biodiversity: A Biological Indicator for the Monitoring and Assessment of Water Quality

Abid Ali Ansari[1]*, Shalini Saggu[1], Sulaiman Mohammad Al-Ghanim[1], Zahid Khorshid Abbas[1], Sarvajeet Singh Gill[2], Fareed A. Khan[3], Mudasir Irfan Dar[3], Mohd Irfan Naikoo[3] and Akeel A. Khan[4]

[1]*Department of Biology, Faculty of Science, University of Tabuk, Tabuk, Saudi Arabia;* [2]*Stress Physiology and Molecular Biology Lab, Centre for Biotechnology, MD University, Rohtak, Haryana, India;* [3]*Department of Botany, Aligarh Muslim University, Aligarh, Uttar Pradesh, India;* [4]*GF College affiliated to MJP Rohilkhand University, Shahjahanpur, U.P., India*

Abstract

The present chapter covers the concept of water pollution, its cause and effects on plant diversity of an aquatic ecosystem. An alteration in the diversity of plants and the disappearance of aquatic plants has been noted in the majority of the world's water bodies as a result of water pollution. In polluted aquatic ecosystems, plant diversity was studied as a strong bioindicator of water quality. Numerous studies on phytoplankton and aquatic macrophytes have been incorporated in this chapter as an indicator of water pollution. Water pollution leads to changes in species composition, declines in overall plant species diversity and the loss of rare and uncommon species. A perturbation in aquatic ecosystems causes succession of macrophytes with complete loss of submerged vegetation and dominance of phytoplanktons and weeds. Monitoring, assessment and measurement of plant diversity through density, frequency, abundance and diversity indices are also integrated in this chapter.

11.1 Introduction

Aquatic ecosystems are very rich sources of biodiversity. Macrophyte richness is a vital parameter directly related to altitude, nutritional status and water quality (Murphy, 2002). Mechanisms regulating species diversity in aquatic ecosystems have been studies very little, compared to terrestrial and benthic ecosystems. The species diversity of various water bodies has been explored, where the water quality parameters showed a direct impact on the dynamics of plant diversity (Murphy *et al.*, 2003). Anthropogenic forcing supports a higher diversity of non-indigenous species as compared to the indigenous species (Occhipinti and Savini, 2003).

When there is degradation of an ecosystem by pollution or over-exploitation, the dominant species are eliminated or debilitated, which shows a causal connection between loss in biodiversity and changes in ecosystem structure and function (Grime, 1998). Signs of biodiversity change were reported from the Undasa wetlands of Ujjain (India) as a result of excessive industrial pollution (Shrivastava *et al.*, 2003). The anthropogenic stresses and open water habitat increase the abundance of exotic species and lead to vegetation removal and plant invasion, causing perturbation in the aquatic ecosystem (Detembeck *et al.*, 1999).

Many physical processes like nutrient cycling, dynamics in temperature, pH, dissolved oxygen,

*E-mail: aansari@ut.edu.sa; aaansari40@gmail.com

 Plant Biodiversity: Monitoring, Assessment and Conservation (eds A.A. Ansari, S.S. Gill, Z.K. Abbas and M. Naeem)

carbon dioxide and light are the important limiting factors for the growth and development of aquatic flora (Shen-Dong and Shen-Dong, 2002). The plant species composition showed significant correlations with nutrient concentrations, water level and depth of water body. Aquatic plant diversity is directly influenced by pH, the concentrations of Cl^-, organic carbon and NH_4^+, and water temperature (Best *et al.*, 1995). Studies from 39 drainage streams in Victoria (Australia) revealed that diatom diversity and richness was found and strongly correlated with irrigation practices and dry land farming (Blinn *et al.*, 2001). Diatom assemblages studied in the Kathmandu Valley reflect physico-chemical characteristics of water. The diversity showed an increase with silica, sodium, and phosphates but a decrease with pH, calcium and magnesium (Juttner *et al.*, 2003).

Algal diversity was studied in different aquatic ecosystems of Bulgaria. The distribution of different groups of algae was influenced by thermal pollution (Stoyneva, 2003). Also, deforestation on the coastal area due to anthropogenic activities promotes algal biodiversity due to an increase in availability of nutrients, light and mixing depth (Nicholls *et al.*, 2003). Algal diversity was studied in Najafgarh drain (River Sahibi) in Delhi, India. The drain receives agricultural, industrial and domestic effluents. The species recorded from the drain were found to be highly tolerant to water pollution caused by these organic compounds and have been suggested as suitable for bio-monitoring and phytoremediation (Sinha, 2001). The growth responses of common duckweeds *Lemna minor* and *Spirodelapolyrrhiza* were reported to be directly correlated with the aquatic environment (Ansari and Khan, 2008; Ansari and Khan 2009).

Aquatic biodiversity is an underestimated component of global biodiversity and its estimation acts as a model for basic researches in evolutionary and experimental biology and ecological studies. Therefore, aquatic organisms signal water quality, and can thus be used for monitoring and assessment of different kinds of water pollution and its influence on ecosystem health (Xu et al., 2007).The emerged, free-floating and submerged macrophytes have always been considered important in long-term monitoring and assessment of water quality in lentic and lotic aquatic ecosystems, as they are very sensitive to the water pollution. However, the submerged macrophytes are reported to be the most suitable for the assessment of lotic water ecosystems (Tremp *et al.*, 1995).

11.2 Water Pollution

The contamination of water bodies results in the degradation of the aquatic environment when pollutants are directly or indirectly discharged. The entire biosphere, plants and animals living in these water bodies are affected by the water pollution. In most cases, the effect is harmful, not only to individual species and population, but also to the natural biological communities. The most common sources of water pollution are nutrients, fertilizers (industrial and household), acid rain, heavy metals, pesticides, oil, and many other industrial chemicals. One of the major forms of water pollutants comes through run-offs as a result of rain. The nitrates and phosphates in the water result in eutrophication, causing algal blooms.

As a result of the increase in algae, a reduced amount of sunlight is able to reach other aquatic plants, leading to their death. The algae themselves also eventually die and bacteria then begin to break down the dead plant material and algae, using up oxygen, resulting in the death of fish and other organisms. The composition and total amount of algae species can indicate the amount of organic pollution. In recent decades, a significant effort has been put forward around the globe for environmental monitoring and assessment of water pollution based not only on chemical parameters (nutrients, metals, pesticides, etc.), but also on biological indicators. It is difficult to draw a conclusion in an intermediate situation about the quality of water without further evaluation. The European Water Framework Directive (2000) has a simple philosophy, very easy to understand but difficult to put in practice: an ecosystem should be evaluated.

11.3 Biological Indicators

The species whose presence, absence or abundance reflects a specific environmental condition such as habitat or community and health of an ecosystem or pollution, is considered an indicator species. An indicator species provides a forecast for any changes in environmental conditions as they are the most sensitive species of an ecosystem. The biological indicators respond very rapidly to environmental changes and are widely used for the assessment and monitoring of different types of normal and perturbed ecosystems. For the management and monitoring of pollution levels, the bioindicators are one of the important tools. Plant species of different groups serve as a reliable index for biological monitoring of pollution load in an aquatic habitat.

The aquatic macrophytes are well known for their great potential to indicate quality of water. They can tolerate and accumulate high concentration of toxic substances in their tissues, and on their biochemical analysis, the presence of any environmental contaminant can be determined, even when in very low concentration (Caffrey *et al.*, 2006). In several susceptible species, morphological and structural changes induced by metal may also be indicative of changes which are specific to metals. The nutrient fortification effect is indicated by the disappearance of vulnerable species, leading to the change of species composition. These may be effectively used as ecological indicators (bioindicators) for assessing and predicting environmental changes (Chandra and Sinha, 2000).

To determine the concentration of pollutants, especially heavy metals in rivers and oceans, biological markers have attracted a great deal of scientific interest. The main idea behind the biomarker approach is to perform the analysis of an organism's metal content and comparison with background metal levels (Rashed, 2001). Indicator species represent a particular environment, and they are easy to adapt and harvest. In view of the remarkable potential of biological systems, rigorous efforts are now needed to identify bioindicators for assessing the state of aquatic systems and working out suitable remedial methods to restore unbalanced or polluted aquatic ecosystems (Chandra and Sinha, 2000).

11.4 Biological Indicators and Biomonitoring Programme

Biological monitoring, or biomonitoring, is the use of biological responses to assess changes in environmental conditions. Biomonitoring programmes, which may be qualitative, semiquantitative or quantitative, are an important tool for assessing water quality in all types of monitoring programmes. There are two types of biomonitoring: first, surveillance before and after a project is complete, or before and after a toxic substance enters the water; second, biomonitoring to ensure compliance with regulations or guidelines, or to ensure maintenance of water quality. Biomonitoring involves the use of indicator species/communities: aquatic plants are used for various pollutants and nutrients (Feminella and Flynn, 1999).

Stream organisms have been reported as the base for pollution detection and water quality monitoring. In the early 1900s, Europeans first adopted this strategy to identify organic pollution in large rivers. In the United States, the use of stream organisms as biological indicators, or 'sentinels', has become widespread. Agencies such as the Environmental Protection Agency (EPA), the Natural Resources Conservation Service (NRCS), and the US Geological Survey (USGS) are now implementing biomonitoring procedures to monitor the health of streams and rivers across the country (Philadelphia Water, 2015).

Predation, competition or geographic barriers are some of the other reasons which may account for the absence of any species. Nonexistence of many species of different orders with similar tolerance levels that were present previously at the same site is more indicative of pollution than absence of a single species. When the pollutant type is well known, certain indicators are more effectively used or are less expensive. When stressors are unknown and/or less is known about species tolerance levels, multiple-level assessment and more intensive and expensive studies will be required. Behavioural changes of organisms may also be considered as indicators for assessment and monitoring.

Biochemical, genetic, morphological, physiological and behavioural changes, populations and community dynamics could also be studied through an indicator species. Some other attributes or changes are also considered as indicators for the health of an aquatic ecosystem; for example: genetic mutations, reproductive success, physiology metabolism, oxygen consumption, photosynthetic rate, disease resistance, tissue/organ damage, bioaccumulation, population survival/mortality, sex ratio abundance/ biomass, behaviour (migration), predation rates, population decline/increase in community abundance ('evenness') of an organism, etc. (Feminella and Flynn, 1999).

11.5 Plant Diversity as a Biological Indicator

Any structural and functional change in species, population, community, relative density, frequency and abundance may indicate pollution. Presence or absence of a particular species may indicate favourable or unfavourable changes to its ecosystems. The response of a single indicator species may not provide sufficient data for the assessment of an ecosystem's health, and further investigations may be required to reach any conclusion. Many migratory species can also indicate the environmental condition: fish and algal species associated with the same food chain have been used in stream biological monitoring and assessment programmes (Feminella and Flynn, 1999). There is the relative influence of local- and landscape-level habitat on aquatic plant diversity, but local habitats are more influential than the landscape (Rooney and Bayley, 2011). Several plants from different divisions of the plant kingdom, such as chlorophyta, bryophyta, pteridophyta and angiosperms, can be used for the monitoring and assessment of water quality in freshwater or marine, lentic or lotic, wetlands and coastal aquatic ecosystems (Table 11.1).

Table 11.1. Aquatic plants: biological indicators for the monitoring and assessment of water quality.

Sl. No.	Species	Taxonomic Group/Family	Ecological Habitat
1	*Microcystis* sp.	Cyanobacteria/Microcystaceae	Free Floating
2	*Chara* sp.	Chlorophyta/Characeae	Rooted Submerged
3	*Nitella* sp.	Chlorophyta/Characeae	Rooted Submerged
4	*Scendesmus* sp.	Chlorophyta/Scendesmaceae	Free Floating
5	*Chlorella* sp.	Chlorophyta/Chlorellaceae	Free Floating
6	*Hydrodictyon* sp.	Chlorophyta/Hydrodictiaceae	Free Floating
7	*Cladophora* sp.	Chlorophyta/Cladophoraceae	Amphibious
8	*Stegeoclonium* sp.	Chlorophyta/Chaetophoraceae	Submerged
9	*Sphagnum acutifolia*	Bryophyta/Sphagnaceae	Amphibious
10	*Fontinalis* sp.	Bryophyta/Fontinalaceae	Submerged
11	*Azollafiliculoides*	Pteridophyta/Salvinaceae	Free Floating
12	*Salvinia* sp.	Pteridophyta/Salvinaceae	Free Floating
13	*Vallisneria* sp.	Angiosperm/Hydrocharitaceae	Rooted Submerged
14	*Hydrilla* sp.	Angiosperm/Hydrocharitaceae	Rooted Submerged
15	*Myrophyllum*	Angiosperm/Haloragaceae	Rooted Submerged
16	*Sagittaria montevidensis*	Angiosperm/Alismataceae	Amphibious/Emergent
17	*Alternanthera philoxiroides*	Angiosperm/Amaranthaceae	Amphibious/Emergent
18	*Centella asiatica*	Angiosperm/Apiaceae	Amphibious/Emergent
19	*Lemna minor*	Angiosperm/Araceae	Free Floating
20	*Spirodella polyrrhiza*	Angiosperm/Araceae	Free Floating
21	*Wolffia brasiliensis*	Angiosperm/Araceae	Free Floating
22	*Wolffiella oblonga*	Angiosperm/Araceae	Free Floating
23	*Hydrocotyl* sp.	Angiosperm/Araliaceae	Amphibious/Emergent
24	*Mikania periplocifolia*	Angiosperm/Asteraceae	Amphibious
25	*Enhydra anagallis*	Angiosperm/Asteraceae	Amphibious/Emergent
26	*Cyprus* sp.	Angiosperm/Cyperaceae	Amphibious/Emergent
27	*Androtrichum trigynum*	Angiosperm/Cyperaceae	Amphibious/Emergent
28	*Scirpus giganteus*	Angiosperm/Cyperaceae	Amphibious/Emergent
29	*Vigna luteola*	Angiosperm/Fabaceae	Amphibious/Emergent
30	*Erythrina crista-galli*	Angiosperm/Fabaceae	Amphibious/Emergent
31	*Utricularia* sp.	Angiosperm/Lentibulariaceae	Submerged/Free
32	*Nymphoides indica*	Angiosperm/Menyanthaceae	Rooted/Leaves Floating
33	*Ludwigia* sp.	Angiosperm/Onagraceae	Amphibious/Emergent
34	*Bacopa monnieri*	Angiosperm/Plantaginaceae	Amphibious/Emergent
35	*Luziola peruviana*	Angiosperm/Poaceae	Amphibious/Emergent
36	*Polygonum* sp.	Angiosperm/Polygonaceae	Amphibious/Emergent
37	*Eichhornia crassipes*	Angiosperm/Pontederiaceae	Free Floating
38	*Potamogeton pectinatus*	Angiosperm/Potamogetonaceae	Rooted Submerged
39	*Salix humboldtiana*	Angiosperm/Salicaceae	Amphibious/Emergent
40	*Typhadomingensis*	Angiosperm/Typhaceae	Amphibious/Emergent
41	*Xyris jupical*	Angiosperm/Xyridaceae	Amphibious/Emergent

11.6 Macrophytes as an Indicator of Water Pollution

Aquatic plants such as *Ottelia alismodes, Fimbristylis pauciflora* and *Blyxa malayana* are reported sensitive to zinc, copper, cadmium, lead, nickel, and iron and chromium concentration. Metal enrichment was found to be dependent on the plant species and metal type. All plants showed their potential to remove copper, zinc and cadmium more rapidly than chromium and nickel. These plants were found to be sensitive to the aquatic environment and may be used as biological indicators of water pollution (She *et al.*, 1984). *Lemna minor*, free-floating duckweed, is highly sensitive to a number of environmental factors, and has been determined as an indicator of water quality (Ansari and Khan, 2002). Various *Potamogeton* species have also been reported as indicators of water pollution, typically *P. filliforms* and *P. polygonifolius*. Many other species like *P. alpinus, P. lucens, P. praelongus, P. zooterifolius, P. aqualifolius, P. colourates* (*P. coloratus*), *P. densus* and *P. ratilus* are slow growing but also act as water quality indicators (Sand-Jensen *et al.*, 2000).

Aquatic macrophytes are photosynthetic and large enough to see with the naked eye. They grow actively submerged or free floating through the water surface. There are seven plant divisions: Cyanobacteria, Chlorophyta, Rhodophyta, Xanthophyta, Bryophyta, Pteridophyta and Spermatophyta for the aquatic macrophytes. Species richness, composition and distribution of aquatic plants in the more primitive divisions are less known than vascular macrophytes in the divisions Pteridophyta and Spermatophyta, of which 2614 aquatic species have been reported and represent only a small fraction (1%) of the total number of vascular plants. They are included in 33 orders, 88 families and 412 genera. Aquatic macrophytes play an important role in the structure and function of aquatic ecosystems. Many of the threats to fresh water result in reduced macrophyte diversity, which, in turn, affects the faunal diversity of aquatic ecosystems and supports the introduction of exotic species at the expense of native species (Chambers *et al.*, 2008).

Pereira *et al.* (2012) studied the potential of aquatic macrophyte communities as bioindicators in six small shallow lakes and water temperature, dissolved oxygen, pH, conductivity, total alkalinity, chlorophyll-*a*, suspended matter, total nitrogen (Nt) and total phosphorus (Pt) were measured. Significant different species diversity was recorded among the 43 species of the studied site. A canonical correspondence analysis showed that the concentration of nutrients, chlorophyll-*a*, suspended matter, dissolved oxygen and pH were the most important predictors of the distribution of macrophytes (Pereira *et al.*, 2012).

For metal pollution, aquatic vascular plants are considered strong indicators. Many sensitive aquatic vascular plants have been identified as indicators of metals because of their potential to accumulate them substantially. Various macrophytes such as *Potamogeton sp.* have been considered strong indicators of Cu, Pb and Zn pollution in aquatic systems. Similarly, *Pontedaria cordata* has been found ideal for detecting copper and lead; *Ceratophyllum demersum, Elodea densa* and *Eichhornia* for mercury contamination; *Limnanthemum cristatum* for chromium concentration and *Wolffia globosa* for cadmium contamination in aquatic systems (Chandra and Sinha, 2000). The downstream water quality has been found to be related to the impact of purple Loosestrife (*Lythrum salicaria*) in the North American wetland, which affected various other species (Emery and Perry, 1996).

11.7 Algae as a Biological Indicator

Most algae that grow on the surfaces of rocks in a slippery layer are called periphyton and if they remain suspended in the water are called phytoplankton. The algal population increases at a faster rate when the water is enriched with nutrients that algae need for their growth (Philadelphia Water, 2015). Lower plants are considered as strong indicators of pollution and algae have been extensively used as indicators of water quality. *Scenedesmus obliquus* has been reported as an indicator of cadmium and lead pollution. Brown algae have also been used as biological indicators of heavy metal pollution.

Variation in species diversity of phytoplankton strongly indicates pollution in aquatic ecosystems. Morphological changes, cellular malformation, chlorosis and significant increase in heterocyst frequency in *Anabaena cylindrica* indicate cadmium pollution. An induction of abnormally long filaments and loss of cellular contents has been reported in *A. inequalis* as a result of cadmium pollution. A notable decrease in zygospore germination in *Chlamydomonas* has also been reported in response to some herbicides and insecticides (Chandra and Sinha, 2000).

Eutrophication can profoundly change rocky shore communities. These changes often cause the replacement of perennial, canopy-forming algae such as *Fucus* spp. with annual, bloom-forming algae such as *Enteromorpha* spp. (Worm and Heike, 2006). Dominance of Chlorophytes and Cryptomonade (Phytoplankton) has been reported at low concentration of P in aquatic ecosystems. The diatoms also attained substantial proportions at lower P loads (Sand-Jensen *et al.*, 2000). Any change in the diversity of phytoplankton reflects water quality (Mayer *et al.*, 1997).

11.8 Diatoms: An Indicator of Water Quality

Diatoms are a single-celled, microscopic algal species with a siliceous covering called a frustule. They occur in a variety of shapes, and are either planktonic (living suspended on the water), benthic (growing associated to a substrate), or both planktonic and benthic. The algal species that develop in an area are very sensitive to different environmental factors like salinity, temperature, pH, water velocity, shading, depth, substrate availability and water chemistry. Thus, the species that can be found in a water body indicate the physico-chemical properties of the water. Algae are considered good indicators for the monitoring of water quality. Diatoms have the benefit of being easily identifiable at the species level, and are very simple to collect and store due to their hard frustule.

The ecological requirements of many diatom species are known and, therefore, many diatom-based indexes of water quality have been developed. The diatoms may remain in their location until or unless there is a disturbance in the aquatic ecosystem: they reflect the characteristics of the water from the area in which they live. Protocols must be followed at the time of sampling, treatments and analysis of diatoms to avoid the risk of using the indexes alone, as there are some situations which can lead to incorrect conclusions about the health of an ecosystem (Martín and Fernández, 2012).

In the most organically polluted sites of the Karasu river basin (Turkey), three diatom species: *Gomphonemaparvulum*, *Nitzschiapalka* and *Naviculacryptocephala*, were found in high densities. The COD, BOD, and concentrations of nutrients ($NO3^{-}$-N, $NO2^{-}$-N, $NH4^{+}$-N and PO^{-3}-P) were co-related negatively with DO (Gurbuz and Kivrak, 2002). The diatoms have been recognized as bioindicators in temperate streams in the Kathmandu Valley, Middle Hills of Nepal and northern India. In the Kathmandu Valley, the richness and diversity of diatoms was found correlated with K^{+}, Cl^{-}, $SO4^{-3}$, NO^{-2}, Al^{+3}, Fe^{+3} surfactant and phenol (Broderson *et al.*, 2001).

11.9 Measurements and Evaluation of Biodiversity

The diversity index of a community is the ratio between the number of species and the number of individuals in that community. The species richness index, Palmer's generic index and Margalef's index are useful for the assessment of water quality parameters (Hariprasad and Ramkrishnan, 2003). The diversity and density index of planktons are generally applied for the monitoring of water pollution by organic contaminants (Zhao *et al.*, 2000). The zooplankton diversity index, (Lougheed and Chow, 2002), Shannon–Weaver Index, Whilm–Dorris diversity index and the saprobity index of Watanabe are diversity indices useful in environmental monitoring (Sadusso and Morana, 2002). The species richness, number of distinct taxa, such as orders, families, species at monitoring site, evenness, the relative abundance of different taxonomic groups, are also important parameters for the measurement of plant diversity.

Comparison or similarity indices are also used for the assessment of community structure using the feeding group (herbivores, detrivores, carnivores) rather than taxa (species, genus, family) (Danilov and Ekelund, 1999; North Carolina State University (NCSU) Water Quality Group, n.d.). There are three indices that provide an overall indication of the ecological health of an aquatic ecosystem: a Native Condition Index (extent and diversity of native plants); an Invasive Condition Index (extent and impact of alien weeds) (both of these are generated from scores allocated to carefully selected vegetation features); and an integrated Lake SPI Index (largely derived from components of the other two indices). The macrophyte index, based on nutrient loads, and the saprobic index, based on organic pollution, are used to indicate the water quality of Lake Geneva (Switzerland). The saprobic index was found sensitive to small-scale changes in species composition as a result of change in water quality (Lehmann and Lachavanne, 1999). GI (Generic Index) was found directly correlated with COD, BOD, DO and nutrients of an aquatic ecosystem (Gurbuz and Kivrak, 2002).

To assess the effect of stressors on (aquatic) populations and communities, diverse indices have been used. Biotic indices are generally specific to the type of pollution or the geographical area; they are used to categorize the degree of pollution by determining the tolerance of an indicator organism to a pollutant (Feminella and Flynn, 1999). The aquatic plant diversity can be measured with the help of frequency, density and abundance of the plant communities within an ecosystem. The diversity index is a useful parameter for the study of biodiversity in an ecosystem. The phosphorus enrichment in the marsh and slough area of the Northern Everglades (USA) caused distinct changes in species frequency. Density represents the numerical strength of a species in the community. It gives the degree of competition in an ecosystem. The abundance is the number of individuals of any species per sampling unit of occurrence (Sharma, 2014).

Population dynamics is the study of how and why population size changes over a time period. Plant populations can be estimated by their size, density, structure, number of individuals of different age groups. To understand the patterns of population dynamics, observations, experiments and mathematical models are used by plant population ecologists. Plant populations are controlled by density-dependent forces, interspecies and intraspecies competition for various resources. Demographic differences among individuals also affect the potential of contribution to population dynamics. Transition matrix models are the most important mode commonly used to study plant populations and guide the management of harvested populations and species of conservation concern (Bierzychudek, 2014).

For predicting environmental perturbations, plants are increasingly being used as highly effective and sensitive tools. Due to industrialization and urbanization, water pollution has become an acute problem. Aquatic plants are progressively being used to provide information on the status of the aquatic environment because they do not migrate from one place to another and they quickly attain equilibrium with their ambient environment. Also, they provide cumulative information of preceding and present environmental conditions, as compared with chemical analytical methods, which provide only the current status of the environment. Although a lot of information is available on the bioindicators of air pollution, information on pollution-indicator plants in the aquatic environment is rather inadequate (Chandra and Sinha, 2000).

11.10 Discussion

Water is one of the most significant natural resources. Skilful management of our water resources is immediately required for diverse domestic and industrial purposes. Irrigation of crop fields, transport, recreation purposes, sports, fisheries, power generation and waste disposal are among the most common uses of water. Water pollution is a worldwide problem, most commonly associated with the discharge of effluents containing various contaminants from sewers or sewage treatment plants, drains, factories and households. Various types of water bodies are under the direct threat of water pollution (Rashed, 2001). Plants are the cheapest monitoring systems or bioassays provided by nature for the estimation and indication of levels and types of air, soil and water pollution in a specific region. Diverse damage symptoms are produced in plants because such symptoms are indicative of specific pollutants and concentrations.

There are a few safety measures to be followed during the use of plants as pollution indicators. The damage symptoms in plants should preferably be studied in the local native species and not in cultivated or introduced species. Species that are sensitive to any type of pollution should first be identified locally, then used for pollution monitoring. Tolerant species should not be used for environmental monitoring purposes. The presence of the pollutant in the area may be cross-checked by studying damage symptoms of a particular plant species to a particular pollution. Damage symptoms of morphological, anatomical, ultra-structural physiological and biochemical types should also be studied in more than one sensitive plant species for confirmation purposes. A number of samples should be taken from different sites in the area of study.

Any possibility of damage symptoms occurring due to some reason other than pollution (e.g. pathogen, environmental factors, nutritional deficiency/enrichment) should be checked thoroughly. The plant species used to monitor pollution in an area should have certain important features for more accurate results. The plant species used for environmental monitoring should be easily identifiable in the field and also be easy to handle for damage analysis. The species must have a wide geographical distribution, so that it can be used in various study sites with different environmental conditions.

The indicator species should be very sensitive to various types of pollutants, so that it can be used to monitor different types of pollutions in that area. Plants commonly used as pollution indicators are lichens, mosses, plankton algae, aquatic ferns and angiosperms, terrestrial ferns and angiosperms, conifers, oaks and many crop plants. Mosses, lichens, ferns, algae and aquatic plants are generally more useful in environmental pollution monitoring (Garg, 2008). Algae, a vital group of bacteria and plants in aquatic ecosystems, are an important component of biological water-quality monitoring programmes. They are the most suitable tool for the water-quality assessment because of their rapid nutrient uptake, rapid multiplication and reproduction, and completion of their life cycles in a very short period of time. Algae are important indicators of ecosystem health as they respond very quickly both in species composition and densities to a wide range of aquatic environmental factors.

An increase in water acidity influences lake pH levels, as well as heavy metals discharged from industrial areas, which may affect the composition of tolerant species. Analysis of water-quality parameters of water bodies determines the diversity and density of algal species and provides a forecast of deteriorating conditions (Bruun, 2012). Plant diversity is a strong bioindicator for aquatic ecosystems of different types and responses (Lorenz *et al.*, 2003). The chironomid taxa (non-biting midges) were found to be the best indicator of highly productive Danish lakes lacking abundant submerged vegetation (Broderson *et al.*, 2001).

Acknowledgements

The authors would like to acknowledge financial support for this work, from the Deanship of Scientific Research (DSR), University of Tabuk, Tabuk, Saudi Arabia, under the no. S-0021-1436. The authors would also like to thank the Department of Biology, Faculty of Sciences, Saudi Digital Library and University Library for providing the facility for literature survey and collection.

References

Ansari, A.A. and Khan, F.A. (2002) Nutritional status and quality of water of a waste water pond in Aligarh showing blooms of *Spirodela polyrrhiza* (L.) Shield. *Journal of Ecophysiology and Occupational Health* 2, 185–189.

Ansari, A.A. and Khan, F.A. (2008) Remediation of eutrophic water using *Lemna minor* in a controlled environment. *African Journal of Aquatic Science* 33, 275–278.

Ansari, A.A. and Khan, F.A. (2009) Remediation of eutrophic water using *Spirodela polyrrhiza* L. Shield in controlled environment. *Pan-American Journal of Aquatic Science* 4, 52–54.

Best, E.P.H., Vander, S.S., Oomes, M.J.M. and Vander, S.S. (1995) Responses of restored grassland ditch vegetation to hydrological changes, 1989–1992. *Vegetation* 116, 107–122.

Bierzychudek, P. (2014) Plant biodiversity and population dynamics. In: Monson R.K. (ed.) (2014) *Ecology and the Environment* (*The Plant Sciences*, Vol. 8). Springer Media, New York, USA, pp. 29–65. DOI: 10.1007/978-1-4614-7501-9_15.

North Carolina State University (NCSU) Water Quality Group (n.d.) Biomonitoring. Available at: www.water.ncsu.edu/watershedss/info/biomon.html (accessed 27 May 2016).

Blinn, D.W., Bailey, P.C.E. and Herbst, D.B. (2001) Land-use influence on stream water quality and diatom communities in Victoria, Australia: a response to secondary salinization. *Hydrobiologia* 499, 231–244.

Broderson, K.P., Odgaard, B.V. and Anderson, N.J. (2001) Chironomid stratigraphy in the shallow and eutrophic Lake Søbygaard, Denmark: chironomid–macrophyte co-occurrence. *Freshwater Biology* 46, 253–267.

Bruun, K. (2012) Algae can function as indicators of water pollution. Nostoca Algae Laboratory, Washington State Lake Protection Association (WALPA), Seattle, Washington. Available at: www.walpa.org/waterline/june-2012/algae-can-function-as-indicators-of-water-pollution (accessed 27 May 2016).

Caffrey, J.M., Dutartre, A., Haury, J., Murphy, K.M. and Wade, P.M. (2006) *Macrophytes in Aquatic Ecosystems: From Biology to Management, Developments in Hydrobiology.* Springer, New York, USA, pp. 263.

Chambers, P.A., Lacoul, P., Murphy, K.J. and Thomaz, S.M. (2008) Global diversity of aquatic macrophytes in freshwater. *Hydrobiologia* 595, 9–26. DOI: 10.1007/s10750-007-9154-6.

Chandra, P. and Sinha, S. (2000) Plant bioindicators of aquatic environment. *EnviroNews Archives* 6(1) (Millennium Issue).

Danilov, R. and Ekelund, N.G.A. (1999) The efficiency of seven diversity and one similarity indices based on phytoplankton data for assessing the level of eutrophication in lakes in central Sweden. *Science of the Total Environment* 234(1), 15–23.

Detembeck, N.E., Galatowitsch, S.M., Atkinson, J. and Ball, H. (1999) Evaluating perturbations and developing restoration strategies for inland wetland in the great lakes basin. *Wetlands* 19, 789–820.

Emery, S.L. and Perry, J.A. (1996) Decomposition rates and phosphorus concentrations of purple Loosestrife (*Lythrum salicaria*) and Cattail (*Typha* spp.) in fourteen Minnesota Wetlands. *Hydrobiologia* 323(2), 129–138.

Feminella, J.W. and Flynn, K.M. (1999) The Alabama Watershed Demonstration Project: Biotic Indicators of Water Quality. ANR-1167. Alabama Cooperative Extension System, Alabama A&M University and Auburn University, Auburn, Alabama, USA. Available at: www.aces.edu/pubs/docs/A/ANR-1167/index2.tmpl (accessed 27 May 2016).

Garg, P.K. (2008) Plants as indicators and monitors of pollution. *Environment of Earth*, 4 August (blog). Available at: environmentofearth.wordpress.com/tag/pollution (accessed 27 May 2016).

Grime, J.P. (1998) Benefits of plant diversity to ecosystems: immediate, filter and founder effects. *Journal of Ecology* 86, 902–910.

Gürbüz, H. and Kivrak, E. (2002) Use of epilithic diatoms to evaluate water quality in the Karasu River of Turkey. *Journal of Environmental Biology* 23(3), 239–246.

Hariprasad, P. and Ramakrishnan, N. (2003) Algal assay used for the determination of organic pollution level in freshwater body at Tiruvannamalai, India. *Journal of Ecotoxicology and Environmental Monitoring* 13(4), 241–248.

Jüttner, I., Sharma, S., Dahal, B.M., Ormerod, S.J., Chimonides, P.J. and Cox, E.J. (2003) Diatoms as indicators of stream quality in the Kathmandu Valley and Middle Hills of Nepal and India. *Freshwater Biology* 48, 2065–2084.

Lehmann, A. and Lachavanne, J.B. (1999) Changes in the water quality of Lake Geneva indicated by submerged macrophytes. *Freshwater Biology* 42, 457–466.

Lorenz, C.M., Markert, B.A., Breure, A.M. and Zechmeister, G.H. (2003) Bioindicators for ecosystem management, with special reference to freshwater system. In: Markert, B.A., Breure, A.M. and Zechmeister, H.G. (eds) *Bioindicators and Biomonitors: Principles Concepts and Applications*. Elsevier, Kidlington, UK, pp. 123–152.

Lougheed, V.L. and Chow, F.P. (2002) Development and use of a zooplankton index of wetland quality in the Laurentian Great lakes basin. *Ecological Applications* 12, 474–486.

Martín, G. and Fernández, M.R. (2012) Diatoms as indicators of water quality and ecological status sampling, analysis and some ecological remarks. In: Voudouris, K. (ed) *Ecological Water Quality – Water Treatment and Reuse*. InTech, Rijeka, Croatia, pp. 183–204. Available at: cdn.intechopen.com/pdfs-wm/36803.pdf (accessed 27 May 2016).

Mayer, J., Dokulil, M.T., Salbrechter, M., Berger, M., Posch, T., Pfister, G., Kirschner, A.K.T., Velimirov, B., Steitz, A. and Ulbricht, T. (1997) Seasonal succession and trophic relations between phytoplankton, zooplankton cilliate and bacteria in a hypertrophic shallow lake in Vienna, Austria. *Hydrobiologia* 342–343, 165–174.

Murphy, K.J. (2002) Plant communities and plant diversity in softwater lakes of northern Europe. *Aquatic Botany* 73, 287–324.

Murphy, K.J., Dickinson, G., Thomaz, S.M., Bini, L.M., Dick, K. *et al*. (2003) Aquatic plant communities and predictors of diversity in a subtropical river flood plain: the upper Rio Parana Brazil. *Aquatic Botany* 77, 257–276.

Nicholls, K.H., Steedman, R.J. and Carney, E.C. (2003) Changes in phytoplankton communities following logging in the drainage basins of three boreal forests lakes in northern western Ontario (Canada), 1991–2000. *Canadian Journal of Fisheries and Aquatic Science* 60, 43–54

Occhipinti, A.A. and Savini, D. (2003) Biological invasions as a component of global changes in stressed marine ecosystems. *Marine Pollution Bulletin* 46, 542–551.

Pereira, S.A., Trindade, C.R., Albertoni, E.F. and Palma-Silva C. (2012) Aquatic macrophytes as indicators of water quality in subtropical shallow lakes, Southern Brazil. *Acta Limnologica Brasiliensia* 1–13.

Philadelphia Water (2015) Algae. Available at: www.phillywatersheds.org/what_were_doing/waterways_assessment/algae (accessed 27 May 2016).

Rashed, M.N. (2001) Biomarkers as indicator for water pollution with heavy metals in rivers. *Seas and Oceans* 2001, 1–13.

Rooney, R.C. and Bayley, S.E. (2011) Relative influence of local- and landscape-level habitat quality on aquatic plant diversity in shallow open-water wetlands in Alberta's boreal zone: direct and indirect effects. *Landscape Ecology* 26, 1023–1034, DOI: 10.1007/s10980-011-9629-8.

Sadusso, M.M. and Morana, L.B. (2002) Comparison of biotic indexes utilized in the monitoring of lotic systems of North East Argentina. *Revista de Biologia Tropical* 50, 327–336.

Sand-Jensen, K., Riis, T., Vestergaard, O. and Larsen, S.E. (2000) Macrophyte decline in Danish lakes and streams over the past 100 years. *Journal of Ecology* 88, 1030–1040.

Sharma, P.D. (2014) *Ecology and Environment*, 12th edn. Rastogi, Meerut, India.

She, L.K., Kheng, L.C. and Hain, T.C. (1984) Selected aquatic vascular plants as biological indicators for heavy metal pollution. *Pertanika* 7(1), 33–47.

Shen-Dong, S. and Shen-Dong, S. (2002) Study on limiting factors of water eutrophication of the network of rivers. *Journal of Zhejeng University of Agriculture and Life Science* 28, 94–97.

Shrivastava, S., Shukla, A.N. and Roa, K.S. (2003) Biodiversity of undasa wetland Ujjain (India) with special reference to its conservation. *Journal of Experimental Zoology* 6, 125–135.

Sinha, A. (2001) Study of water pollution of Najafgarh drain, Delhi on the basis of algae pollution indices. *Journal of Ecology, Taxonomy and Botany* 25, 339–345.

Stoyneva, M.P. (2003) Survey on green algae of Bulgarian thermal springs. *Biologia Bratislava* 58, 563–574.

Tremp, H., Kohler, A., Tremolieres, M. and Muller, S. (1995) The usefulness of macrophytes monitoring systems, exemplified on eutrophication and acidification of running waters. *Acta Botanica Gallica* 142, 541–550.

Worm, B. and Heike, K.L. (2006) Effects of eutrophication, grazing, and algal blooms on rocky shores. *Limnology and Oceanography* 51, 569–579.

Xu, Z., Yan, B., He, Y. and Song, C. (2007) Nutrient limitation and wetland botanical diversity in Northeast China: can fertilization influence on species richness? *Soil Science* 172, 86–93.

Zhao, W., Dong, S., Zhen, W., Gang, Z.Z., Zhao, W., Dong, S.L., Zheng, W.G. and Zhang, Z.Q. (2000) Effects of Nile Tilapia on plankton in enclosures with different treatment in saline-alkaline ponds. *Zoological Research* 21, 108–114.

12 Gymnosperm Diversity of the Kashmir Himalayas

Mohd Irfan Naikoo[1]*, Mudasir Irfan Dar[1], Fareed Ahmad Khan[1], Abid Ali Ansari[2], Farha Rehman[1] and Fouzia Nousheen[3]

[1]*Environmental Botany Laboratory, Aligarh Muslim University, Aligarh, Uttar Pradesh, India;* [2]*Department of Biology, Faculty of Science, University of Tabuk, Tabuk, Saudi Arabia;* [3]*Department of Botany, Women's College, Aligarh Muslim University, Aligarh, Uttar Pradesh, India*

Abstract

The Kashmir is rich in biodiversity and is known as the biomass state of India (Lawrence, 1895). Phytogeographically located at the Holarctic and Paleotropical intersection in the North-Western Himalaya, this bio-region harbours luxurious treasures of plant diversity. The Kashmir region is rich in gymnosperm diversity, which forms an important component, floristically, ecologically and socio-economically: it is known as the green gold of the state of Jammu and Kashmir. Gymnosperms harbour a rich diversity of flora and fauna under their canopies. They are the rich source of diverse economic and medicinal products, providing innumerable products, including timber, fuel, gums, resins, medicines and many more useful products, besides acting as effective wind-breaks, especially the evergreen species, which also slow soil erosion and protect watersheds. The single giant sequoia (*Sequoiadendron giganteum*), the state tree of California, which grows at Yarikhah Drug Farm (Tangmarg) in Kashmir Valley is the lone representative in the India subcontinent. Due to their immense importance, the gymnosperms have been overexploited by the human population. Sustainable management and conservation of these gymnosperms is urgently required. Anthropogenic activities should be checked and the stake holders educated about the proper harvesting of gymnosperm flora for different uses.

12.1 Introduction

The Kashmir Himalayas are world famous for their picturesque appearance, and are known as paradise on earth (Vigne, 1842). Kashmir, endowed with rich biodiversity, is referred to as the biomass state of India (Lawrence, 1895). Phytogeographically, this bio-region, situated between the Holarctic and Paleotropical intersections in the North-Western Himalaya, harbours luxurious treasures of floristic diversity.

Geological evidence (Puri, 1943; Puri, 1947; Vishnu-Mittre, 1963) reveals that the once-tropical flora of Kashmir Himalaya changed into a mix of subtropical-temperate-alpine flora after the Little Ice Age. Gradually flora from a large number of places found entry into the Kashmir Himalayas through natural dispersion or by anthropogenic means; the existing flora is therefore a mix of adjacent phytogeographical regions with a proportion of endemics (Dar and Khuroo, 2013).

Plant exploration in the Kashmir Himalayas dates back to 1812, when a (non-botanist) British veterinary surgeon William Moorcroft collected a bundle of plants. The first botanical collection was made in 1831 by Victor Jacquemont, who was sent by the National Museum of France (Stewart, 1979). Since then a number of botanists have explored and

*E-mail: mdirfanmsbo@gmail.com

documented the plant wealth growing along a variety of vast topographical gradients of this region. This plant life shows a huge diversity, ranging from minute algae to huge plants such as *Eucalyptus*.

The gymnosperms form an important floristic component, due to their extreme ecological and socio-economic value (Dar and Dar, 2006a). Despite this, the taxonomy of gymnosperms has been neglected in India, particularly in the Kashmir Himalayas. Even Hooker (1888), who was the first to study gymnosperms in India, showed little interest in this region. Since then, Lambert (1933), Dhar (1966, 1975), Dhar (1978), Stewart (1972), Javeid (1970, 1979), Singh and Kachroo (1976), Sahni (1990), Dar *et al.* (2002), Dar and Dar (2006a, 2006b), Dar and Christensen (2003), and Dar (A.R. Dar, Taxonomic study on gymnosperms in Kashmir. Department of Botany, University of Kashmir, Srinagar, Jammu and Kashmir, India, 2004, unpublished thesis) have made strenuous efforts to fill the gap. The sum of their work is detailed in this chapter.

12.2 Gymnosperms – the Green Gold of Jammu and Kashmir

In the state of Jammu and Kashmir, the Kashmir valley has the largest area of forest cover (about 50% of its geographical area is under forests). These are mainly temperate and occur between 1600 and 2700 m. The temperate forests comprise mainly the conifers, such as Himalayan Deodar (*Cedrus deodara*), Blue Pine (*Pinus wallichiana*), Silver Fir (*Abies pindrow*), Spruce (*Picea smithiana*) and the Himalayan Yew (*Taxus wallichiana*), associated with some broad-leaved trees, herbs and shrubs. Beyond the temperate forests, between 2700 and 3500 m, are the subalpine forests. The Silver Fir is dominant in the subalpine forests in the lower reaches, and the Birch (*Betula utilis*) can be found above 3200 m, and forms the tree line in the Kashmir Himalaya. Above the tree line is alpine scrub vegetation, comprising mainly species of *Juniperus*, *Salix*, *Cotoneaster*, *Lonicera* and *Rhododendron*. The mountain slopes at the alpine and subalpine altitudes have lush green meadows (Nayyeh or Bahaks) with characteristic herbaceous plants species such as *Aconitum*, *Iris*, *Ranunculus*, *Pedicularis*, *Aquilegia*, *Gentiana*, and *Potentilla*.

Gymnosperms dominate in the forests of the Kashmir Himalayas, and because of their high economic importance are known as the green gold of the state of Jammu and Kashmir. They are the source of diverse economic and medicinal products, providing innumerable products including timber, fuel, resins, gums, important medicines and many more useful products, besides acting as effective wind-breaks, especially the evergreen species, and are important for soil-erosion control and protection of watersheds. The cones of some gymnosperms are a source of food for wildlife and some gymnosperms are used as ornamental plants in gardens and parks.

The gymnosperm species, though only a few grow wild, dominate the coniferous forests of the Kashmir Himalayan region. Due to the low number of gymnosperm species, species richness is well documented and described for this region (Dar and Dar, 2006a). A huge single tree of the gigantic sequoia (*Sequoiadendron giganteum*) growing at Yarikhah, Tangmarg in Kashmir is the lone representative of the state tree of California in the Indian subcontinent.

There have been extensive studies carried out by various workers to collect information about the diversity of gymnosperms in Kashmir Valley. The summary the sum up of their findings follows.

A total of 20 species belonging to six families (Pinaceae, Cupressaceae, Taxodiaceae, Taxaceae, Ginkgoaceae and Ephedraceae) of gymnosperms have been so far reported from the Kashmir Himalayas. These species have been arranged alphabetically within their families, with their correct name, synonyms, English name, vernacular/local names, short description, distribution, where specimens were examined and economic importance (Lone, 2013).

12.3 Pinaceae

The Pinaceae family is represented by six genera in India: Abies, Cedrus, Larix, Picea, Pinus and Tsuga, including 14 species. In Kashmir, seven species belonging to four genera of this family have been reported so far.

12.3.1 *Abies pindrow*

Royle III. Bot. Himal.T.86, pp: 350–351 1839.

English name: West Himalayan Silver Fir.

Vernacular/local name: Kashmiri: Badul; Urdu: budlu.

Evergreen trees (45–55 m tall) with a narrow cylindrical crown of horizontally drooping branches; bark greyish-brown with longitudinal fissures. Shoot monomorphic. Leaves linear, flattened, spirally spread, two greyish bands of stomata on either side of shallow midrib. Cones erect or cylindrical.

Distribution: Jhelum, Lolab Valley and Sindh Valley.

Specimens examined: Forest slopes of Thajwas, 2900 m by A.R. Dar and G.H. Dar (2002); Narang forests, 1900 m by A.R. Dar (2002); Forests of Akhal, 2250 m by A.R. Dar and G.H. Dar (2002) and Prang Forests, 1950 m by G.H. Dar (1993).

Economic importance: Wood soft, easily worked, good surface finishing, very polishable and paintable, suitable for joinery and furniture, building construction and for packing cases. Branches are burnt as fuel for cooking. Fresh leaves, ground and mixed with equal quantity of honey, are used for curing cough and cold. Bark used by nomads as a tea substitute.

12.3.2 *Abies spectabilis*

(D. Don) Spach. Hist. Nat. Veg. Phan. 2: 422.1842

Pinus spectabilis D.Don.

Abies webbiana Lindl.

English name: Himalayan Silver Fir, Webb Fir.

Vernacular/local name: Kashmiri: Reia Badul; Hindi: Bang.

Tall evergreen tree (50–60 m tall) with dense cylindrical crown of pendulous branches; bark dark-brown and less deeply fissured than *Abies pindrow*. Shoots monomorphic and hairy. Leaves are flattened, linear and arranged in three or four rows on each side of branchlet. Cones always erect and shorter than *Abies pindrow.* Male cones are solitary and cylindrical. Female cones are solitary, ovoid-oblong and sub-sessile. Seeds are winged.

Distribution: Hills of Kashmir, Aru Valley Pahalgam

Specimens examined: Dachigam (Kashmir), 2700 m by Gurcharan Singh (1971).

Economic importance: Wood is used for ceilings, framing, decking, panelling, mill work, furniture parts, floor boards and packing cases. The dried leaves are used to cure indigestion and to stop bleeding. A powder, made from the leaves along with juice of *Adhatoda zeylanica* and honey, is used for asthma. The juice of fresh leaves is given to infants suffering from fever and chest infection.

12.3.3 *Cedrus deodara*

(Roxb ex. Lamb) G. Don in London. Hort. Brit. 3888.1830.

Pinus deodara Roxb.

Cedrus libani Barr.

English name: Deodar, Himalayan cedar.

Vernacular/local name: Kashmiri: deodar; Urdu: Deodar.

Large evergreen pyramidal (40–60 m tall) tree with horizontal branches and massive trunk, 3–5 m in girth; bark brown often greyish or reddish, deeply furrowed with irregular or oblong plates. Shoots dimorphic. Leaves needle-like, glaucous green and in clusters, stiff, acuminate with stomatal lining on both sides. Flowers monoecious. Male cones solitary, cylindrical, composed of spirally overlapping microsporophylls. Female cones reddish-brown, solitary, barrel-shaped. Seeds winged.

Distribution: Lolab Valley, Kralpora forests.

Specimens examined: Fourbay forest slopes, Ganderbal, 1800 m; Kashmir university campus, 1600 m; Shankaracharya forest slopes, 1650 m; The forest slopes of Chandanwari (Uri), 1500 m by A.R Dar and G.H. Dar (2002).

Economic importance: Deodar is one of the most important assets of the Kashmir Himalaya. Being the strongest among Indian coniferous woods, it produces valuable timber, which is in high demand. Wood is used for construction of buildings, railway sleepers, bridges, furniture, packing cases, electrical poles for carrying high-tension lines and several other purposes. It is also used for construction of boats, as it is fairly water-resistant. Reddish-brown oil extracted from its wood is used for treating skin disorders, ulcers and rheumatic pain. It is also used as an antidote for snake bites, and is applied to animals' bodies to repel insects, tics and mites.

12.3.4 Picea smithiana

(Wall.) Boiss. Fl. Or 5: 700. 1881.

Piceae morinda Link.

Pinus smithiana Wall.

English name: Western Himalayan Spruce.

Vernacular/local name: Kashmiri: Kachul, Rayal.

Large evergreen pyramidal tree (50–55 m tall) with drooping branches and tapering trunk 5–8 m in girth; bark reddish- or pale grey, rough, exfoliating in thin woody plates. Shoots monomorphic. Leaves needle-like, single, spirally spread around the branches. Male cones solitary, short-stalked and ovoid. Female cones solitary, long-stalked, cylindrical, pendulous and dark brown. Seeds winged.

Distribution: Jehlum, Lolab, Sindh Valley.

Specimens examined: The forests of Gulmarg, 2800 m; Narang forests, 1900 m; Akhal forests, 2250 m by G.H. Dar and A.H. Dar (2002); Ganderbal forests, 1850 m by G.H. Dar (1982).

Economic importance: The main use of *Picea smithiana* is for pulpwood. Wood is used for building construction, general mill works, framing material and for boxes. Oleoresin is applied on wounds and cracked heels. Leaves used as bath salts; leaf oil used as deodorants and room sprays. Leaves are also used as manure and litter for cattle.

12.3.5 *Pinus halepensis*

Miller, Gard. Dict. Ed. 8:8.1768

Pinus arabica Sieber ex Spreng. 1826

Pinus maritima Aiton 1813 *non* Mill. 1768

English name: Aleppo pine.

Vernacular/local name: Kashmiri: Yaer.

Small to medium-sized tree, 15–25 m tall, with a straight trunk (60–100 cm in diameter). Bark thick and reddish-brown and linearly fissured. Shoots dimorphic. Leaves are needle-like, in fascicles of two, twisted, edges minutely serrate and yellowish-green in colour. Stomatal lining on both surfaces. Male cones in clusters, short-stalked, ovate or cylindrical, containing spirally overlapping microsporophylls. Female cones are solitary, or 2–3 together, short-stalked, Ovate, conical, dark-brown. Seeds winged (Dallimore *et al.* 1967).

Pinus halepensis exists only in cultivation in Kashmir.

Distribution: Srinagar, Gulmarg.

Specimens examined: Hill slopes of Shankaracharaya, 1650 m by A.R. Dar and G.H. Dar (2003); Gulmarg, 2800 m by Manju Kapoor.

Economic importance: Wood, being poor in quality, resinous and coarse-grained, is not recommended for construction purposes, but is used for making inferior joinery, boxes, etc., and as fuel. A good-quality resin is obtained from it. The bark, being highly resistant, is used for tanning. The trees are planted to reduce soil erosion and as wind-breaks.

12.3.6 *Pinus roxburghii*

Sargent, Silva.N.Amer.11.1897.

Pinus longifolia Roxb.

English name: Chir pine.

Vernacular/local name: Kashmiri: Kairoo; Hindi: Dhup; Urdu: Chir.

Large evergreen resinous tree (45–55 m tall) with straight trunk, 1–2 m in girth and spreading umbrella shaped crown. Bark brownish-red and deeply furrowed. Shoots dimorphic. Leaves are needle-like, in fascicles of three, very slender, persistent (1–3 years) and bright green.

Male cones in clusters, ovoid conic with numerous, stalkless, spirally overlapping microsporophylls. Female cones are solitary or 2–5 together, oval, short-stalked and brownish-red. Seeds winged. Scales woody with a curved 'beak'.

Distribution: Shankaracharya hill, Kashmir University.

Jehlum, Lidder, Lolab and Sindh Valleys.

Specimens examined: Botanical Garden, University of Kashmir, 1600 m; Lalpul, Uri forests, 1200 m; Hill slopes of Shankacharaya by G.H. Dar and A.H. Dar (2003).

Economic importance: Wood is moderately hard, heavy and resinous. It is used in the construction of buildings, huts, etc. It is also used for making boxes and as firewood. Soft bark is used for tanning. Resin obtained from *P. roxburghii* is of substantial commercial importance. Resin yield is high and the plants are systematically tapped to get resin, which is applied to cracked heels, and is also used for skin diseases and to soothe inflammation. It is also used for wounds, burns, sores, boils, etc. The turpentine extracted from the resin is diuretic, antiseptic, vermifuge and rubefacient, a valuable remedy for kidney and bladder ailments, diseases of the mucous membranes and respiratory disorders such as influenza, coughs, colds and TB. Raw or cooked seeds are eaten as emergency food.

12.3.7 *Pinus wallichiana*

A.B Jackson in Kew Bull: 85. 1938.

Pinus excelsa Wall.

English name: Himalayan blue pine.

Vernacular/Local name: Kashmiri: kaeur; Urdu: Kairo; Hindi: Chilla.

Graceful evergreen tree (35–50 m tall) with straight trunk, 2–4 m in girth and downward-curving branches. When solitary, the larger branches form a dome-like crown. Bark grey-brown, initially smooth, but becomes shallowly fissured over time, forming oblong or ovoid plates. Shoots dimorphic. Leaves needle-like, in fascicles of five, persistent (3–4 years), 7–14 cm long, glaucous, green on abaxial side and multiple bluish-white stomatal lines on ventral side. Male cones in clusters, ovoid conic and short-stalked with numerous, stalkless, spirally overlapping microsporophylls. Female cones solitary or 2–3 together, terminal or subterminal in position, cylindrical, about 5 cm stalked, light brown. Seeds winged. Scales woody with slight, obtuse tips.

Distribution: Lolab Valley, Gulmarg, Sonamarg, Aharbal, Lidder Valley.

Specimens examined: Aharbal, 2850 m; Sonamarg forests, 2800 m; Pahalgam forests, 2134 m; Boniyar forests, 1550 m; by A.R. Dar and G.H. Dar (2002); Narang forests, 1900 m by A.R. Dar (2002)

Economic importance: Wood is moderately hard, durable and rich in resin. It is used in building construction, furniture, railway sleepers, bridges, carpentry, paper pulp and firewood. Cones are also used as firewood, and the residue as charcoal. The resinous wood is splintered and used as torches. Oleoresin obtained from blue pine is applied on cracked heels. In rural areas of Kashmir, a dark-brown, viscous substance called *kellum* is obtained from the wood and applied by farmers to their limbs to protect against insect bites (Khanze is the local/Kashmiri term for insect bites) while transplanting paddy in water lodged fields. Kellum is later removed with kerosene oil.

12.4 *Cupressaceae*

The family *Cupressaceae* comprises about 20 genera with 130 species, of which 11 are found in the Indian subcontinent. In Kashmir, seven species belonging to three genera of this family have been reported.

12.4.1 *Cupressus cashmeriana*

Royle ex Carriere Trait Gen. Conif. ed. 2: 161.1867.

Cupressus assamica

Cupressus darjeelingensis

English name: Kashmir Cypress.

Vernacular/local name: Kashmiri: Sarva.

Graceful, pyramidal evergreen tree (15–20 m tall) with straight trunk, 1–1.5 m in girth and tapering gradually. Bark is fissured superficially and light brown in colour. The bark peels off in longitudinal strips. The leaves are opposite decussate, tightly appressed, scale-like and bluish-green in colour. Male cones are solitary, ovoid in shape, terminal in position and composed of 5–10 pairs of microsporophylls. Female cones are solitary, globular or spherical, faint brown with obvious white bloom. The cones open and shed their seeds on maturity.

Distribution: Srinagar, lower Lidder Valley forests, Ganderbal.

Specimens examined: Behama, Ganderbal forests, 1800 m; Rangil rocky slopes, 1850 m; by A.R. Dar and G.H. Dar (2002). University of Kashmir Campus, 1600 m by A.R. Dar and G.H. Dar (2003).

Economic importance: *Cupressus cashmeriana* is widely grown as an ornamental plant. It is an avenue tree planted in parks and gardens. In addition to its glaucous blue beauty, it is very aromatic and hangs in pendulous, flat sprays.

12.4.2 *Cupressus sempervirens*

Linn. Sp.Pl. 1002. 1753.

Cupressus sempervirens Pyramidalis.

Cupressus sempervirens stricta.

English name: Mediterranean Cypress.

Vernacular/local name: Kashmiri: Sarva; Urdu: saro

Medium-sized coniferous evergreen tree (10–20 m tall) with a conic crown. The dark-green foliage grows in dense sprays. Bark is shallowly fissured and greyish-brown in colour. The leaves are scale-like, opposite decussate, tightly appressed, margin entire and incurving and dark green in colour. The cones are ovoid or oblong, green at first and brown at maturity. The male cones are solitary, ovate cylindrical in shape, terminal in position and composed of 5–10 pairs of microsporophylls. Female cones are solitary or in 2–3 together, globular or spherical, deep brown, containing 6–8 pairs of fused bract and ovuliferous scales. Seeds winged.

Distribution: Srinagar, lower Lidder Valley forests, Ganderbal.

Specimens examined: Rangil forest slopes, 1850 m by A.R. Dar and G.H. Dar (2002); Kashmir University Campus, 1600 m by A.R. Dar and G.H. Dar (2003).

Economic importance: *Cupressus sempervirens* is an ornamental plant and known for its durable, fragrant wood. Wood is used for building construction and furniture purposes. The doors of St Peter's Basilica in Vatican City are made from this wood. The wood is repellent to insects and is used to keep them away from clothes and other articles. Essential oil extracted from leaves and shoots is used in cosmetics. It is used as a fragrance, and for anti-seborrhoeic, anti-dandruff, astringent and anti-ageing purposes.

12.4.3 *Cupressus torulosa*

D. Don. Prodr. Fp. Nep. 55. 1825.

English name: Himalayan Cypress.

Vernacular/local name: Kashmiri: Sarva; Hindi: Surai.

A large evergreen tree (23–35 m tall) with trunk 1–2 m in girth and pyramidal crown with drooping branches. Bark is thick, fissured superficially and pale brown or reddish-brown in colour. The bark peels off in long, thin strips. The leaves are dark green, scale-like, opposite and decussate, triangular closely appressed with margins entire and slightly curved. Male cones are solitary, subglobular in shape, terminal in position and composed of 6–8 pairs of imbricate microsporophylls. Female cones are solitary or grouped on very short stalks, elliptic or globose, green when young, later turning dark brown and composed of 6–10 pairs of opposite and decussate bracts and ovuliferous scales. Seeds pale brownish and winged.

Distribution: Srinagar, Ganderbal, Lidder Valley.

Specimens examined: Beehama forest slopes, Ganderbal 1800 m by A.R. Dar and G.H. Dar (2002) and Botanical Garden, University of Kashmir, 1600 m by A.R. Dar and G.H. Dar (2003).

Economic importance: Wood of *Cupressus torulosa* is moderately hard and very durable. It is a prime quality timber with straight grain and fine texture, resistant to termites and insects. It is considered as equivalent to deodar. It is used for general construction, railway sleepers, window frames, fine art articles, doors, cabinet work, office furniture, ceilings, etc. The wood is aromatic, especially the root. The essential oil extracted from *Cupressus torulosa* has a medicinal importance, and is used to cure inflammatory wounds and as an antiseptic. It is also used for cosmetics. It is the best timber for making pencils. Traditionally, it has been used for the construction of Buddhist temples and religious wood carving. The plant is burnt as incense.

12.4.4 *Juniperus communis*

Linn. Sp. Pl. 1040. 1753.

English name: Common juniper, red cedar

Vernacular/local name: Kashmiri: Yathur, Vaitro; Hindi: Jhora, Billa, Bhitaru; Urdu: Bhentri.

Juniperus communis is a small coniferous evergreen tree (2–4 m tall, but occasionally 10 m), multi-stemmed, decumbent or rarely upright, with spreading or ascending branches. Bark is brown, fibrous, that of small branchlets (4–10 mm diameter) is smooth, exfoliating in strips, or in plates for larger branches. The leaves are green, when glaucous appearing silvery, spreading, in whorls of three, joined at base, needle-like to narrowly lance-shaped, subulate, with a single white stomatal band on the inner surface, apex acute to obtuse and mucronate. Male cones are yellow, solitary, catkin-like and cylindrical in shape and composed of 6–8 pairs of opposite and decussate microsporophylls. They fall after shedding their pollen. Female cones are solitary, berry-like, initially green and purple-black when ripe, globose and composed of 3–4 pairs of opposite and decussate bracts and ovuliferous scales, each scale with a single seed.

Distribution: Lolab Valley, Dachigam, Lidder Valley, Jehlum and Sindh Valley.

Specimens examined: Fourbay, Ganderbal forests, 1800 m by A.R. Dar and G.H. Dar (2002); Sonamarg, 2800 m by A.R. Dar (2002); Khilanmarg, Gulmarg, 3000 m by U. Dhar (1980) and Sangam slopes, 3600 m by U. Dhar (1982).

Economic importance: Wood is mainly used as fuel. Juniper leaves and twigs are burnt as a fumigant and incense. The essential oil extracted from it is used in perfumes. The ripe fruits are aromatic, antiseptic, diaphoretic, strongly diuretic, rubefacient, and tonic. Juniper fruits are used in herbal medicine and are useful in the treatment of digestive disorders, stomach cramping, piles, diarrhoea, dysentery, and infections of the urinary tract, bladder, kidneys and prostate. The berries of juniper have anti-inflammatory properties and are used for relieving pain and inflammation in chronic arthritis, gout and rheumatism. Berries, wood and oil of juniper are reported to be used in remedies for cancer, indurations, polyps, swellings, tumours and warts.

12.4.5 *Juniperus semiglobosa*

Regel. Trudy Imp. S-Peterbursk. Bot. Sada. 6(2): 487–488. 1879.

Juniperus drobovii Sumnev.

English name: Himalayan pencil juniper, cedar

Vernacular/local name: Kashmiri: shir, challai; Hindi: Dhup.

Juniperus semiglobosa is a medium-sized evergreen coniferous shrub (6–15 m tall), usually monopodial, but sometimes multi-stemmed. Bark is reddish-brown to grey-brown, fibrous, longitudinally furrowed and peeling off in long strips. The leaves are light green to yellowish-green in colour, scale-like, appressed, accuminate and opposite decussate in arrangement. Male cones are solitary, terminal or subterminal, cylindrical, initially green, turning yellowish, and composed of 6–10 microsporophylls. Female cones are solitary, terminal or subterminal on ultimate branchlets, when young, they are stellate or spheroid and dark green. Mature cones are semiglobose to triangular and brown to blackish-blue in colour. 1–2 fertile seeds per cone.

Distribution: Lidder Valley, Kokernag, Sonamarg, Gulmarg, Lolab Valley, Ganderbal

Specimens examined: Chittergul forests, 1900 m; Sonamarg forests, 2800 m; Gagangir slopes, 2200 m; Kangan, 2300 m by A.R. Dar and G.H. Dar (2002).

Economic importance: the wood of *Juniperus semiglobosa* is very hard and is used in furniture and pencil making. The wood is fragrant and is burnt as incense in monasteries. It is also used as fuel and charcoal.

12.4.6 *Juniperus squamata*

Buch-Han ex. D.Don in Lambert, Genus *Pinus*, II: 17. 1824.

Juniperus recurva Buch-Ham

English name: Weeping blue juniper.

Vernacular/local name: Kashmiri: Yathu, Vaitro.

Juniperus squamata is an evergreen shrub or small tree (2–10m tall), multi-stemmed with prostrate to irregularly conical crown. Bark is scaly, flaky, brown and peels off in longitudinal strips. Leaves are sharply pointed, awl-shaped, bifacial, acuminate and often strongly glaucous blue-green in colour. Flowers monoecious or dioecious. Male cones are solitary, axillary in position, short-stalked and globose in shape. Female cones are also solitary, axillary in position, ovoid and composed of opposite and decussate bracts and ovuliferous scales. Each cone contains one seed.

Distribution: Lolab Valley, Dachigam, Keran, Gulmarg, Sonamarg.

Specimens examined: Apharwat slopes, 3700 m and Sonamarg forests, 2800 m by A.R. Dar (2002); Razdani Open Slopes by Dhar, Yousf, Gupta (1981); Nilnai rocky mountain slopes, 3650–3900 m and Zojila mountain slopes, 3500 m by G.H. Dar (1983).

Economic importance: The wood of *Juniperus squamata* is essentially used as fuel in alpine ranges, as no other arboreals grow there. The female cone is of great medicinal importance. Berries are used for treating stomach disorders. It is also burnt as incense.

12.4.7 *Thuja orientalis*

Linn., Sp. Pl.: 1002.1753; ed. 2; 1422.1763.

English name: White cedar

Vernacular/local name: Kashmiri: Sarva

Thuja orientalis is a beautiful oval or pyramid-shaped evergreen shrub 2–10 m tall. The bark is smooth, brownish, superficially fissured, with papery scales. The terminal shoots are divided into a spray of branchlets covered by dark green, closely appressed, acute, opposite and decussate leaves. Male cones are usually solitary and terminal in position, short-stalked and ovoid and bear 4–6 pairs of microsprophylls arranged decussately. Female cones are either solitary or arranged in groups of two or more, reddish-brown and bear 3–5 pairs of opposite and decussate, fused bract and ovuliferous scales. Seeds are wingless (Biswas and Johri, 2004).

Distribution: Local gardens and parks of Kashmir.

Specimens examined: University of Kashmir, Botanical garden, 1600 m by A.R. Dar (1983) and G.H. Dar; Kashmir University Campus, 1600 m by G.N. Javeid (1970).

Economic importance: *Thuja orientalis* is an exotic and is used as an ornamental plant in home gardens and parks.

12.5 *Taxodiaceae*

The family *Taxodiaceae* comprises nine genera and 16 species distributed widely in the temperate and subtropical regions of the world. In Kashmir, two exotic species belonging to two genera of *Taxodiaceae* are found.

12.5.1 *Cryptomeria japonica*

D. Don in Trans. Linn. Soc. London 18.167t 13.f.I 1841.

English name: Japanese cedar

Vernacular/local name: Hindi: Suji

Cryptomeria japonica is a large evergreen tree (25–30 m tall) with a conical crown and a straight, slender trunk 1–1.5 m in girth. Bark is reddish-brown to dark gray in colour, fibrous, superficially and linearly fissured and peeling off in strips. Leaves are bright bluish-green in colour, needle-like, spirally arranged in five ranks, acute-acuminate with margins entire. Male cones are in clusters of 8–20, ovoid or ovoid-ellipsoid, and bear many spiral microsporophylls. Female cones are solitary or occasionally aggregated, terminal, sessile, globose or subglobose, reddish-brown in colour and composed of 20–30 bracts and ovuliferous scales in 4–5 whorls. Seeds wingless or with rudimentary wings.

Distribution: Srinagar

Specimens examined: Botanical garden, University of Kashmir, 1600 m by A.R. Dar and G.H. Dar (2002); A.R. Naqash and G.N. Dar, 1600 m (1979).

Economic importance: Wood is light, fragrant and fine-grained. It is often used for construction of buildings and furniture in Japan. The wood is highly rot-resistant and easily worked, and is used for construction of buildings, bridges, furniture, ships, lamp posts, utensils, and paper manufacture. The essential oil and/or a resin from the plant is depurative and used for the treatment of gonorrhoea. The leaves are aromatic and are used as incense sticks.

12.5.2 *Sequoiadendron giganteum*

(Lindley) Buchholz in Am. J. Bot. 1939, xxxvi 536.

English name: Giant redwood

Sequoiadendron giganteum is the lone living species of the genus *Sequoiadendron* and occurs naturally only in groves on the western slopes of the Sierra Nevada mountains of California. The *Sequoiadendron* is the most massive and tallest known tree, reaching more than 100 m in height and 11–12 m in girth. In Kashmir, it is represented by a single tree that grows in Yarikah Drug Farm, Tangmarg. It is a large evergreen tree (22–25 m tall) with a trunk of 2–3 m in girth. The bark is dark-brown in colour, spongy and furrowed. The leaves are evergreen, awl-shaped, acute, margins entire, appressed at the base and spreading towards the tip. Both male and female cones are solitary and terminal in position and oval-ellipsoidal in shape. Female cones contain many spirals, bracts and ovuliferous scales.

Distribution: Tangmarg.

Specimens examined: Yarikhah, Tangmarg, 2154 m by G.L. Dhar (1975).

Economic importance: The wood is not durable, as it is fibrous, brittle and highly resistant to decay, therefore unsuitable for construction. It is an avenue tree and is a very popular ornamental.

12.6 *Taxaceae*

The family *Taxaceae* is represented by two species in Kashmir. *Taxus wallichiania* and *Taxus baccata.*

12.6.1 *Taxus wallichiania*

Zucc. In Abhandl. Bayer. Acad. Classe. Math. Phys. 3: 805. T. 5. 1843.

Taxus bacata Linn.

English name: Himalayan yew.

Vernacular/local name: Kashmiri: Pastul; Hindi: birmi.

Himalayan Yew is a medium-sized evergreen coniferous tree (10–20 m tall) with fluted stem. Bark is thin, smooth and reddish-brown. The leaves are linear, flattened, slightly sickle-shaped, coriaceous, distichous, dark green and shining above, pale or rusty below.

Male cones are solitary, axillary and subglobose. Female cones are solitary, few imbricate scales around an erect ovule. A membranous, cup-shaped and bright red disc surrounds the ovule at its base, which is succulent. Seeds olive-green.

Distribution: Lidder Valley, Jehlum Valley, Uri, Lolab Valley, Guraiz.

Specimens examined: Chitragul forests, Anantnag, 1900 m and Gulmarg forest slopes, 2800 m by A.R. Dar (2002); Limber, Uri forests by Zahid A. Dar (2002).

Economic importance: Wood is used for cabinet work and for making ploughs and fancy articles, such as knife handles and cutlery. A well-known anti-cancer drug, taxol, is obtained from *Taxus*. *Taxus wallichiana* has medicinal importance. The tincture prepared from the young shoots is used for the treatment of headache, dizziness and diarrhoea. It is also used to treat fever and muscular pain (Schippmann, 2001).

12.6.2 *Taxus baccata*

Linn. Sp. Pl. 1040. 1753.

English name: European yew

Vernacular/local name: Kashmiri: Bermi; Hindi: Thunner.

Taxus baccata is a small to medium-sized evergreen tree (10–20 m tall) with a trunk up to 2 m in girth. The bark is thin, dark reddish-brown, often scaly, coming off in small flakes or peelings. The leaves are lanceolate, spirally arranged on the stem, dark green and shiny above, pale green below. The leaves are highly poisonous. The male cones are globose and small. Female cones are partly surrounded by a modified scale, developing into a soft, bright red berry-like structure called an aril, which is gelatinous and very sweet in taste. Each cone contains only one seed.

Distribution: Srinagar, Anantnag.

Specimens examined: Botanical garden, University of Kashmir, 1600 m by A.R. Dar (2003).

Economic importance: Wood is durable, hard, elastic, takes a good finish, and is used in cabinet making. It is also used for making tool handles, ploughs and bows. It is also used as firewood. The plant shoots contain Taxol, a potential anti-cancer drug, used particularly in the treatment of ovarian cancers. All parts of the plant, except the fleshy fruit, are antispasmodic, cardiotonic, and diaphoretic, emmenagogue, expectorant, narcotic and purgative (Duke, 1992). The leaves are used in the treatment of asthma, bronchitis, hiccups, indigestion, rheumatism and epilepsy. The fruit is relished by thrushes.

12.7 Ginkgoaceae

The Ginkgoaceae, a family of gymnosperms, appeared during the Mesozoic Era, of which the only extant representative is *Ginkgo biloba*. In Kashmir, it is exotic and grown in some gardens.

12.7.1 *Ginkgo biloba*

Linn. Mant. Pl. 2: 313. 1771.

English name: Maiden hair tree, Silver fruit.

Vernacular/local name: Chinese: yin-kuo

Ginkgos are large trees (20–35 m tall) with trunk 1.5 m in girth, having long and angular crown. Young trees are often tall, slender and sparsely branched. The crown becomes broader as the tree ages. The leaves are fan-shaped with veins radiating out into the leaf blade, sometimes bifurcating, variously lobed, and venation is conspicuously dichotomous. The leaves are bright green in summer and golden yellow in autumn. Male cones are pendant, catkin like and borne on short shoots in the axil of leaves. Each cone comprises 40–50 microsporophyll. Female cones are borne in groups at the apex of the dwarf shoot.

Distribution: Gardens of Srinagar.

Specimens examined: Lal Mandi Floriculture park, Srinagar, 1600 m by R.A. Qazi, M.Y. Baba, N.A. Dar; Botanical garden, Chesma Shahi, Srinagar, 1700 m by A.R. Dar (2003).

Economic importance: Wood is light, brittle and limited in supply. In China and Japan it is used as firewood and for making chess boards. It is an important Chinese herbal medicine and can be traced back 5000 years. It has immense medicinal properities and is widely used in herbal medicine. Tea made from the leaves is used for memory loss (Le Bars *et al.*, 1997). The leaf extract is used in Europe to reduce symptoms of cognitive disorders due to its antioxidant properties (Diamond *et al.*, 2000). These extracts are given to the persons suffering from cerebrovascular and peripheral circulatory problems. Oedema is completely prevented by the administration of *Ginkgo* extracts (Kleijnen and Knipschild, 1992). The female cones are used to treat respiratory disorders. The seeds of *Ginkgo biloba* are roasted and eaten by the people of China and Japan.

12.8 Ephedraceae

The family Ephedraceae is represented by one species, *Ephedra gerardiana*, in Kashmir.

12.8.1 *Ephedra gerardiana*

Wall.ex Stapf. I.C. 75 emand Florin, Kungl. Sv. Vetensk. Handl. Ser. 3, 12 (1): 21, 1933.

English name: joint pine, joint fir.

Vernacular/local name: Kashmiri: Asmani bhutti; Hindi: Somalata; Chinese: Tse.

Epherda giardiadiana is a perennial dense, tufted, evergreen shrubby plant (20–120 cm tall). It is nearly leafless and has slender, cylindrical, yellow-green branches arising from a woody base and underground runners. Male cones are solitary or in groups of 2–3. Female cones are solitary, sessile or shortly pedunculate, subglobose and red, resembling berries at maturity.

Distribution: Uri, Upper Lidder Valley

Specimens examined: Noorkhan, Uri, Sedimentary rock cervices, 1450 m by A.R. Dar (2002) and Z.S. Khan (2002).

Economic importance: Wood is used locally as a fuel. It is grazed by sheep and goats. An important drug, ephedrine, obtained from *Ephedra geredиana*, has been widely used for the treatment of cold, sinusitis, hay fever, bronchial asthma, allergies and rheumatism. Fruits of *Ephedra gerediana* are edible and sweet. They are effective in hepatic diseases, used as a blood purifier and for cleaning of teeth. The juice of the berries cures respiratory infections.

12.9 Conclusion

This chapter summarizes that gymnosperms are predominant in the Kashmir Valley. Although only a few species of gymnosperms exist in the Kashmir Himalayas, they constitute the major portion of the floristric diversity. Due to their vast socio-economic importance, they have been referred to as the green gold of the state of Jammu and Kashmir. In the Kashmir Himalayas, gymnosperms are represented by 20 species belonging to the five families. Most of these dominate the wild forests and a few species are exotic and are cultivated in gardens and parks. A substantial single and gigantic tree, sequoia (*Sequoiadendron giganteum*) growing at Yarikhah Drug Farm (Tangmarg) in Kashmir Valley is the lone representative of this state tree of California in the Indian subcontinent. Besides providing firewood and timber for various construction purposes, the gymnosperms are the repository of a vast array of medicines for mankind. Gymnosperms harbour a rich diversity of flora and fauna under their canopies. Due to their immense importance, they have been over-exploited by the general population. Sustainable management, along with conservation, are very important: anthropogenic activities should be checked and the stake holders should be educated about the proper harvesting of gymnosperm flora for different uses.

References

Biswas, C. and Johri, B.M. (2004) *The Gymnosperms.* Narosa, New Delhi, India.

Dallimore, W., Jackson, A.B. and Harrison, S.G. (1967) *A Handbook of Coniferae and Ginkgoaceae*, 4th edn. St Martin's Press, New York, USA.

Dar, A.R. and Dar, G.H. (2006a) The wealth of Kashmir Himalaya – Gymnosperms. *Asian Journal of Plant Sciences* 5(2), 251–259.

Dar, A.R. and Dar, G.H. (2006b) Taxonomic appraisal of conifers of Kashmir Himalaya. *Pakistan Journal of Biological Sciences* 9(5), 859–867.

Dar, G.H. and Christensen, K.I. (2003) Gymnosperms of the western Himalaya. The Genus Juniperus (Cupressaceae). *Pakistan Journal of Botany* 35, 283–311.

Dar, G.H. and Khuroo, A.A. (2013) Floristic diversity in the Kashmir Himalaya: progress, problems and prospects. *Sains Malaysiana* 42(10), 1377–1386. Available at: www.ukm.my/jsm/pdf_files/SM-PDF-42-10-2013/05%20G.H.%20Dar.pdf.

Dar, G.H., Bhagat, R.C. and Khan, M.A. (2002) *Biodiversity of the Kashmir Himalaya.* Valley Book House, Srinagar, Jammu and Kashmir, India.

Dhar, G.L. (1966) The distribution of coniferales in India Part I. Pinaceae. *Kashmir Science* 3, 33–42.

Dhar, G.L. (1975). Sequuiadendron giganteum – A report from Kashmir. *Indian Forester* 101, 562–564.

Dhar, U. (1978) Phytogeographic studies on the Alpine flora of Kashmir Himalaya. PhD thesis, University of Kashmir, Srinagar, Jammu and Kashmir, India.

Diamond, B.J., Schiffet, S.C., Feiwel, N., Matheis, R.J., Noskin, O., Richards, J.A. and Schoenberger, N.E. (2000) Ginkgo Biloba extract: mechanism and clinical indications. *Archives of Physical Medicine and Rehabilitation* 81, 669–678.

Duke, J.A. (1992) *Handbook of Phytochemical Constituents of GRAS Herbs and Other Economic Plants.* CRC Press, Boca Raton, Florida, USA.

Hooker, J.D. (1888) *The Flora of British India.* Vol. 5. L. Reeve and Co. Ltd, London, UK.

Javeid, G.N. (1970) Flora of Srinagar, a phytogeographic and taxonomic study of the flowering plants of Srinagar. Vol II. PhD thesis, Kashmir University, Srinagar, Jammu and Kashmir, India.

Javeid, G.N. (1979) Forest flora of Kashmir, a check list-II. *Indian Forester*, 105(2), 148–170.

Kleijnen, J. and Knipschild, P. (1992) Ginkgo biloba for cerebral insufficiency. *British Journal of Clinical Pharmacology* 34, 352–358. DOI: 10.1111/j.1365-2125.1992.tb05642.x PMID:1457269.

Lambert, W.J. (1933) List of trees and shrubs for the Kashmir and Jammu forest circles, Jammu and Kashmir State. *Forest Bulletin* 80, Dehradun, India.

Lawrence, W.R. (1895) *The Valley of Kashmir*. Chinar Publishing House, Srinagar, Jammu and Kashmir, India.

Le Bars, P.L., Katz, M.M., Berman, N., Itil, T.M., Freedman, A.M. and Schatzberg, A.F. (1997) A placebo-controlled, double-blind, randomized trial of an extract of Ginkgo biloba for dementia. North American EGb Study Group. *Journal of the American Medical Association* 278(16), 1327–1332. DOI: 10.1001/jama.1997.03550160047037 PMID:9343463.

Lone, V. (2013) Tree diversity and economic importance of forest trees of Kashmir (Jammu and Kashmir), India. *International Journal of Fundamental and Applied Sciences* 2(4), 56–63.

Puri, G.S. (1943) The occurrence of Woodfordia fruticosa (L.) S. Kurz in the Karewa deposits of Kashmir with remarks on changes of altitude and climate during the Pleistocene. *Journal of Indian Botanical Society* 22, 125–131.

Puri, G.S. (1947) Fossil plants and the Himalayan uplift. *Journal of Indian Botanical Society* (M.O.P. Iyenger Commemoration Vol.) 25, 167–184.

Sahni, K.C. (1990) *Gymnosperms of India and Adjacent Countries*. Bishen Singh and Mahendra Pal Singh, Dehradun, India.

Schippmann, U. (2001) *CITES Medicinal Plants Significant Trade Strategy*. Project S 109, German Federal Agency for Nature Conservation, Bonn, Germany.

Singh, G. and Kachroo, P. (1976) *Forest Flora of Srinagar and Plants of Neighbourhood*. Bishen Singh and Mahendra Pal Singh, Dehradun, India.

Stewart, R.R. (1972) *An Annotated Catalogue of the Vascular Plants of West Pakistan and Kashmir*. Fakhri Press Karachi, Pakistan.

Stewart, R.R. (1979) The first plant collectors in Kashmir and the Punjab. *Taxon* 28, 51–61.

Vigne, G.T. (1842) *Travels in Kashmir, Ladak and Iskardo (1835–1839)*. Henry Colburn, London, UK.

Vishnu-Mittre (1963) Oaks in the Kashmir Valley with remarks on their history. *Grana Palynologica* 4, 306–312.

13 Diversity of Plant Parasitic Nematodes in Pulses

TARIQUE HASSAN ASKARY*

Division of Entomology, Sher-e-Kashmir University of Agricultural Sciences and Technology, Srinagar, Jammu and Kashmir, India

Abstract

Pulses, such as chickpea (*Cicer arietinum*), pigeonpea (*cajanus cajan*), common bean (*Phaseolus vulgaris*), mung bean (*Vigna radiata*), urd bean (*Vigna mungo*) and lentil (*Lens culinaris*), are an excellent source of dietary protein as well as forming part of a cholesterol-free diet for millions of people around the world. Numerous plant parasitic nematodes attack pulse crops and the prominent among them are *Meloidogyne* spp., *Heterodera* spp. and *Paratylenchus* spp., the endoparasites, *Rotylenchulus* spp., the semi-endoparasites, and *Tylenchorhynchus* spp. and *Helicotylenchus* spp., the ectoparasites. These nematodes have diverse methods of attack. *Meloidogyne* causes galling on roots, accompanied by a change in cell morphology, leading to the formation of giant cells in the cortical region of root upon which they feed; *Heterodera* form syncytia in the steler region and have a pearly appearance on the root; *Pratylenchus* form necrotic lesions on the root; *Rotylenchulus reniformis* cause dirty root disease and the mature females attached to the roots have a kidney-like appearance (seen under a microscope), hence the name *reniformis*. *Tylenchorhynchus* and *Helicotylenchus* are ectoparasites, but are considered of lesser importance than endoparasites or semi-endoparasites. They cause mechanical injury by feeding primarily on the epidermal cells of roots. Host range, length of life cycle and soil temperature and moisture for survival vary with nematode genera and species. Economic Threshold Level (ETL) also varies from crop to crop, as for chickpea it is estimated 1–2 J_2 of *M. incognita*/cm^3 of soil or 0.031 *P. thornei*/cm^3 of soil or 1 egg of *H. ciceri*/cm^3 of soil or 1 premature female of *R. reniformis*/gm of soil, whereas in the case of pigeonpea it is 1 J_2 of *M. javanica*/gm of soil or 1–2 J_2 of *H. cajani*/gm of soil or 1.4 premature female of *R. reniformis*/gm of soil. During parasitism, nematodes have an adverse effect on the normal physiology, growth and development of the host plant. Ultimate harm caused to pulse crops are in the form of reduction in yield and quality. Average yield losses caused due to plant parasitic nematodes on a worldwide basis have been estimated as 13.7% in chickpea, 13.2% in pigeonpea and 10.9% in common bean. Different approaches for the management of these nematode pests include cultural, biological, botanical, host plant resistance and chemicals. These approaches need to be tested by research workers to develop integrated nematode management (INM) strategies at farm level in order to achieve maximum yield and better quality of crop.

13.1 Introduction

Pulses are an excellent source of dietary protein for millions of people. They are a nutritious feed for livestock and a cholesterol-free food for all kinds of consumers. They are rich in calcium, iron and some essential amino acids. Supplementation of cereals with pulses provides the best solution to alleviate protein-calorie malnutrition. Pulse crops have the capacity to fix huge amounts of nitrogen through symbiosis and thus minimize dependency on inorganic fertilizers. That is why growing pulses in cereal-based cropping systems has been viewed as a component of integrated nutrient management.

Plant parasitic nematodes are one of the main biotic constraints in reducing the quantity and quality of pulse crops. Due to their microscopic size, hidden habitats and lack of visible symptoms on aerial plant parts, they are often considered as a hidden enemy of crops. A range of parasitic nematodes attack pulse crops, however: root-knot nematode,

*E-mail: tariq_askary@rediffmail.com

Meloidogyne spp., is predominant and widespread throughout the pulse-growing regions of the world. The two species of root-knot nematode, *Meloidogyne incognita* and *M. Javanica*, are the causal agent of root-knot disease in chickpea, pigeonpea, lentil, common bean, field pea, mung bean, urd bean and also some of the minor pulses like lathyrus, horse gram and rice bean (Ali and Askary, 2005). Besides *Meloidogyne*, other important endoparasitic and semi-endoparasitic nematodes of pulses are root-lesion nematode, *Pratylenchus* spp., reniform nematode, *Rotylenchulus* spp., cyst nematode, *Heterodera* spp. Among the ectoparasitic nematodes, stunt nematode, *Tylenchorhynchus* spp. and spiral nematode, *Helicotylenchus* spp. infect pulse crops (Ali and Askary, 2001a), but in most cases these ectoparasites are considered much less significant than endoparasitic nematodes. Nematodes cause infection by piercing the host plant tissue, which leads to mechanical damage. This injury to the plant provides entry site for other pathogens, such as bacteria and fungi, which in many cases results in complex diseases. During the process of feeding by nematodes, salivary juices are injected into the host plant, which results in hydrolysis of host components and altered host metabolism. Root-knot nematode induces the formation of giant cells in the cortical region, accompanied by cell wall lyis, cellular hypertrophy and damage to the cells of epidermis, cortical and vascular regions (Sankaranarayanan and Hari, 2013).

Vascular disorders take place as the conductive tissues responsible for translocation of water and nutrients are blocked at the site of infestation. Thus, the infested plant is deprived of nutrients and water, necessary for its growth and maintenance. The nematode infestation also reduces root penetration of the soil profile, increasing the negative impact that moisture stress exerts on plant health (Sikora and Greco, 1990). Other disorders observed in pulse crops are reduction in rhizobium root nodulation and nitrogen-fixing activities.

Average yield losses caused by plant parasitic nematodes on a worldwide basis are estimated at 13.7% in chickpea, 10.9% in common bean and 13.2% in pigeonpea (Sasser and Freckman, 1987). India is the largest producer of pulses in the world and the avoidable yield losses due to root-knot nematodes in pulses range from 20 to 35% (Gaur *et al.*, 2001). In India, pigeonpea cyst nematode, *Heterodera cajani*, causes an estimate yield loss of 16–30% in different pulse crops (Saxena and Reddy, 1987; Gaur *et al.*, 2001), while reniform nematode *Rotylenchulus* sp. causes 23% yield loss in mung bean, 14–29% in pigeonpea and 11% in chickpea (Gaur *et al.*, 2001).

In this chapter, I have tried to deal with the important plant parasitic nematodes of pulses, their mode of infection, highlighting the diversity of attack on pulse crops. Under each section, while describing a particular nematode, I have briefly reviewed their management strategies adopted by different workers.

13.2 Chickpea/Bengal gram (*Cicer arietinum*)

Chickpea (*Cicer arietinum*), an ancient pulse crop, was first grown in Turkey in about 7,000 BC and is now traditionally grown in the semi-arid zones of India and in the Mediterranean region. It is an important pulse crop, constituting a major source of protein for the large vegetarian population. The major producers of chickpea in the world are India, Pakistan, Turkey, Australia and Mexico.

13.2.1 Nematode pests of chickpea

Chickpea is one of the most important pulse crops of the world, and can be attacked by a range of nematode pests. Plant parasitic nematodes are reported to cause 13.7% yield losses in chickpea (Greco, 1987; Sharma and McDonald, 1990; Abd-Elgawad and Askary, 2015). Around 100 species of plant parasitic nematodes are reported to be associated with chickpea, of which root-knot nematode (*Meloidogyne* spp.), lesion nematode (*Pratylenchus* spp.), reniform nematode (*Rotylenchulus* spp.), cyst nematode (*Heterodera* spp.), spiral nematode (*Helicotylenchus* spp.), stunt nematode (*Tylenchorhynchus* spp.) and lance nematode (*Hoplolaimus* spp.) are common. Besides these, needle nematode (*Longidorus* spp.) and dagger nematode (*Xiphinema* spp.) act as a vector in transmitting viruses to the plant.

13.2.1.1 Occurrence and distribution

Surveys conducted in various countries of the world have revealed evidence of the role of nematodes in chickpea crop damage. However, on a global basis, the most severe problem is caused by endoparasites: root-knot nematode (*Meloidogyne* spp.), cyst nematode (*Heterodera* spp.) and reniform nematode (*Rotylenchulus* spp.). *Helicotylenchus* spp. and *Tylenchorhynchus* spp. are ectoparasites and generally epidermal feeders. Greco *et al.* (1988) reported association of *Heterodera ciceri* with chickpea, causing yield loss. Maqbool (1980) reported association of *Meloidogyne* and *Heterodera* with chickpea in Pakistan; Yassin (1987) reported association of *Aphelenchus avenae*, *Aphelenchoides graminis* and *Ditylenchus myceliophagus* with chickpea in Sudan; Hashim (1979) reported *Pratylenchus thornei* in the rhizosphere of chickpea from Jordan; Di Vito *et al.* (1994a) reported *M. artiellia*, *P. mediterraneus*, *P. thornei*, *P. penetrans*, *Helicotylenchus* spp., *Tylenchus* spp., *Tylenchorhynchus* spp., *Rotylenchulus* spp., *Rotylenchus* spp., *Hoplolaimus* spp. and *Paralongidorus* spp. from the rhizosphere of chickpea in Tunisia. According to Ali (1995), out of 64 nematode species recorded on chickpea from India, 57 belong to the order Tylenchida, five belong to the order Dorylaimida and two to the order Aphelenchina. Ali and Sharma (2003) conducted a random survey in chickpea growing areas of Rajasthan. During the survey, it was observed that *M. incognita*, *M. javanica, Pratylenchus thornei* and *Pratylenchus* spp. were the predominant nematodes infesting chickpea in most of the areas. The cyst nematode *Heterodera swarupi* was reported for the first time on chickpea in some of the areas surveyed.

13.2.2 Root-knot nematode (*Meloidogyne* spp.)

Four species of root-knot nematode: *Meloidogyne arenaria* (Neal) Chitwood, *M. artiella* (Franklin), *M. incognita* (Kofoid and White) Chitwood and *M. javanica* (Treub) Chitwood. are recognized as the most important pests for chickpea in Brazil, Ethopia, Ghana, Italy, India, Malawi, Nepal, Pakistan, Spain, Syria, USA, Zambia and Zimbabwe (Ali, 1995). However, *M. arenaria* is reported on chickpea from India only (Mathur *et al.*, 1969). Severe infestations on chickpea caused by *M. artiella* has been reported from Italy (Greco, 1984), Spain (Tobar-Jiménez, 1973), Egypt (Oteifa, 1987) and Syria (Mamluk *et al.*, 1983; Greco *et al.*, 1984). *M. javanica* causes severe problems to chickpea in Malawi, Zimbabwe, Brazil, Nepal, India and Pakistan, whereas *M. incognita* is one of the major problems for the chickpea crop in India, Pakistan, Brazil, Ethopia, Bangladesh and Nepal (Sharma and McDonald, 1990; Ali, 1993).

13.2.2.1 Economic importance and threshold level

Under field conditions, a study conducted by Ali (2009) reported 25.6% chickpea yield loss in a field heavily infested with *M. javanica*, as compared to a non-infested field. Chickpea plants infested with nematodes bear seeds with reduced grain protein content (Greco and Sharma, 1990). Rehman *et al.* (2012) inoculated chickpea cv. Avarodhi with 1000 second-stage juveniles (J_2) of *M. incognita*. The results showed the highest reduction in plant growth characters such as shoot and root length, fresh and dry weight of plant, number of flowers and pods, total chlorophyll content and nitrate reductase activity. The root-knot index also increased in the nematode-treated plants. A reduction in shoot length and plant weight was observed when 1 J_2/gm soil was used to inoculate chickpea seedlings grown in pots (Ahmad and Husain, 1988). Mani and Sethi (1984) reported 2 J_2/gm soil being the damaging threshold level in chickpea cv. Pusa 209. However, according to Sharma and McDonald (1990), the damage threshold of *M. incognita* on chickpea ranges between 1 to 2 J_2/gm or cm^3 of soil.

13.2.2.2 Symptoms and nature of damage

The common symptoms which can be observed under field conditions are poor and uneven growth of chickpea plants in patches (Fig. 13.1). Heavy infestations lead to stunted growth of plants with lesser branches and pale green leaves. Below-ground symptoms can be seen on roots bearing knots or galls (uneven swellings). Galls are produced mostly on root tips and on their vicinity, but all along the roots the beaded appearance of multiple galls (the result of coalescing of adjacent galls) is also common (Fig. 13.1). Plants which are heavily galled show wilting symptoms under field conditions. A severe infestation by root-knot nematode finally results in poor pod formation with deformed seeds that are less developed, and smaller in size and weight (Ali, 1995). The roots of infested plants bear fewer rhizobium nodules in comparison to healthy ones. Interestingly, bacterial nodules of heavily galled root were also found infested with root-knot nematode.

Fig. 13.1. Sparse, uneven stunted growth of chickpea infested with root-knot nematode.

13.2.2.3 Biology and life cycle

The second-stage juveniles (J_2) carry the infection, which moves towards the root tip and invades roots, penetrates the epidermal cells and cortex, becomes sedentary, and moults three times. The female becomes saccate, whereas the male remains vermiform and free living. By pressing against surrounding cells, the nematodes interrupt the function of xylem and phloem and translocation of water and nutrients to the top of plant are blocked. Females establish themselves inside the vascular region and induce formation of giant cells, on which they feed. Giant cells usually develop in pericycle, phloem parenchyma, cambium and xylem parenchyma. As the nematode reaches maturity, giant cells show a gradual increase of protein in their cytoplasm, but the protein decreases during degeneration of the giant cells. DNA is concentrated in the nuclei of parenchyma, adjacent to the giant cell and nematodes, where RNA and ascorbic acid granules are mainly localized (Ali, 1995). Gall formation takes place within 48 hours after infection. Adjacent galls coalesce and more than one nematode may be embedded in the same gall. The size of galls is influenced by soil temperature and susceptibility of chickpea genotypes (Ali, 1995). Primary galls show a decrease in insoluble polysaccharides including starch. Lipid contents were more in galled roots. As the size of galls increase, their lipid contents also increase (Sarna, 1984). On average 200–500 eggs are laid in a gelatinous matrix by fertilized females. At an optimum temperature of 25–30°C, *M. incognita* and *M. javanica* complete their life cycle in 27–32 days, and more than one life cycle is completed by nematode during a cropping season. However, in the case of *M. artiella*, eggs after hatching require a cool temperature for complete development (Greco, 1987); hence, only one generation in a year is usual (Tobar-Jiménez, 1973).

13.2.2.4 Interaction with other micro-organisms

Nematodes not only attack the plants directly, but also interact with other soil-borne micro-organisms such as fungi and bacteria and under such conditions cause considerable damage to devastate the crop. However, interactions between plant-parasitic nematodes and fungi can vary considerably over plant genotypes, cultivars and lines (Uma Maheswari *et al.*, 1997; Castillo *et al.*, 2003). Khan and Hosseini-Nejad (1991) reported that wilting in chickpea cvs. Pusa 212, Pusa 240, Pusa 209 and Pusa 261 was caused by fungus *Fusarium oxysporum* and aggravated in the presence of root-knot nematode *M. javanica. M. javanica* was found to cause injury to the roots, which facilitated the entry of *F. oxysporum* f. sp. *ciceri* in wilt-resistant chickpea cv. Avarodhi, thus making the plant susceptible to the fungus (Ali and Gurha, 1995). In India, several researchers have found by experiments that

M. incognita or *M. Javanica*, in presence of the wilt fungus *F. oxysporum* f. sp. *Ciceri*, can break down resistance in wilt-resistant genotypes of chickpea (Mani and Sethi, 1987; Uma Maheswari *et al.*, 1995; Krishna Rao and Krishnappa, 1996; Uma Maheswari *et al.*, 1997). A deleterious and synergistic effect on root and shoot length as well as on wilt incidence of chickpea was recorded when both the pathogens were inoculated simultaneously or sequentially. Palomares-Rius *et al.* (2011) revealed key aspects of chickpea *Fusarium oxysporum* f. sp. *ciceris* race 5 and *M. artiellia* interactions studying the fungal- and nematode-induced changes in root proteins, using chickpea lines CA 336.14.3.0 and ICC 14216 K. These two chickpea lines were resistant to *Fusarium oxysporum* f. sp. *ciceris* race 5 but susceptible to *M. artiellia*. However, in the presence of the nematode alone or both the pathogens, the two chickpea lines showed differential responses. *F. solani* caused black root-rot disease in chickpea. *M. javanica* in combination with *F. solani* increased severity of the disease, with significant reduction in the fresh shoot and number of nodules in plants (Dalal and Bhatti, 1985). Another fungal pathogen, *Rhizoctonia bataticola*, which is the causal agent of dry root rot disease in chickpea, when inoculated with *M. javanica* in the rhizosphere of chickpea seedlings reduced plant growth parameters more than a single inoculation of any of the pathogens (Goel and Gupta, 1985). Interaction between *M. javanica*) pathotype 1 and *F. oxysporum* f. sp. *ciceri* was studied on chickpea cv. Dahod Yellow. Simultaneous inoculation of both the organisms resulted in maximum decrease in plant growth characters. The wilting was 40% in the chickpea plant when fungus was inoculated alone; however, the disease severity increased to 75% in the presence of both the organisms (Patel *et al.*, 2000).

Chickpea is a leguminous crop and therefore has the capacity to fix nitrogen in association with rhizobia by using solar energy collected through the process of photosynthesis. Infestation of root-knot nematodes on the chickpea plant causes a significant reduction in rhizobium root nodulation, thereby causing indirect damage to plants (Ali, 1995). Upadhyay and Dwivedi (1987) found greatest reduction in root nodulation when chickpea cv. K 850 was inoculated with *M. javanica* @ 500 J_2/plant. Similar was the observation of Mani and Sethi (1984) when they inoculated *M. incognita* @ 0.5, 1.0, 2.0, 4.0 or 8.0 J_2/g soil to chickpea cv. 207. There was an adverse affect on rhizobium nodulation of chickpea at all the inoculum levels of nematode used in the experiment. In an experiment conducted in greenhouse conditions, it was observed that *M. incognita* infected more than 25% of nodules induced by Mesorhizobium in chickpea cv. UC 648 (Vovlas *et al.*, 1998).

13.2.2.5 Management

Nematode problems are reported on chickpea crops from different parts of the world. There are different approaches to manage root-knot nematode, including cultural, biological, botanical, host-plant resistance and chemicals. These approaches are being tested by research workers to develop integrated nematode management (INM) strategies at farm level.

Cultural

Cultural practices are one of the effective methods of suppressing root-knot nematode populations in soil. An effective control of root-knot nematode on chickpea can be obtained by the inclusion of cereal or grasses in the cropping system. Crop rotation for 2–3 years with a non-host crop like sesame, mustard and winter cereals may be useful in reducing the population of root-knot nematodes *M. incognita* and *M. javanca* in chickpea (Sharma *et al.*, 1992). Weeds are often an excellent host for root-knot nematode and, therefore, while adopting a rotation programme under non-host or fallow conditions, measures should also be adopted to control weeds (Sikora and Greco, 1990).

Biological

Biological control aims to maintain the pest population below the economic threshold level, rather than eliminating them as done by chemicals. It is an eco-friendly approach, wherein beneficial microbes are used as biological nematicides. These beneficial microbes are also called biocontrol agents. (Askary and Martinelli, 2015). Technically, biocontrol is the reduction in pest population accomplished through the introduction of antagonists or manipulation of the environment to make it congenial for the activity of naturally occurring antagonists. In the last two decades, fungal biocontrol agents such as *Aspergillus niger*, *Trichoderma harzianum*, *Paecilomyces lilacinus* and *Pochonia chlamydosporia* have been used commercially and very successfully against endoparasitic nematodes

(Askary, 2015a). Hussain *et al.* (2001) tested ten cultivars of chickpea against *M. javanica* using filtrates of fungal biocontrol agents *Aspergillus niger*, *A. flavus*, *A. temarii*, *A. nidulans*, *A. terreus* and *A. fumigatus*. The most promising results, in terms of significantly reducing the number of galls per root system, was obtained with *A. niger*, followed by *A. nidulans*, *A. flavus*, *A. terreus* and *A. fumigates*, as compared with the untreated control. Pant *et al.* (2004) evaluated some biocontrol agents: *Paecilomyces lilacinus*, *Trichoderma harzianum*, *A. niger* and *Glomus fasciculatum* (VAM fungus), at different concentrations, for the management of *M. incognita*. The best performance was shown by *Glomus fasciculatum* (VAM fungus) followed by *P. lilacinus*, *T. harzianum* and *A. niger*. Siddiqui and Akhtar (2009) conducted an experiment to evaluate plant-growth-promoting rhizobacteria (*Pseudomonas putida* MTCC No. 3604 and *Pseudomonas alcaligenes* MTCC No. 493) and fungal biocontrol agents (*Pochonia chlamydosporia* KIA and *Paecilomyces lilacinus* KIA) alone and in combination with *Rhizobium* sp. (charcoal commercial culture) on the growth of chickpea in presence of *M. javanica*. Individual application of rhizobacteria or fungal biocontrol agents increased the plant growth characters and reduced the multiplication of nematodes; however, the combined application (*P. lilacinus* KIA + *Rhizobium*) was superior to all the other treatments in reducing the root galls and nematode multiplication.

Botanical

Botanicals are materials or products made or derived from plants. Botanicals for the management of nematodes are observed to be easy in application, free from environmental pollution, and improve soil health structurally and nutritionally (Mishra, 2007). A neem seed coating on chickpea seeds before sowing resulted in a decrease in root-knot nematode *M. incognita* infecting chickpea (Mojumder and Mishra, 1991). A seed dressing with the latex of *Calotropis procera* has been found to have potential for the management of *M. incognita*. The treatment also improved the plant growth characters such as root length, number of pods per plant, chlorophyll content of leaves, water absorption capacity of root and rhizobium nodules of roots in chickpea cv. K 850 (Anver and Alam, 1992). Rehman *et al.* (2012) conducted a glasshouse experiment to control *M. incognita* on chickpea cv. Avarodhi by using different concentrations of leaf extract of Persian lilac, *Melia azedarach*. The plants treated with higher concentrations showed the least impact of *M. incognita*. In another experiment, Rehman *et al.* (2013) evaluated the effect of flower extracts of five plants: *C. procera*, *Tagetes erecta*, *Lantana camara*, *Thevetia peruviana* and *Nerium indicum* against *M. incognita* infecting chickpea cv. Avarodhi. All the treatments significantly reduced the infection of *M. incognita* in chickpea roots, but the highest potential impact was shown by *C. procera*.

Host plant resistance

Plant resistance is one of the most desirable components in nematode management. Growing resistant cultivars has the advantage of preventing nematode reproduction, making long-term rotation unnecessary: the crop can be grown on nematode-infested land. Hussain *et al.* (2001) tested ten cultivars of chickpea against *M. javanica*. According to the host response index, all the tested cultivars were moderately resistant. However, gall formation on the root system was lowest on cv. Nes 95004. Haseeb *et al.* (2006) evaluated 32 chickpea accessions for their resistance to *M. incognita* and *F. oxysporum* f. sp. *ciceris*. None of the cultivars were highly resistant to either of the pathogens; however, Phule G-00108, Phule G-00109, Phule G-94259, Phule G-96006 and PDG-84-16 were found resistant and Phule G-00110, Phule G-94091, H-82-2, IPCK-256, IPC-2001-02 and HR-00-299 were moderately resistant to both *M. incognita* and *F. oxysporum*.

Biochemical parameters, such as peroxidase level, which is an indicator of susceptibility or resistance in the host plant, has been exploited to identify nematode resistance in chickpea. An increase in the level of peroxidase shows a positive correlation in the degree of resistance to nematodes. During screening of 20 genotypes of chickpea, on the basis of increase in peroxidase activity, some tolerant genotypes of chickpea were identified: IC 4941, IC 4942 and IC 4944 (Siddiqui and Husain, 1992). Chawla and Pankaj (2007) reported increased peroxidase activity in root-knot nematode-resistant genotypes of chickpea.

Chemicals

Nematicides have been found to provide quick and demonstrable control of root-knot nematodes;

however, their application at higher doses are uneconomical; additionally, they are hazardous to the environment and cause residue problems. Despite having numerous adverse effects, nematicides are the most effective means of disease management (Johnson, 1985). In Zimbabwe, Phenamiphos 20 kg a.i/ha was found effective in reducing the population of root-knot nematode (Sharma *et al.*, 1992). Seed treatment with carbofuran and phenamiphos at 1,2 and 4% reduced the number of root galls on chickpea 42 days after sowing (Kaushik and Bajaj, 1981).

Integrated management approach

Management of nematodes require long-term protection and that is likely to be impractical and uneconomic; therefore, a better approach is integrated management (Askary, 2015b). Integration of two or more different approaches judiciously can stabilize target nematode population at acceptable levels, resulting in long-term socio-economic and eco-friendly consequences. Research workers have found a number of pesticides to be compatible with biocontrol agents and botanicals (Singh *et al.*, 2012; Tapwal *et al.*, 2012). Pandey *et al.* (2005) studied integrated effects of biological control agents and neem cake (alone and in combination) on *M. incognita* infesting chickpea cv. H-208. The treatments comprised neem cake, *Trichoderma harzianum*, *T. viride*, *Paecilomyces lilacinus*, *Aspergillus niger* and *Verticillium chlamydosporium*. The best treatment was a combined application of *P. lilacinus* and neem cake, which improved the plant growth characters and chlorophyll content, besides causing a reduction in root galling. An integrated management approach by seed treatment with different neem-based products and soil application of systemic nematicides indicated that neem seed powder @ 10% w/w along with carbofuran @ 1 Kg. a.i./ha was the most effective management package against *Meloidogyne incognita* infecting chickpea (Chakrabarti and Mishra, 2001). In a pot experiment, it was observed that combined application of leaf powder of *Cassia tora* and *P. lilacinus* successfully managed the root-knot nematode, suppressing the gall formation on the chickpea roots and nematode population in the soil (Azam *et al.*, 2009). In a chickpea field experiment, combined application of farmyard manure (FYM) (5 ton/ha), neem cake (1.5q/ha) and fungal biocontrol agents, *P. lilacinus* and *A. niger*, proved most effective in suppressing the soil population of root-knot nematode, *M. incognita* (Singh *et al.*, 2011).

13.2.3 Root-lesion Nematode (*Pratylenchus* spp.)

Root-lesion nematodes, *Pratylenchus* spp., are a major constraint in chickpea production all over the world. The nematode has a wide host range and can occur in almost any climate. Three species of root-lesion nematode: *P. penetrans*, *P. thornei* and *P. vulnus*, have been frequently noticed on the chickpea crop in the Mediterranean region (Ali, 1995). Greco *et al.* (1984) reported *P. thornei* from 74% of the root samples of chickpea from northern Syria collected during a survey. Di Vito *et al.* (1992) found infection of *Pratylenchus* spp. in 82% of the chickpea samples collected in the province of Aleppo and Idlib and 100% from Izrae, Es Seweldiye, Tartus, Latakia and north-east Syria. In Italy, *P. thornei* is considered as the most important plant parasitic nematode of chickpea (Greco, 1987). *Pratylenchus* spp. has also been reported as an important nematode problem in Zimbabwe and USA (Sharma and McDonald, 1990). In India, surveys were conducted to find out plant parasitic nematodes associated with chickpea in the Bundelkhand region of Uttar Pradesh and Madhya Pradesh. Altogether, 14 nematode genera: *Pratylenchus*, *Tylenchorhynchus*, *Hoplolaimus*, *Tylenchus*, *Helicotylenchus*, *Filenchus*, *Basiria*, *Aphelenchus*, *Rotylenchulus*, *Scutellonema*, *Boleodorus*, *Basiriolaimus*, *Hemicycliophora* and *Paratylenchus* were found associated with chickpea. *Pratylenchus* was the most prominent, with a prominence value of 424.4 (Singh and Jagadeeswaran, 2014).

13.2.3.1 Economic importance and threshold level

Walia and Seshadri (1985) reported a significant reduction in seed germination of chickpea when inoculated with *P. thornei* @ 4000/kg soil. In a field trial conducted in Syria, the threshold limit of *P. thornei* on chickpea was calculated to be 0.031/cm^3 soil. Yield loss of 58% in chickpea was recorded at 2 nematodes/cm^3 soil (Di Vito *et al.*, 1992).

13.2.3.2 Symptoms and nature of damage

Root-lesion nematodes are migratory endoparasites that cause dark brown to black necrotic lesions on the epidermal, cortical and endodermal cells of chickpea roots. With the passage of time, these necrotic spots coalesce, leading to the development of necrosis in the entire root. In young seedlings, the most visible above-ground symptoms observed in the field are stunted growth and pale green foliage, which becomes more prominent as the plant ages (Fig. 13.2).

13.2.3.3 Biology and life cycle

The root-lesion nematodes, *Pratylenchus* spp., are migratory endoparasites and the entire life cycle is completed inside the root. Reproduction may be sexual or parthenogenetic and the female lays eggs in the root cortex (Walia and Bajaj, 2003). The nematodes move between cells and inside cells, within the cortical tissues. Under suitable environmental conditions, the life cycle is completed in about one month. Four moultings take place: the first within the egg and three outside. All the life stages except the J_1 are parasitic (Askary *et al.*, 2012). *Pratylenchus* is an endoparasite nematode and therefore its population densities are typically much greater in plant roots than in the rhizosphere. However, initially the population of nematodes is greater in the soil, but during the vegetative stage of the crop, the nematode population builds up enormously and becomes much greater than the soil population. At the time of crop maturity, the nematode leaves the root and moves into the soil, thus again increasing the soil population (Walia and Bajaj, 2003). Under warm and dry soils, *P. thornei* is more active than in cool, moist soils with in-season rainfall or with irrigation (Castillo *et al.*, 1995). With an increase in soil temperature from 15 to 20°C, reproductive rate, hatching and penetration of *P. thornei* on chickpea roots also increased (Castillo *et al.*, 1996a, b). Reports from Israel (Glazer and Orion, 1983) and southern Spain (Talavera and Valor, 2000) revealed that *P. thornei* has the ability to survive under stress conditions, i.e. a dry fallow season of soil, by undergoing an anhydrobiotic state.

Fig. 13.2. Lesion nematode infesting chickpea roots: arrows show necrosis.

13.2.3.4 Interaction with other micro-organisms

Inoculation of wilt fungus *F. oxysporum* f. sp. *ciceri* @ 2g with *P. thornei* at 1000, 2000 and 3000 level resulted in significant damage to chickpea plants, reducing the shoot and root length and rhizobial nodulation. Moreover, combined inoculation not only increased the disease incidence but also severity of the disease, as compared to single inoculation of fungus. The wilt symptoms first appeared 60 days after inoculation and the multiplication of the nematode was adversely affected by the fungus (Devi, 1995). Tiyagi and Parveen (1992) reported a reduction in rhizobial nodules on chickpea plants as compared to healthy ones when inoculated with 500 *P. thornei*/kg soil. Baghel and Singh (2013) reported that the rhizobium nodules significantly decreased in chickpea plants inoculated with 10,000 *P. thornei*, but was statistically identical with 100 and 1000 inoculum level.

13.2.3.5 Management

Cultural

In Syria, *P. thornei* has been found to reproduce better on winter and summer crops and therefore a crop rotation programme is a good alternative to manage this nematode on chickpea (Greco and

Di Vito, 1987). There are some weeds which are good hosts for the root-lesion nematode, *Pratylenchus* spp., therefore weed management practices may be adopted so as to reduce the multiplication of nematode and subsequent damage to crops (Kornobis and Wolny, 1997; Vanstone and Russ, 2001).

Biological

Pseudomonas fluorescens, a bacterial biocontrol agent and *Trichoderma viride*, a fungal biocontrol agent were used as soil and seed treatment against *Pratylenchus thornei* infecting chickpea. When both the organisms were applied in combination @ 2.5 kg/ha as soil treatment or 5 g/kg seed, the soil population of nematode decreased by 34.8% and 22.3%, respectively. Yield of the host crop was increased from 25.0% to 29.5% over untreated control (Dwivedi *et al.*, 2008).

Botanical

Sebastian and Gupta (1996a) treated chickpea seeds with latex of *Calotropis procera*, *Euphorbia pulcherrima* and *Carica papaya*. The treated seeds were sown in field microplots infested with *P. thornei*. All three treatments resulted in bringing down the nematode population in soil. Also, the growth parameters of chickpea increased with all the treatments. In another experiment, Sebastian and Gupta (1996b) evaluated oil seed cakes of groundnut, linseed, mustard and *Azadirachta indica* for the control of *P. thornei* on chickpeas under field conditions. The results showed that mustard cake was most effective in reducing the root populations of *P. thornei* as well as significantly increasing the shoot and root mass of chickpea, *A. indica* cakes also proved effective in significantly increasing the shoot and root mass of chickpea.

Host plant resistance

Tiwari *et al.* (1992) tested chickpea accessions against root-lesion nematode *P. thornei*. The chickpea lines highly resistant to *P. thornei* were GNG 543, GF 88428 and PKG-24. The nematode was observed confined to the cortex of the plant and showed no adverse effect on seed germination.

Chemicals

Seed treatment of chickpea with aldicarb, carbofuran, fensulfothion or phorate at three doses (1–2% a.i. seed weight) significantly reduced the soil populations of *P. thornei*. The greatest nematode reduction was recorded at the highest dose of chemicals (Walia and Seshadri, 1985). The soil application of carbofuran and phorate @ 2 kg a.i/ha resulted in reducing the population of *P. thornei* on chickpea grown in plots (Sebastian and Gupta, 1997). However, phorate gave better control of *P. thornei* as compared to carbofuran. Greco *et al.* (1988) treated soil with aldicarb @ 5–10 kg a.i/ha and found suppression in the root invasion of chickpea by *P. thornei*. The treatment also increased the chickpea yield; however, seed treatments proved ineffective.

Integrated management approach

Combined application of *Paecilomyces lilacinus* was with neem cake under field conditions. This was found to be promising in reducing the population of root-lesion nematode in chickpea (Tiyagi and Ajaz, 2004).

13.2.4 Cyst Nematode (*Heterodera* spp.)

Cyst-forming nematodes, *Heterodera* spp., are of great importance in chickpea-growing areas throughout the world. Greco (1987) reported *Heterodera trifolii*, *H. cajani*, *H. goettingiana* and *H. vigna* reproducing on chickpea plants. *H. ciceri* were reported infecting chickpea in Syria (Vovlas *et al.*, 1985; Greco and Di Vito, 1987), *H. cajani* and *H. swarupi* in India (Ali, 1995; Sharma *et al.*, 1998; Ali and Sharma, 2003) and *H. goettingiana* in Morocco and Algeria (Di Vito *et al.*, 1994a). *H. ciceri* is considered the most damaging chickpea nematode in Syria and Turkey, where it causes considerable yield loss (Greco *et al.*, 1988; Di Vito *et al.*, 1994b).

13.2.4.1 Economic importance and threshold level

The tolerance limit of chickpea to *H. ciceri* has been estimated at 1 egg/cm^3 (Greco *et al.*, 1988). A complete loss in yield was reported when the fields were infested with 32 eggs of *H. ciceri*/cm^3 soil (Greco and Di Vito, 1987). Greco and Sharma (1990) reported complete failure of the crop at 64 eggs of *H. ciceri*/cm^3 soil.

13.2.4.2 Symptoms and nature of damage

Stunted growth, yellowing of the foliage, poor flowering and podding (containing small or no seeds in case of heavy infestation), extensive necrosis of the roots and early senescence of the plants are the symptoms observed on chickpea plants when infested by *H. ciceri*. Symptoms are more apparent when the infested plants are at the flowering stage (Castillo *et al.*, 2008). The roots of such plants are poorly developed, lacking in nitrogen-fixation nodules. The cysts of adult nematode females are found attached to the roots; initially these are white, creamy or pale yellow in colour and are visible with the naked eye giving a pearly appearance on the root, hence the name 'pearly root disease'. At a later stage, when the plant attains maturity, the cysts turn brown in colour, are detached from the root and fall into the soil.

13.2.4.3 Biology and life cycle

The second-stage juveniles (J_2) are infective and, after hatching from eggs, emerge from the cysts. J_2 migrate through soil and their migration towards the chickpea roots is directed by host stimuli in the form of root exudates. J_2 enter any area of the chickpea roots, but prefer the region just behind the root tip. They move inter- and intracellularly in the root cortex before settling permanently near the steler region. Soon the feeding starts, which stimulates the plant to produce special feeding cells called syncytia. The cells of syncytia are hypertrophic with dense cytoplasm and large nuclei. The nematodes continue feeding, become established and become sedentary. Two more moultings result in the formation of J_3 and J_4. Males and females can be differentiated at this stage. The adult male emerges after the fourth moult. It may be noted that development of male through third and fourth stage proceeds without feeding. The adult male moves into the soil, survives for a few days and then dies: males are required only for mating (Walia and Bajaj, 2003). Females continue to grow and become saccate, feeding at each stage. Swollen or saccate females rupture root tissues with the posterior end of their bodies, which then protrude, with only the neck region inside the root tissue, and start laying eggs. Females retain eggs within the body. A small gelatinous matrix may also be protruded by females, but void of eggs (Kaloshian *et al.*, 1986b). When all the eggs are laid, the female dies and the cuticle becomes thicker and hardens. Due to the process of chromogenesis, the cuticle turns from white to yellow, brown and finally black, and the dead female now is termed as a 'cyst'. The cyst containing eggs detaches from the root, falls into the soil and persists in the soil until the next host crop is available. Egg hatching within the cyst is favoured by root diffusate of host crops, suitable soil moisture and a favourable temperature (Kaloshian *et al.*, 1986b; Greco *et al.*, 1992a, b). Invasion on chickpea roots generally takes place at 8°C, but development of the nematode only occurs at ≥ 10°C (Kaloshian *et al.*, 1986a).

13.2.4.4 Management

Cultural

Crop rotation is an effective practice to manage *H. ciceri* because of the narrow host range of this nematode. In Syria, crop rotation for a period of four years by including barley and wheat as a non-host crop of *H. ciceri* resulted in an increase in the yield of chickpea as compared to the yield without a non-host crop (Saxena *et al.*, 1992). The soil population of nematodes were recorded to be at high levels in plots planted with host crops every year or every other year (maximum 63 eggs/g soil), but declined to 43–63% and 12–16% when non-host crops were cultivated for 1 or 2–3 consecutive years, respectively. The rate of nematode multiplication decreased as the soil population of nematode increased and it was 17 at 2 eggs/g soil and 3 at 14 eggs/g soil.

The practice of fallowing is common in Turkey and Syria. The field when left fallow for some period proves helpful in checking the nematode population, as the nematodes die from starvation. Due to fallowing, a reduction of 35–50% per year in the soil population densities of the nematode has been recorded (Greco *et al.*, 1988).

Host plant resistance

Chickpea germplasm lines ILC 10765 (Reg. no. GP-226, PI 629017) and ILC 10766 (Reg. no. GP-227, PI 629018) were released in the USA for resistance to *H. ciceri*. The germplasm lines were developed as a result of crossing between a wild progenitor of chickpea belonging to the species *Cicer reticulatum* and a cultivated pea accession (Malhotra *et al.*, 2002). Di Vito *et al.* (1988) screened 2001 lines of chickpea for resistance to *H. ciceri* and found only 20 lines having low infestation level of *H. ciceri*. Two chickpea germplasm lines, FLIP 2005- 8C (Reg. No. GP-273, PI 645462), and FLIP 2005-9C (Reg. No. GP-274, PI 645463), resistant to chickpea cyst nematode (CCN; *Heterodera ciceri*), were jointly developed by the International Center for Agricultural Research in the Dry Areas (ICARDA), Syria, and the Istituto per la Protezione delle Piante, Consiglio Nazionale delle Richerche (IPP-CNR), Italy, in August 2006. The lines FLIP 2005-8C and FLIP 2005-9C, were developed from the cross ILWC 292/ILC 482. Two lines, FLIP 2005-8C and FLIP 2005-9C, were rated 2–4 on a 1–9 scale (1=resistant, 9=susceptible), based on symptoms visible in aerial plant parts due to nematode attack on the roots (Malhotra *et al.*, 2008).

13.2.5 Reniform Nematode (*Rotylenchulus reniformis*)

R. reniformis, a semi-endoparasitic nematode was first reported by Linford and Oliveira (1940) on cowpea roots in Hawaii. Later on this nematode was reported on several other crops including chickpea and thus categorized as a polyphagous pest. The common name 'reniform' refers to the kidney-shaped body of the sedentary mature female protruding from infected roots. *R. reniformis* is one of the most economically important species of reniform nematodes, which has been reported on chickpea from India (Rashid *et al.*, 1973; Ali, 1993), Tunisia (Di Vito *et al.*, 1994a), Ghana (Edwards, 1956) and Mediterranean countries (Oteifa, 1987).

13.2.5.1 Economic importance and threshold level

The damaging threshold level of *R. reniformis* was recorded at 1 premature female/1 g soil (Darekar and Jagdale, 1987), but on the basis of a study conducted in greenhouse conditions, it was estimated at 0.5–1.0 nematode/g soil and with the presence of 10 nematodes/g soil, the loss in yield in chickpea was recorded at up to 80% (Mahapatra and Pahdi, 1986).

13.2.5.2 Symptoms and nature of damage

Infected plants are distributed in patches and show stunted growth and early senescence. The leaves of such plants are pale green in colour. The plants show uneven growth when at the younger stage, but no visual symptoms appear when the crop is at maturity. Infection caused by *R. reniformis* on chickpea root causes a reduction in the number of rhizobium nodules as compared with the uninfected ones (Darekar and Jagdale, 1987; Tiyagi and Parveen, 1992). In gently uprooted infected plants, the presence of soil particles adhering to the root at different intervals and in the vicinity of bacterial nodules indicated the presence of *R. reniformis* (Ali, 1995). The presence of premature females on rootlets is pronounced after a gentle wash and this is the only remarkable symptom of infection caused by reniform nematode on chickpea plants.

13.2.5.3 Biology and life cycle

The life cycle of *R. reniformis* takes about 25–30 days from egg to egg. The thermal limits for reproduction range between 15 and 36°C, with an optimum of 30°C (Castillo *et al.*, 2008). The three stages of this nematode, J_2, J_3 and J_4, are non-parasitic. After the fourth moult, approximately equal numbers of vermiform females and males emerge. Females are parasitic, whereas males are non-parasitic. The young vermiform females are infective and cannot develop further without feeding. They penetrate the epidermis and cortical parenchyma of roots, become sedentary and establish a permanent feeding site on a single endodermal cell (Castillo *et al.*, 2008). The body of the nematode is embedded within the root anteriorly, while posteriorly, the body protrudes from the root surface. As the reproductive system matures, the posterior portion of the nematode body swells to assume a kidney-like shape. Egg laying starts at the 9–14th day after infection, while the gelatinous matrix is secreted on the 12th day (Ali, 1995). Copulation starts as the females begin to enlarge. A single female lays 60–200 eggs into the gelatinous matrix, which flows out from the vulva (Sivakumar and Seshadri, 1971). The cellular modifications in the host that follow as a result of feeding on roots by *R. reniformis* are hypertrophy, hyperplasia, thickening of cell walls, dense granular cytoplasm, enlarged nuclei, formation of a feeding site (syncytium) and feeding tube formation within syncytium. However, the type and degree of cellular modification vary with the host and variety involved (Varaprasad, 1986).

13.2.5.4 Interaction with other micro-organisms

Association of *R. reniformis* with *Fusarium oxysporum* f. sp. *ciceri*, the causal agent of wilt disease, and *Rhizoctonia solani*, the causal agent of root-rot disease, have been reported from Kanpur, India (Ali, 1995). Siddiqui and Mahmood (1994) conducted a pot experiment to study the effects of *R. reniformis*, and *F. oxysporum* f. sp. *ciceri* on the disease complex of chickpea. Inoculation of *F. oxysporum* alone caused more reduction on the growth of chickpea than *R. Reniformis*; however, combined inoculation of both the pathogens resulted in more deleterious effects on the plant.

13.2.5.5 Management

Cultural

Management of *R. reniformis* by adopting the practice of crop rotation is very difficult and, therefore, while going for an alternative non-host crop, it must be tested before recommendation. Other practices such as changing the date of crop sowing may also prove successful, as it lowers the soil population densities of *R. Reniformis*, as has been seen in the case of paddy cultivation (Haidar *et al.*, 2001).

Soil solarization

In India in 1984 and 1985, solarization by covering the soil with transparent polythene sheets during summer months (April, May, June) resulted in a significant reduction in population densities of *R. reniformis*, parasitic to chickpea. The reduction in population density of *R. reniformis* was achieved with a 100% success rate in 1984 (Sharma and Nene, 1990a).

Biological

Anver and Alam (1999) tested the efficacy of *P. lilacinus* against *R. reniformis* infesting chickpea. The nematodes caused less damage to chickpea plants in presence of *P. lilacinus*. The multiplication rate of the test nematode also reduced in the presence of *P. lilacinus* as compared to the absence of *P. lilacinus*.

Botanical

Oilseed cakes of *Azadirachta indica*, *Ricinus communis*, *Arachis hypogaea*, *Linum usitatissimum*, *Helianthus annuus* and *Glycine max* were evaluated against some plant parasitic nematodes including *R. reniformis* infesting chickpea. All the treatments were found effective in reducing the multiplication of soil nematodes (Tiyagi and Ajaz, 2004). Akhtar (1998) tested various products prepared from *A. indica* (neem), such as leaf powder, sawdust, oilseed cake and also urea against plant parasitic nematodes, including *R. reniformis*, a predatory nematode (*Dorylaimus elongatus*) and free-living nematodes in a field where chickpea was grown. Under field conditions, chickpea seeds treated with 20% w/w powdered formulation of neem, i.e. seed kernel, seed coat, deoiled cake, Achook and 5% v/w liquid formulations (Neemark and Nimbecidine), resulted in a significant reduction in the soil populations of *M. incognita*, *R. reniformis*, *Tylenchorhynchus mashhoodi*, *Helicotylenchus indicus* and *Hoplolaimus indicus*. On the other hand, multiplication of saprophytic nematodes increased using these treatments. All the treatments resulted in increasing the grain yield of chickpea (Mojumder, 1999). Anver and Alam (1992) reported a significant control of *R. reniformis* when seeds of chickpea cv. K850 were coated with latex of *Calotropis procera*. The treatment also resulted in increasing the number of pods, chlorophyll content of leaves, water absorption capacity of roots, and root nodulation in chickpea plants.

Host plant resistance

Twenty cultivars of chickpea were screened for their resistence against *R. reniformis* and wilt fungus, *F. solani*. No tested cultivars were found resistant or moderately resistant against either of the pathogen used in the study. However, a cultivar IC-4927 gave tolerant reaction against *F. solani* and six cultivars against *R. reniformis*. Others were either susceptible or highly susceptible (Lakshminarayan *et al.*, 1990).

Chemicals

Two split dosages of carbofuran (@ 1 kg a.i./ha), one at the time of sowing of chickpea and the second 40 days after seed germination, reduced the population density of *R. reniformis* by up to 87.3%. The treatment also caused a significant increase in the yield of chickpea (Ali, 1988). Meher *et al.* (2010) evaluated the nematicides carbosulfan, cadusafos, phorate and triazophos for their efficacy and persistence in soil to devise nematode management decisions on chickpea. Cadusafos, then triazophos, proved most effective against *R. reniformis*. The concentration of nematicide residue was always higher in roots than in shoots. The residue persisted beyond 90 days in both. The final recommendation was to apply cadusafos (Rugby 10 G) or, alternatively, a spray application of triazophos (Hostathion 40 EC) in planting furrows, both at 1.0 kg a.i/ha, followed by light irrigation.

Integrated management approach

In India, an experiment was conducted to test the efficacy of *Trichoderma viride* (2.0 kg/ha), *Lantana camara* (2.0 quintal/ha) and phorate 10G (1.0 kg a.i./ha) singly or in combination against *R. reniformis* infesting chickpea cv. Pusa 267. Leaves and twigs of *L. camara* were applied to the soil one week before sowing. *T. viride* (powder formulation) and phorate 10G were applied during sowing and one week after seed germination, respectively. The nematode population decreased significantly 30 days after treatment and harvest (Prasad, 2006). Under glasshouse conditions, the application of fruit wastes for the management of *R. reniformis* have proved successful. However, better results have been obtained when fruit wastes were applied in combination with fungal biocontrol agents. Ashraf and Khan (2008) found fruit wastes of papaya when applied in combination with the fungal biocontrol agent *Paecilomyces lilacinus* @ 2g (mycelium+spores)/plant resulted in an increase in plant growth of chickea and population reduction of *R. reniformis*.

13.2.6 Stunt Nematode (*Tylenchorhynchus* spp.)

These are obligatory root ectoparasitic nematodes reported on chickpea from Syria (Greco and Di Vito, 1987), Morocco, Tunisia (Di Vito *et al.*, 1994a), Netherlands (Nene and Reddy, 1987; Nene *et al.*, 1989), India (Nene *et al.*, 1989), Pakistan (Maqbool, 1987) and Spain (Alcala *et al.*, 1970). However, they are of less importance than the endoparasitic nematodes described above. In this section, besides *Tylenchorhynchus*, I have included the nematodes *Merlinius*, *Bitylenchus* and *Amplimerlinus*, as these genera had previously been included in *Tylenchorhynchus* group nematodes (Ali, 1995).

13.2.6.1 Economic importance and threshold level

T. vulgaris has a wide host range and chickpea is among the 29 weak good hosts (Upadhyay and Swarup, 1972). Castillo *et al.* (1991) reported *Amplimerlinus magnistylus* infesting chickpea in Spain. *Bitylenchus vulgaris* (Upadhyay and Swarup, 1972; Ali, 1993), *B. brevilineatus* and *Merlinius brevidens* (Sitaramaiah *et al.*, 1971) have been reported on chickpea from India. Mahapatra and Das (1979) observed the multiplication rate of *Tylenchorhynchus* to be lower in chickpea even at 1–4 times inoculum level and they considered chickpea as a fair host of this nematode. Tiyagi *et al.* (1986) reported that *T. brassicae* significantly reduced chickpea growth and the infected roots markedly inhibited the absorption capability of chickpea plants. Tiyagi and Alam (1990), on the basis of a pot trial reported that *T. brassicae* was more pathogenic than *R. reniformis* but less pathogenic than *M. incognita* when inoculated on three-week-old seedlings of chickpea cv. K 850. *T. brassicae* was inoculated at 100, 1000 and 10,000 nematodes/plant. Infection of nematodes led to a significant reduction in plant growth, water absorption and chlorophyll content, even at low inoculum level. The reproduction rate of *T. brassicae* was highest at 100 nematode/plant.

13.2.6.2 Symptoms and nature of damage

The chickpea plants infested with stunt nematodes showed stunted growth with a retarded root system. Since these nematodes are ectoparasites, they feed primarily on epidermal cells of roots in the region of elongation, but occasionally they are observed partly or totally embedded in host tissue (Ali, 1995).

13.2.6.3 Interaction with other micro-organisms

Tylenchorhynchus sp., in presence of wilt fungus, has been found to cause wilt symptom and has an adverse affect on the root shoot rate of chickpea on dry weight basis. The emergence of chickpea seedlings was also delayed in the presence of wilt fungus (Sobin *et al.*, 1979). Tiyagi and Alam (1990) reported suppression in rhizobium nodulation on chickpea cv. K 850, when the plant was inoculated with *T. brassicae* under pot culture condition. There was a 72% reduction in rhizobium nodules on chickpea roots when the inoculum level of nematode was 10/g soil.

13.2.6.4 Management

Cultural

The low population of *Tylenchorhynchus* sp. was recorded in cropping sequences of groundnut, sesame, soybean and tomato in *kharif* season and wheat, mustard, chickpea and tomato in *rabi* season (Sharma *et al.*, 1971). In Pakistan, the population of *Tylenchorhynchus annulatus* declined in cotton wheat and barley in all cropping sequences, resulting in 10–15% increase in yield in subsequent cropping of chickpea (Maqbool, 1987).

Soil solarization

The population of *Tylenchorhynchus* sp., parasitic to chickpea cvs. ICCV 1 and JG 74, was reduced significantly in pre-sowing in a solarized plot of chickpea as compared to a non-solarized chickpea plot (Sharma and Nene, 1990a).

Chemical

Maqbool (1987) tested the efficacy of two nematicides, aldicarb and carbofuran @ 1 kg a.i./ha against *T. annulatus* infesting chickpea. The results indicated up to 20% increase in the yield of chickpea with the application of these nematicides.

13.3 Pigeonpea/Red gram/Congo pea/ No-eyed pea (*Cajanus cajan*)

Pigeonpea (*Cajanus cajan*) is considered one of the important pulse crops in the world with almost all production confined to developing countries in Asia. It is considered an important grain legume in the semi-arid tropics with about 90% of the world production in the Indian subcontinent, principally in northern, central and eastern India (Nene and Sheila, 1990). In Asia, pigeonpea is grown in an area of 4.33 million ha with an annual production of 3.8 million tons, of which India shares an area of 4.09 million hectares, with a production of 3.27 million tonnes (Vanisree *et al.*, 2013). Besides Asian countries, it is also grown in Kenya, Uganda, Malawi in eastern Africa and in the Dominican Republic and Puerto Rico in Central America.

13.3.1 Nematode pests of pigeonpea

Pigeonpea is reported to be attacked by more than 100 pathogens (Nene *et al.*, 1989), although only a few cause losses of economic value (Kannaiyan *et al.*, 1984). Plant parasitic nematodes cause significant damage to the pigeonpea crop and may directly affect the physiological functioning of the plants to the extent of producing a lower yield. Many species of plant parasitic nematodes have been found associated with pigeonpea (Nene *et al.*, 1996) and of these *Heterodera cajani*, *Meloidogyne* spp., and *Rotylenchulus reniformis* are considered important (Sharma *et al.*, 1992). Ali and Askary (2001b) reported that along with *Heterodera cajani*, *M. incognita, M. javanica* and *R. reniformis* are serious pests of pigeonpea in India. On a global basis, the annual yield loss to pigeonpea caused by plant parasitic nematodes has been estimated at 13.2% (Abd-Elgawad and Askary, 2015).

13.3.1.1 Occurrence and distribution

Plant parasitic nematodes are generally reported to occur in all pigeonpea-growing areas; however, the crop is severely affected in warm regions where the crop stands in the field for long duration. A high temperature causes an increase in the number of generations per season, resulting in a higher population of nematodes (Ali and Singh, 2005). Although the occurrence of a particular nematode depends upon the prevailing cropping system in the area, the most common nematode parasites of pigeonpea are pigeonpea cyst nematode, root-knot nematode and reniform nematode. *H. cajani* is the most widely distributed cyst nematode of pigeonpea in India (Koshy and Swarup, 1971a; Varaprasad *et al.*, 1997), though this nematode has also been reported from Pakistan, Myanmar and Egypt (Aboul-Eid and Ghorab, 1974; Maqbool, 1980; Shahina and Maqbool, 1995; Myint *et al.*, 2005). A study of the distribution of plant parasitic nematodes in different agroecological zones representing the major pigeonpea-producing regions in north-eastern Kenya revealed that *M. javanica*, *Scutellonema unum* and *Rotylenchulus parvus* are potentially important species associated with pigeonpea (Sharma *et al.*, 1993a). *M. incognita* and *M. javanica* have been reported to attack pigeonpea in Australia, India, Pakistan, Malawi, Nepal, Trinidad and the USA, *M. javanica* in Brazil, India, Pakistan, Peurto Rico, Zambia and Zimbabwe (Reddy *et al.*, 1990).

13.3.2 Pigeonpea Cyst Nematode (*Heterodera cajani*)

H. cajani was first described by Koshy (1967); however, Swarup *et al.* (1964) were the first to record it on pigeonpea. Since then, 21 plant species have been reported as hosts of *H. Cajani*, of which 19 belong to the family *Leguminosae*. The females of the species are lemon-shaped, possessing a short neck and terminal cone. Initially, the cysts are pale yellow but as the time advances, the cysts become hard-walled and turn brown then black in colour. Mature females are quite distinct from other species, having a large egg sac sometimes almost double the size of the cyst itself (Ali and Singh, 2005).

13.3.2.1 Economic importance and threshold level

H. cajani have been found to cause reduction in growth parameters and yield of pigeonpea in various experiments carried out by research workers, but no information on actual yield loss has been reported so far. A reduction in plant growth parameters by 27.6% and grain yield loss of 30.1% has been reported under pot conditions (Saxena and Reddy, 1987). In a greenhouse experiment, 14–22% yield loss was recorded due to infestation on pigeonpea plants caused by *H. cajani* (Sharma *et al.*, 1993b). Sharma and Nene (1988) reported a significant reduction in plant growth parameters at the level of 500 and 1000 J_2/500 cc soil. Zaki and Bhatti (1986a) reported a significant reduction in growth parameters of pigeonpea at an inoculum level of 100 J_2/kg soil under pot conditions.

13.3.2.2 Symptoms and nature of damage

The symptoms of nematode injury appear in the form of stunted plant growth, yellowing of leaves and reduced size of pods. Although symptoms on leaves generally may not appear even in heavily infested soils, a reduction in plant height and vigour may be observed in infected plants (Sharma, 1993). Below-ground symptoms are a pearly white or lemon-shaped female attached with roots at seedling stage (30–40 days) of plant. Such infested plants show a reduction in rhizobial nodulation, which is responsible for nitrogen fixation.

13.3.2.3 Biology and life cycle

Heterodera cajani is sedentary in habitat and can reproduce sexually as well as parthenogenetically. The life cycle involves six stages: egg, J_1, J_2, J_3, J_4 and adult. J_2 is the infective stage. Adult females are lemon-shaped, semi-endoparasitic on plant roots, and lay eggs in an egg sac. The first-stage juvenile (J_1) undergoes the first moulting inside the egg and formation of second-stage juvenile (J_2) takes place. J_2 is vermiform (worm-like) and after hatching it penetrates the epidermal cells of the root and migrates through the cortex to a permanent feeding site in the steler region. The nematode becomes established there, becomes sedentary and feeds via specialized trophic cells (syncytia) formed in the steler region of the root. The formation of syncytia takes place as a result of the host plant response to secretions from the nematode. Gradually the nematode grows obese, undergoes a second moult to become a third-stage juvenile (J_3) and a third moult to become a fourth-stage juvenile (J_4). After the final moulting (fourth moult), J_4 becomes an adult female or male. Adult females are lemon-shaped, semi-endoparasitic and remain in position within the root cortex, but the vermiform male leaves the root and moves into the soil to mate with the female. However, males are not always necessary for reproduction as the females can reproduce without them (Sharma and Swarup, 1984). The adult female dies and its cuticle forms a brown cyst. The cyst retains the eggs and can remain viable in the soil for years. Generally one life cycle is completed in 16 days at 29°C; however, when the temperature is cooler (10–25°C) it takes 45–80 days to complete one life cycle (Koshy and Swarup, 1971b).

13.3.2.4 Interaction with other micro-organisms

H. cajani not only infects pigeonpea plants directly, but also indirectly, in association with other micro-organisms such as wilt fungus, *Fusarium udum*. The synergistic effect of both pathogens together is greater on plants in terms of growth parameters and yield when compared to their single effect. The pathogenicity of *F. udum* is enhanced in the presence of *H. cajani* in a pigeonpea wilt-susceptible genotype ICP 2376 (Sharma and Nene, 1989). Inoculation of *H. cajani* (@ 500 J_2/seedling) and *F. udum* resulted in a greater reduction in plant growth of pigeonpea when both nematodes were inoculated simultaneously (Siddiqui and Mahmood, 1999). It has also been observed that simultaneous inoculation of *H. cajani* and *F. udum* resulted in greater reduction in the growth of pigeonpea cv. Bahar and also the multiplication of *H. cajani* than either of the pathogens inoculated alone (Singh *et al.*, 1993).

13.3.2.5 Management

Cultural

Crop rotation is an acceptable method for the management of *H. Cajani*, as the nematode is a non-host of cereal crops. A rotation of 2–3 years with maize, rice, pearl millet, sorghum, groundnut, castor and cotton may suppress the multiplication of nematode and thereby the harmful effect on the pigeonpea crop (Sharma *et al.*, 1992). Growing sorghum continuously in the rainy season and safflower in the post-rainy season has been found to reduce the population density of *H. cajani* (see CAB International's Invasive Species Compendium at: www.cabi.org/isc/datasheet/27023).

Biological

Wilt disease complex of pigeonpea caused by *H. cajani* and *F. udum* was suppressed with the application of biocontrol agents *Bacillus subtilis*, *Bradyrhizobium japonicum* and *Glomus fasciculatum* when applied alone or in combination. The treatments also increased the plant growth characters, and decreased the nematode multiplication and wilting index (Siddiqui and Mahmood, 1995a). Fungal biocontrol agents *P. lilacinus*, *Verticillium chlamydosporium*, *Gigaspora margarita* (Siddiqui and Mahmood, 1995b), *Trichoderma harzianum* and *Glomus mosseae* (Siddiqui and Mahmood, 1996) have also been reported as suppressing the deleterious effect of *H. cajani* on pigeonpea plants.

Botanical

Organic cakes such as neem cake, sesame cake, mustard cake, cotton cake and castor cake were evaluated in soil at two different doses, 3 and 5 g/pot, against *H. cajani* infesting pigeonpea. The results showed a significant reduction in the nematode population and an increase in growth parameters of pigeonpea plant. Neem cake was most effective followed by sesame cake, mustard cake, cotton cake and castor cake (Meena *et al.*, 2009). In another experiment, pigeonpea seeds were coated with neem seed powder, neem seed kernel and neem seed cake for evaluating their efficacy on *H. cajani* infesting pigeonpea. Maximum reduction in penetration of *H. cajani* on pigeonpea plants were observed when seed was coated with neem seed kernel @ 20% w/w (Mojumder and Mishra, 2001). Anver and Alam (1999) and Haseeb and Shukla (2004) also found neem cake effective in suppressing the population of *H. cajani* in the field.

Host plant resistance

Growing resistant cultivars is one of the most suited nematode management strategies as it is economic, easy to recommend and highly practicable. Sharma *et al.* (1991) developed the greenhouse procedure to screen pigeonpea genotypes for resistance to *H. cajani*. The experiments were conducted in pots. White cysts of *H. cajani* were counted on the roots of *H. cajani* susceptible genotype ICPL 87 at 15, 30 and 45 days after emergence of seedlings which were grown in soil infested with varying inoculums levels of *H. cajani*. Altogether 60 genotypes of pigeonpea were screened and all of them were highly susceptible.

Chemical

Application of chemicals for the management of *H. cajani* does not always prove effective because the adult female after death transforms into a leathery tough brown sac containing eggs. However, in some cases, chemicals have been found promising against *H. cajani*. Zaki and Bhatti (1986b) treated pigeonpea seeds with Fensulfothion. The chemical was found effective in reducing the nematode populations at higher concentrations and at a long soaking period. In a glass-house experiment, efficacy of two nematicides (phorate and carbofuran) and three organic amendments (linseed cake, mustard cake and neem cake), each at 5 q/ha, was evaluated against *H. cajani* on pigeonpea cv. Bahar. The seedlings of pigeonpea at seven days old were inoculated with freshly hatched 500 J_2 of *H. cajani*. All the treatments significantly reduced the nematode population on test crop; however, the reduction in population of cysts, males, egg sacs, eggs and juveniles of *H. cajani*/pot was maximum in carbofuran-amended soil, followed by phorate (Singh, 2004). Velayutham (1988) performed seed soaking with dimethoate, acephate and quinalphos, but did not find an effective control of *H. cajani* on pigeonpea plants.

Integrated management approach

Two different management schedules: (a) soil application of neem seed powder @ 50 kg/ha + soil solarization (transparent polythene sheet of 400 gauge thickness for a period of 4 weeks) + VAM @ 100 kg/ha; and (b) seed treatment with neem seed powder @ 10% w/w + soil solarization + VAM @

100 kg/ha were applied to evaluate their effectiveness against *H. cajani* infesting pigeonpea under field conditions. Similar results were obtained with both the management schedules. The cost–benefit ratio was 1:2.54 and 1:2.60 for treatments (a) and (b), respectively, which indicates that both the treatment schedules were economically viable (Nageswari and Mishra, 2005).

13.3.3 Root-knot Nematode (*Meloidogyne* spp.)

Root-knot nematodes are considered a serious pest of pigeonpea all over the world. Four species of root-knot nematodes, *M. incognita*, *M. javanica*, *M. hapla* and *M. Arenaria*, have been reported to attack pigeonpea (Sharma *et al.*, 1992). Rodriguez-Kabana and Ingram (1978) found pigeonpea plants highly susceptible to the populations of *M. arenaria*. Infestation of *M. incognita* (Askary, 2008) and *M. javanica* (Askary, 2012; Askary and Ali, 2012) have also been observed on different genotypes of pigeonpea under pot and field conditions.

13.3.3.1 Economic importance and threshold level

Root-knot nematodes, *Meloidogyne* spp., are one of the major limiting factors to pulse production. *M. incognita* and *M. javanica* are reported to cause 14–29% yield loss in pigeonpea (Ali, 1997; Sharma *et al.*, 1993a). Bridge (1981) has estimated 8–35% yield loss to pigeonpea due to root-knot nematodes. In India, the avoidable yield loss in pigeonpea cv. Pusa Ageti due to *Meloidogyne* spp. was assessed as 14.2% in Gujarat state, particularly in a field infested with a mixed population of *M. javanica* and *M. incognita* (Patel and Patel, 1993). Root-knot nematode also has negative effects on plant-growth parameters, such as length and weight of plants, number of pods, bulk density of woody stem, chlorophyll content of leaves, root nodulation and water-absorption capacity of roots (Alam *et al.*, 1991). Askary (2011) studied the pathogenic levels of root-knot nematode *M. javanica* on short duration pigeonpea cv. UPAS 120. It was observed that reduction in different plant-growth characters were directly proportional to the initial inoculum level of nematodes (Fig. 13.3).

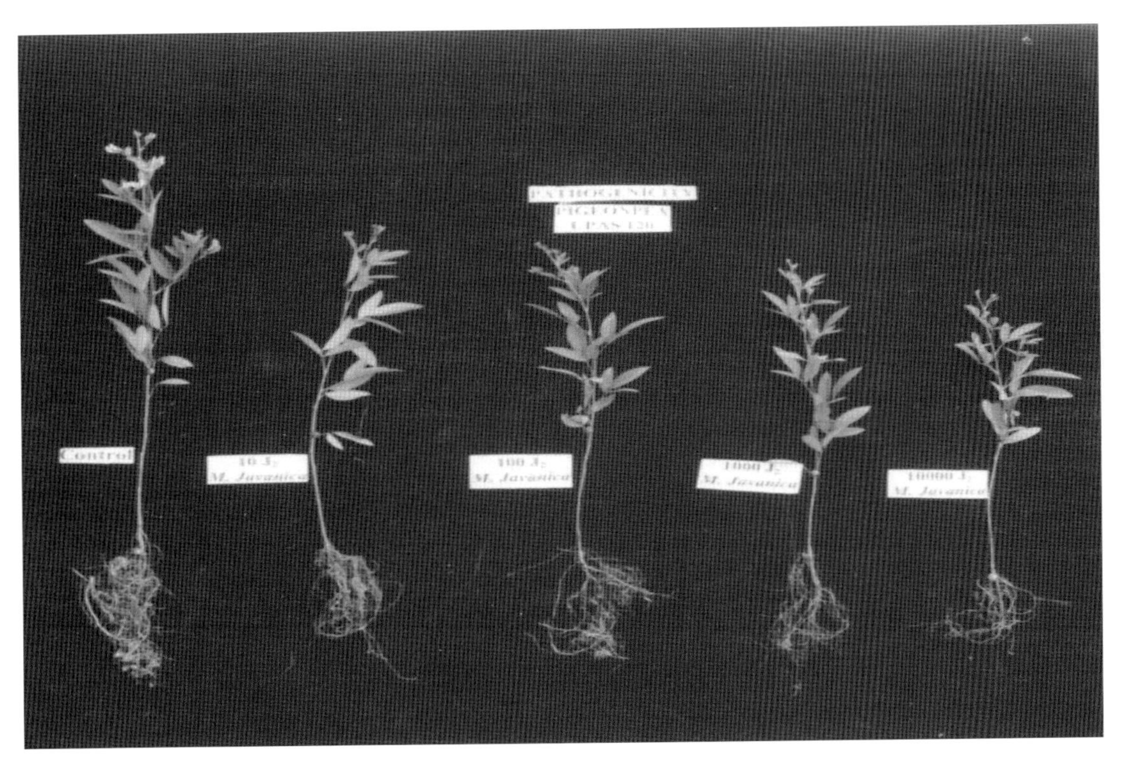

Fig. 13.3. Effect of different inoculum levels [0 (control), 10, 100, 1000 and 10,000] of second-stage juveniles (J_2) of *Meloidogyne javanica* on seedlings of pigeonpea cv. UPAS 120.

A significant reduction in plant growth characters was observed at 1000 J_2/kg soil. However, maximum reduction in shoot length, fresh and dry root and shoot weight of plants were observed at 10,000 levels. Root infection in terms of galls and egg masses were also highest at this inoculum level.

13.3.3.2 Symptoms and nature of damage

The discernible and most characteristic symptom of root-knot nematode infestation is formation of gall on the plant roots (Bird, 1972). The shape and size of galls vary, depending upon the nematode species and population density. A reduction in the number and size of rhizobial nodules were observed on plant roots. It was found that rhizobium does not interfere with the activity of nematodes, but the nematode was causing harmful effect on rhizobium (Taha, 1993). The above-ground symptoms caused by the root-knot nematodes appear as stunting of foliage, smaller size of leaves and pods, chlorosis and patchy growth of plants in a field. Pods may ripen and dry prematurely and remain partially filled and undersized (Reddy *et al.*, 1990). Poor emergence and death of young seedlings may occur in heavily infested soil, but the death of full-grown plants is rare unless there is an association of fungus or bacteria to form a disease complex (Fracl and Wheeler, 1993).

13.3.3.3 Biology and life cycle

Root-knot nematode, *Meloidogyne* spp., is a sedentary endoparasite and second-stage juveniles (J_2) are the infective stage of the nematode. Reproduction is parthenogenetic and when the conditions are favourable, life cycle from egg to egg completes in around three weeks. J_2 does not feed until it is completely inside living plant tissues. They feed on the epidermal cells of roots and penetrate the newly formed tissues such as the meristematic zone. Due to feeding, the cell contents are liquefied and semi-digested extra-corporeally, with the help of hydrolytic enzymes secreted by oesophageal glands of the nematode. The nematode enzymes induce excessive conversion of tryptophan into indoleacetic acid (Dasgupta and Gaur, 1986). This results in enlargement (hypertrophy) of infected root tissue and formation of multinucleated giant cells (2–6) around the nematode's head in steler tissue (Bird, 1962; Haung, 1985; Pasha *et al.*, 1987). The giant cells are formed due to repeated endomitosis without cytokinensis of primary phloem cells or adjacent parenchyma and pericycle cells (Hussey, 1989; Wiggers *et al.*, 1990). The cortical parenchymatous cells around the giant cell undergo excessive multiplication (hyperlesia), giving rise to tiny swellings on the roots or primary galls (Loewenberg *et al.*, 1960), several of which may merge into big multiple galls. J_2 continues feeding on giant cells for a period of several weeks and start to grow obese. The second moult takes place, giving rise to J_3, followed by a quick third moult and the development of J_4. Differentiation in sex occurs after the third moult, as females acquire a V-shaped genital primordium, whereas males acquire an I-shaped genital primordium. The J_3 and J_4 retain their old cuticles as a result of the superimposed moults, and in these two stages nematodes do not feed, as they lack stylets. The pointed tail of J_2 is still visible, and hence are also called spike-tailed stages. After the fourth or final moult the adult stage is reached. The stylet reappears in the adult female, which is sac-like with a fully developed reproductive system and a clear vulval opening. Females at maturity lay on average 200–500 eggs in a gelatinous matrix collectively known as an egg mass. Vermiform (worm-like) adult males are coiled inside the J_4 cuticle, emerge and migrate out of root into the soil. They are short-lived and die after a few days. The length of life cycle and number of generations depends on host health and temperature. Generally, the life cycle of *M. incognita* or *M. javanica* completes in 21–25 days at an optimum temperature of 26–27°C, but it may take as long as 80 days when the temperature is low (14–16°C).

13.3.3.4 Interaction with other micro-organisms

The root-knot nematode, *Meloidogyne* spp., when in combination with wilt fungus *Fusarium udum*, can cause severe damage to pigeonpea crop. The wilting of pigeonpea plants caused by *F. udum* is aggravated in the presence of *M. javanica* (Singh *et al.*, 2004; Askary and Ali, 2012). In a pot trial, the combined effect of *M. javanica* and *F. udum* in ten wilt-resistant accessions of pigeonpea was evaluated. Presence of *M. javanica* with *F. udum* caused 8–33% wilting in KPL 44, 15–60% in AWR 74/15, 15–50% in ICP 12745 and 25–50% in ICP 8859 and ICPL 89049 (Singh *et al.*, 2004). In another study, five wilt-resistant accesions of pigeonpea, DPPA 85-5, DPPA 85-11, DPPA 85-14, Banda

Palera and Sujata, were evaluated against *M. javanica* and *F. udum*. The presence of *M. javanica* with *F. udum* applied to the roots of pigeonpea seedlings caused a susceptible reaction to fusarium wilt in all the five accessions used in the study. Wilt severity was highest in DPPA 85-14 (46.67%) followed by DPPA 85-11 (40.0%). Greatest reduction in root nodulation was observed when plants were inoculated simultaneously with *M. javanica* and *F. Udum*, and it was followed by *M. javanica* prior to *F. udum* and *F. udum* prior to *M. javanica* (Askary and Ali, 2012).

13.3.3.5 Management

Cultural

Leaving the field without vegetation after ploughing helps to expose the nematodes to sunlight, meaning that they die due to starvation and drought (Jonathan *et al.*, 2005). A fallow period of 2–3 years can reduce the root-knot nematode population in pigeonpea (Agrawal, 2003). For the management of root-knot nematode, ploughing at intervals of 2–4 weeks during the dry season could kill the eggs and juveniles in root debris and deeper soil layers and increase the yield of subsequent susceptible crops (Peacock, 1957). Application of green manure crops like *Crotalaria juncea* against *Meloidogyne* spp. (Agrawal, 2003) has been reported on pigeonpea.

Biological

Biocontrol agents have been found quite effective in the management of root-knot nematode disease on pigeonpea. Siddiqui and Shakeel (2006) evaluated 20 isolates of fluorescent Pseudomonads isolated from pigeonpea fields against *M. incognita–Fusarium udum* wilt-disease complex of pigeonpea under pot conditions. Isolate Pf605 emerged the best as it caused the greatest reduction in nematode multiplication and reduced wilt index. The results obtained by the isolates Pf604, Pf611 and Pa616 were also significant. Siddiqui *et al.* (2008) tested six isolates of *Bacillus* and *Pseudomonas* under both pot and field conditions for the management of wilt-disease complex of pigeonpea caused by *Meloidogyne* spp. and *F. udum*. The best two isolates under both pot and field condition were Pa324 and B18 causing greater reduction in multiplication of *M. incognita* and reducing the wilting in plants. Askary (2012) conducted a field experiment to find an economically suitable option through a pre-sowing seed coating of different chemicals, bioagents and botanicals in the management of *M. javanica* on short-duration pigeonpea cultivar UPAS 120. *P. lilacinus* and *A. niger* were found significantly effective in reducing nematode infection on plants and increasing the plant growth characters and yield as compared to the untreated control.

Botanical

Among various neem-based materials such as neem seed powder, Achook, Neemark, Neemgold, Nimbecidine and Field Marshal, neem seed powder was found most effective in suppressing populations of *M. incognita, H. cajani* and *R. reniformis* infecting pigeonpea, followed by Neemgold and Neemark. Achook and Field Marshal were less effective (Das and Mishra, 2000; Das and Mishra, 2002; Das and Mishra, 2003). Seed dressing of pigeonpea and chickpea with latex of *Calotropis procera* resulted in significant control of *M. incognita* and *R. Reniformis*, with a corresponding increase in plant growth, chlorophyll content, water-absorption capacity of roots and root nodulation (Mojumder and Mishra, 1991). In a field experiment, various neem products, neem oil (2%), Neemarin (2%), neem seed powder (5%) and crude neem leaf extract (5%), were used as a seed treatment against root-knot nematode, *M. javanica* in pigeonpea cv. Narendra Arhar 1. At 60 days after sowing, the gall index and soil population of the nematode was minimal and significantly lower than the control (Singh, 2009). Askary (2012) also reported a significant reduction in root galling caused by *M. javanica* on pigeonpea cv. UPAS 120 grown from the seeds treated with neem seed powder @ 5% or latex of *C. procera* @ 1%, as compared to the untreated control. The plant-growth characters also increased due to these two treatments (Fig. 13.4).

Host plant resistance

In India, Simon and Dass (2010) evaluated 70 pigeonpea accessions against *M. incognita*. The accessions highly resistant to *M. incognita* were JBP001, ICPL 151, LRG 38. HY3C, TV-1, WRP 1 and ICP 8863. In some of the genotypes of pigeonpea, only egg masses are produced, not the galls (Sharma *et al.*, 1994), so galling and reproduction can be considered for plant reaction to root-knot

Fig. 13.4. Effect of seed treatment with neem seed powder (5%) and latex of *Caloropis procera* (1%) on pigeonpea cv. UPAS 120 in field infested with root-knot nematode, *Meloidogyne javanica*.

nematodes. The resistant plant acts as a barrier for nematodes in a number of ways: it retards the rate of reproduction, delays the maturity of the parasite, hampers production of more males, making nematodes unable to complete development. Several pigeonpea breeding lines have shown to be highly resistant to both *M. incognita* and *M. javanica*. In a screening test out of 47 lines of pigeonpea, 36 were rated as resistant against *M. Incognita*, based only on galling under field conditions (Ravichandra *et al.*, 1988). Patel *et al.* (1987) reported pigeonpea lines 18-1 and 77-1 as highly resistant to mixed population of *M. incognita* and *M. javanica*.

Chemical

Nursery-bed treatments with nematicides like carbofuran, aldicarb or fenamiphos are effective in reducing the root galling by 27–49% against *M. javanica* (Jain and Bhatti, 1991). Seed treatment with dimethoate @ 8 ml/kg seed, chlorpyriphos @10 ml/kg seed and triazophos @ 1% are reported to control efficiently the infestation of root-knot nematode, thereby reducing the population build-up of nematodes in soil, increasing the plant growth parameters and grain yield (Mishra *et al.*, 2003). Carbosulfan @ 1–3% has also been found effective against *Meloidogyne* spp. on different pulse crops, including pigeonpea (Mishra, 2007). Seed treatment with chlorpyriphos and dimethoate @ 1ml/kg seed resulted in causing a significant reduction in soil population of *M. javanica*, and increasing the plant-growth characters and yield of pigeonpea cultivar UPAS 120 as compared to untreated control under field conditions (Askary, 2012).

Integrated management approach

Askary (2008) applied *Paecilomyces lilacinus* + neem seed powder + Dimethoate as a seed dressing on two pigeonpea cultivars UPAS 120 and CO 6 grown under pot and microplot conditions where the soil was inoculated with second-stage juveniles (J_2) of *M. incognita*. The treatment resulted in a significant reduction in the development of galls on plant roots, production of egg masses and soil population of *M. incognita* as compared to the untreated control. Treatment constituting farmyard manure (FYM), oilseed cake of *Pongamia pinnata* and Vesicular Arbuscular Mycorrhiza (VAM) has been found to reduce the disease incidence caused by root-knot nematode–*Fusarium* complex in pigeonpea to a significant level and improve the plant-growth parameters (Goswami *et al.*, 2007).

13.3.4 Reniform Nematode (*Rotylenchulus reniformis*)

R. reniformis, a semi-endoparasitic nematode is considered an important pathogen of pigeonpea, causing considerable damage to the crop. This nematode has been reported on pigeonpea from

India, Jamaica, Trinidad and Fiji (Sharma and McDonald, 1990).

13.3.4.1 Economic importance and threshold level

Several workers have reported yield loss in pigeonpea due to infestation by *R. reniformis*. Ali (1996) reported yield loss of 19% in pigeonpea due to infestion by *R. reniformis*. However, the loss may vary from 14 to 29%, depending on the initial population level of nematode and duration of the crop (Ali and Singh, 2005). The damaging level of *R. reniformis* in a susceptible variety of pigeonpea was found to be 1000/700 g soil, whereas in the case of resistant varieties it was 10,000/700 g soil (Thakar and Yadav, 1985). Mohanty and Padhi (1987) found a significant reduction in root length, dry weight of shoot and number of rhizobial nodules when the plants were inoculated with 500 nematodes/pot.

13.3.4.2 Symptoms and nature of damage

Pigeonpea plants infested by *R. reniformis* show the symptoms such as yellowing of new leaves, die-back of twigs and also main stem. Sometimes the premature death of plants has also been reported due to nematode infestation (Hutton and Hammerton, 1975). The aerial plant parts do not show explicit symptoms; however, patches of stunted plant growth can be observed in the field and the number of such patches increases under water-stress conditions (Ali and Singh, 2005). The most common below-ground symptoms are the presence of soil-covered egg masses attached to the roots that are dirty in appearance, as the soil particles adhered to it do not easily dislodge by shaking or gently washing the root with water (Sharma *et al.*, 1992); hence the disease caused by this nematode is often known as dirty root disease (Ali and Askary, 2005).

13.3.4.3 Biology and life cycle

Rotylenchulus reniformis are semi-endoparasitic (partially inside roots) nematodes and their life cycle consist of egg, four vermiform juvenile stages (J_1, J_2, J_3 and J_4) and adult. After the second-stage juvenile (J_2) hatches, the third and fourth moults take place. After the fourth moult, the young female (pre-adult) emerges, which is the infective stage of the nematode. The pre-adult nematode prefers to penetrate roots in the zone of elongation, penetrating the root cortex, establishing a permanent feeding site in the steler region of the root, and becoming sedentary or immobile. The female feeds permanently on a single cell in the endodermis, leading to the formation of the feeding site, or syncytia. Females of *R. reniformis* feed continuously and start to attain maturity, become enlarged on the ventral side around the vulval region to form a reniform (kidney) shape within five days after infection. The egg laying starts after 7–8 days. Males and females are normally in the ratio of 1:1. The nematode reproduces by cross-fertilization; however, some populations of reniform nematodes reproduce parthenogenetically, i.e. produce eggs without fertilization. The eggs are fertilized with sperm, and deposited into a gelatinous matrix. One egg mass of *R. reniformis* contain up to 150 eggs. The life cycle is completed in 24–29 days in females, and 16–20 days in males. *R. reniformis* can survive without any host for more than 300 days without losing its infectivity (Sharma and Nene, 1992).

13.3.4.4 Interaction with other micro-organisms

The reniform nematode, *R. reniformis*, in the presence of wilt fungus can cause more harm to pigeonpea plants. The nematode has also been reported to break the wilt resistance of pigeonpea plants. Sharma and Nene (1990b) reported that the presence of *R. reniformis* along with *F. udum* accelerated the death of wilt-susceptible pigeonpea cultivars. Jain and Sharma (1996) observed that wilt-resistant pigeonpea cultivar ICPL 270 loses its resistance in presence of *R. reniformis*.

13.3.4.5 Management

Cultural

A non-host crop such as groundnut and sesame, when intercropped with pigeonpea for the management of *R. Reniformis*, has been found successful in bringing down the soil population of nematodes, reducing the number of galls on pigeonpea plants and ultimately increasing the yield of pigeonpea (Upadhyay *et al.*, 1997).

Biological

Soil application of *P. lilacinus* for the management of *R. reniformis* on pigeonpea plants resulted in reducing the multiplication rate of *R. reniformis* and significant reduction in plant damage (Anver and Alam, 1997).

Botanical

Efficacy of some products of *Azadirachta indica*, for example, achook, neemark, neemgold and nimbecidine, were tested against *R. reniformis* on pigeonpea under pot conditions. All the treatments resulted in lowering the soil population of *R. reniformis* and thereby reducing the nematode infestation on pigeonpea plants (Das and Mishra, 2003).

Host plant resistance

Forty-six medium maturity (mature in 151–200 days at Patancheru, India) pigeonpea genotypes were evaluated for resistance and tolerance to *R. reniformis* in greenhouse and field conditions, over the period 1990–1997. Pigeonpea genotypes C 11, ICPL 87119 and ICPL 270 were used as nematode-susceptible checks. The reniform nematode-tolerant lines were ICP 16329, ICP 16330, ICP 16331, ICP 16332, and ICP 16333 (Sharma *et al.*, 2000). Anver and Alam (2001) screened 37 accessions of pigeonpea against *R. reniformis*. On each three-week-old plant of pigeonpea, 5000 immature females of *R. reniformis* were inoculated. The accession KM-137 was found resistant to *R. reniformis*. In Brazil, ten pigeonpea accessions, g5-94, g59-95, g66-95, g8-95, g109-99, g124-95, g3-94, g127-97, g58-95 and g18-95 were tested against *R. Reniformis*, but all were found susceptible to this nematode (Filho *et al.*, 2010).

Chemical

An application of carbofuran 3G @ 6 kg a.i/ha in a soil heavily infested with *R. reniformis* where pigeonpea plants were grown resulted in a significant reduction in nematode densities and an improvement in plant growth.

Integrated management approach

A combined application of *Paecilomyces lilacinus* with *Zea mays* and *Sesbania aculeate* as green manuring proved highly effective for the management of *R. reniformis* on pigeonpea (Mahmood and Siddiqui, 1993).

13.4 Common bean/haricot bean/red kidney bean/snap bean/rajmash (*Phaseolus vulgaris*)

Among pulses, common bean (*Phaseolus vulgaris*) is one of the most widely distributed crops in the world, grown in the Americas, Africa, Asia and Europe. Asia is the largest producer of common bean, contributing almost half of the world's production. The crop is generally sown in the warm season and therefore often becomes the host of nematodes active at this temperature range.

13.4.1 Nematode pests of common bean

Numerous plant parasitic nematodes belong to genera *Meloidogyne*, *Pratylenchus*, *Heterodera*, *Scutellonema*, *Helicotylenchus*, *Tylenchorhynchus*, *Tylenchus*, *Criconemella*, *Aphelenchus*, *Hemicyliophora* and *Trichodorus* have been found associated to common bean; however, the most damaging are *Meloidogyne* spp. and to some extent *Pratylenchus* spp.

13.4.2 Root-knot Nematode (*Meloidogyne* spp.)

The two most common species of root-knot nematode which are reported to cause damage to common bean in the Americas, Africa and Asia are *M. incognita* and *M. javanica* (Sikora and Greco, 1990).

13.4.2.1 Economic importance and threshold level

Kimenju *et al.* (1999) found a significant reduction in rhizobium nodulation of common bean when infested with *M. incognita*. Sharma (1981) reported a significant growth reduction in common bean when soil was infested with *M. javanica* at 1 egg/g soil. The yield loss can occur up to 60% when beans are grown on soils heavily infected with *Meloidogyne* spp. (Parisi *et al.*, 2004).

13.4.2.2 Symptoms and nature of damage

The plants of common bean infested with *Meloidogyne* spp. show above-ground and below-ground symptoms, including yellow stunted plants, early defoliation, secondary root and stem rots, and presence of root galls (Parisi *et al.*, 2004).

13.4.2.3 Interaction with other micro-organisms

Root-knot nematode has been found to increase the disease severity in the presence of certain micro-organisms such as fungus. Wilting in common bean plants caused by *Fusarium solani* f. sp. *phaseoli* increased in the presence of *M. arenaria* and *M. javanica* (Hutton *et al.*, 1972).

13.4.2.4 Management

Biological

A reduction in plant growth of common bean was observed when seedlings were inoculated with *M. incognita* alone and 15 days prior to inoculation of *Pochonia chlamydosporia*. On the other hand, application of fungus prior to nematode inoculation resulted in improving the plant growth and reducing the number of galls and egg masses per root system (Sharf *et al.*, 2014a). In another experiment, Sharf *et al.* (2014b) observed that a combined application of two biofertilizers (*Trichoderma viride* and *P. chlamydosporia*) and the nitrogenous fertilizer (urea) improved the growth parameters and biochemical parameters (chlorophyll, protein, nitratereductase, nitrogen and phosphorus contents) of red kidney bean infested with *m. incognita* as compared to control (*M. incognita*) alone.

Chemical

Marwoto (1988) conducted a field study in Indonesia to evaluate the effects of systemic nematicides, carbofuran and oxamyl at different doses against *Meloidogyne* spp. on kidney bean. The best result were obtained from oxamyl @ 1.2 kg a.i/ha and carbofuran 1.8 kg a.i/ha; however, oxamyl proved more effective than carbofuran at all the doses used in the experiment.

Host resistance

In a greenhouse study conducted in Venezuela to evaluate the response of some selections of *P. vulgaris* to *M. incognita* race 1 and *M. Enterolobii*, only the cultivar Alabama #l was found resistant to both nematode species (Crozzoli *et al.*, 2011). In Belgium, on the sandy soils of the provinces of Antwerp and Limburg, ten bean cultivars were screened against root-knot nematodes, *M. chitwoodi* , *M. fallax* and *M. hapla*. The tested cultivars were poor to good hosts for *M. chitwoodi*, non-hosts or bad hosts for *M. fallax* and excellent hosts for *M. hapla*. Significantly, fewer *M. fallax* were found in the roots, and their development was delayed. Penetration of *M. hapla* took place over a longer period than that of *M. chitwoodi* and *M. fallax* (Wesemael and Moens, 2012). Pereira *et al.* (2013) conducted a study in Brazil to select cultivars of common bean for resistance to *M. incognita*, race 2. Altogether, 11 common bean cultivars (BRS Requinte, BRS Pontal, CNFC 10470, IPR Tangará IPR, IPR Colibri, Princesa, IPR Siriri, Aporé, Engopa 202 Rubi, IPR Juriti, BRS Majestoso) were screened. On the basis of root-knot reproduction index, the BRS Requinte, BRS Pontal, CNFC 10470, IPR Tangará, IPR Colibri, Princesa, IPR Siriri and Aporé were classified as slightly resistant genotypes and Engopa 202 Rubi, IPR Juriti and BRS Majestoso as susceptible genotypes to root-knot nematode *M. incognita*, race 2.

Integrated management approach

The nematodes parasitic on common bean can be maintained at levels below economic threshold by applying *B. subtilis* in combination with cow manure, an integration which also demonstrates conservation of the nematode diversity (Huising and Okoth, 2011).

13.4.3 Root-lesion Nematode (*Pratylenchus* spp.)

Among the lesion nematodes, *Pratylenchus scribneri* (Thomason *et al.*, 1976) and *P. penetrans* (Elliot and Bird, 1985) have been found to cause reduction in plant growth of common bean when the nematode population in soil exceeds 0.5 nematode/cm^3. In Bulgaria, imazethapyr, a herbicide, was evaluated against some plant parasitic nematodes, *Pratylenchus* spp., *Helicotylenchus* spp., *Paratylenchus* spp., *Trichodorus* spp. and *Xiphinema* spp. on common bean. Except for *Pratylenchus* spp., the herbicide did not cause any significant effect on the population dynamics of nematodes (Trifonova and Peneva, 2003). In Turkey, 15 cultivars and genotypes of *P. vulgaris* (Musica, Şelale, Nadide, Gina, Serra, Karabağ, Funda, Lepus, Zülbiye, Kınalı, Perla, Kwintus, Özayşe, Tokat and Sırık) were tested for host suitability to four root lesion nematodes, *Pratylenchus thornei*, *P. crenatus*, *P. neglectus*, and *P. penetrans*, under controlled conditions. The results indicated differences in host suitability against the tested nematode species. *P. penetrans* reproduced on Gold Nectar, Karabag˘, Zülbiye, Kınalı, Nadide and Funda; however, Perla, Kwintus, Gina, Musica and Özayşe showed poor host reaction to *P. penetrans*. *Pratylenchus thornei*, multiplied on Zülbiye, s¸elale, Kınalı, Lepus, Perla and Nadide. Özays¸e, Tokat Sırık, Gold Nectar, Kwintus, Serra, Gina, Funda and Musica, showed significantly reduced *P. thornei* populations. *P. neglectus* and *P. crenatus* showed extremely low reproduction indices in all bean cultivars (Sög˘üt *et al.*, 2014).

13.5 Mung Bean/Green Gram/Golden Gram (*Phaseolus aureus* Roxb, syn. *Vigna radita* (L.) Wilczek var. *radiata*)

Mung bean (*Phaseolus aureus* Roxb, syn. *Vigna radita* (L.) Wilczek var. *radiata*) is a native to the Indian subcontinent (Vavilov, 1951) but is now also cultivated in China, Southeast Asia, hot dry regions in Africa, Southern Europe and South America. It is an important short-duration pulse crop, which is rich in protein and carbohydrates.

13.5.1 Nematode pests of mung bean

Plant parasitic nematodes belonging to genera *Meloidogyne*, *Rotylenchulus* and *Heterodera* have been reported to cause deleterious effects on mung bean cultivation. *M. javanica* has been reported causing damage to mung bean in the Philippines (Castillo, 1975) and India (Singh, 1972), to *Heterodera glycines* in the United States (Epps and Chambers, 1959) and to *Rotylenchulus reniformis* in the Philippines (Bajet and Castillo, 1974). According to Bridge (1981), *Meloidogyne* spp. are a serious threat to the crop in India, the Philippines, Thailand and the USA.

13.5.2 Root-knot nematode (*Meloidogyne* spp.)

The major nematode pathogen of mung bean is root-knot nematode and all the four major species of this nematode have been reported to parasitize mung bean (Sikora and Greco, 1990). A significant reduction in plant growth, nodulation and nitrogen content of the shoot and root has been observed in mung bean infested with *M. incognita* (Hussaini and Seshadri, 1975; Inderjit Singh *et al.*, 1977). Castillo *et al.* (1977) recorded a yield loss of 28% in mung bean grown in a field infested with mixed population of *M. incognita* and *R. reniformis*.

13.5.2.1 Management

Neem seed powder @ 50mg/kg soil has been reported as highly effective against *M. incognita–Fusarium oxysporum* disease complex of mung bean cv. ML-1108 (Haseeb *et al.*, 2005). Wani and Bhat (2012a) amended soil with urea, coated with nimin [neem (*Azadirachta indica*)-based product with neem-triterpenes] and oils of neem (*A. indica*), castor (*Ricinus communis*), and rocket salad (*Eruca sativa*) for testing their efficacy against *M. incognita* infesting mung bean. The best results in terms of improving plant-growth characteristics and chlorophyll content of plant leaves was obtained with urea coated with nimin followed by neem oil, castor oil and rocket salad. The efficacy of three biocontrol agents, *Trichoderma viride*, *T. harzianum*, Biofor-pf (a combination of *T. harzianum* and *Pseudomonas fluorescens*), neem cake and carbofuran were tested against *M. incognita* infesting mung bean. Biofor-pf @ 100 kg/ha exhibited the best result, followed by neem cake @ 2 tons/ha and carbofuran @ 1 kg a.i./ha in reducing the galls, egg masses and final soil population of nematode (Sumita and Das, 2014).

13.6 Black Gram/Urd Bean/Mash (*Vigna mungo* L.)

Black gram (*Vigna mungo* L.) is a drought-tolerant pulse crop commonly grown in America, Africa and Asia. It is also grown intercropped with maize or sorghum. It is nutritious and, like other pulses, is recommended for diabetes patients.

13.6.1 Nematode pests of black gram

Numerous nematodes have been found associated with black gram. In India, *Meloidogyne* (Nadakal, 1964), *Rotylenchulus reniformis*, *Heterodera cajani* and *Tylenchorhynchus mashhoodi* (Sitaramaiah, 1984), in Brazil, *M. incognita* and *M. javanica* (Freire *et al.*, 1972) and in Puerto Rico, *R. reniformis* (Ayala and Ramirez, 1964) have been reported to infest black gram. However, the most potent nematode from the point of view of damaging this crop is *Meloidogyne* spp.

13.6.2 Management

Several methods have been reported to manage nematodes infesting black gram. Plants inoculated with bacterial biocontrol agents, *Psedomonas fluorescens* or *Bacillus subtilis* and second-stage juveniles of *M. incognita* resulted in improving plant-growth parameters and decreasing the number of root galls, as compared to the plants inoculated with nematode alone (Akhtar *et al.*, 2012). The leaf extract of goat weed plant, *Ageratum conyzoides*, proved effective against *M. incognita* infecting black gram (Pavaraj *et al.*, 2010). In India, Goel (2004) conducted laboratory experiments to screen different urd bean genotypes for their reaction against *M. javanica* and *M. incognita*. The black gram genotypes used against *M. javanica* were KU-300, KU-303, UG-562, WBU-109, Pusa-2, Part 11-19, UPU-95, UPU-95-2, UG-218, UG-737, DPU-91-7, AKU-7, KU-304, KU-305, KU-308, KU-309, whereas against *M. incognita*, the genotypes were TPU-94-2, KU-315, K-98 and UPU-97-10. A total of 1000 freshly hatched nematode juveniles were inoculated in earthen pots ten days after seed germination, and the plants were uprooted after 45 days for gall count. KU-300 and KU-303 were the only genotypes moderately resistant to *M. Javanica*, while TPU-94-2, KU-315 and K-98 were the only moderately resistant genotypes to *M. incognita*. The rest of the genotypes screened were either susceptible or highly susceptible to *M. javanica* and *M. incognita*. Bhat *et al.* (2012), on the basis of an experiment, reported that treatment of *Bradyrhizobium* and *P. lilacinus* significantly reduced the damage to plant growth of black gram than the plants treated with *Bradyrhizobium* at the time of inoculation with root-knot nematode, *M. incognita,* followed by plants where *Bradyrhizobium* and *P. lilacinus* were applied ten days after nematode inoculation. In all the treatments, *Bradyrhizobium* caused a significant increase in the nitrogen content of root and shoot.

13.7 Lentil (*Lens culinaris*)

Lentil (*Lens culinaris*) originated in the region of the Middle East (Webb and Hawtin, 1981), and is commonly grown in India, Turkey, Iran, Pakistan, Bangladesh, Syria, Ethiopia, Morocco, Spain and Chile. It is a cool-season crop, moderately resistant to low temperature and drought and generally rotated with cereals. Lentil is mainly consumed by people as a supplement to their diet, in the form of soup or for baking flour. The straw is used as animal feed.

13.7.1 Nematode pests of lentil

Root-knot nematode, *Meloidogyne incognita* (Singh *et al.*, 2013), and *M. javanica* in India (Siddiqui *et al.*, 2007), stem and bulb nematode, *Ditylenchus dipsaci*, in Syria (Greco and Di Vito, 1987), cyst nematode, *Heterodera ciceri* in North Syria (Sikora and Greco, 1990), reniform nematode, *Rotylenchulus reniformis* in India (Fazal *et al.*, 1992) and root-lesion nematode, *Pratylenchus thornei*, *P. penetrans* and *P. mediterraneus* in Turkey (Di Vito *et al.*, 1994b) have been reported to cause infestation on lentil plants.

13.7.1.1 *Economic importance and threshold level*

M. incognita at an inoculum level of 1000 J_2/kg soil was found to cause damage to the lentil plant (Singh *et al.*, 2013). The tolerance limit of *H. ciceri* on lentil has been estimated at 2.5 eggs/cm^3 of soil (Greco *et al.*, 1988). When the field was infested with 20 eggs/cm^3 of soil, the yield loss was 20% but at 64 eggs/cm^3 the loss in yield was up to 50%. Ali (2009) recorded 15% yield losses in lentil caused by *M. javanica* under field conditions.

13.7.1.2 *Interaction with other micro-organisms*

Wilt fungus, *Fusarium oxysporum* f. sp. *Lentis*, in the presence of root-knot nematode *M. incognita*, proved more aggressive, as the concomitant inoculation caused greater reduction in growth parameters of lentil plant when compared to individual inoculation of either of the pathogens (Fazal *et al.*, 1994).

13.7.1.3 *Management*

Application of neem cake on lentil plants grown from seed and soaked in leaf extract of rocket salad (*Eruca sativa*) resulted in the highest reduction in root-knot development and increase in root nodulation, plant growth and chlorophyll content (Wani and Bhat, 2012b). Combined applications of *rhizobium* and the bacterial biocontrol agent, *Pseudomonas putida*, caused significant reduction in galling and multiplication of *M. javanica* infecting lentil plants (Siddiqui *et al.*, 2007).

13.8 Conclusion and Future Thrusts

This chapter has described the role of plant parasitic nematodes as constraints in the production of pulse crops. In the past, surveys have only been conducted to learn the incidence of nematodes on some major pulse crops such as chickpea, pigeonpea and common bean, but nematodes of minor pulses such as urd bean, mung bean and lentil have not been studied extensively. Hot spots of major nematode pests: *Meloidogyne*, *Heterodera*, *Pratylenchus*, *Rotylenchulus*, have to be identified in pulse-growing areas on a regional basis through extensive surveys. In the above literature, it is very clear that these nematode pests differ in a number of ways: method of attack, symptoms produced in pulse crops, length of life cycle and host range. *Meloidogyne* form giant cells and cause galling on roots; *Heterodera* form syncytia in the steler region and have a pearly appearance on the root; *Pratylenchus* are migratory endoparasites and cause necrotic root lesions; *Rotylenchulus* are semi-endoparasites and cause dirty root disease. Ectoparasites such as *Tylenchorhynchus* and *Helicotylenchus* cause mechanical injury by feeding primarily on the epidermal cells of roots. There are numerous reports of nematodes infesting pulse crops, but in most cases these records are only up to genera and not species level. Therefore, for an effective and precise management of plant parasitic nematodes attacking pulse crops, a prerequisite step should be correct diagnoses of the nematode up to species level and also their pathogenic variants. The conventional identification of nematodes based on their morphological features is hindered by the lack of expertise of diagnosticians, but now there is a need to use newly developed diagnosis technologies, based on biochemical and DNA analyses, which can be of greater help in the management of nematode pathogens (Castillo *et al.*, 2008). Yield loss of different pulse crops under different conditions caused by these nematodes have not yet been determined on a global basis. Mapping of major pulse nematode pests in pulse-growing regions on a global basis is essential for real assessment of yield losses due to nematodes. This would help to plan future strategies in the management of pulse nematodes. A major drawback in the field of plant nematology is that visual symptoms on plants that are due to nematode infestation are very similar to fungal symptoms or nutritional disorders, and so have generally been confused. This often creates difficulties for policy makers and administrators, as they lack expertise. Therefore, while making policies to combat nematode diseases in pulses, trained nematologists with expertise in this field should be involved in discussions.

Nematode pricks on the host plant tissue open ways for other pathogens like bacteria or fungi to cause infection on the host plant. A plant, resistant to a certain fungi or bacteria but grown in the field infested with nematode, can lose its resistance due to early infestation by nematodes. Therefore, while using a breeding programme to produce a variety resistant to fungi or bacteria, breeders should also keep in mind that it should also be resistant to nematodes. Currently, very few reports are available of pulse crops resistant to both nematode and wilt fungus. There is also a big research gap related to nematode–rhizobium and nematode–virus interactions. Furthermore, genotypes of pulse crops having high-level resistance against nematode pests need to be sought out. However, while screening germplasm, breeding materials and land races for the identification of resistant genotypes, confirmation should be made under both pot and field conditions. The multilocational trial of a genotype can help to strengthen the confirmation, which will facilitate identifying donors to be used in the breeding programme. Ali and Singh (2005) are of the view that different races of *M. incognita* and *M. javanica* may have different reactions to different genotypes of pigeonpea and, therefore, the genotype should be screened against specific races.

People are aware of the hazardous effect of chemicals and therefore a substitute for chemicals with other effective methods is required (Moosavi and Askary, 2015). There is a need to integrate various management techniques in order to prepare an integrated nematode management (INM) module to disseminate the nematode management technology to the pulse growers by conducting multilocational trials at farm level (Askary, 2008; Askary, 2015a). The management of nematodes must be based on strategies that integrate the use of available host-resistant or host-tolerant cultivars with crop rotation or inclusion of a non-host crop in cropping systems. However, it should be remembered that crop rotation has certain limitations, particularly in those areas where land is insufficient and choice of crop is limited. The host range of *Meloidogyne* spp. is wide, but that of *Heterodera* spp. is narrow. Under such circumstances, crop rotation for the nematode with a wide host range may not prove advantageous in suppressing nematode population. Therefore, while planning for crop rotation or including a crop in the pulse-based cropping systems, only that crop should be included which may prevent the adverse affects of nematodes on the main crop. In addition, the crop used in crop rotation or included in a cropping system must have some market value and provide quick monetary return to the growers.

Acknowledgements

I am indebted to my teacher, the late Dr S.S. Ali, Emeritus Scientist, Indian Institute of Pulses Research (IIPR), Kanpur, India for providing me with the opportunity to work on pulses. I am also thankful to him for supplying Figs 13.1 and 13.2. Thanks are also due to all the staff at the Central Library, Sher-e-Kashmir University of Agricultural Sciences and Technology of Kashmir (SKUAST-K), Srinagar, India, who were of great help during the collection of information related to the topic.

References

Abd-Elgawad, M.M.M. and Askary, T.H. (2015) Impact of phytonematodes on agriculture economy. In: Askary, T.H. and Martinelli, P.R.P. (eds) *Biocontrol Agents of Phytonematodes.* CAB International, Wallingford, UK, pp. 3–49.

Aboul-Eid, H.Z. and Ghorab, A.I. (1974) Pathological effects of *Heterodera cajani* on cowpea. *Plant Disease Reporter* 58(12), 1130–1133.

Agrawal, S.C. (2003) *Diseases of Pigeonpea*. Concept Publishing Company, New Delhi, India. 352 pp.

Ahmad, S. and Husain, S.I. (1988) Effect of root-knot nematodes on qualitative and quantitative characters of chickpea. *International Nematology Network Newsletter* 5(1), 12–13.

Akhtar, M. (1998) Biological control of plant-parasitic nematodes by neem products in agricultural soils. *Applied Soil Ecology* 7(3), 219–223.

Akhtar, A., Hisamuddin and Abbasi (2012) Interaction between *Meloidogyne incognita*, *Pseudomonas fluorescens* and *Bacillus subtilis* and its effect on plant growth of black gram (*Vigna mungo* L.). *International Journal of Plant Pathology* 3(2), 66–73.

Alam, M.M., Anver, S. and Yadav, A. (1991) Effect of *Meloidogyne incognita* on plant growth and bulk density of plant residues of pigeonpea *Cajanus cajan*. *Afro-Asian Journal of Nematology* 1(1), 73–76.

Alcala, V.J., Tobar-Jimenez, A. and Munoz, M.J.M. (1970) Losiones causadas y racciones provocados por algunos nematodes en las raices de ciertas plantas. *Revista Ibérica de Parasitología* 30(4), 547–566.

Ali, S.S. (1988) *Investigations on plant parasitic nematodes associated with pulse crops. Annual Report*. Directorate of Pulses Research, Indian Council of Agricultural Research (ICAR), Kanpur, India, pp. 77–79.

Ali, S.S. (1993) Prevalence of plant parasitic nematodes associated with chickpea in Gwalior district of Madhya Pradesh. *International Chickpea Newsletter* 28, 11.

Ali, S.S. (1995) *Nematode Problems in Chickpea*. Pawel Graphics Private Limited, Kanpur, India. 184 pp.

Ali, S.S. (1996) Estimation of yield losses in pigeonpea due to reniform nematodes. *Indian Journal of Pulses Research* 9(2), 209–210.

Ali, S.S. (1997) Status of nematode problems and research in India. In: *Diagnosis of Key Nematode Pests of Chickpea and Pigeonpea and Their Management. Proceedings of a Regional Training Course, 25 –30 November, 1996*. ICRISAT, Patancheru, Hyderabad, India. pp. 74–82.

Ali, S.S. (2009) Estimation of unavoidable yield losses in certain *rabi* pulse crops due to the root-knot nematode, *Meloidogyne javanica*. *Trends in Biosciences* 2(2), 48–49.

Ali, S.S. and Askary, T.H. (2001a) Taxonomic status of phytonematodes associated with pulse crops. *Current Nematology* 12(1–2), 75–84.

Ali, S.S. and Askary, T.H. (2001b) Taxonomic status of nematodes of pulse crops. In: Jairajpuri, M.S. and Rahaman, P.F. (eds) *Nematode Taxonomy, Concepts and Recent Trends*. Impressions Quality Printers, Hyderabad, India, pp. 197–216.

Ali, S.S. and Askary, T.H. (2005) Dynamics of nematodes in pulses. In: Singh, G., Sekhon, H.S. and Kolar, J.S. (eds) *Pulses*. Agrotech Publishing Academy, Udaipur, India, pp. 519–532.

Ali, S.S. and Gurha, S.N. (1995) Role of *Meloidogyne javanica* in fusarium wilt of chickpea. *Indian Journal of Pulses Research* 8, 201–203.

Ali, S.S. and Sharma, S.B. (2003) Nematode survey of chickpea production areas in Rajasthan, India. *Nematologia Mediterranea* 31, 147–149.

Ali, S.S. and Singh, B. (2005) Nematodes of pigeonpea and their management. In: Ali, M. and Kumar, S. (eds) *Advances in Pigeonpea Research*. D.K. Agencies Pvt Limited, New Delhi, India, pp. 284–314.

Anver, S. and Alam, M.M. (1992) Effect of latex seed dressing on interacting root-knot and reniform nematodes. *Afro-Asian Journal of Nematology* 2(1–2), 17–20.

Anver, S. and Alam, M.M. (1997) Control of *Meloidogyne incognita* and *Rotylenchulus reniformis* singly and concomitantly on pigeonpea with *Paecilomyces lilacinus*. *Indian Journal of Nematology* 27(2), 209–213.

Anver, A. and Alam, M.M. (1999) Control of *Meloidogyne incognita* and *Rotylenchulus reniformis* singly and concomitantly on chickpea and pigeonpea. *Archives of Phytopathology and Plant Protection* 32(2), 161–172.

Anver, S. and Alam, M.M. (2001) Reaction of pigeonpea accessions to root-knot nematode *Meloidogyne incognita* and reniform nematode *Rotylenchulus reniformis*. *International Chickpea and Pigeonpea Newsletter* 8, 41–42.

Ashraf, M.S. and Khan, T.A. (2008) Biomanagement of reniform nematode, *Rotylenchulus reniformis* by fruit wastes and *Paecilomyces lilacinus* on chickpea. *World Journal of Agricultural Sciences* 4(4), 492–494.

Askary, T.H. (2008) Studies on root-knot nematode infesting pigeonpea and its integrated management. PhD thesis, Department of Plant Protection, Faculty of Agricultural Sciences, Aligarh Muslim University, Aligarh, India.

Askary, T.H. (2011) Effect of different inoculum levels of root-knot nematode *M. javanica* on pigeonpea (*Cajanus cajan* L. Millsp.) cv. UPAS 120. *Indian Journal of Ecology* 38(2), 284–285.

Askary, T.H. (2012) Management of root-knot nematode *M. javanica* in pigeonpea through seed treatment. *Indian Journal of Ecology* 39(1), 151–152.

Askary, T.H. (2015a) Nematophagous fungi as biocontrol agents of phytonematodes. In: Askary, T.H. and Martinelli, P.R.P. (eds) *Biocontrol Agents of Phytonematodes.* CAB International, Wallingford, Oxfordshire, UK, pp. 81–125.

Askary, T.H. (2015b) Limitations, research needs and future prospects in the biological control of phytonematodes. In: Askary, T.H. and Martinelli, P.R.P. (eds) *Biocontrol Agents of Phytonematodes.* CAB International, Wallingford, Oxfordshire, UK, pp. 446–454.

Askary, T.H. and Ali, S.S. (2012) Effect of *Meloidogyne javanica* and *Fusarium udum* singly and concomitantly on wilt resistant accessions of pigeonpea. *Indian Journal of Plant Protection* 40(3), 167–170.

Askary, T.H. and Martinelli, P.R.P. (2015) *Biocontrol Agents of Phytonematodes.* CAB International, Wallingford, Oxfordshire, UK. 470 pp.

Askary, T.H., Banday, S.A., Iqbal, U., Khan, A.A., Mir, M.M. and Waliullah, M.I.S. (2012) Plant parasitic nematode diversity in pome, stone and nut fruits. In: E. Lichtfouse (ed.) *Agroecology and Strategies for Climate Change*. Springer, Heidelberg, Germany. pp. 237–268.

Ayala, A. and Ramirez, C.T. (1964) Host range, distribution and bibliography of the reniform nematode, *Rotylenchulus reniformis* with special reference to Puerto Rico. *Journal of Agriculture – University of Puerto Rico*, 48, 142–161.

Azam, T., Hisamuddin and Singh, S. (2009) Efficacy of plant leaf powder and *Paecilomyces lilacinus* alone and in combination for controlling *Meloidogyne incognita* on chickpea. *Indian Journal of Nematology* 39(2), 152–155.

Baghel, K.S. and Singh, R. (2013) Alarming population of *Pratylenchus* spp. in chickpea growing areas in Rewa and its vicinity and its effect on plant growth and nodulation. *Environment and Ecology* 31(1A), 328–333.

Bajet, N.B. and Castillo, M.B. (1974) effects of *Rotylenchulus reniformis* inoculations on mung bean, soybean and peanut. *Philippine Phytopathology* 1, 50–55.

Bhat, M.Y., Wani, A.H. and Fazal, M. (2012) Effect of *Paecilomyces lilacinus* and plant growth promoting rhizobacteria on *Meloidogyne incognita* inoculated black gram, *Vignamungo* plants. *Journal of Biopesticides* 5(1), 36–43.

Bird, A.F. (1962) The inducement of giant cells of *Meloidogyne javanica*. *Nematologica* 8, 1–10.

Bird, A.F. (1972) Quantitative studies on the growth of syncytia induced in plants by root-knot nematodes. *International Journal of Parasitology* 2, 157–170.

Bridge, J. (1981) Nematodes. In: Ward, A., Mercer, S.L. and Howe, V. (eds) *Pest Control in Tropical Grain Legumes*. Centre for Overseas Pest Research, London, pp. 111–125.

Castillo, M.B. (1975) Plant parasitic nematodes associated with mung bean, soybean and peanut in the Philippines. *Philippines Agriculturist* 59, 91–99.

Castillo, M.B., Alejar, M.S. and Litsinger, J.A. (1977) Pathological reactions and yield loss of mung bean to known populations of *Rotylenchulus reniformis* and *Meloidogyne acrita. Philippine Agriculturist* 61, 12–24.

Castillo, P., Gomez-Barcina, A., Vovlas, N. and Navas, A. (1991) Some plant parasitic nematodes associated with cotton and chickpea in southern Spain with description of *Amplimerlinius magnistylus* sp.n. *Afro-Asian Journal of Nematology* 1(2), 195–203.

Castillo, P., Jiménez Díaz, R.M., Gomez-Barcina, A. and Vovlas, N. (1995) Parasitism of the root-lesion nematode *Praylenchus thornei* on chickpea. *Plant Pathology* 44, 728–733.

Castillo, P., Gómez-Barcina, A. and Jiménez Díaz, R.M. (1996a) Plant parasitic nematodes associated with chickpea in southern Spain and effect of soil temperature on reproduction of *Pratylenchus thornei. Nematologica* 42, 211–219.

Castillo, P., Trapero-Casas, J.L. and Jiménez Díaz, R.M. (1996b) The effect of temperature on hatching and penetration of chickpea roots by *Pratylenchus thornei. Plant Pathology* 45, 310–315.

Castillo, P., Navas-Cortés, J.A., Gomar Tinoco, D., Di Vito, M. and Jiménez-Díaz, R.M. (2003) Interactions between *Meloidogyne artiellia*, the cereal and legume root-knot nematode, and *Fusarium oxysporum* f. sp. *ciceris* race 5 in chickpea. *Phytopathology* 93, 1513–1523.

Castillo, P., Navas-Cortés, J.A., Landa, B.B., Jiménez-Díaz, R.M. and Vovlas, N. (2008) Plant-parasitic nematodes attacking chickpea and their in planta interactions with rhizobia and phytopathogenic fungi. *Plant Disease* 92(6), 840–853.

Chakrabarti, U. and Mishra, S.D. (2001) Seed treatment with neem products for integrated management of *Meloidogyne incognita* infecting chickpea. *Current Nematology* 12(1/2), 15–19.

Chawla, G. and Pankaj (2007) Biochemical basis of resistance in chickpea varieties against the root-knot nematode, *Meloidogyne incognita. Indian Journal of Nematology* 37(1), 105.

Crozzoli, R., Seguro, M., Perichi, G. and Pérez, D. (2011) Response of selections of legumes to *Meloidogyne incognita* and *Meloidogyne enterolobii* (Nematoda; Meloidogynidae). *Fitopatología Venezolana* 24(2), 56–57.

Dalal, M.R. and Bhatti, D.S. (1985) Effect of interaction between *Meloidogyne javanica* and *Fusarium solani* on chickpea. *Indian Journal of Nematology* 5, 287.

Darekar, K.S. and Jagdale, G.B. (1987) Pathogenicity of *Rotylenchus reniformis* to chickpea. *Current Research Reporter* 3(2), 89–91.

Das, D. and Mishra, S.D. (2000) Effect of neem seed powder and neem based formulations as seed coating against *Meloidogyne incognita*, *Heterodera cajani* and *Rotylenchulus reniformis* infecting pigeonpea. *Current Nematology* 11(1, 2), 13–23.

Das, D. and Mishra, S.D. (2002) Seed soaking technology through neem seed powder and neem based formulations for the management of Meloidogyne incognita, Heterodera cajani and Rotylenchulus reniformis infecting pigeonpea. *Current Nematology* 13 (1, 2), 7–17.

Das, D. and Mishra, S.D. (2003) Effect of neem seed powder and neem based formulations for the management of *Meloidogyne incognita*, *Heterodera cajani* and *Rotylenchulus reniformis* infecting pigeonpea. *Annals of Plant Protection Sciences* 11 (1), 110–115.

Dasgupta, D.R. and Gaur, H.S. (1986) The root-knot nematodes *Meloidogyne* spp. in India. In: Swarup, G. and Dasgupta, D.R. (eds) *Plant Parasitic Nematodes of India: Problems and Progress*. Nematological Society of India, Indian Agricultural Research Institute (IARI), New Delhi, India, pp. 139–171.

Devi, S. (1995) Pathogenicity of *Rotylenchulus reniformis*, *Meloidogyne incognita* and *Pratylenchus thornei* and their interaction effect with *Fusarium oxysporum* f.sp. *ciceri* on chickpea. *Indian Journal of Nematology* 25(1), 118–119.

Di Vito, M., Greco, N., Singh, K.B. and Saxena, M.C. (1988) Response of chickpea germplasm lines to *Heterodera ciceri* attack. *Nematologia Mediterranea* 16, 17–18.

Di Vito, M., Greco, N. and Saxena, M.C. (1992) Pathogenicity of *Pratylenchus thornei* on chickpea in Syria. *Nematologia Mediterranea* 20, 71–73.

Di Vito, M., Greco, N., Halila, H.M., Mabsoute, L., Labdi, M., Beniwal, S.P.S., Saxena, M.C., Singh, K.B. and Solh, M.B. (1994a) Nematodes of cool-season food legumes in North Africa. *Nematologia Mediterranea* 22, 3–10.

Di Vito, M., Greco, N., Oreste, G., Saxena, M.C., Singh, K.B. and Kusmenoglu, I. (1994b) Plant parasitic nematodes of legumes in Turkey. *Nematologia Mediterranea* 22, 245–251.

Dwivedi, K., Upadhyay, K.D., Verma, R.A. and Ahmad, F. (2008) Role of bioagents in management of *Pratylenchus thornei* infecting chickpea. *Indian Journal of Nematology* 38(2), 138–140.

Edwards, E.E. (1956) Studies on resistance to the root-knot nematode of the genus *Meloidogyne* Goeldi, 1887. *Proceedings of the Helminthological Society of Washington* 23, 112–118.

Elliot, A.P. and Bird, G.W. (1985) Pathogenicity of *Pratylenchus penetrans* to navy bean (*Phaseolus vulgaris* L.). *Journal of Nematology* 17, 81–85.

Epps, J.M. and Chambers, A.Y. (1959) Mung bean (*Phaseolus aureus*), a host of the soybean cyst nematode (*Heterodera glycines*). *Plant Disease Reporter* 43, 981–982.

Fazal, M., Siddiqui, Z.A. and Imran, M. (1992) Effect of pre, post and simultaneous inoculations of *Rhizobium* sp., *Rotylenchulus reniformis* and *Meloidogyne incognita* on lentil. *Nematologia Mediterranea* 20, 159–161.

Fazal, M., Khan, M.I., Raza, M.M.A. and Siddiqui, Z.A. (1994) Interaction between *Meloidogyne incognita* and *Fusarium oxysporum* f.sp. *lentis* on lentil. *Nematologia Mediterranea* 22, 185–187.

Filho, A.J.V., Inomoto, M.M., Godoy, R. and Feraz, L.C.C.B. (2010) Host response of Brazilian pigeonpea lines to *Rotylenchulus reniformis* and *Pratylenchus zeae*. *Nematologia Brasileir* 34(4), 204–210.

Fracl, L.J. and Wheeler, T.A. (1993) Interaction of plant parasitic nematodes with wilt inducing fungi. In: Khan, M.W. (ed.) *Nematode Interactions*. Chapman and Hall, London, UK, pp. 78–103.

Freire, F.C. das O., Diogenes, A.M. and da Ponte, J.J. (1972) Nematodes das galhas *Meloidogyne javanica* e *M. incognita*, parasitando leguminosas forrageiras. *Revista da Sociedade Brasileira de Fitopatologia* 5, 27–32.

Gaur, H.S., Singh, R.V., Kumar, S., Kumar, V. and Singh, J.V. (2001) Search for nematode resistance in crops. In: *AICRP on Plant Parasitic Nematodes with Integrated Approach for Their Control*. IARI, New Delhi, India.

Glazer, J. and Orion, D. (1983) Studies on anhydrobiosis of *Pratylenchus thornei*. *Journal of Nematology* 15, 333–338.

Goel, S.R. (2004) Reaction of certain urdbean (*Vignamungo* L.) genotypes to root-knot nematode, *Meloidogyne javanica* and *Meloidogyne incognita*. *Annals of Agricultural Research* 25(4), 626–627.

Goel, S.R. and Gupta, D.C. (1985) Interaction of *Meloidogyne javanica* and *Rhizoctonia bataticola* on chickpea (*Cicer arietinum*). *Indian Journal of Nematology* 16(1), 133–134.

Goswami, B.K., Pandey, R.K., Goswami, J. and Tewari, D.D. (2007) Management of disease complex caused by root knot nematode and root wilt fungus on pigeonpea through soil organically enriched with Vesicular Arbuscular Mycorrhiza, karanj (*Pongamia pinnata*) oilseed cake and farmyard manure. *Journal of Environmental Science and Health* 42(8), 899–904.

Greco, N. (1984) *Meloidogyne artiellia* on chickpea in Italy. *Nematologia Mediterranea* 12(2), 235–238.

Greco, N. (1987) Nematodes and their control in chickpea. In: Saxena, M.C. and Singh, K.B. (eds) *The Chickpea*. CAB International, Wallingford, Oxfordshire, UK, pp. 271–281.

Greco, N. and Di Vito, M. (1987) The importance of plant parasitic nematodes in food legume production in the

Mediterranean region. In: Saxena, M.C., Sikora, R.A. and Srivastava, J.P. (eds) *Nematodes Parasitic to Cereals and Legumes in Temperate Semi-Arid Regions. Proceedings of Workshop Held at Larnaca, Cyprus, 1–5 March.* ICARDA, Aleppo, Syria, pp. 29–45.

Greco, N. and Sharma, S.B. (1990) Progress and problems in the management of nematode diseases. In: *Chickpea in the Nineties. Proceedings of Second International Workshop on Chickpea Improvement.* ICRISAT Centre, Patancheru, India, pp. 135–137.

Greco, N., Di Vito, M., Reddy, M.V. and Saxena, M.C. (1984) A preliminary report of survey of plant parasitic nematodes of leguminous crops in Syria. *Nematologia Mediterranea* 12, 87–93.

Greco, N., Di Vito, M., Saxena, M.C. and Reddy, M.V. (1988) Effect of *Heterodera ciceri* on yield of chickpea and lentil and development of this nematode in chickpea in Syria. *Nematologica* 34(1), 98–114.

Greco, N., Di Vito, M. and Nombela, G. (1992a) The emergence of juveniles of *Heterodera ciceri. Nematologica* 38, 514–519.

Greco, N., Vovlas, N. and Inserra, R.N. (1992b) The chickpea cyst nematode, *Heterodera ciceri. Nematology Circular*. Florida Department of Agriculture, Tallahassee, Florida, USA. 198 pp.

Haidar, M.G., Askary, T.H. and Nath, R.P. (2001) Nematode population as influenced by paddy based cropping sequences. *Indian Journal of Nematology* 31, 68–71.

Haseeb, A. and Shukla, P.K. (2004) Management of *Heterodera cajani, Meloidogyne incognita* and *Fusarium* wilt on pigeonpea with some chemicals, biopesticides and bio-agents. *Nematologia Mediterranea* 32, 217–222.

Haseeb, A., Sharma, A. and Shukla, P.K. (2005) Studies on the management of root-knot nematode, *Meloidogyne incognita*-wilt fungus, *Fusarium oxysporum* disease complex of green gram, *Vigna radiate* cv ML-1108. *Journal of Zhejiang University Science B* 6(8), 736–742.

Haseeb, A., Sharma, A., Abuzar, S. and Kumar, V. (2006) Evaluation of resistance in different cultivars of chickpea against *Meloidogyne incognita* and *Fusarium oxysporum* f. sp. *cicero* under field conditions. *Indian Phytopathology* 59(2), 234–236.

Hashim, Z. (1979) A preliminary report of the plant parasitic nematodes in Jordan. *Nematologia Mediterranea* 7, 177–186.

Haung, G.S. (1985) Formation, anatomy and physiology of giant cells induced by root-knot nematodes. In: Sasser, J.N. and Carter, C.C. (eds) *An Advanced Treatise on Meloidogyne*, Vol. 1, *Biology and Control.* North Carolina State University Graphics, Raleigh, North Carolina, USA. pp. 155–164.

Huising, J. and Okoth, P. (2011) Effect of soil fertility management practices and Bacillus subtilis on plant parasitic nematodes associated with common bean, *Phaseolus vulgaris. Tropical and Subtropical Agroecosystems* 13(1), 29–36.

Hussaini, S.S. and Seshadri, A.R. (1975) Interrelationships between *Meloidogyne incognita* and *Rhizobium* sp. on mung bean (*Phaseolus aureus*). *Indian Journal of Nematology* 5, 189–199.

Hussain, S., Zareen, A., Zaki, M.J. and Abid, M. (2001) Response of ten chickpea (*Cicer arietinum* L.) cultivars against *Meloidogyne javanica* (Treub) Chitwood and disease control by fungal filtrates. *Pakistan Journal of Biological Sciences* 4(4), 429–432.

Hussey, R.S. (1989) Disease including secretions of plant-parasitic nematodes. *Annual Review of Phytopathology* 27, 123–141.

Hutton, D.G. and Hammerton, J.L. (1975) Investigating the role of *Rotylenchulus reniformis* in a decline of pigeonpea. *Nematropica* 5(2), 24.

Hutton, D.G., Wilkinson, R.E. and Mai, W.F. (1972) Effect of two plant parasitic nematodes on *Fusarium* dry root rot of beans. *Phytopathology* 63, 749–751.

Inderjit Singh, V.P.S., Chahal, P.K., Sakhuja, P.K. and Chohan, J.S. (1977) Effect of different levels of *Meloidogyne incognita* in the presence or absence of *Rhizobium phaseoli* on *Phaseolus aureus. Indian Journal of Nematology* 7, 172–174.

Jain, R.K. and Bhatti, D.S. (1991) Evaluation of the effective integrated methods for the control of root-knot nematode, *Meloidogyne javanica* in tomato. *Indian Journal of Nematology* 21(2), 107–112.

Jain K.C. and Sharma, S.B. (1996) Loss of *Fusarium* wilt resistance in a pigeonpea line ICPL 270 in reniform nematode infested soil at ICRISAT Asia Centre. *International Chickpea and Pigeonpea Newsletter* 3, 90.

Johnson, A.W. (1985) The role of nematicides in nematode management. In: Sasser, J.N. and Carter, C.C. (eds) *Advanced Treatise on Meloidogyne. Vol. 1. Biology and Control*. North Carolina State University Graphics, Raleigh, North Carolina, USA, pp. 249–268.

Jonathan, E.I., Cannayane, I., Devrajan, K., Kumar, S. and Ramakrishnan, S. (2005) *Agricultural Nematology*. Sri Sakthi Promotional Litho Process, Coimbatore, Tamil Nadu, India. 260 pp.

Kaloshian, I., Greco, N. and Saad, A.T. (1986a) Hatching of cysts and infectivity of *Heterodera ciceri* on chickpea. *Nematologia Mediterranea* 14, 129–133.

Kaloshian, I., Greco, N., Saad, A.T. and Vovlas, N. (1986b) Life cycle of *Heterodera ciceri* on chickpea. *Nematologia Mediterranea* 14, 135–145.

Kannaiyan, J., Nene, Y.L., Reddy, M.V., Ryan, J.G. and Raju, T.N. (1984) Prevalence of pigeonpea diseases and associated crop losses in Asia, Africa and the Americas. *Tropical Pest Management* 30(1), 62–71.

Kaushik, H.D. and Bajaj, H.K. (1981) Control of root-knot nematodes *Meloidogyne javanica* and crop yield. *Haryana Agricultural University Journal of Research* 22(1), 40–45.

Khan, M.W. and Hosseini-Nejad, S.A. (1991) Interaction of *Meloidogyne javanica* and *Fusarium oxysporum* f.sp. *ciceri* on some chickpea cultivars. *Nematologia Mediterranea* 19, 61–63.

Kimenju, J.W., Karanja, N.K. and Macharia, I. (1999) Plant parasitic nematodes associated with common bean in Kenya and the effect of *Meloidogyne* infection on bean nodulation. *African Crop Science Journal* 7(4), 503–510.

Kornobis, S. and Wolny, S. (1997) Occurrence of plant parasitic nematodes on weeds in agrobiocenosis in the Wielkopolska region in Poland. *Fundamental and Applied Nematology* 20, 627–632.

Koshy, P.K. (1967) A new species of *Heterodera* from India. *Indian Phytopathology* 20, 272–274.

Koshy, P.K. and Swarup, G. (1971a) Distribution of *Heterodera avenae*, *H. zeae*, *H.cajani*, and *Anguina tritici* in India. *Indian Journal of Nematology* 1, 106–111.

Koshy, P.K. and Swarup, G. (1971b) Investigations on the life history of the pigeonpea cyst nematode, *Heterodera cajani*. *Indian Journal of Nematology* 1, 44.

Krishna Rao, V. and Krishnappa, K. (1996) Interaction of *Fusarium oxysporum* f. sp. *ciceri* with *Meloidogyne incognita* on chickpea in two soil types. *Indian Phytopathology* 49, 142–147.

Lakshminarayan, S., Husain, S.I. and Siddiqui, Z.A. (1990) Evaluation of twenty chickpea varieties against *Rotylenchulus reniformis* and *Fusarium solani*. *New Agriculturist* 1(1), 29–39.

Linford, M.B. and Oliveira, J.M. (1940) *Rotylenchulus reniformis*, nov. gen., n. sp., a nematode parasite of roots. *Proceedings of the Helminthological Society of Washington* 7(1), 35–42.

Loewenberg, J.R., Sullivan, T. and Schuster, M.L. (1960) Gall induction by *Meloidogyne incognita* by surface feeding and factors affecting the behaviour patterns of the second stage larva. *Phytopathology* 50, 322–323.

Mahapatra, B.C. and Das, S.N. (1979) Host range and pathogenicity of *Tylenchorhynchus mashhoodi* Siddiqi and Basir, 1949 on Maize (*Zea mays*). *Indian Journal of Nematology* 9(1), 64.

Mahapatra, B.C. and Padhi, N.N. (1986) Inoculum potential of the reniform nematode in relation to growth of chickpea. *International Chickpea Newsletter* 15, 15.

Mahmood, I. and Siddiqui, Z.A. (1993) Integrated management of *Rotylenchulus reniformis* by green manuring and *Paecilomyces lilacinus*. *Nematologia Mediterranea* 21(2), 285–287.

Malhotra, R.S., Singh, K.B., Vito, M. di, Greco, N. and Saxena, M.C. (2002) Registration of ILC 10765 and ILC 10766 chickpea germplasm lines resistant to cyst nematode. *Crop Science* 42(5), 1756.

Malhotra, R.S., Greco, N., Vito, M. di, Singh, K.B., Saxena, M.C. and Hajjar, S. (2008) Registration of FLIP 2005–8C and FLIP 2005–9C, chickpea germplasm lines resistant to chickpea cyst nematode. *Journal of Plant Registrations* 2(1), 65–66.

Mamluk, O.F., Augustin, B. and Bellar, M. (1983) New record of cyst and root-knot nematodes on legume crops in the dry areas of Syria. *Phytopathologia Mediterranea* 22, 80.

Mani, A. and Sethi, C.L. (1984) Plant growth of chickpea as influenced by initial inoculums levels of *Meloidogyne incognita*. *Indian Journal of Nematology* 14(1), 41–44.

Mani, A. and Sethi, C.L. (1987) Interaction of root-knot nematode, *Meloidogyne incognita* with Fusarium oxysporum f. sp. ciceris and *F. solani* on chickpea. *Indian Journal of Nematology* 17, 1–6.

Maqbool, M.A. (1980) Occurrence of eight cyst nematodes on some agricultural crops in Pakistan. *Journal of Science – Karachi University* 8, 103–108.

Maqbool, M.A. (1987) Present status of research on plant parasitic nematodes in cereals and food and forage legumes in Pakistan. In: Saxena, M.C., Sikora, R.A. and Srivastava, J.P. (eds) *Nematodes Parasitic to Cereals and Legumes in Temperate Semi-Arid Regions*. *Proceedings of Workshop Held at Larnaca, Cyprus, March 1–5*. ICARDA, Aleppo, Syria, pp. 173–180.

Marwoto, B. (1988) Control of root knot nematodes (*Meloidogyne* spp.) on kidney bean (*Phaseolus vulgaris* L.) using systemic nematicides. *Buletin Penelitian Hortikultura* 16(2), 5–10.

Mathur, B.N., Handa, D.K. and Singh, H.G. (1969) Note on the occurrence of *Meloidogyne arenaria* as a serious pest of *Cicer arietinum*. *Madras Agriculture Journal* 56, 744.

Meena, P., Nehra, S. and Trivedi, P.C. (2009) Efficacy of decomposed organic cakes against *Heterodera cajani* infecting *Cajanus cajan*. *Asian Journal of Experimental Sciences* 23(1), 181–184.

Meher, H.C., Gajbhiye, V.T., Singh, G., Kamra, K. and Chawla, G. (2010) Persistence and nematicidal efficacy of carbosulfan, cadusafos, phorate, and triazophos in soil and uptake by chickpea and tomato crops under tropical conditions. *Journal of Agricultural and Food Chemistry* 58(3), 1815–1822.

Mishra, S.D. (2007) Economic management of phytonematodes in pulse crops. In: Rajvanshi, I. and Sharma, G.L. (eds) *Ecofriendly Management of Phytonematodes*. Oxford Book Company, Jaipur, India, pp. 112–121.

Mishra, S.D., Dhawan, S.C., Tripathi, M.N. and Nayak, S. (2003) Field evaluation of biopesticides, chemicals and bioagents on plant parasitic nematodes infesting chickpea. *Current Nematology* 14(1, 2), 82–92.

Mohanty, C.B. and Padhi, N.N. (1987) Pathogenic effect of reniform nematode at varying levels of inoculation of pigeonpea. *International Nematology Network Newsletter* 4(3), 15–16.

Mojumder, V. (1999) Effect of seed treatment of chickpea with crude neem products and neem-based pesticides on nematode multiplication in soil and the grain yield. *International Journal of Nematology* 9(1), 76–79.

Mojumder, V. and Mishra, S.D. (1991) Nematicidal efficacy of some plant products and management of *Meloidogyne incognita* in pulse crops by soaking seeds in their aqueous extracts. *Current Nematology* 2(1), 27–32.

Mojumder, V. and Mishra, S.D. (2001) Effect of neem products as seed coating of pigeonpea seed on *Heterodera cajani. Pesticide Research Journal* 13(1), 103–105.

Moosavi, M.R. and Askary, T.H. (2015) Nematophagous fungi: commercialization. In: Askary, T.H. and Martinelli, P.R.P. (eds) *Biocontrol Agents of Phytonematodes.* CAB International, Wallingford, Oxfordshire, UK, pp. 187–202.

Myint, Y.Y., Lwin, T., Thwe, K.H., Min, Y.Y., Aye, S.S., Lin, M., Kyi, P.P., Maung, Z.T.Z. and Than, P.P. (2005) New record on the occurrence of cyst nematode, *Heterodera cajani* Koshy, 1967 on sesame, *Sesamum indicum* in Myanmar. In: Thaung, M., Win, K.K. and Hom, N.H. (eds) *Proceedings of the Fourth Agricultural Research Conference in Commemoration of the Ruby Jubilee of Yezin Agricultural University, Yezin Agricultural University, Nay Pyi Taw, Myanmar, 17–18 February 2005.* Yezin Agricultural University, Nay Pyi Taw, Myanmar, pp. 59–67.

Nadakal, A.M. (1964) Studies on plant parasitic nematodes of Kerala. Iii. An additional list of plants attacked by root-knot nematode, *Meloidogyne* sp. (Tylenchoidea: Heteroderidae). *Journal of Bombay Natural History Society* 61(2), 467–469.

Nageswari, S. and Mishra, S.D. (2005) Integrated nematode management schedule incorporating neem products, VAM and soil solarisation against *Heterodera cajani* infesting pigeonpea. *Indian Journal of Nematology* 35(1), 68–71.

Nene, Y.L. and Reddy, M.V. (1987) Chickpea diseases and their control. In: Saxena, M.C. and Singh, K.B. (eds) *The Chickpea.* CAB International, Wallingford, Oxfordshire, UK, pp. 233–270.

Nene, Y.L. and Sheila, VK. (1990) Pigeonpea: geography and importance. In: Nene, Y.L., Hall, S.D. and Sheila, V.K. (eds) *The Pigeonpea.* CAB International and ICRISAT, Wallingford, Oxfordshire, UK and Patancheru, India, pp. 1–14.

Nene, Y.L., Sheila, V.K. and Sharma, S.B. (1989) A world list of chickpea (*Cicer arietinum* L.) and pigeonpea (*Cajanus cajan* (L.) Millsp.) pathogens. *Legumes Pathology Progress Report-7.* ICRISAT, Patancheru, India, 23 pp.

Nene, Y.L., Sheila, V.K. and Sharma, S.B. (1996) *A World List of Chickpea and Pigeonpea Pathogens.* ICRISAT, Patancheru, India, 27 pp.

Oteifa, B.A. (1987) Nematode problems of winter season cereals and food legume crops in the Mediterranean region. In: Saxena, M.C., Sikora, R.A. and Srivastava, J.P. (eds) *Nematodes Parasitic to Cereals and Legumes in Temperate Semi-arid Regions. Proceedings of Workshop Held at Larnaca, Cyprus, 1–5 March.* ICARDA, Aleppo, Syria, pp. 199–209.

Palomares-Rius, J.E., Castillo, P., Navas-Cortés, J.A., Jiménez-Díaz, R.M. and Tena, M. (2011) A proteomic study of in-root interactions between chickpea pathogens: the root-knot nematode *Meloidogyne artiellia* and the soil-borne fungus *Fusarium oxysporum* f. sp. *ciceris* race 5. *Journal of Proteomics* 74(10), 2034–2051.

Pandey, R.K., Pant, H., Yadav, S., Varshney, S., Pandey, G. and Dwivedi, B. (2005) Application of biocontrol agents and neem cake for the management of *Meloidogyne incognita* on chickpea (*Cicer arietinum* L.). *Pakistan Journal of Nematology* 23(1), 57–60.

Pant, H., Pandey, G. and Shukla, D.N. (2004) Effect of different concentrations of bio-control agents on root-knot disease of chick pea and its rhizosphere microflora. *Pakistan Journal of Nematology* 22(1), 103–109.

Parisi, B., Baschieri, T., del bianco, F., di vito, M., Ranalli, P. and Carboni, A. (2004) Breeding for nematodes resistance in common bean (*Phaseolus vulgaris* L.). *Proceedings of the XLVIII Italian Society of Agricultural Genetics – SIFV-SIGA Joint Meeting Lecce, Italy,* 15–18 September 2004.

Pasha, M.J., Siddiqui, Z.A., Khan, M.W. and Qureshi, S.I. (1987) Histopathology of egg plant roots infected with root-knot nematode, *Meloidogyne incognita. Pakistan Journal of Nematology* 5, 27–34.

Patel, G.A. and Patel, D.J. (1993) Avoidable yield losses in pigeonpea cv. Pusa Ageti due to *Meloidogyne javanica. International Pigeonpea Newsletter* 17, 26–27.

Patel, B.A., Chavda, J.C. Patel, S.T. and Patel, D.J. (1987) Susceptibility of some pigeonpea lines to root-knot nematodes (*Meloidogyne incognita* and *M. javanica*). *International Pigeonpea Newsletter* 6, 55–57.

Patel, B.A., Patel, D.J. and Patel, R.G. (2000) Effect of interaction between Meloidogyne *javanica* pathotype 1 and wilt inducing fungus, *Fusarium oxysporum* f. sp. *ciceri* on chickpea. *International Chickpea and Pigeonpea Newsletter* 7, 17–18.

Pavaraj, M., Karthikairaj, K. and Rajan, M.K. (2010) Effect of leaf extract of *Ageratum conyzoides* on the biochemical profile of blackgram *Vignamungo* infected by root-knot nematode, *Meloidogyne incognita. Journal of Biopesticides* 3(1), 313–316.

Peacock, F.C. (1957) Studies on root-knot nematodes of the genus *Meloidogyne* in the Gold coast, Part II. *Nematologica* 2, 114–122.

Pereira, P.R., Fidelis, R.R., Santos, M.M. dos, Santos, G.R. dos and Nascimento, I.R.do (2013) Tolerance of the common bean genotypes *Meloidogyne incognita* race 2 under high temperature conditions. *Revista Verde de Agroecologia e Desenvolvimento Sustentável* 8(4), 202–207.

Prasad, D. (2006) Integrated management of *Rotylenchulus reniformis* infecting chickpea. *Annals of Plant Protection Sciences* 14(2), 518–519.

Rashid, A., Khan, F.A. and Khan, A.M. (1973) Plant parasitic nematodes associated with vegetables, fruits, cereals and crops following infection by *Rotylenchulus reniformis*. *Indian Journal of Nematology* 3, 8–23.

Ravichandra, N.C., Krishnappa, K., Anil Kumar, T.B. and Saifulla, M. (1988) Reaction of pigeonpea lines to the root-knot nematode (*Meloidogyne incognita*). *International Pigeonpea Newsletter* 7, 31–32.

Reddy, M.V., Sharma, S.B. and Nene, Y.L. (1990) Pigeonpea disease management. In: Nene, Y.L., Susan, D.H. and Sheila, Y.K. (eds) *The Pigeonpea*. CAB International, Wallingford, Oxfordshire, UK, pp. 303–347.

Rehman, B., Ganai, M.A., Parihar, K., Siddiqui, M.A. and Usman, A. (2012) Management of root-knot nematode, *Meloidogyne incognita* affecting chickpea, *Cicer arietinum* for sustainable production. *Biosciences International* 1(1), 1–5.

Rehman, B., Ahmad, F., Babalola, O.O., Ganai, M.A., Parihar, K. and Siddiqui, M.A. (2013) Usages of botanical extracts for the management of root-knot nematode, *Meloidogyne incognita* in chickpea. *Journal of Pure and Applied Microbiology* 7(3), 2385–2388.

Rodriguez-Kabana, R. and Ingram, E.G. (1978) Susceptibilidad del gandul a species de fitonematodos en Alabama. *Nematropica* 8, 32–34.

Sankaranarayanan, C. and Hari, K. (2013) Biomanagement of root-knot nematode *Meloidogynejavanica*in sugarcane by combined application of arbuscularmycorrhizal fungi and nematophagous fungi. *Journal of Sugarcane Research* 3(1), 62–70.

Sarna, N.T. (1984) Studies on histopathology and histochemistry of root galls incited by *M. incognita* in *Cicer arietinum*. PhD thesis, Rajasthan Agricultural University, Rajasthan, India.

Sasser, J.N. and Freckman, D.W. (1987) A world prospective in nematology: the role of society. In: Veech, J.A. and Dickson, D.W. (eds.) *Vistas on Nematology, a Commemoration of the Twenty-Fifth Anniversary of the Society of Nematologists*. Society of Nematologists Inc., Hyattsville, Maryland, USA, pp. 7–14.

Saxena, R. and Reddy, D.D.R. (1987) Crop losses in pigeonpea and mungbean by pigeonpea cyst nematode, *Heterodera cajani*. *Indian Journal of Nematology* 17(1), 91–94.

Saxena, M.C., Greco, N. and Di Vito, M. (1992) Control of Heterodera ciceri by crop rotation. *Nematologia Mediterranea* 20(1), 75–78.

Sebastian, S. and Gupta, P. (1996a) Evaluation of three plant latices as seed treatment against Pratylenchus *thornei* on chickpea. *International Chickpea and Pigeonpea Newsletter* 3, 39–40.

Sebastian, S. and Gupta, P. (1996b) Evaluation of some oil-cakes against Pratylenchus thornei on chickpea. *International Chickpea and Pigeonpea Newsletter* 3, 40–41.

Sebastian, S. and Gupta, P. (1997) Crop loss trial of chickpea infested with *Pratylenchus thornei*. *Indian Journal of Nematology* 27(1), 142–143.

Shahina, F. and Maqbool, M.A. (1995) *Cyst Nematodes of Pakistan (Heteroderidae)*. National Nematological Research Centre, University of Karachi, Pakistan.

Sharf, R., Shiekh, H., Syed, A., Akhtar, A. and Robab, M.I. (2014a) Interaction between *Meloidogyne incognita* and *Pochonia chlamydosporia* and their effects on the growth of *Phaseolus vulgaris*. *Archives of Phytopathology and Plant Protection* 47(5), 622–630.

Sharf, R., Hisamuddin, Abbasi and Akhtar, A. (2014b) Combined effect of biofertilizers and fertilizer in the management of *Meloidogyne incognita* and also on the growth of red kidney bean (*Phaseolus vulgaris*). *International Journal of Plant Pathology* 5(1), 1–11.

Sharma, R.D. (1981) Pathogenicity of *Meloidogyne javanica* to bean (*Phaseolus vulgaris* L.) *Sociedade Brasilerira de Nematologia* 5, 137–144.

Sharma, S.B. (1993) Pearly root of pigeonpeas caused by *Heterodera cajani*. *Indian Journal of Nematology* 21, 169.

Sharma, S.B. and McDonald, D. (1990) Global status of nematode problems of groundnut, pigeonpea, chickpea, sorghum and pearl millet and suggestions for future work. *Crop Protection* 9, 453–458.

Sharma, S.B. and Nene, Y.L. (1988) Effect of *Heterodera cajani*, *Rotylenchulus reniformis* and *Hoplolaimus seinhorsti* on pigeonpea biomass. *Indian Journal of Nematology* 18(2), 273–278.

Sharma, S.B. and Nene, Y.L. (1989) Interrelationship between *Heterodera cajani* and *Fusarium udum* in pigeonpea. *Nematropica* 19(1), 21–28.

Sharma, S.B. and Nene, Y.L. (1990a) Effects of soil solarization on nematodes parasitic to chickpea and pigeonpea. *Journal of Nematology* 22(4S), 658–664.

Sharma, S.B. and Nene, Y.L. (1990b) Effect of *Fusarium udum* alone and in combination with *Rotylenchulus reniformis* or *Meloidogyne* spp. on wilt incidence, growth of pigeonpea and multiplication of nematodes. *International Journal of Tropical Plant Diseases* 8, 95–101.

Sharma, S.B. and Nene, Y.L. (1992) Spatial and temporal distribution of plant parasitic nematodes on pigeonpea in alfisols and vertisols. *Nematropica* 22, 13–20.

Sharma, S.B. and Swarup, G. (1984) *Cyst Forming Nematodes of India*. Cosmos Publications, New Delhi, India. 152 pp.

Sharma, S.K., Singh, I. and Sakhuja, P.K. (1971) Effect of different cropping sequences on the population of root-knot nematodes, *Meloidogyne incognita*. *Indian Journal of Nematology* 9(1), 57–58.

Sharma, S.B., Kumar, A. and MacDonald, D. (1991) A greenhouse technique for resistance to *Heterodera cajani*. *Annals of Applied Biology* 118(2), 351–356.

Sharma, S.B., Smith, D.H. and McDonald, D.I. (1992) Nematode constraints of chickpea and pigeonpea production in the semi-arid tropics. *Plant Disease* 76(9), 868–874.

Sharma, S.B., Ali, S.S., Patel, D.J., Patel, H.V., Patel, B.A. and Patel, S.K. (1993a) Distribution and importance of plant parasitic nematodes associated with pigeonpea in Gujarat state, India. *Afro-Asian Journal of Nematology* 3(1), 55–59.

Sharma, S.B., Nene, Y.L., Reddy, M.V. and McDonald, D. (1993b) Effect of *Heterodera cajani* on biomass and grain yield of pigeonpea on vertisol in pot and field experiments. *Plant Pathology* 42(2), 163–167.

Sharma, S.B., Mohiuddin, M., Jain, K.C. and Remanandan, P. (1994) Reaction of pigeonpea cultivars and germplasm accessions to the root-knot nematode, *Meloidogyne javanica*. *Journal of Nematology* 26(4), 644–652.

Sharma, S.B., Siddiqi, M.R., Rahaman, P.F., Ali, S.S. and Ansari, M.A. (1998) Description of *Heterodera swarupi* sp. n. (Nematoda Heteroderidae), a parasite of chickpea in India. *International Journal of Nematology* 8, 111–116.

Sharma, S.B., Jain, K.C. and Lingaraju, S. (2000) Tolerance to reniform nematode (*Rotylenchulus reniformis*) race A in pigeonpea (*Cajanus cajan*) genotypes. *Annals of Applied Biology* 136(3), 247–252.

Siddiqui, Z.A. and Akhtar, M.S. (2009) Effect of plant growth promoting rhizobacteria, nematode parasitic fungi and root-nodule bacterium on root-knot nematodes *Meloidogyne javanica* and growth of chickpea. *Biocontrol Science and Technology* 19(5), 511–521.

Siddiqui, Z.A. and Husain, I. (1992) Response of 20 chickpea cultivars to *Meloidogyne incognita* race 3. *Nematologia Mediterranea* 20(1), 33–36.

Siddiqui, Z.A. and Mahmood, I. (1994) Interactions of *Meloidogyne javanica*, *Rotylenchulus reniformis*, *Fusarium oxysporum* f. sp. *ciceri* and *Bradyrhizobium japonicum* on the wilt disease complex of chickpea. *Nematologia Mediterranea* 22(2), 135–140.

Siddiqui, Z.A. and Mahmood, I. (1995a) Biological control of *Heterodera cajani* and *Fusarium udum* by *Bacillus subtilis*, *Bradyrhizobium japonicum* and *Glomus fasciculatum* on pigeonpea. *Fundamentals of Applied Nematology* 18(6), 559–566.

Siddiqui, Z.A. and Mahmood, I. (1995b) Some observations on the management of the wilt disease complex of pigeonpea by treatment with a vesicular arbuscular fungus and biocontrol agents for nematodes. *Bioresource Technology* 54(3), 227–230.

Siddiqui, Z.A. and Mahmood, I. (1996) Biological control of *Heterodera cajani* and *Fusarium udum* on pigeonpea by *Glomus mosseae*, *Trichoderma harzianum*, and *Verticillium chlamydosporium*. *Israel Journal of Plant Sciences* 44(1), 49–56.

Siddiqui, Z.A. and Mahmood, I. (1999) Effect of *Heterodera cajani* and *Meloidogyne incognita* with *Fusarium udum* and *Bradyrhizobium japonicum* on the wilt disease complex of pigeonpea. *Indian Phytopathology* 52(1), 66–70.

Siddiqui, Z.A. and Shakeel, U. (2006) Use of fluorescent *Pseudomonads* isolates for the biocontrol of wilt disease complex of pigeonpea in green house assay and under pot condition. *Plant Pathology Journal* 5(1), 99–105.

Siddiqui, Z.A., Baghel, G. and Akhtar, M.S. (2007) Biocontrol of *Meloidogyne javanica* by rhizobium and plant growth-promoting rhizobacteria on lentil. *World Journal of Microbiology and Biotechnology* 23(3), 435–441.

Siddiqui, Z.A., Shakeel, U. and Siddiqui, S. (2008) Biocontrol of wilt disease complex of pigeonpea by fluorescent *Pseudomonads* and *Bacillus* spp. under pot and field conditions. *Acta Phytopathologica et Entomologica Hungarica* 43, 79–94.

Sikora, R.A. and Greco, N. (1990) Nematode parasites of food legumes. In: Luc, M., Sikora, R.A. and Bridge, J. (eds) *Plant Parasitic Nematodes in Subtropical and Tropical Agriculture*. CAB International, Wallingford, Oxfordshire, UK, pp. 181–235.

Simon, L.S. and Dass, S. (2010) Screening of chickpea, field pea, lentil and pigeonpea against root-knot nematode, *Meloidogyne incognita*. *Indian Journal of Nematology* 40(2), 231–233.

Singh, R.N. (1972) Root-knot disease of urd and mung in India. *Indian Journal of Mycology and Plant Pathology* 2, 87.

Singh, V.K. (2004) Management of *Heterodera cajani* on pigeonpea with nematicides and organic amendments. *Indian Journal of Nematology* 34(2), 213–214.

Singh, B. (2009) Management of root-knot nematode *Meloidogyne javanica* by seed treatment with neem based products in pigeonpea. *Journal of Food Legumes* 22(1), 71–72.

Singh, B. and Jagadeeswaran, R. (2014) Phytoparasitic nematodes associated with chickpea crop in Bundelkhand region of Uttar Pradesh and Madhya Pradesh. *Indian Journal of Agricultural Sciences* 84(10), 1284–1287.

Singh, V.K., Rai, P.K. and Singh, K.P. (1993) Effect of simultaneous and sequential inoculation of *Heterodera cajani* and *Fusarium udum* on pigeonpea. *Current Nematology* 4(2), 129–133.

Singh, B., Ali, S.S., Naimuddin and Askary, T.H. (2004) Combined effect of *Fusarium udum* and *Meloidogyne javanica* on wilt resistant accessions of pigeonpea. *Annals of Plant Protection Sciences* 12(1), 130–133.

Singh, S., Bhagawati, B. and Goswami, B.K. (2011) Biomanagement of root-knot disease of chickpea caused by *Meloidogyne incognita*. *Annals of Plant Protection Sciences* 19(1), 159–163.

Singh, V.P., Srivastava, S., Srivastava, S.K. and Singh, H.B. (2012) Compatibility of different insecticides with *Trichoderma harzianum* under *in vitro* condition. *Plant Pathology Journal* 11(2), 73–76.

Singh, S., Abbasi and Hisamuddin (2013) Histopathological response of *Lens culinaris* roots towards root-knot nematode, *Meloidogyne incognita*. *Pakistan Journal of Biological Sciences* 16, 317–324.

Sitaramaiah, K. (1984) *Plant Parasitic Nematodes of India*. Today and Tomorrow's Printers and Publishers, New Delhi, India. 292 pp.

Sitaramaiah, K., Singh, R.S., Singh, K.P. and Sikora, R.A. (1971) Plant parasitic soil nematodes of India. *Experimental Station Bulletin No. 3*. UP Agricultural University, Pantnagar and US Department of Agriculture. 69 pp.

Sivakumar, C.V. and Seshadri, A.R. (1971) Life history of the reniform nematode, *Rotylenchulus reniformis* Linford and Oliveira, 1940. *Indian Journal of Nematology* 1, 7–20.

Sobin, N., Nema, K.G. and Dave, G.S. (1979) The possible interrelationship between plant parasitic nematode (*Tylenchorhynchus*) and a root-rot fungus from gram (*Cicer arietinum* L.). In: Agarwal, G.P. and Bilgrami, K.S. (eds) *Physiology and Parasitism*, Vol. 7. *Current Trends in Life Sciences*. Today and Tomorrow WIS Printers and Publishers, New Delhi, India, pp. 451–456.

Söğüt, M.A., Göze, F.G., Önal, T., Devran, Z. and Tonguç, M. (2014) Screening of common bean (*Phaseolus vulgaris* L.) cultivars against root-lesion nematode species. *Turkish Journal of Agriculture and Forestry* 38(4), 455–461.

Sumita, K. and Das, D. (2014) Management of root-knot nematode, *Meloidogyne incognita* on green gram through bioagents. *International Journal of Plant, Animal and Environmental Sciences* 4(4), 287–289.

Swarup, G., Prasad, S.K. and Raski, D.J. (1964) Some *Heterodera* species from India. *Plant Disease Reporter* 48, 235.

Taha, A.H.Y. (1993) Nematode interactions with root-nodule bacteria. In: Khan, M.W. (ed.) *Nematode Interactions*. Chapman and Hall, London, UK, pp. 175–202.

Talavera, M. and Valor, H. (2000) Influence of the previous crop on the anhydrobiotic ability of *Pratylenchus thornei* and *Merlinius brevidens*. *Nematologia Mediterranea* 28, 77–81.

Tapwal, A., Kumar, R., Gautam, N. and Pandey, S. (2012) Compatibility of *Trichoderma viride* for selected fungicides and botanicals. *International Journal of Plant Pathology* 3(2), 89–94.

Thakar, N.A. and Yadav, B.S. (1985) Comparative pathogenicity of the reniform nematode, *Rotylenchulus reniformis*, on susceptible and resistant varieties of pigeonpea. *Indian Journal of Nematology* 15(2), 167–169.

Thomason, I.J., Rich, J.R. and Omelia, F.C. (1976) Pathogenicity and histopathology of *Pratylenchus scribneri* infecting snap bean and lima bean. *Journal of Nematology* 8, 347–352.

Tiwari, S.P., Vadhera, I., Shukla, B.N. and Bhatt, J. (1992) Studies on the pathogenicity and relative reactions of chickpea lines to *Pratylenchus thornei* (Filipjev, 1936) Sher & Allen, 1953. *Indian Journal of Mycology and Plant Pathology* 22(3), 255–259.

Tiyagi, S.A. and Alam, M.M. (1990) Effect of root-knot, reniform and stunt nematodes on plant growth, water absorption capability and chlorophyll content of chickpea. *International Chickpea Newsletter* 22, 40–42.

Tiyagi, S.A. and Parveen, M. (1992) Pathogenic effect of root-lesion nematode *Pratylenchus thornei* on plant growth, water absorption capability, and chlorophyll content of chickpea. *International Chickpea Newsletter* 26, 18–20.

Tiyagi, S.A. and Ajaz, S. (2004) Biological control of plant parasitic nematodes associated with chickpea using oil cakes and *Paecilomyces lilacinus*. *Indian Journal of Nematology* 34(1), 44–48.

Tiyagi, S.A., Bano, M., Anver, S. and Alam, M.M. (1986) Water absorption capability in chickpea in the relation to nematode infection. *National Academy of Sciences, India* 9, 191–192.

Tobar Jiménez, A. (1973) Nematodos de los 'secanos' de la comarca de 'Alhama'. 1. Niveles de población y cultivos hospedadores. *Revista Ibérica de Parasitología* 33, 525–566.

Trifonova, Z. and Peneva, A. (2003) Influence of imazethapyr on the population densities of plant parasitic nematodes on bean (*Phaseolus vulgaris* L.). *Bulgarian Journal of Agricultural Science* 9(4), 515–520.

Uma Maheshwari, T., Sharma, S.B., Reddy, D.D.R. and Haware, M.P. (1995) Co-infection of wilt resistant chickpeas by *Fusarium oxysporum* f. sp. *ciceri* and *Meloidogyne javanica*. *Journal of Nematology* 27, 649–653.

Uma Maheswari, T., Sharna, S.B., Reddy, D.D.R. and Haware, M.P. (1997) Interaction of *Fusarium oxysporum* f. sp. *ciceris* and *Meloidogyne javanica* on *Cicer arietinum*. *Journal of Nematology* 29, 117–126.

Upadhyay, K.D. and Dwivedi, K. (1987) Analysis of crop losses in pea and gram due to *Meloidogyne incognita*. *International Nematology Network Newsletter* 4(4), 6–7.

Upadhyay, K.D. and Swarup, G. (1972) Culturing, host range and factors affecting multiplication of *Tylenchorhynchus vulgaris* on maize. *Indian Journal of Nematology* 2, 139–145.

Upadhyay, K.D., Dwivedi, K. and Srivastava, S.K. (1997) Effect of intercropping on pigeonpea infested with *Meloidogyne incognita*. *Indian Journal of Nematology* 27(2), 270.

Vanisree, S., Sreedhar, M. and Raju, C.S. (2013) Studies on genetic characteristics of pigeonpea and determination of selection criteria with path co-efficient analysis. *International Journal of Applied Biology and Pharmaceutical Technology* 4(2), 223–226.

Vanstone, V.A. and Russ, M.H. (2001) Ability of weeds to host the root lesion nematodes *Pratylenchus neglectus* and *P. thornei* I. Grass weeds. Australas. *Plant Pathology* 30, 245–250.

Varaprasad, K.S. (1986) *Rotylenchulus reniformis* Linford and Oliveira 1940. A comprehensive account of systematics, biology and management. In: Swarup, G. and Dasgupta, D.R. (eds) *Plant Parasitic Nematodes, Problems and Progress*. Indian Agricultural Research Institute, New Delhi, India, pp. 194–210.

Varaprasad, K.S., Sharma, S.B. and Loknathan, T.R. (1997) Nematode constraints to pigeonpea and chickpea in Vidarbha region of Maharashtra in India. *International Journal of Nematology* 7(2), 152–157.

Vavilov, N.I. (1951) The origin, variation and immunity and breeding of cultivated plants. *Chronica Botanica* 13, 1–364.

Velayutham, B. (1988) Efficacy of nematicidal seed treatment in the control of the pigeonpea cyst nematode, *Heterodera cajani* Koshy, 1967 affecting pigeonpea, *Cajanus cajan* L. *Indian Journal of Nematology* 18(2), 365–366.

Vovlas, N., Greco, N. and Di Vito, M. (1985) *Heterodera ciceri* sp. n. (Nematoda: Heteroderidae) on *Cicer arietinum* L., from northern Syria. *Nematologia Mediterranea* 13, 239–252.

Vovlas, N., Castillo, P. and Troccoli, A. (1998) Histology of nodular tissue of three leguminous hosts infected by three root-knot nematode species. *International Journal of Nematology* 8, 105–110.

Walia, R.K. and Bajaj, H.K. (2003) *Textbook on Introductory Plant Nematology*. Directorate of Information and Publications of Agriculture, ICAR, New Delhi, India. 227 pp.

Walia, R.K. and Seshadri, A.R. (1985) Chemical control of *Pratylenchus thornei* on chickpea through seed treatment. *International Chickpea Newsletter* 13, 32–34.

Wani, A.H. and Bhat, M.Y. (2012a) Control of root-knot nematode, *Meloidogyne incognita* by urea coated with Nimin or other natural oils on mung, *Vignaradiata* (L.) R. Wilczek. *Journal of Biopesticides* 5, 255–258.

Wani, A.H. and Bhat, M.Y. (2012b) Management of *Meloidogyne incognita* on lentil by plant extracts. *Annals of Plant Protection Sciences* 20(1), 201–204.

Webb, C. and Hawtin, G. (1981) Introduction. In: *Lentils*. Commonwealth Agricultural Bureaux, London, UK and ICARDA, Aleppo, Syria, pp. 1–5.

Wesemael, W.M.L. and Moens, M. (2012) Screening of common bean (Phaseolus vulgaris) for resistance against temperate root-knot nematodes (*Meloidogyne* spp.). *Pest Management Science* 68(5), 702–708.

Wiggers, R.J., Starr, J.C. and Price, H.J. (1990) DNA content and variation in chromosome number in plant cells affected by the parasitic nematodes *Meloidogyne incognita* and *M. arenaria*. *Phytopathology* 80, 1391–1395.

Yassin, A.M. (1987) The status of research on plant nematology in cereals and food and fodder legumes in the Sudan. In: Saxena, M.C., Sikora, R.A. and Srivastava, J.P. (eds) *Nematodes Parasitic to Cereals and Legumes in Temperate Semi-arid Regions. Proceedings of Workshop Held at Larnaca, Cyprus.* ICARDA, Aleppo, Syria, pp. 181–192.

Zaki, F.A. and Bhatti, D.S. (1986a) Effect of pigeonpea cyst nematode, *Heterodera cajani* Koshy, 1967 on macro and micro nutrients of pigeonpea and moth. *Indian Journal of Nematology* 16(1), 103–105.

Zaki, F.A. and Bhatti, D.S. (1986b) Control of pigeonpea cyst nematode, *Heterodera cajani* Koshy, 1967 by chemical seed treatment. *Indian Journal of Nematology* 16(1), 106–108.

14 The Influence of Soil Microbes on Plant Diversity

Mohammad Mobin*

Department of Biology, Faculty of Science, University of Tabuk, Tabuk Saudi Arabia

Abstract

The ecological consequences of biodiversity loss have sparked a serious debate during recent times. However, significant advances have been made in exploring the relationship between plant diversity and ecosystem processes, unravelling the underlying mechanisms. Soil microbes have key roles in nutrient cycling, soil formation and plant interactions. These roles are vital to ecosystem processes and biodiversity. Therefore, this chapter aims to discuss the relation of soil microbes in determining biodiversity and ecosystem processes.

14.1 Introduction

In recent times, one of the major challenges to the ecosystem is the disappearance of biodiversity. Ecosystem and biodiversity are strongly linked to each other where growing anthropogenic alteration of the ecosystem has slowly converted them into an almost defunct system. This unparalleled loss of biodiversity has created unease over the ramifications for the precise functioning of the ecosystem (Sala *et al.*, 2000; Loreau *et al.*, 2001; Jenkins, 2003; Millennium Ecosystem Assessment, 2005). As producers, plants assimilate mineral nutrients from inorganic sources that are mainly provided by decomposers, whereas decomposers (chiefly soil micro-organisms) take up carbon from organic resources that are delivered by autotrophic organisms. According to Naeem *et al.* (2000), ecosystem processes related to producer–consumer interactions are greatly disturbed by the loss of biodiversity. Plant-based organic resources infiltrate the soil via root exudation, rhizodeposition and leaf and root litters (Naeem *et al.*, 2000; Wardle *et al.*, 2004). In as much as the biochemical constituents of plant species vary greatly, alterations in plant diversity may affect these organic reserves both quantitatively and qualitatively. Consequently, the composition and metabolic functions of soil microbial communities are modified (Zak *et al.*, 2003; Nilsson *et al.*, 2008).

Soil microbes have a prodigious impact on ecosystem dynamics that is closely linked to the ecosystem processes such as nutrient acquisition, nitrogen and carbon recycling, and soil formation (Tiedje, 1988; Smith and Read, 1997; Hogberg *et al.*, 2001; Kowalchuk and Stephen, 2001; Sprent, 2001; Rillig and Mummey, 2006). Moreover, in spite of the significant role played by soil microbes in plant community composition and/or diversity, very little information is available on the response of soil microbes to altered plant diversity. This chapter focuses on the role of soil microbes in modulating the productivity and diversity of plant communities.

14.2 Role of Soil Characteristics in Shaping Up Microbial Diversity

Soil is a dynamic, highly diverse and indispensable living matrix, which is vital to the maintenance of most life processes. It serves as a habitat for an array of microbial communities.

The magnitude of soil microbes in terms of number and mass can easily be envisaged: one gram of

*E-mail: mhasa@ut.edu.sa

 Plant Biodiversity: Monitoring, Assessment and Conservation (eds A.A. Ansari, S.S. Gill, Z.K. Abbas and M. Naeem)

soil is estimated to contain 10^{10}–10^{11} bacteria (Horner-Devine *et al.*, 2003), 6000–50,000 bacterial strains (Curtis *et al.*, 2002), and up to 200 million fungal hyphae (Leake *et al.*, 2004). However, in temperate grassland soil, Killham (1994) recorded 1–2 and 2–5 t ha^{-1} of bacterial and fungal biomass respectively.

It is well recognized that soil ecology is governed by several environmental variables. Some of the important factors which could significantly alter the ecological processes are the available mineral nutrients, moisture content, temperature, electromagnetic radiation, pH, cation-exchange capacity and positive/negative interactions between microorganisms, among numerous others. Nonetheless, these environmental variables are in a flux that creates unique soil microhabitats. Various factors may converge to play a critical role in the soil microbial community. However, among abiotic factors, the soil pH has been found to significantly regulate the relative diversity of the bacterial community in the soil (Fierer and Jackson, 2006). To demonstrate the effect of pH, Rousk *et al.*, (2010) collected soil samples from a long-term liming experiment, where the pH ranged from 4.0 to 8.3 and the other experimental variables were kept constant. They observed a strong correlation between soil pH and the diversity and composition of bacterial communities across biomes. It was concluded that variable responses to microbial diversity are due to the pH sensitivity of bacterial cells, as bacterial populations display a relatively lower pH growth tolerance.

14.3 Soil Microbes and Plant Productivity

Plant productivity is largely governed by soil microbes, which collaborate with plants in either beneficial or harmful propinquity. Beneficial effects include root-associated mutualism or free-living microbes' modulated nutrient supply and partitioning of resources.

14.4 Beneficial Interactions

Root-zone microbial activity is vital to the availability of nutrients, which greatly influences plant productivity. There is ample evidence available to corroborate the claim that the soil microbes play a fundamental role in enriching the nutrient-scarce soil with important mineral nutrients. In this regard, nitrogen-fixing bacteria are of considerable importance as they are inherently linked to the N-pool of the ecosystem. Furthermore, soil–microbe interactions evolve into intimate symbiotic associations with plants and can vitalize plant productivity. One of the most commonly explored examples is the symbiotic association of plants and nitrogen-fixing bacteria that converts atmospheric N into ammonium-N. Most notably, approximately 20% of the total N comes from the N-fixing bacterial symbionts inhabiting the grassland, tropical savannah and tropical forests, where legume vegetation has the dominant presence (Cleveland *et al.*, 1999; Van der Heijden *et al.*, 2006a). However, not only legumes but other plant species have been reported to exist in symbiotic associations.

Approximately 150 cycads, nearly 400 shrubs and several unidentified plant species have been recorded as having associations with actinomycetes (Bond, 1983), cyanobacteria (Rai *et al.*, 2000) and endophytes to fix the atmospheric nitrogen, respectively.

Actinorhizal plants such as *Casuarina*, *Myrica*, *Hippophae* and *Alnus* symbiotically establish an association with cyanobacteria, which has momentous implications for ecosystem functioning. Vitousek and Walker (1989) noted a remarkable improvement in the N-profile of the soil and plant productivity when N-limited forests of Hawaii were invaded by actinorhizal shrubs.

In addition to the group of rhizospheric bacteria, the mycorrhizal fungi can also form a symbiotic association with plants. It was estimated that more than 80% of surviving plant families are capable of forming a mycorrhizal association (Smith and Read, 1997). The ability of forming mycorrhizal association is not exclusively limited to the extant plant families – fossil evidence suggested that these symbiotic associations were in existence about 400 million years ago, when the earth surface started to be colonized by land plants for the first time (Remy *et al.*, 1994). The importance of mycorrhizal fungi in sustainable agriculture is as an agent of nutrient transport and nutrient availability. In the rhizospheric fungal community, arbuscular mycorrhizal (AM) fungi, the ecto-mycorrhizal (EM) fungi and the ericoid mycorrhizal (ERM) fungi constitute an extremely significant category related to the efficient functioning of ecosystems. They are known to enhance resource complementarity by transforming the unavailable mineral nutrients into a usable form and facilitating the acquisition of inaccessible nutrients to the plant roots. Significant improvement in the nutrient profile of soil by mycorrhizal association was noted by Marschner and Dell (1994), who found that 80% of the plant's P and 25% of the plant's N were delivered by the association of the plant with the external hyphae of AM fungi. The ectomycorrhizal associations have also been found to be highly beneficial. Bowen (1973) recorded 3.2 times more P and 1.8 times more N in ectomycorrhizal roots of pine than non-mycorrhizal roots.

The relative abundance of mycorrhizal groups is spatially distinct. AM fungi symbiotically associate with the grassland, savannah and tropical forests (Read and Perez-Moreno, 2003). EM fungi associate with temperate and boreal forests and in some tropical forests in abundance (Alexander and Lee, 2005). ERM fungi are reported to be in symbiotic association with the members of the family *Ericaceae*, especially plentiful in heath land (Smith and Read, 1997).

Among different groups of mycorrhizal fungi, AM fungi, more than any other group of fungi, have been found to enhance the availability of scarce or soil minerals when associated with plants. By increasing the availability of nutrients and stabilizing the soil aggregates, they can act as a determining factor for the plant community structure (Grime *et al.*, 1987). They can also bring about a significant change in the course of succession as being observed by Gange *et al.* (1990). It has been noted that AM fungus/host-plant predilection varies in different plant species within natural communities (Vandenkoornhuyse *et al.*, 2003). They have been found to be essential for sustainable growth and plant productivity in vegetation as diverse as the grassland, the shrubs, and the rainforests (Hartnett *et al.*, 1993; Koide *et al.*, 1994).

Boreal and temperate forests are usually characterized by nutrient-poor soil, where nearly all nutrients are trapped in litter and humus in unavailable organic forms. The existence of EM fungi in such an ecosystem contributes hugely towards the sustainable plant diversity by facilitating the acquisition of nitrogen (Read and Perez-Moreno, 2003). Approximately 80% of all plant N in boreal forests is obtained from EM fungi as confirmed by pot (Simard *et al.*, 2002) and field trials (Hobbie and Hobbie, 2006).

Soil microbes in the form of free-living microbes can also modulate the ecosystem-related processes such as plant productivity indirectly. Free-living microbes decompose the soluble and insoluble organic matter into inorganic minerals. The nitrogen budget of some terrestrial ecosystems largely depends upon free-living N-fixing bacteria, which fix < 3 kg N ha^{-1} $year^{-1}$ (Cleveland *et al.*, 1999). A large proportion of the soil N (96–98%) exists in the form of dead organic matter, where the polymers of proteins, nucleic acids and chitin are disintegrated into dissolved organic N (DON) by extracellular enzymes secreted by soil microbes (Schimel and Bennett, 2004). The DON has one of two possible fates; either it is absorbed by free-living microbes or it is mineralized by the microbial biomass when carbon is a limiting factor and releases inorganic nitrogen into soil. In lieu of this, the plants may assimilate DON in the form of amino acids directly from the soil, thereby omitting the microbial mineralization step. This omission of microbial mineralization steps is particularly conspicuous in the N-deprived ecosystems of the Arctic (Nordin *et al.*, 2004), alpine tundra (Raab *et al.*, 1999), boreal (Nasholm *et al.*, 1998) temperate forest (Finzi and Berthrong, 2005) and low-productivity grassland (Bardgett *et al.*, 2003).

14.5 Harmful Interactions

Soil microbes may bring about a great change in ecosystem processes when they affect the keystone species. A whole spectrum of dominant forest trees such as Oak, Acacia and Eucalyptus have been prone to pathogenic attacks of *Phytophthora*, *Fusarium* and *Pythium* strains (Burdon *et al.*, 2006). Evidently, soil microbes compete with plants for nutrient acquisition. However, the competitive challenges posed by the soil microbes to the plant roots for the acquisition of nutrients are extreme in the nutrient-poor Arctic and alpine tundra region (Nordin *et al.*, 2004). In some instances, the nitrogen content of microbial and plant pools were comparable with each other (Jonasson *et al.*, 1999). Furthermore, the abundance of carbon activated the microbial growth, but ecosystem N pool and plant productivity declined, as observed by Dunn *et al.* (2006).

The soil microbes can also lower the ecosystem N availability by transforming the N to nitrate in a bacterial process of nitrification (Kowalchuk and Stephen, 2001). As we know, nitrate is highly mobile and can easily percolate into ground water and surface run-off (Scherer-Lorenzen *et al.*, 2003). Denitrification is another microbial process that may substantially reduce the N pool of ecosystems. Under anaerobic conditions, bacteria and some fungi can transform the plant's available nitrate into a gaseous form of nitrogen. According to estimates,

105–185 tons of N are lost per annum through denitrification worldwide (Tiedje, 1988), which is nearly 7% of the worldwide terrestrial productivity (Schlesinger, 1997). Houlton *et al.* (2006) estimated that up to 50% of available soil can be lost by denitrification from tropical forests. Taken together, these observations signify the relevance of changes in microbial biomass and community structure for regulating N-acquisition.

14.6 Soil Microbes and Plant Diversity

There is indisputable evidence indicating that the soil microbes as symbiont play a significant role in shaping up the plant diversity. N-fixing symbionts may determine the vegetation succession (Vitousek and Walker, 1989), plant productivity (Spehn *et al.*, 2002), plant invasibility (Parker *et al.*, 2006), plant community composition (Van der Heijden *et al.*, 2006a) and plant diversity (Van der Heijden *et al.*, 2006a). In the Fynbos area of South Africa, a biodiversity hotspot, several hectares of the unique and fragile vegetation was ravaged by *Acacia* spp. The successful colonization of acacia was linked by Sprent and Parsons (2000) with their symbiotic association of N-fixing bacteria.

Soil fungi also play a crucial role in ecosystem processes, including plant diversity. They aid the establishment of seedlings and the competitive ability of subordinate plant species (Grime *et al.*, 1987; Van der Heijden *et al.*, 2006b). Based on the results of several observations, it has been concluded that AM fungi enhances plant diversity in European grassland by as much as 30% (Grime *et al.*, 1987). However, the plant diversity in the tall grass prairie, which relies heavily on mycorrhizal association, may be reduced by AM fungi (Hartnett and Wilson, 1999). Correspondingly, ectomycorrhizal associations encourage the dominance of certain tree species in the tropical rainforests by enabling them to thwart pathogenic attacks and efficiently transforming the nutrients into an available form (Connell and Lowman, 1989).

Contemporary work in the field of plant diversity involving soil pathogens has revealed that they contribute significantly to spatial and temporal patterns in natural plant communities by negative plant–soil feedback mechanisms. Packer and Clay (2000) have observed that the survival of *Prunus serotina* seedlings was higher when they were not near the adult plants, while those seedlings which were below the adult plant could not grow, due to the accumulation of the soil fungus *Pythium*. It suggests that soil pathogens may enhance spatial variation in plant communities (Van der Putten, 2003). Bever *et al.* (1997) extrapolated a predictive model based on the presumption that the negative effects of soil pathogens on plant species increase with plant abundance. They noted that the soil pathogens maintain plant diversity by subduing the growth of dominant plant species. It was further ascertained by Bell *et al.* (2006) that the seedling mortality of a neotropical tree (*Sebastiana longicuspis*) was severely inhibited by soil pathogenic fungi when the plant density was highest. However, antithetical to this, Klironomos (2002) suggested that the soil pathogens dictate the rarity of the plant and reduce plant diversity. He observed that the rare plant species is strongly influenced by negative plant–soil feedback in comparison to the abundant plants.

The role of soil pathogens in plant life histories and their effect on successional dynamics has attracted much attention among plant ecologists (Jarosz and Davelos, 1995; Bever *et al.*, 1997; Clay and Van der Putten, 1999; Olff *et al.*, 2000; Packer and Clay, 2000; Van der Putten, 2001). According to Bever (1994), in negative soil feedback processes, soil pathogens can disturb the growth and destroy young plants (Packer and Clay, 2000; Reinhart *et al.*, 2005; Bell *et al.*, 2006; Van der Heijden *et al.*, 2008). It has been noted that the early successional species from ruderal plant communities are more sensitive to soil pathogens than species from more stable communities (Kardol *et al.*, 2006). However, successional change (Van der Putten *et al.*, 1993; Kardol *et al.*, 2006) or the maintenance of diversity (Janzen, 1970; Connell, 1971; Bever, 1994; Bever *et al.*, 1997; Bonanomi and Mazzoleni, 2005; Bradley *et al.*, 2008) have been reported to be based upon competitive interactions (Alexander and Holt, 1998; Kardol *et al.*, 2007; Petermann *et al.*, 2008) carried out by density-dependent pathogen attack (Mills and Bever, 1998; Packer and Clay, 2000; Bell *et al.*, 2006). Soil pathogens can also alter the invasion dynamics of exotic species (Wolfe and Klironomos, 2005; Reinhart and Callaway, 2006). Van Grunsven *et al.* (2007) compared the invasive behaviour of six species (three native species and three invading species), and found that invading plant species were less affected by soil pathogens than native species. Similar observations were also made by Mangla *et al.* (2008), who noted that the root exudates from invasive tropical weed, *Chromolaena odorata*, triggered the abundance of the soil pathogen *Fusarium* spp., which favoured the seedling growth of invasive species but inhibited the growth of one naturalized and one native species.

14.7 Conclusions

There are many direct and indirect mechanisms by which plant–soil microbe interactions can affect plant diversity. Further studies should be conducted in the biogeographical context, which might provide insight into the relative importance of the different groups of organisms and underlying mechanisms favouring or repulsing the invading species. Considering the extraordinary diversity of both fungal and bacterial communities in natural environments, the significance of soil pathogen effects on plant community composition deserves more attention.

References

Alexander, H.M. and Holt, R.D. (1998) The interaction between plant competition and disease. *Perspectives in Plant Ecology Evolution and Systematics* 1(2), 206–220.

Alexander, I.J. and Lee, S.S. (2005) Mycorrhizas and ecosystem processes in tropical rain forest: implications for diversity. In: Burslem, D.F.R.P., Pinard, M.A. and Hartley, S.E. (eds) *Biotic Interactions in the Tropics: Their Role in the Maintenance of Species Diversity*. Cambridge University Press, Cambridge, UK, pp. 165–203.

Bardgett, R.D., Streeter, T. and Bol, R. (2003) Soil microbes compete effectively with plants for organic nitrogen inputs to temperate grasslands. *Ecology* 84, 1277–1287.

Bell, T., Freckleton, R.P. and Lewis, O.T. (2006) Plant pathogens drive density-dependent seedling mortality in a tropical tree. *Ecological Letters* 9, 569–574.

Bever, J.D. (1994) Feedback between plants and their soil communities in an old field community. *Ecology* 75, 1965–1977.

Bever, J.D., Westover, K.M. and Antonovics, J. (1997) Incorporating the soil community into plant population dynamics: the utility of the feedback approach. *Journal of Ecology* 85, 561–573.

Bonanomi, G. and Mazzoleni, S. (2005) Soil history affects plant growth and competitive ability. *Community Ecology* 6, 23–28.

Bond, G. (1983) Taxonomy and distribution of non-legume nitrogen-fixing systems. In: Gordon, J.C. and Wheeler, C.T. (eds) *Biological Nitrogen Fixation in Forest Ecosystems: Foundations and Applications*. Martinus Nijhoff/Dr W. Junk Publishers, The Hague, The Netherlands, pp. 7–53.

Bowen, G.D. (1973) Mineral nutrition of mycorrhizas. In: Marks, G.C. and Kozlowski, T.T. (eds) *Ectomycorrhizas*. Academic Press, New York, USA, pp. 151–201.

Bradley, D.J., Gilbert, G.S. and Martiny, J.B.H. (2008) Pathogens promote plant diversity through a compensatory response. *Ecology Letters* 11, 461–469.

Burdon, J.J., Thrall, P.H. and Ericson, L. (2006) The current and future dynamics of disease in plant communities. *Annual Review of Phytopathology* 44, 19–39.

Clay, K. and Van der Putten, W.H. (1999) Pathogens and plant life histories. In: Vuorisalo, T.O. and Mutikainen, P.K. (eds) *Life History Evolution in Plants*. Kluwer Academic Publishers, Dordrecht, The Netherlands, pp. 275–301.

Cleveland, C.C., Townsend, A.R., Schimel, D.S. and Fisher, H. (1999) Global patterns of terrestrial biological nitrogen (N-2) fixation in natural ecosystems. *Global Biogeochemical Cycles* 13, 623–645.

Connell, J.H. (1971) On the role of natural enemies in preventing competitive exclusion in some marine animals and in rain forests. In: Den Boer, P.J. and Gradwell, G.R. (eds) *Dynamics in Populations*. Centre for Agricultural Publishing and Documentation, Wageningen, The Netherlands, pp. 298–312.

Connell, J.H. and Lowman, M.D. (1989) Low-diversity tropical rain forests – some possible mechanisms for their existence. *The American Naturalist* 134, 88–119.

Curtis, T.P., Sloan, W.T. and Scannell, J.W. (2002) Estimating prokaryotic diversity and its limits. *Proceedings of National Academy of Sciences USA* 99, 10494–10499.

Dunn, R., Mikola, J., Bol, R. and Bardgett, R.D. (2006) Influence of microbial activity on plant–microbial competition for organic and inorganic nitrogen. *Plant Soil* 289, 321–334.

Fierer, N. and Jackson, R.B. (2006) The diversity and biogeography of soil bacterial communities. *Proceedings of National Academy of Sciences USA* 103, 626–631.

Finzi, A.C. and Berthrong, S.T. (2005) The uptake of amino acids by microbes and trees in three cold-temperate forests. *Ecology* 86, 3345–3353.

Gange, A.C., Brown, V.K. and Farmer, L.M. (1990) A test of mycorrhizal benefit in an early successional plant community. *New Phytologist* 115, 85–91.

Grime, J.P., Mackey, J.M.L., Hillier, S.H. and Read, D.J. (1987) Floristic diversity in a model system using experimental microcosms. *Nature* 328, 420–422.

Hartnett, D.C. and Wilson, W.T. (1999) Mycorrhizae influence plant community structure and diversity in tall grass prairie. *Ecology* 80, 1187–1195.

Hartnett, D.C., Hetrick, B.A.D., Wilson, G.W.T. and Gibson, D.J. (1993) Mycorrhizal influence on intra- and inter-specific neighbour interactions among co-occurring prairie grasses. *Journal of Ecology* 81, 787–795.

Hobbie, J.E. and Hobbie, E.A. (2006) N-15 in symbiotic fungi and plants estimates nitrogen and carbon flux rates in Arctic tundra. *Ecology* 87, 816–822.

Hogberg, P., Nordgren, A., Buchmann, N. and Taylor, A.F.S. (2001) Large-scale forest girdling shows that current photosynthesis drives soil respiration. *Nature* 411, 789–792.

Horner-Devine, M.C., Leibold, M.A., Smith, V.H. and Bohannan, B.J.M. (2003) Bacterial diversity patterns along a gradient of primary productivity. *Ecological Letters* 6, 613–622.

Houlton, B.Z, Sigman, D.M. and Hedin, L.O. (2006) Isotopic evidence for large gaseous nitrogen losses from tropical rainforests. *Proceedings of National Academy of Sciences USA* 103, 8745–8750.

Janzen, D.H. (1970) Herbivores and the number of tree species in tropical forests. *American Naturalist* 104, 501–528.

Jarosz, A.M. and Davelos, A.L. (1995) Effects of disease in wild plant populations and the evolution of pathogen aggressiveness. *New Phytology* 129, 371–387.

Jenkins, M. (2003) Prospects for biodiversity. *Science* 302, 1175–1177.

Jonasson, S., Michelsen, A. and Schmidt, I.K. (1999) Coupling of nutrient cycling and carbon dynamics in the Arctic: integration of soil microbial and plant processes. *Applied Soil Ecology* 11, 135–146.

Kardol, P., Bezemer, T.M. and Van der Putten, W.H. (2006) Temporal variation in plant–soil feedback controls succession. *Ecological Letters* 9, 1080–1088.

Kardol, P., Cornips, N.J., Van Kempen, M.L., Bakx-Shotman, J.M. and Van der Puttenm, W.H. (2007) Microbe-mediated plant–soil feedback causes historical contingency effects in plant community assembly. *Ecological Monograph* 77, 147–162.

Killham, K. (1994) *Soil Ecology*. Cambridge University Press, Cambridge, UK.

Klironomos, J.N. (2002) Feedback with soil biota contributes to plant rarity and invasiveness in communities. *Nature* 417, 67–70.

Koide, R.T., Shumway, D.L. and Mabon, S.A. (1994) Mycorrhizal fungi and reproduction of field populations of *Abutilon theophrasti* Medic. (Malvaceae). *New Phytologist* 126, 123–130.

Kowalchuk, G.A. and Stephen, J.R. (2001) Ammonia-oxidizing bacteria: a model for molecular microbial ecology. *Annual Reviews in Microbiology* 55, 485–529.

Leake, J.R., Johnson, D., Donnelly, D.P., Muckle, G.E., Boddy, L. and Read, D.J. (2004) Networks of power and influence: the role of mycorrhizal mycelium in controlling plant communities and agroecosystem functioning. *Canadian Journal of Botany* 82, 1016–1045.

Loreau, M., Naeem, S., Inchausti, P. and Bengtsson, J. (2001) Biodiversity and ecosystem functioning: current knowledge and future challenges. *Science* 294, 804–808.

Mangla, S., Inderjit and Callaway, R. (2008) Exotic invasive plant accumulates native soil pathogens which inhibit native plants. *Journal of Ecology* 96, 58–67.

Marschner, H. and Dell, B. (1994) Nutrient uptake in mycorrhizal symbiosis. *Plant and Soil* 159, 89–102.

Millennium Ecosystem Assessment (2005) *Ecosystems and Human Well-Being: Biodiversity Synthesis*. World Resources Institute, Washington, DC, USA.

Mills, K.E. and Bever, J.D. (1998) Maintenance of diversity within plant communities: soil pathogens as agents of negative feedback. *Ecology* 79, 1595–1601.

Naeem, S., Hahn, D.R. and Schuurman, G. (2000) Producer–decomposer co-dependency influences biodiversity effects. *Nature* 403, 762–764.

Nasholm, T., Ekblad, A., Nordin, A., Giesler, R., Hogberg, M. and Hogberg, P. (1998) Boreal forest plants take up organic nitrogen. *Nature* 392, 914–916.

Nilsson, M.C., Wardle, D.A. and DeLuca, T.H. (2008) Belowground and aboveground consequences of interactions between live plant species mixtures and dead organic substrate mixtures. *Oikos* 117, 439–449.

Nordin, A., Schmidt, I.K. and Shaver, G.R. (2004) Nitrogen uptake by arctic soil microbes and plants in relation to soil nitrogen supply. *Ecology* 85, 955–962.

Olff, H., Hoorens, B., De Goede, R.G.M., Van der Putten, W.H. and Gleichman, J.M. (2000) Small-scale shifting mosaics of two dominant grassland species: the possible role of soil-borne plant pathogens. *Oecologia* 125, 45–54.

Packer, A. and Clay, K. (2000) Soil pathogens and spatial patterns of seedling mortality in a temperate tree. *Nature* 404, 278–281.

Parker, M.A., Malek, W. and Parker, I.M. (2006) Growth of an invasive legume is symbiont limited in newly occupied habitats. *Diversity and Distribution* 12, 563–571.

Petermann, J.S., Fergus, A.I., Turnbull, L.A. and Schmid, B. (2008) Janzen–Connell effects are widespread and strong enough to maintain diversity in grasslands. *Ecology* 89, 2399–2406.

Raab, T.K., Lipson, D.A. and Monson, R.K. (1999) Soil amino acid utilization among species of the *Cyperacea*: plant and soil processes. *Ecology* 80, 2408–2419.

Rai, A.N., Soderback, E. and Bergman, B. (2000) Cyanobacterium plant symbioses. *New Phytology* 147, 449–481.

Read, D.J. and Perez-Moreno, J. (2003) Mycorrhizas and nutrient cycling in ecosystems – a journey towards relevance? *New Phytology* 157, 475–492.

Reinhart, K.O. and Callaway, R.M. (2006) Soil biota and invasive plants. *New Phytology* 170, 445–457.

Reinhart, K.O., Royo, A.A., Van Der Patten, W.H. and Clay, K. (2005) Soil feedback and pathogen activity in *Prunus serotina* throughout its native range. *Journal of Ecology* 93, 890–898.

Remy, W., Taylor, T.N., Hass, H. and Kerp, H. (1994) Four hundred-million-year-old vesicular arbuscular mycorrhizae. *Proceedings of the National Academy of Sciences of the USA* 91, 11841–11843.

Rillig, M.C. and Mummey, D.L. (2006) Mycorrhizas and soil structure. *New Phytology* 171, 41–53.

Rousk, J., Baath, E., Brookes, P.C., Lauber, C.L., Lozupone, C., Caporaso, J.G., Knight, R. and Fierer, N. (2010) Soil bacterial and fungal communities across a pH gradient in an arable soil. *ISME Journal* 4, 1340–1351.

Sala, O.E., Chapin, F.S., Armesto, J.J. and Berlow, E. (2000) Global biodiversity scenarios for the year 2100. *Science* 287, 1770–1774.

Scherer-Lorenzen, M., Palmborg, C., Prinz, A. and Schulze, E.D. (2003) The role of plant diversity and composition for nitrate leaching in grasslands. *Ecology* 84, 1539–1552.

Schimel, J.P. and Bennett, J. (2004) Nitrogen mineralization: challenges of a changing paradigm. *Ecology* 85, 591–602.

Schlesinger, W.H. (1997) *Biogeochemistry: An Analysis of Global Change*, 2nd edn. Academic Press, San Diego, California, USA.

Simard, S.W., Jones, M.D. and Durall, D.M. (2002) Carbon and nutrient fluxes within and between mycorrhizal plants. In: Van der Heijden, M.G.A. and Sanders, I.R. (eds) *Mycorrhizal Ecology*. Ecological Studies. Springer, Heidelberg, Germany, pp. 33–74.

Smith, S.E. and Read, D.J. (1997) *Mycorrhizal Symbiosis*, 2nd edn. Academic Press, London, UK.

Spehn, E.M., Scherer-Lorenzen, M., Schmid, B. and Hector, A. (2002) The role of legumes as a component of biodiversity in a cross-European study of grassland biomass nitrogen. *Oikos* 98, 205–218.

Sprent, J.I. (2001) *Nodulation in Legumes*. Royal Botanical Gardens, Kew, London, UK.

Sprent, J.I. and Parsons, R. (2000) Nitrogen fixation in legume and non-legume trees. *Field Crops Research* 65, 183–196.

Tiedje, J.M. (1988) Ecology of denitrification and dissimilatory nitrate reduction to ammonium. In: Sehnder, A.J.B. (ed.) *Biology of Anaerobic Microorganisms*. Wiley, New York, USA, pp. 179–244.

Van der Heijden, M.G.A., Bardgett, R.D. and Van Straalen, N.M. (2008) The unseen majority: soil microbes as drivers of plant diversity and productivity in terrestrial ecosystems. *Ecology Letters* 11, 296–310.

Van der Heijden, M.G.A., Bakker, R., Verwaal, J. and Scheublin, T.R. (2006a) Symbiotic bacteria as a determinant of plant community structure and plant productivity in dune grassland. *FEMS Microbiological Ecology* 56, 178–187.

Van der Heijden, M.G.A., Streitwolf-Engel, R., Riedl, R., Siegrist, S., Neudecker, A. and Ineichen, K. (2006b) The mycorrhizal contribution to plant productivity, plant nutrition and soil structure in experimental grassland. *New Phytology* 172, 739–752.

Van der Putten, W.H. (2003) Plant defense belowground and spatiotemporal processes in natural vegetation. *Ecology* 84, 2269–2280.

Van der Putten, W.H., Van Dijk, C. and Peters, B.A.M. (1993) Plant specific soil borne diseases contribute to succession in foredune vegetation. *Nature* 362, 53–56.

Van der Putten, W.H., Vet, L.E.M., Jarvey, J.A. and Wäckers, F.L. (2001) Linking above- and belowground multitrophic interactions of plants, herbivores and their antagonists. *Trends in Ecology and Evolution* 16, 547–554.

Van Grunsven, R.H.A., Van der Putten, W.H., Bezemer, T.M., Tamis, W.L.M., Berendse, F. and Veenendaal, E.M. (2007) Reduced plant–soil feedback of plant species expanding their range as compared to natives. *Journal of Ecology* 95, 1050–1057.

Vandenkoornhuyse, P., Ridgway, K.P., Watson, I.J., Fitter, A.H. and Young, J.P.W. (2003) Co-existing grass species have distinctive arbuscular mycorrhizal communities. *Molecular Ecology* 12, 3085–3095.

Vitousek, P.M. and Walker, L.R. (1989) Biological invasion by *Myrica-faya* in Hawaii – plant demography, nitrogen-fixation, ecostystem effects. *Ecological Monography* 59, 247–265.

Wardle, D.A., Bardgett, R.D. and Klironomos, J.N. (2004) Ecological linkages between aboveground and belowground biota. *Science* 304, 1629–1633.

Wolfe, B.E. and Klironomos, J.N. (2005) Breaking new ground: soil communities and exotic plant invasion. *Bioscience* 55, 477–487.

Zak, D.R., Holmes, W.E., White, D.C., Peacock, A.D. and Tilman, D. (2003) Plant diversity, soil microbial communities, and ecosystem function: are there any links? *Ecology* 84, 2042–2050.

15 Plant-associated Endophytic Plethora as an Emerging Source of Antimicrobials

Syed Baker[1], P. Azmath[1], H.C. Yashavantha Rao[1], D. Rakshith[1], K.S. Kavitha[1] and S. Satish[1&2*]

[1]Herbal Drug Technological Laboratory, Department of Studies in Microbiology, University of Mysore, Mysore, Karnataka, India;
[2]Department of Plant Pathology, University of Georgia, Athens, Georgia, USA

Abstract

An ongoing strategy to isolate unique metabolites with antimicrobial activity from myriad natural niches is one of the top research priorities among scientific communities, owing to rapid expansion of multi-drug-resistant microbes. Prospecting of medicinal plants for various biological activities can be traced back to the ancient era. Before scientific knowledge was widespread, plants served as an immortal resource of structurally diverse phytocomponents. A large number of antimicrobial metabolites have been successfully isolated from medicinal plants. But harvesting of endangered plant species may pose a risk and cause an imbalance in plant diversity; hence, finding an alternative feasible source of bioactive compounds has been an area of interest in recent decades. Among which the endophytic plethora has revealed the diverse chemistry of metabolites bearing therapeutic properties, resulting in the rapid expansion of research on endophytes across the globe, with various valuable compounds of pharmaceutical importance constantly being explored. Hence, this chapter envisages the antimicrobial potentials of endophytic origin which can give an insight into the isolation of potent antimicrobial agents to combat life-threatening infections caused by microbes.

15.1 Introduction

An antimicrobial agent refers to a substance that kills or inhibits microbial growth by interpreting or binding to vital components responsible for metabolism, which in turn suppress the synthesis of functional biomolecules or impede normal cellular activities. In 1910 the first antimicrobial agent salvarsan was synthesized by Ehrlich to treat syphilis, and since then research on antimicrobials has gained momentum. The greatest step in antimicrobial research was the discovery of penicillin by Alexander Fleming, which saved thousands of lives during the Second World War (Powers, 2004). Antibiotics discoveries have generated remarkable advances in the past, with a large number of antimicrobial agents being constantly explored from myriad sources, among which plants and microbes play major roles. From plants can be extracted biologically active metabolites in the form of antimicrobials, antioxidants, anticancerous products, antidiabetics, etc. These have an immense impact, but the exploitation of endangered plant species may pose a risk to plant diversity; hence, bioactive compounds of microbial origin have gained importance. Microbial symbionts residing inside the plants are called endophytes (Baker and Satish, 2012a; Rakshith and Satish, 2011). The term was first coined by de Bary (1866) and is reported to be one of the richest sources of natural products (Strobel *et al.*, 2004). Interaction of plants and their microbial symbionts

*E-mail: satish.micro@gmail.com and satish@uga.edu

have been a subject of interest among scientific communities in recent decades. Endophytic plethora inhabit unique niches in their host and are capable of secreting bioactive compounds of agricultural, pharmaceutical and industrial importance (Fig. 15.1). The interest in endophytes began with the isolation of fungal endophytes from grasses and the subsequent synthesis of alkaloids that are toxic to pests. This report was the foundation for screening and isolation of secondary metabolites from endophytes responsible for bioactivity, which can be used as biopesticides in agriculture. Since then, research into endophytes' potential for bioactive compounds has been taken up by researchers across the globe (Baker and Satish, 2012a).

The rapid expansion of multidrug-resistant microbes against existing antibiotics is a major area of concern. The hunt for antimicrobial compounds from all possible natural resources has increased the expectations of mankind in the prevention of drug-resistant pathogens, which are growing at alarming rate. Approximately 20,000 bioactive secondary metabolites have been isolated from microbial origins, and are used in various sectors such as agriculture, pharmaceutical and industrial (Berdy, 2005; Brady and Clardy, 2000; Ebrahim *et al.*, 2012; Gordien *et al.*, 2010; Harper *et al.*, 2003; Mitchell *et al.*, 2010; Musetti *et al.*, 2006; Pongcharoen *et al.*, 2008; Randa *et al.*, 2010; Syed and Satish, 2012, You *et al.*, 2009).

Among the huge microbial diversity, endophytic symbionts can imitate the host chemistry and can synthesize bioactive metabolites or their derivatives similar to their host: in some cases, they can improve on their host (Owen and Hundley, 2004). Secretion of functional bioactive metabolites by endophytes is due to independent evolution, which in turn leads to the transformationn of genetic information from the host for adaptation and protects the host from invading pathogenic pests and microorganisms (Strobel and Daisy, 2003; Strobel *et al.*, 2004; Baker and Satish, 2012b). Therefore endophytes are an extremely rich, untapped reservoir of bioactive compounds and the most logical sources for screening completely new classes of metabolites with wide-ranging applications (Bacon and White, 2000). This chapter explores the antimicrobial potential of endophytes. Table 15.1 represents the myriad of antimicrobial metabolites secreted by different endophytes. To date, only a few potent endophytes have been studied. In recent years there has been a dramatic explosion of interest in endophytes for commercial exploitation. During plant–endophyte interactions, the plant supplies nutrients and shelter for the endophyte. In return,

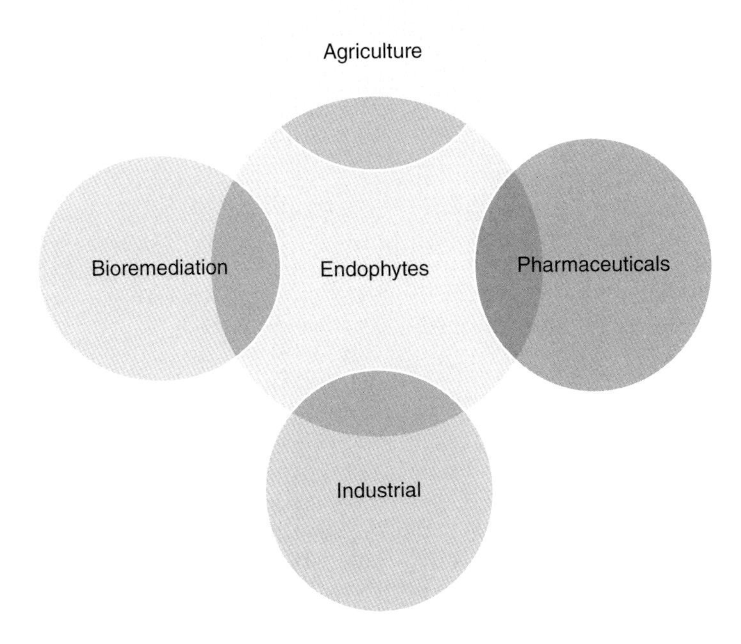

Fig. 15.1. Endophytes in various sectors.

Table 15.1. Representing endophytes and their biological antimicrobial activity.

Endophytes	Host	Bioactive compound	Bioactivity	References
Antimicrobial				
Aspergillus niger IFB-E003	*Cyndon dactylon*	Rubrofusarin B	Antimicrobial	(Song *et al.*, 2004)
Pseudomonas viridiflava	Wild grasses	Ecomycins B and C	Antimicrobial	(Miller *et al.*,1998)
Hypericum perforatum	Unidentified fungus	Hypericin	Antimicrobial	(Kusari *et al.*, 2008)
Chaetomium globosum	*Ginkgo biloba*	Chaetoglobosins A and C	Antimicrobial	(Qin *et al.*, 2009)
Talaromyces sp.	*Kandelia candel*	Secalonic acid A	Antimicrobial	(Liu *et al.*, 2010b)
Penicillium Commune	*Hibiscus tiliaceus*	1-O-(2,4-dihydroxy-6-methylbenzoyl)-glycerol	Antimicrobial	(Yan *et al.*, 2010)
Pestalotiopsis Pauciseta	*Tabebuia pentaphylla*	Anticancer Taxol	Antimicrobial	(Vennila *et al.*, 2010)
Phoma medicaginis	*Medicago* sp.	Phomol	Antimicrobial	(Weber *et al.*, 2004)
Phoma sorghina	*Tithonia diversifolia*	1,7-dihydroxy-3-methyl-9,10-anthraquinone, 1,6-dihydroxy-3-methyl-9,10-anthraquinone, 1-hydroxy-3-methyl-9,10-anthraquinone, 1,7-dihydroxy-3-hydroxymethyl-9,10-anthraquinone	Antimicrobial	(Borges and Pupo, 2006)
Nigrospora sp.	*Aegle marmelos*	Ethyl acetate fraction	Antimicrobial	(Gond *et al.*, 2007)
Cylindrocarpon sp.	*Saussurea involucrate*	Ethyl acetate fraction	Antimicrobial	(Yali *et al.*, 2010)
Streptomyces griseus	*Kandelia candel*	p-Aminoacetophenonic acids	Antimicrobial	(Guan *et al.*, 2005)
Streptomyces NRRL 30562	*Kennedia nigriscans*	Munumbicins Munumbicin D	Antimicrobial	(Castillo *et al.*, 2002)
Pichia guilliermondii Ppf9	*Paris polyphylla* var. *yunnanensis*	Ergosta-5,7,22-trienol, helvolic acid	Antimicrobial	(Zhao *et al.*, 2010)
Phomopsis sp. ZSU-H76	*E. agallocha*	Phomopsin metabolites	Antimicrobial	(Liu *et al.*, 2010a)
Antibacterial				
Rhizoctonia sp.	*Cynodon dactylon*	Rhizotonic acid	Antibacterial	(Ma *et al.*, 2004)
Alternaria sp.	*Vinca minor* and *Eonymus europaeus*	Altersetin	Antibacterial	(Hellwig *et al.*, 2002)
Periconia sp.	*Taxus cuspidate*	Periconicins A and B	Antibacterial	(Kim *et al.*, 2004)
Guignardia sp.	*Spondias mombin*	Guignardic acid	Antiviral, antibacterial	(Rodrigues-Heerklotz *et al.*, 2001)
Muscodor albus	*Cinnamomum zeylanicum*	Volatile antibiotics (alcohols, esters, ketones, acids and lipids)	Antifungal, antibacterial	(Strobel *et al.*, 2001)
Pichia guilliermondii	*Paris polyphylla*	Ergosta-5,7,22-trienol, 5α,8α-epidioxyergosta-6,22-dien-3β-ol, Ergosta-7,22-dien-3β,5α,6β-triol, helvolic acid	Antibacterial activity	(Zhao *et al.*, 2010)
Phoma sp.	*Saurauia scaberrinae*	Phomodione	Antibacterial	(Hoffman *et al.*, 2008)
Streptomyces SUK 06	*Thottea grandiflora*	Anti-methicillin-resistant *Staphylococcus aureus*	Antibacterial	(Ghadin *et al.*, 2008)

Continued

Table 15.1. Continued.

Endophytes	Host	Bioactive compound	Bioactivity	References
Penicillium janthinellum	*Melia azedarach*	Polyketide citrinin	Antibacterial	(Marinho *et al.*, 2005)
Diaporthe phaseolorum	*L. racemosa*	3-Hydroxypropionic acid	Antibacterial agent	(Sebastianes *et al.*, 2012)
Antifungal				
Cryptosporiopsis quercina	*Tripterigeum wilfordii*	Cryptocandin A	Antifungal	(Strobel *et. al.*, 1999b)
Pestalotiopsis microspora	Tropical rainforest plants	Ambuic acid	Antifungal	(Li *et al.*, 2001)
Pestalotiopsis jester	*Fragraea bodenii*	Jesterone	Antifungal	(Li and Strobel, 2001)
Fusarium	*Selaginella pallescens*	Pentaketide	Antifungal	(Brady and Clardy, 2000)
Aspergillus fumigatus CY018	*Cynodon dactylon*	Asperfumoid	Antifungal	(Liu *et al.*, 2004)
Acremonium zeae	Maize (*Zea mays*)	Pyrrocidines A and B	Antifungal	(Wicklow and Poling, 2009)
Curvularia geniculata	*Catunaregam tomentosa*	Curvularide B	Antifungal	(Wiyakrutta *et al.*, 2010)
Alternaria alternata	*Vitis vinifera*	Anti-sporulation compounds	Antifungal	(Musetti *et al.*, 2007)
Paecilomyces sp.	*Torreya grandis*	Brefeldin A	Antifungal	(Huang *et al.*, 2001)
Pestalotiopsis microspora	*Terminalia morobensis*	1,3-dihydro isobenzofuran	Antifungal	(Harper *et al.*, 2003)
Botryosphaeria rhodina	*Bidens pilosa*	Anti-sporulation compounds	Antifungal	(Abdou *et al.*, 2010)
Serratia marcescens	*Rhyncholacis penicillata*	Oocydin A	Antifungal	(Strobel *et al.*, 1999a)
Streptomyces sp.	*Monstera* sp.	Coronamycin	Antimalarial, antifungal	(Ezra *et al.*, 2004)
Bacillus pumilus	*Cassava* sp.	Pumilacidin	Antifungal	(De Melo *et al.*, 2009)
Antiviral				
Cytonaema sp.	*Quercus* sp.	Cytonic acids A	Antiviral	(Guo *et al.*, 2000)
Penicillium chrysogenum	Unidentified plant	Xanthoviridicatin E	Antiviral	(Singh *et al.*, 2003)
Xylaria mellisii	Unidentified medicinal plant of Kaeng Krachan National Park, Phetchaburi	Mellisol	Antiviral	(Pittayakhajonwut *et al.*, 2005)
Pullularia sp.	*Sonneratia caseolaris*	Pullularin A	Antiviral	(Isaka *et al.*, 2007)
Guignardia sp.	*Spondias mombin*	Guignardic acid	Antiviral, antibacterial	(Rodrigues-Heerklotz *et al.*, 2001)
Cytonaema sp.	*Quercus* sp. 103	Cytonic acids A and D	Antiviral	(Guo *et al.*, 2000)
Antimycobacterial				
Muscodor crispans	*Ananas ananassoides*	Volatile antimicrobials propanoic acid, 2-methyl-, methyl ester; propanoic acid, 2-methyl-; 1-butanol, 3-methyl-; 1-butanol, 3-methyl-, acetate; propanoic acid, 2-methyl-, 2-methylbutyl ester; ethanol	Antimyco-bacterial	(Mitchell *et al.*, 2010)

Continued

Table 15.1. Continued.

Endophytes	Host	Bioactive compound	Bioactivity	References
Phomopsis sp. PSU-D15	*Garcinia dulcis*	Phomoenamide, phomonitroester, deacetylphomoxanthone B	Antimycobacterial	(Rukachaisirikul *et al*., 2008)
Cladonia arbuscula, Empetrum nigrum, Juniperis communis, Calluna vulgaris, Vaccinium myrtillus	Fungal endophytes	Anti-TB compounds	Antimycobacterial	(Gordien *et al*., 2010)
Fusarium sp.	Mangrove medicinal plant	Metal complexes of fusaric acid	Antimycobacterial	(Pan *et al*., 2011)
Periconia sp.	*Piper longum*	Piperine	Antimycobacterial	(Verma *et al*., 2011)
Antimalarial				
Streptomyces sp.	*Monstera* sp.	Coronamycin	Antimalarial, Antifungal	(Ezra *et al*., 2004)
Streptomyces NRRL 30562	*Kennedia nigriscans*	Munumbicins, munumbicin D	Antibiotic, antimalarial	(Castillo *et al*., 2002)
Colletotrichum gloeosporioides	*Artemisia annua*	Artemisinin-related compounds	Antimalarial	(Tan and Zou, 2001)

the endophyte secretes secondary metabolites, which act as plant growth promoters and play a vital role in plant protection (Arnold *et al*., 2003).

15.2 Selection of Plant Material

Plants with an ethno-botanical history form important criteria for the screening of endophytes. Endophytes inhabit unique niches within the plant and perform myriad functions. Selecting a plant with ethnobotanical property can lead to isolation of novel endophytes, which in most cases are capable of secreting novel secondary metabolites. Normally, endophytes can also mimic the host chemistry in secreting better metabolites compared to their host (Mittermeier, 1999). Several criteria underlie the selection of plants for endophyte screening.

1. Medicinal property: Plants with proven medicinal properties and with a history of use in traditional medicine should be selected first.
2. Plant native to a biodiverse region: Plants that are native to a richly biodiverse area may harbour diverse endophytes.
3. Geographical area: Plants inhabiting a unique environment play an important role, as these plants have biological mechanisms to survive various environmental stresses. For example, Strobel *et al.* (1999a) reported on *Rhyncholacis penicillata*, which exists in a harsh aquatic environment: upon endophytic screening the researchers isolated *Serratia marcescens*, which secreted a Oocydin A, exhibiting a potent antifungal activity (Strobel *et al*., 1999a).
4. Selection of healthy plant materials: Young plant tissue is more suitable for isolation of endophytes, especially fungal, rather than bacterial. In older tissues, there may be many additional fungal inhabitants, which makes isolation difficult. The collected plant samples are stored at 4°C until the isolation procedure is carried out, and isolation should be as soon as possible after collection to avoid contamination by air microspora (Bacon and White, 1994).

15.3 Host–endophyte interaction

Host–endophyte interaction underlies the dynamic existing between host and microbes, which may result in both harm and benefit, depending on their mode of interaction. Endophytes adapt themselves inside the host plant and occupy a myriad of niches resulting in highly integrated and specialized symbioses. The complex host–endophyte interactions can range from mutualistic commensalism to parasitism. The mutual relationship benefits the endophytes through provision of energy, nutrients, shelter and protection from environmental stress. At the same time, endophytes benefit their host by secreting unique secondary metabolites and enzymes, which influence plant growth, adaptation of plants to environmental stresses, and defence mechanism against disease-causing pathogenic micro-organisms, insects and nematodes (Kogel *et al.*, 2006). According to Schulz and Boyle, as long as endophytic virulence and plant defence are balanced, the interaction remains asymptomatic. Once the host–endophyte interaction becomes unbalanced, either disease arises in the host plant or the plant defence machinery kills the pathogenic endophytic fungus (Baker and Satish, 2013). Whether the interaction is balanced or unbalanced depends on the general status of the partners, the virulence of the endophytes and the defences of the host, both virulence and defence being variable and influenced by environmental factors, nutritional status and developmental stages of the partners. Hence, commensalism and mutualism require a sophisticated balance between the defence responses of the plant and the nutrient demand of the endophyte. The process of interaction unveils and facilitates architectural, genetic interference, in which endophytes mimic the host chemistry, protecting the plant from invading pests, pathogens and environmental stress (Schulz and Boyle, 2005)

15.4 Methods of Endophyte Isolation

Plants materials are thoroughly washed under running water followed by sterile water to remove soil particles. Later the samples are washed with sterile water and cut into segments of 2–3 cm, followed by immersion into different surfactants, such as sodium hypochlorite, mercuric chloride and 70% of alcohol. After surface sterilization, the plant segments are placed onto the microbiological media, depending on the selection of endophytes, such as fungal or bacterial (Fig. 15.2).

To confirm successful sterilization and verify that the sample is free from biological contamination, sterility checks were carried out for each sample and 0.1 ml from the final rinse was plated out on nutrient synthetic agar media. The absence of bacteria and fungi after six days of incubation in the sterile environment was taken as confirmation that surface sterilization protocol was accurate (Strobel and Daisy, 2003).

15.5 Antimicrobial Compounds Secreted from Different Endophytes

Perusal of studies so far has resulted in the isolation and evaluation of myriad antimicrobial metabolites from endophytes. Some of the important bioactive metabolites with antimicrobial activity are explored in the present chapter. A study reports isolation of *A. niger* IFB-E003, inhabiting *Cyndon dactylon*, which secreted four known compounds displaying significant

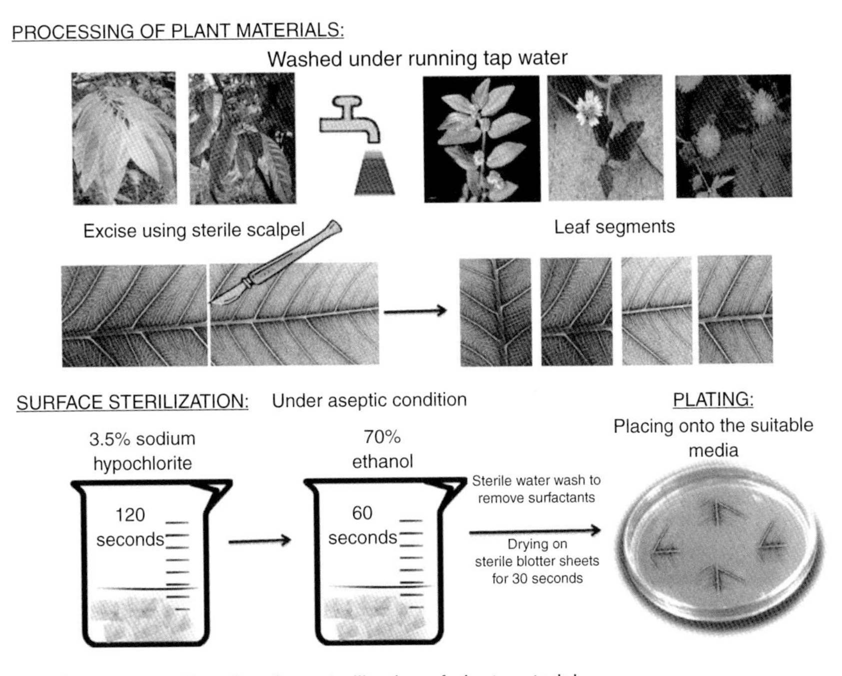

Fig. 15.2. Schematic representation of surface sterilization of plant material.

activity against test pathogens. These compounds were further characterized as naphtho-gamma-pyrones rubrofusarin B, fonsecinone A, asperpyrone B and aurasperone A (Song *et al.*, 2004). Ecomycins B and C, antifungal agents secreted from endophyte *Pseudomonas viridiflava*, which was isolated from *Grass*. Ecomycins displayed significant activities against human and plant pathogenic fungi (Miller *et al.*, 1998). Similarly Hypericin and Emodin were isolated from an endophytic fungus inhabiting an Indian medicinal plant. These compounds exhibited significant antimicrobial activity against a panel of test pathogens (Kusari *et al.*, 2008). Similarly, an endophytic fungus *Chaetomium globosum* screened from the leaves of *Ginkgo biloba* secreted a novel metabolite, chaetomugilin D. The isolation and purification was carried out by bioassay-guided fractionation with ethyl acetate as solvent. The structure was elucidated by analyses of spectroscopic methods, with 2D-NMR. The isolated compounds displayed significant growth-inhibitory activity against the brine shrimp (*Artemia salina*) and *Mucor miehei* (Qin *et al.*, 2009). Phomol, secreted from an endophyte inhabiting *Tithonia diversifolia*, displayed potent antimicrobial and anti-inflammatory activity (Weber *et al.*, 2004). Dendryol derivatives were secreted from *Phoma sorghina*, associated with *Tithonia diversifolia*, and expressed a broad spectrum of antimicrobial activity (Borges and Pupo, 2006).

p-Aminoacetophenonic acids were elucidated from *Streptomyces griseus* associated with *Kandelia candel*, which is known to possess antimicrobial activity against significant microbial pathogens (Guan *et al.*, 2005). When endophytes associated with *Kennedia nigriscans* were screened, this resulted in the production of munumbicin and its derivatives. These compounds are known to support antimicrobial potential pathogenic micro-organisms including the malarial parasite *Plasmodium falciparum* (Castillo *et al.*, 2002). Endophytic *Pichia guilliermondii* Ppf9 isolated from *Paris polyphylla* var. *yunnanensis* exhibited antimicrobial activity by secreting steroids and triterpenoid (Zhao *et al.*, 2010). Similarly rhizoctonic acid was secreted by endophytic *Rhizoctonia* sp. (Cy064), which was purified by bioassay-guided fractionation techniques. The metabolite structure was elucidated by ^{13}C-NMR, which displayed activity against significant pathogens including *Helicobacter pylori* strains (Ma *et al.*, 2004). When periconicins A and B were extracted from the endophytic fungus *Periconia* sp., associated with branches of *Taxus cuspidate*, upon evaluation they were revealed to have potent antibacterial activity. The compound was purified and characterized via combined spectroscopic methods (Kim *et al.*, 2004). An endophytic *Penicillium janthinellum*, isolated from *Melia azedarach*, secreted the polyketide citrinin, which upon evaluation revealed significant antibacterial activity, characterized by classical chromatographic procedures and identified by spectroscopic techniques (Marinho *et al.*, 2005). Similarly 3-hydroxypropionic acid (3-HPA) was isolated from *Diaporthe phaseolorum*, associated with *Laguncularia racemosa*, and on evaluation the compound displayed antibacterial activities against both *S. aureus* and *S. typhi* (Sebastianes *et al.*, 2012). Cryptocandin, isolated from endophytic *Cryptosporiopsis* cf. *quercina*, exhibited activity against pathogenic fungi such as *Trichophyton mentagrophytes, Candida albicans, Botrytis cinerea, Trichophyton rubrum,* and *Sclerotinia sclerotiorum* (Li *et al.*, 2001). Similarly, ambuic acid was purified and elucidated from endophytes of the *Pestalotiopsis* species and *Monochaetia* species associated with many plants of the tropical rainforest. Isolated ambuic acid displayed potent antifungal activity (Li *et al.*, 2001). Similarly, *Aspergillus fumigatus* CY018, inhabiting the leaf of *Cynodon dactylon*, secreted five secondary metabolites: fumigaclavine C, asperfumoid, fumitremorgin C, helvolic acid and physcion, and these metabolites were found to inhibit *Candida albicans* (Liu *et al.*, 2004). Coronamycin, secreted by endophytic species of *Streptomyces*, exhibited antifungal activity along with the malaria parasite *Plasmodium falciparum* (Ezra *et al.*, 2004). When endophytic bacteria *B. pumilus* MAIIM4a, associated with cassava, was isolated, it was found to secrete pumilacidin, which showed antifungal activity against *Rhizoctonia solani*, *Pythium aphanidermatum* and *Sclerotium rolfsii* (De Melo *et al.*, 2009). The endophytic *Xylarial* species from *Ginkgo biloba* L. was reported to secrete 7-amino-4-methyl coumarin, which exhibited broad-spectrum activity against food-borne pathogens known to produce toxins: these include *S. aureus, E. coli, S. typhia, S. typhimurium, S. enteritidis, A. hydrophila, Yersinia* sp., *V. anguillarum, Shigella* sp., *V. parahaemolyticus, C. albicans, P. expansum,* and *A. niger* (Liu *et al.*, 2008). Similarly 2,4-dihydroxy-5,6-dimethylbenzoate and phomopsilactone, secreted from the endophytic fungus *Phomopsis cassia*, inhabiting *Cassia spectabilis*, displayed significant antifungal activity against *Cladosporium cladosporioides* and *C. sphaerospermum* (Silva *et al.*, 2005). Brefeldin A, secreted from the endophyte *Cladosporium* sp., displayed significant antifungal activity against an array of microbial pathogens (Wang *et al.*, 2007). The endophytic fungus *Pestalotiopsis adusta* was reported to secrete pestalachloride derivatives A and B, which displayed significant antifungal activity against selected pathogens, including *Fusarium culmorum, Gibberella zeae,* and *Verticillium alboatrum*. The endophytic *Streptomyces* species inhabiting *Grevillea pteridifolia* was capable of secreting kakadumycin A and echinomycin, which displayed broad antibacterial activity against gram-positive bacteria and impressive antimalarial activity against *Plasmodium falciparum* (Castillo *et al.*, 2003).

Apart from antibacterial and antifungal activities, metabolites from endophytic flora are also capable of secreting antiviral metabolites, according to various studies. One such study including endophytic fungal species of *Cytonaema* sp. was capable of secreting the metabolites cytonic acid A and B, which exhibited antiviral activity against human cytomegalovirus protease inhibitors (Guo *et al.*, 2000). Similarly, polyketides such as xanthoviridicatins E and F were secreted from endophytic *Penicillium chrysogenum* and represented anti-HIV-1 activity via integrase inhibition, thus forming a promising drug molecule for antiretroviral therapy. Similarly, inhibition of cleavage reaction of HIV-1 integrase was observed with xanthoviridicatins E and F with IC50 values 6 and 5 respectively (Singh *et al.*, 2003). Endophytic metabolites are also reported to secrete antimycobacterial metabolites; for instance, *Muscodor crispans*, isolated from *Ananas ananassoides*, secreted a mixture of VOCs displaying activity against *Mycobacterium tuberculosis*, along with antibiotic activity against and array of human and phytopathogens. *Fusarium* sp. endophyte, upon chemical analysis, led to characterization of fusaric acid displaying inhibitory activity against the *M. bovis* BCG and *M. tuberculosis* H37Rv (Silva *et al.*, 2005).

15.6 Production, Extraction and Chemical Profiling of Antimicrobial Metabolites from Endophytes

Production, extraction and characterization of antimicrobial metabolites secreted from the endophytes is a complex process (Fig. 15.3). Antimicrobial metabolites are plentifully secreted as secondary metabolites during idiophase (the stationary phase of cell growth) in the culture media inoculated with endophytes. The production of secondary metabolites is influenced by various parameters such as nutrients in the media, multiplication rate and precursor

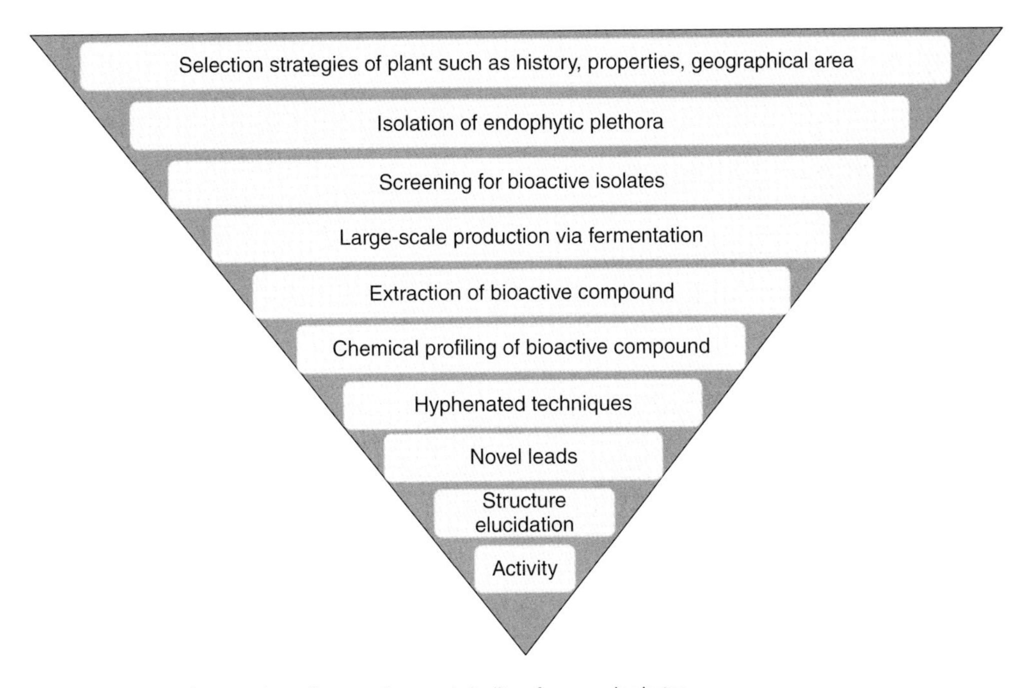

Fig. 15.3. Isolation and extraction of secondary metabolites from endophytes.

molecules (Rao *et al.*, 2015). Endophytes have metabolic diversity and synthesis of secondary metabolites is encoded by cluster genes. Considerable research efforts have been devoted in the last two decades to isolate and characterize antimicrobial metabolites secreted from endophytes.

To extract potent antimicrobial metabolites from endophytes, different approaches and rational strategies play a very important role. To increase the yield of the desired antimicrobial metabolite, different microbial media are used to culture the endophytes at a large scale under optimized variables. The majority of scientific studies report liquid–liquid extraction as the most popular and best choice of extraction process. In order to separate the antimicrobial metabolite to obtain a high purity, different techniques are employed via bioassay-guided fractionation techniques, which involve combinations of chromatographic techniques with bioassay; for instance, bioautography, which can minimize the impurities associated with the antimicrobial metabolite (Rao and Satish, 2015). It is desirable to employ chromatographic techniques, such as preparative HPLC.

15.7 TLC-Bioautography and Column Chromatography

Extraction and purification of secondary metabolites with antimicrobial activity is a multidisciplinary endeavour, which involves myriad techniques to obtain pure metabolites (Fig. 15.4). Biophysical

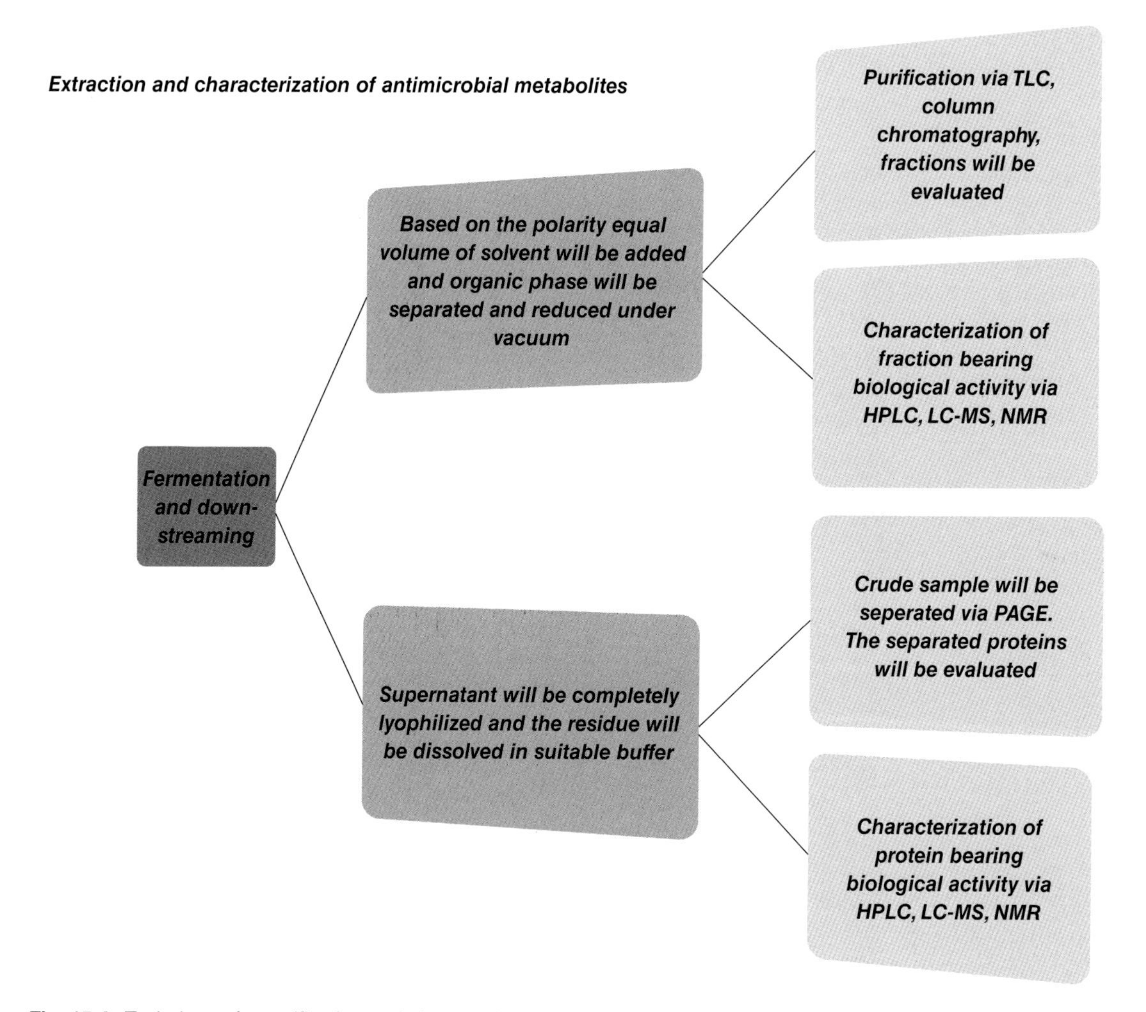

Fig. 15.4. Techniques for purification and characterization of antimicrobial metabolites.

characterization of antimicrobial metabolites can be achieved using various hyphenated techniques. The characterization mainly depends on the quality and purity of the antimicrobial metabolites. Spectroscopic techniques, such as UV-visible spectrophotometry, Raman spectroscopy, FTIR, LC-MS, GC-MS, NMR and XRD, are widely employed in characterization. If the metabolite is already reported, then it becomes easier to compare the spectroscopic data with the existing studies. If the metabolite is novel, then some of the advanced spectroscopic techniques; for instance, 2D-NMR and HRESI-MS, are employed in order to elucidate the structure.

15.8 Future Prospects for Microbial Endophytes

Over the years, natural products have provided fertile ground for researchers because of their enormous structural diversity and complexity. These products are naturally derived metabolites and byproducts of micro-organisms, plants or animals (Wrigley *et al.*, 2000). Considerable research efforts have been devoted in recent decades to the discovery of new biological active compounds from natural resources, but the most unexplored source is the endophytic plethora. In spite of an increasing amount of research into endophytes, the number of compounds extracted and purified from endophytes entering clinical trials is scanty. Some future avenues for endophytic microbe research are as follows.

1. Isolation of endophytes from unique niches: Isolation of microbial endophytes from unexplored biological niches is crucially important, but this parameter limits the scope of the isolates and novel endophytes.
2. Preliminary screening: In isolation of endophytes, the primary screening of endophytes is essential, and improved technologies such as metagenomics can be valuable in designing the primer of biological activity which may be employed for the identification of the potential endophyte.
3. Screening of secondary metabolites: As the goal is to obtain the widest possible screening for bioactive compounds, different types of bioactivity for secondary metabolites are tested; for instance, antimicrobial, anticancerous, and antioxidant activity.
4. Extraction: Bioactive compounds are extracted by the process of fermentation, and various solvents such as ethyl acetate, methanol, chloroform etc., in combination or as individual solvents, may be used to get crude extracts, which should then be evaluated for biological activity.
5. Purification: When bioactivity of the crude extract is confirmed, purification of the bioactive compound is the next step. By using hyphenated techniques, the crude extract is purified and subjected to characterization to determine the nature and type of compounds. After purification and characterization of the bioactive compound, pharmacological screening is employed. *In vitro* and *in vivo* experiments are carried out to determine the efficiency of the compound.
6. Industrial development of natural bioactive products: Collaborative research programmes combine various industries for large-scale production of natural bioactive compounds. This is the usual route by which purified bioactive compounds are developed into desirable formulations to originate into a product.

15.9 Conclusion

Ongoing efforts to combat multidrug-resistant micro-organisms have created a huge impact across the globe, with research delving into the traditional medicine records, and improving scientific strategies to discover bioactive compounds with antimicrobial properties. Due to problems obtaining treatment for life-threatening microbial infections in developing countries, funding has increased six-fold, with the result that many countries have been making progress in preventing new infections. Considerable research efforts have been devoted in the last two decades to the discovery of novel bioactive compounds from natural resources such as plants. But the slow growth rate and harvesting of rare endangered species pose a risk to biodiversity; in addition, the chemical synthesis of biomolecules is not economically feasible for developing countries. Hence microbes form a feasible alternative source of bioactive compounds, and endophytes form potential sources for the discovery of bioactive compounds.

The discovery of bioactive compounds is a multidisciplinary area to search for new and potent pharmaceutical agents from diverse sources. Endophytes are one such source: upon isolation from unique and underexplored ecological niches they secrete unique secondary metabolites. The recent developments and execution of new technologies has led to a new area of endophytic research, which focuses on its emerging potential as an untapped source of antimicrobial agents; this has been outlined in the present chapter.

References

Abdou, R., Scherlach, K., Dahse, H.M., Sattler, I. and Hertweck, C. (2010) Botryorhodines A–D, antifungal and cytotoxic depsidones from *Botryosphaeria rhodina*, an endophyte of the medicinal plant *Bidens pilosa*. *Phytochemistry* 71(1), 110–116.

Arnold, A.E., Mejia, L.C., Kyllo, D., Rojas, E.I., Maynard, Z., Robbins, N. and Herre, E.A. (2003) Fungal endophytes limit pathogen damage in a tropical tree. *Proceedings of the National Academy of Sciences* 100, 15649–15654.

Bacon, C.W. and White, J.F. (1994) *Biotechnology of Endophytic Fungi of Grasses*. CRC Press Inc., Boca Raton, Florida, USA.

Bacon, C.W. and White, J.F. (2000) *Microbial Endophytes*. Marcel Dekker, New York, USA, pp. 341–388.

Baker, S. and Satish, S. (2012a) Endophytes: natural warehouse of bioactive compounds: *Drug Invention Today* 4(11), 548–553.

Baker, S. and Satish, S. (2012b) Antimicrobial activity and biosynthesis of nanoparticles by endophytic bacterium inhabiting *Coffee arabica* L. *Scientific Journal of Biological Sciences* 1(5), 107–113.

Baker, S. and Satish, S. (2013) Bioprospecting of endophytic bacterial plethora from medicinal plants. *Plant Science Feed* 3(3), 42–45.

Berdy, J. (2005) Bioactive microbial metabolites. *Journal of Antibiotechnology* 58, 1–26.

Borges, W.S. and Pupo, M.T. (2006) Novel anthraquinone derivatives produced by *Phoma sorghina*, an endophyte found in association with the medicinal plant *Tithonia diversifolia* (Asteraceae). *Journal of the Brazilian Chemical Society* 17, 929–934.

Brady, S.F. and Clardy, J. (2000) CR377, a new pentaketide antifungal agent isolated from an endophytic fungus. *Journal of Natural Products* 63, 1447–1448.

Castillo, U., Strobel, G.A., Ford, E.J., Hess, W.M., Jensen, J.B. and Albert, H. (2002) Munumbicins, wide spectrum antibiotics produced by *Streptomyces munumbi*, endophytic on *Kennedia nigriscans*. *Microbiology* 148, 2675–2685.

Castillo, U.F., Harper, J.K., Strobel, G.A., Sears, J., Alesi, K. and Ford, E. (2003) Kakadumycins, novel antibiotics from *Streptomyces* NRRL 30566, an endophytic of *Grevillea pteridifolia*. *FEMS Microbiology Letters* 224, 183–190.

De Melo, F.M.P., Fiore, M.F., De Moraes, L.A.B., Silva Stenico, M.F., Scramin, S. and de Araujo Teixeria, M. (2009) Antifungal compound produced by the cassava endophyte *Bacillus pumilus* MAIIIM4A. *Science Agricola* 66, 583–592.

Ebrahim, W., Kjer, J., El Amrani, M., Wray, V., Lin, W. and Ebel, R. (2012) Pullularins E and F, two new peptides from the endophytic fungus *Bionectria ochroleuca* isolated from the mangrove plant *Sonneratia caseolaris*. *Marine Drugs* 10(5), 1081–1091.

Ezra, D., Castillo, U.F., Strobel, G.A., Hess, W.M., Jensen, J.B. and Condron, M.A. (2004) *Coronamycins,* peptide antibiotics produced by a *verticillate Streptomyces* sp. (MSU-2110) endophytic on *Monstera* sp. *Microbiology* 150, 785–793.

Ghadin, N., Zin, N.M., Sabaratnam, V., Badya, N., Basri, D.F. and Lian, H.H. (2008) Isolation and characterization of a novel endophytic *Streptomyces* SUK 06 with antimicrobial activity from Malaysian plant. *Asian Journal of Plant Science* 7, 189–194.

Gond, S.K., Verma, V.C., Kumar, A. and Kumar, V. (2007) Study of endophytic fungal community from different parts of *Aegle marmelos* Correae (Rutaceae) from Varanasi (India). *World Journal of Microbiology and Biotechnology* 23, 1371–1375.

Gordien, A.Y., Gray, A., Ingleby, K., Franzblau, S.G. and Seidel, V. (2010) Activity of Scottish plant, lichen and fungal endophyte extracts against *Mycobacterium aurum* and *Mycobacterium tuberculosis*. *Phytothermal Research* 24, 692–698.

Guan, S., Grabley, S., Groth, I., Lin, W., Christner, A., Guo, D. and Sattler, I. (2005) Structure determination of germacrane-type sesquiterpene alcohols from an endophyte *Streptomyces griseus* subsp. *Magnetic Resonance in Chemistry* 43, 1028–1031.

Guo, B., Dai, J., Ng, S., Huang, Y., Leong, C. and Ong, W. (2000) Cytonic acids A and B: novel tridepside inhibitors of hCMV protease from the endophytic fungus *Cytonaema species*. *Journal of Natural Products* 63, 602–604.

Harper, J.K., Ford, E.J., Strobel, G.A., Arif, A., Grant, D.M. and Porco, J. (2003) Pestacin: a 1,3-dihydro isobenzofuran from *Pestalotiopsis microspora* possessing antioxidant and antimycotic activities. *Tetrahedron* 59, 2471–2476.

Hellwig, V., Grothe, T., Bartschmid, M.A., Endermann, R., Geschke, F.U., Henkel, T. and Stadler, M.A. (2002) Altersetin, a new antibiotic from cultures of endophytic *Alternaria* spp.: taxonomy, fermentation, isolation, structure elucidation and biological activities. *Journal of Antibiotics* 55, 881–892.

Hoffman, A.M., Mayer, S.G., Strobel, G.A., Hess, W.M., Sovocool, G.W. and Grange, A.H. (2008) Purification, identification and activity of phomodione, a furandione from an endophytic *Phoma* species. *Phytochemistry* 69, 1049–1056.

Huang, Y., Wang, J., Li, G., Zheng, Z. and Su, W. (2001) Antitumor and antifungal activities in endophytic fungi isolated from pharmaceutical plants *Taxus mairei*, *Cephalataxus fortunei* and *Torreya grandis*. *FEMS Immunology and Medical Microbiology* 31, 163–167.

Isaka, M., Berkaew, P., Intereya, K., Komwijit, S. and Sathitkunanon, T. (2007) Antiplasmodial and antiviral cyclohexadepsipeptides from the endophytic fungus *Pullularia* sp. BCC 8613. *Tetrahedron* 63(29), 6855–6860.

Kim, S., Shin, D.S., Lee, T. and Oh, K.B. (2004) Periconicins, two new fusicoccane diterpenes produced by an endophytic fungus *Periconia* sp. with antibacterial activity. *Journal of Natural Products* 67, 448–450.

Kogel, K.H., Franken, P. and Huckelhoven, R. (2006) Endophyte or parasite – what decides? *Current Opinion in Plant Biology* 9, 358–363.

Kusari, S., Lamshoft, M., Zuhlke, S. and Spiteller, M. (2008) An endophytic fungus from *Hypericum perforatum* that produces hypericin. *Journal of Natural Products* 71, 159–162.

Li, J.Y. and Strobel, G.A. (2001) Jesterone and hydroxy-jesterone antioomycetes cyclohexenenone epoxides from the endophytic fungus *Pestalotiopsis jesteri*. *Phytochemistry* 57, 261–265.

Li, J.Y., Harper, J.K., Grant, D.M., Tombe, B.O., Bashyal, B. and Hess, W.M. (2001) Ambuic acid, a highly functionalized cyclohexenone with antifungal activity from *Pestalotiopsis* spp. and *Monochaetia* sp. *Phytochemistry* 56, 463–468.

Liu, J.Y., Song, Y.C., Zhang, Z., Wang, L., Guo, Z.J. and Zou, W.X. (2004) *Aspergillus fumigatus* CY018, an endophytic fungus in *Cynodon dactylon* as a versatile producer of new and bioactive metabolites. *Journal of Biotechnology* 114, 279–287.

Liu, X., Dong, M., Chen, X., Jiang, M., Lv, X. and Zhou, J. (2008) Antimicrobial activity of an endophytic *Xylaria* sp. YX-28 and identification of its antimicrobial compound 7-amino-4-methylcoumarin. *Applied Microbiology and Biotechnology* 78, 241–247.

Liu, A.R., Chen, S.C., Lin, X.M., Wu, S.Y., Xu, T. and Cai, F.M. (2010a) Endophytic *Pestalotiopsis* species associated with plants of Palmae, Rhizophoraceae, Planchonellae and Podocarpaceae in Hainan, China. *African Journal of Microbiology Research* 4, 2661–2669.

Liu, F.I., Cai, X.L., Yang, H., Xia, X.K., Guo, Z.Y. and Yuan, J. (2010b) The bioactive metabolites of the mangrove endophytic fungus *Talaromyces* sp. ZH-154 isolated from *Kandelia candel* (L.) Druce. *Planta Medica* 76(2), 185–189.

Ma, Y., Li, M., Liu, Y., Song, J.Y. and Tan, R.X. (2004) Anti-*Helicobacter pylori* metabolites from *Rhizoctonia* sp. Cy064, an endophytic fungus in *Cynodon dactylon*. *Fitoterapia* 75, 451-456.

Marinho, A.M.R., Rodrigues-Filho, E., Moitinho, M.D. and Santos, L.S. (2005) Biologically active polyketides produced by *Penicillium janthinellum* isolated as an endophytic fungus from fruits of *Melia azedarach*. *Journal of the Brazilian Chemical Society* 16, 280–283.

Miller, C.M., Miller, R.V., Garton-Kenny, D., Redgrave, B., Sears, J. and Condron, M.M. (1998) Ecomycins, unique antimycotics from *Pseudomonas viridiflava*. *Journal of Applied Microbiology* 84, 937–944.

Mitchell, A.M., Strobel, G.A., Moore, E., Robison, R. and Sears, J. (2010) Volatile antimicrobials from *Muscodor crispans*, a novel endophytic fungus. *Microbiology* 156, 270–277.

Mittermeier, R. (1999) *Hotspot: Earth's Biologically Richest and Most Endangered Ecoregions*. CEMEX Conservation International, University of Chicago Press, Washington, DC, USA.

Musetti, R., Vecchione, A., Stringher, L., Borselli, S., Zulini, L. and Marzani, C. (2006) Inhibition of sporulation and ultrastructural alterations of grapevine downy mildew by the endophytic fungus *Alternaria alternate*. *Phytophathology* 96, 689–698.

Musetti R., Polizzotto, R., Vecchione, A., Borselli, S., Zulini, L., D'Ambrosio, M., Toppi, L.S.D. and Pertot, I. (2007) Antifungal activity of diketopiperazines extracted from *Alternaria alternata* against *Plasmopara viticola*: an ultrastructural study. *Micron* 38, 643–650.

Owen, N.L. and Hundley, N. (2004) Endophytes – the chemical synthesizers inside plants. *Science Progress* 87, 79–99.

Pan, J.H., Chen, Y., Huang, Y.H., Tao, Y.W., Wang, J. and Li, Y. (2011) Antimycobacterial activity of fusaric acid from a mangrove endophyte and its metal complexes. *Archives of Pharmacology Research* 34, 1177–1181.

Pittayakhajonwut, P., Suvannakad, R., Thienhirun, S., Prabpai, S., Kongsaeree, P. and Tanticharoen, M. (2005) An anti-herpes simplex virus-type 1 agent from *Xylaria mellisii* (BCC 1005). *Tetrahedron Letters* 46, 1341–1344.

Pongcharoen, W., Rukachaisirikul, V., Phongpaichit, S., Kuhn, T., Pelzing, M. and Sakayaroj, J. (2008) Metabolites from the endophytic fungus *Xylaria* sp. PSU-D14. *Phytochemistry* 69, 1900–1902.

Powers, J.H. (2004) Antimicrobial drug development – the past, the present, and the future. *Clinical Microbiology Infection* 10, 23–31.

Qin, J.C., Zhang, Y.M., Gao, J.M., Bai, M.S., Yang, S.X., Laatsch, H. and Zhang, A.L. (2009) Bioactive metabolites produced by *Chaetomium globosum*, an endophytic fungus isolated from *Ginkgo biloba*. *Bioorganic Medical Chemistry Letters* 19, 1572–1574.

Rakshith, D. and Satish, S. (2011) Endophytic mycoflora of *Mirabilis jalapa* L., and studies on antimicrobial activity of its endophytic *Fusarium* sp. *Asian Journal of Experimental Biology* 2(1), 75–79. Available at: eprints.uni-mysore.ac.in/15112/1/13.pdf (accessed 27 May 2016).

Randa, A., Kirstin, S., Hans-Martin, D., Isabel, S. and Christian, H. (2010) Botryorhodines A–D, antifungal and cytotoxic depsidones from *Botryosphaeria rhodina*, an endophyte of the medicinal plant *Bidens pilosa*. *Photochemistry* 71, 110–116.

Rao, H.C.Y. and Satish, S. (2015) Genomic and chromatographic approach for the discovery of polyketide antimicrobial metabolites from an endophytic *Phomopsis liquidambaris* CBR-18. *Frontiers in Life Science* 8(2), 200–207. DOI: 10.1080/21553769.2015.1033768.

Rao, H.C.Y., Santosh, P., Rakshith, D. and Satish, S. (2015) Molecular characterization of an endophytic *Phomopsis liquidambaris* CBR-15 from *Cryptolepis buchanani* Roem. and impact of culture media on biosynthesis of antimicrobial metabolites. *3 Biotech* 5, 165–173. DOI: 10.1007/s13205-014-0204-2.

Rodrigues-Heerklotz, K.F., Drandarov, K., Heerklotz, J., Hesse, M. and Werner, C. (2001) Guignardic acid, a novel type of secondary metabolite produced by endophytic fungus *Guignardia* sp.: isolation, structure elucidation and asymmetric synthesis. *Helvica Chimica Acta* 84, 3766–3771.

Rukachaisirikul, V., Sommart, U., Phongpaichit, S., Sakayaroj, J. and Kirtikara, K. (2008) Metabolites from the endophytic fungus *Phomopsis* sp. PSU-D15. *Phytochemistry* 69, 783–787.

Schulz, B. and Boyle, C. (2005) The endophytic continuum. *Mycological Research* 109, 661–686.

Sebastianes, F.L.S., Lacava, P.T., Favaro, L.C.L., Rodrigues, M.B.C., Araujo, W.L. and Azevedo, J.L. (2012) Genetic transformation of *Diaporthe phaseolorum*, an endophytic fungus found in mangrove forests, mediated by *Agrobacterium tumefaciens*. *Current Genetics* 58, 21–33.

Silva, G.H., Teles, H.L., Trevisan, H.C., da S Bolzani, V., Young, M.C.M. and Pfenning, L.H. (2005) New bioactive metabolites produced by *Phomopsis cassiae*, an endophytic fungus in *Cassia spectabilis*. *Journal of the Brazilian Chemical Society* 16, 1463–1466.

Singh, S.B., Zink, D.L., Guan, Z., Collado, J., Pelaez, F. and Felock, P.J. (2003) Isolation, structure, and HIV-1 integrase inhibitory activity of xanthoviridicatin E and F, two novel fungal metabolites produced by *Penicillium chrysogenum*. *Helvetica Chimica Acta* 86(10), 3380–3385.

Song, Y.C., Li, H., Ye, Y.H., Shan, C.Y., Yang, Y.M. and Tan, R.X. (2004) Endophytic naphthopyrone metabolites are co-inhibitors of xanthine oxidase, SW1116 cell and some microbial growths. *FEMS Microbiology Letters* 241, 67–72.

Strobel, G. and Daisy, B. (2003) Bioprospecting for microbial endophytes and their natural products. *Microbiology and Molecular Biology Review* 67, 491–502.

Strobel, G.A., Li, J.Y., Sugawara, F., Koshino, H., Harper, J. and Hess, W.M. (1999a) Oocydin A, a chlorinated macrocyclic lactone with potent anti-oomycete activity from *Serratia marcescens*. *Microbiology* 145, 3557–3564.

Strobel, G.A., Miller, R.V., Teplow, D.W. and Hess, W.M. (1999b) Cryptocandins, a potent antimycotic from the endophytic fungus *Cryptosporiopsis* cf. quercina. *Microbiology* 145, 1919–1926.

Strobel, G.A., Dirkse, E., Sears, J. and Markworth, C. (2001) Volatile antimicrobials from *Muscodor albus*, a novel endophytic fungus. *Microbiology* 147, 2943–2950.

Strobel, G., Daisy, B., Castillo, U. and Harper, J. (2004) Natural products from endophytic microorganisms. *Journal of Natural Products* 67, 257–268.

Syed, B. and Satish, S. (2012) Endophytes: toward a vision in synthesis of nanoparticle for future therapeutic agents. *International Journal of Bio-Inorganic Hybrid Nanomaterials* 1, 67–77.

Tan, R.X. and Zou, W.X. (2001) Endophytes: a rich source of functional metabolites. *Natural Product Reports* 18, 448–459.

Vennila, R., Thirunavukkarasu, S.V. and Muthumary, J. (2010) Evaluation of fungal taxol isolated from an endophytic fungus *Pestalotiopsis pauciset* AVM1 against experimentally induced breast cancer in Sprague Dawley rats. *Research Journal of Pharmacology* 4, 38–44.

Verma, V.C., Lobkovsky, E., Gange, A.C., Singh, S.K. and Prakash, S. (2011) Piperine production by endophytic fungus *Periconia* sp. isolated from *Piper longum* L. *Journal of Antibiotics* 64, 427–431.

Wang, F.W., Jiao, R.H., Cheng, A.B., Tan, S.H. and Song, Y.C. (2007) Antimicrobial potentials of endophytic fungi residing in *Quercus variabilis* and Brefeldin A obtained from *Cladosporium* sp. *World Microbiology and Biotechnology* 23, 79–83.

Weber, D., Sterner, O., Anke, T., Gorzalczancy, S., Martino, V. and Acevedo, C. (2004) Phomol, a new anti inflammatory metabolite from an endophyte of the medicinal plant *Erythrina cristagalli*. *Journal of Antibiotics* 57, 559–563.

Wicklow, D.T. and Poling, S.M. (2009) Antimicrobial activity of pyrrocidines from *Acremonium zeae* against endophytes and pathogens of maize. *Phytopathology* 99, 109–115.

Wiyakrutta, C.P., Aree, S., Sriubolmas, T., Ngamrojanavanich, N., Mahidol, N., Ruchirawat, C. and Kittakoop, S. (2010) Prasat. Curvularides A–E: antifungal hybrid peptide-polyketides from the endophytic fungus *Curvularia geniculata*. *Chemistry* 16, 11178–11185.

Wrigley, S.K., Hayes, M.A., Thomas, R., Chrystal, E.J.T., Nicholson, N., Baker, D., Mocek, U. and Garr, C. (2000) *Biodiversity: New Leads for Pharmaceutical and Agrochemical Industries*. The Royal Society of Chemistry, Cambridge, UK, pp. 66–72.

Yali, L.V., Fusheng, Z., Juan, C., Jinlong, C., Yongmei, X., Xiangdong, L. and Shunxing, G. (2010) Diversity and antimicrobial activity of endophytic fungi associated with the alpine plant *Saussurea involucrate*. *Biology and Pharmcology Bulletin* 33, 1300–1306.

Yan, H., Gao, S., Li, C., Li, X. and Wang, B. (2010) Chemical constituents of a marine-derived endophytic fungus *Penicillium commune* G2M. *Molecules* 15, 3270–3275.

You, F., Han, T., Wu, J.Z., Huang, B.K. and Qin, L.P. (2009) Antifungal secondary metabolites from endophytic *Verticillium* sp. *Biochemistry and Systematic Ecology* 37, 162–165.

Zhao, J., Mou, Y., Shan, T., Li, Y., Zhou, L., Wang, M. and Wang, J. (2010) Antimicrobial metabolites from the endophytic fungus *Pichia guilliermondii* isolated from *Paris polyphylla* var. *yunnanensis*. *Molecules* 15, 7961–7970.

16 Biodiversity, Bioindicators and Biogeography of Freshwater Algae

MARTIN T. DOKULIL*

EX Research Institute Mondsee, Mondsee, Austria

Abstract

Biodiversity and the problems associated with it are outlined and then the focus turns to continental waters. Levels and factors affecting biodiversity are described and listed. Measures of diversity focus on species richness and the Shannon index using abundances and biovolumes. Spatial scales of measuring and monitoring of biodiversity are listed and explained. Diversity of algae is then discussed in detail, including phytoplankton and algae on substrates, and a discussion of the paradox of the plankton and the 'everything is everywhere' hypothesis. The biogeographic distributions of algal groups are mentioned, with a particular focus on endemism, followed by a brief consideration of trophic interactions. Biodiversity of algae is described and evaluated from five case studies using long-term data on phytoplankton from three lakes, a length profile of plankton from the River Danube and epilithic algae from an artificial system. The results and findings are critically discussed with respect to advantages and drawbacks in using single indices. The use of algae and their diversity as biological indicators (biomarkers) in environmental assessments is finally outlined, followed by concluding remarks.

16.1 Introduction

The huge number of diverse organisms living on Earth is usually referred to as biodiversity. These species and the different ecosystems contain an enormous variety of genes, adding a biogeographical context to biological diversity (Hubbell, 2011). The biodiversity debate is dominated by discussions about strategies of conservation and sustainable development (Maczulak, 2010). Since it still remains unclear to a large extent how many distinct species inhabit the planet, proper research and conservation are hampered (May, 2011). The nonexistence of a unified species concept accentuates the problem. For example, Wilkins (2002) lists 26 species concepts. As a consequence, estimates of the total number of organisms vary from 5 to 50 million species (May, 2011). Using a predictive approach, Mora *et al.* (2011) calculated a figure of 8.7 ± 1.3 million species, which has been questioned by Caley *et al.* (2014). Appeltans *et al.* (2012) estimate that about 80% non-marine species and 75% marine species are yet undescribed.

Functioning of ecosystems is significantly affected by taxonomic and functional traits included in biodiversity. Species number or richness are hence often used as measures of the health of biological systems. The present status for biodiversity in natural ecosystems has a dim future. Global change already disturbs several regions on the globe, as predicted by Sala *et al.* (2000). However, diversity has increased to the north in Siberian Rivers affected by air temperature, ice-free periods and trophic variables (Barinova *et al.*, 2014). Declines estimated for tropical rainforests are often similar or less than those of northern freshwater ecosystems (Heino *et al.*, 2009). The reason for this rapid change in diversity is the disproportionate level of biodiversity in freshwater ecosystems compared to their spatial coverage. Fresh waters cover only 0.8% of the Earth's surface area and constitute only 0.01% of the world's water, but supports at least 6% of the estimated 1.8 million described species (Dudgeon *et al.*, 2006). The biodiversity in fresh waters will certainly largely increase when ground-water systems (Gilbert and Deharveng,

*E-mail: martin.dokulil@univie.ac.at

2002), and especially micro-organisms, are studied more intensively (Dudgeon *et al.*, 2006).

Since biodiversity includes all aspects of life and its processes, it is more than species richness or species number. Generally, three levels of biodiversity are accepted which are distinct but interrelated.

1) Genetic diversity – variety of genes within a species; determine how organisms develop, their traits and abilities. This level of diversity can differ by alleles, by entire genes, or by units larger than genes, such as chromosomal structure.
2) Species diversity – variety or richness of species; important for conservation, species extinction and the recognition of 'invasive species'.
3) Ecosystem diversity – ecosystem or habitat richness; dealing with species distributions and community patterns, the role and function of key species, and combines species functions and interactions.

A number of factors affecting species diversity have been proposed which were summarized in Valiela (1995).

1) Time Hypothesis – More time means more diversity.
2) Spatial Heterogeneity Hypothesis – More ecological niches mean more species.
3) Competition Hypothesis – More biological competition means more specialization.
4) Environmental Stability Hypothesis – More stability allows more diversity, due to adaptation.
5) The Productivity Hypothesis – More productivity means more species.
6) The Predation Hypothesis – Lowers competition, increases speciation due to less selection.
7) Time–Stability Hypothesis – Stressful, shallow habitats are less diverse than environmentally constant habitats.

16.2 Measures of Diversity

The number of different species is a simple measure of species richness in a habitat, ecosystem or region. This measure does not account for abundance.

More commonly, diversity is estimated using the popular Shannon Index, often cited as Shannon Wiener, Shannon–Weiner or Shannon Weaver Index, which considers both species number and abundance (Shannon, 1948, Shannon and Weaver, 1949, Spellerberg and Fedor, 2003). The Index H' is calculated for an assemblage of N individuals with abundance n_i from S different species using relative frequency of the species p_i from:

$$H' = -\Sigma p_i * \ln p_i n_i \qquad \text{mit } p_i = \frac{ni}{N}$$

The natural logarithm $^e\log$ or ln is often replaced by the binary logarithm $^2\log$. The Shannon Index is equivalent to entropy in information theory. Several other approaches and indices have been applied to the measure of diversity (see e.g. compilation by Magurran, 2004). Applying principal component analyses, a multivariate index can be derived describing the relative influence of biodiversity on ecosystem functioning by Naeem (2002).

Microorganisms vary in cell size by several orders of magnitude. Numerical based diversity is here replaced by the biomass contributed by each species. If the term H_n refers to numerical diversity and H_b to biomass diversity with b_i and B replacing n_i and N respectively, the corresponding equations using binary logarithms are as follows (Bürgi and Stadelmann, 2002):

$$H_n = -\Sigma\left(n_i * N^{-1*2}\log\left(n_i * N^{-1}\right)\right)$$

$$H_b = -\Sigma\left(b_i * B^{-1*2}\log\left(b_i * B^{-1}\right)\right)$$

The index of evenness can be calculated according to Krebs (1993) from:

$$En = (H_n - H_{nmin})(H_{nmax} - H_{nmin})^{-1}$$

with H_{nmin} = minimal possible numerical-based diversity (totally uneven abundance), and H_{nmax} = maximal possible numerical-based diversity (all species have the same abundance).

Measurements and monitoring of biodiversity can be performed at different spatial scales which may differ among habitats (Whittaker, 1972; Whittaker *et al.*, 2001).

Alpha diversity – diversity within a particular area or ecosystem; usually expressed by the number of species or species richness. For different approaches to calculate alpha diversity, consult Whittaker *et al.* (2001) and Tuomisto (2010a, b).

Beta diversity – change in species diversity between ecosystems; comparison of the total number of species that are exclusive to each of the ecosystems. For details of calculation, consult Whittaker (1972), Whittaker *et al.* (2001) or Tuomisto (2010a).

Gamma diversity – a measure of the overall diversity for the different ecosystems within a region; may be called 'geographic-scale species diversity'. Several different definitions and ways of calculation of gamma diversity exist (see e.g. Jost, 2007 and references therein; Tuomisto, 2010a, b). Gamma diversity is not very much used at present.

Epsilon diversity – proposed as a measure of regional diversity; rarely used (see Whittaker *et al.*, 2001).

16.3 Diversity of Algae

Knowledge of how many species of algae exist is hampered by the lack of an appropriate species concept, as mentioned above (Guiry, 2012). Estimates range from 30,000 to over 1 million species. A conservative approximation by Guiry (2012) yields 72,500 algal species. Part of the uncertainty originates from the fact that algae are phylogenetically hard to define, since they are a heterogenous polyphyletic mixture spanning two kingdoms and several phyla (Falkowski *et al.*, 2004, Keeling, 2004).

Plankton diversity thrives in the tension between the slogans 'plankton paradox' and 'everything is everywhere'. The paradox of plankton, first formulated by Hutchinson (1961), is based on the question of how a large number of phytoplankton species can coexist in a seemingly isotropic or unstructured environment all competing for the same resources. On the other hand, if morphology is used to define taxonomic units, organisms are potentially cosmopolitan, as the 'everything is everywhere' (EiE) hypothesis states (Fenchel and Finlay, 2004; Fenchel and Finlay, 2006). Based on a large number of observations, most authors conclude that everything is everywhere but the environment selects. The environment, however, does not simply act as a filter: micro-organisms adapt to specific habitats (Weisse, 2006). Experimental results show that fluctuations in environmental variables maintain phytoplankton diversity (e.g. Sommer, 1984). Models of plankton biodiversity have given controversial results. Huisman and Weissing (1999, 2000) concluded that internal system feedbacks inducing oscillations or chaos result in more coexisting species than resources available. This phenomenon was called 'supersaturated coexistence' by Schippers *et al.* (2001), who concluded from the model that it is rare and unlikely to occur in nature. Introducing evolution into such models results in destabilisation of previously stable species communities (Shoresh *et al.*, 2008). Overviews of the mechanism proposed to resolve the paradox were compiled by Wilson (1990) and Roy and Chattopadhyay (2007).

A new paradox of the plankton was suggested by Passy and Legendre (2006). Analysing a large algal data set from streams in the United States including phytoplankton and phytobenthos, the authors concluded that diversification at higher taxa are greater in phytoplankton than in the benthos. The array of ecological niches necessary poses a new paradox of plankton.

16.4 Biogeography

Although freshwater phytoplankton species are considered to be cosmopolitan, a considerable number are local or regional endemic algal taxons (review by Padisák, 2003). Moreover, the absence of certain phytoplankton species from whole continents cannot be explained by the lack of suitable habitats. True biogeography has been demonstrated for e.g. the freshwater heterokont algae *Synura petersenii*. This morphotaxon contains both cosmopolitan and regionally endemic cryptic species, as revealed by molecular techniques (Boo *et al.*, 2010).

Elucidation of prokaryotic biogeography is largely limited by the species concept as applied to prokaryotes (Fontaneto, 2011). Recognition of diversity and abundances of species are essential in community and population studies (Lacap *et al.*, 2011). Since the traditional definition of a species is in general not applicable to prokaryotes, an ecological species concept has been proposed, where ecotypes can arise in different niches (Cohan, 2002).

The biogeography of various algal groups has been summarised by Kristiansen (1994). For the cyanobacteria, Hoffmann (1996) concluded that temperature is a main controlling factor limiting species to particular climatic zones besides the distribution of ecological niches. Since cyanobacterial groups are associated with climatic domains, composition may shift to more frequent occurrence of microcystin-producing taxa (Pitois *et al.*, 2014). A comprehensive overview of cyanobacterial diversity was compiled by Whitton and Potts (2000). Species with a restricted regional distribution (endemics) cannot be ruled out. Although dispersal is a problem in chrysophytes, they are distributed worldwide and quite a large portion are endemic (Kristiansen, 2008). Endemism and rarity has also been stressed for diatoms by Mann and Droop (1996). The global distribution of diatom species has recently been shown to range from cosmopolitanism to narrow endemic. Geographical factors independent of environmental conditions structure diatom communities and influence their diversity (Vanormelingen *et al.*, 2008). Endemism may also be relevant in desmids, since their distribution seems to be microclimatologically determined (Coesel, 1996). Endemism of freshwater algal taxa in the Australian region was discussed by Tyler (1996). Several species from the chloro-, chryso-, and dinophyta were designated a 'robust' endemics, such as e.g. the chrysophyte *Chrysonephele palustris* from Tasmania. Tell *et al.* (2011) were able to show that diversity of the group Chlorococcales among the Chlorophyta was directly related to the trophic status of lakes and inversely correlated with latitude when analysing water bodies along a transect in Patagonia, Argentina. Dinoflagellates are highly versatile in habitat distribution and trophic diversity. Continental waters contain about 10% of the described species and are dominated by plastid-containing free-living forms, mainly in the phytoplankton, while benthic species are less common (Gómez, 2012). From the number of living species, only one explicitly benthic species has been described so far (Taylor *et al.*, 2008). Dinoflagellate biogeographical distribution has been termed 'modified latitudinal cosmopolitanism', as an indication that the same morphospecies occur within similar climatic zones. Few endemics are known (Taylor *et al.*, 2008).

16.5 Trophic Interaction

Effects of predation and competition on size diversity of freshwater and marine sites were analysed for phytoplankton, zooplankton and fish by Quintana *et al.* (2015). Results for phytoplankton differed markedly from the other levels of the food web. Diversity of phytoplankton size classes increased when nutrient availability was high and zooplankton grazing low. Effects of grazing on freshwater benthic algal communities were summarized by Steinmann (1996). Intensity and frequency of disturbance is associated with a decline in species richness and diversity, as most studies indicate.

The functional role of biodiversity in ecosystems requires an integration of diversity within trophic levels, termed 'horizontal diversity', and across trophic levels, termed 'vertical diversity' by Duffy *et al.* (2007). Horizontal diversity of producers and consumers can increase similarly with biomass and the usage of resources. Changes in vertical diversity among predators can cascade down to affect plant biomass. Both diversities also interact (Duffy *et al.*, 2007).

Eight fundamental questions were articulated by Cardinale *et al.* (2011) about how efficiency of resource use and biomass production in ecosystems is influenced by diversity of primary producers such as algae. One of their conclusions was that the effects of diversity are driven by complementarity in aquatic ecosystems.

16.6 Case Studies on Algal Biodiversity

In the following section, algal diversity will be demonstrated using five different water bodies as examples. Long-term data have been suggested as potentially important for diversity dynamics (Magurran *et al.*, 2010). Therefore, long-term data from a deep, stratifying lake in Austria, typical of the alpine region (Dokulil, 2005), are used to demonstrate the effects of oligotrophication on phytoplankton species richness and diversity (Fig. 16.1). The oligotrophic lake underwent eutrophication starting in the late 1960s and culminating in the mid-1970s, when the annual average total phosphorus (TP) concentration peaked at 36 µg L^{-1} TP (Dokulil and Teubner, 2005). Associated with eutrophication, the filamentous cyanobacterium *Planktothrix rubescens* appeared, became dominant and declined (Dokulil and Teubner, 2012). In-lake nutrient concentrations began to decline after 1978 when sewage diversion became effective.

Phytoplankton composition and biomass is not well-enough documented for the early eutrophication period (1969 to 1982) to allow evaluation of biodiversity. Annual average TP, species number (richness) and biomass-based diversity in Mondsee was calculated for the 23-year period 1982 to 2004 (Fig. 16.1). Phosphorus concentrations varied around an average of 10 µg L^{-1} with the lowest values in the mid-1990s (Fig. 16.1(A)). The species number in the 534 data points ranged from 7 to 37, with annual averages from 15.3 to 29.0 and mean 25.2 and median 25.0 for all (Fig. 16.1(B)). Annual mean Shannon diversity indices increased until 2002 at a rate of 0.001 per decade ($r^2 = 0.77$, $n = 21$, $F = 63.4$, $p<0.001$). Annual averages varied considerably from year to year in the 1980s and became less variable in the 1990s (Fig. 16.1(C)). Yearly mean evenness (not shown) varied around 0.6 in the 1980s and increased to around 0.75 in 1991 and after. This evolvement reflected the progression in oligotrophication and the alteration of species composition. The decline in 2003 and 2004 was probably caused by changes in the analytical technique and/or the observer. Using the 5%, 95% percentiles and the median as diagnostic plot, as

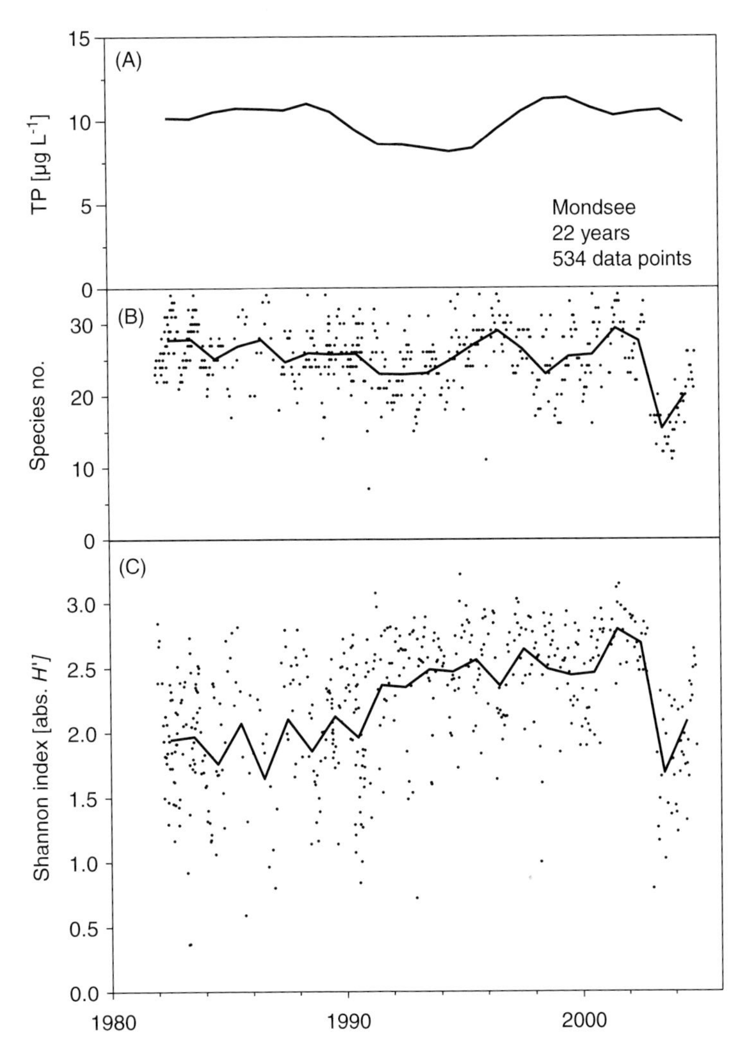

Fig. 16.1. Long-term data from Mondsee, Austria, 1982–2004. (A) Annual average total phosphorus (TP) concentrations in the epilimnion. Smoothed with Lowess function. (B) Species number. (C) Shannon diversity index (abs. H') calculated from phytoplankton biovolumes. Dots = individual data integrated for the top 20 m. Line = Annual average.

suggested by Straile *et al.* (2013), ruled out inconsistencies in expertise or detection limits. Long-term data (38 years) from Neusiedler See, a large shallow alkaline turbid lake (320 km^2, max. depth 1.8 m) bordering Austria and Hungary (Dokulil, 1979; Dokulil and Padisák, 1994), demonstrate the progress of phytoplankton diversity in a polymictic, non-stratifying lake (Fig. 16.2).

Increasing mean TP-concentrations indicate eutrophication in the 1970s peaking at 164 µg L^{-1} in 1982 (Fig. 16.2(A)). In-lake concentrations gradually decreased thereafter, approaching a minimum of 45 µg L^{-1} TP in 1999. The early years of the 21st century were characterized by rising concentrations (max. 117 µg L^{-1} in 2004), perhaps due to low water levels. During the eutrophication period (1968–1982) and most of the oligotrophication period until 1993, species richness remained around an average number of 10.4 (5.7–14.2) as shown in Fig. 16.2(B). Spatial variation among stations varied largely from 2 to 25 taxa while H' was between 0.09 and 2.56. Average annual H' was 1.35 for 1968–1993, ranging from 0.66 to 1.83 (Fig. 16.2(C)). Yearly mean evenness (not shown) varied from

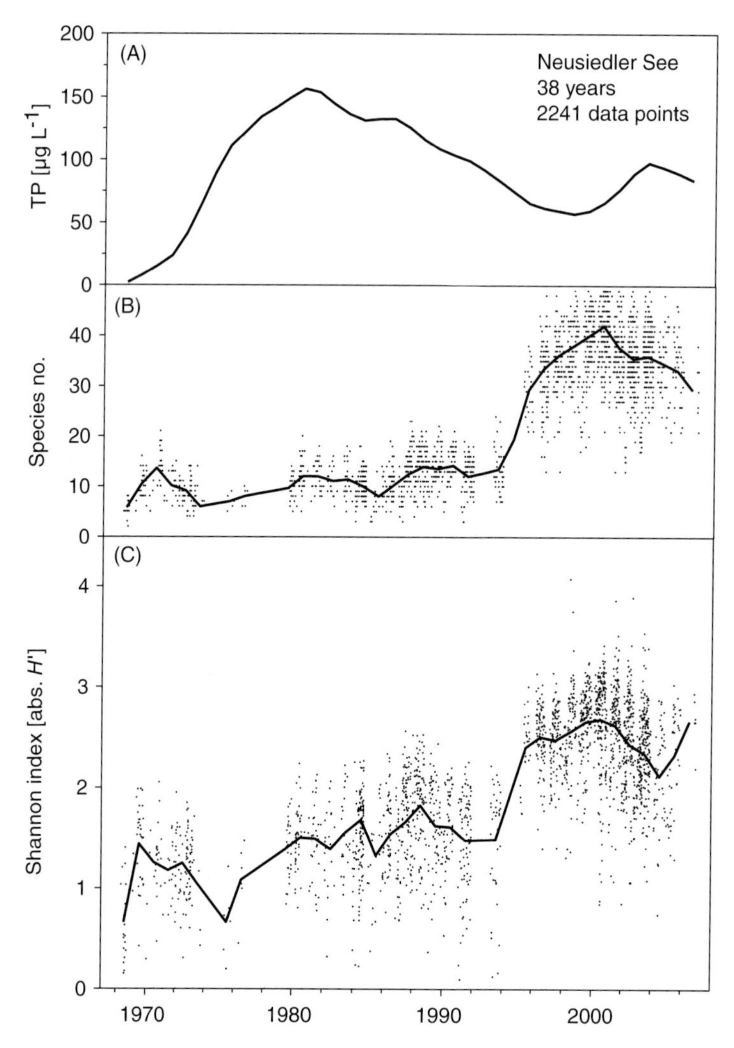

Fig. 16.2. Long-term data from Neusiedler See, Austria, 1968–2006. (A) Annual average total phosphorus (TP) concentrations for the water column. Smoothed with Lowess function. (B) Species number. (C) Shannon diversity index (abs. H') calculated from phytoplankton biovolumes. Dots = individual data for the water column. Line = Annual average.

year to year, but tended to steadily increase from about 0.5 to 0.7 over the investigation period. Calculating species number and diversity from data collected in 1958 by Ruttner (unpublished) revealed 12 and 1.84 respectively, values which are close to those mentioned above.

When TP reached minimum values, species number and diversity indices significantly increased to annual averages of 35 (29–42) and 2.48 (2.12–2.68) respectively. As the observer changed at approximately that time, data were checked by a diagnostic plot, according to Straile *et al.* (2013). No inconsistencies, however, were detected. Correspondence between the trophic status of the lake and phytoplankton biodiversity is rather weak and is not statistically significant.

The shallow urban seepage Lake Alte Donau (AD) in Vienna switched from macrophyte domination to algal domination in 1992/1993 and was restored thereafter (Dokulil *et al.*, 2006; Dokulil *et al.*, 2011). As a consequence of dilution and particularly P-flocculation, mean in-lake TP concentrations declined from 66 in 1993 to 19.1 mg L^{-1}in 1996 Concentrations remained around an average of 19.1

TP until 2003, and then dropped further to a mean of 11.4 µg L^{-1} (Fig. 16.3(A)), associated with the return of the underwater vegetation. These events were reflected by species richness and diversity (Fig. 16.3(B),(C)).

In the two highly eutrophic years 1993 and 1994, the mean annual Shannon diversity was 1.54 and 1.60, while species numbers were 14 and 12 respectively. During restoration (1994/95) both increased to 2.36 H' and 31 species. Species number fluctuated from year to year around an average of 24 until 2004. Mean H' was 2.0 in this period, increasing to 2.4 after 2004, when species numbers reached an average level of 38 (Fig. 16.3(B),(C)). Annual average evenness (not shown) varied around the long-term mean of 0.64 with greater yearly variation at the beginning of the observation period. After restoration, evenness became much more invariable. Spatial variability of diversity is demonstrated using data on river phytoplankton from a length profile of the river Danube from the city of Regensburg (river km

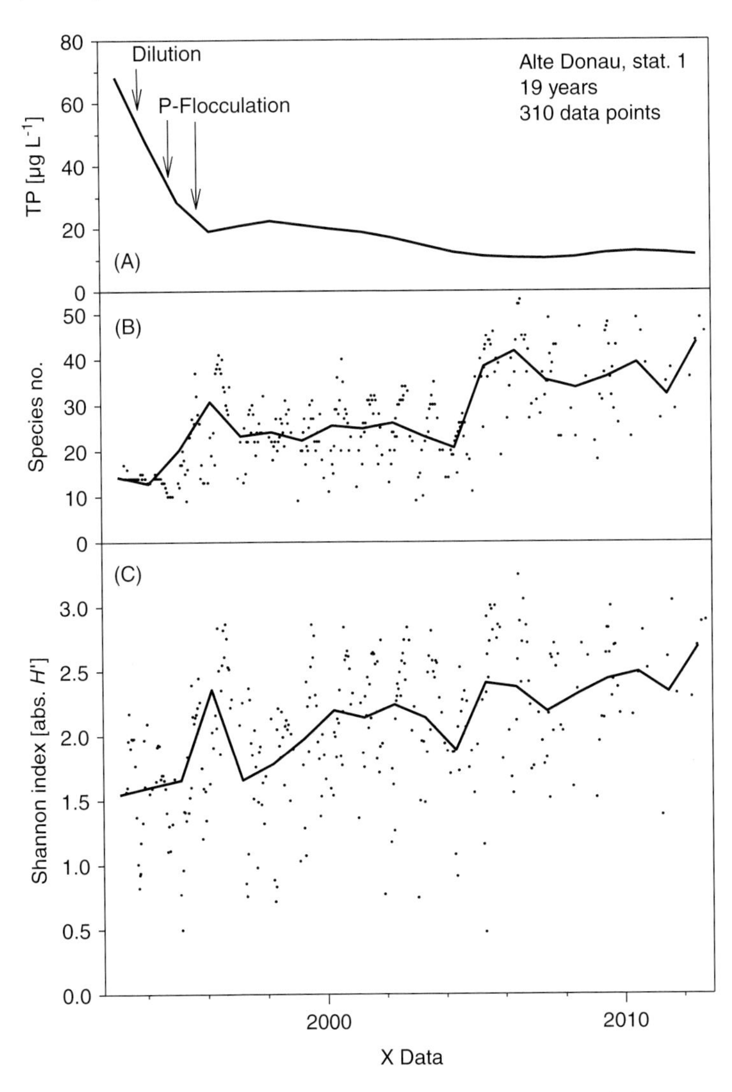

Fig. 16.3. Long-term data from Alte Donau, Vienna, 1993–2013. (A) Annual average total phosphorus (TP) concentrations for the water column. Smoothed with Lowess function. (B) Species number. (C) Shannon diversity index (abs. H') calculated from phytoplankton biovolumes. Dots = individual data for the water column. Line = Annual averages. Dilution and P-flocculation as restoration measures (Dokulil *et al.*, 2011).

2600) to the delta at the Black Sea (river km 0) sampled in 2007 (Dokulil and Kaiblinger, 2008). Data on major tributaries are provided for comparison (Fig. 16.4).

Soluble reactive phosphorus (SRP) is used here to indicate the nutritional status of the river (Fig. 16.4(A)). In general, SRP concentrations remained below 10 μg L^{-1}. Concentrations were largely discharge-related and not correlated with the number of species or diversity. Species richness and diversity peaked at 43 and 2.9 respectively in the middle part of the river (km 1600–1200), where SRP was low (Fig. 16.4(B),(C)). Both diversity indicators showed a clear spatial variability not significantly influenced by the inputs from the tributaries. The decline of H' in the lower stretch of the river (Fig. 16.4(C)) was paralleled by an almost equal decline in evenness. This decline of species diversity and species evenness is

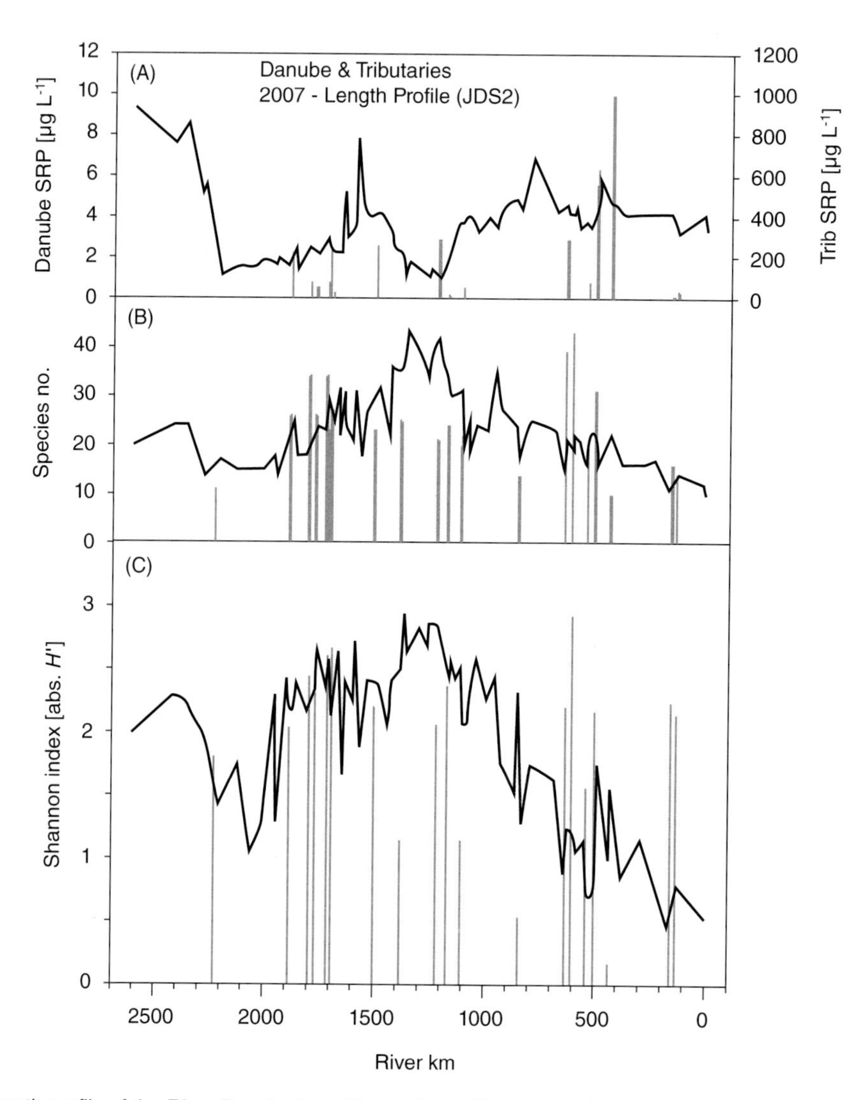

Fig. 16.4. Length profile of the River Danube from Regensburg, Germany to the delta at the Black Sea (river km 0) using JDS2 results from September 2007. (A) Soluble reactive phosphorus (SRP) concentrations. Grey bars indicate concentrations in main tributaries (note different y-axis scale). (B) Species number. (C) Shannon diversity index (abs. H') calculated from phytoplankton biovolumes. Grey bars indicate main tributaries. All samples are surface samples. For further details, see Dokulil and Kaiblinger (2008).

likely to have been caused by turbidity. The ups and downs of H' were significantly related to species richness, explaining 50% of the variability (r^2 = 0.50, n = 74, F = 73.2, p < 0.001).

The diversity of epilithic diatoms is evaluated from samples taken in an artificial flood water by-pass of the River Danube in Vienna in 1988 (Dokulil, 1979; Dokulil and Janauer, 2000). Because the system is operated by sluices, it is stagnant for most of the year under normal operation, but is river-like during floods when sluices are opened (Fig. 16.5).

Yearly average SRP concentrations were 88 μg L^{-1} in the Danube and between 12.5 and 7.3 μg L^{-1} decreasing from up- to downstream stations (Fig. 16.5a). Although SRP concentrations were so dissimilar between the river and the impoundment,

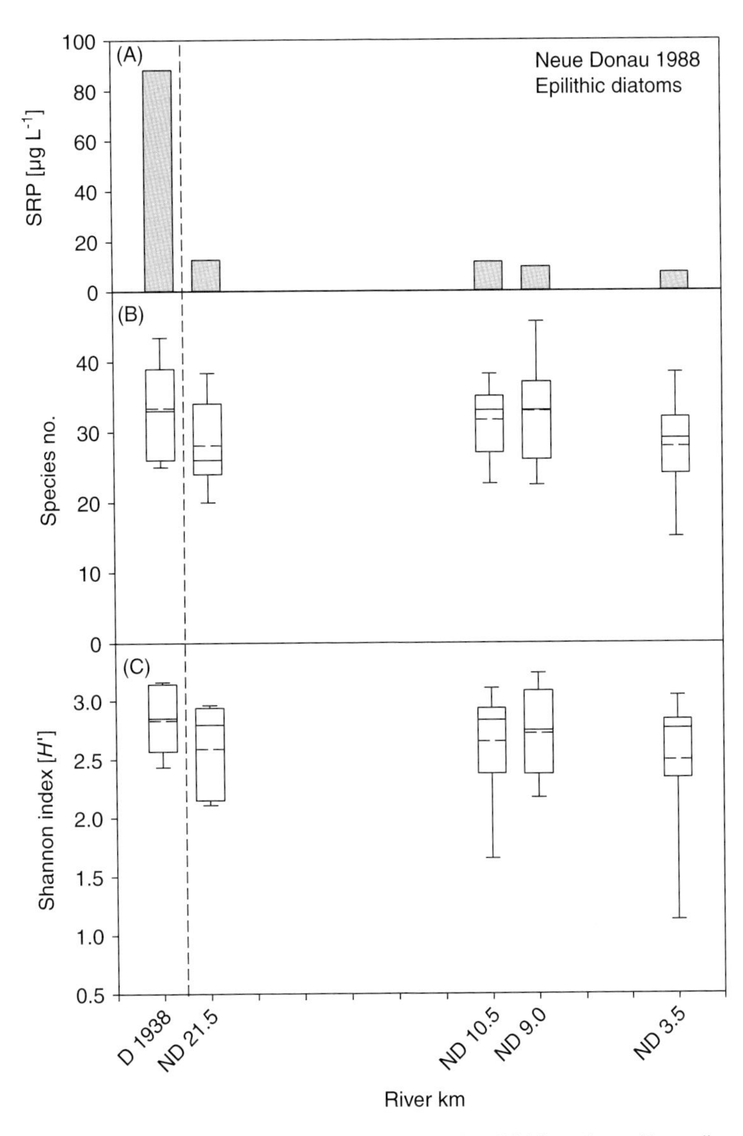

Fig. 16.5. Epilithic diatom data from Neue Donau, Vienna for 1988. Box-Whisker plots with median (solid line) and mean (dashed line). A reference station in the Danube at km 1938 was used (D 1938). The stations in the artificial system Neue Donau are indicated by ND followed by the respective km from upstream (21.5) to downstream (3.5). Sample interval was monthly. For further details, see the text and Dokulil *et al.* (1997).

epilithic diatom assemblages did not statistically differ in either species richness or diversity. There was also no significant differentiation among the stations, except that variability became slightly larger at the downstream station ND 3.5 (Fig. 16.5(B),(C)).

16.7 Critical Discussion of Algal Diversity

The examples described above may serve to discuss problems associated with biodiversity calculation for algae, particularly of phytoplankton. Diversity indices should be used cautiously for several reasons. Using common methodology, species richness is poorly estimated because it usually relates more to the counting effort than to ecological indications. Richness largely depends on the level of taxonomic separation. More detailed identification leads to more taxa and hence higher richness. The probability for species detection in counts of a community is considered to be less than one. Consistent effort, sample volume and reliable observers are the prime requirements for proper evaluation of algal biodiversity. Only minute proportions of algae in habitats are examined and many rare taxa are not recognized, which nevertheless might be important. As a consequence, taxa numbers in most counts or sample scans do not represent true species richness. Assuming such estimates to be a measure of the number of taxa in a sample or habitat is therefore highly suspect. Since evenness of abundances strongly affects numbers observed in counts, the assumption that observed counts are proportional to the number of species in samples is invalid (Stevenson, 2014). In summary, species richness is influenced by incomplete detection and identification during analysis procedure. Empirical estimates must therefore be negatively biased.

The type of sample, the sampling technique and the method used for identification and enumeration largely affect correct diversity estimation. As for species richness, problems of detection, missed species or inadequate identification apply. Estimating changes of diversity over time or space adopts the unrealistic assumption that detection probability is constant. The probability to detect a species is also a function of the number of individuals and the difficulty to detect each individual. Identification of small flagellates is especially difficult in phytoplankton assemblages. Molecular techniques can possibly help in this field, as demonstrated by Luo *et al.* (2011). A comparison of traditional microscopic and molecular approaches revealed remarkable differences. Autotrophic organisms predominated in microscopic analyses, whereas the molecular method resulted in higher diversity of heterotrophic flagellates. Genetic diversity has significant effects on many ecological processes at the population, community and ecosystem levels (Hughes *et al.*, 2008).

Statistical artefacts are minimized if diversity is expressed as a function of maximum diversity possible for a given sample. Doing so, a scale of 0–1 is created, which might be more easily understood. The similarity of samples at the genus or species level can be estimated from presence/absence data. Expressing the detailed analysis of a sample analysis as a single number, however, oversimplifies community features and may not be reliable by itself. Indices may have their advantage for the management of water, since an index amalgamates several influential factors. Important details may be lost or overlooked, however. The level of biodiversity index may, for instance, suggest that moderate changes have occurred. In fact, internal species composition may have completely altered, replacing species which were previously present by others due to changes in water quality. Larger spring diatom species were replaced by small centric diatoms during oligotrophication and *Planktothrix rubescens* gradually declined, disappeared and recently re-appeared to some extent. Biodiversity indices can thus not fully substitute for detailed analysis of species composition.

A further problem associated with the estimation of algal diversity is the question of whether to use numerical or biovolume data for calculation. Considering the huge size differences of algal species, many investigators use biovolume-based estimates of diversity, although differences seem moderate, as Bürgi and Stadelmann (2002) concluded. All data shown above are biovolume-based estimates of biodiversity.

An excellent example of using alpha, beta and gamma diversity in algal assemblages is provided by Zheng and Stevenson (2006). Results indicated that restored wetlands had higher alpha, beta, and gamma diversity than existing wetlands. Non-diatom and diatom diversity among habitats were quite different. Alpha, beta, and gamma diversity of non-diatom epiphyton and phytoplankton were greater than epipelon in all wetlands. Sediment diatoms had higher alpha and gamma diversity, but lower beta diversity, than epiphytic diatoms. Algal beta and gamma diversity in stream ecosystems was significantly higher in geomorphically stable reaches than in degraded reaches (Passy and Blanchet, 2007). Alpha diversity and cell density were significantly higher in unstable reaches. Stream restoration influenced algal beta and gamma diversity, but the stable reference reaches retained highest beta diversity.

16.8 Environmental Assessment and the Water Framework Directive (WFD)

Algal diversity is a good indicator of human disturbance, and is therefore highly relevant for management. For that reason, algal species or communities are often used as biological indicators (Bellinger and Sigee, 2015). Both phytoplankton and algae on substrates can be used in lakes and streams (Dokulil, 2003; Poulíčková *et al.*, 2004; Janauer and Dokulil, 2006; Soltani *et al.*, 2012).

The composition and abundance of phytoplankton and phytobenthos is needed to determine the ecological status of inland waters, according to the WFD (European Commission, 2000), or, in other words, biological diversity as a biomonitor of the ecosystems of the world (UNESCO, 2000). Considerable progress has been made in the implementation of the directive already (Phillips, 2014). Reference conditions and class boundaries have been established for phytoplankton biomass in alpine lakes on the basis of algal taxa lists (Wolfram *et al.*, 2009; Wolfram *et al.*, 2014). The ecology of epipelic algae and their use in biodiversity and biomontoring has been summarized by Poulíčková *et al.* (2008). Observed species decreases should be related very cautiously to biodiversity loss. Blanco *et al.* (2012), for instance, discourage use of diatom diversity indices after their evaluation.

16.9 Concluding Remarks

Various aspects of biodiversity have recently attracted attention, possibly because of greater awareness of their functional role. Environmental influences, climatic change or spatial and temporal patterns of diversity can certainly be investigated and analysed from long-term data sets with advantage. The case studies shown above and the discussion associated with them indicate, however, that detailed conclusions and comparisons must be drawn with caution. Diversity largely depends on the resolution of taxa identification. Biodiversity should not be used in isolation or as a substitute for more detailed analyses. Understanding of ecosystems will hopefully improve when biodiversity is carefully evaluated, perhaps with the help of genetic diversity. More information on higher-level biodiversity (beta to epsilon diversity) seems necessary. Diversities in different ecosystems within a region, or on a wider regional scale, need to be compared to improve our understanding of algal biogeography.

Acknowledgements

Very many individuals have contributed to the case studies. The logistics of the Institute of Limnology in Mondsee (University of Innsbruck, formerly the Austrian Academy of Sciences) have provided a solid basis for all activities throughout the years. Thanks are particularly extended to Katrin Teubner, Christina Kaiblinger, Claudia Skolaut and Lotte Eisl for help in the field and extended laboratory analysis of the samples. Additional phosphorus data were supplied by Albert Jagsch from the Institute for Aquatic Ecology, Fish Biology and Lake Science (Federal Bureau of Water Resources). The Biological Station Neusiedler See, Illmitz (State Burgenland) has provided logistic data over many years on Neusiedler See. Special thanks go to Alois Herzig for continuous support, Judit Padisák and Harald Krisa for phytoplankton data. Phytoplankton data for Alte Donau were provided by the DWS-Hydro-Ecology, Vienna (Georg Wolfram, Karl Donabaum). The continuous support of the municipality of Vienna since 1986 is kindly acknowledged for both studies on Alte Donau and Neue Donau. Danube data were collected during the Joint Danube Survey 2 (JDS2) in August to October 2007. The contract of the leading organisation ICPDR (International Commission for the Protection of the Danube Region) is kindly acknowledged. Special thanks are extended to the JDS2 team, the boat crew and Christina Kaiblinger for sampling and laboratory analyses.

References

Appeltans, W., Ahyong, S.T., Anderson, G., Angel, M.V. and Artois, T. (2012) The magnitude of global marine species diversity. *Current Biology* 22, 2189–2202.

Barinova, S., Gabyshev, V. and Gabysheva, O. (2014) Climate impact on freshwater biodiversity: general patterns in extreme environments of North-Eastern Siberia (Russia). *British Journal of Environment and Climate Change* 4(4), 423–443.

Bellinger, E.G. and Sigee, D.C. (2015) *Freshwater algae: identification and use as bioindicators*, 2nd edn. Wiley, Chichester, UK. 296 pp.

Blanco, S., Cejudo-Figueiras, C., Tudesque, L., Becares, E., Hoffmann, L. and Ector, L. (2012) Are diatom diversity indices reliable monitoring metrics? *Hydrobiologia* 695, 199–206.

Boo, S., Kim, H.S., Shin, W., Boo, G.H., Cho, S.M. and Jo, B.Y. (2010) Complex phylogeographic patterns in the freshwater alga *Synura* provide new insights into ubiquity vs. endemism in microbial eukaryotes. *Mono Lake Ecology* 19, 4328–4338.

Bürgi, H.R. and Stadelmann, P. (2002) Change of phytoplankton composition and diversity in Lake Sempach before and during restoration. *Hydrobiologia* 469, 33–48.

Caley, M.J., Fisher, R. and Mengersen, K. (2014) Global species richness estimates have not converged. *Trends in Ecology and Evolution* 29(4), 187–188.

Cardinale, B.J., Matulich, K.L., Hooper, D.U., Byrnes, J.E. and Duffy, E. (2011) The functional role of producer diversity in ecosystem. *American Journal of Botany* 998(3), 572–592.

Coesel, P.F.M. (1996) Biogeography of desmids. *Hydrobiologia* 336, 41–53.

Cohan, F.M. (2002) What are bacterial species? *Annual Reviews of Microbiology* 56, 457–487.

Dokulil, M.T. (1979) Seasonal pattern of phytoplankton. In: Löffler, H. (ed.) *Neusiedlersee: The Limnology of a Shallow Lake in Central Europe*. Dr W. Junk bv Publishers, The Hague, The Netherlands, pp. 203–231.

Dokulil, M.T. (2003) Algae as ecological bio-indicators. In: Markert, B.A., Breure, A.M. and Zechmeister, H.G. (eds) *Bioindicators and Biomonitors*. Elsevier Science Ltd, Amsterdam, The Netherlands, pp. 285–327.

Dokulil, M. (2005) European alpine lakes In: O'Sullivan, P. and Reynolds, C.S. (eds) *The Lakes Handbook*, Vol. 2. Blackwell Science, Oxford, UK, pp. 159–178.

Dokulil, M.T. and Janauer, G.A. (2000) Alternative stable states of macrophytes versus phytoplankton in two interconnected impoundments of the New Danube (Vienna, Austria). *Archiv für Hydrobiologie* Suppl. 135 (Large Rivers 12), 75–83.

Dokulil, M.T. and Kaiblinger, C. (2008) *Phytoplankton. Full Report + Annexes. Results of the Joint Danube Survey 2*. ICPDR, Vienna, Austria. Available at: www.icpdr.org/JDS (accessed 27 May 2016).

Dokulil, M.T. and Padisák, J. (1994) Long-term compositional response of phytoplankton in a shallow, turbid environment, Neusiedlersee, Austria. *Hydrobiologia* 275/276, 125–137.

Dokulil, M.T. and Teubner, K. (2005) Do phytoplankton assemblages correctly track trophic changes? Assessment from contemporary and palaeolimnological data. *Freshwater Biology* 50, 1594–1604.

Dokulil, M.T. and Teubner, K. (2012) Deep living *Planktothrix rubescens* modulated by environmental constraints and climate forcing. *Hydrobiologia* 698, 29–46.

Dokulil, M.T., Schmidt, R. and Kofler, S. (1997) Benthic diatom assemblages as indicators of water quality in an urban flood-water impoundment, Neue Donau, Vienna, Austria. *Nova Hedwigia* 65, 273–283.

Dokulil, M.T., Donabaum, K. and Pall, K. (2006) Alternative stable states in floodplain ecosystems. *Ecohydrology and Hydrobiology* 6, 37–42.

Dokulil, M.T., Donabaum, K. and Pall, K. (2011) Successful restoration of a shallow lake: a case study based on bistable theory. In: Ansari, A.A., Gill, S.S., Lanza, G.R. and Rast, W. (eds) *Eutrophication: Causes, Consequences and Control*. Springer, Dordrecht, The Netherlands, pp. 285–294.

Dudgeon, D., Arthington, A.H., Gessner, M.O., Kawabata, Z.I., Knowler, D.J. and Leveque, C. (2006) Freshwater biodiversity: importance, threats, status and conservation. *Biological Reviews* 81, 163–182.

Duffy, J.E., Cardinale, B.J., France, K.E., McIntyre, P.B., Thébault, E. and Loreau, M. (2007) The functional role of biodiversity in ecosystems: incorporating trophic complexity. *Ecology Letters* 10, 522–538.

European Commission (2000) European Commission Directive 2000/60/EC of the European Parliament and of the Council of 23rd October 2000 of establishing a framework for community action in the field of water policy. *The Official Journal of the European Commission* 22(12).

Falkowski, P.G., Katz, M.E., Knoll, A.H., Quigg, A., Raven, J.A., Schofield, O. and Taylor, F.J.R. (2004) The evolution of modern eukaryotic phytoplankton. *Science* 305, 354–360.

Fenchel, T. and Finlay, B.J. (2004) The ubiquity of small species: patterns of local and global diversity. *BioScience* 54, 777–784.

Fenchel, T. and Finlay, B.J. (2006) The diversity of microbes: resurgence of the phenotype. *Biological Science* 361, 1965–1973.

Fontaneto, D. (2011) *Biogeography of Microscopic Organisms: Is Everything Small Everywhere?* Cambridge University Press, Cambridge, UK.

Gilbert, J. and Deharveng, L. (2002) Subterranean ecosystems: a truncated functional biodiversity. *BioScience* 50, 653–666.

Gómez, F. (2012) A quantitative review of the lifestyle, habitat and trophic diversity of dinoflagellates (Dinoflagellata, Alveolata). *Systematics and Biodiversity* 10(3), 267–275.

Guiry, M.D. (2012) How many species of algae are there? *Journal of Phycology* 68, 1057–1063.

Heino, J., Virkkala, R. and Toivonen, H. (2009) Climate change and freshwater biodiversity: detected patterns, future trends and adaptations in northern regions. *Biological Review* 84, 39–54.

Hoffmann, L. (1996) Geographic distribution of freshwater blue-green algae. *Hydrobiologia* 336, 33–40.

Hubbell, S.P. (2011) *The Unified Neutral Theory of Biodiversity and Biogeography*. Princeton University Press, Princeton, New Jersey, USA.

Hughes, A.R., Inouye, B.D., Johnson, M.T.J., Underwood, N. and Vellend, M. (2008) Ecological consequences of genetic diversity. *Ecology Letters* 11, 609–623.

Huisman, J. and Weissing, F.J. (1999) Biodiversity of plankton by species oscillations and chaos. *Nature* 402, 407–410.

Huisman, J. and Weissing, F.J. (2000) Coexistence and resource competition. *Nature* 407, 694.

Hutchinson, G.E. (1961) The paradox of the plankton. *American Naturalist* 95, 137–145.

Janauer, G. and Dokulil, M.T. (2006) Macrophytes and algae in running waters. In: Ziglio, G., Siligardi, M. and Flaim, G. (eds) *Biological Monitoring of Rivers: Applications and Perspectives*. Wiley, Chichester, UK, pp. 89–109.

Jost, L. (2007) Partitioning diversity into independent alpha and beta components. *Ecology* 88, 2427–2439.

Keeling, P.J. (2004) Diversity and evolutionary history of plastids and their hosts. *American Journal of Botany* 91(19), 1481–1493.

Krebs, C.J. (1993) *Ecology: The Experimental Analysis of Distribution and Abundance*. Harper Collins, New York, USA.

Kristiansen, J. (1994) Biogeography of freshwater algae. *Developments in Hydrobiology* 118, 1–161.

Kristiansen, J. (2008) Dispersal and biogeography of silica-scaled chrysophytes. *Biodiversity and Conservation* 17, 419–426.

Lacap, D.C., Lau, M.C.Y. and Pointing, S.B. (2011) Biogeography of prokaryotes. In: Fontaneto, D. (ed.) *Biogeography of Microscopic Organisms: Is Everything Small Everywhere?* Cambridge University Press, Cambridge, UK, pp. 35–40.

Luo, W., Bock, C., Li, H.R., Padisák, J. and Krienitz, L. (2011) Molecular and microscopic diversity of planktonic eukaryotes in the oligotrophic Lake Stechlin (Germany). *Hydrobiologia* 661, 133–143.

Maczulak, A.E. (2010) *Biodiversity: Conserving Endangered Species*. Infobase, New York, USA.

Magurran, A.E. (2004) *Measuring Biological Diversity*. Blackwell, Malden, Massachusetts, USA.

Magurran, A.E., Baillie, S.R., Buckland, S.T., Dick, J. Mc.P. and Elston, D.A. (2010) Long-term datasets in biodiversity research and monitoring: assessing change in ecological communities through time. *Trends in Ecology and Evolution* 25(10), 574–582.

Mann, D.G. and Droop, J.M. (1996) Biodiversity, biogeography and conservation of diatoms. *Hydrobiologia* 336, 19–32.

May, R.M. (2011) Why worry about how many species and their loss? *PLoS Biology* 9(8), e1001130.

Mora, C., Tittensor, D.P., Adl, S., Simpson, A.G.B. and Worm, B. (2011) How many species are there on earth and in the ocean? *PLoS Biology* 9(8), e1001127.

Naeem, S. (2002) Disentangling the impacts of diversity on ecosystem functioning in combinatorial experiments. *Ecology* 83(10), 2925–2935.

Padisák, J. (2003) Phytoplankton. In: O'Sullivan, P. and Reynolds, C.S. (eds) *The Lakes Handbook*, Vol 1. Blackwell Science, Oxford, UK, pp. 251–308.

Passy, S.I. and Blanchet, F.G. (2007) Algal communities in human-impacted stream ecosystems suffer beta-diversity decline. *Diversity and Distribution* 13, 670–679.

Passy, S.I. and Legendre, P. (2006) Power law relationships among hierarchical taxonomic categories in algae reveal a new paradox of the plankton. *Global Ecology and Biogeography* 15, 528–535.

Phillips, G. (2014) Progress towards the implementation of the European Water Framework Directive (2000–2012). *Aquatic Ecosystem Health & Management* 17(4), 424–436.

Pitois, F., Thoraval, I., Baurès, E. and Thomas, O. (2014) Geographical patterns in Cyanobacteria distribution: climate influence at regional scale. *Toxins* 6, 509–522.

Poulíčková, A., Duchoslav, M. and Dokulil, M.T. (2004) Littoral diatom assemblages as bioindicators for lake trophy: a case study from alpine and pre-alpine lakes in Austria. *European Journal of Phycology* 39, 143–152.

Poulíčková, A., Hašler, P., Lysákova, M. and Spears, B. (2008) The ecology of freshwater epipelic algae: an update. *Phycologia* 47(5), 437–450.

Quintana, X.D., Arim, M., Badosa, S.A., Blanco, J.M. and Boix, D. (2015) Predation and competition effects on the size diversity of aquatic communities. *Aquatic Science* 77, 45–57.

Roy, S. and Chattopadhyay, J. (2007) Towards a resolution of 'the paradox of the plankton': a brief overview of the proposed mechanisms. *Ecological Complexity* 4, 26–33.

Sala, O.E., Chapin, F.S., Armesto, J.J., Berlow, E., Bloofield, J. and Dirzo, R. (2000) Global biodiversity scenarios for the year 2100. *Science* 287, 1770–1774.

Schippers, P., Verschoor, A.M., Vos, M. and Mooij, W.M. (2001) Does 'supersaturated coexistence' resolve the 'paradox of the plankton'? *Ecology Letters* 4, 404–407.

Shannon, C.E. (1948) A mathematical theory of communication. *Bell System Technical Journal* 27, 379–423.

Shannon, C.E. and Weaver, W. (1949) *The Mathematical Theory of Communication*. University of Illinois at Urbana–Champaign, Illinois, USA.

Shoresh, N., Hegreness, M. and Kishony, R. (2008) Evolution exacerbates the paradox of the plankton. *Proceedings of the National Academy of Sciences* 105(34), 12365–12369.

Soltani, N., Khodaei, K., Alnajar, N., Shahsavari, A. and Ardalan, A.A. (2012) Cyanobacterial community patterns as water quality bioindicators. *Iranian Journal of Fisheries Sciences* 11(4), 876–891.

Sommer, U. (1984) The paradox of the plankton: fluctuations of phosphorus availability maintain diversity of phytoplankton in flow-through cultures. *Limnology and Oceanography* 29(3), 633–636.

Spellerberg, I.F. and Fedor, P.J. (2003) A tribute to Claude Shannon (1916–2001) and a plea for more rigorous use of species richness, species diversity and the 'Shannon–Wiener' Index. *Global Ecology Biogeography* 12, 177–179.

Steinmann, A.D. (1996) Effects of grazers on freshwater benthic algae. In: Stevenson, R.J., Bothwell, M.L. and Lowe, R.L. (eds) *Algal Ecology. Freshwater Benthic Ecosystems*. Academic Press, San Diego, California, USA, pp. 341–373.

Stevenson, J. (2014) Ecological assessment with algae: a review and synthesis. *Journal of Phycology* 50, 437–446.

Straile, D., Jochimsen, M.C. and Kümmerlin, R. (2013) The use of long-term monitoring data for studies of planktonic diversity: a cautionary tale from two Swiss lakes. *Freshwater Biology* 58, 1292–1301.

Taylor, F.J.R., Hoppenrath, M. and Saldarriaga, J.F. (2008) Dinoflagellate diversity and distribution. *Biodiversity and Conservation* 17, 407–418.

Tell, G., Izaguirre, I. and Allende, L. (2011) Diversity and geographic distribution of *Chlorococcales* (*Chlorophyceae*) in contrasting lakes along a latitudinal transect in Argentinean Patagonia. *Biodiversity and Conservation* 20, 703–727.

Tuomisto, H. (2010a) A diversity of beta diversities: straightening up a concept gone awry. Part 1. Defining beta diversity as a function of alpha and gamma diversity. *Ecography* 33, 2–22.

Tuomisto, H. (2010b) A consistent terminology for quantifying species diversity? Yes, it does exist. *Oecologia* 4, 853–860.

Tyler, P.F. (1996) Endemism in freshwater algae. *Hydrobiologia* 336, 127–135.

UNESCO (2000) *Solving the Puzzle: The Ecosystem Approach and Biosphere Reserves*. UNESCO, Paris, France. Available at: http://unesdoc.unesco.org/images/0011/001197/119790eb.pdf (accessed 27 May 2016).

Valiela, I. (1995) *Marine Ecological Processes*, 2nd edn. Springer, New York, USA.

Vanormelingen, P., Verleyen, E. and Vyverman, W. (2008) The diversity and distribution of diatoms: from cosmopolitanism to narrow endemism. *Biodiversity and Conservation* 17, 393–405.

Weisse, T. (2006) Biodiversity of freshwater microorganisms – achievements, problems, and perspectives. *Polish Journal of Ecology* 54(4), 652–662.

Whittaker, R.H. (1972) Evolution and measurement of species diversity. *Taxon* 21, 213–251.

Whittaker, R.J., Willis, K.J. and Field, R. (2001) Scale and species richness: towards a general, hierarchical theory of species diversity. *Journal of Biogeography* 28, 453–470.

Whitton, B.A. and Potts, M. (2000) *The Ecology of Cyanobacteria: Their Diversity in Time and Space*. Kluwer Academic, Dordrecht, The Netherlands.

Wilkins, J.S. (2002) Summary of 26 species concepts. Available at: http://researchdata.museum.vic.gov.au/forum/wilkins_species_table.pdf (accessed 7 January 2015).

Wilson, J.B. (1990) Mechanisms of species coexistence: twelve explanations for Hutchinson's 'paradox of the plankton': evidence from New Zealand plant communities. *New Zealand Journal of Ecology* 13, 17–42.

Wolfram, G., Argillier, C., de Bortoli, J., Buzzi, F., Dalmiglio, A. and Dokulil, M.T. (2009) Reference conditions and WFD compliant class boundaries for phytoplankton biomass and chlorophyll-a in Alpine lakes. *Hydrobiologia* 633, 45–58.

Wolfram, G., Buzzi, F., Dokulil, M., Friedl, M., Hoehn, E. and Laplace-Trytue, C. (2014) Alpine Lake Phytoplankton ecological assessment methods. In: Poikane, S. (ed.) *Water Framework Directive Intercalibration Technical Report*, Part 2. Publications Office, Joint Research Centre, EuroPCom, Luxembourg.

Zheng, L. and Stevenson, R.J. (2006) Algal assemblages in multiple habitats of restored and extant wetlands. *Hydrobiologia* 561, 221–238.

17 Quantitative Description of Upper Storey Vegetation at a Foothill Forest in Indian Eastern Himalayas

Gopal Shukla[1]*, Nazir A. Pala[1], Saikat Gantait[2] and Sumit Chakravarty[1]

[1]*Department of Forestry, Uttar Banga Krishi Viswavidyalaya, West Bengal, India;* [2]*AICRP on Groundnut, Directorate of Research, Bidhan Chandra Krishi Viswavidyalaya, Kalyani, Nadia, West Bengal, India*

Abstract

Studies on biological diversity are usually descriptive and concentrated at higher spatial scales and rarely concentrated at local landscapes. The present study is an attempt to describe and quantify the upper-storey vegetation of a forest at Terai Duars in West Bengal state, India. Stratified random nested quadrat sampling was adopted to mark 57 quadrats of size 20 x 20 m to describe the quantitative characters of upper-storey vegetation composition in the forest using number of individuals, occurrence of a species in quadrats, diameter, status of distribution and Importance Value Index (IVI). Upper-storey vegetation comprised 131 tree species, represented by 43 families and 92 genera. On the basis of total number of individuals recorded from all the quadrats, species were categorized as few, medium and high. The distribution status of species was described as frequent, rare and abundant in occurrence. The abundant species can be regarded as the prominent upper-storey species in the forest. This type of species occurrence is expected in typical species-rich tropical forests that reflect heterogeneity or distribution of dominance across the species. The tree species found in the forest were also described as very low diversity, low diversity, diverse and highly diverse, to indicate an overall picture of a species and their importance in the plant community. On the basis of diameter, the species were grouped as small, intermediate and large. This indicates that neither younger trees nor mature trees dominated in the forest. The quantitative characters described in this study could well act as indicators of change and susceptibility to anthropogenic stressors among various vegetation categories and their formation which could be further interpreted as a distinct wildlife habitat.

17.1 Introduction

Tropical forests are one of the most structurally and functionally complex systems on earth. The diversity of forest trees provides resources and habitat for almost all other life forms of the forest (Huston, 1994; Cannon *et al*., 1998). Tree species diversity in tropical areas varies greatly from place to place, mainly due to variation in biogeography, habitat and disturbance (Whittaker, 1972). The diversity of biological communities has never been more appreciated globally, and particularly in Asia, than they are now, as they become increasingly threatened by environmental crises. Because of their remarkably high species richness and endemism, the rapid loss and degradation of forests is likely to have serious impacts on Asian biodiversity (Laurance, 2007). The scientific management of the forest in India, especially in the foothills of Eastern Himalaya, has become difficult due to lack of proper understanding of the structural and functional relationships of the forest ecosystem (Pande, 2001).

*E-mail: gopalshukla12@gmail.com

Previous studies of biological diversity have been descriptive and concentrated on larger areas, i.e. regional, national and global (Raven, 1987; May, 1988; Groombridge, 1992; Daily, 1995; Heywood, 1995). However, the focus of such studies has now changed from higher spatial scales to locally convenient landscapes, at which land-use decisions and management policies are frequently implemented (Ricklefs and Schluter, 1993; Nagendra and Gadgil, 1999; Negi, 1999). This chapter is an attempt to describe and quantify the upper-storey vegetation of a forest at Terai Duars in West Bengal state in India, which lies within the Indo-Malayan Biodiversity Hot Spot (Myers and Mittermeier, 2000).

17.2 Materials and Methods

The study was conducted at Chilapatta Reserve Forest, located in the subhumid tropical foothills of Indian Eastern Himalaya. The study area is located at 26° 32.85' N latitude and 89° 22.99' E longitude with an altitude of 47 m above mean sea level. Using stratified random nested quadrat sampling, 57 quadrats 20 x 20 m in size were laid down on the forest floor for analysing the upper-storey vegetation composition. Quantitative characters of upper-storey vegetation in the forest were described by number of individuals, occurrence of a species in quadrats out of total quadrats, diameter, status of distribution and Importance Value Index (IVI). The species were categorized on the basis of the total number of individuals recorded in all the quadrats as few (fewer than 20 individuals), medium (20–100 individuals) and high (more than 100 individuals). The distribution status of the species was described on the basis of its occurrence or presence in the quadrats (out of a total of 57 quadrats) as rare occurrence/presence (species found in fewer than five quadrats), frequent occurrence or presence (species found in 5–25 quadrats) and abundant occurrence/presence (species found in more than 25 quadrats), following Johnson (F.D. Johnson, Glossary of ecological terms. College of Fish and Wildlife Resources, University of Idaho, Moscow, Idaho, USA, 1983, unpublished). The tree species were classified into three diameter (at breast height) classes, based on their population mean: small (0–40 cm), intermediate (40–80 cm) and large (above 80 cm). IVI of each species was also estimated (Curtis, 1959; Phillips, 1959; Mishra, 1968; Kershaw, 1973; Cintron and Novelli, 1984) and was categorized according to their estimated IVI value as very low diversity (less than 0.5), low diversity (0.5–1.5), diverse (1.5–2.5) and highly diverse (above 2.5) to indicate an overall picture of a species and their importance in the plant community.

17.3 Results and Discussion

Upper-storey vegetation in the Chilapatta Reserve Forest comprised 131 tree species, represented by 43 families and 92 genera (Tables 17.1 and 17.2).

The dominating families in term of genera and species richness were Lauraceae and Fabaceae, each represented by eight genus with 16 and 15 species, respectively (Table 17.2). In the forest, the family Lauraceae was recorded with the maximum number of individuals (818) followed by Verbenaceae (481), Sterculiaceae (447) and the lowest number of only three individuals was recorded for the family Poaceae (Table 17.1). The upper-storey species richness recorded in this study is comparable with other studies for tropical forests in Southeast Asia and other parts of the country (Whitmore, 1984; Khan *et al.*, 1997; Hajra *et al.*, 1999; Jamir and Pandey, 2002; Padalia *et al.*, 2004; Sukumaran *et al.*, 2005; Kumar *et al.*, 2006; Reddy *et al.*, 2006; Dash *et al.*, 2009; Chandra *et al.*, 2010). The species richness recorded in this study, however, was lower than the Amazon rainforest, the most species-rich plant community on earth: in the Amazon rainforest, tree richness was 244 tree genera over an area of 300 km^2 (Laurance *et al.*, 2004). Species diversity in tropical areas is rich (Odum, 1971) and varies greatly from place to place, mainly due to variations in biogeography or local landscape conditions (e.g. soil type, topography), habitat and disturbance (Whitmore, 1998; Fajardo and Alaback, 2005).

On the basis of total number of individuals recorded from all the quadrats, 39 species were categorized as few, 87 species as medium and five species as high. The high number of species in the medium band indicates that dominance was shared by many species, i.e. no single species is dominant over the others. This distribution of dominance among the upper-storey vegetation in the forest was further established by their distribution status, as high-number species were categorized as frequent occurrence (108 species), i.e. present in 5–25 quadrats, followed by rare occurrence (14 species) present in five or fewer quadrats, and least were the abundant species (nine species), present in more than 25 quadrats. The species categorized as abundant can be regarded as the prominent upper-storey

Table 17.1. Quantitative description of upper-storey vegetation.

	1	2	3	4	5
	Anacardiaceae				
1.	*Drymicarpus racemosus* (Roxb) Hk. f.	M	Fr	I	LD
2.	*Saurindia madagascarensis*	M	Fr	I	LD
	Annonaceae				
3.	*Uvaria hamiltonii* Hook. f. & Thomson	M	Fr	S	LD
4.	*Polyalthia simiarum* Benth. & Hk. f.	M	Fr	S	LD
	Apocynaceae				
5.	*Alstonia scholaris* (R. BR.)	M	Fr	S	LD
6.	*Wrightia arborea* (Dennst.) Mabb	M	Fr	I	LD
7.	*Wrightia tomentosa* (Roxb.) Roem. & Schult.	M	Fr	I	LD
	Arecaceae				
8.	*Pinangra gracilis* (Roxb.) Blume	F	Fr	S	VLD
9.	*Calamus erectus* Roxb.	F	R	S	VLD
10.	*Calamus guruba* Buch-Ham ex martius	F	R	S	VLD
	Bignoniaceae				
11.	*Oroxylum indicum* Vent.	M	Fr	I	LD
	Bombacaceae				
12.	*Bombax ceiba* L.	F	Fr	I	LD
	Boraginaceae				
13.	*Cordia obliqua* Willd.	F	Fr	S	VLD
	Burseraceae				
14.	*Canarium sikkimense* King	M	Fr	S	LD
	Capparidaceae				
15.	*Capparis multiflora* Hook f. & Thomson	M	Fr	S	LD
16.	*Capparis sikkimensis* Kurz	M	Fr	S	VLD
17.	*Stixis suaveolens* (Roxb.) Pierre	M	Fr	S	LD
	Combretaceae				
18.	*Termenalia bellirica* (Gaertner) Roxb.	M	FR	I	LD
19.	*Terminalia chebula* Retz.	M	Fr	I	LD
20.	*Terminalia crenulata* Roth.	F	Fr	I	LD
21.	*Terminalia myriocarpa* Heurck & Muell.	M	Fr	I	LD
22.	*Termenalia tomentosa*	M	Fr	I	D
	Dilleniaceae				
23.	*Dillenia indica* Linn.	M	Fr	I	LD
24.	*Dillenia pentagyna* Roxb.	M	Fr	S	LD
25.	*Tetracera sarmentosa* (L.) Vahl	M	Fr	S	LD
	Dipterocarpaceae				
26.	*Shorea robusta* Gaerth f.	H	Fr	L	HD
	Ehretiaceae				
27.	*Ehretia acuminata* R. Br.	M	Fr	S	LD
	Elaeocarpaceae				
28.	*Sloania Sterculiaceae*	M	Fr	S	LD
29.	Elaeocarpus lanceifolius (Roxb.)	M	Fr	S	LD
	Euphorbiaceae				
30.	*Baliospermum montanum* (Willd.) Mueller	F	Fr	S	LD
31.	*Croton caudatus* Geiseler	M	Fr	S	LD
32.	*Macaranga denticulata* (Blume) Mueller	M	Fr	I	D
33.	*Macaranga indica* Wight	M	Fr	I	LD
34.	*Mallotus philippensis* (Lam.) Mueller	M	Fr	I	D
35.	*Phyllanthus reticulatus* Poiret	M	Fr	S	LD
36.	*Sapium baccatum* Roxb.	M	Fr	S	LD
37.	*Trewia nudiflora* L.	M	Fr	L	LD

Continued

Table 17.1. Continued.

	1	2	3	4	5
	Fabaceae				
38.	*Acacia concenna* (Willd.) DC.	F	R	S	VLD
39.	*Albizia lebbeck* Linn.	M	Fr	I	LD
40.	*Albizia procera* (Roxb.) Benth.	F	R	I	VLD
41.	*Bridelia sikkimensis*	M	Fr	S	LD
42.	*Dalbergia latifolia* Roxb.	M	Fr	I	LD
43.	*Albizia lucidior*	M	Fr	I	D
44.	*Bauhinia variegata* L.	M	Fr	I	LD
45.	*Bauhinia scandens* L.	F	Fr	I	LD
46.	*Dalbergia stipulacea* Roxb.	M	Fr	I	LD
47.	*Pterocarpus marsupium* (Roxb.)	M	Fr	I	LD
48.	*Butea parviflora* Roxb.	F	Fr	S	VLD
49.	*Tephrosia candida* D.C.	M	Fr	I	LD
50.	*Acacia lenticularis Ham.*	M	Fr	I	LD
51.	*Albizia chinensis (Osbeck) Merr.*	F	Fr	I	LD
52.	*Bauhinia vahlii* W. and A.	M	Fr	I	LD
	Fagaceae				
53.	*Quercus Castanopsis*	F	Fr	L	VLD
54.	*Castanopsis indica (Roxb.) A. Dc.*	M	Fr	S	LD
55.	*Castanea indica*	M	Fr	I	LD
	Flacourtiaceae				
56.	*Gynocardia odorata* R. Br.	F	Fr	S	LD
	Hypoxidaceae				
57.	*Curculigo orchiodes*	M	Fr	S	LD
58.	*Curculigo* sp.	F	Fr	S	VLD
	Lauraceae				
59.	*Actinodaphne obovata* (Ness) Blume	M	Fr	I	D
60.	*Actinodaphne sikkimensis* Meisner	M	Fr	I	D
61	Beilschmiedia dalzellii (Meisner) Koster.	F	Fr	S	LD
62.	*Cinnadenia paniculata* (Hook f.) Koster.	F	Fr	I	LD
63.	*Cinnamomum bejolghota* (Hamilton) Sweet	M	Fr	I	LD
64.	*Litsea cubeba* (Lour.) Persoon	M	Fr	I	LD
65.	*Litsea glutinosa* (Lour.) Robinson	M	Fr	I	D
66.	*Litsea hookeri* (Meisner) Long	M	Fr	I	D
67.	*Litsea lacta* (Nees) Hook f.	M	Fr	I	LD
68.	*Litsea monopetala* (Roxb.) Persoon	M	Fr	S	D
69.	*Litsea panamanja* (Ness) Hook f.	M	Fr	S	D
70.	*Litsea salicifolia* (Ness) Hook f.	M	Fr	S	LD
71.	*Machilus villosa* Hk. F.	H	A	S	D
72.	*Persea glaucescens*(Nees) Long	M	Fr	S	D
73.	*Persea odoratissima* (Nees) Kastermans	M	Fr	S	LD
74.	*Phoebe lanceolata* (Nees) Nees	M	Fr	S	LD
	Lecythidaceae				
75.	*Careya arborea* Roxb.	M	15	I	D
	Lythraceae				
76.	*Lagerstroemia parviflora* Roxb.	H	A	I	HD
77.	*Lagerstroemia speciosa* (Linn.) Pers.	M	Fr	I	D
78.	*Duabanga sonneratioides* Ham.	F	Fr	I	VLD
	Magnoliaceae				
79.	*Michelia champaca* L.	M	A	L	D
	Menispermaceae				
80.	*Tinospora cordiofolia* Miers (Willd.)	M	Fr	S	LD

Continued

Table 17.1. Continued.

	1	2	3	4	5
	Meliaceae				
81.	*Cedrela toona* Roxb.	M	Fr	I	LD
82.	*Chukrasia tabularis* A. Juss.	M	Fr	I	D
83.	*Amoora wallichii* King.	M	Fr	I	LD
84.	*Switenia mahogany*	M	Fr	I	LD
	Moraceae				
85.	*Artocarpus chama* Hamilton	M	Fr	I	LD
86.	*Artocarpus chaplasha* Roxb.	M	Fr	I	LD
87.	*Artocarpus lakoocha* Wall.	M	Fr	I	LD
88.	*Ficus neriifolia* Smith	F	R	L	VLD
89.	*Ficus religiosa* L.	F	R	L	VLD
90.	*Streblus asper* Lour.	F	R	S	VLD
91.	*Morus laevigata* Wall	F	R	S	VLD
92.	*Ficus elastica* Roxb.	F	R	I	VLD
	Myrsinaceae				
93.	*Ardisia thyrsiflorus*	F	Fr	S	VLD
	Myrtaceae				
94.	*Myristica eratica*	F	Fr	I	VLD
95.	*Myristica longifolia* Wall.	F	Fr	I	VLD
96.	Syzygium operculatum Gamble	M	Fr	I	LD
97.	*Syzygium balsameum*(Wight) Walpers	M	Fr	I	LD
98.	*Syzygium claviflorum* (Roxb.) Cowan & Cowan	M	Fr	I	LD
99.	*Syzygium cumunii* L.	M	Fr	I	LD
100.	*Syzygium formosum* (Wall.) Masamune	M	Fr	I	LD
101.	*Eugenia cumini* L.	F	Fr	S	VLD
	Phyllanthaceae				
102.	*Bischofia javanica* Bl.	F	Fr	I	LD
	Poaceae				
103.	*Neomicrocalamus andropogonifolius* (Griff)	F	R	S	LD
	Rhamnaceae				
104.	*Ziziphus rugosa* Lamarck	M	Fr	S	LD
105.	*Ziziphus mauritiana* Lam.	M	Fr	S	VLD
	Rubiaceae				
106.	*Haldina cordifolia* (Roxb.) Ridsdale	M	Fr	I	LD
107.	*Hyptianthera stricta* (Smith) Wight & Arnott	M	Fr	I	LD
108.	*Ixora javanica* (Blume) DC.	M	Fr	I	LD
109.	*Ixora nigricans*	M	Fr	I	LD
110.	*Morinda angustifolia* Roxburgh	M	Fr	I	LD
111.	*Mussaenda roxburghii* Hook. f.	M	Fr	I	LD
112.	*Neolamarikia kadamba* (Roxb.) Bosser	M	Fr	L	D
113.	*Psychotria calocarpa* Kurz.	M	A	S	LD
	Rutaceae				
114.	*Zanthoxylum rhetsa* (Roxb.) DC.	M	Fr	I	LD
115.	*Aegle marmelos* (L.) Corr.	F	R	I	VLD
	Sabiaceae				
116.	*Meliosma pinnata* (Roxb.) Maximowicz	M	Fr	S	LD
	Sapindaceae				
117.	*Aesculus assamica* Griff.	M	Fr	I	LD
	Sapotaceae				
118.	*Mimusops elengi* L.	M	Fr	I	LD
	Simaroubaceae				
119.	*Ailanthus integrifolia* Lamk.	F	Fr	I	LD
120.	*Ailanthus grandis* Prain	F	Fr	I	LD

Continued

Table 17.1. Continued.

	1	2	3	4	5
	Solanaceae				
121.	*Ardisia solanacea*	M	A	S	LD
	Sterculiaceae				
122.	*Pterygota alata* (Roxb.) R. Brown	H	A	I	HD
123.	*Sterculia vilosa* Roxb.	M	Fr	S	HD
	Tetraceraceae				
124.	*Tetracera sarmentosa*	M	Fr	I	LD
	Theaceae				
125.	*Eurya cerasifolia*	F	Fr	S	VLD
126.	*Schima wallichii* (DC.) Korth.	M	Fr	I	D
127.	*Eurya japonica* Thumb.	F	R	S	VLD
	Ulmaceae				
128.	*Trema orientalis* Blume	F	R	I	LD
	Verbenaceae				
129	*Gmelina arborea* Roxb.	M	Fr	I	HD
130.	*Tectona grandis* L. f.	H	A	L	HD
131.	*Vitex quinata* (Lour.)	F	R	I	VLD

Notes
1= Scientific Name; **2** = total number of individuals in a quadrat [F: few (fewer than 20), M: medium (20–100), H: high (more than 100)]; **3** = number of quadrats out of total quadrats in which a species occurred [R: rare occurrence/presence (less than 5), Fr: frequent (5–25), A: abundant (more than 25)]; **4** = diameter class [S: small (0–40 cm), I: intermediate (40–80 cm), L: large (above 80 cm)]; **5** = IVI [VLD: very less diverse (less than 0.5), LD: less diverse (0.5–1.5), D: diverse (1.5–2.5), HD: highly diverse (more than 2.5)]

species of Chilapatta Reserve Forest. This type of species occurrence is expected in a typical species-rich/diverse tropical/subtropical forest that reflects the higher heterogeneous nature of the forest (Odum, 1971).

IVI helps in understanding the ecological significance of the species irrespective of vegetation types, and the higher the IVI value the more ecologically significant is a species in that particular ecosystem. The high number of upper-storey species (83 species) was estimated with IVI values of 0.5–1.5 and were categorized as low diversity, followed by very low diversity with 25 species (IVI values less than 0.5), diverse (1.5–2.5) with 17 species, and least with six species categorized as highly diverse (above 2.5), indicating an overall picture of species and their importance in the plant community. This clearly denotes the ecological importance of highly diverse upper-storey species in the forest, as the species were more in number and had higher occurrence. This may be because of the high species richness/diversity of the forest, leading to a nearly equal chance of all the species occurring with less deviation from each other due to favourable/optimum climatic and edaphic conditions for all the species, or not a single species highly dominating over the other. This indicates that the upper-storey vegetation in the forest was diverse and was present in sufficient numbers, i.e. neither too high nor too low. Further, 57.25% of species were categorized as small, with a mean diameter of less than 40 cm, followed by the intermediate class with 36.64% of species having a mean diameter of 40–80 cm, and the remaining 6.11% of species were categorized as large, i.e. more than 80 cm in diameter. This indicates optimum growth and stock distribution across the tree species in the forest from seedling/pole stage to maturity, with neither younger trees nor mature trees dominating. This normal distribution of upper-storey species can be attributed to a lack of disturbance, owing to its protection from the law as a Reserve Forest or Protected Forest for wildlife.

17.4 Conclusion

It is useful to rank tropical and subtropical forests through a quantitative description of upper-storey vegetation. The rarer tree species with poor representation in the quadrats need proper attention from plant biologists to determine their conservation status and key functions. Mapping the areas where these species are concentrated, and further study of their key ecological and cultural functions,

Table 17.2. Families, genera and number of individuals of upper-storey vegetation.

Sl No.	Family	No. of species	No. of genera	No. of individuals
1.	Anacardiaceae	2	2	50
2.	Annonaceae	2	2	81
3.	Apocynaceae	3	2	77
4.	Arecaceae	3	2	11
5.	Bignoniaceae	1	1	26
6.	Bombacaceae	1	1	18
7.	Boraginaceae	1	1	9
8.	Burseraceae	1	1	27
9.	Capparidaceae	3	2	74
10.	Combretaceae	5	1	238
11.	Dilleniaceae	2	2	130
12.	Dipterocarpaceae	1	1	176
13.	Elaeocarpaceae	2	2	61
14.	Ehretiaceae	1	1	34
15.	Euphorbiaceae	8	7	370
16.	Fabaceae	15	8	427
17.	Fagaceae	3	3	76
18.	Flacourtiaceae	1	1	16
19.	Hypoxidaceae	2	1	39
20.	Lauraceae	16	8	818
21.	Lecythidaceae	1	1	76
22.	Lythraceae	3	2	412
23.	Magnoliaceae	1	1	78
24.	Menispermaceae	1	1	35
25.	Meliaceae	4	4	199
26.	Moraceae	8	4	121
27.	Myrsinaceae	1	1	16
28.	Myrtaceae	8	3	184
29.	Phyllanthaceae	1	1	39
30.	Poaceae	1	1	3
31.	Rhamnaceae	2	1	37
32.	Rubiaceae	8	7	303
33.	Rutaceae	1	2	59
34.	Sabiaceae	1	1	47
35.	Sapindaceae	1	1	42
36.	Sapotaceae	1	1	49
37.	Simaroubaceae	2	1	36
38.	Solanaceae	1	1	40
39.	Sterculiaceae	2	2	447
40.	Tetraceraceae	1	1	50
41.	Theaceae	3	2	76
42.	Ulmaceae	1	1	6
43.	Verbenaceae	3	3	481

would help to identify locations for conservation action and determine which wildlife species may depend on them. Forest managers can use such information about rare and common species to help manage wildlife habitats. These species also provide cultural resources, food, animal fodder, etc. for the tribal peoples dwelling in and around the forest area. The quantitative characters described in this study could well act as indicators of change and susceptibility to anthropogenic stressors among various vegetation categories, which could be further interpreted as a distinct wildlife habitat (Kumar *et al.*, 2006). The knowledge of distribution patterns of several species would be of prime importance in deciding the management options for specific host populations of native wildlife, which otherwise may face local extinction due to human activity around the Chilapatta Reserve Forest, leading to destruction of the original forest conditions.

References

Cannon, C.H., Peart, D.R. and Leighton, M. (1998) Tree species diversity in commercially logged Bornean rain forest. *Science* 28, 769–788.

Chandra, J., Rawat, V.S., Rawat, Y.S. and Ram, J. (2010) Vegetational diversity along an altitudinal range in Garhwal Himalaya. *International Journal of Biodiversity and Conservation* 2, 14–18.

Cintron, G. and Novelli, Y.S. (1984) Methods for studying mangrove structure. In: Snedaker, C. and Snedaker, G. (eds) *The Mangrove Ecosystem: Research Methods*. UNESCO, Paris, France.

Curtis, J.T. (1959) *Vegetation of Wisconsin*. Wisconsin Press, Madison, Wisconsin, USA.

Daily, G.C. (1995) Restoring value to the world's degraded lands. *Science* 269, 350–354.

Dash, P.K., Mohapatra, P.P. and Rao, Y.G. (2009) Diversity and distribution pattern of tree species in Niyamgiri hill ranges, Orissa, India. *Indian Forester* 135, 927–942.

Fajardo, A. and Alaback, P. (2005) Effects of natural and human disturbances on the dynamics and spatial structure of *Nathofagus glauca* in south-central Chile. *Journal of Biogeography* 32, 1811–1825.

Groombridge, B. (1992) *Global Biodiversity: Status of the Earth's Living Resources*. Chapman and Hall, London, UK. 585 pp.

Hajra, P.K., Rao, P.S.N. and Mudgal, V. (1999) *Flora of Andaman and Nicobar Islands*. Botanical Survey of India, Kolkata, India.

Heywood, V.H. (1995) *Global Biodiversity Assessment*. Cambridge University Press, Cambridge, UK.

Huston, M.A. (1994) *Biological Diversity: The Coexistence of Species in Changing Landscapes*. Cambridge University Press, Cambridge, UK.

Jamir, S.A. and Pandey, H.N. (2002) Status of biodiversity in the sacred groves of Jaintia hills, Meghalaya. *Indian Forester* 128, 738–744.

Kershaw, K.A. (1973) *Quantitative and Dynamic Plant Ecology*. Edward Arnold Ltd, London, UK.

Khan, N.L., Menon, S. and Bawa, K.S. (1997) Effectiveness of the protected area network in biodiversity conservation: a case study of Meghalaya state. *Biodiversity and Conservation* 6, 853–868.

Kumar, A., Marcot, B.G. and Saxena, A. (2006) Tree species diversity and distribution patterns in tropical forests of Garo hills. *Current Science* 91, 1370–1381.

Laurance, W.F. (2007) Forest destruction in tropical Asia. *Current Science* 93, 1544–1550.

Laurance, W.F., Oliveria, A.A., Laurance, S.G., Condit, R., Nascimento, H.E.M. *et al.* (2004) Pervasive alteration of tree communities in undisturbed Amazonian forests. *Nature* 428, 171–175.

May, R.M. (1988) How many species are there on earth? *Science* 241, 1441–1449.

Mishra, R. (1968) *Ecology Work Book*. Oxford and IBH Publishing Co., New Delhi, India.

Myers, N. and Mittermeier, R.A. (2000) Biodiversity hotspots for conservation priorities. *Nature* 403, 853–854.

Nagendra, H. and Gadgil, M. (1999) Biodiversity assessment at multi scale: linking remotely sensed data with field information. *Proceedings of the National Academy of Sciences* 96, 9154–9158.

Negi, H.R. (1999) Co-variation in diversity and conservation value across taxa: a case study from Garhwal Himalaya. PhD thesis, Indian Institute of Science, Bangalore, India, pp. 1441–1449.

Odum, E.P. (1971) *Fundamentals of Ecology*, 3rd edition. W.B. Saunders Co., Philadelphia, Pennsylvania, USA.

Padalia, H., Chauhan, N., Porwal, M.C. and Roy, P.S. (2004) Phytosociological observations on tree species diversity of Andaman Islands, India. *Current Science* 87, 799–806.

Pande, P.K. (2001) Litter nutrient dynamics of *Shorea robusta* Gaertn. plantation at Doon Valley (Uttaranchal), India. *Indian Forester* 127, 980–994.

Phillips, E.A. (1959) *Methods of Vegetation Study*. Henry Holt and Co. Inc., New York, USA.

Raven, P.H. (1987) *We Are Killing Our World. The Global Ecosystem Crisis*. MacArthur Foundation, Chicago, Illinois, USA.

Reddy, C.S.H., Pattanik, C., Dhal, N.K. and Biswal, A.K. (2006) Vegetation and floristic diversity of Bhitarkanika National Park, Orissa, India. *Indian Forester* 132, 664–680.

Ricklefs, R.E. and Schluter, D. (1993) Species diversity regional and historical influences. In: Ricklefs, R.E. and Schluter, D. (eds) *Species Diversity in Ecological Communities: Historical and Geographical Perspectives*. University of Chicago Press, Chicago, Illinois, USA, pp. 350–363.

Sukumaran, S., Balasingh, R.G.S., Kavitha, A. and Raj, A.D.S. (2005) The floristic composition of sacred groves – a functional tool to analyse the miniforest ecosystem. *Indian Forester* 131, 773–785.

Whitmore, T.C. (1984) A new vegetation map of Malaysia at the scale 1:5 million with commentary. *Journal of Biogeography* 11, 961–971.

Whitmore, T.C. (1998) *An Introduction to Tropical Forests*, 2nd edition. Clarendon Press, Oxford, UK and University of Illinois Press, Urbana, Illinois, USA.

Whittaker, R.H. (1972) Evolution and measurement of species diversity. *Taxonomy* 21, 213–251.

18 Significance of Permanent Sample Plots (PSPs) Established in Different Forest Ecosystems in Monitoring Ecological Attributes and Conservation of Biodiversity: A Review

JYOTI K. SHARMA*

Environmental Sciences & Natural Resource Management, School of Natural Sciences, Shiv Nadar University, Village Chithera, Dadri, Gautam Budh Nagar, India

Abstract

Many forest areas in the tropics are undergoing rapid, wide-ranging changes in land cover, due to deforestation on account of anthropogenic pressure and diversion of land for agriculture and infrastructure development. As a consequence the tropical biodiversity is getting depleted at a faster rate. In India, at least 27 mammals have become rare and threatened with possibility of extinction and over 800 plant species are either extinct or threatened with extinction. Conservation efforts for these species are often hampered by the absence of basic information on which to build conservation strategies, and reliable alternatives to get rid of uncontrolled and probably dangerous developmental activities. The dynamics and patterns of biodiversity, which is measured in terms of species richness, distribution of organisms, and their ecological roles in a defined, physical environment, are influenced by the characteristics of both the immediate physical environment and the surrounding landscape. For generating basic data, a clear understanding of the dynamics of all biophysical parameters in the forest ecosystem is needed. To monitor biodiversity over a period of time, use of long-term permanent sample plots is considered the best method. This chapter reviews various sampling techniques and highlights the importance of permanent sample plots in monitoring various ecological attributes and biodiversity in the tropical forest ecosystem.

18.1 Introduction

Forests are major stores of species, habitat, and genetic diversity (Noble and Dirzo, 1997). In recent years anthropogenic activities on forest lands have made a significant impact on local, regional and global diversity, and the health and function of natural ecosystems (Kimmins, 1997). Many forests which are under great anthropogenic pressure require management intervention to maintain the overall biodiversity, productivity and sustainability (Kumar *et al.*, 2002). Understanding the ecosystem dynamics, species diversity and distribution patterns is important to evaluate the complexity and resources of these forests (Kumar *et al.*, 2006) and to plan appropriate biodiversity conservation strategies.

Biodiversity profiles encompass biotic composition, structure, function and interrelationships of resident organisms in the defined sampling unit. The dynamics and patterns of biodiversity are influenced by the characteristics of both the immediate

*E-mail: jyoti.sharma@snu.edu.in and jyotikumarsharma@gmail.com

physical environment and the surrounding landscape. Biodiversity is measured in terms of species richness, distribution of organisms, and their ecological roles in a defined, physical environment. This multi-taxon approach to measuring biodiversity of a defined sampling unit provides a comprehensive description of species assemblages and community structure within an ecosystem. Ecosystem profile data can be used to examine interspecific relationships of resident species with various ecological functions. The relationship between biodiversity and ecosystem functioning cannot be revealed by one-time ecological studies of communities that focus on the structure and behaviour of species and populations at a location. What is needed in addition are studies that address the flux of energy and matter through the ecosystem in time and space. Hence, for generating basic data, a clear understanding of dynamics of all biophysical parameters in the forest ecosystem and their role in providing various ecosystem services long-term replicated studies is needed.

Ecological dynamics of a forest ecosystem is a function of interconnected relationships among biodiversity, ecosystem processes and landscape dynamics. Floristic inventory is a necessary prerequisite for such fundamental research into tropical community ecology, such as modelling patterns of species diversity or understanding species distributions (Phillips *et al.*, 2003). Many floristic diversity studies have been conducted in different parts of the world and the majority of these studies focused on inventory (Whittaker and Niering, 1965; Risser and Rice, 1971; Gentry, 1988; Linder *et al.*, 1997; Chittibabu and Parthasarathy, 2000; Sagar *et al.*, 2003; Padalia *et al.*, 2004; Appolinario *et al.*, 2005). Apart from inventory, disturbance intensity on regeneration (Kennard *et al.*, 2002; Denslow, 1995), phenological assessment (Frankie *et al.*, 1974), comparison of tree species diversity (Pitman *et al.*, 2002), biodiversity monitoring (Sukumar *et al.*, 1992), species area and species individual relationship (Condit *et al.*, 1996) have also been studied through floristic analysis.

Thus, it is evident that floristic and biodiversity studies have been undertaken by researchers worldwide, at different levels and following a variety of sampling and measurement techniques based on their objectives. A comparable inventory data set is required to locate areas for *in situ* conservation and to efficiently allocate available scarce resources. However, it is very difficult to use and to compare all the data that are available through inventory and diversity studies, because the sampling techniques and their advantages and pitfalls, sample size and measurements taken in the fields vary considerably between studies.

Insight into the underlying mechanisms in the functioning of ecosystems is the best guarantee for long-term survival of these ecosystems. This knowledge of the ecosystem allows a flexibility in the necessary management practices for preservation. Similarly, inventorying and monitoring biodiversity in the forest ecosystem yield valuable information on its status, which has practical application in planning conservation strategies. To obtain this knowledge on vegetation dynamics and biodiversity status, long-term permanent plot monitoring is considered to be an appropriate research tool.

This review discusses the commonly followed methods and measurement techniques in biodiversity monitoring and floristic inventory and biodiversity studies, especially in permanent sampling plots and briefly discusses the strategic planning of projects involving PSPs in India.

18.2 Biodiversity Monitoring

Biodiversity monitoring is the repeated observation or measurement of biological diversity to determine its status and trend. Monitoring thus contrasts with surveys, in which biodiversity is measured at a single point of time, e.g. to determine the current distribution of a species. To understand the causes for change in status and trends, biodiversity monitoring must also cover measurements of environmental pressures, including anthropogenic factors.

Biodiversity monitoring is an obligatory component in many international agreements. The Convention on Biological Diversity obliges each contracting party:

> as far as possible and as appropriate [to] identify components of biological diversity important for its conservation and sustainable use ... [to] monitor, through sampling and other techniques, the components of biological diversity identified ... [and to] identify processes and categories of activities which have or are likely to have significant adverse impacts on the conservation and sustainable use of biological diversity, and monitor their effects through sampling and other techniques.

Biodiversity, its status and trends, have been of regulatory, social and management interest for decades; however, despite this interest and a fair amount of research related to its ecological importance and measurement, few efforts have been made to develop and implement standard methods for the assessment and monitoring of biodiversity at local to national spatial scales.

Although a variety of approaches have been investigated for measuring components of biodiversity and estimating the number of species, most theoretical work has been associated with species–area relationships arising from, in part, island biogeography theory. Some studies have taken both a retrospective view, attempting to assess past biodiversity patterns and a prospective view, in trying to understand the potential long-term evolutionary effects on numbers of species due to human land-use change. Recent empirical studies have attempted to incorporate gradients in habitat quality to the construction and interpretation of species–area curves that the earlier theoretical work did not address. Other approaches have attempted to use species–habitat relationships without invoking any particular theoretical foundation for how many species a particular area will sustain. Tools are needed that, while addressing remaining theoretical questions associated with species biodiversity status measurement and forecasting, are practical for resource managers to use and are scalable across multiple spatial scales. Moreover, such tools should be capable not only of accurately assessing and monitoring native species biodiversity at different spatial scales, they also should enable managers and decision makers to determine the effects on such biodiversity of different land use and management actions. Although a variety of approaches have been investigated for measuring components of biodiversity, different approaches associated with estimating the number of species belonging to different groups of flora and fauna are discussed below.

18.2.1 Biodiversity monitoring approach

The initial phase in biodiversity surveys is estimating diversity at one point in time and location (in other words, knowing what species or communities are present). The second phase, monitoring biodiversity, is estimating diversity at the same location at more than one time period for drawing inferences about change (Gaines *et al.*, 1999). Wilson (1996) identified various attributes of biodiversity that can be assessed at each level of ecological organization. At the landscape level, attributes that could be monitored include the identity, distribution and proportions of each type of habitat, and the distribution of species within those habitats. At the ecosystem level, richness, evenness and diversity of species, guilds and communities are important. At the species level, abundance, density and biomass of each population may be of interest. And, at the genetic level, genetic diversity of individual organisms within a population is important. It is best to assess and interpret biodiversity across all these levels of organization by using various approaches at several spatial and temporal scales (Noss and Cooperrider, 1994).

Gaines *et al.* (1999) proposed a three-phase approach to monitoring biodiversity: (i) identify monitoring questions derived from regional, provincial, or watershed assessments; (ii) identify monitoring methods; and (iii) analyse and interpret information for integration into management strategies. This includes identifying and refining biodiversity-monitoring questions, determining data needs to address the questions, and prioritizing monitoring questions and data needs. Prioritizing monitoring questions is important because the resources available to accomplish monitoring are likely to be limited. Identifying monitoring questions is a critical and difficult step. It could be accomplished through an interdisciplinary process with experts knowledgeable of the issues at the appropriate level (e.g. landscape, ecosystem, species, genetic, etc.) and should be considered an iterative process that is adapted as new information becomes available. Monitoring questions could be derived from published information available.

The sampling methods selected for monitoring biodiversity depend on the required outcome. A management objective of maintaining species viability would involve different monitoring methods from an objective of restoring inherent disturbance regimes. Selecting the appropriate biodiversity monitoring approach includes identifying methods that will provide answers to specific monitoring questions. A wide range of methods are available, and a selection of methods should be made based on costs, available resources and statistical constraints. It might be helpful, if not absolutely necessary, to consult a statistician at this stage to determine sampling sizes, strategies and statistical design. Periodically, data collected from monitoring would be analysed and integrated into management strategies based on the knowledge gained. If monitoring data reveal that adjustments need to be made, then this monitoring approach needs to be revisited for any correction/modification required.

18.2.2 Monitoring questions

Examples of monitoring questions at this level could include: What is the current level of landscape diversity and its richness and how does it compare with previous sustainable levels? What are the trends in habitats or populations of a particular species and its distribution pattern? What are the trends in landscape features, such as the amount of edge, patch size, forest interior, etc.?

18.2.3 Monitoring methods

There are several approaches to assessing biological diversity at a landscape scale. The need for biodiversity assessment and monitoring is explicitly recognized by policy processes such as Convention on Biological Diversity (CBD) and Agenda 21. However, as biodiversity is a highly complex concept, uncertainty exists among practitioners about how biodiversity can be assessed in practice, including issues such as the selection of variables for measurement, definition of appropriate measurement techniques, approaches for sampling and data analysis, and the selection and use of indicators.

In this review, various methods which can be applied for biodiversity assessment and monitoring are discussed. The process of undertaking a biodiversity assessment and monitoring can be conceived as a series of stages.

- Identification of information needs and monitoring questions.
- Identification of different biodiversity values.
- Assessment of existing information, and identification of information gaps.
- Definition of what to measure, and how to measure it.
- Development and implementation of a sampling programme on a long-term basis.
- Analysis and presentation of results.

18.2.4 Sampling techniques

The sampling protocols and procedures for the ecosystem profile include two approaches to inventories.

1) Ecosystem profile inventory – The purpose of this inventory is to describe the composition, structure, function and interrelationships of biotic and abiotic components of a plot unit that can be repeated. Precise protocols have been developed to sample specific taxa from soil level to tree canopies, both plants and animals or any one of the biodiversity components.
2) Stand-based inventory – The purpose of this inventory is to illustrate the pattern of biodiversity within the larger, stand-based unit and to compliment data from the ecosystem profile inventory. The sampling protocols can be used to answer general or taxon-specific research questions by expanding the sampling effort. In that general context, two complementary approaches are often proposed in order to estimate the ongoing changes:

a. using remote sensing, mapping and geographic information systems as means to detect, represent and assess the global changes in forest area and the flows from one forest type to another

b. installing permanent plots for assessing the local changes in forest biomass and structure, in tree species richness and diversity, and for describing and understanding the processes involved in forest dynamics and maintenance or loss of biodiversity.

2. There is an excellent review by Jayakumar *et al.* (2011), who have reviewed in detail various sampling techniques used in floristic surveys. The sampling methods used for floristic and biodiversity assessment could be broadly classified into:

a. plot sampling method
b. line transect method
c. *k*-tree or fixed tree count sampling method
d. *ad-hoc* method.

18.2.4.1 Plot sampling method

Due to the rapid degradation of forest area in qualitative terms, replacing 'natural' closed forests by disturbed open forests, the focus of most forest inventories has shifted from the one-time assessment of forest diversity to the continuous monitoring of their changes over a period of time. Hurst and Allen (2007a) listed two approaches to monitor long-term changes in forests: 1) spatially extensive small plots that sample landscapes (classically used in forest inventory and forest health studies in Scandinavia, Central Europe, North America and New Zealand); and 2) large plots that sample small-scale diversity (exemplified by 50-ha plots in tropical rainforests). The majority of the biodiversity inventory and floristic diversity studies have been done following the plot sampling method which has three distinct types: (a) permanent plot technique; (b) random plot technique; and (c) stratified random plot technique.

18.2.4.2 Random plot (RP) technique

Gordon and Newton (2006b) recommended that randomized selection of site for sampling would ideally assess the diversity in any locality. In Random Plot Technique, plots are laid in the field randomly to represent the entire floristic region in order to avoid bias sampling (Magurren, 1988; Zhang and Wei, 2009; Zhang, 2010, 2011). As compared to big permanent plot studies, literatures on random plot studies are very limited.

Rarely, a circular plot of 20 m radius has also been used (Linder *et al.*, 1997). However, the reason for the circular plot and its significance was not given. As far as the shape of the plot is concerned, Condit *et al.* (1996) confirm that very narrow rectangular plots, 1000 x 1 m, were more diverse (18% to 27%) than square plots.

18.2.4.3 Stratified random plot (SRP) technique

Stratified random sample is a technique by which the floristic area is first divided into homogenous vegetation groups based on type or density (forest cover/type classification) using satellite data, and then samples are distributed to each vegetation group proportionately, based on their aerial extent. The SRP method has been used for floristic studies by only a very few studies, as is evident from the literature. In the early 1970s, Paijmans (1970) estimated the floristic diversity of tropical rainforest in New Guinea following the SRP technique, where aerial photos were used to classify the forests. In India, Behera *et al.* (2002, 2005), in the eastern Himalayas, and Balaguru *et al.* (2006), in the Eastern Ghats, have estimated the floristic diversity by the SRP method, using 20 x 20 m plots. Padalia *et al.* (2004) conducted a floristic study in the Andaman Islands of India adopting the SRP technique, with 462 samples of various plot sizes covering an area of 12.52 ha. Killeen *et al.* (1998) assessed the tropical semi-deciduous forest in the Chiquitania region of Santa Cruz, Bolivia following the SRP method.

18.2.4.4 Line transect method

In this method, a line transect is formed in the field to a specific distance and then those species which are touching the line are measured. This method, for floristic inventory, is not popular among ecologists (Zhang and Wei, 2009). Frankie *et al.* (1974) adopted this technique in the Lowlands of Costa Rica. For phenological variation between wet and dry forest 17 line transects of 200 m each using aerosol paint were used. This method is more popular in animal diversity studies. For example, Shahabuddin and Kumar (2006) studied the bird communities in Tropical Dry Forest of Sariska Tiger Reserve in Rajasthan, India in connection with vegetation structure and anthropogenic disturbance using a 500-m line transect. In Tanzania, Rovero and Marshall (2005) studied the forest antelopes using two line transects of about 3100 m long. Although the line transect is easy to follow, the inability of the resulting datasets to estimate various diversity indices based on area make it unpopular.

18.2.4.5 Temporary plots (transects)

This method is useful for rapid estimation of diversity and to identify trends in species richness and floristic diversity, and also to identify the crucial areas when sampled along a continuum (Elouard and Krishnan, 1999). Transects can be used in different ways depending on the objectives, resources and time available. A rapid assessment of the vegetation type and the major species within an area can be done by transects: a rope is laid over a certain distance (usually 1 km) and all forms of biodiversity nearest the rope are identified and trees measured (girth and height). Using this method, estimates can be completed within a short period (i.e. in a few hours).

18.2.4.6 Stand-based inventories: vegetation plots

In this method, plots are established along the transect (Mahan *et al.*, 1998). Russell (E.W.B. Russell, Terrestrial vegetation monitoring with permanent plots: a procedural manual. National Park Service Chesapeake/Allegheny Clusters, Philadelphia, Pennsylvania, USA, 1996, unpublished report) established a minimum of 20 plots of 0.04 ha (400 m^2) along the four transversing transects in each forest stand (five plots/transect). These 20 plots were established at equally spaced intervals along the transects to allow for adequate coverage of the stand (Orwig and Abrams, 1993). Plots should be located at least 25 m apart along the transects, and vegetation plots at least 50 m from the edge of a particular forest stand to minimize any edge effects (Myers and Irish, 1981; Orwig and Abrams, 1993).

18.2.4.7 k-tree sample/fixed tree count

k-tree or sample/fixed tree count plot is synonymous with point samples or variable area plot, requiring no plot demarcation but only a fixed number of trees to be measured per point sample (Gordon and Newton, 2006a; Sheil *et al.*, 2003). The advantage of the *k*-tree sampling method is that it is plot-less (Engeman *et al.*, 1994) and the same number of trees is measured in all samplings. Kleinn and Vilcko (2006) conducted floristic diversity studies with the *k*-tree sampling method in the Miombo woodlands of Northern Zambia. Gordon and Newton (2006a) in Southern Mexico have also used *k*-tree sampling at eight sites. Hall (1991) attempted to effectively survey montane forest in Africa adopting fixed tree counts, but provided no direct comparison of its efficiency with other methods. Condit *et al.* (1996) have supported fixed tree count methods on statistical grounds. They argued that by comparing equal numbers of stems, the resulting diversity indices would be unbiased. As the number of studies following this method is very limited, comparison of results with other studies is also very limited.

18.2.4.8 Ad-hoc method

This method is deliberately informal, the sampling being started from the perimeter and circling inwards until the team decides, subjectively, that no more new species are likely to be found (Jayakumar *et al.*, 2011). Gordon and Newton (2006a) adopted this method in their study and concluded that the *ad-hoc* protocol is the most efficient in accumulating new species during sampling. They also claimed that this method is highly efficient and simple, where the resources are limited and statistical analysis is not considered. However, Nelson *et al.* (1990) argued that this method is of limited value to ecologists, as it is subject to various forms of sampling bias.

In conclusion, careful matching of inventory purpose to method has always been important for ecologists, and is especially so now in the tropical context of rapid environmental changes. The need for efficient sampling is a dominant factor determining methodological decisions, but comparative analysis of efficiency has been lacking in the tropical eco-floristic literature.

18.3 Permanent Sample Plots (PSPs)

Permanent sample plots are permanently defined areas of forest that are periodically but regularly remeasured to provide data on stock, stand density, regeneration, tree dimensions and volume, plus ecological aspects, including biodiversity. Information on changes in the composition, structure and growth of a forest over time can be derived from PSPs. With the emphasis shifting from the one-time assessment of forest dynamics and biodiversity to the continuous monitoring of their changes, the question of installing permanent plots has become increasingly acute across the world (Elouard and Krishnan, 1999). The main objective for laying permanent plots is to answer questions that only long-term monitoring can provide: about ecosystem dynamics and temporal changes in the floristic and biodiversity composition. Permanent sample plots established in forests have been used extensively to track ecological processes, changes in forest habitat, biodiversity measurement and climate change effects in the US, Canada, European countries and Australia.

It is clear that for long-term replicated studies we need to establish PSPs in different forest ecosystems. They may be laid down either as a network of sampling plots, passively sampling existing forest management practices; or as measurement plots within an experimental design.

Permanent plot observations are not only important because they enable the description of the effect of external causes, but also because of the possibility of generating hypotheses on internal causes and mechanisms of species replacement during vegetation succession. Hubbell and Foster (1992) advocated for establishment of PSPs in natural forests with the aim of documenting and monitoring plant diversity, and obtaining long-term data on ecosystem structures, dynamics and properties.

Some of the long-term studies of tree population dynamics in permanent plots revealed that in relatively undisturbed forests, the community maintains stability in spite of fluctuations in the mortality and recruitment rates of some populations (Phillips, 1996; Kohyama *et al.*, 2003; Sheil *et al.*, 2000). Long-term studies in permanent plots established in disturbed forests also helped to evaluate the changes in tree species diversity, composition and dynamics in relation to the degree and frequency of disturbance (Coates, 2002; Fashing *et al.*, 2004). Observations of vegetation dynamics in permanent plots not only provide insight into several ecological patterns that are not predicted, but also form a scientific base for evaluating existing management approaches and their impacts on forest ecosystem (Sheil *et al.*, 2000).

Even when well-designed replicated field experiments have been established, without long-term observations the results obtained may be insufficient or even misleading. Tilman (1989) and Inouye and Tilman (1995) provided examples where the more frequently used (3–5 year) period of observations and manipulations was not long enough to capture the main trends in vegetation change. Longer periods of observation are needed when the community under investigation is rich in species. Tilman (1989) also found that among the papers published in the journal *Ecology*, only 7% used material from experiments which lasted for five or more field seasons. This percentage was higher for observational studies. Historically, these studies have been difficult to maintain because of short-term funding, the lack of continuity of human resource, and short-term experimentation to suit the researchers (Gocz, 1998).

It needs a great deal of discipline to maintain a series of permanent plots and analyse them yearly or at periodic intervals over a period long enough to answer relevant questions on dynamics of the ecosystem and biodiversity. This raises the question: How long do 'long-term observations' have to last? The series of permanent plots in the Park Grass Experiment at Rothamsted started in 1856 (Johnston, 1991; Silvertown *et al.*, 1994, 2006) and most probably holds the world record.

The permanent sample plots from which data are repeatedly collected over long periods of time have been widely recognized as invaluable for monitoring and documenting natural processes and long-term changes in habitat composition and condition (Alder and Synnott, 1992; Scott, 1998; Dyrness and Acker, 1999, 2000; Ahlstrand *et al.*, 2001; Smits *et al.*, 2002; J. Henderson and R. Lesher, Survey protocols for benchmark plots (permanent intensive ecoplots) for western Washington, version 1.22. US Department of Agriculture, Forest Service, Pacific Northwest Research Station, Portland, Oregon, USA, 2002, unpublished report). Permanent sample plots result in precise estimates of change (Scott, 1998), giving much greater statistical power to detect change than would a series of temporary sampling units in the same habitats. Long-term data on ecosystems obtained from PSPs are now recognized as crucial for our understanding of environmental dynamics in the forest ecosystem and monitoring biodiversity.

In the US, the Long Term Ecological Research (LTER) Network was launched in 1980, a network programme at 18 sites involving about 1400 scientists and students with the following objectives.

- Promote and enhance understanding of long-term ecological phenomena across national and regional boundaries.
- Facilitate interaction among participating scientists and disciplines.
- Promote comparability of observations and experiments, the integration of research and monitoring.
- Promote comparative analysis across sites.
- Enhance training and education in comparative long-term ecological research.

With LTER still going on, in more than 30 years it has identified a number of lessons that may be of value to other researchers in undertaking such long-term ecological studies. Here are a few of them.

- Short-term studies can be misleading.
- Studies at only one scale can be misleading.
- Single-species studies can be misleading.
- Multiple-factor control dynamics: complexity is the rule.
- Make measurements at comparable, temporal and spatial aggregation scales.
- Spatial variation is greater than interannual (temporal) variations.
- Biotic data exhibit more variability than abiotic data.
- The greater the variety of systems considered, the more general the metric of comparison.
- Synthesis requires interdisciplinary approaches.
- Long-term studies should be coordinated by an agency which has long-term research goals, committed funding and can attract a team of scientists from various institutions to work in a network mode.

In the Netherlands since the 1930s, more than 6000 PSPs have been established, many in forested habitats (Smits *et al.*, 2002). They provide insight into vegetation succession, fluctuations within plant communities over time, and the effects of changes in the environment on the vegetation. PSPs have also been widely used in the US forests of the Pacific Northwest, which established 3097 PSPs in western Washington during the 1980s and is continuing to monitor them (Henderson *et al.*, 1989). Details of other PSPs in the US are given below.

- Thirty-eight PSPs were established in or near the H.J. Andrews Experimental Forest in the late 1970s, and have been revisited at 5–6-year intervals (Dyrness and Acker, 1999).
- Eighteen PSPs were established in Mount Rainier National Park in 1976–1997, which have been utilized in a number of research studies, including old-growth forest characterization, how forest characteristics change through time, and soil carbon and nitrogen in old growth (Dyrness and Acker, 2000). The National Park Service will soon be establishing PSPs in both the North Cascades and Olympic National Parks under their Long-term Ecological Monitoring Program (Freet, 2001). Data from these PSPs will help document the distribution of vegetative assemblages, forest structure and fuel loading, and monitor how these characteristics change through time.
- Twenty-one PSPs were established in second-growth forest in the Cedar River Watershed (CRW) in 1946–1979. They have been re-measured from one to ten times, most recently in 1986, and include some sites that have been thinned. Whenever possible, the data is being gathered from these previously established plots in conjunction with data from the proposed grid-based PSPs. Unfortunately, the previous PSPs are limited in number, located in spatially restricted areas, inconsistent in size, and used various sampling protocols, making many types of comparison difficult.

During the past two decades many studies (Harms *et al.*, 2001; Chittibabu and Parthasarathy, 2000; Phillips *et al.*, 2003; Fashing *et al.*, 2004; Proctor *et al.*, 1983) have followed the permanent-plot sampling technique for floristic diversity analysis where the size of the permanent plot varied from 1 ha to 50 ha.

Studies such as Proctor *et al.* (1983) in Gunung Mulu National Park, Sarawak; Parthasarathy and Karthikeyan (1997) in the Coromandel Coast, India; Kadavul and Parthasarathy (1999) in the Eastern Ghats, India; Aldrich *et al.* (2002) in east-central Indiana; Grau *et al.* (1997) in Tucuman, Argentina; Mani and Parthasarathy (2006) in Shevaroys, India; and Bhat *et al.* (2000) in Uttara Kannada, India, have used plots smaller than 10 ha to estimate floristic diversity. Phillips *et al.* (2003) in Amazonian Peru, and Pitman *et al.* (2002) in Ecuador and Peru, have estimated using plots larger than 10 ha but smaller than 50 ha. Harms *et al.* (2001) in Barro Colorado, Panama Island, Nath *et al.* (2006) in Mudumalai Wildlife Sanctuary, India, and He *et al.* (1996) in Negeri Sembilan, Malaysia, used 50-ha plots in their studies.

For convenience, large plots were divided into subplots of 10 x 10 m (Chittibabu and Parthasarathy, 2000; Mani and Parthasarathy, 2005; Venkateswaran and Parthasarathy, 2005), 20 x 20 m (Harms *et al.*, 2001; He *et al.*, 1996; Grau *et al.*, 1997), 5 x 5 m (Franklin and Rey, 2007), or 2 x 50 m and 6 x 50 m (Gordon and Newton, 2006a). Occasionally the plot shape was also changed to circular (20 m radius) in Linder *et al.* (1997) or rectangular (10 x 500 m) in Shankar (2001).

Additionally, PSPs in tropical forests have been utilized to predict the impacts of environmental change on vegetation-related processes. Tropical forests contain about 40% of the carbon stored as terrestrial biomass (Dixon *et al.*, 1994) and represent a substantial fraction of the world's forest net primary production (NPP) (Melillo *et al.*, 1993; Field *et al.*, 1998). Field estimates of NPP for tropical ecosystems are important to assess carbon-recycling rates in the face of global change, and to validate global-scale ecosystem models. Field studies for estimating forest NPP involves measurements of separated below- and above-ground components, which include the increments and losses of forest biomass throughout the time period (Clark *et al.*, 2001b). Above-ground biomass increment or stand mass increment (dry mass increment, or DMI) (Chambers *et al.*, 2001; Clark *et al.*, 2001a, b), is an important component of NPP, and is defined as the change in mass of surviving trees in an inventory plot over time. In a recent detailed review of available data for estimating total NPP from 39 old-growth tropical forest sites, Clark *et al.* (2001a) found that only 18 of these sites had reliable information to accurately quantify DMI, which varied from 0.3 to 3.8 Mg C ha^{-1} per year. Some results from PSPs indicate that mature neotropical forests are a net sink for atmospheric carbon, due to increasing forest biomass (Phillips *et al.*, 1998, 2002; Baker *et al.*, 2004; Lewis *et al.*, 2004). Measurements over shorter time scales using the eddy covariance method in tropical forests also suggested a net carbon sink (Malhi *et al.*, 1998).

In India, various researchers have established large research plots as PSPs in different forest ecosystems to study the stand/structural dynamics, species richness, vegetation structure and floristic composition, biodiversity inventory, etc. (Table 18.1). In most cases, only one-time data have been collected from these plots and results published. This defeats the very purpose of establishing the PSPs, and means that all the efforts were in vain. The main reason for these one-time studies in PSPs is lack of strategic planning and scientific policies, long-term goals pertaining to scientific research in the country and hence the paucity of adequate funding support on a long-term basis.

18.3.1 Size of permanent sample plots

Sample size plays a major role in determining the total diversity of a region. Whether we should go for a small number of large-sized plots or for a large number of small-sized plots. The most efficient plot area is a compromise between the total forest area sampled and the physical effort needed to establish, maintain and regularly re-measure plots. The aim is to minimize the plot edge to area ratio. The most efficient plot size in any particular forest will depend on inventory objectives, the level of precision required, forest variability and the costs of PSP establishment, maintenance and re-measurement.

However, an important point that is still unresolved to date is the size of the permanent plots (Elouard and Krishnan, 1999). More precisely, the debate is centred on the question: Do several small plots reflect diversity better than a single large plot? Several studies have shown that estimates of species diversity and richness are higher when many small plots are sampled, as compared to a few but larger plots (Parsons and Cameron, 1974; Routledge, 1979; Whitmore, 1984; Whitmore *et al.*, 1985). However, large plots also have their advantages: a more diverse representation of life forms; the gradient of species commonality and rarity become apparent; site monitoring of dynamic changes in the ecosystem can be undertaken continuously (Bakker *et al.*, 1996a, b; Herben, 1996).

In most of the Indian studies the size of the plot has varied considerably from 0.5 ha to 50 ha (Table 18.1). Elsewhere, in other studies the size of the plot/quadrat varied from 1 x 1 m to 20 x 50 m (Zhang and Barrion, 2006; Zhang *et al.*, 2008). Gordon and Newton (2006b) have conducted random plot analysis in Huatulco, Mexico with 2 x 50 and 6 x 50m size. Knight (1975) investigated the floristic diversity in Barro Colorado Island, Panama with 10 x 20 m plots. In Santa Catalina Mountain, Arizona, Whittaker and Niering (1965), Swamy *et al.* (2000) in Agasthyamalai hills of South India, Sagar and Singh (2006) in Vindhyan Dry Tropical Forest of India have estimated the diversity of forest using 10 x 10 m plots. Ramanujam and Kadamban (2001) in South eastern coastal of India have used 25 plots of 20 x 20 m. Huang *et al.* (2003) in the East Usambara Mountains Arc Africa and Kalacska *et al.* (2004) in Parque National Santa Rosa in the Province of Guanacaste conducted floristic studies with 20 x 50 m random plots. Bazzaz (1975) in Southern Illinois used 40 plots of 2 x 1 m size and 25 plots of 4 x 4 m size in his diversity studies.

The permanent 20 × 20 m plot method has been fully described in detail by Hurst and Allen (2007a, b) in two versions, expanded and field. The expanded version includes more information about survey design and sampling than the more compact field version.

However, the sample size varies considerably among studies involving PSPs, as evident from an exhaustive table given by Jayakumar *et al.* (2011). For example, in PP, RP and SRP the ranges of sample size based on the literature survey are 0.54–54 ha, 0.2–50 ha and 3.2–12.52 ha, respectively. These variations support the assumption that the sample size is determined mainly by time, money and human resources (Phillips *et al.*, 2003). These drawbacks hinder the biodiversity studies proportionately. All the diversity studies aim to bring out the species diversity of any region, but it is doubtful that they estimate full diversity, because the sample size may be insufficient in many studies to derive the floristic pattern of a region. For example, the study at Jaú National Park in Central Amazonia, Brazil, with 4 ha of sample size (Ferreira and Prance, 1998) concluded that it is difficult to elucidate major floristic patterns at regional and local scales with this sample size. The authors also suggested that the sampled area should be 1 ha or more and the sampling should be from independent replicated samples. Biodiversity studies may underestimate the species richness in tropical forests

Table 18.1. Permanent sample plots established in various forest types in different regions of India (compiled by Dr U.M. Chandrashekara, Kerala Forest Research Institute, Kerala, India).

	Study area	Targeted forest type/s	Year of establishment	Plot size and number of plots	Period of study	Reference
A	Western Ghats, India					
	Uttara Kannada, Karnataka	Evergreen and moist deciduous forests	1984	Eight plots: 1 ha each	1984–1994	Bhat *et al.*, 2000
	Mudumalai Wildlife Sanctuary, Tamil Nadu	Dry deciduous forest	1988	One plot: 50 ha	1988–1993	Sukumar *et al.*, 1998
	Uppangala, Karnataka	Evergreen forest	1990	One plot: 28 ha		Pascal and Pelissier, 1996
	Kerala	Shola, evergreen forest, moist deciduous and dry deciduous forest	1996	Four plots: 1 ha each in a forest type	Once only: 1996–1998	Chandrashekara *et al.*, 1998
	Anamalai Hills, Tamil Nadu	Evergreen forest	1998	One plot: 30 ha	Once only: 1998–1999	Ayyappan and Parthasarathy, 1999
	Kerala	Shola, evergreen forest, moist deciduous and dry deciduous forest	2001	60 plots: 0.5 ha each in Shola (one plot), evergreen (14 plots),semi-evergreen (21 plots), moist deciduous (20 plots) and dry deciduous forests (one plot)	Once only: 2001–2002	Chandrashekara and Jayaraman, 2002
	Bhadra Sanctuary, Karnataka	Dry deciduous forest	2004	One plot: 2 ha each	Once only	Krishnamurthy *et al.*, 2010
B	Eastern Ghats, India					
	Kolli Hills, Tamil Nadu	Evergreen forest	1996	Four plots: 2 ha each	Once only: 1996–1999	Chittibabu and Parthasarathy, 2000
C	Central India					
	Vindhyan Hill Ranges	Dry deciduous forests	1998	15 plots: 1 ha each	Once only: 1998–2000	Sagar *et al.*, 2003
D	Western Himalaya, India					
		Birch–fir forest, Mixed coniferous forest, Birch–rhododendron forest	2009	Four plots: 1 ha each	Once only: remote sensing studies	Chawla *et al.*, 2012

when the stem counts are <1000 (Condit *et al.*, 1996). However, it is important for a study to make ample biodiversity measurement, to determine the sample size based on species area curve/asymptotic curve. These methods are widely accepted among biodiversity measurements, but are available in few studies to check whether the sample size of respective studies is sufficient to represent the true biodiversity profile of the region. Therefore, every biodiversity study should assign more importance to the sample size before concluding the study. Otherwise the results and conclusions made in a study may not be a true representation and, moreover, may give a false impression of biodiversity status.

Permanently marked 20 × 20 m (400 m^2) plots have emerged as the standard plot size and are currently the most widely applied of all vegetation plot methodologies used. The purpose of this plot is to reveal changes in structure and composition of vegetation communities. Stems of all individuals are identified and counted within the plot. Growth, mortality and recruitment of individuals are derived from repeated measurements of the forest overstorey (tree stems) and understorey (saplings, seedlings). Plots are divided into 16 subplots of 5 × 5 m to increase measurement efficiency and minimize counting errors. Trees are defined as stems that have reached a threshold 2.5 cm diameter at 1.35 m high (diameter breast height, DBH). Growth and basal area of each tree species is calculated from stem diameter measurements. Understorey stem density is determined from complete counts of saplings within a plot. All woody seedlings are counted within 24 (0.75 m^2) permanently marked understorey subplots and categorized into height tiers. Data obtained from seedling subplots is used to calculate seedling densities within the height tiers.

Floristic diversity analysis of a 1-ha plot is very popular and is suited to a variety of additional purposes, such as monitoring forest dynamics, as well as phenological and ethnobotanical research (Condit *et al.*, 1996). Elouard and Krishnan (1999) also recommend that the minimum size for a permanent plot should be 1 hectare. Species diversity and dynamic processes are poorly estimated in plots smaller than this. Phillips *et al.* (2003) have compared the suitability of 0.1-ha and 1-ha plots for floristic studies. They have concluded that the 0.1-ha inventory method achieves more in floristic knowledge and understanding and in detecting significant habitat–species associations than the 1-ha inventory method. Moreover, the 1-ha method also demands more time and human resources than the 0.1-ha plot. The permanent plots require regular monitoring and assessment. However, in many studies the plots are abandoned after yielding the inventory data. For example, in western Amazonia the failure rate is estimated as >50% (Phillips *et al.*, 2003). In India, there is a similar trend where the so-called permanent plots are abandoned after the inventory. The reasons could be inadequate funds to re-census, the impossibility of relocating the plot's position, disturbance from local residents in terms of commercial logging and removal of tags, natural disturbance such as rapid radial tree growth 'swallowing' tags, and inaccessibility due to over-growth of liana. Moreover, conversion to permanent plot status is expensive, time-consuming and uncertain. Temporary and inadequate funding is the main reason that most 1-ha plots remain simply temporary floristic samples. Therefore, Phillips *et al.* (2003) suggest that it is advisable to install 1-ha plots only with long-term funding programmes. Condit *et al.* (1996) verified that small plots tend to have fewer species than larger plots with the same number of individuals, but the differences were slight and sometimes nonexistent. In many cases, larger plots did not have more species. They also suggested that rectangular plots recorded 10% more species than square plots.

One hectare square-shaped plots are widely used in many tropical forests, with the following advantages.

- They are easily subdivided into 25 subplots of 0.04 ha, or 100 subplots of 0.01 ha, for subsampling within a plot.
- They represent values per hectare for stock and other measured parameters, thereby avoiding conversion problems.
- They provide good representation of species diversity.

Square plots have the following advantages.

- They have shorter boundaries than an equal area of strips, thus reducing the effort and cost of plot establishment and maintenance. The shorter boundaries will reduce errors which might be caused by trees located on plot boundaries.
- Square plots are easier to locate than strips without introducing bias and are less likely to

be interfered with by tracking or road-building. Edges and boundaries of square plots are easier to relocate than circular plots.

- Square plots are more practical to establish and maintain in many tropical forests than are circular plots. It is impractical to set out large circular plots accurately in dense vegetation, because trees near the plot boundary cannot be seen from the centre. Circular plots cannot easily be subdivided into subplots.

Studies need good sampling designs in order to evaluate the drivers of change, and designs vary in accordance with the study objectives.

18.3.1.1 Inventory efficiency – plot size

Phillips *et al.* (2002) worked out the inventory efficiency and reported that 1-ha samples on average record more species than 0.1-ha samples, but require much more effort, in both Loreto and Madre de Dios. As a result, 0.1-ha inventories were substantially more efficient in terms of floristic data gained per effort invested. The crude inventory efficiency of 0.1-ha samples is three to four times that of 1-ha samples in Loreto and in Madre de Dios. The inventory efficiency results showed that the 0.1-ha protocol is still about twice as efficient as the 1-ha protocol in shrub-rich forests and about three times as efficient as in shrub-poor forests. When only tree species are considered in the sample and the target flora, then the 0.1-ha protocol is more than three times as efficient (tree inventory efficiency) as the 1-ha protocol, regardless of the assumption made about the richness of shrub species in the flora. Although these results suggest that the inventory method itself was an important factor in determining effort and efficiency, they do not prove it conclusively. The apparent difference between methods could be driven by co-varying differences in species richness, plant density or the number of field assistants.

18.3.1.2 Protocol for laying of PSPs

Elouard and Krishnan (1999) have described various protocols for laying of PSPs. For example, to lay a plot of 1 ha (100 m x 100 m), a compass is used to fix the direction while drawing the boundaries of the square plot. The direction is continuously checked with the compass while the boundary is extended. Poles (pegs) or stones are placed every 10 m using a measuring tape. The total 1-ha plot is progressively divided into subplots of 10 m x 10 m and directions for setting the subplots are changed to avoid measuring errors. The plot is then subdivided into 10-m subplots (or quadrats) with rope and pegs, using measuring tape and compass. If necessary for the study (seedlings count, ground cover, etc.), quadrats of 5 x 5 m are laid within the 10 x 10 m subplots.

18.3.1.2.1 EXAMPLE OF PSP PROTOCOL This protocol was designed to monitor forest diversity and dynamics in the Agastyamalai region, Southern Western Ghats, South India. It involved the following.

1) Establishment of a 1-ha plot (100 x 100 m), with three replicates for each vegetation type, to study the spatial and floristic structure and dynamics of the forest.

2) Establishment of several smaller plots of 0.01 ha (10 x 1 m) located around each 1-ha permanent plot, in order to capture the structural and floristic variations of the major types. The satellite plots are laid in floristically different or transition types. The methodology is as follows.

a) The 1-ha plot is divided into 10 x 10 m subplots:
   - All individuals with DBH ≥0 cm are measured for height and girth (1 ha).
   - In addition, all individuals with DBH ≥1 cm are measured for height and girth, in 14 subplots (0.14 ha).
   - In the same subplots, another smaller plot measuring 5 x 5 m is laid and all individuals with DBH ≤1 cm are measured for height; heights are classified into 20-cm class intervals (0.035 ha).
   - For grasses, *Strobilanthes*, bamboos and reeds, the percentage of the 5 x 5 m area covered is estimated.
   - The slopes of the plot are measured and all individuals with DBH ≥10 cm diameter are specially mapped in the 1-ha plot.

b) Measurements are taken in the satellite plots as follows.
   - All individuals with DBH ≥10 cm are identified and measured for girth (100 m^2).
   - In the central 5 x 5 m quadrat of each plot, all individuals with DBH ≥1 cm are identified, measured for girth and counted. This protocol is applicable to all the floristic variations in a continuum and for all the forest types encountered in the Southern Western

Ghats. Further, it also addresses the need to monitor transition forest zones with the help of several satellite plots.

The time frame, human resources and budget for such an exercise have to be worked out for each situation based on the working conditions there.

18.3.1.3 Parameters to be sampled: field measurements

In all sampling techniques, the important measurement made in the field with reference to trees, lianas and other plants are very crucial for monitoring the biodiversity and floristic of the given region (Elouard and Krishnan, 1999). For tree growth, in addition to the inventory, girth at breast height (GBH) (Chittibabu and Parthasarathy, 2000; Shankar, 2001; Kumar *et al.*, 2006), which is also called circumference at breast height (CBH) (Padalia *et al.*, 2004); diameter at breast height (DBH) (Pitman *et al.*, 2002; Knight, 1975; Gillespie *et al.*, 2000) or diameter at base trunk (DBT) (Appolinário *et al.*, 2005) are also measured. Variations are noted in many studies in measurements, irrespective of GBH, CBH, DBH and DBT. From the literature survey, the GBH measurement is taken at 10 and 30 cm, the CBH is measured at 10, 17 and 30 cm, DBH is measured from 1 to 10.2 cm and DBT is measured at 5 cm. Because tree stems taper, 'breast height' (BH), the height above ground where diameter is measured, should influence DBH recorded (Brokaw and Thompson, 2000). Using a BH of 130 cm was customary in continental Europe (Robbie, 1955), whereas the seemingly odd value of 137 cm (4.5 feet) was the usual BH where English units were employed (Grubb *et al.*, 1963). The present literature review and the study made by Brokaw and Thompson (2000) reveal two important items of information: 1) more than 50% of the studies did not mention the height at which the measurement was taken, i.e. breast height; and 2) of those that did report BH used, the range was 120 cm to 160 cm.

In many studies (Pitman *et al.*, 2002; Parthasarathy, 1999; Linder *et al.*, 1997; Kalacska *et al.*, 2004) values for basal area, biomass and growth of standing trees and sometimes lianas (Reddy and Parthasarathy, 2003; Parthasarathy and Sethi, 1997) are given on a per hectare basis. As basal area and biomass are exponential functions of diameter, consistency should be maintained when measuring DBH, because using different values of BH may lead to erroneous comparisons of diameter-class distributions, growth, basal area or biomass (Brokaw and Thompson, 2000).

Another important issue with reference to the girth measurement is that it is responsible for the inclusion and exclusion of species in a study. For example, if the girth threshold for a tree is fixed as ≥30 cm, then only tree stems with a girth equal to and above the threshold will be considered as trees. In such cases, tree stems with a girth below this threshold will not be considered as trees. Those species which failed to reach 30 cm girth will be omitted from the sampling, resulting in poor species richness and stand density.

Elouard and Krishnan (1999) argue that this will affect the total number of species, stand density and basal area of the study. For example, in the inventory study by Shankar (2001) in the Eastern Himalaya, apart from adult trees (≥30 cm girth), the saplings and seedlings (<30 cm girth) were also enumerated. As a result, the number of species, stand density and basal area increased considerably.

As evidenced from the literature, there is no strict rule or threshold value for the girth of tree stems. The girth threshold varies considerably from ≥1 cm to 30 cm. The girth threshold set in each study is at the author's discretion. In inventory studies, all the species inside the plot should be enumerated, as in Guariguata *et al.* (1997). For better understanding, the authors classified the stems into: 1) trees (stems ≥10 cm DBH); 2) treelets (stems ≥5 cm DBH and <10 cm DBH); 3) saplings (stems ≥1 m tall and <5 cm DBH: stem diameter at 1.3 m); and 4) seedlings (stems ≥0.2 m tall and <1 m tall). According to this grouping, all the stems will be enumerated. Ferreira and Prance (1998) suggest that the minimum recommended DBH for a tree is 10 cm, as it is becoming standard for quantitative inventories for many ecologists; but if resources permit, a smaller minimum DBH should be included as a subsample in any floristic inventory.

For monitoring of forest dynamics with respect to floral diversity, the following should be taken into account.

Numbering of the trees: All trees more than 30 cm GBH should be enumerated and numbered. The numbering of trees should be done preferably using a metal label, which may be attached to the tree using a single nail. Although painting is cheap initially, it has to be renewed every year, requiring additional expenditure and manpower.

Spatial location of the trees: The trees should be located, marked and mapped within each plot, using a measuring tape to get the *x* and *y* coordinates. The coordinates are measured from one corner of the plot, facing north, for each 10 m quadrat and then adjusted for the whole plot. The mapping of trees is necessary for studies involving stand structure, species and tree distribution, fruit production, seed dispersal and germination, regeneration and colonization in canopy gaps.

Species identification: As far as possible, species should be identified in the field and herbarium specimens collected and processed from the unidentifiable species.

Girth: Girth at breast height (130 cm) is measured with a measuring meter or with microdendrometers (metal ring permanently fixed on the tree trunk). The first method is cheaper, valid only for a one-time measurement of the girth. The second method is preferable if regular monitoring of the tree's growth is required (long-term studies). Where there is a buttress at 130 cm, the measurement should be taken above the buttresses.

Height: Depending on the tree, height is measured with a slope meter at a defined distance from the tree (15, 20 or 30 m).

Crown diameter: To understand spatial development of trees at canopy level, growth of canopy and canopy gap ecology, it is necessary to measure and record the crown diameter of trees.

Recruitment and regeneration:

- Saplings below 10 cm and above 1 cm GBH are identified, measured (exact measurement or grouped into different height classes, e.g., 25 or 50 cm class intervals) and counted during recruitment studies.
- Seedlings (<1 cm DBH) are identified, measured (20 or 50 cm class intervals) and counted (number of seedlings of each species within a 5 x 5 m subplot) for regeneration studies.

Intervals between measurements: The time interval between the measurements depends on the objectives of the study. For assessing the forest dynamics the following aspects should be considered.

- Long-term changes (forest structure, recruitment, floristic changes): this requires monitoring at low periodicity, e.g. five- or ten-year intervals. The measurements are then made for trees >30 cm GBH.

- Survey of mortality and regeneration: this requires monitoring at a high periodicity, e.g. every year or two years. All trees (≥10 cm dbh), saplings (≥lcm DBH) and seedlings (below 1 cm DBH) are measured. Saplings and seedlings should be monitored on a partial area of the plot in 10 x 10 m subplots for saplings and 5 x 5 m subplots for seedlings, randomly selected within the 1-ha plot. The turnover for recruitment can be studied yearly or on a long-term basis, depending on the objectives: assessment of global changes over a long period or a continuous survey of regeneration processes and mortality rates.

18.3.1.3.1 MONITORING AND ASSESSING SPECIES DIVERSITY The main problem is to clearly state at which level species diversity is to be assessed. Let us take two contrasted examples.

1) If the diversity is to be estimated at the stand or community level (α-diversity and, possibly, β-diversity): the choice will be between a single large plot (say between 5 and 50 ha), a set of a few medium-size plots (say 3–5 plots of 1 ha), or a cluster of several small plots (say 20–30 plots of 20 trees).
2) If the diversity is to be estimated at the level of a forest type (γ-diversity): it will be necessary to sample various plots located in different forest patches/types and stands and the choice will concentrate on the compromise between the number of plots and their size.

In both cases, it is necessary to state at which scale the estimates of diversity are provided: 1 ha or the total community; 1 ha or the total vegetation type. An important point which actually concerns most operational forest inventories is that they use some form of variable-size plots, where larger trees have a higher probability to be sampled. Let $p(d)$ be the probability of a tree of size d to be sampled. The estimates of the standing stock easily take this factor into account, by weighting the observations by the inverse of the probability to be observed. The situation is a bit different for diversity.

1) For richness, this procedure may result in a bias as soon as there is any form of correlation between species composition and size structure (e.g. some species yielding only small individuals).
2) For species frequency, the procedure described above can be applied, without any adverse consequence except that rare small-size species have more chance of being missed.

18.3.1.3.2 MONITORING SPECIES RICHNESS AND DIVERSITY – CHANGES IN SPECIES RICHNESS The first step in assessing changes in species richness over time is fairly similar to the first step in assessing the changes in the standing stock: the estimate of the overall balance is to be estimated; this estimate is more precise if permanent plots are used than if temporary plots are used. Then comes the question of estimating the components of this balance: how many new species have appeared and how many have disappeared? This can actually be carried out even if temporary plots are used, which is a major difference from the assessment of the components of changes in the standing stock. But, of course, the estimates may not be reliable when the plots are temporary. Other important points are:

1) that the scale-dependence of diversity may yield paradoxical results: for example, changes occurring at plot scale may no longer appear when all plots are pooled together
2) that the duration of richness and diversity changes is usually longer than the duration of biomass changes: of course, this is not the case during some special event such as colonization of open land, tree-fall due to heavy storms or massive dieback.

It is worth noting that in such cases, the standing stock also changes dramatically. More important than numbers of species, the lists of the species (those which remained, those which disappeared and those which appeared) aid our understanding of the nature of the changes: for example, whether the species that appeared are light-demanding or tolerant, to which strata they belong, whether they are indicators of disturbance, whether they are deciduous or evergreen, exotic or indigenous.

18.3.1.3.3 MONITORING SPECIES RICHNESS AND DIVERSITY – CHANGES IN SPECIES DIVERSITY Changes in diversity indices can be estimated simply and more precisely if permanent plots are used. However, these indices being composite, it is difficult to break the changes into components. Monitoring the changes in diversity means assessing the changes not only in species richness but also in species frequency. In that respect, this is fairly comparable to monitoring changes in the size structure of a stand. For example, comparing two successive species composition can be achieved by a chi-squared (c^2) test. Methods discussed for assessing the changes in the overall standing stock can also be applied to single species: for instance, estimate of mortality, recruitment and growth for each species in a stand. One major difficulty in tropical forests is that the high number of species combined with the usual constraints on sampling intensity result in few individuals sampled per species. So, such a species-based approach is rarely feasible.

18.3.1.3.4 DATA ANALYSIS, INTERPRETATION AND REPORTING The approach to data analysis depends on the objectives of the monitoring programme. Always seek statistical advice from a biometrician or suitably experienced person prior to undertaking any analysis. The time and resources that are needed to undertake analysis of permanent plot data are substantial, but they are routinely underestimated. Advanced data handling and analytical skills are necessary to process and interpret this data. Inadequate training in the analysis packages is thought to be an impediment to routine analysis of plot data.

Before any analyses are undertaken, it is crucial that data errors are identified and corrected.

For permanent plot data, analysis programs exist that have been specifically tailored to analyse overstorey data (tree stems) using PC-DIAM (Hall, 1994a) and understorey data (saplings and seedlings) using PC-USTOREY (Hall, 1994b). Like any analysis package, these programs require training and expertise to use proficiently. If these programs are to be used, then data must be obtained in the appropriate file format. Manuals for the PC packages (Hall, 1994a, b) can be obtained free of charge from Landcare Research, and these outline the file formats.

18.4 Significance of Permanent Sample Plots: Advantage and Disadvantage

Phillips *et al.* (2002) have outlined the advantages and disadvantages of each of the sampling methods with respect to the typical range of purposes in tropical forest ecology and biodiversity monitoring. In conclusion, careful matching of inventory purpose to method has always been important for ecologists, and is especially so now in the tropical context of rapid environmental changes. The need for efficient sampling is a dominant factor determining methodological decisions, but comparative analysis of efficiency has been lacking in the tropical eco-floristic literature. Phillips *et al.* (2002) showed for the first time that conventional approaches to tropical floristic inventory vary greatly in their relative inventory efficiencies. These preliminary findings suggest that the urgent need for extensive plot-based floristic assessment in remote areas of the tropics can be addressed most simply by sampling small size-classes in narrow transects, but do not imply that this is the optimum approach for all inventory research. Further comparative analyses are needed using simulated and empirical results to explore how assessment techniques perform under different conditions.

18.4.1 Advantages – suitability for inventory and monitoring

The major advantages of setting permanent plots rather than having successive independent surveys are that permanent plots: 1) reduce the imprecision in the estimates of change; and 2) provide a means to break global estimates of changes into their components (mortality, growth, recruitment). The PSP method does provide data on species distribution (e.g. presence of rare, endangered and threatened plants), but the cost and time associated with the laying and undertaking measurements of plots at an adequate spatial scale makes it unsuitable for the purposes of inventory. RECCE plots (which are carried out in association with permanent plots) are more appropriate for inventory objectives (Hurst and Allen, 2007c).

Permanent 20 × 20 m plots are accepted as the best method available for monitoring structural and compositional change in shrublands and forests over long time periods in New Zealand. Long-term permanent plot data sets with greater than three re-measurement periods permit the analysis of trends at a range of spatial and temporal scales. Temporal trends are best interpreted in combination with other covariate monitoring data on pest animal abundances (e.g. faecal pellet counts) and other habitat and site condition assessments.

18.4.2 Disadvantages – transversal sampling using age instead of time

One problem with successive surveys, be they permanent or temporary plots, is that it takes several years before changes can be assessed: thus, the idea to use plots sampled at the same date but situated in different stands ordinate along an age gradient. The main difficulty with this approach lies in the fact that the age gradient may be correlated to some ecological factors, which have themselves a strong influence on biodiversity or standing stock. This is a very general problem, which also applies to biodiversity studies.

Houllier *et al.* (1998) reviewed some of the problems that arise when permanent plots are set up with the aim of monitoring forest biodiversity and/or forest dynamics. The major advantages of setting permanent plots rather than having successive independent surveys are that this 1) reduces the imprecision in the estimates of change and 2) provides a means to investigate the components of the changes. However, there are several technical issues which always come to the surface and have to be discussed and solved according to the specific context and objectives.

- Whether the plots should be physically demarcated in the light of the diffusion of global positioning systems (GPS).
- The choice between several alternative sampling designs: minimum size of the plot, cluster of plots vs a single connected plot, network of medium-size plots vs a few large plots.
- The minimum size of the sample trees.
- The way changes are estimated if the duration is long.

Phillips *et al.* (2002) have outlined the reasons why 1-ha plots should not be used as monitoring sites, including: 1) inadequate funds to re-census; 2) impossibility of relocating the plot's position; 3) removal of aluminium labels and nails by local residents; 4) forest disturbance by residents or commercial interests; 5) changing research interests of principal investigators; 6) rapid radial tree growth 'swallowing' tags; 7) liana or bamboo growth discouraging access; and 8) re-location or death of the principal investigator. Clearly, not all these factors can be anticipated, but they illustrate the need for a realistic appraisal of the risks and benefits before conducting any 1-ha inventory: conversion to permanent plot status is expensive, time-consuming and uncertain.

Temporary and inadequate funding is the main reason that most 1-ha plots remain simply temporary floristic samples. We suggest that installation of 1-ha plots for monitoring purposes may only be worthwhile when long-term funding programmes are identified from the start. However, as well as their key (but often unrealized) role in long-term studies, 1-ha plots may still be an appropriate e-method in some studies where the primary research purpose concerns a floristic inventory of trees, and their principal attraction arguably lies in their reasonable suitability for many purposes.

In a geographic area, the permanent plots can be established in two ways (CBD, 2001).

1) Small number of large-sized plots in major ecosystems of the region

Here, we can select the major forest ecosystems in the state. For example, in Kerala, shola forest, tropical wet evergreen, tropical moist deciduous and tropical dry deciduous can be selected. A suitable locality for each type of ecosystem can be identified and a large plot of a minimum of 10 ha can be established.

Advantages:

- Since it covers a larger area, the plant and animal (particularly small animals) diversity, their dynamics, interaction, etc. can be studied.
- Here, an ecosystem-level study is possible, considering abiotic and biotic factors and their interaction.
- Multidisciplinary research is possible in such large plots.
- Since the number of plots is lower, further management and monitoring will also be comparatively easier.
- Comparatively less expensive, less laborious and less time-consuming.

Disadvantages:

- The study will lead only to site-specific results. Even in a given type of forest, the results may not be applicable to those areas located in altitudinal/longitudinal/latitudinal gradients and those forest patches having different species composition.
- Very often, finding a suitable locality of larger size with rare human intervention is difficult, as in most of the places in the country, forests are highly fragmented.
- Initial establishment of larger plots is tougher than that of small-sized plots.

2) Large number of small-sized plots to cover all diverse ecosystems (subtypes of given type of forest)

Here we can select all major forest types and their subtypes. For instance, in Kerala there are around 100 subtypes of forests (Chandrasekharan, 1962; Pascal, 1988). In each subtype, small-sized plots (ranging from 0.01 to 0.1 ha, depending on the vegetation type) with replicates can be established considering the altitudinal/latitudinal/longitudinal/disturbance/successional gradient.

Advantages:

- Since it covers all subtypes of different types of forests, studies in replicate plots will provide valuable information on vegetation structure, composition, successional patterns and dynamics of a large number of plots.
- It will help to undertake a long-term monitoring programme of the entire forest ecosystem in a given state.
- Locating suitable localities for establishing plots is comparatively easy, as each plot will cover a small area.

Disadvantages:

- Since the plot size is small (ranging from 0.01 to 0.1 ha, depending on the vegetation type), studies may have to be restricted to the plant community only. Thus multidisciplinary studies may not be possible.
- Within the state, there will be a large number of plots. A permanent set-up and sustained financial support to maintain and monitor plots are required.
- Extensive field work will be necessary to identify suitable localities for different forest types and subtypes and establish permanent plots.
- Establishment and management of a large number of plots are more expensive (logistics, travel costs, etc.) and labour-intensive.
- A GIS integration is necessary to identify/relocate for further management.

18.5 Discussion

Floristic and biodiversity assessments are carried out at local and regional levels to understand the present status and to make effective management strategies for conservation. In this regard, various sampling techniques and measurement methods are followed, based on the objectives of the study. In the majority of the studies, the availability of time, money and human resources is the major constraint. Several issues are discussed in this review in relation to sampling techniques, measurements, sample size, etc. The sampling methods should satisfy the objectives of the study and also bring out the inherent diversity status of a region of investigation. The measurement of stem size in the field is the major issue in diversity studies, where a unanimous decision should be reached in relation to the girth of stem considered to be a tree and the height from the ground at which the measurement is to be taken. Sample size in floristic diversity study is an important issue, which determines the success and failure of a diversity study in establishing the true diversity status. Much attention should be paid to this issue in determination of the sample size and distribution of the samples.

This review has emphasized the usefulness of permanent plots in terms of precision of the estimates of change as well as in terms of the possibility of tracing and estimating the components of the changes. It has also shown that species richness and diversity, as measures of heterogeneity and of variability, are of a very different nature from volume, biomass, basal area, etc., and that this has several implications on the sampling and estimation processes.

In the field of classical forest inventories, many empirical as well as theoretical studies have shown that, for the same global sample size, the most efficient sampling designs are those comprising numerous very small plots. When cost and time are taken into account (e.g. cost of travelling from one plot to another), it appears, in temperate forests, that many small plots of, say, 0.01–0.10 ha provide better estimates than several larger plots. This general guideline can be extended to biodiversity assessment (Gimaret *et al.*, 1996).

A point which is more open to dispute is whether the plots should have a fixed size, contain a fixed number of trees, or have a variable size depending on the size of the trees. In the field of classical forest inventories, it is widely accepted that the emphasis should be put on the larger trees which have a higher economic value. However, this results in plots of variable size (e.g. relascope plots) and poses some technical problems when it comes to assessing changes or estimating species richness and diversity. Furthermore, the question of plot size takes another twist when it comes to monitoring and understanding the dynamic processes that take place. Such an understanding often requires locating the trees in the space, to consider their environment (neighbours, local ecological conditions). In that case, it is preferable and more efficient to set up large plots where the ratio of perimeter to area is small (i.e. there are relatively few trees whose neighbours are outside the plot). It is impossible to propose a unique framework for monitoring biodiversity, but some guidelines may be proposed.

- For research purposes, large permanent plots (from 1 to 50 ha) are the best solution.
- For operational purposes, clusters (or tracts) of small fixed-area permanent plots are a good solution.

18.5.1 Strategic planning of the project

Since permanent sample plots (PSPs) are long-term commitments with projects running for decades, there should be a strategic planning for implementing such projects, considering all the bottlenecks outlined above. Before establishing a permanent plot, the objectives and planning have to be clearly defined: the nature of studies to be carried out in the plot (e.g. biodiversity, phenology, regeneration, recruitment, mortality, decolonization processes, litter fall and litter decomposition, primary production, architecture, forest mosaic) and the limitations (funds, human resources, time allocated for these studies). The size of the plot, monitoring period and method are then decided based on these criteria. Hence, the projects involving PSPs need to be planned in such a way that importance is given to the system in place and, of course, the necessary institutional arrangements, the most important being the commitment for long-term funding, without which no long-term planning is possible for studies involving PSPs. For this, the following planning is suggested.

1. Establishment of a coordination cell in the funding agency/government department/ministry: to maintain the continuity of the project, it is appropriate that overall coordination is done through the government department/ministry.
2. Identification of key institutions/organizations, preferably research institutions with a commitment to undertake long-term ecosystem studies as part of the coordinated project at the national level, and signing of a memorandum of understanding with them.
3. Identification of the main/key researchers in various institutions/organizations with appropriate research background and commitment to being part of this long-term project and coordinating research in a specific field in a network mode.
4. Identification of researchers in different organizations/institutions in different areas/disciplines with appropriate research background.

5. Formulating uniform methodology (size of plot, replications, parameters to be monitored, etc.) for all the components of the project.
6. Different forest types should be identified for establishing PSPs and if there are already established PSPs where baseline data have been generated, those should be preferred for the study.
7. The project should be taken in phases; each phase of the project should be for five years' duration.
8. Guidelines and format for periodical review of progress of the project; interim reports, review meetings, site visits, etc.
9. At the end of each phase, the results should be analysed and reports prepared and discussed in a national workshop to get feedback on the quality of work. The workshop should also take stock of positive and negative aspects of the project implementation and corrective measures taken for the implementation of the next phase; the workshop may also be used for identifying new researchers to join the project, if required.
10. It should be mandatory for the researchers to publish papers in national and international journals; all the papers published as a result of this project should bear the words: '*Research Paper No. ... ; work presented forms part of the National Project on ... supported by ...* '; the publication paper number should be allotted by the coordination cell in the government department/ministry after the paper is accepted or approved for publication in the journal. This way there will be a record of all the papers published as a result of this project.

Acknowledgement

I wish to acknowledge with gratitude the excellent publications and reviews of Houllier *et al.* (1998), Mahan *et al.* (1998), Gaines *et al.* (1999), Elouard and Krishnan (1999), Phillips *et al.* (2002), Jayakumar *et al.* (2011) which I consulted besides several others, and the material used liberally in the preparation of this review paper.

References

Alder, D. and Synnott, T. J. (1992) Permanent sample plot techniques for mixed tropical forests. *Tropical Forestry Papers* No. 25. Oxford Forestry Institute, Oxford, UK.

Aldrich, P.R., Parker, G.R. and Ward, J.S. (2002) Spatial dispersion of trees in an old-growth temperate hardwood forest over 60 years of succession. *Forest Ecology and Management* 180, 475–491.

Ahlstrand, G.J., Hadfield, J., Holmes, R., Kailin, J., Reeberg, P. and Schreiner, E. (2001) *Long-term Ecological Monitoring Conceptual Plan for Terrestrial Vegetation*. Cited in: Munro, D., Nickelson, S., Chapin, D., Joselyn, M., Lackey, B., Paige, P., Beedle, D., Boeckstiegel, L., Sammarco, W. and Anderson, C. (eds) *A Proposal for the Establishment of Permanent Sample Plots in the Cedar River Municipal Watershed*. 24 January 2003. Available at: www.seattle.gov/Util/cs/groups/public/@spu/@ssw/documents/webcontent/spu02_015244.pdf (accessed 27 May 2016).

Appolinário, V., Filho, A.T.O. and Guilherme, F.A.G. (2005) Tree population and community dynamics in a Brazilian tropical semi-deciduous forest. *Revista Brasil* 28, 347–360.

Ayyappan, N. and Parthasarathy, N. (1999) Biodiversity inventory of trees in a large-scale permanent plot of tropical evergreen forest at Varagalaiar, Anamalais, Western Ghats, India. *Biodiversity Conservation* 8, 1533–1554.

Baker, T.R., Phillips, O.L., Malhi, Y., Almeida, S., Arroyo, L. *et al.* (2004) Increasing biomass in Amazonian forest plots? *Philosophical Transactions of Royal Society of London* 359, 353–365.

Bakker, J.P., Willems, J.H. and Zobel, M. (1996a) Long term vegetation dynamics: introduction. *Journal of Vegetation Science* 7, 146–147.

Bakker, J.P., Willems, J.H. and Zobel, M. (1996b) Why do we need permanent plots in the study of long term vegetation dynamics? *Journal of Vegetation Science* 7, 147–156.

Balaguru, B., Britto, S.J. and Nagamurugan, N. (2006) Identifying conservation priority zones for effective management of tropical forests in Eastern Ghats of India. *Biodiversity and Conservation* 15, 1529–1543.

Bazzaz, F.A. (1975) Plant species diversity in old-field successional ecosystems in southern Illinois. *Ecology* 56, 485–488.

Behera, M.D., Kushwaha, S.P.S. and Roy, P.S. (2002) High plant endemism in an Indian hotspot – eastern Himalaya. *Biodiversity and Conservation* 11, 669–682.

Behera, M.D., Kushwaha, S.P.S. and Roy, P.S. (2005) Rapid assessment of biological richness in a part of Eastern Himalaya: an integrated three-tier approach. *Forest Ecology and Management* 207, 363–384.

Bhat, D.M., Naik, M.B. and Patagar, S.G. (2000) Forest dynamics in tropical rain forests of Uttara Kannada district in Western Ghats, India. *Current Science* 79, 975–985.

Brokaw, N. and Thompson, J. (2000) The H for DBH. *Forest Ecology and Management* 129, 89–91.

CBD (2001) *Global Biodiversity Outlook*. CBD Secretariat, Montreal, Canada.

Chambers, J.Q., dos Santos, J., Ribeiro, R.J. and Higuchi, N. (2001) Tree damage, allometric relationships, and

above-ground net primary production in central Amazon forest. *Ecological Management* 152, 73–84.
Chandrashekara, U.M. and Jayaraman, K. (2002) Stand structural diversity and dynamics in natural forests of Kerala. *KFRI Research Report No. 232*. Kerala Forest Research Institute, Peechi, India.
Chandrashekara, U.M., Menon, A.R.R., Nair, K.K.N., Sasidharan, N. and Swarupanandan, K. (1998) Evaluating plant diversity in different forest types of Kerala by laying out permanent sample plots. *KFRI Research Report No. 156*. Kerala Forest Research Institute, Peechi, Kerala.
Chandrasekharan, C. (1962) Forest types of Kerala State (1). *Indian Forester* 88, 660–674.
Chawla, A., Yadav, P.K., Uniyal, S.K., Kumar, A., Vats, S.K., Kumar, S. and Ahuja, P.S. (2012) Long-term ecological and biodiversity monitoring in the western Himalaya using satellite remote sensing. *Current Science* 102, 1143–1156.
Chittibabu, C.V. and Parthasarathy, N. (2000) Attenuated tree species diversity in human-impacted tropical evergreen forest sites at Kolli hills, Eastern Ghats, India. *Biodiversity and Conservation* 9, 1493–1519.
Clark, D.A., Brown, S., Kicklighter, D.W., Chambers, J.Q., Thomlinson, J.R. and Ni, J. (2001a) Measuring net primary production in forests: concepts and field methods. *Ecological Applications* 11, 356–370.
Clark, D.A., Brown, S., Kicklighter, D.W., Chambers, J.Q., Thomlinson, J.R., Ni, J. and Holland, E.A. (2001b) Net primary production in tropical forests: an evaluation and synthesis of existing field data. *Ecological Application* 11, 371–384.
Coates, K.D. (2002) Tree recruitment in gaps of various size, clear cuts and undisturbed mixed forest of interior British Columbia, Canada. *Forest Ecology and Management* 155, 387–398.
Condit, R., Hubbell, S.P. and Lafrankie, J.V. (1996) Species–area and species–individual relationship for tropical trees: a comparison of three 50-ha plots. *Ecology* 84, 549–562.
Dixon, R.K., Brown, S., Houghton, R.A., Solomon, A.M., Trexler, M.C. and Wsniewski, J. (1994) Carbon pools and flux of global forest ecosystems. *Science* 263, 185–190.
Denslow, J.S. (1995) Disturbance and diversity in tropical rain forests: the density effect. *Ecological Applications* 5, 962–968.
Dyrness, C.T. and Acker, S.A. (1999) Reference Stands in or near the H.J. Andrews Experimental Forest. Permanent Plots of the Pacific Northwest. Report Number 2. Available at: http://andrewsforest.oregonstate.edu (accessed 27 May 2016).
Dyrness, T. and Acker, S. (2000) Permanent plots in Mount Rainier National Park. Permanent Plots of the Pacific Northwest. Report Number 4. Available at: http://andrewsforest.oregonstate.edu (accessed 27 May 2016).
Elouard, C. and Krishnan, R.M. (1999) Assessment of forest biological diversity. A FAO training course. 2. Case study in India. In: *Pondy Papers in Ecology, Vol. 5*. French Institute of Pondicherry, Pondicherry, India.
Engeman, R.M., Sugihara, R.T., Pank, L.F. and Dusenberry, W.E. (1994) A comparison of plotless density estimators using Monte Carlo simulation. *Ecology* 75, 1769–1779.
Fashing, P.J., Forrestel, A., Scully, C. and Cords, M. (2004) Long-term tree population dynamics and their implications for the conservation of the Kakamega Forest, Kenya. *Biodiversity and Conservation* 13(4), 753–771.
Field, C.B., Behrenfeld, M.J., Randerson, J.T. and Falkowski, P. (1998) Primary production of the biosphere: integrating terrestrial and oceanic components. *Science* 281, 237–240.
Ferreira, L.V. and Prance, G.T. (1998) Species richness and floristic composition in four hectares in the Jaú National Park in upland forests in central Amazonia. *Biodiversity and Conservation* 7, 1349–1364.
Frankie, G.W., Baker, H.G. and Opler, P.A. (1974) Comparative phenological studies of trees in tropical wet and dry in the lowlands of Costa Rica. *Journal of Ecology* 62, 881–919.
Franklin, J. and Rey, S.J. (2007) Spatial patterns of tropical forest trees in Western Polynesia suggest recruitment limitations during secondary succession. *Journal of Tropical Ecology* 23, 1–12.
Freet, B. (2001) *North Cascades National Park Service Complex Long-Term Ecological Monitoring Conceptual Plan*. Cited in: Munro, D., Nickelson, S., Chapin, D., Joselyn, M., Lackey, B., Paige, P., Beedle, D., Boeckstiegel, L., Sammarco, W. and Anderson, C. (eds) *A Proposal for the Establishment of Permanent Sample Plots in the Cedar River Municipal Watershed*. 24 January 2003. Available at: http://www.seattle.gov/Util/cs/groups/public/@spu/@ssw/documents/webcontent/spu02_015244.pdf (accessed 27 May 2016).
Gaines, L.W., Richy, J.H. and John, F.L. (1999) *Monitoring Biodiversity: Quantification and Interpretation*. General Technical Report PNW-GTR-443. United States Department of Agriculture, Forest Service, Pacific Northwest Research Station, Portland, Oregon, USA.
Gentry, A.H. (1988) Tree species richness of upper Amazonian forests. *Proceedings of the National Academy of Sciences* 85, 156–159.
Gillespie, T.W., Grijalva, A. and Farris, C.N. (2000) Diversity, composition, and structure of tropical dry forests in Central America. *Plant Ecology* 147, 37–47.
Gimaret, C., Pélissier, R. and Pascal, J.P. (1996) Estimation et variations de la richesse et de la diversité spécifiques en forêt sempervirente humide. Poster, Symposium Biodiversité et fonctionnement des écosystèmes (12–14/06/96). Ecole normale supérieure, Paris, France.

Gocz, R.J. (1998) International long-term international ecological research priorities, opportunities and lessons learned. In: Iwakuma, T. (ed.) *Long-Term Ecological Research in the East Asia-Pacific Region: Biodiversity and Conservation of Terrestrial and Freshwater Ecosystems*. Environment Agency of Japan, National Institute for Environmental Studies, Japan, pp. 9–14.

Gordon, J.E. and Newton, A.C. (2006a) Efficient floristic inventory for the assessment of tropical tree diversity: a comparative test of four alternative approaches. *Forest Ecology and Management* 237, 564–573.

Gordon, J.E. and Newton, A.C. (2006b) The potential misapplication of rapid plant diversity assessment in tropical conservation. *Journal for Nature Conservation* 14, 117–126.

Grau, H.R., Arturi, M.F. and Brown, A.D. (1997) Floristic and structural patterns along a chronosequence of secondary forest succession in Argentinean subtropical montane forests. *Forest Ecology and Management* 95, 161–171.

Grubb, J., Lloyd, J.R. and Pennington, T.D. (1963) A comparison of montane and lowland forests in Ecuador. I. The forest structure, physiognomy, and floristic. *Journal of Ecology* 51, 567–601.

Guariguata, M.R., Chazdon, R.L. and Denslow, J.S. (1997) Structure and floristics of secondary and old-growth forest stands in lowland Costa Rica. *Plant Ecology* 132, 107–120.

Hall, J.B. (1991) Multiple-nearest-tree sampling in an ecological survey of Afromontane catchment forest. *Forest Ecology and Management* 42, 245–299.

Hall, G.M.J. (1994a) *PC-DIAM: Stem Diameter Data Analysis*. Manaaki Whenua – Landcare Research, Lincoln, New Zealand.

Hall, G.M.J. (1994b) *PC-USTOREY: Seedling and Sapling Data Analysis*. Manaaki Whenua – Landcare Research, Lincoln, New Zealand.

Harms, K., Condit, R., Hubbell, S.P. and Foster, R.B. (2001) Habitat associations of trees and shrubs in a 50-ha neotropical forest plot. *Journal of Ecology* 89, 947–959.

He, F., Legendre, P. and Lafrankie, J.V. (1996) Spatial pattern of diversity in a tropical rain forest in Malaysia. *Journal of Biogeography* 23, 57–74.

Henderson, J.A., Peter, D.H., Lesher, R.D. and Shaw, D.C. (1989) *Forested Plant Associations of the Olympic National Forest*. Ecological Technical Paper 001-88. US Department of Agriculture, Forest Service, Pacific Northwest Research Station, Portland, Oregon, USA.

Herben, T. (1996) Permanent plots as tools for plant community ecology. *Journal of Vegetation Science* 7, 195–202.

Houllier, F., Krishnan, R.M. and Elouard, C. (1998) *Assessment of Forest Biological Diversity, A FAO Training Course. 1 – Lecture Notes*. Pondy Papers in Ecology. French Institute of Pondicherry, Pondicherry, India, 102 pp.

Huang, W., Pohjonen, V. and Johansson, S. (2003) Species diversity, forest structure and species composition in Tanzanian tropical forests. *Forest Ecology and Management* 173, 11–24.

Hubbell, S.P. and Foster, R.B. (1992) Short term dynamics of a neotropical forest: why ecological research matters to tropical conservation and management. *Oikos* 63, 48–61.

Hurst, J.M. and Allen, R.B.A (2007a) *Permanent Plot Method for Monitoring Indigenous Forests – Expanded Manual.* Version 4. Landcare Research Contract Report (LC0708/028). Manaaki Whenua – Landcare Research, Lincoln, New Zealand.

Hurst, J.M. and Allen, R.B. (2007b) *The RECCE Method for Describing New Zealand vegetation – Expanded Manual*. Version 4. Landcare Research Contract Report (LC0708/029). Manaaki Whenua – Landcare Research, Lincoln, New Zealand.

Hurst, J.M. and Allen, R.B. (2007c) *The RECCE Method for Describing New Zealand Vegetation – Field Protocols*. Manaaki Whenua – Landcare Research, Lincoln, New Zealand.

Inouye, R.S. and Tilman, D. (1995) Convergence and divergence of old-field vegetation after 11 yr of nitrogen addition. *Ecology* 76(6), 1872–1887.

Jayakumar, S., Seong, S.-K. and Joon, H. (2011) Floristic inventory and diversity assessment – a critical review. *Proceedings of the International Academy of Ecology and Environmental Sciences* 1(3–4), 151–168.

Johnston, A.E. (1991) Benefits from long-term ecosystem research: some examples from Rothamsted. In: Risser, P.G. (ed.) *Long-Term Ecological Research*. Wiley, Chichester, UK, pp. 89–113.

Kadavul, K. and Parthasarathy, N. (1999) Plant biodiversity and conservation of tropical semi-evergreen forest in the Shervarayan hills of Eastern Ghats, India. *Biodiversity and Conservation* 8, 419–437.

Kalacska, M., Sanchez-Azofeifa, G.A. and Calvo-Alvarado, J.C (2004) Species composition, similarity and diversity in three successional stages of a seasonally dry tropical forest. *Forest Ecology and Management* 200, 227–247.

Kennard, D.K., Gould, K. and Putz, F.E. (2002) Effect of disturbance intensity on regeneration mechanisms in a tropical dry forest. *Forest Ecology and Management* 162, 197–208.

Killeen, T.J., Jardim, A. and Mamani, F. (1998) Diversity, composition and structure of a tropical semideciduous forest in the Chiquitanía region of Santa Cruz, Bolivia. *Journal of Tropical Ecology* 14, 803–827.

Kimmins, J.P. (1997) Biodiversity and its relationship to ecosystem health and integrity. *Forest Chronicle* 73, 229–232.

Kleinn, C. and Vilcko, F. (2006) A new empirical approach for estimation in k-tree sampling. *Forest Ecology and Management* 237, 522–533.

Knight, D.H. (1975) A phytosociological analysis of species-rich tropical forest on Barro Colorado Island, Panama. *Ecological Monographs* 45, 259–284.

Kohyama, T., Suzuki, E., Partomihardjo, T., Yamada, T. and Kubo, T. (2003) Tree species differentiation in growth, recruitment and allometry in relation to maximum height in a Bornean mixed dipterocarp forest. *Journal of Ecology* 91, 797–806.

Krishnamurthy, Y.L., Prakasha, H.M., Nanda, A., Krishnappa, M., Dattaraj, H.S. and Suresh, H.S. (2010) Vegetation structure and floristic composition of a tropical dry deciduous forest in Bhadra Wildlife Sanctuary, Karnataka, India. *Tropical Ecology* 51, 235–246.

Kumar, A., Gupta, A.K. and Marcot, B.G. (2002) *Management of Forests in India for Biological Diversity and Forest Productivity, a New Perspective. Volume IV: Garo Hills Conservation Area (GCA).* Wildlife Institute of India – USDA Forest Service Collaborative Project Report, Wildlife Institute of India, Dehradun, India.

Kumar, A., Marcot, B.G. and Saxena, A. (2006) Tree species diversity and distribution patterns in tropical forests of Garo Hills. *Current Science* 91, 1370–1381.

Lewis, S.L., Phillips, O.L., Baker, T.R., Lloyd, J., Malhi, Y. *et al.* (2004) Concerted changes in tropical forest structure and dynamics: evidence from 50 South American long-term plots. *Philosophical Transactions of The Royal Society of London Series B: Biological Sciences* 359(1443), 421–436.

Linder, P., Elfving, B. and Zackrisson, O. (1997) Stand structure and successional trends in virgin boreal forest reserves in Sweden. *Forest Ecology and Management* 98, 17–33.

Magurren, A.E. (1988) *Ecological Diversity and Its Measurement*. Princeton University Press, Princeton, New Jersey, USA.

Mahan, C., Sullivan, K.K., Ke, C., Yahner, R. and Abrams, M. (1998) *Ecosystem Profile Assessment of Biodiversity: Sampling Protocols and Procedures*. Center for Biodiversity Research Environmental Resources Research Institute, The Pennsylvania State University, US Department of Interior, National Park Service, University Park, Pennsylvania, USA.

Malhi, Y., Nobre, A.D., Grace, J., Kruijt, B., Pereira, M.G.P., Culf, A. and Scott, S. (1998) Carbon dioxide transfer over a Central Amazonian rain forest. *Journal of Geophysical Research* 103, 593–612.

Mani, S. and Parthasarathy, N. (2005) Biodiversity assessment of trees in five inland tropical dry evergreen forests of peninsular India. *Systematic and Biodiversity* 3, 1–12.

Mani, S. and Parthasarathy, N. (2006) Tree diversity and stand structure in inland and coastal tropical dry evergreen forests of peninsular India. *Current Science* 90, 1238–1246.

Melillo, J.M., McGuire, A.D., Kicklighter, D.W., Moore, B. III, Vorosmarty, C.J. and Schloss, A.L. (1993) Global climate change and terrestrial net primary production. *Nature* 363, 234–240.

Myers, W.L. and Irish, R.R. (1981) *Vegetation Survey of Delaware Water Gap National Recreation Area*. Final report. USDI, National Park Service, Mid-Atlantic Regional Office, Philadelphia, Pennsylvania.

Nath, C.D., Dattaraja, H.S., Suresh, H.S., Joshi, N.V. and Sukumar, R. (2006) Patterns of tree growth in relation to environmental variability in the tropical dry deciduous forest at Mudumalai, southern India. *Journal of Biosciences* 31, 651–669.

Nelson, B.W., Ferreira, C. and Da Silva, M. (1990) Endemism centers, refugia and botanical collection density in Brazilian Amazonia. *Nature* 345, 714–716.

Noble, I. and Dirzo, R. (1997) Forests as human-dominated ecosystems. *Science* 277, 522–525.

Noss, R.F. and Cooperrider, A. (1994) *Saving Nature's Legacy: Protecting and Restoring Biodiversity*. Island Press, Washington, DC, USA.

Orwig, D.A. and Abrams, M.D. (1993) Land-use history (1720–1992), composition, and dynamics of oak-pine forests within the Piedmont and Coastal Plain of northern Virginia. *Canadian Journal for Research* 24, 116–125.

Padalia, H., Chauhan, N. and Porwal, M.C. (2004) Phytosociological observations on tree species diversity of Andaman Islands, India. *Current Science* 87, 799–806.

Paijmans, K. (1970) An analysis of four tropical rain forest sites in New Guinea. *Journal of Ecology* 58, 77–101.

Parsons, R.F. and Cameron, D.G. (1974) Maximum plant species diversity in terrestrial communities. *Biotropica* 6, 202–203.

Parthasarathy, N. (1999) Tree diversity and distribution in undisturbed and human-impacted sites of tropical wet evergreen forest in southern Western Ghats, India. *Biodiversity and Conservation* 8, 1365–1381.

Parthasarathy, N. and Karthikeyan, R. (1997) Plant biodiversity inventory and conservation of two tropical dry evergreen forests on the Coromandel coast, south India. *Biodiversity and Conservation* 6, 1063–1083.

Parthasarathy, N. and Sethi, P. (1997) Trees and liana species diversity and population structure in a tropical forest in south India. *Tropical Ecology* 38, 19–30.

Pascal, J.P. (1988) *Wet Evergreen Forests of the Western Ghats of India: Ecology, Structure, Floristic Composition and Succession*. French Institute of Pondichery, Pondicherry, India.

Pascal, J.P. and Pelissier, R. (1996) Structure and floristic composition of a tropical evergreen forest in southwest India. *Journal of Tropical Ecology* 12, 191–214.

Phillips, O.L. (1996) Long-term environmental change in tropical forests: increasing tree turnover. *Environmental Conservation* 23, 235–248.

Phillips, O.L., Malhi, Y., Higuchi, N., Laurance, W.F., Nuñez Vargas, P. and Vásquez, M. (1998) Changes in the carbon balance of tropical forest: evidence from long-term plots. *Science* 282, 439–442.

Phillips, O.L, Malhi, Y., Vinceti, B., Baker, T., Lewis, S. and Higuchi, N. (2002) Changes in growth of tropical forests: evaluating potential biases. *Ecological Applications* 12, 576–587.

Phillips, O.L., Vasquez Martinez, R., Nuñez Vargas, P., Monteagudo, A.L., Chuspe Zans, M., Galiano Sanchez, W., Pena Cruz, A., Timana, M., Yli-Halla, M. and Rose, S. (2003) Efficient plot-based floristic assessment of tropical forests. *Journal of Tropical Ecology* 19, 629–645.

Pitman, N.C.A., Terborgh, J.W. and Silman, M.R. (2002) A comparison of tree species in two upper Amazonian forests. *Ecology* 83, 3210–3224.

Proctor, J., Anderson, J.M. and Chai, P. (1983) Ecological studies in four contrasting lowland rain forests in Gunung Mulu national park, Sarawak: I. Forest environment, structure and floristics. *Journal of Ecology* 71, 237–260.

Ramanujam, M.P. and Kadamban, D. (2001) Plant biodiversity of two tropical dry evergreen forests in the Pondicherry region of South India and the role of belief systems in their conservation. *Biodiversity and Conservation* 10, 1203–1217.

Reddy, M.S. and Parthasarathy, N. (2003) Liana diversity and distribution in four tropical dry evergreen forests on the Coromandel coast of south India. *Biodiversity and Conservation* 12, 1609–1627.

Risser, P. and Rice, E.L. (1971) Diversity in tree species in Oklahoma upland forest. *Ecology* 52, 876–880.

Robbie, T.A. (1955) *Teach Yourself Forestry*. English Universities Press, London, UK.

Routledge, R.D. (1979) Diversity indices: which ones are admissible? *Journal of Theoretical Biology* 76, 503–515.

Rovero, F. and Marshall, A.R. (2005) Diversity and abundance of diurnal primates and forests antelopes in relation to habitat quality: a case study from the Udzungwa mountains of Tanzania. In: Huber, B.A., Sinclair, B.J. and Lampe, K.-H. (eds) *African Biodiversity: Molecules, Organisms, Ecosystems. Proceedings of 5th International Symposium on Tropical Biology, Koenig Museum, Bonn.* Springer Verlag, Berlin, Germany, pp. 297–304.

Sagar, R., Raghubanshi, A.S. and Singh, J.S. (2003) Tree species composition, dispersion and diversity along a disturbance gradient in a dry tropical forest region of India. *Forest Ecology and Management* 186, 61–71.

Sagar, R. and Singh, J.S. (2006) Tree density, basal area and species diversity in a disturbed dry tropical forest of northern India: implications for conservation. *Environmental Conservation* 33(3), 256–262.

Scott, C.T. (1998) Sampling methods for estimating change in forest resources. *Ecological Applications* 8(2), 228–233.

Shahabuddin, G. and Kumar, R. (2006) Influence of anthropogenic disturbance on bird communities in a tropical dry forest: role of vegetation structure. *Animal Conservation* 9, 404–413.

Shankar, U. (2001) A case of high tree diversity in a sal (*Shorea robusta*)-dominated lowland forest of Eastern Himalaya: floristic composition, regeneration and conservation. *Current Science* 81, 776–786.

Sheil, D., Jennings, S.B. and Savill, P. (2000) Long-term permanent plot observations of vegetation dynamics in Budongo, a Ugandan rain forest. *Journal of Tropical Ecology* 16, 765–800.

Sheil, D., Ducey, M.J. and Samsoedin, I. (2003) A new type of sample unit for the efficient assessment of diverse tree communities in complex forest landscapes. *Journal of Tropical Forestry Science* 15(1), 117–135.

Silvertown, J., Dodd, M.E., McConway, K., Potts, J. and Crawley, M. (1994) Rainfall, biomass variation, and community composition in the Park Grass Experiment. *Ecology* 75, 2430–2437.

Silvertown, J., Paul P., Edward J., Grant E., Matthew, H. and Pamela, M.B. (2006) The Park Grass Experiment 1856–2006: its contribution to ecology. *Journal of Ecology* 94, 801–814.

Smits, N.A.C., Schaminee, J.H.J. and Duuren, L.V. (2002) 70 years of permanent plot research in The Netherlands. *Applied Vegetation Science* 5(1), 121–126.

Sukumar, R., Dattaraja, H.S. and Suresh, H.S. (1992) Long-term monitoring of vegetation in a tropical deciduous forest in Mudumalai, southern India. *Current Science* 62, 608–613.

Sukumar, R., Suresh, H.S., Dattaraja, H.S. and Joshi, N.V. (1998) Dynamics of tropical deciduous forest: population changes (1988 through 1993) in a 50-ha plot at Mudumalai, Southern India. In: Dallmeier, F. and Comiskey, J.A. (eds) *Forest Biodiversity Research, Monitoring and Modelling: Conceptual Background and Old World Case Studies*. Parthenon Publishing, Paris, France, pp. 529–540.

Swamy, P.S., Sundarapandian, S.M. and Chandrasekar, P. (2000) Plant species diversity and tree population structure of a humid tropical forest in Tamil Nadu, India. *Biodiversity and Conservation* 9, 1643–1669.

Tilman, D. (1989) Ecological experimentation: strengths and logical experimentation: strengths and conceptual problems. In: Likens, G.E. (ed.) *Long Term Studies in Ecology. Approaches and Alternatives.* Springer, New York, USA, pp. 137–157.

Venkateswaran, R. and Parthasarathy, N. (2005) Tree population changes in a tropical dry evergreen forest of south India over a decade (1992–2002). *Biodiversity and Conservation* 14, 1335–1344.

Whittaker, R. and Niering, W.A. (1965) Vegetation of the Santa Catalina Mountains, Arizona: a gradient analysis of the south slope. *Ecology* 46, 429–452.
Whitmore, T.C. (1984) Plant species diversity in tropical rain forests. *Biology International* Special Issue 6, 5–7.
Whitmore, T.C., Peralla, R. and Brown, K. (1985) Total species count in a Costa Rican tropical rain forest. *Journal of Tropical Ecology* 1, 375–378.
Wilson, D.E., Cole, F.R., Nichols, J.D., Rudran, R. and Foster, M.S. (1996) *Measuring and Monitoring Biological Diversity: Standard Methods for Mammals*. Smithsonian Institution Press, Washington, DC, USA.
Zhang, W.J. (2010) *Computational Ecology: Artificial Neural Networks and Their Applications*. World Scientific, Singapore.
Zhang, W.J. (2011) Simulation of arthropod abundance from plant composition. *Computational Ecology and Software* 1(1), 37–48.
Zhang, W.J. and Barrion, A.T. (2006) Function approximation and documentation of sampling data using artificial neural networks. *Environmental Monitoring and Assessment* 122, 185–201.
Zhang, W.J. and Wei, W. (2009) Spatial succession modeling of biological communities: a multi-model approach. *Environmental Monitoring and Assessment* 158, 213–230.
Zhang, W.J., Zhong, X.Q. and Liu, G.H. (2008) Recognizing spatial distribution patterns of grassland insects: neural network approaches. *Stochastic Environmental Research and Risk Assessment* 22(2), 207–216.

19 Effects of Harvesting Plans on Tree Species Diversity: An Evaluation of Two Logged Forest Compartments

Saiful Islam*

Forest Department Headquarters, Ban Bhaban, Agargaon, Dhaka, Bangladesh

Abstract

Two adjacent forest stands or compartments were logged with different harvesting plan and intensity. There was no striking variation in elevation and terrain structure between these compartments. Also, both the compartments did not vary significantly in number of trees or volume felled. The difference in mean number of tree species in both compartments was not significant before logging. However, conventional logging reduced the species richness in heavily damaged Compartment 29 as compared to Compartment 28. The similarity of species composition between the compartments also changed from the original level. Logging has had effect on Compartment 29 due to harvesting layout with high frequency of skid tracks and their positioning. On the contrary, Compartment 28 was lightly logged with low frequency of skid tracks. Like this study, some studies also concluded that logging methods were more responsible for damage to the residual stands but not the number of trees harvested or volume felled. The study strongly suggests integration of biodiversity with the existing management system for sustainable forest management.

19.1 Introduction

Peninsular Malaysia is still fortunate to have a high percentage of forest cover and richness in tree species diversity (Whitmore, 1984; Berger, 1990; Whitmore, 1990; Ali Budin and Salleh, 1993; Bidin and Latiff, 1995). The natural forests in Malaysia are estimated to cover 58.6% of the country's total land area (Anon., 1992). As per National Forestry Policy 1978, 14.06 million ha of forest have been allocated as Permanent Forest Estate (PFE) which included Production Forest for timber production in perpetuity. Nevertheless, the current method of timber exploitation is not compatible with sustainable forest management in terms of maintenance of structure and species diversity (Wyatt-Smith, 1987; Saiful, 2002).

There is a dearth of knowledge on the biodiversity of production forest that has been subjected to logging operations; however, very few studies are available on reduced impact logging (e.g. Hendrison, 1990; Crome *et al.*, 1992; Shamsudin and Chong, 1999).

There are some studies (e.g. Jonkers, 1987; Verissimo *et al.*, 1992; Johns *et al.*, 1996; Pinard and Putz, 1996) that found no correlation between the harvesting intensity and the amount of damage. By contrast, Nicholson (1958, 1979) and Hendrison (1990) have demonstrated the relationship between logging damage and the basal area extracted. According to Whitmore (1990), the amount of damage to the forest depends more on how many trees are felled than on timber volume extracted. Incidental loss of stems as a result of severe crown and bole injury varied between different studies. Nicholson (1979) reported higher incidental losses in Sabah, Borneo, and Johns (1988) in Pahang, Malaysia, with losses of 58% and 50.9%, respectively. The extent of tree mortality has also been shown to be different among local habitat types, and found to be greatest on the ridge crest and the log landing (Johns, 1988).

*E-mail:sislam47@hotmail.com

With indiscriminate damage by logging, all rare species might be susceptible to depletion, including commercially valuable timber species (Johns, 1992; Appanah, 1999), and the logged forest may look different in species composition (Johns, 1992; Saiful, 2002). Based on available studies, there is also a lack of specific research to enable comparisons of loss of species diversity under different methods of logging.

The aim of this study was to evaluate tree species diversity of two adjacent logged compartments under different harvesting layouts and intensities in dipterocarp forests on a hill in Peninsular Malaysia. The evaluation results in recommendations for improvement of the existing management system.

19.2 Study Site and Methodology

19.2.1 Study site

The study area was situated at Sungai Weng Catchment of Ulu Muda Forest Reserve, Kedah, Peninsular Malaysia (5°50′N; 100°55′E) (Figs 19.1 and 19.2). The field survey was done in the two adjacent compartments, C28 and C29, comprising 45.0 ha of primary forest.

The elevation of the study site ranges from 340 to 550 m above sea level (Saiful, 2002) characterized by hilly and undulating terrain up to 45° slope. The vegetation is primarily mixed hill dipterocarp forest (Whitmore, 1984). The climate is consistently hot, averaging about 25°C, with

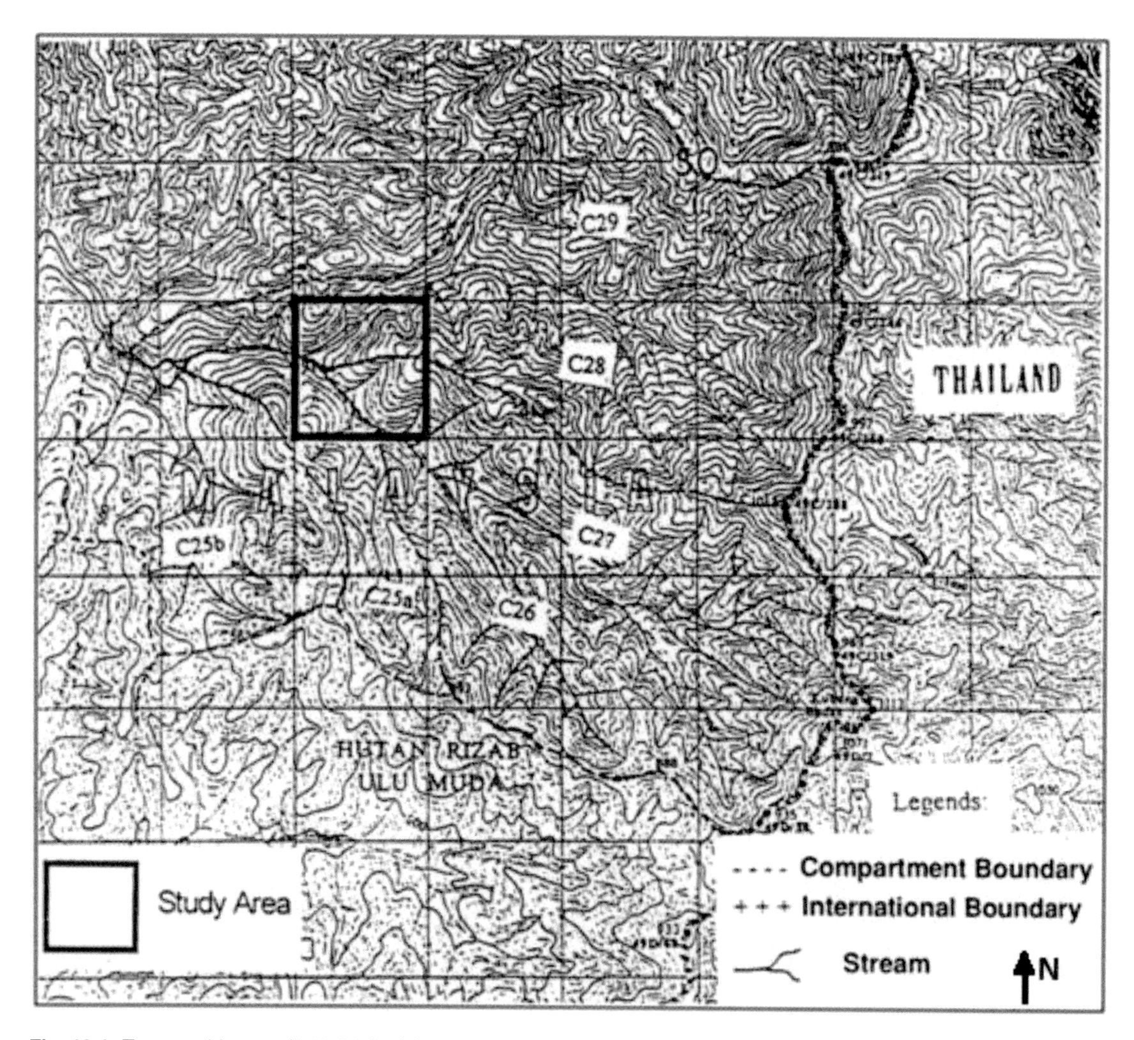

Fig. 19.1. Topographic map (1:50,000) of the study area.

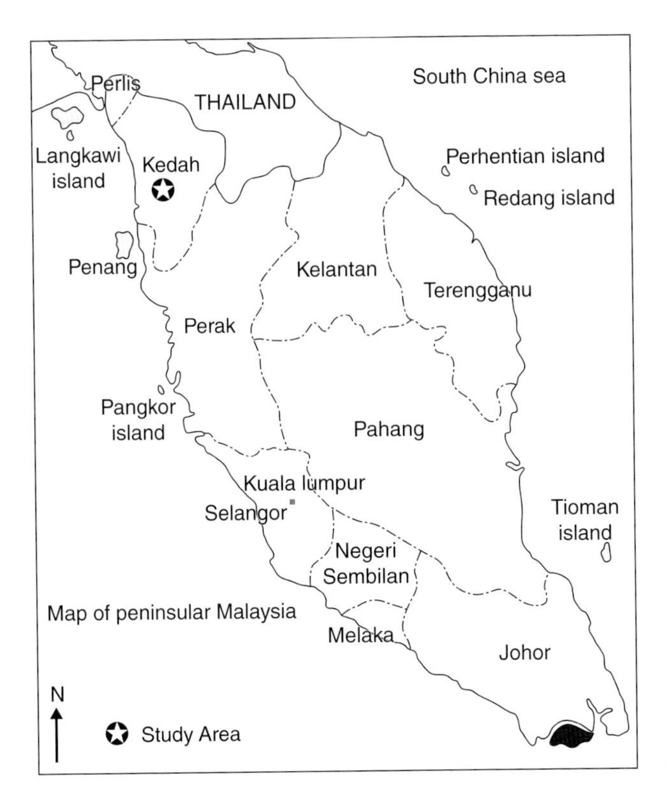

Fig. 19.2. Map of peninsular Malaysia viewing study place.

mean annual rainfall averaging 2869 mm. The parent material is primarily made up of quartzite and sandstone (RRIM, 1988). The soil is red-yellow podzolic with pH ranging from 3.2 to 4.5 (Saiful, 2002).

19.2.2 Methodology

19.2.2.1 Sampling design

A regular sampling was adopted along the gradient-directed transect (Gillison and Brewer, 1985; Austin and Heyligers, 1989) established in both compartments, and each transect originated at the stream bank, followed the centre of the ridge and finally ended at the ridge crest (Fig. 19.3).

The lateral transects had also been positioned systematically at a 40-m distance on either side of the main transect. Study plots were located at a 40-m distance on the transect line. Trees ≥20 cm diameter at breast height (dbh), poles (5.0 to <20 cm dbh) and saplings (1.5 m ht to <5.0 m dbh) were measured in different sizes of plots. Trees were measured in the main plot of 30 × 30 m or 20 × 45 m size and two kinds of nested subplot of 10 × 10 m and 5 × 5 m were distributed inside the main plot for poles and saplings, respectively.

19.2.2.2 Data collection

Trees of ≥20.0 cm dbh were measured with a diameter tape at 1.3 m above ground level or just above the buttress. A tree sample was tagged with a separate identification code. The individual tree position inside the plot was mapped. Voucher specimens were oven-dried and identified up to species level, using keys and descriptions of Malaysian flora (Whitmore, 1972; Whitmore, 1973; Ng, 1978, 1989). Specimens were also verified with the collections preserved in the herbarium of the Forest Research Institute Malaysia (FRIM).

After six months to one year of logging, resurveying of the same sample plots was carried out to detect the effects of selective logging. Tree measurements of all residual trees for different diameter classes were taken to relocate the trees. For presence of a tree, the soundness (i.e. no injury) or any type of injury was recorded as per damage assessment

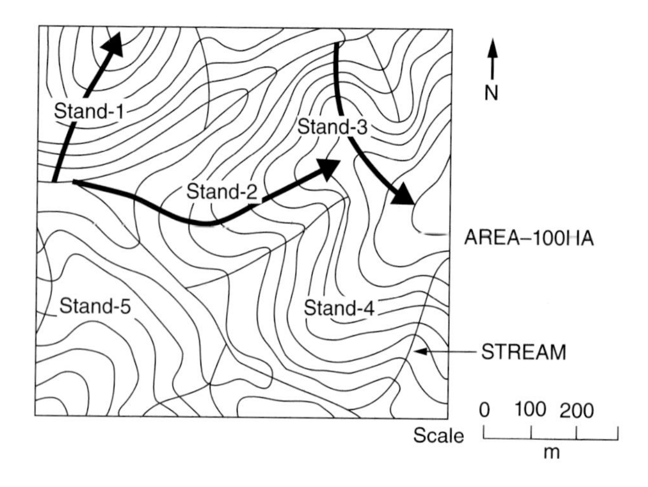

Fig. 19.3. Direction of survey transects (indicated by arrow) along elevation gradient of the study site. Stands 1, 2 and 3 are selectively logged, covering 45 hectares.

classification provided elsewhere (Johns *et al.*, 1996; Saiful, 2002).

19.2.2.3 Methods of logging

Bulldozers (D6C caterpillar) were engaged for construction of roads, skid tracks and log landings, as well as to get timber out of the stump site. Tree felling was done by a power-saw and followed the felling method of directing the fall of the tree. Ground vegetation and top soil were removed during construction of the logging tracks. However, the layout of skid tracks and their frequency varied between study compartments. In Compartment 29, skid tracks were laid out along the centre of the ridge and ridge-tops with lateral branches towards the hillside. On the contrary, in Compartment 28, the skid track was sited on the hillside and never along the centre of the ridge. The frequency of skid tracks in C28 was low compared to C29.

19.2.2.4 Data analysis

By using a histogram, the distribution of the data set was examined. To prevent a skewed data set, data were log-ten transformed for calculation of mean values. Using two-sample t-test, mean comparisons of the variables were determined, and the statistical significance levels were established at $P < 0.05$. Analysis was performed using the Minitab statistical package. Species richness and the Shannon diversity index were used to measure diversity of the logging compartments. Sorensen's similarity index was used to indicate similarity or dissimilarity of species composition.

19.3 Results and Discussion

19.3.1 Harvest levels

Tree removal in the two study compartments C28 and C29 differed (31.6 and 23.0 trees/ha, respectively), but it was not statistically significant (two-sample t-test: $t = 1.45$, $P = 0.16$, $df = 25$). No significant difference was observed in removal of the basal area and mean volume of these two compartments (Basal area: $t = 0.15$, $P = 0.88$, $df = 25$; Volume removal: $t = 0.15$, $P = 0.89$, $df = 25$). However, the overall extraction level in 2.52 ha of sampled area was 27.0 trees/ha, while the removal of basal area and volume was 19.39 m^2 and 73.7 m^3/ha, respectively. As per the existing cutting limit, the harvest levels signify more rigorous logging, and were in the higher range as reported elsewhere (Saiful, 2002).

19.3.2 Intensity of logging damage

Damage assessment was recorded after six months to one year of logging, to allow the injured trees to re-sprout or survive. The mean number of stems that were injured by logging was significantly higher in C29 than in C28 for all three size-classes, i.e. timber ($t = 2.15$, $df = 20$, $P < 0.05$); poles ($t = 2.21$, $df = 22$, $P < 0.05$), and saplings ($t = 2.82$, $df = 17$, $P < 0.05$). This was also reflected across all stems ≥5 cm diameter, and the proportion of stems smashed was much higher in C29 than in C28. In other words, the number of intact stems was much higher in Compartment 28. This variation in logging injury to trees was due mainly to the harvesting layout.

The logged forest floor area inside the study plots affected by the logging tracks was mapped (Figs 19.4 and 19.5).

The mean damaged area for all kinds of logging tracks together was significantly higher in C29 than in C28 ($t = 3.41$, $df = 17$, $P < 0.01$). Across all types of logging tracks, skid tracks alone occupied 56.6% of the total area affected, followed by log landing (27.2%) and roads (16.3%). The positioning of a high percentage of skid tracks along the centre of the ridge and ridge tops was responsible for the high proportion of ground damage in C29, where most exploitable trees dominated. In contrast, the positioning of skid tracks away from the centre of the ridge as well as their low frequency was possibly the key factor in reducing damage in C28. There are other studies (e.g. Jonkers, 1987; Verissimo *et al.*, 1992; Johns *et al.*, 1996; Pinard and Putz, 1996) that, like this study, found no correlation between the harvesting intensity and amount of damage, and concluded that logging methods were more important for damage to the residuals but not the harvesting intensity.

19.3.3 Species richness

At ≥5.0 cm dbh threshold level, there were totals of 110 and 131 tree species before logging in C28 and C29, respectively. Nevertheless, conventional logging depleted the number of species and the depletion was much higher in C29 than C28 (Table 19.1).

Skorupa (1986) documented 6.6% depletion of species in the plots of lightly logged, while the reduction was 28.9% in the heavily logged plots. The Queensland study (Crome *et al.*, 1992) revealed less damage and no loss of species.

Likewise, pole-sized tree species (≥5.0 cm to <20.0 cm dbh) did not vary between the compartments before logging (two-sample t-test: C28, $N = 16$, $Mean = 8.0$; C29, $N = 13$, $Mean = 6.77$, $P > 0.05$), and after logging, significant variation was observed with higher retention in C28 (two-sample t-test: C28, $N = 16$, $Mean = 6.13$; C29, $N = 13$, $Mean = 2.08$, $P < 0.001$). That is, logging has an effect on C29 due to the higher frequency of logging tracks and their position mainly along the centre of ridge and ridge-top, as already described.

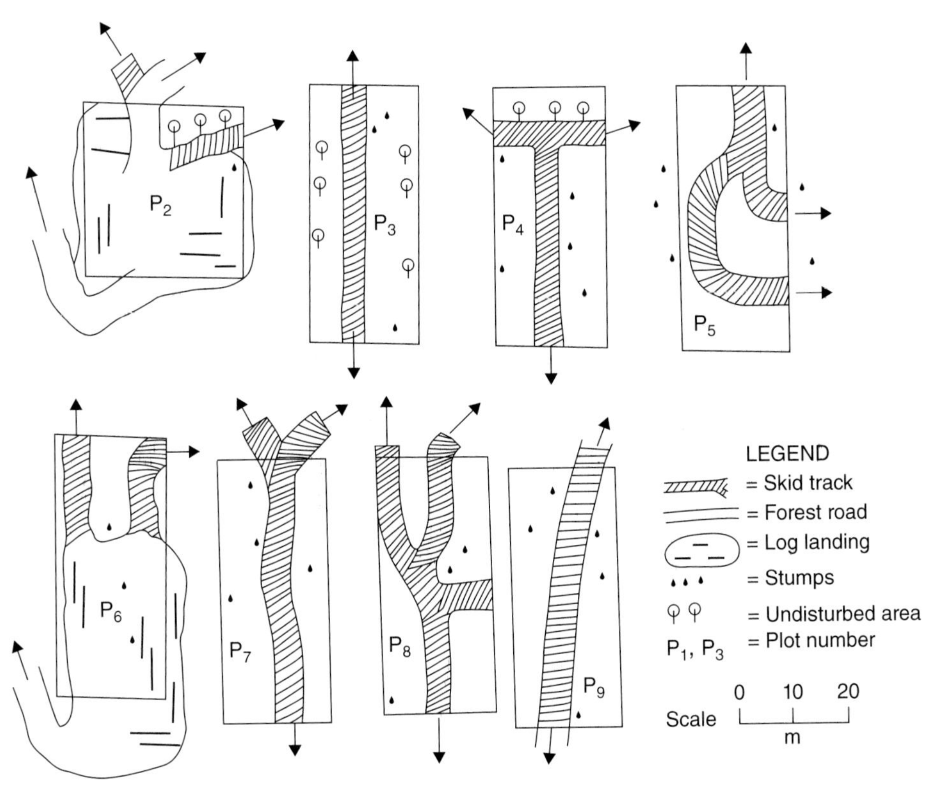

Fig. 19.4. Harvesting layout inside study plots showing positioning of logging tracks and their pattern in Compartment 29. Harvested tree stumps and undisturbed plot area are shown (see legend).

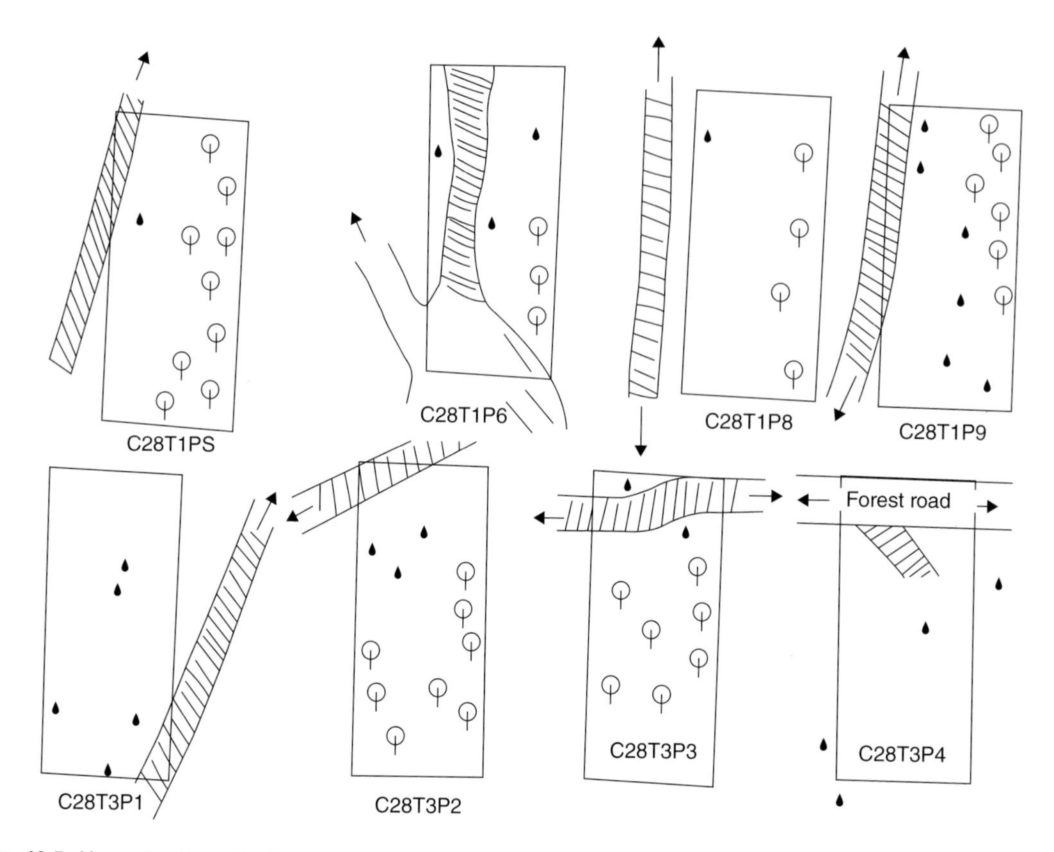

Fig. 19.5. Harvesting layout in Compartment 28 showing position of logging tracks passing at a distance from the study plots, causing minimum damage to trees. See Fig. 19.4 for legend.

Table 19.1. Effects of logging on tree species richness between the study compartments, Ulu Muda Forest Reserve, Kedah, Malaysia.

	Before logging	After logging	% Reduction in richness
Compartment 28 (Trees >5.0 cm dbh)	No of species 110	No of species 92	16.4
Compartment 29 (Trees >5.0 cm dbh)	No of species 131	No of species 69	47.33

Also, at size-class level, the mean species for trees (≥20.0 cm dbh) did not differ significantly before logging (two-sample t-test: C28, N = 15, Mean = 13.3; C29, N = 10, Mean = 16.7, P > 0.05) but much higher reduction was shown in C29 from the original level than in C28 (Table 19.2).

19.3.4 Similarity measures

Sorensen's index of similarity between the compartments also changed from 0.49 to 0.42 along with the severe reduction of the number of common species from the original level (Table 19.3). In fact, C29 was comparatively richer in number of species than C28 (i.e. 131 versus 110) as well as better off in terms of species–individual ratio (i.e. 0.45 versus 0.38) before logging, but the compartment was most affected by the logging design.

19.3.5 Species diversity

The effect of logging intensity in reduction of Shannon's species diversity was more pronounced in C29 than C28; variation in Shannon Evenness (E) was also not as prominent as in C29 (Table 19.4). A similar observation was reported by Skorupa (1986)

Table 19.2. T-test for comparison of mean species of trees (≥20.0 cm dbh) following logging in the study compartments, Ulu Muda Forest Reserve, Kedah.

Compartment	Sample size	Plot mean (species richness)	% Reduction in richness	T-test
C29				
Before logging	10	16.7	57	T-value = 4.58
After logging	10	7.20		P-value = 0.000
C28				
Before logging	15	13.33	30	T-value = 3.30
After logging	15	9.33		P-value = 0.003

Table 19.3. Sorensen's index of similarity (Cs) of species composition following selective logging for trees >5.0 cm dbh.

	Compartment 28		Compartment 29		No. species common in C28 & C29	Sorensen Cs
Before logging	No. species 110	No. stems 290	No. species 131	No. stems 289	59	0.49
After logging	No. species 92	No. stems 192	No. species 69	No. stems 109	34	0.42

Table 19.4. Effects of logging on species diversity under different logging intensity at Ulu Muda Forest Reserve, Kedah. Means are for trees ≥20.0 cm dbh.

Compartment	Sample size	Plot mean (Shannon's H)	% Reduction of Shannon's (H)	Evenness (E)
Heavily damaged C29				
– Before logging	10	2.72	36.03	0.98
– After logging	10	1.74		0.89
Lightly damaged C28				
– Before logging	15	2.46	12.61	0.97
– After logging	15	2.15		0.98

in which both Shannon and Hills evenness declined in the lightly and heavily logged study plots.

19.3.6 Effects on rare and commercial species

Due to the high percentage reduction of species in C29 (Table 19.2), the loss of rare species with one individual was also much higher than C28 (36% versus 14%, respectively). Ten families including two commercially important (i.e. Anacardiaceae and Leguminosae) lost more or less 50% of the total species. *Koompassia excelsa* and *Dialium platysepalum* of family Leguminosae have been totally lost from the species list (for trees >1.5 m ht), and the former was only one individual in C28, but logged. Conversely, among nine individuals of *Dialium platysepalum*, six were from C29, three logged and three (>20 cm dbh) were fatally damaged. Johns (1988) also reported a similar trend, and in his study some valuable legumes (*Intsia palembanica* and *Koompassia malaccensis*) were heavily harvested. Johns (1992) documented vulnerability of infrequent species to depletion from the moderately logged areas.

19.4 Conclusions and Management Concern

As the harvesting pattern and intensity were found to be the main controlling aspects, this study strongly emphasizes the proper placing of logging tracks in areas (e.g. ridge tops) where logging is subjected to special care to minimize damage due to high density of exploitable timber. Therefore, a logging layout plan should be drawn up in advance, to avoid unnecessary logging tracks in such areas.

To prevent biodiversity loss from the concession area, however, some improvements in the existing operational sequence designed for Selective Management System (SMS) in Malaysia are important. For example, a pre-felling inventory using systematic line-plots should determine the status of tree species composition and rarity using an appropriate level of sampling intensity. Besides, instead of a general cutting limit, determination of a site-specific or habitat-based cutting regime would certainly reduce the logging damage in areas with a large number of exploitable trees. In this regard, logging roads and skid tracks must be constructed, taking account of trees that should be retained, for example, 'mother trees' for seed collection or trees of rare species. This study stresses training for both forest workers and logging operators as an important contributing factor in reducing damage during operation. Incentives for best practice in logging are very important, in addition to penalties for unnecessary felling damage.

Acknowledgements

The author is obliged to IRPA Project Grant No 08-02-02-0009 for financial assistance for this study. Also to the State Forestry Department, Kedah, Malaysia, for their cooperation. Thanks are also due to the director and curator of the herbarium of FRIM for their permission to examine tree specimens.

References

Ali Budin, K. and Salleh, S. (1993) Status and approaches towards biodiversity conservation in Peninsular Malaysia. *Proceedings of the ASEAN Seminar on the Management and Conservation of Biodiversity*. Kuala Lumpur, Malaysia (mimeograph).

Anon. (1992) *Progress Report towards Sustainable Management of National Tropical Forest in Malaysia*. 13th Session of ITTC, Yokohama, Japan.

Appanah, S. (1999) Trends and issues in tropical forest management: setting the agenda for Malaysia. Paper presented at the conference on Tropical Forest Harvesting: New Technologies Examined. Forest Research Institute of Malaysia and Asian Strategy & Leadership Institute, Kuala Terengganu, Malaysia (mimeograph).

Austin, M.P. and Heyligers, P.C. (1989) Vegetation survey design for conservation: gradsect sampling of forests in north-eastern New South Wales. *Biological Conservation 50, 13–32*.

Berger, R. (1990) *Malaysia's Forests: A Resource without Future?* Packard Publishing Limited, Chichester, West Sussex, UK.

Bidin, A.A. and Latiff, A. (1995) The status of terrestrial biodiversity in Malaysia. In: Zakri, A.H. (ed.) *Prospects in Biodiversity Prospecting*. Genetics Society of Malaysia, Kuala Lumpur, Malaysia, pp. 59–73.

Crome, F.H.J., Moore, L.A. and Richards, G.C. (1992) A study of logging damage in upland rainforest in North Queensland. *Forest Ecology Management* 49, 1–29.

Gillison, A.N. and Brewer, K.R.W. (1985) The use of gradient directed transects or gradsects in national resource survey. *Journal of Environmental Management* 20, 103–127.

Hendrison, J. (1990) *Damage-controlled Logging in Managed Tropical Rain Forest in Suriname*. Agricultural University, Wageningen, The Netherlands.

Johns, A.D. (1988) Effects of 'selective' timber extraction on rain forest structure and composition and some consequences for frugivores and folivores. *Biotropica* 20, 31–37.

Johns, A.D. (1992) Species conservation in managed tropical forests. In: Whitmore, T.C. and Sayer, J.A. (eds) *Tropical Deforestation and Species Extinction*. Chapman and Hall, New York, USA, pp. 15–53.

Johns, J.S., Barreto, P. and Uhl, C. (1996) Logging damage during planned and unplanned logging operations in the eastern Amazon. *Forest Ecological Management* 89, 59–77.

Jonkers, W.B.J. (1987) *Vegetation Structure, Logging Damage and Silviculture in a Tropical Rain Forest in Suriname. Ecology and Management of Tropical Rain Forest in Suriname 3*. Wageningen Agricultural University, Wageningen, The Netherlands.

Ng, F.S.P. (ed.) (1978) *Tree Flora of Malaya*, Vol. 3. Longman Publishers, London, UK.

Ng, F.S.P. (ed.) (1989) *Tree Flora of Malaya*, Vol. 4. Longman Publishers, London, UK.

Nicholson, D.I. (1958) An analysis of logging damage in tropical rain forest, North Borneo. *Malayan Forester* XXI, 235–245.

Nicholson, D.I. (1979) *The Effects of Logging and Treatment on the Mixed Dipterocarp Forests of Southeast Asia*. Report FO. MISC/79/8. Food and Agriculture Organization of the United Nations, Rome, Italy.

Pinard, M.A. and Putz, F.E. (1996) Retaining forest biomass by reducing logging damage. *Biotropica* 28, 278–295.

RRIM (1988) *Training Manual on Soil, Management of Soils and Nutrition of Hevea*. Rubber Research Institute of Malaysia, Kuala Lumpur, Malaysia.

Saiful, I. (2002) *Effects of selective logging on tree species diversity, stand structure and physical environment of tropical hill dipterocarp forest of Peninsular Malaysia*. PhD thesis, National University of Malaysia, Bangi, Selangor, Malaysia.

Shamsudin, I.I.H. and Chong, P.F (1999) Long haulage ground cable system as an alternative technique of harvesting in hill forests of Malaysia. *Paper presented at the conference on Tropical Forest Harvesting: New Technologies Examined*. Forest Research Institute of Malaysia and Asian Strategy & Leadership Institute, Kuala Terengganu, Malaysia (mimeograph).

Skorupa, J.P. (1986) Responses of rainforest primates to selective logging in Kibale Forest, Uganda: a summary report. In: Benirschke, K. (ed.) *Pimates: The Road to Self-sustaining Populations*. Springer, New York, USA, pp. 57–70.

Verissimo, A., Barreto, P., Mattos, M., Tarifa, R. and Uhl, C. (1992) Logging impacts and prospects for sustainable forest management in an old Amazonian frontier: the case of Paragominas. *Forest Ecology and Management* 55, 169–199.

Whitmore, T.C. (1972) *Tree Flora of Malaya*, Vol. 1. Longman Publishers, Kuala Lumpur, Malaysia and London, UK.

Whitmore, T.C. (1973) *Tree Flora of Malaya*, Vol. 2. Longman Publishers, Kuala Lumpur, Malaysia and London, UK.

Whitmore, T.C. (1984) *Tropical Rain Forests of the Far East*, 2nd edition. Clarendon Press, Oxford, UK.

Whitmore, T.C. (1990) *An Introduction to Tropical Rain Forests*. Clarendon Press, Oxford, UK.

Wyatt-Smith, J. (1987) The management of tropical moist forest for the sustainable production of timber: some issues. *IUCN/IIED Tropical Forest Policy Paper* 4, 20.

20 Diversity of Angiospermic Flora of West Bengal, India

SUNIT MITRA[1] AND SOBHAN K. MUKHERJEE[2*]

[1]*Department of Botany, Ranaghat College, Ranaghat, Nadia, West Bengal, India;* [2]*Department of Botany, University of Kalyani, Kalyani, Nadia, West Bengal, India*

Abstract

West Bengal, 87,616 km^2 in size, is the 13th-largest state in India, situated at 21° 10′ to 27° 35′ N latitude and 85° 50′ to 89° 52′ E longitude. For administration purposes, the state is divided into 20 districts comprising more than 400 blocks and 4300 villages. Forests of West Bengal state can be categorized broadly in to six major types. Total recorded forest land in the state is 11,879 km^2, which constitutes 13.38% of the geographical area of the state. Of the 11,879 km^2 forested area, 7054 km^2 is Reserve Forest, 3772 km^2 is Protected Forest and 1053 km^2 is Unclassified Forest. Floristic composition of the state is still unknown to us and the preparation of information about the flora of the state is a long-time demand. In the present study, an attempt has been made to study the flora of the state, which reveals that the state possesses 4493 taxa of Angiosperms under 1578 genera belonging to 232 families. West Bengal has a wetland area of about 3,44,527 ha: 22 natural wetlands and one man-made cover more than 100 hectares of area. Besides these, there are about 57,000 small ponds with individual land area of less than 2.25 ha. This huge aquatic land area possesses 770 taxa of aquatic plants which belong to 456 genera under 159 families. There are 35 species belonging to 34 genera under 22 families, which are confined to the political boundary of West Bengal and regarded as the endemic plants of the state. Beside these, there are another 25 species of Angiosperm belonging to 22 genera under 13 families, which are endemic to West Bengal and are also found in other Indian states. In total, the state West Bengal contains 57 species of endemic plants belonging to 53 genera under 30 families. In the present study, after a critical perusal of the literature it has been observed that the state possesses 489 exotic taxa, which is 11.03% of the total floristic element of the state, and they belong to 18 different floristic regions of the world. The flora of West Bengal state are under immense pressure due to several biotic and abiotic factors, as a result of which the floristic diversity of the state leads towards many species being threatened with extinction. The flora of the Darjeeling district, part of the eastern Himalayan Hotspots region, face destruction due to tourist activities and increasing need for land development, for establishment of tea gardens, for industry, etc.; this has led to the absence of several floristic elements from the list of the state flora. In this connection, *Aldrovanda vesiculosa* is noteworthy. Different *ex-situ* and *in-situ* conservation strategies have been considered at government level to protect the flora of the state. As of 2007, there are five national parks in the state, 18 wild life sanctuaries, two biosphere reserves and one UNESCO World Heritage site, which play a part in the conservation process of the state flora and fauna. Beside these, there is one Ramsar site (wetland of international importance: East Calcutta Wet Land) which works for conservation of aquatic flora and fauna.

20.1 Introduction

The state of West Bengal was founded on 15 August 1947, with the independence of India from British Rule, at the cost of partition of the country. During the process of partition of British India, Bengal Province was portioned in an East–West direction by the Radcliffe Line. The western part of this arbitrary divisional line, comprising one-third of the original Bengal province, was incorporated into India as a state and renamed as West Bengal. The eastern half was incorporated into Pakistan and is now Bangladesh.

*E-mail: sobhankr@yahoo.com

 Plant Biodiversity: Monitoring, Assessment and Conservation (eds A.A. Ansari, S.S. Gill, Z.K. Abbas and M. Naeem)

20.1.1 Geographical location and political boundary of West Bengal

West Bengal, 87,616 km^2 in size, is the 13th-largest state in India, situated at 21° 10′ to 27° 35′ N latitude and 85° 50′ to 89° 52′ E longitude, covering an area of 87,616 km^2. The state is bounded by the rolling waves of the Bay of Bengal at its southern part, at the eastern side by Bangladesh, at the north by Assam, Sikkim, Nepal and Bhutan, and at the west by Bihar, Jharkhand and Orissa (Fig. 20.1).

For administration purposes, the state is divided into 20 districts, comprising more than 400 blocks and 4300 villages. A conspectus of these 20 districts can be found in Table 20.1.

20.1.2 Geology

Geologically, the major land mass of the state is a thick alluvial deposition, apart from the Himalayan sector of the Darjeeling district and spurs on the western boundaries.

20.1.3 Soil

The predominant part of the state (28,921.3 km^2) is made up of the alluvial type of soil. West Bengal has five broad categories of soil types, which are as follows:

1. Alluvial type
2. Lateritic type
3. Red soil
4. Terai soil
5. Tidal soil.

20.1.4 Climate

The climate of the state is tropical in the plains area and subtropical to temperate in the Terai–Himalayan region of the state. In the winter season, the night temperature falls to subzero levels at different part of the Darjeeling districts and there is snow in areas such as Ghoom, while other parts of the Darjeeling district experienced a minimum temperature of 2–3°C at night and 9–12°C in the daytime. During the summer, the temperature may rise to 30°C and the mean average summer temperature is 15–21°C. Darjeeling district is hotter than the other districts of West Bengal, and the temperature varies from 8–22°C during the winter to 21–39°C in the summer. The western part of the state has the hottest summer and the coldest winter, but in other parts of the state the temperature variation is not so wide.

The state has a high level of rain in the northern part, and is drier at its western fringe. The annual rainfall of the state ranges between 150–200 cm. Some parts of the Darjeeling district have an average rainfall of up to 450 cm per annum, and in parts of the Purulia district the annual average rainfall is 60–80 cm.

20.1.5 Geographical/botanical division

Geographically, the state is divided into five broad geographical regions (Chakravarty *et al.*, 1999), which are as follows:

1. The Darjeeling Himalayan Region
2. Terai–Duars region
3. Western undulating high land and plateau
4. North and South Bengal plains
5. Gangetic delta (Sundarbans).

Prain (1903) in an introductory note described 11 botanical zones, of which the present West Bengal possesses four.

20.1.6 Vegetation of the state

Total recorded forest land in the state is 11,879 km^2, constituting 13.38% of the geographical area of the state. Of the 11,879 km^2 forested area, 7054 km^2 is Reserve Forest, 3772 km^2 is Protected Forest and 1053 km^2 is Unclassified Forest.

Forests of West Bengal state can be categorized broadly as six major types. These six types of forests, their areas in the state, and the major species found in these forests are given in Table 20.2 below.

In the present study, forests are classified into ten broad classes, based on the forest classification system of Champion and Seth (1968) with some modification (Figs 20.2 and 20.3).

20.1.7 Grass land

Besides the abovementioned forest types, another important type of vegetation is grassland. It is the characteristic feature of the riverine areas, which tend to flood during the rainy seasons, but remain dry for the rest of the year. The moist habitat and the nutrient-rich soil of this region support the luxuriant growth of tall grasses like *Saccharum* spp., *Themeda spp.*, *Narenga* spp., *Erianthus* spp., *Sclerostachya* spp., *Phragmites* spp. and *Paspalum* spp. Along with the grass species, others are *Acacia*

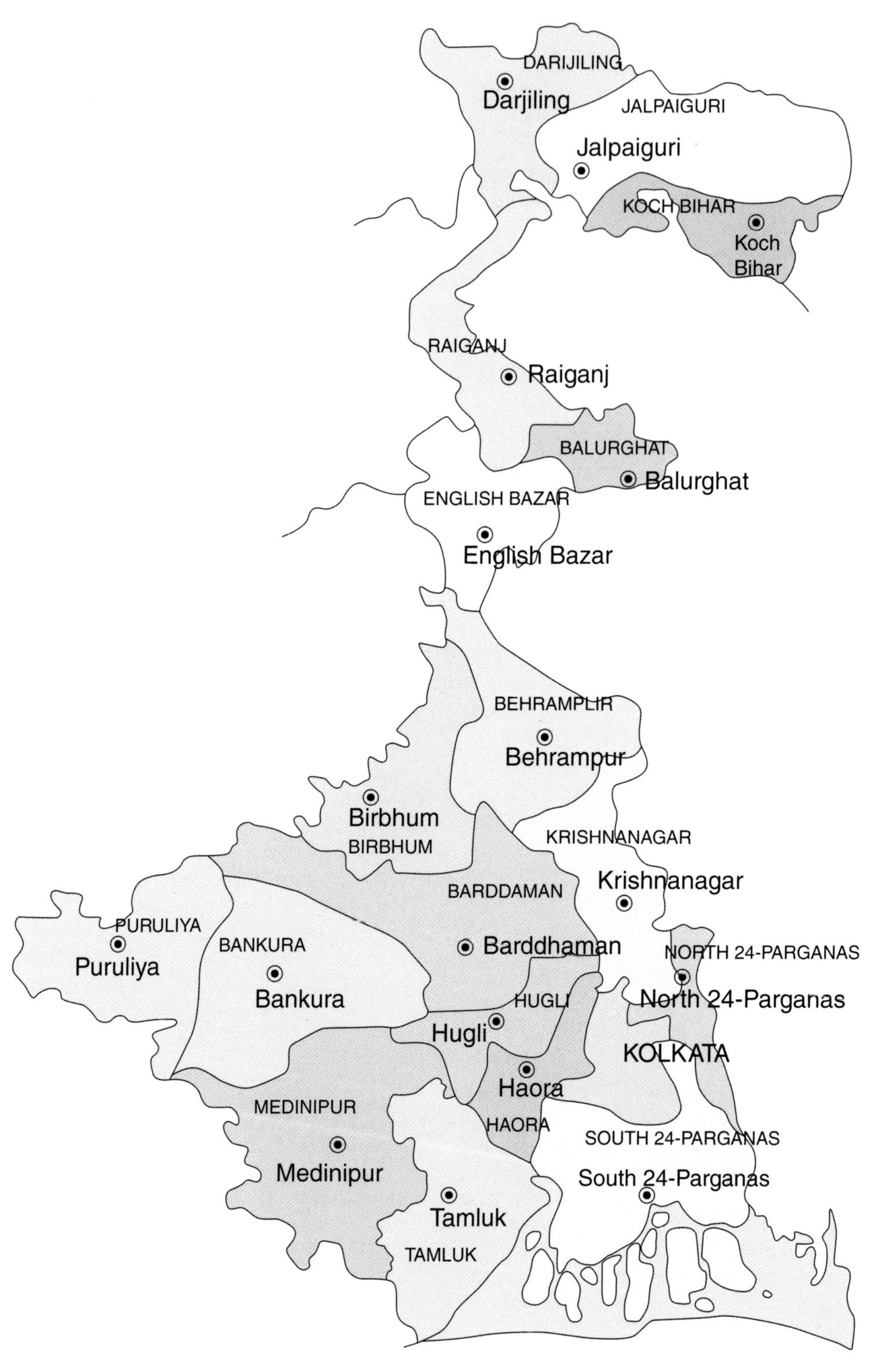

Fig. 20.1. Political map of West Bengal State (excluding newly formed Alipurduar District).

Table 20.1. Statistical conspectus of the districts of West Bengal.

Sl no.	Name of the district	Area (km²)	District headquarters	Population (as per 2011 census report)*			Population density/ km²*
				Male	Female	Total	
1	Darjeeling	3,149	Siliguri	934,796	907,238	1,842,034	585
2	Jalpaiguri	6,227	Jalpaiguri	1,980,068	1,889,607	3,869,675	621
3	Cooch Bihar	3,387	Cooch Behar	1,453,590	1,369,190	2,822,780	833
4	Uttar Dinajpur	3,140	Raiganj	1,550,219	1,450,630	3,000,849	956
5	Dakshin Dinajpur	2,219	Balurghat	855,104	815,827	1,670,931	753
6	Maldah	3,733	Malda	2,061,593	1,936,377	3,997,970	1,071
7	Murshidabad	5,324	Baharampur	3,629,595	3,472,835	7,102,430	1,334
8	Birbhum	4,545	Seuri	1,791,017	1,711,370	3,502,387	77
9	Barddhaman	7,024	Bardhaman	3,975,356	3,748,307	7,723,663	1,100
10	Nadia	3,927	Krishnanagar	2,655,056	2,513,432	5,168,488	1,316
11	North Twenty Four Parganas	3,927	Barasat	5,172,138	4,910,714	10,082,852	2,463
12	Hoogli	3,149	Chinsurah	2,819,100	2,701,289	5,520,389	1,753
13	Bankura	6,882	Bankura	1,840,504	1,755,788	3,596,292	523
14	Puruliya	6,259	Purulia	1,497,656	1,430,309	2,927,965	468
15	Hawrah	1,467	Hoerah	2,502,453	2,339,185	4,841,638	3,300
19	Kolkata	185	Kolkata	2,362,662	2,124,017	4,486,679	24,252
17	South Twenty Four Parganas	9,960	Alipore	4,182,758	3,970,418	8,153,176	819
18	Paschim Medinipur	9,345	Kontai	3,032,630	2,910,670	5,943,300	636
19	Purba Medinipur	4,736	Tamluk	2,631,094	2,463,144	5,094,238	1,076
20	Alipurduar	3,383	Alipurduar	NA	NA	About 17 lakh	22

*Source: State population report, 2011.

Table 20.2. Conspectus of forests of West Bengal. (Source: State forest report 2011–2012.)

Sl no.	Forest types	Area covered in the state (million hectares)	Area covered in india (million hectares)	Major species
1	Tropical moist deciduous forests	0.459	23.245	*Shorea robusta, Michelia champaca, Lagerstroemia parviflora, Terminalia belerica, Chukrasia velutina*
2	Tropical dry deciduous forests	0.43	29.149	*Shorea robusta, Anogeissus latifolia, Boswellia serrata, Terminalia belerica, T. tomentosa*
3	Sub-tropical broadleaved hill forests	0.005	0.287	*Schima wallichi, Castanopsis indica, Phoebe attenuata, Castanopsis tribuloides*
4	Montane wet temperate forests	0.005	1.613	*Quercus* spp., *Acer* spp., *Machilus* spp., *Michelia* spp.
5	Littoral and swamp forests	0.279	0.671	*Ceriops* spp., *Avicennia* spp., *Rhizophora candelaria*
6	Alpine forests	0.005	1.79	*Tsuga brunoniana, Picea* spp., *Abies densa, Quercus* spp., *Juniperus* spp., *Rhododendron* spp., *Betula utilis*

ferrugiana, Adina cordifolia, Leea crispa, L. macrophylla, Bauhinia malaqbarica, Bombax ceiba, Ziziphus mauritiana, etc.

20.2 Origin of the Research Problem

Inventorying of plant diversity is a great need for development of a country and to benefit its citizens.

Fig. 20.2. 1 - Kanchanjangha, highest peak of the state. 2 - View of Teesta River. 3 - A tea garden worker on the way to work. 4 - A panoramic view of the western plateau of the state.

West Bengal is biologically very rich, as it contains Darjeeling district within its political boundary, one of the Eastern Himalaya Hotspots of the Indian subcontinent. But, until a century or so ago, it was unfortunate in not possessing any floristic account of its own.

David Prain (1903) performed a herculean task. He published the floristic account of the then Bengal province under British rule in his publication *Bengal Plants* in two volumes. In his study, Prain included 2771 species under 1124 genera belonging to 151 families. To date, it is the only list of plants of the state. But in the intervening century, the political and geographical position, and the location of the present West Bengal are very different to the Bengal province in David Prain's day. With the partition of India during its independence, a major portion of Bengal province was separated from India and at present it exists as an independent country, Bangladesh. Present West Bengal state consists of the four botanical zones of the 11 botanical zones as considered by Prain (1903) in his floristic conspectus. Besides these, Prain himself considered that his flora was not a complete account of the plants of Bengal because he kept apart the flora of Darjeeling hills and the Sundarban regions of the state. Later on, Prain (1905) studied and published an account of the flora of the Sundarbans in *Records of the Botanical Survey of India*, but the flora of Darjeeling district remains untouched.

An account of the plants of Bengal state is badly needed, because neighbouring Bangladesh possesses its own flora, with up-to-date nomenclature (Pasha and Uddin, 2013), but still the state of West Bengal lack its own floristic documentation.

Fig. 20.3. 5 - Sal forest of Bankura district. 6 - *Datura stramonium*, a neglected plant having medicinal properties. 7 - *Lagerstroemia speciosa*, a common ornamental avenue tree of some parts of Bengal.

20.3 Scrutiny of Literature

Efforts were made by several workers starting with Clushaw (1952) to study the floristic composition of West Bengal and a long list of other publications followed. A bibliographic account of the state has been published by Mitra *et al.* (2010), after which several important articles on the flora and vegetation of West Bengal state were published, including Singh *et al.* (2010), Sujay *et al.* (2010), Chowdhury and Das (2010), Chowdhury and Das (2011a, b), Chowdhury *et al.* (2011), Chowdhury and Das (2013), Biswas *et al.* (2013), Rai *et al.* (2011), Rao *et al.* (2013), Krishna *et al.* (2014), Shukla *et al.* (2014), Yonzone *et al.* (2014) and Moktan and Das (2014). Perusal of these revealed that, with the exception of *Flora West Bengal*, Vol. I, there is no consolidated publication about the flora of West Bengal. However, more or less all flora of all the districts of the state have been studied in the assorted publications.

Recently, Chakravarty *et al.* (1999) has published an account of the statistical analysis of the flora of West Bengal, with 3589 species under 1333 families under 165 orders. But that work was an incomplete enumeration of the plants of Bengal, so still we are in the dark regarding the statistics of the flora of West Bengal. Keeping this in mind, this present work has been taken in to account to provide a complete statistical picture of the floristic composition of the state.

20.4 Materials and Methods

The present work is entirely based on screening of the collected specimens deposited at different state herbariums, such as the Central National Herbarium of Botanical Survey of India (CAL), Herbarium of Indian Museum Industrial Section (BSIS), Herbarium of Calcutta University, Herbarium of Eden Gardens Calcutta, Herbarium of Lloyd Botanical Garden, Herbarium deposited at the University of Kalyani (KAL), Herbarium deposited at K.N. College Murshidabad, some stray collections deposited at the Carrey Museum of Serampore College, etc. and scrutiny of the published literature and the field observations carried out by the authors for the last 20 years in different parts of the state.

20.4.1 Statistical analysis of the state flora

The state of West Bengal shows a spectacular plant diversity, with flora ranging from the littoral forest of the Sundarbans to the temperate vegetation of the Darjeeling district, along with the dry deciduous scrub vegetation of the Western stretch of the state (Figs 20.4 and 20.5).

In the present study it has been observed that the state possesses 4493 taxa belonging to 1578 genera under 232 families. The previous study of Chakravarty *et al.* (1999) has reported 3580 taxa under 1333 genera belonging to 200 families. The present statistical conspectus of the floristic account of West Bengal is given in Table 20.3.

A comparative list of ten dominant families, based on the number of genera, is enumerated in Table 20.4, along with the same as observed by Chakravarty *et al.* (1999). A list based on the numbers of species (including infra-specific taxa) is given in Table 20.5 below, along with the same as observed in the previous study by Chakravarty *et al.* (1999). This shows that the top position in this list is occupied by the family Orchidaceae, which was at seventh position in the previous list.

Fig. 20.4. 8 - Mountain ranges of Bankura. 9 - Dry deciduous forest of Duars. 10 - *Ipomoea capparis*, a mangrove plant from the Sundarban region. 11 - *Grangea maderaspatana*, an uncommon plant of Asteraceae.

Fig. 20.5. 12 - *Ficus benghalensis*, a common sacred plant of West Bengal. 13 - *Ficus religiosa*, a common sacred plant of West Bengal. 14 - *Sterculia foetida*, a plant with pungent flowers. 15 - *Saraca asoca*, a common medicinal plant of therapeutic value.

Table 20.3. Analysis of the floristic account of West Bengal.

Present study				Observation made by Chakravarty *et al.* (1999)			
Rank of the taxa	Magnoliopsida	Liliopsida	Total	Rank of the taxa	Dicot	Monocot	Total
Family	188	44	232	Family	165	35	200
Genera	1176	399	1578	Genera	1014	319	1333
Species	3147	1346	4493	Species	2641	939	3580

From the table it has been observed that the ratio of the family, genera and species belongs to the group Magnoliopsida and Liliopsida is 4.27:1, 2.94:1 and 2.34:1 respectively, in comparison to 5:1 in the family level and 3:1 in genera and species level, as observed by Chakravarty *et al.* (1999), in Indian context the ratio of the dicot and monocot genera is 3.25:1 and dicot to monocot species ratio is 3:1, respectively (Figs 20.6 and 20.7).

The family Gramineae is second, from its first position in the previous list.

One aspect of studying the floristic diversity is to study the monotypic families and monotypic genera of the floristic components of any region. In the floristic composition of West Bengal, there are 15 monotypic families, comprising 6.43% of the total family numbers of the state. Of these, ten (5.31%) belong to the group Magnoliopsida and five (11.36%) to the group Liliopsida. Of the 1575

Table 20.4. Dominant families of the state based on generic diversity.

Ten dominant family based on generic diversity under present study			Ten dominant family based on generic diversity as recorded by Chakravarty *et al.* (1999)		
Sl No.	Name of the family	Number of genera	Sl No.	Name of the family	Number of genera
1	Poaceae	133	1	Poaceae	150
2	Orchidaceae	111	2	Fabaceae (*s.l.*)	97
3	Asteraceae	99	3	Asteraceae	79
4	Fabaceae (*s.l.*)	98	4	Acanthaceae	46
5	Acanthaceae	49	5	Rubiaceae	44
5a	Rubiaceae	49			
6	Lamiacaea	39	6	Orchidaceae	43
7	Euphorbiaceae	37	7	Euphorbiaceae	37
8	Scrophulariaceae	31	8	Asclepiadaceae	29
9	Asclepiadaceae	29	9	Scrophulariaceae	24
9a	Cucurbitaceae	29			
10	Apocynaceae	23	10	Apocynaceae	23

Table 20.5. Dominant families of the state based on species diversity.

Ten dominant family based on species diversity under present study			Ten dominant family based on species diversity as recorded by Chakravarty *et al.* (1999)		
Sl No.	Name of the family	No. of species	Sl No.	Name of the family	No. of species
1	Orchidaceae	467	1	Poaceae	433
2	Poaceae	393	2	Fabaceae (*s.l.*)	324
3	Fabaceae (*s.l.*)	331	3	Asteraceae	133
4	Asteraceae	229	4	Cyperaceae	125
5	Cyperaceae	139	5	Acanthaceae	120
6	Acanthaceae	131	6	Rubiaceae	116
7	Rubiaceae	119	7	Orchidaceae	112
8	Euphorbiaceae	116	8	Euphorbiaceae	111
9	Lamiaceae	90	9	Scrophulariaceae	64
10	Scrophulariaceae	71	10	Convolvulaceae	57

genera, 59 genera (3.74%) are monotypic genera reported from the state. A list of the monotypic families and 17 representing monotypic genera are given in Tables 20.6 and 20.7 respectively (Figs 20.8, 20.9 and 20.10).

20.4.2 Exotic flora of the state

Plant invasion is a natural phenomenon. With the expansion of civilization, numerous exotic species are introduced into a new area, perhaps as an economically important cultivated species, as an ornamental plant, or as a weed. Later on these species become a part of the floristic elements of that area and are recognized as naturalized species. There are some exotic plants, which, after entering a new region gradually destroy the natural vegetation and become the chief floristic composition of that region: these are known as invasive species. In the Biodiversity Convention 1992, these Invasive Alien species are considered as severe a threat to biodiversity as that posed by habitat destruction.

Much work has been done on the exotic flora of the state: recently Das (2002), and Das and Ghosh (2011) recognized 114 species of plants as naturalized or semi-naturalized from the Darjeeling region, of which 99 species belong to the dicotyledon group and the remaining 15 to the monocotyledon group.

Bhattacharyya (1997) in *Introductory Essay on Flora of West Bengal* listed the exotic weeds of the state in eight categories, whereas Chakravarty *et al.* (1999) listed seven types.

Fig. 20.6. 16 - *Phoenix sylvestris*, a common plant used for its sugar yield in villages. 17 - *Trewia nudiflora*, used for matchstick and matchbox production. 18 - *Moringa oleifera*, widely cultivated for its fruits. 19 - *Melia azedarach*, an uncommon plant with termite-resistant properties.

In the present study, after a critical perusal of the literature, it has been observed that the state possesses 489 exotic taxa, which is 11.03% of the total floristic element of the state, and they belong to 18 different floristic regions of the world. A conspectus of these 18 groups of the floristic elements, their representative taxon and their representative percentage, is given in Table 20.8.

20.5 Endemic Flora of West Bengal

Observing the large numbers of exotic plants in the composition of the Indian flora, Hooker (1907) commented that India did not possess any flora of its own. Later on, Chatterjee (1940) and Chatterjee (1962) estimated that India has about 60% of floristic elements which are Indian in origin, of which about 40% are endemic to the country.

As mentioned earlier, the endemic taxa of a region indicate its floristic richness and also the biogeographic character of that region. In this connection, there is now a need to make a proper documentation of the endemic flora of West Bengal.

The word 'endemic' was first coined by de Candolle (1855), and means a taxonomic unit of any rank or taxa of any organism, which is restricted in distribution and remains confined to a particular area, isolated from its surrounding region through geographical or temporal barriers.

According to Engler (1882), there are two types of endemics – one is based on the preservation of the ancient forms and the other one is the development of the entirely new forms. Various types of endemics that are popularly used in phytogeography have been described by Richardson (1978), Nayar (1996) and Ahmedullah (2000).

Fig. 20.7. 20 - *Carissa carandus*, a thorny shrub with edible fruits. 21 - *Gmelina arborea*, commonly used for construction purposes. 22 - *Ranunculus sceleratus*, an aquatic weed of the primitive family Rarunculaceae. 23 - *Limonia acidissima*, an uncommon fruiting plant of Bengal.

Normally, it is the trend that endemism is high in those regions where the climatic condition is extreme or the place is isolated from the adjoining areas by geographical barriers. In West Bengal, the climatic conditions are not extreme in any sense, nor is it separated from the adjoining areas by any natural barriers. So it is predictable that this place possesses a very low level of endemism in its floristic composition. But an exception to this is in the hilly tract of the Darjeeling district, part of the Eastern Himalaya and considered as one of the hotspots of Indian biodiversity. This is a centre of plant speciation and endemic flora.

Prain (1903) did not comment on the endemic flora of Bengal province. The earliest available publication on any part of the Eastern Himalaya was the list of plants published by Don (1821). Soon after came the publication of *Prodromus Florae Nepalensis* (Don, 1825). Hooker's (1875–1897) *The Flora of British India* also covered this region. Later on, a large number of publications appeared on the flora of this region, including Gamble (1896), Hooker (1849), King and Pantling (1898), Cowan and Cowan (1929), Biswas (1966), Hara (1966), Hara (1971), Ohashi (1975), Hara *et al.* (1978), Hara and William (1979), Hara *et al.* (1982), Grierson and Long (1983), Grierson and Long (1984), Grierson and Long (1987), Grierson and Long (1991), Grierson and Long (1999), Grierson and Long (2001), Noltie (1994), Noltie (2000), Pearce and Cribb (2002), A.P. Das (On the floristic and palynological survey of Darjeeling and adjoining places, University of Calcutta, Kolkata, West Bengal, India, 1986, unpublished thesis, Vols 1 and 2),

Table 20.6. Conspectus of the monotypic families of West Bengal.

Sl No.	Name of the family	Generic diversity			Species diversity		
Magnoliopsida (dicotyledons)		State	India	World	State	India	World
1	Aquifoliaceae	1	1	1	6	30	400
2	Cardipteridaceae	1	1	1	1	1	2
3	Grossulariaceae	1	X	1	1	X	200
4	Moringaceae	1	1	1	2	2	13
5	Nelumbonaceae	1	1	1	1	1	2
6	Nyctanthaceae	1	1	Treated within Oleaceae	1	1	Treated within Oleaceae
7	Nymphyeaceae	1	1	Treated within Nelumbonaceae	4	7	Treated within Nelumbonaceae
8	Sphenocleaceae	1	X	1	1	X	2
9	Stachyuraceae	1	1	1	1	1	6
10	Tropeolaceae	1	1	1	1	1	90
Liliopsida (monocotyledons)							
11	Aponogetonaceae	1	1	1	2	8	45
12	Cannaceae	1	1	1	1	4	10
13	Flagellariaceae	1	1	1	1	1	4
14	Ruppiaceae	1	1	1	1	2	7
15	Typhaceae	1	1	1	2	3	12

Table 20.7. Conspectus of the monotypic genera of West Bengal.

Sl No.	Name of the monotypic genus of state	Family
1	*Amherstia**	Fabaceae (*s.l.*)
2	*Blumeopsis*	Asteraceae
3	*Casesulia*	Asteraceae
4	*Cavea*	Asteraceae
5	*Chrysanthellum*	Asteraceae
6	*Eleutheranthera*	Asteraceae
7	*Goniocaulon*	Asteraceae
8	*Hemistipta*	Poaceae
9	*Hygrorhyza*	Poaceae
10	*Lablab*	Fabaceae (*s.l.*)
11	*Ochthocloa*	Poaceae
12	*Tamarindus*	Fabaceae (*s.l.*)
13	*Thysaenolaena*	Poaceae
14	*Foeniculum*	Apiaceae
15	*Bischofia*	Euphorbiaceae
16	*Ricinus*	Euphorbiaceae
17	*Trewia*	Euphorbiaceae

*This is a cultivated species, so it is not incorporated in the counting of the floristic elements of the state.

Das and Chanda (1986), R.B. Bhujel (Studies on the dicotyledonous flora of Darjeeling district. University of North Bengal, Darjeeling, West Bengal, India, 1996, unpublished thesis) and A.K. Samanta (Taxonomical and phytosociological studies on the Angiospermic climbers of Darjeeling and Sikkim Himalayas, University of North Bengal, Darjeeling, West Bengal, 1998, unpublished thesis). In fact, more and more people are now taking an interest in detailed floristic studies of this region.

Besides these, some other recent publications of this region which were used to prepare the present account of the endemic taxa of the state are Ball (1869), Bhattacharyya (1997), Bremkamp (1959), Chakravarty *et al.* (1999), A.P. Das (On the floristic and palynological survey of Darjeeling and adjoining places, University of Calcutta, Kolkata, West Bengal, India, 1986, unpublished thesis, Vols 1 and 2), Das (2004), Das and Chanda (1986) and Das and Lahiri (1997).

Similar studies are not available for other parts of Eastern Himalaya. However, north-eastern states became centres of exploration. Kanjilal *et al.* (1934–1940) and Bor (1940) published *Flora of Assam*, covering an area including Assam, Meghalaya and Arunachal Pradesh, in *Flora of Tripura*, *Flora of Manipur*, *Flora of Mizoram*, *Materials for the Flora of Arunachal Pradesh*, *Flora of Jowai*, *Forest Flora of Meghalaya*, *Flora of Namdapha*, *Orchids of Arunachal Pradesh*. Das and Lama (1992) and Bhujel and Yonzone (1994), in their works on the floristic composition of the Darjeeling hills, show a good number of endemic plants from this region. In

Fig. 20.8. 24 - *Alternanthera tenella*, an obnoxious exotic weed from tropical America. 25 - *Calotropis gigantea*, a common weed of drier parts of the state. 26 - *Rivinia humilis*, an exotic uncommon weed from tropical America. 27 - *Pterospermum acerifolium*, an indigenous plant cultivated for its aromatic flowers and medical value.

the recent past, Chakravarty *et al.* (1999) claim that the state has eight endemic taxa to its credit, although other studies have cast doubt on this figure.

Bhujel and Das (2002) and Das (2004) estimate that about 21.26% dicot plants of the floristic composition of the Darjeeling districts are endemic. Studying the endemic flora is best done by concentrating on endemic plants in particular categories. Normally, categorization is based on the extent of distribution of the endemic plants.

In the present study, the endemic plants of West Bengal state have been studied in three categories, which are:

1. plants restricted within the political boundary of the West Bengal state only (true endemics)
2. plants endemic to India as well as in West Bengal state
3. plants previously considered as endemic to the state but which have lost their endemic status as they are found in adjoining countries.

It has been found that at present there are 35 species belonging to 34 genera under 22 families which are confined to the political boundary of West Bengal. Hence, these 35 species of Angiosperm are endemic to West Bengal. Besides these, there are another 25 species of Angiosperm belonging to 22 genera under 13 families, which are endemic to West Bengal but also found to be present in other Indian states. So only in a broad sense are they considered endemic to West Bengal. In total, the state of West Bengal contains 57 species of endemic plants belonging to 53 genera under 30 families. It has also been observed that 23 species belonging to

Fig. 20.9. 28 - *Haldina cordifolia*, an uncommon woody plant with yellow wood. 29 - *Hygroryza aristata*, a common floating grass of Bengal. 30 - *Rauvolfia serpentine*, a valuable root and rhizome drug of the state. 31 - *Tectona grandis*, a valuable timber-yielding plant.

15 genera have lost their endemic status because of their extended distribution, or because they have changed their taxonomic status by merging with other taxa.

20.5.1 Aquatic vegetation

West Bengal possesses about 3,44,527 ha of wetland. In the state, there are about 22 natural wetlands and one man-made, covering more than 100 ha. Besides these, there are about 57,000 small ponds with individual land area of less than 2.25 ha. All these 22 natural lakes are given in Table 20.9 below.

These aquatic bodies of the state harbour about 770 aquatic taxa belonging to 456 genera under 159 families, which is about 17.13% of the total taxa of the state and about 73% of the Indian aquatic vascular plants mentioned by Cook (1996).

20.5.2 Threats to the flora

Threats to the flora of the West Bengal state are same as with other flora, as follows:

- Depletion of vegetation due to high population pressure
- Requirement of land for expansion of agricultural fields
- Excessive collection of Non Timber Forest Products (NTFP)
- Excessive collection of medicinal, aromatic and other plants with economic potential
- Uncontrolled grazing in the forests and clearing of the forest floor, preventing sapling formation and reducing humus deposition
- Wrong plantation programme like the plantation of the *Cryprtomeria* sp. in Darjeeling or

Table 20.8. Conspectus of exotic floristic elements of West Bengal.

Sl No.	Floristic region	Total no. of taxa present	Percentage of representation (%)
1	African Elements	32	6.54
2	Asian elements	23	4.70
3	Australian elements	24	4.89
4	Central American elements	44	8.99
5	Central Asian elements	18	3.68
6	Chinese elements	21	4.29
7	Eurasian elements	17	3.47
8	European elements	31	6.33
9	Japanise elements	15	3.06
10	Malayasian elements	21	4.29
11	Mediterranean elements	29	5.93
12	Mexican elements	19	3.88
13	North American elements	21	4.29
14	North Temperate elements	15	3.06
15	Pacific Island elements	15	3.06
16	Sino – Japanese elements	20	4.80
17	South American elements	67	13.70
18	Tropical American elements	57	11.65
TOTAL		**489**	**100**

Table 20.9. List of the wet lands of the state of West Bengal.

Sl No.	Name of the district and area	Locality	Name of the wetland	Area in ha	Type of wetland
1	Malda	Pachia	Adh Soi Beel	140	Natural
2	Malda	Kasimpur	Ashi Dob Beel	280	Natural
3	Malda	Jatra Danga	Balotali Beel	120	Natural
4	Malda	Arai – Danga	Barbila Talao	120	Natural
5	Malda	Bowalia	Bochamari Beel	120	Natural
6	Malda	Bajnana	Goal Bod Beel	120	Natural
7	Malda	Mobarakpur	Hazar Takla	140	Natural
8	Malda	Kupur Ganj	Konar Beel	136	Natural
9	Malda	Malda Town	Madhupur Beel	100	Natural
10	Malda	Chandipur	Sanak Beel	200	Natural
11	Malda	Chandipur	Singsar Beel	140	Natural
12	Jalpaiguri	Madarihat	Jalua para Beel	6096	Natural
13	Jalpaiguri	Odalbari	Kathambari Beel	136	Natural
14	Cooch Behar	Haldibari	Buxiganj Nijiarap Beel	1400	Natural
15	Cooch Behar	Toofan Ganj	Rasik Beel	9952	Natural
16	Cooch Behar	Makheli Ganj	Teesta Nadi Beel	1800	Natural
17	Uttar DinajPur	Sitagram	Chalua Beel	140	Natural
18	24 – Parganahs	Namkhana	Talao – Holiday – Island	350	Natural
19	24 – Parganahs	Namkhana	Lothian	3800	Natural
20	24 – Parganahs	Gosaba	Sajna Khali	36,236	Natural
21	24 – Parganahs	Kolkata	East Calcutta Wet Land	3000	Natural
22	Birbhum	Lang Thata	Land Tata Beel	2000	Natural

the plantation of different species of *Acacia*, *Eucalyptus*, etc. In social forestry, the plantation programme in South Bengal prevents the growth of natural vegetation.

20.6 Rare Threatened and Endangered Flora of the State

The flora of West Bengal state are under immense pressure due to several biotic and abiotic factors,

Fig. 20.10. 32 - *Podophyllum hexandrum*, an endangered medicinal plant of Darjeeling district. 33 - *Bombax ceiba*, red silk-cotton used for stuffing purposes. 34 - *Phoenix rupicola*, an uncommon palm.

seriously endangering floristic diversity. This is particularly so in the case of the flora of the Darjeeling district, part of the eastern Himalayan Hotspots region, which faces destruction due to tourist activities and excessive collection by undergraduate and postgraduate students.

Das (1998) prepared a list of 222 plants from the Darjeeling district of the state which are in need of preservation.

A list of some RET plants available in the state, their IUCN status and distribution in the state, is given in Table 20.10.

Besides the abovementioned plants, there are others that have become rare in the wild due to overexploitation, such as *Aconitum bisma, Anemone rupicola, Bauhinia scandens, Heracleum wallichi, Mahonia napaulensis, Michelia doltspora* and *M. velutina*. *Aponogiton natans, Potamogeton* sp., *Ottelia* sp. and *Vallineria* sp. have also become rare, due to habitat destruction.

20.7 Taxa Missing from the State Flora

The abovementioned factors, singly or in combination, demonstrate the different negative effects on the flora and vegetation of the state. For example, some species are now extinct in the wild. In this connection, *Aldrovanda vesiculosa*, formerly abundant at Salt Lake, and members of the *Orobanchae* family, present in the Sonarpur region (Datta and Majumdar, 1966), are now absent in their former locality. Similarly, the Darjeeling district, part of the Eastern Himalaya Hotspots zone, was explored by several workers at different times from the late 19th century onwards. Several

Table 20.10. List of RET listed plants of the state.

Sl No.	Family	Sl No.	Name of the plant	RET category	Distribution in the state
1	Aceraceae	1	*Acer hookeri*	En	Darjeeling at 600–1500 m
		2	*Acer osmastonii*	En	Darjeeling, endemic in Birch Hills
2	Apiaceae	3	*Pimpinella tongloensis*	En	Darjeeling, endemic in Singaleela Range
3	Apocynaceae	4	*Rauvolfia serpentine*	Vu	Distributed in the plains and in the Duars region of the state
4	Arecaceae	5	*Phoenix rupicola*	Ra	Duars region at an elevation of 450 m
5	Begoniaceae	6	*Begonia scutala*	Ra	Darjeeling at an elevation of 1000–1500 m
6	Berberidaceae	7	*Podophyllum hexandrum*	En	Phalut region of Darjeeling
7	Campanulaceae	8	*Codonopsis affinis*	Ra	Darjeeling at an elevation of 1830–2000 m
8	Dioscoreaceae	9	*Dioscorea deltoidea*	En	Locality not precisely mentioned at an elevation of 1800 m
9	Orchidaceae	10	*Bulleyia yunnanensis*	Ra	Darjeeling hills
		11	*Cypripedium himalaicum*	En	Darjeeling (Dumsong and in Munsang)
		12	*Diplomeris hirsute*	Vu	Darjeeling at an elevation of 2000 m
10	Rubiaceae	13	*Hedyotis brunonis*	Ra	Locality not precisely known
		14	*Hedyotis scabra*	Ra	Locality not precisely known
		15	*Ophiorrhiza lurida*	Ra	Darjeeling at an elevation of 1500 m

En = Endangered; Ra = Rare; Th = Threatened; Vu = Vulnerable

species were reported from that region but are missing from the currently available checklists of plants of the Darjeeling district. Gamble (1896), and Cowan and Cowan (1929) mentioned the presence of *Hibiscus scandens* Roxb., *Euonymus macrocarpus* Gamble, *Mucuna monosperma* DC, *Dunbaria grandiflora* (Backer) Maesen, *Derris monticola* (Kurtz) Prain, *D. polystachya* Bentham, *Argyreia atroperpurea* (Wall.) Raizada, *A. argentea* (Roxb.) Choisy, *A. thomsonii* (Clarke) Babu, *A. sikkimensis* (Clarke) Ooststroom and others from the Darjeeling districts and in the plains of North Bengal, which are absent from the modern floristic account of the district in the work of Samanta and Das (1995), R.B. Bhujel (Studies on the dicotyledonous flora of Darjeeling district. University of North Bengal, Darjeeling, West Bengal, India, 1996, unpublished thesis) and A.K. Samanta (Taxonomical and phytosociological studies on the Angiospermic climbers of Darjeeling and Sikkim Himalayas, University of North Bengal, Darjeeling, West Bengal, 1998, unpublished thesis). In addition, some plants, such as *Aconitum bisma* (Hamilton) Rapaics, *Anemone rupicola* Cambess, *Bauhinia scandens* L., *Heracleum wallichi* DC, *M. velutina* DC, *Nardostachys grandiflora* DC and *Rauvolfia serpentina* (L.) Kurz, have become very rare in the hills of Darjeeling due to overexploitation (Figs 20.11 and 20.12).

20.8 Conservation Strategies

The Rio convention in 1992 provided the opportunity to realize and think over the importance of biodiversity as well as conservation strategies. The diversity associated with pre-existing forests, vast natural water bodies and natural vegetation tracts have been engulfed by enormous development activities. The silent hill tracts are now vibrating with the sound of the car, and virgin forests have vanished to meet the need of developing human societies. The large mangrove vegetation of the Sundarbans, which formed a natural wall to protect the southern part of the state from the cyclonic wind of the Bay of Bengal, has been severely reduced.

To meet the need of industry and the demand for house building materials and furniture, the sal tree (*Shorea robusta*) and teak in the western and northern parts of the state (Purulia, Paschim Medinipur, Jalpaiguri, Darjeeling, Uttar Dinajpur) have been lost forever.

The dense forest of the sub-Himalayan region, known as the Duars region, has become thin due to regular collection of plants for food, fodder, timber and dye, and destruction of the natural forest to build tea gardens and tourist bungalows.

All these developmental activities of human society cause regular damage to the biodiversity of the state. The Ministry of Environment and Forests at

Fig. 20.11. 35 - *Clerodendrum viscosum*, a very common weed of the state. 36 - *Cleome rutidosperma*, a common exotic plant found on railway tracks. 37 - *Cuscuta reflexa*, a common parasitic plant of variable host plants. 38 - *Drosera burmannii*, an uncommon insectivorous plant.

both state and national levels are now conscious of this loss of biodiversity. So, different *ex-situ* and *in-situ* conservation strategies have been considered by the government. Up to the period of 2007 in the state, there are five national parks, 18 wildlife sanctuaries, two biosphere reserves and one UNESCO World Heritage site, which play a part in the conservation process of the state flora and fauna. Besides these, there is one Ramsar site (wetland of international importance: East Calcutta Wet Land) which works for conservation of aquatic flora and fauna.

One more aspect of the conservation strategy is conservation through religious beliefs. The state has a significant number of sacred groves in all 20 districts of the state, especially concentrated in the northern and western parts.

20.9 Concluding Remarks

This chapter provides a sketch of the present status of the flora of West Bengal, a century after the publication of *Bengal Plants* by David Prain (1903). It is true that the complete flora of the state is yet to be properly known, as there are some parts of the extreme northern corner of the state which are still out of the reach of the botanist due to their remoteness. Parts of the Darjeeling district are included within the Eastern Himalaya Hotspots region, where speciation is a continuous process, creating neo-endemics and enriching the state flora by adding new species to the floristic composition. More or less every phyto-sociological condition exists, from the mangrove vegetation of the Sundarban region to the temperate-to-alpine vegetation of the Sandakphu and Phalut region of the

Fig. 20.12. 39 - *Vanda tessellate*, a common epiphytic plant of the mango tree. 40 - *Tiliacora acuminata*, a primitive dicot plant with trimerous flowers. 41 - *Erythrina variegata*, a plant of Papilionaceae having extra floral nectarines (EFN). 42 - *Vernonia cinerea*, a common plant of Asteraceae in wild habitat.

state, including the semi-desert vegetation of the western half of the state. But due to lack of proper conservation strategies, high pressure of development, huge demand of the land for agriculture, industry and tourism, the naturalness of the state flora is on the verge of extinction.

Several species have already become extinct, either due to loss of habitat or over-exploitation and a lack of effective conservation strategies.

So it is time to take immediate action to document biodiversity for proper conservation and maintenance.

Acknowledgements

The authors are grateful to the Director and the entire staff of the Botanical Survey of India for their enormous help during screening of the herbarium specimens. We are also grateful to the Curator of the herbarium of Lloyd Botanic Garden, Darjeeling for his help during a visit to the garden. Last, but not least, we thank all the foresters of different Forest Circles during the field trips for their help during field collection, as well as providing data regarding forest utility patterns and the forest community.

References

Ahmedullah, M. (2000) Endemism in the Indian flora. In: Singh, N.P. and Vohra, J.N. (eds) *Flora of India* (introductory volume), Part II. Botanical Survey of India, Kolkata, India, pp. 246–265.

Ball, V. (1869) Notes on the flora of Manbhum. *Journal of Asiatic Society of Bengal* 33, 112–124.

Bhattacharyya, U.C. (1997) *Introductory Essay on Flora of West Bengal*, Vol. I. Botanical Survey of India, Kolkata, India.

Bhujel, R.B. and Das, A.P. (2002) Endemic status of the dicotyledonous flora of Darjeeling district. In: Das, A.P. (ed.) *Perspective of Plant Biodiversity*. Bishen Singh and Mahendra Pal Singh Publications, Dehradun, India, pp. 593–609.

Bhujel, R.B. and Yonzone, G.S. (1994) A new variety of *Baliospermum calycianum* Muell. – Arg. from Darjeeling. *Journal of Economic Taxonomy and Botany* 18(3), 613–614.

Biswas, K.P. (1966) *Plants of Darjeeling and Sikkim Himalaya*. Government of West Bengal, India.

Biswas, R., Das, A.P. and Paul, T.K. (2013) Floristic diversity of Rasik Beel and its adjoining areas in Cooch Behar district of West Bengal, India. *Pleione* 7(2), 501–507.

Bor, N.L. (1940) *Flora of Assam*, Vol. V. Government of Assam, Shillong, Meghalaya, India.

Bremkamp, C.R.B. (1959) New Ixora species from Bengal, Burma and Nicobar Island. *Indian Forester* 85, 371–375.

Champion, H.G. and Seth, S.K. (1968) *A Revised Survey of the Forest Types of India*. Government of India, New Delhi, India.

Chakravarty, R.K., Srivastava, R.C., Mitra, S., Bandyopadhyay, S. and Bandyopadhyay, S. (1999) West Bengal. In: Mugal, V. and Hajra, P.K. (eds) *Floristic Diversity and Conservation Strategies in India*. Botanical Survey of India, Kolkata, India, pp. 1575–1630.

Chatterjee, D. (1940) Studies on the Endemic Flora of India and Burma. *Journal of the Royal Asiatic Society (Sc.)* 5, 19–67.

Chatterjee, D. (1962) Floristic patterns of the Indian vegetation. In: Maheswari, P., Johri, B.M. and Vasil, I.K. (eds) *Proceedings of the Summer School of Botany*, held at Darjeeling, 1960, pp. 32–42.

Chowdhury, M. and Das, A.P. (2010) Hydrophytes of different wet lands in Maldah district of West Bengal, India. *Environmental Biology Conservation* 15, 22–28.

Chowdhury, M. and Das, A.P. (2011a) A note on the distribution and association of *Rosa clinophylla* Thory var. *glabra* (Lindley ex Prain) C. Ghora and G. Panigrahi (Rosaceae) at the Maldah district of West Bengal. *Pleione* 5(1), 196–197.

Chowdhury, M. and Das, A.P. (2011b) Macrophytic diversity and community structure of Adh Soi wet land of Maldah district of Paschimbanga, India. In: Ghosh, C. and Das, A.P. (eds) *Recent Studies in Biodiversity Traditional Knowledge in India*. Gour Mahavidyalaya, Malda, India, pp. 109–115.

Chowdhury, M. and Das, A.P. (2013) Biodiversity and present status of Gangetic wetlands of Maldah district of West Bengal, India. NBU. *Journal Plant Science* 7(1), 29–34.

Chowdhury, M., Chowdhury, A., Sikdar, A., Das, A.P. and Paul, T.K. (2011) Occurrence of *Solvia anthemifolia* (A. Jussieu) R. Brown (Asteraceae) in Eastern India. *Pleione* 5(2), 352–356.

Clushaw, J.C. (1952) Some West Bengal plants. *Journal of Bombay Natural History Society* 49, 188–196.

Cook, C.D.K. (1996) *Aquatic and Wetland Plants of India*. Oxford University Press, New York, USA. 385 pp.

Cowan, A.M. and Cowan, J.M. (1929) *The Trees of North Bengal Including Shrubs, Woody Climbers, Bamboos, Palms, and Tree Ferns Being a Revision of the List by Gamble*. Bengal Secretariat Press, Calcutta, India.

Das, A.P. (1998) Endemic and endangered angiosperms of Darjeeling Hills: an assessment. *Paper presented at the Regional Conference on the Assessment and Conservation of Biodiversity, Corbett National Park, January 1998*.

Das, A.P. (2002) Survey of naturalized exotics in the flora of Darjeeling Hills, West Bengal (India). *Journal of Economic Taxonomy and Botany* 26(1), 31–37.

Das, A.P. (2004) Floristic studies in Darjeeling hills. *Bulletin of Botanical Survey of India* 43(1–4), 1–18.

Das, A.P. and Chanda, S. (1986) Notes on some naturalised exotics in the flora of Darjeeling Hills, West Bengal (India). *Indian Botany Report* 5(2), 144–147.

Das, A.P. and Ghosh, C. (2011) Plant wealth of Darjeeling and Sikkim Himalaya *vis a vis* conservation. *NBU Journal of Plant Science* 5(1), 25–33.

Das, A.P. and Lahiri, A.K. (1997) Phytosociological studies of the ground flora in different types of vegetation on Tiger Hills, Darjeeling, West Bengal. *Indian Forester* 123(12), 1176–1187.

Das, A.P. and Lama, D. (1992) *Liparis breviscapa* A.P. Das et Dorjy – a new species of Orchidaceae from the Darjeeling Hills, West Bengal, India. *Journal of Economic Taxonomy and Botany* 16(1), 226–227.

Datta, S.C. and Majumdar, N.C. (1966) Flora of Calcutta and vicinity. *Bulletin of the Botanical Society of Bengal* 20, 16–120.

de Candolle, A.P. (1855) *Géographie botanique raisonnée*. V. Masson, Paris, France and Geneva, Switzerland.

Don, D. (1821) Description of several taxa from the Kingdom of Nepal, taken from specimens preserved in the herbarium of A.B. Lambert. *Memoirs of the Wernerian Natural History Society* 3, 407–415.

Don, D. (1825) *Prodromus Florae Nepalensis*. London, UK.

Engler, A. (1882) *Versuch einer Entwicklungsgeschichte der Pflanzenwelt*. W. Engelmann, Leipzig, Germany.

Gamble, J.S. (1896) *List of Trees, Shrubs, and Climbers Found in the Darjeeling District, Bengal*, 2nd edn. Government Press, Calcutta, India.

Grierson, A.J.C. and Long, D.G. (1983, 1984, 1987) *Flora of Bhutan*, Vol. 1. Parts 1, 2, and 3. Royal Botanic Garden Edinburgh and London, UK.

Grierson, A.J.C. and Long, D.G. (1991, 1999, 2001) *Flora of Bhutan*, Vol. 2. Parts 1, 2, and 3. Royal Botanic Garden Edinburgh and London, UK.

Hara, H. (1966, 1971) *Flora of Eastern Himalaya*, 1st and 2nd report. Tokyo University Press, Tokyo, Japan.

Hara, H. and William, L.H.J. (1979) *An Enumeration of Flowering Plants of Nepal*, Vol. II. Trustees of British Museum, London, UK.

Hara, H., Steam, W.T. and William, L.H.J. (1978) *An Enumeration of Flowering Plants of Nepal*, Vol. I. Trustees of British Museum, London, UK.

Hara, H., Charter, A.Q. and William, L.H.J. (1982) *An Enumeration of Flowering Plants of Nepal*, Vol. III. Trustees of British Museum, London, UK.

Hooker, J.D. (1849) Notes Chiefly botanical made during the excursion from Darjeeling to Tonglu. *Journal of Asiatic Society of Bengal* 18, 419–449.

Hooker, J.D. (1907) Botany of British India. *Imperial Gazette of India* 1, 157–212.

Kanjilal, U.N., Kanjilal, P.C., Das, A. and De, R.N. (1934–1940) *Flora of Assam*, Vols I–IV. Government of Assam, Shillong, Meghalaya, India.

King, G. and Pantling, R. (1898) *The Orchids of Sikkim-Himalaya*. Royal Botanic Garden, Calcutta, India.

Krishna, G., Maity, D. and Venu, P. (2014) Three new records for the flora of West Bengal, India. *Pleione* 8(1), 193–198.

Moktan, S. and Das, A.P. (2014) Diversity and composition of tree stand in tropical forests of the Darjeeling part of Eastern Himalaya. *International Journal of Current Research* 6(12), 10446–10451.

Mitra, S., Bandyupadhyay, S. and Mukherjee, S.K. (2010) *Bibliography of Flora and Ethnobotany of West Bengal*. East Himalaya Society for Spermatophyte Taxonomy. Siliguri, Darjeeling, West Bengal, India.

Nayar, M.P. (1996) *Hot Spots of Endemic Plants of India, Nepal, Bhutan*. Tropical Botanic Garden and Research Institute, Thiruvananthapuram, Kerala, India.

Noltie, H.J. (1994, 2000) *Flora of Bhutan*, Vol. 3, parts 1 and 2. Royal Botanic Garden, Edinburgh and London, UK.

Ohashi, H. (1975) *Flora of Eastern Himalaya*. Tokyo University Press, Tokyo, Japan.

Pasha, M.K. and Uddin, S.B. (2013) *Dictionary of Plant Names of Bangladesh (Vascular Plants)*. Janokalyan Prokashani. Chittagong, Dhaka, Bangladesh, pp. 1–434.

Pearce, N.R. and Cribb, P.J. (2002) *The Orchids of Bhutan, Flora of Bhutan*, Vol. 3. Royal Botanic Garden, Edinburgh and London, UK.

Prain, D. (1903) *Bengal Plants*, Vols I and II. Bishen Singh and Mahendra Pal Singh Publications, Dehradun, India.

Prain, D. (1905) The vegetation of the districts Hooghly, Howrah and 24-Parganas. *Records of the Botanical Survey of India* 3, 143–339.

Rai, U., Das, A.P. and Singh, S. (2011) Understanding the forest types in lower hills of the Darjeeling Himalaya using satellite and ground truth data. In: Ghosh, C. and Das, A.P. (eds) *Recent Studies in Biodiversity and Traditional Knowledge in India.* Sarat Book House, Kolkata, India, pp. 203–213.

Rao, P.S., Sijatha, B., Lakshminarayana, K. and Ratnam, S.V. (2013) A study on phytosociology, soil conservation and socio-economic aspects in red sand dunes near Bhimili of Vishakapattanam. *Archives of Applied Science and Research* 5(1), 45–46.

Richardson, I.B.K. (1978) Endemic taxa and the taxonomist. In: Street, H.E. (ed.) *Essays in Plant Taxonomy*. Academic Press, London, UK, pp. 243–262.

Samanta, A.K. and Das, A.P. (1995) Angiospermic climbers of Darjeeling Hills. In: Pandey, A.K. (ed.) *Taxonomy and Biodiversity*. CBS Publications, New Delhi, UK, pp. 139–147.

Shukla, G., Biswas, R., Das, A.P. and Chakravarty, S. (2014) Plant diversity in Chilapatta Reserve Forest of Tarai Duars in sub-humid tropical foothills of Indian Eastern Himalayas. *Journal of Forest Research* 25(3), 591–596.

Singh, K.P., Shukla, A.N. and Singh, J.S. (2010) State level inventory of the invasive alien plants, their source region and use of the potential. *Current Science* 99(1), 107–114.

Sujay, Y.H., Sattagi, H.N. and Patil, R.K. (2010) Invasive alien insects and their impact on agro-ecosystems. *Karnataka Journal of Agricultural Science* 23(1), 26–34.

Yonzone, R., Dorjay, L., Bhujel, R.B. and Rai, S. (2014) Present status, diversity and distribution of *Goodyera* R. Brown (Orchidaceae) in Darjeeling part of the Eastern Himalaya. *Pleione* 8(1), 89–91.

21 Status of Invasive Plants in Tamil Nadu, India: Their Impact and Significance

S.M. Sundarapandian* and K. Subashree

Department of Ecology and Environmental Sciences, Pondicherry University, Puducherry, India,

Abstract

Alien invasive plants always pose a major risk to native biodiversity and human welfare. Numerous species have been introduced into India, particularly in Tamil Nadu and particularly during the British administration. At present, there are 279 alien, invasive taxa in Tamil Nadu and about 69% of these are herbs. About 61% of the invasive taxa have migrated to Tamil Nadu from tropical America. Most of the exotic, invasive flora of Tamil Nadu belong to the families Fabaceae and Asteraceae. About 30.8% of the invasive plants are prevalent in all the districts of Tamil Nadu. Nilgiri district has the maximum number of invasive plants (146 taxa). A lot of research has been done on the invasive flora of Tamil Nadu and the numbers are expected to rise in the future. A brief account on the top ten invaders of Tamil Nadu is provided. Several factors facilitate invasion of alien plants such as globalization, global warming, human migration, land-use change, etc. Although several control measures (physical, chemical, biological and cultural) are in practice, the best way to suppress the growth of invasive plants is by exploitation and adopting an integrated approach.

21.1 Introduction

European invasion in Africa and Asia has led to the deliberate or inadvertent introduction of alien flora and fauna for food, fodder, medicinal, ornamental and recreational purposes (Pyšek and Richardson, 2008; Khuroo *et al.*, 2012a; Kannan *et al.*, 2013), that have later become notorious invaders in the new environment. The International Union for Conservation of Nature and Natural Resources (IUCN) defines 'alien invasive species' as an alien species which becomes established in natural or semi-natural ecosystems or habitat as an agent of change and threatens native biological diversity (Srivastava *et al.*, 2014). The Convention on Biological Diversity (CBD, 1992) recognizes 'biological invasion of alien species as the second worst threat after habitat destruction'. In terms of ecology, any species that has been introduced to an ecosystem that is beyond its home range that successfully establishes, naturalizes, proliferates and affects the environment, ecosystem and economy is regarded as an invasive species (Williamson, 1996; Singh *et al.*, 2010).

Invasive plants are known to have faster rates of growth and accumulation of biomass than native flora and are characterized with high competitive ability by copious seed production, efficient dispersal mechanisms, prolific vegetative reproduction, swift establishment, great tolerance and allelopathic potential, besides other traits that help them to adapt in new diverse habitats (Sharma *et al.*, 2005; Simberloff *et al.*, 2005; Huang *et al.*, 2009). Invasive species compete with the native species for light, water, nutrients and space (Mack *et al.*, 2000; Pimental *et al.*, 2000; Ehrenfeld, 2003; Udayakumar *et al.*, 2014), and with their better adaptive features, they gradually outnumber and dislodge the native flora, coercing them to extinction. Although

*E-mail: smspandian65@gmail.com

Darwin (1872) and Wallace (1902) observed the extinction of local flora due to alien plant invasion centuries ago, the problem of biological invasion has received due attention only very recently (Rao and Sagar, 2012). Nowadays, alien invasive flora are often attributed to decline in the numbers of endemic, endangered and threatened species, as they alter vegetation structures and modify ecosystem functioning by attaining high densities and biomass (Denslow, 2007; Surendra *et al.*, 2013).

According to Kohli *et al.* (2012), India is very susceptible to invasion due to three main causes: 1) more migration of humans across the country leading to easy spread of seeds and propagules of the invasive plants; 2) availability of fragmented and disturbed habitats, which provide conducive niches for the ecesis of alien flora; and 3) prevalence of favourable environmental and climatic conditions for invasive plants to flourish. Moreover, as India is a mega-diverse nation with ever-increasing trade and travel by land, air and water, it faces an even higher risk of plant invasion (Saxena, 1991; Sharma *et al.*, 2005; Khuroo *et al.*, 2012a). It has been reported by Nayar (1997) that alien flora comprise 18% of the total Indian flora. There are 173 alien invasive plants in India, which belong to 117 genera and 44 families, and about 80% of them were of neotropical origin (Reddy, 2008). Of late, the Botanical Survey of India, the Forest Department, many researchers and non-profit organizations have made tremendous efforts to document the alien invasive flora, both regionally and state-wise, to preserve native biodiversity as well as to design better management strategies. It is crucial to generate and more importantly to update the baseline data on the taxonomic information, distribution, patterns and pathways of the invasion of exotic flora.

21.2 A Brief Profile of Tamil Nadu

Tamil Nadu is located in the south-east of the Indian Peninsula (8° 04'–13° 34' N and 76° 14'–80° 21' E). It spans 130,058 sq. km and is divided into 30 administrative districts (ibcn.in/wp-content/uploads/2015/05/Tamil-Nadu.pdf). In general, Tamil Nadu enjoys a tropical climate, except in the coastal zone where maritime climate prevails. The temperature ranges from 28°C to 40°C during summer and 18°C to 26°C during winter. The temperature may drop as low as 3°C during winter in the upper mountains. The normal average annual rainfall is 945 mm (www.dcmsme.gov.in/publications/traderep/sptnadu.pdf).

Physiographically, Tamil Nadu is divided into three regions – the eastern coastal region, the western hilly region and the plains. The Western Ghats constitute the main hilly areas in the state, with an average elevation of 1220 m and highest point of 2440 m. Some major rivers that traverse the state are Cauvery, Palar, Bhavani, Vaigai and Tamiraparani. Forests constitute 17.40% of the total land area of the state (ibcn.in/wp-content/uploads/2015/05/Tamil-Nadu.pdf).

The forest types of Tamil Nadu comprise three major groups and nine type groups. The three major groups are tropical forests, montane subtropical forests and montane temperate forests. Of these, the major group tropical forests comprises seven type groups:, 1) tropical wet evergreen forests; 2) tropical semi evergreen forests; 3) tropical moist deciduous forests; 4) littoral and swamp forests; 5) tropical dry deciduous forests; 6) tropical thorn forests; and 7) tropical dry evergreen forests (www.forests.tn.nic.in/forestatglance/TN_forest_type.html).

21.3 Diversity and Distribution of Alien Invasive Flora in Tamil Nadu

The number of exotic invasive plants has been increasing alarmingly and many species are prevalent in almost all the districts. Of late, since tackling invasive flora has demanded considerable attention, more studies have documented them and explored management options of the highly problematic invasive plants. According to the Alien Species Database, hosted by ENVIS Centre of Tamil Nadu, there are 1274 alien species in Tamil Nadu, 998 of which are under cultivation. Alien flora in Tamil Nadu were largely introduced by the British during their administration in Chennai, Nilgiri and Palani Hills. Of this great number of introduced exotic flora, only 276 taxa have turned out to be invasive and are naturalized (tnenvis.nic.in/tnenvis_old/IASintamilnadu.htm). However, with some new additions, there are 279 invasive alien taxa in Tamil Nadu and the numbers are expected to increase with further research. Of these, most of them are terrestrial and only very few are aquatic: *Eichhornia crassipes*, *Kappaphycus alvarezii* (an alga) and *Alternanthera philoxeroides*, while species like *Ipomoea carnea* are semi-aquatic.

21.3.1 Life forms of invasive taxa

Invasive flora are represented in all the life forms. About 69% of the invasive taxa belong to the herbaceous community in Tamil Nadu (Fig. 21.1). Herbs comprise a major share of the invasive flora in many other parts of India like Kashmir (Khuroo *et al.*, 2012b), Uttar Pradesh (Singh *et al.*, 2010; Srivastava *et al.*, 2014), and Maharashtra (Deshmukh *et al.*, 2012), as well as in China (Huang *et al.*, 2009; Randall, 2012). Reddy (2008) has listed 151 invasive alien herbs from India. Shrubs contribute the next major proportion (18%) and it is followed by climbers (5%), trees and undershrubs (4% each). *Kappaphycus alvarezii* is a marine alga that has been documented as invasive in the marine biosphere reserve, Gulf of Mannar.

21.3.2 Nativity of invasive taxa

About 61% of the invasive taxa in Tamil Nadu have their origin in America, followed by the Mediterranean region (15%), Africa (8%), Australia and Eurasia (4% each), Europe and Asia (3% each) and North Temperate region (2%) (Fig. 21.2). Only one species from the Gulf States has been reported to be invasive in Tamil Nadu, *Solanum carolinense* (Chinnusamy *et al.*, 2012a). America is reported to have contributed the largest number of noxious invasive plants in China (Xu and Qiang, 2004; Huang *et al.*, 2009).

Reddy (2008) has reported that 74% of the invasive alien flora in India have originated from tropical America, particularly in Uttar Pradesh (Singh *et al.*, 2010; Srivastava *et al.*, 2014), Andhra Pradesh (Surendra *et al.*, 2013) and Maharashtra (Deshmukh *et al.*, 2012).

Apart from foreign species, few invaders have migrated to Tamil Nadu from other parts of India. *Filago arvensis*, *Persicaria nepalensis*, *Polygonum molle* and *Prinsepia utilis* have been reported to have migrated from the Himalayan states, while *Melica scaberrima* has been recorded to have come from north-west India.

21.3.3 Family-wise distribution of invasive taxa

The majority of invasive plants in Tamil Nadu belong to Fabaceae and Asteraceae (25% each; Fig. 21.3) followed by Poaceae (21%) and Solanaceae (17%). Brassicaceae and Caryophyllaceae represent 6% of invasive flora in Tamil Nadu. Fabaceae has always been noted for being a notorious contributor to the local flora worldwide (Binggeli, 1996; Pyšek, 1998; Wu *et al.*, 2003). Asteraceae has been reported to be a dominant family among the exotic flora of India, China and South Africa (Heywood, 1989; Rao and Murugan, 2006; Huang *et al.*, 2009). Large families could contribute more species to the invasive flora in Tamil Nadu as reported by Singh *et al.* (2010) in Uttar Pradesh, India.

21.3.4 District-wise distribution of invasive taxa

About 30.8% (86 taxa) of alien invasive flora are prevalent in all the districts of Tamil Nadu (Fig. 21.4).

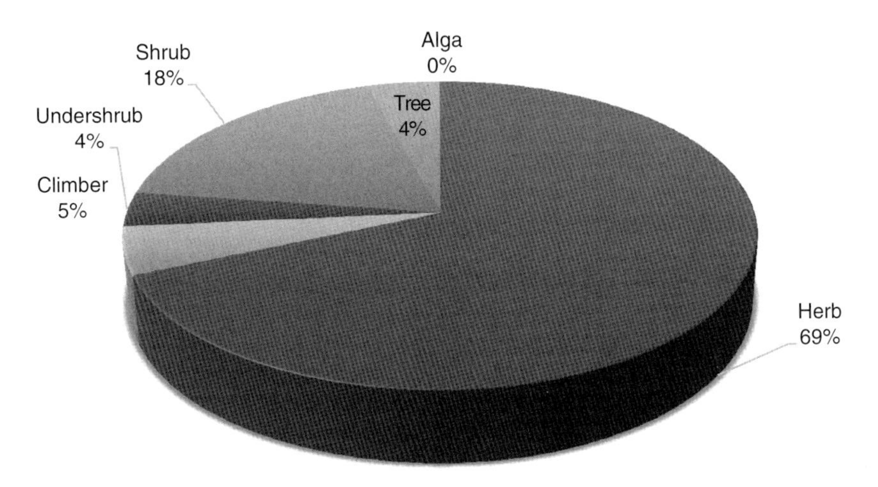

Fig. 21.1. Life forms of invasive taxa in Tamil Nadu, India.

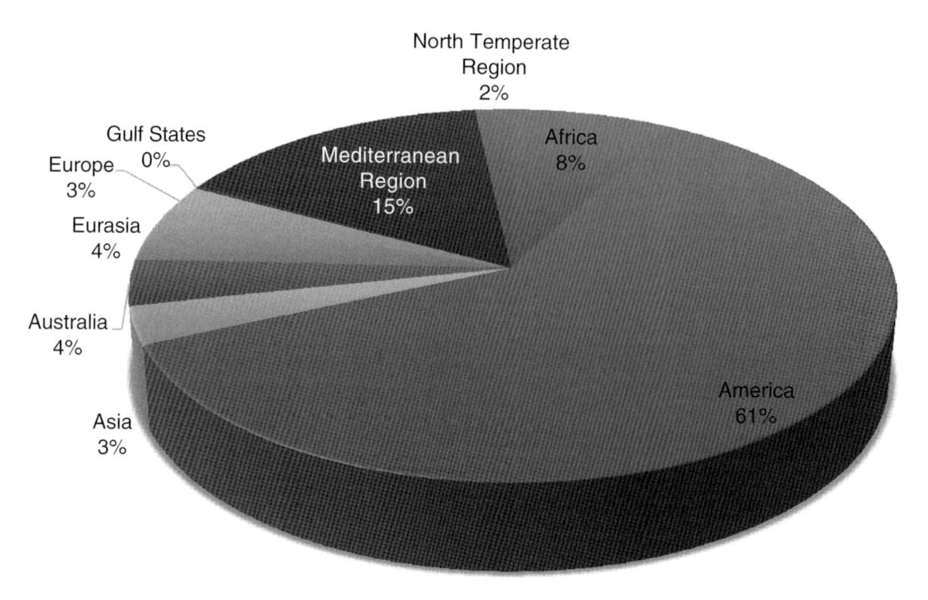

Fig. 21.2. Nativities of invasive plants of Tamil Nadu, India.

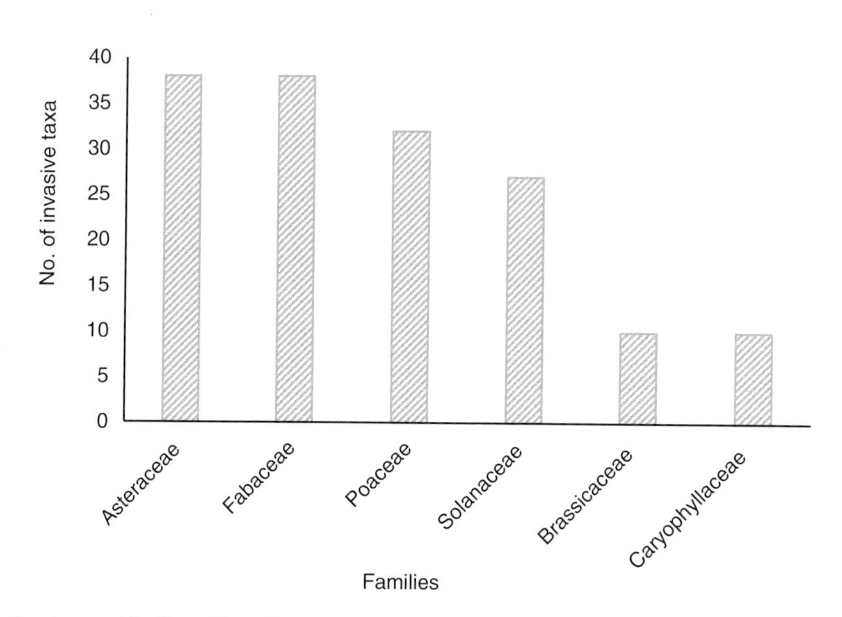

Fig. 21.3. Family-wise contribution of invasive taxa in Tamil Nadu, India.

Nilgiri district has the highest number of invasive taxa (146). This is followed by Dindigul (98), Salem (49), Coimbatore (48) and Theni (29) districts. Since a cool climate prevails in Nilgiris, similar to the temperate zone, it favours the flourishing and ecesis of several invasive exotic flora.

21.4 Studies in Tamil Nadu

Several studies have been carried out by various researchers on different aspects of alien plant invasion, such as documentation, impact assessment, utilitarian value, etc. Chandrasekaran *et al.* (1997)

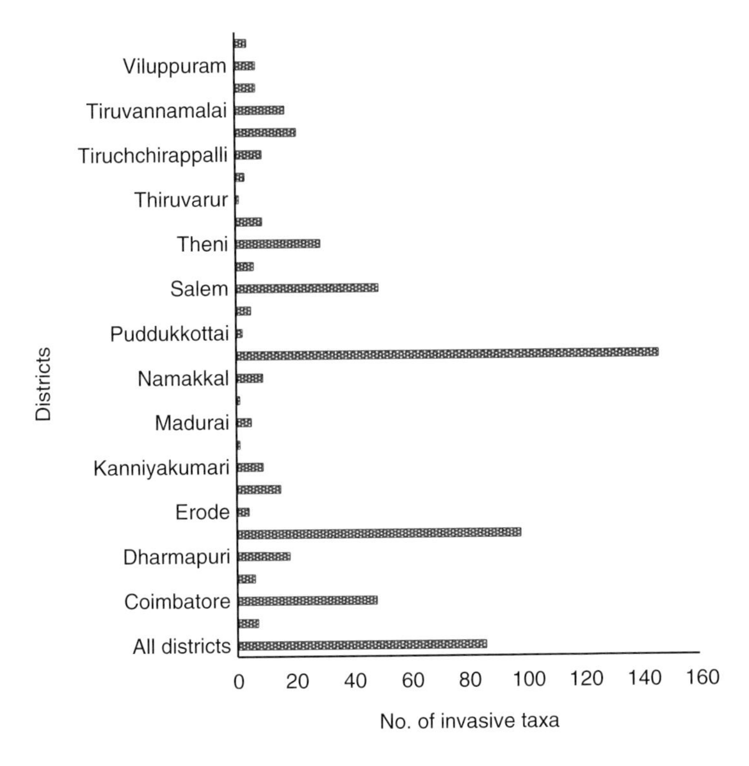

Fig. 21.4. Distribution of invasive taxa in Tamil Nadu, India.

and Sundarapandian *et al.* (2004) have stated that exotic weeds, i.e. *Ageratum conyzoides*, *Chromolaena odorata*, *Mesosphaerum suaveolens*, *Lantana camara* and *Synedrella nodiflora*, form important components of the plant community in young, successional fallows for up to five years in Kodayar in the Western Ghats, Tamil Nadu. Sundarapandian and Swamy (2000) have reported that *Chromolaena odorata* and *Ageratum conyzoides* dominate open and disturbed natural ecosystems in Kodayar. Chandrasekaran *et al.* (2001) have compared the exotic plant invasions in a natural forest, a savannah and human-modified ecosystems such as plantations and wasteland from Kodayar. Chandrasekaran and Swamy (2002) have observed that exotic plants like *Lantana camara* and *Chromolaena odorata* accumulate greater biomass in a natural secondary forest and also in human-modified ecosystems in Kodayar. Sundarapandian *et al.* (2000, 2012) have stated that exotic weeds, such as *Ageratum conyzoides*, *Chromolaena odorata*, *Imperata cylindrica*, *Lantana camara*, *Prosopis juliflora*, *Mimosa pudica*, *Ipomoea carnea* and *Synedrella nodiflora*, are distributed differently in various disturbance regimes.

Rameshkumar *et al.* (2005) have conducted phytochemical and antibacterial studies of two exotic plants of Tamil Nadu, *Chromolaena odorata* and *Mimosa pudica*. According to Ramachandran (cited in Kumar, 2009), 223 angiosperms belonging to 56 families have naturalized in the Nilgiri Biosphere Reserve alone. Joshi *et al.* (2009) studied the edge effects and disturbance in the forest structure of tropical rainforest fragments of the Anamalai Hills in the Western Ghats, due to the invasion by cultivated coffee species, *Coffea arabica* and *Coffea canephora*. Narasimhan *et al.* (2010) have presented a comprehensive review on the invasive alien plant species of Tamil Nadu. *Ipomoea carnea*, *Pistia stratiotes* and *Eichhornia crassipes* have invaded five ephemeral lakes in Thiruvallur district (Udayakumar and Ajithadoss, 2010). Parthasarathy *et al.* (2012) have studied the invasive alien plants in six hill complexes of the south-eastern Ghats of Tamil Nadu. Chinnusamy *et al.* (2012a) have reported the occurrence of an alien invasive weed (*Solanum carolinense*) in eight districts of Tamil Nadu. Jaikumar (2012) have reviewed and suggested the utilization of *Eichhornia crassipes* for phytoremediation in Velachery lake. Kuberan *et al.* (2012) have reported that extracts of *Parthenium hysterophorus* and *Ageratum conyzoides* suppress the growth of *Glomerella cingulata*, which causes brown blight disease of tea; hence they could be used as biocontrol agents.

Frezina (2013) has assessed and explored in detail the utilization options of *Eichhornia crassipes* for effective management in the water bodies of Tamil Nadu. Pushpakaran and Gopalan (2013) have emphasized the great risk posed by *Lantana camara* and *Chromolaena odorata* to the health of seasonally waterlogged grassland ecosystems of Mudumalai Tiger Reserve. Sundarapandian and Karoor (2013) have reported that exotic invaders such as *Lantana camara*, *Chromolaena odorata* and *Ageratum conyzoides* dominate the disturbed areas and open habitats of Periyar Wildlife Sanctuary in the Western Ghats. Understorey vegetation of Sathanur reserve forest, which forms a part of the Eastern Ghats, is dominated by *Lantana camara* (Gandhi and Sundarapandian, 2014). Udayakumar *et al.* (2014) have recorded 136 alien invasive species from Thiruvallur district. Aravindhan and Rajendran (2014) have documented 90 invasive alien plants from the Boluvampatti forest range in the south-western Ghats of Tamil Nadu. Arun *et al.* (2014) studied the chemical constituents of essential oils of the alien weed *Delairea odorata* during flowering and post-flowering periods.

21.5 Top Ten Worst Invasive Species of Tamil Nadu

21.5.1 Lantana camara L.

Lantana camara, the world's most notorious plant invader, is a native of tropical America. It was introduced for ornamental purposes by the British into Calcutta Botanical Gardens in 1809 (Kohli *et al.*, 2006). It is prevalent in all the districts of Tamil Nadu. Because of its great allelopathic potential, this Verbenaceae member quickly kills and displaces the native flora and has now invaded a majority of the grasslands, dry forests, scrub jungles, roadsides and other disturbed and deforested habitats. The invasion success of *Lantana camara* could be attributed to its adaptation to the local climatic and environmental conditions (Singh, 2012). Muthumperumal and Parthasarathy (2010) have reported that *Lantana camara* contributed to 24.04% of all the invasive lianas in six major hill complexes of the Eastern Ghats. Ramaswami and Sukumar (2014) have observed that the invasion of *Lantana camara* is higher along the streams and the area of invasion showed a positive trend with increase in annual rainfall in the tropical dry forests of Mudumalai. Moreover, the attractive umbels and fruits attract pollinators and frugivorous animals that aid in their dispersal. Physical control of *Lantana camara* is a widely sought control measure, although it is very expensive and labour-intensive.

Kumar and Maneemegalai (2008) investigated the larvicidal properties of extracts of *Lantana camara* on the larvae of *Aedes aegypti* and *Culex quinquefasciatus* mosquitoes. Ganjewala *et al.* (2009) compared the biochemical composition and antibacterial activity of *Lantana camara* plants with differently coloured flowers. Thamotharan *et al.* (2010) examined the anti-ulcer activity of the leaf extracts of *Lantana camara* in rats. Venkatachalam *et al.* (2011) conducted both physicochemical and phytochemical studies on the fruits of *Lantana camara*. Kanagavalli (2011) conducted *in-vitro* studies on the antiviral properties of the flower extracts of *Lantana camara* against polio virus type I. Kalita *et al.* (2012) published a detailed review on the medicinal properties of *Lantana camara*. Mariajancyrani *et al.* (2014) have isolated terpenes and assessed the antimicrobial activity of *Lantana camara* leaf extracts.

21.5.2 *Chromolaena odorata* (L.) R. M. King and H. Rob.

Chromolaena odorata (Asteraceae) is an obnoxious shrub that originated in Central America and was introduced as an ornamental elsewhere. It has been regarded as a serious invader in many countries and is predominant in all the districts of Tamil Nadu, threatening the health of forest and grassland ecosystems. This naturalized weed thrives well in dry sites and disturbed habitats, also affecting plantations, pastures, gardens, roadsides and fringes of human settlements (McFadyen, 2002; Orapa *et al.*, 2004; Tripathi *et al.*, 2012a). The ecesis of this invasive plant is due to its high regenerative and biotic potential; its prolific seed production has longer viability (Kohli *et al.*, 2012). Chandrasekaran and Swamy (2010) have studied the growth and allocation patterns of biomass and nutrients of *Chromolaena odorata* populations in varied human-modified ecosystems of Kodayar. Chandrasekaran *et al.* (1998) have observed high levels of nitrogen in the form of ammonia and nitrates in the soil of regions dominated by *Chromolaena odorata* in Kodayar. *Chromolaena odorata* was distributed mostly in disturbed and open habitats of both the Eastern Ghats and Western Ghats during the late 1990s, but in the recent past (2005 onwards), it was found to establish on roadsides and wastelands of even coastal districts, particularly Sivagangai, Ramanathapuram, Pudukkottai and Thanjavur. Physical control by stem cutting and root digging is the most commonly used control measure for successful eradication. In Kanyakumari district, *Chromolaena odorata* is used as a green manure on farmland: it has been mechanically removed from plantations (mostly rubber) and associated ecosystems for this use, which further supports its vigorous proliferation in these regions (Sundarapandian *et al.*, 2012).

Ayyanar and Ignacimuthu (2009) have highlighted the use of this weed in ethnomedicinal systems for wound healing by the Kani tribes of Tirunelveli hills. Exploration of medicinal properties, i.e. secondary metabolites and antibacterial activity, of *Chromolaena odorata* was carried out by Doss *et al.* (2011). Rajmohan and Logankumar (2011) studied the insecticidal properties of *Chromolaena odorata* against the mosquito *Aedes aegypti*, and Suriyavathana *et al.* (2012) have studied its biochemical and antimicrobial properties. Natheer *et al.* (2012) evaluated antibacterial activity of *Chromolaena odorata*, and Parameswari and Suriyavathana (2012) have assessed antioxidant activity. The antifungal activity of few medicinal plants which also includes *Chromolaena odorata* was studied by Sharanya *et al.* (2013). Archana *et al.* (2014) have studied the antimalarial activity of compounds derived from *Chromolaena odorata* and have opined that falcarinol could be developed as a powerful drug for malaria.

21.5.3 Ipomoea carnea Jacq.

Ipomoea carnea (Convolvulaceae) is a serious invader that infests aquatic and semi-aquatic ecosystems in all the districts of Tamil Nadu. It exhibits vigorous growth and thrives well in river beds, canals and other waterlogged areas. It quickly chokes aquatic ecosystems by prolific ramet and seed production. Invasion of *Ipomoea carnea*, along with other invasive plants in different regimes of forest and human-made ecosystems of Tamil Nadu, were studied by Sundarapandian *et al.* (2012). This invasive semi-aquatic weed is often controlled by mechanical removal followed by chemical control using herbicides (2,4-D and glyphosate) in the water reservoirs.

Ganesh *et al.* (2008) studied the production of methane gas by anaerobic digestion of *Ipomoea carnea*, Sahayaraj and Ravi (2008) conducted a preliminary phytochemical analysis of the leaves. and Anand *et al.* (2013) have investigated the anticancerous effects of the hydroalcoholic extract of *Ipomoea carnea* Ehrlich ascites carcinoma. Pandian *et al.* (2013) have found that acid-activated *Ipomoea carnea* could be used as an adsorbent to remove Cu (II) ions from polluted water. *Ipomoea carnea* can also be used for phytoremediation because of its tendency to accumulate significant levels of cadmium and mercury (Kadirvel and Jegadeesan, 2014; Kavitha and Jegadeesan, 2014). *Ipomoea carnea* leaves are used by local people for making garlands, and stems are used to make baskets as a means of positive utilization.

21.5.4 Mikania micrantha Kunth

Mikania micrantha (Asteraceae), a native of South America and a notorious invader of all the districts of Tamil Nadu, was brought into the country for camouflaging the army during the Second World War (Kohli *et al.*, 2012) and also as ground cover for tea plantations (Tripathi *et al.*, 2012b). In Tamil Nadu, *Mikania micrantha* distribution was first noted by Swamy and his research group in teak plantations adjacent to Periyar wildlife sanctuary in 1998, but it has now colonized the entire Western Ghats of Tamil Nadu and a few places in Palani Hills. The infestation of this plant has been recently reported even from the Cauvery delta and beyond. It has been declared as one of the ten most invasive plants of the world (Lowe *et al.*, 2001). It is widely prevalent in all tropical countries and affects almost all types of ecosystems, such as forests, grasslands, plantations, disturbed and fragmented habitats. The high invasion success of *Mikania micrantha* could be attributed to its prolific ramet and rosette production, efficient seed dispersal mechanisms and allelopathy. Elavazhagan *et al.* (2009) have segregated and described the characteristic features of endophytic bacteria from this weed. Prabu *et al.* (2014) studied the ecophysiological traits of *Mikania micrantha* in the Western Ghats, Tamil Nadu.

21.5.5 Prosopis juliflora (Sw.) DC

Prosopis juliflora (Fabaceae), a native of Central and South America was introduced in India mainly for fuel. Now, it has widely spread across the country colonizing all the districts of Tamil Nadu. Adaptation to dry, xeric habitats, aggressive growth and dispersal by cattle, have paved the way for ecesis of *Prosopis juliflora*. Saraswathi *et al.* (2012) have studied the significance of *Prosopis juliflora* as a livelihood support to the rural economy of Sivagangai district. According to Sato (2013), *Prosopis juliflora* is being utilized as an energy source and boosts the economy of rural livelihood in semi-arid districts of Tamil Nadu such as Ramanathapuram, Sivagangai, Virudhunagar and Pudukkottai. Diversity and the nesting success of wetland birds at Vettangudi Bird Sanctuary have been reported to be adversely affected by the aggressive establishment of *Prosopis juliflora* (Chandrasekaran *et al.*, 2014).

Nandagopalan *et al.* (2014) analysed the uses and commercial prospects that could be derived from *Prosopis juliflora* in Pudukkottai district. Shakila and Sukumar (2014) conducted pharmacological studies on *Prosopis juliflora*. Sivakumar *et al.* (2009) isolated phytochemicals from the bark and assessed their anti-inflammatory potential. Selvi *et al.* (2010) isolated the hydrophilic mucilage from seeds and studied its possible use as a tablet binder. Sharmila *et al.* (2013) studied the biodegrading ability of *Prosopis juliflora* extracts to treat tannery effluents. Manual cutting of stems for firewood and excavation of roots is the preferred way of eradicating this plant.

21.5.6 Parthenium hysterophorus L.

Parthenium hysterophorus (Asteraceae), a native of tropical America, has long been regarded not only as a notorious invader threatening ecosystems, but also for causing respiratory and skin diseases in humans. It has been supposed to have gained entry via contaminated cereal grains or pasture seeds that were imported from the USA (GISD, 2010). This herb quickly establishes itself in an introduced habitat owing to its great allelopathic potential and large seed production that is easily disseminated through wind, water and transport (Haseler, 1976). In addition, this weed is able to thrive in a wide range of environmental conditions (Batish *et al.*, 2012). *Parthenium hysterophorus* is largely controlled by mechanical removal and application of weedkillers and, of late, a considerable decline in the populations of this noxious weed has been observed. Pandiselvi *et al.* (1998) studied the growth, resource-allocation patterns and reproductive effort of *Parthenium hysterophorus* in a semi-arid environment in Madurai. Ramanujam *et al.* (2011) studied the usage of *Parthenium hysterophorus* as a bio-control agent to protect tomato plants from *Alternaria alternate*, which causes leaf-spot disease. Jeyalakshmi *et al.* (2011) studied the occurrence and distribution of soil microbes under *Parthenium hysterophorus* from all the districts of Tamil Nadu. They have isolated 13 micro-organisms from the *Parthenium*-infested soils from Tamil Nadu and, among them, *Aspergillus niger* and *Rhizopus stolonifer* were present in all the districts. Anbalagan *et al.* (2012) experimented on the biomanagement of *Parthenium hysterophorus* by vermicomposting using an earthworm, *Eisenia fetida*. Gnanavel (2013) published a detailed review on the habitat, morphology, biology, allelopathic characteristics and control measures of *Parthenium hysterophorus* and its adverse effects on agriculture, and human and animal health. Krishnavignesh *et al.* (2013) conducted a phytochemical screening and assessed the antimicrobial activity of *Parthenium hysterophorus*, and Malarkodi and Manoharan (2013) studied its antifungal activity against *Aspergillus niger, Candida albicans* and *Candida kefyr*. Kalaiselvi *et al.* (2013) synthesized silver nanoparticles from the leaf extracts of this weed and also analysed its antibacterial and antioxidant potential.

21.5.7 Mesosphaerum suaveolens (L.) Kuntze

Mesosphaerum suaveolens (formerly called *Hyptis suaveolens*, Lamiaceae), a native of central and south America, has been reported to dominate most of the ecosystems in all the districts of Tamil Nadu. It is a vigorous invader of rail tracks, roadsides, backyards, forest clearings, wastelands, dry localities and fragmented or disturbed habitats. The aggressive growth of this ruderal weed is due to its large seed production; seeds are widely disseminated, along with vegetative propagation and regeneration. Besides these, it also secretes allelochemicals that kill the neighbouring native plants. Moreover, the unpleasant aroma that emanates from this herb protects it from being eaten by cattle. This weed is often eradicated with physical or chemical control methods. The inhibitory effect of the leaf extracts on fish pathogens was studied by Malar *et al.* (2012). Tennyson *et al.* (2012) tested the extracts of the aerial parts of the plant for larvicidal activity against *Aedes aegypti*, which is a dengue and chikungunya vector. Elumalai *et al.* (2013) evaluated the biological activity of plant extracts against *Culex quinquefasciatus* mosquito. Priyadharshini and Sujatha (2013) conducted antioxidant and cytotoxic studies of the extracts of the leaves. Muthukrishnan *et al.* (2014) studied the inhibition of mild steel corrosion by the aqueous leaf extracts. Rajarajan *et al.* (2014) conducted a phytochemical analysis and analysed the antifungal properties of plant extracts on aflatoxin-producing fungi such as *Aspergillus flavus* and *Aspergillus parasiticus*.

21.5.8 Ageratum conyzoides L.

Ageratum conyzoides (Asteraceae) is a well-noted invader in all the districts of Tamil Nadu. A native of tropical America, it has become naturalized in several parts of the country. This fast-growing herb exhibits strong allelopathy and also aggressive stolon formation. In addition, it has a high regenerative and reproductive potential, producing a large number of minute, lightweight seeds alongside pappi for easy dissemination by wind (Kohli *et al.*, 2012). Moreover, it has a long flowering period (Kaur *et al.*, 2012) which helps in successful colonization within a short time. This serious and highly adapted weed often prefers to grow in shady habitats, cleared lands, disturbed, fragmented and abandoned wastelands, grasslands and scrublands. Physical and chemical methods are commonly used to subdue the growth of this weed. The potential use of *Ageratum conyzoides* as a biopesticide to control the population of *Spodoptera litura* was studied by Renuga and Sahayaraj (2009). The potential use of leaf extracts of *Ageratum conyzoides* as a bionematicide to control the root-knot nematode *Meloidogyne incognita* that infects *Vigna mungo* was experimentally observed by Pavaraj *et al.* (2010). Varadharajan and Rajalingam (2011) have observed anticonvulsant activity of the methanolic extracts of *Ageratum conyzoides*. The wound healing activity of the herbal extracts of various medicinal plants which also includes *Ageratum conyzoides* was listed by Kumarasamyraja *et al.* (2012). The adverse effects of the allelochemicals of *Ageratum conyzoides* on green and black grams were experimentally studied by Jayaraman and Ramalingam (2014). Natarajan *et al.* (2014) have studied the inhibitory effects of the extracts of this plant on germination and growth of *Sesamum indicum*. The phytochemical constituents of the leaves of *Ageratum conyzoides* was assessed by Anbarasan and Prabhakaran (2014).

21.5.9 Eichhornia crassipes (Mart.) Solms

Eichhornia crassipes (Pontederiaceae), commonly known as water hyacinth, is a native of South America that has become widely naturalized in most of the wetlands in all the districts of Tamil Nadu. It is often referred to as the world's worst aquatic weed, as it affects water flow, blocks sunlight from reaching other aquatic plants and deprives the water of oxygen, often leading to the death of fish and other aquatic animals (Frezina, 2013). Its unique physiological growth and reproductive strategies are the main reasons for its ecesis in aquatic ecosystems. Although physical, chemical, biological and cultural control methods are in practice, most of them have only negative impacts due to the fast regrowth of water hyacinth (Frezina, 2013). The impact of *Neochetina eichhorniae* as a biological control agent to suppress the population of this aquatic invader in Singanallur pond, Coimbatore, was studied by Yasotha and Lekeshmanaswamy (2012). Chinnusamy *et al.* (2012b) evaluated the effect of post-emergence herbicide management on water hyacinth. *Eichhornia crassipes* has been recognized as a main contributor of eutrophication in Veeranam lake (Saravanakumar and Prabhakaran, 2013). Deivasigamani (2013) assessed the bio-efficacy of three herbicides to control water hyacinth and the influence of these herbicides on fish mortality. Of late, several initiatives have been taken to control this weed by utilizing it for economic and eco-friendly purposes.

Sivaraj *et al.* (2010) prepared activated carbon from *Eichhornia crassipes*, which was assessed for its usage as an adsorbent to remove dyes. Jayanthi and Lalitha (2011) assessed the reducing potential of the solvent extracts of *Eichhornia crassipes*, to develop antioxidants. Acute toxicity studies of extracts were carried out by Lalitha *et al.* (2012), and Rani *et al.* (2014) have experimented on the usage of *Eichhornia crassipes* as oil sorbent to treat oil spillage. Firdhouse and Lalitha (2014) studied using it as a reductant to produce graphene from graphene oxide.

21.5.10 *Kappaphycus alvarezii* (Doty) Dotyex P.C. Silva

Kappaphycus alvarezii (Solieriaceae) is a red macroalga, native to the Philippines. It was introduced for mariculture into the Gulf of Mannar in 2000 (Chandrasekaran *et al.*, 2008). Since then, it has spread widely and has become a serious invader of the coral reefs of the Ramanathapuram, Pudukottai, Thoothukudi, Thanjavur and Kanyakumari districts in Tamil Nadu. The bioinvasion of this alga on corals was reported for the first time in Gulf of Mannar, India, by Chandrasekaran *et al.* (2008). It rapidly invades and successfully establishes itself due to vegetative propagation, its ability to grow in low and high tides, phenotypic plasticity, rapid proliferation and chemical defence mechanisms against plant feeders (for ecological success of alien/invasive algae in Hawaii, see Chandrasekaran *et al.*, 2008). Athithan (2014) studied the growth performance of *Kappaphycus alvarezii* when it is cultivated in a floating bamboo cage in Thoothukudi coast. It is usually cultivated by a raft method, where seeds are spread on bamboo rafts, which are anchored in the sea. It is mainly utilized for its yield of kappa-carrageenan and for its use as a liquid fertilizer. Eswaran *et al.* (2005) obtained a patent (EP 1534757 A1) for developing an integrated method for maximum utilization of fresh biomass that can be crushed to release the sap which is used as a liquid fertilizer, while the residue can be used to extract kappa-carrageenan. Rajasree and Gayathri (2014) conducted a study on the employment potential of fisherwomen engaged in *Kappaphycus alvarezii* farming and the subsequent economic benefits in Kanyakumari. Kamalakannan *et al.* (2014) described the unsuccessful attempt and negative outcome of the eradication of *Kappaphycus alvarezii* from the coral reef ecosystem of Kurusadai Island, Gulf of Mannar.

Thirumaran and Anantharaman (2009) studied the seasonal variation in daily growth rate of *Kappaphycus alvarezii* among different seedling densities in the Vellar estuary. Johnson and Gopakumar (2011) have discussed at length the status and restraints faced by the farming of *Kappaphycus alvarezii* in the coastal zones of Tamil Nadu. Venkatesh *et al.* (2011) conducted preliminary studies on antixanthomonas activity and also screening of phytochemicals and antimicrobial compounds from *Kappaphycus alvarezii*. Ganesan *et al.* (2013) studied the ability of *Kappaphycus alvarezii* to reduce silver nitrate and thereby to synthesize silver nanoparticles. Alwarsamy and Ravichandran (2014) studied the *in-vitro* antibacterial activity of the extracts of *Kappaphycus alvarezii* that were collected from Rameswaram. Pushparaj *et al.* (2014) has also studied antibacterial activity of *Kappaphycus alvarezii* extracts against six human pathogens. Ranganayaki *et al.* (2014) conducted a phytochemical exploration of the aqueous extract of *Kappaphycus alvarezii* and tested the metabolites for anti-inflammatory activity.

21.6 Factors Facilitating Invasion

Several factors facilitate further invasion of exotic plants that are at different levels of invasion. Some of them are urbanization, increase in CO_2 concentration, coupled with global warming resulting in hotter, drier climates, human migration across the world, modernization leading to deforestation and degradation of natural habitats, and canopy opening in thick forests, land-use change and accidental introductions. Moreover, similarities in climate with that of native range and absence of natural enemies in the new, introduced ecosystem also paves the way for the successful growth and establishment of invasive weeds. Moreover, the notorious invaders are naturally endowed with unique, specialized adaptive features that are generally lacking in the native flora, leading to the quick extinction of the latter, and thus enhancing the chances for successful survival of the former.

21.7 Eradication by Exploitation

Invasive plants are largely controlled by physical methods (such as hand pulling, stem cutting, root digging, mowing, prescribed burning) that are labour-intensive and expensive; chemical methods (applying herbicides via foliar sprays, bark application and cut-stem treatment); time-consuming biological methods (using microbes, insects, etc.); or cultural methods (mulching, canopy closure, soil solarization). Often, an integrated approach combining two or more methods has proven to be more successful in the eradication. However, since invasive plants are generally highly resistant and with a good regenerative potential, they pop out very soon with even greater vigour after implementation of any control method. Therefore, a wiser way of effective removal of the invasive plants would be by exploitation for economic and ecological needs. For example, the stems of *Prosopis juliflora* are used for timber and firewood, *Lantana camara* for making baskets and furniture, *Eichhornia crassipes* for phytoremediation and biogas production, *Chromolaena odorata* for green manure, *Ageratum conyzoides* as a natural herbicide for weed control in paddy fields (Xuan *et al.*, 2004), *Mikania micrantha* for drug production and enrichment of soil fertility (Chen *et al.*, 2009), *Kappaphycus alvarezii* for carrageenan production, and so on. Undoubtedly, more options for effective management by exploitation would open up further research and field studies.

21.8 Conclusion

Since prevention is always better than cure, i.e. restoring the invaded habitats, it is recommended that more stringent quarantine measures be adopted that would check the entry of the infesting, invasive weeds. As for the existing invasive weeds, more studies on the distribution and the stage of invasion ought to be carried out to plan successful management strategies. These apart, a general awareness should be raised among the public on the common invaders, warning against planting them. Overall, it could be concluded that invasive plants could be better tackled with a smart, optimistic way of handling using an integrated approach.

Acknowledgements

The authors thank Dr S. Chandrasekaran, Associate Professor and Dr C. Muthumperumal, Kothari Postdoctoral Fellow, Department of Plant Sciences, School of Biological Sciences, Madurai Kamaraj University, for their support.

References

Alwarsamy, M. and Ravichandran, R. (2014) In-vitro antibacterial activity of *Kappaphycus alvarezii* extracts collected from Mandapam Coast, Rameswaram, Tamil Nadu. *International Journal of Innovative Research in Science, Engineering and Technology* 3(1), 8436–8440.

Anand, G., Sumithira, G., Raja, R.C., Muthukumar, A. and Vidhya, G. (2013) In vitro and in vivo anticancer activity of hydroalcoholic extract of *Ipomoea carnea* leaf against Ehrlich Ascites Carcinoma cell lines. *International Journal of Advance Pharmacology and Genetic Research* 1(1), 39–54.

Anbalagan, M., Manivannan, S. and Prakasm, B.A. (2012) Biomanagement of *Parthenium hysterophorus* (Asteraceae) using an earthworm, *Eisenia fetida* (Savigny) for recycling the nutrients. *Advances in Applied Science and Research* 3(5), 3025–3031.

Anbarasan, R. and Prabhakaran, J. (2014) Leaf phytochemistry of invasive alien species 'Billy goat weeds' (*Ageratum conyzoides* L.). *Asian Journal of Science and Technology* 5(11), 634–638.

Aravindhan, V. and Rajendran, A. (2014) Diversity of invasive plant species in Boluvampatti Forest Range, the Southern Western Ghats, India. *American-Eurasian Journal of Agricultural & Environmental Sciences* 14(8), 724–731.

Archana, C.M., Harini, K., Showmya, J.J. and Geetha, N. (2014) *Insilico* docking study of *Chromolaena odorata* derived compounds against antimalarial activity. *International Journal of Pharmacological Science and Business Management* 2(9), 42–58.

Arun, K.D., Prabu, N.R., Stalin, N. and Swamy, P.S. (2014) Chemical composition of essential oils in the leaves of *Delairea odorata* during flowering and post-flowering periods. *Chemistry of Natural Compounds* 49, 1158–1159.

Athithan, S. (2014) Growth performance of a seaweed, *Kappaphycus alvarezii* under lined earthen pond condition in Tharuvaikulam of Thoothukudi coast, South East of India. *Research Journal of Animal, Veterinary and Fishery Sciences* 2(1), 6–10.

Ayyanar, M. and Ignacimuthu, S. (2009) Herbal medicines for wound healing among tribal people in Southern India: ethnobotanical and scientific evidences. *International Journal of Applied Research in Natural Products* 2(3), 29–42.

Batish, D.R., Kohli, R.K., Singh, H.P. and Kaur, G. (2012) Biology, ecology and spread of the invasive weed *Parthenium hysterophorus* in India. In: Bhatt, J.R., Singh, J.S., Singh, S.P., Tripathi, R.S. and Kohli, R.K. (eds) *Invasive Alien Plants: An Ecological Appraisal for the Indian Subcontinent*. CAB International, Wallingford, Oxfordshire, UK, pp. 10–18.

Binggeli, P. (1996) A taxonomic, biogeographical and ecological overview of invasive woody plants. *Journal of Vegetation Science* 7, 121–124.

CBD (1992) The Convention on Biodiversity website. Available at: www.cbd.int/convention (accessed 27 May 2016).

Chandrasekaran, S. and Swamy, P.S. (2002) Biomass, litterfall and aboveground net primary productivity of herbaceous communities in varied ecosystems at Kodayar in the Western Ghats of Tamil Nadu. *Agricultural Ecosystems and Environment* 88, 61–71.

Chandrasekaran, S. and Swamy, P.S. (2010) Growth patterns of *Chromolaena odorata* in varied ecosystems at Kodayar in the Western Ghats, India. *Acta Oecologica* 36, 383–392.

Chandrasekaran, S., Sundarapandian, S.M. and Swamy, P.S. (1997) Contribution of exotic weeds to plant community structure and primary production in successional fallows at Kodayar in Western Ghats of Tamil Nadu. *International Journal of Ecology and Environmental Science* 23, 381–388.

Chandrasekaran, S., Sundarapandian, S.M. and Swamy, P.S. (1998) Influence of introduced species composition on soil physic-chemical characteristics in natural and man-modified ecosystems at Kodayar in the Western Ghats of Tamil Nadu. *Proceedings of International Conference on Conservation of tropical species, communities and ecosystems*, TBGRI, Thiruvananthapuram, India.

Chandrasekaran, S., Sundarapandian, S.M., Chandrasekar, P. and Swamy, P.S. (2001) Exotic plant invasions in disturbed and man-modified forest ecosystems in the Western Ghats of Tamil Nadu. In: Varma, R.V., Bhat, K.V., Muralidharan, E.M. and Sharma J.K. (eds) Tropical forestry research: challenges in the new millennium. *Proceedings of the International Symposium*, Peechi, Kerala, India, pp. 32–39.

Chandrasekaran, S., Nagendran, N.A., Pandiaraja, D., Krishnankutty, N. and Kamalakannan, B. (2008) Bioinvasion of *Kappaphycus alvarezii* on corals in the Gulf of Mannar, India. *Current Science* 94(9), 1167–1172.

Chandrasekaran, S., Saraswathi, K., Saravanan, S., Kamaladhasan, N. and Nagendran, N.A. (2014) Impact of *Prosopis juliflora* on nesting success of breeding wetland birds at Vettangudi Bird Sanctuary, South India. *Current Science* 106(5), 676–678.

Chen, B.M., Peng, S.L. and Ni, G.Y. (2009) Effect of the invasive plant Mikania micrantha H.B.K. and soil nitrogen availability through allelopathy in South China. *Biological Invasions* 11, 1291–1299.

Chinnusamy, C., Nandhakumar, M.R., Govindarajan, K. and Muthukrishnan, P. (2012a) Incidence of quarantine invasive weed *Solanum carolinense* L. in different ecosystems of Tamil Nadu. *Pakistan Journal of Weed Science and Research* 18, 749–755.

Chinnusamy, C., Janaki, P., Arthanari, P.M. and Muthukrishnan, P. (2012b) Effects of post-emergence herbicides on water hyacinth (*Eichhornia crassipes*) – tank culture experiment. *Pakistan Journal of Weed Science and Research* 18, 105–111.

Darwin, C. (1872) *On the Origin of Species*, 6th edn. John Murray, London, UK.

Deivasigamani, S. (2013) Influence on certain herbicides for the control of water hyacinth (*Eichhornia crassipes* (Mart.) Solms) and its impact on fish mortality. *Journal of Biofertilizers and Biopesticides* 4, 138.

Denslow, J.S. (2007) Managing dominance of invasive plants in wild lands. *Current Science* 93(11), 1579–1586.

Deshmukh, U.B., Shende, M.B. and Rathor, O.S. (2012) Invasive alien angiosperms of Chandrapur district of Maharashtra (India). *Bionanotechnology Frontier* 5(2), 100–103.

Doss, A., Parivuguna, V., Vijayasanthi, M. and Surendran, S. (2011) Antibacterial evaluation and phytochemical analysis of certain medicinal plants, Western Ghats, Coimbatore. *Journal of Research in Biology* 1, 24–29.

Ehrenfeld, J.G. (2003) Effects of exotic plant invasions on soil nutrient cycling processes. *Ecosystems* 6, 503–523.

Elavazhagan, T., Jayakumar, S., Balakrishnan, V. and Chitravadivu, C. (2009) Isolation of endophytic bacteria from the invasive alien weed, *Mikania micrantha* and their molecular characterisation. *American-European Journal of Science and Research* 4(3), 154–158.

Elumalai, D., Kaleena, P.K., Fathima, M. and Muthappan, M. (2013) Evaluation of the biological activity of *Hyptis suaveolens* L. Poit and *Leucas asper* a (Wild) against *Culex quinquefasciatus* (Say). *International Journal of Bioscience Research* 2(6), 1–6. Available at brtpublishers.com/manuscriptfinal/IJBR-12-0126_IJBR-2012-0124-%20Kaleena_final.pdf (accessed 27 May 2016).

Eswaran, K., Ghosh, P.K., Mehta, A.S., Mody, K.H., Pandya, J.B. *et al.* (2005) Method for production of carrageenan and liquid fertilizer from fresh seaweeds. EP 1534757 A1 (text from WO2004016656A1). Available at: www.google.com/patents/EP1534757A1?cl=en (accessed 27 May 2016).

Firdhouse, M.J. and Lalitha, P. (2014) Phyto-reduction of graphene oxide using the aqueous extract of *Eichhornia crassipes* (Mart.) Solms. *International Nano Letters* 4, 103–108.

Frezina, N.C.A. (2013) Assessment and utilization of water hyacinth in the water bodies of Tamil Nadu. *International Journal of Scientific Research and Reviews* 2(1), 58–77.

Gandhi, D.S. and Sundarapandian, S.M. (2014) Diversity and distribution pattern of understory vegetation in tropical dry forests of Sathanur Reserve Forest in Eastern Ghats, India. *International Journal of Science and Nature* 5(3), 452–461.

Ganesh, P.S., Sanjeevi, R., Gajalakshmi, S., Ramasamy, E.V. and Abbasi, S.A. (2008) Recovery of methane-rich gas from solid-feed anaerobic digestion of ipomoea (*Ipomoea carnea*). *Bioresource Technology* 99, 812–818.

Ganesan, V., Devi, A.J., Astalakshmi, A., Nima, P. and Thangaraja, A. (2013) Eco-friendly synthesis of silver nanoparticles using a sea weed, *Kappaphycus alvarezii* (Doty) Doty ex P.C. Silva. *International Journal of Engineering and Advanced Technology* 2(5), 559–563.

Ganjewala, D., Sam, S. and Khan, K.H. (2009) Biochemical compositions and antibacterial activities of *Lantana camara* plants with yellow, lavender, red and white flowers. *EurAsian Journal of BioSciences* 3, 69–77.

GISD (2010) Global Invasive Species Database. Available at: www.issg.org/database/species/search.asp?st=100ss (accessed 27 May 2016).

Gnanavel, I. (2013) *Parthenium hysterophorus* L.: a major threat to natural and agro eco-systems in India. *Science International* 1(5), 124–131.

Haseler, W.H. (1976) *Parthenium hysterophorus* L. in Australia. *PANS* (*Pest Articles and News Summaries*) 22, 515–517.

Heywood, V.H. (1989) Patterns, extents and modes of invasions. In: Drake, J.A. (ed.) *Biological Invasion: A Global Perspective*, John Wiley, New York, USA, pp. 31–60.

Huang, Q.Q., Wu, J.M., Bai, Y.Y., Zhou, L. and Wang, G.X. (2009) Identifying the most noxious invasive plants in China: role of geographical origin, life form and means of introduction. *Biodiversity Conservation* 18, 305–316.

Jaikumar, M. (2012) A review on water hyacinth (*Eichhornia crassipes*) and phytoremediation to treat aqua pollution in Velachery Lake, Chennai – Tamil Nadu. *International Journal of Recent Science Research* 3(2), 95–102.

Jayanthi, P. and Lalitha, P. (2011) Reducing power of the solvent extracts of *Eichhornia crassipes* (Mart.) Solms. *International Journal of Pharmacy and Pharmaceutical Science* 3(3), 126–128.

Jayaraman, P. and Ramalingam, A. (2014) Allelopathy potential of invasive Alien species *Ageratum conyzoides* L. on growth and developmental responses of green gram (*Vigna radiata* (L.) R. Wilczek) and black gram (*Vigna mungo* (L.) Hepper). *International Journal of Advanced Pharmacy and Biological Chemistry* 3(2), 437–442.

Jeyalakshmi, C., Doraisamy, S. and Valluvaparidasan, V. (2011) Occurrence of soil microbes under *Parthenium* weed in Tamil Nadu. *Indian Journal of Weed Science* 43(3,4), 222–223.

Johnson, B. and Gopakumar, G. (2011) *Farming of the Seaweed* Kappaphycus alvarezii *in Tamil Nadu Coast – Status and Constraints*. Marine Fisheries Information Service T and E Ser., No. 208. Mandapam, Tamil Nadu, India. Available at: eprints.cmfri.org.in/8882/1/208-1.pdf (accessed 27 May 2016).

Joshi, A.A., Mudappa, D. and Raman, T.R.S. (2009) Brewing trouble: coffee invasion in relation to edges and forest structure in tropical rainforest fragments of the Western Ghats, India. *Biological Invasions* 11, 2387–2400.

Kadirvel, K.K. and Jegadeesan, M. (2014) Mercury and cadmium accumulation in selected weed plants: implications for phytoremediation. *Asian Journal of Plant Science Research* 4(5), 1–4.

Kalaiselvi, M., Subbaiya, R. and Selvam, M. (2013) Synthesis and characterization of silver nanoparticles from leaf extract of *Parthenium hysterophorus* and its anti-bacterial and antioxidant activity. *International Journal of Current Microbiology and Applied Science* 2(6), 220–227.

Kalita, S., Kumar, G., Karthik, L. and Rao, K.V.B. (2012) A review on medicinal properties of *Lantana camara* Linn. *Research Journal of Pharmacy and Technology* 5(6), 711–715.

Kamalakannan, B., Jeevamani, J.J.J., Nagendran, N.A., Pandiaraja, D. and Chandrasekaran, S. (2014) Impact of removal of invasive species *Kappaphycus alvarezii* from coral reef ecosystem in Gulf of Mannar, India. *Current Science* 106(10), 1401–1408.

Kanagavalli, R. (2011) In-vitro studies on the antiviral property of *Lantana camara* flowers. *International Journal of Pharmacy Research and Development* 3(10), 111–116.

Kannan, R., Shackleton, C.M. and Shaanker, R.U. (2013) Playing with the forest: invasive alien plants, policy and protected areas in India. *Current Science* 104(9), 1159–1165.

Kaur, S., Batish, D.R., Kohli, R.K. and Singh, H.P. (2012) *Ageratum conyzoides*: an alien invasive weed in India. In: Bhatt, J.R., Singh, J.S., Singh, S.P., Tripathi, R.S. and Kohli, R.K. (eds) *Invasive Alien Plants: An Ecological Appraisal for the Indian Subcontinent*. CAB International, Wallingford, Oxfordshire, UK, pp. 57–76.

Kavitha, K.K. and Jegadeesan, M. (2014) Phytoremediation of soil mercury and cadmium by weed plants, *Trianthema Portulocastrum* L., *Saccharum Spontaneum* L. and *Ipomoea Carnea* Jacq. *International Journal of Scientific and Research Publications* 4(10), 1–3. Available at: www.ijsrp.org/research-paper-1014/ijsrp-p3435.pdf (accessed 27 May 2016).

Khuroo, A.A., Reshi, Z.A., Malik, A.H., Weber, E., Rashid, I. and Dar, G.H. (2012a) Alien flora of India: taxonomic composition, invasion status and biogeographic affiliations. *Biological Invasions* 14, 99–113.

Khuroo, A.A., Reshi, Z.A., Dar, G.H. and Hamal, I.A. (2012b) Plant invasions in Jammu and Kashmir State, India. In: Bhatt, J.R., Singh, J.S., Singh, S.P., Tripathi, R.S. and Kohli, R.K. (eds) *Invasive Alien Plants: An Ecological Appraisal for the Indian Subcontinent*. CAB International, Wallingford, Oxfordshire, UK, pp. 216–226.

Kohli, R.K., Batish, D.R., Singh, H.P. and Dogra, K.S. (2006) Status, invasiveness and environmental threats of three tropical American invasive weeds (*Parthenium hysterophorus* L., *Ageratum conyzoides* L., *Lantana camara* L.) in India. *Biological Invasions* 8, 1501–1510.

Kohli, R.K., Batish, D.R., Singh, J.S., Singh, H.P. and Bhatt, J.R. (2012) Plant invasion in India: an Overview. In: Bhatt, J.R., Singh, J.S., Singh, S.P., Tripathi, R.S. and Kohli, R.K. (eds) *Invasive Alien Plants: An Ecological Appraisal for the Indian Subcontinent*. CAB International, Wallingford, Oxfordshire, UK, pp. 1–9.

Krishnavignesh, L., Mahalakshmipriya, A. and Ramesh, M. (2013) *In vitro* analysis of phytochemical screening and antimicrobial activity of *Parthenium hysterophorus* L. against pathogenic microorganisms. *Asian Journal of Pharmacy and Clinical Research* 6(5), 41–44.

Kuberan, T., Balamurugan, A., Vidhyapallavi, R., Nepolean, P., Jayanthi, R., Beulah, T. and Premkumar, R. (2012) *In vitro* evaluation certain plant extracts against *Glomerella cingulata* causing brown blight disease of tea. *World Journal of Agricultural Science* 8(5), 464–467.

Kumar, B.A. (2009) Invasive species, a big threat to biodiversity of Nilgiris. *Times of India*, 8 June. Available at: timesofindia.indiatimes.com/city/chennai/Invasive-species-a-big-threat-to-biodiversity-of-Nilgiris/articleshow/4629187.cms (accessed 27 May 2016).

Kumar, M.S. and Maneemegalai, S. (2008) Evaluation of larvicidal effect of *Lantana Camara* Linn against mosquito species *Aedes aegypti* and *Culex quinquefasciatus*. *Advances in Biological Research* 2(3–4), 39–43.

Kumarasamyraja, D., Jeganathan, N.S. and Manavalan, R. (2012) A review on medicinal plants with potential wound healing activity. *International Journal of Pharmacy Science* 2(4), 105–111.

Lalitha, P., Sripathi, S.K. and Jayanthi, P. (2012) Acute toxicity studies of extracts of *Eichhornia crassipes* (Mart.) Solms. *Asian Journal of Pharmacy and Clinical Research* 5(4), 59–61.

Lowe, S., Browne, M., Boudjelas, S. and Depoorter, M. (2001) *100 of the World's Worst Invasive Alien Species. A Selection from the Global Invasive Species Database*. IUCN/SSC Invasive Species Specialist Group (ISSG), Auckland, New Zealand.

Mack, R.N., Simberloff, D., Lonsdale, W.M., Evans, J., Clout, M. and Bazzaz, F.A. (2000) Biotic invasions: causes, epidemiology, global consequences, and control. *Ecological Applications* 10, 689–710.

Malar, R.J.J.T., Sushna, S.L., Johnson, M., Janakiraman, N. and Ethal, R.J.J.T. (2012) Bio-efficacy of the leaves extracts of *Hyptis suaveolens* (L.) Poit against the fish pathogens. *International Journal of Life Science and Pharmacy Research* 2(1), 128–133.

Malarkodi, E. and Manoharan, A. (2013) Antifungal activity of *Parthenium hysterophorus* L. *Journal of Chemical and Pharmaceutical Research* 5(1), 137–139.

Mariajancyrani, J., Chandramohan, G. and Ravikumar, S. (2014) Terpenes and antimicrobial activity from *Lantana camara* leaves. *Research Journal of Recent Sciences* 3(9), 52–55.

McFadyen, R.E.C. (2002) *Chromolaena* in Asia and the Pacific: spread continues but control prospects improve. In: Zachariades, C. and Strathie, L. (eds) *Proceedings of the Fifth International Workshop on Biological Control and Management of Chromolaena odorata*. ARC – Plant Protection Research Institute, Durban, South Africa, pp. 13–18.

Muthukrishnan, P., Jeyaprabha, B. and Prakash, P. (2014) Mild steel corrosion inhibition by aqueous extract of *Hyptis suaveolens* leaves. *International Journal of Industrial Chemistry* 5, 5.

Muthumperumal, C. and Parthasarathy, N. (2010) A large-scale inventory of vine diversity in tropical forests of South Eastern Ghats, India. *Systematics and Biodiversity* 8, 289–300.

Nandagopalan, V., Doss, A. and Anand, S.P. (2014) Uses and commercial prospects of *Prosopis juliflora*, in Pudukkottai district, South India. *International Journal of Phototherapy* 4(4), 162–166.

Narasimhan, D., Arisdason, W., Irwin, S.J. and Gnanasekaran, G. (2010) Invasive Alien plant species of Tamil Nadu. *Proceedings of National Seminar on 'Invasive Species of Tamil Nadu'*. Department of Environment, Government of Tamil Nadu, Chennai, India, pp. 29–38. Available at: www.researchgate.net/publication/268522142_Invasive_Alien_Plant_Species_of_Tamil_Nadu (accessed 27 May 2016).

Natarajan, A., Elavazhagan, P. and Prabhakaran, J. (2014) Allelopathic potential of Billy Goat Weed *Ageratum conyzoides* L. and *Cleome viscosa* L. on germination and growth of *Sesamum indicum L. International Journal of Current Biotechnology* 2(2), 21–24.

Natheer, S.E., Sekar, C., Amutharaj, P., Rahman, M.S.A. and Khan, K.F. (2012) Evaluation of antibacterial activity of *Morinda citrifolia*, *Vitex trifolia* and *Chromolaena odorata*. *African Journal of Pharmacy and Pharmacology* 6(11), 783–788.

Nayar, M.P. (1997) Changing patterns of the Indian flora. *Bulletin of Botanical Survey of India* 19, 145–154.

Orapa, W., Englberger, K. and Lal, S.D. (2004) *Chromolaena* and other weed problems in the Pacific Islands. In: Day, M.D. and McFadyen, R.E.C. (eds) *Chromolaena* in the Asia-Pacific Region. ACIAR Technical Report No. 55, Australian Centre for International Agricultural Research, Canberra, Australia, p. 13.

Pandian, P., Arivoli, S., Marimuthu, V., Regis, A.P.P. and Prabaharan, T. (2013) Adsorption of Cu (II) ions from aqueous solution by activated *Ipomoea carnea*: equilibrium isotherms and kinetic approach. *International Journal of Engineering and Innovative Technology* 3(1), 156–162.

Pandiselvi, R., Sundarapandian, S.M. and Swamy, P.S. (1998) Growth and allocation strategy of *Parthenium hysterophorus* L. Under semi-arid environment at Madurai. *Proceedings of the Eighty-Fifth Session of the Indian Science Congress*, Hyderabad, India.

Parameswari, G. and Suriyavathana, M. (2012) *In-vitro* antioxidant activity of *Chromolaena odorata* (L.) King & Robinson. *International Research Journal of Pharmacy* 3(11), 187–192.

Parthasarathy, N., Pragasan, L.A. and Muthumperumal, C. (2012) Invasive alien plants in tropical forests of the South-eastern Ghats, India: ecology and management. In: Bhatt, J.R., Singh, J.S., Singh, S.P., Tripathi, R.S. and Kohli, R.K. (eds) *Invasive Alien Plants: An Ecological Appraisal for the Indian Subcontinent*. CAB International, Wallingford, Oxfordshire, UK, pp. 162–173.

Pavaraj, M., Karthikairaj, K. and Rajan, M.K. (2010) Effect of leaf extract of *Ageratum conyzoides* on the biochemical profile of black-gram *Vigna mungo* infected

by root-knot nematode, *Meloidogyne incognita*. *Journal of Biopesticides* 3(1), 313–316.
Pimental, D., Lach, L., Zuniga, R. and Morrison, D. (2000) Environmental and economic costs of nonindigenous species in the United States. *BioScience* 50, 53–63.
Prabu, N.R., Stalin, N. and Swamy, P.S. (2014) Ecophysiological attributes of *Mikania micrantha*, an exotic invasive weed, at two different elevations in the tropical forest regions of the Western Ghats, South India. *Weed Biology and Management* 14, 59–67.
Priyadharshini, S.D. and Sujatha, V. (2013) Antioxidant and cytotoxic studies on two known compounds isolated from *Hyptis suaveolens* leaves. *International Journal of Pharmacy and Pharmaceuticals Science* 5(4), 283–290.
Pushpakaran, B. and Gopalan, R. (2013) Study on the seasonally waterlogged grasslands of Mudumalai Tiger Reserve. *International Journal of Science Research Publications* 3(2), 1–6. Available at: www.ijsrp.org/research-paper-0213/ijsrp-p1471.pdf (accessed 27 May 2016).
Pushparaj, A., Raubbin, R.S. and Balasankar, T. (2014) Antibacterial activity of *Kappaphycus alvarezii* and *Ulva lactuca* extracts against human pathogenic bacteria. *International Journal of Current Microbiology and Applied Sciences* 3(1), 432–436.
Pyšek, P. (1998) Is there a taxonomic pattern to plant invasions? *Oikos* 82, 282–294.
Pyšek, P. and Richardson, D.M. (2008) Invasive plants. In: Jorgensen, S.E. and Brian, D.F. (eds) *Ecological Engineering Vol. [3] of Encyclopedia of Ecology*. Elsevier, Amsterdam, The Netherlands, pp. 2011–2020.
Rajarajan, P.N., Rajasekaran, K.M. and Devi, N.K.A. (2014) Antifungal and phytochemical screening of *Hyptis suaveolens* (L. Poit) lamiaceae on aflatoxin producing Fungi. *Scholars Academic Journal of Pharmacy* 3(1), 50–52.
Rajasree, S.R.R. and Gayathri, S. (2014) Women enterprising in seaweed farming with special references fisherwomen widows in Kanyakumari District Tamil Nadu. *Indian Journal of Coast Development* 17(1), 383. DOI: 10.4172/1410-5217.1000383.
Rajmohan, D. and Logankumar, K. (2011) Studies on the insecticidal properties of *Chromolaena odorata* (Asteraceae) against the life cycle of the mosquito, *Aedes aegypti* (Diptera: culicidae). *Journal of Biology Research* 4, 253–257.
Ramanujam, J.R., Kulothungan, S., Anitha, S. and Deepa, K. (2011) A study on compatibility of *Pseudomonas fluorescenes* L. and *Parthenium hysterophorus* L. as a biocontrol agent to leaf spot by *Alternaria alternata* f.sp. *lycopercisi* in tomato. *South Asian Journal of Biological Sciences* 1(2), 71–86.
Ramaswami, G. and Sukumar, R. (2014) *Lantana camara* L. (Verbenaceae) invasion along streams in a heterogeneous landscape. *Journal of Biosciences* 39, 717–726.
Rameshkumar, C., Murugesan, S. and Sundarapandian, S.M. (2005) Phytochemical and antibacterial studies of two important exotic medicinal plants. In: *Proceedings of National Symposium on Scope and Opportunities in Medicinal Plant Research*. J.J. College of Arts and Science, Pudukkottai, Tamil Nadu, India.
Randall, R.P. (2012) *A Global Compendium of Weeds*, 2nd edn. Department of Agriculture and Food, Western Australia.
Ranganayaki, P., Susmitha, S. and Vijayaraghavan, R. (2014) Study on metabolic compounds of *Kappaphycus alvarezii* and its *in-vitro* analysis of anti-inflammatory activity. *International Journal of Current Research and Academic Reviews* 2(10), 157–166.
Rani, M.J., Murugan, M., Subramaniam, P. and Subramanian, E. (2014) A study on water hyacinth *Eichhornia crassipes* as oil sorbent. *Journal of Applied Natural Sciences* 6(1), 134–138.
Rao, R.R. and Murugan, R. (2006) Impact of exotic adventives weeds on native biodiversity in India: implications for conservation. In: Rai, L.C. and Gaur, J.P. (eds) *Invasive Alien Species and Biodiversity in India*. Banaras Hindu University, Varanasi, Uttar Pradesh, India, pp. 93–109.
Rao, R.R. and Sagar, K. (2012) Invasive Alien weeds in the tropics: the changing pattern in the herbaceous flora of Meghalaya in North-east India. In: Bhatt, J.R., Singh, J.S., Singh, S.P., Tripathi, R.S. and Kohli, R.K. (eds) *Invasive Alien Plants: An Ecological Appraisal for the Indian Subcontinent*. CAB International, Wallingford, Oxfordshire, UK, pp. 189–198.
Reddy, C.S. (2008) Catalogue of invasive alien flora of India. *Life Science Journal* 5(2), 84–89.
Renuga, F.B. and Sahayaraj, K. (2009) Influence of botanicals in total head protein of *Spodoptera litura* (Fab.). *Journal of Biopesticides* 2(1), 52–55.
Sahayaraj, K. and Ravi, C. (2008) Preliminary phytochemistry of *Ipomea carnea* Jacq. and *Vitex negundo* Linn. leaves. *International Journal of Chemical Science* 6(1), 1–6.
Saraswathi, K., Nagendran, N.A., Karthigaicham, R. and Chandrasekaran, S. (2012) Livelihood support from an invasive species *Prosopis juliflora*. *International Journal of Current Science* 2012, 31–36.
Saravanakumar, K. and Prabhakaran, J. (2013) Aquatic floral populations in Veeranam Lake Command area, Tamil Nadu, India. *International Journal of Current Biotechnology* 1(7), 18–26. Available at: ijcb.mainspringer.com/1_7/704.html (accessed 27 May 2016).
Sato, T. (2013) Beyond water-intensive agriculture: expansion of *Prosopis juliflora* and its growing economic use in Tamil Nadu, India. *Land Use Policy* 35, 283–292.
Saxena, K.G. (1991) Biological invasion in the Indian subcontinent: review of invasion by plants. In: Ramakrishnan, P.S. (ed.) *Ecology of Biological Invasion in the Tropics*. International Scientific Publications, New Delhi, India, pp. 53–73.

Selvi, R.S., Gopalakrishanan, S., Ramajayam, M. and Soman, R. (2010) Evaluation of mucilage of *Prosopis juliflora* as tablet binder. *International Journal of Pharmacy and Pharmaceutical Science* 2(3), 157–160.

Shakila, K. and Sukumar, D. (2014) Pharmacological studies on *Prosopis juliflora*. *Indian Journal of Applied Research* 4(12), 80–82.

Sharanya, M., Oviya, I.R., Poornima, V. and Jeyam, M. (2013) Antifungal susceptibility testing of few medicinal plant extracts against *Aspergillus* spp. and *Microsporum* sp. *Journal of Applied Pharmacy Science* 3(8), S12–S16.

Sharma, G.P., Singh, J.S. and Raghubanshi, A.S. (2005) Plant invasions: emerging trends and future implications. *Current Science* 88, 726–734.

Sharmila, S., Rebecca, J.L. and Saduzzaman, M. (2013) Biodegradation of tannery effluent using *Prosopis juliflora*. *International Journal of Chemical Technology Research* 5(5), 2186–2192.

Simberloff, D., Parker, I.M. and Windle, P.M. (2005) Introduced species policy, management and future research needs. *Frontiers in Ecology and Environment* 3, 12–20.

Singh, I. (2012) Control of *Lantana* and restoration of biodiversity in reserve forests of Chandigarh: a case study. In: Bhatt, J.R., Singh, J.S., Singh, S.P., Tripathi, R.S. and Kohli, R.K. (eds) *Invasive Alien Plants: An Ecological Appraisal for the Indian Subcontinent*. CAB International, Wallingford, Oxfordshire, UK, pp. 292–298.

Singh, K.P., Shukla, A.N. and Singh, J.S. (2010) State-level inventory of invasive alien plants, their source regions and use potential. *Current Science* 99(1), 107–114.

Sivakumar, T., Srinivasan, K., Rajavel, R., Vasudevan, M., Ganesh, M., Kamalakannan, K. and Mallika, P. (2009) Isolation of chemical constituents from *Prosopis juliflora* bark and anti-inflammatory activity of its methanolic extracts. *Journal of Pharmacy Research* 2(3), 551–556.

Sivaraj, R., Venckatesh, R. and Gowri Sangeetha, G. (2010) Activated carbon prepared from *Eichhornia crassipes* as an adsorbent for the removal of dyes from aqueous solution. *International Journal of Engineering Science Technology* 2(6), 2418–2427.

Srivastava, S., Dvivedi, A. and Shukla, R.P. (2014) Invasive alien species of terrestrial vegetation of North-Eastern Uttar Pradesh. *International Journal of Forest Research* 2014, 1–9. Article ID 959875. Available at: www.hindawi.com/journals/ijfr/2014/959875 (accessed 27 May 2016).

Sundarapandian, S. and Karoor, P.J. (2013) Edge effects on plant diversity in tropical forest ecosystems at Periyar Wildlife Sanctuary in the Western Ghats of India. *Journal of Forest Research* 24(3), 403–418.

Sundarapandian, S.M. and Swamy, P.S. (2000) Forest ecosystem structure and composition along an altitudinal gradient in the Western Ghats, South India. *Journal of Tropical Forest Science* 12(1), 104–123.

Sundarapandian, S.M., Chandrasekaran, S. and Swamy, P.S. (2000) Vegetation structure and composition under different landscape elements at Kodayar in Western Ghats of Tamil Nadu. In: Khan, M.A. and Farooq, S. (eds) *Environment, Biodiversity and Conservation*. APH Publishing Corporation, New Delhi, India, pp. 197–224.

Sundarapandian, S.M., Natarajan, K.K., Murugesan, S. and Swamy, P.S. (2004) Effect of anthropogenic perturbation on plant biodiversity in tropical forests in the Western Ghats of Tamil Nadu. In: Muthuchelian, K. (ed.) *Proceedings of the National Workshop on Biodiversity Resource Management and Sustainable Use*. Madurai Kamaraj University, Madurai, Tamil Nadu, India, pp. 18–24.

Sundarapandian, S.M., Chandrasekaran, S. and Swamy, P.S. (2012) Disturbance regimes and plant invasiveness in Southern Tamil Nadu, India. In: *Proceedings of 4th International Ecosummit: Ecological Sustainability Restoring the Plants Ecosystem Services (Eco Summit 2012) 30 September–5 October*. Elsevier/Ohio State University, Columbus, Ohio, USA.

Surendra, B., Muhammed, A.A., Temam, S.K. and Solomon, R.A.J. (2013) Invasive Alien plant species assessment in urban ecosystem: a case study from Andhra University, Visakhapatnam, India. *International Research Journal of Environmental Science* 2(5), 79–86.

Suriyavathana, M., Parameswari, G. and Shiyan, S.P. (2012) Biochemical and antimicrobial study of *Boerhavia erecta* and *Chromolaena odorata* (L.) King and Robinson. *International Journal of Pharmacy Science Research* 3(2), 465–468.

Tennyson, S., Ravindran, K.J. and Arivoli, S. (2012) Bioefficacy of botanical insecticides against the dengue and chikungunya vector *Aedes aegypti* (L.) (Diptera: Culicidae). *Asian Pacific Journal of Tropical Biomedicine* 2(3), S1842–S1844.

Thamotharan, G., Sekar, G., Ganesh, T., Sen, S., Chakraborty, R. and Kumar, N.S. (2010) Anti-ulcerogenic effects of *Lantana camara* Linn. leaves on *in vivo* test models in rats. *Asian Journal of Pharmacy and Clinical Research* 3(3), 57–60.

Thirumaran, G. and Anantharaman, P. (2009) Daily growth rate of field farming seaweed *Kappaphycus alvarezii* (Doty) Doty ex P. Silva in Vellar Estuary. *World Journal of Fish and Marine Sciences* 1(3), 144–153.

Tripathi, R.S., Yadav, A.S. and Kushwaha, S.P.S. (2012a) Biology of *Chromolaena odorata*, *Ageratina adenophora* and *Ageratina riparia*: a review. In: Bhatt, J.R., Singh, J.S., Singh, S.P., Tripathi, R.S. and Kohli, R.K. (eds) *Invasive Alien Plants: An Ecological Appraisal for the Indian Subcontinent*. CAB International, Wallingford, Oxfordshire, UK, pp. 43–56.

Tripathi, R.S., Khan, M.L. and Yadav, A.S. (2012b) Biology of Mikania micrantha H.B.K.: a review. In: Bhatt, J.R., Singh, J.S., Singh, S.P., Tripathi, R.S. and Kohli, R.K. (eds) *Invasive Alien Plants: An Ecological Appraisal for the Indian Subcontinent*. CAB International, Wallingford, Oxfordshire, UK, pp. 99–107.

Udayakumar, M. and Ajithadoss, K (2010) Angiosperms, hydrophytes of five ephemeral lakes of Thiruvallur District, Tamil Nadu, India. *Check List* 6(2), 270–274.

Udayakumar, M., Bharathidasan, E. and Sekar, T. (2014) Invasive Alien Flora of Thiruvallur District, Tamil Nadu, India. *Scholars Academic Journal of Bioscience* 2(4), 295–306.

Varadharajan, R. and Rajalingam, D. (2011) Anticonvulsant activity of methanolic extracts of *Ageratum conyzoides* L. *International Journal of Innovative Drug Discovery* 1(1), 24–28.

Venkatachalam, T., Kumar, V.K., Selvi, P.K., Maske, A.O. and Kumar, N.S. (2011) Physicochemical and preliminary phytochemical studies on the *Lantanacamara* (L.) fruits. *International Journal of Pharmacy and Pharmaceutical Science* 3(1), 52–54.

Venkatesh, R., Shanthi, S., Rajapandian, K., Elamathi, S., Thenmozhi, S. and Radha, N. (2011) Preliminary study on antixanthomonas activity, phytochemical analysis, and characterization of antimicrobial compounds from *Kappaphycus alvarezii*. *Asian Journal of Pharmacy and Clinical Research* 4(3), 46–51.

Wallace, A.R. (1902) *Island Life*, 3rd edn. Macmillan, New York, USA.

Williamson, M. (1996) *Biological Invasion*. Chapman and Hall, London, UK.

Wu, S.H., Chaw, S.M. and Rejmanek, M. (2003) Naturalized Fabaceae (Leguminosae) species in Taiwan: the first approximation. *Botanical Bulletin of Academia Sinica*, 44, 59–66.

Xu, H.G. and Qiang, S. (2004) *Inventory of Invasive Alien Species in China*. China Environmental Science Publications, Beijing, China.

Xuan, D., Shinkichi, T., Hong, N.H., Khanh, T.D. and Min, C. (2004) Assessment of phytotoxic action of *Ageratum conyzoides* L. (billy goat weed) on weeds. *Crop Protection* 23, 915–922.

Yasotha, D. and Lekeshmanaswamy, M. (2012) Impact of *Neochetina eichhorniae* Warner. on Biological Control of Water Hyacinth (*Eichhornia crassipes* (Mart.) Solms.) of Singanallur pond, Coimbatore, Tamil Nadu, India. *Indian Journal of Natural Science* 2(10), 721–726.

22 Patterns of Plant Endemism and Forest Regeneration Processes in Northern Western Ghats

Pundarikakshudu Tetali[1] and Sujata Tetali[1]*

[1]*Temple Rose Livestock Farming Pvt. Ltd 201, Shivaji Nagar, Model Colony, Pune, Maharashtra, India;* [2]*Agharkar Research Institute, Pune, Maharashtra, India*

Abstract

Western Ghats is the backbone of the Peninsular Indian economy and a means of sustenance for more than 400 million people. It has been in the limelight for several good reasons throughout history; however, its recent attraction is biodiversity. Among the different regions of Western Ghats, the biodiversity of the Northern Western Ghats (NWG) is more exploited and less understood. The NWG is a home for 23 endemic genera and 815 Indian endemic flowering plant taxa, 162 of which are local endemics. A great number of these flowering plant endemic taxa are distributed largely in three types of macrohabitats: 1) plateaus; 2) sacred groves; and 3) hill forts. The four-month south-west monsoon and a prolonged dry period shape the basic forest processes in NWG. In this chapter, the influence of natural processes in forest regeneration and the patterns of flowering plant endemism of Maharashtra are discussed. Herb domination, the greater number of monocot taxa, underground storage organs, the domination of Poaceae are the highlights of the NWG flowering plant endemism. The secondary forest regeneration processes in NWG are prominently supported by non-invasive species like *Carissa spinarum*. Their presence indicates positive signs of secondary forest regeneration processes.

22.1 Introduction

The Western Ghats is a major component of the Indian landscape (Fig. 22.1; Subramanyam and Nayar, 1974). In a global context, it is viewed as a biodiversity-rich region and a hotspot that is represented by a small area of primeval forest (Myers *et al.*, 2000; Anon., 2013a). In addition, it is considered as an eco-sensitive region that, over time, has developed into a complex entanglement of living systems. Throughout history, it has successfully acted as a backbone of the Peninsular Indian economy, supporting more than 400 million people (Molur *et al.*, 2011). For all the people within this ecosystem, it is vital.

The Western Ghat mountain range runs for nearly 1490 km (Gadgil, 2011) parallel to the western edge of Peninsular India (21° 16′ N to 8° 19′ N), virtually uninterrupted, except for the Palakkad gap. Geographically, the hill range passes through six states of India. Geologically and floristically, it has similarities with Sri Lanka, near the southern tip of India. The states of Maharashtra and Gujarat are located towards the northern tip of the Western Ghats. In a true sense, the Western Ghats is not considered a mountain range. It is believed to be an elevated region of the Deccan plateau (Krishnan, 1953; Radhakrishna, 1967; Gunnel and Radhakrishna, 2001).

The Western Ghats forms a continuous strip of hills from Tapti to Kanyakumari, and shows much dissimilarity within different geographical gradients. The natural landscape is characterized by steep slopes and deep valleys. It is not a dynamic geological formation like the Himalayas, but is ecologically sensitive and is one of the eight 'hottest hotspots' of biological diversity in the world. Primary vegetation covers roughly 7% of the land (Myers *et al.*, 2000).

*E-mail:sujatatetali@gmail.com

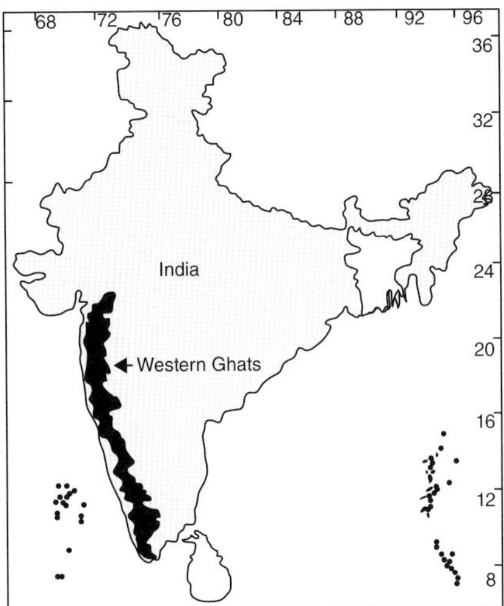

Fig. 22.1. The Northern Western Ghats of Maharashtra.

Although generally referred to as Sahyadri, the Western escarpment in Puranic times was marked as two distinct parts: 1) the Sahya Parvata (the Northern range); and 2) the Malaya Parvata (the Southern range). The mountain range was inhabited by human societies even in the prehistoric period and, as a result, human interactions, particularly trading of forest produce, took place through the serpentine trails known as Ghats, popularly considered synonymous with the mountain region itself.

The word 'jungle', generally meaning secondary forests, denotes more unpredictability and structural complexity. The Western Ghats occupy 129,037 km^2, of which protected forests cover only 15% (Gadgil, 2011, 8). Within the forest category, 75% of the land is managed as reserved forest category, while 9% is managed under protected forest category (Bhat *et al.*, 2001). It is important to mention here that floristic diversity and endemism gradually decrease from Southern to Northern Western Ghats.

The Western Ghats is a cradle of evolution for many economically useful plant and animal species and is a genetic treasure trove. The Western Ghats comprise about 5800 species of flowering plants of which 2100 are endemic (Rao, 2013). It has sustained the shifting demands of India from time to time, as well as the basic needs of Peninsular India, such as water, food, shelter and medicine.

While the growing population demands more and more food, water, health care and energy, India, as a nation, turns once more to the Western Ghats. This naturally leads to debates about conservation approaches and how to harvest resources (Gadgil, 2011; Kasturirangan *et al.*, 2013). These debates do not fall within the purview of the present discourse and are therefore omitted. However, the authors certainly believe that the crucial ecological role that the Western Ghats can play and the structure required to accommodate India's growing demands should go hand in hand. The rule is very simple – utilization of the forests of the Western Ghats is possible only when we encourage the natural processes that help regenerate forests. For this, we need a larger area than currently exists under government protection.

To understand forests better, it is necessary to realize that forests are not just one species or variety. They are nature's unique compositions, formed as a result of complex relationships. Various natural forest processes, such as regeneration, formation, recovery, structure and composition, are expressions of these relationships. Further, the secondary forests, complex and disorderly in nature, and embodied in the word 'jungle', borrowed from the Sanskrit, pose an entirely different scenario. Unfortunately, the NWG has failed to attract as much research attention as the biologically rich Southern Western Ghats.

In this chapter, the historical background of forest utilization and its management; flowering plant endemism and patterns; importance of natural processes in forest establishment and forest regeneration studies in the Northern Western Ghats of Maharashtra state have been considered.

22.2 Methodology

For the purpose of study, endemic flowering plant species/taxa found in the Northern Western Ghats of Maharashtra state were classified into two categories, i.e. those species endemic to Indian region that are distributed in the state (IEn), and those known to be found within the geographical boundaries of the state or the locally endemic taxa (MEn). Checklists of endemics to Maharashtra state have been prepared in the past (Ahmedullah and Nayar, 1986; Yadav 1997; Tetali *et al.*, 2000; Mishra and Singh, 2001; Gaikwad *et al.*, 2014). Our earlier publication, a checklist of endemics of Maharashtra state, was revised by verifying all the recently published data. (The number of endemic species of any region changes from time to time because of new discoveries, due to identification of new distribution localities of species and merging of taxa during taxonomic revisions.) The following information was collected to determine the patterns of endemism. From a conservation point of view, endemic species are more important as their area of occupation is comparatively less than that of the total number of species.

- Distribution of endemic taxa in each of the four biogeographical regions: 1) Konkan; 2) Madhya Maharashtra; 3) Marathwada; 4) Vidarbha (for the purpose of the study, 3) and 4) are not included in the present work).
- Distribution of endemic taxa in each biogeographical region according to habit
  - ➢ Herbs
  - ➢ Shrubs and undershrubs
  - ➢ Climbers, lianas, twiners
  - ➢ Trees.
- Distribution of endemic taxa according to macrohabitat
  - ➢ Closed forests and sacred groves
  - ➢ Open forests and fallow lands
  - ➢ Grasslands
  - ➢ Marshy land and water bodies
  - ➢ Plateaus (table lands)
  - ➢ Rice fields
  - ➢ Hill forts.
- Status of visible adaptations
  - ➢ Modifications of underground parts, such as formations of tubers, corms, rhizomes, bulbs
  - ➢ Leaf modifications, such as development of glandular hairs, thorns, prickles
  - ➢ Stem modifications, such as formation of different types of barks
  - ➢ Epiphytic habit, such as epiphytic parasites, hemi-epiphytes, saprophytes.

Plant association index: the Tanaji Sagar Dam catchment area near Pune from NWG was selected for the association study. Fallow lands of 6–12 years old located on gentle hill slopes (<25°) were selected in the Shirkoli village. Twelve quadrats were randomly selected, with each measuring 20 x 20 m where clump formation was observed. The number and variety of species of each clump were separately counted. The plant association index was measured by the frequency with which 12 commonly found shrub, tree and climber species were associated. The data were analysed by x^2 test with Yates's correction. The association index was calculated based on total number of clumps, number of *C. spinarum* clumps and presence or absence of these species within the clumps (P. Tetali, Studies on the status of people and forests in the Tanaji Sagar Dam Catchment area: the ecological impacts due to construction of a dam with alternative strategies for redevelopment, University of Poona, Poona, India, Vol. 1, 1991, unpublished thesis, Vol. 1).

22.3 Results and Discussion

22.3.1 The study area (Northern Western Ghats of Maharashtra)

The NWG or Sahyadris spreads from Tapi or Tapti River (in Gujarat) to Panaji (in Goa) between 72° 61′ to 74° 40′ E and 15° 60′ to 20° 75′ N.

Geographically, the NWG are found between the coastal belt and the Deccan plateau (Wadia, 1919). The hill ranges cover a distance of about 750 km from north to south, with an average width of 80 km. On the western side, they subside suddenly into a small stretch of near flat land along the Arabian coast known as Konkan (Aparanta, during Puranic times) and on the Eastern side the Ghats are gradually diminished into an extensive plateau, which is occasionally interrupted by small ranges known as Deccan (Dandakaryana, during Puranic times). Due to its fantastic forms, elevation, rainfall and weather, the NWG are naturally variable. The basic feature of the NWG is basaltic trap (Medlicott and Blanford, 1893). The basaltic rocks are mostly composed of pyroxene, plagioclase and iron oxides. The precipitous Ghats rise to an elevation of 1000 m, while their summits occasionally reach to 1500 m high.

Kalsubai is the highest peak in the NWG, located at a height of 1646 m/ASL. The hills are normally flat- or table-topped rocky outcrops.

The NWG experiences a tropical climate. The rainfall in the region gradually decreases from west to east and from south to north. Unlike the southern part, the NWG solely depends upon the southwest monsoon. Because of the uniqueness of the mountain spurs, the heavily loaded monsoon clouds empty their water on the cliffs and as a result create rain-shadow regions in the immediate eastern parts or in its fringe regions.

The NWG of Maharashtra state are flanked on the southern side by the humid tropical forests of Goa and Karnataka, and in the northern frontiers, it shares a border with the drier Gujarat state, which in turn is a neighbour of Rajasthan, the desert state of India. Because of its symbolic position, the region is represented by common as well as endemic flora, which can fit in its unique state of environmental factors and ecological conditions.

Legally, the forests of Maharashtra are classified into three categories. They are 1) Reserved Forests (51,549 km^2), 2) Protected Forests (6727 km^2), and 3) Unclassified Forests (3082 km^2) (Anon., 2013b).

In terms of forest-canopy density classification, Maharashtra has 8739 km^2 of dense forest, 20,834 km^2 of moderately dense forest, and 21,077 km^2 of open forests. The state lost 8 km^2 of closed forest in 2012, which is alarming (Anon., 2013b).

22.4 A Historical Background of Forest Utilization and Its Management

Historically, the Northern Western Ghats were an acknowledged source of a variety of plant-based resources, including medicinal plants, bamboos, spices, charcoal, wild edible fruits and vegetables. These resources were regularly brought for trading in well-established urban markets, located mostly in the plateau region or desh. Notable among these were Pune, Lonanad, Junnar, Aundh and Kolhapur. Although not documented, timber, firewood and charcoal were considered as major forest products in the pre-colonial period, while Bamboo, Karvi, Hirda, Shikakai, Gulvel, Shatavari; Kadipatta, Tamal patra, Dhayati, Apta, Tendu, Chavani leaves, Mango, Kokkam, Jack fruit, Shilar bark and gum were recognized as minor forest products (Table 22.1). Minor forest products comprised an important revenue generation activity for tribal and peasant communities. In the past too, protection was assured for economically important trees through *Adnya patra* (government orders of Chatrapati Shivaji). In addition, evergreen forests were not used for any commercial purpose. Forests were harvested mainly for teak, black-wood and sandalwood. Forests in general were managed prudently.

The Western Ghats forest ecosystems were harvested initially by three groups of people – the hunter gatherers, the shifting cultivators and the settled cultivators (Gadgil, 1990). The mode of ecosystem utilization and harvesting has changed from time to time, since the times of Asoka (269 BCE to 232 BCE), mainly because of spreading Buddhism and later by Satavahanas (230 BCE to 220 CE). The peak of it took place during the expansion of the Maratha Empire. As a result, many of the inaccessible regions were colonized and forts were built to protect the local kingdoms from invaders and plunderers. The consequence of this development is that out of 350-odd forts that are present in the Maharashtra state, more than 200 are located in the Northern Western Ghats. This had, for the first time, brought a general impact on the forest regeneration and harvesting procedures. Many forts were strategically developed into complexes to support food production for the army. Gentle hill slopes at the base were cleared for shifting agriculture, while the lower plains along the river banks were converted for rice cultivation. Many of those practices were sustainable even at that point of time. Forest history information clearly reveals four categories of land existed in almost all villages and they were mostly managed by village communities (Tetali *et al.*, 1993). They were as follows.

- **Devrai** (sacred groves): Forest patches and other common natural resources such as water sources, medicinal plants dedicated to local deities (Gadgil and Vartak, 1975; Tetali and Gunale, 1990; Waghchaure *et al.*, 2006).
- **Inam** (gifted forests)**:** Donated lands; kept reserved for cultural activities, income generated was exclusively utilized for cultural and religious activities.
- **Gaoran** (village pasture lands)**:** Common lands exclusively kept for cattle grazing.
- **Kuran:** Kurans were originally reserved for timber and fuel. They were mostly composed of grasslands, *Acacia nilotica* (Babul kurans) and *Tectona grandis* (Sag kurans). They have occasionally been used for other purposes.

Table 22.1. Important major and minor forest products from the Northern Western Ghats.

Bamboo

Bassia latifolia seeds

Butea frondosa leaves

No.	Category	Plant species
1	Fruits	*Garcinia* sp. (Kokum), *Mangifera indica* (Mango), *Artocarpus* (Jack fruit), *Meyna laxiflora* (Aloo), *Carissa spinarum* (Karvand), *Cucumis* sp. (Meki), *Citrallus* sp., *Elaegnus conferta* (Amgul), *Diospyros* sp.
2	Medicinal plants	**Alternative systems of medicine**
		Terminalia chebula (Hirda), *Terminalia bellerica* (Behada), *Acacia concinna* (Shikakai), *Asparagus racemosus* (Shatavari), *Gloriosa superba* (Kallavi), *Tinospora cordifolia* (Gulvel); *Gymnema sylvestre* (Bedkichi pan); *Pueraria tuberosa* (Vidarikand); *Carraluma fimbriata* (Makad-sheng); *Embelia tsjerium-cottam* (Vavding); *E. ribes* (Vavding); *Helicters isora* (Murudsheng); *Rauwolfia serpentina* (Sarpagandha); *Mimosa pudica* (Lajalu); *Phyllanthus emblica* (Amla), *Pterocarpus marsupium* (Bivla); *Mucuna pruriens* (Kajkuili); *Cyperus rotundus* (Lavala)
		Modern medicine
		Nothapodytes foetida (Narkia), *Iphigenia stellata, I. pallida, I. indica, I. magnifica, Gloriosa superba* (Kallavi), *Bosewila serrata* (Salaiguggul)
3	Edible tubers	*Dioscorea* sp., *Canavalia ensiformis, Flemingea tuberosa, Ceropegia* sp., *Brachystelma* sp.
4	Spices	*Zanthoxylum rhetesa* (seeds), *Cinnamomum tamala* (Tamalpatra); *Curcuma* sp.
5	Timber and furniture	*Tectona grandis, Mangifera indica, Melia composita* (Limbara), *Careya arborea* (Kumba), *Adina cordifolia* (Hedu), *Gmelina arborea* (Shivan), *Bambusa bamboos* (Kalak)
6	Honey	Mixed kind
7	Bamboos	*Pseudoxytenanthera ritcheyi, P. stocksii, Dendrocalmus* sp., *Bambusa bamboos* (Kalak)
8	Charcoal	*Terminalia chebula* (Hirda), *Lagerstroemia lanceolata* (Nana)
9	Plant-supporting sticks	*Strobilanthus callosus* (Karvi)
10	*Bidi* leaves	*Bauhinia racemosa* (Apta), *Diospyros melanoxylon* (Tendu)
11	Housing and furniture	*Tectona grandis* (Teak)
12	Oils	*Actinodapne hookeri* (Pisadi seed oil); *Pongamia pinnata* (Karanj); *Santalum album* (Chandan)

Continued

Table 22.1. Continued.

13	Moss	Many species of bryophytes
14	Fibre	*Sterculia* sp.
15	Biopesticides	*Azadirachta indica* (Kadilimb), *Vitex negundo* (Nirgudi), *Bassia latifolia* (Mohava)
16	Leaf plates	*Ensete superba* (Chavani) , *Musa* sp. (Keli), *Anthocephalus cadamba* (Kadmaba), *Butea frondosa* (Palas)

More or less similar forest practices were encouraged during the Peshwa's administration, with some additional aspects in forest management, i.e. plantation of economically useful timber species such as *Tectona grandis* (teak) on degraded hill slopes under the supervision of Nana Fadnavis (1741–1800) in the Wai and Mahabaleswar region (Garland, 1934). The most destructive period for forests was, however, during colonial times. According to estimates prepared by Gadgil and Guha (1992), the forest exploitation was 65% higher during the Second World War than the pre-war period. The colonial masters during the second phase were not only responsible for degradation of the community or privately owned forests, but they also steadily destroyed the organizational set-up that sustained them. The British adopted a twofold policy to access India's forest resources: the state take-over of community lands; and the claim over all teak trees as the property of British East India Company (Gadgil, 2008).

Bhat *et al.* (2001) described the degradation of primary forests to secondary forests in three phases: 1) the pre-colonial period, when dependence on forests was extensive with less impact; 2) the colonial administration and post-colonial period, where intense use of forests was recorded; and 3) recover and rehabilitation stage during recent years.

Even in the third phase, or the so-called recovery phase, it remains a misnomer (Table 22.2). The private forests demonstrated utter policy neglect. In the NWG, they were more victimized than the state-owned forests. Declaration of SEZs (Special Economic Zones) in the NWG is a visible demonstration of this destructive policy behaviour.

22.5 Wants, Needs and the Victims

The degradation cycle, like any other tropical forests, began here too with excessive logging which reduced the forest to a less commercial or a minor forest producing resources. Depending upon the income-generating capacity, privately owned degraded forests of the NWG were often converted to shifting agriculture. Unsustainable agriculture practices, such as reducing the fallow period, made many forested regions unproductive and commercially less lucrative. When it was coupled with government pursuits of constructing dams for command area agriculture and electricity, the disaster levels were reported to be very high. The status of forests in general is discouraging.

The TOF inventory of the state government, which was carried out between 2002 and 2008 revealed 3.08% tree cover, comprising only 9.466 km^2 of the total geographical area. This estimate clearly explains the extent of tree cover degradation in quantitative terms (Forest Survey of India, 2009).

Many activities had in fact impacted the natural forest processes and their regeneration in the NWG. The status of private forests is a serious concern. The private forests which were under shifting cultivation in the past are now facing a serious threat from the SEZs, satellite townships, resorts, horticulture projects and tourist attractions with changing wants, rather than changing needs. Most of these lands are already sold to builders and other kinds of vested interests. The scenario is depressing.

Among the different groups of plants for which the NWG was known, medicinal plants suffered a lot due to bad planning. Incredible global demand and lucrative income at local levels further adds to this problem. We provide two recent examples below on how things went wrong in the NWG. *Terminalia chebula* (Myrobylan tree), which is popularly known as *Hirda*, was one of the most common trees on gentle hill slopes of the NWG. The fruits were extensively used in medicine and the local tanning industry. In addition, the quality charcoal produced from the wood was in great demand in nearby townships. Human intervention had over

Table 22.2. The Northern Western Ghats – a historical perspective of changing disturbance regimes.

No.	Threat type/category	Intensity: Pre-Independence	Intensity: Post-Independence	Intensity: Recent (2000 onwards)
1	Dam submersions	Minor	Major/destructive	Moderate
2	Residential complexes and other construction activities	Minor	Minor	Major/destructive
3	Mining	Unknown	Major	Major/destructive
4	Querying	Minor	Moderate	Moderate
5	Forest fire	Moderate	Major/destructive	Major/destructive
6	Wildlife	Minor	Minor	Minor
7	Livestock grazing	Minor	Major/destructive	Minor
8	Charcoal preparation	Minor	Major	Minor
9	Fuel wood	Minor	Moderate	Minor
10	Forest harvesting	Moderate	Major/destructive	Moderate
11	Medicinal plant industry	Moderate	Major	Major/destructive
12	Shifting agriculture	Moderate	Major	Minor
13	Windmills	Nil	Nil	Moderate
14	Tourism and recreation activities	Nil	Minor	Major/destructive
15	Industrial zones	Nil	Moderate	Most destructive
16	Satellite city development	Nil	Minor	Most destructive
17	Fort construction	Major	Nil	Nil
18	Exotic species invasion	Unknown	Minor	Major
19	Road preparation	Minor	Moderate	Major/destructive
20	Botanical collections/Research studies	Unknown	Moderate	Moderate

centuries selectively improved the local germplasm to suit the requirements of medicine and the leather industry. Among the many variants that were available throughout the state, the *Devna* and *Raireshwar* variants fetched a better price because of their colour and obovate-shaped fruit with elongated apex (Fig. 22.2). The nuts once contributed heavily to the coffers of the British Empire. Due to lack of political vision and good research support, the Myrobalan industry in the NWG had collapsed by 1980 (Tetali and Amre, 2003). Consequently, many mature trees from private forests were cut down for preparing charcoal as a short-term income source. The Panshet dam (Tanaji Sagar Dam) construction story is an example (Gadgil, 1979; Gole and Tetali, 1985; Brahme and Tetali, 1987).

In recent years the forests, both private and government, were destructively harvested for another medicinal plant, *Nothapodytes nimmoniana*, whose stem and root bark were known to produce an alkaloid, Camptothecin. Tragically, the NWG trees were reported to have yielded a high percentage of Camptothecins and enjoyed global demand. The results were disastrous (Tetali *et al.*, 2003). *Nothapodytes* soon vanished from many parts of the NWG until it became a threatened species in the CAMP exercises (Tetali, 2002).

22.6 Floristic Abundance of Maharashtra – How Rich Is the NWG?

The state of Maharashtra is well known for flowering plants. The most recent status of flowering plants of the state is available from three different publications on state flora. Notably, the *Flora of Maharashtra*, published by the Botanical Survey of India, is most comprehensive (Singh and Karthikeyan, 2000, 2001). However, for more taxonomic details, Almeida's *Flora of Maharashtra* (1996–2009) and Naik's *Flora of Marathwada* (1998) are helpful. The number of flowering plant taxa found in the Maharashtra state varies from 3873 to 4400 (Lakshminarasimhan, 1996). Wild flowering-plant flora of the Maharashtra state comprise only 3042 species, 22 subspecies, 148 varieties, one subvariety and two formas. Roughly 90% of these plants are represented in the NWG. According to *Flora of Maharashtra* (Almeida, 1996–2009) dicots dominate the state flora. About 70% of the wild flowering plants are dicots. The monocot to dicot ratio is 1:2.36.

Fig. 22.2. Nuts of *Terminalia chebula* Retz. One of the most important medicinal plants of India: the dried fruits form one of the most frequently used ingredients in Ayurvedic formulations; variation in fruit shapes and sizes is the combined result of human and natural selection. Two selections – the Raireshwar and the Devana fetch more money in the commercial trading market because they possess three important grading points – golden-yellow colour, long beak and small seeds (Tetali and Amre, 2003).

Number of dicots: 2131 species + other taxa – 18 sub species + 108 varieties + 1 forma

Number of monocots: 913 species + other taxa – 3 subspecies + 40 varieties + 1 subvariety + 1 forma

However, this data is almost 15 years old. Floristic explorations in the state, meanwhile, have added about 75 taxa to the existing flora.

The NWG pass through 11 districts of the state and all these districts display fabulous flowering plant diversity as compared to the other regions of the state. Botanically the region is well explored.

22.7 Forest and Vegetation Types of NWG

Ramesh and Pascal in 1997 identified four major forest types and 23 subforest types from the Western Ghats. The NWG of Maharashtra is represented by three major types of forests as per Champion's classification (Champion, 1936; Champion and Seth, 1968) (Table 22.3). The natural regeneration process of each forest type is locally modified by different vegetation types that determine the dominance, structure and composition. And these forests naturally blended with a large number of species. Even so, relatively few species dominate.

22.8 NWG – A Treasure Trove for Endemic Genera

Forest ecosystems are not simple structures of dominant and recessive taxa. They are also recognized by the unique taxa they represent. Taxa that are restricted to certain geographical areas are

Table 22.3. Major forest types of the Northern Western Ghats and their structure

No.	Forest type	Vegetation type	Description
1)	**Moist tropical forests**		
1a)	Tropical semi-evergreen forests (=west coast semi-evergreen forests)	*Albizzia-Bridelia-Terminalia-Mangifera-Actinodaphne*	Evergreens predominate; distributed mainly in the higher elevations between 450 and 1050 msl; average annual rainfall varies from 150cm to 300cm; trees over 25 m high.
2)	**Tropical moist deciduous forests (=subtropical semi-deciduous lowland forest) Dominants mainly deciduous; top canopy rarely dense; average annual rainfall varies from 150 to 250 cm; trees up to 25 m high.**		
2a)	Tropical moist teak forests	*Tectona-Albizzia-Terminalia-Bassia-Dalbergia-Acacia*	Found mostly in the low altitudes; teak is the most dominant species; from a commercial point of view, these are the most important and valuable forests of the state.
2b)	Southern moist mixed deciduous forests	*Bridelia-Syzygium-Terminalia-Ficus*	These forests are found mostly along the eastern side of the Western Ghats, where the mean annual rainfall ranges from 130 to 180 cm.
3)	**Montane subtropical forests**		
3a)	Subtropical hill forests (=Western subtropical hill forests)	*Memecylon-Syzygium-Terminalia-Ficus*	Found in the upper reaches; usually over 900 msl; trees are generally represented by broad-leaved species of low to medium; receive high mean annual rainfall, above 300 cm; tree height ranging between 5 and 15 m.

known as endemics. Areas with high endemism are valued for their biological wealth. Among the different biogeographical regions of India, the Western Ghats shows a phenomenal representation of endemic genera (Rao, 2007; Irwin and Narasimhan, 2011). Out of the 49 endemic genera, 39 are represented in the Western Ghats. Twenty-three of these genera are found in the NWG (Table 22.4). It means that the majority of the monotypic genera that are found in the Southern Western Ghats are also found in the NWG and it also explains that at the generic level there are no major differences. At the same time, the NWG maintains its uniqueness with the *Frerea*-like succulent genus.

22.9 Patterns in Flowering Plant Endemism

The list of endemic flowering plants of Maharashtra has been revised from time to time. The latest revision carried out by us indicates the presence of a total of 796 species (+ 98 subspecies/varieties), i.e. 16% of Indian endemic species (IEn) and interspecific taxa in the Maharashtra state. The occurrence of 58 endemic taxa in the state from the prepared list is unresolved. These species are either reported locally without proper support or provided with insufficient information in relation to location of collection. The absence of authentic herbarium specimens complicates the process further. The total number of species and intraspecific taxa that are strictly endemic to Maharashtra state (MEn) is 177, which comprises only 3% of the total flowering-plant taxa present in the state. Table 22.5 provides the list of MEn taxa, their families, habitats and the type of underground storage.

Why only endemic taxa? And is there any difference between the flora of the NWG of Maharashtra and its counterpart, the Southern Western Ghats? What controls the regeneration processes, and how? All these are difficult questions to answer (Fig. 22.3).

Endemic plants comprise an important aspect of biodiversity and are important from the conservation point of view. The listing of taxa and understanding of the patterns of endemism are integral parts of modern conservation management strategies, since more than 90% of the threatened plants of any country are endemic taxa (Gaston, 1998; Gaston, 2005; Hughes *et al.*, 2014).

22.10 Is the South-West Monsoon the Major Determinant of Forest Processes in NWG?

All flora and fauna of the NWG are impacted by the south-west monsoon (Raghvan, 1964). It is the elixir of life for them (Patwardhan and Asnani, 2000). The precipitation gradient, coupled with other factors like habitat and relative humidity, shows a pronounced effect on forest types, natural vegetation and their

Table 22.4. Endemic genera of the Northern Western Ghats.

Erinocarpus	*Seshagiria*	*Glyphocloa*

	Name	Habit	Family
Dicots			
1	*Erinocarpus* Nimmo ex J. Graham	Tree	Malvaceae
2	*Hardwickia* Roxb.	Tree	Fabaceae
3	*Pinda* P.K.Mukh. & Const.	Herb	Apiaceae
4	*Polyzygus* Dalz.	Herb	Apiaceae
5	*Frerea* Dalz.	Herb	Apocynaceae
6	*Seshagiria* M.Y. Ansari	Climber	Apocynaceae
7	*Haplanthodes* Kuntz.	Herb	Acanthaceae
8	*Bonnayodes* Blatt. & Hallib.	Herb	Scropulariaceae
9	*Helicanthus* Danser	Herb; epiphytic parasite	Loranthaceae
10	*Adenon* Dalz.	Herb	Asteraceae
11	*Lamprachaenium* Benth.	Herb	Asteraceae
12	*Leucoblepharis* Arn.	Herb	Asteraceae
13	*Nanothamnus* Thomson	Herb	Asteraceae
Monocots			
14	*Smithsonia* C.J. Saldanha	Epiphyte	Orchidaceae
15	*Bhidea* Stapf ex Bor.	Herb	Poaceae
16	*Danthonidium* E.E. Hubb.	Herb	Poaceae
17	*Glyphocloa* Clayton	Herb.	Poaceae
18	*Hubbardia* Bor.	Herb	Poaceae
19	*Indopoa* Bor.	Herb	Poaceae
20	*Lophopogon* Hack.	Herb	Poaceae
21	*Nanothamnus* Thomson	Herb	Poaceae
22	*Pogonachne* Bor.	Herb	Poaceae
23	*Trilobachne* Schenck ex Henrard	Herb	Poaceae
24	*Triplopogon* Bor.	Herb	Poaceae

survival strategies in the NWG. The NWG experiences the monsoon season for a period of four months. The monsoon sets in around 6 June and recedes in October. The coastal region receives a rainfall of 190–250cm per annum. The higher elevations of the Ghat region receive an annual rainfall of 250–440cm. Amboli, located in the southern end, receives the highest rainfall (up to 7477 mm) in the NWG. Because of this unique position, the NWG is also the source region for two of the most important rivers of India, the Godavari and the Krishna, and also a myriad of other rivers that join them. Due to the torrential rains, many high altitudinal regions of the NWG region are inaccessible during the monsoon period. The question is whether high rainfall in the crest region, particularly for a limited period of four months, is a boon to native flora. Studies elsewhere show that wind

Table 22.5. Endemic flowering plants of the Northern Western Ghats.

Abutilon ranadei

Pinda concanensis

Ceropegia maccanii

No.	Botanical name	H	S	C	T	Family	Habitat/Adaptation
1	***Delphinium malabaricum*** (Huth) Munz var. malabaricum var. ghaticum Billore	+	–	–	–	Ranunculaceae	Open moist mixed deciduous forests; hill slopes with grasses and small shrubs; hairy; roots thick
2	***Thalictrum obovatum*** Blatt.	+	–	–	–	Ranunculaceae	Open moist mixed deciduous forests; degraded rocky slopes with sparse shrub cover
3	***Abutilon ranadei*** Woodr. et Stapf	–	+	–	–	Malvaceae	Protected moist semi-evergreen forest hill slopes and hill forts; with *Strobilanthus callosus* and small trees
4	***Salacia brunoniana*** Wight & Arn.	–	+	–	–	Celastraceae	Open forests, scandent shrub
5	***Ventilago madraspatana*** Gaertn a) var. fructifida	–	+	–	–	Rhamnaceae	Ravines of moist deciduous forests, closed forests, ferruginous
6	***Ziziphus rugosa*** Lam a) var. glabra Bhandari & Bhansali	–	+	–	–	Rhamnaceae	Moist deciduous forests; fulvous tomentose; spiny
7	***Rotala floribunda*** (Wight) Koehne	+	–	–	–	Lythraceae	Hill slopes of higher elevations; on wet rock of streams
8	***Rotala sahyadrica*** Gaikwad, Sardesai & Yadav	+	–	–	–	Lythraceae	High-altitude plateaus; aquatic, in marshy lands and puddles
9	****Alysicarpus luteo-vexillatus*** Naik & Pokle	+	–	–	–	Fabaceae	Gravelly hill slopes
10	***Alysicarpus narimanii*** S.M. Almeida & M.R. Almeida	+	–	–	–	Fabaceae	Open hill slopes
11	****Alysicarpus salim-ali*** S.M. Almeida & M.R. Almeida	+	–	–	–	Fabaceae	Open rocky hill slopes
12	***Alysicarpus sanjappae*** S. Chavan, Sardesai & Pokle	+	–	–	–	Fabaceae	Open dry and waste places; along with graminoids
13	***Alysicarpus tetragonolobus*** Edgew var. pashanensis S.M. Almeida & M.R. Almeida	+	–	–	–	Fabaceae	Open hills and rocky places

Continued

Table 22.5. Continued.

No.	Botanical name	H	S	C	T	Family	Habitat/Adaptation
14	***Crotalaria decasperma** Naik	+	–	–	–	Fabaceae	Along roadsides; bulbous hairy
15	**Flemingia rollae** (Billore & Hemadri) A. Kumar	+	–	–	–	Fabaceae	Open rocky hill forests, plateaus; clothed with white or yellowish hairs
16	**Galactia tenuifolia** (Klein ex Willd.) Wight & Arn. var. minor Baker	–	–	+	–	Fabaceae	On rocky hill slopes, grey silky hairs
17	**Indigofera deccanensis** Sanj.	+	–	–	–	Fabaceae	In deciduous forests; open gravelly hill slopes; covered with white tomentum
18	**Indigofera santapaui** Sanj.	+	–	–	–	Fabaceae	On exposed hill slopes of hill fort; hairy, gland-dotted
19	**Indigofera triata** L. var. purandharensis Sanj.	–	+	–	–	Fabaceae	Open slopes of dry places on hill forts; covered with grey pubescence
20	**Mucuna sanjappae** Aitwade & Yadav	–	–	+	–	Fabaceae	In open moist to dry deciduous forests; on hill slopes
21	**Smithia agharkarii** Hemadri	+	–	–	–	Fabaceae	In disturbed areas; open places on rocky plains and plateaus
22	**Spenostylia bracteatus** (Baker) Gillett.	–	–	+	–	Fabaceae	Near waterfalls, in protected moist mixed deciduous forests; roots thick
23	**Cassia kolabensis** Kothari, Moorthy & Nayar	+	–	–	–	Caesalpinaceae	Among grasses in open places and road sides of moist mixed deciduous forests; whitish tomentose
24	**Begonia phixophylla** Blatt. & McCann	+	–	–	–	Begoniaceae	In subtropical hill forests; open, steep and wet hill slopes; crevices of large boulders with graminoids; also epiphytic on tree; densely bristly; roots rhizomatous
25	***Heracleum dalgadianum** Almeida	+	–	–	–	Apiaceae	Open hill slopes; tuberous root, softly hairy
26	***Pimpinella rollae** Billore & Hemadri	+	–	–	–	Apiaceae	Unresolved name; open hill slopes with graminoids; white ciliate
27	**Pimpinella tomentosa** (Dalz. & Gibs.) Clarke	+	–	–	–	Apiaceae,	Open protected areas along the hedges of forests; with graminoids; tomentose; roots tuberous
28	**Pinda concanense** (Dalz.) Mukherjee & Constance	+	–	–		Apiaceae	Hedges of forests and as forest undergrowth in protected moist, semi-evergreen forest hill slopes, hill forts and plateaus; roots tuberous

Continued

Table 22.5. Continued.

No.	Botanical name	H	S	C	T	Family	Habitat/Adaptation
29	***Neanotis sahyadrica*** Billore & Mudaliar	+	–	–	–	Rubiaceae	On high elevations; open, steep, wet and shady hill slopes; forms colonies; greyish, hairy
30	***Blumea venkatramanii*** Rolla Rao & Hemadri	+	–	–	–	Asteraceae	Bunds on cultivated fields; glandular, hairy
31	***Cythocline purpurea*** (Buch-Ham. ex D. Don)).Ktze. a) var. *alba* Sant. b) var. *bicolor* Sant.	+	–	–	–	Asteraceae	Open places of moist, semi-evergreen forest; marshy lands; rice fields, plateaus; glandular, hairy
32	***Phyllocephalum hookeri*** (C.B.Cl.) Uniyal	+	–	–	–	Asteraceae	Data deficient; pubescent
33	***Brachystelma malwanense*** Yadav & Singh	+	–	–	–	Apocynaceae	Open hill slopes; gravelly localities and crevices in laterite with graminoids and small shrubs; roots tuberous
34	***Brachystelma naorojii*** P.Tetali *et al.*	+	–	–	–	Apocynaceae	Open degraded hill slopes; near boulders; with graminoids; roots tuberous
35	***Ceropegia anjanerica*** Malpure, Kamble & Yadav	+	–	–	–	Apocynaceae	Open places of moist, semi-evergreen forest; among grasses; roots tuberous
36	***Ceropegia anantii*** Yadav, Sardesai & Gaikwad	+	–	–	–	Apocynaceae	Open places of moist, semi-evergreen forest; among grasses and shrubs; roots tuberous
37	***Ceropegia concanensis*** Kamble, Chandore & S.R. Yadav	+	–	–	–	Apocynaceae	On lateritic plateaus; along with graminoids
38	***Cerpoegia evansii*** McCann	–	–	+	–	Apocynaceae	In secondary forests of moist, semi-evergreen forests and hill forts; in scrub land; roots tuberous
39	***Ceropegia huberi*** Ansari	–	–	+	–	Apocynaceae	Protected, open rocky places and courses of streams in moist, semi-evergreen forests; with *Euphorbia* sp. and other similar shrubs; roots tuberous
40	***Ceropegia jainii*** Ansari & Kulkarni	+	–	–	–	Apocynaceae	Lateritic plateaus; in crevices of rocks with graminoids; roots tuberous
41	***Ceropegia karulensis*** Punekar *et al.*	+	–	–	–	Apocynaceae	In secondary forests of moist, semi-evergreen forests; roots tuberous
42	***Ceropegia lawii*** Hook.f.	+	–	–	–	Apocynaceae	In secondary forests of moist, semi-evergreen and mixed deciduous forests and hill forts; roots tuberous

Continued

Table 22.5. Continued.

No.	Botanical name	H	S	C	T	Family	Habitat/Adaptation
43	***Ceropegia macanii*** Ansari	+	–	–	–	Apocynaceae	In secondary forests of moist, semi-evergreen and mixed deciduous forests and hill forts; roots tuberous
44	***Ceropegia mahabalei*** Hemadri & Ansari Var. *mahabalei* Var. *hemalatae* S.S. Rahangdale & S.R. Rahangdale	+	–	–	–	Apocynaceae	In secondary forests of moist, semi-evergreen and mixed deciduous forests ; on steep hill slopes with grasses and small shrubs; roots tuberous
45	***Ceropegia maharashtrensis*** Punekar *et al.*	+	–	–	–	Apocynaceae	In secondary forests of moist, semi-evergreen forests; roots tuberous
46	***Ceropegia media*** (Huber) Ansari	–	–	+	–	Apocynaceae	In secondary forests of moist, semi-evergreen forests and plateaus; roots tuberous
47	***Ceropegia mohanramii*** Yadav, Gavade & Sardesai	+	–	–	–	Apocynaceae	In secondary forests of moist, semi-evergreen forests and plateaus; roots tuberous
48	***Ceropegia noorjahaniae*** Ansari	+	–	–	–	Apocynaceae	In protected secondary forests and scrub lands of moist, semi-evergreen forests; roots tuberous
49	***Ceropegia oculata*** Hook. var. satpudensis Punekar *et al.*	–	–	+	–	Apocynaceae	Open hill slopes; roots tuberous
50	***Ceropegia panchganensis*** Blatt. & McCann	+	–	–	–	Apocynaceae	In open places; at the hedges of subtropical hill forests; in rocky crevices with graminoids; roots tuberous
51	***Ceropegia rollae*** Hemadri	+	–	–	–	Apocynaceae	Open hill slopes of high altitude regions and forts; in moist, semi-evergreen forests among grasses; roots tuberous
52	***Ceropegia sahyadrica*** Ansari & Kulkarni	+	–	–	–	Apocynaceae	In forts and protected secondary forests and scrub lands of moist, semi-evergreen forests; roots tuberous
53	***Ceropegia santapaui*** Wadhwa & Ansari	–	–	+	–	Apocynaceae	In open places of moist, semi-evergreen forests; among grasses
54	***Ceropegia vincaefoloia*** Hook.f.	–	–	+	–	Apocynaceae	In secondary forests of moist, semi-evergreen and mixed deciduous forests and plateaus; roots tuberous

Continued

Table 22.5. Continued.

No.	Botanical name	H	S	C	T	Family	Habitat/Adaptation
55	***Frerea indica*** Dalz.	+	–	–	–	Apocynaceae	Mountain cliffs and hill forts; associated with *Euphorbia nerifolia*; succulent
56	***Dregea lanceaolata*** (Cooke) Sant. & Wagh	–	–	+	–	Apocynaceae	Data deficient
57	*****Canscora diffusa*** (Vahl) R.Br. ex Roem & Schult. var. tetraptera Naik & Pokle	+	–	–	–	Gentianaceae	Exposed in rocky areas
58	***Canscora khandalensis*** Sant.	+	–	–	–	Gentianaceae	In secondary forests of semi-evergreen forests; on rocky crevices
59	****Canscora devendrae***	+	–	–	–	Gentianaceae	Goliguddae Plateau
60	***Argyreia boseana*** Sant. & Patel	–	–	+	–	Convolvulaceae	In secondary forests of semi-evergreen forests; open situations; softy pubescent
61	***Ipomea clarkei*** C.B.Cl	–	–	+	–	Convolvulaceae	On gravelly and rocky hill slopes
62	***Ipomea salsettensis*** Sant. & Patel	–	–	+	–	Convolvulaceae	Fringes of mixed deciduous or moist, semi-evergreen forests; more along the coastal belt.
63	***Operculina tansaensis*** Sant. & Patel	–	–	+	–	Convolvulaceae	Marshy land, puberelous tomentose
64	***Bonnayodes limnophiloides*** Blatt. & Hallib.	+	–	–	–	Scrophulariaceae	Insufficient data; marshy land; bottom of a dried lake
65	***Lindernia quinqueloba*** (Blatt. & Hallib.) Mukh.	+	–	–	–	Scrophulariaceae	Marshy lands; rice fields
66	***Rotala floribunda*** (Wight) Koehne	+	–	–	–	Lythraceae	Higher elevations; on wet rocks along Ghats; clump making
67	***Rotala sahyadrica*** Gaikwad, Sardesai& Yadav	+	–	–	–	Lythraceae	High-elevation plateaus; ponds or puddles; clump making
68	***Utricularia janarthanamii*** Yadav, Sardesai & Gaikwad	+	–	–	–	Lentibulariaceae	Marshy lands; in open places with graminoids; clump making
69	***Utricularia naikii*** Yadav, Sardesai & Gaikwad	+	–	–	–	Lentibulariaceae	Marshy land; in open places with graminoids
70	***Dicliptera ghatica*** *Sant.*	+	–	–	–	Acanthaceae	In secondary forests; open situations; forest clearance; woolly tomentose
71	***Dicliptera nasikensis*** Lakshminarasimhan & Sharma	+	–	–	–	Acanthaceae	In secondary forests of semi-evergreen forests; fallow lands, open situations; hairy
72	***Hypoestes lanata*** Dalz.	–	+	–	–	Acanthaceae	Degraded hill slopes and secondary forests; flowers woolly tomentose, bract glandular hairy
73	***Lepidagathis bandraemis*** Blatt.	+	–	–	–	Acanthaceae	On gravelly hill slopes of degraded hill slopes; among grasses; velvety pubescent

Continued

Table 22.5. Continued.

No.	Botanical name	H	S	C	T	Family	Habitat/Adaptation
74	***Nilgirianthus reticulatus*** (Stapf.) Bremek.	–	+	–	–	Acanthaceae	Higher elevations; exposed slopes or rocky plateaus; with grasses; short and strigose
75	***Synnema anomalum*** (Blatt.) Sant.	–	–	+	–	Acanthaceae	Strongly ciliate
76	***Leucas deodikarlii*** Billore & Hemadri	+	–	–	–	Lamiaceae	In protected forests of moist, semi-evergreen and mixed deciduous forests; as undergrowth or in open situations; greyish strigose hairs
77	***Nepeta bombainsis*** Dalz.	+	–	–	–	Lamiaceae	Data deficient
78	***Scurrula stocksii*** Hook.f.) Danser	+	–	–	–	Loranthaceae	Epiphytic parasite; moist, deciduous or semi-evergreen forest; buff tomentose
79	***Euphorbia katrajensis*** Gage	+	–	–	–	Euphorbiaceae	Open forest
80	***Euphorbia khandalensis*** Blatt. & Hallib.	+	–	–	–	Euphorbiaceae	Higher elevations; open exposed rocky ground or plateaus; rhizomatous
81	***Euphorbia panchganensis*** Blatt. & McCann	+	–	–	–	Euphorbiaceae	Higher elevations; open exposed rocky ground or plateaus; rhizomatous
82	***Jatropha nana*** Dalz.	–	+	–	–	Euphorbiaceae	Secondary forests; on open gravely hill slopes of moist, deciduous forests
83	*****Habenaria caranjensis*** Dalz.	+	–	–	–	Orchidaceae	Possibly extinct; tuberous
84	***Habenaria panchganiensis*** Sant. & Kapadia	+	–	–	–	Orchidaceae	Higher elevations; in open situations on hill forts and plateaus with graminoids; roots tuberous
85	***Curcuma inodora*** Blatt.	+	–	–	–	Zingiberaceaae	Higher elevations; at the edges of forests as undergrowth; shady areas in hilly forest; roots rhizomatous
86	***Curcuma purpurea*** Blatt.	+	–	–	–	Zingiberaceaae	Higher elevations; at the edges of forests as undergrowth; shady areas in hilly forest; roots rhizomatous
87	***Hitchenia caulina*** (Grah.) Baker	+	–	–	–	Zingiberaceaae	Higher elevations; on plateaus and at the edges of forests; in open situations; roots rhizomatous
88	***Crinum elenorae*** Blatt. & McCann a) var. *elenorae* b) var. *purpurea* Blatt. & McCann	+	–	–	–	Amaryllidaceae	Higher elevations; open forest slopes. roots bulbous

Continued

Table 22.5. Continued.

No.	Botanical name	H	S	C	T	Family	Habitat/Adaptation
89	***Crinum solapurense** S.P. Gaikwad, K.U. Garad & R.D. Gore	+	–	–	–	Amaryllidaceae	Along river banks; swampy regions; roots bulbous
90	***Crinum woodrowii*** Baker	+	–	–	–	Amaryllidaceae	Higher elevations; at the edges of forests; open situations; roots bulbous
91	***Pancratium sanctae-mariae*** Blatt. & Hallib.	+	–	–	–	Amaryllidaceae	Open forest; roots bulbous
92	***Camptorrhiza indica*** Yadav, Singh & Mathew	+	–	–	–	Colchicaceae	Along coastal plateaus; in marshy places with graminoids; clumping habit; roots bulbous
93	***Iphigenia stellata*** Blatt.	+	–	–	–	Colchicaceae	High-altitude plateaus; in marshy places with graminoids; clumping habit; roots bulbous
94	***Chlorophytum glaucoides*** Blatt.	+	–	–	–	Asparagaceae	In secondary forests of moist, semi-evergreen and mixed deciduous forests; as undergrowth; thick roots
95	***Chlorophytum gothanensis*** Malpure & Yadav	+	–	–	–	Asparagaceae	High-altitude plateaus; in open rocky regions; roots tuberous
96	***Chlorophytum kolhapurense*** Sardesai, Gaikwad & Yadav	+	–	–	–	Asparagaceae	In secondary forests as undergrowth; roots tuberous
97	***Dipcadi concanense*** (Dalz.) Baker	+	–	–	–	Asparagaceae	Open rocky plateaus; roots bulbous
98	***Dipcadi maharashtrensis*** Deb & Dasgupta	+	–	–	–	Asparagaceae	High-altitude plateaus; marshy places in rocky areas; roots bulbous
99	***Dipcadi minor*** Hook.f.	+	–	–	–	Asparagaceae	Sandy plateau; bulbous roots
100	***Dipcadi saxorum*** Blatt.	+	–	–	–	Asparagaceae	Degraded hill slopes; rocky areas; roots bulbous
101	***Dipcadi ursulae*** Blatt.	+	–	–	–	Asparagaceae	On plateaus with graminoids; roots bulbous
102	***Drimia razii*** Ansari	+	–	–	–	Asparagaceae	Higher elevations; on open gravelly or rocky hill slopes with sparse shrub cover and grasses; roots bulbous
103	***Protoasparagas karthikeyanii*** Kamble	–	–	+	–	Asparagaceae	Data deficient; roots tuberous
104	***Scilla viridis*** Blatt. & Hallib.	+	–	–	–	Asparagaceae	Data deficient; roots bulbous
105	***Amorphophallus mysorensis** E. Branes & C.E.C. Fisch. var. *bhandarensis* (S.R. Yadav, Khalkar & Bhuskate) Sivadasan & Jaleel	+	–	–	–	Araceae	Unresolved name; roots bulbous
106	***Arisaema caudatum*** Engler	+	–	–	–	Araceae	Open forest as undergrowth; roots bulbous

Continued

Table 22.5. Continued.

No.	Botanical name	H	S	C	T	Family	Habitat/Adaptation
107	***Arisaema murrayii*** a) var. *murrayii* b) var. *sonubenii* Tetali, Punekar & Lakshmin.	+	-	-	-	*Araceae*	Higher elevations; in protected moist, evergreen forests as undergrowth; roots bulbous
108	***Arisaema sahyadricum*** Yadav, Patil & Bhachulkar var. *sahyadricum* var. *ghaticum* Sardesai, Gaikwad & Patil	+	–	–	–	Araceae	Higher elevations; in protected moist, evergreen forests as undergrowth; roots bulbous
109	***Arisaema sivadasanii*** Yadav, Patil & Janardhanam	+	–	–	–	Araceae	Higher elevations; in protected moist, evergreen forests as undergrowth; roots bulbous
110	***Cryptocoryne cognata*** Schott	+	–	–	–	Araceae	Aquatic; in monsoon-swollen streams; thick roots
111	***Aponogeton bruggenii*** Yadav & Govekar	+	–	–	–	Aponogetonaceae	Along the river bank, rice field; roots bulbous
112	***Aponogeton satarensis*** Sundraraghavan, Kulkarni & Yadav	+	–	–	–	Aponogetonaceae Bulbous	High-altitude plateaus; in monsoon puddles and marshy lands; roots bulbous
113	***Aponogeton nateshii*** Yadav et al.	+	–	–	–	Aponogetonaceae	Coastal plateaus; in monsoon puddles; roots bulbous
114	***Ericaulon bolei*** Bole & Almeida	+	–	–	–	Erocaulaceae	Open hill slopes
115	***Eriocaulon apetalum*** Punekar, Malpure & Lakshmin.	+	–	–	–	Eriocaulaceae	Higher elevations; in marshy lands
116	*****Eriocaulon baramaticum*** V.B. Shimple *et al.*	+	–	–	–	Eriocaulaceae	Grass lands; wet places in rain shadow regions
117	***Eriocaulon cookie*** Punekar, Malpure & Lakshmin.	+	–	–	–	Eriocaulaceae	Higher elevations; in marshy lands
118	***Eriocaulon epedunculatum*** Potdar *et al.*	+	–	–	–	Eriocaulaceae	Higher elevations; in marshy lands
119	***Ericarulon kolhapurense*** S.P. Gaikwad, Sardesai & S.R. Yadav	+	–	–	–	Eriocaulaceae	Grass lands; higher elevations; in marshy lands
120	***Eriocaulon konkanense*** Punekar, Malpure & Lakshmin.	+	–	–	–	Eriocaulaceae	Higher elevations; in marshy lands
121	***Eriocaulon koynense*** Punekar, Mungikar & Lakshmin.	+	–	–	–	Eriocaulaceae	Higher elevations; in marshy lands
122	***Eriocaulon maharashtrense*** Punekar & Lakshmin.	+	–	–	–	Eriocaulaceae	Higher elevations; in marshy lands
123	***Eriocaulon ratnagiricus*** S.R. Yadav, S.P. Gaikwad & Sardesai	+	–	–	–	Eriocaulaceae	Plateaus; in marshy lands
124	***Eriocaulon rouxianum Steud***	+	–	–	–	Eriocaulaceae	Higher elevations; in marshy lands
125	***Eriocaulon sahyadricum*** Punekar, Malpure & Lakshmin.	+	–	–	–	Eriocaulaceae	Higher elevations; in marshy lands
126	***Eriocaulon santapauii*** Mold.	+	–	–	–	Eriocaulaceae	Higher elevations; in marshy lands along streams

Continued

Table 22.5. Continued.

No.	Botanical name	H	S	C	T	Family	Habitat/Adaptation
127	***Eriocaulon sharmae*** Ansari & Balakr.	+	–	–	–	Eriocaulaceae	Plateaus; higher elevations; open places; with graminoids
128	***Eriocaulon tuberiferum*** A.R.Kulkarni & Desai	+	–	–	–	Eriocaulaceae	High-altitude plateaus; open places; in marshy lands and puddles with graminoids
129	***Cyperus decumbens*** Govind.	+	–	–	–	Cyperaceae	Marshy lands; rhizomatous
130	***Cyperus diwakarii*** W. Khan & S. Solanki	+	–	–	–	Cyperaceae	Marshy lands
131	***Cyperus yadavi*** W. Kahn, Chavan & Solanki	+	–	–	–	Cyperaceae	Marshy lands
132	***Cyperus pentabracteatus*** Govind & Hemadri	+	–	–	–	Cyperaceae	Marshy lands, rhizomatous
133	***Eleocharis lankana*** T. Koyama ssp. mohamadi W. Khan	+	–	–	–	Cyperaceae	Stagnant water; rhizomatous
134	***Fimbrystylis ambavanensis*** V.P. Prasad & N.P. Singh	+	–	–	–	Cyperaceae	Open areas; marshy lands
135	***Fimbristylis dichotoma*** (L.) Vahl. var. *poladpurensis* D.P. Chavan	+	–	–	–	Cyperaceae	Marshy lands
136	***Fimbristylis naikaii*** W. Khan & Lakshmin.	+	–	–	–	Cyperaceae	Marshy lands
137	****Fimbrystylis nagpurensis** V.P. Prasad & N.P. Singh	+	–	–	–	Cyperaceae	Open areas; marshy lands
139	****Fimbrystylis poklii** W. Khan, Bhuskute & Kahalkar	+	–	–	–	Cyperaceae	Marshy lands
140	***Fimbrystylis ratnagirica*** V.P. Prasad & N.P. Singh	+	–	–	–	Cyperaceae	Plateaus; marshy lands
141	***Fimbrystylis unispicalaris*** Govind & Hemadri	+	–	–	–	Cyperaceae	Plateaus; marshy lands
142	***Mariscus blatteri*** Blatt.	+	–	–	–	Cyperaceae	Open marshy lands
143	***Mariscus konkanensis*** (Cooke) Sedgw	+	–	–	–	Cyperaceae	Rocky soils
144	***Pycreus bolei*** S.M. Almeida	+	–	–	–	Cyperaceae	Paddy fields, riverbank
145	***Pycreus lancelotii*** S.M. Almeida	+	–	–	–	Cyperaceae	Paddy fields, riverbank
146	***Pycreus malabaricus*** C.B.Cl.	+	–	–	–	Cyperaceae	Open gravelly hill slopes
147	****Scirpus naikianus** W.Khan	+	–	–	–	Cyperaceae	Along the margins of tanks
148	***Scirpus poklii*** W. Khan	+	–	–		Cyperaceae	Grass lands, paddy fields, forest clearance
149	***Chrysopogon castneus*** Veldkamp & Salunke	+	–	–	–	Poaceae	Plateaus, open grass lands
150	***Danthonidium gammiei*** (Bhide) C.E.Hubb.	+	–	–	–	Poaceae	Plateaus, muddy plains
151	***Dichanthium armatum*** (Hook.f.) Blatt. & McC.	+	–	–	–	Poaceae	Open hill slopes; plateaus
152	***Dichanthium compressum*** (Hook.f.) Jain & Deshpande	+	–	–	–	Poaceae	Open hill slopes
153	***Dicanthium jainii*** (Deshpande & Hemadri) Deshpande	+	–	–	–	Poaceae	Open hill slopes
154	***Dicanthium macanii*** Blatt.	+	–	–	–	Poaceae	Open hill slopes

Continued

Table 22.5. Continued.

No.	Botanical name	H	S	C	T	Family	Habitat/Adaptation
155	***Dicanthium panchganiensis*** Blatt. & McC.	+	–	–	–	Poaceae	High-altitude plateaus; with small herbs
156	***Dicanthium woodrowii*** (Hook.f.) Jain & Deshpande	+	–	–	–	Poaceae	In the plains of low rainfall areas
157	***Dimera blatteri*** Bor.	+	–	–	–	Poaceae	Open areas, plateaus
158	***Eulalia shrirangii*** Salunke & Potdar	+	–	–	–	Poaceae	Plateaus; with graminoids
159	***Glyphocloa ratnagirica*** (Kulkarni et Hemadri) W.D. Clayton	+	–	–	–	Poaceae	Plateaus; with graminoids
160	***Glyphochloa santapaui*** (Jain & Deshpande) Clayton	+	–	–	–	Poaceae	Coastal plateaus; in open situations
161	***Isachne bicolor*** Naik & Patunkar	+	–	–	–	Poaceae	Plateaus
162	***Isachne borii*** Hemadri	+	–	–	–	Poaceae	Plateaus and hill forts
163	***Ischaemum bolei*** M.R. Almeida	+	–	–	–	Poaceae	Grassland, open forest
164	***Ischaemum bombaiense*** Bor.	``	–	–	–	Poaceae	Grassland, open forest
165	***Ischaemum diplopogon*** Hook.f.	+	–	–	–	Poaceae	Open gravelly hill slopes
166	***Ischaemum huegllii*** Hack.	+	–	–	–	Poaceae	Abandoned fields in secondary forest (Ratnagiri)
167	***Ischaemum impressum*** Hack.	+	–	–	–	Poaceae	Open gravelly hill slopes
168	***Jansenella neglecta*** Yadav, Chivalkar & Gosavi	+	–	–	–	Poaceae	Open gravelly hill slopes
169	***Mnesithea veldkampii*** Potdar, Gaikwad, Salunkhe & Yadav	+	–	–	–	Poaceae	High-altitude plateaus; with graminoids
170	***Ophiuros bombaiensis*** Bor.	+	–	–	–	Poaceae	Data deficient
171	***Panicum johnii*** Almeida					Poaceae	Paddy fields (Sindhugurg-Sateli)
172	***Panicum painanum*** Niak & Patunkar var. *minor* Naik & Patunkar	+	–	–	–	Poaceae	Gulleys, open grasslands
173	*****Panicum phoinicládos*** Naik & Patunkar	+	–	–	–	Poaceae	Stagnant water poles
174	***Pogonachne racemosa*** Bor.	+	–	–	–	Poaceae	Open gravelly hill slopes
175	***Sacciolepis indica*** (L.) var. *intermedia* Almeida					Poaceae	In moist places (Sindhugurg)
176	***Themeda pseudotrimula*** Potdar *et al.*	+	–	–	–	Poaceae	Open gravelly hill slopes
177	*****Triplopogon polyanthus*** Naik & Patunkar	+	–	–	–	Poaceae	Wetlands, margins of tanks

**Taxa found in locations other than the Northern Western Ghats
*Unresolved taxa

current is an important abiotic factor that affects the survival of seedlings, particularly in seasonal forests where wet and dry seasons alternate (Delissio and Primack, 2003). Unseasonal rains during the dry season may induce seed germination, but these seedlings die, as ephemeral conditions cannot support their growth and establishment (Skoglund, 1992). The flowering plants in general and the endemic taxa of the NWG in particular tackle this problem in three ways by adapting themselves.

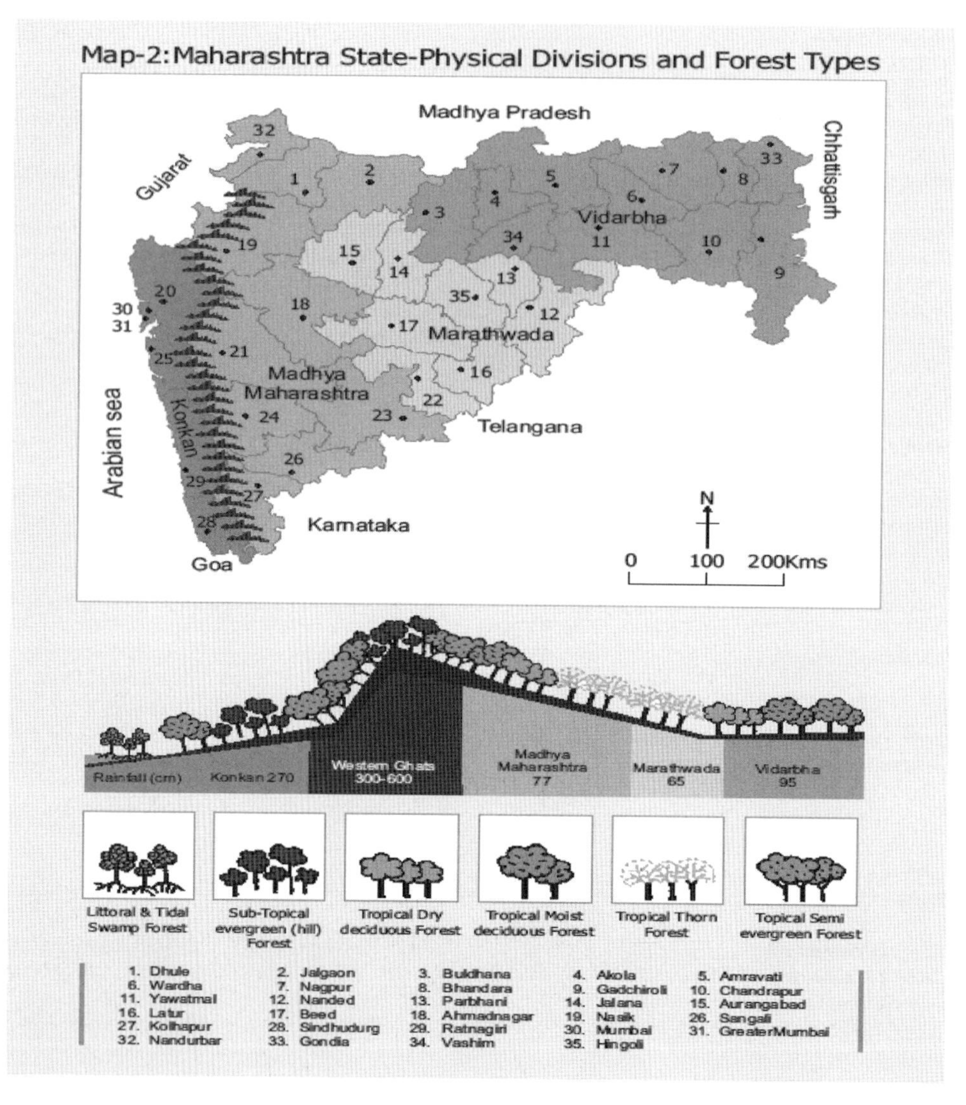

Fig. 22.3. Maharashtra State – physical divisions and forest types.

1) By phenological adaptations of flowering as to suit seasonal monsoon,
2) Developing underground storage organs, and
3) More preference for herbaceous habit or graminoids.

All these three features reflect patterns of endemism.

22.11 Phenological Adaptations and the Rise of Monocots and the Rule of Herbs

An impressive character of the NWG is that it is dominated by herbaceous flora, in particular the endemics. The higher percentage of monocot endemism could also be the result of the same ecological conditions. This can be easily noticeable when a comparison is made between the flora of the state and those of the endemic taxa. The monocot ratio is far higher than the dicots in number of endemic taxa (Table 22.6). Many monocot genera do show more radiations, resulting in more endemism. Families like Poaceae, Eriocaulaceae, Aponogetonaceae, Cyperaceae, Zingiberaceae, Araceae, Antheriaceae from monocots display more radiations and are represented by more Indian and Maharashtra endemic taxa. At the same time, we also notice the fall in the numbers of endemic Leguminaceae taxa in MEn (Table 22.7). Obviously, nitrogen-fixing ability may be good for general dominance. There are hardly any local (NWG) endemics belonging to nitrogen-fixing families. This point clearly indicates that this type of adaptive feature does not make a significant contribution in the evolution of newer taxa in NWG (Fig. 22.4).

Much of the endemic flora of the NWG completes their life cycle, chiefly the reproductive part, during four months of monsoon. Both the inaccessibility due to torrential rains and monsoon-dependent reproduction to a large extent help to protect the seed bank, present both on trees as well as in the soil and their germination and mortality.

Water is the limiting factor that shapes the natural processes to a large extent and the south-west monsoon in particular has a major role to play in patterns of plant endemism in the NWG.

The monsoon impact is so great that all of the 188 IEn reported to occur on plateaus (Watve, 2008; Watve, 2013) display monsoonal flowering. A very high level of flowering adaptation for monsoon is observed even in MEn. Of the total 177 MEn, 159 flower during the monsoon season, including pre- and post-monsoon periods.

Table 22.6. Changing flowering plant ratio of monocots to dicots in Maharashtra state – general flora versus endemism and the rise of endemic monocot ratio.

Category	General state flora	Indian endemics (IEn) in Maharashtra	Maharashtra state endemics (MEn)
Monocot to dicot ratio	1:2.36	1:1.62	1:0.86

Table 22.7. Patterns in flowering plants – the top five families in Maharashtra.

	Family	No. of taxa
Floristic richness – no. of wild taxa		
1	Poaceae	458
2	Leguminaceae	375
3	Cyperaceae	174
4	Acanthaceae	149
5	Asteraceae	116
Endemism – Indian endemics		
1	Poaceae	139
2	Leguminaceae	110
3	Acanthaceae	57
4	Orchidaceae	46
5	Apocynaceae	47
Local endemism – Maharashtra state only		
1	Poaceae	35
2	Apocynaceae	25
3	Cyperaceae	19
4	Eriocaulaceae	15
5	Leguminaceae	15

22.12 Poaceae is the King of NWG Forest Ecosytems

Among the different families that are present in the state, Poaceae (Graminaceae or grass family), is undisputedly the most dominant family by virtue of number of taxa as well as area of occupation (Deshpande and Singh, 1986; Potdar *et al.*, 2012). Grasslands cover a larger part of the state. The grass family is represented by 415 species, of which 139 taxa are endemic. Almost 15% (34 taxa) of the total endemic taxa of the state are grasses. This is in accordance with the general pattern in the Western Ghats. Poaceae has the highest number of endemic genera (Daniels, 2001; Irwin and Narasimhan, 2011). Three grass genera from the NWG, *Dicanthium*, *Isachne* and *Glyphocloa*, show supreme levels of endemism and are expressed by a large number of species and varieties. From the evolutionary point of view, Poaceae is the most successful family of the Maharashtra state, as well as in the NWG (Fig. 22.5).

Besides this, other important families showing more endemism are Leguminaceae, Acanthaceae, Orchidaceae, Eriocaulaceae and Asclepiadaceae. However, when the endemism-rich families Eiocaulaceae, Colchicaceae, Aponogetonaceae and Isachineae are compared with the Poaceae, Poaceae tops the list.

22.13 Functional Adaptations – Survival by Hiding (in Soil) Through Developing Underground Storage Parts

Plant adaptations in the NWG are generally of two types: those for water-limited environments and those for defence. Adaptation to four months of heavy rainfall followed by a water-limited environment for eight months is a serious challenge to the

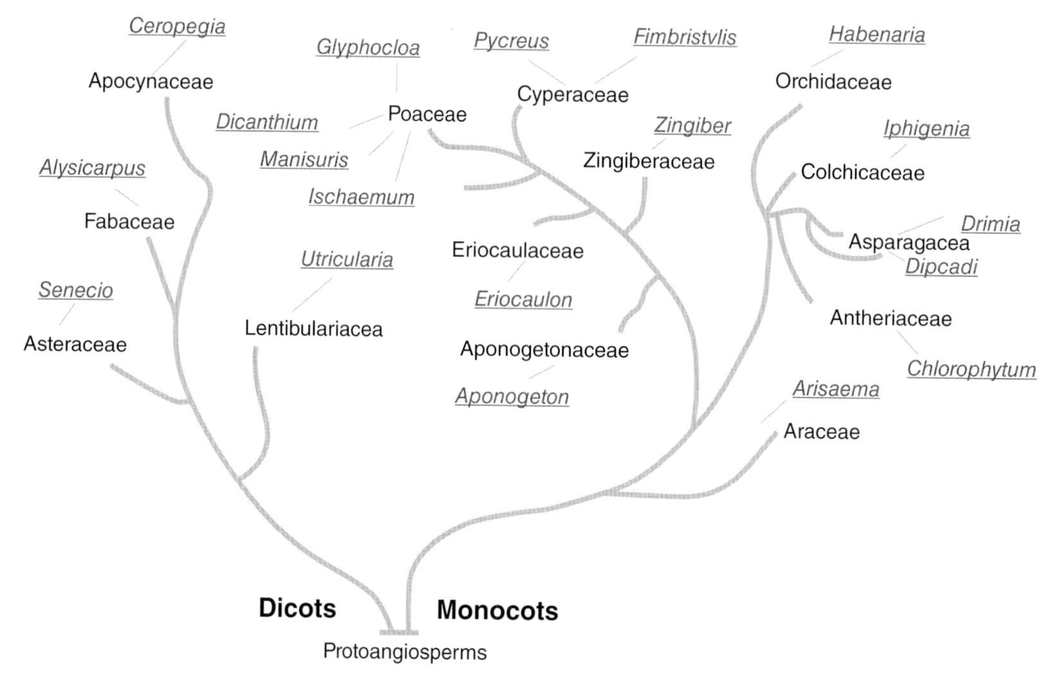

Fig. 22.4. Flowering plants of the Northern Western Ghats.

Fig. 22.5. Grasslands of the Northern Western Ghats near Amba Ghat.

NWG endemics. Another interesting part of their survival strategy in the NWG flora is that many herbaceous endemics develop underground storage parts, such as tubers, rhizomes, bulbs, corms, pseudobulbs and fleshy roots. As a result they do not die: they live underground and become dormant immediately after the rains.

As many as 72 IEn and 59 MEn have underground storage parts and become dormant after the rains. Although the genus Eriocaulons, which is adapted to marshy lands and has a fibrous root system like its close relative, the grass family, only one species in the entire family, *E. tuberiferum*, a local endemic found on the Kas plateau, has tuberous roots. This certainly indicates the direction of evolution and adaptation.

The data shows that the state is represented by a large number of taxa adapted for underground modifications, and the percentage increases in MEn: 34%, as opposed to 19% in IEn. The study further highlights that there is a greater number of monocots than dicots with root modifications in IEn. An insignificant number of taxa represent the other types of adaptation. For example, *Frerea indica* is one such unique monotypic genus evolved as a succulent, considered as palaeo-endemic. The phenomenon of modification of leaves for the water-limited environment is negligible in the NWG. In the case of defensive adaptations, we observed a low percentage of thorny taxa. The only MEn taxon known with such features is *Zizyphus rugosa* var. *glabra*. However, the absence of type specimens in Herbaria doubtfully endorses its existence.

22.14 How Do Forests Regenerate?

Natural forest regeneration is the process in which lands are filled up by trees that develop from seeds and other propagules that are naturally available. Regeneration processes vary in type, intensity, frequency and scale and they are triggered by the disturbance regime. In forest ecosystems, the disturbance regime unfolds in many ways and no single factor determines the system. A number of contingencies work for regeneration. Foremost among them is the presence of seed-bearing plants, i.e. the seed bank. Several other conditions, such as altitude and latitude gradients; rainfall and rainy days; soil types and soil cover, nutrients in the soil, animal and human interactions – all of these play roles in regeneration (Ramachandra *et al.*, 2012).

The forests of the Western Ghats have a dual-edged problem. Regeneration in many Indian forests, including the Western Ghats, is inadequate (Sukumar *et al.*, 1991). The situation is further complicated because many of the sites suitable for seed germination may not be suitable for the growth and establishment of seedlings (Shivanna and Tandon, 2014).

Seed germination in undisturbed sites may not be a serious problem. Studies on the Gap Phase regeneration of primary (*Pallaquium ellipticum*), late secondary (*Actinodaphne malabarica*), and early secondary (*Macaranga peltata*) tree species of different successional status in undisturbed sites and natural gaps have revealed that primary successional species complete their germination quickly, and all the categories of species showed higher establishment and growth under gaps. The survival of seedlings increased with gap size but sharply declined with gap age in the humid tropical forests of Kerala (Chandrashekara and Ramakrishnan, 1993).

The regeneration status in dry and moist deciduous forests is different. In the tropical deciduous forests of Madumalai Wildlife Sanctuary in Tamil Nadu, only 28.8% of the 104 tree species showed good regeneration. In the case of dry deciduous forests, regeneration is slightly higher. Of 86 tree species, 33.7% showed good regeneration (Reddy and Prachi, 2008).

22.15 Regeneration of Secondary Forests in NWG

Back in the 1980s, forest regeneration processes in the Tanaji Sagar Dam Catchment area in the NWG was studied. The catchment was represented mainly by three types of forest types that are common to the NWG: 1) moist evergreen forest; 2) moist mixed deciduous forest; and 3) broad-leaved hill or valley forest. The hill or valley type of forest is restricted to sacred groves and riverbanks.

The catchment once had luxuriant forests (Garland, 1934), but by the 1960s, it had lost much primary forest belonging to government as well as private organizations, due to dam construction, agriculture, and illicit cutting of trees for charcoal and forest fires. The natural forest cover lost in the decade 1973 to 1982 was estimated to be 4.25%, with a simultaneous rise of 6.19% in agricultural land (Patil *et al.*, 1986; Tetali *et al.*, 1993). As a result, all land except for sacred groves but including forest departments, turned to secondary forests. However, the level of depredations varied with the intensity of land use or forest exploitation. Shrub and small tree cover, intermingled with grasses, dominated all these secondary forests.

The hill slopes showed two prominent features. The hill tops, notably the steep hill slopes, were dominated by *Strobilanthus callosus* and *Pseudo-xytenanthera* (bamboo) species, while the middle hill slopes showed more frequency, density and abundance of *Carissa congesta, Lantana camara, Allophyllus cobbe, Gnidia glauca, Woodfordia fruticosa, Glochidion hohenackeri, Flacourtia indica, Colebrookea oppositifolia, Holarrhaena antidysentrica, Casearia graveolens, Securingea leucopyrus, Mynea laxiflora, Xeromphis spinosa* and *Osyris lanceolata*, depending upon the length of the fallow period. Among them, *Strobilanthus callosus* and *Pseudo-oxytenathera* are endemic to the Western Ghats and *Lantana camara* is an exotic weed. *Themeda quadrilocularis* and *Heteropogon contortus* were the two grass species that were predominant in the secondary forests, more prominent in degraded soils.

All these species are typical of secondary forests of the NWG in the initial forest-regeneration stages. They are all pyrophytes and adapted to the above-mentioned anthropogenic pressures. Each one of them has their distinct characteristic behaviour. For example, *Holarrhaena antidysentrica* spreads more with the help of root suckers in freshly cleared fallows than through seed germination. *Lantana camara* is more common near boulders and other shrubs. *Flacourtia* is thorny but its leaves and fruits are edible to wild animals. Birds are its natural dispersal agents. *Strobilanthus callosus* (*Karvi*) is a gregarious shrub, deciduous, strongly hairy, unpalatable and glandular. It flowers once in seven years and produces a tremendous number of seeds before its death. It forms dense and abundant thickets at such a height that a patch of *Karvi* thicket of 20 x 20m area may consist of 450–600 plants, with an average of 525 plants (based on stem count). The percentage of seed germination is very high (above 96%) and specific habitat adaptations support its monopoly on steep hill slopes. Except for *Lantana camara*, no germinating seed or seedlings of any of the common shrub or tree species were observed in freshly abandoned fallow land. The case is similar with some climbers such as *Tylophora dalzelli*, *Hemidesmus indicus*, and *Cocculus hirsutus*. Among the more frequently occurring climber species of the secondary forests, germinating seeds or seedlings of *Smilax zeylanica*, *Jasminum malabaricum* were recorded. These are the species of primary succession. On the other hand, the secondary forests showed seedlings of tree species which occur less frequently, such as *Bombax ceiba*, *Erythrina stricta* and *Terminalia species*. Their seedlings occur in such a large quantity that sometimes they cover the whole ground. In addition to these species, we also recorded the occasional presence of *Bridelia retusa* and *Syzygium cumini* seedlings. But survival of all these seedlings of different species after the monsoon is extremely rare. In an open situation, their survival percentage is almost nil.

During the vegetation studies, it was revealed that species of fallow lands were not totally destroyed after slash-and-burn agriculture and other anthropogenic activities. At Shirkoli village, where fallow land was monitored before and after slash-and-burn agriculture, the species count revealed all tree species, and roughly 90% of the shrub species of fallow lands did not die, but sprouted immediately during the monsoon. These sprouts after two to six years developed into strong clumps, as many other species started growing in their vicinity. These clumps expanded their size gradually and merged with other clumps to form a jungle.

These clumps harboured more species. In general, the larger the clumps, the higher the number of species. The clump expansion process was also studied in the secondary forests, particularly the fallow lands. Among the species that thrived in these fallow lands of secondary forests, five different categories were observed, based on their ability to make relationships with other species.

1. **Colony makers:** These species form strong colonies or thickets of their own species (e.g. *Carvia callosa, Holarrhena antidysentrica, Casearia graveolens, Gnidia glauca*).
2. **Loners:** These species prefer to grow alone and are randomly distributed (e.g. *Flacourtia indica*).
3. **Stoic species:** These have no specificities or interests (e.g. *Heterophragma quadrilocularis*).
4. **Clump or cluster makers:** These allow many other species to grow in their vicinity and as a result form clumps of many species (e.g. *Carissa spinarum, Zizyphus rugosa*).
5. **Opportunist exotics (weeds):** These are exotics, ready to grow with or without any other species. They also have tendency to make colonies (e.g. *Lantana camara*).

Further, among these different species categories, the clump or cluster makers were thought to be important for secondary forest regeneration process and clump formation was more distinctly visible in land kept fallow for 6–12 years. The clump formation among different species was quantified. The quadrat study clearly indicated that *Carissa spinarum* is the most common clump-making shrub found in all quadrats of middle hill slopes. The degree of association between 12 different herb, shrub and tree species that were commonly found in the secondary forests kept fallow in the catchment area for 6–12 years with *C. spinarum* was studied. The data were analysed by x^2 test with Yates's correction. An association index was calculated, based on total number of clumps, number of *C. spinarum* clumps and presence or absence of these species within the clumps. Overall, 12 quadrats each measuring 20 x 20 m in Shirkoli village were studied (P. Tetali, Studies on the status of people and forests in the Tanaji Sagar Dam Catchment area: the ecological impacts due to construction of a dam with alternative strategies for redevelopment, University of Poona, Poona, India, 1991, unpublished thesis, Vol. 1).

Statistical analysis showed that among seven tree species that were studied for association with *C. spinarum*, three species, *Bridelia retusa*, *Syzygium cumini* and *Osyris lanceolata*, showed positive association. *Terminalia elliptica* showed weak association. Among all the tree species, *Osyris*, an evergreen tree from the sandalwood family, showed the maximum association. Both the shrub species *Lantana camara* and *Zizyphus rugosa* showed an insignificant or a very weak relationship with *C. spinarum* (Fig. 22.6). The association index of all three climbing species with *C. spinarum* was naturally very high, ranging from 20 to 80. Of all the species studied, the highest positive association was observed with *Cocculus hirsutus* and *Smilax zeylanica* (Brahme and Tetali, 1991).

22.16 Why Only *C. spinarum*?

Association of two species in nature is possible only because of certain common features which are either attractive or essential to both. The survival of species in any given environment is dependent on a number of contingencies and *Carissa* inadvertently provides many of them. Here, *C. spinarum*'s congenial nature is more because of its attractive biological characters, not because of dependence.

C. spinarum is a large evergreen climbing shrub. It is a deep-rooted species and a pyrophyte, surviving both fire and axe. The branches are armed with twin stout, sharp, horizontal spines, which protects anything that germinates and grows in its vicinity. This is one of the reasons why tree species that are characteristic to secondary succession showed more positive association with it. Leaves and branches are unpalatable to domestic animals such as cattle and sheep, and this feature makes it a prominent species that occupies fallow lands more frequently. The fruits are edible and are sold by local communities in the urban markets. This means that it is an economically important plant and therefore it naturally gets some level of protection. It flowers and fruits profusely during the summer months and the fruits form an important source of food for wild animals, particularly birds and muntjacs, which are natural seed-dispersal agents.

This simple quantification study of species association clearly highlights that forest-regeneration processes in the NWG are highly dynamic. Like any other forest ecosystems, the processes are blended with both positive and negative interactions, or controlling mechanisms. Association of species varies from species to species. The study clearly concludes that secondary forests regenerate through the process of clump formation, which expand in the number of species and size, finally joining up to form a jungle. Species like *C. spinarum* make positive interactions and are essential for these secondary forests to develop into a jungle and to enhance biodiversity (Fig. 22.7).

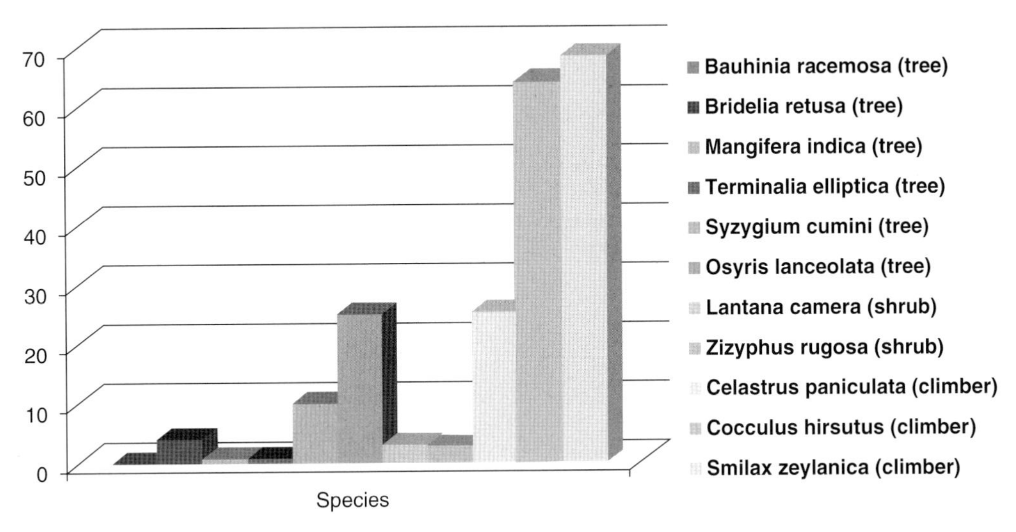

Fig. 22.6. Association index of 12 flowering plant species with *Carissa spinarum* in Tanaji Sagar Dam Catchment area of the Northern Western Ghats.

Fig. 22.7. Study area: Tanaji Sagar Dam Catchment area (Panshet Catchment) (A) Eight-year-old burned fallow land showing expanding *Carissa spinarum* clumps. (B) Single clump close-up with associated species. (C) Study area: Shirkoli village.

22.17 What is the status of NWG (Maharashtra) plant conservation programmes?

Among the different biogeographical regions of the state, Konkan is represented by the highest number of IEn taxa, closely followed by Madhya Maharashtra. Vidharbha and Marathwada are represented by fewer endemics. In the case of MEn, Madhya Maharashtra has more local endemics than Konkan districts. The difference is obviously due to the high-altitude plateaus, which increase the MEn percentage. Henceforth, flowering-plant conservation planning must reflect this scenario.

In terms of area, Maharashtra Forest Department controls 21% of the total area, i.e. 63,993 km^2. Dense forests occupy only 26,177 km^2 (11.75%), while open forests occupy 17,767 km^2. The wildlife of the state is protected through six national parks and 22 sanctuaries. Three of the national parks are tiger reserves, and three of the sanctuaries are exclusively preserved for bustard, black buck and peacock, with one wetland sanctuary at Jaikwadi for water fowl. There are no biosphere reserves in the state. Protected areas cover 15,300 km^2 or 4.97% of the total area of the state. These figures may be misleading, since 8496 km^2 (849,644 hectares) of the total protected area belongs to the Great Indian Bustard sanctuary. As compared to the diversity, protected areas in the NWG are far less than what actually is required.

Although recent, the UNESCO-sponsored MAB programme is one of the agents that recognizes biodiversity heritage sites for which India is a signatory. Thirty-nine places in the Western Ghats have been declared as world heritage sites so far under the UNESCO–MAB programme. NWG has four such sites: 1) Kaas plateau; 2) Koyna Wildlife Sanctuary; 3) Chandoli National Park; and 4) Radhanagari Wildlife Sanctuary. Each one is unique and represents the rich biodiversity heritage of the NWG.

22.18 Where Do Endemic Flowering Plants Occur and How Should We Conserve Them?

Habitats are important factors that contribute to rich species diversity. They are vital components of biodiversity conservation and planning programmes. The NWG is a mesohabitat of the Western Ghats and comprises many macro- and microhabitats of considerable importance. Many endemic taxa are habitat-specific. The role of microhabitats in endemic plant distribution and their survival is well explained by Lekhak and Yadav (2012), Bhattarai *et al.* (2012) and Watve (2008). Fortunately we have excellent research work with good ground-research data on important flowering-plant habitats. Nonetheless, it is practically impossible to protect all these habitats. Existing studies indicate that three types of macrohabitats are especially valuable for endemic flowering plant conservation (a brief account of their status is presented in Table 22.8).

1) Plateaus (Watve, 2013)
2) Hill forts (Ghanekar, 2007; Datar, 2011)
3) Sacred groves (Gadgil and Vartak, 1975; Deshmukh, 1999).

22.19 Conclusion

The NWG match the Southern Western Ghats in flowering-plant richness. However, they have not been widely recognized, due to the predominant herbal endemism. Study of endemism is important since these species do not occur anywhere else and, in a sense, it reflects the unique biological wealth of the geographical region. The NWG of Maharashtra is a source of many economically important plant species, 23 endemic genera and 162 endemic taxa. Furthermore, the NWG is enriched with two major groups: grasses and tuberous plants. In addition, two important rivers of Peninsular India, Godavari and Krishna, which sustain the Peninsular Indian economy, originate here. Historically and traditionally, the local communities are heavily dependent on the forest and its produce.

Water availability, particularly the monsoon, shapes the evolution, survival and success of the species that are unique to this region. The endemic species data shows a clear advantage for herbaceous plants, monocots, tuberous plants and species that can finish their life cycle during the monsoon or those that can withstand a prolonged dry period. The resultant conditions provide an excellent opportunity for the family Poaceae. As a result, it is represented by a very high percentage of flowering-plant endemism, both at Indian and Maharashtra levels. The absence of endemic tree species is unique and a distinguishing feature of the NWG of Maharashtra. Consideration of the nature of the south-west monsoon and its impact on floristic diversity, vegetation and the patterns of endemism it shapes is essential in conservation programmes.

Table. 22.8. Flowering plant rich localities (habitats) of Maharashtra

Kas Plateau

Purandhar fort

Shirkai sacred grove

	Macro habitat	Number	Number of endemic & threatened species
1	Plateaus	67	188
2	Hill forts	350	300
3	Sacred groves	5000	List incomplete

Forest-regeneration processes in the NWG are highly dynamic, with both positive and negative interactions, or controlling mechanisms. The role of *Carissa spinarum* is very important as it plays a beneficial role in forest-regeneration processes in the secondary forests.

As we expect more and more from the Western Ghats, we must accept the fact that we know little about the natural forces that sustain these ecosystems. This chapter therefore addressed some of the natural processes and patterns of flowering-plant endemism, such as species distribution in correspondence with habit, habitat, biogeographical locations and adaptations. This study, we believe, will provide insights into the NWG forest ecosystems and help in providing guidelines to design and implement conservation programmes. All this data calls for 'ecological thinking' in addressing the problems and planning of the so-called 'growth oriented development programmes' in the NWG of Maharashtra.

22.20 Acknowledgements

The authors would like to thank Mr V.M. Crishna, Director of the Naoroji Godrej Centre for Plant Research for necessary support. Thanks are also due to the Department of Science and Technology for funding the Women Scientist Project (WOS-A) and Mrs Sulabha Brahme of Shankar Brahme Samaj Vidnyan Granthalaya, Pune for funding part of the work. Authors are also thankful to Dr Sitaramam for going through the manuscript and providing useful suggestions.

References

Ahmedullah, M. and Nayar, M.P. (1986) *Endemic Plants of the Indian Region, Vol. 1, Peninsular India*. Botanical Survey of India, Calcutta, India.

Almeida, M.R. (1996–2009) *Flora of Maharashtra (Vols. 1–5A) Mumbai*. Oriental Press, Mumbai, India.

Anon. (2013a) Western Ghats website. Available at: whc.unesco.org/en/list/1342 (accessed 31 May 2016).

Anon. (2013b) *A Statistical Outline: Current Salient Forest Statistics 2013*. Government of Maharashtra, Forest Department, Nagpur, India. Available at: www.indiaenvironmentportal.org.in/files/file/maharashtra%20forest%20statistics%202013.pdf (accessed 31 May 2016).

Bhat, D.M., Murali, K.S. and Ravindranath, N.H. (2001) Formation and recovery of secondary forests of India: a particular reference to Western Ghats in South India. *Journal of Tropical Forest Science* 13(4), 601–620.

Bhattarai, U., Tetali, P. and Kelso, S. (2012) Contributions of vulnerable hydrogeomorphic habitats to endemic plant diversity on the Kas Plateau, Western Ghats. Available at: springerplus.springeropen.com/articles/10.1186/2193-1801-1-25 (accessed 31 May 2016).

Brahme, S and Tetali, P. (1987) *Economic and Ecological Impacts of Panshet Reservoir – A Study in Problem and Eco Development Alternatives*. Shankar Brahme Samaj Vidnyan Granthalaya, Pune, India.

Brahme, S and Tetali, P. (1991) *Experimentation for the Regeneration of Vegetation of Panshet Reservoir Catchment Area*. Shankar Brahme Samaj Vidnyan Granthalaya, Pune, India.

Champion, H.G. (1936) *A Preliminary Survey of the Forest Types of India and Burma*. Indian Forest Records (New Series). Government of India Press, New Delhi, India.

Champion, H.G. and Seth, S.K. (1968) *A Revised Survey of the Forest Types of India*. Government of India Press, Nasik, India.

Chandrashekara, V.M. and Ramakrishnan, P.S. (1993) Gap phase regeneration of tree species of different successional status in a humid tropical forest of Kerala. *Indian Journal of Biosciences* 18(2), 279–290.

Daniels, R.J.R. (2001) *National Biodiversity Strategy and Action Plan: Western Ghats Eco-region*. Report submitted to Ministry of Environment and Forests, Government of India, New Delhi, India.

Datar, M.N. (2011) *Gad Killyanvaril Vanaspati* (in the Marathi language). Snehal Prakashan, Pune, India.

Delissio, L.J. and Primack, R.B. (2003) The impact of drought on the population dynamics of canopy-tree seedlings in an aseasonal Malaysian rain forest. *Journal of Tropical Ecology* 19, 489–500.

Deshmukh, S.V. (1999) *Final Report of the World Bank Aided Maharashtra Forestry Project 'Conservation and Development of Sacred Groves in Maharashtra' Submitted to the Development of Sacred Groves in Maharashtra*. Bombay Natural History Society, Mumbai, India.

Deshpande, U.R. and Singh, N.P. (1986) *Grasses of Maharashtra – An Annotated Inventory*. Mittal Publications, New Delhi, India.

Forest Survey of India (2009) *India State of Forest Report, 2009*. Available at: http://fsi.nic.in/details.php?pgID=sb_61 (accessed 10 October 2016).

Gadgil, M. (1979) Hills, dams and forests; some field observations from Western Ghats. *Proceeding of the Indian Academy of Science* C2(3), 291–303.

Gadgil, M. (1990) State subsidies and resource use in a dual society. *The Economic Times*, 5 April, 24. New Delhi, India.

Gadgil, M. (2008) *Let Our Rightful Forests Flourish*. National Centre for Advocacy Studies, Pune, India.

Available at: www.indiaenvironmentportal.org.in/files/wp27.pdf (accessed 31 May 2016).
Gadgil, M. (2011) *Report of WGEEP (Western Ghats Ecology Experts Panel) Submitted to Ministry of Environment and Forests*. Government of India, India.
Gadgil, M. and Vartak, V.D. (1975) The sacred groves of Western Ghats in India. *Economic Botany* 30(2), 152–160.
Gadgil, M. and Guha, R. (1992) *This Fissured Land, an Ecological History of India*. Oxford University Press, New Delhi, India.
Gaikwad, S.D., Gore, R., Garad, K. and Gaikwad, S. (2014) Endemic flowering plants of northern Western Ghats (Sahyadri Ranges) of India: a checklist. *The Journal of Biodiversity Data* 10(3), 462–472. DOI: dx.doi.org/10.15560/10.3.461.
Garland, E.A. (1934) *Poona District Forest Working Plan*. Forest Department, Government of Maharashtra, Poona, India.
Gaston, K.J. (1998) Rarity as double jeopardy. *Nature* 394, 229–230.
Gaston, K.J. (2005) Biodiversity and extinction: species and people. *Progress in Physical Geography* 29(2), 239–247.
Ghanekar, P.K. (2007) *Sad Shyadrichi Bhatkanti Killyanchi* (in the Marathi language). Snehal Prakashan, Pune, India.
Gole, P. and Tetali, P. (1985) *An Inquiry into the Status of Animals and Plants in Critical Areas of Western Ghats in Order to Evolve a Plan to Conserve Biological Diversity*. Ecological Society, Pune, India.
Gunnel, Y. and Radhakrishna, B.P. (2001) *Sahyadri – The Great Escarpment of the Indian Subcontinent*, Vol. 1 of *Patterns of Landscape Development in the Western Ghats*. Geological Society of India, Bangalore, India.
Hughes, T.P., Bellwood, D.R., Connolly, S.R., Cornell, H.V. and Karlson, R.H. (2014) Double jeopardy and global extinction risk in corals and reef fishes. *Current Biology* 24(24), 2946–2951.
Irwin, S.J. and Narasimhan, D. (2011) Endemic genera of Angiosperms in India. A review. *Rheedea* 21(1), 87–105.
Kasturirangan, K., Babu, C.R., Mauskar, J.M., Chopra,K., Jagdish, K. *et al.* (2013) *Report of the High Level Working group on Western Ghats*, Vol. 1. Ministry of Environment and Forests, Government of India, New Delhi, India.
Krishnan, M.S. (1953) Structural and tectonic history of India. *Memoirs of the Geological Survey of India* 81, 41.
Lakshminarasimhan, P. (1996) Monocotyledons. In: Singh, N.P. and Sharma, B.D. (eds) *Flora of Maharashtra State*. Botanical Survey of India, Calcutta, India, pp. 1–794.
Lekhak, M.M. and Yadav, S.R. (2012) Herbaceous vegetation of threatened high altitude lateritic plateau ecosystems of Western Ghats, Southwestern Maharashtra, India. *Rheedea* 22, 39–61.
Medlicott, H.B. and Blanford, W.T. (1893) *A Manual of the Geology of India*, 2nd edn, revised by R.D. Oldham. Office of the Superintendent of Government Printing, Calcutta, India.
Molur, S., Smith, K.G. Daniel, B.G. and Darwall, W.R.T. (comp.) (2011) *The Status and Distribution of Fresh Water Biodiversity in the Western Ghats, India*. The IUCN Species Survival Commission, Cambridge UK and Gland, Switzerland. Available at: portals.iucn.org/library/efiles/documents/RL-540-001.pdf (accessed 31 May 2016).
Mishra, D.K. and Singh, N.P. (2001) *Endemic and Threatened Flowering Plants of Maharashtra*. Botanical Survey of India, Calcutta, India.
Myers, N., Mittermeier, R.A., Mittermeier, C.G., da Fonseca, G.A.B. and Kent, J. (2000) Biodiversity hotspots for conservation priorities. *Nature* 403, pp. 853–858. DOI: 10.1038/35002501.
Naik, V.N. (1998) *Flora of Marathwada*, Vols 1 and 2. Amrut Prakashan, Aurangabad, India.
Patil, D.N., Sawant, S.B. and Dey, B. (1986) Changes in natural vegetation pattern of some river basins of Western Maharashtra, India: a study based on remote sensing techniques. *Proceedings of the International Seminar on Phytogeography and Remote Sensing for Developing Countries*, Vol. 1. New Delhi, India, pp. T.4–P/5.1–5.13.
Patwardhan, S.K. and Asnani, G.C. (2000) Meso-scale distribution of summer monsoon rainfall near the Western Ghats (India). *International Journal of Climatology* 20, 575–581.
Potdar, G.G., Salunkhe, C.B. and Yadav, S.R. (2012) *Grasses of Maharashtra*. Shivaji University, Kolhapur, Maharashtra, India.
Radhakrishna, B.P. (1967) *The Western Ghats of the Indian Peninsula: Proceedings of the Seminar on Geomorphological Studies in India, 1965*. University of Sagar, Sagar, Madhya Pradesh, India, pp. 5–14.
Raghvan, K. (1964) Influence of Western Ghats on monsoon rainfall. *Indian Journal of Meteorology and Geophysics* 15(4), 617–620.
Ramachandra, T.V., Subash Chandran, M.D., Joshi, N.V., Sooraj, N.P., Rao, G.R. and Vishnu, M. (2012) *Ecology of sacred Kan forests in Central Western Ghats. ENVIS Technical Report: 41*. Available at: wgbis.ces.iisc.ernet.in/biodiversity/pubs/ETR/ETR41/index.htm (accessed 31 May 2016).
Ramesh, B.R. and Pascal, J.P. (1997) *Atlas of the Endemics of the Western Ghats (India): Distribution of Tree Species in the Evergreen and Semievergreen Forests*, Vol. 38. French Institute of Pondicherry, Puducherry, India.
Rao, N.A. (2007) *Forest Ecology in India – Colonial Maharashtra (1850–1950)*. Cambridge University Press, Cambridge, UK.

Rao, R.R. (2013) *Floristic Diversity in Western Ghats: Documentation, Conservation and Bioprospection – A Priority Agenda for Action*. Available at: wgbis.ces.iisc.ernet.in/biodiversity/sahyadri_enews/newsletter/issue38/article/index.htm (accessed 31 May 2016).

Reddy, C.S. and Prachi, U. (2008) Survival threat to the flora of Mudumalai Wildlife sanctuary, India: an assessment based on regeneration status. *Nature and Science* 6(4), 42–54. Available at: www.sciencepub.net/nature/0604/08_0468_mudumalai_regeneration_science.pdf (accessed 31 May 2016).

Shivanna, K.R. and Tandon, R. (2014) *Reproductive Ecology of Flowering Plants: A Manual*. Springer, New Delhi, India.

Singh, N.P. and Karthikeyan, S. (2000) *Flora of Maharashtra State (dicotyledons)*, Vol. 1. Botanical Survey of India, Calcutta, India.

Singh, N.P. and Karthikeyan, S. (2001) *Flora of Maharashtra State (dicotyledons)*, Vol. II. Botanical Survey of India, Calcutta, India.

Skoglund, J. (1992) The role of seed banks in vegetation dynamics and restoration of dry tropical ecosystems. *Journal of Vegetation Science* 3, 357–360.

Subramanyam, K. and Nayar, M.P. (1974) Vegetation and Phytogeography of the Western Ghats. In: Mani, M.S. (ed.) *Ecology and Biogeography of India*. The Hague, Netherlands.

Sukumar, R., Dattaraja, H.S., Suresh, H.S., Radhakrishna, J., Vasudeva, R., Nirmala, S. and Joshi, N.V. (1991) Long term monitoring of vegetation in a tropical deciduous forest in Mudumalai, Southern India. *Current Science* 62, 608–616.

Tetali, P. (2002) Threatened medicinal plants of Maharashtra state, India. In: Bhattarai, N. and Karki, M. (eds) *Sharing Local and National Experience in Conservation of Medicinal and Aromatic Plants in South Asia. Proceedings of the Regional Workshop Held at Pokhara, Nepal, 21–23 January 2001.* Archiv Karki no. 120935. International Development Research Centre (IDRC), Ottawa, Canada, pp. 139–140. Available at:idl-bnc.idrc.ca/dspace/bitstream/10625/29240/1/120935.pdf (accessed 31 May 2016).

Tetali, P. and Amre, R. (2003) *Grading and Trading Patterns of Hirda (Terminalia chebula Retz.) in Maharashta State*. Naoroji Godrej Centre for Plant Research, Shindewadi, Maharashtra, India.

Tetali, P. and Gunale, V.R. (1990) Status of sacred groves in the Western Ghats – Tanaji Sagar Dam catchment area. *Biology India* 1(1), 9–16.

Tetali, P.D., Patil, N., Gunale, V.R. and Brahme, S. (1993) Changing patterns of vegetation in the Mula-Mutha River basin with special reference to the Panshet Dam Catchment Area, Maharashtra, India. *Biology India* 4(1,2), 39–52.

Tetali, P, Tetali, S., Kulkarni, B.G., Prasanna, P.V., Lakshminarasimhan, P., Lale, M., Kumbhojkar, M.S., Kulkarni, D.K. and Jagatap, A.P. (2000) *Endemic Plants of India (A Status Report of Maharashtra State), Satara, India*. Naoroji Godrej Centre for Plant Research, Shindewadi, Maharashtra, India.

Tetali, P., Tetali, S. and Munde, P. (2003) *Status of Nothapodytes nimmoniana (Grah.) Mabb. in the Maharashtra State*. Naoroji Godrej Centre for Plant Research, Shindewadi, Maharashtra, India.

Wadia, D.N. (1919) *Geology of India*. Macmillan Publishers, London, UK.

Waghchaure, C.K., Tetali, P., Gunale, V.R., Antia, N.H. and Birdi, T.J. (2006) Sacred groves of Parinche Valley of Pune District of Maharashtra, India and their importance. *Anthropology and Medicine* 13(1), 55–76.

Watve, A. (2008) Special habitats: rock outcrops in northern Western Ghats. Special habitats and threatened plants of India, wildlife and protected areas. *ENVIS* 11(1), 147–153.

Watve, A. (2013) Status review of rocky plateaus in the northern Western Ghats and Konkan region of Maharashtra, India with recommendations for conservation and management. *Journal of Threatened Taxa* 5, 3935–3962.

Yadav, S.R. (1997) Endemic plants of peninsular India with special reference to Maharashtra. In Pokle, D.S., Kanir, S.P. and Naik, V.N. (eds) *Proceedings, VII IAAT Annual Meet and National Conference,* Aurangabad. Regency Publications, New Delhi, India, pp. 31–51.

23 DNA Barcoding as a Molecular Tool for the Assessment of Plant Biodiversity

Subrata Trivedi[1]*, Abid Ali Ansari[1], Hasibur Rehman[1], Shalini Saggu[1], Zahid Khorshid Abbas[1] and Sankar K. Ghosh[2]

[1]*Department of Biology, Faculty of Science, University of Tabuk, Tabuk, Saudi Arabia;* [2]*Department of Biotechnology, Assam University, Silchar, Assam, India*

Abstract

Increasing population and pollution are resulting in the reduction of biological diversity. In addition, several human activities are adversely affecting the natural habitat, thereby reducing biodiversity. For proper biodiversity assessment, accurate species identification is very important. Disagreement among traditional taxonomists often makes species identification based on morphological characters a difficult task. With increasing sophistication, standardized molecular techniques are proving to be an effective means of species identification. Global food security is highly dependent on plant genetic resources. Attacks by invasive alien species (IAS) can have a devastating effect on native biodiversity, thereby causing severe loss of biodiversity. In this chapter, we discuss DNA barcoding in plants, including the super-barcode concept.

23.1 DNA Barcoding and Plant Biodiversity: An Introduction

Our planet sustains a huge variety of plants, animals and microbes. Plants can be as tiny as microscopic phytoplanktons or as large as perennial trees. The proper assessment and conservation of this massive biodiversity is not an easy task. Plants, being autotrophic organisms, trap energy from the non-living environment and in turn make it available to other living organisms. On the other hand, animals, being heterotrophic, are either directly or indirectly dependent on plants. The rapid pace of industrialization and deforestation is resulting in a negative impact on plant biodiversity. The disappearance of one plant species may pave the way for extinction of many animal species. Therefore the identification of each plant species is a vital aspect of plant biodiversity assessment. An IUCN report revealed that 12–52% of species from the higher taxa, such as vertebrates and vascular plants, are facing extinction. Hence, there is an urgent need for assessment and conservation of biodiversity. In this chapter, we discuss DNA barcoding as a modern molecular tool for the assessment of plant biodiversity.

23.2 The Concept of DNA Barcoding

For the proper assessment of biodiversity, accurate identification of species is of prime importance. Species identification using traditional morphological methods often evokes disagreement among taxonomists. To overcome this situation, modern standardized molecular identification methods

*E-mail:strivedi@ut.edu.sa, trivedi.subrata@gmail.com

 Plant Biodiversity: Monitoring, Assessment and Conservation (eds A.A. Ansari, S.S. Gill, Z.K. Abbas and M. Naeem)

emerged for identification of species. In 2003, Professor Paul Hebert and his research group working at the University of Guelph (Canada) published a paper entitled 'Biological identifications through DNA barcodes'. They proposed a new system of identification of species using a short standardized segment of DNA, which gave rise to the unique idea of DNA barcoding (Hebert *et al.*, 2003). A brief standardized region from the 5' end of mitochondrial cytochrome c oxidase subunit 1 (COI) was put forward as the candidate barcode gene (Fig. 23.1). In 2004, the DNA Barcode of Life Data System (BOLD, http://www.boldsystems.org) was developed and was officially established in 2007 (Ratnasingham and Hebert, 2007).

Subsequently, the COI gene was used for species identification for varied animal groups, such as mosquitoes (Cywinska *et al.*, 2006), spiders (Greenstone *et al.*, 2005), springtails (Hogg and Hebert, 2004), shrimps (Trivedi *et al.*, 2011), oysters (Trivedi *et al.*, 2012; Trivedi *et al.*, 2013; Trivedi *et al.*, 2015b) and fish (Trivedi *et al.*, 2014a). Biodiversity of massive and diverse marine ecosystems can be assessed by DNA barcoding (Bucklin *et al.*, 2011; Ghosh *et al.*, 2009; Smriti *et al.*, 2010; Radulovici *et al.*, 2010; Trivedi *et al.*, 2013; Trivedi *et al.*, 2014b; Trivedi *et al.*, 2016b; Trivedi *et al.*, 2015a). Figure 23.2 shows the gel image after successful PCR amplification of the COI region of eight samples and Fig. 23.3 shows the position of the COI gene in the mitochondrial genome.

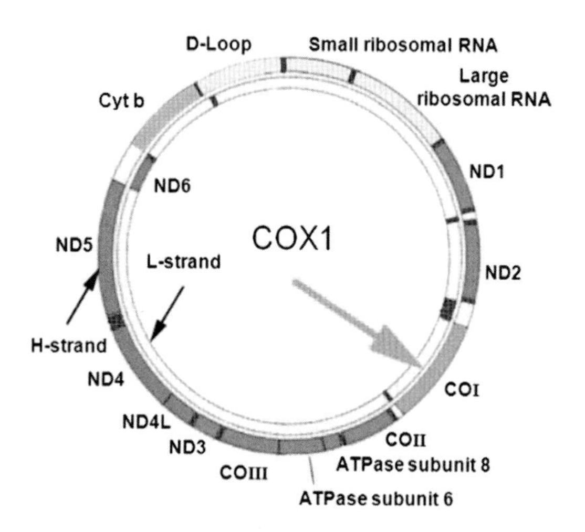

Fig. 23.1. The COI region in the mitochondrial genome.

Although analysis of the COI gene proved to be efficient as a barcode in different animal groups, it has not produced fruitful and desirable results in most plant groups. Low divergence and rapid genome rearrangements in this gene are the reasons for such low success in plants. The flowchart of the DNA barcoding procedure is shown in Fig. 23.2.

Yao *et al.* (2010) analysed several plant and animal sequences and subsequently suggested ITS2 as a universal DNA barcode for identifying plant species.

The CBOL Plant Working Group (2009) studied the performance of seven candidate plastid DNA regions (*atpF–atpH*spacer, *matK*gene, *rbcL* gene, *rpoB*gene, *rpoC1* gene, *psbK–psbI*spacer, and *trnH–psbA*m spacer) and recommend the two-locus combination of *rbcL_matK* as the barcode for land plants.

23.3 Potential Use of DNA Barcoding

Biological invasions by invasive alien species (IAS) are regarded as serious threats to global biodiversity. Invasive alien species can cause serious loss of biodiversity. The lower estimate of economic loss caused by IAS in Europe every year is €12 billion. In 2011, the IUCN reported that more than a quarter of critically endangered European native species are threatened by IAS, which is likely to result in their local extinction. They are also capable of inflicting socio-economic damage (Pimentel *et al.*, 2005). This technique can effectively distinguish the IAS from noninvasive counterparts (Ghahramanzadeh *et al.*, 2013).

Pryer *et al.* (2010) advocated the use of this modern technique in the horticulture industry. The enigmatic and cryptic species can be effectively and promptly identified by DNA barcoding. The study of cryptic species is an important aspect of biodiversity assessment.

Species causing Harmful Algal Bloom (HAB) can be authenticated by DNA barcoding, thereby proving its role in abatement of marine pollution. The different plants used as ingredients in herbal products like fruit juice, hair oil, herbal medicine, food pastes, etc. can be authenticated by applying this technique (Mankga *et al.*, 2013). It also finds use in prompt and accurate identification of agricultural pests and disease-causing organisms. For DNA barcoding, voucher specimens are required, which are regarded as an important asset for plant biodiversity studies. Finally, this method is effective in the discovery of new species, which will refine our past biodiversity records.

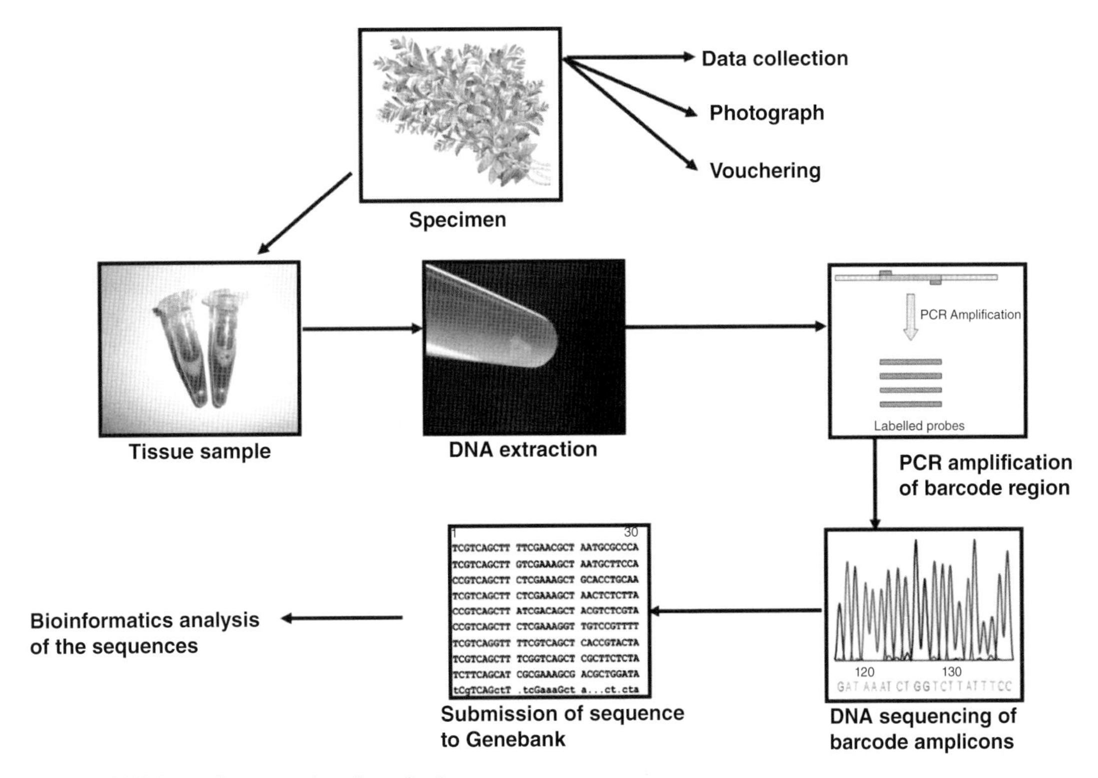

Fig. 23.2. DNA barcoding procedure flow chart.

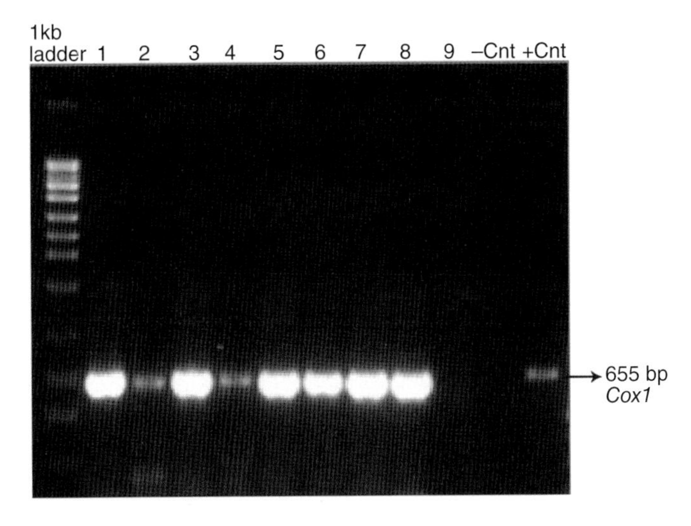

Fig. 23.3. Gel image of PCR amplified DNA of COI region (eight samples). Last lane shows the COI (also known as Cox 1) marker of 655 bp.

23.4 DNA Barcoding of Thallophytes

Different genes/regions were targeted by different researchers to barcode the algal species. They are COI (Ruennes *et al.*, 2010), 18S rDNA (Alemzadeh *et al.*, 2014), COI and UPA (Clarkston and Saunders, 2010), COI, UPA and ITS (Xiaobo *et al.*, 2013), 16S rRNA and 23S rRNA (Ahmad *et al.*, 2013). Several primer pairs were suggested for barcoding red algae (Saunders and Moore, 2013). A study conducted in Canadian subarctic algal population established the power of DNA barcoding in algal biodiversity research (Saunders and McDevit, 2013).

23.5 DNA Barcoding of Bryophytes

The bryophytes are represented by liverworts, hornworts and mosses. Mosses are the most diverse among bryophytes (Shaw, 2009). Accurate identification of mosses by morphological techniques is difficult and often misleading. Recent studies have shown that DNA barcoding using ITS1 locus can distinguish closely related moss species (Stech *et al.*, 2013).

Climate change is one of the most important environmental concerns of the present era. Bryophytes are regarded as model organisms to study climate change especially in extremely cold environments (Gordon *et al.*, 2001; Krab *et al.*, 2008). Lang *et al.* (2014) successfully used DNA barcoding to elucidate the molecular phylogeny of bryophytes belonging to the genus *Dicranum*.

23.6 DNA Barcoding of Pterodophytes

Pteridophytes have well-differentiated roots, stems and leaves, but are flowerless and seedless plants. *Huperzia* species – a pteridophyte medicinal plant – is used to treat Alzheimer's disease. A study conducted on medicinal pteridophytes in China suggested that *psbA-trnH* intergenic region can identify medicinal pteridophytes (Ma *et al.*, 2011). In this study, samples representing 51 species from 24 families were analysed with five DNA markers (*psbA-trnH* intergenic region, *rbcL*, *rpoB*, *rpoC1*, and *matK*). By using *rbcL*, *atpA*, and *trnG-R* sequence data, it was revealed that a fern which was marketed as *Cheilanthes wrightii* (endemic to the south-western USA and northern Mexico) in the horticultural trade was actually *Cheilanthes distans* (endemic to Australia and adjacent islands) (Pryer *et al.*, 2010).

23.7 DNA Barcoding of Gymnosperms

Gymnosperms are fruitless flowering plants that have naked ovules. Li *et al.* (2011) used *rbcL* and *matK* primers to study more than 50 gymnosperm species of China. DNA barcoding was carried out for the native flowering plants and conifers of Wales using two plant barcode markers, *rbcL* and *matK*. In this study, both herbarium and fresh samples were used, comprising 1143 species. This provided a valuable database of flowering plants in Wales that has a various range of applications (de Vere *et al.*, 2010).

23.8 DNA Barcoding of Angiosperms

Angiosperms are the most diverse among the plants. It is suggested that *matK* is the best target gene for angiosperms. An efficient *matK* primer pair for angiosperms was put forward by Yu *et al.* (2011). Alternative target genes for barcoding angiosperms were also proposed by different researchers (Kress *et al.*, 2005). Enan and Ahmed (2014) barcoded several date palm trees.

Although seagrasses have a very high ecological value, particularly for their role in reducing the effect of climate change (Ghosh *et al.*, 2015), they are depleting globally at a very alarming rate (Waycott *et al.*, 2002). Identification of seagrasses is challenging for traditional taxonomists. Lucas *et al.* (2012) showed that molecular identification of seagrasses is possible by DNA barcoding.

23.9 DNA Barcoding of Mangroves

Mangroves are composed of varied and unique halophytic plants that dwell in the coastal regions. These halophytic species have very significant economic and ecological significance. Every year the mangrove forests produce services valued at billions of dollars (Costanza *et al.*, 1997; Field *et al.*, 1998). The high level of nutrients in the mangrove ecosystem attracts a huge number of varied species thereby making these habitats biodiversity rich. The rapid destruction of mangroves worldwide is a matter of grave concern. Therefore, the remaining mangrove vegetation should be properly conserved.

The Sundarbans (located in India and Bangladesh) is a huge mangrove forest. The Sundarbans mangrove cover is shrinking in order to provide settlements for the rapidly increasing population. Increasing water pollution is aggravating the fragile situation. Several publications reflect the water quality and heavy metal accumulation in mangrove and mangrove-associated biota of this region (Chakraborty *et al.*, 2014; Pramanick *et al.*, 2014; Trivedi *et al.*, 1995; Mitra *et al.*, 1994a, b; Mitra *et al.*, 1995; Mitra *et al.*, 1996; Trivedi and Ansari, 2015). Recent studies suggested that with time, the salinity profile is changing in the different sections of Sundarbans. This shift in salinity profile is causing a corresponding shift in the mangrove zonation and distribution (Trivedi *et al.*, 2016a). DNA barcoding of these mangroves is essential for framing proper management strategies for mangrove restoration, as suggested by Daru *et al.* (2013). Phytoplanktons of this region were barcoded using the *rbcL* subunit of the RuBisCO enzyme (Bhattacharjee *et al.*, 2013).

23.10 DNA Barcoding Plants in the Biodiversity Hotspots

Several ecological studies are centered on the biodiversity hotspots. A biodiversity hotspot has a significant reservoir of biodiversity at the same time they are facing destruction. Ecologists generally agree that conservation of these regions is vital for the conservation of biodiversity. A study conducted on thousands of specimens from two biodiversity hotspots, Mesoamerica and Maputaland–Pondoland–Albany in southern Africa, revealed *matK* as the barcode gene. Barcoding can effectively identify cryptic and endangered species (Lahaye *et al.*, 2008).

23.11 DNA Barcoding for Identification of Toxic Plants

Food poisoning by toxic plants is not only a serious concern for human beings but also for domestic and wild animals. Prompt identification of these plants is necessary for treatment as well as providing public awareness. Morphological identification of the toxic plants is not possible, particularly for poisoning caused by fruit juice, jelly, pastes, etc. DNA barcoding is used for poisonous plant identification (Bruni *et al.*, 2010; Federici *et al.*, 2014; Xie *et al.*, 2014). Important articles on DNA barcoding in plants are listed in Table 23.1.

Table 23.1. Some important publications on DNA barcoding of plants.

Sl No.	Title of Article	Reference
1	A proposal for a standardised protocol to barcode all land plants	Chase *et al.* (2007)
2	Challenges in the DNA barcoding of plant material	Cowan and Fay (2012)
3	DNA barcoding in plants: taxonomy in a new Perspective	Vijayan and Tsou (2010)
4	Diatoms and DNA barcoding: a pilot study on an environmental sample	Jahn *et al.* (2007)
5	DNA barcodes for biosecurity: invasive species identification	Armstrong and Ball (2005)
6	DNA barcoding and genetic divergence in the giant kelp *Macrocystis* (Laminariales)	Macaya and Zuccarello (2010)
7	DNA barcoding detects contamination and substitution in North American herbal products	Newmaster *et al.* (2013)
8	DNA barcoding exposes a case of mistaken identity in the fern horticultural trade	Pryer *et al.* (2010)
9	DNA barcoding of Arctic bryophytes: an example from the moss genus *Dicranum* (Dicranaceae, Bryophyta)	Lang *et al.* (2014)
10	DNA barcoding of select freshwater and marine red algae (Rhodophyta)	Ruennes (2010)
11	DNA barcoding the native flowering plants and conifers of Wales	de Vere *et al.* (2012)
12	Efficient distinction of invasive aquatic plant species from non-invasive related species using DNA barcoding	Ghahramanzadeh *et al.* (2013)
13	DNA barcoding methods for land plants	Fazekas *et al.* (2012)
14	High universality of *matK* primers for barcoding gymnosperms	Li *et al.* (2011)
15	High-level diversity of dinoflagellates in the natural environment, revealed by assessment of mitochondrial *cox1* and *cob* genes for dinoflagellate DNA barcoding	Lin *et al.* (2009)
16	Molecular species delimitation in the *Racomitriumcanescens* complex (Grimmiaceae) and implications for DNA barcoding of species complexes in mosses	Stech *et al.* (2013)
17	Multiple multilocus DNA barcodes from the plastid genome discriminate plant species equally well	Fazekas *et al.* (2008)
18	New universal *matK* primers for DNA barcoding angiosperms	Yu *et al.* (2011)
19	DNA barcoding the floras of biodiversity hotspots	Lahaye *et al.* (2008)

Continued

Table 23.1. Continued.

Sl No.	Title of Article	Reference
20	A DNA barcode for land plants	CBOL Plant Working Group (2009)
21	Refining the DNA barcode for land plants	Hollingsworth (2011a)
22	Refinements for the amplification and sequencing of red algal DNA barcode and RedToL phylogenetic markers: a summary of current primers, profiles and strategies	Saunders and Moore (2013)
23	Species identification of medicinal pteridophytes by a DNA barcode marker, the chloroplast *psbA-trnH* intergenic region	Ma *et al.* (2010)
24	The changing epitome of species identification – DNA barcoding	Ali *et al.* (2014)
25	The genus *Lactarius s. str.* (*Basidiomycota, Russulales*) in Togo (West Africa): phylogeny and a new species described	Maba *et al.* (2014)
26	The promise of DNA barcoding for taxonomy	Hebert and Gregory (2005)
27	Use of DNA barcodes to identify flowering plants	Kress *et al.* (2005)
28	Use of ITS2 region as the universal DNA barcode for plants and animals	Yao *et al.* (2010)
29	Role of DNA barcoding in marine biodiversity assessment and conservation: an update	Trivedi *et al.* (2016b)
30	Identification of poisonous plants by DNA barcoding approach	Bruni *et al.* (2010)
31	Rapid plant identification using species- and group- specific primers targeting chloroplast DNA	Wallinger *et al.* (2012)
32	Choosing and using a plant DNA barcode	Hollingsworth *et al.* (2011b)
33	Molecular identification of Malaysian pineapple cultivar based on internal transcribed spacer region	Hidayat *et al.* (2012)
34	Unveiling distribution patterns of freshwater phytoplankton by a next generation sequencing based approach	Eiler *et al.* (2013)
35	Phytoplanktons and DNA barcoding: characterization and molecular analysis of phytoplanktons on the Persian Gulf	Alemzadeh *et al.* (2014)
36	History, applications, methodological issues and perspectives for the use of environmental DNA (eDNA) in marine and freshwater environments	Díaz-Ferguson and Moyer (2014)
37	DNA barcoding unmasks overlooked diversity improving knowledge on the composition and origins of the Churchill algal flora	Saunders and McDevit (2013)
38	Plant DNA barcoding: from gene to genome	Li *et al.* (2015)
39	DNA barcoding of arid wild plants using *rbcL* gene sequences	Bafeel *et al.* (2012)
40	DNA barcoding based on plastid *matK* and RNA polymerase for assessing the genetic identity of date (*Phoenix dactylifera* L.) cultivars	Enan and Ahmed (2014)
41	DNA barcoding for minor crops and food traceability	Galimbert *et al.* (2014)
42	DNA barcoding: a tool for species identification from herbal juices	Mahadani and Ghosh (2013)
43	Identification of Amazonian trees with DNA barcodes	Gonzalez *et al.* (2009)
44	How effective are DNA barcodes in the identification of African rainforest trees?	Parmentier *et al.* (2013)
45	Prospects and problems for identification of poisonous plants in China using DNA barcodes	Xie *et al.* (2014)
46	Food forensics: using DNA technology to combat misdescription and fraud	Woolfe and Primrose (2004)
47	A rapid diagnostic approach to identify poisonous plants using DNA barcoding data	Federici *et al.* (2014)
48	Efficacy of the core DNA barcodes in identifying processed and poorly conserved plant materials commonly used in South African traditional medicine	Mankga *et al.* (2013)
49	A successful case of DNA barcoding used in an international trade dispute	Jian *et al.* (2014)
50	DNA markers for food products authentication	Scarano and Rao (2014)

23.12 Use of Super-barcodes for Plant Identification

It is noteworthy that plants do not have a universal barcode region like the animals. A new super-barcode concept has emerged in recent years to address this issue. Some researchers have suggested that the super-barcode in plants is the chloroplast genome. Super-barcodes in conjugation with single-locus barcodes is the most efficient method of identification in plants (Li *et al.*, 2015) (Fig. 23.4).

Acknowledgements

We gratefully acknowledge the support provided to Dr Subrata Trivedi from the Deanship of Scientific Research (DSR), University of Tabuk, Ministry of Higher Education, Kingdom of Saudi Arabia, for the projects numbers S-1434-0106, S-1435-0112 and S-1436-0252. The assistance of the Saudi Digital Library is also acknowledged.

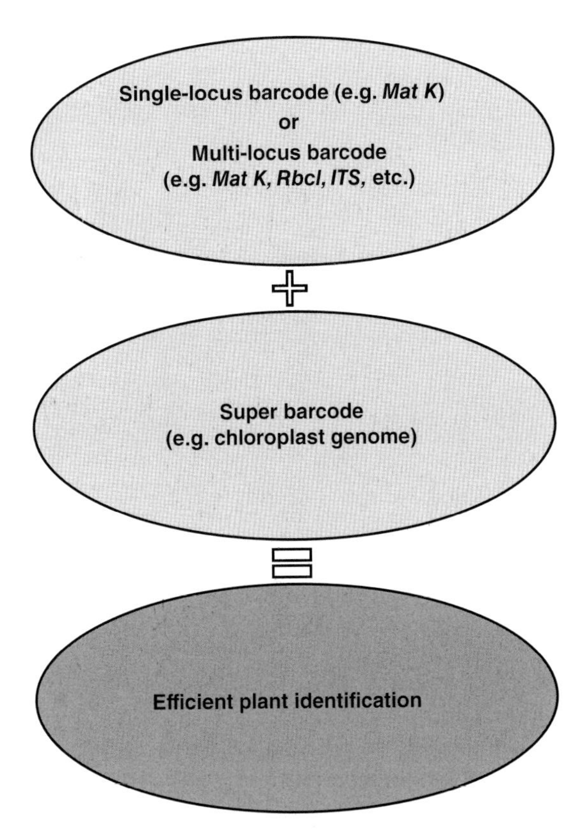

Fig. 23.4. Use of super-barcode in conjugation with traditional barcodes for efficient plant identification.

References

Ahmad, I., Fatima, Z., Yazdani, S.S. and Kumar, S. (2013) DNA barcode and lipid analysis of new marine algae potential for biofuel. *Algal Research* 2(1), 10–15. DOI: 10.1016/j.algal.2012.10.003.

Alemzadeh, E., Haddad, R. and Ahmadi, A.R. (2014) Phytoplanktons and DNA barcoding: characterization and molecular analysis of phytoplanktons on the Persian Gulf. *Iranian Journal of Microbiology* 6(4), 296–302.

Ali, M.A., Gyulai, G., Hidvégi, N., Kerti, B., Al Hemaid, F.M.A., Pandey, A.K. and Lee, J. (2014) The changing epitome of species identification – DNA barcoding. *Saudi Journal of Biological Sciences* 21, 204–231.

Armstrong, K.F. and Ball, S.L. (2005) DNA barcodes for biosecurity: invasive species identification. *Philosophical Transactions of the Royal Society of London* 360, 1813–1823.

Bafeel, S.O., Arif, I.A., Bakir, M.A., Al-Homaidan, A.A., Al Farhan, A.H. and Khan, H.A. (2012) DNA barcoding of arid wild plants using *rbcL* gene sequences. *Genetics and Molecular Research* 11(3), 1934–1941.

Bhattacharjee, D., Samanta, B., Danda, A.A. and Bhadury, P. (2013) Temporal succession of phytoplankton assemblages in a tidal creek system of the Sundarbans mangroves: an integrated approach. *International Journal of Biodiversity*. Article ID 824543. DOI: 10.1155/2013/824543.

Bruni, I., Mattia, F.D., Galimberti, A., Galasso, G., Banfi, E., Casiraghi, M. and Labra, M. (2010) Identification of poisonous plants by DNA barcoding approach. *International Journal of Legal Medicine* 124(6), 595–603. DOI: 10.1007/s00414-010-0447-3.

Bucklin, A., Stienke, D. and Blanco-Bercial, L. (2011) DNA barcoding of marine metazoa. *Annual Review of Marine Science* 3, 471–508.

CBOL Plant Working Group (2009) A DNA barcode for land plants. *Proceedings of the National Academy of Sciences USA* 106(31), 12794–12797.

Chakraborty, S. Trivedi, S., Fazli, P., Zaman, S. and Mitra, A. (2014) *Avicennia alba*: an indicator of heavy metal pollution in Indian Sundarban estuaries. *Journal of Environmental Science, Computer Science and Engineering and Technology* 3(4), 1796–1807.

Chase, M.W., Cowan, R.S., Hollingsworth, P.M., Van den Berg, C., Madriñán, S., Petersen, G., Seberg, O., Jørgsensen, T., Cameron, K.M. and Carine, M. (2007) A proposal for a standardised protocol to barcode all land plants. *TAXON* 56(2), 295–299.

Clarkston, B.E. and Saunders, G.W. (2010) A comparison of two DNA barcode markers for species discrimination in the red algal family Kallymeniaceae (Gigartinales, Florideophyceae), with a description of *Euthoratimburtonii* sp. *Botany* 88, 119–131. DOI: 10.1139/B09-101.

Costanza, R., D'Arge, R., de Groot, R., Farber, S., Grasso, M. *et al.* (1997) The value of the world's ecosystem services and natural capital. *Nature* 387, 253–260. Available at: www.esd.ornl.gov/benefits_conference/nature_paper.pdf (accessed 31 May 2016).

Cowan, R.S. and Fay, M.F. (2012) Challenges in the DNA barcoding of plant material. In: Sucher, N.J. (ed.) *Plant DNA Fingerprinting and Barcoding*. Humana Press, New York, USA, pp. 23–33.

Cywinska, A., Hunter, F.F. and Hebert, P.D.N. (2006) Identifying Canadian mosquito species through DNA barcodes. *Medical and Veterinary Entomology* 20, 4213–424.

Daru, B.H., Yessoufou, K., Mankga, L.T. and Davies, T.J. (2013) A global trend towards the loss of evolutionarily unique species in mangrove ecosystems. *PLoS One* 8(6), e66686. DOI: 10.1371/journal.pone.0066686.

de Vere, N., Rich, T.C.G., Ford, C.R., Trinder, S.A., Long, C. *et al.* (2012) DNA barcoding the native flowering plants and conifers of Wales. *PLoS One* 7(6), e37945. DOI: 10.1371/journal.pone.0037945.

Díaz-Ferguson, E.E. and Moyer, G.R. (2014) History, applications, methodological issues and perspectives for the use of environmental DNA (eDNA) in marine and freshwater environments. *International Journal of Tropical Biology* 62(4), 1273–1284.

Eiler, A., Drakare, S., Bertilsson, S., Pernthaler, J., Peura, S. *et al.* (2013) Unveiling distribution patterns of freshwater phytoplankton by a next generation sequencing based approach. *PLoS One* 8(1), e53516. DOI: 10.1371/journal.pone.0053516.

Enan, M.R. and Ahmed, A. (2014) DNA barcoding based on plastid *mat*K and RNA polymerase for assessing the genetic identity of date (*Phoenix dactylifera* L.) cultivars. *Genetics and Molecular Research* 13(2), 3527–3536.

Fazekas, A.J., Burgess, K.S., Kesanakurti, P.R., Graham, S.W., Newmaster, S.G. *et al.* (2008) Multiple multilocus DNA barcodes from the plastid genome discriminate plant species equally well. *PLoS One* 3(7), e2802. DOI: 10.1371/journal.pone.0002802.

Fazekas, A.J., Kuzmina, M.L., Newmaster, S.G. and Hollingsworth, P.M. (2012) DNA barcoding methods for land plants. *Methods in Molecular Biology* 858, 223–252. DOI: 10.1007/978-1-61779-591-6_11.

Federici, S., Fontana, D., Galimberti, A., Bruni, I., deMattia, F. *et al.* (2014) A rapid diagnostic approach to identify poisonous plants using DNA barcoding data. *Plant Biosystems* 149(3), 537–545. DOI: 10.1080/11263504.2014.941031.

Field, C.B., Osborn, J.G., Hoffman, L.L., Polsenberg, J.F., Ackerly, D.D. *et al.* (1998) Mangrove biodiversity and ecosystem function. *Global Ecology and Biogeography Letters* 7, 3–14.

Galimbert, A., Labra, M., Sandionigi, A., Bruno, A., Mezzasalma, V. and De Mattia, F. (2014) DNA barcoding for minor crops and food traceability. *Advances in Agriculture*.Article ID 831875. DOI: 10.1155/2014/831875.

Ghahramanzadeh, R., Esselink, G., Kodde, L.P., Dustermaat, H., van Valkenburg, J.L. *et al.* (2013) Efficient distinction of invasive aquatic plant species from non-invasive related species using DNA barcoding. *Molecular Ecology Resources* 13, 21–31.

Ghosh, S.K., Ghosh, P.R., Trivedi, S., Das, P.J., Chetry, A.J. *et al.* (2009) Mitochondrial genome: the biomarker for Indian biodiversity. In: *Proceedings of the 3rd International Barcode of Life Conference*. Mexico City, Mexico.

Ghosh, R., Trivedi, S., Pramanick, P., Zaman, S. and Mitra, A. (2015) Seagrass: a store house of carbon. *Journal of Energy, Environment and Carbon Credits* 5(2), 23–29.

Gonzalez, M.A., Baraloto, C., Engel, J., Mori, S.A., Pétronelli, P. *et al.* (2009) Identification of Amazonian trees with DNA barcodes. *PLoS One* 4(10), e7483. DOI: 10.1371/journal.pone.0007483.

Gordon, C., Wynn, J.M. and Woodin, S.J. (2001) Impacts of increased nitrogen supply on high Arctic heath: the importance of bryophytes and phosphorus availability. *New Phytology* 149, 461–471. DOI: 10.1046/j.1469-8137.2001.00053.x.

Greenstone, M.H., Rowley, D.H., Heimbach, U., Lundgren, J.G., Pfannenstiel, R.S. and Rehner, S.A. (2005) Barcoding generalist predators by polymerase chain reaction: carabids and spiders. *Molecular Ecology* 14(10), 3247–3266.

Hebert, P.D.N. and Gregory, T.R. (2005) The promise of DNA barcoding for taxonomy. *Systemic Biology* 54(5), 852–859.

Hebert, P.D., Cywinska, A. and Ball, S.L. (2003) Biological identifications through DNA barcodes. *Proceedings of The Royal Society B: Biological Sciences* 270(1512), 313–321. DOI: 10.1098/rspb.2002.2218.

Hidayat, T., Abdullah, F.I., Kuppusamy, C., Samad, A.A. and Wagiran, A. (2012) Molecular identification of Malaysian pineapple cultivar based on internal transcribed spacer region. *APCBEE Procedia* 4, 146–151. DOI: 10.1016/j.apcbee.2012.11.025.

Hogg, I.D. and Hebert, P.D.N. (2004) Biological identification of springtails (Collenbola: Hexapoda) from Canadian Arctic, using mitochondrial DNA barcodes. *Canadian Journal of Zoology* 82, 1–6.

Hollingsworth, P.M. (2011a) Refining the DNA barcode for land plants. *Proceedings of the National Academy of Sciences USA* 108(49), 19451–19452. DOI: 10.1073/pnas.1116812108.

Hollingsworth, P.M., Graham, S.W. and Little, D.P. (2011b) Choosing and using a plant DNA barcode. *PLoS One* 6(5), e19254. DOI: 10.1371/journal.pone.0019254.

Jahn, R., Zetzsche, H., Reinhardt, R. and Gemeinholzer, B. (2007) Diatoms and DNA barcoding: a pilot study on an environmental sample. In: Kusber, W.-H. and Jahn, R. (eds) *Botanic Garden and Botanical Museum Berlin-Dahlem, Proceedings of the 1st Central European Diatom Meeting*. Freie Universität, Berlin, Germany, pp. 63–68.

Jian, C., Deyi, Q., Qiaoyun, Y., Jia, H., Dexing, L., Xiaoya, W. and Leiqing, Z. (2014) A successful case of DNA

barcoding used in an international trade dispute. *DNA Barcodes* 2, 21–28.
Krab, J., Cornelissen, J.H.C., Lang, S.I. and Logtestijn, R.S.P. (2008) Amino acid uptake among wide-ranging moss species may contribute to their strong position in higher-latitude ecosystems. *Plant Soil* 304, 199–208. DOI: 10.1007/s11104-008-9540-5.
Kress, W.J., Wurdack, K.J., Zimmer, E.A., Weigt, L.A. and Janzen, D.H. (2005) Use of DNA barcodes to identify flowering plants. *Proceedings of the National Academy of Sciences USA* 102(23), 8369–8374.
Lahaye, R., van der Bank, M., Bogarin, D., Warner, J., Pupulin, F. *et al*. (2008) DNA barcoding the floras of biodiversity hotspots. *Proceedings of the National Academy of Sciences USA* 105(8), 923–2928. DOI: 10.1073_pnas.0709936105.
Lang, A.S., Kruijer, J.D. and Stech, M. (2014) DNA barcoding of Arctic bryophytes: an example from the moss genus *Dicranum* (Dicranaceae, Bryophyta). *Polar Biology* 37, 1157–1169.
Li, Y., Gao, L.M., Poudel, R.C., Li, D.Z. and Forrest, A. (2011) High universality of *matK* primers for barcoding gymnosperms. *Journal of Systematics and Evolution* 49(3), 169–175.
Li, X., Yang, Y., Henry, R.J., Rossetto, M., Wang, Y. and Chen, S. (2015) Plant DNA barcoding: from gene to genome. *Biological Reviews* 90, 157–166. DOI: 10.1111/brv.12104.
Lin, S., Zhang, H., Hou, Y., Zhuang, Y. and Miranda, L. (2009) High-level diversity of dinoflagellates in the natural environment, revealed by assessment of mitochondrial *cox1* and *cob* genes for dinoflagellate DNA barcoding. *Applied and Environmental Microbiology* 75(5), 1279–1290.
Lucas, C., Thangaradjou, T. and Papenbrock, J. (2012) Development of a DNA barcoding system for sea grasses: successful but not simple. *PLoS One* 7(1), e29987. DOI: 10.1371/journal.pone.0029987.
Ma, X.Y., Xie, C.X., Liu, C., Song, J.Y., Hui Yao, H. *et al*. (2010) Species identification of medicinal pteridophytes by a DNA barcode marker, the chloroplast *psbA-trnH* intergenic region. *Biology and Pharmacy Bulletin* 33(11), 1919–1924.
Maba, D.L., Guelly, A.K., Yorou, N.S., Kesel, A.D., AnnemiekeVerbeken, A. and Agerer, R. (2014) The genus *Lactarius s. str.* (*Basidiomycota, Russulales*) in Togo (West Africa): phylogeny and a new species described. *IMA Fungus* 5(1), 39–49.
Macaya, E.C. and Zuccarello, G.C. (2010) DNA barcoding and genetic divergence in the giant kelp *Macrocystis* (Laminariales). *Journal of Phycology* 46, 736–742.
Mahadani, P. and Ghosh, S.K. (2013) DNA barcoding: a tool for species identification from herbal juices. *DNA Barcodes* 1, 35–38. DOI: 10.2478/dna-2013-0002.
Mankga, L.T., Yessoufou, K., Moteetee, A.M., Daru, B.H. and van der Bank, M. (2013) Efficacy of the core DNA barcodes in identifying processed and poorly conserved plant materials commonly used in South African traditional medicine. In: Nagy, Z.T., Backeljau, T., De Meyer, M. and Jordaens, K. (eds) *DNA Barcoding: A Practical Tool for Fundamental and Applied Biodiversity Research*. *ZooKeys* 365 (special issue), 215–233. DOI: 10.3897/zookeys.365.5730.
Mitra, A., Trivedi, S. and Choudhury, A. (1994a) Interrelationship between trace metal pollution and physico-chemical variables in the frame work of Hooghly estuary. *Indian Ports* 10, 27–35.
Mitra, A., Trivedi, S. and Choudhury, A. (1994b) Interrelationship between gross primary production and metal accumulation by *Crassostrea cucullata*. *Pollution Research* 13(4), 391–394.
Mitra, A., Trivedi, S., Gupta, A., Choudhuri, A., Bag, M., Ghosh, I. and Choudhury, A. (1995) Balanus balanoides as an indicator of heavy metals. *Indian Journal of Environmental Health* 37(1), 42–45.
Mitra, A. Trivedi, S., Gupta, A., Choudhuri, A. and Choudhury, A. (1996) Distribution of heavy metals in the sediments from Hooghly Estuary, India. *Pollution Research* 15(2), 137–141.
Newmaster, S.G., Grguric, M., Shanmughanandhan, D., Ramalingam, S. and Ragupathy, S. (2013) DNA barcoding detects contamination and substitution in North American herbal products. *BMC Medicine* 11, 222.
Parmentier, I., Duminil, J., Kuzmina, M., Philippe, M., Thomas, D.W. *et al*. (2013) How effective are DNA barcodes in the identification of African rainforest trees? *PLoS One* 8(4), e54921. DOI: 10.1371/journal.pone.0054921.
Pimentel, D., Zuniga, R. and Morrison, D. (2005) Update on the environmental and economic costs associated with alien-invasive species in the United States. *Ecological Economics* 52, 273–288.
Pramanick, P., Jana, H.K., Zaman, S., Bera, D., Trivedi, S. and Mitra, A. (2014) Microbial status of mangrove fruit (*Sonneratia apetala*) jelly. *International Journal of Universal Pharmacy and Bio Sciences* 3(4), 221–227.
Pryer, K.M., Schuettpelz, E., Huiet, L., Grusz, A.L., Rothfels, C.J., Avent, T., Schwartz, D. and Windham, M.D. (2010) DNA barcoding exposes a case of mistaken identity in the fern horticultural trade. *Molecular Ecology Resources* 10, 979–985.
Radulovici, A.E., Archambault, P. and Dufresne, F. (2010) DNA barcodes for marine biodiversity: moving fast forward? *Diversity* 2(4), 450–472.
Ratnasingham, S. and Hebert, P.D.N. (2007) BOLD: the barcode of life data system. *Molecular Ecology Notes* 7, 355–364.
Ruennes, J. (2010) DNA barcoding of select freshwater and marine red algae (Rhodophyta). *Cryptogamie, Algologie* 31(4), 377–386.
Saunders, G.W. and McDevit, D.C. (2013) DNA barcoding unmasks overlooked diversity improving knowledge on the composition and origins of the Churchill algal flora. *BMC Ecology* 13(9). Available at: www.biomedcentral.com/1472-6785/13/9 (accessed 31 May 2016).

Saunders, G.W. and Moore, T.E. (2013) Refinements for the amplification and sequencing of red algal DNA barcode and RedToL phylogenetic markers: a summary of current primers, profiles and strategies. *Algae* 28(1), 31–43.

Scarano, D. and Rao, R. (2014) DNA markers for food products authentication. *Diversity* 6, 579–596. DOI: 10.3390/d6030579.

Shaw, A.J. (2009) Bryophyte species and speciation. In: Goffinet, B. and Shaw, A.J. (eds) *Bryophyte Biology*, 2nd edn. Cambridge University Press, Cambridge, UK.

Smriti, S., Shrivastava, A., Tarafdar, A. and Trivedi, S. (2010) DNA barcoding of some economic and endangered species from Sunderbans, India. *National Conference on Glimpses of Medical, Molecular and Marine Biotechnology*. Institute of Technology and Marine Engineering, West Bengal, India.

Stech, M., Veldman, S., Larraín, J., Muñoz, J., Quandt, D., Hassel, K. and Kruijer, H. (2013) Molecular species delimitation in the *Racomitrium canescens* complex (Grimmiaceae) and implications for DNA barcoding of species complexes in mosses. *PLoS One* 8(1), e53134. DOI: 10.1371/journal.pone.0053134.

Trivedi, S. and Ansari, A.A. (2015) Molecular mechanisms in the phytoremediation of heavy metals from coastal waters. In: Ansari, A.A., Gill, S.S., Gill, R., Lanza, G.R. and Newman, L. (eds) *Phytoremediation: Management of Environmental Contaminants, Volume* 2. Springer International Publishing, Cham, Switzerland, pp. 219–231. DOI: 10.1007/978-3-319-10969-5_19.

Trivedi, S., Mitra, A., Bag, M., Ghosh, I. and Choudhury, A. (1995) Heavy metal concentration in mud-skipper Boleophthalmus boddaerti of Nayachara Island, India. *Indian Journal of Environmental Health* 37(2), 120–122.

Trivedi, S., Ghosh, S.K. and Choudhury, A. (2011) *Cytochrome c oxidase* subunit 1 (COI) sequence of *Macrobrachium rosenbergii* collected from Sunderbans, India. *Journal of Environment and Sociobiology* 8(2), 69–172.

Trivedi, S., Ghosh, S.K. and Choudhury, A. (2012) Mitochondrial DNA sequence of *Cytochrome c oxidase subunit 1* (COI) region of an oyster *Crassostrea cuttakensis* collected from Sunderbans. *Journal of Environment and Sociobiology* 9(1), 13–16.

Trivedi, S., Affan, R., Alessa, A.H.A., Dhar, B., Mahadani, P. and Ghosh, S.K. (2013) DNA barcoding of fishes collected from Red Sea coastal waters of Tabuk, Saudi Arabia. *Proceedings of the 5th International Barcode of Life Conference*. Kunming, China.

Trivedi, S., Affan, R., Alessa, A.H.A., Ansari, A.A., Dhar, B., Mahadani, P. and Ghosh, S.K. (2014a) DNA barcoding of Red Sea fishes from Saudi Arabia – the first approach. *DNA Barcodes* 2, 17–20.

Trivedi, S., Aloufi, A., Ansari, A.A., Mitra, A. and Ghosh, S.K. (2014b) DNA barcoding in marine perspective. *Proceedings of the Aqaba International Conference on Marine and Coastal Environment, Status and Challenges* in Arab World. Aqaba, Jordan.

Trivedi, S., Aloufi, A., Ansari, A.A. and Ghosh, S.K. (2015a) DNA barcoding as a molecular tool to assess biodiversity in the Red Sea. *Proceedings of the Conference on Economics of the Red Sea and Their Development*. Tabuk, Saudi Arabia.

Trivedi, S., Aloufi, A., Ansari, A.A. and Ghosh, S.K. (2015b) Molecular phylogeny of oysters belonging to the genus *Crassoatrea* through DNA barcoding. *Journal of Entomology and Zoology Studies* 3(1), 21–26.

Trivedi, S., Zaman, S., Ray Choudhury, T., Pramanick, P., Pardis, F., Gahul, A. and Mitra, A. (2016a) Inter-annual variation of salinity in Indian Sundarbans. *Indian Journal of Geo-Marine Sciences* 45(3), 410–415.

Trivedi, S., Aloufi, A., Ansari, A.A. and Ghosh, S.K. (2016b) Role of DNA barcoding in marine biodiversity assessment and conservation: an update. *Saudi Journal of Biological Sciences* 23(2), 161–171. DOI: 10.1016/j.sjbs.2015.01.001.

Vijayan, K. and Tsou, C.H. (2010) DNA barcoding in plants: taxonomy in a new perspective. *Current Science* 99(11), 1530–1541.

Wallinger, C., Juen, A., Staudacher, K., Schallhart, N., Mitterrutzner Steiner, E.E.M., Thalinger, B. and Traugott, M. (2012) Rapid plant identification using species- and group-specific primers targeting chloroplast DNA. *PLoS One* 7(1), e29473. DOI: 10.1371/journal.pone.0029473.

Waycott, M., Freshwater, D.W., York, R.A., Ainsley, C. and Judson, K.W. (2002) Evolutionary trends in the seagrass genus *Halophila* (thouars): insights from molecular phylogeny. *Bulletin of Marine Science* 71, 1299–1308.

Woolfe, M. and Primrose, S. (2014) Food forensics: using DNA technology to combat misdescription and fraud. *Trends in Biotechnology* 22(5), 222–226.

Xiaobo, Z., Shaojun, P., Tifeng, S. and Feng, L. (2013) Applications of three DNA barcodes in assorting intertidal red macroalgal flora in Qingdao, China. *Journal of Oceanic University of China* 12(1), 139–145. Available at: dx.doi.org/10.1007/s11802-013-2052-9.

Xie, L., Wang, Y.W., Guan, S.Y., Xie, L.J., Long, X. and Sun, C.Y. (2014) Prospects and Problems for Identification of Poisonous Plants in China using DNA Barcodes. *Biomedical and Environmental Sciences* 27(10), 794–806.

Yao, H., Song., J., Liu., C., Luo, K., Han, J. *et al.* (2010) Use of ITS2 region as the universal DNA barcode for plants and animals. *PLoS One* 5(10), e13102. DOI: 10.1371/journal.pone.0013102.

Yu, J., Xue, J.H. and Zhou, S.L. (2011) New universal *matK* primers for DNA barcoding angiosperms. *Journal of Systematics and Evolution* 49(3), 176–181.

24 Onion and Related Taxa: Ecogeographical Distribution and Genetic Resources in the Indian Subcontinent

ANJULA PANDEY[1]*, K. PRADHEEP[1] AND K.S. NEGI[2]

[1]*Plant Exploration and Germplasm Collection Division, National Bureau of Plant Genetic Resources, New Delhi, India;* [2]*Regional Station, Bhowali, Niglat, Uttarakhand, India*

Abstract

This chapter is based on a study of the ecogeography and distribution of the onion and allied taxa of the Indian subcontinent. A field- and herbarium-based study of this group helped to classify them at the infrageneric level. The chapter also includes information on genetic resources and domestication trends of lesser-known, locally important and wild economic species from the Indian subcontinent.

24.1 Introduction

The genus Allium has more than 800 taxa occurring in different parts of the world (Fritsch *et al.*, 2010; Fritsch and Abbasi, 2013). The most important economic species in this genus which are cultivated include onion, garlic, shallot, leek, chives, Chinese chives and giant onion (Burba and Galmarini, 1994; Fritsch and Friesen, 2002). Among these, the common onion (*Allium cepa* L.) is one of the oldest domesticated species.

The genus *Allium* is highly variable in the characteristics of bulb/rhizome, roots, leaves, flowers, seeds and method of propagation. Variation is also reported in growth form (annual to perennial) and dormancy in winter season, biochemical composition and ploidy levels (Xu *et al.*, 1998; Zhou *et al.*, 2007).

In India, the genus is represented by commonly cultivated taxa, *A. cepa* L. (onion) and and *A. sativum* L. (garlic), as well as some locally important species occurring in selected pockets (Negi, 2006; Pandey *et al.*, 2005a,b). Two distinct centres of diversity occur in India for the genus *Allium* – the western Himalaya and the eastern Himalaya. The former region records the higher concentration (over 85%) of diversity in wild species; the eastern Himalaya has a lower figure (6%). About 35–40 species occur in temperate and alpine regions of Himalaya (Kachroo *et al.*, 1977; Polunin and Stainton, 1984; Pandey *et al.*, 2008), including the major cultivated species in India. The alpine-subtemperate region (2500–4500 m) of the western Himalaya, with about 25 taxa, has been the zone of concentration of species diversity (Gohil, 1992; Pandey *et al.*, 2008).

Studies of the distribution pattern of taxa, their ecogeography and genetic resource value, have helped in assessing biodiversity in general and agribiodiversity in particular. For the Indian region assessment of biodiversity of the genus *Allium*, species distribution pattern and use has helped in identifying areas of concentration of diversity *vis-à-vis*

*E-mail: anjuravinder@yahoo.com

germplasm collected and conserved under the national genetic resource programme (Pandey *et al.*, 2005a, b; Pandey *et al.*, 2008; Verma *et al.*, 2008; Pandey *et al.*, 2014). Meagre information on the pattern of species distribution and diversity, especially on the status (rare, threatened) of indigenous taxa, has necessitated concerted efforts in this direction (Nayar *et al.*, 1992; Negi and Pant, 1992; Singh and Rana, 1994; Pandey *et al.*, 2005b; Khosa *et al.*, 2014).

This chapter highlights the study of the ecogeography, systematics and distribution of the genus *Allium* and infers infrageneric classification for species in India. Characters of subgenus and section with habitat specificity are recorded. Some species like *A. gilgiteum* E.J. Wang and Tang, *A. tenuicaule* Regel and *A. barsczewskii* Lipsky reported from Kashmir (Gohil, 1992; Dasgupta, 2006), *A. caesioides* Wendelbo from Gilgit, Ladakh (Murti, 2001) and others of minor ornamental importance, such as *A. neopolitanum* Cirillo from Kashmir, have not been dealt with in detail due to lack of availability of material or information for study. Additionally, information on some economically important species of *Allium* in India is also included, along with trends of domestication in the Indian subcontinent. The authors have tried to present the above information for the benefit of users working in this area.

24.2 Ecogeographic Distribution

The genus *Allium* is widespread in warm temperate to temperate regions of the northern hemisphere, with the Irano-Turanian biogeographical region as a primary centre and the Mediterranean Basin, south-western and central Asia and western North American regions as secondary centres of origin (Fritsch and Friesen, 2002; Fritsch and Abbasi, 2013) (Fig. 24.1).

Diversity-rich regions with a high rate of endemism include west Asia (total native species 363; endemics 47.66 %), Middle Asia (244; endemics 30.74%), south-eastern Europe (131; 46.71%) and China (145; 24.48 %) (WCSP, 2015). A few species occur in mountains or highlands within the subtropics and tropics (Sri Lanka, Ethiopia, Central America). With increased exploratory and floristic studies undertaken to hitherto unexplored or underexplored areas of diversity-rich regions, new taxa of *Allium*, including some endemic and rare species, are being described and added (Khassanov and Fritsch, 1994; Khassanov and Tojibaev, 2010; Tojibaev and Karimov, 2012).

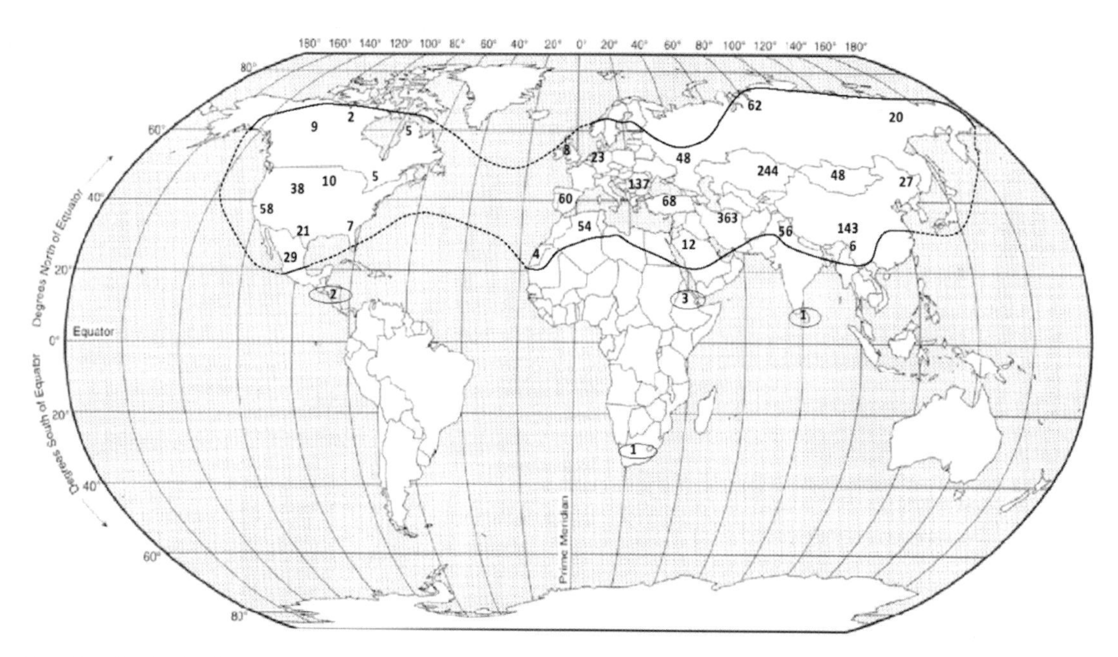

Fig. 24.1. World distribution of wild species in the genus *Allium* (richness of species diversity to a region designated by number indicated) (WCSP, 2015).

The majority of species are distributed in wide ecological conditions, such as open grassland, stony mountain slopes, sunny, dry sites in steppes and semi-deserts, rocky crevices and coastal cliffs. High polymorphism and ploidy variation has helped in acquiring traits for wider adaptability for diverse ecological conditions and high endemism in the genus *Allium*.

The diversity of *Allium* in the Indian subcontinent in the Himalayas is represented by about 7.5% of species, including two or three endemic taxa. This region forms a part of the extreme southern range of the global distribution, with Pakistan to the west (see Flora of Pakistan at: www.efloras.org/florataxon.aspx?flora_id=5&taxon_id=101086) and China in the eastern part (see Flora of China at: www.efloras.org/florataxon.aspx?flora_id=2&taxon_id=101086). In the Indian context, *Allium* species diversity is mostly concentrated in the high mountain ranges of western and eastern Himalaya, with endemic species in the former part (Pandey *et al.*, 2008).

24.3 Taxonomy and Infrageneric Classification of *Allium*

Allium taxonomy remains a matter of confusion. Taxonomic studies based on inadequate or incomplete material, close morphological similarities among related taxa, polymorphism and over-reliance on dried (herbarium) specimens have created a proliferation of synonyms. In addition, disagreements on the importance of specific morphological traits in taxonomic delimitation and subdivision of the genus have further complicated the situation (Hanelt *et al.*, 1992; Gregory *et al.*, 1998).

Infrageneric classification of *Allium* from the time of Linnaeus (1753; see Linnaeus, 1957–1959) to the present day has shown a constant increase in infrageneric categories: from 30 species classified in three groups to 262 species into six sections (Don, 1832); nine sections categorizing 228 species of the former USSR (Vvedensky, 1935); three subgenera, 36 sections and subsections grouping about 600 species (Traub, 1968); six subgenera, 44 sections and subsections (Kamelin, 1973); and three subgenera and 12 sections (Stearn, 1992).

Recent phylogenetic studies have helped in better understanding intrageneric classification in the *Allium* (Friesen *et al.*, 2006). Modern *Allium* systematics begin with a reassessment of the genus, based on molecular data (von Berg *et al.*, 1996). Evidence from morphology, anatomy, cytology, distributional ecology, developmental biology, serology and isozyme data using living collections and a herbarium-based study revealed a new taxonomic categorization of the genus *Allium* into six subgenera and 50 sections and subsections for 600–700 species (Hanelt *et al.*, 1992), which was escalated to 14 subgenera with about 60 taxonomic groups (Fritsch and Friesen, 2002). Based on phylogeny, Friesen *et al.* (2006) proposed a new classification, consisting of 15 subgenera and 72 sections for about 780 species.

Recent classifications have shown relationships among the infrageneric categories based on data from chloroplast and nuclear (ITS) genome (von Berg *et al.*, 1996; Samoylov *et al.*, 1999; Friesen *et al.*, 2006). Data from studies of macro- and micromorpholgy, anatomy, cytology and molecular science have facilitated an understanding of character development in the genus across the wide geographical range (Choi and Cota-Sánchez, 2010; Choi *et al.*, 2012). With continued work on phylogeny and biogeography of a region, the existing classification is likely to evolve further with better understanding of endemic taxa (Choi and Cota-Sánchez, 2010).

Among the distinguishable vegetative characters used in classification of *Allium*, bulb morphology is one of the important characteristics. The bulb membrane, particularly cell patterns on the inner surface of the outer bulb coat, have been used in taxonomic delimitation and species determination at infraspecific level (Wheeler *et al.*, 2013; Rola, 2014). The ultrastructure of the bulb membrane of subgenera *Anguinum*, *Butomissa*, *Rhizirideum* and *Reticulatobulbosa* is reticulate and fibrous, but in subgenus *Allium* it has hexagonal-elongated cell patterns; subgenus *Cepa* has thick membranous rectangular-elliptic cells, and a specific pattern has been observed in subgenera *Melanocrommyum* and *Polyprason*.

Allium species from India, China and Japan were classified into two series and seven groups, on the basis of taxonomy and systematic studies based the filament, bulb, spathe and direction of perianth segment in the expanded flower (Baker, 1874). Based on bulb and leaf characters, 27 Indian taxa were classified into three sections (*Schoenoprasum*, *Rhizirhideum* and *Molium*) (Hooker, 1892). Taxonomic studies based on both the field and herbarium data have been carried out for a few common Indian taxa (Gohil, 1992; Pandey and

Pandey, 2005; Pandey *et al.*, 2013; Pandey *et al.*, 2014). Revisionary works on the genus have relied on data from regional floristic records and the material available in national herbaria (Dasgupta, 2006).

Study of newer records or doubtful taxa could not be established due to poor access to the original range of distribution/occurrence. In some Himalayan taxa, such as *A. fedtschenkoanum* Regel and *A. seminovii* Regel, *A. jacquemontii* Kunth and *A. przewalskianum* Regel, *A. griffithianum* Boiss. and *A. roylei* Stearn., and *A. griffithianum* Boiss and *A. jacquemontii* Kunth, the problems associated with resolving the identity of allied taxa across different taxonomic groups were solved using numerical taxonomy (Yousaf *et al.*, 2004). Due to the lack of availability of information on rare and endemic *Allium* species and meagre species representation across the distributional range, it is difficult to review the complete ecogeographical and distributional study. Most of the phylogenetic studies on world *Allium* taxa have not taken into account Indian taxa, due to the lack of data on some significant characters such as bulb or rhizome morphology, leaf anatomy, micro-morphology of ovary, seed macro- and micro-morphology, and ecogeography.

24.3.1 Classification

Morphology of the bulb is one of the important characteristics used in the phylogeny of *Allium* (Fritsch and Friesen, 2002). The phylogeny of *Allium* was discussed under three main evolutionary lines, grouped on the basis of rhizome/bulb morphology and molecular study (Friesen *et al.*, 2006). The subgenus Amerallium belongs to the most ancient evolutionary line which consists of only bulbous plants, which rarely produce a notable rhizome. The second evolutionary line comprises subgenera *Anguinum* and Melanocrommyum that consist of small rhizomes and no bulb. The third evolutionary line, with subgenera *Butomissa*, *Rhizirideum*, *Allium*, *Cepa*, *Reticulatobulbosa* and *Polyprason*, are characterized by a well-developed rhizome and bulb, which are often much reduced or modified.

This chapter includes infrageneric classification based on morphological, distributional data and herbarium study and groups recognized taxa into ten subgenera and 22 sections (Fig. 24.2). We have broadly followed the infrageneric classification by Friesen *et al.* (2006). Subgenus and sectional details are briefly given below with the salient characters and ecological features.

24.3.2 First evolutionary line

a. *Allium* subgen. *Amerallium* Traub. Sect. *Bromatorrhiza* Ekberg.: Subgenus *Amerallium* is a large subgenus containing 120–140 species that occur in North America, Europe, north Africa, Ethiopia, the Caucasus, northern Iran, south-east Tibet, and south-west China (Hanelt *et al.*, 1992). Morphologically and ecologically distinct groups (isolated geographically) comprise *Allium* species: 1) native to North America (New World); 2) species from the Mediterranean region; and 3) species from eastern Asia (Old World). Subgenus *Amerallium* has characteristic tepals (one nerved), leaf vascular bundles (one row), absence of leaf palisade parenchyma (if present, it is secondarily evolved from spongy mesophyll), and the presence of subepidermal leaf laticifers (Fritsch, 1988); species have serological affinities and $x = 7$ as basic chromosome number. All this evidence strongly supports its distinct status.

Subgenus *Amerallium* is one of the most primitive groups in the evolutionary line. Taxa belonging to section *Bromatorrhiza* have no bulb formation, and thick, fleshy, swollen and cylindrical roots, sometimes modified into tubers or fasciculate structure. Four taxa belonging to this section are distributed mainly in the eastern Himalaya (Sikkim, Bhutan, Arunachal Pradesh) of India. *A. wallichii* Kunth widely occurs in high altitude (2300–4800 m) regions, forest clearings and shrubberies, through Indian Himalayas and the north-eastern region, extending to Pakistan and south-west China. Among other species, *A. fasciculatum* Rendle is distributed in the dry slopes, meadows and sandy locations in Bhutan, Nepal and Sikkim, to eastern Himalaya and China (2200–5400 m), whereas *A. macranthum* Baker and *A. hookeri* Thwaites are available at a lower altitudinal range (about 1400 m) in forest margins, moist places, meadows, river banks and damp areas at higher altitude (4200 m). Distribution of *A. hookeri* is very interesting; this species is mainly distributed in the Khasi Hills of Meghalaya, India, extending eastward to western China. *A. hookeri* was first reported from Pedurutalagala in Sri Lanka and its disjunct geographic distribution in this area is still unresolved. The authors have observed

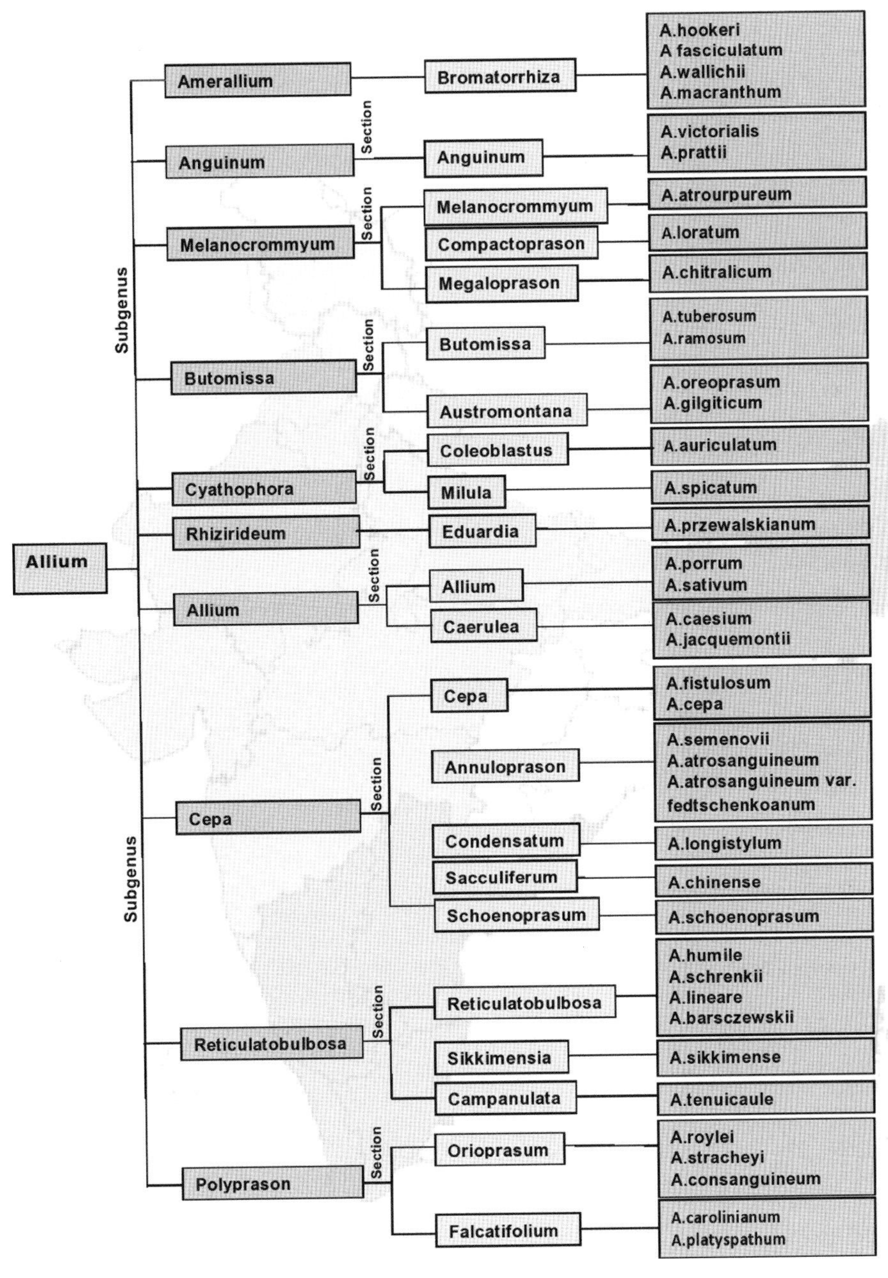

Fig. 24.2. Pictorial representation of infrageneric classification of genus *Allium* into subgenera, sections and species (adopted from Friesen *et al.*, 2006).

sporadic occurrence of this taxon in the tribal-dominated area of the north-eastern region (Palamau hills), with disjunct distribution, which needs further study on nativity in this region. In the region of its distribution in India, it is commonly reported in the wild and under cultivation, as was evident through field study by the authors and floristic data (Fig. 24.3). Intraspecific differentiation among the taxa belonging to this section is not reported.

Fig. 24.3. Different species of *Allium* grown in India. (A) *Allium macranthum* – a wild species grown in home gardens in Sikkim. (B) *Allium hookeri* – leaves harvested from plants grown in home gardens and being sold in the local market of Sikkim. (C) *Allium cepa* var. *aggregatum* (shallot) – less common cultivated species in hills of Himalaya. (D) *Allium cepa* var. *aggregatum* (multiplier onion) – commonly cultivated in south India and the eastern region. (E) *Allium sativum* (common garlic) – types with a pinkish-white bulb are less common. (F) *Allium sativum* – a white-coloured bulb is the more commonly cultivated type. (G) *Allium tuberosum* (Chinese chive) – commonly cultivated home garden species in hills. (H) *Allium chinense* (rokkyo) – lesser-known cultivated species in the north-eastern hills.

A. macranthum was first described from upper Sikkim, and is now reported to have a natural distribution eastwards to Yunnan (China). It is an uncommon species in the wild as well as under cultivation in the tribal areas of Arunachal Pradesh (Devi *et al.*, 2014). This species has a morphological resemblance to an American species, *A. cernuum* Roth, in having distinct pink-purple flowers arranged in drooping inflorescences, and with many North American species in the shape and characteristics of the ovary. Section *Bromatorrhiza* formed a sister clade with sections *Narkissoprason* Kam. Briseis (Salisb.) Stearn, *Arctoprasum* Kirschl. and *Molium*, which are distributed mainly in south-eastern Europe, west Asia, northern Africa and Macronesia (Nguyen *et al.*, 2008). On the basis of chromosome morphology, *A. fasciculatum* has shown a close relationship with *A. spicatum*, an endemic species to the Tibetan Plateau (subgen. *Cyathophora* sect. *Milula* (Prain) N. Friesen), also found in India (Sikkim, western Himalaya) (Tang *et al.*, 2005).

24.3.3 Second evolutionary line

b. *Allium* subgen. *Anguinum* (G.Don ex Koch) N.Friesen. The subgenus *Anguinum* is disjunct in distribution, ranging from hilly regions of south-western Europe to east Asia and North America

(Fritsch and Friesen, 2002). Based on distinct morphological, anatomic and cytological characters, this subgenus has been found to be a very specialized group. It forms the sister clade with subgenera *Caloscordum*, *Vvedenskya*, *Porphyroprason* and *Melanocrommyum*. The members of this subgenus are slow-growing plants characterized by a prominent ascending rhizome with developed lateral shoots (of first order only) and two or three glabrous leaves, broadly elliptical or lanceolate with distinct petiole.

Section *Anguinum* is represented by two distinct groups, one of Eurasian-American distribution (*A. victorialis* L.) and the other of east Asian distribution (*A. pratti* Wight). Only two species are distributed in cool and humid conditions in India; a common species, *A. victorialis*, from temperate Himalayan region and the other species, *A. prattii* C. H. Wright ex Hemsley, with sporadic distribution. *A. victorialis* occurs in moist habitats in forests, shady and moist slopes, pasture land and streamside (600–2500 m), from temperate Himalaya from Kashmir to Sikkim. *A. prattii*, from west Nepal to south-west China (2000–4500m), is distributed in shady and damp forests, thickets, scrub, stream banks and meadows. The two species differ distinctly in flower colour, general habit and leaf character. The taxa belonging to this section are serologically related to those of subgenus *Butomissa*.

c. *Allium* subg. *Melanocrommyum* (Webb & Berth.) Rouy.: represents the second evolutionary line with largest taxa belonging to this group. Taxa are morphologically very distinct. The subgenus is concentrated in the eastern parts of the Mediterranean region, south-west and Central Asia (Fritsch *et al.*, 2010; Fritsch and Abbasi, 2013) and further spreads to the Canary Islands, Kazakhstan, China and Pakistan. Species are distinct in having a tunicated bulb, showy flowers, nectaries without excretory tubes (a simple character state), as compared to those in the subgenus *Reticulobulbosa* and *Cepa*. Nectaries are located only in the lower half of the ovaries and excretion is through spurs rather than short tubes and excretory tubes bent downwards.

Taxa belonging to this group are strongly xerophilous and heliophilous and completely lack adaptability to mesic environmental conditions. They occur in dry steppes, deserts, arid hilly slopes and scrub. Some popular ornamental species include *A. aflatunense* B. Fedtsch., *A. giganteum* Regel and *A. aschersonianum* Barbey. In India, it is represented by three sections, each with one species: sect. *Melanocrommyum* Webb & Berth. (*A. atropurpureum* Waldst. & Kit.); sect. *Megaloprason* (*A. chitralicum* Wang & Tang); and sect. *Compactoprason* (*A. loratum* Baker), all three of which are mainly confined to the high altitudes of western Himalaya and extending to Afghanistan. The first two species are common and the third one is rare in occurrence.

24.3.4 Third evolutionary line

The third evolutionary line is complex and comprises heterogeneous taxa.

d. *Allium* subgn. *Butomissa* (Salisb.) N.Friesen: The taxa are characterized by simple habit and chromosome morphology. Plants have characteristic elongated cylindrical or conic bulb, outer coat brown, reticulate or fibrous and inner one membranous white and stout rhizome. They occur in steppes of Siberian-Mongolian-north China. Taxa of this subgenus have shown close affinity with those of the subgenus *Anguinum* based on morphological and serological study. Molecular data indicate mid position of this subgenus between subgenus *Rhizirideum* and subgenus *Cythophora*.

In India, two sections occur. Sect. *Butomissa* (Salisb.) Kamelin has two species, *A. ramosum* L. and *A. tuberosum* Rottler ex Spreng.; the latter is a less variable but widely occurring species in the Himalayan range and cultivated in plains (true wild types reported from the Himalaya). Occurrence of *A. ramosum* is doubtful and needs more studies to confirm it in the Indian region of western Himalaya.

Sect. *Austromontana* has clustered bulbs covered with reticulate membrane, well-developed horizontal rhizome, linear flat leaves aggregated on the basal scape part. *A. oreoprasum* Schrenk and *A. gilgiticum* Wang & Tang occur in the hilly region of eastern and central Asia extending up to the adjoining eastern Mediterranean region. *A. oreoprasum* Schrenk occurs in sunny slopes and stony rivers shores of the Ladak Himalaya (1200–2700 m). Another species *A. gilgiticum* is rare and endemic species reported from Gilgit (type locality) and suspected to be extinct (Gohil, 1992). *A. oreoprasum* is allied to the members of Sect. *Butomissa* *A. ramosum* and *A. tuberosum*) but differs in seed testa ornamentation having verrucose periclinal walls (Friesen *et al.*, 2006).

e. *Allium* subgn. *Cyathophora* (R.M. Fritsch) R.M. Fritsch. It is a small subgenus with distribution in Tibet and the Himalayas with two sections; sect. *Coleoblastus* (*A. auriculatum* Kunth.) is a widely occurring in western Himalaya but sect. *Milula* (*A. spicatum* (Prain) N. Friesen syn. *Milula spicata* Prain) is restricted to the central Himalayan region. Based on a molecular study and anatomical investigations of leaf characters, taxa from this group (two sections: section *Coleoblastus* and section *Milula*) have shown closeness with taxa belonging to section *Cyathophora* of Asian temperate origin (Friesen *et al.*, 2000). Specialized morphological features of taxa belonging to both *Coleoblastus* and *Milula* mean they are highly adaptable to the hardy habitat of western Himalaya.

A. auriculatum has a distinctive elongated, narrow bulb, covered with a reticulate fibrous tunic of orange-brown to brown. It grows in the rock crevices on rocky soil near streams. *A. spicatum* is morphologically very distinct from other members of the subgenus in having inflated, thickened leaf bases, less-specialized roots, absence of storage leaves and bulb, and densely flowered spiklet. It grows in thickets, grassy slopes and sandy grasslands (2900–4800 m) in Nepal Himalaya and adjoining areas. Molecular data have supported the inclusion of a closely related genus *Milula* in the *Allium* genus (Friesen *et al.*, 2006).

Evidence based on geography, morphology and molecular data have supported phylogenetic position of subgenus *Cyathophora* closer to subgenus *Amerallium* and subgenus *Butomissa* near *A. tuberosum* and *A. ramosum* (Huang *et al.*, 2014). *A. cyathophorum*, a related species from neighbouring areas (China), has more advanced characteristics (growth form, foliation and structures of nectar and excretory tube) than other related taxa in this subgenus in India (Fritsch, 1992).

f. *Allium* subgen. *Rhizirideum* (G.Don ex Koch) Wendelbo s.s.: sect. *Eduardia* N.Friesen shows its distribution from the western Himalaya, Pakistan and central Nepal, Tibet, south-west China. Taxa occur on scrub, dry or rocky areas, and are characterized by narrow, ovoid-cylindric bulbs usually covered with a common reticulate membrane and leaves channelled shorter than scape, and filaments slightly longer than perianth segments. In India, only one species, *A. przewalskianum* Regel, occurs in diverse habitats in scrub, dry slopes, ravines and rocky crevices (2000–4500 m), and is morphologically and genetically distinct from its close relatives in subgenus *Rhizirideum* (Zhuo *et al.*, 2007). It differs from its related species, *A. eduardii* Stearn, in the stamen, and is distributed on dry slopes and plains of the adjoining regions of Mongolia and Russia. Beyond Indian boundaries, this subgenus is well represented in China, where over 20 taxa occur.

g. *Allium* subgen. *Allium* L. subgen. *Allium* is the largest among all the subgenera, comprising 40% of the total number of species in the genus. This group is distributed from Europe to north-west China and Japan and its richness of diversity is concentrated in the Mediterranean and Irano-Turanian region (Hanelt *et al.*, 1992). Subgenus *Allium* is supposed to be an actively evolving group in which the determination of species is often difficult (Hanelt *et al.*, 1992). Taxa are characterized by the presence of ovoid or subglobose well-developed bulbs, with a membranous inner coat. Popular cultivars belonging to this subgenus are: *A. sativum* (garlic); *A. porrum*/*A. ampeloprasum* (leek, kurrat, pearl onion and elephant garlic) and ornamentals (*A. atroviolaceum* Boiss. and *A. sphaerocephalon* L.). Plants are ecologically restricted to an open dry habitat, where they face little competition with other plants.

Taxa in the *Allium* section are commonly cultivated as spice and condiments; in *A. sativum* there is not much variability in bulb characteristics, except bold-sized cloves and colour variation in the bulb membrane from pink, mauvish and white. *A. porrum* (leek) is under limited cultivation in the Himalayan and foothill regions, where flowers and cloves/bulbils are equally important for use as condiment and medicine.

Members of sect. *Caeruleum* and *Avulsea* occur wild in western Himalaya and adjoining areas of west and central Asia in the desert regions, dry pasture land and dry hills/slopes (700–2000 m). In India, the wild taxa *A. caesium* Schrenk is a less commonly available species in the deserts and dry fields of western Himalaya, and extending to central Asia (up to 2000 m). A much more common species, *A. griffithianum* Boiss., is distributed in the western Himalaya and westwards to Central Asia, Afghanistan and Pakistan.

h. *Allium* subgn. *Cepa* (Mill.) Radić. This is represented by five sections, including cultivated taxa of exotic origin. In India, subgenus *Cepa* is the largest, with 11 taxa and seven species (four cultivated and three wild). Taxa belonging to the sect. *Cepa* have a short vertical rhizome, cylindrical-globose bulb(s) developed on short rhizomes, a bulb covered with coriaceous outer layer and membranous inner layer, cylindrical fistulose leaves, often covering the scape, which is also fistulose, and ovaries with distinct nectariferous pores and two ovules per locule.

In sect. *Cepa* (Mill.) Prokh., *A. cepa* var. *cepa*, *A. cepa* var. *aggregatum* G.Don (aggregate onion), *A. cepa* var. *proliferum* or *A. x proliferum* (Moench) Regel (tree onion, occasionally cultivated in Asia and probably a taxon of hybrid origin) and shallot (*A. fistulosum* L.) are some of the important cultivated taxa in western Himalaya. *A. fistulosum* is a variable cultivated species of primary importance in east Asia, grown only in the hilly areas of Himalaya and in Nilgiris. Sect. *Seculiferum* P.P.Gritz has *A. chinense* G.Don, which is a commonly cultivated species in the north-eastern hill region. Two types are commonly reported: narrow bulbs with wide-channelled (fistulose) leaves, and narrow bulb with narrow triquetrous leaves.

Section *Cepa* is morphologically closely related to the sect. *Schoenoprasum* Dumort. and other species of the subgenus *Rhizirideum*. *A. schoenoprastum* L. has high ecological and morphological adaptability across different geographical regions of the world (Friesen and Blattner, 2000). In India, *A. schoenoprasum* occurs in the meadows, valleys, damp slopes, along streams and on rocky alpine slopes (2000–3000 m) in the western Himalaya (Uttarakhand, Himachal Pradesh, Kashmir, Drass) extending to south-west Asia, north and east Europe and also in North America.

Taxa of the sect. *Annuloprason* T.V.Egorova occur in meadows, high mountain bogs, near streams and in moist places (2400–5000 m). Two closely related taxa, *A. atrosanguineum* subsp. *atrosanguineum* Schrenk and *A. atrosanguineum* subsp. *fedtschenkoanum* Regel, differ in perianth colour and fistulose leaves, and are both distributed in meadows, high mountain bogs and near streams (2400–3500 m) in the Himalayan ranges extending to Afghanistan, China, Kazakstan, Kyrgyzstan, Pakistan, Tajikistan and Uzbekistan. *Allium atrosanguineum* var. *tibeticum* (Regel) G. Zhu & Turland is endemic to China (Zhu and Turland, 2000).

A. semenovii Regel is a less common species, found in forest margins, open slopes, damp slopes and meadows (2000–4000 m) in parts of western Himalaya (Himachal Pradesh, Jammu and Kashmir) to middle Asia and China. It has characteristic stout, hollow leaves and a cylindrical bulb covered with old leaves/sheath. A rare species, *A. longistylum* Baker, of sect. *Condensatum*, occurs on the slopes of western Himalaya (1500–3000 m). Taxa belonging to this section were reported to have closer affinity to the subgenus *Allium* than the subgenus *Cepa* on the basis of seed testa structure. This has the general habit of *A. stracheyi*, but differs in leaf placement and longer reproductive parts (Baker, 1874).

i. *Allium* subgenus *Reticulatobulbosa* (Kamelin) N.Friesen has three sections, *Reticulatobulbosa*, *Sikkimensia* and *Campanulata*. *Reticulatobulbosa* includes two taxa, *A. humile* Kunth, an endemic species of western Himalaya, a rare taxon, and *A. schrenkii* Regel, a species widely distributed from the Himalayan Mountains to Siberia on high mountains and stony, open alpine slopes (3000–4000 m). The former species is found in high altitudes in north-west Himalaya (including Pakistan), and the latter is endemic to the Sikkim Himalaya, as far as Siberia.

A. sikkimense Baker of sect. *Sikkimensia* is a Chinese-Himalayan species distributed across Sikkim, Tibet, Bhutan, Nepal, the south-western and southern parts of China, occurring in forest margins, scrub, slopes and meadows. It grows in meadows and on the edges of forests, scrub, slopes, meadows (2400–5000 m). *A. tenuicaule* Regel of sect. *Campanulata* is reported from high mountain regions (over 2500 m) in the Kashmir area (Gohil, 1992; Murti, 2001) with wider distribution in the Hindu Kush and Pamir-Alai mountains. It is distinguished from other, related taxa of the section in scape, rhizome and ovary characteristics (Memariani *et al.*, 2007).

j. *Allium* subgenus *Polyprason* Radić. This subgenus shares features of scape anatomy with subgen. *Rhizirideum* sect. *Rhizirideum* (leathery bulb tunic, breaking into strips in the upper part), but growth forms are advanced as subgenus *Reticulobulbosa*. In India, subgenus *Polyprason* has section *Oreiprason* F.Herm., a narrow endemic species, *A. stracheyi* Baker and *A. roylei* Stearn, occurring in rock crevices of the western Himalaya and *A. consanguineum* Kunth, commonly occurring in western (Kashmir) and central Himalaya. Of these, *A. roylei* is an Indian species which is a source of powdery mildew and other diseases. Second sect. *Falcatifolium* N.Friesen. has three species: *A. blandum* Wall., *A. carolinianum* DC. (syn. *A. thomsonii* Baker), and *A. platyspathum* Schrenk. These are hardy species highly adaptable to wide environmental conditions of the western Himalaya. They have characteristic coriaceous tunic, falcate-linear thick leaves and a short, vertical, sturdy rhizome enabling the species to survive on the steep slopes of the subalpine mountain.

24.4 Factors Responsible for Diversification of Genus *Allium*

Diverse climatic factors (from arid to warm temperate), mostly characterized by differences in precipitation, temperature and evapotranspiration, have played a pivotal role in developing species diversity in the genus *Allium*. Geographic isolation, habitat heterogeneity and topography are the major factors that together shape and maintain high species diversity in *Allium* (Huang *et al.*, 2014). Wide geographical conditions have helped in developing morphological and ecological diversity among the Asian and European subgroups in *A. schoenoprastum* (Friesen and Blattner, 2000).

Ecology and molecular work on *Allium* species has strongly supported the fundamental study on evolutionary origin, adaptation and divergence of polyploids within the species complex by Kui *et al.* (2008). A discrete distributional pattern was evident in populations of diploid-polyploidy complex of *A. przewalskianum* that was associated with variable distributional capacity, colonization potential and growth forms (Xu *et al.*, 1998; Yao *et al.*, 2011).

In the common leek (*A. porrum*) domestication-related traits like adaptation to horticultural conditions and yield were found to be associated to heterozygosity rather than to polyploidy (Guenaoui *et al.*, 2013).

24.5 *Allium* Genetic Resources

Among the cultivated species of *Allium*, onion (*A. cepa* L.), garlic (*A. sativum* L.), leek (*A. porrum* L.), and aggregate onion (*A. cepa* var. *aggregatum* syn. A. *ascalonicum* auct. Non L.) are well-known vegetable crops widely grown in different parts of India. Many introduced species like onion and garlic have been under cultivation for a long time and have developed tropical-subtropical types. Regional variability has developed on the basis of agricultural pattern, climate variables, weather conditions, drought, flood and other biotic factors (Khar *et al.*, 2011; Shaaf *et al.*, 2014). Some significant types of cultivated and semi-wild taxa of *Allium* in India are given in Figure 24.3.

In *A. sativum*, mostly white or pinkish bulb types are grown in the hilly tracts but lilac-pink garlic is more popular in the north-eastern region and hills of Kashmir (Fig. 24.3). Among the regionally important cultivars, aggregate onion (*A. cepa* var. *aggregatum*) is extensively grown in south India with variability in membrane colour from copper, lilac, violet-brown, including a rare type with white membrane. The hill types of *A. cepa* var. *Aggregatum*, locally called 'doona', are also reported under cultivation in the Himalayas. A lesser-known species, *A. chinense*, is under cultivation from the majority of states of the north-eastern hills of India, having variation in plant type (leaves and bulb size). Among other less-commonly cultivated crops, *A. porrum/ ampeloprasum* var. *ampeloprasum* (locally known as hargandh, sidhum), and *A. tuberosum* are grown as backyard cultigen in tribal-dominated tracts of Indian Himalaya and north-eastern hill regions. Commonly sold under the name 'Chinese garlic' or 'Vilayati lahsun', this cultivar has gradually gained popularity over the common garlic (*A. sativum*) in many parts of India. The regional importance of *A. tuberosum* has been realized at the national level and efforts to promote large-scale cultivation in the north-eastern hills are in progress (Pandey *et al.*, 2014).

Generally, the majority of alliums are eaten as vegetables for all plant parts except perhaps the seeds. Many of the wild *Allium* species are consumed as a vegetable, spice, condiment, or used for their medicinal properties in the hills of India (Negi *et al.*, 2006). Mature leaves and bulbs are consumed as a vegetable, and when young the plant is used as a spice or condiment, or in soups in the north-eastern hill region. Several species (*A. carolinianum, A. humile, A. consanguineum, A. wallichi, A. stracheyi*) are locally important traditional spices that are regionally grown for consumption as well as for trade (K.C. Bhatt, Ethnobotanical studies of some remote areas of Pithoragarh and Chamoli District, HNB Garhwal University, Srinagar, Uttarakhand, India, 1990, unpublished thesis; Pandey *et al.*, 2008). Freshly harvested leaves or bulbs of wild alliums are occasionally sold in village markets (Fig. 24.3). Use of *Allium* species for commercial ornamental purposes, especially *A. chinense*, *A. hookeri*, *A. macranthum* and *A. tuberosum* is yet to be recognized in India (Pandey *et al.*, 2008; Devi *et al.*, 2014). *A. roylei* has been the donor for genes for powdery mildew and leaf blight resistance to *A. cepa* (van der Meer and de Vries, 1990; de Vries *et al.*, 1992; Khrustaleva and Kik, 1998; Khrustaleva and Kik, 2000; Kohli and Gohil, 2009).

Many local or vernacular names are used for *Allium* species in different parts of the country. The local name 'Zimu' is used for a species which has the odour of onion and the morphology of common garlic, and is presumed to be a interspecific hybrid between the two. Field- and herbarium-based studies have helped in resolving the identity-related taxonomical validation of the species as *A. tuberosum* (Pandey *et al.*, 2014). In Nepal, the local name 'Jimu' is used for other wild species of *Allium* used for edible purposes, such as *A. tuberosum* (R.C. Nepal, Status, use and management of jimbu (*Allium* spp.): a case study from Upper Mustang, Nepal, Norwegian University of Life Sciences, Oslo, Norway, 2006, unpublished thesis).

Cytological and karyotype studies are part of the genetic resource programme. In recent studies, accessions of *A. roylei* from Jammu and Kashmir were observed to be translocation heterozygotes (Sharma and Gohil, 2003; Sharma and Gohil, 2008) when compared to material used by de Vries *et al.* (1992), which was a diploid with $2n = 2x = 16$ chromosomes. This species, with highly unstable cytology of chromosomes under diverse habitat conditions, has revealed scope for use of the genes for broadening adaptability (Sharma and Gohil, 2003; Sharma and Gohil, 2008; Kohli and Gohil, 2009; Sharma and Gohil, 2013).

24.5.1 Domestication trends in Allium

Domestication of wild alliums started millennia ago, and is still continuing. Among recent domesticates, the wild species *A. komarovianum* Vved. (*A. sacculiferum* Maxim.) has entered into commercial cultivation in North Korea, as has *A. canadense* L. in Cuba (Fritsch and Friesen, 2002). Other wild species of *Allium* having an onion- or garlic-like odour and flavour are in use at local levels. Examples of some of the potentials for domestication are Chinese or Japanese garlic (*A. macrostemon*), Naples garlic (*A. neapolitanum* Cirillo), ramsons (*A. ursinum* L.), long-rooted garlic (*A. victorialis*) and Canada garlic (*A. canadense*), which have been taken up for cultivation.

From the Indian Himalaya and adjoining regions, many wild species of *Allium* are used (Gohil, 1992; Negi and Pant, 1992; Yao *et al.*, 2011; Devi *et al.*, 2014; Shah, 2014). In the tribal areas, domestication trends are apparent through occurrence of wild populations of these species growing in nearby habitations and also in the protected areas near human habitations, especially in gardens or field bunds. During exploration and germplasm collection in these areas by the authors, some of the species, *A. macranthum*, *A. hookeri*, *A. stracheyi*, *A. przewalskianum*, *A. victorialis* and *A. consanguineum*, were observed to have traditional use and have been found under cultivation in the homestead. Many of these are sold regionally, fetching good prices for the farmers (R.C. Nepal, Status, use and management of jimbu (Allium spp.): a case study from Upper Mustang, Nepal, Norwegian University of Life Sciences, Oslo, Norway, 2006, unpublished thesis; Kohli and Gohil, 2009).

24.6 Future Thrust

Study of ecogeographical distribution and genetic resources in India are based on herbarium material and other published literature. Herbarium material lacks important characteristics that are lost during drying (flower, membrane/tunic characters, perianth colour, and orientation of anther in open flower), which has led to an incomplete information base.

More emphasis on the neglected characteristics for infraspecific differentiation, especially for problem taxa, would be desirable to segregate variation among taxa across the groups. Some characters, like membrane texture of bulb/rhizome, roots in sect. *Bromatorrhiza*, foliage characters in subgenus *Cepa*, other micro-morphological characters of flower (anther, ovary, nectaries) and seed (seed coat, shape, testa), can help in delimiting allied taxa. The presence, shape and orientation of ovarian crests are important for species identification (Wheeler *et al.*, 2013).

Study of a diverse collection of taxa from subgenera *Cepa* and *Allium*, especially the wild species, is desirable, particularly of their morphology and karyology variability across the region of distribution. Poor representation of study material, especially of subgenus *Polyprason* section *Oreiprason*, that includes rare and endemic species (*A. consanguineum* Kunth, *A. roylei* Stearn, *A. stracheyi* Baker), need appropriate attention.

Validation of the taxonomic identity of locally important taxa (common names, mistaken identity), doubtful/dubious species and nomenclatural status need properly addressing in the context of Indian taxa. New records of species variants, newer areas of distribution/occurrence (such as *A. hookeri* from the eastern region and *A. ramosum* from India), status of taxa (endangered/rarity of *Allium stracheyi* Baker) (Nayar and Sastry, 1987–1988; Hajra, 1983), popular but lost crops of a region (Shah, 2014), and the extinction of *A. gilgitum* need investigating. Access to species for study using live material from hitherto unexplored or difficult terrains and habitats, and extended distribution to adjoining regions, should be made a priority. The molecular taxonomy of *Allium*, in reference to the Indian taxa, needs to be properly addressed to provide answers for the above issues.

Acknowledgements

The authors express their sincere thanks to the Director of the National Bureau of Plant Genetic Resources (NBPGR) for valuable guidance and support in completing this work.. Acknowledgements are also due to Drs E.R. Nayar, K.C. Bhatt, Ms Rita Gupta, Mr O.P. Dhariwal for providing help in all respects. Thanks are also due to my other colleagues, especially Mr Shashi Kant Sharma, for preparing the map and flow chart for inclusion in this chapter.

References

Baker, J.G. (1874) On the *Allium* of India, China and Japan. *Journal of Botany* 12, 289–295.

Burba, J.L. and Galmarini, C.R. (1994) *First International Symposium on Edible Alliaceae*. International Society for Horticultural Science, Mendoza, Argentina.

Choi, H.J. and Cota-Sánchez, J.H. (2010) A taxonomic revision of *Allium* (Alliaceae) in the Canadian Prairie Provinces. *Botany* 88, 787–809.

Choi, H.J., Liliana, M.G., Chang, G.J., Byoung, U.O. and Cota-Sánchez, J.H. (2012) Systematics of disjunct north eastern Asian and northern North American *Allium* (Amaryllidaceae). *Botany* 90, 491–508.

Dasgupta, S. (2006) Alliaceae. In: Singh, N.P. and Sanjappa, M. (eds) *Fascicles of Flora of India, no. 23*. Botanical Survey of India, Calcutta, India, pp. 1–48.

Devi, A., Rakshit, K., and Sarania, B. (2014) Ethnobotanical notes on *Allium* species of Arunachal Pradesh, India. *Indian Journal of Traditional Knowledge* 13(3), 606–612.

de Vries, J.N., Wietsma, W.A. and Jongerius, M.C. (1992) Introgression of characters from *Allium* roylei Stearn into *A. cepa* L. In: Hanelt, P., Hammer, K. and Knüpffer, H. (eds) *The Genus Allium: Taxonomic Problems and Genetic Resources*. Proceedings of an international symposium held at Gatersleben, Germany, 11–13 June 1991. Institut für Pflanzengenetik und Kulturpflanzenforschung (IPK), Gatersleben, Germany, pp. 321–325.

Don, G. (1832) A monograph of the genus *Allium*. *Memoirs of the Wernerian Natural History Society* 6, 1–102.

Friesen, N. and Blattner, F.R. (2000) RAPD Analysis reveals geographic differentiations within *Allium schoenoprasum* L. (Alliaceae). *Plant Biology* 2(3), 297–305.

Friesen, N., Fritsch, R.M., Pollner, S. and Blattner, F.R. (2000) Molecular and morphological evidence for an origin of the aberrant genus *Milula* within Himalayan species of *Allium* (Alliaceae). *Molecular Phylogenetics and Evolution* 17, 209–218.

Friesen, N., Fritsch, R.M. and Blattner, F.R. (2006) Phylogeny and new intrageneric classification of *Allium* (Alliaceae) based on nuclear ribosomal DNA sequences. *Aliso* 22, 372–395.

Fritsch, R.M. (1988) Anatomische Untersuchungen an der Blattspreite bei *Allium* L. (Alliaceae). I. Arten mit einer einfachen Leitbündelreihe. *Flora: Morphologie, Geobotanik, Oekophysiologie* 181, 83–100.

Fritsch, R.M. (1992) Septal nectaries in the genus *Allium* L. In: Hanelt, P., Hammer, K. and Knüpffer, H. (eds) *The Genus Allium: Taxonomic Problems and Genetic Resources.* Proceedings of an international symposium held at Gatersleben, Germany, 11–13 June 1991. InstitutfürPflanzengenetikundKulturpflanzenforschung (IPK), Gatersleben, Germany, pp. 77–85.

Fritsch, R.M. and Abbasi, M. (2013) A taxonomic review of *Allium* subg. *Melanocrommyum* in Iran. Institut für Pflanzengenetik und Kulturpflanzenforschung (IPK), Gatersleben, Germany. Available at: www.ipk-gatersleben.de/fileadmin/content-ipk/content-ipk-ressourcen/Download/IrMeRevAllN.pdf (accessed 31 May 2016).

Fritsch, R.M. and Friesen, N. (2002) Evolution, domestication and taxonomy. In: Rabinowitch, H.D. and Currah, L. (eds) *Allium Crop Science: Recent Advances*. CAB International, Wallingford, Oxfordshire, UK, pp. 5–30.

Fritsch, R.M., Blattner, F.R. and Gurushidze, M. (2010) New classification of *Allium* L. subg. *Melanocrommyum* (Webb & Berthel) Rouy (Alliaceae) based on molecular and morphological characters. *Phyton* 49, 145–220.

Gohil, R.N. (1992) Himalayan representatives of alliums. In: Hanelt, P., Hammer, K. and Knüpffer, H. (eds) *The Genus Allium: Taxonomic Problems and Genetic Resources*. Proceedings of an international symposium held at Gatersleben, Germany, 11–13 June 1991. InstitutfürPflanzengenetikundKulturpflanzenforschung (IPK), Gatersleben, Germany, pp. 335–340.

Gregory, M., Fritsch, R.M., Friesen, N., Khassanov, F.O. and Mcneal, D.W. (1998) *Nomenclator Alliorum: Allium Names and Synonyms – A World Guide*. Royal Botanic Gardens, Kew, Surrey, UK.

Guenaoui, C., Mang, S., Figliuolo, G. and Neffati, M. (2013) Diversity in *Allium ampeloprasum*: from small and wild to large and cultivated. *Genetic Resources and Crop Evolution* 60, 97–114.

Hajra, P.K. (1983) Plants of northwestern Himalayas with restricted distribution – a census. In: Jain, S.K. and Rao, R.R. (eds) *An Assessment of Threatened Plants of India*. Botanical Survey of India, Calcutta, India.

Hanelt, P., Schulze-Motel, J., Fritsch, R., Kruse, J., Maaß, H.I., Ohle, H. and Pistrick, K. (1992) Infrageneric grouping of *Allium* – the Gatersleben approach. In: Hanelt, P., Hammer, K. and Knüpffer, H. (eds) *The Genus Allium: Taxonomic Problems and Genetic Resources. Proceedings of an international symposium held at Gatersleben, Germany, 11–13 June 1991*. Institut für Pflanzengenetik und Kulturpflanzenforschung (IPK), Gatersleben, Germany, pp. 107–123.

Hooker, J.D. (1892; rev. 1973) *Liliaceae: Allium L. Flora of British India*, Vol. VI (repr. Edn). Bishen Singh Mahendra Pal Singh, Dehradun, Uttarakhand, India and Periodical Experts, New Delhi, India, pp. 337–345.

Huang, D.Q., Yang, J.T., Zhou, C.J., Zhou, S.D. and He, X.J. (2014) Phylogenetic reappraisal of *Allium* subgenus *Cyathophora* (Amaryllidaceae) and related taxa, with a proposal of two new sections. *Journal of Plant Research* 127(2), 275–286.

Kachroo, B.L., Sapru, B.L. and Dhar, U. (1977) *Flora of Ladak – An Ecological and Taxonomic Appraisal*. Bishen Singh Mahendra Pal Singh, Dehradun, Uttarakhand, India.

Kamelin, R.V. (1973) *Florogeneticheskij analiz estestvennoj flory gornoj Srednej Azii*. Nauka, Leningrad, Russia.

Khar, A., Lawande, K.E. and Negi, K.S. (2011) Microsatellite marker based analysis of genetic diversity in short day tropical Indian onion and cross amplification in related *Allium* spp. *Genetic Resources and Crop Evolution* 58, 741–752.

Khosa, J.S., Dhatt, A.S. and Negi, K.S. (2014) Morphological characterization of *Allium* spp. using multivariate analysis. *Indian Journal of Plant Genetic Resources* 27(1), 24–27.

Khassanov, F.O. and Fritsch, R.M. (1994) New taxa in *Allium* L. subg. *Melanocrommyum* (Webb & Berth.) Rouy from Central Asia. *Linzer biologische Beiträge* 26, 965–990.

Khassanov, F.O. and Tojibaev, K.S. (2010) Two more new *Allium* L. species from the Fergana depression (Central Asia). *Stapfia* 92, 27–28.

Khrustaleva, L.I. and Kik, C. (1998) Cytogenetical studies in the bridge cross *Allium cepa* x (*A. fistulosum* x *A. roylei*). *Theoretical Applications of Genetics* 96, 8–14.

Khrustaleva, L.I. and Kik, C. (2000) Introgression of *A. fistulosum* into *A. cepa* mediated by *A. roylei. Theoretical Applications of Genetics* 100, 17–26.

Kohli, B. and Gohil, R.N. (2009) Need to conserve *Allium roylei* Stearn: a potential gene reservoir. *Genetic Resources and Crop Evolution* 56, 891–893.

Kui, X.-K., Ao, C.-Q., Zhang, Q., Chen, L.-T. and Liu, J.-Q. (2008) Diploid and tetraploid distribution of *Allium prezwalskianum* Regel. (Liliaceae) in the Qinghai–Tibetan Plateau and adjoining regions. *Caryologia* 61(2), 192–200.

Linnaeus, C.V. (1957–1959) *Species Plantarum*, Vol. 1. *Allium*. Ray Society, London, UK, pp. 294–302. (Facsimile of 1753 edition published by Laurentiis Salvii, Stockholm, Sweden.)

Memariani, F., Joharchi, M.R. and Khassanov, F.O. (2007) *Allium* L. subgen. *Rhizirideum* sensu lato in Iran, two new records and a synopsis of taxonomy and phytogeography. *Iranian Journal Botany* 13(1), 12–20.

Murti, S.K. (2001) *Flora of cold deserts of western Himalaya. Vol. 1 (Monocotyledons)*. Botanical Survey of India, Calcutta, India.

Nayar, M.P. and Sastry, A.R.K. (1987–1990) *Red Data Book of Indian Plants. Vols. I–III*. Botanical Survey of India, Calcutta, India.

Nayar, N.M., Ahmedulla, M. and Singh, R. (1992) *Allium* in South Asia: importance, ethnobotanical uses, genetic resources, enumeration of species and distribution. In: Hanelt, P., Hammer, K. and Knüpffer, H. (eds) *The Genus Allium: Taxonomic, Problems and Genetic Resources*. Proceedings of an international symposium held at Gatersleben, Germany, 11–13 June 1991. Institut für Pflanzengenetik und Kulturpflanzenforschung (IPK), Gatersleben, Germany, pp. 205–213.

Negi, K.S. (2006) *Allium* species in Himalayas and their uses with special reference to Uttaranchal. *Ethnobotany* 18, 53–66.

Negi, K.S. and Pant, K.C. (1992) Less-known wild species of *Allium* L. (Amaryllidaceae) from mountain region of India. *Economic Botany* 46(1), 112–114.

Nguyen, N.H., Heather, E.D.I. and Chelsea, D.S. (2008) A molecular phylogeny of the wild onions (*Allium*; Alliaceae) with a focus on the western North American centre of diversity. *Molecular Phylogenetics and Evolution* 47, 1157–1172.

Pandey, A. and Pandey, R. (2005) Wild useful species of *Allium* in India – key to identification. *Indian Journal of Plant Genetic Resources*, 18(2), 180–182.

Pandey, U.B., Kumar, A., Pandey, R. and Venkateswaran, K. (2005a) Bulbous crops – cultivated Alliums. In: Dhillon, B.S., Tyagi, R.K., Saxena, S. and Randhawa, G.J. (eds) *Plant Genetic Resources of Vegetable Crops*. Narosa Publishing House Pvt Ltd, New Delhi, India, pp. 108–120.

Pandey, A., Pandey, R. and Negi, K.S. (2005b) Wild *Allium* species in India: biodiversity distribution and systematic studies. *Abstracts from the first National Conference on Allium*, 24–25 February, Banaras Hindu University, Varanasi, Uttar Pradesh, India, p. 44.

Pandey, A., Pandey, R., Negi, K.S. and Radhamani, J. (2008) Realizing value of genetic resources of *Allium* in India. *Genetic Resources and Crops Evolution* 55, 985–994.

Pandey, A., Pradheep, K. and Gupta, R. (2013) Seed morphological study on *Allium* (Amaryllidaceae): an aid in taxonomic identification. In: *Proceedings of National Seed Seminar, Indian Society of Seed Technology XIII, 8–10 June 2013*. University for Agricultural Sciences, Bengaluru, Karnataka, India.

Pandey, A., Pradheep, K. and Gupta, R. (2014) Chinese chives (*Allium tuberosum* Rottler ex Spreng.) – a home garden species or a commercial crop in India. *Genetic Resources and Crops Evolution* 61, 1433–1440.

Polunin, O. and Stainton, A. (1984) *Flowers of the Himalaya*. Oxford University Press, New Delhi, India, pp. 413–416.

Rola, K. (2014) Cell pattern and ultra sculpture of bulb tunics of selected *Allium* species (Amaryllidaceae) and their diagnostic value. *Acta Biologica Cracoviensia Series Botanica* 56(1), 28–41.

Samoylov, A., Friesen, N., Pollner, S. and Hanelt, P. (1999) Use of chloroplast polymorphisms for the phylogenetic study of *Allium* subgenera *Amerallium* and *Bromatorrhiza* (Alliaceae) II. *Feddes Report* 110, 103–109.

Shaaf, S., Sharma, R., Kilian, B., Walther, A., Özkan, H., Karami, E. and Mohammadi, B. (2014) Genetic structure and eco-geographical adaptation of garlic landraces (*Allium sativum* L.) in Iran. *Genetic Resources and Crops Evolution* 61, 1565–1580.

Shah, N.C. (2014) Status of cultivated and wild *Allium* species in India: a review. *The Science Technology Journal* 1(1), 28–36.

Sharma, G. and Gohil, R.N. (2003) Cytology of *Allium roylei* Stearn. Meiosis in a population with complex interchanges. *Cytologia* 68(2), 115–119.

Sharma, G. and Gohil, R.N. (2008) Intrapopulation karyotypic variability in *Allium roylei* Stearn – a threatened species. *Botany Journal of Linneus Society* 158, 242–248.

Sharma, G. and Gohil, R.N. (2013) Origin and cytology of a novel cytotype of *Allium tuberosum* Rottl. ex Spreng (2n=48). *Genetic Resources and Crops Evolution* 60, 503–511.

Singh, B.P. and Rana, R.S. (1994) Collection and conservation of *Allium* genetic resources: an Indian perspective. *Acta Horticulturae* 358, 181–190. DOI: 10.17660/ActaHortic.1994.358.29.

Stearn, W.T. (1992) How many species of *Allium* are known? *Kew Magazine* 9, 180–182.

Tang, H., Lihuab, M., Shiqimga, A. and Jianquaanab Liu (2005) Origin of the Qinghai–Tibetan Plateau endemic *Milula* (Liliaceae): further insights from karyological comparisons with *Allium*. *Caryologia* 58(4), 320–331.

Tojibaev, K.S. and Karimov, F.I. (2012) Endemic monocotyledonous geophytes of Ferghana valley flora (in Russian). *Rastitelniy mir Asiatskoy Rossii* 1(9), 55–59.

Traub, H.P. (1968) The subgenera, sections and subsections of *Allium* L. Pl. *Life* 24, 147–163.

van der Meer, Q.P. and de Vries, J.N. (1990) An interspecific cross between *Allium roylei* Stearn and *Allium cepa* L., and its backcross to *A. cepa*. *Euphytica* 47, 29–31.

Verma, V.D., Pradheep, K., Khar, A., Negi, K.S and Rana, J.C. (2008) Collection and characterization of *Allium* species from Himachal Pradesh. *Indian Journal of Plant Genetic Resources* 21, 225–228.

von Berg, G.L., Samoylov, A., Klaas, M. and Hanelt, P. (1996) Chloroplast DNA restriction analysis and the infrageneric grouping of *Allium* (Alliaceae). *Plant Systematics and Evolution* 200, 253–261.

WCSP (2015) World Checklist of Selected Plant Families. Facilitated by the Royal Botanic Gardens, Kew, Richmond, Surrey, UK. Available at: apps.kew.org/wcsp (accessed 31 May 2016).

Wheeler, E.J., Mashayekhi, S., McNeal, D.W., Columbus, J.T. and Pires, J.C. (2013) Molecular systematic of Allium subgenus Amerallium (Amaryllidaeae) in North America. *American Journal of Botany* 100(4), 701–711.

Xu, J.M., Yang, L. and He, X.J. (1998) A study on karyotype differentiation of *Allium fascicultum* (Liliaceae). *Acta Phytotaxonomica Sinica* 36, 346–352.

Yao, B., Deng, J. and Liu, J. (2011) Variations between diploids and tetraploids of *Allium przewalskianum*, an important vegetable and/or condiment in the Himalayas. *Chemical Biodiversity* 8(4), 686–691.

Yousaf, Z., Shinwari, Z.K. and Aleem, R. (2004) Can complexity of the genus *Allium* be resolved through some numerical techniques? *Pakistan Journal of Botany* 36(3), 487–501.

Zhou, S.D., He, X.J., Yu, Y. and Xu, J.M. (2007) Karyotype studies on twenty-one populations of eight species in *Allium* section *Rhiziridium*. *Acta Phytotaxonomica Sinica* 45, 207–216.

Zhu, G. and Turland, N.J. (2000) Two new combinations in Central Asian and Chinese *Allium* (Alliaceae). *Novon* 10(2), 181–182.

25 Traditional Ecological Knowledge and Plant Biodiversity Conservation in a European Transfrontier Landscape

José Antonio González[1]*, Ana Maria Carvalho[2] and Francisco Amich[1]

[1]*Grupo de Investigación de Recursos Etnobiológicos del Duero-Douro (GRIRED), Department of Botany, Faculty of Biology, University of Salamanca, Salamanca, Spain;* [2]*Mountain Research Centre (CIMO), School of Agriculture, Polytechnic Institute of Bragança, Campus de Santa Apolónia, Bragança Portugal*

Abstract

On the basis of ecological and cultural characterization of the landscape, combined with empirical local knowledge and historic data, this chapter focuses on traditional plant-use in two adjoining natural protected areas at the northern tip of the Spanish–Portuguese border (Iberian Peninsula). The river Douro, one of the major Iberian rivers, known for its impressive canyons, flows across north-central Spain and Portugal to its mouth in the Atlantic Ocean. For a 120-km stretch, the Douro ravines define the frontier between the two countries and create a singular, mosaic-like landscape, shaped by ages-long interaction between natural processes and human activities. Despite the physical evidence of a natural border, there is an important shared cultural heritage, with particular significance and similarity in plant-use and local management of habitats and territory. This study shows that to preserve traditional ecological knowledge about plant-use within a transfrontier landscape it is fundamental to overcome difficulties associated with natural resources and territorial management. Moreover, several ideas, tools, skills and effective measures for plant biodiversity conservation are needed, as well as intergovernmental cooperation, which is of crucial importance in an internationally shared landscape.

25.1 Introduction

Common European rural landscapes are culturally relevant and a consequence of the integration of natural biophysical processes and human activities that has led to a multiplicity of sustainable landscapes (Council of Europe, 2000; Antrop, 2005; Council of Europe, 2006). Intrinsically, such landscapes were and still are under human influence, as besides natural values, they also combine individual and communitarian values and significant socio-economic aspects. Moreover, the effects of farming and forestry systems and land-use patterns have had a greater impact on rural landscape than ecological features and processes within the last five decades. In Europe, the long and complex history of land-use has promoted diverse natural landscapes that have been shaped by local practices, beliefs and specific purposes, and maintained in those areas where physical, socio-economic and political constraints have prevented modernization

*E-mail: ja.gonzalez@usal.es

and change in farming systems until recently (Vos and Meekes, 1999; Antrop, 2005; Plieninger *et al.*, 2006; Calvo-Iglesias *et al.*, 2009; Carvalho *et al.*, 2010; Gullino and Larcher, 2013; Knight and Harrison, 2013; Lourenço-Gomes *et al.*, 2014).

Therefore, Traditional Ecological Knowledge (TEK) about the use of many natural resources, both wild and domesticated, is of critical importance. TEK plays a decisive role in the management and conservation of natural habitats and ecosystems, and encompasses a wide range of behavioural and agricultural representations and practices. However, TEK is not changeless but instead evolves from knowledge acquired over time and is transformed through experimenting and learning. It is currently adapting to temporal changes, thereby favouring the integrated management of the available natural resources. Thus, TEK is crucial to improve cooperation and communication among local populations and managers of different protected areas. Participation and involvement of local communities should provide the basis for creating a more focused management system to protected areas (Berkes *et al.*, 2000; Drew and Henne, 2006; Nazarea, 2006; Pardo-de-Santayana *et al.*, 2012).

In this sense, ethnobiology, the interdisciplinary field defined as 'the scientific study of dynamic relationships among peoples, biota, and environments' (see Society of Ethnobiology website: ethnobiology.org) gives important approaches, methods and tools. As Wolverton *et al.* (2014) have stated recently, ethnobiologists (as well as ethnobotanists) are important mediators between local people and decision makers, ensuring that programmes targeting biodiversity and conservation will be also engaged with the human community's needs.

Nevertheless, these assumptions face practical obstacles in the case of European transfrontier landscapes. Wascher and Pérez-Soba (2004) use the term 'transfrontier landscape' to describe such 'a piece of land where natural and cultural characteristics form recognizable coherent entities which are divided by national or sub-national administrative boundaries, resulting in two or more areas of sovereignty or jurisdiction'.

Among other difficulties associated with this kind of landscape, the different (sometimes conflicting) laws and the differences in the authority given to managers and staff may reduce the effectiveness of transboundary cooperation (Wascher and Pérez-Soba, 2004; Prieur, 2006).

Along the border between Portugal and Spain (Iberian Peninsula), the official line separating the two countries matches, in most of its extent, to important protected areas established more than 20 years ago. Those borders were defined according to political and administrative policies, allowing, in theory, a greater contact between nature, people, ways of life and ideas. Some examples are Peneda Gerês National Park and Baixa Limia Sierra do Xurés, Montesinho and Sanabria Lake Natural Parks. In other situations, national boundaries correspond to physiographical features, such as a river that flows into the two countries; for example, Douro Internacional and Arribes del Duero Natural Parks (henceforth Douro International or DI). Even though river canyons impose a physical barrier between the two countries, both sides evolved together a long time ago, maintaining a high level of biological diversity, unique natural features and associated cultural heritage, conforming to a 'cultural landscape' (see whc.unesco.org/en/culturallandscape).

Several ethnographic and ethnobotanical studies conducted in DI (Frazão-Moreira *et al.*, 2009; Carvalho *et al.*, 2010; Carvalho and Frazão-Moreira, 2011; Carvalho, 2012a, b; González *et al.*, 2013) have shown that the particular landscape of such protected areas is deeply related to the biogeographical and historical events, which, together with human activity over many generations, have provided a mosaic of many different plant communities and crops, where seasonality is clearly marked by the vegetation chromatic contrasts and the various tasks performed on arable fields. The main strength of DI territory is its landscape heterogeneity: agricultural and forest interacting patches and geomorphologic dominating areas (Mata Olmo and Sanz Herráiz, 2004).

This chapter aims to discuss and demonstrate that, in order to preserve both biodiversity and traditional ecological knowledge about plant-use in a transfrontier landscape, it is fundamental to overcome difficulties associated with natural resources and territorial management. Ecosystem services and plant biodiversity in a European transfrontier landscape are expected to benefit from synergies between TEK and science, namely conservation biology.

25.2 Study Area

DI territory matches the political and natural boundary between Spain and Portugal for 120km

(40° 50′–41° 35′ N, 6° 00′–7° 05′ W; Fig. 25.1). These Portuguese and Spanish lands represent a high-quality connecting protected area (Natural Parks of the Douro Internacional and the Arribes del Duero), declared as Sites of Community Importance (European Commission Habitats Directive 92/43/EEC) with singular floristic, ecological and geomorphological traits. The morphological homogeneity of surrounding peneplain, whose altitude is 700–800 m above sea level, is abruptly interrupted by deep gorges. The river Douro and some of its main tributaries (Tormes, Águeda, Huebra, Uces, Sabor and Côa) run through an extraordinary labyrinth of canyons (*arribas* or *arribes*) (see Sanz *et al.*, 2003; García Feced *et al.*, 2007; Carvalho *et al.*, 2010), with rocky cliffs that often exceed 400 m.

The climate is characterized by fairly mild annual temperatures (11°C mean for the whole area), rare periods of frost during early spring, and rainfall of about 700 mm/year, characteristic of a Mediterranean mesoclimate (Calonge-Cano, 1990).

Biogeographically, this area is included in the Carpetan-Leonese subprovince (Mediterranean West Iberian province), more precisely, in the Lusitan Duriensean sector (Rivas-Martínez *et al.*, 2002). Being within the Mediterranean and Eurosiberian worlds, DI is floristically rich and complex, with many endemic and subendemic plant species, some of them reflecting in their scientific name their particular chorology (e.g. a specific epithet such as 'duriensis' or 'lusitanica').

Some important examples are *Allium schmitzii* Cout., *Anarrhinum duriminium* (Brot.) Pers., *Anthyllis sampaioana* Rothm., *Antirrhinum lopesianum* Rothm., *Armeria transmontana* (Samp.) Lawrence, *Delphinium fissum* subsp. *sordidum* (Cuatrec.) Amich, E. Rico & J. Sánchez, *Digitalis purpurea* subsp. *amandiana* (Samp.) Hinz, *Epipactis duriensis* Bernardos *et al.*, *Epipactis lusitanica* Tyteca, *Holcus annuus* subsp. *duriensis* (P. Silva) Franco & Rocha Afonso, *Isatis platyloba* Link ex Sted., *Linaria intricata* Coincy, *Paradisea lusitanica* (Cout.) Samp., *Scrophularia valdesii* Ortega Olivencia & Devesa, *Silene boryi* subsp. *duriensis*

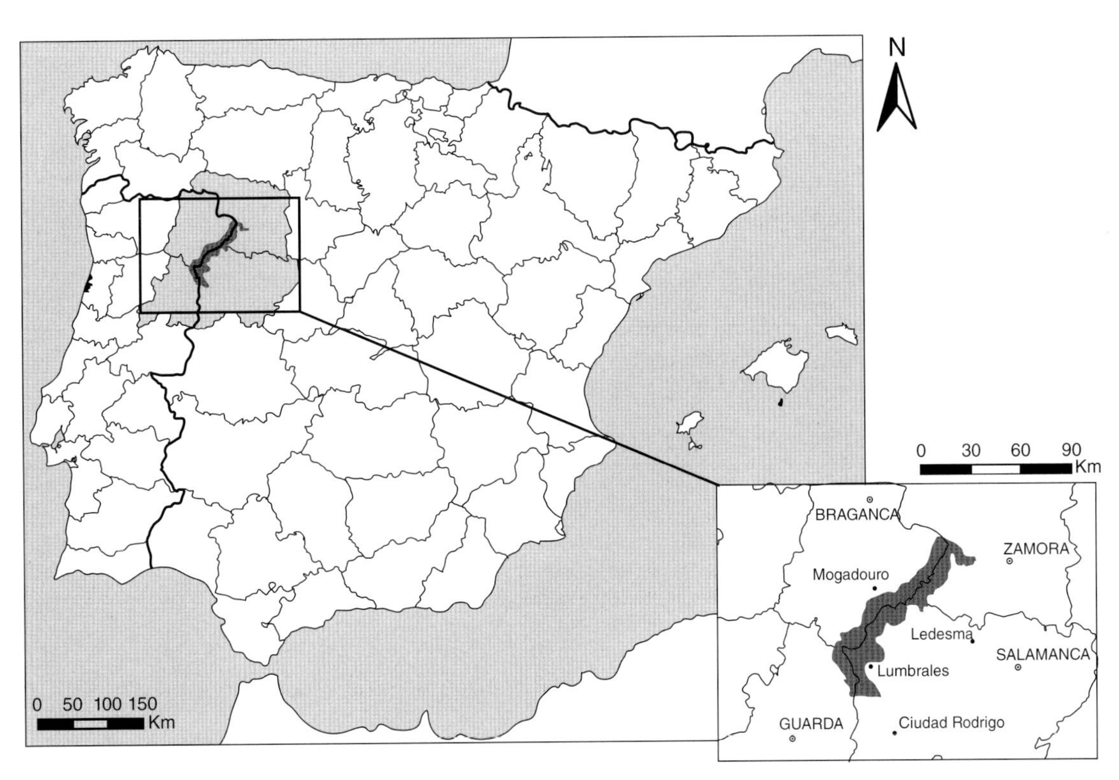

Fig. 25.1. Geographical location of the Douro International area. The edges of the Douro Internacional Natural Park (Portugal) and the Arribes del Duero Natural Park (Spain) are shaded.

(Samp.) Cout., *Silene marizii* Samp. and *Trigonella polyceratia* subsp. *amandiana* (Samp.) Amich & Sánchez (e.g. Amich *et al.*, 2004; Ramírez-Rodríguez and Amich, 2014; Rocha *et al.*, 2014).

The main vegetation types include: oak acidophilous or neutral-acidophilous forests dominated by *Quercus pyrenaica* Willd. (*Quercenion pyrenaicae* Rivas Goday ex Rivas-Martínez 1964 suballiance), with *Genista falcata* Brot., *Lonicera periclymenum* subsp. *hispanica* (Boiss. & Reuter) Nyman, *Ranunculus ollisiponensis* subsp. *carpetanus* (Boiss. & Reuter) Rivas Mart.; cork-oak subhumid to humid forest of *Quercus suber* L. (*Quercenion broteroi* Rivas-Martínez 1987 suballiance), with *Arisarum simorrhinum* Durieu, *Epipactis lusitanica* Tyteca, *Physospermum cornubiensis* (L.) DC., *Quercus faginea* Lam., *Sanguisorba hybrida* (L.) Font Quer; holm-oak semicontinental and continental dry to subhumid forest of *Quercus rotundifolia* Lam. (*Paeonio broteroi-Quercenion rotundifoliae* Rivas-Martínez in Rivas-Martínez, Costa & Izco 1986 suballiance), with *Genista hystrix* Lange, *Lonicera etrusca* G. Santi, *Lonicera implexa* Aiton, *Paeonia broteroi* Boiss. & Reuter, *Pistacia terebinthus* L., *Ruscus aculeatus* L.; nano to microphanerophytic communities, on acid or neutral soils (*Ericenion arboreae* Rivas-Martínez 1975 suballiance, with two endemic association: *Erico arboreae-Buxetum sempervirentis* Aguiar, Esteves & Penas 1999 and *Cytiso grandiflori-Arbutetum unedonis* Monteiro-Henriques *et al.*, 2012) with *Arbutus unedo* L., *Cytisus grandiflorus* (Brot.) DC., *Erica arborea* L., *Phyllirea angustifolia* L.; scrub vegetation with Cistaceae (*Cistus* L., *Halimium* (Dunal) Spach, *Helianthemum* Mill.), Fabaceae (*Cytisus* Desf., *Echinospartum* (Spach) Fourr., *Genista* L.), Liliaceae (*Asparagus* L., *Ruscus* L.), Oleaceae (*Olea* L., *Phillyrea* L.), Rhamnaceae (*Rhamnus* L.), Rosaceae (*Prunus* L.); and riparian vegetation with *Alnus glutinosa* (L.) Gaertn., *Celtis australis* L., *Clematis campaniflora* Brot. (endemic association *Clematido campaniflorae-Celtidetum australis* Monteiro-Henriques *et al.*, 2012), *Fraxinus angustifolia* Vahl, *Populus* L. sp. pl. and *Salix* L. sp. pl. (see Rivas-Martínez, 2011; Costa *et al.*, 2012).

The territory is also rich in endemic communities, such as *Phagnalo saxatilis-Antirrhinetum lopesianii*, *Rumici indurati-Anarrhinetum durimini*, *Rusco aculeati-Juniperetum lagunae*, and *Sileno duriensis-Aphyllanthetum monspeliensis*, among others (see Aguiar *et al.*, 2003; Bernardos *et al.*, 2004).

The floristic catalogue of the territory comprises more than 1100 species of vascular plants, corresponding to around 120 botanical families (see Bernardos *et al.*, 2004; González *et al.*, 2013).

This territory has been altered to such an extent that few plant communities remain pristine. Although human management of vegetation and habitats has been performed for centuries by local people, DI territory is still considered a reservoir of plant diversity and of *in-situ* conservation of many species and crops, providing an interesting pool of endogenous resources, inextricably linked with particular TEK and cultural perceptions. These characteristics were the essential basis for the legal establishment of this territory as a natural protected area (Carvalho and Frazão-Moreira, 2011).

In former times, DI surroundings were intensively explored by locals whose daily routines were largely dependent on natural resources, due to geographical isolation, self-subsistence-oriented choices and the limited potential of regional soils. Pastoralism, cattle transhumance, small-scale animal breeding, extensive cropping and small farming systems were the main activities contributing to a self-sufficient economy, sometimes complemented by extra income from trading and selling surpluses or from temporary job opportunities (e.g. mining and smuggling) (García Feced *et al.*, 2007; Luis Calabuig, 2008; Carvalho and Frazão-Moreira, 2011).

Different ecosystems (e.g. woods, scrubland, grassland, riverside) and natural vegetation types provided for medicinal and edible plants, mushrooms, berries, pasture, fodder, and different sorts of raw materials (e.g. fibres, fuel, domestic tools, and building). Those spaces as well as arable crops were essential to meet the basic needs of households and animals. Moreover, their management also contributed to satisfy many religious and symbolic purposes, which were socially relevant to the communities (Carvalho *et al.*, 2010; Carvalho and Frazão-Moreira, 2011; González *et al.*, 2013).

A very long history of human occupation and adaptive management within a rural society resulted in a communitarian agropastoralism, promoted TEK, influenced landscape characteristics, and controlled land-use. The particular arrangement of settlements and gardens, taking into account slope and soil fertility gradients and water proximity, as well as multifunctional, productive and diverse landscapes, enabled certain species and habitats to remain relatively stable over time (García Feced *et al.*, 2007; Luis Calabuig, 2008; Aguiar *et al.*, 2009; Carvalho *et al.*, 2010).

The continuous movement of people and exchanges between the countries and several other regions of the Iberian Peninsula, Europe and even America has introduced species that have adapted to the prevailing growing conditions and have become part of the local diversity of plants and crops. An impressive assortment of useful species (native, semi-domesticated and cultivated) is linked to certain uses and practices. Some of them played important roles in local economies (households were essentially self-sufficient and subsistence-oriented) and were praised and valued in folklore, gastronomy and local culture. Many of them are still used or the seeds are preserved and kept by the farmers. For instance, Barbela wheat (a well-adapted landrace of common wheat *Triticum aestivum* L.), flax (*Linum usitatissimum* L.), sorghum (*Sorghum bicolor* (L.) Moench), peppers (varieties of *Capsicum annuum* L.), olive tree (*Olea europaea* L.), turpentine tree (*Pistacia terebinthus* L.), sumach (*Rhus coriaria* L.), cade (*Juniperus oxyicedrus* L.), wild asparagus (*Asparagus acutifolius* L.), or the roman chamomile (*Chamaemelum nobile* (L.) All.).

The landscape of DI reflects cultural heritage, local people's perceptions and representations, in line with highly significant values and ideals related to dwellings, neighbourliness, intergenerational relationships, tradition, history and landmarks.

25.3 Research Methods

Ethnobotanical information was obtained from a review of the authors' previous research and published works (e.g. Carvalho, 2012a, b; González *et al.*, 2013). In all cases, data were acquired using consented and semi-structured interviews with informants with a sound TEK of useful plants who were born in the study area, largely older people. Open questions were asked about useful species which sought to ascertain knowledge of past and present use. In addition, during the interviews, informants' opinions regarding different issues were reported. For instance, comments about new agricultural and livestock-raising practices, current lifestyles within DI territory, the continuing deterioration in the oral transmission of local knowledge, and how all this affects the preservation of natural heritage and current use of plant resources.

25.4 Results and Discussion

25.4.1 Primary ethnobotanical data

Taking into consideration the data relating to wild plants (including naturalized) traditionally used as medicinal plants and/or wild food plants (wild vegetables and fruits, seasoning, beverages), and those plant species used for technological purposes (building, furniture and household utensils, farming implements and tools, plant fibres, dyeing, leather tanning, fuel), the ethnofloristic catalogue of DI comprises 153 and 181 wild plant species for the Portuguese and Spanish territories respectively. The number of common useful species, used in both areas, is 125. Table 25.1 lists the 75 plant species with the broadest importance of use in the area, belonging to 39 botanical families.

Of the 75 species considered, a total of 68 (91%) are native to the study area, whereas seven (9%) are allochthonous plants, although many of them (such as *Opuntia maxima*, *Portulaca oleracea*, *Rhus coriaria*) have become naturalized and currently form part of DI vegetation and landscape. Taking the classical biotype classification from Raunkiaer (1934), the group with the highest proportion of representation was hemicryptophytes (28, or 37%), followed by phanerophytes (18, 24%), nanophanerophytes (14, 19%) and chamaephytes (7, 9%). Regarding the biogeographical spectrum, analysing the distribution of these species according to the criteria of Rivas-Martínez *et al.* (2002) and Rivas-Martínez (2011), a total of 38 (51%) can be considered Mediterranean elements (late-Mediterranean, circum-Mediterranean, Eurosiberian Mediterranean, Euroasiatic Mediterranean Macaronesian Islands). Twenty-four species (32%) are cosmopolitan or subcosmopolitan (including those distributed on two or more continents and in two or more biogeographic regions) and there is also a considerable number of endemic Iberian plants (including Iberian-North African and Franco-Iberian species), with a total of 13, representing 17% of the considered ethnoflora.

In relation to the habitat category to which each selected taxon belonged, for a better analysis with fewer categories, the results are grouped in broad groups of vegetation (Table 25.1). The high percentage (37%) of plant species collected from forest, woodland, dwarf scrub and scrub vegetation should be noted. It is also important that many aromatic and medicinal shrubs are collected in serial vegetation communities and climacic forests. This also shows that the area has natural zones that are still well

Table 25.1. List of the main useful wild plant species in the Douro International area. Chorology: WR = wide-range elements; MD = Mediterranean elements; IB = Iberian elements. Vegetation group: SYN = synanthropic vegetation; GRM = grassland and meadow vegetation; FWS = forest, woodland, scrub vegetation; CHS = chasmophytic and scree vegetation; AMP = amphibious vegetation.

Families and species	Use categories	Chorology	Biotype	Vegetation group
CONIFEROPSIDA				
Cupressaceae				
Juniperus oxycedrus L.	Medicine, technology	MD	Phanerophyte	FWS
MAGNOLIOPSIDA				
Aceraceae				
Acer monspessulanum L.	Technology	MD	Phanerophyte	FWS
Anacardiaceae				
Pistacia terebinthus L.	Technology	MD	Phanerophyte	FWS
Rhus coriaria L.	Technology	MD	Nanophanerophyte	FWS
Apiaceae				
Apium nodiflorum (L.) Lag.	Food	MD	Hemicryptophyte	AMP
Ferula communis L.	Technology	MD	Hemicryptophyte	GRM
Foeniculum vulgare Mill.	Food, medicine	MD	Hemicryptophyte	SYN
Asteraceae				
Chamaemelum nobile (L.) All.	Food, medicine	WR	Chamaephyte	GRM
Chondrilla juncea L.	Food, medicine, technology	WR	Hemicryptophyte	SYN
Taraxacum officinale Weber ex F. H. Wigg.	Food, medicine	WR	Hemicryptophyte	SYN
Brassicaceae				
Rorippa nasturtium-aquaticum (L.) Hayek	Food, medicine	WR	Hemicryptophyte	AMP
Cactaceae				
Opuntia maxima Mill.	Food, medicine	WR	Phanerophyte	SYN
Caprifoliaceae				
Sambucus nigra L.	Food, medicine, technology	MD	Phanerophyte	SYN
Caryophyllaceae				
Paronychia argentea Lam.	Medicine	MD	Chamaephyte	SYN
Chenopodiaceae				
Chenopodium ambrosioides L.	Food, medicine	WR	Therophyte	SYN
Cistaceae				
Cistus ladanifer L.	Medicine, technology	IB	Nanophanerophyte	FWS
Crassulaceae				
Umbilicus ruprestris (Salisb.) Dandy	Medicine	MD	Hemicryptophyte	CHS
Cucurbitaceae				
Bryonia dioica Jacq.	Food, medicine	MD	Geophyte	SYN
Ericaceae				
Arbutus unedo L.	Food, medicine, technology	MD	Phanerophyte	FWS
Fabaceae				
Adenocarpus complicatus (L.) J. Gay in Durieu	Technology	MD	Nanophanerophyte	FWS
Cytisus multiflorus (L'Hér.) Sweet	Medicine, technology	IB	Nanophanerophyte	FWS
Cytisus scoparius (L.) Link	Medicine, technology	MD	Nanophanerophyte	FWS
Cytisus striatus (Hill) Rothm.	Medicine, technology	IB	Nanophanerophyte	FWS
Fagaceae				
Castanea sativa Mill.	Food, medicine, technology	MD	Phanerophyte	FWS

Continued

Table 25.1. Continued.

Families and species	Use categories	Chorology	Biotype	Vegetation group
Quercus faginea Lam.	Medicine, technology	IB	Phanerophyte	FWS
Quercus rotundifolia Lam.	Food, medicine, technology	MD	Phanerophyte	FWS
Quercus pyrenaica Willd.	Medicine, technology	MD	Phanerophyte	FWS
Guttiferae				
Hypericum perforatum L.	Medicine	WR	Hemicryptophyte	SYN
Juglandaceae				
Juglans regia L.	Food, medicine, technology	WR	Phanerophyte	SYN
Lamiaceae				
Lavandula pedunculata (Mill.) Cav.	Food, medicine, technology	IB	Chamaephyte	FWS
Melissa officinalis L.	Food, medicine	MD	Hemicryptophyte	SYN
Mentha aquatica L.	Food, medicine	WR	Hemicryptophyte	AMP
Mentha cervina L.	Food, medicine	IB	Hemicryptophyte	AMP
Mentha pulegium L.	Food, medicine	MD	Hemicryptophyte	AMP
Mentha spicata L.	Food, medicine	WR	Hemicryptophyte	AMP
Mentha suaveolens L.	Food, medicine	MD	Hemicryptophyte	AMP
Rosmarinus officinalis L.	Food, medicine	MD	Nanophanerophyte	FWS
Salvia verbenaca L.	Medicine	MD	Hemicryptophyte	GRM
Thymus mastichina (L.) L.	Food, medicine	IB	Chamaephyte	FWS
Thymus zygis Loefl. ex L.	Food, medicine	IB	Chamaephyte	FWS
Lauraceae				
Laurus nobilis L.	Food, medicine	MD	Phanerophyte	SYN
Lythraceae				
Lythrum salicaria L.	Medicine	WR	Hemicryptophyte	AMP
Malvaceae				
Malva sylvestris L.	Food, medicine	WR	Hemicryptophyte	SYN
Oleaceae				
Fraxinus angustifolia Vahl	Medicine, technology	MD	Phanerophyte	FWS
Olea europaea var. *sylvestris* (Miller) Lehr	Food, medicine, technology	MD	Phanerophyte	FWS
Papaveraceae				
Chelidonium majus L.	Medicine	WR	Hemicryptophyte	SYN
Plantaginaceae				
Plantago lanceolata L.	Medicine	WR	Hemicryptophyte	SYN
Polygonaceae				
Rumex acetosa L.	Food	WR	Hemicryptophyte	SYN
Rumex induratus Boiss. & Reut.	Food	IB	Chamaephyte	CHS
Portulacaceae				
Montia fontana L.	Food	WR	Therophyte	AMP
Portulaca oleracea L.	Food	WR	Therophyte	SYN
Rosaceae				
Crataegus monogyna Jacq.	Food, medicine, technology	WR	Phanerophyte	FWS
Prunus spinosa L.	Food, medicine, technology	MD	Nanophanerophyte	FWS
Rosa canina L.	Food, medicine	WR	Nanophanerophyte	FWS
Rubus ulmifolius Schott	Food, medicine, technology	MD	Nanophanerophyte	FWS
Rutaceae				
Ruta montana (L.) L.	Medicine	MD	Chamaephyte	FWS
Salicaceae				
Salix atrocinerea Brot.	Technology	MD	Phanerophyte	FWS
Salix fragilis L.	Technology	MD	Phanerophyte	FWS
Santalaceae				
Osyris alba L.	Technology	MD	Nanophanerophyte	FWS

Continued

Table 25.1. Continued.

Families and species	Use categories	Chorology	Biotype	Vegetation group
Scrophulariaceae				
Digitalis purpurea L.	Medicine	MD	Hemicryptophyte	FWS
Digitalis thapsi L.	Medicine	IB	Hemicryptophyte	CHS
Odontitella virgata (Link) Rothm.	Technology	IB	Te	GRM
Verbascum pulverulentum Vill.	Medicine	MD	Hemicryptophyte	SYN
Verbascum thapsus L.	Medicine	IB	Hemicryptophyte	SYN
Thymelaeaceae				
Daphne gnidium L.	Technology	MD	Nanophanerophyte	FWS
Ulmaceae				
Ulmus minor Mill.	Technology	WR	Phanerophyte	FWS
Urticaceae				
Parietaria judaica L.	Medicine	MD	Hemicryptophyte	SYN
Urtica dioica L.	Food, medicine	WR	Therophyte	SYN
Urtica urens L.	Food, medicine	WR	Therophyte	SYN
Verbenaceae				
Verbena officinalis L.	Medicine	WR	Hemicryptophyte	SYN
LILIOPSIDA				
Cyperaceae				
Cyperus longus L.	Technology	WR	Geophyte	AMP
Dioscoreaceae				
Tamus communis L.	Food, medicine	MD	Hemicryptophyte	FWS
Liliaceae				
Asparagus acutifolius L.	Food	MD	Nanophanerophyte	FWS
Ruscus aculeatus L.	Medicine, technology	MD	Nanophanerophyte	FWS
Poaceae				
Stipa gigantea L.	Technology	IB	Hemicryptophyte	GRM

conserved and that they are known and used by the inhabitants of the DI area. The following groups in importance are synanthropic vegetation zones, modified by human activity and formed by nitrophilous annual or perennial herbs, mostly in ruderal environments (31%), and amphibious vegetation (13%). Most of the ethnobotanical resources can be catalogued as being present in anthropically influenced environments, usually from nitrophilous or subnitrophilous plant communities. This strengthens the idea that many resources are collected in the readily accessible proximities of inhabited zones, since many of the individuals who collect them are older people (because for them wild plant resources are culturally much more important than for younger people). It also confirms that rural communities have used different management strategies, above all in anthropogenic environments outside the plants' original habitat. Some relevant examples, with uses belonging to different use-categories, are: *Chondrilla juncea, Sambucus nigra, Bryonia dioica, Malva sylvestris, Portulaca oleracea, Urtica dioica* and *U. urens*.

Future studies should analyse what the local inhabitants of DI consider the most important plants, in an attempt to implement appropriate strategies for better conservation and sustainable development. The cultural importance of plant resources is usually related to the role they play in any given community, reflected in aspects such as the ease with which the resource can be gathered or the period of time during which it has been used. Works addressing the significance of plants in human cultures have increased steadily in recent years (e.g. Reyes-García *et al.*, 2006; Thomas *et al.*, 2009). Although none of DI's ethnobotanical resources are included in the Red Lists of Vascular Plants (Moreno, 2010; Bilz *et al.*, 2011), all the wild plants used in DI represent an important natural (and cultural) resource for all the inhabitants of the area. Thus, their inclusion in inventories that record and document overall TEK is crucial to help both their conservation and management. Compilation of these inventories is crucial.

25.4.2 Scenarios of plant diversity and plant genetic resources conservation

Ethnobotanical inventories carried out for the last decades within the DI territory have shown that plant diversity is highly correlated with agricultural practices, particularly with animal farming (Frazão-Moreira *et al.*, 2009; Carvalho *et al.*, 2010; Carvalho and Frazão-Moreira, 2011). Meeting the different needs of people and animals involved many available wild species and growing staple products that were used for food and fodder. Moreover, these surveys have also documented that TEK and plant-use are mostly the elders' domain because these people have experienced geographic isolation and socio-economic constraints that have induced creative sustainable management of natural resources, benefiting as much as possible from it.

Critical changes in rural societies, mainly agriculture abandonment, globalization, ageing and a gap in generational knowledge transmission, are seen as key factors affecting plant-use and endangering species diversity, habitats and agroecosystems.

Therefore, in such a context, dominant scenarios involve important losses of TEK and skills that usually supported diversity, arable species conservation and rather specific site management. Apparently, conservation strategies should focus on not only the biological patrimony but intangible heritage as well.

25.4.2.1 Non-prevailing agricultural practices affect plant resources diversity

People living in DI have kept their own distinct cultural identity and deep sense of belonging to a unique region. For generations of users and consumers, local knowledge about agricultural practices and sustainable management made it possible to maintain a pool of both wild and domesticated plant genetic resources that are still of great interest when considering different approaches, such as ecosystems services, *in-situ* conservation programmes, food security and the safeguard of biocultural heritage.

Specific interactions between wilderness, arable crops, gardens and human behaviour and aptitudes were responsible for providing minimal resources for subsisting all year round. Agriculture, animal husbandry, as well as fishing, hunting, beekeeping or crafting, were fundamental interrelated activities that profited from natural environment management, along with arable crops, and had implicit acquired expertise and skills. When some of these skills disappear or become obsolete in a global context, many goals established to sustain daily life are not reached and may not last, which threatens all other interconnected components of the system.

Cattle transhumance, communitarian management, multipurpose species, selection of landraces, crop rotation, mixed crops, manure and compost, partial harvest, digging crop residues into the soil, opening clearings in the woods and preparing firewood and charcoal, are some examples of ancient practices of great direct or indirect consequence for plant resources diversity.

Previous ethnobotanical surveys carried out in DI territories (Frazão-Moreira *et al.*, 2009; Carvalho *et al.*, 2010; Carvalho and Frazão-Moreira, 2011) have reported several cases of such connectivity.

The use of well-adapted landraces of grain and old fodder species (e.g. wheat, rye, pulses and vetches) was related to numbers of livestock and draft animals, because such crops provided for different sources of fodder (e.g. grains, straw, green biomass, pods and stubbles). However, some of their byproducts supplied raw materials for weaving, bedding, manure and biofertilizers. The number of plots assigned to fodder and grain production depended on livestock and on the availability of local varieties of seed. Under extensive management and because of their rusticity, these crops had a good field performance. They also had associations with a useful adventitious flora that was used to feed pigs and chickens or to prepare some homemade remedies to treat sick humans and animals.

The continuous presence in the territory of cattle, sheep and goats and the extensive semi-natural grazing systems were an interesting contribution to species protection and enabled the natural balance between the different vegetation types. Grazing controlled flora life forms, avoiding a tallest stratum with increased risk of wildfires. Furthermore, this balance improved plant diversification conforming to more productive, aesthetic and attractive landscapes. Flocks and cattle itineraries made possible the use of several sites and gardens and access to other types of fodder (e.g. woods, scrubland, fallows, stubble, crop residues, leaves from riparian trees, ruderal species, home-garden surpluses). These sites also profited from animal waste, and some emblematic medicinal species in meadows benefited too (e.g. *Thymus pulegioides* L. and *Mentha pulegium* L.); animals had a trampling effect on the soil that favoured plant propagation. Grazing or cutting hay prevented many useful species from being shaded out, which would affect their availability and use (Carvalho *et al.*, 2010).

It seems that if the strong connection existing between natural resources and TEK is broken, plant genetic resources diversity is affected as well, because purposes, values and skills (which are some of the components of TEK) may separate into meaningless pieces without being transmitted or preserved. The lack of resources management and practices provide important changes in flora composition and plant communities. For instance, herbaceous communities in unexploited meadows are rapidly replaced by scrubland species.

25.4.2.2 Perennial species take the place of seasonal crops

Characteristic land forms of the DI territory are peneplain and canyon. Taking advantage of rolling plains, farming systems formerly included arable crops, meadows and grasslands that, as mentioned earlier, were fundamental to subsistence-oriented households. Annual crops and perennials managed on an annual basis were grown in these lands. Besides grains (rye, barley, oats), some other examples are different kinds of cabbages and pumpkins, turnips, rutabaga, beetroots and sorghum.

Extensive management of arable crops and annual species required flexible labour inputs and particular productive strategies, reflecting both local (traditional) techniques and a more general agronomic technology. Agricultural labour was seasonal, mostly manual and time consuming. During recent decades, agricultural policies along with the socio-economic context already described have decreased farmers' initiatives and caused a decline in arable crop production and cattle raising (Carvalho and Frazão-Moreira, 2011).

According to local people, forestation was the best choice to avoid abandonment and to keep productive lands while maintaining their market and sentimental value (income versus a patrimony for future generations). Traditional food and fodder crops were suddenly substituted by broadleaves and mixed forests. Autochthonous species, imported varieties of native species and exotic ones were introduced into arable lands and meadows, for timber, pulpwood and fruit production.

Nowadays, the DI peneplains have a different character; the extensive arable lands have almost disappeared and scattered forested patches combine with the remaining annual crops, meadows and scrubland.

Some of these perennial plots depend on particular techniques of tillage (much more intensive than reduced and conservation tillage) that may interfere with flora diversity, removing natural cover and plant communities in edges. It can be argued that the replacement of seasonal crops by perennial wood species seems to lead to a loss of flora diversity, since the processes used to maintain such perennial plots are very intensive and demanding technical resources.

Traditional land-use for annual crops included three-field rotation. The land was divided into three parts. One plot was planted in the autumn with winter wheat or rye; crops for fodder or grazing were grown in the second plot (e.g. turnips in winter and peas, lentils or mixed peas, oats, barley in spring); the third field was left fallow. To some extent, these practices made it possible to restore or to maintain a productive soil, favoured the diversity of adventitious flora, and spared the edges and vegetation islands. For instance, in those arable fields were often kept isolated holm oak trees, olive trees or oak trees, under whose canopy different plant communities were able to survive.

The decline of agriculture and recent demographic trends has also generated new approaches to home gardens. They used to be less diverse because other agricultural activities, such as forestry, grain production and animal husbandry, were fundamental for the household economy. Today, besides staple products, people enjoy growing exotic herbaceous and woody ornamentals. Many of these plants took the place of wild species previously harvested from the forest by women and children. Seasonal crops once only cultivated in arable lands may be found currently in home gardens (e.g. cabbages, pumpkins, vetches and grains).

Home gardens and allotments have become reservoirs of flora diversity and places for innovation. The flora composition of home gardens and new green spaces inside the settlements are also signs of transformation in land-use and may be perceived as new structural components of local landscapes (Fig. 25.2).

25.4.3 Douro International: plant diversity and future scenarios

DI territories are ideal for exploring relationships between landscape, physiography and culture. Physical geography and land-use history of this region have strongly affected the processes and patterns of human activity, and landscape is its expression. European and Iberian socio-economic transformations have induced important changes that stand out in rural contexts. Many issues are viewed and discussed at a more global level, which interfere with local knowledge, endogenous resources, sustainable management and the systems of plant-use and local beliefs. Modern and different insights into rural lifestyles, generational differences and environmental problems may provide more technical then sociocultural approaches endangering cultural landscapes, agro-ecosystems biodiversity and the flora and fauna habitats, for whose conservation protected areas were created.

The focus is that knowledge and abilities gained through time are being rapidly superseded. New attitudes and values are adopted without having been previously tested. This may lead to important changes with serious long-term repercussions that may influence the status of such protected areas and, in addition, plant biodiversity (Frazão-Moreira *et al.*, 2009; Carvalho *et al.*, 2010; Carvalho and Frazão-Moreira, 2011; Carvalho and Morales, 2013).

Based on empirical evidence from the study area and several data from ethnobotanical surveys, variability in size, shape and configuration of vegetation patches seems to increase landscape heterogeneity that corresponds to plant diversity (richness and abundance of herbaceous species, including species

Fig. 25.2. Some examples of home gardens and allotments in the Douro International area.

with conservation value). Considering that traditional land-use and traditional crops and products are some of the most important parameters used to recognize cultural landscape (Gullino and Larcher, 2013), it seems that local land-use and plant-use are key components of cultural landscapes that may influence plant diversity. Moreover, there is an adaptive knowledge transmitted through oral traditions that is not merely of academic or historical interest, but is fundamental to maintaining cultural continuity and identity and, possibly, could play a role in achieving sustainable use of plant resources in the future.

25.4.4 Management strategies

In this transfrontier context, cooperation becomes even more important. Numerous different variables correlate with cooperation but, not surprisingly, the most important ones are distinctly human variables relating to human relationships. Thus, the conservation projects depend on inclusion of all stakeholders (see Zbicz, 2003). Bilateral cooperation is dealt with on different levels (Zbicz, 2003) and in different sectors, which for the DI area are described in Table 25.2.

To meet such an objective and to provide effective plant-diversity conservation of an important biocultural patrimony, several ideas, tools and skills may be necessary.

1. Awareness of the importance of local knowledge and cultural heritage.
2. Cultural landscapes legacy, built over generations of experimentation and observation, provides ideas and opportunities for sustainable and multipurpose use of resources and offers contemporary strategies for preserving cultural and ecological diversity.
3. Much of aesthetic, historical or cultural value of rural landscapes remains to be inventoried and recorded before it disappears permanently.
4. Farming and rural lifestyles must be recognized as important subjects for conservation and sustainable development.
5. Legal processes and governmental policies should involve users and communities in both countries (Portugal and Spain), as well as local knowledge systems, because local perceptions and conceptions can be considered important tools for landscape conservation and management.
6. Participatory approaches should be implemented in order to adopt measures that will develop more outstanding natural features of protected areas, respecting landscaping, socio-economic and cultural trends and being more favourable to people living in these areas.
7. Environmental education and other educational efforts at different levels (children, young people and adults) and scales.
8. Awareness of the increased value of biodiversity and cultural heritage in order to provide heritage safeguarding and resources availability.
9. The importance of assigning a value to cultural landscapes, land-use and plant-use.
10. Rural development policies enabling people to live satisfactorily in such particular areas.

International contributions are important, but ultimate sustained success depends on the day-to-day involvement and efforts of those at the local level who must interact. Changes in land-use (e.g. marked abandonment of cultivated lands and traditional

Table 25.2. Stakeholder groups in the natural area of Douro International (Portugal–Spain).

	Countries	
Administrative scale	Portugal	Spain
National	Nature Conservation and Forest Institute by delegation of the Ministry of Agriculture and the Sea	Ministry of Agriculture, Food and Environmental Affairs, and Ministry of Industry, Energy and Tourism
Regional	Northern Department of Nature Conservation and Forests. Tourism Institute of Portugal	Government and administration of the Autonomous Region of Castile and León
Sub regional (provincial)	Northern Department of Nature Conservation and Forests Regional Services of PNDI	Diputaciones Provinciales of the provinces of Salamanca and Zamora
Local	Four municipalities: Figueira de Castelo Rodrigo; Freixo de Espada à Cinta; Miranda do Douro; Mogadouro	37 municipalities

crops, a huge decrease in livestock numbers) imply that strategies aimed at the management of ruderal communities in the study zone (e.g. the maintenance of slopes and of the separation between plots and smallholdings) are now essential, because in general they cause few disturbances to the landscape. Another important management strategy in the DI area is the conversion of traditional home gardens and allotments into areas of *ex-situ* and *in-situ* conservation, both of wild species and of local and territorial landraces of domesticated plants; for both nostalgic and pragmatic reasons.

25.5 Final Remarks

The TEK amassed in the rural community of DI, constructed over generations of experience and observation, may provide ideas and opportunities for the sustainable and polyvalent use of plants. However, in order for this to be successful, the active participation of people is necessary and merely documenting and validating local knowledge is insufficient. The informants indicated that such participation should be set up to adopt measures aimed at a type of management that will improve and develop the most outstanding characteristics of the protected area, respecting the socio-economic trends and favouring the people who live in the area. Implementing the goals of preserving the plant biodiversity of the region depends on networks and on cooperation at different levels, both horizontally (e.g. cooperation among farmers) and vertically (e.g. cooperation of nature conservation organizations).

Thus, transfrontier cooperation represents a real opportunity to explore landscape conservation strategies, by rewarding the economy of the local communities, having in mind the application of the European Landscape Convention (article 9, Transfrontier Landscapes): to give added value to traditional culture and solve conflicts, in order to show people that the protected area can help and sustain their activities. Finally, it is also necessary to increase the level of cooperation: a full cooperation, i.e. fully integrated, ecosystem-based planning, with common goals and joint decision making by an international committee, sometimes even involving joint management. All of this will create new opportunities: high landscape resources, increase of rural tourism or promotion of high-quality agricultural and cattle products (e.g. beef, milk, cheese, butter, wool, manure).

References

Aguiar, C., Costa, J.C., Capelo, J., Amado, A., Honrado, J., Espírito Santo, M.D. and Lousã, M. (2003) Aditamentos à vegetação de Portugal continental. Nota 34 en Notas do Herbario da Estação Florestal Nacional (LISFA). Fasc. XVII. *Silva Lusitana* 11(1), 101–111.

Aguiar, C., Rodrigues, O., Azevedo, J. and Domingos, T. (2009) In: Pereira, H.M., Domingos, T., Vicente, L. and Proença, V. (eds) *Ecossistemas e Bem-Estar Humano: Avaliação para Portugal do Millennium Ecosystem Assessment*. Escolar Editora, Lisbon, Portugal, pp. 293–337.

Amich, F., Bernardos, S., Aguiar, C., Fernández-Diez, J. and Crespi, A.L. (2004) Taxonomic composition and ecological characteristics of the endemic flora of the lower Duero Basin (Iberian Peninsula). *Acta Botanica Gallica* 151(4), 341–352.

Antrop, M. (2005) Why landscapes of the past are important for the future. *Landscape and Urban Planning* 70, 21–34.

Berkes, F., Colding, J. and Folke, C. (2000) Rediscovery of traditional ecological knowledge as adaptive management. *Ecological Applications* 10, 1251–1262.

Bernardos, S., Amado, A., Aguiar, C., Crespí, A.L., Castro, A. and Amich, F. (2004) Aportaciones al conocimiento de la flora y vegetación del centro-occidente ibérico (CW de España y NE de Portugal). *Acta Botanica Malacitana* 29, 285–295.

Bilz, M., Kell, S.P., Maxted, N. and Lansdown, R.V. (2011) *European Red List of Vascular Plants*. Publications Office of the European Union, Luxembourg.

Calvo-Iglesias, M.S., Fra-Paleo, U. and Díaz-Varela, R.A. (2009) Changes in farming system and population as drivers of land cover and landscape dynamics: the case of enclosed and semi-openfield systems in Northern Galicia (Spain). *Landscape and Urban Planning* 90, 168–177.

Calonge-Cano, G. (1990) La excepcionalidad climática de los Arribes del Duero. *Ería* 21, 45–59.

Carvalho, A.M. (2012a) *Etnobotânica da Terra de Miranda. Projeto Cultibos, Yerbas i Saberes*. FRAUGA – Associação para o Desenvolvimento Integrado de Picote and Instituto Politécnico de Bragança, Bragança, Portugal.

Carvalho, A.M. (2012b) *Etnoflora da Terra de Miranda. Projeto Cultibos, Yerbas i Saberes*. FRAUGA – Associação para o Desenvolvimento Integrado de Picote and Instituto Politécnico de Bragança, Bragança, Portugal.

Carvalho, A.M. and Frazão-Moreira, A. (2011) Importance of local knowledge in plant resources management and conservation in two protected areas from Trás-os-Montes, Portugal. *Journal of Ethnobiology and Ethnomedecine* 7, 36.

Carvalho, A.M. and Morales, R. (2013) Persistence of wild food and wild medicinal plant knowledge in a Northeastern region of Portugal. In: Pardo-de-Santayana, M., Pieroni, A. and Puri, R.K. (eds) *Ethnobotany in the New Europe: People, Health and Wild Plant Resources*. Berghahn Books, Oxford, UK, pp. 147–170.

Carvalho, A.M., Ramos, M.T. and Frazão-Moreira, A. (2010) Connecting landscape conservation and management with traditional ecological knowledge: does it matter how people perceive landscape and nature? In: Azevedo, J.C., Feliciano, M., Castro, J. and Pinto, M.A. (eds) *Proceedings of the IUFRO Landscape Ecology Working Group International Conference*. IUFRO – IPB, Bragança, Portugal, pp. 474–479.

Costa, J.C., Neto, C., Aguiar, C., Capelo, J., Espírito Santo, M.D. and Honrado, J. (2012) Vascular plant communities in Portugal (continental, the Azores and Madeira). *Global Geobotany* 2, 1–180.

Council of Europe (2000) *European Landscape Convention. ETS No. 176*. Council of Europe Publishing, Strasbourg, France.

Council of Europe (2006) *Landscape and Sustainable Development: Challenges of the European Landscape Convention*. Council of Europe Publishing, Strasbourg, France.

Drew, J.A. and Henne, A.P. (2006) Conservation biology and traditional ecological knowledge: integrating academic disciplines for better conservation practice. *Ecological Society* 11(2), 34. Available at: www.ecologyandsociety.org/vol11/iss2/art34 (accessed 31 May 2016).

Frazão-Moreira, A., Carvalho, A.M. and Martins, M.E. (2009) Local ecological knowledge also 'comes from books': cultural change, landscape transformation and conservation of biodiversity in two protected areas in Portugal. *Anthropological Notebooks* 15(1), 27–36.

García Feced, C., Escribano Bombín, R. and Elena Rosselló, R. (2007) Comparación de la estructura de los paisajes en Parques Naturales fronterizos: Arribes del Duero *versus* Douro Internacional. *Montes* 91, 8–14.

González, J.A., García-Barriuso, M., Ramírez-Rodríguez, R., Bernardos, S. and Amich, F. (2013) Ethnobotanical resources management in the Arribes del Duero Natural Park (Central Western Iberian Peninsula): relationships between plant use and plant diversity, ecological analysis, and conservation. *Human Ecology* 41(4), 615–630.

Gullino, P. and Larcher, F. (2013) Integrity in UNESCO World Heritage Sites. A comparative study for rural landscapes. *Journal of Cultural Heritage* 14(5), 389–395.

Knight, J. and Harrison, S. (2013) 'A land history of men': the intersection of geomorphology, culture and heritage in Cornwall, southwest England. *Applied Geography* 42, 186–194.

Lourenço-Gomes, L., Costa Pinto, L.M. and Rebelo, J.F. (2014) Visitors' preferences for preserving the attributes of a World Heritage Site. *Journal of Cultural Heritage* 15(1), 64–67.

Luis Calabuig, E. (2008) *Arribes del Duero: Guía de la naturaleza*. Edilesa, León, Spain.

Mata Olmo, R. and Sanz Herráiz, C. (2004) Case study II – Atlantic Mountains: the Arribes del Duero/Arribes do Douro (Spain/Portugal). In: Wascher, D.M. and Pérez-Soba, M. (eds) *Learning from European Transfrontier Landscapes – Project in Support of the European Landscape Convention*. Alterra report 964. Alterra Wageningen UR, Wageningen, The Netherlands, pp. 18–20.

Moreno, J.C. (2010) *Lista Roja de la Flora Vascular Española*. Ministerio de Medio Ambiente y Medio Rural y Marino – Sociedad Española de Biología de la Conservación de Plantas, Madrid, Spain. Available at: www.magrama.gob.es/es/biodiversidad/temas/inventarios-nacionales/inventario-especies-terrestres/inventario-nacional-de-biodiversidad/lista_roja_flora.aspx (accessed 31 May 2016).

Nazarea, V.D. (2006) Local knowledge and memory in biodiversity conservation. *Annual Review of Anthropology* 35, 317–335.

Pardo-de-Santayana, M., Morales, R., Aceituno, L., Molina, M. and Tardío, J. (2012) Etnobiología y biodiversidad: el Inventario Español de los Conocimientos Tradicionales. *Ambienta* 99, 6–24.

Plieninger, T., Höchtl, F. and Spek, T. (2006) Traditional land-use and nature conservation in European rural landscapes. *Environmental Science & Policy* 9(4), 317–321.

Prieur, M. (2006) Landscapes and policies, international programmes and transfrontier landscapes. In: Council of Europe (ed.) *Landscape and Sustainable Development: Challenges of the European Landscape Convention*. Council of Europe Publishing, Strasbourg, France, pp. 141–161.

Ramírez-Rodríguez, R. and Amich F. (2014) Notes on rare and threatened flora in western-central Iberia. *Lazaroa* 35, 221–226.

Raunkiaer, C. (1934) *The Life Forms of the Plants and Statistical Plant Geography*. Oxford University Press, Oxford, UK.

Reyes-García, V., Huanca, T., Vadez, V., Leonard, W. and Wilkie, D. (2006) Cultural, practical, and economic value of wild plants: a quantitative study in the Bolivian Amazon. *Economic Botany* 60(1), 62–74.

Rivas-Martínez, S. (2011) Mapa de series, geoseries y geopermaseries de vegetación de España. *Itinera Geobotanica* 18(1–2), 800–805.

Rivas-Martínez, S., Díaz, T.E., Fernández-González, F., Izco, J., Loidi, J., Lousã, M. and Penas, A. (2002)

Vascular plant communities of Spain and Portugal. *Itinera Geobotanica* 15(1–2), 5–922.
Rocha, J., Almeida da Silva, R., Amich, F., Martins, A., Almeida, P. and Aranha, J.T. (2014) Biogeographic trends of endemic and subendemic flora in the western Iberian Peninsula under scenarios of future climate changes. *Lazaroa* 35, 19–35.
Sanz, C., Mata, R., Gómez, J., Allende, F., López, N., Molina, P. and Galiana, L. (2003) *Atlas de los Paisajes de España*. Ministerio de Medio Ambiente, Madrid, Spain.
Thomas, E., Vandebroek, I., Sanca, S. and Van Damme, P. (2009) Cultural significance of medicinal plant families and species among Quechua farmers in Apillapampa, Bolivia. *Journal of Ethnopharmacology* 122(1), 60–67.
Vos, M. and Meekes, H. (1999) Trends in European cultural landscape development: perspectives for a sustainable future. *Landscape and Urban Planning* 46, 3–14.
Wascher, D.M. and Pérez-Soba, M. (2004) *Learning from European Transfrontier Landscapes – Project in Support of the European Landscape Convention*. Alterra report 964. Alterra Wageningen UR, Wageningen, The Netherlands.
Wolverton, S., Nolan, J.M. and Ahmed, W. (2014) Ethnobiology, political ecology, and conservation. *Journal of Ethnobiology* 34(2), 125–152.
Zbicz, D.C. (2003) Imposing transboundary conservation: cooperation between internationally adjoining protected areas. *Journal of Sustainable Forestry* 17(1–2), 21–37.

26 Cryoconservation Methods for Extended Storage of Plant Genetic Resources

Saikat Gantait[1,2*], Uma Rani Sinniah[2], Gopal Shukla[3] and Narayan Chandra Sahu[4]

[1]AICRP on Groundnut, Directorate of Research, Bidhan Chandra Krishi Viswavidyalaya, Kalyani, Nadia, West Bengal, India; [2]Department of Crop Science, Faculty of Agriculture, Universiti Putra Malaysia, Serdang, Selangor, Malaysia; [3]Department of Forestry, Uttar Banga Krishi Viswavidyalaya, Pundibari, Coochbehar, West Bengal, India; [4]Sasya Shyamala Krishi Vigyan Kendra, Ramakrishna Mission Vivekananda University, Arapanch, Sonarpur, Kolkata, West Bengal, India

Abstract

This chapter seeks to deliver an elucidation of the diverse technological attributes identifying the vastly effectuated and evolving technique of cryoconservation, a biotechnology developed to enable the prolonged storage of diversified flora. The foremost cryogenic methodologies and the pivotal phases for their effective adjustment to varied kinds of germplasms are expounded. Herein, numerous examples of cryopreservation of plant species are mentioned, to illustrate the incredible breakthrough that has been made, along with its additional roles in supporting genetic breeding programmes and in eliminating systemic plant pathogens by means of cryotherapy; thus making it an effective substitute for the purpose of conservation of germplasm.

26.1 Introduction

Seeds of a variety of food plants often face dehydration at the point of maturation; however, they can be preserved at low moisture content and low temperature for extended lengths of time due to their ability to withstand extreme aridness. Seeds suitable for this treatment are mostly found in the temperate quarters of our planet, and are known as orthodox (Roberts, 1973). On the other hand, no occurrence of maturation-related drying is noticed in the seeds of, for instance, oil palm, coffee, cacao, rubber, coconut, and some forest and fruit trees that originate in the tropics; rather, the seeds are shed off at a moderately high moisture content (Chin, 1988). Seeds like these are incapable of enduring extensive desiccation and since they persistently display a vulnerability to chilling, it becomes almost impossible to preserve them using traditional seed storage methods. Seeds of this nature are labelled either as 'recalcitrant' (Roberts, 1973; Chin and Roberts, 1980) or as 'intermediate' (Ellis *et al.*, 1990, 1991). To maintain the viability of this type of seeds, the factors that determine the level of dehydration sensitivity have to be considered and the seeds preserved in a moist and moderately warm environment; nonetheless, even under such favourable conditions, the lifespan of these seeds is restricted to days and infrequently months.

*E-mail: saikatgantait@yahoo.com

Conservation of many plants such as edible bananas and plantain in seed form cannot be achieved due to their seedless nature. Others are propagated through vegetative means particularly to maintain clonal properties such as several allogamous root and tuber crops, for example, potato (Solanum tuberosum), sweet potato (Ipomoea batatas), yams (Dioscorea spp.), sugarcane (Saccharum spp.) and cassava (Manihot esculenta), hence conservation is executed by maintaining the standard vegetative technique in field genebanks.

One of the advantages of the field genebanks, utilized as the *ex-situ* storage medium (Engelmann and Engels, 2002), is that the genetic assets laid out for conservation can easily be examined or retrieved if necessary and analysed in detail. However, there are certain encumbrances: for instance, the exposure of the plants to pests, diseases, weather damage, drought and vandalism, which impede its effectiveness and jeopardize its safety (Withers and Engels, 1990; Engelmann, 1997). Moreover, germplasm exchange becomes risky, as the odds of disease via the swapping of vegetative material are high. Besides, the land, management, material and labour that are indispensible to facilitate a field genebank to maintain and preserve a wide diversity, are restricted because of the high costs.

A number of significant factors were considered about how to mitigate the issues associated with the conservation of plant genetic resources at field genebanks. As a result, the concept of storage using liquid nitrogen at an ultra-low temperature (–196°C) was conceived with the aim of extending the storage period, and termed as 'cryoconservation' or 'cryopreservation'. The significance of this ultra-low temperature is that the metabolic activities and cell divisions can be paused and plant materials maintained for a prolonged period, escaping microbial contaminations or cumbersome maintenance expenses, unlike the field genebank. The other important aspect of this storage system is the range of planting materials that can be cryopreserved falls under a wide range of planting materials consisting of both *in vivo* and *in vitro* explants such as seeds, dormant buds, cell suspensions, shoot tips, pollens, somatic and zygotic embryos. In this chapter, we strive to explicate a variety of the latest cryoconservation routines that have evolved based on the materials at hand, along with the optimistic prospect of extensive preservation of plant genetic resources.

26.2 Cryoconservation Principles

Plant materials such as pollen and orthodox seeds which go through the task of typical dehydration do not require any treatment preceding cryopreservation; although plant materials, like embryos, shoot tips, cell suspensions and calluses, that preserve volumes of water, have to be artificially dehydrated in order to safeguard them from the injury they may suffer if the intracellular water crystallizes into ice (Mazur, 1984). With reference to the contradistinction that continues to exist between the conventional and modern cryopreservation routines (Withers and Engelmann, 1998); in the former method, the plant samples are dehydrated prior to and during the ongoing process of cryopreservation (freeze-induced dehydration), whereas in the latter, one dehydration only is executed prior to cryopreservation. Dehydration is the process wherein cellular water is sufficiently eliminated, in order to allow vitrification to occur, whereby water from its liquid state is converted into an amorphous state, avoiding crystalline ice formation (Fahy *et al.*, 1984).

26.3 Contemporary Techniques of Cryoconservation

In the contemporary procedure of cryopreservation, the dehydration mechanism, which is completed at the pre-cryopreservation stage, involves the process by which the plant samples are exposed to concentrated cryoprotectant solutions in order to expel all freezable cellular water, followed by vitrification of the aqueous chamber, with no generation of intracellular ice throughout the process. Usually, samples are dipped straight away in liquid nitrogen during dehydration, although a course of action based on cooling speed still exists which aids in reducing the temperature, via three modes: ultra-fast, fast or slow cooling. Additionally, with a view to retaining a cooling environment, a programmable freezing apparatus is also used (Engelmann and Dussert, 2013). In the conventional method, the plant samples are initially treated with cryoprotective solutions (which comprise a single or a combination of colligative chemical substances). The same is then cooled gradually (0.5–2.0°C min^{-1}) up to a pre-freezing temperature which usually stays close to –40°C and thereafter, leading up to a quick dip in liquid nitrogen (Gonzalez-Arnao *et al.*, 2008).

The objective of the freeze-dehydration mechanism, which is characterized by a slow freezing system, that also forms a part of the traditional process of cryopreservation, is to advocate the development of amorphous solids. As noted in the research by Benson *et al.* (2006), as long as the quantity of residual intracellular water stays small, the aqueous chamber undergoes a rapid vitrification as soon as the samples are dipped in the liquid nitrogen. Well-regulated labour-saving yet high-priced procedures, for instance, like the one in controlled cooling (Ashmore, 1997), mean that handfuls of cryovials can be handled all at once right through a cooling run; the same mechanism gaining ascendancy over other standard vitrification methods where it becomes compulsory to cautiously handle the cryovials manually, lifting one at a time.

The mentioned advanced procedures are usually meant to be applied for complex plant materials like shoot tips, embryos or embryonic axes. Furthermore, the same advanced method exhibits another of its features, halting ice formation, and is technically uncomplicated enough to have an edge over the normal cooling methods (e.g. advanced cryopreservation procedures do not necessitate the usage of controlled freezers). This element is especially vital for cryoconservation of tropical plant germplasm that can be effectuated in several tropical countries, as long as the basic tissue culture facilities, along with a reliable source of liquid nitrogen, are available (Engelmann and Dussert, 2013). On a positive note, provided the plant samples are subjected to dehydration prior to cryopreservation, and to satisfactorily low water contents, only then can a major improvement in the survival rate be expected, in contrast to the controls which do not experience dehydration (Engelmann, 2009). Following are some noteworthy cryopreservation methods: dehydration, preculture-dehydration, encapsulation-dehydration, vitrification, encapsulation-vitrification, and droplet-vitrification (see Fig. 26.1).

26.4 Dehydration

To protect the samples from the detrimental effects of freezing during direct exposure to liquid nitrogen, it is necessary to ensure the removal of free water which is responsible for ice crystal formation and rupturing of cells while in the ultra-cooling phase. The simplest procedure to remove the free water prior to freezing treatment is dehydration. Generally, the samples passed through the dehydration phase are exposed to liquid nitrogen immediately, with the exception of oily seeds, which need to be pre-cooled gradually, prior to cryoexposure. Since dehydration is the basic procedure of pre-freezing treatment and is harsh for desiccation-sensitive tissue, this method is chiefly applicable for whole seeds, embryonic axes or zygotic embryos. However, distinct methodologies are followed for individual types of samples while carrying out the dehydration job, i.e. specific protocols are followed for seed, embryonic axes or zygotic embryos, based on their tissue nature. A number of recalcitrant or intermediate tropical seeds have been conserved in liquid nitrogen following dehydration (Engelmann, 1997). A saturated salt-assisted controlled drying usually follows to dehydrate the seeds, whereas the usage of laminar airflow remains the conventional practice in the case of embryonic axes or zygotic embryos.

The simple procedure of dehydration has been evolved and improved with the passage of time and the interventions of compressed and sterile dry airflow, silica gel chamber or desiccation solution saturated with salts are taken into account to ensure a high success rate of dehydration, especially in tropical countries with high humidity. The flow of compressed and sterile air ensuring accelerated dehydration with least desiccation-injury has been developed (Berjak *et al.*, 1989). However, the most critical part in the process of free-water removal is the optimization of water content within the cells to survive the freezing-injury with assured post-cryo regrowth. In understanding the nature and amount of water to be removed, especially for recalcitrant or intermediate seeds, usage of Differential Scanning Calorimetry is necessary (Dussert *et al.*, 2001; Hor *et al.*, 2005).

26.5 Preculture-dehydration

It has been observed that the juvenile explants or samples that are sensitive to dehydration usually face the detrimental impact of direct dehydration.

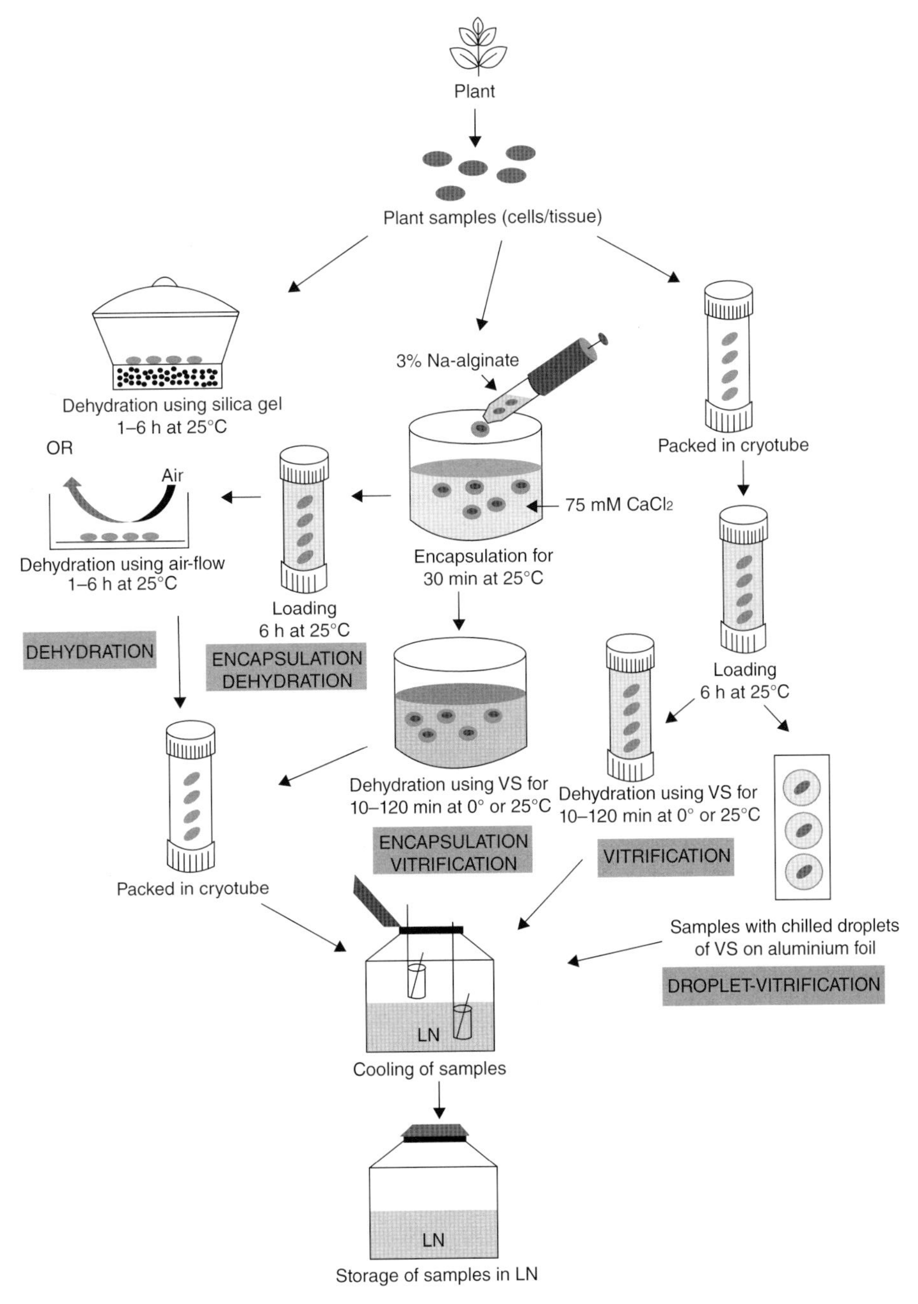

Fig. 26.1. A schematic representation of competent cryoconservation procedures comprising dehydration, encapsulation-dehydration, vitrification, encapsulation-vitrification, and droplet-vitrification for long-term storage of plant germplasm.

Therefore, it is necessary to precondition those sensitive explants prior to their exposure to laminar airflow or the silica gel chamber. To resolve this issue, preculture-desiccation plays a basic yet pivotal role, wherein the desiccation-sensitive samples such as somatic embryo, zygotic embryonic axes, or meristematic segments are treated with cryoprotectant solutions, followed by conventional dehydration and liquid nitrogen exposure. There are plenty of examples of preculture-desiccation of asparagus (Uragami *et al.*, 1990), coconut (Assy-Bah and Engelmann, 1992), oil palm (Dumet *et al.*, 1993), and coriander (Popova *et al.*, 2010).

26.6 Encapsulation-dehydration

The term 'encapsulation-dehydration' defines the conjugal approach of 'encapsulation' and 'dehydration' techniques. Instead of direct exposure to physical or chemical dehydration, the samples are protected with a spherical layer of calcium alginate. Explants emerged in sodium alginate are dropped in aliquots into calcium chloride solution to form an encapsulated structure (also known as synthetic or artificial seed). The encapsulated samples are usually exposed to sucrose containing a liquid medium to accomplish their pregrowth for a week, following which they are dehydrated either by air in a laminar air flow chamber or in a desiccator using silica gel. The conventional dehydration requirement reported is around 20% of water content on a fresh weight basis. Once the targeted dehydration is achieved, the samples are directly exposed to liquid nitrogen for ultra-freezing, following which an increased level of regrowth, avoiding the callogenesis phase, occurs. An ample number of plant species from temperate and tropical zones have been conserved through this encapsulation-dehydration approach, aided with tissue culture, cell culture and somatic embryogenesis-based techniques (Gonzalez-Arnao and Engelmann, 2006; Engelmann *et al.*, 2008). Later on, a unique yet simple approach came into play where preconditioning of samples is achieved using an osmoticum like glycerol and sucrose blended within the solution of encapsulation agent (Bonnart and Volk, 2010). Hence, during the time of encapsulation, the samples become acclimatized for instant dehydration either by silica gel or by laminar airflow.

26.7 Vitrification

Treatment of samples with relatively higher levels of solutions of cryoprotectants prior to their exposure to liquid nitrogen is the key approach of the vitrification technique. The process of vitrification starts with the exposure of samples to a solution usually comprised of 2 M glycerol + 0.4 M sucrose having transitional concentration level, termed as loading (Matsumoto *et al.*, 1994). A series of loading solution-based approaches, for sensitive species like chrysanthemum, has been conceived by Kim *et al.* (2009a). This technique has been exceedingly effective for freezing sensitive plant species.

Subsequent to the loading treatment, the samples pass through chemical desiccation with an increased level of vitrification solutions known as plant vitrification solutions (PVS). Widely used PVS solutions are PVS2 and PVS3, which have been developed by Sakai *et al.* (1990) and Nishizawa *et al.* (1993), respectively. These PVS are comprised (*w/v*) of 15% ethylene glycol + 30% glycerol + 15% DMSO + 13.7% sucrose and 50% glycerol + 50% sucrose, respectively. An advancement of conventional vitrification with the employment of PVS2- and PVS3-derived solutions was developed by Kim *et al.* (2009b), to further the survival and recovery frequency in comparison to the traditional vitrification solution. However, following any mode of vitrification exposure, the explants are usually emerged in 0.5–1.0 ml vitrification solution taken in cryovials. Then the vials are swiftly exposed to liquid nitrogen for freezing.

For post-cryo treatment, the liquid nitrogen-treated samples are placed in a hot water bath at 37–40°C. Following this thawing, the samples are treated with concentrated sucrose solution (0.8–1.2 M) to remove the vitrification-associated toxicity, and gradual desiccation is achieved. The success of vitrification is estimated through the revival of samples following ultra-freezing exposure. For the subsequent regeneration of the samples, an ambient environment and medium is required, which might be in the standard nutrient medium but is maintained without illumination for an initial seven days. Later on, the surviving samples are conventionally regrown following normal *in-vitro* culture conditions. The success of cryoconservation through the vitrification approach has been reported with numerous tropical and temperate plant species (Sakai and Engelmann, 2007; Sakai *et al.*, 2008; Suranthran *et al.*, 2012; Sinniah *et al.*, 2014).

26.8 Encapsulation-vitrification

Encapsulation-vitrification is a process combining encapsulation-dehydration and vitrification. The mechanism involves the samples being encapsulated in alginate beads, following a regular vitrification protocol, as discussed earlier. The major advantage of this technique is that encapsulated explants are not subjected to contact directly with the condensed vitrification solutions, in this manner lessening their noxiousness. Wide ranges of plant species belonging to tropical and temperate regimes responded positively to the encapsulation-vitrification technique, and underwent successful cryoconservation (Sakai and Engelmann, 2007; Sakai *et al.*, 2008).

26.9 Droplet-vitrification

Droplet-vitrification is the most up-to-date system yet developed (Panis *et al.*, 2005). The method suggests a distinctive vitrification technique with an exceptional step. Prior to rapid dipping in liquid nitrogen, the explants are positioned individually on aluminum strips and covered with droplets of vitrification solution. Some of the key innovations of this approach are the direct and complete exposure of the explants with liquid nitrogen during freezing, as well as with solution while thawing. These steps eventually endorse improved freezing and thawing. Droplet-vitrification, a relatively new approach to cryoconservation, has been employed for a number of plant species and the numbers are increasing steadily (Sakai and Engelmann, 2007; Gantait *et al.*, 2015).

26.10 Conclusion

An inadequate amount of plant germplasm has been employed for cryoconservation in a consistent manner. Nevertheless, the advancement of the original method based on vitrification has established its utility to a wider array of plant species. An extremely significant benefit of these innovative systems is their ease of use, important for tropical subcontinents, where the major share of plant germplasm of delinquent species is found, and where, on occasion, conditions are rudimentary. Another benefit is their specific significance to the long-term storage of wild species: the huge genetic divergence necessitates cryoconservation. In this chapter, we have endeavoured to elucidate the ways in which cryoconservation has improved, to the extent of being established as a biotechnological imperative, with the realistic aim of extended storage for plant biodiversity. In plant cryopreservation, the most significant improvement has been the application of cryogenic techniques to various species which formerly could not endure freezing in liquid nitrogen. Furthermore, cryopreservation now offers solutions for the eradication of systemic pathogens, and renders plant germplasm safe and suitable for breeding programmes.

Acknowledgements

The authors are appreciative to the Department of Crop Science, Faculty of Agriculture, Universiti Putra Malaysia, Malaysia and the Bidhan Chandra Krishi Viswavidyalaya, West Bengal, India for providing the research and library facilities, respectively.

References

Ashmore, S.E. (1997) *Status Report on the Development and Application of In Vitro Techniques for the Conservation and Use of Plant Genetic Resources*. International Plant Genetic Resources Institute, Rome, Italy.

Assy-Bah, B. and Engelmann, F. (1992) Cryopreservation of mature embryos of coconut (*Cocos nucifera* L.) and subsequent regeneration of plantlets. *Cryo Letters* 13, 117–126.

Benson, E.E., Johnston, J., Muthusamy, J. and Harding, K. (2006) Physical and engineering perspectives of *in vitro* plant cryopreservation. In: Gupta S. and Ibaraki, Y. (eds) *Plant Tissue Culture Engineering*, Vol. 6. Springer, Berlin, Germany, pp. 441–476.

Berjak, P., Farrant, J.M., Mycock, D.J. and Pammenter, N.W. (1989) Homoiohydrous (recalcitrant) seeds: the enigma of their desiccation sensitivity and the state of water in axes of *Landolphia kirkii* Dyer. *Planta* 186, 249–261.

Bonnart, R. and Volk, G.M. (2010) Increased efficiency using the encapsulation-dehydration cryopreservation technique for *Arabidopsis thaliana*. *Cryo Letters* 31, 95–100.

Chin, H.F. and Roberts, E.H. (1980) *Recalcitrant Crop Seeds*. Tropical Press Sdn Bhd, Kuala Lumpur, Malaysia.

Chin, H.F. (1988) *Recalcitrant Seeds: A Status Report*. International Plant Genetic Resources Institute, Rome, Italy.

Dumet, D., Engelmann, F., Chabrillange, N. and Duval, Y. (1993) Cryopreservation of oil palm (*Elaeis guineensis* Jacq.) somatic embryos involving a desiccation step. *Plant Cell Reproduction* 12, 352–355.

Dussert, S., Chabrillange, N., Roquelin, G., Engelmann, F., Lopez, M. and Hamon, S. (2001) Tolerance of coffee (*Coffea* spp.) seeds to ultra-low temperature exposure in relation to calorimetric properties of tissue water, lipid composition and cooling procedure. *Physiologia Plantarum* 112, 495–505.

Ellis, R.E., Hong, T. and Roberts, E.H. (1990) An intermediate category of seed storage behaviour? I. Coffee. *Journal of Experimental Botany* 41, 1167–1174.

Ellis, R.E., Hong, T., Roberts, E.H. and Soetisna, U. (1991) Seed storage behaviour in *Elaeis guineensis*. *Seed Science Research* 1, 99–104.

Engelmann, F. (1997) Importance of desiccation for the cryopreservation of recalcitrant seed and vegetatively propagated species. *Plant Genetic Resources Newsletter* 112, 9–18.

Engelmann, F. (2009) Use of biotechnologies for conserving plant biodiversity. *Acta Horticulturae* 812, 63–82.

Engelmann, F. and Dussert S. (2013) Cryopreservation. In: Normah, M.N., Chin, H.F. and Reed, B.M. (eds.), *Conservation of Tropical Plant Species*. Springer, New York, USA, pp. 107–119.

Engelmann, F. and Engels, J.M.M. (2002) Technologies and strategies for ex situ conservation. In: Engels, J.M.M., Rao, V.R., Brown, A.D.H. and Jackson, M.T. (eds) *Managing Plant Genetic Diversity*. CAB International, Wallingford, Oxfordshire, UK/International Plant Genetic Resources Institute, Rome, Italy, pp. 89–104.

Engelmann, F., Gonzalez-Arnao, M.T., Wu, W.J. and Escobar, R.E. (2008) Development of encapsulation-dehydration. In: Reed, B.M. (ed.) *Plant Cryopreservation: A Practical Guide*. Springer, Berlin, Germany, pp. 59–76.

Fahy, G.M., MacFarlane, D.R., Angell, C.A. and Meryman, H.T. (1984) Vitrification as an approach to cryopreservation. *Cryobiology* 21, 407–426.

Gantait, S., Sinniah, U.R., Suranthran, P., Palanyandy, S.R. and Subramaniam, S. (2015) Improved cryopreservation of oil palm (*Elaeis guineensis* Jacq.) polyembryoids using droplet-vitrification approach and assessment of genetic fidelity. *Protoplasma* 252, 89–101.

Gonzalez-Arnao, M.T. and Engelmann, F. (2006) Cryopreservation of plant germplasm using the encapsulation-dehydration technique: review and case study on sugarcane. *Cryo Letters* 27, 155–168.

Gonzalez-Arnao, M.T., Panta, A., Roca, W.M., Escobar, R.H. and Engelmann, F. (2008) Development and large scale application of cryopreservation techniques for shoot and somatic embryo cultures of tropical crops. *Plant Cell, Tissue and Organ Culture* 92, 1–13.

Hor, Y.L., Kim Y.J., Ugap, A., Chabrillange, N., Sinniah, U.R., Engelmann, F. and Dussert, S. (2005) Optimal hydration status for cryopreservation of intermediate oily seeds: *Citrus* as a case study. *Annals of Botany* 95, 1153–1161.

Kim, H.H., Lee, Y.G., Ko, H.C., Park, S.U., Gwag, J.G., Cho, E.G. and Engelmann, F. (2009a) Development of alternative loading solutions in droplet-vitrification procedures. *Cryo Letters* 30, 291–299.

Kim, H.H., Lee, Y.G., Shin, D.J., Kim, T., Cho, E.G. and Engelmann, F. (2009b) Development of alternative plant vitrification solutions in droplet-vitrification procedures. *Cryo Letters* 30, 320–334.

Matsumoto, T., Sakai, A. and Yamada, K. (1994) Cryopreservation of *in vitro* grown apical meristems of wasabi (*Wasabia japonica*) by vitrification and subsequent high plant regeneration. *Plant Cell Reproduction* 13, 442–446.

Mazur, P. (1984) Freezing of living cells: mechanisms and applications. *American Journal of Physiology* 247, 125–142.

Nishizawa, S., Sakai, A., Amano, A.Y. and Matsuzawa, T. (1993) Cryopreservation of asparagus (*Asparagus officinalis* L.) embryogenic suspension cells and subsequent plant regeneration by vitrification. *Plant Science* 91, 67–73.

Panis, B., Piette, B. and Swennen, R. (2005) Droplet vitrification of apical meristems: a cryopreservation protocol applicable to all *Musaceae*. *Plant Science* 168, 45–55.

Popova, E., Kim, H.H. and Paek, K.Y. (2010) Cryopreservation of coriander (*Coriandrum sativum* L.) somatic embryos using sucrose preculture and air desiccation. *Scientia Horticulturae* 124, 522–528.

Roberts, H.F. (1973) Predicting the viability of seeds. *Seed Science and Technology* 1, 499–514.

Sakai, A. and Engelmann, F. (2007) Vitrification, encapsulation-vitrification and droplet-vitrification: a review. *Cryo Letters* 28:151–172.

Sakai, A., Kobayashi, S. and Oiyama, I.E. (1990) Cryopreservation of nucellar cells of navel orange (*Citrus sinensis* Osb. *var. brasiliensis* Tanaka) by vitrification. *Plant Cell Reproduction* 9, 30–33.

Sakai, A., Hirai, D. and Niino, T. (2008) Development of PVS-based vitrification and encapsulation-vitrification protocols. In: Reed, B.M. (ed.) *Plant Cryopreservation: A Practical Guide*. Springer, Berlin, Germany, pp. 33–58.

Sinniah, U.R., Gantait, S. and Suranthran, P. (2014) Cryopreservation technology for conservation of selected vegetative propagules. *Journal of Crop and Weed* 10, 10–13.

Suranthran, P., Gantait, S., Sinniah, U.R., Subramaniam, S., Sarifa, S.R.S.A. and Roowi, S.H. (2012) Effect of loading and vitrification solutions on survival of cryopreserved oil palm polyembryoids. *Plant Growth Regulators* 66, 101–109.

Uragami, A., Sakai, A. and Magai, M. (1990) Cryopreservation of dried axillary buds from plantlets of *Asparagus officinalis* L. grown *in vitro*. *Plant Cell Reports* 9, 328–331.

Withers, L.A. and Engelmann, F. (1998) *In vitro* conservation of plant genetic resources. In: Altman, A. (ed.) *Biotechnology in Agriculture*. Marcel Dekker, New York, USA, pp. 57–88.

Withers, L.A. and Engels, J.M.M. (1990) The test tube genebank – a safe alternative to field conservation. *IBPGR Newsletter for Asia, the Pacific and Oceania* 3, 1–2.

27 Interspecific Chemical Differentiation within the Genus *Astragalus* (Fabaceae) Based on Sequential Variability of Saponin Structures

Abir Sarraj-Laabidi[1,2] and Nabil Semmar[2,3*]

[1]Université de Tunis El Manar (University of Tunis El Manar), Faculty of Sciences of Tunis, Tunis, Tunisia; [2]Université de Tunis El Manar (University of Tunis El Manar), Laboratory of Bioinformatics, Biomathematics and Biostatistics (BIMS), Institut Pasteur de Tunis, Tunis, Tunisia; [3]Medical School of Marseilles, Aix-Marseille Université, Marseilles, France

Abstract

Phytochemistry represents an emergent field aimed at the characterization of plants from their secondary metabolites' structures. Identification of original molecules from a wide metabolic pool leads to chemically differentiated individual species within their genus. Research into the chemical fingerprints specific to different biological taxa is associated with chemotaxonomy; it provides an efficient way for analysis of complex biodiversity through flexible, hierarchical and precise analytical parameters. In this chapter, an illustrative study case is presented for the genus *Astragalus*, which was widely studied for its synthesized saponins (triterpenic metabolites). From a pool of 240 metabolites analysed in 53 species of *Astragalus*, 46 species showed chemical differentiations according to five hierarchical metabolic processes: (1) biosynthesis of rare aglycones (chemical skeletons); (2) increasing of desmosylation levels in common aglycones; (3) substitutions of atypical chemical groups; (4) substitutions of common groups at atypical positions of aglycones; and (5) formation of unusual sequences from common chemical groups substituted at common positions. Remaining saponins (the most usual structures) allowed identification of a chemical backbone at the generic scale, leading to recognition of deep saponin traits in the genus *Astragalus*. Finally, a geographical trend was highlighted for desmosylation level in the saponins of *Astragalus*.

27.1 Introduction

The plant world is botanically classified in a hierarchy through which organisms are identified as biological species or varieties belonging to higher taxonomic levels (genus, family, class). Such botanical classifications are traditionally based on morphological traits and morphometrical measures completed by histological sections.

More precision is provided by genetic factors, which help to:

- confirm/check the botanical groups with heterogeneous botanical aspects (e.g. Liliaceae) (APG, 2003)
- analyse links between different plant taxa
- highlight biological (genetic) features of individual plants that can serve as fingerprints for the traceability of biodiversity from analysis of single biological tissues.

In relation to this last point, phytochemistry is an efficient way of characterizing individual plants at specific or varietals levels by means of secondary metabolites. Secondary metabolites, including phenolic compounds, terpenes and nitrogen-containing compounds (e.g. alkaloids), show high structural diversity compared to the universally constitutive

*E-mail: nabilsemmar@yahoo.fr

primary metabolites (sugars, fatty acids, amino acids). This high diversity of chemical structures consists of more than 8000 phenolics, 40,000 terpenes and 12,000 alkaloids (Croteau *et al.*, 2000; Roberts, 2007; Withers and Keasling, 2007; Dai and Mumper, 2010). Secondary metabolites show chemotaxonomical usefulness in the sense that they are more precise than botanical parameters and more easily accessible than genetic ones. Moreover, secondary metabolites occupy a key intermediate position between genome and environmental conditions, making them informative on both plant states and their living conditions.

On the basis of these characteristics, secondary metabolites can be used for:

- chemical differentiation between different plant taxons
- analysis of variation trends of secondary metabolites' contents in relation to intrinsic or extrinsic factors.

This chapter focuses on the flexible usefulness of secondary metabolites for chemical differentiation of plant species using different structural criteria. The secondary metabolites explored here are saponins, which are widely present in the plant kingdom and particularly well produced by some families and genera. The chemotaxonomical use of saponins will be illustrated by the genus *Astragalus* (Fabaceae), which has been widely researched for its saponins content (more than 110 published articles).

The genus *Astragalus* L. belongs to the Fabaceae family (also named Leguminosae or Papilionaceae). It is the largest genus in this family and one of the largest genera of vascular plants. It comprises about 2500–3000 herb or shrub species which are mostly perennial and widely distributed through the temperate region of the world (Aytaç, 2000). About 2000 species are distributed in the northern temperate regions and tropical African mountains (Radwan *et al.*, 2004a). Of these, 445 are found in Turkey, 372 in North America and 133 in Europe (Aytaç, 2000; Davis, 1982; Rios and Waterman, 1997).

The plants of *Astragalus* genus proved to be rich in saponins, particularly those based on a cycloartane skeleton (or aglycone). Saponins belonging to the family of triterpenes demonstrated high chemical variability due to diversity of aglycones (including cycloartane, 9,10-secocycloartane, oleanane), chemical groups substituting for aglycones, substitution positions and chaining sequences of such groups.

The high diversity of saponin structures in the terrestrial plant kingdom and their production by different plant species make them efficient biochemical markers for monitoring and assessment of plant biodiversity. More generally, chemical characterizations of plant taxa from their secondary metabolites provide efficient ways to define chemical fingerprints at specific levels, so bringing complementary and confirmative tools into the field of biodiversity research.

27.2 Material and Method

27.2.1 Background presentation of saponins

Saponins are secondary metabolites with a molecular weight ranging from 600 to 2000 Da (Hostettmann and Marston, 1995; Böttger and Melzig, 2011). They are based on a hydrocarbon skeleton (called aglycone or sapogenin) which is of triterpenic (C30) or steroidal (C27) type. Sapogenin is metabolized into saponin by substituting one or more sugar moieties (sapogenin – sugar = saponin). The different substituted sugar moieties (called glycosyls) can be attached on one, two, three or four carbons of the aglycone, leading to different desmosylation levels in saponins: mono-, bi-, tri- and tetradesmosides, respectively. Desmosidic linkages can be of different length and branching. Apart from saccharides, several other substitutions are observed including hydroxyl, methyl acyl, etc. (Fig 27. 1).

The name 'saponin' is derived from the Latin word sapo which means 'soap', because saponin molecules form stable soap-like foams in water solutions (Hostettmann and Marston, 1995). This characteristic trait is caused by the amphiphilic nature of saponins due to linkage of the lipophilic sapogenin to hydrophilic saccharide side chains. This soap-like attribute of saponins enables them to interfere with cellular membranes, inducing cytotoxicity even at low concentrations.

Occurrence of saponins is widespread among plants belonging to the division of Magnoliophyta, covering both dicotyledons and monocotyledons. However, dicotyledons have been shown to produce more saponins than monocotyledons (Vincken *et al.*, 2007). The important distribution and variability of saponins in the terrestrial plant world mean that these secondary metabolites can be used for determination of chemical signatures or fingerprints enabling better understanding and monitoring of biodiversity.

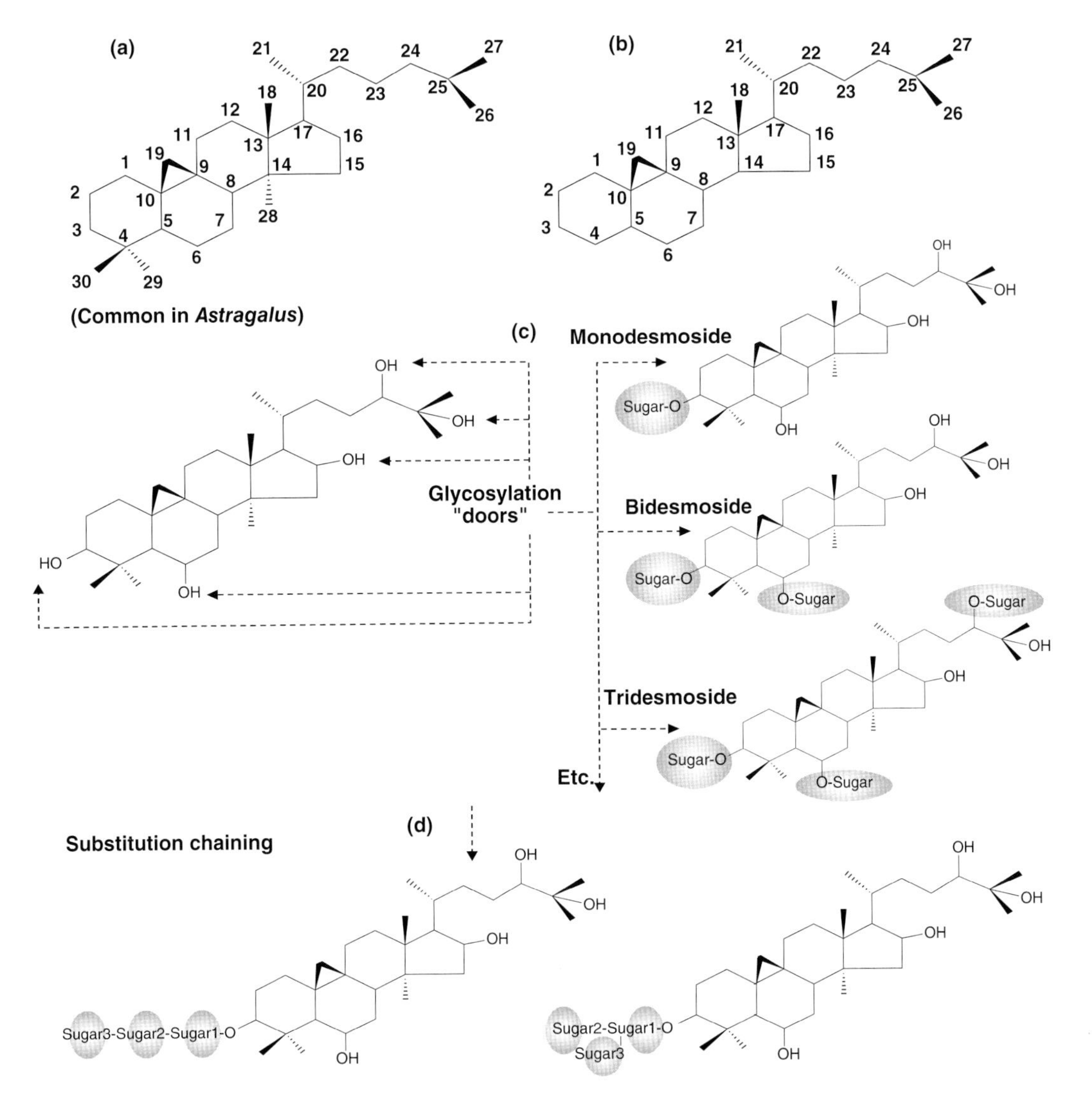

Fig. 27.1. General presentation of saponin diversification processes in the plant world. (a, b) basic aglycone structures of triterpenes (a) and steroids (b); (c) desmosylation leading to different numbers of glycosylated carbons on aglycone; (d) variation of substitution chaining by different arrangements between substituted units.

Apart from their chemotaxonomical values, saponins (and more generally, secondary metabolites) are well known to be sensitive to variations in environmental conditions. Although saponins are generally accumulated as part of normal plant development, their accumulation is also known to be influenced by several environmental factors, such as nutrients, water availability, light irradiation and seasonal fluctuations (Szakiel *et al.*, 2011). This multifactorial sensitivity of saponins to abiotic conditions means that they can be used to identify plant metabolic trends associated with ecological gradients, leading to better understanding of biodiversity distribution at regional and global scales. Beyond physical chemical environmental factors, variations in saponin distribution and levels in different plant tissues have been suggested as representing varying needs for protection against specific herbivores and pests (Ndamba *et al.*, 1994; Papadopoulou *et al.*, 1999; Howe and Jander, 2008; Yendo *et al.*, 2010).

27.2.2 Presentation of structural criteria for characterization of *Astragalus* species

Chemical differentiations of *Astragalus* species from their synthesized saponins were carried out by identifying several rare structural traits which are specific to the different plants. Chemical and botanical pools covered 240 molecules synthesized by 53 species (Fig. 27.2).

Structural traits were identified in the 240 molecules on the basis of five hierarchical (sequential) criteria, as follows.

1. Aglycone type, i.e. the type of hydrocarbon skeleton on the basis of which the synthesis of saponin is achieved.
2. Desmosylation level in aglycone, i.e. the number of carbons of aglycone on which sugars are

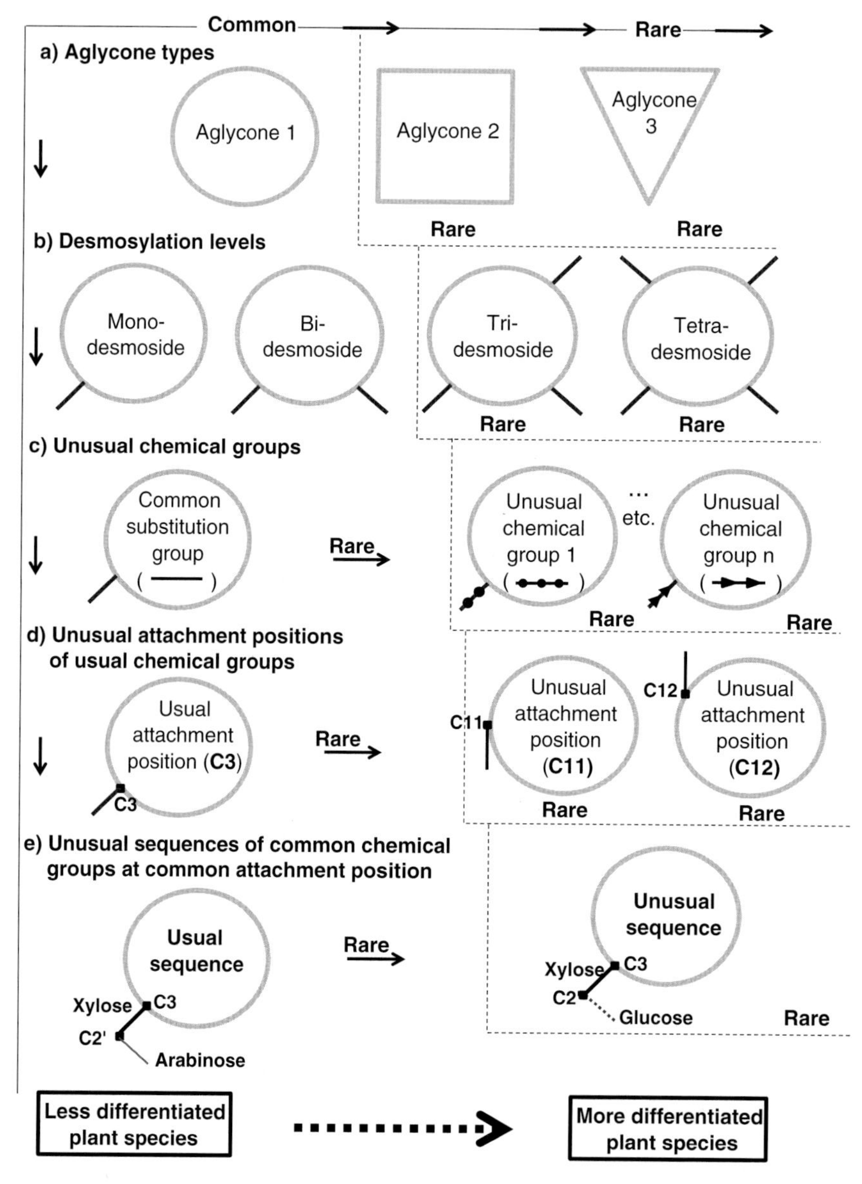

Fig. 27.2. Diagram illustrating five structural criteria (a–e) used to identify original saponins characterizing different species of *Astragalus* genus.

attached (number of glycosylated carbon of aglycone).
3. Types of chemical groups substituted on aglycone. This criterion concerns the type of substitutions, including hydroxylation, glycosylation, acetylation, etc.
4. Positions of chemical groups on aglycone, i.e. carbons on which chemical groups are attached.
5. Substitution sequences resulting from chaining substituted chemical groups.

The different modalities of each criterion showed differential frequencies leading to definition of usual (common) and unusual (original) chemical structures at the scale of the *Astragalus* community (Fig. 27.2).

1. Aglycone types included cycloartane, oleanane, 9,10-secocycloartane and lanostane. Also, cycloartane was divided into different subgroups, depending on its epoxylation level and type.
2. Desmosylation levels varied from one to four; rare features among these four cases were indicative of original saponin according to the desmosylation level.
3. Chemical groups were diversified, including hydroxyl, methyl, acetyl, acids, sugars and, cetone, with greater or fewer occurrence levels according to the aglycone. Rare chemical groups allowed identification of original saponins for the aglycone type.
4. When chemical groups were frequent, they contributed to defining original saponins from some unusual attachment positions on the aglycone.
5. Finally, when chemical groups are frequent and attached on usual positions, they can contribute to defining original saponins from some unusual chaining sequences.

27.3 Analysis of Structural Diversity in *Astragalus* Genus

27.3.1 Structural diversity based on aglycone (sapogenin) type

Among 240 triterpene derivatives (221 saponins and 19 sapogenins) extracted from 53 species of *Astragalus*, 80.8% were revealed to be based on the cycloartane type aglycone (Figs 27.3 and 27.4).

This percentage showed that the genus *Astragalus* is a high producer of cycloartane-based saponins. The 19.2% of remaining saponins were shared between 14.6% of oleanane, 2.9% of 9,10-secocycloartane, 0.8% (1 saponin and 1 sapogenin) of lanostane and 0.8% (1 saponin and 1 sapogenin) of β-sitosterol (Figs 27.3 and 27.4).

The rarity of these four aglycone types (more particularly lanostane, β-sitosterol and 9,10-secocycloartane) means that they are interesting chemotaxonomical markers of productive plant species.

A lanostane-based saponin, named orbigenin, was found in *A. orbiculatus* (epigeal parts) (Mamedova *et al.*, 2003); the plant species showed the occurrence of both free orbigenin and its 3-xylosylated form. This finding represents a unique lanostane occurrence in the genus *Astragalus*, leading to a strong chemical characterization of the species *A. orbiculatus* among all the other species of *Astragalus*. Also, rare aglycones included β-sitosterol, which was found in *A. dissectus* (roots and stems), *A. coluteocarpus* (epigeal parts), *A. amarus*, *A. sieversianus* (roots) and *A. orbiculatus* (roots) in free and 3-glucosylated forms (Sukhina and Isaev, 1995; Sukhina *et al.*, 2000; Iskenderov *et al.*, 2008b) (Fig. 27.5).

Among the rare triterpenic aglycones in *Astragalus*, the 9, 10-secocycloartane was reported in only two species: *A. membranaceus* (leaves) and *A. macropus* (roots) (Fig. 27.6) (Kuang *et al.*, 2009; Isaev *et al.*, 2010a; Kuang *et al.*, 2011). In *A. membranaceus*, six saponins were reported named huangqiyenin E–J, whereas the species *A. macropus* showed the occurrence of a single aglycone, named secomacrogenin B.

On the basis of these findings, *A. membranaceus* and *A. macropus* can be distinguished by their ability to synthesis 9,10-secocycloartane, uncommon to the family Fabaceae, and well reported in two families: Ranunculaceae and Buxaceae (Kadota *et al.*, 1995; Atta-ur-Rahman *et al.*, 1999; Choudhary *et al.*, 2003).

Oleanane was reported in eight *Astragalus* species, among which *A. tauricolus* (whole plant) showed ten saponins (Fig. 27.7) followed by *A. sinicus* (seeds) (nine saponins) (Fig. 27.8), *A. complanatus* (seeds) and *A. flavescens* (roots) (six saponins) (Figs 27.9 and 27.10), *A. caprinus*, *A. membranaceus*, *A. trojanus* (roots) and *A. hareftae* (whole plant) (one saponin) (Fig. 27.11) (Kitagawa *et al.*, 1983e; Kitagawa *et al.*, 1983b; Cui *et al.*, 1992a; Cui *et al.*, 1992b; Bedir *et al.*, 1999; Mitaine-Offer *et al.*, 2006; Avunduk *et al.*, 2008; Horo *et al.*, 2012; Gülcemal *et al.*, 2013).

It is worth noting that the phytochemical investigation of *A. tauricolus* (whole plant) revealed the occurrence of only oleanane-type triterpene glycosides without cycloartane-type glycosides (the main constituents of most *Astragalus* species). This peculiar feature was observed in a limited group of *Astragalus*

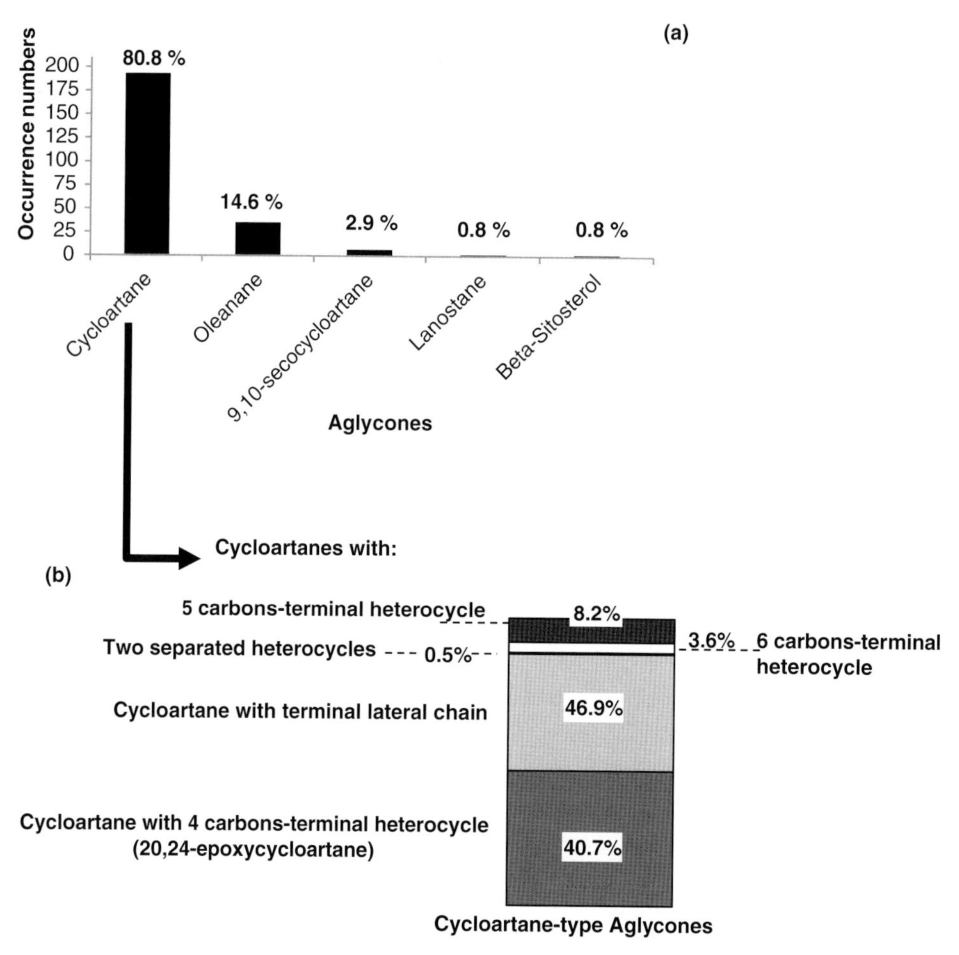

Fig. 27.3. (a) Distribution of relative frequencies of different triterpenic aglycones in 240 triterpenic molecules extracted from 53 *Astragalus* species. (b) Relative frequencies of four types of cycloartane characterized by different terminal structures.

spp., including *A. hamosus*, *A. complanatus*, *A. sinicus*, and *A. corniculatus* (Ionkova, 1991; Cui *et al.*, 1992a; Cui *et al.*, 1992b; Krasteva *et al.*, 2006; Krasteva *et al.*, 2007). Moreover, this category of saponin was extracted from several leguminous plants: from the seeds of *Glycine max* (Kitagawa *et al.*, 1976), the roots of *Sophora flavescens* (Yeshikawa *et al.*, 1981), the roots of *Hedysarum fruticosum* (Yeshikawa *et al.*, 1981) and the seeds of *Vigna angularis* (Kitagawa *et al.*, 1983a).

Other original triterpenes found in *Astragalus* consisted of cycloartanes containing two or one terminal heterocycles with four to six carbons. When the heterocycle contained six carbons, this was due to the presence of two enclosed heterocycles. These rare aglycone structures represent 12.3% of the whole set of cycloartane-based saponins. The other 87.6% of cases showed either a terminal chain (40.7%) or a lateral heterocyle with four carbons (46.9%) (Fig. 27.3).

The 12.3% of less frequent cycloartanes concerned structures of five different epoxylated terminal heterocycles (Figs 27.12, 27.13 and 27.14).

1. Heterocycle with six carbons showing a 16, 24-epoxyl-bridge enclosing a second epoxylated heterocycle in C16, C23 (Figs 27.3 and 27.12).
2. Heterocycle with six carbons showing a 16, 24-epoxyl-bridge enclosing a second heterocycle epoxylated in C20, C24 (Figs 27.3 and 27.13).
3. Two successive heterocycles with five and four carbons, respectively showing three epoxyls in 16-O-23, 23-O-26, 24-O-25 (Figs 27.3 and 27.15).

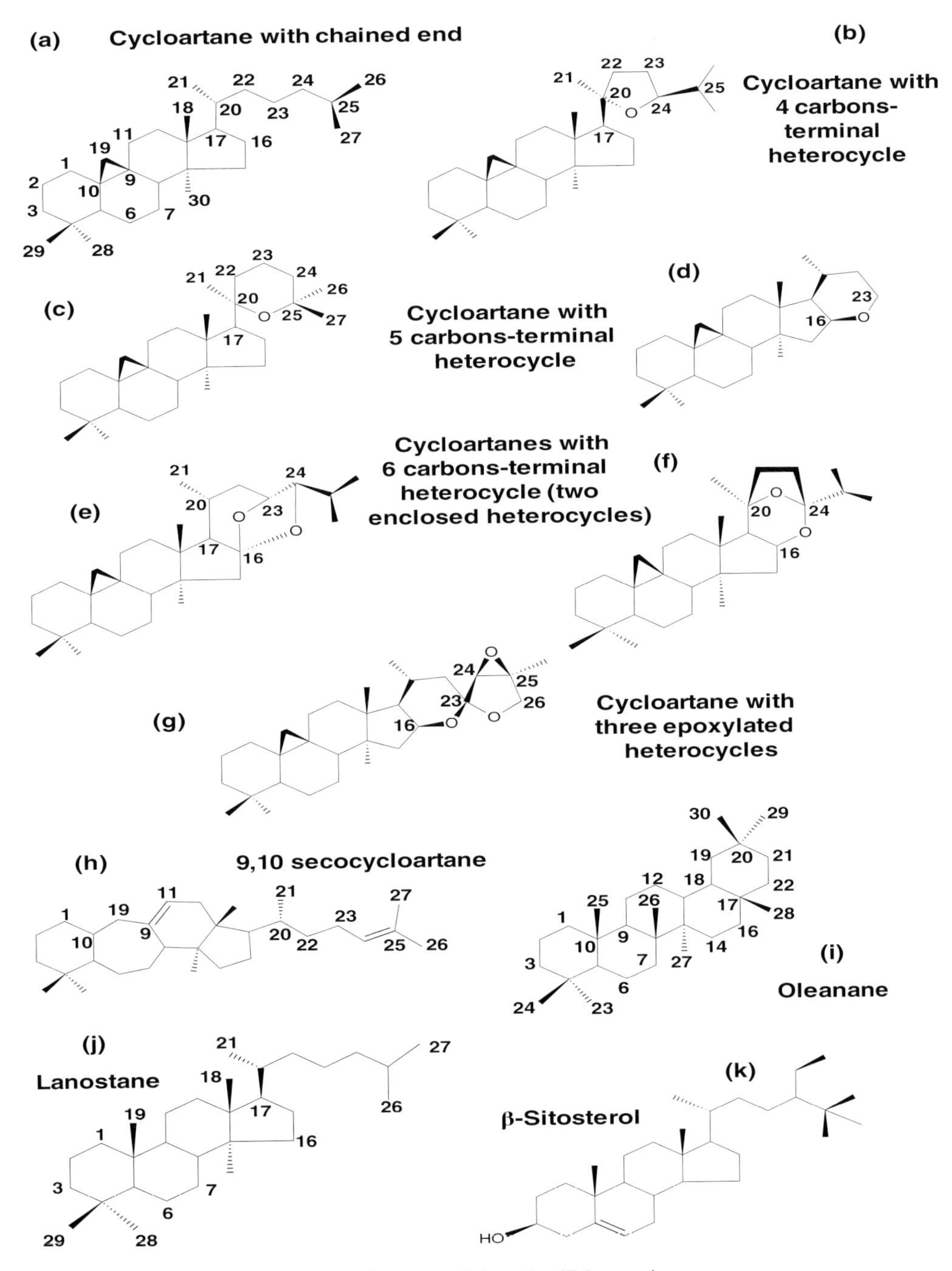

Fig. 27.4. Different sapogenins observed in the genus *Astragalus* (Fabaceae).

4. Heterocycle with five carbons epoxylated in C16, C23 (Figs 27.3 and 27.14).
5. Heterocycle with five carbons epoxylated in C20, C25 (Figs 27.3 and 27.16).

Diepoxylated cycloartanes concerned six saponins shared between *A. orbiculatus* (epigeal parts and roots; seven cases), *A. campylosema* (roots; one case) and *A. alopecurus* (aerial parts; one case)

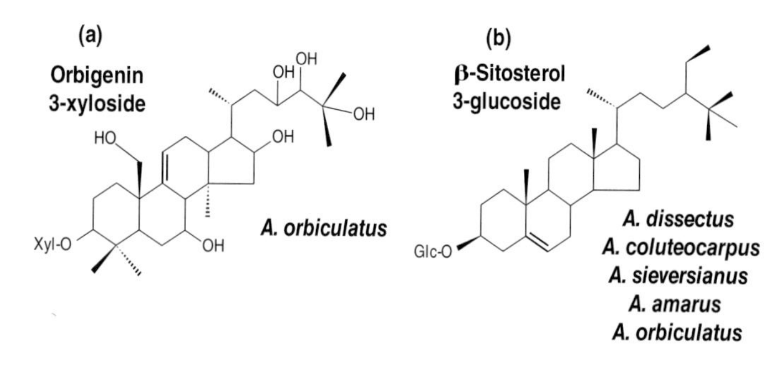

Fig. 27.5. Chemical structures of orbigenin and β-sitosterol 3-glucoside, two rare saponins found in the genus *Astragalus* and based on lanostane and β-sitosterol aglycones.

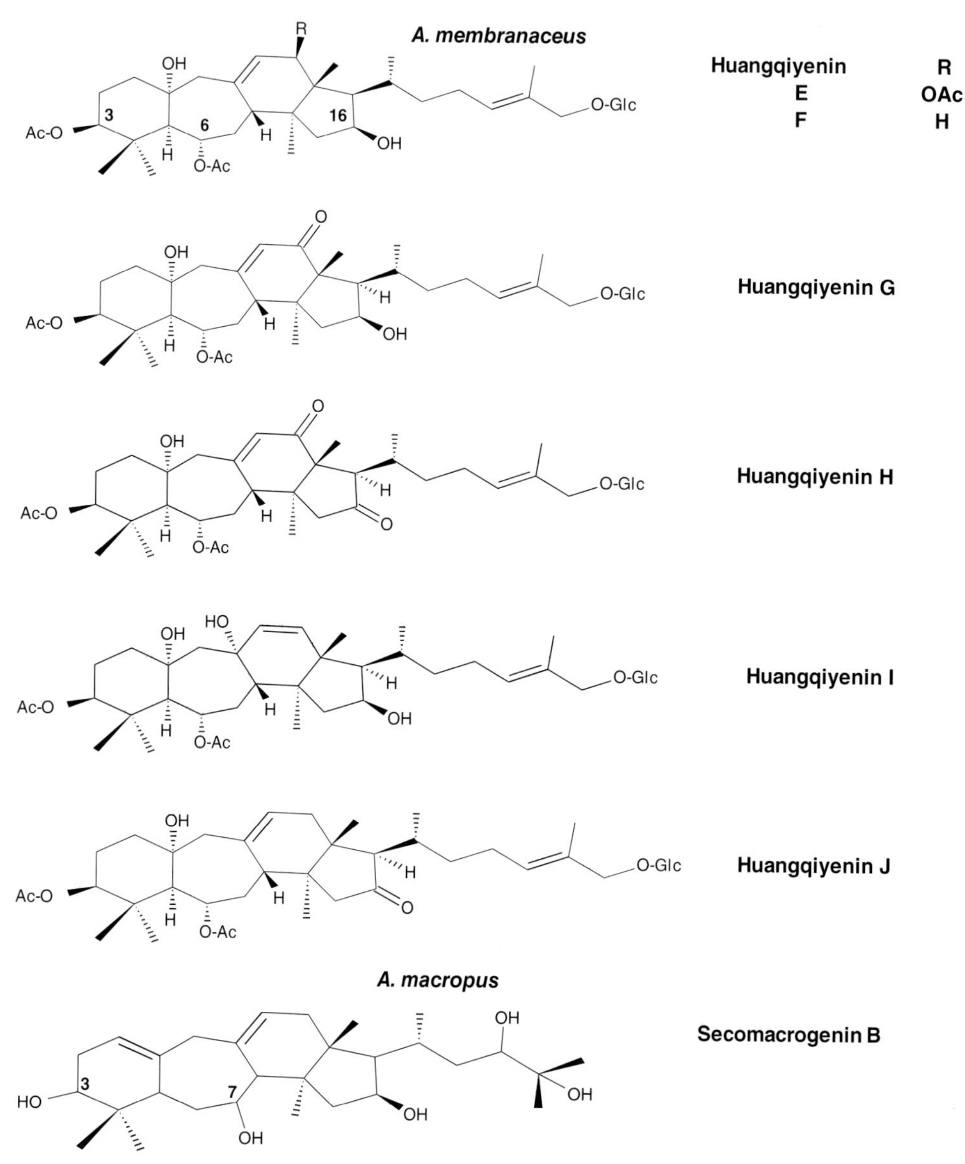

Fig. 27.6. Different 9,10-secocycloartane-based saponins found in the genus *Astragalus*. Legend: Ac = acetyl; Glc = glucose

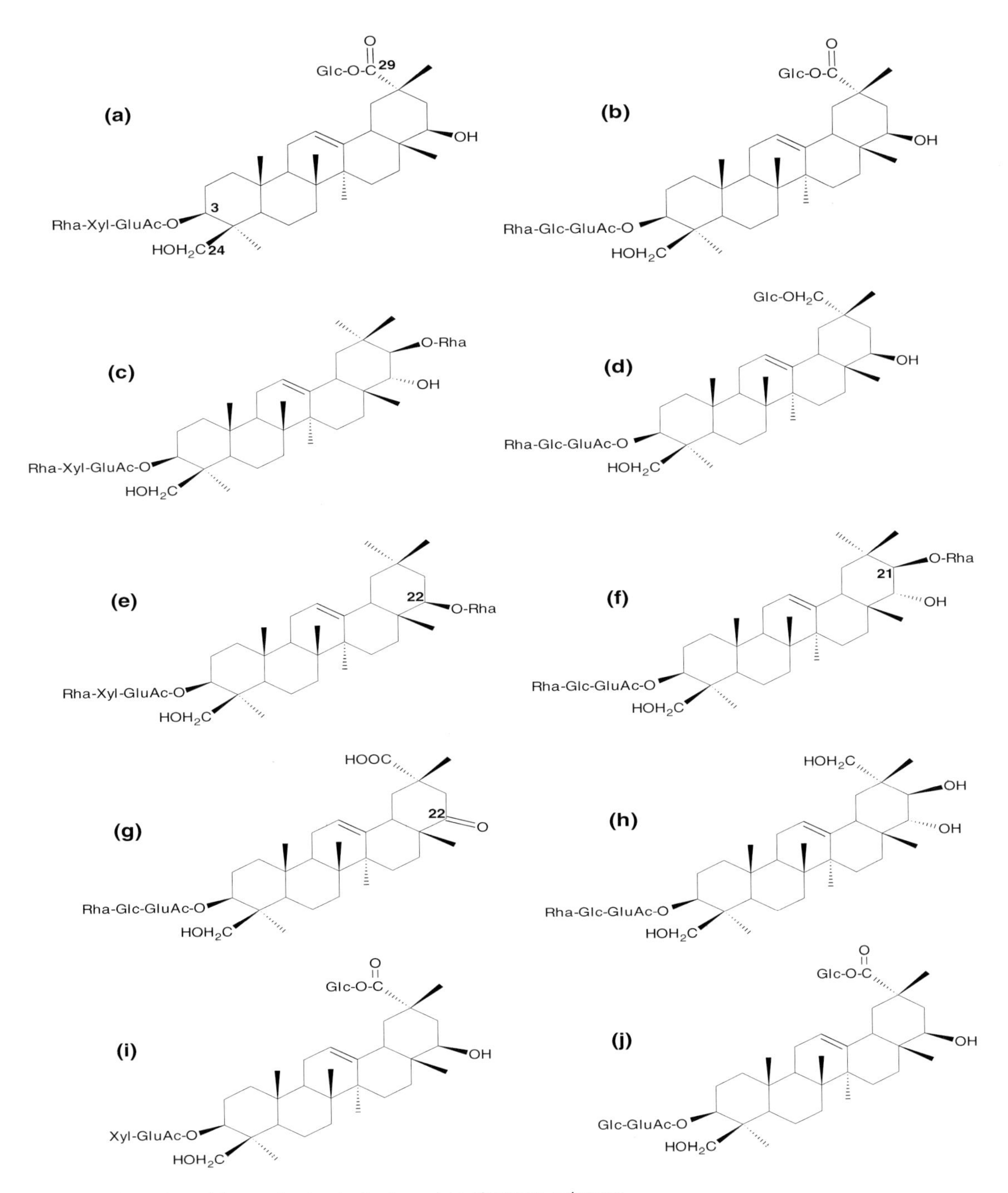

Fig. 27.7. Saponins of *Astragalus tauricolus* based on oleanane aglycone.

(Agzamova *et al.*, 1987a; Agzamova *et al.*, 1987b; Agzamova *et al.*, 1988; Agzamova *et al.*,1990; Agzamova and Isaev, 1995a; Agzamova and Isaev, 1997, Agzamova and Isaev, 1998a; Mamedova *et al.*, 2002a; Mamedova *et al.*, 2002b; Çalis *et al.*, 2008b; Isaev *et al.*, 2010b) (Figs 27.12 and 27.13). Diepoxylated structures in (C16, C23), (C16, C24) concerned four saponins: cycloorbicosides A, B, C and G in

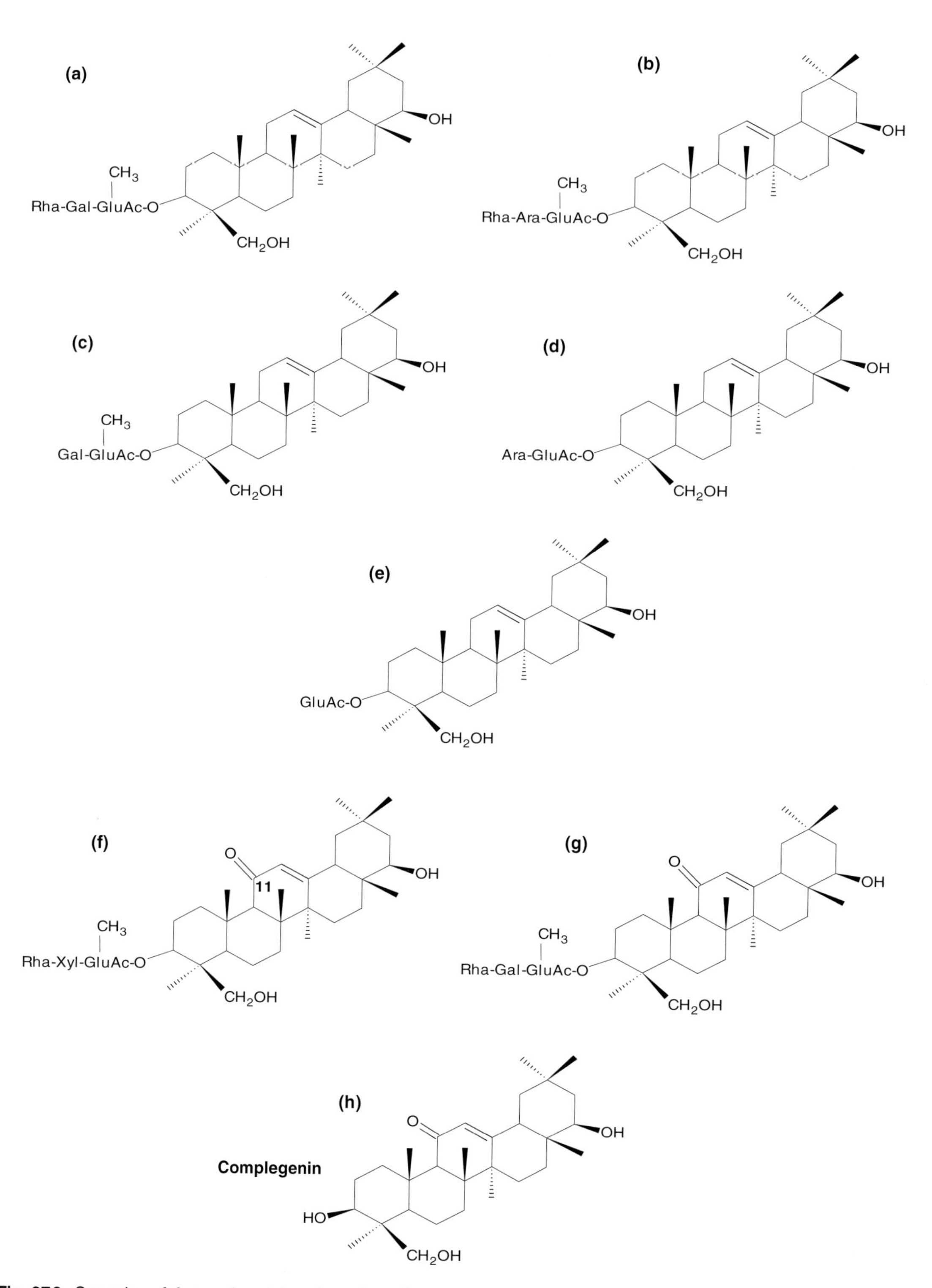

Fig. 27.8. Saponins of *Astragalus sinicus* based on oleanane aglycone.

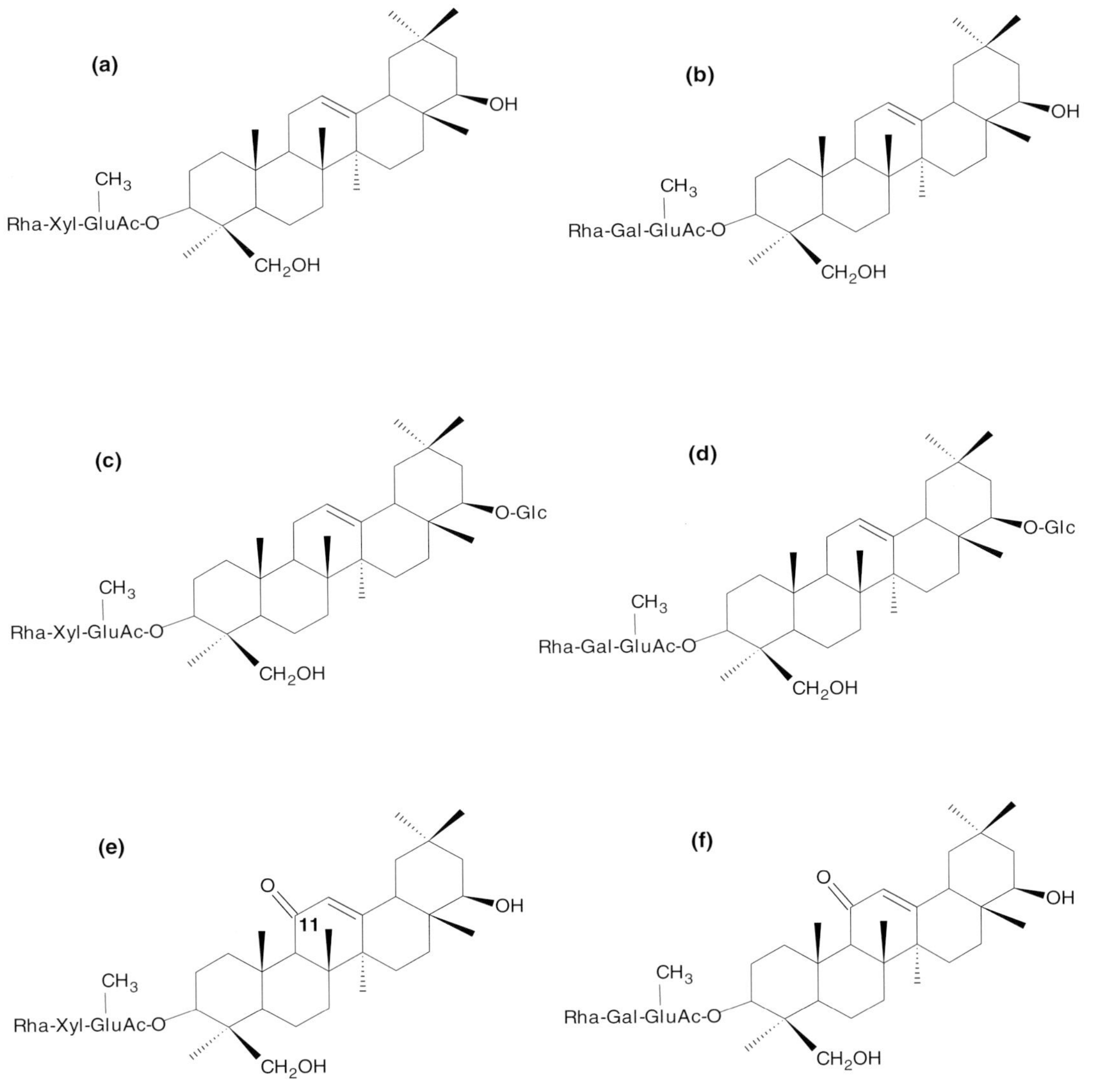

Fig. 27.9. Saponins of *Astragalus complanatus* based on oleanane aglycone.

A. orbiculatus; these saponins were based on original aglycones (cycloorbigenin A, B, dihydrocycloorbigenin A) which were highlighted by hydrolysis (Fig. 27.12). Similar chemical structures based on (16β, 23), (16α, 24)-diepoxycycloartanes were reported to be characteristics of *Cimicifuga* genus (Ranunculaceae) (Isaev *et al*., 1985b), whereas in *Astragalus* genus, they have been found in only *A. orbiculatus*. Concerning *A. campylosema* and *A. alopecurus*, they were characterized by unique diepoxy structures linking C24 to both C16 and C20 (Fig. 27.13) (Agzamova and Isaev, 1995a; Çalis *et al*., 2008b). Similar chemical structures (very unusual in the plant kingdom) were reported in *Souliea vaginata* and *Beesia calthaefolia* belonging to the family of Ranunculaceae (Sakurai *et al*., 1990).

Finally, the rarity of these diepoxylated triterpenes was increased by the fact that they showed two enclosed cycles containing the epoxyls: smaller cycle (C20-O-C24) or (C16-O-C23) with four or five carbons, respectively, within bigger cycle (C16-O-C24) with six carbons (Fig. 27.12 and 27.13).

Other types of epoxylated cycloartanes (C16-O-C23) were reported in two *Astragalus* species:

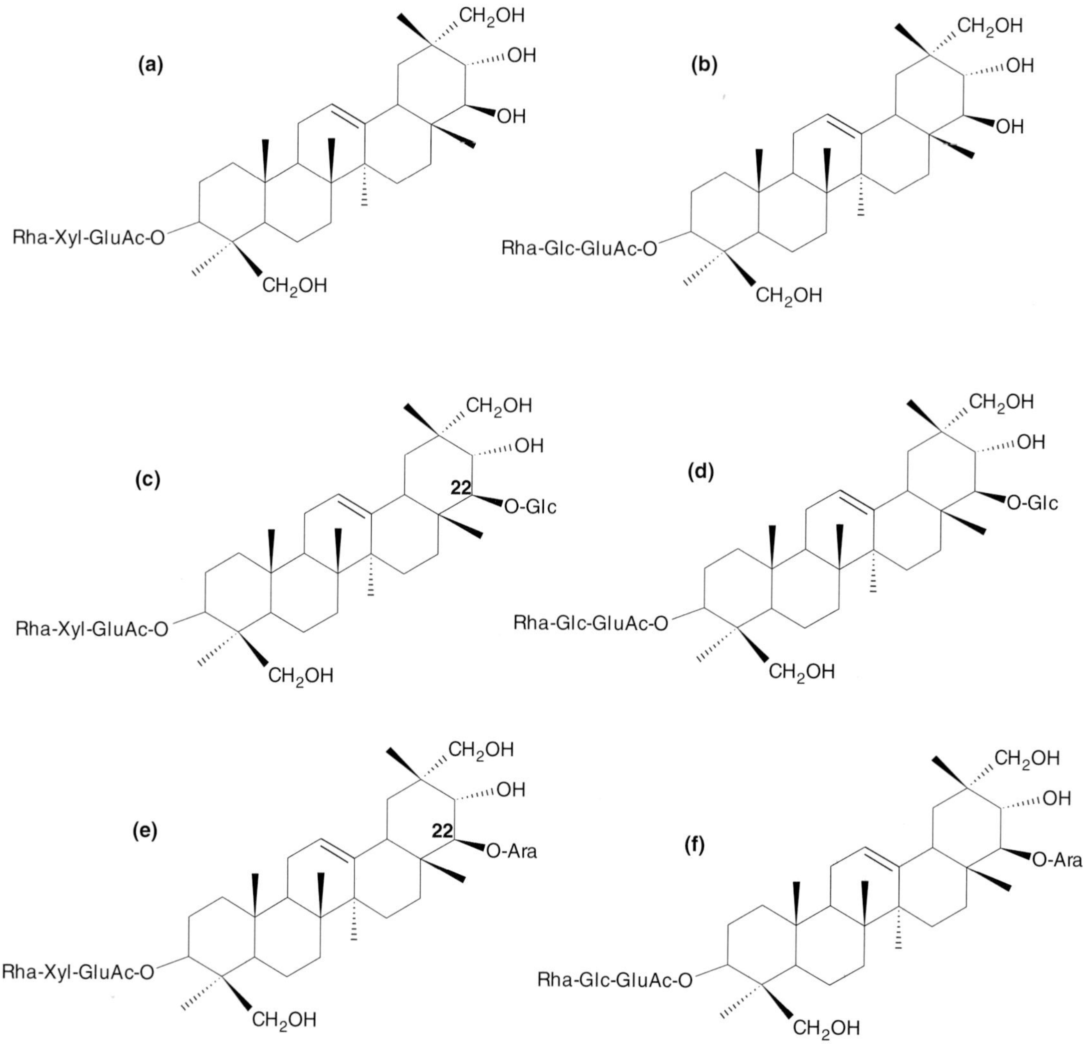

Fig. 27.10. Saponins of *Astragalus flavescens* based on oleanane aglycone.

A. bicuspis (whole plant; two saponins) and *A. tomentosus* (aerial parts; four saponins) (Radwan *et al.*, 2004a; Choudhary *et al.*, 2008). Concerning these two species (*A. tomentosus*, *A. bicuspis*), more chemical differentiations of the 23,16-epoxycycloartanes are observed at substitution levels (see 27.3.3.3.3 below). In addition to 16,23-epoxylation, the originality of the two saponins of *A. bicuspis* (bicusposides A, B) and the four of *A. tomentosus* was increased by the fact that the aglycone showed a number of carbons equal to 26 (<30) (Fig. 27.14); this suggests that four carbons would have been lost for aglycone biosynthesis. Apart from the two 16, 23-monoepoxylated saponins (bicusposides A, B), *A. bicuspis* showed a unique tri-epoxylated structure (bicusposide C) at (C16, C23), (C23, C26) and (C24, C25) (Fig. 27.15).

Concerning 20, 25-epoxylyated cycloartane, it was reported in seven *Astragalus* species including: *A. aureus* (whole plant; three saponins), *A. dissectus* (roots and stems), *A. icmadophilus* (whole plant; two saponins), *A. zahlbruckneri*, *A. caprinus*, *A. microcephalus* (roots) and *A. hareftae* (whole plant; one saponin) (Fig. 27.16) (Bedir *et al.*, 1998a; Sukhina *et al.*, 1999; Sukhina *et al.*,

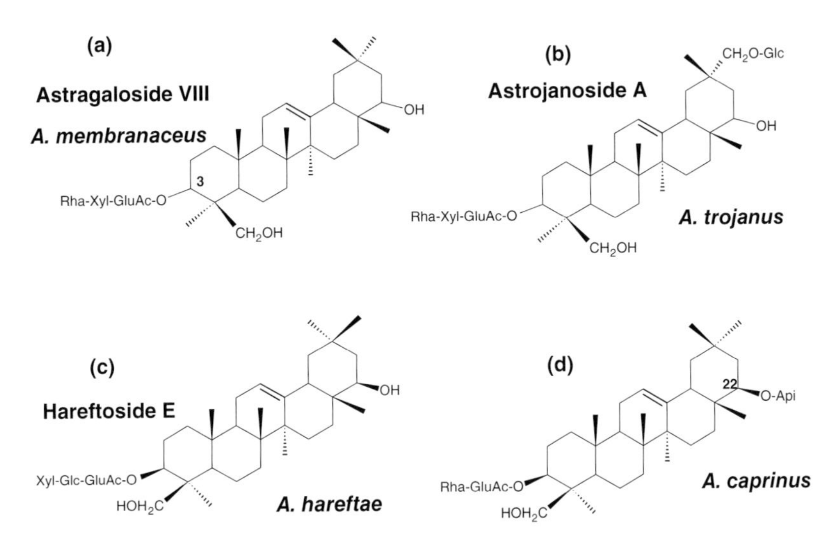

Fig. 27.11. Oleanane-based saponins found in *Astragalus membranaceus*, *A. trojanus*, *A. hareftae* and *A. caprinus*.

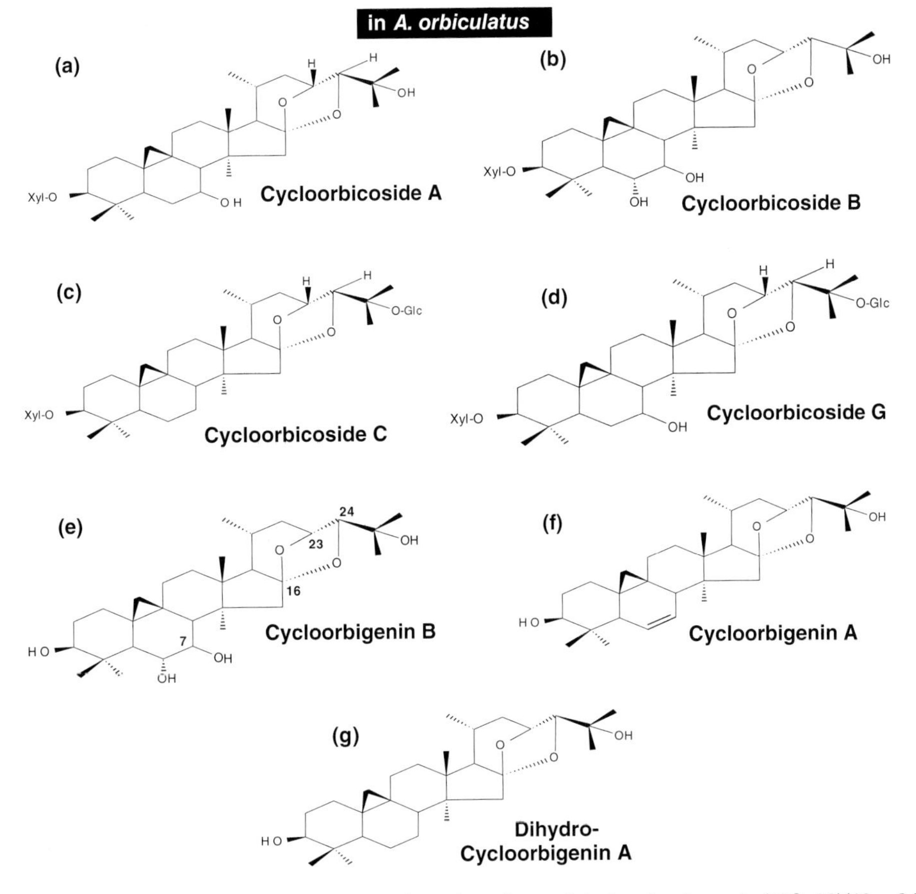

Fig. 27.12. Saponins in three *Astragalus orbiculatus* based on diepoxylated cycloartanes in (16β, 23)(16α, 24).

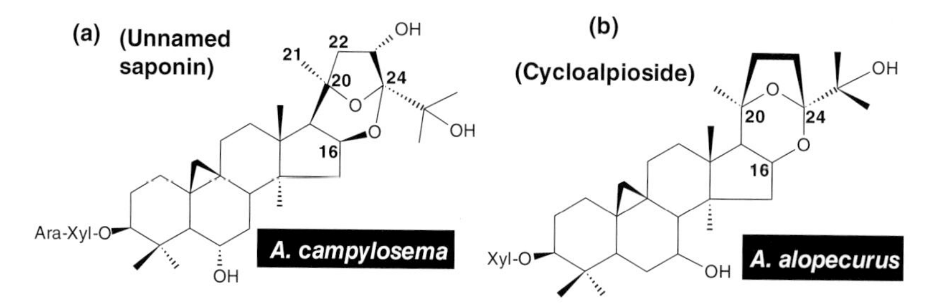

Fig. 27.13. Saponins in three *Astragalus campylosema* and *A. Alopecurus* showing diepoxylated cycloartane in (C16, C24)(C20, 24).

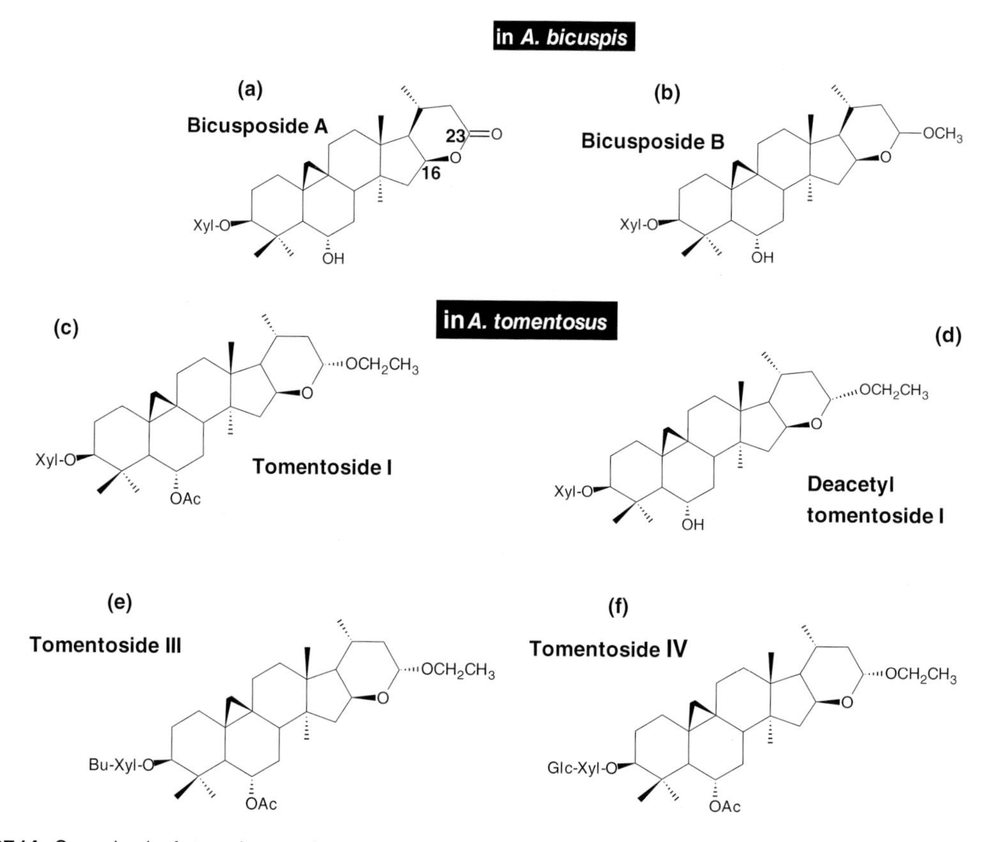

Fig. 27.14. Saponins in *Astragalus* species containing cycloartane with 16,23-epoxy terminal heterocycle.

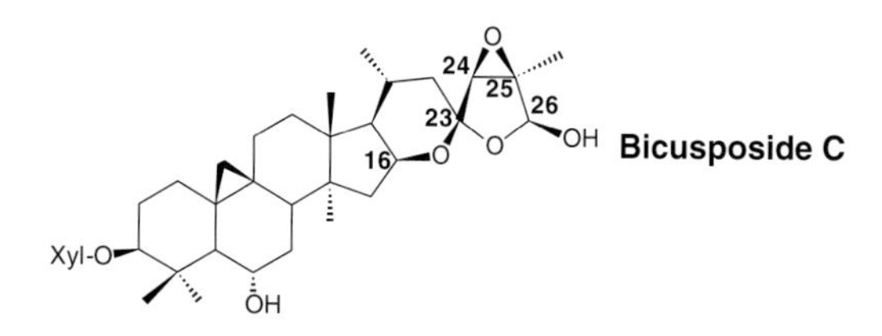

Fig. 27.15. Tri-epoxylated cycloartane-based saponin in *Astragalus bicuspis*, a unique triterpenic structure in the genus *Astragalus*.

2000; Çalis *et al.*, 2001; Semmar *et al.*, 2001; Sukhina *et al.*, 2007; Horo *et al.*, 2010; Horo *et al.*, 2012; Gülcemal *et al.*, 2011).

27.3.2 Chemical differentiation linked to desmosylation levels

Desmosylation returns to the number of ramifications showing glycosylation on aglycone. Using this chemical differentiation criterion, original saponins

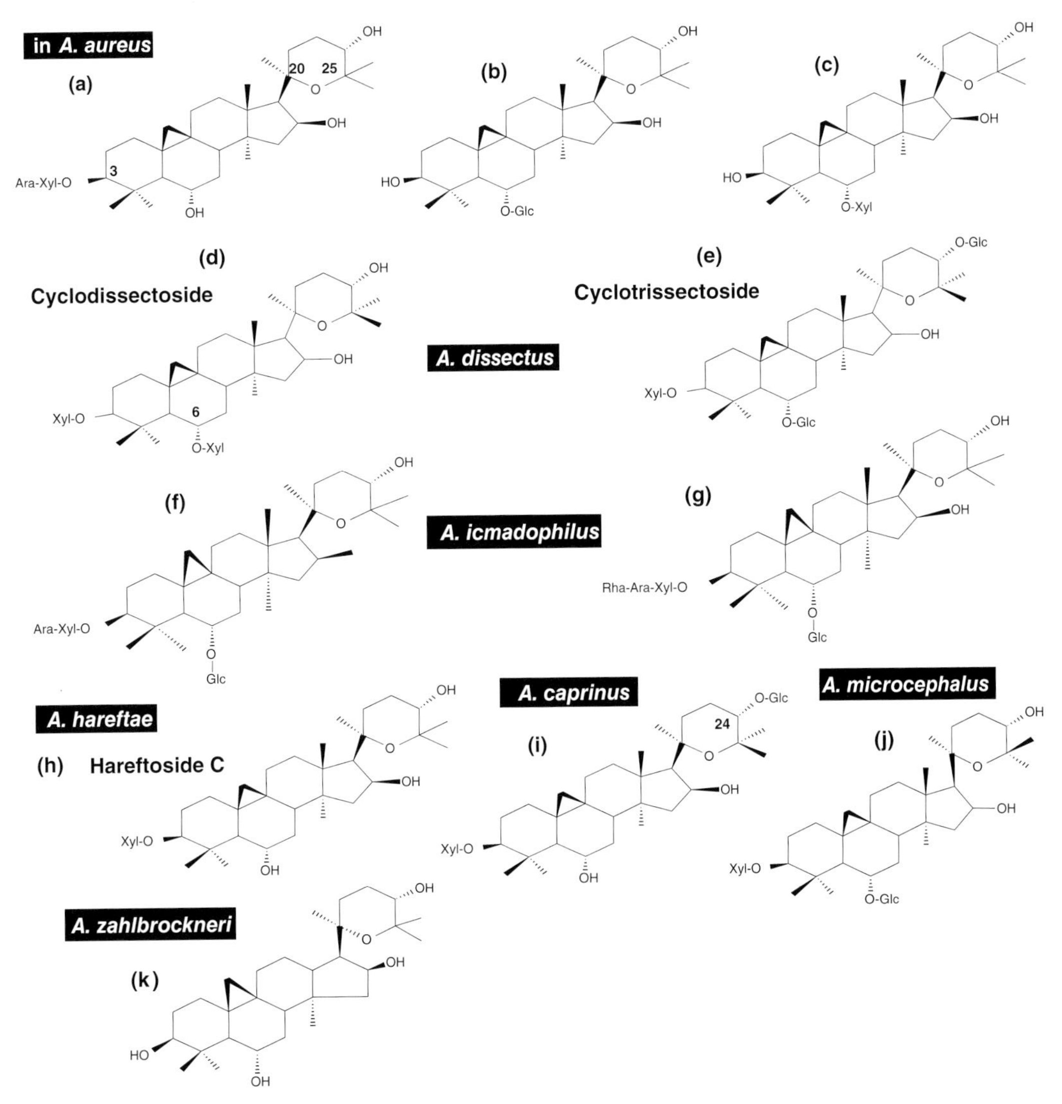

Fig. 27.16. Chemical structures of 20,25-epoxycycloartane-based saponins (a–j) and sapogenin (k) found in *Astragalus* species.

can be identified as molecules showing unusual desmosylation levels in the *Astragalus* genus. For that, the numbers of saponins with different desmosylation levels were graphically analysed by separately considering the three aglycones: 9,10-secocycloartane, oleanane and cycloartane. Plots helped to identify whether there were rare desmosylation levels for each of these three triterpene aglycones (Fig. 27.17).

Thus, it can be revealed that:

- The six saponins based on 9,10-secocycloartane showed only monodesmosidic structures.
- The 34 saponins based on oleanane were shared between two moieties including monodesmosides (18 saponins) and bidesmosides (16 saponins). Because there were equal moieties, no unusual cases were observed.
- Finally, the 179 saponins based on cycloartane were shared between 63, 93, 22 and 1 mono-, bi- tri- and tetradesmosides, respectively. From such a distribution, the 23 tri- and tetradesmosides represented 12.9% of cases and can be considered as unusual structures characterizing the corresponding *Astragalus* species.

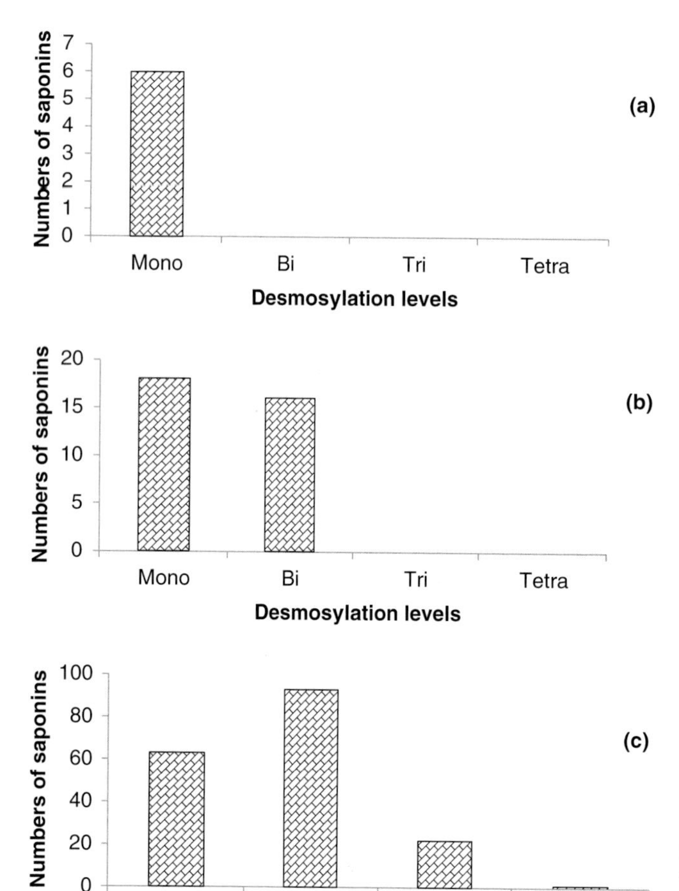

Fig. 27.17. Plots showing numbers of saponins with mono-, di-, tri- or tetradesmosylation levels for three sapogenins produced by *Astragalus* species: (a) 9,10-secocycloartane; (b) oleanane; (c) cycloartane.

After highlighting unusual cycloartane-based saponins due to high desmosylation levels, a graphical analysis of desmosylation levels was carried out by searching tri- and tetradesmosides within different types of cycloartane: (i) lateral chain cycloartane; (ii) 20,24-epoxycyclortane; (iii) other epoxylated cycloartanes (Fig. 27.18).

27.3.2.1 Chemical differentiation linked to desmosylation levels in lateral chain cycloartane

Of 74 saponins based on lateral chain cycloartane, 14 and 1 proved to be tri- and tetradesmosides, respectively (Fig. 27.18). The unique tetradesmoside concerned ciceroside A produced by *A. cicer* (Fig. 27.19) (Linnek *et al.*, 2011).

Apart from this unique tetradesmoside, 14 tridesmosides were found among the unusual saponins of *Astragalus*. These saponins were distinguished from the others by different glycosylation positions. Six tridesmosylation features were observed with the following decreasing rarity order: (C6, C16, C25) and (C6, C16, C24); (C3, C16, C24); (C3, C24, C25); (C3, C6, C25); (C3, C6, C24).

The (C6, C16, C25) and (C6, C16, C24) tridesmosylation features were unique in that they did not show glycosylation at C3 (quasi-glycosylated in *Astragalus* saponins). They were represented by two saponins characterizing *A. amblolepis* (Polat *et al.*, 2009).

Tridesmosylation at C3, C16, C24 concerned only one saponin, cephalotoside A, which was produced by *A. cephalotes* (Çalis *et al.*, 1999). Originality of this saponin increased by glycosylation

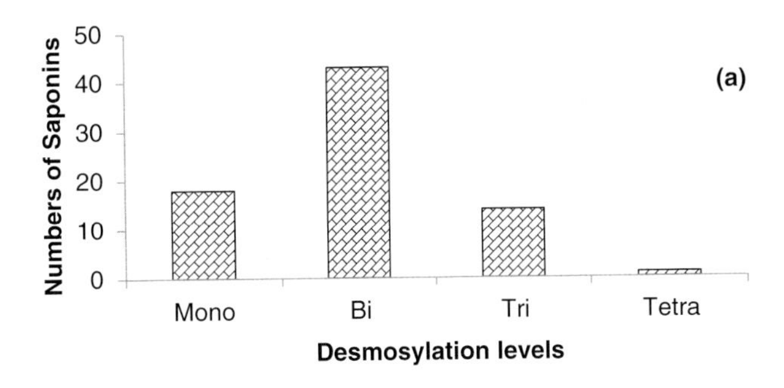

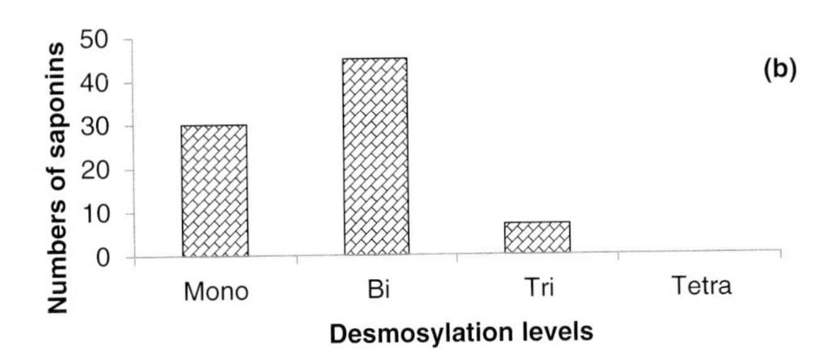

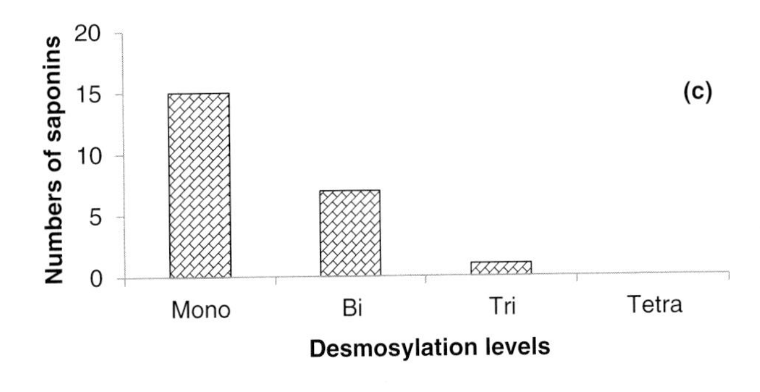

Fig. 27.18. Plots showing the numbers of saponins with mono-, di-, tri- or tetradesmosylation levels for three cycloartane types produced by *Astragalus* species: (a) lateral chain cycloartane; (b) 20,24-epoxycycloartane; (c) other epoxylated cycloartanes.

at C16 which proved to be less frequent than glycosylations at C3, C6 and C24.

The (C3, C24, C25) feature showed originality within tridesmosides, due to glycosylation at C25, which is less frequent than those occurring at C3, C6, C24. This feature was observed twice: in *A. amblolepis* (unnamed saponin) and *A. cicer* (cicero-side B) (Polat *et al.*, 2009; Linnek *et al.*, 2011).

Tridesmosylation at C3, C6, C25 concerned only two saponins characterizing *A. aureus* (Gülcemal *et al.*, 2011). This tridesmosidic feature revealed to be significantly less frequent than the (C3, C6, C24) one.

The (C3, C6, C24)-tridesmosidic feature was relatively more frequent than the five previous ones, but it contained rare saponins linked to additional structural criteria: the most unusual case concerned two saponins of *A. wiedemannianus*, which proved to be tridesmosides and tetraglycosides in addition to their acetylation at C16 (Polat *et al.*, 2010). Two other tridesmosidic saponins showed a rare occurrence because of their tetraglycosylation levels. They consisted of trojanosides E and F produced by *A. trojanus* (Bedir *et al.*, 1999). Finally, three other saponins showed the (C3, C6, C24) tridesmosylation feature and consisted of brachyoside C, hareftoside B and trojanoside D produced by *A. brachypterus*, *A. hareftae* and *A. trojanus*, respectively (Fig. 27.19) (Bedir *et al.*, 1998b, Bedir *et al.*, 1999; Horo *et al.*, 2012).

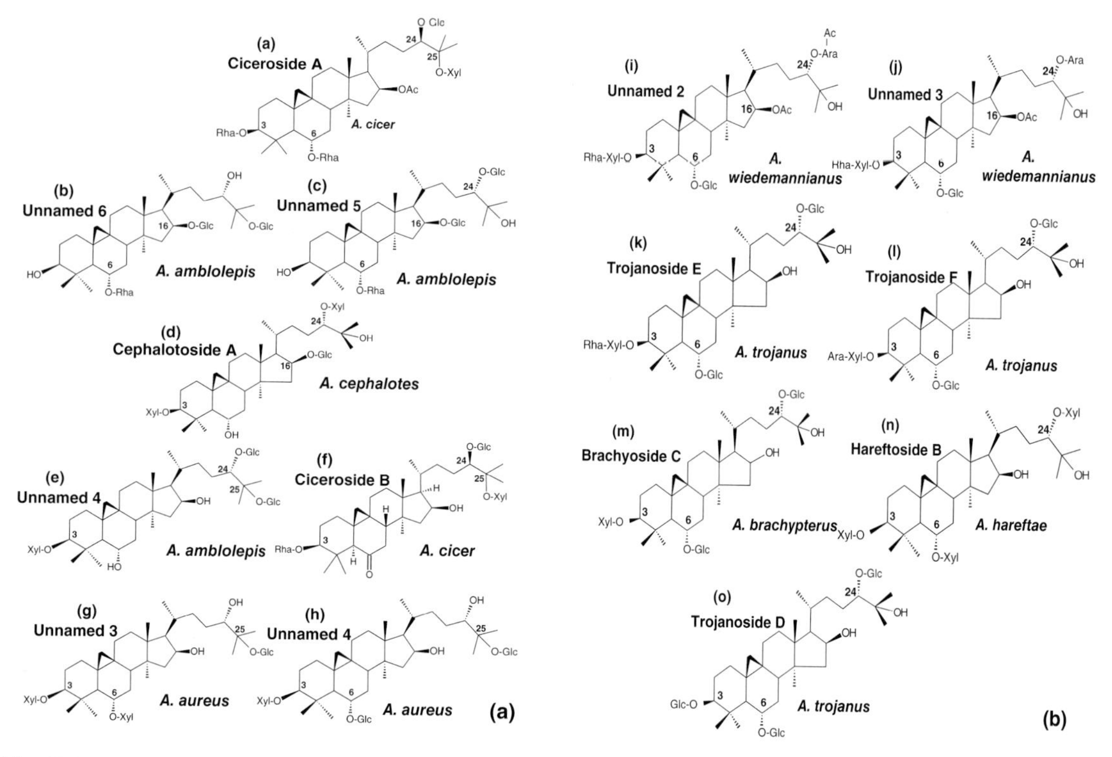

Fig. 27.19. Lateral chain cycloartane-based saponins found in *Astragalus* species and showing structural originality due to tetradesmosylation (a) or tridesmosylation (b–o). Legend: Ara = arabinose; Glc = glucose; Rha = rhamnose; Xyl = xylose.

27.3.2.2 Chemical differentiation linked to desmosylation levels in 20,24-epoxycycloartane

In the 20,24-epoxycycloartane, tridesmosylation was concerned with two attachment carbons' features: (C3, C6, C25) and (C3, C6, C16). The latter feature was significantly less frequent than the former.

- Tridesmosylation at C3, C6, C16 concerned only one saponin, trojanoside K, which characterized *A. trojanus* (Fig. 27.20) (Bedir *et al.*, 2001).
- The (C3, C6, C25) tridesmosylation feature concerned six saponins which were divided into two types, according to attached sugars. The first tridesmoside type consisted of (3-Xyl, 6-Glc, 25-Glc) and concerned three saponins: agroastragalosides III and IV found in *A. membranaceus*, (Zhou *et al.*, 1995), and astragaloside VII found in *A. membranaceus, A. ptilodes, A. trojanus, A. oldenburgi* and *A. dissectus* (Kitagawa *et al.*, 1983e; Bedir *et al.*, 2001; Sukhina *et al.*, 2007; Linnek *et al.*, 2011; Naubeev and Isaev, 2012). The second type consisted of (3-Xyl, 6-Xyl, 25-Glc) and concerned three saponins: trojanoside B characterizing *A. trojanus* and armatosides I and II found in *A. armatus* (Fig. 27.20) (Bedir *et al.*, 1999; Semmar *et al.*, 2010).

27.3.3 Analysis of chemical substitution types and levels in different sapogenins in *Astragalus* population

Beyond the aglycone type and desmosylation level, chemical substitutions provide a new dimension to identify original saponins, leading to more chemical differentiation of *Astragalus* species. Such a chemical originality is directly identified from some chemical groups showing low frequencies at the population scale. In a preliminary step, a plot of relative frequencies was established for all the substitutions whatever the aglycone in the whole studied *Astragalus* community (Fig. 27.21). This helped to identify rare substitutions at this global scale. Then, frequency distributions of different substitutions were analysed in detail by reference to each aglycone type.

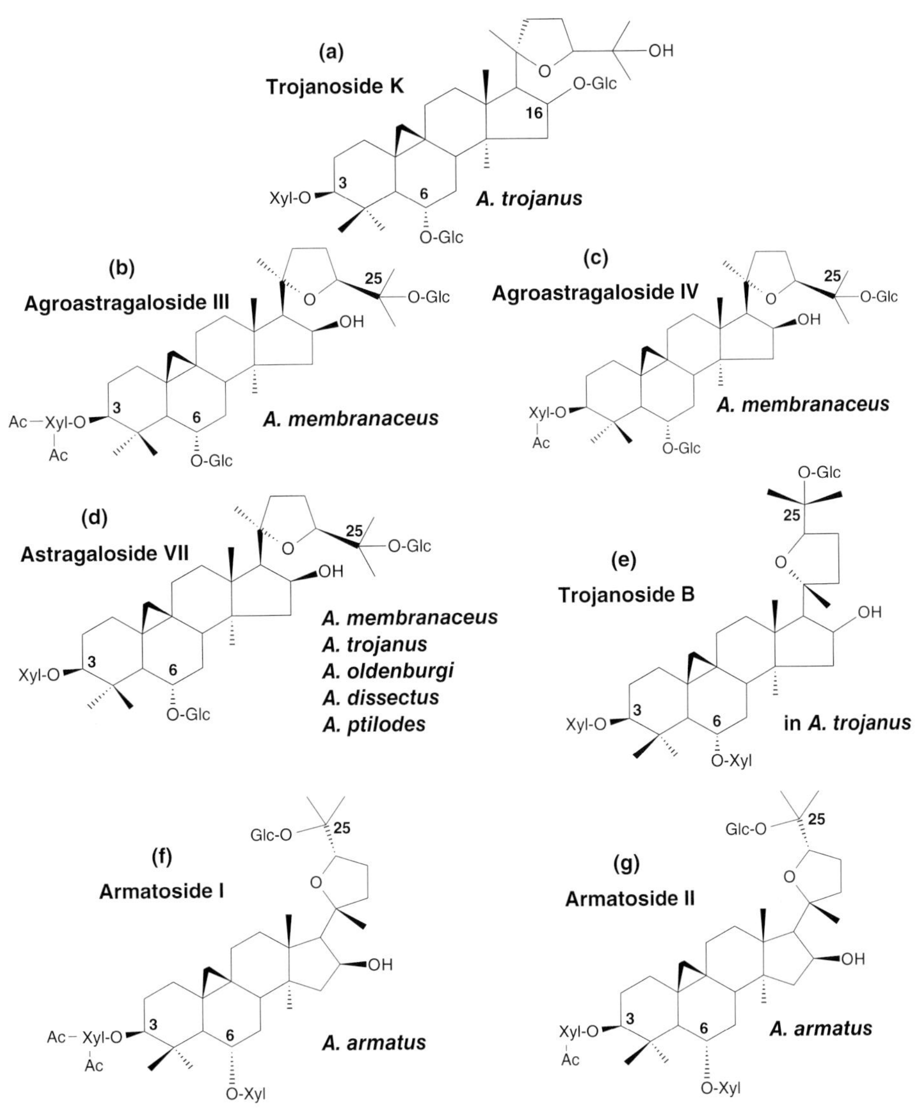

Fig. 27.20. 20,24-epoxycycloartane-based saponins found in *Astragalus* species and showing structural originality due to tridesmosylation (a–g). Legend: Ara = arabinose; Glc = glucose; Rha = rhamnose; Xyl = xylose.

According to their relative frequencies, the different chemical substitutions can be ranged into four occurrence levels: highly frequent (>20%), commonly frequent (10≤<20%), few frequent (1≤<10%) and rare (<1%) (Fig. 27.21).

- The highest occurrence levels concerned hydroxyl with relative frequency close to 40%. Free hydroxyls are known to be attached in the upstream of metabolism to favour other types of substitutions in the following steps (Sawai and Saito, 2011; Turgut-Kara and Ari, 2011.
- Frequent substitutions concerned epoxyl (9.1%), glucose (12. 9%) and xylose (13.1%).
- Few or less frequent substitutions included acetyl (6.2%), rhamnose (5.0%), glucuronic acid (2.7%), arabinose (3.2%) and ketone (1.6%).
- Rare chemical substitutions included methyl (0.9%), glacatose (0.5%), methanoic acid (0.4%), apiose (0.2%), ethyl (0.2%), butenoyl (0.1%), hydroxyacetoxyl (0.1%) and isopropylidenedioxyl (0.1%).

These rare chemical groups provide additional dimensions for more chemotaxonomical characterization

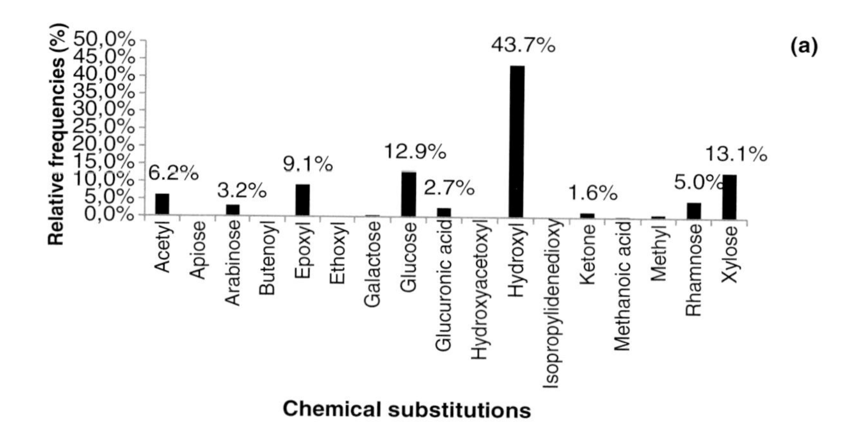

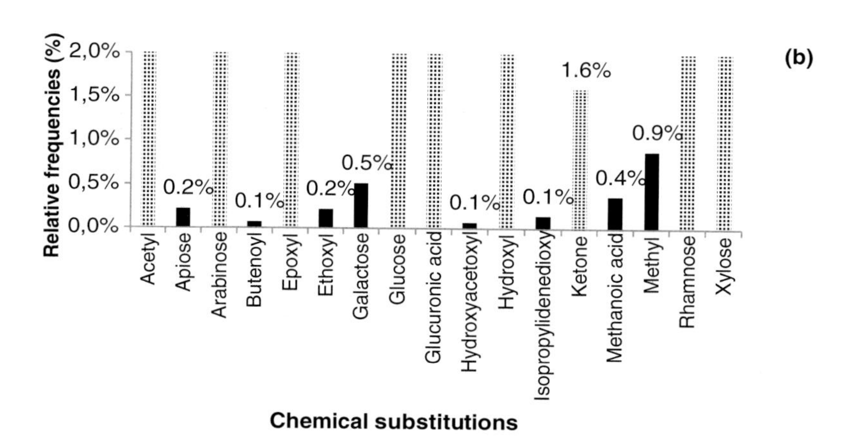

Fig. 27.21. Relative frequencies of different chemical substitutions in 240 saponins found in 53 species of the genus *Astragalus* chemically studied between 1981 and 2014. (a) Major substitutions; (b) minor substitutions.

of productive *Astragalus* plants. Distribution analysis of chemical substitutions through different saponins was carried out by separately considering different types of aglycones. Using this method, some chemical groups seemed to be common or specific to some sapogenins (Fig. 27.22).

27.3.3.1 Analysis of chemical substitution types in 9,10 secocycloartane

The 9,10 secocycloartane showed only four substitution types (acetyl, ketone, glucose, hydroxyl), among which acetyl represented 34.2% vs 6.5% and 0% (absence) in cycloartane and oleanane, respectively. Also, ketone showed relatively high frequency (10.5%) in 9,10 secocycloartane, compared to its relative occurrence in cycloartane (1%) and oleanane (3.3%) (Fig. 27.22). Saponins showing more occurrences of acetyl and ketones were observed in *A. membranaceus*, including huangqiyen in E–J (Fig. 27.6) (Kuang *et al.*, 2009; Kuang *et al.*, 2011). Finally, hydroxyls occurred widely and comparably in the three sapogenin types, with 39.5%, 45.2% and 35.4% in 9,10 secocycloartane, cycloartane and oleanane, respectively (Fig. 27.22). The most illustrative chemical structure is given by the secomacrogenin B of *A. macropus* followed by huangqiyenins E–J of *A. membranaceus* (Fig. 27.6) (Kuang *et al.*, 2009; Isaev *et al.*, 2010a; Kuang *et al.*, 2011).

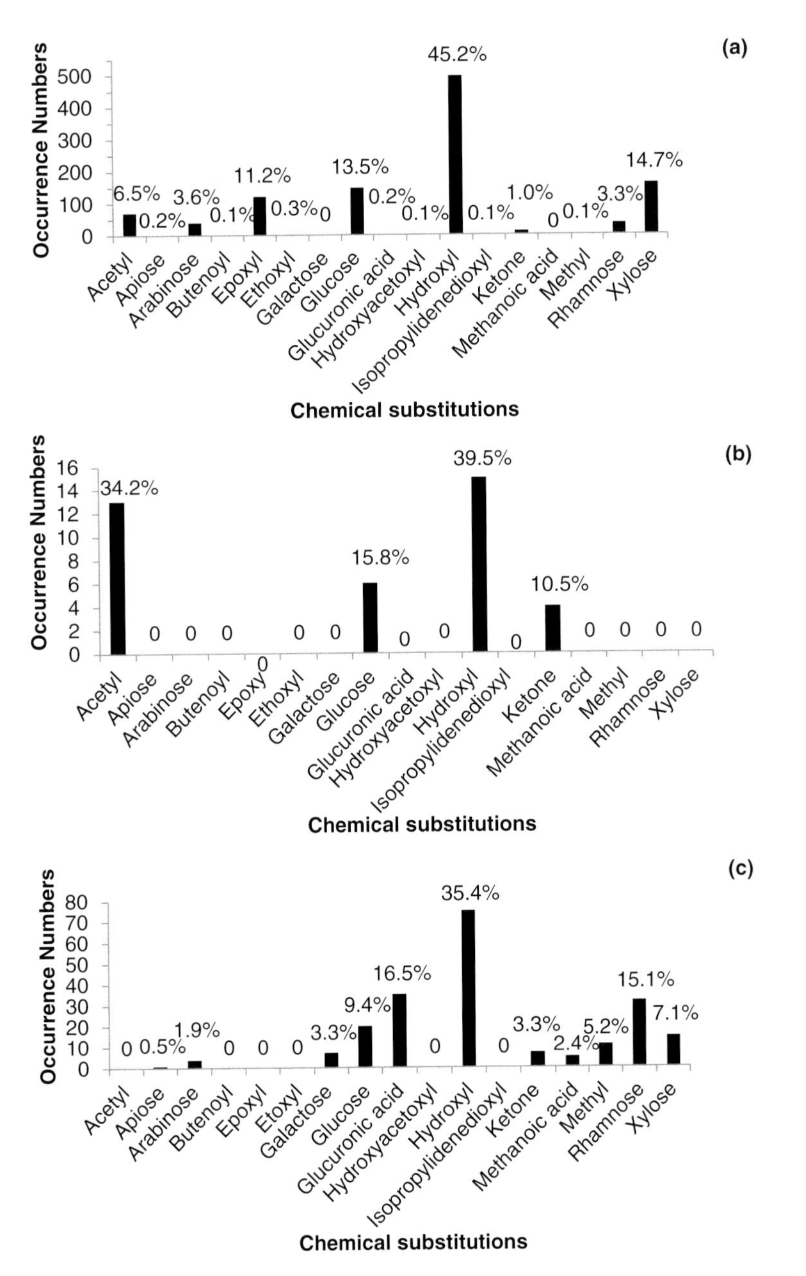

Fig. 27.22. Occurrence numbers and relative frequencies of different chemical substitutions in three different sapogenins more or less observed in the genus *Astragalus*: (a) Cycloartanes; (b) 9,10-secocycloartane; (c) Oleanane.

27.3.3.2 Analysis of chemical substitution types in oleanane

Oleanane-based saponins were generally characterized by glucuronic acid (16.5%), rhamnose (15.1%), methyl (5.2%) and galactose (3.3%), which was significantly less present or entirely absent in cycloartane and 9,10 secocycloartane (Fig. 27.22). The general distributions of glucuronic acid and rhamnose in oleanane-based saponins of *Astragalus* are illustrated by the saponins of *A. tauricolus* (Fig. 27.7), *A. sinicus* (Fig. 27.8), *A. complanatus* (Fig. 27.9), *A. flavescens* (Fig. 27.10), *A. membranaceus*,

A. trojanus and *A. caprinus* (Fig. 27.11) (Kitagawa *et al.*, 1983e; Cui *et al.*, 1992a; Cui *et al.*, 1992b; Bedir *et al.*, 1999; Mitaine-Offer *et al.*, 2006; Avunduk *et al.*, 2008; Gülcemal *et al.*, 2013). When substitution by methyl was observed, it occurred on glucuronic acid; this feature concerned saponins of *A. complanatus* (Fig. 27.9) and in *A. sinicus* (Fig. 27.8). Galactose was directly attached on glucuronic acid of different saponins in both *A. sinicus* and *A. complanatus* (Fig. 27.8 and 27.9) (Cui *et al.*, 1992a; Cui *et al.*, 1992b).

By considering the high occurrence level of glucuronic acid and rhamnose in oleanane-based saponins, absence of such chemical substituents could provide an interesting criterion to identify original saponins. Absence of glucuronic acid was not observed until now, whereas that of rhamnose concerned two saponins of *A. tauricolus* (Fig. 27.7), three saponins of *A. sinicus* (Fig. 27.8) and one saponin of *A. hareftae* (hareftoside E) (Fig. 27.11) (Cui *et al.*, 1992b; Horo *et al.*, 2012; Gülcemal *et al.*, 2013).

Besides the most frequently occurring chemical groups, some substitutions occurred with low frequencies in oleanane, leading to the specific characterization of some productive plant species. They concerned methanoic acid (2.4%) and apiose (0.5%) (Fig. 27.22). Methanoic acid characterized *A. tauricolus* and was directly substituted in the 29-position of five saponins (Fig. 27.7) (Gülcemal *et al.*, 2013). Apiose in oleanane was exclusively observed in *Astragalus caprinus* (Fig. 27.11) (Mitaine-Offer *et al.*, 2006).

27.3.3.3 Analysis of rare chemical substitutions in cycloartane

Compared to 9,10 secocycloartane and oleanane, cycloartane showed several substitutions with low frequencies (≤1%), most of them being absent in 9,10 secocycloartane and oleanane (Fig. 27.22). Rare substitutions include ketone (1%), ethoxyl (0.3%), hydroxyacetoxyl (0.1%), isopropylidenedioxyl (0.1%) and butenoyl (0.1%), in addition to apiose (0.2%) which was revealed to be also rarely observed in oleanane (0.5%). Analysis of chemical substitutions in cycloartane was carried out by separately considering the different forms of this sapogenin (see Fig. 27.23), including:

- Cycloartane with terminal lateral chain (Fig. 27.4)
- Cycloartane with terminal 20,24-epoxylated heterocycle (Fig. 27.4)
- Less frequent or rare forms of cycloartane (Fig. 27.4).

27.3.3.3.1 ANALYSIS OF RARE CHEMICAL SUBSTITUTIONS IN LATERAL CHAIN CYCLOARTANE In cycloartane containing terminal lateral chain, five saponins and two sapogenins were revealed to be highly original because of the occurrence of glucuronic acid, isopropylidenedioxyl and ketone which represented 0.5%, 0.2% and 0.9% of the whole substitution set, respectively (Fig. 27.23).

- Glucuronic acid occurred in *A. taschkendicus* (roots; askendoside K) and *A. amblolepis* (roots; unnamed saponin) (Fig. 27.24) (Polat *et al.*, 2009; Isaev and Isaev, 2011a).
- Ketone was observed in *A. cicer* (aerial parts; ciceroside B), *A. membranaceus* (leaves; huangqiyenin B) and *A. taschkendicus* (cycloasgenin B, 3-dehydrocycloasgenin C) (Fig. 27.24) (Isaev *et al.*, 1984; Isaev *et al.*, 1985b; Ma *et al.*, 1997; Linnek *et al.*, 2011).
- Isopropylidenedioxyl was observed in *A. macropus* (roots; cyclomacroside A) (Fig. 27.24) (Iskenderov *et al.*, 2009d).

Less rare chemical groups in lateral chain cycloartane included acetyl, which represented about 4% among all the observed substitutions (Fig. 27.23). Acetyls were generally substituted on sugars, which are directly attached to the aglycone (Fig. 27.25). Nine *Astragalus* species showed occurrence of acetyl on lateral chain cycloartane: *A. kahiricus*, *A. cicer*, *A. mongholicus* (aerial parts), *A. icmadophilus*, *A. wiedemannianus* (whole plant), *A. macropus*, *A. oleifolius*, *A. membranaceus* (roots) and *A. tragacantha* (stem) (Isaev *et al.*, 1992; Zhu *et al.*, 1992; Hirotani *et al.*, 1994; Çalis *et al.*, 1996; Radwan *et al.*, 2004b; Iskenderov *et al.*, 2009c; Horo *et al.*, 2010; Polat *et al.*, 2010; Linnek *et al.*, 2011). Among the 14 acetylated saponins, those showing acetylation at C1, C6 and C7 seemed to be less frequent than those acetylated at other positions (particularly C3) (Fig. 27.25).

27.3.3.3.2 ANALYSIS OF RARE CHEMICAL SUBSTITUTIONS IN 20,24-EPOXYCYCLOARTANE In cycloartane containing 20,24-epoxylated cycle, three chemical groups showed the lowest relative frequencies: hydroxyacetoxyl (0.2%), apiose (0.5%), and ketone (1.4%), followed by arabinose (2.5%) and rhamnose (3.2%) (Fig. 27.23).

- Apiose was observed in cycloaraloside C (also named astrailienin A) produced by *A. amarus*, *A. illiensis*, *A. unifoliolatus* (roots) and *A. verrucosus*

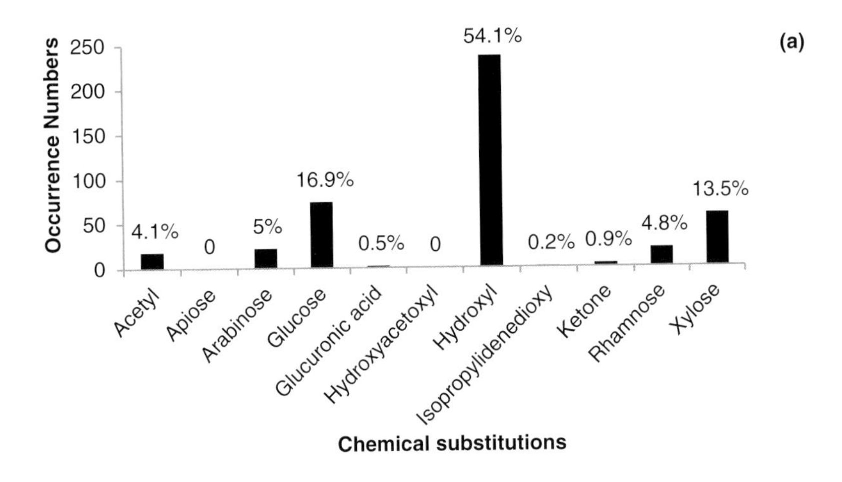

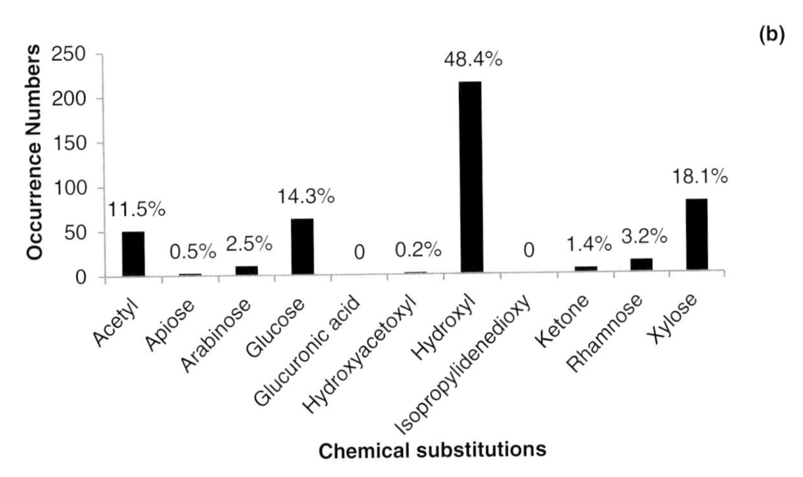

Fig. 27.23. Occurrence numbers and relative frequencies of different chemical substitutions in saponins of the genus *Astragalus* based on the two most frequent forms of cycloartane. (a) Cycloartane with terminal chain (79 saponins, 24 plant species); (b) 20,24-epoxycycloartane (91 saponins, 33 plant species).

(aerial parts) as well as cycloaraloside F, found in *A. amarus* (roots) and *A. villosissilmus* (Fig. 27.26) (Isaev and Abubakirov, 1990b; Yu-Qun *et al.*, 1990; Isaev and Imomnazarov, 1991; Kucherbaev *et al.*, 2002; Pistelli *et al.*, 2003).

- Hydroxyacetoxyl was observed in *A. campylosema* (roots; unnamed saponin) (Fig. 27.26) (Çalis *et al.*, 2008b).
- Ketone was observed in cycloartanes of *A. alopecurus* (aerial parts; cycloalpigenin A, cycloalpioside A), *A. pycanthus* (seeds; cyclopycnanthogenin), *A. membranaceus* (leaves; Huangqiyenin A), *A. taschkendicus* (cycloasgenin A), *A. zahlbrukneri* (roots; unnamed saponin) (Fig. 27.26) (Isaev *et al.*, 1982a, Agzamova and Isaev, 1994a; Ma *et al.*, 1997; Agzamova and Isaev, 1998b; Çalis *et al.*, 2001).
- Arabinose was found in nine saponins shared between *A. campylosema* (three saponins), *Astragalus armatus*, *A. elongatus*, *A. trigonus*, *A. taschkendicus* (two saponins), *A. icmadophilus*, *A. halicacabus* and *A. trojanus* (one saponin) (Fig. 27.27) (Isaev *et al.*, 1983a; Isaev *et al.*, 1983b; Isaev, 1991; Gariboldi *et al.*, 1995; Çalis *et al.*, 2008a; Çalis *et al.*, 2008b; Horo *et al.*, 2010; Semmar *et al.*, 2010; Djimtombaye *et al.*, 2013).
- Rhamnose was found in 15 saponins synthesized by *Astragalus amarus*, *A. coluteocarpus*, *A. wiedemannianus*, *A. oldenburgii*, *A. alexandrinus* (one saponin), *A. trojanus*, *A. ernestii* (three saponins), *A. verrucosus* (four saponins) and *A. sieversianus* (five saponins) (Figs 27.28 and 27.29) (Li-Xian *et al.*, 1986, Wang *et al.*, 1989; Imomnazarov and Isaev, 1992; Orsini

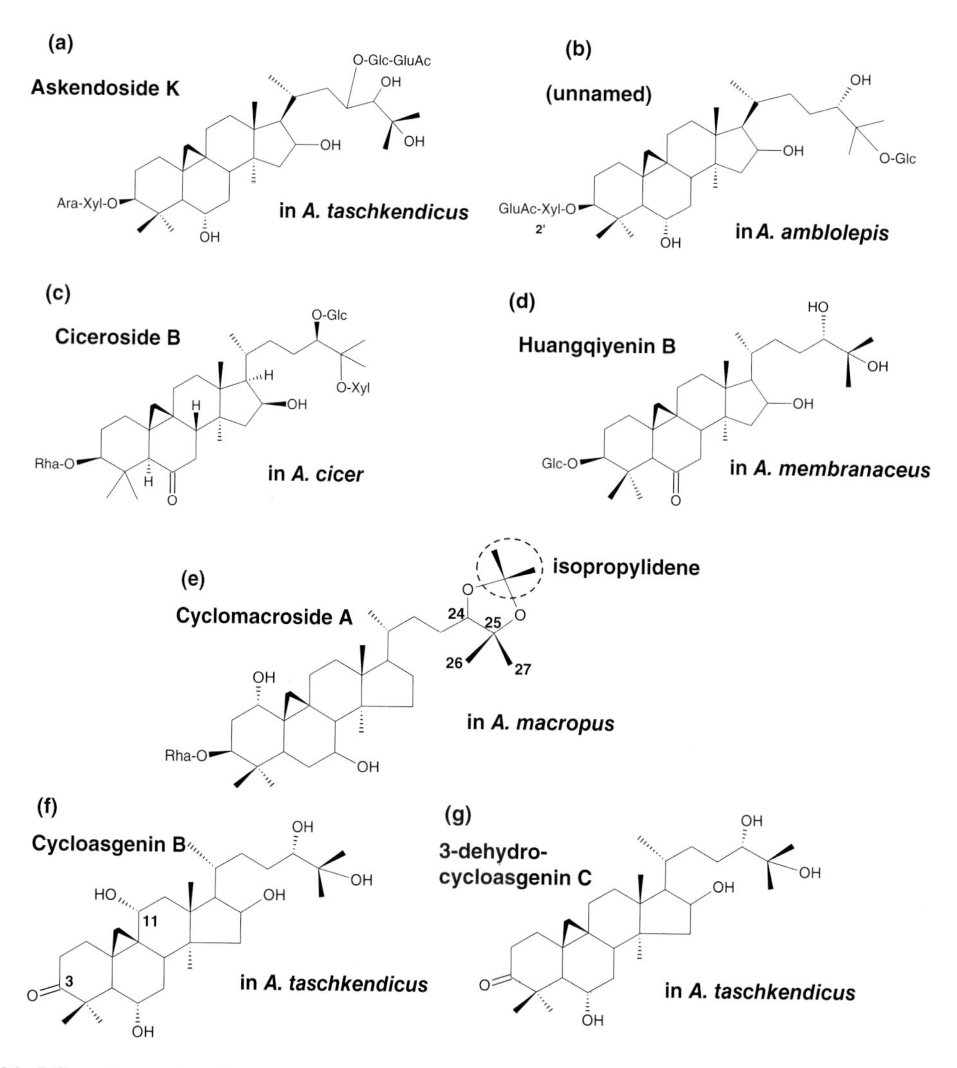

Fig. 27.24. Different saponins of *Astragalus* species showing original structures due to some rare substitutions in lateral chain cycloartane: occurrence of glucuronic acid (GluAc) (a, b), ketone (c, d, f, g), isopropylidenedioxyl (e).

et al., 1994; Pistelli *et al.*, 1998; Bedir *et al.*, 2001; Pistelli *et al.*, 2003; Iskenderov *et al.*, 2008b; Polat *et al.*, 2010; Naubeev and Isaev, 2012). The low relative frequence of rhamnose seemed to be linked to its quasi-exclusive occurrence in carbon 3, except cyclocarposide B of *A. coluteocarpus* (Fig. 27.28) (Imomnazarov and Isaev, 1992).

27.3.3.3.3 ANALYSIS OF RARE CHEMICAL SUBSTITUTIONS IN OTHER CYCLOARTANE TYPES

Apart from the most frequent cycloartanes (with lateral chain or with 20,24-epoxylated cycle), original chemical substitutions concerned also rare cycloartane forms including 16, 23-epoxylated, 20, 25-epoxylated and di-epoxylated cycloartanes (Fig. 27.30).

In 16,23-epoxylated cycloartane, ethoxyl, methoxyl and butenoyl occurred at relative frequencies of 17.6, 5.9 and 5.9%, respectively, whereas they were absent in the other cycloartane forms (Fig. 27.30); also, ketone (5.9%) occurred in 16, 23-epoxylated cycloartane among the rarely observed chemical substitutions in cycloartane forms.

- Ethoxyl (OC_2H_5) was observed in four saponins of *A. tomentosus* (aerial parts; tomentoside I,

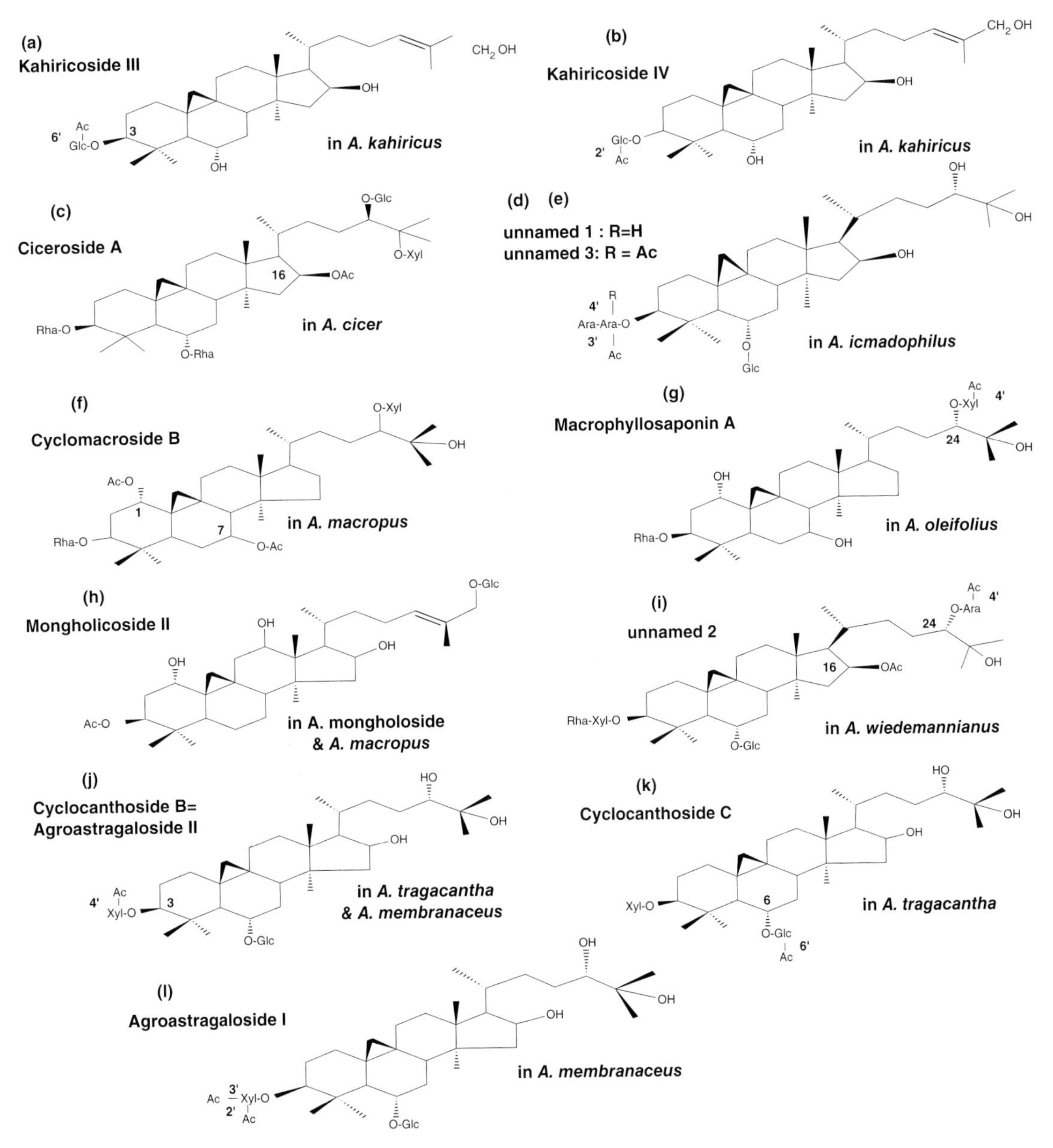

Fig. 27.25. Original saponins of *Astragalus* species due to acetylation of lateral chain cycloartane. Legend: Ac = acetyl, Xyl = xylosyl, Glc = glucosyl, Rha = rhamnosyl, Ara = arabinosyl.

deacetyltomentoside I, tomentoside III, tomentoside V) (Fig. 27.31) (Radwan *et al.*, 2004a).

- Butenoyl (Bu) occurred in tomentoside III of *A. tomentosus* (Fig. 27.31) (Radwan *et al.*, 2004a).
- Methoxyl (OCH_3) was observed in bicusposide B produced by *A. bicuspis* (whole plant) (Fig. 27.31) (Choudhary *et al.*, 2008).
- Ketone (=O) was also observed in *A. bicuspis* through the saponin bicusposide A (Fig. 27.31) (Choudhary *et al.*, 2008).

In 20, 25-epoxycycloartane, chemical substitutions with relatively low frequencies included arabinose (6.3%) and rhamnose (2.1%) (Fig. 27.30).

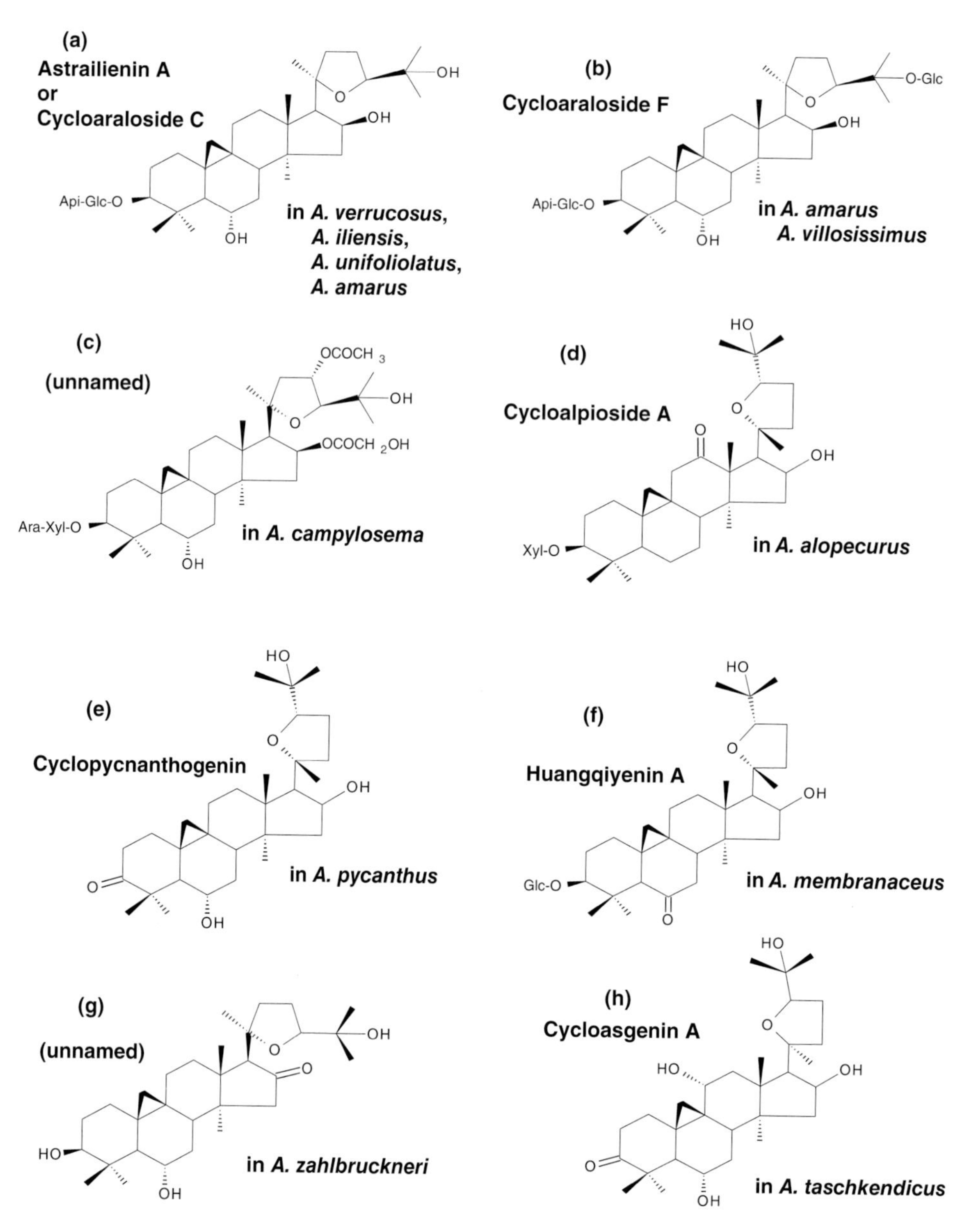

Fig. 27.26. Saponins of species of *Astragalus* based on 20,24-epoxycycloartane and showing rare chemical substitutions including apiose (a, b), hydroxyacetoxyl (c) and ketone (d–h). Legend: Api = apiose; Glc = glucose; Ara = arabinose; Xyl = xylose.

- Arabinose occurred in two saponins of *A. icmadophilus* (whole plant) and one saponin of *A. aureus* (whole plant) (Fig. 27.32) (Horo *et al.*, 2010; Gülcemal *et al.*, 2011).
- Rhamnose occurred in one saponin of *A. icmadophilus* (Fig. 27.32) (Horo *et al.*, 2010).

27.3.4 Chemotaxonomical analysis at attachment carbon scale

After highlighting of original saponins based on: (1) rare aglycones; (2) desmosylation levels; or (3) rare chemical substitutions, another chemotaxonomical dimension can be brought in by considering

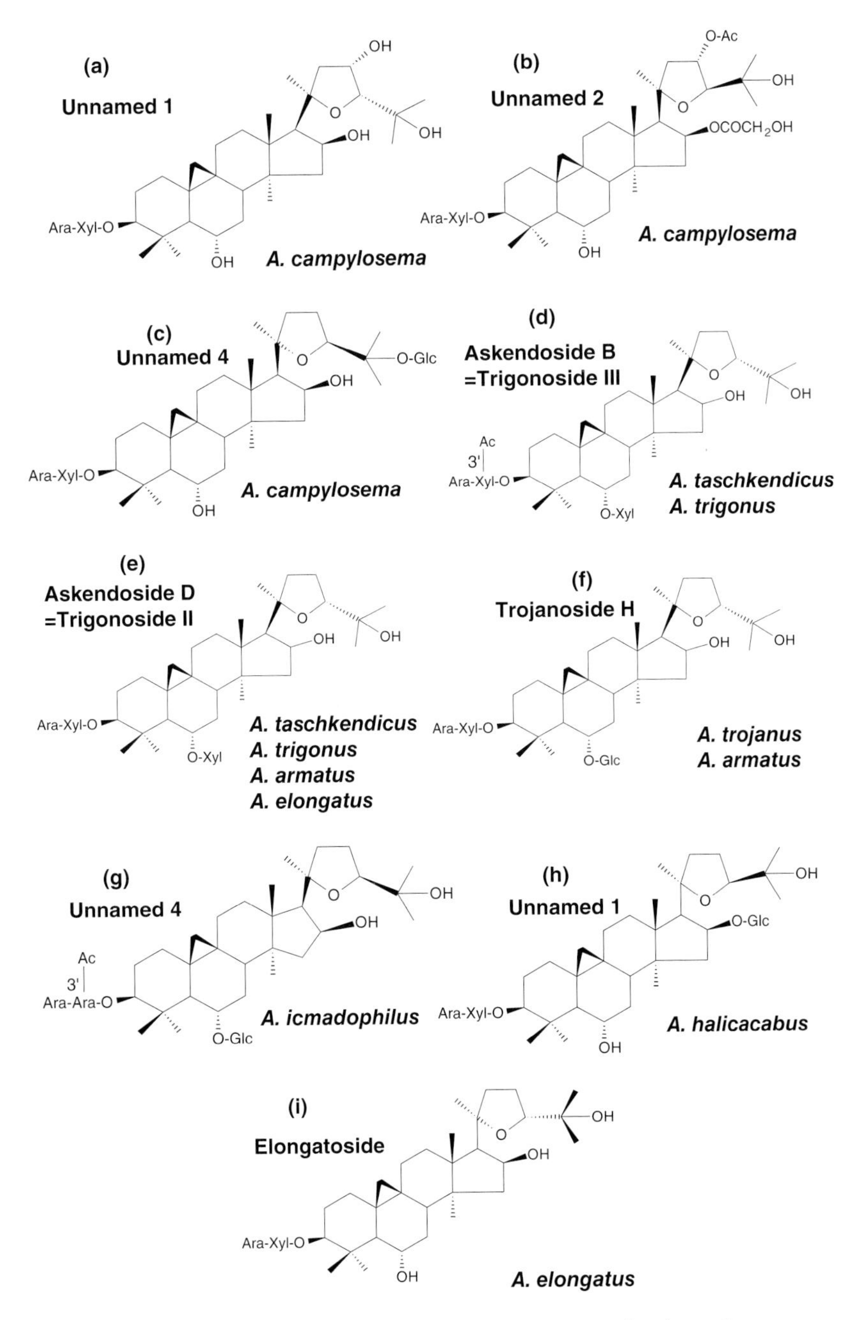

Fig. 27.27. Original 20,24-epoxycycloartane-based saponins of *Astragalus* species due to the occurrence of arabinose (unusual sugar observed in nine saponins among 91 based on 20,24-epoxylated cycloartane).

uncommon attachment positions of common chemical groups in common aglycones. This fourth criterion enables the characterization of further saponins for chemical differentiation of more *Astragalus* species.

By plotting the frequencies of different chemical groups found in cycloartane (all confounded forms), hydroxyl (45.2%), epoxyl (11.2%), glucose (13.5%) and xylose (14.7%) were revealed to be the most frequent (Fig. 27.33). The percentage of

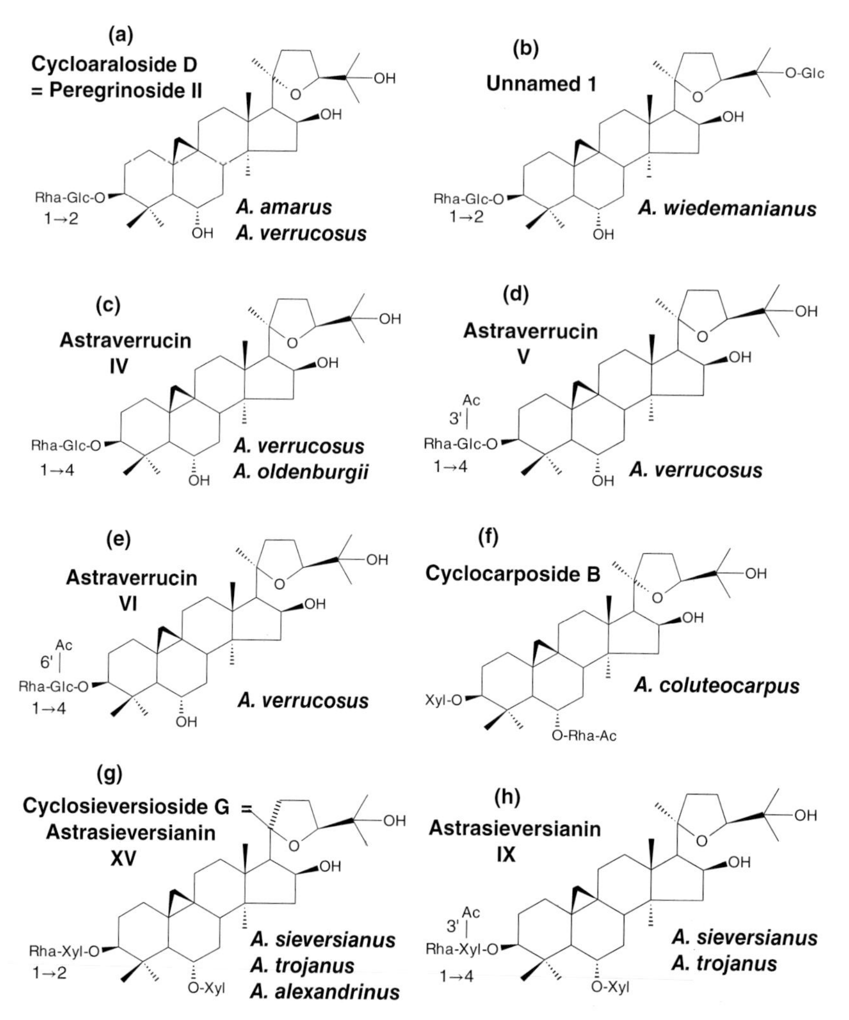

Fig. 27.28. (a–h) 20,24-epoxycycloartane-based saponins showing original structures due to the occurrence of rhamnose (a sugar observed in 15 saponins among 91 based on 20,24-epoxycycloartane).

each chemical group was calculated by adding together its occurrences in all the saponins, then dividing by the number of substitutions (observed in all the carbons of all the saponins). Epoxyl, naturally attached on two carbons, was counted once.

Epoxyl was not considered in this section, because it was shown to play a role at aglycone level, leading to previously discussed heterocyclic cycloartane forms. By considering xylose, glucose and hydroxyl, original saponins were identified from rare attachment positions of these very frequent chemical groups. Rare attachment positions were deduced from low percentages calculated by dividing the number of occurrences of a given chemical group at a given carbon (in all the saponins) by the number of its occurrences in all the carbons of all the saponins (Fig. 27.34).

- Xylose revealed to be rarely substituted on C25 (1.2%) (Fig. 27.34).
- Glucose was rarely observed on C23 (1.3%) and C27 (2%) (Fig. 27.34).
- Hydroxyl showed several rare positions including C11 (0.4%) and C26 (0.2%), C20, C23 and C27 (0.6%), C12 (1%) (Fig. 27.34).

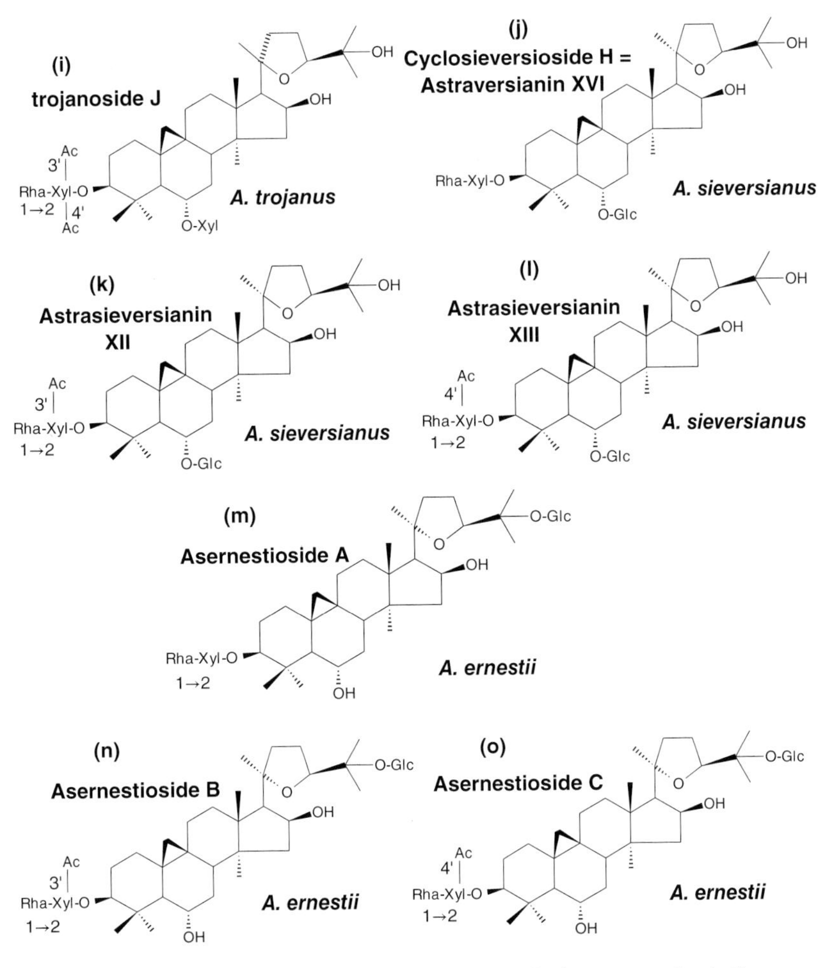

Fig. 27.29. (i –o) 20,24-epoxycycloartane-based saponins showing original structures due to the occurrence of rhamnose (a sugar observed in 15 saponins among 91 based on 20,24-epoxycycloartane).

Other less frequent chemical groups showed also rare occurrence at some carbons compared to other carbons.

The acetyl group was mainly attached on C3 (82.6%) and was less frequent in C16 (7.2%), C6 (2.9%), C24 (2.9%) and C1, C7, C23 (1.4%) (Fig. 27.35). Moreover, acetylation frequency was revealed to be highly dependent on cycloartane type: it was significantly more present in 20,24-epoxycycloartane than lateral chain cycloartane. Acetylated structures of lateral chain cycloartane were discussed in section 27.3.3.3.1, showing original saponins based on rare substitutions. However, in the case of 20,24-epoxycycloartane where acetyl is more frequent, original saponins will be identified on the basis of rare acetylation positions, i.e. C6 (2.9%), C16 (7.2%) and C23 (1.4%) (Fig. 27.35).

Rare substitution positions in cycloartane-based saponins were analysed with more precision by separately considering the two most frequent forms of cycloartane.

- Lateral chain cycloartane (Fig. 27.36).
- 20,24-epoxycycloartane (Fig. 27.40).

27.3.4.1 Analysis of rare substitution positions in lateral chain cycloartane

Cycloartane with lateral chain showed several original saponins due to rarely substituted carbons by common chemical groups: xylose (13.5%),

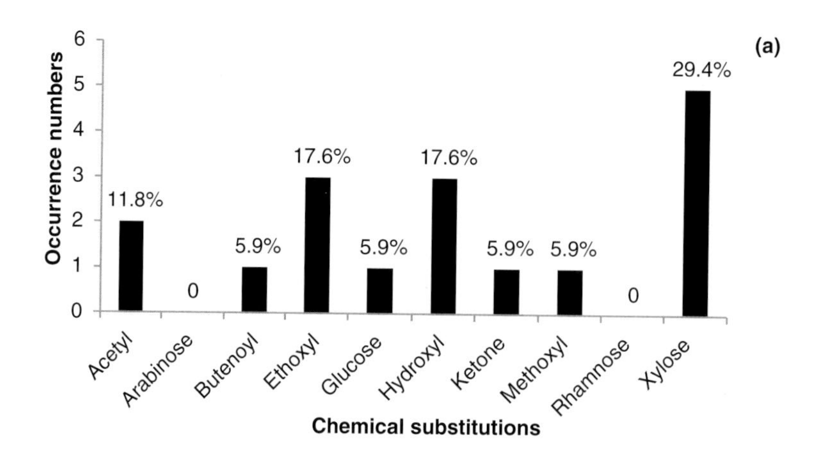

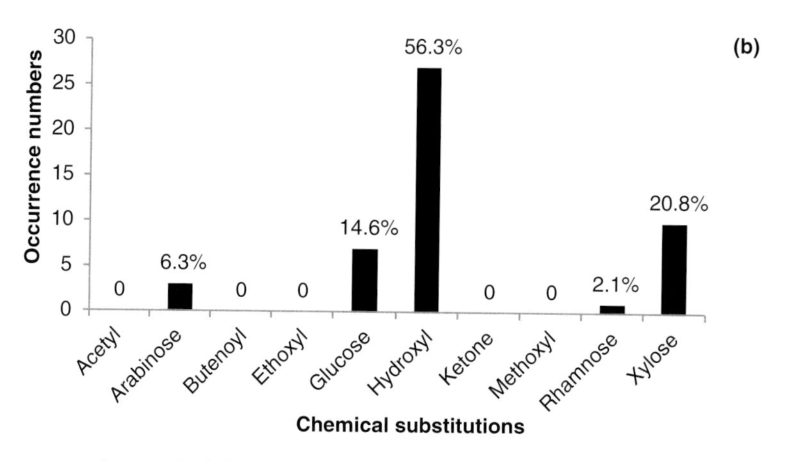

Fig. 27.30. Occurrence numbers and relative frequencies of different chemical substitutions in saponins of *Astragalus* based on epoxylated cycloartanes at the carbons C16, C23 (a) and C20, C25 (b).

glucose (16.9%) and hydroxyl (54.1%) in the aglycone (Figs 27.36 and 27.37).

- Hydroxyl showed five rare substitution positions including C11, C12 and C23 (0.4%), C20 and C27 (1.3%) (Fig. 27.36).
  - Hydroxylation at C11 was concerned with a single saponin characterizing *A. taschkendicus* (cycloasgenin B) (Fig. 27.24) (Isaev *et al.*, 1984).
  - The 12-OH substitution was observed in mongholicoside II of *A. mongholicus* (Fig. 27.38) (Zhu *et al.*, 1992).
  - Hydroxylated C23 was observed in cycloorbicoside D of *A. orbiculatus* (epigeal parts) (Fig. 27.39) (Mamedova *et al.*, 2005).
  - The 20-OH concerned three unnamed saponins of *A. stereocalyx* (roots) (Fig. 27.39) (Yalçın *et al.*, 2012).
  - Hydroxylated C27 was observed in kahiricosides II–IV of *A. kahiricus* (Fig. 27.39) (Radwan *et al.*, 2004b).
- Xylose was rarely substituted on C25 (3.4%) (Fig. 27.36). This structural feature concerned two saponins, cicerosides A and B, produced by *A. cicer* (aerial parts) (Fig. 27.38) (Linnek *et al.*, 2011).
- Glucose was rarely observed on C23 (two saponins) and C27 (three saponins) (Fig. 27.36). The 23-Glc structure concerned askendoside K and askendoside H produced by *A. tacshkendicus* (roots), whereas 27-Glc structures were observed in kahiricoside V of *A. kahiricus* (aerial parts) and in mongholicosides I and II produced by *A. mongholicus* (aerial parts) (Fig. 27.38) (Zhu *et al.*, 1992; Radwan *et al.*, 2004b, Isaev and Isaev, 2011a; Isaev and Isaev, 2011b).

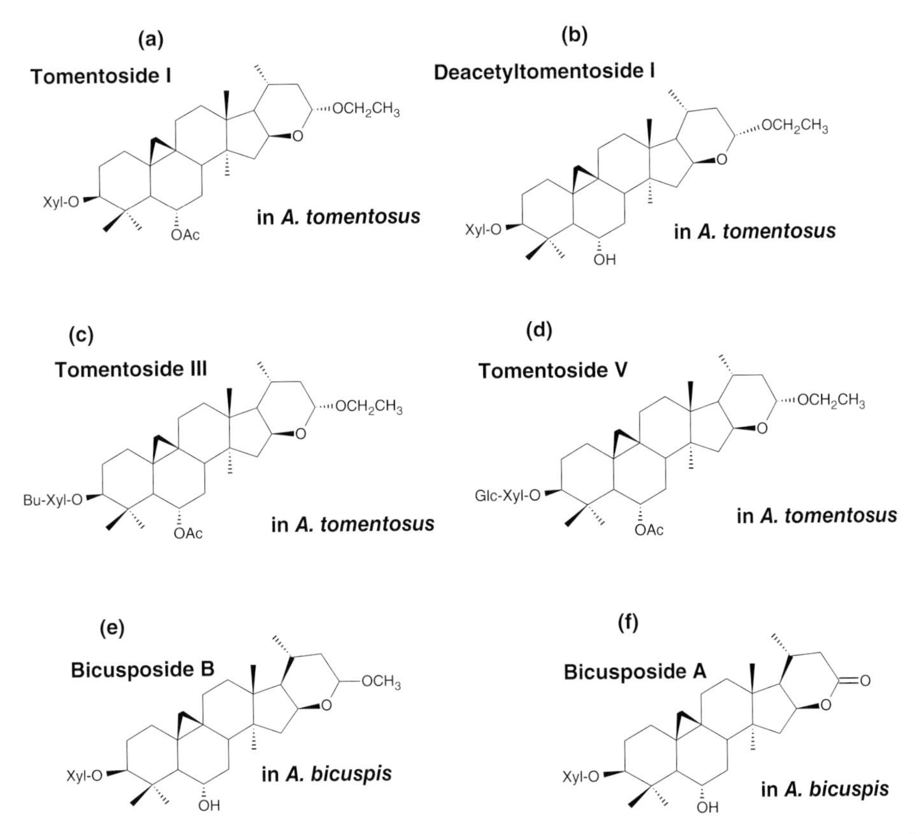

Fig. 27.31. Saponins of species *Astragalus* based on 16,23-epoxycycloartane and showing original chemical substitutions including ethoxyl (a–d), methoxyl (e) and ketone (f).

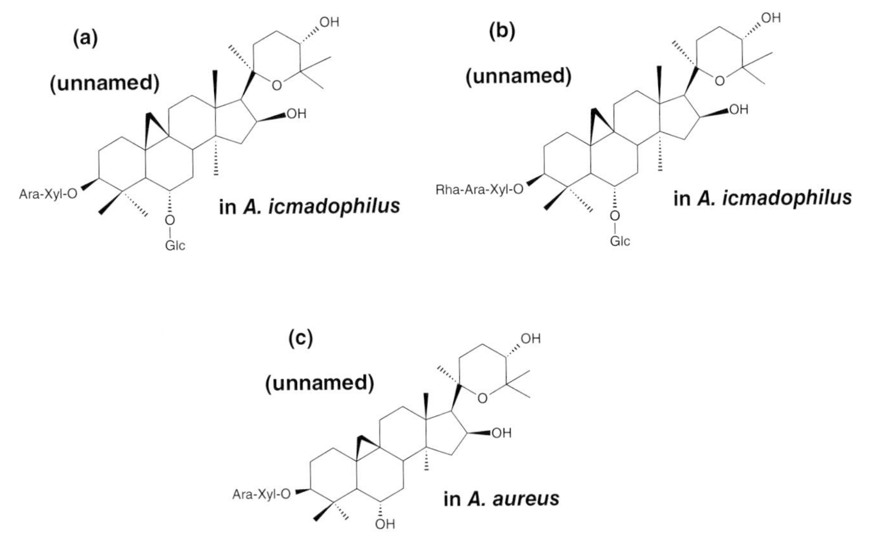

Fig. 27.32. Saponins of species *Astragalus* based on 20,25-epoxycycloartane and showing original chemical substitutions including arabinose (a–c) and rhamnose (b).

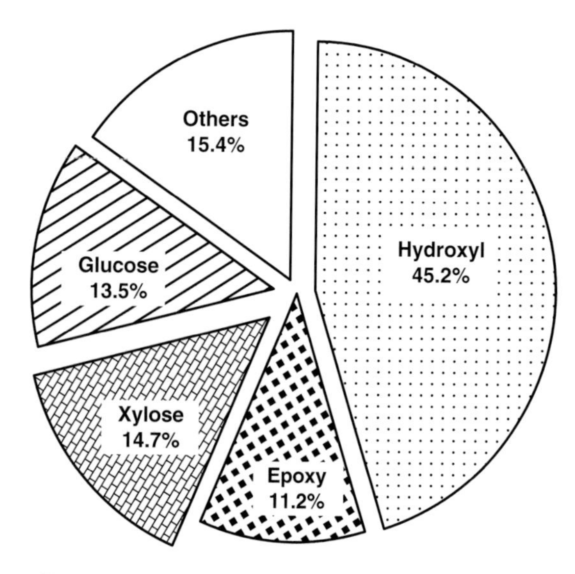

Fig. 27.33. Relative frequencies of different chemical substitutions on cycloartane-based saponins in population of *Astragalus* species.

27.3.4.2 Analysis of rare substitution positions in 20,24-epoxycycloartane

The 20,24-epoxycycloartane showed some rare attachment positions for hydroxyl, glucose and acetyl but not for xylose (Fig. 27.40). These four most frequent chemical groups had occurrence percentages of 48.4, 14.3, 11.5 and 18.1, respectively (Fig. 27.41). These values were calculated without consideration of the occurrence number of epoxyl group, which was systematically present in all the 20,24-epoxycyloartane-based saponins.

Rare attachment positions of glucose in 20,24-epoxycycloartane concerned Carbon 16 (Fig. 27.40). C16-glucosylation was observed in two saponins, trojanoside K and hareftoside D, characterizing *A. trojanus* (aerial parts) and *A. hareftae* (whole plant), respectively (Bedir *et al.*, 2001; Horo *et al.*, 2012). Concerning hareftoside D, it is worth noting that its structure increased by the absence of glycosylation at C3 (a rare feature in saponins of *Astragalus*) (Fig. 27.42).

For hydroxyl, rare positions in 20,24-epoxycycloartane concerned C11 (0.4%), C7 (0.9%), C23 (0.9%) and C12 (1.7%) (Fig. 27.40).

- Hydroxylation at C11 was concerned with a single sapogenin (cycloasgenin A) found in *A. taschkendicus* (roots) (Fig. 27.42) (Isaev *et al.*, 1982a).
- Hydroxylation at C23 was concerned with two unnamed saponins found in *A. campylosema* (roots) (Fig. 27.13 and Fig. 27.42) (Çalis *et al.*, 2008b). It is worthy to note that the diepoxylated saponin presented in Fig. 27.13 could be derived from that of Fig. 27.42 by cyclization between C16 and C24 favoured by the 16-OH.
- Hydroxylation at C7 was concerned with one saponin, including cycloalpioside D and corresponding aglycone (cycloalpigenin D) of *A. alopecurus* (aerial parts) (Fig. 27.42) (Agzamova and Isaev, 1991). Beside this 20,24-epoxycyclortane-based saponin, it is worth noting that the 7-OH seemed to be more present in rare saponins based on diepoxylated cycloartanes, including cycloorbicosides A, B, G of *A. orbiculatus* (roots and epigeal parts) (Fig. 27.12) (Mamedova *et al.*, 2002a; Isaev *et al.*, 2010b), as well as a cycloalpioside of *A. alopecurus* (Fig. 27.13) (Agzamova and Isaev, 1995a).
- Hydroxylated 20,24-epoxycycloartane at C12 was observed in two saponins (cycloalpiosides B and C) as well as their respective aglycones (cycloalpigenins B, C) characterizing *A. alopecurus* (Fig. 27.42) (Agzamova and Isaev, 1994b; Agzamova and Isaev, 1995b). These two saponins structurally differed only by 12-OH configuration, which was of α- and β-types in cycloalpiosides B and C, respectively. This shows the importance of spatial atomic configurations for further structural characterization of saponins when aglycone, substitution types and attachment positions are not sufficient for chemical discrimination between molecules.

In addition to xylose, glucose and hydroxyl, acetyl was commonly observed in 20,24-epoxycycloartanes, and particularly at C3. However, its attachments to C6, C16 and C23 were relatively rare: among all the invested saponins based on 20,24 epoxycycloartane, four were revealed to be original due to their acetylation on C16 (two cases), C6 (one case), and C23 (one case).

- 6-acetylation was observed in cyclocarposide B of *A. coluteocarpus* (Fig. 27.43) (Immomnazarov and Isaev 1992).
- 16-acetylation was observed in trojanosides A and I of *A. trojanus* (Fig. 27.43) (Bedir *et al.*, 1999; Bedir *et al.*, 2001; Çalis *et al.*, 2008b).
- 23-acetylation was observed in one saponin (unnamed) of *A. campylosema* (Fig. 27.43).

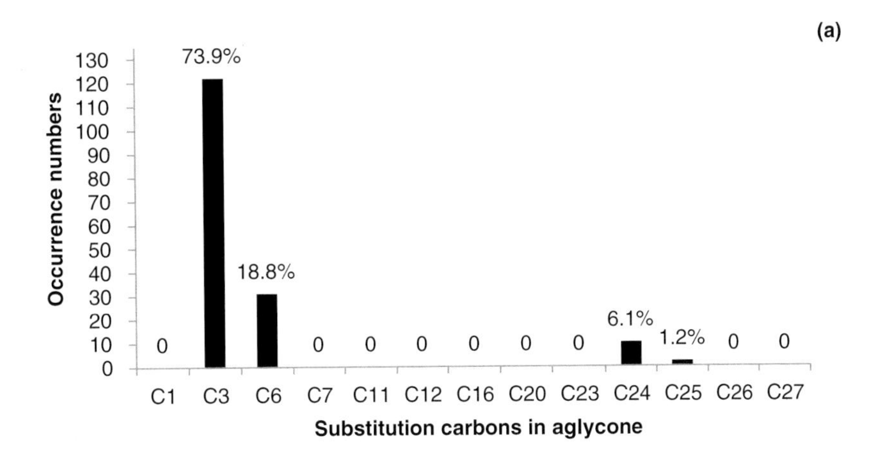

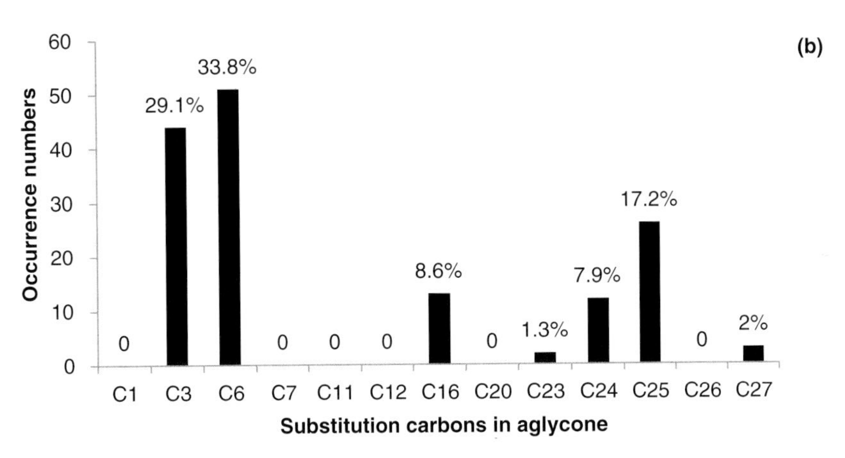

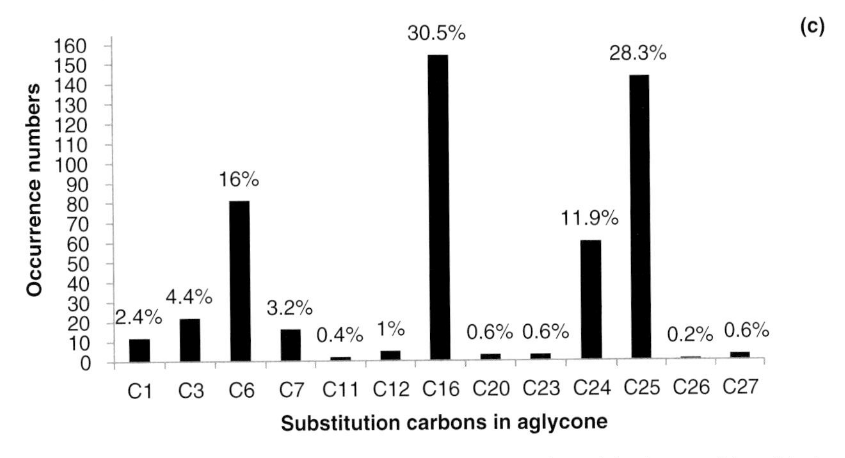

Fig. 27.34. Distribution of the most frequent chemical substitutions: xylose (a), glucose (b) and hydroxyl (c), through different carbons of cycloartanes in *Astragalus* species. The percentages of occurrences of considered chemical group at different attachment carbons were calculated relative to the sum of its occurrences on all the carbons.

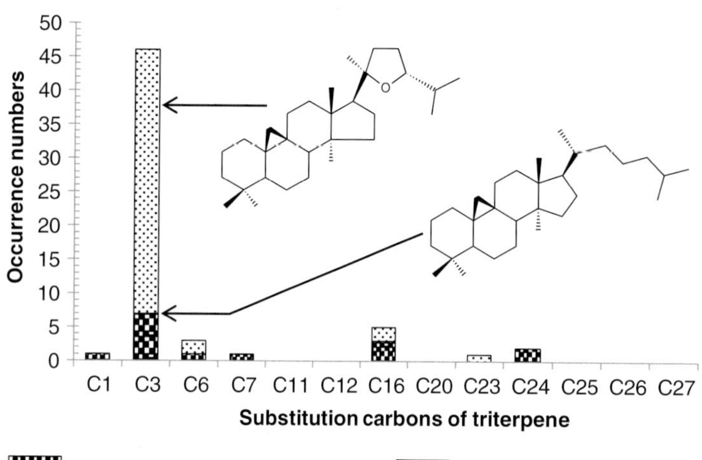

Fig. 27.35. Distribution of acetyl (a relatively frequent chemical group) through different carbons of lateral chain cycloartanes and 20,24-epoxycycloartanes produced by the genus *Astragalus*. Occurrence percentages of acetyl at each carbon were calculated relative to the sum of its occurrences on all the carbons.

27.3.5 Chemotaxonomical analysis based on rare substituted chaining sequences

Saponins in *Astragalus* showed high structural diversity due to different types of (i) sapogenins, (ii) desmosylation levels as well as (iii) substituted chemical groups which were more or less distributed through (iv) different carbons of aglycones. In addition to these four chemotaxonomical criteria, a fifth differentiation dimension is brought in by considering chaining sequences of chemical groups at a same carbon of sapogenin. Among the carbons of sapogenins, the C3 benefited from more glycosylations compared to the other carbons which were more hydroxylated than glycosylated. Therefore, the C3 offered a high number of glycosylation sequences, leading us to easily identify the most unusual ones.

In lateral chain cycloartane, glycosylated sequences starting by 3-O-arabinosyl seemed to be less frequent than 3-O-glucosyl, 3-O-rhamnosyl or 3-O-xylosyl. Among the 79 lateral chain cycloartane-based saponins substituted at C3 (see Fig. 27.44), these four cases represented percentages of:

- 2.53% (two saponins of 79) for 3-O-Ara-R; (R representing following chemical substitutions attached to the first sugar, i.e. arabinose in this case)
- 13.92% (11 saponins) for 3-O-Rha-R
- 20.25% (16 saponins) for 3-O-Glc-R
- 50.63% (40 saponins) for 3-O-Xyl-R.

Acetyl and glucuronic acid were discussed in section 27.3.3.3.1, which concerns rare chemical groups in lateral chain cycloartane. Taking into account the attachment positions, rare acetylations were observed at C2', C3' and C4' of 3-O-xylosyl (Fig. 27.25). Moreover, acetylation was observed at C2' and C6' of 3-O-glycosyls among the unusual substitutions occurring in lateral chain cycloartane (Fig. 27.25). Concerning glucuronic acid, only *A. amblolepis* showed a saponin where this chemical group was attached at carbon 2' of 3-O-xylosyl (Fig. 27.24).

In 20,24-epoxycycloartane, glycosylated sequences at C3 were initiated by different sugars including arabinose, glucose and xylose. In 3-O-diglycosyls, the second sugar was arabinose, rhamnose, apiose or glucose. Among these sugars, apiose, arabinose and rhamnose have already been considered in section 27.3.3.3.2 above, concerning rare chemical groups in 20,24-epoxycycloartanes. This led to the highlighting of original saponins due to the occurrence of these less frequent sugars. Therefore, next original saponins will be highlighted from unusual sequences combining xylose, glucose and acetyl (the three most frequent substitution types in 20,24-epoxyxycloartane apart from hydroxyl; Fig. 27.23).

27.3.5.1 Original substitution sequences in lateral chain cycloartane

In lateral chain cycloartane-based saponins, some saponins showed rare glycosidic sequences due to the unusual relative position of enchained sugars. Five unusual glycosidic sequences can be reported on the C3 of lateral chain cycloartane-based saponins (Fig. 27.45):

- 3-O-xylose-3'-xylose
- 3-O-xylose-2'-glucose

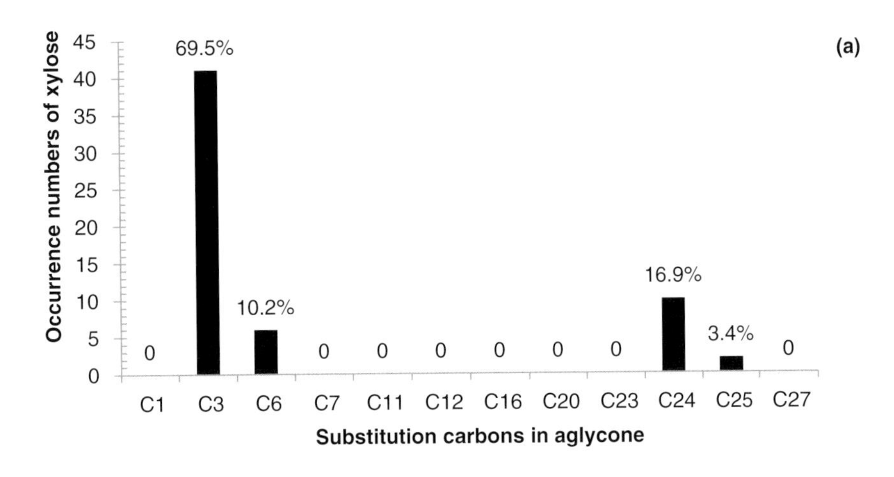

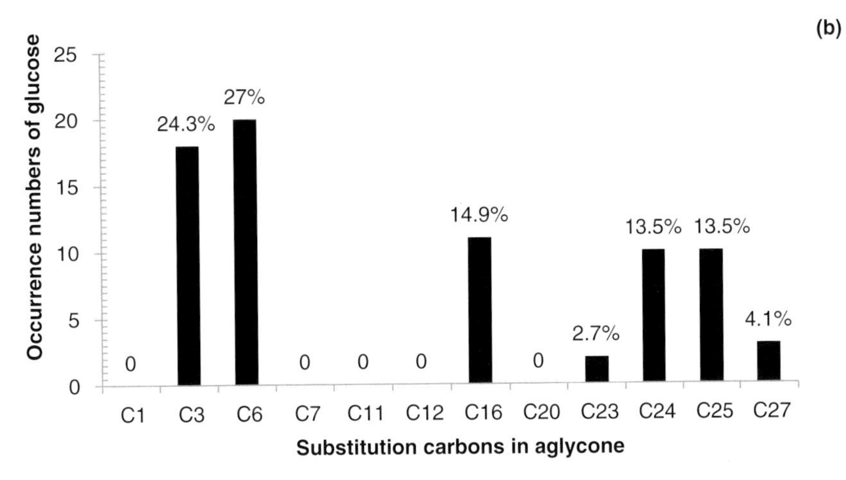

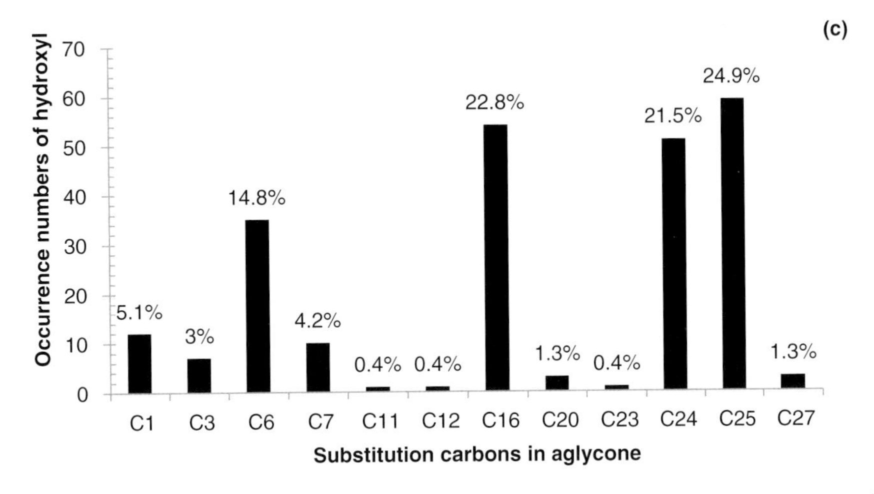

Fig. 27.36. Distribution of the most frequent chemical substitutions: xylose (a), glucose (b) and hydroxyl (c), through different carbons of lateral chain cycloartane in *Astragalus* species.

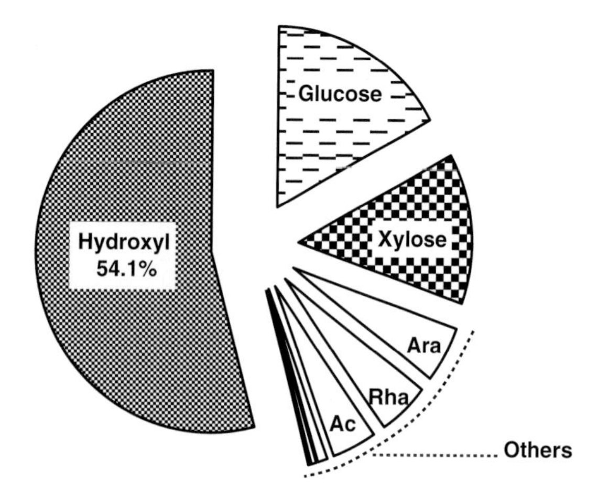

Fig. 27.37. Plot showing the percentages of the most frequent chemical substitutions in lateral chain cycloartane-based saponins of *Astragalus*. Legend: Ara = arabinose; Rha = rhamnose; Ac = acetyl.

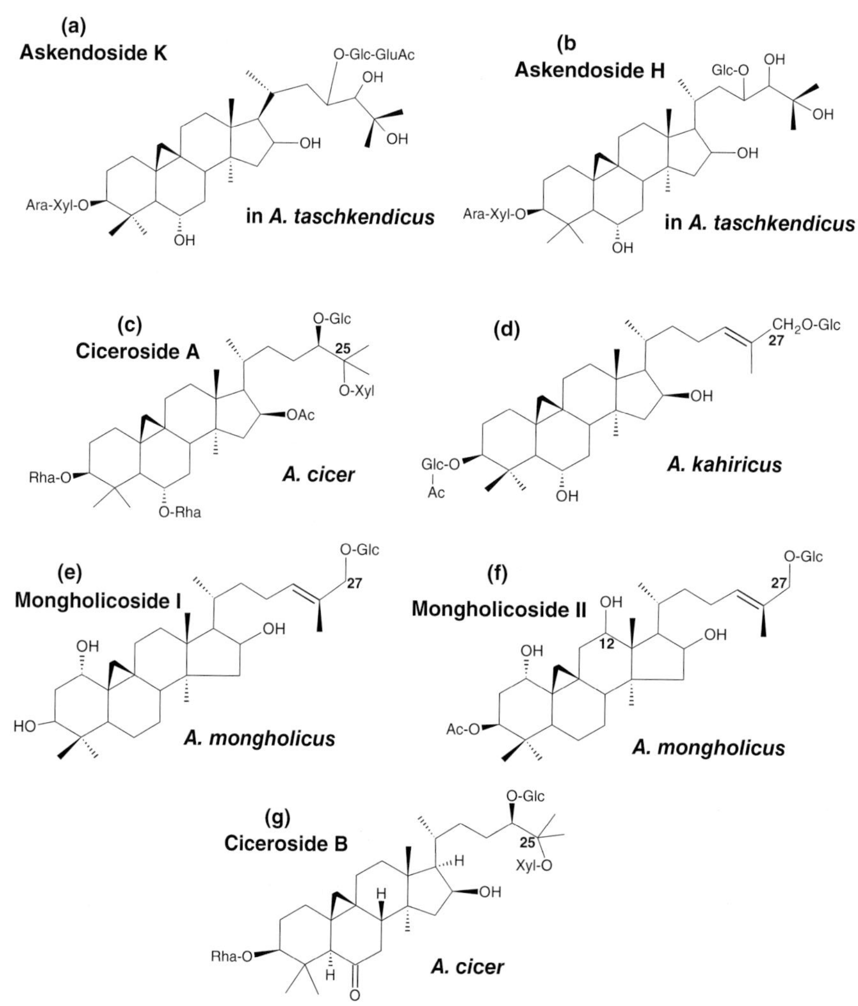

Fig. 27.38. Saponins based on cycloartane with lateral chain and showing original structures due to rare occurrences of xylose on C25 (c, g), glucose on C23 (a, b), glucose on C27 (d–f) and hydroxyl on C12 (f).

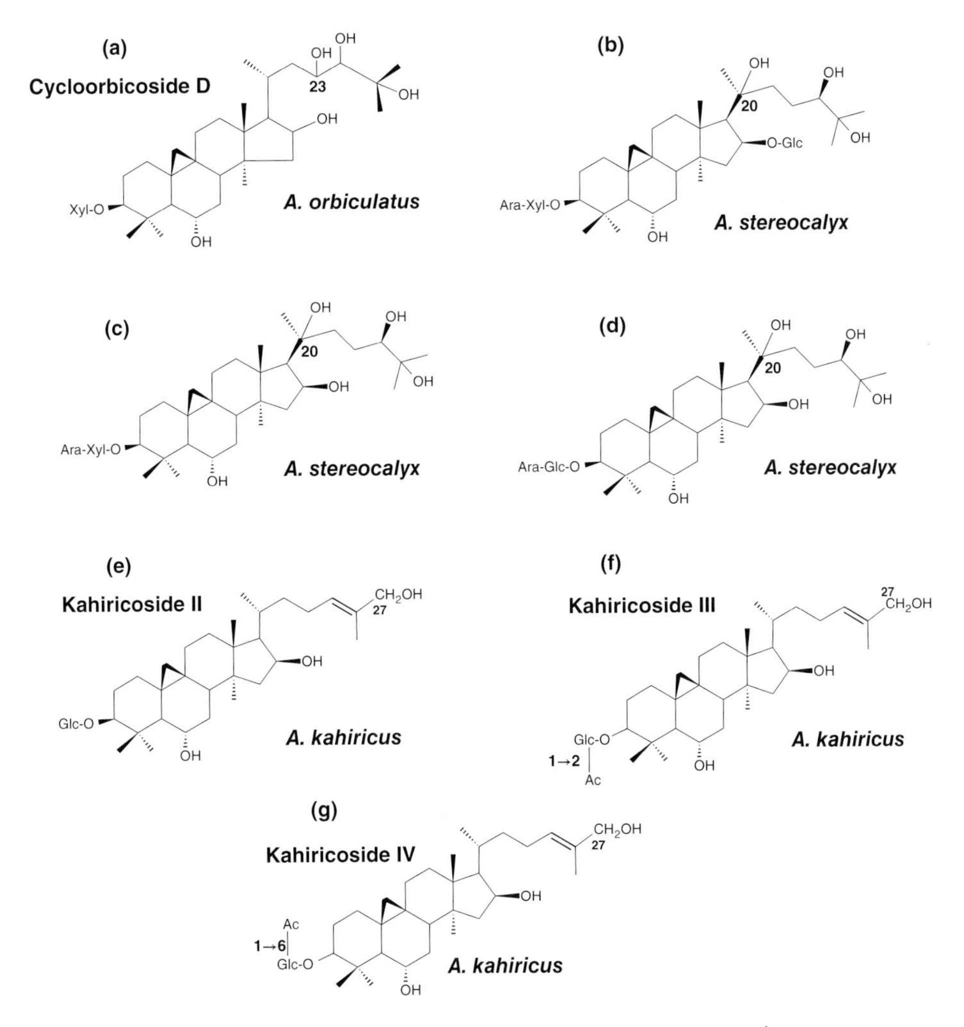

Fig. 27.39. Saponins based on lateral chain cycloartane and showing original structures due to rare occurrences of hydroxyl on C23 (a), C20 (b–d) and C27 (e–g).

- 3-O-glucose-(2'-arabinose, 4'-rhamnose)
- 3-O-xylose-2'-arabinose-2"-rhamnose
- 3-O-arabinose-(2'-arabinose,3'-acetyl) and 3-O-arabinose-(2'-arabinose,3', 4'-diacetyl).

Among these five unusual sequences, the first one (3-O-Xyl-3'-Xyl) was distinguished by the occurrence of a second xylose attached to the first one, whereas in all the other saponins, the second sugar attached to xylose was arabinose (in 14.1% cases), rhamnose (5.06%) or glucose (2.5%). The 3-O-Xyl-3'-Xyl sequence was observed in brachyoside A produced by *A. brachypterus* (Bedir *et al.*, 1998b). The second sequence (3-O-Xyl-2'-Glc) showed originality because of the rarity of glucose on the C2' of xylose. This feature concerned two saponins (2.5% cases): astramembranoside B of *A. membranaceus* and cyclocanthoside G of *A. tragacantha* (Fig. 27.45) (Isaev *et al.*, 1992; Kim *et al.*, 2008).

The third sequence: 3-O-Glc-(2'-Ara,4'-Rha), was original for several reasons (Fig. 27.45): (1) rhamnose was rarely observed at carbon 4' of 3-O-glucose; (2) triglycosylation revealed to be not frequent in cycloartane-based saponins of *Astragalus*; (3) from the 3-O-glucose, the two sugars (arabinose and rhamnose) showed ramified postions. This rare glycosylated sequence was observed in one saponin of *A. stereocalyx* (Yalçın *et al.*, 2012).

The fourth glycosidic sequence (3-O-Xyl-2'-Ara-2"-Rha) was original because of its triglycosylation

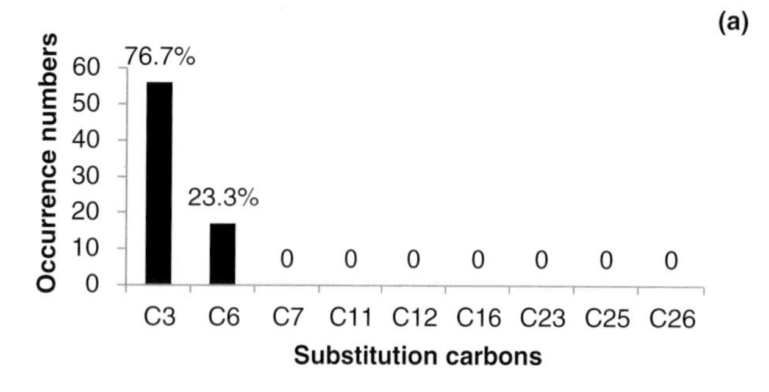

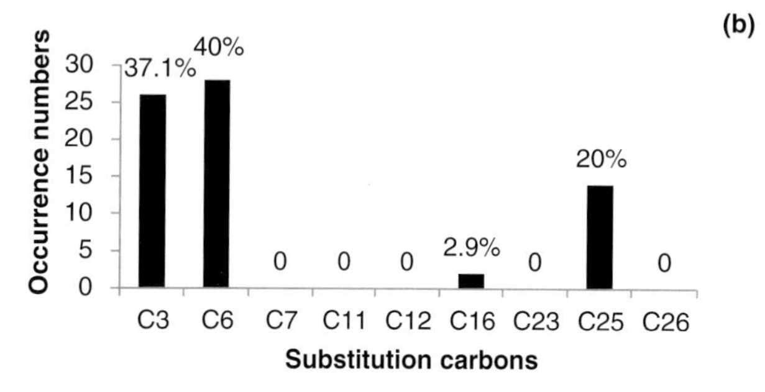

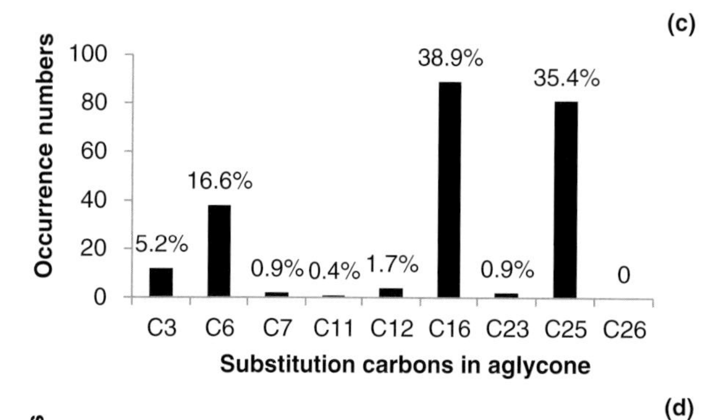

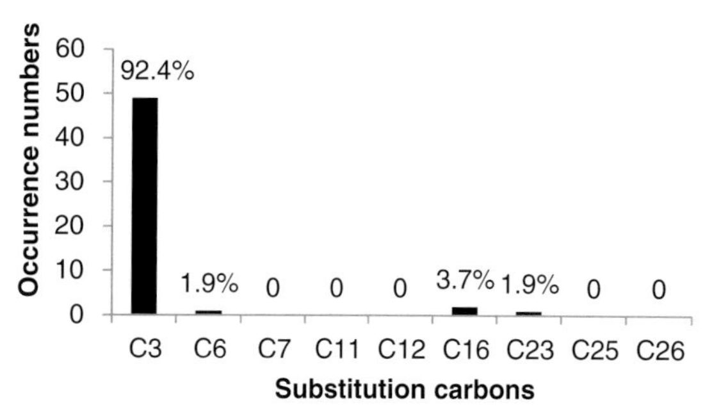

Fig. 27.40. Distribution of the most frequent chemical substitutions: xylose (a), glucose (b), hydroxyl (c) and acetyl (d) through different carbons of 20,24-epoxycycloartane in *Astragalus* species.

enchaining, whereas diglycosylated sequence (3-O-Xyl-2'-Ara) proved to be the most frequent feature among all the diglycosylated sequences (12.7% cases). This triglycoside feature was observed in two saponins produced by *A. aureus* and *A. icmadophilus* (Fig. 27.45) (Gülcemal *et al*., 2011; Horo *et al*., 2010).

The fifth sequence (3-O-Ara-2'-Ara) was original because of its double arabinose enchaining. Each of the two arabinoses brought some originality to the other:

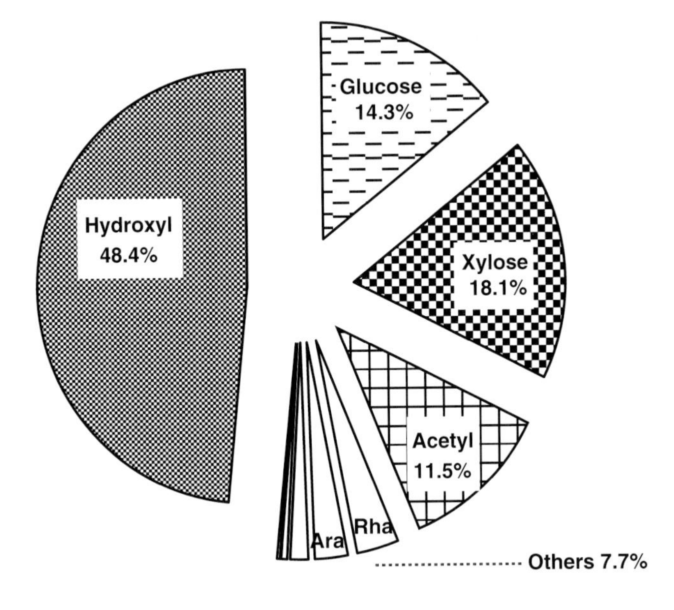

Fig. 27.41. Plot showing the percentage of the most frequent chemical substitutions in 20,24-epoxycycloartane-based saponins of *Astragalus*. Legend: Ara = arabinose; Rha = rhamnose.

the first originality came from the fact that arabinose directly attached to C3 (3-O-Ara-) was significantly less frequent than the second arabinose (3-O-sugar-Ara). This originality was reinforced by the rarity of double arabinosylation due to the second arabinosyl. This structural feature was observed in two saponins of *A. icmadophilus* (Fig. 27.45) (Horo *et al.*, 2010). These saponins were also shown to be original by their acetylation (see section 27.3.3.3.1 above).

27.3.5.2 Original substitution sequences in 20,24-epoxycycloartane

According to rare glycosidic sequences at C3, the 20,24-epoxycycloartane showed one original feature consisting of 3-O-xylose-2'-glucose (Fig. 27.44): originality was directly due to glucose which proved to be significantly less substituted to 3-O-xylose (4.9%) compared to attachments of rhamnose (14.8%), arabinose (14.8%) or acetyl (65.6%) to 3-O-Xyl. This feature was observed in astragalosides III, V and VI of *A. membranaceus* (Fig. 27.46) (Kitagawa *et al.*, 1983d).

Other glucosylated sequences, such as 3-O-glucose-2'-glucose, showed some low frequencies (Fig. 27.44) but they can't be considered as original for two reasons: (i) the second sugar attached to the carbon 2' of 3-O-glucose was more frequently glucose rather than rhamnose or apiose; (ii) the set of glycosylated sequences starting by a 3-O-glucose were significantly lower than those starting by a 3-O-xylose leading to lesser confirmations of rarity of 3-O-Glc-R sequences (R other substitutions).

By analysing acetylations of different sugars, the feature 3-O-glucose-2'-acetyl showed the lowest frequency, compared to other acetylations at different carbons of different sugars (Fig. 27.44). This original acetylated feature concerned astraverrucin II produced by *A. verrucosus* (Fig. 27.46) (Pistelli *et al.*, 1997).

27.4 Synthesis of Chemical Differentiation Levels of Different Original Saponins

Synthesis of chemical differentiation between *Astragalus* species was carried out taking into account the five considered structural criteria consisting of (i) rare aglycones, (ii) rare desmosylation levels, (iii) rare substitutions, (iv) uncommon substitution positions and (v) rare substituted chaining sequences (Fig. 27.2). For each *Astragalus* species, the five criteria were separately counted, taking into account the saponins produced. This resulted in five calculated (AS, DS, SS, PS, CS) for each *Astragalus* species, corresponding to rare aglycones, desmosylation levels, substitutions, positions and chaining sequences, respectively. These scores helped to classify plants according to the five chemical differentiation criteria. Also, the five scores were

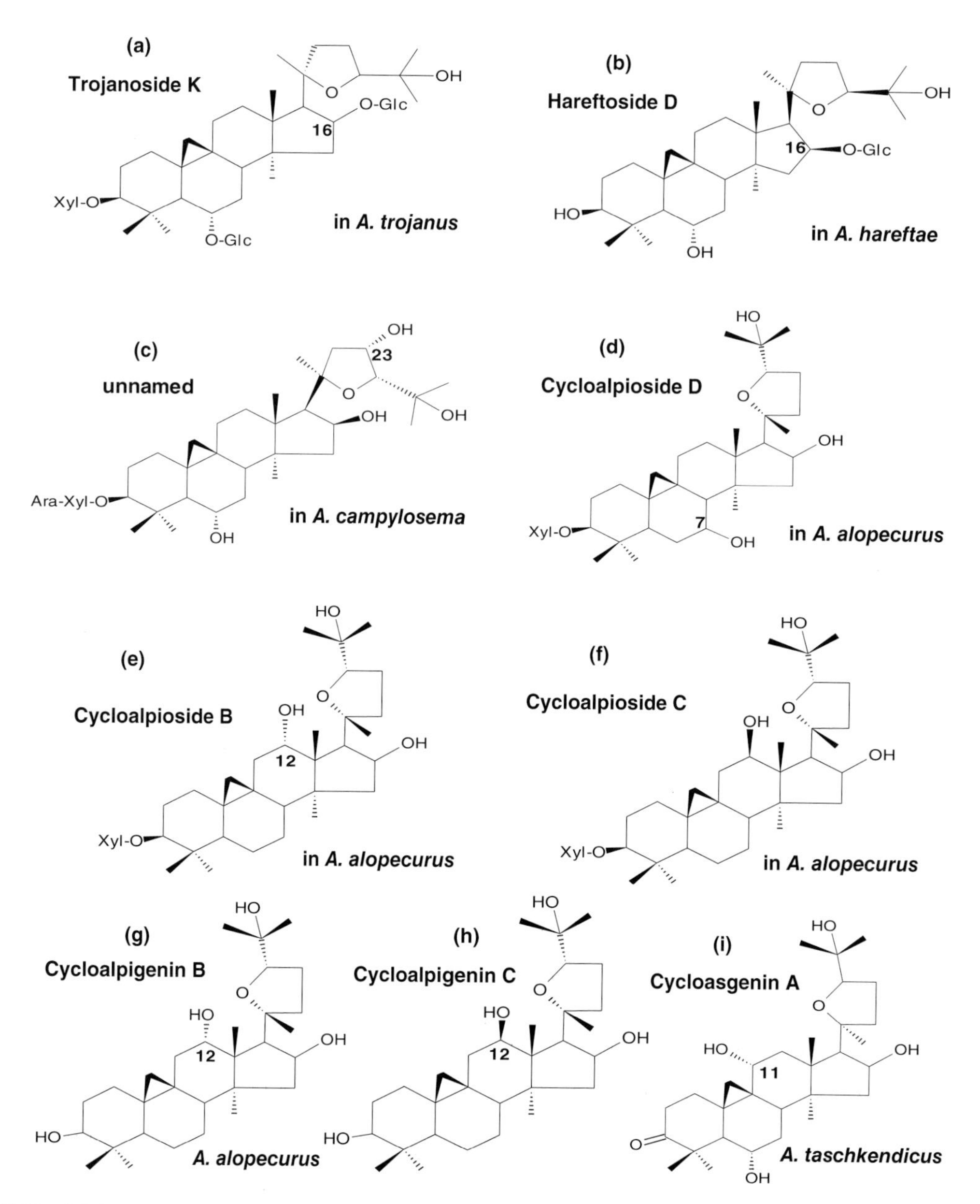

Fig. 27.42. Original saponins in *Astragalus* species due to the occurrence of glucose and hydroxyl (common chemical substitutions) on unusual attachment positions among carbons of 20,24-epoxycycloartane.

cumulated (summed) to obtain a global score (GS) helping to evaluate global chemical differentiation of each *Astragalus* species within the invested plant community.

Among 46 differentiated *Astragalus* species, nine showed the highest GS (Fig. 27.47): *A. membranaceus* (GS=18), *A. tauricolus* (GS=15), *A. trojanus* (GS=14), *A. icmadophilus* (GS=11), *A. tomentosus*, *A. taschkendicus*, *A. sinicus*, *A. alopecurus* and *A. orbiculatus* (GS=9).

Immediately lower GS values concerned 18 species: *A. campylosema* (GS=8), *A. sieversianus* and *A. aureus* (GS=7), *A. verrucosus*, *A. Kahiricus*, *A. flavescens*, *A. complanatus* and *A. cicer* (GS=6), *A. wiedemannianus*,

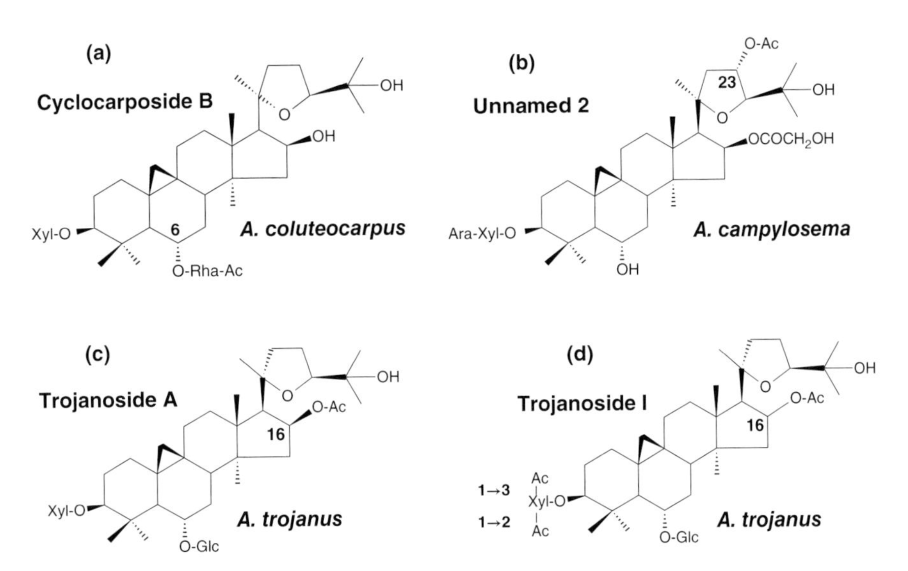

Fig. 27.43. Original saponins in *Astragalus* species due to the occurrence of acetyl on uncommon attachment positions (C6, C16, C23) among carbons of 20,24-epoxycycloartane.

A. dissectus, *A. bicuspis* and *A. amarus* (GS=5), *A. hareftae*, *A. mongholicus*, *A. stereocalyx*, *A. coluteocarpus*, *A. armatus* and *A. amblolepis* (GS=4) (Fig. 27.47).

Lower GS (=3 and 2) concerned eight *Astragalus* species: *A. tragacantha*, *A. macropus*, *A. ernestii* and *A. caprinus* (GS=3), *A. zahlbrukneri*, *A. oldenburgi*, *A. elongatus* and *A. brachypterus* (GS=2) (Fig. 27.47).

Finally, the lowest GS values (GS=1) concerned 11 plants, eight of which were original by a rare substitution unit (GS=SS=1) (Figs 27.47 and 27.48): *A. villosissimus*, *A. trigonus*, *A. iliensis*, *A. oleifolius*, *A. pycanthus*, *A. unifoliatus*, *A. halicacabus* and *A. alexandrinus*. Two plants (*A. ptilodes*, *A. cephalotes*) were original by a rare desmosylation levels (GS=DL=1). The other one (*A. microcephalus*) had a GS=1 due to an original aglycone (GS=AS=1) (Fig. 27.49).

Analysis of the species with scores relatively close to each other revealed four metabolic strategies for chemical differentiation of different *Astragalus* species (Fig. 27.50).

1. The first strategy was represented by relatively high AS values compared to the other scores. This indicates a strategy by which plants chemically differentiate in upstream steps of saponin metabolism. The extreme case is illustrated by the species *A. sinicus* which showed chemical differentiation exclusively due to original aglycone biosynthesis (GS=AS=9). The list of concerned (less extreme) plants includes *A. orbiculatus*, *A. dissectus*, *A. tauricolus*, *A. bicuspis*, *A. caprinus*, *A. zahlbruckneri*, *A. tomentosus*, *A. amarus*, *A. membranaceus*, *A. aureus*, *A. macropus*, *A. hareftae*, *A. flavescens*, *A. complanatus*, *A. coluteocarpus* and *A. sieversianus* (Fig. 27.50).

2. The second metabolic strategy was represented by relatively high DS values vs low AS ones. This indicated that species which don't synthesize unusual aglycones can early differentiate by increasing the desmosylation level (number of glycosylated carbons) on common sapogenins. This strategy concerned the species *A. trojanus*, *A. amblolepis*, *A. ptilodes*, *A. oldenburgi*, *A. armatus*, *A. cicer*, *A. cephalotes*, *A. brachypterus* and *A. wiedemannianus* (Fig. 27.50).

3. The third metabolic strategy dominated in plants showing relatively high SS values with non-null CS values occasionally. This strategy was based on intermediate or terminal metabolic steps consisting of (i) substituting atypical chemical groups or (ii) arranging common chemical groups into atypical substitution sequences. The extreme case concerned three species: *A. icmadophilus* (SS=6 vs GS=11), *A. sieversianus* (SS=5 vs GS=7) and A. verrucosus (SS=5 vs GS=6). Note that the species *A. sieversianus* was also classified in strategy 1 because it was strongly distinguished by the biosynthesis ability of β-sitosterol; however, its

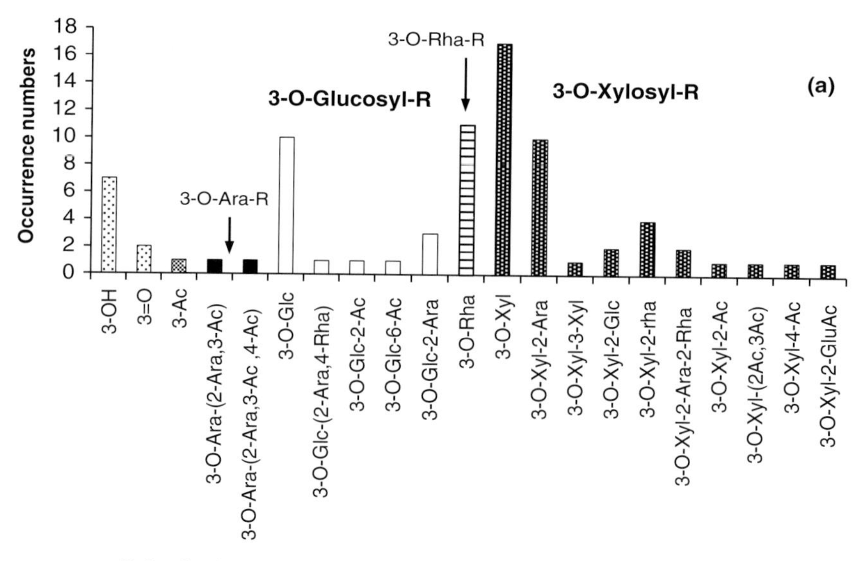

Substitution sequences at C3 of lateral chain cycloartane

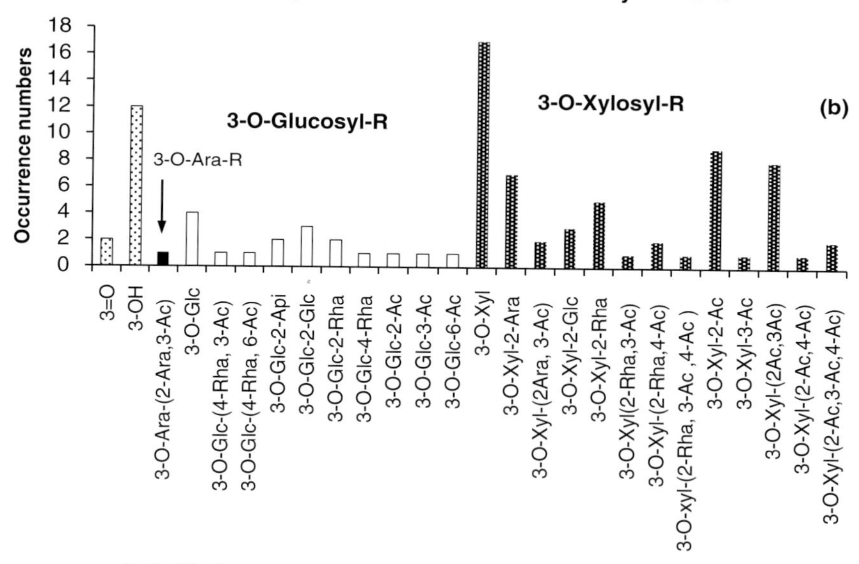

Substitution sequences at C3 of 20,24-epoxycycloartane

Fig. 27.44. Plot showing occurrence levels of different chemical substitutions attached on the first sugar (xylose, glucose or arabinose) substituted at carbon C3 of saponin. (a) Lateral chain cycloartane-based saponins; (b) 20,24-epoxycycloartane-based saponins. Legend: Ac = acetyl; Ara = arabinose; Glc = glucose; GluAc = glucuronic acid; Rha = rhamnose; Xyl = xylose.

cycloartane-based saponins were differentiated by some original substitutions. Apart from these two species, this metabolic differentiation strategy concerned less extreme plants showing generally a low global score (GS ≤3): *A. alexandrinus*, *A. elongatus*, *A. ernestii*, *A. halicacabus*, *A. ilensis*, *A. oleifolius*, *A. pycanthus*, *A. tragacantha*, *A. trigonus*, *A. unifoliatus* and *A. villosissimus* (Fig. 27.50).

4. The fourth metabolic strategy was represented by relatively high PS values compared to the other scores. This indicated that plants producing common aglycone with usual desmosylation levels and

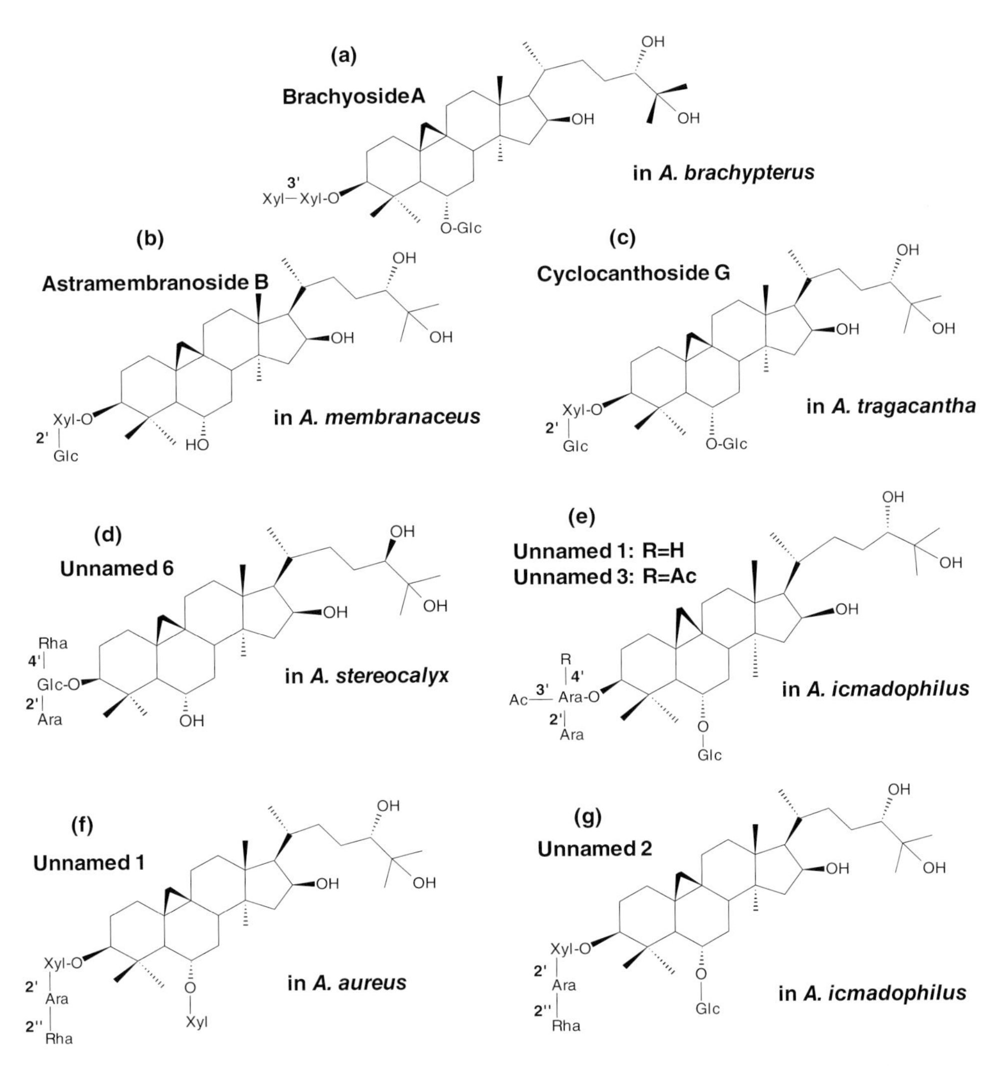

Fig. 27.45. Original lateral chain cycloartane-based saponins due to rare glycosylated and acetylated sequences. Legend: Ac = acetyl; Ara = arabinose; Glc = glucose; Rha = rhamnose; Xyl = xylose

substitution types can be chemically differentiated by attaching common chemical groups on unusual positions of sapogenin. This strategy was dominant in *A. alopecurus*, *A. kahiricus*, *A. campylosema*, *A. mongholicus*, *A. taschkendicus* and *A. stereocalyx* (Fig. 27.50).

The synthesis of these four metabolic trends highlighted the hierarchical role of aglycone biosynthesis in the chemical differentiation of a majority of plants followed by substitution type, desmosylation level and substitution position strategies. The rare substitution sequence did not appear as a distinct differentiation strategy for two reasons: (i) the number of saponins concerned is relatively low compared to the numbers of other original saponins; (ii) *Astragalus* species having original saponins with unusual sequences showed other original saponins due to the four other criteria. More saponins with original sequences need to be studied (by future research) to interpret the existence of a fifth metabolic diversification strategy in the *Astragalus* genus.

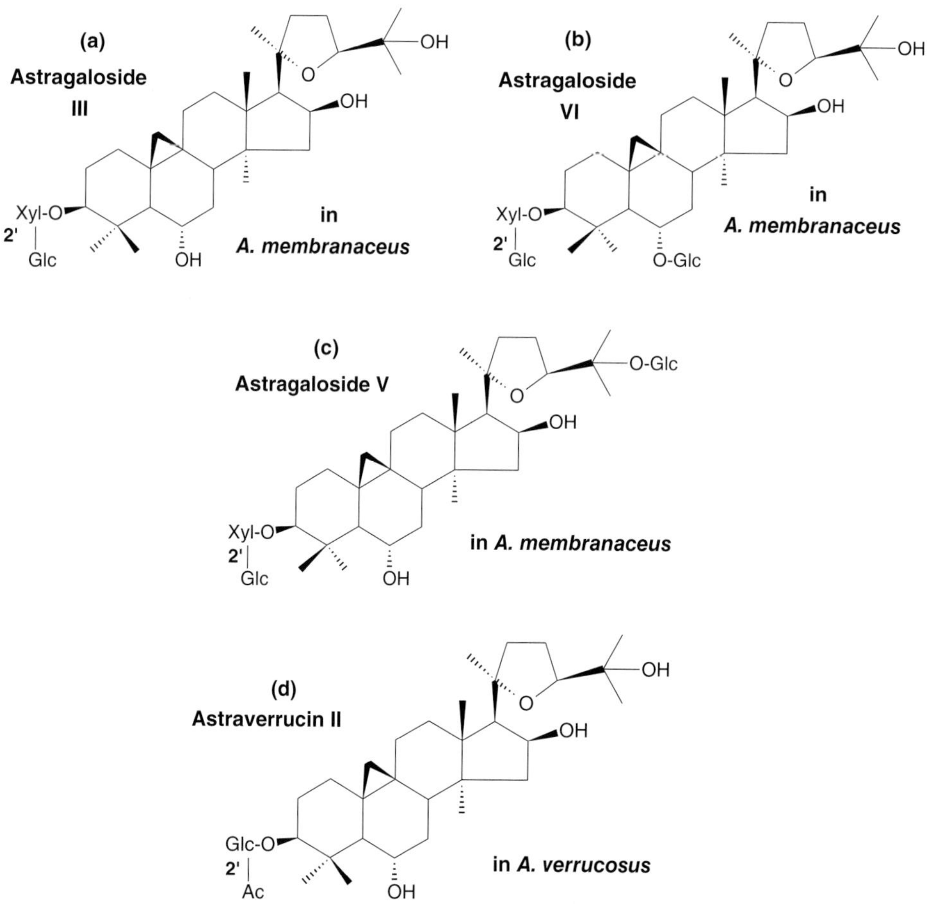

Fig. 27.46. Original 20,24-epoxycycloartane-based saponins due to rare glycosylated (a-c) and acetylated sequences (d). Legend: Ac = acetyl; Glc = glucose; Rha = rhamnose; Xyl = xylose.

27.5 General Chemical Features in *Astragalus* Community: Most Common Features of Saponins

After highlighting the original saponins in different *Astragalus* species based on rare aglycones, desmosylation level, chemical groups, substitution positions and glycosidic sequences, the remaining molecules were considered to find the most common structural traits in this plant community. This analysis helped to extract a general feature of saponins at generic level. Global results showed that the most substituted positions concerned C3, C16, C24 and C25 in both lateral chain cycloartane and 20,24-epoxycycloartane. Other carbons, including C1, C7 and C20, showed aglycone type-dependent substitutions: C1 and C7 were substituted in lateral chain cycloartane but were generally free in 20,24-epoxycycloartane; C20 was naturally substituted in 20,24-epoxycycloartane vs no substitution in later chain aglycone (Fig. 27.51).

Qualitatively, the C3 showed high diversity of substitution groups compared to the other carbons in the two cycloartane types (Fig. 27.51). These chemical groups included xylose (>35%), hydroxyl (7–21%), glucose (>19%), rhamnose (≤ 14%), arabinose (≤ 11%).

Comparison between lateral chain cycloartane and 20,24-epoxycycloartane showed that the former was more familiar with rhamnose (14%) and arabinose (11%) at C3, whereas the latter was particularly distinguished by a high occurrence of acetyl

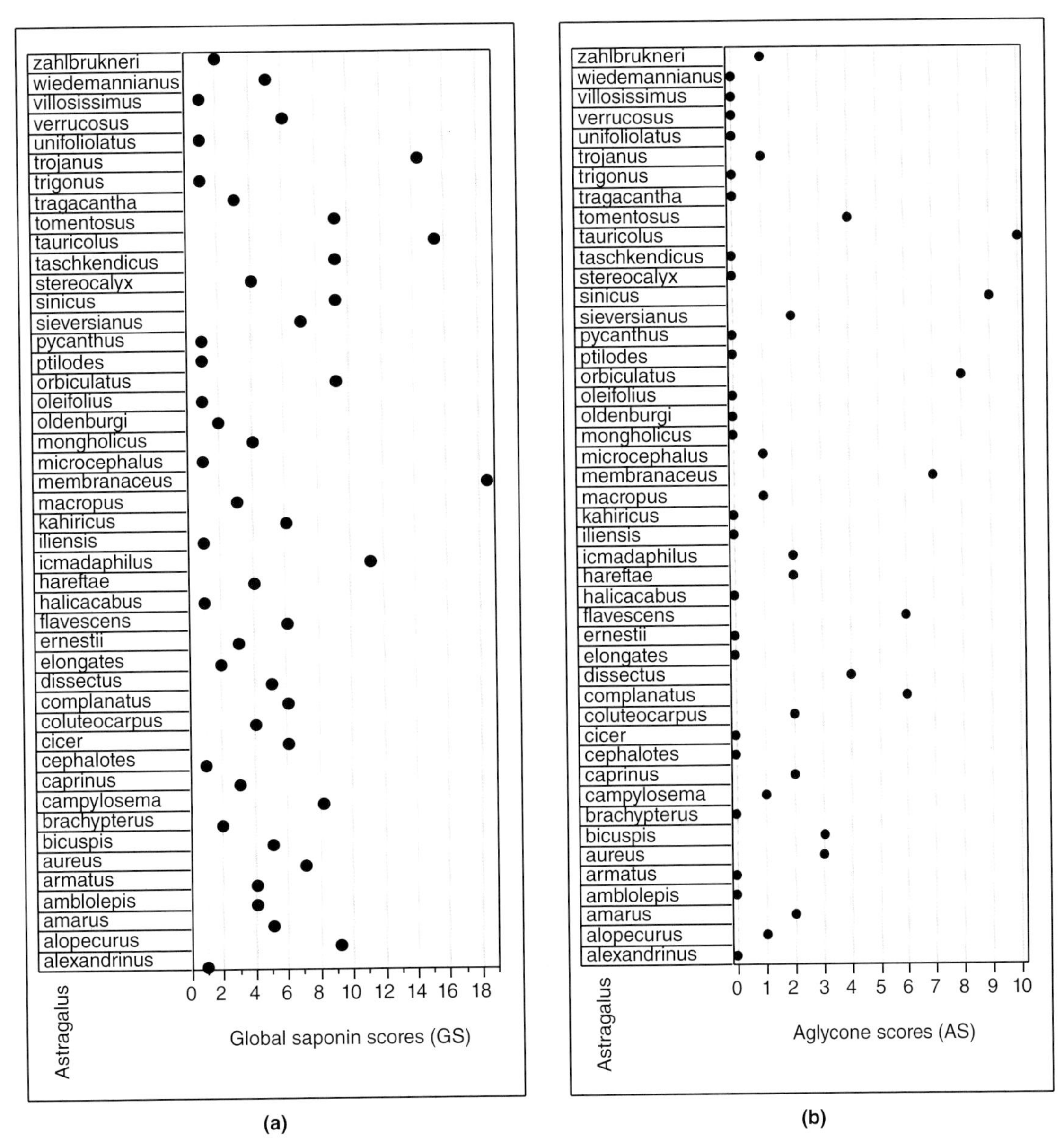

Fig. 27.47. Calculated scores for original saponins found in different *Astragalus* species. (a) Global scores (GS) representing the sum of original structural features due to rare aglycones, rare desmosylation levels, rare substitution types, rare substitution positions and rare substitution sequences in all the saponins of considered species; (b) rare aglycone scores (AS) representing the numbers of rare aglycones found in considered species.

(27/5%) on C3, exclusively hydroxylated C16 and practically not substituted at C1 and C7.

- Acetyl generally occurred on a sugar and showed a substitution percentage of 33%, especially in 20,24-epoxycycloartane, relative to the sum of substitutions in C3 (Fig. 27.51).
- Glucosylation at C16 was more frequent in lateral chain cycloartane, whereas it concerned rare cases in 20,24-epoxycycloartane, which were

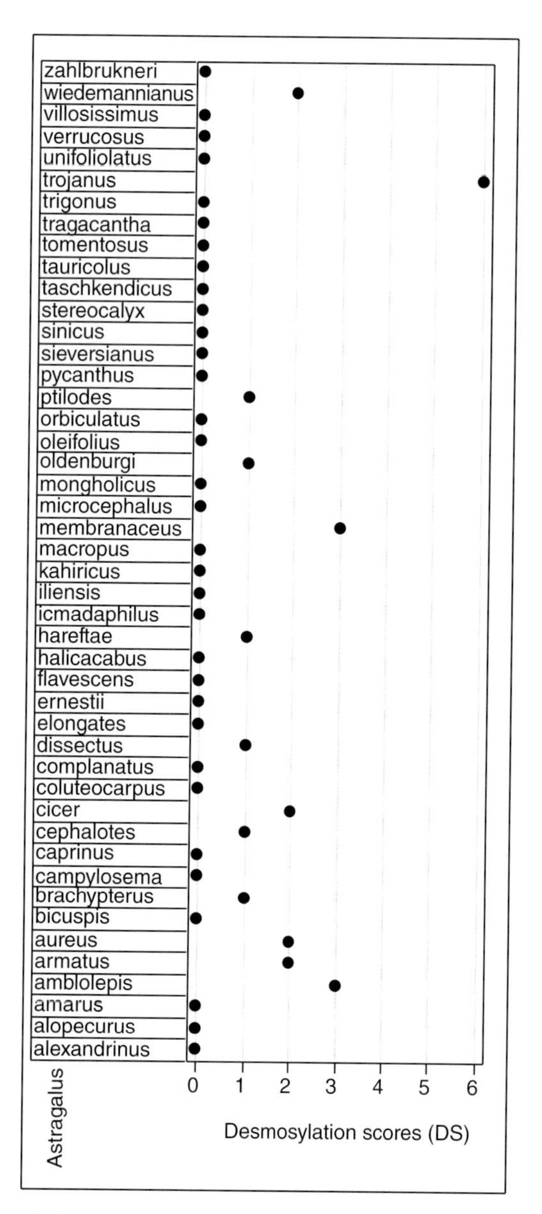

(a) Rare desmosylation level scores

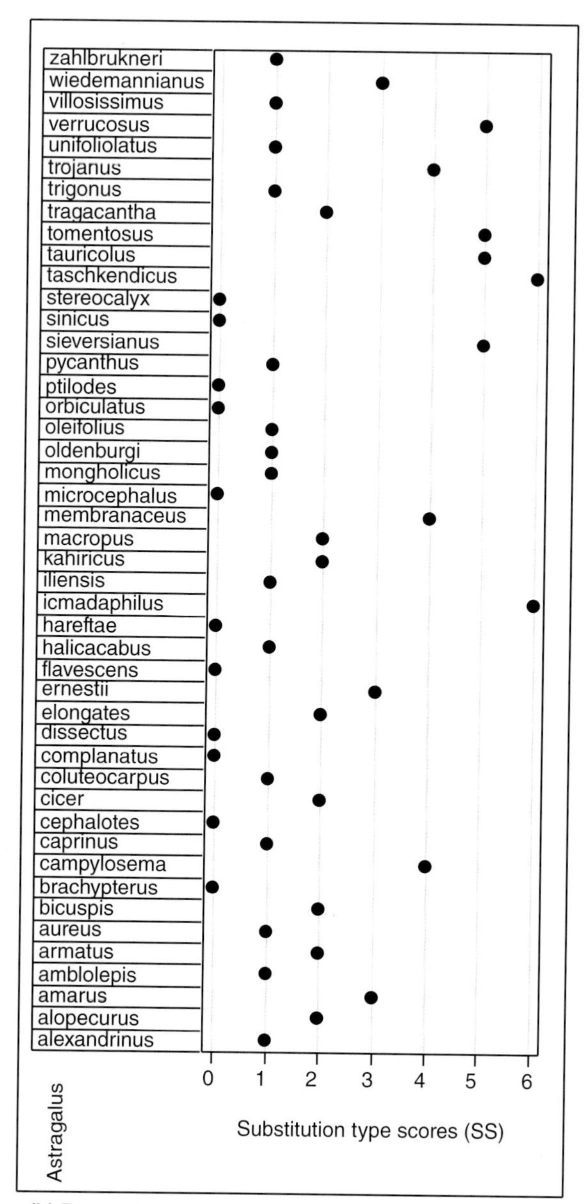

(b) Rare substitution type scores

Fig. 27.48. Calculated scores for original saponins found in different *Astragalus* species. (a) Rare desmosylation level scores (DS) corresponding to the number of tri- or tetradesmosides found in *Astragalus* species. (b) Rare substitution types scores (SS) representing the number of rare chemical groups attached in saponins of *Astragalus* species.

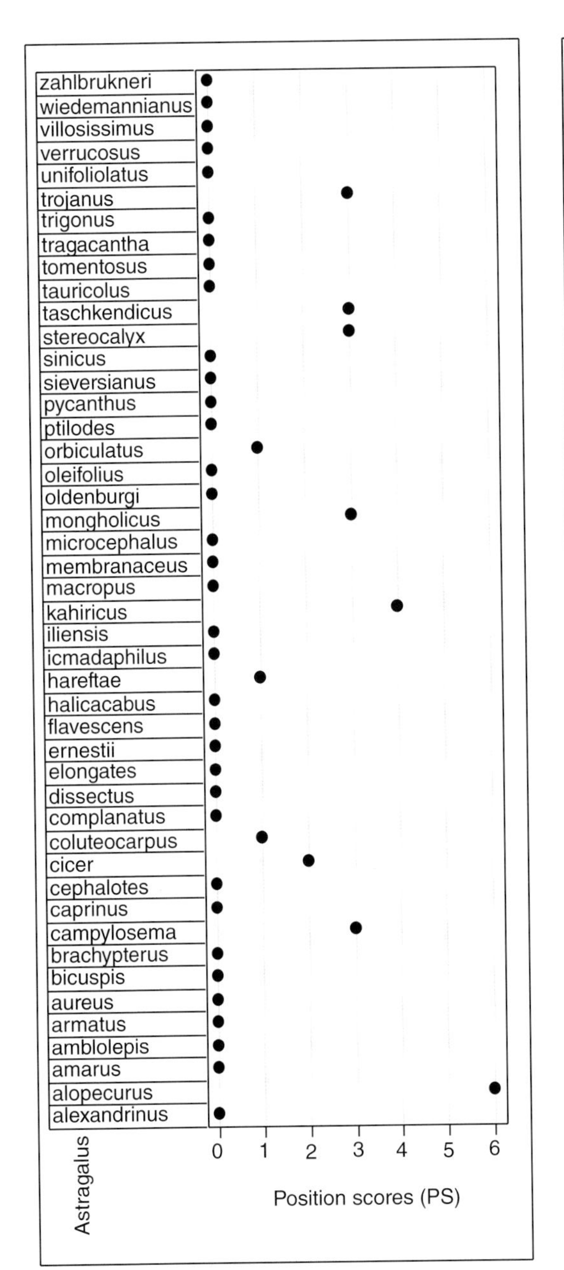

(a) Rare substitution position scores

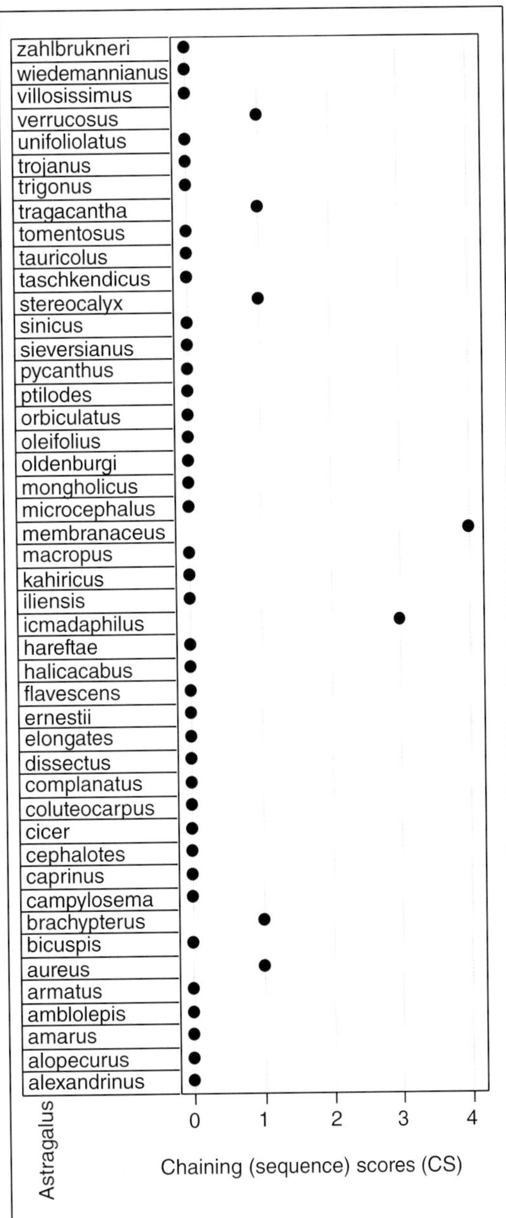

(b) Rare substitution sequence scores

Fig. 27.49. Calculated scores for original saponins found in different *Astragalus* species. (a) Rare substitution position scores (PS) corresponding to the number of times where frequent chemical groups were attached at atypical positions on the sapogenin. (b) Rare substitution chaining scores (CS) representing the number of rare sequences made by usual substitution units in saponins of each *Astragalus* species.

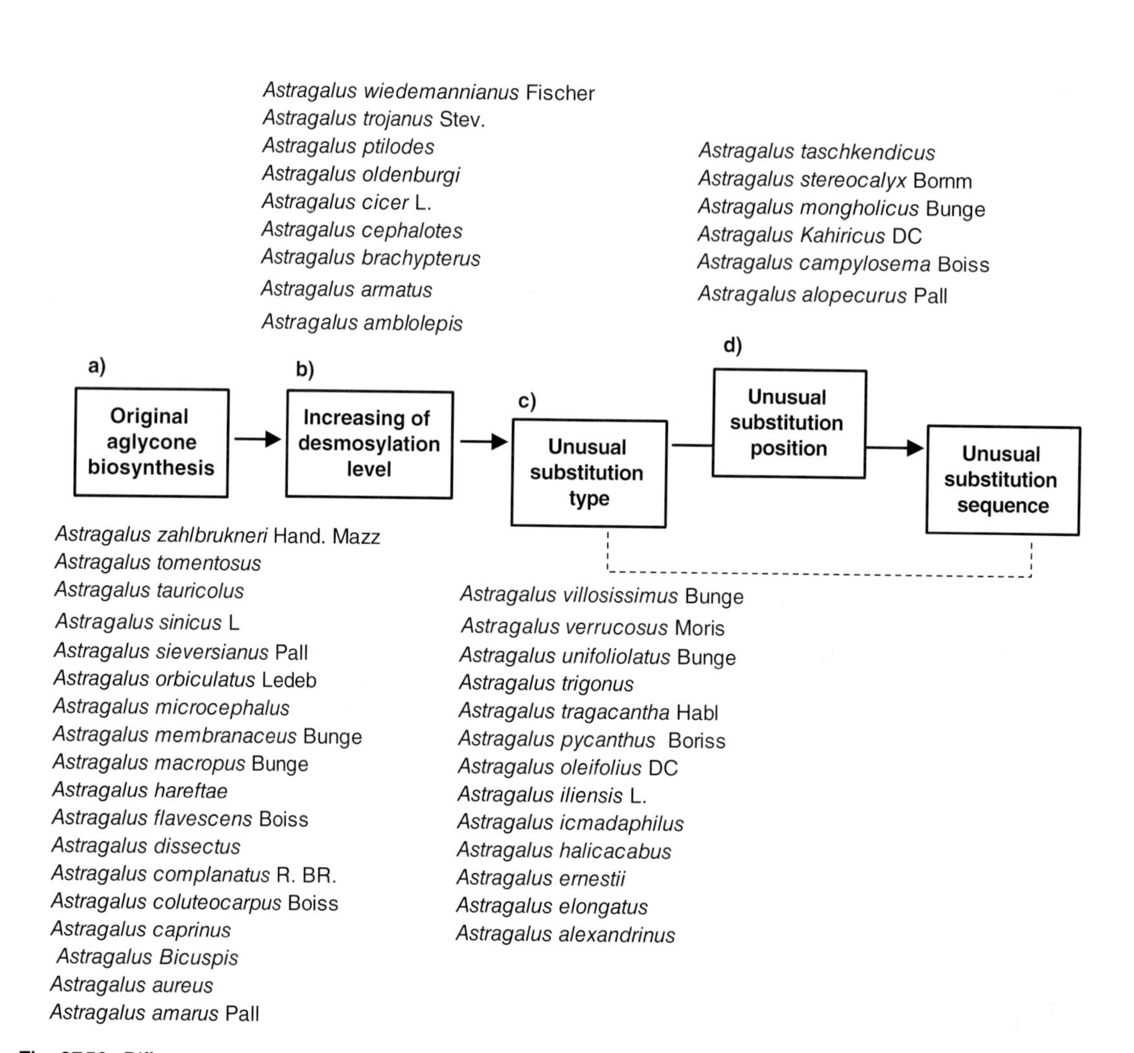

Fig. 27.50. Different metabolic strategies governing the chemical differentiation of different *Astragalus* species.

classified among original saponins in the previous section (see section 27.3.4.2).

- Lateral chain cycloartane was distinguished by its ability to attach hydroxyls on both C1 and C7 (Fig. 27.51).
- Finally, rhamnose and arabinose were found in 20,24-epoxycycloartane, but with relatively lower percentages compared with lateral chain cycloartane (Figs 27.23 and 27.51).

In quantitative terms, C3 was the most substituted in 20,24-epoxycycloartane, whereas the highest substitutions concerned the carbons 24 and 25 in lateral chain cycloartane (Fig. 27.51). In this latter case, C24 and C25 were essentially hydroxylated (80–91%) and glucosylated (8–11%). Also, the C24 showed some ability to be xylosylated. Although C1 and C7 were substituted in lateral chain cycloartane in contrast to 20,24-epoxycycloartane (not substituted), they were significantly less substituted than the other carbons of this sapogenin.

Finally, the C6 was substituted by glucose, hydroxyl and xylose in the two types of cycloartanes (Fig. 27.51). The difference between the two aglycone types was essentially quantitative: in lateral chain cycloartane, hydroxyl most frequently occurred in C6 (61.4%) followed by glucose (25.7%) and xylose (12.9%). In 20,24-epoxycycloartane, hydroxyl and glucose showed comparable substitution percentages on C6: 38% vs 24% for xylose. These quantitative

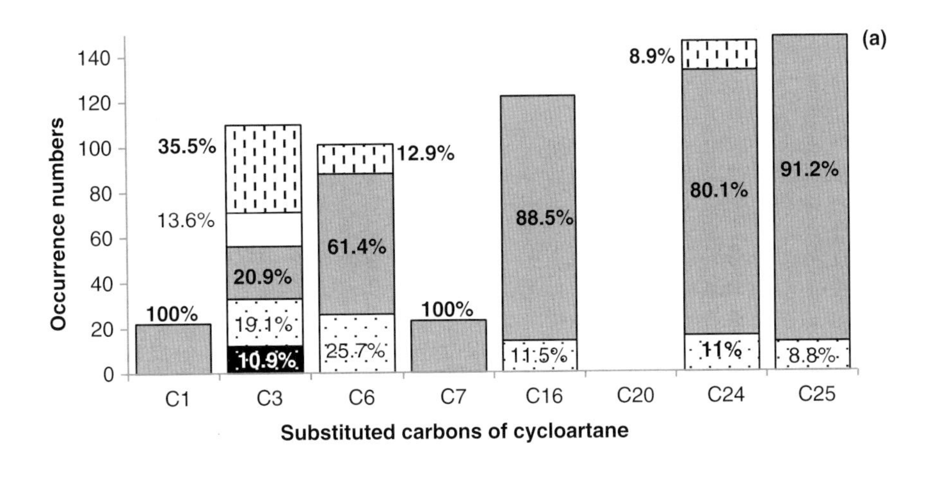

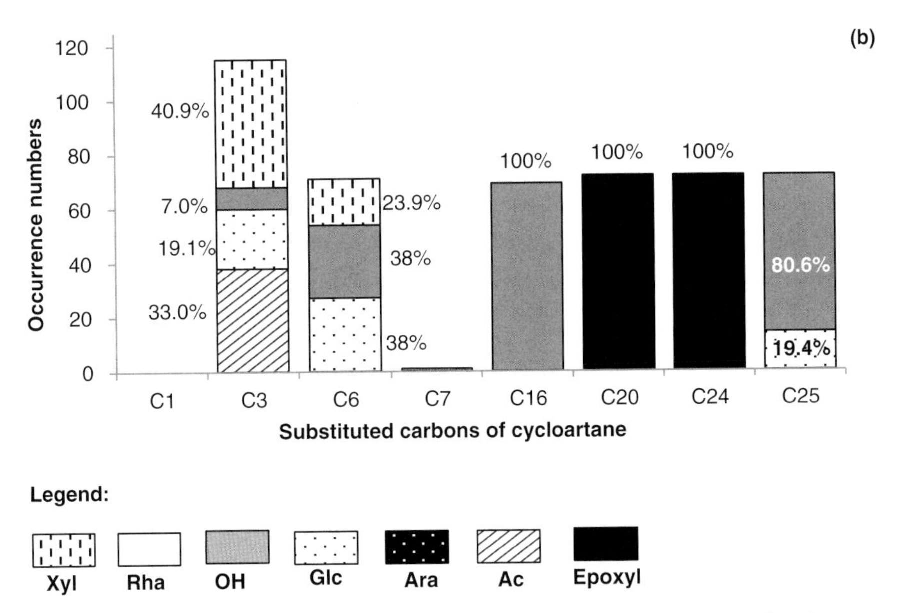

Fig. 27.51. Distribution of the most frequent substitutions through the most substituted carbons of cycloartanes containing terminal lateral chain (a) and 20,24-epoxylated cycle (b). Percentages of chemical groups in each carbon were calculated using the set of the 'most ordinary' saponins, i.e. saponins which were not differentiated by the five structural criteria. These saponins are representative of the most common structural features observed in the genus *Astragalus*.

features highlight some aglycone type-dependent substitution trends in the *Astragalus* community.

Concerning chemical substitution sequences, lateral chain cycloartane showed systematic glycosylation at C3, which was revealed to be generally initiated by xylose (3-O-Xyl) in 51.9% of all the lateral chain cycloartane-based saponins (Fig. 27.52). In other cases, glucose (3-O-Glc), rhamnose (3-O-Rha) and arabinose (3-O-Ara) were directly linked to C3 in 20.25%, 13.9% and 1.3% of cases, respectively. Beyond the first sugar (directly linked to C3), one or two chained sugars can be observed leading to di- or triglycosylated lateral chain cycloartane-based saponins at C3 (Fig. 27.52): the most frequent 3-diglycosylated chain was 3-O-Xyl-Ara (12.7%), followed by 3-O-Xyl-Rha (5.1%), then 3-O-Xyl-Glc (2.5%). Triglycosylation at C3 was rarely observed (3.8%). Apart from glycosylation,

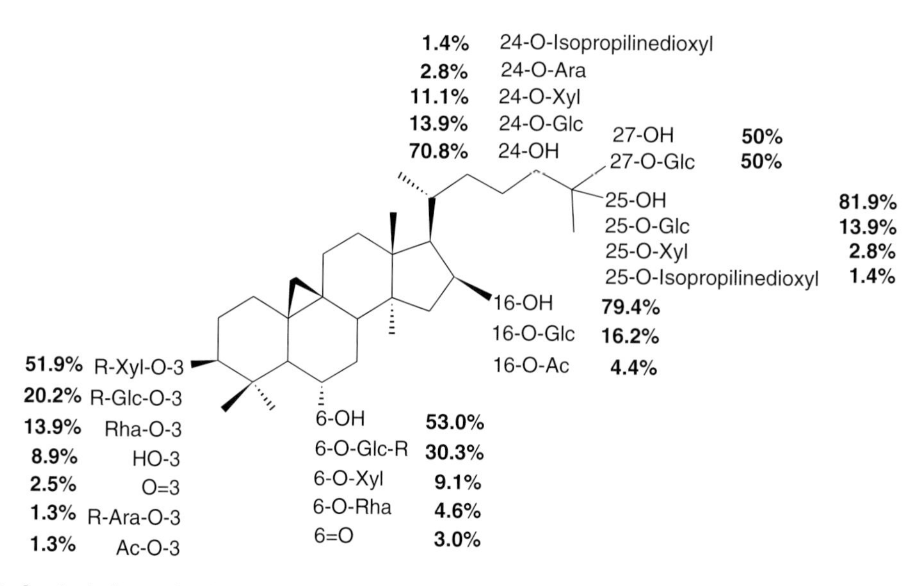

Fig. 27.52. Synthetic figure showing different substitution sequences at different carbons of lateral chain cycloartane in *Astragalus* genus. Arrows show chemical groups from most to least frequent. Legend: Ac = acetyl; Ara = arabinose; Glc = glucose; Rha = rhamnose; Xyl = xylose; R = chemical group attached to first sugar.

acetylation was found at C3. In this case, the acetyl group in lateral chain cycloartane was revealed to be directly attached on the first sugar (generally xylose) (Fig. 27.52): acetyl attached on carbon 2', 3', 4' of xylose showed percentages of 2.52, 1.26 and 1.26% of cases, respectively.

For 20,24-epoxycycloartane, most of the saponins were either mono- or diglycosylated at C3, except some cases where C3 showed a free OH (Fig. 27.53). The C3 was revealed to be directly substituted by xylose in 64.8% of all the 20,24-epoxycycloartane-based saponins. Beyond this general structural feature, less frequent cases were observed through 3-O-monoglycosyls, including 3-O-glucosyl (18.7%) and 3-O-arabinosyl (1.1%). Diglycosylations were also observed, among which the most frequent was the substitution chain 3-O-Xyl-Ara (7.6%). Less frequent diglycoside sequences included 3-O-Xyl-Rha (5.43%) and 3-O-Xyl-Glc (3.26%).

In summary, it is worth noting that C3 was the only carbon to generally attach xylose as the first sugar, whereas all the other glycosylated carbons initially attached glucose.

At carbon 6 of lateral chain cycloartane, hydroxylation was the most frequent feature, with 53% of cases. When glycosylation occurred, it was generally initiated by glucose (30.3%) (Fig. 27.52). In less frequent cases, C6-O-glycosyl was concerned with xylose (9.1%) and rhamnose (4.6%).

For 20,24-epoxycycloartane, when substitution occurred at C6, it consisted generally of free hydroxyl (41.0% of cases) (Fig. 27.53). When 6-O-glycosylation occurred, it was generally initiated by either glucose (28.9%) or xylose (27.7%).

The highest hydroxylation rates in lateral chain cycloartane were observed at C16 (79.4%), C24 (70.8%) and C25 (81.9%). Glycosylation of these three carbons was generally initiated by glucose in 16.2%, 13.9% and 13.9% of all the substitutions occurring at C16, C24, C25, respectively. 24-O-Xylose was also relatively frequent, with a comparable percentage to that of glucose, i.e. 11.1% vs 13.9% (Fig. 27.52).

In 20,24-epoxycycloartane, the C25 and C16 were generally hydroxylated in 84.6% and 93.4% of cases, respectively. Glucosylation was rare at C16 (2.2%) but relatively frequent at C25 (15.4%).

Beyond these general aspects, other substitutions rarely occurred leading to original saponins. They included: acetylation at C16 (2.2%) and C23 (1.1%), free hydroxylation at C7 (2.2%), C11

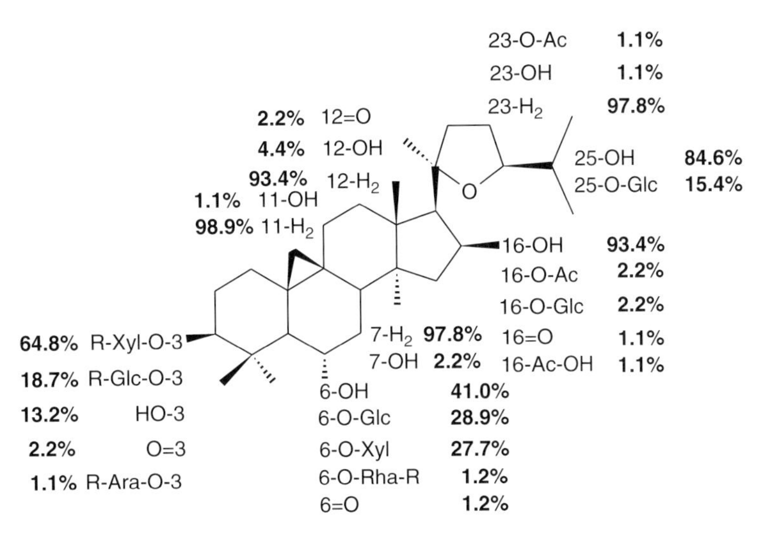

Fig. 27.53. Synthetic figure showing different substitution sequences at different carbons of 20,24-epoxycycloartane in *Astragalus* genus. Arrows show chemical groups from most to least frequent. Legend: Ac = acetyl; Ara = arabinose; Glc = glucose; Rha = rhamnose; Xyl = xylose; R: chemical group attached to first sugar.

(1.1%), C12 (4.35%) and C23 (1.1%), as well as ketone at C12 (2.2%) (Fig. 27.53).

General structural features concerning the most common saponins in *Astragalus* genus are summarized in Tables 27.1 and 27.2. Chemical structures illustrating general saponin features in *Astragalus* are presented in Figs 27.54 and 27.55.

27.6 Analysis of Environmental Meaning of Structural Variations of *Astragalus* Saponins

After characterization of *Astragalus* species by several original saponins based on different structural criteria, it is interesting to analyse their geographical distributions along north–south and east–west axes. This helps to highlight a chemical ecological gradient, enabling the attribution of ecological meaning to some structural criteria.

By considering latitude and longitude of studied plant samples, statistical link analysis showed a significant effect of this geographical factor on desmosylation levels of saponins belonging to different aglycones (Fig. 27.56).

- In lateral chain cycloartane, the average desmosylation level tended to be significantly higher in east–west direction, i.e. in the direction of decreasing longitudes. Significance of such a trend was given by a p-value of 0.0021 for calculated Fisher F ratio statistics.
- In 20,24-epoxycycloartane, desmosylation level tended to be significantly higher in north–south direction, i.e. with decreasing latitude (p-value = 0.0002).

These two desmosylation gradients proved to be associated with inverse hydroxylation gradients: in fact, hydroxylation showed increasing levels with longitude and latitude in lateral chain cycloartane and 20,24-epoxycycloartane, respectively. Such inversion between desmosylation and hydroxylation levels could be explained by the fact that OH (O-H) initially occur on sapogenin to be substituted by O-glycosylation in next metabolic steps. The result is that the next glycosylations of initially hydroxylated carbons increase desmosylation levels at the expense of free OH. Inversely, as long as glycosylation did not occur, the free OH level remains relatively higher.

Moreover, glycosylation levels were revealed to be positively linked to desmosylation levels (Fig. 27.57). Geographical trends were also found to be aglycone type-dependent: higher glycosylation levels tended to be found in southern locations for 20,24-epoxycycloartane-based saponins (higher

Table 27.1 Summary of general (most frequent) structural features of saponins in *Astragalus* genus, based on lateral chain cycloartane with chemical substitutions indicated in Fig. 27.51.

Sp. No.	*Astragalus* species (Sp)	Saponin names	Plant tissue	References
1	*A. hareftae*	Hareftoside A	Whole plant	Horo *et al.*, 2012
2	*A. alexandrinus* (Boiss)	Alexandroside I	Aerial plant and roots	Orsini *et al.*, 1994
3	*A. amblolepis* (Fischer)	unnamed 1	Roots	Polat *et al.*, 2009
		unnamed 2	Roots	Polat *et al.*, 2009
4	*A. aureus* (Willd)	unnamed 2	Whole plant	Gülcemal *et al.*, 2011
		unnamed 4	Whole plant	Gülcemal *et al.*, 2011
		unnamed 5	Whole plant	Gülcemal *et al.*, 2011
5	*A. chivensis* (Bunge)	Aleksandroside I	Aerial plant	Naubeev *et al.*, 2007
		Cyclochivinoside B	Aerial plant	Naubeev *et al.*, 2007
		Cyclochivinoside C	Aerial plant	Naubeev and Uteniyazov, 2007
6	*A. macropus* (Bunge)	Cyclomacroside E	Roots	Iskenderov *et al.*, 2010
		Cyclomacrogenin B	Roots	Iskenderov *et al.*, 2008a
		Cyclomacroside C	Roots	Iskenderov *et al.*, 2009a
		Cyclomacroside D	Roots	Iskenderov *et al.*, 2009b
7	*A. oleifolius* (DC)	Cyclocanthoside E	Lower stem	Özipek *et al.*, 2005
		Macrophyllosaponins B, C, D	Roots	Çalis *et al.*, 1996
		Macrophyllosaponin E	Roots	Bedir *et al.*, 2000
		Oleifoliosides A, B	Lower stem	Özipek *et al.*, 2005
8	*A. pycnanthus* (Boriss)	Cyclopycoanthoside	Seed	Agzamova and Isaev, 1998c
9	*A. stereocalyx*	unnamed 4	Roots	Yalçin *et al.*, 2012
		unnamed 5	Roots	Yalçin *et al.*, 2012
10	*A. taschkendicus* (C.Bge)	Cycloasgenin C	Roots	Isaev *et al.*, 1982b
11	*A. tragacantha* (Habl)	Cyclocanthoside D	Stem	Fadeev *et al.*, 1988
		Cyclocanthoside E	Stem	Isaev *et al.*, 1992
		Cyclocanthogenin	Stem	Isaev *et al.*, 1992
		Cyclocanthoside A	Stem	Isaev *et al.*, 1992
12	*A. trigonus* (DC)	unnamed 1	Roots	Verotta *et al.*, 1998
13	*A. trojanus* (Stev)	Trojanoside C	Roots	Bedir *et al.*, 1999

Table 27.2. Summary of general (most frequent) structural features of saponins in *Astragalus* genus based on 20,24-epoxycycloartane with chemical substitutions indicated in Fig. 27.51.

Sp. No.	*Astragalus* species	Saponin names	Plant tissue	References
1	*Astagalus brachypterus*	Brachyoside B	Roots	Bedir *et al.*, 1998b
2	*A. alexandrinus* (Boiss)	Astraversianins VI, X, XV	Aerial parts, roots	Orsini *et al.*, 1994
3	*A. amarus* (Pall)	Cycloaraloside A	Roots	Isaev *et al.*, 1989
		Cycloaraloside E	Roots	Isaev and Abubakirov, 1990a
		Cyclosieversigenin	Roots	Isaev *et al.*, 1989
4	*A. dissectus* (Fedtsch)	Cyclosieversioside E	Roots, stem	Sukhina *et al.*, 1999
		unnamed 6	Roots, stem	Sukhina *et al.*, 2000
5	*A. exilis* (A. Kor)	Cycloexoside	Roots	Imomnazarov and Isaev, 1992
		Cycloexoside B	Roots	Mamedova *et al.*, 2002c

Continued

Table 27.2. Continued.

Sp. No.	*Astragalus* species	Saponin names	Plant tissue	References
6	*A. galegiformis* (L.)	Cyclogaleginoside A	Flowers	Alaniya *et al.*, 1985
		Cyclogaleginoside B	Flowers	Alaniya *et al.*, 1985
7	*A. lehmannianus* (Bunge)	Cyclolehmanoside C	Aerial parts	Zhanibekov *et al.*, 2013
8	*A. membranaceus* (Bunge)	Astramembranoside A	Roots	Kim *et al.*, 2008
		Acetylastragaloside I	Roots	Kitagawa *et al.*, 1983c
		Astragalosides I, II, IV	Roots	Kitagawa *et al.*, 1983c
		Isoastragalosides I, II	Roots	Kitagawa *et al.*, 1983c
9	*A. microcephalus* (Willd)	Brachyoside B	Roots	Bedir *et al.*, 1998b
		Cyclocephaloside II	Roots	Bedir *et al.*, 1998b
10	*A. oldenburgii* (Fedtsch)	Cyclosieversioside F	Aerial parts	Naubeev and Isaev, 2012
11	*A. pycnanthus* (Boriss)	Cyclosieversioside F	Seed	Agzamova and Isaev, 1998c
12	*A. sieberi*	Sieberosides I, II	Aerial parts	Verotta *et al.*, 1998
13	*A. sieversianus* (Paull)	Cyclosieversiosides A, B, F	Roots	Iskenderov *et al.*, 2008b
14	*A. trigonus* DC	Trigonoside I	Roots	Gariboldi *et al.*, 1995
15	*A. unifoliolatus* (Bunge)	Cyclounifolioside B	Aerial parts	Kucherbaev *et al.*, 2002
16	*A. verrucosus* (Moris)	Astraverrucins I, III	Aerial parts	Pistelli *et al.*, 1997
		Astraverrucin VII	Aerial parts	Pistelli *et al.*, 2003

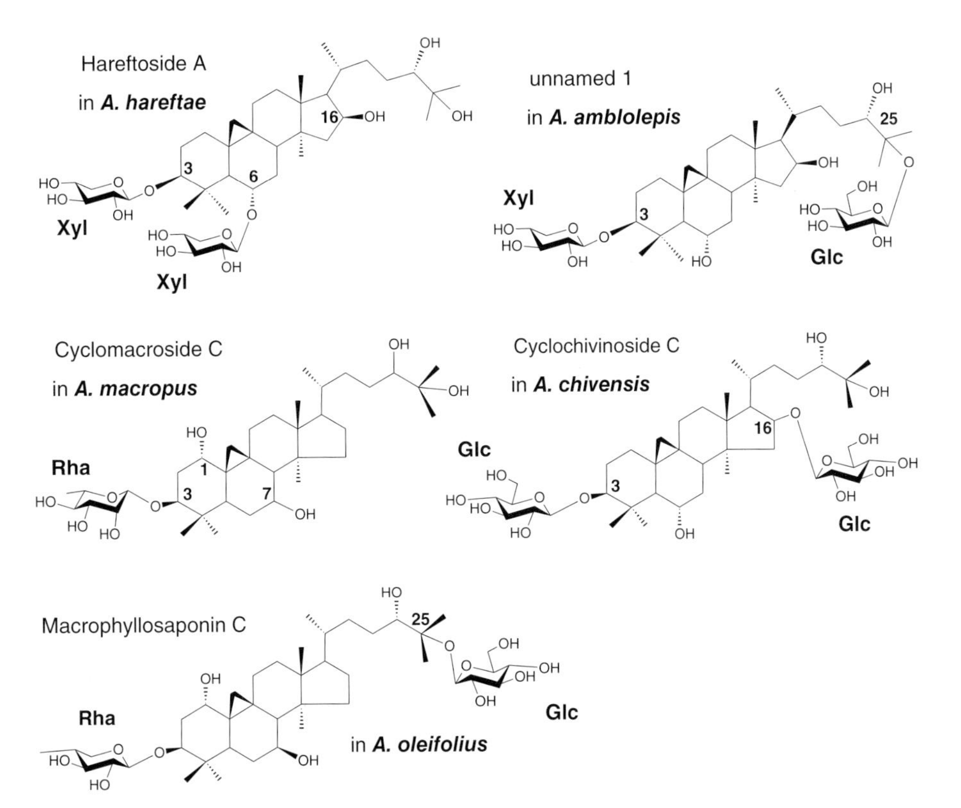

Fig. 27.54. Most common saponin structures in *Astragalus* community based on lateral chain cycloartane.

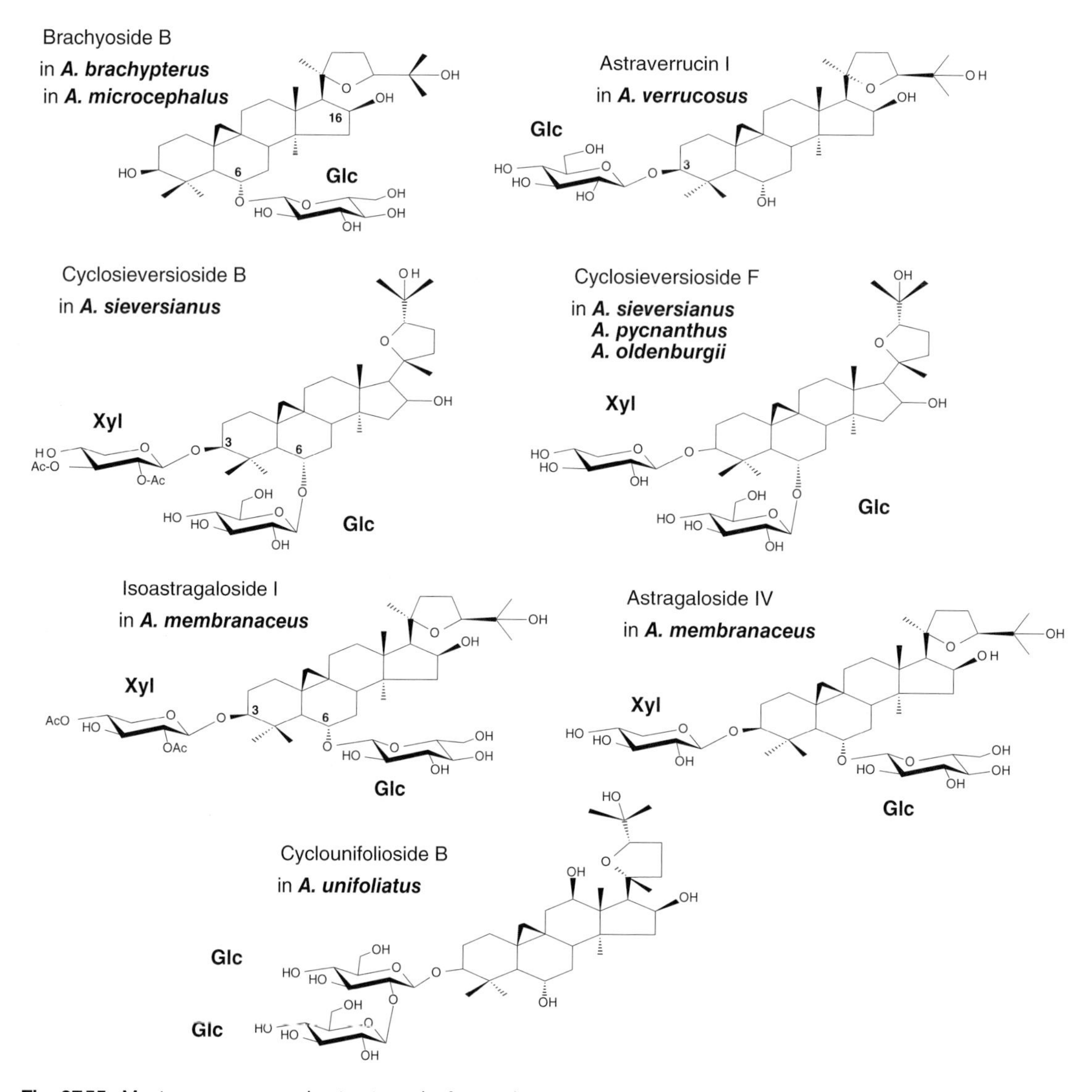

Fig. 27.55. Most common saponin structures in *Astragalus* community based on 20,24-epoxycycloartane.

glycosylated saponins in 37° N than 42° N) (Fig. 27.57). Also, higher glycosylation levels were found in western locations (more in 30° E than 50° E) for lateral chain cycloartane-based saponins (Fig. 27.57). These observations aligned with those of desmosylation levels (Fig. 27.56).

These geographical gradients of desmosylation-hydroxylation levels could be indicative of metabolic responses of *Astragalus* species to gradual variations in the environmental conditions, including regional climate as well as local soil and co-occurring biological populations. More precise responses on environmental factors governing these metabolic gradients could be provided through experiments carried out under controlled abiotic and biotic conditions.

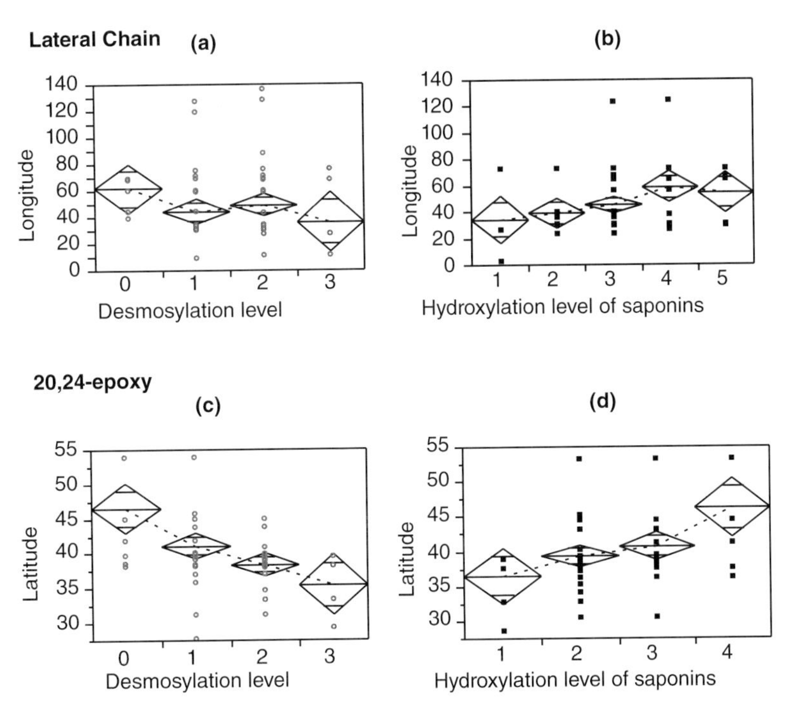

Fig. 27.56. Different geographical trends of desmosylation (a, c) and hydroxylation levels (b, d) in saponins of *Astragalus* species. Longitude gradient of desmosylation (a) and hydroxylation (b) levels in lateral chain cycloartane. Latitude gradient of desmosylation (c) and hydroxylation (d) levels in 20,24-epoxycycloartane.

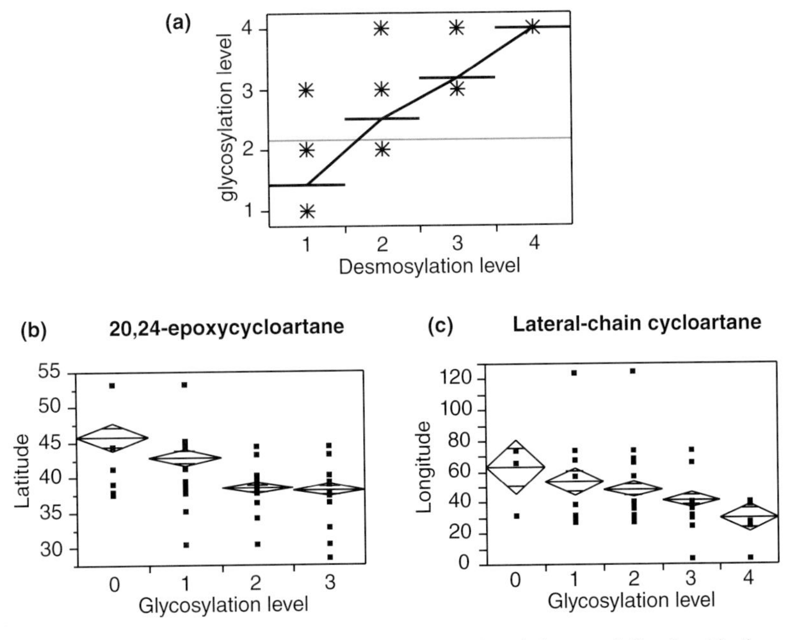

Fig. 27.57. (a) Plot showing postive trend between glycosylation level and desmosylation level in the saponins of *Astragalus* plants. (b, c) Plots showing geographical trends of glycosylation level along latitude (b) and longitude (c) directions for 20,24-epoxycycloartane- and lateral chain cycloartane-based saponins, respectively.

References

Agzamova, M. and Isaev, M. (1991) Triterpene glycosides of astragalus and their genins. XXXVIII. Cycloalpigenin D and cycloalpioside D from *Astragalus alopecurus*. *Chemistry of Natural Compounds* 27(3), 326–332.

Agzamova, M. and Isaev, M. (1994a) Triterpene glycosides of *Astragalus* and their genins. XLVII. Structures of cycloalpigenin a and cycloalpioside A. *Chemistry of Natural Compounds* 30(3), 346–351.

Agzamova, M. and Isaev, M. (1994b) Triterpene glycosides of *Astragalus* and their genins. XLVIII. The structures of cycloalpigenin B and cycloalpioside B. *Chemistry of Natural Compounds* 30(4), 474–479.

Agzamova, M. and Isaev, M. (1995a) *Astragalus* triterpene glycosides and their genins. I. Structures of cycloalpigenin and cycloalpioside. *Chemistry of Natural Compounds* 31(5), 589–595.

Agzamova, M. and Isaev, M. (1995b) Triterpene glycosides of *Astragalus* and their genins XLIX. Structures of cycloalpigenin C and cycloalpioside C. *Chemistry of Natural Compounds* 31(1), 70–75.

Agzamova, M. and Isaev, M. (1997) Triterpene glycosides of *Astragalus* and their genins LV. Structure of cycloorbigenin A. *Chemistry of Natural Compounds* 33(6), 663–664.

Agzamova, M. and Isaev, M. (1998a) Triterpene glycosides of *Astragalus* and their genins LVIII. The structure of dihydrocycloorbigenin A. *Chemistry of Natural Compounds* 34(4), 477–479.

Agzamova, M. and Isaev, M. (1998b) Triterpene glycosides of *Astragalus* and their genins LVII. The structure of cyclopycnanthogenin. *Chemistry of Natural Compounds* 34(4), 474–476.

Agzamova, M. and Isaev, M. (1998c) Triterpene glycosides of *Astragalus* and their genins. LVI. Cyclopycnanthoside – A new cycloartane glycoside. *Chemistry of Natural Compounds* 34(2), 155–159.

Agzamova, M., Isaev, M., Gorovits, M. and Abubakirov, N. (1987a) Triterpene glycosides of *Astragalus* and their genins. XXII. Cycloorbicoside A from *Astragalus orbiculatus*. *Chemistry of Natural Compounds* 22(6), 671–676.

Agzamova, M., Isaev, M., Gorovits, M. and Abubakirov, N. (1987b) Triterpene glycosides of *Astragalus* and their genins. XX. Cycloorbigenin from *Astragalus orbiculatus*. *Chemistry of Natural Compounds* 22(4), 425–429.

Agzamova, M., Isaev, M., Gorovits, M. and Abubakirov, N. (1988) Triterpene glycosides of *Astragalus* and their genins. XXIV. Cycloorbicoside G from *Astragalus orbiculatus*. *Chemistry of Natural Compounds* 23(6), 696–700.

Agzamova, M., Isaev, M., Gorovits, M. and Abubakirov, N. (1990) Triterpene glycosides of *Astragalus* and their genins. XXXI. Cycloorbigenin B from *Astragalus orbiculatus*. *Chemistry of Natural Compounds* 25(6), 688–690.

Alaniya, M.D., Isaev, M.I., Gorovits, M.B., Abdullaev, N.D., Kemertelidze, E.P. and Abubakirov, N.K. (1985) Triterpene glycosides of *Astragalus* and their genins. XVI. Cyclogaleginosides A and B from *Astragalus galegiformis*. *Chemistry of Natural Compounds* 20(4), 451–454.

APG (Angiosperm Phylogeny Group) (2003) An update of the Angiosperm Phylogeny Group classification for the orders and families of flowering plants: APG II. *Botanical Journal of the Linnean Society* 141, 399–436.

Atta-ur-Rahman, A., Ata, A., Naz, S., Choudhary, M.I., Sener, B. and Turkoz, S. (1999) New steroidal alkaloids from the roots of *Buxus sempervirens*. *Journal of Natural Products* 62, 665.

Avunduk, S., Mitaine-Offer, A.C., Alankuş-Çalişkan, Ö., Miyamoto, T., Şenol, S.G. and Lacaille-Dubois, M.A. (2008) Triterpene glycosides from the roots of *Astragalus flavescens*. *Journal of Natural Products* 71(1), 141–145.

Aytaç, Z. (2000) *Astragalus* L. In: Güner A, Özhatay N, Ekim T and Başer, K.H.C. (eds) *Flora of Turkey and the East Aegean Islands*, vol. 11. Edinburgh University Press, Edinburgh, UK, pp. 79–88.

Bedir, E., Çalis, I., Zerbe, O. and Sticher, O. (1998a) Cyclocephaloside I: a novel cycloartane-type glycoside from *Astragalus microcephalus*. *Journal of Natural Products* 61(4), 503–505.

Bedir, E., Çalis, I., Aquino, R., Piacente, S. and Pizza, C. (1998b) Cycloartane triterpene glycosides from the roots of *Astragalus brachypterus* and *Astragalus microcephalus*. *Journal of Natural Products* 61(12), 1469–1472.

Bedir, E., Çalis, I., Aquino, R., Piacente, S. and Pizza, C. (1999) Secondary metabolites from the roots of *Astragalus trojanus*. *Journal of Natural Products* 62(4), 563–568.

Bedir, E., Çalis, I. and Khan, I.A. (2000) Macrophyllosaponin E: a novel compound from the roots of *Astragalus oleifolius*. *Chemistry and Pharmaceutical Bulletin* 48(7), 1081–1083.

Bedir, E., Tatli, I.I., Çalis, I. and Khan, I.A. (2001) Trojanosides I–K: new cycloartane-type glycosides from the aerial parts of *Astragalus trojanus*. *Chemistry and Pharmaceutical Bulletin* 49(11), 1482–1486.

Böttger, S. and Melzig, M.F. (2011) Triterpenoid saponins of the Caryophyllaceae and Illecebraceae family. *Phytochemistry Letters* 4, 59–68.

Çalış, I., Zor, M., Saracoğlu, İ., Işımer, A. and Rüegger, H. (1996) Four novel cycloartane glycosides from *Astragalus oleifolius*. *Journal of Natural Products* 59(11), 1019–1023.

Çalis, I., Yusufoglu, H., Zerbe, O. and Sticher, O. (1999) Cephalotoside A: a tridesmosidic cycloartane type glycoside from *Astragalus cephalotes* var. Brevicalyx. *Phytochemistry* 50(5), 843–847.

Çalis, I., Gazar, H.A., Piacente, S. and Pizza, C. (2001) Secondary metabolites from the roots of *Astragalus zahlbruckneri*. *Journal of Natural Products* 64(9), 1179–1182.

Çalis, I., Barbic, M. and Jürgenliemk, G. (2008a) Bioactive cycloartane-type triterpene glycosides from *Astragalus elongatus*. *Zeitschrift für Naturforschung C – A Journal of Biosciences*, 63(11/12), 813–820.

Çalis, I., Dönmez, A.A., Perrone, A., Pizza, C. and Piacente, S. (2008b) Cycloartane glycosides from *Astragalus campylosema* Boiss. ssp. *campylosema*. *Phytochemistry* 69(14), 2634–2638.

Choudhary, M.I., Shahnaz, S., Parveen, S., Khalid, A., Ayatollahi, S.A.M. and Atta-ur-Rahman, P.M. (2003) New triterpenoid alkaloid cholinesterase inhibitors from *Buxus hyrcana*. *Journal of Natural Products* 66(6), 739–742.

Choudhary, M.I., Jan, S., Abbaskhan, A., Musharraf, S.G., Samreen Sattar, S.A. and Atta-ur-Rahman, P.M. (2008) Cycloartane triterpenoids from *Astragalus bicuspis*. *Journal of Natural Products* 71(9), 1557–1560.

Croteau, R., Kutchan, T.M. and Lewis N.G. (2000) Natural products (secondary metabolites). In: Buchanan, B., Gruissem, W. and Jones, R. (eds) *Biochemistry and Molecular Biology of Plants*. American Society of Plant Physiologists, Rockville, Maryland, pp. 1250–1318.

Cui, B., Sakai, Y., Takeshita, T., Kinjo, J. and Nohara T. (1992a) Four new oleanene derivatives from the seeds of *Astragalus complanatus*. *Chemistry* and *Pharmaceutical Bulletin* 40(1), 136–138.

Cui, B., Inoue, J., Takeshita, T., Kinjo, J. and Nohara, T. (1992b) Triterpene glycosides from the seeds of *Astragalus sinicus* I. *Chemistry* and *Pharmaceutical Bulletin* 40(12), 3330–3333.

Dai, J. and Mumper, R.J. (2010) Plant henolics: extraction, analysis and their antioxidant and anticancer properties. *Molecules* 15, 7313–7352.

Davis, A.M. (1982) Crude protein, crude fiber, tannin, and oxalate concentrations of 33 *Astragalus* species. *Journal of Range Management* 35, 32–34.

Djimtombaye, B.J., Alankus çalıskan, O., Gülcemal, D., Khan, I.A., Anıl, H. and Bedir, E. (2013) Unusual secondary metabolites from *Astragalus halicacabus* Lam. *Chemistry and Biodiversity* 10(7), 1328–1334.

Fadeev, Yu M., Isaev, M.I., Akimov, Yu A., Kintya, P.K., Gorovits, M.B. and Abubakirov, N.K. (1988) Triterpeneglycosides of *Astragalus* and their genins. XXV. Cyclocanthoside D from *Astragalus tragacantha*. *Chemistry of Natural Compounds* 24(1), 62–65.

Gariboldi, P., Pelizzoni, F., Tatò, M., Verotta, L., El-Sebakhy, N., Asaad, A.M., Abdallah, R.M and Toaima, S.M. (1995) Cycloartane triterpene glycosides from astragalus trigonus. *Phytochemistry* 40(6), 1755–1760.

Gülcemal, D., Alankuş-Çalışkan, Ö., Perrone, A., Özgökçe, F., Piacente, S. and Bedir E. (2011) Cycloartane glycosides from *Astragalus aureus*. *Phytochemistry* 72(8), 761–768.

Gülcemal, D., Masullo, M., Napolitano, A., Karayıldırım, T., Bedir, E., Alankuş-Çalışkan, Ö. and Piacente, S. (2013) Oleanane glycosides from *Astragalus tauricolus*: isolation and structural elucidation based on a preliminary liquid chromatography-electrospray ionization tandem mass spectrometry profiling. *Phytochemistry* 86, 184–194.

Hirotani, M., Zhou, Y., Rut, H. and Furuya, T. (1994) Cycloartane triterpene glycosides from the hairy root cultures of *Astragalus membranaceus*. *Phytochemistry* 37(5), 1403–1407.

Horo, I., Bedir, E., Perrone, A., Özgökçe, F., Piacente, S. and Alankuş-Çalışkan, Ö. (2010) Triterpene glycosides from *Astragalus icmadophilus*. *Phytochemistry* 71(8), 956–963.

Horo, I., Bedir, E., Masullo, M., Piacente, S., Özgökçe, F. and Alankuş-Çalışkan, Ö. (2012) Saponins from *Astragalus hareftae* (NAB.) *SIRJ*. *Phytochemistry* 84, 147–153.

Hostettmann, V.K. and Marston, U.A. (1995) *Saponins*. Cambridge University Press, Cambridge, UK.

Howe, G.A. and Jander, G. (2008) Plant immunity to insect herbivores. *Annual Reviews in Plant Biology* 59, 41–66.

Imomnazarov, B. and Isaev, M. (1992) Triterpene glycosides of *Astragalus* and their genins XL. Cyclocarposide B from *Astragalus coluteocarpus*. *Chemistry of Natural Compounds* 28, 195–198.

Ionkova, I. (1991) Producing triterpene saponins by conventional and genetically transformed cultures of *Astragalus hamosus* L. (Fabaceae). *Problems Farmakol Farmat* 5, 32–38.

Isaev, M.I. (1991) Triterpene glycosides and their genins from *Astragalus*. XXXIX. Cycloaraloside D from *Astragalus amarus*. *Chemistry of Natural Compounds* 27(4), 457–459.

Isaev, M.I. and Abubakirov, N.K. (1990b) Triterpene glycosides of *Astragalus* and their genins XXXV. Cycloaraloside C from *Astragalus amarus*. *Chemistry of Natural Compounds* 26(6), 667–670.

Isaev, M. and Abubakirov, N. (1990a) Triterpene glycosides of *Astragalus* and their genins XXXIV. Cycloaraloside E from *Astragalus amarus*. *Chemistry of Natural Compounds* 26(5), 559–561.

Isaev, M.I. and Imomnazarov, B.A. (1991) Triterpene glycosides of *Astragalus* and their genins XXXVII. Cycloaraloside F from *Astragalus amarus* and *Astragalus villosissimus*. *Chemistry of Natural Compounds* 27(3), 323–326.

Isaev, I. and Isaev, M. (2011a) Triterpene glycosides from astragalus and their genins. XC. Askendoside K from *Astragalus taschkendicus*. *Chemistry of Natural Compounds* 47(4), 587–591.

Isaev, I. and Isaev, M. (2011b) Triterpene glycosides from *Astragalus* and their genins. LXXXIX. Askendoside H from *Astragalus taschkendicus*. *Chemistry of Natural Compounds* 47, 411–414.

Isaev, M., Gorovits, M., Abdullaev, N., Yagudaev, M. and Abubakirov, N. (1982a) Triterpene glycosides of *Astragalus* and their genins. III. Cycloasgenin A from *Astragalus taschkendicus*. *Chemistry of Natural Compounds* 17(5), 411–418.

Isaev, M.I., Gorovits, M.B., Abdullaev, N.D. and Abubakirov, N.K. (1982b) Triterpene glycosides of *Astragalus* and their genins. VI. Cycloasgenin C from *Astragalus taschkendicus*. *Chemistry of Natural Compounds* 18(4), 424–430

Isaev, M.I., Gorovits, M.B., Abdullaev, N.D. and Abubakirov, N.K. (1983a) Triterpene glycosides of *Astragalus* and their genins. XII. Askendoside B from *Astragalus taschkendicus*. *Chemistry of Natural Compounds* 19, 428–431.

Isaev, M.I., Gorovits, M.B., Abdullaev, N.D. and Abubakirov, N.K. (1983b) Triterpene glycosides of *Astragalus* and their genins. IX. Askendoside D from *Astragalus taschkendicus*. *Chemistry of Natural Compounds* 19, 170–174.

Isaev, M., Gorovits, M., Abdullaev, N. and Abubakirov, N. (1984) Triterpene glycosides of *Astragalus* and their genins. XVII. Cycloasgenin B from *Astragalus tashkendicus*. *Chemistry of Natural Compounds* 20, 691–694.

Isaev, M., Gorovits, M.B. and Abubakirov, N.K. (1985a) Triterpenoids of the cycloartane series. *Chemistry of Natural Compounds* 21, 399–447.

Isaev, M., Umarova, R., Gorovits, M. and Abubakirov, N. (1985b) Triterpene glycosides of *Astragalus* and their genins. XVIII. 3-Dehydrocycloasgenin C from *Astragalus taschkendicus*. *Chemistry of Natural Compounds* 21, 203–208.

Isaev, M., Gorovits, M. and Abubakirov, N. (1989) Triterpene glycosides of *Astragalus* and their genins XXX. Cycloaraloside A from *Astragalus amarus*. *Chemistry of Natural Compounds* 25(6), 684–687.

Isaev, M., Imomnazarov, B., Fadeev, Y.M. and Kintya, P. (1992) Triterpene glycosides of *Astragalus* and their genins XLII. Cycloartanes of *Astragalus tragacantha*. *Chemistry of Natural Compounds* 28(3–4), 315–320.

Isaev, I.M., Iskenderov, D.A. and Isaev, M.I. (2010a) Triterpene glycosides and their genins from *Astragalus*. LXXXIV. Secomacrogenin B, a new 9,10-seco-cycloartane. *Chemistry of Natural Compounds* 46(1), 36–38.

Isaev, I., Mamedova, R., Agzamova, M. and Isaev, M. (2010b) Triterpene glycosides from *Astragalus* and their genins. LXXXVI. Chemical transformation of cycloartanes. VIII. Syntheses based on cycloorbicoside A. *Chemistry of Natural Compounds* 46(3), 400–406.

Iskenderov, D., Isaev, I. and Isaev, M. (2008a) Triterpene glycosides from *Astragalus* and their genins. LXXVII. Cyclomacrogenin B, a new cycloartane triterpenoid. *Chemistry of Natural Compounds* 44(5), 621–624.

Iskenderov, D.A., Keneshov, B.M. and Isaev, M.I. (2008b) Triterpene glycosides from *Astragalus* and their genins. LXXVI. Glycosides from *A. sieversianus*. *Chemistry of Natural Compounds* 44(3), 319–323.

Iskenderov, D., Isaev, I. and Isaev, M. (2009a) Triterpene glycosides from *Astragalus* and their genins. LXXIX. Structure of cyclomacroside C. *Chemistry of Natural Compounds* 45(1), 132–134.

Iskenderov, D.A., Isaev, I.M. and Isaev, M.I. (2009b) Triterpene glycosides from *Astragalus* and their genins. LXXX. Cyclomacroside D, a new bisdesmoside. *Chemistry of Natural Compounds* 45(1), 55–58.

Iskenderov, D.A., Isaev, I.M. and Isaev, M.I. (2009c) Triterpene glycosides from *Astragalus* and their genins. LXXXII. Cyclomacroside B, a new glycoside. *Chemistry of Natural Compounds* 45(4), 511–513.

Iskenderov, D.A., Isaev, I.M. and Isaev, M.I. (2009d) Triterpene glycosides from *Astragalus* and their genins. LXXXIII. Structure of cyclomacroside A. *Chemistry of Natural Compounds* 45(5), 656–659.

Iskenderov, D., Isaev, I. and Isaev, M. (2010) Triterpene glycosides from *Astragalus* and their genins. LXXXV. Structure of cyclomacroside E. *Chemistry of Natural Compounds* 46(2), 250–253.

Kadota, S., Li, J.X., Tanaka, K. and Namba, T. (1995) Constituents of cimicifugae rhizoma II.Isolation and structures of new cycloartenol triterpenoids and related compounds from *Cimicifuga foetida* L. *Tetrahedron* 51, 1143.

Kim, J.S., Yean, M.H., Lee, E.J., Jung, H.S., Lee, J.Y., Kim, Y.J. and Kang, S.S. (2008) Two new cycloartane saponins from the roots of *Astragalus membranaceus*. *Chemistry and Pharmaceutical Bulletin* 56(1), 105–108.

Kitagawa, I., Yoshikawa, M. and Yosioka, I. (1976) Saponin and sapogenol. XIII. structures of three soybean saponins: Soyasaponin I, Soyasaponin II, and Seyasaponin III. *Chemistry and Pharmaceutical Bulletin* 24, 121.

Kitagawa, I., Wang, H.K., Saito, M. and Yoshikawa, M. (1983a) Saponin and sapogenol. XXXIII: Chemical constituents of the seeds of *Vigna angularis* (Willd.) Ohwi et Ohashi. (3): Azukisaponins V and VI. *Chemistry and Pharmaceutical Bulletin* 31(2), 683–688.

Kitagawa, I., Hui Kang, W., Takagi, A., Fuchida, M., Miura, I. and Yoshikawa, M. (1983b) Saponin and sapogenol. XXXIV: Chemical constituents of astragali radix, the root of *Astragalus membranaceus* Bunge. (1): Cycloastragenol, the 9,19-cycloanostane-type aglycone of astragalosides, and the artifact aglycone astragenol. *Chemistry and Pharmaceutical Bulletin* 31(2), 689–697.

Kitagawa, I., Wang, H., Saito, M., Takagi, A. and Yoshikawa, M. (1983c) Saponin and sapogenol. XXXV. Chemical constituents of Astragali radix, the root of *Astragalus membranaceus* Bunge. (2). Astragalosides I, II and IV, acetylastragaloside I and isoastragalosides I and II. *Chemistry and Pharmaceutical Bulletin* 31(2), 698–708.

Kitagawa, I., Wang, H., Saito, M. and Yoshikawa M. (1983d) Saponin and sapogenol. XXXVI. Chemical constituents of astragali radix, the root of *Astragalus membranaceus* Bunge. (3). Astragalosides III, V, and VI. *Chemistry and Pharmaceutical Bulletin* 31(2), 709–715.

Kitagawa, I., Wang, H. and Yoshikawa, M. (1983e) Saponin and sapogenol. XXXVII. Chemical constituents of astragali radix, the root of *Astragalus membranaceus* Bunge. (4). Astragalosides VII and VIII. *Chemistry and Pharmaceutical Bulletin* 31(2), 716–722.

Krasteva, I., Nikolov, S., Kaloga, M. and Mayer, G. (2006) Triterpenoid saponins from *Astragalus corniculatus*. *Zeitschrift für Naturforschung B* 61, 1166–1169.

Krasteva, I., Nikolov, S., Kaloga, M. and Mayer, G. (2007) A new saponin lactone from *Astragalus corniculatus*. *Natural Products Research* 21, 941–945.

Kuang, H., Okada, Y., Yang, B., Tian, Z. and Okuyama, T. (2009) Secocycloartane triterpenoidal saponins from the leaves of *Astragalus membranaceus* Bunge. *Helvetica Chimica Acta* 92(5), 950–958.

Kuang, H.X., Wang, Q.H., Yang, B.Y., Wang, Z.B., Okada, Y., Okuyama, T. and Huangqiyenins, G.J. (2011) Four new 9,10-secocycloartane (=9,19-cyclo-9,10-secolanostane) triterpenoidal saponins from *Astragalus membranaceus* Bunge leaves. *Helvetica Chimica Acta* 94(12), 2239–2247.

Kucherbaev, K.D., Uteniyazov, K., Kachala, V., Saatov, Z., Shashkov, A., Uteniyazov, K. and Khalmuratov, P. (2002) Triterpene glycosides from *Astragalus*. Structure of cyclounifolioside B from *Astragalus unifoliolatus*. *Chemistry of Natural Compounds* 38(1), 62–65.

Li-Xian, G., Xiano-Bing, H. and Yu-Qun, C. (1986) Astrasieversianins IX, XI and XV, cycloartane derived saponins from *Astragalus sieversianus*. *Phytochemistry* 25(10), 1437–1441.

Linnek, J., Mitaine-Offer, A.C., Miyamoto, T., Tanaka, C., Paululat, T., Avunduk, S., Alankuş-Çalişkan, Ö. and Lacaille-Dubois, M.A. (2011) Cycloartane glycosides from three species of *Astragalus* (Fabaceae). *Helvica Chimica Acta* 94(2), 230–237.

Ma, Y., Tian, Z., Kuang, H., Yuan, C., Shao, C., Ohtani, K., Kasai, R., Tanaka, O., Okada, Y. and Okuyama, T. (1997) Studies of the constituents of *Astragalus membranaceus* Bunge. III. Structures of triterpenoidal glycosides, huangqiyenins A and B, from the leaves. *Chemistry and Pharmaceutical Bulletin* 45(2), 359–361.

Mamedova, R., Agzamova, M. and Isaev, M. (2002a) Triterpene glycosides of *Astragalus* and their genins. LXV. Cycloartane and lanostane triterpenoids of *Astragalus orbiculatus*. *Chemistry of Natural Compounds* 38(4), 354–355.

Mamedova, R.P., Agzamova, M.A. and Isaev, M.I. (2002b) Triterpene glycosides of *Astragalus* and their genins. LXVI. Cycloorbicoside C, a new bisdesmoside. *Chemistry of Natural Compounds* 38(6), 570–573.

Mamedova, R., Agzamova, M. and Isaev M. (2002c) Triterpene glycosides of *Astragalus* and their genins. LXVII. Structure of cycloexoside B. *Chemistry of Natural Compounds* 38(6), 579–582.

Mamedova, R.P., Agzamova, M.A. and Isaev, M.I. (2003) Triterpene glycosides and their genins from *Astragalus*. LXIX. Orbigenin, the first lanostanoid from *Astragalus* plants. *Chemistry of Natural Compounds* 39(5), 475–478.

Mamedova, R., Agzamova, M. and Isaev, M. (2005) Triterpene glycosides from *Astragalus* and their genins. LXXI. Cycloorbicoside D, a new glycoside from *Astragalus orbiculatus*. *Chemistry of Natural Compounds* 41(4), 429–431.

Mitaine-Offer, A.C., Miyamoto, T., Semmar, N., Jay, M. and Lacaille-Dubois, M.A. (2006) A new oleanane glycoside from the roots of *Astragalus caprinus*. *Magnetic Resonance in Chemistry* 44(7), 713–716.

Ndamba, J., Lemmich, E. and Mølgaard P. (1994) Investigation of the diurnal, ontogenetic and seasonal variation in the molluscicidal saponin content of *Phytolacca dodecandra* aqueous berry extracts. *Phytochemistry* 35, 95–99.

Naubeev, T.K. and Uteniyazov, K. (2007) Structure of cyclochivinoside C from *Astragalus chivensis*. *Chemistry of Natural Compounds* 43(5), 560–562.

Naubeev, T.K. and Isaev, M.I. (2012) Triterpene glycosides of *Astragalus* and their genins. XCII. Cycloartane glycosides from *A. oldenburgii*. *Chemistry of Natural Compounds* 48(4), 704–705.

Naubeev, T.K., Uteniyazov, K., Kachala, V. and Shashkov, A. (2007) Cyclochivinoside B from the aerial part of *Astragalus chivensis*. *Chemistry of Natural Compounds* 43(2), 166–169.

Naubeev, T.K., Uteniyazov, K.K. and Isaev, M.I. (2011) Triterpene glycosides from *Astragalus* and their genins. LXXXVIII. Cycloascidoside A, a new bisdesmoside of cycloasgenin C. *Chemistry of Natural Compounds* 47(2), 250–253.

Orsini, F., Verotta, L., Barboni, L., El-Sebakhy, N.A., Asaad, A.M., Abdallah, R.M. and Toaima, S.M. (1994) Cycloartane triterpene glycosides from *Astragalus alexandrinus*. *Phytochemistry* 35(3), 745–749.

Özipek, M., Dönmez, A.A., Çalış, İ., Brun, R., Rüedi, P. and Tasdemir, D. (2005) Leishmanicidal cycloartane-type triterpene glycosides from *Astragalus oleifolius*. *Phytochemistry* 66(10), 1168–1173.

Papadopoulou, K., Melton, R.E., Leggett, M., Daniels, M.J. and Osbourn, A.E. (1999) Compromised disease resistance in saponin-deficient plants. *Proceedings of National Academy of Science USA* 96, 12923–12928.

Pistelli, L., Pardossi, S., Flamini, G., Bertoli, A. and Manunta, A. (1997) Three cycloastragenol glucosides from *Astragalus verrucosus*. *Phytochemistry* 45(3), 585–587.

Pistelli, L., Pardossi, S., Bertoli, A. and Potenza, D. (1998) Cycloastragenol glycosides from *Astragalus verrucosus*. *Phytochemistry* 49(8), 2467–2471.

Pistelli, L., Giachi, I., Lepori, E. and Bertoli, A. (2003) Further saponins and flavonoids from *Astragalus*

verrucosus Moris. *Pharmaceutical Biology* 41(8), 568–572.
Polat, E., Caliskan-Alankus, O., Perrone, A., Piacente, S. and Bedir, E. (2009) Cycloartane-type glycosides from *Astragalus amblolepis*. *Phytochemistry* 70(5), 628–634.
Polat, E., Bedir, E., Perrone, A., Piacente, S. and Alankus-Caliskan, O. (2010) Triterpenoid saponins from *Astragalus wiedemannianus* fischer. *Phytochemistry* 71(5), 658–662.
Radwan, M.M., Farooq, A., El-Sebakhy, N.A., Asaad, A.M., Toaima, S.M. and Kingston D.G.I. (2004a) Acetals of three new cycloartane-type saponins from Egyptian collections of *Astragalus tomentosus*. *Journal of Natural Products* 67(3), 487–490.
Radwan, M.M., El-Sebakhy, N.A., Asaad, A.M., Toaima, S.M. and Kingston, D.G.I. (2004b) Kahiricosides II–V, cycloartane glycosides from an Egyptian collection of *Astragalus kahiricus*. *Phytochemistry* 65(21), 2909–2913.
Rios, J.L and Waterman, P.G. (1997) A review of the pharmacology and toxicology of *Astragalus*. *Phytotherapy Research* 11, 411–418.
Roberts, S.C. (2007) Production and engineering of terpenoids in plant cell culture. *Nature Chemical Biology* 3, 387–395.
Sakurai, N., Goto, T., Nagai, M., Inoue, T. and Xiao, P. (1990) Studies on the constituents of *Beesia calthaefolia*, and *Souliea vaginata* III. Beesioside IV, a cyclolanostanol xyloside from the rhizomes of *B. calthaefolia* and *S. vaginata*. *Heterocycles* 30, 897–904.
Sawai, S. and Saito, K. (2011) Triterpenoid biosynthesis and engineering in plants. *Frontiers of Plant Science* 2, 1–8.
Semmar, N., Fenet, B., Lacaille-Dubois, M.A., Gluchoff-Fiasson, K., Chemli, R. and Jay, M. (2001) Two new glycosides from *Astragalus caprinus*. *Journal of Natural Products* 68, 656–658.
Semmar, N., Tomofumi, M., Mrabet, Y. and Lacaille-Dubois, M.A. (2010) Two new acylated tridesmosidic saponins from *Astragalus armatus*. *Helvica Chimica Acta* 93(5), 870–876.
Sukhina, I.A. and Isaev, M.I. (1995) Triterpene glycosides of *Astragalus* and their genins. LI. Isoprenoids of *Astragalus dissectus*, *A. epheromotorum*, and *A. kulabensis*. *Chemistry of Natural Compounds* 31(5), 639–640.
Sukhina, I., Agzamova, M. and Isaev, M. (1999) Triterpene glycosides of *Astragalus* and their genins IX. Cyclodissectoside – a new dixyloside of cyclocephalogenin. *Chemistry of Natural Compounds* 35(4), 442–444.
Sukhina, I., Agzamova, M., Imomnazarov, B. and Isaev, M. (2000) Triterpene glycosides of *Astragalus* and their genins. LXI. Occurrence of cycloartane 6-O-monodesmosides. *Chemistry of Natural Compounds* 36(4), 373–376.
Sukhina, I., Mamedova, R., Agzamova, M. and Isaev, M. (2007) Triterpene glucosides of *Astragalus* and their genins. LXXIV. Cyclotrisectoside, the first trisdesmoside of cyclocephalogenin. *Chemistry of Natural Compounds* 43(2), 159–161.
Szakiel, A., Pączkowski, C. and Henry, M. (2011) Influence of environmental abiotic factors on the content of saponins in plants. *Phytochemistry Reviews* 10, 471–491.
Turgut-Kara, N. and Ari, S. (2011) Analysis of elicitor inducible cytochrome P450 induction in *Astragalus chrysochlorus* cells. *Plant Omics Journal* 4(5), 264–269.
Verotta, L., Orsini, F., Tatò, M., El-Sebakhy, N.A. and Toaima, S.M.A (1998) Cycloartane triterpene 3β, 16β diglucoside from *Astragalus trigonus* and its non natural 6-hydroxy epimer. *Phytochemistry* 49(3), 845–852.
Vincken, J.P., Heng, L., de Groot, A. and Gruppen, H. (2007) Saponins, classification and occurrence in the plant kingdom. *Phytochemistry* 68, 275–297.
Wang, H.K., He, K., Ji, L., Tezuka, Y., Kikuchi, T. and Kitagawa, I. (1989) Asernestioside C, a new minor saponin from the roots of *Astragalus ernestii* comb.; first example of negative nuclear Overhauser effect in the saponins. *Chemistry and Pharmaceutical Bulletin* 37(8), 2041–2046.
Withers, S.T. and Keasling, J.D. (2007) Biosynthesis and engineering of isoprenoid small molecules. *Applied Microbiology and Biotechnology* 73, 980–990.
Yalçın, F.N., Piacente, S., Perrone, A., Capasso, A., Duman, H. and Çalış, İ. (2012) Cycloartane glycosides from *Astragalus stereocalyx* Bornm. *Phytochemistry* 73, 119–126.
Yeshikawa, M., Fuchida, M., Tosirisuk, V. and Kitagawa, I. (1981) *Proceedings of 31st Annual Meeting of the Kinki Branch of the Pharmaceutical Society of Japan*, Kobe, Japan.
Yendo, A.C.A., de Costa, F., Gosmann G. and Fett-Neto, A.G. (2010) Production of plant bioactive triterpenoid saponins: elicitation strategies and target genes to improve yields. *Molecular Biotechnology* 46, 94–104.
Yu-Qun, C., Guli, A. and Yong-Rong, L. (1990) Astrailienin A from *Astragalus iliensis*. *Phytochemistry* 29(6), 1941–1943.
Zhanibekov, A., Naubeev, T.K., Uteniyazov, K., Bobakulov, K.M. and Abdullaev, N. (2013) Triterpene glycosides from *Astragalus*. Structure of cyclolehmanoside C from *A. lehmannianus*. *Chemistry of Natural Compounds* 49(3), 475–477.
Zhou, Y., Hirotani, M., Rui, H. and Furuya, T. (1995) Two triglycosidic triterpene astragalosides from hairy root cultures of *Astragalus membranaceus*. *Phytochemistry* 38(6), 1407–1410.
Zhu, Y.Z., Lu, S.H., Okada, Y., Takata, M. and Okuyama, T. (1992) Two new cycloartane-type glucosides, mongholicoside I and II, from the aerial part of *Astragalus mongholicus* Bunge. *Chemistry and Pharmaceutical Bulletin* 40(8), 2230–2232.

28 Implementing Traditional Ecological Knowledge in Conservation Efforts

Meg Trau[1], Robin Owings[2] and Nishanta Rajakaruna[3*]

[1]Natick, Massachusetts, USA; [2]West Blocton, Alabama USA; [3]College of the Atlantic, Bar Harbor, Maine, USA and Unit for Environmental Sciences and Management, North-West University, Potchefstroom, South Africa

Abstract

Many different fields are involved in the complex practice of biodiversity conservation. To be successful, ecologists and other natural scientists must collaborate with experts from disparate fields such as economics, politics, engineering, and anthropology. However, one of the most recently recognized contributors to a successful conservation effort are the community stakeholders who are intertwined with the local ecology. The people who live on the land that is to be restored or protected hold a crucial role in effective conservation efforts. They can provide scientists and conservation biologists with invaluable information known as traditional ecological knowledge (TEK), which can include how the land is and was used, how species composition has changed over time, and what forces they perceive to be affecting the land and its biota, among other observations. Such knowledge can help create an effective, well-rounded conservation plan. Locals are also key enactors of a conservation plan. The success of a biodiversity conservation project hinges upon the cooperation and enthusiasm of local community members, for without the support from local stakeholders, conservation efforts are neither practical nor sustainable. We will explore these concepts through case studies, showing examples of successes and failures, collaborations and disharmony. We will also address the nuances involved in discussing traditional ecological knowledge, including the associated misconceptions and stereotypes, and how these assumptions can affect the understanding and implementation of TEK.

28.1 Introduction

Within the past 400 years, hundreds of plant species have become extinct and approximately 20–25 species of birds and mammals are lost every 100 years, a rate far greater than the estimated average through geological time (CBD, 2001). In addition, traditional cultures across the globe are also becoming extinct at an alarming rate (Davis, 2007; Davis, 2009). These traditional cultures are important from an anthropological perspective as they teach us different ways of being and knowing. Conscientiously incorporating these diverse belief systems and lifestyles into our approach may be crucial for us to survive our current conservation crisis (Linehan and Gross, 1998). Simply eliminating or marginalizing these differences will do us no good; we must find ways to respectfully integrate them into our scientific understanding and conservation policies (Fenstad *et al.*, 2002). A successful conservation effort is an interdisciplinary one. Individuals from many different fields must collaborate to create and enact a plan that is comprehensive in its ecological understanding, respectful of the locals' lifestyles, and sustainable in the long term (Usher, 2000; Nadasdy, 2003; Padilla and Kofinas, 2014). Involving traditional ecological knowledge (TEK) in the planning and management processes can help ensure all three factors. Community stakeholders are intertwined with their local ecology, and their intimate knowledge of their home environment can

*E-mail: nrajakaruna@coa.edu

enrich biological research as well as improve the functionality of management plans. In this chapter, we will explore this integration using case studies, discuss the methods and complexities of co-management, and suggest trajectories for the future role of TEK in biodiversity conservation.

28.2 Defining TEK

28.2.1 Definitions

Traditional ecological knowledge (TEK) as a term has not been consistently defined within the scientific community (Berkes, 1993; Berkes, 1999). Many studies include their own definitions of TEK, which vary slightly depending on how flexibly the term is used. For example, Huntington (2000) describes TEK as 'the knowledge and insights acquired through extensive observation of an area or species'. Charnley *et al.* (2007, 2008) describe 'a cumulative body of knowledge about the relationships living things (including people) have with each other and with their environment, that is handed down across generations through cultural transmission'. In addition, TEK is often considered a holistic way of 'knowing', as opposed to the compartmentalized method of scientific study. It fuses fields labelled as distinct, such as science with economics and religion (Anderson, 2011). A similar term, 'local ecological knowledge'(or LEK), has also been used to describe the knowledge of people who do not necessarily have a long-term relationship with a particular environment, but nevertheless have expertise and practices adapted to the local ecosystem (Ballard and Huntsinger, 2006).

28.2.2 Stereotypes

Many stereotypes and Western biases accompany the concept of TEK (Whyte, 2013). The word 'traditional' in TEK may seem to indicate knowledge of the past, a form which has already been developed and is unchanging. However, viewing TEK as a romantic, static and archaic way of knowing is a false stereotype, which does not provide space for development and adaptive change (Usher, 2000). Like other living systems and bodies of knowledge, TEK has the ability to evolve (and continue to evolve) over time. TEK holders are not a homogeneous group, but diverse in their practices and beliefs (Nadasdy, 1999; Turner *et al.*, 2000). They face the challenge of maintaining their unique ancestral practices while adapting to changes of the present and future (Turner *et al.*, 2000; Padilla and Kofinas, 2014), including the pressure to conform to scientific methods of conservation and land management. Scientists often subject TEK to a process of validation and, although challenging to knowledge holders, this may ultimately prove to be a useful tool for encouraging governments and outside sources to trust the knowledge (Gratani *et al.*, 2011). Governments and agencies are now paying closer attention to TEK and touting the benefits of incorporating it and other interdisciplinary dialogue into global biodiversity conservation work (CBD, 2005; UNESCO, 2005; CEAA, 2012; IPBES, n.d.). The 2005 United Nations Convention on Biological Diversity states in Article 8(j):

> Each contracting Party shall, as far as possible and as appropriate: Subject to national legislation, respect, preserve and maintain knowledge, innovations and practices of indigenous and local communities embodying traditional lifestyles relevant for the conservation and sustainable use of biological diversity and promote their wider application with the approval and involvement of the holders of such knowledge, innovations and practices and encourage the equitable sharing of the benefits arising from the utilization of such knowledge innovations and practices.
>
> (CBD, 2005)

This step forward in prioritizing the use of traditional knowledge should not be ignored. However, while governments and agencies may express a clear desire to incorporate TEK into conservation efforts, there are often no clear expectations of how to do so. At this scale, TEK is in jeopardy of being transformed from a broad and ever-changing body of knowledge into a 'bureaucratic object' (Anderson, 2011). Such an oversimplification makes it easy for governing agencies to further isolate and ignore the knowledge and the people who possess it (Nadasdy, 1999). Additionally, some scientists have been slow to include TEK in their work because there is a 'general resistance to change', specifically the change required by using it (Huntington, 2000). Suspicions may arise about the validity of observation-based TEK in a scientific study. Some researchers view TEK as unsubstantiated (Johannes, 1993), even to the extent of accusing it of being 'simply a political ploy invented by aboriginal people to wrest control of wildlife from "qualified" scientific managers' (Nadasdy, 1999). Any potential distrust can go both ways, as TEK holders may be suspicious of others' motives for wanting to use TEK, believing that 'others will not use it responsibly or in a manner that benefits the knowledge holders' (Charnley *et al.* 2007). TEK holders wishing to aid conservation efforts may find great difficulty in conforming to the methods of outside researchers, while finding little opportunity to speak with any sort of decision-making power. Finally, TEK is not limited to aboriginal/indigenous peoples (Berkes 1999; Usher, 2000), but may include anyone using adaptive instincts and skills to navigate an ecosystem, whether urban or rural (Wavey, 1993; Ballard and Huntsinger, 2006; Bethel *et al.*, 2014).

28.3 Potential Roles in Conservation

There are numerous ways in which TEK can contribute to conservation efforts. From informing research to enacting management plans, TEK can inform all stages of conservation. TEK experts have an understanding of ecology that has been passed down for generations, making them knowledgeable about the past and present of the environment as well as invested in its future. Their knowledge of the ecosystem is a part of life, for making one's living off the land successfully requires an intimate understanding of growing seasons, migrations, and species–habitat relations (Ballard and Huntsinger, 2006; Emery and Barron, 2010; Turner and Turner, 2008). And when one of those crucial details changes – a harvest occurs sooner, a population declines, or species composition changes – a TEK expert is likely to notice (Wavey, 1993; Turner, 2003). On the other hand, scientists are often limited in their observations by time and by geography. Likewise, land managers cannot be everywhere at once. In these situations, TEK provides information to scientists and conservationists about the ecological, social, and historical aspects that science may not capture (Uprety *et al.*, 2012).

TEK can improve the quality of research by filling in gaps between limited scientific data. Locals can provide invaluable observational knowledge that may not be available to scientists, including how the land is and was used, how species composition has changed over time, and what forces they perceive to be affecting the land (Bethel *et al.*, 2014). With the collective memory of elders and the oral record, a community can account for a history of the land that far precedes modern scientific records. They also have a more complete understanding of the land that is not restricted by the limited spatial reach or small time window of scientific surveys (Nadasdy, 2003). As Jackson and Hobbs (2009) note: 'Systematic monitoring of ecosystems, whether deeply degraded or nearly pristine, rarely spans more than the past few decades'. This limitation can pose problems when attempting to set restoration targets for a landscape. While paleoecology and other natural records can provide insight to a site's ecological history (Jackson and Hobbs, 2009), a people's lived experience may provide a deeper understanding of this history as well as a clearer reflection of what the future may hold.

The locals of a conservation area can also be key enactors of a management plan (Johannes, 1993; Huntington, 2000). The most successful conservation plans engage local knowledge holders as key participants. Charnley *et al.* (2007) describes several formats through which TEK experts can be involved in conservation: collaborative species-specific management; co-management projects; integrated scientific panels; formal institutional liaisons; and ecological modelling. Collaborative species-specific management occurs when locals collaborate with scientists to create a management plan for a particular species of concern. This allows for scientists to learn about historic management methods, while locals learn about new technology and research that could enhance or inform these methods. Co-management projects are similar, but do not have a species focus. Here, TEK experts and conservation biologists collaborate to create a management plan to restore an entire landscape. Integrated scientific panels, such as the Intergovernmental Platform on Biodiversity and Ecosystem Services (IPBES.net), may be used during this process to facilitate discussion and knowledge sharing among all parties – from governments and nonprofits to scientists and local peoples. However, longer-term relationships, such as formal institutional liaisons, may be formed to enact a management plan. These groups, such as the Indigenous Peoples Restoration Network, work with the community as well as conservationists to encourage the incorporation of TEK in conservation efforts (SER, n.d.). Ecological modelling allows for TEK experts and scientists to share their knowledge of a resource, landscape or species and discuss their hopes for the future. Examples of many of these formats are explored in the case studies section of this chapter. TEK experts can also contribute to co-management through data gathering, decision making, resource protection, regulation enforcement and plan creation (Pinkerton, 1989).

It is crucial that locals are willingly involved and invested in conservation efforts, for it is their home and their future. When communication is weak and understanding is lacking, it is unlikely that rules will be enforced within the community (Padilla and Kofinas, 2014). On the other hand, if locals are enthusiastically invested in the management plan – because it aligns with their beliefs, hopes, and lifestyle – they will likely be more eager to work with scientists and follow new regulations or guidelines.

28.4 Challenges

Few studies discuss conservation and management efforts that have failed to properly utilize TEK. This does not signify an absence of challenges or failure, but rather that these are rarely described within the scientific community. It is difficult to learn from others' mistakes if they are not honestly addressed. In this section, we will explore several potential challenges of incorporating TEK into conservation efforts.

While studies of integrating conservation biology and TEK are often focused on the technical aspects of the collaboration, political aspects are often ignored (Nadasdy, 1999). The political frameworks in which co-management efforts take place illuminate underlying struggles for power. In most conservation efforts, traditional peoples must conform to the scientific way of defining and understanding 'knowledge' in order to contribute. This ultimately concentrates power in administrative centres instead of with the local stakeholders (Nadasdy, 1999). It is now often politically imperative for scientists, governments and resource managers to include TEK in some form or another into conservation projects (CBD, 2005; UNESCO, 2005; CEAA, 2012.). Ironically, guidelines for how to utilize TEK lack consistency, resulting in an often miniscule use of TEK, which may not ultimately assist these local groups in any way (Usher, 2000). As a result, locals may develop an underlying sense of distrust of the term 'traditional ecological knowledge', the motives for utilizing it, and the odds of a truly successful outcome.

Another challenge inherent to biodiversity conservation is defining what is 'natural' to conserve. Conservationists can no longer ignore the profound influence humans have had on the environment (Kareiva and Marvier, 2012). The concept of 'natural' is understood by some traditional peoples as the integrated and interconnected world we live in, and includes 'active human manipulation' as a necessary component of the ecology of a landscape (Charnley *et al.*, 2007). Conservation biologists are becoming increasingly aware of this notion of 'natural' as inclusive of human activity (Jackson and Hobbs, 2009). This creates a predicament for ecologists and conservation biologists who are tasked to 'save' nature, which is neither 'natural' nor wild (Anderson, 2011). Therefore, ecologists must find effective ways to assess and evaluate ecosystems in varying states of alteration, which will maximize benefits to both people and biodiversity (Jackson and Hobbs, 2009; Kareiva and Marvier, 2012).

28.5 Case Studies

There are many ways by which traditional ecological knowledge can be collected and integrated into conservation policy. The methods differ based on the research team, the locale and its residents, and the aim of the research (Huntington, 2000; Usher, 2000; Carlsson and Berkes, 2005; Charnley *et al.*, 2007; Charnley *et al.*, 2008). The following case studies display a few ways in which TEK has been investigated and used, both successfully and unsuccessfully. The three examples of successes are data collection studies for future implementation in policy making while the failures are reflective studies examining two cases of unsuccessful attempts to incorporate TEK into policy or science. These, plus three more case studies not discussed here, can be found in Table 28.1, which compares the purposes, methods and outcomes of studies conducted around the globe. It is important to note that this is in no way a comprehensive list of studies on TEK, but merely a selection of examples we found illustrative of a particular method or issue.

Information from the data collection studies featured here can be consulted for future conservation efforts or management plans for their studied regions. But it is valuable to acknowledge that positive collaborations are already happening in forms such as the Maidu Land Stewardship project in the Sierra Nevada mountains (which brings the Feather River Land Trust in partnership with the Maidu community; FRLT.org, n.d.) and land management groups such as the Iisaak Forest Resources (which produces timber in Clayquot Sound, British Columbia, while being deeply invested in the well-being of the surrounding community; Iisaak.com).

Table 28.1. Comparison of eight case studies of TEK in conservation.

Authors	Topic	Location	Participants	Aim of study	Methods	Conclusions
Ballard and Huntsinger, 2006	Salal harvesting	Washington state, USA	20 immigrant harvesters	Data collection: to inform land management policies	Interviews, site visits	The harvesters know how to harvest sustainably but do not have the proper access to the land to allow them to put this knowledge to full use.
Bethel *et al.*, 2014	Louisiana coastline	Louisiana bayou, USA	13 local resource users	Data collection: to inform future restoration efforts	Interviews, site visits, maps	Combining TEK with GIS mapping technology can help visually identify sites for future restoration, as well as areas of consensus and conflicts between local stakeholders and policymakers.
Emery and Barron, 2010	Morel harvesting	Mid-Atlantic, USA	41 local harvesters	Data collection: to investigate morel decline, learn about morel ecology	Interviews, site visits	TEK could guide future research on the taxonomy of *Morchella* species, as well as suggest that future research should focus on changes in forest composition and land use.
Nadasdy, 2003	Dall sheep	Yukon Territory, Canada	Ruby Range Sheep Steering Committee	Review: attempted co-management to address decline in sheep population	Observational, reflective	This effort failed due to lack of trust between scientists and First Nation people, as well as political involvement of hunting outfitters.
Padilla and Kofinas, 2014	Caribou leaders	Yukon Territory, Canada	29 First Nation hunters and elders	Review: attempted reduction of highway hunting of caribou population	Interviews, archival review	This effort failed due to lack of understanding of the differences in tribal TEK and the need for rule enforcement within the tribes.
Gómez-Baggethun *et al.*,2012	Environmental extremes	Doñana region, Spain	33 interviews and 3 village focus groups, involving 13 villages	Review: community responses to changes in climate and resource availability	Interviews, focus groups, archival review	TEK helps facilitate a community-wide response to crisis and keeps social–ecological systems alive across generations.
Gratani *et al.*, 2011	Fish poisons	Queensland, Australia	2 Malanbarra Yidinji elders with groups of non-indigenous scientists	Experiment: validating TEK through scientific laboratory testing	Experiments, interviews	Scientific validation of TEK can provide legitimacy in the eyes of scientists and government environmental agencies.
Cardoso and Ladio, 2011	Local forestry practices	Patagonia region, Argentina	28 Mapuche residents	Data collection: analysing local forestry practices for better forestry management	Interviews	The influence of external agricultural agents has created a hybrid system of slow-growing native woods, as well as less sustainable exotic species.

These are just two of many conservation collaborations that are happening now (see SER.org for more examples), and they provide hope that successful co-management is possible.

28.5.1 Successes

28.5.1.1 *Input for future restoration*

In response to a 2012 Coastal Master Plan developed for the Louisiana coast, Bethel *et al.* (2014) conducted a survey of local resource users to get their perspectives on areas of concern and potential restoration approaches. The Master Plan sought to incorporate stakeholder input, but as the authors noted, the public forums used to solicit this input were either too technical for the public or not representative of their ideas. As a result, public engagement with the Master Plan process was limited. Therefore, Bethel and colleagues sought to fill in the gaps and find out what the people who know this area intimately from living on the land and using the resources have to say about the restoration needs. The research team chose their participants carefully, based on multiple iterations of peer referral. Finally, 13 of the most-referred people – fishers, trappers, and hunters – agreed to participate in the study. Each participant took the researchers on a site visit, or data-collection trip, to explore the areas they were most concerned about. The researchers transcribed and coded the interviews conducted during these trips and later reviewed the transcriptions with the participants to confirm that they were accurate. Four dominant themes emerged from the interviews: 1) methods of coastal restoration; 2) issues of freshwater salinity; 3) land loss; and 4) resource use and change (Bethel *et al.*, 2014). The researchers asked their participants about potential responses to each issue, discussing priority areas, restoration tactics and possible conflicts with local interests. These discussions revealed that TEK experts possessed a deep understanding of the local ecological processes and a thoughtfulness that revealed a true concern for the future of the coastal region.

While the researchers' methods were more representative of statistics and biological research than ethnography, their concern with accuracy ensured that they chose well-respected, knowledgeable TEK experts and that their participants agreed with the interpretations of the data. What made this TEK collection unique was their effective use of maps. GPS coordinates were collected throughout the data-collection trips, and the data from each visit were compiled into one map to compare the routes. They also asked each participant to mark on a map which areas they saw as a priority for restoration, as well as the restoration method they believed would be most effective. These maps were also combined to show TEK expert consensus for restoration sites. By sharing the data in a visual map form, policy makers can quickly and easily understand which areas are priorities for local stakeholders. The maps combined with the interview data can help inform policy makers and scientists as they carry out the Coastal Master Plan and determine sites for restoration, considering areas of consensus and conflict and better understanding locals' values and concerns.

28.5.1.2 *Insight for better land management*

Land is often managed by people far removed from the actual property; as a result, policies, while made with good intentions, often are impractical for those using the land on a daily basis. Ballard and Huntsinger's (2006) study sought to find out what people who rely on the land know about the local ecology with the hopes of integrating this knowledge into future land management. While their study did not directly contribute to a management plan, their research revealed several important points. A defining feature of this study was the participant population. They interviewed 20 harvesters of salal (*Gaultheria shallon*; Ericaceae) in the Olympic Peninsula of Washington state; most of the harvesters (17) were recent immigrants to the United States. This defies the popular notion that TEK is only held by Native peoples. Regardless of their ethnicity, these harvesters derived their livelihood from the land and relied on their intimate understanding of the ecology for a successful harvest. The interviews also revealed, as may be suspected, that the harvesters with more experience (eight or more years) had 'distinctly more detailed answers' and a more nuanced perspective on the ecology of the forests they worked in (Ballard and Huntsinger, 2006). Their experience had taught them that harvesting at a low intensity and allowing an area to lie unharvested for a year to recover, or a 'rest-rotation system,' is the ideal approach. However, their temporary access to the land does not allow them to utilize the sustainable practices they would prefer. Instead, a permit allowed a person to harvest only one species or category of species (i.e. 'floral greens'). Many harvesters expressed a desire to stay in one area while harvesting numerous species. But their permits did not guarantee that a 'rest-rotation system' will protect the plants from illegal harvesting by others (Ballard and Huntsinger, 2006). This study shows that the people using the land may possess a better understanding of what works than forest managers and scientists may assume. If these harvesters were allowed more long-term access to a parcel of forest, they could securely practise the sustainable harvesting methods they know from years of experience are best for the plant and the forest as a whole. If policy makers were to collaborate with the people using the land, a more sustainable management plan could be developed, satisfying both human and ecological needs.

28.5.1.3 *Information about under-researched species*

While scientific observation and research has been conducted for many years, limitations of time, geography and funding have resulted in gaps in scientific knowledge. Emery and Barron's (2010) study began at the request of the US National Park Service (NPS) with the hopes of filling in such gaps in information about morel mushrooms (*Morchella* spp.; Morchellaceae). The NPS had been receiving reports of a decline in morels in the US Mid-Atlantic region. Lacking sufficient scientific data on the mushroom, the NPS turned to local morel harvesters to ask for their perspective on the potential decline.

Emery and Barron (2010) conducted interviews and site visits with 41 participants, who had been harvesting morels for less than ten to more than 30 years. Morel harvesters must have a fairly detailed understanding of the genus if they want to harvest the correct species. The aspects they shared with the researchers included types, tree associations, disturbance relationships and seasonality. Because the genus is variable and its range extensive, there is little scientific information about morels, particularly in the region where this study was conducted. Therefore, the harvesters' insight is valuable to mycologists and ecologists who seek to learn more about the *Morchella* species throughout their range.

Regarding the reported decline of morels, the harvesters expressed some concern but also provided alternative explanations for the reduced sightings. Many participants noted that the ideal weather conditions for morel growth are becoming less frequent and the harvesting season has become earlier over several years. The participants agreed with mycologists that climate change and habitat destruction could be contributing to the decline; however, many of them also noted that social reasons could account for the perceived decline in harvests. Many harvesters had observed 'more people hunting morels today than in the past and several noted that what they perceive as a decline in morels may actually be increased competition for them' (Emery and Barron 2010).

Clearly, determining the causes for morel decline – if there is indeed a decline – will require more careful study. This investigation into harvesters' TEK provided more insight into the species in the Mid-Atlantic region, and can guide the direction of future morel research. Harvesters' beliefs about the species decline also suggest that future studies will have to consider social as well as ecological factors. Their observations over time can also contribute to ecologists' understanding of the past and present of the morel population.

28.5.2 Failures

28.5.2.1 *Misunderstanding TEK*

Padilla and Kofinas (2014) analysed the case of a failed attempt to use TEK to guide hunting regulations in the Yukon Territory of Canada. After the opening of a highway through north-western Canada in 1979, First Nation hunters as well as government officials were concerned about the highway's impact on the local Porcupine caribou herd (*Rangifer tarandus granti*; Cervidae). The newly formed Dempster Highway committee of the Yukon Department of Renewable Resources created their first recommendation – a 16 km no-hunting corridor, later reduced to 2 km – without consulting any indigenous hunters. Then, in 1994, the local resource council recommended the government consult TEK as the basis for creating new hunting regulations. A group of First Nations people expressed concern that hunting along the highway would deflect the herd's migration. A new regulation was thus created, based on some First Nation elders' teachings of 'letting the leaders pass', which advised that hunters should not take the leaders of the herd in order to avoid spooking and diverting the others. However, as the Porcupine

Caribou Management Board would soon learn, this sort of TEK was very context-specific as well as tribe-specific. The new regulation created a week-long closure on highway hunting that began upon the arrival of the first caribou in the fall. This closure was based solely on TEK in light of the dearth of scientific studies on caribou herd leaders and the impacts of the highway on the herd. This new regulation was primarily weak because 'local perspectives on the number, sex, and ages of caribou leaders needed to let the leaders pass [...] was not well defined' (Padilla and Kofinas, 2014). A look into the context in which this teaching was shared reveals why it failed to be broadly applied. Until the mid-1900s, hunting was often a community-oriented, cooperative effort, consisting of a group travelling on foot directed by a chief. Here, hunters held each other accountable and followed the guidance of elders. But beginning in the mid-1900s, particularly with the advent of snowmobiles, hunting became more individualized. Rarely was there a leader to oversee the hunting practices. As a result, the hunters lacked an authority who could determine whether the caribou leaders had passed and an overseeing entity in the community to easily share this information with. This new social context, combined with inadequate definitions and a lack of communication, made the new hunting closure regulation confusing and difficult to enforce, and, consequently, it received mixed responses from the community. Finally, in 2007, a young hunter who had violated the closure contested his case in court. He testified that the new regulation did not reflect the teachings of Dawson First Nation elders. During Padilla and Kofina's interviews with nine hunters and elders in Dawson, three also asserted that the 'let the leaders pass' rule did not reflect their elders' teaching, and four others partially disagreed with the regulation. In response, the Yukon Territory Minister of Environment declared the closure voluntary until consensus had been reached.

This case study demonstrates the complex, contextual nature of TEK. The ecological knowledge is intertwined with politics, social mores and spirituality. This complexity does not negate TEK's usefulness; it only means that it must be collected, considered and implemented while being mindful of the human component of the knowledge. This case also illustrates a common misconception: that all indigenous people have the same knowledge and culture. Policy makers had not considered that what an elder says in one First Nation tribe is not necessarily true in another tribe. Here again, the social context of the information must be considered. This implementation of TEK failed due to gaps in communication: disconnect between board members and tribal members; conflict between communities; and a divide between generations. The authors suggest that, in the future, co-managers must be cognizant of the difference between informal hunting customs and formal laws, confirm that all stakeholders fully support any recommendations made by a co-management board, engage elders in education efforts and oversight roles, and host community-wide discussions about the similarities and differences in their hunting traditions. These suggestions can be applied to any co-management situation involving local customs, and the principles of communication, context and complexity seen in this case are relevant to any TEK co-management situation.

28.5.2.2 Mistrusting TEK

A second case of a failed co-management effort also occurred in the Yukon Territory. Nadasdy (2003) reviewed the struggle of the Ruby Range Sheep Steering Committee (RRSSC) as it sought to gather data about populations of Dall sheep (*Ovis dalli dalli*; Bovidae). The RRSSC was created in response to growing concern about a declining Dall sheep population, and its members included scientists, resource managers, hunting outfitters and First Nation people. Ostensibly, the members' mixed backgrounds would bring diverse perspectives to the table, enriching their collective understanding of the situation. However, collaboration was instead replaced with discord. According to Nadasdy (2003), 'virtually the only significant instance of knowledge-integration that occurred during the entire RRSSC process' was between scientists and hunting outfitters. Scientists, who distrusted the results of their 1996 annual aerial survey due to snowy conditions, consulted the results of a hunting outfitter's sheep count conducted a few months later. The outfitter's count included 100 more sheep than the aerial survey, yet the scientists and outfitter 'jointly concluded that the drop in the sheep count represented problems with the survey (e.g., different time of year, snow cover, and 100 moving sheep) rather than a drop in the actual number of sheep' (Nadasdy, 2003). However, the other parties of the RRSSC disagreed about the extent of, reasons for, and responses to the decline of the Dall sheep population, and each seemingly refused to consider the others' perspective. First Nation representatives argued that they have seen the population steadily declining since the 1960s, yet biologists were doubtful the decline was as severe or longstanding. Each considered the others' knowledge as invalid and unreliable; biologists wanted formal research and written reports, while First Nations people were dubious of studies so limited in time and geographic scope. Also at play in this situation was the political power of big game outfitters. With financial incentive for continuing hunting and the political sway of the Yukon Outfitters Association in the government, RRSSC recommendations could have been made irrelevant were they opposed by the outfitters. Ultimately, the RRSSC produced 24 recommendations: '12 dealt with the basically non-contentious issues of harassment, access, education, and predation. An additional six recommendations dealt specifically with how and when to conduct future scientific research. Five recommendations dealt with the contentious issues of hunting' (Nadasdy, 2003). These latter five recommendations were worded such that only First Nation peoples' hunting practices were ultimately affected.

The case of the Ruby Range Sheep Steering Committee demonstrates many concepts that may be overlooked when considering co-management. Successful co-management is more than having different stakeholders in the same room; there are social interactions, cultural differences and political influences at play that can cause conflict and make fruitful conversations scarce. Additionally, the context in which these conversations were taking place was troublesome. As Nadasdy (2003) notes: 'since the RRSSC process was created within the context of (and inserted into) existing systems of state resource management, biologists had no choice but to undervalue the artifacts of TEK vis-a-vis those of biology'. The bureaucracy which ultimately oversees the co-management body can limit the outcome of even the most collaborative co-management efforts. The biologists wanted numbers because they required numbers in order to make recommendations for legislation. TEK was undervalued, not only by the RRSSC, but also by the system within which the RRSSC functioned. If governments are going to integrate TEK with management (CEAA, 2012), then they must truly value TEK as a valid source of data. Otherwise, all of the diverse committees and discussions with indigenous people and local stakeholders are all for the sake of appearances, nothing more.

28.6 Discussion and Recommendations

In light of the information we have today regarding co-management practices, what steps can be taken to better integrate TEK into conservation efforts? First, the dynamics of management planning processes would surely be changed by giving TEK holders equal decision-making power: '"traditional knowledge" cannot truly be "incorporated" into the management process until native elders and hunters have achieved full decision-making authority in that realm' (Nadasdy, 1999). In addition, communities must have the scientific tools and technical support to control and refine their own management methods (Wavey, 1993). Co-management systems will be unable to reach their true potential as long as TEK holders are coerced to validate their lifestyles in the terms of external Western and scientific standards (Gratani *et al.*, 2011). Granted, our current political and management systems, in which power over TEK holders seems almost inherent, may make an idealistic balance seem unachievable. However, the outcome is in the hands of co-managers, who can choose to collaborate and optimize their use of TEK as well as scientific data.

Co-management can be difficult and complex, but Carlsson and Berkes (2005) provide some insight into the types and methods of co-management; they note that 'management processes can be improved by making them adaptable and flexible through the use of multiple perspectives and a broad range of ecological knowledge and understanding'. This recognizes the complexities of both human and ecological systems. If we see co-management as a network, we can acknowledge the many authorities with whom individuals interface throughout the conservation process, as well as the sometimes fragmentary nature of a community (Carlsson and Berkes, 2005). With these concepts in mind, one will be able to approach co-management situations with the understanding of the humanity of the situation. Biodiversity conservation is not solely about the threatened species; conservation plans affect human lives and require human knowledge. On this note, we recognize that many conservation biologists are not social scientists; however, it benefits all to have good ethnographic and anthropological skills (Johannes, 1993). By knowing the area and culture, conducting interviews in the interviewee's preferred language, and being open and honest with participants, a researcher creates an underlying foundation of respect. This goes a long way towards eliminating the inequalities in the dynamic between scientists and local knowledge holders.

28.7 Conclusion

TEK is an indispensable resource to biodiversity conservation efforts. However, Chief Wavey warned in his 1993 keynote address to the International Workshop on Indigenous Knowledge and Community-Based Resource Management: 'traditional ecological knowledge is not another frontier for science to discover' (Wavey, 1993). We must commit not to simply 'discover' and appropriate TEK from others, but to find ways of being active and engaged members of our shared ecological landscapes (Kimmerer, 2002). We should question what science can do to support TEK (such as by providing resources, technical support, supplies, education in methodology, etc.), so that we may cultivate a mutually beneficial relationship with TEK holders (Wavey, 1993). We must also recognize TEK as a way to better promote the value of interacting with and understanding our natural (and even urban) environments in a more holistic way. Because TEK is not transmitted in an institutionalized way like scientific knowledge, acquiring TEK requires participatory learning (Setalaphruk and Price, 2007). Thus, teaching children the skills needed to interact with nature is a way of activating our own type of local ecological knowledge (Kimmerer, 2002). By doing so, we may prevent the development of a cultural apathy for nature, and instead find that our local conservation efforts become enhanced by the participation of newly engaged citizens. Ultimately, integrating different ways of knowing and changing our mindsets could turn the tide on global conservation efforts. By fostering mutual well-being and respect at the core of our knowledge systems, we may allow better systems of co-management to evolve in the future.

References

Anderson, E.N. (2011) Ethnobiology: overview of a growing field. In: Anderson, E.N., Pearsall, D., Hunn, E. and Turner, N. (eds) *Ethnobiology*. Wiley-Blackwell, Hoboken, New Jersey, USA, pp. 1–14.

Ballard, H.L. and Huntsinger, L. (2006) Salal harvester local ecological knowledge, harvest practices and understory management on the Olympic Peninsula, Washington. *Human Ecology* 34, 529–547.

Berkes, F. (1993) Traditional ecological knowledge in perspective. In: Inglis, J.T. (ed.) *Traditional Ecological Knowledge: Concepts and Cases*. International Program on Traditional Ecological Knowledge and International Development Research Centre, Ottawa, Canada, pp. 1–9.

Berkes, F. (1999) *Sacred Ecology: Traditional Ecological Knowledge and Resource Management*. Taylor and Francis, Philadelphia, Pennsylvania, USA.

Bethel, M.B., Brien, L.F., Esposito, M.M., Miller, C.T., Buras, H.S. and Laska, S.B. (2014) Sci-TEK: a GIS-based multidisciplinary method for incorporating traditional ecological knowledge into Louisiana's coastal restoration decision-making processes. *Journal of Coastal Research* 30(5), 1081–1099.

Cardoso, M.B. and Ladio, A.H. (2011) Peridomestic forestation in Patagonia and traditional ecological knowledge: a case study. *Sitientibussérie Ciências Biológicas* 11(2), 321–327.

Carlsson, L. and Berkes F. (2005) Co-management: concepts and methodological implications. *Journal of Environmental Management* 75, 65–76.

CBD (Convention on Biological Diversity) (2001) Status and trends of global biodiversity. In: *Global Biodiversity Outlook* 1. Montreal, Canada. pp. 59–118. Available at: www.cbd.int/doc/publications/gbo/gbo-ch-01-en.pdf (accessed 31 May 2016).

CBD (Convention on Biological Diversity) (2005) Traditional knowledge, innovations and practices of indigenous and local communities. In: *Handbook of the Convention on Biological Diversity Including its Cartagena Protocol on Biosafety*, 3rd edn. Montreal, Canada. pp. 138–151. Available at: www.cbd.int/doc/handbook/cbd-hb-all-en.pdf (accessed 31 May 2016).

CEAA (Canadian Environmental Assessment Act) (2012). S.C. 2012, c. 19, s. 52: 19.3. Available at: laws-lois.justice.gc.ca/PDF/C-15.21.pdf (accessed 31 May 2016).

Charnley, S., Fischer, A.P. and Jones, E.T. (2007) Integrating traditional and local ecological knowledge into forest biodiversity conservation in the Pacific Northwest. *Forest Ecology and Management* 246, 14–28.

Charnley, S., Fischer, A.P. and Jones, E.T. (2008) Traditional and local ecological knowledge about forest biodiversity in the Pacific Northwest. In: *General Technical Report PNW-GTR-751*. Pacific Northwest Research Station, Forest Service, US Department of Agriculture, Portland, Oregon.

Davis, W. (2007) *Light at the Edge of the World: A Journey through the Realm of Vanishing Cultures*. Douglas and McIntyre Ltd, Vancouver, British Columbia, Canada.

Davis, W. (2009) *The Wayfinders: Why Ancient Wisdom Matters in the Modern World*. House of Anansi Brothers Press, Toronto, Ontario, Canada.

Emery, M.R. and Barron, E.S. (2010) Using local ecological knowledge to assess moral decline in the US Mid-Atlantic region. *Economic Botany* 64(3), 205–216.

Fenstad, J.E., Hoyningen-Huene, P., Qiheng, H., Kokwaro, J., Nakashima, D. and Salick, J. (2002) *Science and Traditional Knowledge: Report from the ICSU Study Group on Science and Traditional Knowledge*. International Council for Science, Paris, France. Available at: www.icsu.org/publications/reports-and-reviews/science-traditional-knowledge/Science-traditional-knowledge.pdf (accessed 31 May 2016).

FRLT.org (n.d.) *Maidu Land Stewardship*. Feather River Land Trust, Quincy, California. Available at: www.frlt.org/experience-land/maidu-stewardship (accessed 31 May 2016).

Gómez-Baggethun, E., Reyes-Garcia, V., Olsson, P. and Montes, C. (2012) Traditional ecological knowledge and community resilience to environmental extremes: a case study in Doñana, SW Spain. *Global Environmental Change* 22(3), 640–650.

Gratani, M., Butler, J.R.A., Royee, F., Valentine, P., Burrows, D. and Canendo, W.I. (2011) Is validation of indigenous ecological knowledge a disrespectful process? A case study of traditional fishing poisons and invasive fish management from the wet tropics, Australia. *Ecological Society* 16(3), 25.

Huntington, H.P. (2000) Using traditional knowledge in science: methods and applications. *Ecological Applications* 10(5), 1270–1274.

Iisaak.com. *Iisaak: Wood with Respect*. Iisaak Forest Resources Ltd, Ucluelet, British Columbia, Canada.

IPBES.net (n.d.) Intergovernmental Platform on Biodiversity and Ecosystem Services website. IPBES, Bonn, Germany. Available at: www.ipbes.net (accessed 31 May 2016).

Jackson, S.T. and Hobbs, R.J. (2009) Ecological restoration in the light of ecological history. *Science* 325(5940), 567–569.

Johannes, R.E. (1993) Integrating traditional ecological knowledge and management with environmental impact assessment. In: Inglis, J.T. (ed.) *Traditional Ecological Knowledge: Concepts and Cases*. International Program on Traditional Ecological Knowledge and International Development Research Centre, Ottawa, Canada, pp. 33–39.

Kareiva, P. and Marvier, M. (2012) What is conservation science? *Bioscience* 62, 962–969.

Kimmerer, R.W. (2002) Weaving traditional ecological knowledge into biological education: a call to action. *Bioscience* 52(5), 432–438. Available at: bioscience.oxfordjournals.org/content/52/5/432.full.pdf+html (accessed 31 May 2016).

Linehan, J.R. and Gross, M. (1998) Back to the future, back to basics: the social ecology of landscapes and the future of landscape planning. *Landscape and Urban Planning* 42, 207–233.

Nadasdy, P. (1999) The politics of TEK: power and the 'integration' of knowledge. *Arctic Anthropology* 36(1–2), 1–18.

Nadasdy, P. (2003) Reevaluating the co-management success story. *Arctic* 56(4), 367–380.

Padilla, E. and Kofinas, G.P. (2014) Letting the leaders pass: barriers to using traditional ecological knowledge in comanagement as the basis of formal hunting regulations. *Ecological Society* 19(2), 7.

Pinkerton, E. (1989) Attaining better fisheries management through co-management: prospects, problems, and propositions. In: Pinkerton, E. (ed.) *Co-operative Management of Local Fisheries: New Directions for Improved Management and Community Development.* University of British Columbia Press, Vancouver, Canada, pp. 3–33.

SER.org (n.d.) *Indigenous Peoples Restoration Network*. Society for Ecological Restoration, Washington, DC, USA. Available at: www.ser.org/iprn/iprn-home (accessed 31 May 2016).

Setalaphruk, C. and Price, L.L. (2007) Children's traditional ecological knowledge of wild food resources: a case study in a rural village in Northeast Thailand. *Journal of Ethnobiology and Ethnomedicine* 3, 33. DOI: 10.1186/1746-4269-3-33.

Turner, N.J. (2003) The ethnobotany of 'edible seaweed' (*Porphyraabbottae* and related species; Rhodophyta: Bangiales) and its use by First Nations on the Pacific coast of Canada. *Canadian Journal of Botany* 81(2), 283–293.

Turner, N.J. and Turner, K.L. (2008) Where our women used to get the food: cumulative effects and loss of ethnobotanical knowledge and practice: case study from coastal British Columbia. *Botany* 86(2), 103–115.

Turner N.J., Ignace, M.B. and Ignace, R. (2000) Traditional ecological knowledge and wisdom of aboriginal peoples in British Colombia. *Ecological Applications* 10(5), 1275–1287.

UNESCO (2005) Local & Indigenous Knowledge of the Natural World: An Overview of Programmes and Projects. *International Workshop on Traditional Knowledge, Panama City, Republic of Panama, 21–23 September 2005*. UNESCO, New York, USA, pp. 1–8. Available at:www.un.org/esa/socdev/unpfii/documents/workshop_TK_UNESCO.pdf (accesssed 31 May 2016).

Uprety, Y., Asselin, H., Bergeron, Y., Doyon, F. and Boucher, J. (2012) Contribution of traditional knowledge to ecological restoration: practices and applications. *Ecoscience* 19(3), 225–237.

Usher, P. (2000) Traditional ecological knowledge in environmental assessment and management. *Arctic* 53(2), 183–193.

Wavey, R. (1993) International workshop on indigenous knowledge and community-based resource management: keynote address. In: Inglis, J.T. (ed.) *Traditional Ecological Knowledge: Concepts and Cases*. International Program on Traditional Ecological Knowledge and International Development Research Centre, Ottawa, Canada, pp. 11–16.

Whyte, K.P. (2013) On the role of traditional ecological knowledge as a collaborative concept: a philosophical study. *Ecological Processes* 2(7). Available at: www.ecologicalprocesses.com/content/2/1/7 (accessed 31 May 2016).

29 Conserving Forest Biodiversity

Petros Ganatsas*

Laboratory of Silviculture, Aristotle University of Thessaloniki, Thessaloniki, Greece

Abstract

Forests demonstrate a high degree of biodiversity, being thought to comprise the most diverse ecosystems on land, as most of the terrestrial species in the world dwell there. Although taking actions for the conservation of biodiversity at a local, regional, national and global level has been declared as a universal priority, biodiversity decrease is a crucial and ongoing environmental issue. Among the most common disorders causing biodiversity decrease in forest ecosystems are the introduction of alien species, climate shift, and habitat fragmentation as a result of forest management for timber production purposes. In the past few decades, biodiversity conservation has been assigned a vital part in forestry, and it holds a key role in understanding the dynamism and heterogeneity of natural forests, thus offering drivers for their management. One of the challenges faced by current forestry practice is how to achieve biodiversity conservation and successfully correspond to the demands for timber products. This approach gives rise to the necessity for management of forests and for coming up with silvicultural practices intended for preserving ecosystem integrity, while meeting the demands for wood. Nevertheless, it is universally acknowledged that biodiversity conservation requires separating and managing lands as protected areas. Thus, conservation of forest biodiversity has been oriented towards generating a network of areas under protection. Nonetheless, there is still great controversy over the impact caused by forest management on ecosystem biodiversity, and over which management practice should be implemented for the purpose of achieving biodiversity conservation. It has been held that effective forest management should rely on alternative silvicultural practices emulating natural disturbances of low or intermediate intensity, promoting stand structural characteristics of the later stages of succession and forests of old growth, including the creation of uneven-aged, multi-storey stands, diverse forest types, and the preservation of a few large trees. On the contrary, interventions that result in the simplification of stand structures and composition of species, such as clear-cuttings, should be avoided.

29.1 Introduction – Forest Biodiversity

As the human population on the planet increases, the influences of human societies spread to more and more areas. Thus, disruption of natural ecosystems in many cases is becoming more frequent and intense. Sustaining natural ecosystems and their biodiversity is recognized as a very important priority for human societies worldwide.

Biodiversity is defined as the degree of diversity of all life forms within a species, an ecosystem, a region, or around the globe. Especially, for forest ecosystems, biodiversity includes, among others, all of the stand structural characteristics, such as tree height, density, complexity, age distribution, which determine ecosystems function, value and services. Forests comprise ecosystems with high biodiversity values and they are considered as the most diverse ecosystems on land, because they hold the vast majority of the world's terrestrial species (CIFOR, 2014). They offer shelter habitat to a disproportionately large part of the world's biodiversity (Battles *et al.*, 2001). Based on the latest reports, it is estimated that forests support about 65% of the terrestrial species living in the world (World Commission on Forests and Sustainable Development, 1999; Lindenmayer *et al.*, 2006). Additionally, many important rare or under extinction organisms live within forest environments. Thus, conserving forest biodiversity is a critical issue not only in current ecological science and practice (Aanderaa *et al.*, 1996; Hunter, 1999; Lindenmayer *et al.*, 2006), but also for human societies, as well as for the ecological

*E-mail: pgana@for.auth.gr

stability of the Earth. Thus, national authorities worldwide, and a number of forest management bodies at a national and international level, enter into numerous agreements for the adoption of measures for forest biodiversity conservation (e.g. Commonwealth of Australia, 1998; Montreal Process Liaison Office, 2000; Commonwealth of Australia, 2001; FAO, 2001; MCPFE, 2003; CBD, 2006; MCPFE and PEBLDS, 2006).

Biodiversity conservation has held a critical part in the science of forestry, at least for the past few decades, and it holds a key role in understanding how dynamic and heterogeneous natural forests are, thus offering drivers for their management. Furthermore, diversity in species is increasingly considered crucial to the operation of the ecosystem (Scherer-Lorenzen *et al.*, 2005), while recent international reports have stressed the need to prevent biodiversity loss and enhance sustainable forest management (Parviainen *et al.*, 2007; Paillet *et al.*, 2010).

At a worldwide level, based on the data on forest areas designated primarily for the conservation of biodiversity, over 400 million hectares of forests, which corresponds to 11.2% of the total forest area of the countries reporting, appear to function primarily for the purpose of conserving biodiversity. The largest area is located in South America, succeeded by North America, while the majority of the forests in Central America and Western and Central Africa primarily serve the purposes of biodiversity conservation. The lowest percentage of forests primarily destined for biodiversity conservation is found in Europe and Western and Central Asia (FAO, 2005). However, based on the results of the Expanded programme of work on forest biodiversity by the Convention on Biological Diversity, there is adequate conservation in 10% of each forest type in the world (CBD, 2006).

Although taking actions for the conservation of biodiversity at a local, regional, national and global level has been declared as a universal priority, biodiversity decrease is an ongoing crucial environmental issue. Among the most common disorders causing biodiversity decrease in forest ecosystems are the introduction of alien species, climate shift, and habitat fragmentation as a result of forest management for timber production purposes (Primack, 2000). Interestingly, biodiversity conservation has taken precedence in forest management, together with the effective regeneration of species of forest trees which are economically valuable for many forests undergoing management (Ellum, 2009). Additionally, some early works have highlighted the significance of conservation of biodiversity in the management of forests, and thus relevant questions pertaining to forests being managed have been addressed (e.g. Bratton, 1994; Matlack, 1994; Roberts and Gilliam, 1995).

Forestry currently faces the challenge of how to achieve biodiversity conservation and satisfy the demand for timber products (Battles *et al.*, 2001). In practice, management of forests in many countries has begun to adopt the objective of sustainability for all components of forest ecosystems. Under this approach, this interest in biodiversity conservation generates the need for forest management and silvicultural techniques designed to maintain the integrity of ecosystems, while satisfying society's need for timber resources. Thus, the disturbances arising from exploitation of forests, and primarily by logging, often try to mimic natural disturbances, for the purpose of ensuring that the forest could function as it would have functioned in the absence of human intervention. For example, a major concept involves sustaining forest understorey plant communities, in view of the fact that plant diversity in the majority of forests is to a great extent included therein, and they hold a crucial role with regard to numerous ecosystem functions (Wilson and Puettmann, 2007; Ellum, 2009).

Nevertheless, it is universally acknowledged that biodiversity conservation calls for separating and managing lands as protected areas (FAO, 2005). Accordingly, conservation of forest biodiversity has been primarily oriented towards generating a network of areas under protection at a national or international level (European Economic Community, 1992; CBD, 2006; Lindenmayer *et al.*, 2006; MCPFE and PEBLDS, 2006; Ganatsas *et al.*, 2013). Biodiversity conservation within protected areas encompasses maintaining specific portions of land, excluding any human intervention, and letting the ecosystems operate as they would have operated if they had not been subject to human intervention. In the last few years, nature conservationists and policy makers have supported the concept that new forest reserves should be formed within managed forests (Parviainen *et al.*, 2000), for the purpose of facilitating biodiversity conservation. Thus, any reliable forest biodiversity conservation seems to be oriented towards the creation of natural reserve networks (Norton, 1999; Lindenmayer *et al.*, 2006). Such strategies are based on the speculation that the lack of forest management may be favourable for many forest species.

However, three important questions arise from the implementation of this policy. First, are reserves as such sufficient to sustain forest biodiversity effectively (Sugal, 1997; Daily *et al.*, 2001; Lindenmayer and Franklin, 2002), in the light of the fact that 92% of forests in the world are not within the territory of formally protected areas? Second, what measures should be adopted for the promotion of biodiversity in the remaining 92% of the forest areas? And third, should any prohibition be imposed on silvicultural and management actions within the protected areas, or should the management plans for the protected areas contain specific forest management interventions for the promotion of biodiversity, as opposed to leaving the ecosystem to operate on its own?

29.2 Factors Determining Forest Biodiversity

There are many stand characteristics contributing to forest stand biodiversity capacity. Forest stands vary according to their composition, density, age differentiation and distribution, vertical structure, existence and characteristics of shrub species, herbaceous vegetation and individual spatial heterogeneity (Kerr, 1999).

29.2.1 Stand composition

The composition of stands is either pure or mixed. Pure stands are the stands in which at least 80% of the trees in the main canopy belong to a single species, while in mixed stands, less than 80% of the trees in the canopy belong to a single species. It is generally accepted that mixed stands support higher biodiversity due to their variety of structural characteristics.

29.2.2 Stand density

Stand density depends on the number of trees, the basal area, the volume, or other factors, on an area-specific basis. Stand density affects forest biodiversity to a great extent, since it determines some important habitat features such as canopy closure, amount of light penetration, number and composition of understorey species, space for wildlife, etc.

29.2.3 Age structure

Forest stands are commonly distinguished as even-aged, uneven-aged or all-aged according to tree ages. Thus, stands are viewed on the basis of the age classes comprising them. These are: 1) even-aged stands, i.e. stands exhibiting relatively minor differences in terms of age (up to 20 years for mature stands (Dafis, 1990) between individual trees; 2) uneven-aged stands, i.e. stands with relatively major age differences (over 20 years) between individual trees; and 3) all-aged stands, where all tree age classes are present in a relative small area.

29.2.4 Age distribution – stand developmental stages

Depending on the age of trees, any forest stand follows a series of developmental stages (Oliver and Larson, 1996), through the development of individual trees, which compete, survive and die at different rates, showing great fluctuation among the different forest types. This series usually appears after a natural stand-replacing disturbance or after an intense human forest intervention (e.g. clear-cutting in a wide area) that removes the previous stands. During the natural process, different stages can be described that usually follow the general pattern described by Oliver and Larson (1996).

- **Stand-initiation stage:** The first stage following a disturbance that gives rise to stand replacement. This stage is characterized by new individuals and species occupying the space that was freed as a result of the destruction of the trees that previously existed in the area. Numerous trees, shrubs, herbs and grasses are revived through various mechanisms and take up again the space available, in the first years following the disturbance. This stage expires at the point when the growing space becomes dominated and permanently occupied by some plants, thus limiting the possibilities of regeneration for new plants. The stage usually lasts for a couple of years in cases when species recruitment occupies the available space, or it can last for decades, depending on the type of disturbance, the characteristics of the site and the ecophysiology of the species.
- **Stem-exclusion stage:** This stage succeeds the previous stage, upon the occupation of the whole space. In this stage there are no new individuals, while some of the already existing ones die. Initially, the trees form a stratum; then some trees

grow larger, occupy growing space from other trees, and compete against the others, causing them to decelerate their growth speed, or finally to die. Some species existing in stands of mixed species fall short in terms of height growth, get overtopped, and develop very slowly, dominated by the stronger ones. This process gives rise to a 'stratification' pattern, where large trees take over the upper stratum (overstorey), and small trees take over lower strata (understorey), even though in most cases they all have roughly the same age.

- **Understorey-reinitiation stage:** Later, after many years have passed, forest floor herbs and grasses, shrubs and advance tree regeneration again appear under the canopy of the initial stand, and survive in the understorey, growing at a rather slow speed. This process leads to the development of a vegetative understorey, which gives rise to the understorey stage. The overstorey shade prevents this vegetation from growing tall, and a separate layer of vegetation of one to two metres high can persist for a quite long period of time.
- **Old-growth stage:** Later on, overstorey trees die after their senescence, in an irregular manner, and some of the trees on the understorey begin gradually to grow on the overstorey. This stage appears if no disturbance occurred in a forest stand within the lifespan of the dominant tree species.

Forest biodiversity alters with the passage of time through the transition among the different stages of stand development, since during these stages forest stands present quite different structural features in terms of distribution of sizes, numbers, and species of trees, shrubs, herbs, snags and logs. These structural features play a critical role, as they offer habitats for different species, and different susceptibilities to disturbances such as winds, fires, insects, etc., which affect the wildlife capacity of the forest ecosystem. The transition process directly follows an initial stand-replacing disturbance, and usually lasts for decades (Oliver and Larson, 1996). These alterations are not linear; it has been demonstrated that species wealth fluctuates through a number of peaks and lows during the transition between the various stand development stages (Peet and Christensen, 1980; Halpern and Spies, 1995; Roberts and Gilliam, 1995).

In general, species wealth is inclined to increase through the course of stand initiation, as resources are abundant and new species enter the disturbed site (Ellum, 2009). Then, species richness usually decreases to a minimum degree during the stem-exclusion stage, when stand density is high, and there is massive competition for light, in view of the low availability of resources under the dense tree canopy. In the stand-reinitiation stage, species richness again increases in view of the high availability of light to stand interior, through the canopy openings. Then, it follows a rather stable or slightly falling trend as the stands approach the old-growth stage.

29.2.5 Vertical structure

Vertical stand structure exhibits a broad range of forms, based on the height-based differentiation of the trees comprising the stands. These are commonly described as follows: single-storey stands, two-storied, or multiple-layered stands. Additionally, in the even-aged stands (but also in all types of stands), trees are classified according to the crown position in the canopy, as follows.

29.2.5.1 Crown classification

- **Dominant.** The largest trees in a stand, with crowns that exceed the average stand canopy level, thus getting full light from above and in part from the sides. These, along with the codominant trees, form the stand overstorey.
- **Codominant.** Trees with crowns that form the average level of the stand canopy and receive full light from above and some light from the sides; usually with medium-sized crowns, crowded on the sides to a larger or smaller extent.
- **Intermediate.** Trees shorter compared to the previous two classes; their crowns occupy the space between and under the crowns of dominant and codominant trees, which generally receive some light from above, but no light from the sides, usually with small crowns noticeably crowded on the sides.
- **Suppressed.** Trees with crowns existing below the overstorey, thus receiving only diffused light.

Vertical stand structure and crown characteristics hold a crucial role in forest biodiversity, especially with regard to bird nesting and habitation.

Spatial heterogeneity is a property usually attributed to a landscape or a population. It is associated with the spatial distribution exhibited by various concentrations of each species in an area or a landscape. A landscape demonstrating high spatial heterogeneity exhibits a mixture of concentrations of multiple plant or animal species (biological), or of terrain formations (geological), or environmental features (e.g. rainfall, temperature, wind). In populations with spatial heterogeneity, various concentrations of individuals belonging to this species are distributed across an area in an uneven manner. Plant species richness indicates spatial heterogeneity in an ecosystem. In the light of the fact that vegetation offers food, habitats, etc., if vegetation is limited, populations of animals will be equally limited. Environments having a wide range of habitats, such as different topographies, soil types and climates can provide shelter to a larger number of species.

Three main types of spatial patterns are usually identified: random, aggregated or regular (Begon *et al.*, 1996), while the main patterns of spatial heterogeneity are described as follows: evenly spaced, variably spaced, aggregated pattern or random pattern. Generally, natural forests follow an aggregated or random pattern, depending on the establishment process and the seedling recruitment pattern, while artificial forests (plantations) follow the evenly spaced or variably spaced pattern. It is obvious that the most diverse patterns support higher biodiversity.

29.2.5.2 *Shrub layer characteristics*

Understorey woody species at the shrub layer generally contribute to higher forest diversity. The species that exist in this layer constitute a significant habitat for a large number of bird species, as they provide food such as berries and seeds (Wender *et al.*, 2004) and shelter for small mammals (Martin and McComb, 2002); they contribute to the diversity of stand foliage height, and they constitute an important substrate for lichen growth (Wilson and Puettmann, 2007). Moreover, deciduous shrub components provide shelter to a broad variety of arthropods, which are a significant source of food for many neotropical songbirds (Schowalter, 1995; Muir *et al.*, 2002). A rich and diverse understorey layer is also valuable for numerous animal species, since deciduous shrubs, grasses and forbs are tastier than evergreen trees, and are, therefore, a significant source of food for many small mammals and invertebrates (Muir *et al.*, 2002).

29.2.5.3 *Vegetation at herb layer*

Among the different layers of a forest stand, undoubtedly the herb layer is the richest and most abundant in biodiversity components, in almost all types of forests (Ellum, 2009). Additionally, understorey vegetation greatly affects wildlife diversity; for instance, a number of studies have found that young forest stands exhibiting rich and diverse understorey vegetation offer significant benefits for many wildlife species, particularly in cases of habitat rarity, as in the case of dense unthinned stands (Hayes and Hagar, 2002; Wilson and Puettmann, 2007; L.R. Beggs, Vegetation response following thinning in young Douglas-fir forests of western Oregon: can thinning accelerate development of late successional structure and composition? Oregon State University, Corvallis, Oregon, USA, 2005, unpublished thesis, p. 110). Species at the herb layer are thought to be a vital constituent of forest stand structure, being a significant source of food for many arthropods and small mammals (Muir *et al.*, 2002).

29.3 Estimation of Forest Biodiversity

Many approaches have been proposed for the evaluation of the ecological conditions and the assessment of the conservation status of a forest ecosystem, most of which contain some common indices or indicators. Moreover, there are some estimators, including naturalness, natural regeneration, deadwood volume, and number of large trees, which are commonly used in estimating forest biodiversity, in view of the crucial role they play in critical ecosystem functions, such as nutrient cycling, ecosystem stability, etc.

29.3.1 Species richness and diversity indices

A commonly applied biodiversity estimation approach at the entity or species scale is species richness (S), which estimates diversity ‘as the number of species present per unit land area’, irrespective of any data regarding abundance (Magurran, 1988; Gotelli and Colwell, 2001; Ganatsas *et al.*, 2012). When species richness is combined with information regarding individual abundances of species, richness can be used in calculating indices such as Shannon Index (H') or Evenness Index (E) (Magurran, 1988). Although the Shannon Index and species richness are valuable when it comes to understanding the heterogeneity of diversity at different locations, attention must be paid to their interpretations, as they are unable to supply information about which species are present at any temporal point or spatial extant.

29.3.2 Stand structural diversity

Stand structural diversity is also considered as a crucial component of forest biodiversity. Traditionally, tree species, diameter and height are commonly measured for indicating changes in vertical and horizontal stand structure. However, modified Shannon indices were recently used for the purpose of estimating forest structural diversity (Staudhammer and LeMay, 2001). This involves replacing the number of species present in a stand with the tree basal area distribution per hectare by diameter and height classes or species, thus indicating a decent estimation of stand differentiation in terms of tree diameter, height and age.

29.3.3 Focal and indicator species approach

Forest biodiversity can also be estimated through the implementation of the focal species approach, which is broadly used as a measure of forest management sustainable from an ecological perspective (Hannon and McCallum, 2002; Lindenmayer *et al.*, 2006). Based on this approach, some specific species are used for estimation of biodiversity, namely **focal species**. Focal species are those that are mostly affected by threatening processes; for example, any species that can be 'most limited by dispersal abilities, resources or ecological processes' (Lambeck, 1997). In general, focal species are rather small in number, and their distributions and abundances are known. They are used in conservation planning, they are considered to demonstrate the distribution and abundance of the regional biota, and they encompass indicators and umbrella species (Hannon and McCallum, 2002).

Among the focal species commonly used for ecological and conservational purposes are several types of indicators, species at risk, umbrella species and keystone species. However, it is maintained that single indicators are not adequate in most cases, and that a combination of indicator species is preferable (Landres *et al.*, 1988; McLaren *et al.*, 1998). Usually, population indicators, guild indicators and condition indicators are used in evaluating conservation and management actions, whereas biodiversity and composition indicators are used in determining areas or priorities for conservation using protected areas (Hannon and McCallum, 2002).

Indicator species define a trait of the specific environment. For example, an indicator species may outline an ecoregion or signify a special environmental condition, e.g. a climate shift, a disease outbreak or high pollution. Indicator species usually rank among the most vulnerable species in a region or for a specific environmental factor, and they, therefore, operate occasionally as an early indicator of environmental decline. Indicator species can also function as measured environmental conditions. Many species of flora and fauna have been utilized as indicators for the purpose of gathering data regarding the ecological evaluation of specific areas. For example, lichens serve as pollution indicators, and vertebrates are used as an indicator for population trends, and as a habitat for other species.

29.3.4 Management indicators

Management indicators include species belonging to groups of species where management focuses on resource production, population recovery, population viability or ecosystem diversity (Hannon and McCallum, 2002). This term contains a set of different indicators. as follows.

- **Population indicators** are those species whose status reflects the status or the dynamics held by other species. For instance, a species may be always associated with another species, and the population dynamics of one species may be similar to the population dynamics of another species. Thus, variations in terms of abundances of a species can be correlated with alterations in the abundance of another species.
- **Guild indicators** involve species representing other species in the guild. For example, downy woodpecker can be used in monitoring primary cavity nesters.
- **Condition indicators** involve species that are prone to stress environmental factors and thus, demonstrate quick responses to human interventions in the ecosystem. This is very useful in early warning for detecting environmental changes, and thus considering management measures. Condition indicators may also encompass species whose habitat demands may be threatened

by specific management actions. Condition indicators are often specialists, while sometimes invasive species can be used as indicators of anthropogenic stress (e.g. exotic plant species). According to McLaren *et al.* (1998), selecting a representative number of specialists on different types of habitats, forest seral stages, different types of structures within forests is an effective method for monitoring forest ecological status and conservation.

- **Biodiversity indicators** are species indicating areas of high species richness of other species or taxal groups, e.g. hot spots of bird species richness usually overlap with centres of butterfly species richness. Their biological features should be well known, and they should extend in a wide geographical range (Pearson, 1994).
- **Composition indicators** involve those species that, based on their own traits, can represent specific habitat types or elements, e.g. a spotted owl can be considered as representing old-growth forests.

29.3.4.1 Species at risk

Species at risk include the species listed as threatened or endangered by international or national wildlife bodies.

29.3.4.2 Umbrella species

Umbrella species require large areas, e.g. *Ursus arctos*. Conservation of these species should automatically entail the conservation of a host of other species. Umbrella species can be defined through the use of the following criteria (Lambeck, 1997; Caro and O'Doherty, 1999; Noss, 1999; Fleishman *et al.*, 2000): they are rather large species, with vast habitats, with minimum area requirements that encompass those of the rest of the community (area-limited); they should also be non-migratory, demonstrate low temporal variation and be relatively abundant and widely distributed.

29.3.4.3 Keystone species

Keystone species significantly affect many other species, despite their low abundance and limited biomass. They can be predators, prey, plants, mutualists and habitat modifiers (e.g. beaver). Keystone species may also be umbrella species.

29.3.4.4 Natural regeneration

The natural regeneration percentage of a forest area, compared to other parts of a forest where planting and afforestation are the most common regeneration methods, is a significant factor that is indicative of the self-renewing ability of the ecosystem, and reflects positive future trends.

29.3.4.5 Naturalness

The naturalness degree of a forest involves the intensity and history of human interventions implemented on it. Different intensities of utilization are estimated based on the remaining forest area, as well as on the alterations in the structures and the different species communities within a forested area. In general, forest areas and other wooded lands can be classified as 'undisturbed by humans', 'semi-natural' or 'plantations'.

29.3.4.6 Deadwood volume

This involves the average volume of standing and lying deadwood on forests and other wooded areas. Deadwood is thought to be a significant substrate for numerous forest-dwelling species such as insects, lichens, bryophytes and fungi, while it also functions as a shelter and nesting place for many mammal and bird species (European Environment Agency, 2008). There is considerable difference in the amount of deadwood found in undisturbed and managed forests. Generally, late developmental stages of natural forests exhibit high abundance and diversity of deadwood, while managed forests are poor in deadwood.

29.3.4.7 Number of large trees

Similarly to the previous estimator, the quantity of large trees constitutes an important component and substrate, which can support numerous forest species.

29.3.4.8 The concept of thresholds of vegetation cover

The notion of thresholds of vegetation cover was suggested by some authors for the purpose of estimating the risk for species and population losses. According to With and Crist (1995), thresholds of vegetation cover are 'abrupt, non-linear changes that occur in some measure (such as the rate of loss of species) across a small amount of habitat loss' (Rolstad and Wegge, 1987; Andren, 1994; Enoksson *et al.*, 1995; With and Crist, 1995; Andren, 1999; With and King, 1999). They formed the hypothesis that if a threshold response occurs below a critical amount of habitat cover, species and population loss is greater than what can be predicted based on a smooth relationship with habitat cover alone (Lindenmayer *et al.*, 2005).

However, as Lindenmayer *et al.* (2005) put it, it is hardly likely that any generic rules pertaining to critical change points or threshold levels of vegetation or habitat cover (e.g. 10%, 30%, 70%) may be applied across different landscapes and different biotic groups. When it comes to thresholds established in an empirical manner, they depend on the landscape, the groups or species of interest, and the ecological processes investigated.

29.4 Ecologically Sustainable Forest Management Focusing on Biodiversity Conservation

Silvicultural actions carried out on a stand include all procedures implemented by humans on small or large areas of forests, for the purpose of fulfilling the forestry management objectives, including growth acceleration, maintenance and amelioration of the environmental conditions for trees and stands, and increase of the wood and non-wood values of a forest. Over the past few decades, conservation of biodiversity and sustainability of natural resources on the planet have been acknowledged as a main objective of ecological thought and environmental policy. Accordingly, forest management in many countries has encompassed sustainability of all parts of forest ecosystems as one of the main objectives of management. Therefore, forest management and silvicultural treatments are accordingly reformed in order to keep the integrity of forest ecosystems, and at the same time meet the needs of society for wood resources. The US National Forest Management Act of 1976 was the first to recognize and mandate these objectives, which constitute significant management priorities included in the goals pursued by The Forest Stewardship Council (1993), The Montreal Process (1998), The Sustainable Forestry Initiative (1995), and The Society of American Foresters (Ellum, 2009). World forestry practice follows this trend in several countries; forest managers update the current silvicultural methods in order to include all the components of a forest ecosystem, thus emphasizing the autecology of noncommodity forest species.

For example, patterns of flora diversity at the landscape, stand and plant levels have recently been acknowledged as a specific objective to be included in ecologically sustainable forestry for the management of entire forest ecosystems (Ellum, 2009). This has given rise to the need to enhance understanding of the impact of forest management on floristic diversity patterns (Reader and Bricker, 1992; Gilliam *et al.*, 1995; Meier *et al.*, 1995; Roberts and Gilliam, 1995; Maschinski *et al.*, 1997; Jenkins and Parker, 1999; Lindenmayer, 1999; Rubio *et al.*, 1999; Battles *et al.*, 2001; Brosofske *et al.*, 2001; Johnson *et al.*, 1993; Lindenmayer *et al.*, 2006). Similarly, faunal diversity has been recognized as an important ecosystem component, which should be considered in forest management (DeGraaf and Miller, 1996).

The disturbances caused by forest interventions, and especially the logging procedures, may try to emulate natural disturbances, thus leading to the concept of ecologically sustainable forestry. The term 'ecologically sustainable forestry' can be defined as: 'perpetuating ecosystem integrity while continuing to provide wood and non-wood values; where ecosystem integrity means the maintenance of forest structure, species composition, and the rate of ecological processes and functions within the bounds of normal disturbance regimes' (Lindenmayer et al., 2006). Accordingly, all the silvicultural interventions implemented on a forest could operate under this concept for forest biodiversity conservation purposes.

Traditionally, the most important forestry interventions in the forestry practice are: (i) selection of tree species for the initial stand establishment stage according to site features; (ii) site preparation for the forest stand establishment; (iii) plantings; (iv) fertilization; (v) clearings and removal of understorey and purifications; (vi) thinning; (vii) logging and gap creation (USDA Forest Service, 2013). Selecting plant species for a specific site is one of the initial and highly significant steps in installing a forest. The various woody species have different requirements for nutrients and water, while different sites may vary in availability of the required conditions. The type of forest stand created holds a key role in forest biodiversity; for example, the establishment of a forest with an overstorey of photophilous tree species can support an abundant understorey of shade-tolerant species, as it mitigates the stress caused by temperature and light. Site preparation also changes the specific microclimate conditions, which may lead to favourable tree growth, moisture, control of insects and diseases, and reduction of low vegetation. By controlling growth of competing low vegetation, tree species face lighter competition, thus, being able to grow unhindered and dominate by better exploiting any resources. Pruning concerns removing the lower branches of a tree, while thinning reinforces the health of forest stands by cutting down the amount of trees per hectare (stand density); thus, the remaining trees have access to more resources, thereby increasing in a faster and more effective manner (USDA Forest Service, 2013).

Physical disturbances occurring in a natural forest ecosystem determine the course and the future development of forests, in terms of many levels, including composition, structure and biodiversity (Roberts and Gilliam, 1995). Disturbance plays a key role in many mechanistic models of species diversity (Battles *et al.*, 2001). According to Roberts and Gilliam (1995), the intermediate-disturbance hypothesis is most applicable in forest management. Similarly to natural forest ecosystems, in managed forest ecosystems, non-physical disturbances introduced by man determine the development of the forests (Franklin *et al.*, 2002; Torras *et al.*, 2012). Forest management disrupts the canopy cover, creating gaps and discontinuities in overstorey cover, which in turn results in an alteration of stand microclimatic conditions. It is, therefore, evident that post-disturbance alterations in stand understorey conditions vary, based on the intervention type, the management intensity, and the forest type. Moreover, harvesting may cause damage to the understorey vegetation, in particular, tall shrubs. Apart from the initial effect of logging and site preparation, other management activities, including fertilization, herbicide application and grazing, may affect the composition and quantity of understorey vegetation (Halpern and Spies, 1995; Battles *et al.*, 2001).

Thinning of young forest stands may also affect wildlife in a positive manner, as a result of higher light penetration, which leads to an increase in understorey vegetation cover, and contributes to the composition of diverse plant species (Carey *et al.*, 1999; Suzuki and Hayes, 2003; Hagar *et al.*, 2004). Thus, the heterogeneous nature of understorey vegetation in stands can result in an increased generation of habitats, suitable for many different species.

29.5 How Can Forest Biodiversity Be Improved? Managed Versus Unmanaged Forests

In view of the necessity to eliminate biodiversity losses in all ecosystem types, biodiversity conservation should be considered crucial in setting forest management priorities. Therefore, the relationships between forest manipulations and the alterations they cause to forest ecosystems need to be investigated, as they greatly affect biodiversity (Pastur *et al.*, 2002; Mitchell *et al.*, 2008). It is crucial that the effects of silvicultural treatments on forest ecosystem biodiversity be examined before decisions are adopted about the implementation of forest management oriented towards biodiversity conservation.

Formerly, it was commonly believed that minimizing human interventions on forest ecosystems would lead to the minimization of negative effects on forest biodiversity (Niemela, 1999). However, much research has demonstrated that forest management contributes to biodiversity conservation, by improving the structural characteristics of forest ecosystems for several key species groups (e.g. focal species). On the other hand, it is now commonly acknowledged that intensive forest management for timber purposes intensifies ecosystem unbalance, posing a threat for the survival of many natural forest-dependent species (Bengtsson *et al.*, 2000; Paillet *et al.*, 2010).

From a theoretical perspective, the impact of silvicultural treatments applied in a forest can be seen as a type of disturbance that affects its ecological conditions. Massive attention has been drawn to the fact that disturbances alter the main components of an ecosystem, thus affecting the diversity of species at a local and a landscape level (White, 1979; Petraitis *et al.*, 1989; Franklin *et al.*, 2002). A similar approach was adopted in forest management (Roberts and Gilliam, 1995), and the silvicultural treatments that are commonly applied on forest ecosystems.

With regard to natural disturbance, equilibrium and nonequilibrium models have been proposed as tools for illustrating the effects of disturbances on determining species diversity in various ecosystems (Connell, 1978). Equilibrium models are based on the assumption that as a result of a disturbance, the composition of species dwelling at an ecosystem develops up to a maximum threshold, where it remains until another disturbance gives rise to a new relevant process. According to the nonequilibrium models, 'repeated random and catastrophic disturbance events prevent species composition from ever reaching equilibrium' (Ellum, 2009).

Pursuant to this approach, Roberts and Gilliam (1995) maintain that such a hypothesis expressly requires a disturbance incident to maintain diversity. In managed forest ecosystems, overstorey trees function as key system determinants, since they are the dominant competitors, able to intercept light and moisture and, thus, determine the understorey conditions and communities. With regard to the disturbance type, it is accepted that the main constituents of any natural and anthropogenic disturbance often refer to the frequency, intensity and size of the events (White, 1979; White and Pickett, 1985), while far less attention has been drawn to event timing and its possible effect on plant distribution and patterns of successional ecosystem dynamics (Oliver, 1981; Runkle, 1985; Berger and Puettman, 2004).

Nevertheless, there is still massive controversy over the impact of forest management on ecosystem biodiversity (Siitonen, 2001; Paillet *et al.*, 2010), and over the most appropriate management practice for sustaining forest biodiversity. A few decades ago, it was maintained that, at the local scale, a larger number of species lived in unmanaged forests than in managed forests, which is indicative of higher biodiversity (Okland *et al.*, 2003; Paillet *et al.*, 2010). Today, worldwide research results are quite contradictory, and many studies do not support this; in numerous published works, no differences are found in particular species, such as vascular plants, birds and soil invertebrates (Graae and Heskjaer, 1997; Bobiec, 1998). A number of studies have also shown that forest management has some positive effects on species richness of vascular plants (Schmidt, 2005; Ellum, 2009), fauna species (Carey, 1995; Carey and Johnson, 1995; Barber *et al.*, 2001), beetles (Vaisanen *et al.*, 1993), or carabids (Desender *et al.*, 1999). Battles *et al.* (2001) also reported that single tree selection or group selection of reserve management regimes presented a greater proportion of late-seral vs early-seral species and a lower proportion of introduced exotic species compared to plantations and shelterwoods. Additionally, Martín-Queller *et al.* (2013) report that partial harvesting in both coniferous and broad-leaved Mediterranean forests may result in higher tree species richness, compared to complete harvesting and lack of management.

Setting important forest ecosystems as nature reserves for maintaining forest biodiversity presents some further uncertainties; e.g. in many cases it is anticipated that biodiversity may recover slowly after forest management has ceased and, therefore, the establishment of new forest reserves may temporarily seem ineffective. Thus, an assessment of the time required for biodiversity to recover is of utmost importance for conservation policy. Furthermore, forest management encompasses many different practices, with contradictory effects on biodiversity: in the light of the assumption with regard to the negative effects of forest management, one can assume that the more intense the management, the greater the difference in biodiversity between unmanaged and managed forests (Stephens and Wagner, 2007; Paillet *et al.*, 2010).

In the light of the foregoing, it seems that worldwide literature does not consistently adopt the hypothesis that unmanaged forests exhibit greater species richness than managed forests. However, understanding how forest ecosystems react against disturbances caused by timber harvest is necessary, when silvicultural practices are implemented for conservation of flora. For example, according to Ellum (2009), research in southern New England has shown that the seasonal timing of disturbances involving canopy removal may significantly affect the growth and survival of several forest understorey plants. Moreover, dormant season disturbances enable understorey plants to adapt their anatomy, morphology and physiology via developmental plasticity, leading to positive responses to increased light levels.

29.6 Effects of Silvicultural Treatments on Specific Components of Forest Biodiversity

Silvicultural manipulations create gaps in forest stands, which enhance availability of light, thus influencing numerous factors that support the growth or limit of new species, depending on the degree of intensity of the interventions (Kerr, 1999). In turn, these could increase the understorey diversity of some plant species that would otherwise be dominated only by a few shade-tolerant species (Schuman *et al.*, 2003). However, if the frequency of disturbances is very high, the preserved plant communities will be probably dominated only by early successional species. The levels of biodiversity in those communities seem to be lower (Roberts and Gillam, 1995). By contrast, interventions of lower frequency allow forest communities to absorb the impacts and keep their ecological balance.

Forest areas derived from clear-cutting logging are usually characterized by new feature, homogeneous, environmental conditions, which support the predominance of a few pioneer species that colonize the area quickly after the disturbances. Studies of Brashears *et al.* (2004) and Brokaw and Lent (1999) reinforce this view. However, it has been recently proposed by Schulze *et al.* (2014) that 'clear-felling followed by natural succession may even be superior to the protection of old-growth forests, regarding biodiversity'. In contrast, Mikoláš *et al.* (2014) vehemently rejected this hypothesis, arguing that it is misleading to compare clear-fellings to protected areas dominated by old-growth or primary forests by means of a simplistic measure of biodiversity and without a landscape perspective on the role of different types of habitats (successional stages) in conserving biodiversity over time and space.

As the high-intensity interventions in forest ecosystems have been suggested to inhibit the growth of biodiversity (Battles *et al.*, 2001; Paillet *et al.*, 2010), contemporary forest management should come up with alternative silvicultural methods that promote the growth characteristics of later stages of succession, such as large trees, diverse forest types, multi-storey stands and large logs. Creating uneven-aged forests seems to be in close proximity with the ideal model with respect to smooth degree disorders, while it enhances the diversity of wild fauna species, especially those associated with mature forests (McComb *et al.*, 1993). The even-aged forests treated with high rotation ages provide shelter for some wildlife species which prefer mature forests, and some features of these should be maintained in these large forests' peer age. Forests of this type require special silvicultural manipulations, so that they can acquire such appropriated stand characteristics (O'Hara, 2014).

Torras *et al.* (2012) report that after 11 years of monitoring conducted in undisturbed stands, they noticed that biodiversity was increased particularly in terms of species abundance and diversity of the shrub layer. Biodiversity was also enhanced in managed stands which applied selective thinning and ameliorative treatments, while biodiversity reduction was reported only on areas where extensive interventions have been applied, such as creating large gaps and clear-cutting logging. Selective logging appears to be the most suitable method for the regeneration of forest stands aiming at biodiversity conservation. Also, other appropriate manipulations such as low-intensity thinning and pruning, seem to favor the conservation of biodiversity. These results are also supported by the findings of Anderson and Ostlund (2004), Wen *et al.* (2010) and Fuji *et al.* (2010). They also observed that the impact of forest stand handling during the regeneration of forests on diversity of tree species depends on the length of rotation age, which should be relatively large, but varies between different forest types. Another factor that enhances the stability of biodiversity is the attainment of maturity of the forest (Torras *et al.*, 2012).

It is generally accepted that the simplification of stand structures and species composition leads to systems with low connectivity, rendering them susceptible to insect and mammalian herbivory (Thompson *et al.*, 2003). By contrast, silvicultural prescriptions enhancing variability in the same stand can offer significant habitat characteristics across multiple scales and enhance habitat quality beyond that provided by stand-level prescriptions (Wilson and Puettmann, 2007). However, predicting the response of forest understorey plants to disturbances involving canopy removal should be a primary objective of silviculture practices that consider the integrity of whole forest systems (Ellum, 2009), focusing on the conservation of all components of forest ecosystems.

29.6.1 Effects of silvicultural treatments on floristic diversity

Various plant species may be favoured by alterations in the environmental conditions caused by disturbances and the various forestry operations, while some others are likely to be adversely affected. This causes a change in the composition and abundance of plant species. A typical example is that many species of early stages of succession which are present in gaps resulting from logging (Schuman *et al.*, 2003) are absent in undisturbed ecosystems. Thus, one of the major questions is how the understorey plant responds to disturbance in an effort to inform silvicultural practices intended to conserve plant diversity in forests undergoing management. According to Grime (1977), based on a general theory that describes plant survival strategies under different levels of resource stress and disturbance conditions, strategies can be divided into three groups: plants that persist in zones exhibiting low stress and high disturbance (ruderals); plants that persist in zones exhibiting high stress and low disturbance (stress-tolerants); and plants that persist under conditions of low stress and low disturbance (competitive plants).

However, although it is commonly acknowledged that forest management has numerous effects on floristic diversity patterns, the magnitude and trajectory of such effects are still examined with regard to several silvicultural treatments available to managers, or across the multitude of forest types currently under management (Ellum, 2009). According to Roberts and Gilliam (1995), the maximum attainable biodiversity conservation levels are achieved through moderate intensity and frequency of disturbance. Among the major factors determining alterations in stand floristic diversity are the disturbance type (e.g. the type of silvicultural treatment), the stand type and site conditions, the management intensity, and the frequency of the interventions (time intervals between treatments). Significant differences in the effects on biodiversity are identified by Niemela (1999), between application of silvicultural treatments and natural forest disturbances, particularly as regards the frequency and method of carrying out the logging. Other recent studies on the impact of these processes on biodiversity introduced alternative techniques for maintaining conservation of biodiversity in managed forests. It should be noted the vast majority of the published studies apply observational methods to make snapshot assessments of the existing floristic patterns, and compare them to past management records and unmanaged stand conditions (Jenkins and Parker, 1999; Peltzer *et al.*, 2000; Small and McCarthy, 2005). Moreover, the limited information on the physiological, anatomical and morphological plasticity of understorey plants in response to the environmental shifts caused by forest management are to a great extent derived from the tropics (Clearwater *et al.*, 1999). Today there is scant information on the impact of timber management on the understorey flora of numerous forests in the world.

29.6.1.1 Effects on species richness and composition

In the early stages of a forest stand development, a rapid canopy closure occurs a few decades after stand establishment, which greatly affects floristic composition of understorey vegetation during these early phases, and it usually restricts species richness and abundance (Dyrness, 1973; Halpern, 1989; Puettmann and Berger, 2006; Wilson and Puettmann, 2007). During these early phases, the application of any forest manipulation and specific silvicultural treatments alter the stand environmental conditions, usually by providing more light, which in turn positively influences understorey vegetation composition. Early research results (e.g. Halpern, 1989; L.R. Beggs, Vegetation response following thinning in young Douglas-fir forests of western Oregon: can thinning accelerate development of late successional structure and composition? Oregon State University, Corvallis, Oregon, USA, 2005, unpublished thesis, p. 110) show that following thinning, the composition of species is shifted towards an early-seral type independent of the thinning intensity, and is characterized by increasing occurrence and cover of early-seral or invading species. This change in composition usually occurs directly after thinning and persists for at least ten years (Halpern, 1989; Puettmann and Berger, 2006; L.R. Beggs, Vegetation response following thinning in young Douglas-fir forests of western Oregon: can thinning accelerate development of late successional structure and composition? Oregon State University, Corvallis, Oregon, USA, 2005, unpublished thesis, p. 110). By contrast, understorey vegetation in old-growth forests that have been subjected to clearcut or burned, after treatment, exhibited a composition similar to the pre-disturbed stands within 20–40 years (Schoonmaker and McKee, 1988; Halpern and Spies, 1995), thus indicating the trend for understorey vegetation to return to mature forest conditions.

Thinning and creation of gaps through single tree logging or group logging enhance the heterogeneous nature of composition in terms of space. Nevertheless, minor gaps caused by silvicultural interventions with little intensity may not result in a distinct species shift, but they may offer space and benefits for rare late-seral species over a longer period (Spies and Franklin, 1989). However, in relation to thinned stands, vegetation in small gaps (0.1 ha) showed a shift in composition towards the competitive (see Grime, 1977) forest-residual species (Wilson and Puettmann, 2007). Retrospective researches have demonstrated that thinning results in a long-term increase of species richness, and temporarily, in a compositional shift towards early-seral species, primarily because of the light increase to the forest floor. This in turn results in increased grasses, sedges and nitrogen-fixing species. By contrast, sites that have not undergone thinning were found to be similar to adjacent old growth (Bailey *et al.*, 1998). This continued presence may be a crucial consideration in latter thinnings, as many species propagate clonally and can respond vigorously if already established in the stand (Tappeiner *et al.*, 2001), close to near pre-treatment levels. However, it is not straightforward whether these species will be completely lost from the thinned stands even after several decades (Halpern and Spies, 1995; Bailey *et al.*, 1998; Puettmann and Berger, 2006). When thinning is applied in small patches, the influence is less important. Unthinned patches of any size can offer habitats for many species not present in the thinned stands, indicating the value of these patches in terms of within-stand heterogeneity and their potential for propagation into adjacent areas (Wilson and Puettmann, 2007).

Thinning can have positive effects on understorey species abundance and cover, and it is likely to significantly affect the wildlife habitat and other ecosystem functions. On the other hand, the high cover of early-seral invading species in the thinned stands is anticipated to gradually decrease as overstorey canopy closes to near pre-treatment levels. The time required for these species to completely disappear from the thinned stands cannot be accurately estimated, even after several decades have passed from the application of thinning (Halpern and Spies, 1995; Bailey *et al.*, 1998; Puettmann and Berger, 2006).

29.6.1.2 Effects on herbaceous cover

Understorey plants dwelling under the forest canopy have developed certain physiological, anatomical and morphological features, which differ largely from species existing in more open environments (Bazzaz, 1979). Ellum (2009), in his review article on the literature on responses of plants to disturbance – primarily shifts in light environments – suggested that forest interventions typically cause an increase of the light levels to the groundstorey through canopy tree removal, caused by timber harvest. The ability of species to adapt and endure in light-changing environments has been found to be a major criterion when it comes to determining the floristic patterns inherent in a landscape.

While undergoing timber harvest or other silvicultural interventions (thinning, clearings, pruning, etc.), understorey vegetation sustains key changes in terms of physical environment. Some of these changes involve increased light, higher nutrient levels, increased diurnal temperature fluctuations, as well as fluctuations in soil moisture content and relative humidity (Minckler *et al.*, 1973; Denslow, 1985; Phillips and Shure, 1990; Ganatsas, 1993; Ellum, 2009). The most dramatic change involves the increase of the levels of direct photosynthetically active radiation (PAR) compared to the previous relatively shaded environment (Chazdon and Fletcher, 1984; Canham, 1988; Canham *et al.*, 1990; Ganatsas, 1993). The PAR levels under the closed canopy of several forests can reach less than 5% of incident radiation, following canopy removal, due to forest operation to full sun (Anderson, 1964; Ganatsas, 1993). Such changes are not uniformly distributed in the forest stands, but they vary, to a great extent based on the canopy opening, the position of a microsite in relation to the edge or centre of the disturbance area, and the spatial orientation of non-circular gaps (Runkle, 1982; Canham *et al.*, 1990; Ganatsas, 1993; Ellum, 2009).

The diversity of plant species increases with increasing levels of disturbance up to a point, after which it begins to deviate. Based on this hypothesis, many scientists (e.g. Battles *et al.*, 2001; Schuman *et al.*, 2003) assessed the impact of forest management at different intensities in different types and locations, reinforcing it with their results.Diversity of species in areas with clear-cutting logging was also found to be lower than the undisturbed or low-intensity management areas. Both the formation of gaps and the implementation of thinning exhibited statistically significant effects in terms of abundance and diversity of flora species. A number of other studies also reported that the herb cover in thinned stands is much higher (up to twice as high) in unthinned or old-growth stands (Alaback and Herman, 1988; Spies, 1991; Bailey *et al.*, 1998; Hanley, 2005). By contrast, Wilson and Puettmann (2007) found that herbaceous cover was less responsive to density treatments compared to shrub cover, and increasing herb production may pose an important challenge for management. They also report that herb cover responds either neutrally or somewhat negatively to thinning across a broad range of residual densities and spatial patterns.

29.6.1.3 Effects on shrub cover

There are only few studies about the evaluation on the impact of silvicultural treatments on shrub species changes. In general, thinning has the tendency to increase shrub cover and within-stand variability, especially when the shrub cover is limited prior to treatment (Harrington *et al.*, 2005), as in the case of dense stands. Nevertheless, density management has the tendency to slightly homogenize shrub cover when pre-treatment cover is relatively high. The research of Torras and Saura (2008) in the Mediterranean region showed that positive selection thinning highly increased the diversity of shrub and tree species. Pruning was also beneficial to shrubby species diversity. Dense uniform shrub cover of a few species may reduce the diversity of forest function, and, thus, treatments such as variable density thinning or creation of gaps have been proposed for the purpose of increasing spatial variability in shrub cover and stand diversity (Wilson and Puettmann, 2007).

Klinka *et al.* (1996) found that in western British Columbia, unthinned young plantations showed an inverse relationship between overstorey canopy cover and shrub cover. Nevertheless, this is common in plantation ecosystems in the Mediterranean region (Salvatore *et al.*, 2012). This is indicative of the fact that long-term development of shrub layer is primarily dependent on the overstorey cover. However, any distinction between early and late responses to thinning must also take into account the stand dynamics trends, especially in the case where a higher shrub and herb cover appears, as the stands proceed from the stem exclusion to the understorey initiation and old-growth stages (Spies, 1991; Wender *et al.*, 2004).

29.6.1.4 Effects on exotic species

In general, exotic species tend to expand and localize to areas subjected to intense disturbances, such as roads and clear-cuts (Ganatsas *et al.*, 2012), but they are also common in some unmanaged riparian areas (Heckman, 1999; Parendes and Jones, 2000). Thus, these species are rare in natural forest stands, but when they appear they entail serious dangers, since they tend to dominate site resources, which may cause a significant reduction of native species richness (Heckman, 1999). Thinning disturbs vegetation and releases resources, bearing the risk that exotic species may propagate and dominate the understorey vegetation. However, following clear-cutting, the exotic species cover usually decreases rapidly due to the crown closure (DeFerrari and Naiman, 1994; Halpern and Spies, 1995). Heckman (1999) also found that exotics in Pacific Northwest forests in northern America are mainly ruderal herbs occupying a niche of heavily disturbed sites, while they are rather scarce in natural forest stands.

29.6.2 Effects of silvicultural treatments on fauna

Although there is scant long-term information concerning the effects of basic silviculture and intense forest management on forest fauna diversity, studies have reported that the decrease of structural diversity in forest stands usually has a significant common impact, which gives rise to vertebrate species responses: fewer structural characteristics in managed forests than in natural forests entail smaller numbers of fauna species dependent on those structures, such as cavity-using species and species needing large decaying woody debris (Thompson *et al.*, 2003).

Silvicultural treatments affecting stand density management, and especially those incorporating gap creation and other spatially variable density treatments increase within-stand heterogeneity in canopy structure, thus providing suitable habitats for wildlife. According to Barber *et al.* (2001), thinning and gap creation may increase biodiversity, and create different mosaics, which can support the activities of many species (nesting, hunting, habitation), and each species depending on the size of the food habits etc., has different space requirements between the stands in the canopy openings, and type and amount of food (Chambers and McComb, 1997).

29.6.2.1 Effects of silvicultural treatments on bird fauna

As has already been mentioned, forest stands constitute an important habitat for numerous bird species. Thinning intensity and type play an important role in stand capacity for supporting bird fauna. For example, uniform thinning usually generates uniform canopy openings and structures, namely conditions that are usually more favourable for wildlife species with larger home ranges (Wilson and Puettmann, 2007). By contrast, horizontal canopy heterogeneity appears to be common in older, unmanaged stands, where there is high variation among canopy gaps (Spies *et al.*, 1990). In general, an intricate vertical canopy structure should be one of the main management objectives when wildlife biodiversity is addressed, since it has been commonly associated with increased songbird diversity and use (Carey and Johnson, 1995; Carey, 1996; Thompson *et al.*, 2003). Home ranges of many songbirds are generally small, varying between two and five hectares, while wildlife habitats for many of these species often demand contrasting fine-scale habitat characteristics in close juxtaposition within such home ranges (Hayes and Hagar, 2002; Wilson and Puettmann, 2007).

A great foliage height diversity is highly desirable (MacArthur and MacArthur, 1961), or relevant diversity-oriented measures that usually appear under uneven-aged silvicultural systems, or preferably in all-aged mixed forests. Forming variable canopy conditions in the long term is possible through the implementation of suitable silvicultural treatments. Thinning operations or selective cuttings greatly contribute to this direction; the small gaps created, as well as any variable density thinning create tree spatial heterogeneity and increase canopy openness, thus increasing within-stand variability. By contrast, uniform low intensity thinning has a smaller effect on the structural variability of stands, since it does not dynamically affect canopy structures, and, therefore, varied intensities of low thinning do not radically increase diversity of foliage height (L.R. Beggs, Vegetation response following thinning in young Douglas-fir forests of western Oregon: can thinning accelerate development of late successional structure and composition? Oregon State University, Corvallis, Oregon, USA, 2005, unpublished thesis, p. 110).

In general, canopy closure variability in unthinned stands is low, depending on how stands are established, whereas following the application of thinning, the gaps created within evenly spaced thinning treatments cause an up to 79% increase in stand variability (L.R. Beggs, Vegetation response following thinning in young Douglas-fir forests of western Oregon: can thinning accelerate development of late successional structure and composition? Oregon State University, Corvallis, Oregon, USA, 2005, unpublished thesis, p. 110). Similarly, it can result in heavy evenly spaced thinning, which can also increase within-stand variability of canopy closure, but the ecological repercussions are different, as heavy thinning does not provide for strong edge contrasts or large openings (Wilson and Puettmann, 2007).

At stand scale, diversity of foliage height in young, single-storied canopies tends to be low. Thus, appropriate thinning that intentionally creates tree and crown-size diversity may be required when it comes to monoculture stands, with a view to ameliorating habitats for a large number of species (Kerr, 1999; O'Hara, 2014). It has also been noticed that many songbird and bat species usually nest or roost in deciduous trees, and thus, any retention of such trees in thinning operations can offer vital characteristics in an otherwise suitable habitat (Hayes and Hagar, 2002). Furthermore, regeneration cutting of an under- and eventually mid-storey tree layer can increase both horizontal and vertical stand variability, as well as stand foliage height diversity. Shade-tolerant tree species are able to survive and grow under low understorey light levels (Ganatsas, 1993; Gray and Spies, 1996), offering a good illustration of how spatial diversity in the overstorey can offer stand variability in the long term.

However, in many studies, forestry operations, particularly those pertaining to timber removal, seem to be connected with the decrease in wildlife diversity. Chambers *et al.* (1999) demonstrated that as forestry operations increase and the openings in the area become larger, the number of birds keeps falling for up to three years after conclusion of logging. On the other hand, silvicultural operations based on natural disturbance patterns showed that it may be favourable to species associated with the last stages of succession or old forests. Moreover, some of those species may appear in these forests only after application of such operations. Most of the intense forestry treatments may have crucial effects on biodiversity decrease; thus, disturbances of low or intermediate intensity, such as those resulting from the methods of selecting small groups of trees, or single tree management may be more effective in increasing bird fauna diversity (Torras *et al.*, 2012; Dafis, 2013; Martín-Queller *et al.*, 2013).

29.6.2.2 Effect of silvicultural treatments on small mammals

Many studies have investigated the effect of forestry operations on small mammals. Their findings generally support the view that intensive interventions in a forest ecosystem have an adverse impact on the number of species and their abundance (Aubry *et al.*, 1991; Corn and Bury, 1991; Gilbert and Allwine, 1991). According to Waldien (2005), after moderate interventions, the densities and abundances of various species of small mammals tended to be greater in stands 8–12 years after logging, in relation to the control areas, or were identical in all. This can easily be explained as the understorey vegetation responds to excess light entering after canopy openings, providing shelter, complex and varied microclimates and abundance and variety of food for small mammals. Increased available resources, such as food, lead to increased populations of small mammals, until the point when a resource constraint becomes limiting due to its inadequacy (Rosenberg and Anthony, 1993; Carey, 1995; Carey and Johnson, 1995; Hayes *et al.*, 1995; Carey, 2001).

29.6.2.3 *Effect of silvicultural treatments on insects and fungi*

The research findings of many studies, e.g. Hamer and Hill (2000), Holloway *et al.* (1992), Raguso and Llorente-Bousquets (1990), Hamer *et al.* (1997), have shown that species diversity and abundance may vary upon application of forestry operations, but this is largely dependent on the intensity and method of application. According to the research findings of Stork *et al.* (2003) on the effects of forest manipulation on populations of butterflies, the abundance of species was significantly higher in undisturbed areas than in those which had undergone logging. Moreover, the stands that had been subjected to medium-intensity interventions exhibited lower diversity of species than undisturbed stands, but they were evidently larger than the stands that had been cleared. Further monitoring of the areas for a longer period showed no further decrease in the diversity or abundance, but there was stability in the numbers obtained after logging.

Dollin (2004)reported that most species of saproxylic beetles were found in middle- and old-aged stands, while only a few species were found in young stands. The number of species and abundance was also significantly lower in stands that had undergone logging. The saproxylic beetles are a group of Coleoptera that, at some point in their lives, depend on eating disintegrating wood. They need mature forest ecosystems, represent a large proportion of the total species diversity of the stand, and contribute greatly to the decomposition and recycling of nutrients (Probst and Crow, 1991).

Fungi diversity is usually higher in closed forests than in open stands, because of their requirements in specific temperature and humidity conditions, which mostly appear in closed stands. The removal of the largest part of the canopy can cause changes in the forest microclimate, which may negatively affect the abundance and diversity of fungi (Luoma *et al.*, 2003).

29.6.3 The role of forest management intensity

Many research findings corroborate the view that the impact of forest management on forest biodiversity depends on management intensity. It is anticipated that intense forest management leads to greater difference in terms of species richness between unmanaged and managed forests (Paillet *et al.*, 2010). In comparison with unmanaged forests, the greatest difference in species richness was found in forests that underwent clear-cutting and had experienced alterations in tree species in the past (Stephens and Wagner, 2007; Paillet *et al.*, 2010). Conversely, species richness in forests which had undergone clear-cutting, but did not face any change in tree species (natural or artificial regeneration) hardly differed from unmanaged references. Clear-cutting is frequently applied in boreal forests, because it imitates the natural fire disturbance regime. Clear-cutting, however, may significantly diverge from the natural fire regime with respect to disturbance intensity and frequency and effect on habitat features (Niemela *et al.*, 2007; Paillet *et al.*, 2010). It should be stressed at this point that there are massive adverse effects on biodiversity, which appear when native forest types are replaced by fast-growing exotic tree species after clear-cutting. However, in most boreal forests, native tree species are usually planted after logging, and thus, any effects on the related wildlife communities can be noticeably lighter than in plantations of non-native species.

In forest stands exhibiting low biodiversity, such as the monospecific plantations, biodiversity increase can be achieved by promoting age differentiation (Kerr, 1999). Plantations in many countries such as in the United Kingdom are managed under short rotation ages of 35–45 years, reaching only the stem exclusion stage (see Oliver and Larson,

1996). Extending the rotation age and moving to the understorey re-initiation stage is a common method recommended for diversified habitat structure, in favour of forest biodiversity (Kerr, 1999).

The effect of selective cutting, whether close-to-nature management or not, was rather insignificant, and this demonstrates that the implementation of small-scale disturbance does not adversely affect forest biodiversity. However, the analysis carried out by Paillet *et al.* (2010) on the effects of management types for different taxonomic groups did not manage to disclose any clear and statistically significant trends. Their findings indicate that large and intense disturbances succeeded by a change in the composition of tree species have the most detrimental effect on species richness. The remaining management-intensity gradient did not indicate any clear trend. Martín-Queller *et al.* (2013) emphasized the potential possible effects of medium-intensity harvesting for native tree species richness in the Mediterranean region. These findings can have significant implications for forest management.

Current and future developments of forest exploitation may also have significant effects on forest biodiversity in some timber management forests. For example, in the Canadian boreal forests, while mild silvicultural regeneration techniques, such as planting, seeding and tending, have been employed since the 1940s, more intensive techniques (intensive forest management) such as increased area planted, pre-commercial and commercial thinning, extra tending events, fertilizing and short rotations, may soon be implemented (Thompson *et al.*, 2003). This may give rise to more extensive conflicts between conservation of biodiversity and timber production, in view of the major effects that are anticipated from a more intensive silviculture on biodiversity in the long term, compared to natural regeneration after the application of logging or after a stand-replacing natural disturbance. In this case, the concentration of intensive forest management in stands developed on the more productive sites could intensify effects (whether positive or negative), due to the positive relationship between forest productivity and diversity of flora and fauna. Therefore, several species such as black-backed woodpeckers (*Picoides arcticus*) may be reduced over large areas by stand conversion to mixedwoods, stand structural changes and age-class truncation (Thompson *et al.*, 2003).

29.7 Application of Appropriate Forest Management and Silvicultural Practices Aiming at Forest Biodiversity Conservation

In managed forests, the forest structure varies according to several factors, the most important being the applied silvicultural treatment, both on a short-term or long-term basis. Generally, canopy openings resulting from management operations, and forest-floor disturbance associated with tree-fall gaps in old-growth stands are considered crucial for conserving forest biodiversity (Battles *et al.*, 2001; Franklin *et al.*, 2002). According to the literature, silvicultural operations appear to affect the abundance and diversity of flora and fauna species. A general rule is that severe interventions in forest ecosystems can lead to a radical decrease of forest biodiversity. It has been proven that high-intensity interventions in forest ecosystems have adverse effects on biodiversity; thus, effective forest management should encompass alternative silvicultural operations that promote stand features of the later stages of succession and old-growth forests, such as retention of large trees, diverse forest types, multi-storey stands and large logs. The creation of uneven-aged multi-storey stands is closer to this model with regard to mild degree disorders, which promotes diversity of wild fauna species, especially those associated with mature forests (McComb *et al.*, 1993; O'Hara, 2014).

Also, even-aged stands treated with high rotation ages can provide shelter for wildlife species specialists on mature forests, and some specific features of these large aged forests shall be maintained. However, this type of forest requires suitable silvicultural operations, so that it can acquire such stand characteristics. Silvicultural prescriptions that increase within-stand variability can provide significant habitat characteristics across multiple scales and improve habitat quality beyond that provided by stand-level prescriptions (Wilson and Puettmann, 2007). By contrast, simplifying stand structures and species composition may lead to systems with low connectivity, vulnerable to insect and mammalian herbivory (Thompson *et al.*, 2003).

In the last few years, modern biodiversity conservation techniques have been proposed at a worldwide level, oriented towards small-scale interventions, which follow and try to replace, or assist, the natural process and succession occurring in forest stands. Such techniques seek to conserve the forest encountered in the last stages of succession, as it has been demonstrated that such forests may support the highest biodiversity. Thinning, and in general any silvicultural treatment of low to medium intensity, diversifies the microclimate of the stand, and creates more favourable conditions for the development of the understorey flora and fauna. Moreover, the creation of forest gaps appears to be impairing the increase of biodiversity, while their existence can fulfil the demands of certain bird species.

The specific treatments required for each forest stand depend also on the short-term management objectives. For example, with regard to the stand of floristic diversity, a valid model has been proposed according to which minor to moderate silvicultural manipulations enhance diversity, and when they are applied to large strains, there has been a decrease of biodiversity. However, the ability of predicting how forest understorey plants will respond to canopy-removing disturbances should be a major objective of silvicultural practices that focus on the integrity of whole forest systems (Ellum, 2009) and are oriented towards the conservation of all components of forest ecosystems. In areas where tall shrub cover is of special concern, harvesting activities may be designed for the protection of shrub patches. Alternatively, on nutrient-limited sites, fertilization may increase understorey vegetation growth without needing to alter the density management regimes (Prescott *et al.*, 1993). Heavier thinning or gap creation interventions are usually necessary to promote tree flowering and seed production of the overstorey trees when stand regeneration is the short-term management goal (Ganatsas *et al.*, 2008). With respect to stands managed primarily for wildlife conservation purposes, variable density thinning may reduce the risk of the uniform understorey shrub layer which is dominated by a single or few species (Alaback and Herman, 1988; Bailey *et al.*, 1998; Deal and Tappeiner, 2002). Moreover, the gaps caused by the silvicultural interventions can provide space, food sources and favourable microclimatic conditions for a number of arthropod species, and openings for aerial insect capture by songbirds (Hagar *et al.*, 2004). For small mammals, even where the management goal involves production of timber products, rotation ages of <150 yr, the retention of live trees at harvest, combined with coarse woody debris and understorey vegetation enhancement is suggested, since they can be very useful in biodiversity conservation in the managed landscape (Carey and Johnson, 1995).

However, in view of the complexity and diversity of any ecosystem, biodiversity-oriented forest management requires that information be collected on the manner in which various species react against different silvicultural measures and logging systems, which can be achieved by the management objectives.

29.7.1 General principles for conserving forest biodiversity

Based on the above-mentioned analysis of the world literature and practice, some general principles can be set for conserving forest biodiversity in managed forests through the design and application of several appropriate silvicultural treatments. However, these can be used as well to conserve forest biodiversity within the protected areas, when it is needed. These are summarized below (Lindenmayer *et al.*, 2006; Dafis, 2013).

- The need for continuous, repeatable silviculture interventions, based on the specific site conditions, stand characteristics and forest management goals.
- The conservation of forest stand natural composition.
- The conservation of the complex nature of stand structures.
- The setting of the priority for natural regeneration in order to perpetuate all forest genes and genetic resources of an area.
- The maintenance of the spatial heterogeneity of stands.
- The preservation of nutrient cycling and the protection of ecosystems from nutrient losses.
- The preservation of the heterogeneous nature of habitats and landscapes, as well as landscape connectivity.
- The promotion of low-energy harvest machinery and equipments.
- The preservation of aquatic system integrity through the implementation of hydrological and geomorphological processes.
- The need for monitoring and updating operations.
- The promotion of international knowledge exchange.
- The development of common conservation strategy and policy.

- The creation of international ecological networks for nature conservation (e.g. the Natura 2000 network in Europe).

These general principles can be specialized for management and silvicultural measures at stand-level, at region-scale, and at biogeographical-level biodiversity. The stand-level measures include the main conservation measures which are addressed to forest structure and functions (Lindenmayer *et al.*, 2006). At the region-scale, the measures include actions focusing on the conservation of habitat and landscape characteristics, such as the maintenance of connectivity, the wildlife corridors, and the integrity of the different systems. They also consider the special habitat requirements for populations of rare species, and biological hotspots. At biogeographical-level biodiversity, the measures concern mainly policy actions, and national and international cooperation and rules, as well as experience exchange and development of a common conservation policy at greater scales (EU, 2015).

29.7.2 Silvicultural treatments at stand-level biodiversity aiming at the conservation of forest structure and functions

Forest management oriented towards forest conservation can be successfully combined with harvesting, provided that it has been appropriately designed. However, this type of management demands specific silvicultural tools, and an approach to more close-to-nature silviculture, to be effective. Accordingly, more emphasis should be placed on the following (Dafis, 2013).

- The need for continuous, repeatable silviculture interventions.
- The conservation and enhancement of the natural composition of forest stands, incorporating species that definitely pre-existed in the area, but are now absent for a variety of reasons (e.g. forestry practice, repeated wildfires, etc.).
- The adaption of small-scale disturbance and the complete exclusion of clear-cuttings.
- The maintenance of stand structural complexity; this may include (Lindenmayer *et al.*, 2006):
  - use of natural disturbance regimes as a template for logging regimes
  - creation of specific structural complexity through well-designed stand management activities, including the application of novel thinning or silvicultural systems, for the purpose of accomplishing stand-level objectives
  - special treatment of regenerated and existing stands for the purpose of creating special structural conditions, such as through new kinds of thinning activities (Carey *et al.*, 1999) or structural retention during regeneration harvest.
- Always applying tree natural regeneration in order to perpetuate all forest genes and genetic resources of an area.
- The maintenance of stand spatial heterogeneity; this may comprise the consideration of adjacency to other plants and stands, as well as the landscape context.
- The habitat consideration within the management units or stands of a forest.
- The retention of structures and organisms during regeneration harvesting; this may involve maintaining 5–10 old trees (with hollows) per hectare (or mature trees if there are no old trees), for nesting wildlife (e.g. wood-food or bark-food invertebrates), distributed along the stand area.
- The promotion of uneven-aged forest structure with high structural diversity at stand level; this may include the long (lengthened) rotations or cutting cycles (Seymour and Hunter, 1999; Lindenmayer *et al.*, 2006; O'Hara, 2014).
- Removing only wood volume from the forest, and keeping all the rest as biomass materials (foliage, branches, thin wood material, bark etc.).
- The maintenance of nutrient cycling and thus the protection of ecosystems from nutrient losses and degradation.
- The conservation of dense edges of forest stands.
- The increase of stand and lying deadwood by maintaining in the stands some dead trees, a part of lying thick wood materials, and the majority of forest harvest residuals and debris. However, this should be a lower priority in fire-prone forest ecosystems.
- Livestock grazing exclusion from the forests, which, except for the damage to young trees, forest regeneration and soil depressing, also acts as competition to wildlife (herbivory species).
- The consideration of targeted management strategies for particularly important species.
- The maintenance or the scarce creation of small patches with no forest cover (forest gaps) and avoidance of the creation of huge uniform forest stands. This may involve (Lindenmayer *et al.*,

2006) conservation of open areas, and heath and grassland habitats within forests that can be vital for some key elements of biota.
- The implementation of control strategies for unwelcome species (e.g. weed management, feral animal control).
- Promotion of low-energy harvest machinery and equipment.
- Protection from wildfires, especially in fire-prone ecosystems.

Furthermore, Martín-Queller *et al.* (2013) points out that in order for forestry management to be effective in conserving forest diversity at a local scale, it needs to:

1) implement practices tailored according to the region, exhibiting lower intensities in terms of harvest in areas of greater hydric stress
2) extensive implementation of a single silvicultural system over large areas should be avoided
3) maintain an assortment of forests with species richness, which can provide sources of colonizers to enrich the adjacent regenerating stands.

29.7.3 Conservation measures at region-scale biodiversity

At region-scale biodiversity, the measures may include strategies and actions focusing on the conservation of forest habitats and landscape characteristics of a region, placing special emphasis on the maintenance of connectivity, the wildlife corridors and the integrity of the different systems. They may also consider the special habitat requirements for populations of rare species, and biological hotspots. More specifically, some of the main strategies may include the following (Lindenmayer *et al.*, 2006; Dafis, 2013; EU, 2015).

- Measures that ensure effective site protection, management and restoration of a region.
- Establishing extensive ecological reserves.
- Developing landscape-level strategies of conservation within off-reserve forest.
- Protecting habitats in landscape–protected areas at intermediate-spatial scales.
- Maintaining landscape heterogeneity.
- Maintaining connectivity; this involves creating and managing suitable wildlife corridors.
- Preserving the integrity of aquatic systems and riparian buffers by implementing hydrological and geomorphological processes.
- Considering demands in sources and habitats for rare species populations.
- Considering and prioritizing biological hotspots, special habitats such as cliffs, caves, rockslides, springs, seeps, lakes, ponds, wetlands, streams and rivers and related buffers.
- Considering restoration and reconstruction of late-successional (old-growth) forests or other habitat features, as well as residuals of late-successional forests.
- Considering spatial and temporal patterns of timber harvesting systems, as well as the size of harvest units and the rotation lengths (Lindenmayer *et al.*, 2006).
- Considering the transportation systems (e.g. road networks).
- Including management strategies for species of special interest (e.g. rare or threatened species).
- Considering natural disturbance regimes as templates for logging regimes (e.g. identification of natural disturbance refuges as places for logging exemption).
- Control strategies for unwanted species at the region scale.
- Landscape-level objectives for specific structural characteristics (e.g. large trees with hollows).
- Other landscape-level considerations.
- Developing new management insights, (cross-border) stakeholders' cooperation frameworks, networks of specialists and site managers, etc. (EU, 2015).
- Setting the priority of monitoring needs and updating treatments.

However, the final impact of forest management on species richness also depends on the adverse climate conditions of a region, and on the kind and extent of other management practices in the surrounding non-forest landscapes (Martín-Queller *et al.*, 2013).

29.7.4 Conservation measures at biogeographical-level biodiversity

With regard to biogeographical-level biodiversity, the measures primarily involve policy strategies and actions pertaining to national and international collaboration and rules, including exchange of experience, and development of a common conservation policy across different countries and large biogeographical areas. In general, the actions to be taken should focus on:

- the need to establish global priorities in terms of conservation by identifying common objectives, priorities and management activities
- gathering current data on threats and conservation needs for species and habitats (EU, 2015)
- supporting prioritized actions involving management, restoration and improvement of the conservation status of species and habitats of wider interest
- supplying different collaborating countries with the required information, which enables them to reflect conservation or restoration priorities agreed at the biogeographical level (EU, 2015)
- promoting the sharing of expertise and experience and international knowledge through exchanging experiences, case studies and best practices
- the development of common conservation strategy and policy
- pointing out the necessity for collaboration for the purpose of developing common actions and policies for the promotion of forest biodiversity conservation at the biogeographical level
- identifying species and forest habitats whose conservation status needs to be improved in a more widely addressed manner, primarily through the implementation of common policies (EU, 2015)
- the need for monitoring and for the creation of a database at the biogeographical level
- the establishment and development of international ecological networks engaging in nature conservation, with a view to effective site protection, management and restoration (e.g. the Natura 2000 network in Europe)
- developing and reinforcing collaboration between different countries, including international cooperation with regard to the management of ecologically significant forest sites
- setting priorities and suggesting cost-effective ways of achieving a desired conservation status and of coping with climate change (EU, 2015)
- facilitating exchange of information about conservation goals and measures, including best practice
- maintaining connectivity and integration management at the region scale
- identifying possible interactions and advantages of management measures for significant forest ecosystems with other environmental and climate change goals (EU, 2015)
- creating a permanent internet-based platform of communication for exchange of information regarding management of key forest sites
- suggesting that broader sectoral policies be adopted where necessary, so that conservation measures within ecologically important forest sites will be complemented.

29.8 Future Needs

Continuous and comparable data are essential for a broad type of forest ecosystem with regard to the impact of different silvicultural treatments and management operations on different kinds of forest diversity.

The creation of a wide coordinated monitoring network (e.g. regional or biogeographical networks) for comparing biodiversity between unmanaged and managed forests is recommended by many researchers (Parviainen *et al.*, 2000; Larsson, 2001; Meyer, 2005; Paillet *et al.*, 2010), for the purpose of fulfilling the needs for an effective policy on the conservation of forest biodiversity. The creation and operation of such networks will provide information, such as species richness and changes for several taxonomic groups, as well as further fundamental information on the patterns and processes involved in forest biodiversity changes.

Finally, the development and the reinforcement of any existing networking programmes and communication platforms could facilitate more profound cooperation, networking and cooperative actions between different members of the scientific and management communities, thus paving the way for more effective forest biodiversity conservation.

References

Aanderaa, R., Rolstad, J. and Sognen, S.M. (1996) *Biological Diversity in Forests*. Norges Skogeierforbund og A/S Landbruksforlaget, Oslo, Norway.

Alaback, P.B. and Herman, F.R. (1988) Long-term response of understory vegetation to stand density in Picea-Tsuga forests. *Canadian Journal of Forest Research* 18, 1522–1530.

Anderson, M.C. (1964) Studies of the woodland light climate. II. Seasonal variation in the light climate. *Journal of Ecology* 52, 643–663.

Andersson, R. and Östlund, L. (2004) Spatial patterns, density changes and implications on biodiversity for old trees in the boreal landscape of northern Sweden. *Biological Conservation* 118 (4), 443–453.

Andren, H. (1994) Effects of habitat fragmentation on birds and mammals in landscapes with different proportions of suitable habitat: a review. *Oikos* 71, 355–366.

Andren, H. (1999) Habitat fragmentation, the random sample hypothesis and critical thresholds. *Oikos* 84, 306–308.

Aubry, K.B., Crites, M.J. and West, S.D. (1991) Regional patterns of small mammal abundance and community composition in Oregon and Washington. In: Ruggiero, L.F., Aubry, K.B., Carey, A.B. and Huff, M.H. (eds) *Wildlife and Vegetation of Unmanaged Douglas-fir Forests*. United States Department of Agriculture Forest Service General Technical Report PNW-GTR-285. Pacific Northwest Research Station. Portland, Oregon, USA, pp. 285–294.

Bailey, J.D., Mayrsohn, C., Doescher, P.S., St Pierre, E. and Tappeiner, J.C. (1998) Understory vegetation in old and young Douglas-fir forests of western Oregon. *Forest Ecology and Management* 112, 289–302.

Barber, D.R., Martin, T.E., Melchiors, M.A., Thill, R.E. and Wigley, T.B. (2001) Nesting success of birds in different silvicultural treatments in Southeastern U.S. pine forests. *Conservation Biology* 15, 196–207.

Battles, J.J., Shlisky, A.J., Barrett, R.H., Heald, R.C. and Allen-Diaz, B.H. (2001) The Effects of forest management on plant species diversity in a Sierran Conifer Forest. *Forest Ecology and Management* 146, 211–222.

Bazzaz, F.A. (1979) The physiological ecology of plant succession. *Annual Review of Ecology and Systematics* 10, 351–371.

Begon, M., Harper, J.L. and Townsend, C.R. (1996) *Ecology: Individuals, Populations and Communities*, 3rd edn. Blackwell Science, Oxford, UK.

Bengtsson, J., Nilsson, S.G., Franc, A. and Menozzi, P. (2000) Biodiversity, disturbances, ecosystem function and management of European forests. *Forest Ecology and Management* 132, 39–50.

Berger, A.L. and Puettmann, K. (2004) Harvesting impacts on soil and understory vegetation: the influence of season of harvest and within-site disturbance patterns on clear-cut aspen stands in Minnesota. *Canadian Journal of Forest Research* 43, 2159–2168.

Bobiec, A. (1998) Themosaic diversity of field layer vegetation in the natural and exploited forests of Bialowieza. *Plant Ecology* 136,175–187.

Brashears, M.B., Fajvan, M.A. and Schuler, T.M. (2004) An assessment of canopy stratification and tree species diversity following clearcutting in Central Appalachian Hardwood. *Forest Science* 50, 54–64.

Bratton, S.P. (1994) Logging and fragmentation of broadleaved deciduous forests: are we asking the right questions? *Conservation Biology* 8, 295–297.

Brokaw, N.V. and Lent, R.A. (1999) *Vertical Structure. Maintaining Biodiversity in Forest Ecosystems*. Cambridge University Press, Cambridge, UK, pp. 335–361.

Brosofske, K.D., Chen, J. and Crow, T. (2001) Understory vegetation and site factors: implications for a managed Wisconsin landscape. *Forest Ecology and Management* 146, 75–87.

Canham, C.D. (1988) An index for understory light levels in and around canopy gaps. *Ecology* 69, 1634–1638.

Canham, C.D., Denslow, J., Platt, W., Runkle, J., Spies, T. and White, P. (1990) Light regimes beneath closed canopies and tree-fall gaps in temperate and tropical forests. *Canadian Journal of Forest Research* 20, 620–631.

Carey, A.B. (1995) Sciurids in Pacific Northwest managed and old-growth forests. *Ecological Applications* 5, 648–661.

Carey, A.B. (1996) Interactions of Northwest forest canopies and arboreal mammals. *Northwest Science* 70, 72–78.

Carey, A.B. (2001) Experimental manipulation of spatial heterogeneity in Douglas-fir forests: effects on squirrels. *Forest Ecology and Management* 152, 13–30.

Carey, A.B. and Johnson, M.L. (1995) Small mammals in managed, naturally young, and old-growth forests. *Ecological Applications* 5, 336–352.

Carey, A.B., Kershner, J., Biswell, B.L. and de Toledo, L.D. (1999) Ecological scale and forest development: squirrels, dietary fungi, and vascular plants in managed and unmanaged forests. *Wildlife Monograph* 142, 1–71.

Caro, T.M. and O'Doherty, G. (1999) On the use of surrogate species in conservation biology. *Conservation Biology* 13, 805–814.

CBD (2006) Forest biological diversity: implementation of the programme of work. CBD COP 8 Decision VIII/19. Available at: www.cbd.int/decisions/?dec=VIII/19 (accessed February 2014).

Chambers, C. and McComb, W.C. (1997) Effects of silvicultural treatments on wintering bird communities in Oregon Coast Range. *Northwest Science* 71, 298–304.

Chambers, C.L., McComb, W.C. and Tappeiner, J.C. (1999) Breeding bird responses to three silvicultural treatments in the Oregon Coast Range. *Ecological Applications* 9, 171–185.

Chazdon, R.L. and Fletcher, N. (1984) Photosynthetic light environments in a low land tropical rain forest in Costa Rica. *Journal of Ecology* 72, 553–564.

CIFOR (Center for International Forestry Research) (2014) Forests and biodiversity. Available at:www.cifor.org/Publications/Corporate/FactSheet/forests_biodiversity.htm (accessed January 2015).

Clearwater, M.J., Susilawaty, R., Effendi, R. and van Gardingen, P. (1999) Rapid photosynthetic acclimation of *Shorea johorensis* seedling after logging disturbance in Central Kalimantan. *Oecologia* 121, 478–488.

Commonwealth of Australia (1998) *A Framework of Regional (Subnational) Level Criteria and Indicators of Sustainable Forest Management in Australia*. Commonwealth of Australia, Canberra, Australia.

Commonwealth of Australia (2001) *Australia's Forests – The Path to Sustainability*. Commonwealth of Australia, Canberra, Australia.

Connell, J.H. (1978) Diversity in tropical rainforests and coral reefs. *Science* 199,1302–1310.

Corn, P.S. and Bury, R.B. (1991) Small mammal communities in the Oregon Coast Range. In: Ruggiero, L.F., Aubry, K.B., Carey, A.B. and Huff, M.H. (eds) *Wildlife and Vegetation of Unmanaged Douglas-fir Forests*. United States Department of Agriculture Forest Service General Technical Report PNW-GTR-285. Pacific Northwest Research Station. Portland, Oregon, USA, pp. 241–254.

Dafis, S. (1990) *Applied Silviculture*. Giahoudis-Giapoulis Editions, Thessaloniki, Greece.

Dafis, S. (2013) Silvicultural treatments of forests in Natura 2000 network. *Amfivion* 96, 9–10 [published in Greek].

Daily, G.C., Ehrlich, P.R. and Sanchez-Azofeifa, G.A. (2001) Countryside biogeography: use of human-dominated habitats by the avifauna of southern Costa Rica. *Ecological Applications* 11, 1–13.

Deal, R.L. and Tappeiner, J.C. (2002) The effects of partial cutting on stand structure and growth of western hemlock–Sitka spruce stands in southeast Alaska. *Forest Ecology and Management* 159, 173–186.

DeFerrari, C.M. and Naiman, R.J. (1994) A multi-scale assessment of the occurrence of exotic plants on the Olympic Peninsula, Washington. *Journal of Vegetation Science* 5, 247–258.

DeGraaf, R.M. and Miller, R.I. (1996) *Conservation of Faunal Diversity in Forested Landscapes*, Vol. 6 of Goldsmith, F.B. and Duffey, E. (eds) *Conservation Biology*, Springer, London, UK, pp. 389–406.

Denslow, J.S. (1985) Disturbance-mediated coexistence of species. In Pickett, S.T.A. and White, P. (eds) *The Ecology of Natural Disturbance and Patch Dynamics*. Academic Press,San Diego, California, USA, pp. 307–321.

Desender, K., Ervynck, A. and Tack, G. (1999) Beetle diversity and historical ecology of woodlands in Flanders. *Belgian Journal of Zoology* 129, 139–155.

Dollin, P. (2004) *Effects of Stands Age and Silvicultural Treatment on Beetle (Coleoptera) Biodiversity in Coniferous Stands in Southwest Nova Scotia*. Dalhousie University, Halifax, Nova Scotia, Canada.

Dyrness, C.T. (1973) Early stages of plant succession following logging and burning in the western Cascades of Oregon. *Ecology* 54, 57–69.

Ellum, D.S. (2009) Floristic diversity in managed forests: demography and physiology of understory plants following disturbance in Southern New England forests. *Journal of Sustainable Forests* 28, 132–151.

Enoksson, B., Angelstam, P. and Larsson, K. (1995) Deciduous forest and resident birds: the problem of fragmentation within a coniferous forest landscape. *Landscape Ecology* 10, 267–275.

EU (2015) *The Natura 2000 Biogeographical Process*. Available at: ec.europa.eu/environment/nature/natura2000/seminars_en.htm. (accessed January 2016).

European Economic Community (1992) Council Directive 92/43/EEC of 22 May 1992 on the conservation of natural habitats and of wild fauna and flora. *Official Journal L* 206, 7–50.

European Environment Agency (EEA) (2008) *European Forests – Ecosystem Conditions and Sustainable Use*. EEA Report No 3/2008. European Environment Agency, Copenhagen, Denmark.

FAO (2001) *State of the World's Forests*. Food and Agriculture Organization of the United Nations, Rome, Italy. Available at: www.fao.org/docrep/003/y0900e/y0900e00.htm (accessed April 2016).

FAO (2005) Biological diversity. In: *Global Forest Resources Assessment 2005: Progress Towards Sustainable Forest Management*. FAO Forestry Paper 147. Food and Agriculture Organization of the United Nations, Rome, Italy, Chapter 3, pp. 40–43. Available at: ftp://ftp.fao.org/docrep/fao/008/A0400E/A0400E04.pdf (accessed April 2016).

Fleishman, E., Murphy, D.D. and Brussard, P.F. (2000) A new method for selection of umbrella species for conservation planning. *Ecological Applications* 10, 569–579.

Franklin, J.F., Spies, T.A., van Pelt, R., Carey, A., Thornburgh, D., Berg, D.R., Lindenmayer, D.B., Harmon, M., Keeton, W. and Shaw, D.C. (2002) Disturbances and the structural development of natural forest ecosystems with some implications for silviculture. *Forest Ecological Management* 155, 399–423.

Fuji, S., Kubota, Y. and Enoki, T. (2010) Long-term ecological impacts of clear-fell logging on tree species diversity in a Subtropical Forest, Southern Japan. *The Japanese Forest Society* 15, 289–298.

Ganatsas, P. (1993) *Stand Structure and Natural Regeneration of Spruce Forest in Elatia Drama, Northern Greece*. Dissertation, School of Forestry and Natural Environment, Aristotle University of Thessaloniki, Thessaloniki, Greece (in Greek). Available at: thesis.ekt.gr/thesisBookReader/id/3704#page/1/mode/2up (accessed April 2016).

Ganatsas, P., Tsakaldimi, M. and Thanos, C. (2008) Seed and cone diversity and seed germination of *Pinus pinea* in Strofylia Site of the Natura 2000 Network. *Biodiversity Conservation* 17, 2427–2439.

Ganatsas, P., Tsitsoni, T., Tsakaldimi, M. and Zagas, T. (2012) Reforestation of degraded Kermes oak shrublands with planted pines: effects on vegetation cover, species diversity and community structure. *New Forest* 43, 1–11.

Ganatsas, P., Tsakaldimi, M. and Katsaros, D. (2013) Natural resource management in national parks: a management assessment of a Natura 2000 wetlands site in Kotychi-Strofylia, southern Greece. *International Journal of Sustainable Development World* 20, 152–165.

Gilbert, F.F. and Allwine, R. (1991) Small mammal communities in the Oregon Cascade Range. In: Ruggiero, L.F., Aubry, K.B., Carey, A.B. and Huff, M.H. (eds) *Wildlife and Vegetation of Unmanaged Douglas-fir Forests*. United States Department of Agriculture Forest Service General Technical Report PNW-GTR-285. Pacific Northwest Research Station. Portland, Oregon, USA, pp. 257–267.

Gilliam, F.S., Turrill, N. and Adams, M. (1995) Herbaceous-layer and overstory species in clearcut and mature central Appalachian hardwood forests. *Ecological Applications* 5, 947–955.

Gotelli, N.J. and Colwell, R.K. (2001) Quantifying biodiversity: procedures and pitfalls in the measurement and comparison of species richness. *Ecology Letters* 4, 379–391.

Graae, B.J. and Heskjaer, V.S. (1997) A comparison of understorey vegetation between untouched and managed deciduous forest in Denmark. *Forest Ecological Management* 96, 111–123.

Gray, A.N. and Spies, T.A. (1996) Gap size, within-gap position and canopy structure effects on conifer seedling establishment. *Journal of Ecology* 84, 635–645.

Grime, J.P. (1977) Evidence for the existence of three primary strategies in plants and its relevance to ecological and evolutionary theory. *The American Naturalist* 111, 1169–1194.

Hagar, J.C., Howlin, S. and Ganio, L. (2004) Short-term response of songbirds to experimental thinning of young Douglas-fir forests in the Oregon Cascades. *Forest Ecological Management* 199, 333–347.

Halpern, C.B. (1989) Early successional patterns of forest species: interactions of life history traits and disturbance. *Ecology* 70, 704–720.

Halpern, C.B. and Spies, T.A. (1995) Plant species diversity in natural and managed forests of the Pacific Northwest. *Ecological Applications* 5, 913–934.

Hamer, K.C. and Hill, J.K. (2000) Scale-dependent effects of habitat disturbance on species richness in Tropical Forests. *Conservation Biology* 14, 1435–1440.

Hamer, K.C., Hill, J.K., Lace, L.A. and Langan, A.M. (1997) Ecological and biogeographical effects of forest disturbance on tropical butterflies of Sumba, Indonesia. *Journal of Biogeography* 24, 67–75.

Hanley, T.A. (2005) Potential management of young-growth stands for understory vegetation and wildlife habitat in southeastern Alaska. Landsc. *Urban Planning* 72, 95–112.

Hannon, S.J. and McCallum, C. (2002) *Using the Focal Species Approach for Conserving Biodiversity in Landscapes Managed for Forestry*. Sustainable Forest Management Network, University of Alberta, Edmonton, Alberta, Canada. Available at: era.library.ualberta.ca/downloads/5712m722m (accessed April 2016).

Harrington, C.A., Roberts, S.D. and Brodie, L.C. (2005) Tree and understory responses to variable density thinning in western Washington. In: Peterson, C.E. and Maguire, D.A. (eds) *Balancing Ecosystem Values*. US Department of Agriculture, Forest Service, Pacific Northwest Research Station, Portland, Oregon, USA, pp. 97–106.

Hayes, J.P. and Hagar, J.C. (2002) Ecology and management of wildlife and their habitats in the Oregon Coast Range. In: Hobbs, S.D., Hayes, J.P., Johnson, R.L., Reeves, G.H., Spies, T.A., Tappeiner, J.C.II and Wells Gail, E. (eds) *Forest and Stream Management*. Oregon State University Press, Corvallis, Oregon, USA, pp. 99–134.

Hayes, J.P., Horvath, E.G. and Hounihan, P. (1995) Townsend's chipmunk populations in Douglas-fir plantations and mature forests in the Oregon Coast Range. *Canadian Journal of Zoology* 73, 67–73.

Heckman, C.W. (1999) The encroachment of exotic herbaceous plants into the Olympic National Forest. *Northwest Science* 73, 264–276.

Holloway, J.D., Kirk-Spriggs, A.H. and Khen, C.V. (1992) The response of some rainforest insect groups to logging and convertion to plantation. *Philosophical Transactions of the Royal Society* 335, 425–436.

Hunter, M.L. (1999) *Managing Biodiversity in Forest Ecosystems*. Cambridge University Press, Cambridge, UK.

Jenkins, M.A. and Parker, G. (1999) Composition and diversity of ground-layer vegetation in silvicultural openings of southern Indiana forests. *The American Midland Naturalist* 142, 1–16.

Johnson, A.S, Ford, W. and Hale, P. (1993) The effects of clearcutting on herbaceous understories are still not fully known. *Conservation Biology* 7, 433–435.

Kerr, G. (1999) The use of silvicultural systems to enhance the biological diversity of plantation forests in Britain. *Forestry* 72, 191–205.

Klinka, K., Chen, H.Y.H., Wang, Q. and deMontigny, L. (1996) Forest canopies and their influence on understory vegetation in early seral stands on west Vancouver Island. *Northwest Science* 70, 192–200.

Lambeck, R.J. (1997) Focal species: a multi-species umbrella for nature conservation. *Conservation Biology* 11, 849–856.

Landres, P.B., Verner, J. and Thomas, J.W. (1988) Ecological uses of vertebrate indicator species: a critique. *Conservation Biology* 2, 316–328.

Larsson, T.B. (2001) Biodiversity evaluation tools for European forests. *Ecology Bulletins* 50, 1–237.

Lindenmayer, D.B. (1999) Future directions for biodiversity conservation in managed forests: indicator species, impact studies, and monitoring. *Forest Ecological Management* 115, 277–288.

Lindenmayer, D.B. and Franklin, J.F. (2002) *Conserving Forest Biodiversity: A Comprehensive Multiscaled Approach*. Island Press, Washington, DC, USA.

Lindenmayer, D.B., Fischer, J. and Cunningham, R.B. (2005) Native vegetation cover thresholds associated with species responses. *Biological Conservation* 124, 311–316.

Lindenmayer, D.B., Franklin, J.F. and Fischer, J. (2006) General management principles and a checklist of strategies to guide forest biodiversity conservation. *Biological Conservation* 131, 433–445.

Luoma, D.L., Trappe, J.M., Claridge, A.W., Jacobs, K.M. and Cazares, E. (2003) *Relationships among Fungi and Small Mammals in Forested Ecosystems*. Cambridge University Press, Cambridge, UK.

MacArthur, R.H. and MacArthur, J.W. (1961) On bird species diversity. *Ecology* 42,594–598.

Magurran, A.E. (1988) *Ecological Diversity and Its Measurement*. Princeton University Press, Princeton, New Jersey, USA.

Martin, K.J. and McComb, W.C. (2002) Small mammal habitat associations at patch and landscape scales in Oregon. *Forest Science* 48, 255–266.

Martín-Queller, E., Diez, J.N., Ibanez, I. and Saura, S. (2013) Effects of silviculture on native tree species richness: interactions between management, landscape context and regional climate. *Journal of Applied Ecology* 50, 775–785.

Maschinski, J., Kolb, T.E., Smith, E. and Philips, B. (1997) Potential impacts of timber harvesting on a rare understory plant, *Clematis hirsutissima* var. *arizonica*. *Biological Conservation* 80, 49–61.

Matlack, G. (1994) Plant demography, land-use history, and the commercial use of forests. *Conservation Biology* 8, 298–299.

McComb, W.C., Spies, T.A. and Emmingham, W.H. (1993) Douglas-fir forests: managing for timber and mature-forest habitat. *Journal of Forestry* 91, 31–42.

McLaren, M.A., Thompson, I.D. and Baker, J.A. (1998) Selection of vertebrate wildlife indicators for monitoring sustainable forest management in Ontario. *The Forestry Chronicle* 74, 241–548.

MCPFE (2003) *State of Europe's Forests 2003. The MCPFE Report on Sustainable Forest Management in Europe*. Jointly prepared by the MCPFE Liaison Unit Vienna and UNECE/FAO, Vienna, Austria.

MCPFE and PEBLDS (2006) *The Pan-European Understanding of the Linkage between the Ecosystem Approach and Sustainable Forest Management.* Joint position of the MCPFE and the EfE/PEBLDS. Ministerial Conference on the Protection of Forests in Europe, Warsaw, Poland. Available at: www.foresteurope.org/documentos/SFM_EA.pdf (accessed April 2016).

Meier, A.J., Bratton, S. and Duffy, D. (1995) Possible ecological mechanisms for loss of vernal-herb diversity in logged eastern deciduous forests. *Ecological Applications* 5, 935–946.

Meyer, P. (2005) Network of strict forest reserves as reference system for close to nature forestry in Lower Saxony, Germany. *Forest Snow and Landscape Research* 79, 33–44.

Mikoláš, M., Svoboda, V., Pouska, R.C., Morrissey, D.C., Donato, W.S., Keeton, T.A. Nagel, V.D., Popescu, J., Müller, C., Bässler, J., Knorn, L., Rozylowicz, C.M., Enescu, V., Trotsiuk, P., Janda, H., Mrhalová, Z., Michalová, F. and Krumm Kraus, D. (2014) Comment on 'Opinion paper: forest management and biodiversity': the role of protected areas is greater than the sum of its number of species. *Web Ecology* 14, 61–64.

Minckler, L.S., Woerheide, J. and Schlesinger, R. (1973) *Light, Moisture and Tree Reproduction in Hardwood Forest Openings (U.S.F.S Research Paper NC-89)*. US Department of Agriculture, St Paul, Minnesota, USA.

Mitchell, M.S., Reynold-Hogland, M.J., Smith, M.L., Wood, P.B., Beebe, J.A., Keyser, P.D., Loehle, C., Reynolds, C.J., Van Deusen, P. and White, D. Jr (2008) Projected long-term response of southeastern birds to forest management. *Forest Ecology and Management* 256, 1884–1896. Available at: www.umt.edu/mcwru/documents/Mitchell_Publications/FORECO11221.pdf (accessed March 2016).

Montreal Process Liaison Office (2000) *The Montréal Process: Progress and Innovation in Implementing Criteria and Indicators for the Conservation and Sustainable Management of Temperate and Boreal Forests*. Year 2000 Progress Report, Canadian Forest Service. Montréal Process Liaison Office, Canadian Forest Service, Ottawa, Canada.

Muir, P.S., Mattingly, R., Tappeiner, J.C., Bailey, J.D., Elliott, W.E., Hagar, J.C., Miller, J.C. and Peterson, E.B. (2002) *Managing for Biodiversity in Young Douglas-fir Forests of Western Oregon*. Biological Science Report. USGS/BRD/BSR-2002–0006. US Geological Survey, Forest and Rangeland Ecosystem Science Center, Corvallis, Oregon, USA.

Niemela, J. (1999) Management in relation to disturbance in the boreal forest. *Forest Ecology and Management* 115, 127–134.

Niemela, J., Koivula, M. and Kotze, D.J. (2007) The effects of forestry on carabid beetles (Coleoptera: Carabidae) in boreal forests. *Journal of Insect Conservation* 11, 5–18.

Norton, D.A. (1999) Forest reserves. In: Hunter, M.J. Jr (ed.) *Managing Biodiversity in Forest Ecosystems*. Cambridge University Press, Cambridge, UK, pp. 525–555.

Noss, R.F. (1999) Assessing and monitoring forest biodiversity: a suggested framework and indicators. *Forest Ecology and Management* 115, 135–146.

O'Hara, K. (2014) *Multiaged Silviculture: Managing for Complex Forest Stand Structures*. Oxford University Press,Oxford, UK.

Okland, T., Rydgren, K., Økland, R.H., Storaunet, K.O. and Rolstad, J. (2003) Variation in environmental conditions, understorey species number, abundance and composition among natural and managed Picea-abies forest stands. *Forest Ecology and Management* 177, 17–37.

Oliver, C.D. (1981) Forest development in North America following major disturbances. *Forest Ecology and Management* 3, 153–168.

Oliver, C.D. and Larson, B. (1996) *Forest Stand Dynamics*, updated edn.Wiley, New York, USA.

Paillet, Y., Berges, L., Hjalten, J., Odor, P., Avon, C., Bernhardt-Romermann, M., Bijlsma, R.J., De Bruyn, L., Fuhr, M., Grandin, U., Kanka, R., Lun-din, L., Luque, S., Magura, T., Matesanz, S., Meszaros, I., Sebastia, M.T., Schmidt, W., Standovar, T., Tothmeresz, B., Uotila, A., Valladares, F., Vellak, K. and Virtanen, R. (2010) Biodiversity differences between managed and unmanaged forests: meta-analysis of species richness in Europe. *Conservation Biology* 24, 101–112.

Parendes, L.A. and Jones, J.A. (2000) Role of light availability and dispersal in exotic plant invasion along roads and streams in the H.J. Andrews Experimental Forest, Oregon. *Conservation Biology* 14, 64–75.

Parviainen, J., Bucking, W., Vandekerkhove, K., Schuck, A. and Paivinen, R. (2000) Strict forest reserves in Europe: efforts to enhance biodiversity and research on forests left for free development in Europe (EU-COST-Action E4). *Forestry* 73, 107–118.

Parviainen, J., Bozzano, M., Estreguil, C., Koskela, J., Lier, M., Vogt, P. and Ostapowicz, K. (2007) Maintenance, conservation and appropriate enhancement of biological diversity in forest ecosystems. In: Kohl, M. and Rametsteiner, E. (eds) *State of Europe's Forests 2007-MCPFE Report on Sustainable Forest Management in Europe*. Ministerial Conference on the Protection of Forests in Europe, Liaison Unit, Warsaw, Poland, pp. 45–72.

Pastur, M.G., Peri, P.L., Fernandez, M.C., Staffieri, G. and Lencinas, M.V. (2002) Changes in understory species diversity during the *Nothofagus pumilio* Forest Management Cycle. *The Japanese Forest Society* 7, 165–174.

Pearson, D.L. (1994) Selecting indicator taxa for the quantitative assessment of biodiversity. *Philosophical Transactions of the Royal Society of London B* 345, 75–79.

Peet, R.K. and Christensen, N. (1980) Succession: a population process. *Vegetation* 43, 131–140.

Peltzer, D.A., Bast, M., Wilson, S. and Gerry, A. (2000) Plant diversity and tree responses following contrasting disturbances in boreal forest. *Forest Ecology and Management* 127, 191–203.

Petraitis, P.S., Latham, R. and Neisenbaum, R. (1989) The maintenance of species diversity by disturbance. *The Quarterly Review of Biology* 64, 393–418.

Phillips, D.L. and Shure, D. (1990) Patch-size effects on early succession in southern Appalachian forests. *Ecology* 71, 204–212.

Prescott, C.E., Coward, L.P., Weetman, G.F. and Gessell, S.P. (1993) Effects of repeated fertilization on the ericaceous shrub, salal (*Gaultheria shallon*), in two coast Douglas-fir forests. *Forest Ecology and Management* 61, 45–60.

Primack, R.B. (2000) *A Primer of Conservation Biology*. Sinauer Associates, Inc., Sunderland, Massachusetts.

Probst, J.R. and Crow, T.R. (1991) Integrating biological diversity and resource management. *Journal of Forestry* 89, 12–17.

Puettmann, K.J. and Berger, C.A. (2006) Development of tree and understory vegetation in young Douglas-fir plantations in western Oregon. *Western Journal of Applied Forestry* 21, 94–101.

Raguso, R.A. and Llorente-Bousquets, J. (1990) The butterflies (Lepidoptera) of the Tuxtlas Mts., Vera Cruz, Mexico, revised: species richness and habitat disturbance. *Journal of Research on the Lepidoptera* 29, 105–133.

Reader, R.J. and Bricker, B. (1992) Value of selectively cut deciduous forest for understory herb conservation: an experimental assessment. *Forest Ecology and Management* 51, 317–327.

Roberts, M.R. and Gilliam, F.S. (1995) Patterns and mechanisms of plant diversity in forested ecosystems: implications for forest management. *Ecological Applications* 5, 969–977.

Rolstad, J. and Wegge, P. (1987) Distribution and size of capercaillie leks in relation to old forest fragmentation. *Oecologia* 72, 389–394.

Rosenberg, D.K. and Anthony, R.G. (1993) Differences in townsends chipmunk populations between second- and old-growth forests in Western Oregon. *Journal of Wildlife Management* 57, 365–373.

Rubio, A., Gavilan, R. and Escudero, A. (1999) Are soil characteristics and understory composition controlled by forest management? *Forest Ecology and Management* 113, 191–200.

Runkle, J.R. (1982) Patterns of disturbance in some old-growth mesic forests of eastern North America. *Ecology* 63, 1533–1546.

Runkle, J.R. (1985) Disturbance regimes in temperate forests. In: Pickett, S.T.A. and White, P.S. (eds) *The Ecology of Natural Disturbance and Patch Dynamics*. Academic Press, San Diego, California, USA, pp. 17–33.

Salvatore, P., La Mantia, T. and Rühl, J. (2012) The impact of *Pinus halepensis* afforestation on Mediterranean spontaneous vegetation: do soil treatment and canopy cover matter? *Journal of Forestry Research* 23, 517–528.

Scherer-Lorenzen, M., Korner, C. and Schulze, E.D. (2005) *Forest Diversity and Function: Temperate and Boreal Systems*. Springer, Berlin, Germany.

Schmidt, W. (2005) Herb layer species as indicators of biodiversity of managed and unmanaged beech forests. *Forest Snow and Landscape Research* 79, 111–125.

Schoonmaker, P. and McKee, A. (1988) Species composition and diversity during secondary succession of coniferous forests in the western Cascade Mountains of Oregon. *Forest Science* 78, 960–979.

Schowalter, T.D. (1995) Canopy arthropod community responses to forest age and alternative harvest practices in western Oregon. *Forest Ecology and Management* 78, 115–125.

Schuman, M.E., White, A.S. and Witham, J.W. (2003) The effects of harvest-created gaps on plant species diversity, composition and abundance in a Maine Oak-Pine. *Forest Ecology and Management* 176, 543–561.

Schulze, E.D., Bouriaud, L., Bussler, H., Gossner, M., Walentowski, H., Hessenmöller, D., Bouriaud, O. and Gadow, K.V. (2014) Opinion paper: forest management and biodiversity. *Web Ecology* 14, 3–10.

Seymour, R.S. and Hunter, M.L. (1999) Principles of ecological forestry. In: Hunter, M.L. (ed.) *Maintaining Biodiversity in Forest Ecosystems*. Cambridge University Press, Cambridge, UK, pp. 22–61.

Siitonen, J. (2001) Forest management, coarse woody debris and saproxylic organisms: Fennoscandian boreal forests as an example. *Ecology Bulletins* 49, 11–41.

Small, J.C. and McCarthy, B. (2005) Relationship of understory diversity to soil nitrogen, topographic variation, and stand age in an eastern oak forest, USA. *Forest Ecology and Management* 217, 229–243.

Spies, T.A. (1991) *Plant Species Diversity and Occurrences in Young, Mature and Old-Growth Douglas-fir Stands in Western Oregon and Washington*. General Technical Report PNW-GTR-285. US Department of Agriculture, Forest Service, Pacific Northwest Research Station, Portland, Oregon, USA, pp. 111–121.

Spies, T.A. and Franklin, J.F. (1989) Gap characteristics and vegetation response in coniferous forests of the Pacific Northwest. *Ecology* 70, 543–555.

Spies, T.A., Franklin, J.F. and Klopsch, M. (1990) Canopy gaps in Douglas-fir forests of the Cascade Mountains. *Canadian Journal of Forest Research* 20, 649–658.

Stephens, S.S. and Wagner, M.R. (2007) Forest plantations and biodiversity: a fresh perspective. *Journal of Forestry* 105, 307–313.

Staudhammer, C.L. and LeMay, V.M. (2001) Introduction and evaluation of possible indices of stand structural diversity. *Canadian Journal of Forest Research* 31, 1105–1115.

Stork, N.E., Srivastava, D.S., Watt, A.D. and Larsen, T.B. (2003) Butterfly diversity and silvicultural practice in lowland rainforests of Cameroon. *Biodiversity and Conservation* 12, 387–410.

Sugal, C. (1997) Most forests have no protection. *World Watch* 10,9.

Suzuki, N. and Hayes, J.P. (2003) Effects of thinning on small mammals in Oregon coastal forests. *Journal of Wildlife Management* 67, 352–371.

Tappeiner, J.C., Zasada, J.C., Huffman, D.W. and Ganio, L. (2001) Salmonberry and salal annual aerial stem production: the maintenance of shrub cover in forest stands. *Canadian Journal of Forest Research* 31, 1629–1638.

Thompson, I.D., James, A., Baker, J.A. and Ter-Mikaelian, M. (2003) A review of the long-term effects of post-harvest silviculture on vertebrate wildlife, and predictive models, with an emphasis on boreal forests in Ontario, Canada. *Forest Ecology and Management* 177, 441–469.

Torras, O. and Saura, S. (2008) Effects of silvicultural treatments on forest biodiversity indicators in the Mediterranean. *Forest Ecology and Management* 255, 3322–3330.

Torras, O., Gil-Tena, A. and Saura, S. (2012) Changes in biodiversity indicators in managed and unmanaged forests in NE Spain. *The Japanese Forest Society* 17, 19–29.

USDA Forest Service (2013) Silvicultural treatments (North Central Region Forest Management Guides). Available at: www.ncrs.fs.fed.us/fmg/nfmg/fm101/silv/p2_treatment.html (accessed April 2016).

Vaisanen, R., Bistrom, O. and Heliovaara, K. (1993) Subcortical Coleoptera in dead pines and spruces: is primeval species composition maintained in managed forests? *Biodiversity and Conservation* 2, 95–113.

Waldien, D. (2005) *Population and Behavioral Responses of Small Mammals to Silviculure and Downed Wood Treatments in the Oregon Coast Range*. Oregon State University Press, Cornvallis, Oregon, USA.

Wender, B.W., Harrington, C.A. and Tappeiner, J.C. (2004) Flower and fruit production of understory shrubs in western Washington and Oregon. *Northwest Science* 78, 124–140.

Wen, Y., Ye, D., Chen, F., Liu, S. and Liang, H. (2010) The changes of understory plant diversity in continuous cropping system of Eucalyptus plantations, South China. *The Japanese Forest Society* 15, 252–258.

White, P.S. (1979) Pattern, process, and natural disturbance in vegetation. *Botanical Review* 45, 229–299.

White, P.S. and Pickett, S. (1985) Natural disturbance and patch dynamics: an introduction. In: Pickett, S.T.A. and White, P.S. (eds) *The Ecology of Natural Disturbance and Patch Dynamics*. Academic Press San Diego, California, USA, pp. 3–9.

Wilson, D.S. and Puettmann, K.J. (2007) Density management and biodiversity in young Douglas-fir forests: challenges of managing across scales. *Forest Ecology and Management* 246, 123–134.

With, K.A. and Crist, T.O. (1995) Critical thresholds in species' responses to landscape structure. *Ecology* 76, 2446–2459.
With, K.A. and King, A.W. (1999) Extinction thresholds for species in fractal landscapes. *Conservation Biology* 13, 314–326.
World Commission on Forests and Sustainable Development (1999) *Our Forests, Our Future*. Summary Report of the World Commission on Forests and Sustainable Development. Cambridge University Press, Cambridge, UK. Available at: www.iisd.org/sites/default/files/publications/wcfsdsummary.pdf (accessed 31 May 2016).

30 Invasive Alien Weed Species: A Threat to Plant Biodiversity

Disha Jaggi, Mayank Varun, Saurabh Pagare, Niraj Tripathi, Meenal Rathore, Raghwendra Singh and Bhumesh Kumar*

Directorate of Weed Research, Jabalpur, India

Abstract

Biological invasion is defined as the introduction (intentional or non-intentional), successful establishment, and potential spread of species outside their native range of habitat, and recognized as a threat to the economy, environment and biodiversity globally. Together, such species are regarded as 'invasive alien species'. Liberalization of global trade, increase in transport, travel and tourism are the root causes of invasions by alien species that pose a serious threat to biodiversity of different ecosystems, food security, human and animal health. In the recent past, the implications of climate change for biodiversity have been widely recognized, however, the ill effects of invasive alien species on global biodiversity have received little attention. Climate change is itself a major factor responsible for the complete shift of native flora diversity, which may further complicate the puzzle of biological invasions. It has been generally visualized that predicted changes in climatic factors will also modify interaction scenarios between crops and weeds that will make weed management more difficult. And, if this happens, invasive alien weed species will dominate the other flora due to their superior adaptability to new environments. An attempt has been made to review the available literature on invasive alien weed species and their significant impact on different ecosystems, including forestry, agro-forestry, cropped and non-cropped lands in India and elsewhere.

30.1 Introduction

India has a geography with characteristics of the three major biogeographic realms, the Indo-Malayan, the Eurasian and the Afro-tropical, and is considered to be one of the 12 centres of origin and diversity of several plant species (Hooker, 1904). Indian flora has 35% of south-east Asian and Malayan, 8% temperate, 1% steppe, 2% African, and 5% Mediterranean-Iranian elements (Nayar, 1977). India occupies only 2.4% of the world's total land area but contributes about 8% to the world's species diversity. India is an important hub of agri-biodiversity and contributes about 167 species to world agriculture and more than 300 species to wild relatives of crops. The number of reported angiosperm species in India is about 17,000 (Hajra and Mudgal, 1997).

In the recent past, an increase in activities related to travel, tourism and trade led to an unprecedented movement of non-native plant materials across the globe, which, in turn, caused instability in the plant biodiversities of different countries. Non-native weed species, having the ability to establish themselves in a new ecosystem, are regarded as invasive alien species (IAS). Any non-native plant species, introduced either intentionally or non-intentionally to a specific region, dominates or replaces native flora either through direct competition or other natural phenomenon and gradually becomes a threat to the plant biodiversity of that region. In the past few years, introduced species have shown their harmful effects to environments, ecosystem services, and human physical and cultural health. The available literature clearly suggests that invasion by alien

*E-mail: kumarbhumesh@yahoo.com

plant species has increased rapidly throughout the world during this and the previous century and is responsible for the homogenization of flora, which causes a substantial threat to biodiversity and the ecological integrity of native habitats and ecosystems (Booth *et al.*, 2003; Hulme, 2003). Although implications of climate change for biodiversity have been widely recognized, adverse effects of IAS on global biodiversity have received little attention.

Introduced species pose a serious threat to native flora and fauna communities, either directly or by modifying ecosystem process and functions (Vitousek *et al.*, 1997; Gordon, 1998; Hulme, 2006). Due to their potential to change the complete structure of a plant community, invasive alien species top this list, along with climate change. Introduced species establish, spread and subsequently invade in areas where they are not native. According to the World Conservation Union and Convention on Biological Diversity, IAS are the second most significant threat to biodiversity and socio-economic welfare of the planet. In addition, IAS are recognized as one of the most troublesome threats, causing extinctions of numerous species at the global scale (Clavero and García-Berthou, 2005). In an analysis, IAS were regarded to be a severe threat to amphibians, birds and mammals (Vié *et al.*, 2009).

For the last three centuries, invasive alien species have been a contributing factor to the extinction of many species (SCBD, 2006). Because of this, IAS have attracted the attention of ecologists and weed biologists with a focus on plant invasions, their history, impact and to some extent genetics. Government and non-government organizations have initiated considerable projects to manage local and global biodiversity. In this context, the active participation of international agencies has been increased to conserve biodiversity and sustain livelihoods by minimizing the spread and impact of invasive species.

30.2 Invasive Alien Weeds Species (IAWS)

Biological invasions are defined as the introduction, establishment and spread of species outside their native range (Richardson and Pysek, 2006), and are recognized as a major threat to the economy, environment and social issues worldwide. Introduction of the species to the new location can either be accidental or intentional (Enserink, 1999; Van der Putten, 2007). Accidental introduction can happen by way of unwittingly transporting the plant (Hughes, 2003), plant parts (Usher *et al.*, 1988) or propagules, as contaminants in food grains (Shimono and Konuma, 2008), fodder (Panetta and Scanlan, 1995), or attached to vehicles (Carlton and Ruiz, 2005). Intentional introductions are made for purposes like agriculture, horticulture (Reichard and Hamilton, 1997), forestry (Sankaran, 2002) and for other aesthetic reasons. Lack of proper screening protocols and quarantine procedures may prompt movement of a species to a new area where it can impact local flora in various ways (Mack and Lonsdale, 2001).

Indian plant diversity is also suffering from impacts of IAS. Negative impacts have been felt through losses to grazing areas, agricultural production and health hazards for humans and animals. Invasive alien weed species also pose serious problems, as these have enough potential to alter ecological processes and displace native plant and animal species. Through hybridization with native species, such species may also alter gene pools. Weeds also serve as hosts for many plant pathogens and insects, and so considerably impact crops. *Phytophthora capsici* is a difficult-to-control pathogen which utilizes weeds as alternate hosts in the absence of a suitable host crop and causes damage whenever the host crop is available. Human activities and disturbance, such as road and track construction, grazing, and dumping of agricultural waste, also contribute to the rapid spread and proliferation of IAWS. In the absence of adequate regulations for the import of foodstuffs and distribution of imported material, there is always a high risk of non-intentional introduction of IAWS. India is a country lacking stringent polices to tackle such issues, and consequently, Indian native diversity is under the threat of invasion by IAWS.

It is important to study the invasive species' characteristics and their adaptive potential in order to understand the process of invasion in regions of entry (Vermeij, 1996; Lonsdale, 1999; Alpert *et al.*, 2000). A wide range of soil and climate conditions, enormous seed production, easy seed dispersal, seed dormancy, extensive root develoment, shorter lifespan, high seed dispersal rates, photoperiod insensitiveness, early colonizing ability, and high input use efficiency are the characteristic features that contribute to invasiveness of a weed species (Sakai *et al.*, 2001; Monaco *et al.*, 2005; Funk and Vitousek, 2007; Leishman *et al.*, 2007). Due to fast establishment and growing habit, IAWS have an ability to make use of tree-fall gaps, degraded forests and forest fringes that is superior to that of native species (Grotkopp *et al.*, 2002; Burns, 2004; David *et al.*, 2005; Burns, 2006; Rojas *et al.*, 2011). Most of the invasive plant species produce allelochemicals, which prevent native plants growing and establishing in the vicinity, and thus aid colonization (Callaway and Aschehoug, 2000). Another characteristic of IAWS is phenotypic plasticity, which helps them to adapt to a variety of habitats. For example, *Lantana camara* and *Mimosa diplotricha*, which remain shrubs in open lands but turn into climbers when growing under a closed canopy of trees, where there is competition for light (Hulme, 2008; Niklas, 2008). The dual mode of reproduction (sexual and vegetative) in many invasive species makes them able to spread throughout the year, even under adverse climate conditions (Silvertown, 2008). *Parthenium hysterophorus* displays a number of peculiar characteristics and poses a high risk to biodiversity in many countries, including India. It has a considerable degree of phenotypic and morphological plasticity in response to varying environmental conditions. Lack of natural enemies, habitat destruction and competition are the three primary mechanisms which help an introduced plant species to affect the biodiversity of any region. Habitat destruction is one of the factors leading to the decline and/or extinction of species. Introduced species outcompete native flora by either overexploiting resources required by native species for survival or by aggressively excluding native species from space and/or resources (Dickman, 2006; Stokes *et al.*, 2009).

30.3 Invasive Alien Weeds as a Threat to Biodiversity

Invasive alien weed species pose a sizable threat to native plant communities in disturbed habitats globally (D'Antonio *et al.*, 2001); however, only a few alien plant species have the capacity to invade undisturbed habitats (Rejmánek, 1989). Previous studies have demonstrated that the effects of invasive species are very complex and they can sometimes permanently alter the composition of flora of a region (Holway *et al.*, 2002; Carlton, 2003). There is evidence on the impact of invasive alien species on biodiversity and ecosystem functioning (D'Antonio and Vitousek, 1992; Richardson, 1998), shifts of native species (Walker and Vitousek, 1991; Holmes and Cowling, 1997; Kwiatkowska *et al.*, 1997), and dynamics of post-disturbance community (Mack and D'Antonio, 1998). The plant biodiversity of a region is important, as it governs all the necessary bio-functions, distribution and climatic link of the different habitats, hence biodiversity is considered a boon for the biological system. Adverse effects of IAWS on an ecosystem include changes in species richness, abundance and functions (Grice, 2004) directly or indirectly. Direct impacts can be listed as changes in composition of flora, loss of productivity, litter breakdown rates, nutrient dynamics, changes in hydrological cycles and fire regimes (Brooks *et al.*, 2004; Yurkonis *et al.*, 2005). However, indirect impacts include detrimental associations with micro-organisms such as bacteria and mycorrhizae, and changes in population of larger invertebrate and vertebrate fauna (Zedler and Kercher, 2004).

The distribution extent, spreading rate and persistence of IAWS directly influence the plant biodiversity of the invaded region and, therefore, the trend of invasion of IAWS acts as an important indicator of the damage to biodiversity. The Millennium Ecosystem Assessment (2005) lists IAS as one of the five primary drivers of change in ecosystem composition, structure and function. Overall, species extinctions implicated a negative role of IAS in 50% of those extinctions where a cause could be identified or inferred, second only to habitat transformation (Ervin, 2003). Considering the adaptive potential and plasticity of IAWS, global climate change may play a significant role in accelerating the rate of introduction, distribution and spread of IAS into new areas where they were previously absent, or boost their performance by enhancing competitive ability relative to indigenous species. Besides suppressing native biodiversity through direct competition, IAWS, having strong allelopathic potential, may also influence regional plant diversity by replacing a complex plant community with a much simpler one, characterized by the dominance of only a few species that were potentially more competitive and allelopathic in nature. In extreme situations where the local plant communities are severely affected by a single species, this results in dominance of that species; and also has implications for the food-chain length and complexity of the food web.

Unfortunately, present research studies lack full assessment of the impact of invasive plant species on ecosystem processes. Most of the studies are based on observational comparisons of native diversity and composition in invaded and non-invaded locations. On the basis of such observations, it is hard to identify the exact mechanisms and documentation of the affected community structure at invaded and non-invaded areas. However, a few studies have elaborately examined the mechanism(s) underlying impacts on community structure, and the majority of these pointed to 'competition for resources' being responsible for the observed impact on community structure and composition (i.e. Braithwaite *et al.*, 1989; Woods, 1993; Wyckoff and Webb, 1996; Holmes and Cowling, 1997; Kwiatkowska *et al.*, 1997; Lavergne *et al.*, 1999; Martin, 1999). Available evidence, so far, seems to be based on comparative counts, and in most cases, key mechanistic clues are absent.

30.4 Major Invasive Alien Weeds in India

India occupies only 2.4% of the world's land area but contributes about 8% of the world's species diversity. In India, like other parts of the world, invasive weed species have been introduced via different pathways. Most of the alien plant species that are now known to be invasive in India were first introduced as garden ornamentals. Other reasons for the introduction of alien species were to fulfil other human needs like fuel-wood and food requirements, for commercial cultivation and for conservation of the environment, e.g. to prevent desert spread. Cultivation was stimulated during the last half-century as the globalization of trade, tourism and industry has resulted in increased mobility of people and goods, and the associated transport of plants, animals and micro-organisms around the world. The introduction and naturalization of foreign weeds on Indian soil has greatly influenced the shifts and plant community composition of Indian flora. India has a flora of its own; nevertheless, immigration provides a continual source of modifying diversity for a particular region.

In India, 18,000 plant species, 30 mammal species, four bird species, and more than 300 fish species are alien (Pimentel *et al.*, 2001). About 40% of Indian flora is regarded as alien, of which 25% is considered to be invasive (Raghubanshi *et al.*, 2005). Invasive alien weed species already documented as having a significant impact on the agro-ecosystem, as well as on human health, are *Argemone mexicana, Lantana camara, Ageratum* spp., *Chromolaena odorata, Cuscuta campestris, Eichhornia crassipes, Ipomoea carnea, Mikania micrantha, Mimosa* spp., *Parthenium hysterophorus* and *Salvinia molesta*. Threats posed by these alien weeds to agriculture and biodiversity, and their management, have been discussed in many reviews.

Recently, Kour *et al.* (2014) documented 55 aquatic invasive alien plant species belonging to 24 families in the Jammu region with other background information. The majority of noxious invasive plants in the Jammu region were introduced from America, followed by Europe and Africa. Acoording to a comprehensive inventory of the invasive alien flora in the state of Uttar Pradesh, India, 152 species from 109 genera and 44 families are present (Singh *et al.*, 2010). Reddy (2008) prepared a comprehensive list of invasive alien species in India with useful background information on family, habit and nativity; a total of 173 invasive alien species belonging to 117 genera under 44 families were documented. Tropical America (74%) and Tropical Africa (11%) are the main contributors to the invasive alien flora of India. According to Khuroo *et al.* (2012), the alien flora of India amounts to 1599 species, belonging to 842 genera in 161 families, and constitutes 8.5% of the total vascular flora found in the country. The negative impacts of alien species have been felt through losses to grazing lands, decrease in agricultural yield as well as ill effects on human health. Some of the important IAWS of forestry, agro-forestry, wastelands and aquatic systems (excluding most of those of crop lands) which have already had negative impacts on biodiversity are described below in detail.

30.4.1 *Parthenium hysterophorus* L. (Asteraceae)

Parthenium hysterophorus is a branched, annual, short-lived, ephemeral, erect and herbaceous plant that grows 0.5–2.0 m tall. Mature stems are greenish and longitudinally grooved, hairy and well branched at maturity. Leaves are pale green, pubescent, lobed, alternate, sessile and bipinnate, and resemble the leaves of carrot. The lower leaves are relatively large and are deeply divided. Leaves on the upper branches are usually smaller in size and less divided than the lower leaves. Numerous small flower-heads (capitula) are arranged in clusters at the tips of the branches and each capitulum attaches to a stalk of 1–10 mm long. Flower-heads are creamy-white in colour and have five small petals. Numerous white flowers (disc florets) are surrounded by two rows of small green bracts. The colour of flowers changes to light brown when seeds are mature. Flowering starts about a month after germination and continues throughout the year, with maximum flowering in the rainy season. The fruit of *Parthenium* is cypsella. Seed production is enormous and a fully developed plant can produce up to 100,000 seeds during its lifetime. Seeds do not have a dormancy period and are capable of germinating at any time when favourable environmental conditions are met. The tap root system with many secondary roots is an adaptive feature, which helps in efficient absorption of nutrients even in resource-poor soils.

Parthenium hysterophorus is an example of an accidental introduction of an alien species into India that has become invasive. It was first reported in 1951 from Maharashtra and believed to be introduced from Tropical America. Currently, it exists throughout the country. It is an aggressive colonizing alien species of dry and disturbed land, and is widely demonstrated to affect crop production, animal husbandry, human health and biodiversity (Evans, 1997). It has spread to more than 20 countries of Africa, Asia and Oceania (Dhileepan and Strathie, 2009). Recent reports indicated the severe negative impacts of *P. hysterophorus* in countries like India, Ethiopia and South Africa, where many farmers have been forced to abandon grazing and cultivation on land invaded by *P. hysterophorus*.

The attribute of easy seed dispersal also contributes to the invasive habit of *P. hysterophorus*. The seeds are very small and light, with short, wing-like structures that are known to float on the wind or are carried easily by various means (Navie *et al.*, 1996). Evidence indicates long-distance seed dispersal, with vehicles, air, water, movement of livestock, animal dung and grain seeds contributing to the spread (Frew *et al.*, 1996; T. Tamado, Biology and management of *Parthenium hysterophorus* L. in Eastern Ethiopia. School of Graduate Studies of Swedish University of Agricultural Sciences Uppsala, Sweden, 2001, unpublished thesis). High reproductive ability is one of the biological characteristics contributing to the success of *P. hysterophorus* as an aggressive weed (Pandey and Dubby, 1989). Favourable conditions prompt flowering within four weeks of seed germination and plants continue to flower for up to eight months (McFadyen, 1992). All such characteristics contribute to the successful survival and establishment of *P. hysterophorus* and indicate the ability of this weed to pose numerous constraints on sustainable development, economic growth, poverty alleviation and food security (GISP, 2004).

30.4.2 *Lantana camara* L. (Verbenaceae)

Lantana camara is a perennial, erect or prostrate shrub growing 4–8 feet in height, has a tap root and many other shallow secondary roots. Stems are square in cross-section and hairy when young but become cylindrical at maturity. Leaves are ovate, oppositely arranged, leaf blades are serrated, have the feel of fine sandpaper, and release a pungent odour upon being crushed. Flower heads contain 20–40 flowers and are clustered at the tip of stems. Flowers change colour with time from white to pink or lavender, or yellow to orange or red. Typically, mature flowers are darker in colour (lavender and red). The fruit is a berry and seeds will change from green to a deep purple and eventually black, on reaching maturity. *Lantana* exhibits the dual mode of reproduction, i.e. vegetatively as well as by seeds. Vegetative reproduction occurs when stems come into contact with moist soil, initiating root formation at the contact site. It can also regrow from the base of the cut stem. Flowers produce round the year and exhibit both self- and cross-pollination. Seed production is prolific, with approximately 12,000 fruits per plant. Normally seed germination is low; however, when seed passes through the digestive system of an animal, the germination rate is increased.

Lantana camara, considered as one of the ten worst weeds of the world, was introduced into India as an ornamental plant brought to the Calcutta Botanical Garden in 1809 from tropical and subtropical America (Thakur *et al.*, 1992). Now it has reached and spread from the sub-mountainous regions of the outer Himalaya to the southernmost part of India, covering a variety of habitats. It shows tremendous morphological and phenotypic variability and is reported to have about 650 varieties globally. It has already established or is expanding into many regions of the world, often as a result of forest clearing for timber or agriculture, and has had a severe impact on agriculture as well as on natural ecosystems. Plants of *Lantana* can grow individually in clumps or as dense thickets, virtually eliminating many desirable species. In disturbed native forests, it can be a dominant understorey species, leading to disruption of succession and sometimes the complete loss of biodiversity. It has been reckoned as one of the most noxious weeds and has encroached on most of the social areas and reserve lands. In India, the outer Himalayas are almost completely occupied by this weed, resulting in significant impact on the yield of crops and pastures. The costs of manual/mechanical removal of this weed have increased enormously, leading to heavy expenditure for its management.

According to a report, *L. camara* occurrence has doubled across the 540 km^2 of Biligiri Rangaswamy Temple Tiger Reserve (BRT) in India during the last decade (Sundaram and Hiremath, 2012). The increase in spatial extent was accompanied by a disproportionate increase in density. There have been several recent reviews of hypotheses pertaining to species or community characteristics that make them invasive (Catford *et al.*, 2009; Gurevitch *et al.*, 2011; Jeschke *et al.*, 2012). Several alternative mechanisms also appear to contribute to the successful invasion of *L. camara*. The absence of herbivores and pathogens in introduced areas (Keane and Crawley, 2002) and non-palatability of *L. camara* to herbivores are advantages for this species, ensuring unchecked establishment and spread in the target region. It has been shown to exhibit high nutrient use efficiency (Bhatt *et al.*, 1994), which potentially gives it a competitive advantage over other species, especially in nutrient-poor soils. Its ability to flower throughout the year, prolific fruiting and easy dispersal are other factors which contribute to the success of *L. camara*. Allelopathy is another mechanism which can reduce the growth and development, vigour and productivity of associated plant species. Together, all these factors contribute to the invasiveness of *L. camara* and make management of this weed a tough task. In a review on *L. camara*, Bhagwat *et al.* (2012) described its management as a lost battle, suggesting that attempts to eradicate it have failed to a large extent. Utilization of *L. camara* as fuel wood and for other uses to enhance livelihoods can offset some of its management costs and may be one of the few viable options available, as suggested by Shaanker *et al.* (2009).

30.4.3 *Eichhornia crassipes* (Mart.) Solms (Pontederiaceae)

Eichhornia crassipes (water hyacinth) is a free-floating perennial hydrophyte native to South America. It grows generally to 0.25–0.50 m in height and forms dense floating mats. Roots are adventitious, fibrous and purplish-black due to the presence of several pigments including anthocyanins. Leaves are petiolate, thick, waxy, rounded, glossy, broadly ovate to circular, 10–20 cm in diameter, arranged spirally, leaf veins are dense, numerous, fine and longitudinal. Leaf stalks are swollen and form spongy and bulbous structures. Flowers are borne terminally as a spike, with 10–15 flowers on an elongated peduncle. Flowers have six petals, purplish-blue or lavender to pinkish, six stamens (sometimes five or seven), having curved filaments with glandular hairs, and purple-coloured anthers 1.4–2.2 mm long. Fruit is a thin-walled capsule enclosed in a relatively thick-walled hypanthium, and a mature capsule containing up to 450 seeds.

Eichhornia crassipes has been reckoned among the top ten worst weeds worldwide. In India, the plant was first introduced in Bengal during the early 1780s. It is now found in more than 50 countries on five continents. Water hyacinth is a very fast-growing plant with an explosive growth rate and contributes to the eutrophication in water bodies. The dense and thick mat-like canopy of water hyacinth also prevents sunlight and oxygen from reaching the water column, and creates a hindrance for other aquatic living beings (i.e. submerged plants and fish), and dramatically reduces biological diversity in aquatic ecosystems. Heavy infestations of this weed adversely impact commercial, recreational and social activities like navigation, hydroelectric generation, swimming and fishing. In addition, it poses health hazards by harbouring harmful insects and vectors of diseases such as malaria, encephalitis and filariasis. Water hyacinth is also responsible for high evapotranspiration losses and often has been blamed for drying-up of water bodies, which are the mainstay of agriculture in rainfed areas.

30.4.4 *Mikania micrantha* Kunth (Asteraceae)

Mikania micrantha is a perennial herbaceous vine of the Asteraceae family and commonly known as American rope, bittervine, chinese creeper, climbing hempweed and mile-a-minute weed. It is a fast-growing, well-branched vine, with a climbing, twining or creeping habit. Stems can grow up to 6 m long or sometime more. They are slender, ribbed and hairy; lateral stems are as vigorous as the main stem, and sometimes produce adventitious roots at nodes. Leaves are opposite, ovate-deltoid, glabrous on both sides, minutely glandular beneath; petiolate (petiole 3–7 cm long). The inflorescence is axillary panicled corymbs; the capitula is cylindrical with four flowers per capitula; oblong to obovate, green in colour. In south-west India, flowering time is August to January and fruit setting between September and February. It has prolific seed production and a single plant can produce 20,000–40,000 mature seeds in one season. Seeds are minute, black or blackish-brown, with pappus, and elongated in shape. Plants have dual mode of reproduction and can grow vegetatively from the nodes. Growth of young plants is extremely fast (8–9 cm in a day) and can rapidly form a dense cover over trees, posing a serious threat to plantations crops and ornamentals.

Mikania micrantha is thought to have been introduced by the Allied forces during the Second World War to camouflage airfields built along the Indo-Burmese border as a defence against the Japanese forces (Randerson, 2003). Since then, it has spread across the entire region and has become a menace in natural forests, plantations, and agricultural systems in north-east and south-west India. It climbs to the top of the canopy, spreads over it entirely and seriously impacts on the growth of trees, shrubs and herbs. It has been recognized as an aggressive colonizer, due to its extremely fast growth rate and dual mode of reproduction. At flowering, it attracts a large number of pollinators, thus creating competitive pressure on the native species. The easy dispersal ability of the seeds also make this species suitable for fast spread. In addition, after drying out, it increases the risk of forest fires.

30.4.5 *Prosopis juliflora* (Sw.) DC. (Fabaceae)

Prosopis juliflora is deciduous shrub or small tree and grows up to 12 m height with thick bark of brown or blackish colour. Branches are characteristically zig-zag with long paired thorns (modified stipules) at each bend. Leaves are deciduous, pinnately compound with 10–15 pairs of linear-oblong leaflets. Flowers are in long cylindrical spikes; greenish-yellow in colour and lightly scented. Pods are long, pulpy, and yellowish in colour at maturity, 10–20 cm long, cylindrical and contain 10–15 seeds. Seeds have a hard coat, are compressed and oval or elliptic, smooth and brown in colour. Fruits and seed production is very high.

Prosopis juliflora is thought to be native to Central America, northern South America and the Caribbean islands (Burkart, 1976) and has become a major invasive weed species in many regions throughout the world, including southern Africa, the Middle East, Pakistan, India and Hawaii (Pasiecznik *et al.*, 2001). It forms pure stands in its invaded range in forests, wastelands and at the boundaries of crop fields. *Prosopis juliflora* was first introduced to India in 1877 to halt the spread of the Thar desert in Northwest India. Following the initial success, it was planted on a large scale in the western Indian states of Gujarat, Rajasthan and Maharashtra (Tiwari, 1999), where it has become invasive. However, it is also considered as a good source of fuel wood, fodder, charcoal and timber. In 1940, *P. juliflora* was declared a 'Royal Plant', and given special protection in the erstwhile princely kingdom of Jodhpur in Rajasthan (Pasiecznik *et al.*, 2001). Further, *P. juliflora* was also introduced to several protected areas in Rajasthan, such as Keoladeo Ghana (K.R. Anoop, Progress of *Prosopis juliflora* eradication work in Keoladeo National Park, Rajasthan Forest Department, Jaipur, Rajasthan, 2010, unpublished report), and Ranthambore (Dayal, 2007). Currently, it is widely recognized as a problem in Rajasthan and other part of the country; however, *P. juliflora* is still being cited as a successful example of afforestation, especially in Kutch region of Gujarat.

According to Sharma and Dakshini (1996), the presence of thorns, its seed dispersal characteristics and its low palatability, give *P. juliflora* an advantage over native species. In addition, it has been shown to exhibit allelopathic impacts on native vegetation (Kaur *et al.*, 2012). In and around Kumbalgarh wildlife sanctuary, the *P. juliflora* invasion, accompanied by other unpalatable species like *Lantana camara*, has led to the exclusion of important fodder grasses (Robbins, 2001). It has also been reported to encroach on unique habitats for grassland birds such as the Houbara bustard (Tiwari, 1999).

30.4.6 *Ageratum conyzoides* L. (Asteraceae)

Ageratum conyzoides is commonly known as Goatweed, Tropical Whiteweed and Chickweed. It is an erect, branched, soft, aromatic, annual herb with a shallow, fibrous root system. Stems are covered with fine white hairs, leaves are opposite, pubescent with long petioles and glandular trichomes. Inflorescence may contain 30–45 light-pink flowers, arranged as a corymb and are self-incompatible. Fruit is an achene with an aristate pappus and is easily dispersed by wind. Seeds are positively photoblastic, often lost within 12 months. The species has great morphological and phenotypic plasticity and adaptability to different ecological conditions.

Ageratum conyzoides, an alien noxious weed, is a native of tropical America. It has spread in almost all parts of India, especially in forests, fallow lands, roadsides, railway tracks, plantation crops and even in crop fields. Due to its ability to produce vegetatively, the control of this weed species is very difficult. It is considered to be a strong competitor of other native associated vegetation for nutrients, moisture and light, and suppresses their growth and development to a great extent. A visible impact of the invasion by *Ageratum* is shrinkage of grazing area for animals, along with reduction in productivity of grasslands. In general, it is considered a threat to biodiversity in forest areas. In addition, several health hazards like nausea, giddiness, headache, and skin and eye irritation have also been reported by agriculturists. Cases of poisoning in farm animals have also been reported.

30.4.7 *Chromolaena odorata* (L.) King & H.E. Robins (Asteraceae)

Chromolaena odorata belongs to Asteraceae family and commonly known as Siam Weed, Christmas Bush, Devil Weed and Common Floss Flower. In open areas, it spreads into tangled, dense thickets up to 2 m tall if standing alone, and higher when climbing up vegetation. It has usually soft stems but the base may become woody with age. Leaves are arrowhead-shaped with three characteristic veins in a forked pattern. Clusters of 10–35 pale-pink or creamy-white flowers are found at the ends of branches. Seeds are achenes, small and brown with a white feathery 'parachute', photoblastic, normally dispersed with air but can also cling to fur, clothes and farm machinery, enabling long-distance dispersal. A single plant can produce 50,000–100,000 seeds. The root system is fibrous and can reach a depth of 30 cm. The plant can regenerate from the roots in addition to seeds.

Chromolaena odorata is a perennial shrub originating from South and Central America. It is considered as one of the world's worst tropical weeds due to its extremely fast growth rate (up to 2 cm per day), prolific seed production and long lifespan (approximately ten years). It was first introduced to India as an ornamental plant into the Royal Botanic Garden, Calcutta, in 1845. Nowadays it can easily be found on forest edges and paths, in barren fields and grasslands, at building sites, along roads, railways and streams, in plantation crops such as coconuts and rubber, and to some extent in field crops such as tobacco and sugarcane. This weed grows best in tropical and subtropical regions, and prefers well-drained soils and full sun. It constitutes a potential fire hazard in forests, particularly during the dry season but it can survive fires and grows back vigorously following rain, which provides a competitive advantage over other plant species.

30.4.8 *Cuscuta campestris* Yunck. (Convolvulaceae)

Cuscuta campestris, commonly known as golden dodder or field dodder, is an annual stem parasitic plant which lacks normal roots. Stems are usually yellow or orange in colour, glabrous, trailing or twining, and are attached to the host by numerous small connective structures known as haustoria. In place of normal leaves, tiny scales are found arranged alternatively. Inflorescence is corymbiform, flowers subsessile with short pedicel, tetra- or pentamerous, white, membranous, gland-like laticiferous cells mostly in the calyx and less obvious in the corolla, ovary and capsule too. The fruits are a light-brown capsule. Seeds are oval in shape and light brown or brownish in colour. *Cuscuta campestris* is considered as an invasive noxious parasite weed that attaches itself to the stems and leaves of a wide variety of host plant species. After establishment, it draws nutrients from the host plant, leading to a drastic retardation of growth and development of the host plant. *Cuscuta* parasitizes mostly dicots, such as clovers, linseed, onion, sugarbeet, chillies, green gram, black gram, niger, alfalfa and chickpea. High levels of *Cuscuta* in fodder are toxic to cattle and horses if continuously consumed for several weeks.

30.4.9 *Alternanthera philoxeroides* Griseb. (Amaranthaceae)

Alternanthera philoxeroides, commonly known as alligator weed, is a perennial herb found in both aquatic and terrestrial habitats. The stems are often hollow in aquatic situations, where they make a floating mat with upright branches, and may vary in colour. Fibrous roots emerge from nodes of the stem hanging free in water or can penetrate into the sediment/soil, depending upon the depth of the water. Leaves are opposite, entire, and leaf base forms a sheath around the stem at the node. The flower has no petals, but has five white sepals, 10–15 mm in size, silvery-white, papery, axillary or terminal with a short stalk, and numerous flowers form a circular cluster.

Alternanthera philoxeroides is thought to be native to South America. However, it can be found in many parts of the world, infesting rivers, lakes, ponds and irrigation canals, as well as many terrestrial habitats. In India and the Asia-Pacific region, this weed is considered as an invasive exotic in subtropical to temperate regions. The aquatic form of the weed has the potential to become a serious threat to waterways, agriculture and the environment. Management of this weed is expensive and extremely difficult, due to its ability to propagate via vegetative fragmentation and dispersal of vegetative propagules. These attributes also contribute to its success as an invasive species. The terrestrial form of *Alternanthera philoxeroides* grows into a dense mat with a massive underground rhizomatous root system and has the capacity to smother most other herbaceous plant species.

30.4.10 *Salvinia molesta* D. Mitchell (Salviniaceae)

Salvinia molesta is a free-floating aquatic fern. It produces a horizontal rhizome, having two types of fronds (leaves), emergent and submerged, arranged in three whorls. Emergent leaves are green in colour and obovate in shape. The surface of the leaves is covered with many hairs that split and then rejoin at the tips and form a cage-like structure. The purpose of this is to trap air, giving the plant buoyancy in the water. Submerged fronds are brownish in colour and feather-like in appearance, and are sometimes mistaken for roots. The papillae, hairs and upper surface of the plant are water-repellent in comparison to the lower surface, which in fact attracts water. Plants of *Salvinia molesta* exhibits morphological variations to a great extent depending on the prevailing conditions such as space and nutrient availability. Within the submerged leaves, egg-shaped sporocarps are found which accommodate spores.

Salvinia molesta is commonly regarded as one of the worst aquatic weeds in India because of its invasiveness, potential for spread, and economic and environmental impacts. It thrives in slow-moving, nutrient-rich, warm freshwater, and has the potential for long-distance dispersal within water bodies through water currents. Among water bodies, it can be dispersed via animals, human intervention and contaminated equipment such as boats and fishery tools. Optimum water temperature for *Salvinia* ranges between 20 and 30° C. The growth rate of this weed is extremely fast, and in optimal conditions it can double in size within three days. The high growth rate of this weed leads to formation of very dense mats, which reduces flow of water, availability of light at the water surface and causes a deficiency of dissolved oxygen. Together, all these factors result in a stagnant, dark environment, which adversely affects the biodiversity of the habitat. In addition, it can also disturb wetland ecosystems, posing a severe threat to socio-economic activities due to flooding, pollution of drinking water, hindrances in irrigation, water sports, and other recreational activities such as swimming, fishing and boating.

30.4.11 *Ipomoea carnea* Jace. (Convolvulaceae)

Ipomoea carnea belongs to the family Convolvulaceae and is commonly known as Morning Glory. It is a large diffuse shrub and can grow up to a height of 6 m on terrestrial land; however, it remains shorter in the aquatic habitat. Stems are erect, woody, hairy, more or less cylindrical, and light greenish in colour. leaves are simple, petiolate and upper surface of leaf is dull green while the lower surface is pale. Flowers vary in colour (pale rose, pink, light violet, white), axillary or terminal, and pedunculate cymes.

Ipomoea carnea is one of the most dominant and harmful of the weeds in tropical and subtropical regions globally. It is thought to be a native of South America and was introduced into India as a hedge plant. Since then, it has spread rapidly in terrestrial as well as in aquatic habitats. It is a hardy and resilient species and can resist successfully all the control measures, including biological agents and herbicides application. It can flower and fruit throughout the year and has the potential to survive in any environmental conditions and stresses for several months. In addition to reproduction by seeds, its capability of rooting from the stem within a few days provides this species with a robust mode of vegetative propagation, which has been a key factor for its survival and spread in adverse conditions. Due to its wide range of ecological and stress adaptability, and its robustness *Ipomoea* has become a serious problem, affecting fisheries, navigation, agriculture, irrigation and, above all, biodiversity.

30.5 Plant Invasions in Relation to Climate Change

As predicted by numerous models, the impact of climate change on agriculture production is almost certain, with variations depending on the species and region. Climatic factors, such as variation in rainfall, may determine the physical and economic viability for crop production, depending on how sensitive the crop is to climatic changes and how significant those changes are for a given region (FAO, 2008b). Frequently, there are warnings from experts that climate change will be detrimental to food security, especially in developing countries and for the resource-poor population (FAO, 2008b). In some cases, climate change has also been predicted to be beneficial for plant growth in agricultural areas; however, crops are invariably associated with a number of weed species. In the agriculture system, net outcome (productivity) depends on the arithmetic of favourable factors and limiting factors. Thus, consequences of change in climatic factors (e.g. rainfall, temperature, atmospheric CO_2) may determine whether crops or weeds will predominate. Based on the available evidence, weeds, especially the invasive weeds, are expected to be dominant over crops or native flora due to their high adaptability to climatic variability and stress conditions (Nelson *et al.*, 2009). Invasive species (plants, animals, insects and diseases) are already regarded as the largest impediment to global food security and agricultural productivity (FAO, 2008a; Rangi, 2009), and predicted climatic changes would further aggravate negative impacts of invasive species on agricultural production and biodiversity.

Climatic and biogeographic changes have significant implications for both native and non-native species. Different species require different sets of ecological and climatic parameters to complete the various phases of their life cycles. A shift or alteration in climatic variables such as the composition of greenhouse gases, temperature and water availability, can also alter the dominance and competitiveness of the species or even may prompt species to shift (if physically possible) to new areas, where these variables are in favour of their survival. Invasive species are generally considered as highly tolerant species in comparison to native species, thereby providing invaders with a wider array of suitable habitats (Walther *et al.*, 2009). Among different climatic parameters, a shift in temperature, for example, might then have significant impacts on a native species, but little impact on an introduced species, due to differences in their tolerance levels; thereby altering the competitive dynamic among them. In addition, changes in precipitation patterns, rising CO_2 levels and increased nitrogen deposition may play a greater role in deciding plant community dynamics and dominance by the specific plant species (Dukes, 2000; Richardson *et al.*, 2000).

Changes in plant community dynamics by competition or other natural phenomena is not expected to be uniform globally, rather, these may vary in different scenarios and prevalent climatic conditions in different geographic regions. For example, community dynamics may change differently in tropical and temperate regions. Similarly, higher latitudes and altitudes will probably experience a shift of species from an adjacent region with the change in temperature (Parmesan, 2006). However, warming tropical systems are not expected to face the same level of threat as warming temperate systems. At the same time, changes in other variables, such as precipitation pattern, may also influence the impact of other factors in either direction, which in turn would dictate the magnitude of invasion and the vulnerability of native flora to invasive species. In addition to range expansion, range contraction or reduction in overall impact of invasive species are also very much possible (Hellmann *et al.*, 2008). As predicted by different research groups, climate change will also increase the frequency and severity of extreme weather events such as wind storms, floods, cyclones, avalanches, and severe heat or cold. Such extreme events would accelerate the physical shift of invasive species at regional and global scales. For example, in 1984, cyclone 'Demonia' facilitated the introduction of *P. hysterophorus* across the landlocked country Swaziland, which subsequently spread and adversely impacted on agricultural production systems, indigenous hunting areas and wildlife reserves (IRIN, 2010). Authors also opined that extreme weather events may or may not be directly responsible for the spread and shifts of invasive species, but definitely play a facilitating role in the movement of species, and by virtue of their resilient and competitive ability, invasive species can easily colonize disturbed areas.

Reaction to climate change and rate of adoption of mitigation strategies at regional and societal level will also contribute to the potential introduction and spread of invasive species. Resilient species with high growth rate, adaptability to harsh conditions and natural disturbances will be in demand for agriculture, forestry, aquaculture, biofuels and other needs, such as restoration of habitats. Coincidentally, all or most of these traits persist in most of the invasive species, hence, offer a preference for their introduction at regional or societal level (Low and Booth, 2007). In India, *P. juliflora* is an example of intentional introduction in the Thar desert to stabilize ecosystems, which, despite several benefits, has become a threat to the native biodiversity of the entire Kutch region.

The impact of climate change as well as invasive alien weed species on food security has been a focus of ecologists and climatologists for the last two decades. Management of invasive weed species under the regime of climate change is currently a focus of weed scientists. There is evidence suggesting a decline in the efficacy of herbicides on invasive plants with rising atmospheric CO_2 (Ziska, 2005). Similar changes in the success of other management strategies, such as fungicides, insecticides and biological control, are also possible under climate change (Chakraborty *et al.*, 2000). The decline in efficacy of herbicides and other pesticides may warrant increases in dose or frequency of chemical applications or may require a switch to another strategy, leading to incurrence of heavy cost and expertise that may not be accessible to small-scale farmers (Gay *et al.*, 2006). All these factors together may have a definite negative impact food security at regional level or perhaps globally.

It is also critical to emphasize the need to re-examine the efficacy of current management practices, given that climate change may alter ways for movement, distribution and impact of invasive species; as well as the effectiveness of our management strategies (Hellmann *et al.*, 2008). Keeping in mind the extent of threats posed by invasive species, blocking of pathways for the unintentional introduction of such species would be the safest preventive measure. Location-specific risk assessment, development of early detection and rapid response systems, and study of interaction with climate change derivatives are equally important and would be useful tools to develop models for analysis of habitat vulnerability for introduction and spread of non-native weed species, as well as the potential threat posed by introduced species (Sutherst, 2000).

30.6 Impact

30.6.1 Agriculture

Weeds alone cause almost 25% of the losses in agricultural productivity and degrade agricultural fields and food quality. Based on eight major crops, Oerke *et al.* (1994) calculated a 13% loss in the world's agricultural output due to weeds. There are numerous ways in which invasive species affect agriculture. Non-indigenous species often compete with native ones, act as insect/disease vectors or sometimes hybridize with them, resulting in hybridized species that are difficult to control and more adaptable to climate change. Herbicides are commonly being used to manage weeds; however, to some extent, herbicides also degrade the agro-ecosystem and negatively impact on beneficial flora and fauna.

The management of invasive species may demand heavy economic inputs in addition to the direct loss of productivity. Losses in agriculture due to invasive species are enormous throughout the world. According to a global estimate, costs associated with the negative impacts of invasive alien species have been US$ 1.4 trillion per year, which constitutes nearly 5% of global GDP at the time of analysis (Pimentel *et al.*, 2000). In Africa, invasive species of the genus *Striga* have a direct impact on local livelihoods, affecting more than 100 million people and as much as 40% of arable land in the savannahs (UNEP, 2004). The unintentional introduction of invasive alien species into the agro-ecosystem due to liberalized global trade policies may provide major threats and is of major concern to the ecologists, environmentalists, the public and policy makers. Hence, an effective and stringent policy needs to be enforced in developing countries such as India to prevent unintentional/intentional introduction of alien weed species. Accordingly, the list of quarantined weeds must be updated from time to time, keeping in view the coming threats from climate change and free trade. Efforts have been made to summarize the documented impacts of some major IAWS at global scale (Table 30.1) and in India as well (Table 30.2).

Table 30.1. Impact of some major invasive alien weed species at global scale.

Sl. No.	Botanical name	Family	Common name	Origin	Introduced in	Threat reported	Reference
1	*Acroptilon repens*	Asteraceae	Russian knapweed	Russia	North America	Affects seedling emergence and growth of native grasses	Grant *et al*. (2003)
2	*Agrop yron cristatum*	Poaceae	Crested wheatgrass	North America	Asia	Strong competitor, displaces or prevents the establishment of native species	Bakker and Wilson (2001)
3	*Alliaria petiolata*	Brassicaceae	Garlic mustard	Europe	US, Canada	Poses severe threat to native flora	Nuzzo (1999)
4	*Alternanthera philoxeroides*	Amaranthaceae	Alligator Weed	South America	China	Forms dense mat over water and land, threatens the native flora and fauna, reduces crops yields, promotes flooding and blocks ships	Holm *et al*. (1997)
5	*Andropogon gayanus*	Poaceae	Gamba grass	Africa	Brazil	Threat to native species	Fisher *et al*. (1995)
6	*Arundo donax*	Poaceae	Giant Cane	India	California	Displaces native species around the water channels	Dukes and Mooney (2004)
7	*Brachiaria mutica*	Poaceae	California grass	Africa	Humid tropics, subtropics	Threat to native vegetation	Williams and Baruch (2000)
8	*Carduus acanthoides*	Asteraceae	Welted thistle	Europe	North America	High probability of establishment and persistence, and has a wide impact on native species	Jongejans *et al*. (2007)
9	*Carrichtera annua*	Brassicaceae	Ward's weed	--	South Australia	Affects native communities in chenopod shrub lands	Harris and Facelli (2003)
10	*Eragrostis lehmanniana*	Poaceae	Lehmann lovegrass	South Africa	Southwestern USA	Wide range of ecological conditions with little to no genetic variation	Schussman *et al*. (2006)

Continued

Table 30.1. Continued.

Sl. No.	Botanical name	Family	Common name	Origin	Introduced in	Threat reported	Reference
11	*Heracleum persicum*	Apiaceae	Persian Hogweed	Turkey, Iraq, and Iran	Scandinavia, Europe	Threat to native plant flora	Jahodová *et al.* (2007)
12	*Imperata cylindrica*	Poaceae	Japanese bloodgrass	Subtropical and tropical Asia	Southeastern USA	Displaces native plant and animal species and alters fire regimes; adding phosphorus reduces invasion of long-leaf pine savanna	Brewer and Cralle (2003)
13	*Lolium multiflorum*	Poaceae	Italian rye grass	Spain	California, USA	Poses a long-term threat to native alkali biodiversity	Dawson *et al.* (2007)
14	*Lythrum salicaria*	Lythraceae	Purple lythrum	Eurasia	North America	Displaces native plants, clogs waterways, and reduces the quality of habitat	Shadel and Molofsky (2002)
15	*Microstegium vimineum*	Poaceae	Japanese stiltgrass	Japan	USA	Threatens the growth of native species	Redman (1995)
16	*Panicum maximum*	Poaceae	Guinea grass	South Africa	USA, West Indies, Central and South America	Extreme structural changes of the vegetation	Williams and Baruch (2000)
17	*Pennisetum clandestinum*	Poaceae	Kikuyu grass	Eastern and Southern Africa	North Mexico, Southwestern US	Threatens the native flora	Williams and Baruch (2000)
18	*Polygonum perfoliatum*	Polygonaceae	Devil's tail	India, China, Korea, Japan, Bangladesh,	Northeastern USA	Causes ecological problems in invaded areas, as the plant grows rapidly and covers shrubs and other vegetation, dominating in its new community	Ding *et al.* (2007)

19	*Prosopis juliflora*	Fabaceae	Mesquite	Jamaica	India	Aggressive and has not only successfully invaded several habitats but has also caused substratum degradation	Sharma and Dakshini (1998)
20	*Spartina densiflora*	Poaceae	Dense-flowered cordgrass	Chile	Gulf of Cadiz, Spain,	Poses a threat to the biodiversity of southern European marshes	Castillo *et al*. (2000)
21	*Senecio madagascariensis*	Asteraceae	Madagascar ragwort	Afro-Madagascan	Hawaii	Competes strongly with existing pasture flora for light, moisture and soil nutrients, leading to the ultimate deterioration of pastures	Roux *et al*. (2006)
22	*Sorghum halepense*	Poaceae	Johnson grass	Africa	North and South America	Threat to indigenous plants	Williams and Baruch (2000)
23	*Tamarix ramosissima*	Caryophyllaceae	Salt Cedar	Russia	North America	Strongly influences species composition, ecological processes, productivity, and biodiversity	Stohlgren *et al*. (1998)
24	*Tradescantia fluminensis*	Commenlinaceae	River Spiderwort	South America	New Zealand	Affects litter decomposition and nutrient availability	Standish *et al*. (2004)
25	*Ziziphus mauritiana*	Rhamnaceae	Jujube	India	Australia	Threatens native biodiversity	Grice *et al*. (2000)

Note: Information in table is compiled from the sources provided in the final column.

Table 30.2. Impact of some invasive alien weed species in India.

Sl. No.	Botanical name	Family	Common name	Origin	Introduced in	Threat reported	Reference
1	*Acanthospermum hispidum*	Asteraceae	Bristly starbur	South America	Uttar Pradesh, Himachal, Andhra Pradesh	Weed of agricultural ecosystems, invades native rangeland pastures and out-competes more desirable native species, negative impact on the productivity of these pastures and also affects their biodiversity	Chatterjee *et al.* (2009)
2	*Aeschynomene Americana*	Fabaceae	Shyleaf	Tropical America	Uttar Pradesh, Bihar	Paddy crop fields	Chatterjee *et al.* (2009)
3	*A. houstonianum*	Asteraceae	Blueweed	Mexico	Tamil Nadu, Assam, Nilgiri Hills, Sikkim, Dehradun, Madhya Pradesh, Nepal, Bengal	Frequently invades bush land and other natural habitats, resulting in substantial changes in native plant communities; displaces indigenous plants and possibly also native animals; and is particularly invasive along waterways and in riparian vegetation	Chatterjee *et al.* (2009)
4	*Alternanthera polygonoides*	Amaranthaceae	Joyweed	Tropical and subtropical America	Upper Gangetic Plain, West Bengal, Coimbatore, Tamil Nadu,	Inhibits germination and early seedling growth of crops and vegetables	Chatterjee *et al.* (2009)
5	*A. pungens*	Amaranthaceae	Khaki weed	Tropical America	Coimbatore, Bangalore, Chennai, Andhra Pradesh, Mumbai, Orissa, Delhi	Regarded as a weed of lawns, pastures and disturbed sites near habitation and poses severe threat to native flora	Chatterjee *et al.* (2009)
6	*Argemone Mexicana*	Papaveraceae	Mexican poppy	Mexico	Delhi and Madhya Pradesh	Significant effects on cultivated agricultural fields	Chatterjee *et al.* (2009)

7	*Ageratum conyzoides*	Asteraceae	Goat weed	Tropical America	Throughout India	Wastelands, plantations, pastures and all forest types; poses threat to indigenous vegetation in NW Himalaya and many other parts of India	Dogra *et al*. (2009)
8	*Eupatorium adenophorum*	Asteraceae	Crofon Weed	Mexico		Occupies vacant places in teak, rubber and other forest plantations and causes serious threat to forests. In hilly areas of south and north India, it forms dense thickets on grazing lands	National Focal Point for APFISN (2005)
9	*Lantana camara*	Verbenaceae	Wild sage	Tropical America	Himachal Pradesh, Uttar Pradesh, Madhya Pradesh, Karnataka, Maharashtra, Tamil Nadu	Common throughout the country in the forests, plantations, agricultural land, disturbed areas, grass lands, wetlands, riparian and urban areas; a serious threat to grasslands and indigenous medicinal plants	National Focal Point for APFISN (2005); Dogra *et al*. (2009)
10	*Parthenium hysterophorus*	Asteraceae	Carrot grass	Tropical America		Fields, forest areas, grass lands and urban areas; aggressive colonizer of degraded areas with poor ground cover and exposed soil, such as fallow wastelands, roadsides and overgrazed pastures between existing plant cover and native weed density	National Focal Point for APFISN (2005); Dogra *et al*. (2009)

30.6.2 Nutrient cycling

Invasive alien weed species have the potential to alter the soil nutrient dynamics by altering soil nutrients like N, P, K, organic carbon, key parameters like pH and EC, and the activity of bio-fixers. The effects of invasive species on nutrient cycling have been studied in detail (Belnap and Phillips, 2001; Ehrenfeld *et al.*, 2001; Scott *et al.*, 2001). Nitrogen cycling, in particular, was the focus of attention, based on the *Myrica* invasion in Hawaii. The impacts of plant invasions on soil nutrient cycling suggest that introduced plant species differ from native species in their biomass and productivity, tissue chemistry, plant morphology and phenology, and are able to alter soil nutrient dynamics. Introduced plants often increase biomass and net primary production, alter nitrogen availability and nitrogen-fixation rates, produce litter with higher decomposition rates than co-occurring natives (Ehrenfeld, 2003), and in turn may alter structure and dynamics at plant community level.

Pathways by which invasive species alter nitrogen cycles have been evaluated in some studies. Evans *et al.* (2001) found that *B. tectorum* reduced nitrogen-mineralization rates by having greater carbon–nitrogen and lignin–nitrogen ratio than native species; however, similar effects could not be explained in the case of invasive *Hieracium* in New Zealand grasslands (Scott *et al.*, 2001). Mack *et al.* (2001) suggested that the impacts of exotic grasses on nitrogen cycling in Hawaiian woodlands were through their alteration of community structure due to fire, instead of through direct invasion. Many other studies have also suggested that high variations in the nitrogen cycle, etc. can be ascribed to different mechanisms. For example, *Hyparrhenia rufa*, a perennial grass invading secondary pastures in Costa Rica, was found to have lower rates of nitrogen cycling compared with intact uncut forest. On the other hand, exotic perennial *Melinis minutiflora*, invading Hawaiian shrub lands, accelerates the rate of nitrogen cycling. Effects on total standing biomass and litter may also be variable (Dascanio *et al.*, 1994). Ehrenfeld *et al.* (2001) conducted studies on the shrub *Berberis thunbergii* and the grass *Microstegium vimineum*, which have invaded eastern USA deciduous forests. Although both species increase nitrification and pH, it was suggested that the shrub does so through the production of highly decomposable tissue, while the grass does so through low annual production.

Several reports indicate that introduced species in Australia are capable of nitrogen-fixing and can alter nutrient regimes. Being highly competitive, *Cytisus scoparius* in South Australia has the ability to change the nutrient availability in nutrient-rich soils, and this could be an important mechanism for this species to invade native woodlands (Fogarty and Facelli, 1999). Other studies on grasses have also identified the mechanisms that non-nitrogen-fixing species use to alter nutrient cycling (Mack and D'Antonio, 2003; Rossiter *et al.*, 2006). Despite substantial evidence suggesting that introduced invasive species can alter nutrient cycling, the definite mechanism by which such species influence nutrient availability at plant community level is still lacking and even sometimes confusing (Levine *et al.*, 2003). For example, the increased availability of nitrogen following the invasion of nitrogen-fixing species might be an important pathway by which invaders dominate community structure, and possibly provide the ideal conditions for the invasion of more exotic species as suggested by Davis *et al.* (2000). However, in most cases, the nitrogen-fixers reduce local plant diversity by interfering with resident species, and seem to have no effect on the abundance of other exotic plants (Mueller-Dombois and Whiteaker, 1990). The mechanism involved in replacement has not been clearly understood, but probably combines the effects of shading, competition, allelopathy and altered nutrient cycle; however, the relative contribution of these factors may vary with the characteristics of the invader as well as composition of flora at the invaded site. For example, shading seems to be a more crucial factor than altered nitrogen cycling for the impact of *Myrica* and *Acacia* on surrounding plants (Holmes and Cowling, 1997).

30.6.3 Hydrology and aquatic ecosystems

Wetland ecosystems are prone to plant invasions and introduced species can quickly affect the diversity of aquatic environments and hydrological cycles. According to an estimate, less than 6% of the Earth's land mass is wetland while 24% of the total invasive species are wetland species (Zedler and Kercher, 2004). Invasive wetland plant species can change the dynamics of evapotranspiration and run-off in addition to alteration in water composition (Levine *et al.*, 2003). For example, in Australia, introduced willow (*Salix* species) form thickets that completely displace native vegetation and spread their roots throughout the bed of a watercourse, and hence hamper water flow (NLWRA, 2008). At the same time, fallen leaves by willows contribute organic matter during the autumn, reducing water quality and available oxygen that subsequently affects the biodiversity of aquatic fauna and flora.

In general, aquatic weeds block waterways and irrigation channels, restrict light and deplete oxygen, leading to several other adverse effects of aquatic fauna (Waters and Rivers Commission, 2000). However, invasive wetland plants reduce diversity of both fauna and flora, often form monotype vegetation, alter habitat structure, affect nutrient recycling, productivity and food webs (Zedler and Kercher, 2004). *Juncus acutu*, a fast-spreading invasive sedge, is capable of altering the physical structure and hydrological conditions of a site, and subsequently restricting native animals to access waterways, harbouring pest animal species and blocking the flow of water, leading to frequent floods (Parsons and Cuthbertson, 2001; Keighery and Keighery, 2006). An extensive survey was made by Zavaleta (2000) on the impact of invasive species *Tamarix*. Results of the survey suggested that invasion by this species led to increase in evapotranspiration by 300–460 mm per year. Similarly, invasion by *Centaurea solstitialis* in annual grasslands of western North America has increased summer water use by 105–120 mm per year. In another study, Cline *et al.* (1977) noted that invasive species generally have shallower root systems. Similarly, exotic annual grasses in California have been reported to change hydrology, and such change has been ascribed to competitive displacement of the deeper-rooting native perennials (Dyer and Rice, 1999). In all the above cases, an enhanced rate of evapotranspiration is considered as a functional trait of invasive species with increased leaf area for active growth during summer.

30.6.4 Fire

Increase in frequency and intensity of fire hazard associated with the introduction of invasive species has been the focus of numerous studies (D'Antonio and Vitousek, 1992; D'Antonio, 2000; Brooks *et al.*, 2004). Such plants affect fire regimes by invading an area and substantially modifying the structure and composition of the fuel bed (Levine *et al.*, 2003). For example, the grass–fire cycle occurs when an introduced grass species invades a habitat, alters the vegetation structure and provides a substantial fuel load, resulting in an increase in fire frequency (D'Antonio and Vitousek, 1992). In addition, introduced grass species may increase combustibility of the available fuel load, hence subsequently altering fire intensity (Levine *et al.*, 2003; Grice, 2004). In the western USA, invasion of desert shrub lands by annual grasses from the Mediterranean region has resulted in an increase in fire frequency over the past century, and this has resulted in conversion of shrub lands to grasslands with an impact on biodiversity (D'Antonio and Vitousek, 1992). Similarly, in Hawaii, introduced grasses have increased fire frequency, with a sizeable impact on biodiversity (Tunison *et al.*, 2001).

Woody invaders are also responsible for fire hazards, but to a lesser extent compared to the case of introduced grasses. In such cases, frequency may be less but the intensity and duration is expected to be more. Higher temperatures and greater heat release have been recorded due to longer duration and more lengths of flame for a variety of exotic shrubs and trees invading humid grassland, savannah or fire-prone shrub land ecosystems (Lippincott, 2000). After fire, the enhanced chance of invasion by invasive herbaceous plants has been considered a serious threat in Western Australia (Lamont and Markey, 1995). In such cases, fire opens up areas of vegetation and creates a rich ash bed, allowing invasive plants with competitive advantages to rapidly establish in the absence of native vegetation. Milberg and Lamont (1995) documented the invasion of remnant sclerophyll woodland vegetation by exotic species after fire. It was inferred that the population and coverage of exotic plant species increased after fire with a concomitant decrease in the abundance of native species. At the same time, susceptibility of the vegetation to fire may also have increased due to compositional changes as suggested by Milberg and Lamont (1995).

30.6.5 Economy

Like other weeds, invasive weeds also affect the economy in two direct ways. First, losses in production of a system, and second, the huge costs involved in the management of invasive species. In addition to the direct losses, indirect economic losses can also be sizeable through loss of recreational and tourism revenues, loss of biodiversity, adverse societal impacts, including health hazards and soil and water dynamics. In 1993, the Office of Technology Assessment of the US Congress estimated that the 70 most harmful invasive species had caused damage of US$ 97 billion in the USA since 1906. In India, the cost involved in the management of *Parthenium* is huge, and this weed is still on the rampage in many areas, creating havoc. Similarly, management of *Lantana* and *Mikania* demands heavy expenditure, and remains a constraint in the control of such weeds. If we add together the monetary losses due to the extinction of species, loss in biodiversity, and loss of ecosystem services, then economic losses from impacts of invasive species would be increased many times.

30.6.6 Soil dynamics

Soil plays an important role in the establishment of introduced plant species by accelerating invasions and proliferation. The introduction of invasive exotic plant species affects the soil dynamics and ecosystem functions, due to differences between the exotic and native species such as nutrients-cycling and water-use efficiency which, in turn, may lead to big changes in ecosystem composition and functions (Yelenik *et al.*, 2007). For example, introduction of two exotic plants – *Berberis thunbergii* and *Microstegium vimineum* – in the forests of New Jersey has shown a significant increase in pH of soil, available nitrate and net potential nitrification as compared to soils under native shrubs (Kourtev *et al.*, 1999). In north and north-west India, *Prosopis juliflora* has been reported to degrade substratum in the semi-arid and arid zones (Sharma and Dakshini, 1998). Resource availability is also a decisive factor which contributes to invasion of native grassland communities by non-native plants (White *et al.*, 1997; Davis *et al.*, 2000; Kolb *et al.*, 2002). Presence and composition of soil microbes may also play a crucial role in establishment of an invasive species in the invaded region.

30.7 Management of Invasive Alien Weed Species

Effective management practices have been implicated in order to utilize all available resources necessary for the restoration of native vegetation. The weed management strategy may warrant effective prevention, control and restoration of invaded areas and depends on a wide variety of ecological factors (Zimdahl, 1999; Monaco *et al.*, 2002). Management options include the following measures.

30.7.1 Early detection and prevention

Early detection of newly entered species at international, national, provincial and local levels is considered as a basic and important step of management. Whenever a new invasive species enters, it is essential to detect and respond in a proactive manner before the situation gets out of hand. Effective and site-specific scouting around critical points of entry, protected areas, urban and agricultural ecosystems is important. Once the new entry is detected, immediate contingency quarantine measures are essential before the plant starts spreading beyond the limit of management. Immediate actions should be adopted by quarantine authorities, to prevent the spread of the entered species and protect the native biodiversity. Often, it has been emphasized that the best way to control an invasive species is prevention (DiTomaso, 2000). Invasive alien species have the reputation of being able to establish themselves in adverse conditions, to spread and dominate the native flora. By virtue of these attributes, once a species is introduced into a new area, complete eradication of such species is nearly impossible due to several limitations in the control measures to be applied, such as inapplicability, low rate of success, cost of implementation and availability of expertise at local level.

30.7.2 Control

30.7.2.1 Cultural control

Various cultural control practices have been employed in weed management for a long time, and allow the crop to become established without experiencing the negative effects of weed interference. Proper selection of cultivar, sowing time, planting method and geometry, and timing and method of fertilizer application are the factors which can be manipulated to improve the competitive ability of a crop. These are very cost-effective and can be implemented at larger scales. In non-cropped situations, over-grazing with domestic livestock prevents seed-setting of certain weed species (depending on the preference) and weakens the dominance of such weeds. For example, sheep and goats can be used to graze broad-leaved weeds, while cattle prefer grasses and can be used to manage undesirable grasses (Tu *et al.*, 2001). Prescribed or controlled burning consists of planning, setting, and managing fire (CNAP, 2000). The most successful use of fire for invasive species control results from burns that try to mimic or restore natural fire regimes, which otherwise have been disrupted for one reason or another, i.e. land-use changes, suppression practices, or developmental activities (Tu *et al.*, 2001).

30.7.2.2 Manual and mechanical control

Manual and mechanical weed control involves physical disturbance or removal of the weeds, through activities like mowing, cutting or sawing, digging, pulling or ploughing. Mechanical weed control can be done through hand weeding, using small instruments or by using power-driven tools. This strategy is very effective and relies on the prevention of seed production in annual and biennial weeds. In perennial plants, frequent physical disturbance exhausts the root reserves and weakens the overall performance of such weeds, if performed at the proper time, giving competitive advantage to native vegetation. However, for perennial plants with a vegetative mode of reproduction from root parts such as rhizomes, mechanical operations are not expected to provide complete control, yet these can be used as a tool for stressing the plants, making other treatments more effective (Derscheid *et al.*, 1961; Renz and DiTomaso, 1998). On the other hand, as mechanical control is labour- and cost-intensive, and also time-consuming, its adoption may be limited (van Wilgen *et al.*, 2001). The feasibility of mechanical operations due to growth stage, geometry of crops, and unfavourable field moisture levels are the factors which sometimes restrict the practicability of mechanical measures.

30.7.2.3 Chemical control

Herbicides have been used to control unwanted vegetation and considered as a tool of choice for weed management globally. Selective herbicides can kill grasses or broad-leaved species, leaving other plants unharmed, while non-selective herbicides target almost all types of vegetation. At present, a variety of herbicide molecules is commercially available in a range of formulations which make them applicable in different ways. Before application of herbicides, several factors such as soil type, moisture level, crop and crop rotation, composition of weed flora, weed-control, efficiency of herbicide, cost of herbicide, and possible impact on non-target organisms must be kept in mind for economical or sustainable weed management.

There are growing concerns over the excessive use and harmful impact of herbicides in terms of impacts on the environment and on non-target organisms. Although many newly introduced herbicides have been claimed to have comparatively low toxicity and fewer residual effects, issues such as detrimental environmental impact remain in question. Another limitation with the use of herbicides is the need for a relatively high level of skill and training (van Wilgen *et al.*, 2001). Herbicides have also been employed successfully to control invasive weed species. In Florida, water hyacinth was managed by use of the herbicide 2,4-D, combined with mechanical removal (Schardt, 1997). Glyphosate is another herbicide used for controlling invasive species globally; this has a high adoption rate, as it is considered to be a relatively non-toxic chemical with low residual persistence in the soil or environment. However, in the recent past, concerns have also been raised about glyphosate toxicity, leading to a ban on its use. At present, the scientific community is divided in opinion about the safety concerns associated with the use of glyphosate. In this context, authors opine that continuous use of any chemical (be it glyphosate or any other) for a long time may lead to adverse impacts on the environment and on non-target organisms. However, judicious use of herbicides can minimize most of the associated ill-effects.

30.7.2.4 Biological control

Biological control can be defined as the use of living organisms to control pest species (Waage and Greathead, 1988; Watson, 1991). Biological control aims at establishing an equilibrium between biological agent and target species and maintains the population of target species at a level of negligible harm, instead of complete elimination (Bani, 2002). Deliberate introduction of natural enemies of the target species (such as insects and pathogens), and subsequent manipulation of population with the goal of suppressing the pest population are the key factors for biological control (Wilson and Huffaker, 1976). Biological control of weeds has been in practice for many years, with substantial success recorded in many parts of the world, especially in the USA, Australia, South Africa, Canada, New Zealand and India. According to a report, more than 350 species of invertebrates and pathogens have been deliberately introduced into 75 countries for the control of at least 133 weed species (Julien and Griffiths, 1998).

Basically, biological controls explore the hypothesis that the success of many invasive alien weeds is due to the absence of natural enemies when they are introduced into a new regime (Cronk and Fuller, 1995). The introduction of enemy insects or pathogens, usually from the weeds' native habitat, check the establishment and spread of that particular weed, allowing native plants to compete on more equal terms. Biological control of weeds offers certain benefits over the other control measures. Biocontrol agents, once established, can sustain the required population build-up throughout the range of the target weed without external efforts. In addition, it can be easily practised in areas where chemical or mechanical control is not possible. Environmental and health hazards are almost negligible with biological control. Biological control of a weed is continuous, self-sustained, long-lasting and relatively cheap. On the other hand, biological control has a few limitations. Complete eradication of the weed population is not possible with this approach, because as the number of host plants declines, the population of biocontrol agents also declines. Host specificity is always a big concern for the success of biological control. Adverse climatic conditions may hinder the successful establishment of bioagents.

30.7.2.5 Restoration

After successful invasion of an ecosystem by invasive plant species, the process of the recovery becomes degraded, damaged or destroyed (SER, 2004). In such a situation, the restoration of desirable native vegetation requires assistance from outside. Assisting the establishment of desirable vegetation through replanting practices, and at the same time, management of the invasive weed species contributes to the larger goal of restoration (Jacobs *et al.*, 1998). There are always greater risks of encroachment by invasive weed species in a region where monotype vegetation exists. However, during the restoration process, establishment of a diverse community of desirable vegetation can prevent weed encroachment by utilizing all or most of the available resource niches (Sheley *et al.*, 1996). During the process of restoration of an ecosytsem, several location-specific steps may be required, which include seedbed preparation, broadcast seeding, drill seeding, container planting and sprigging live branches (Roundy, 1996). Post-planting care also may be required, at least in the initial stage, in order to ensure desirable results and successful restoration.

30.7.2.6 Integrated weed management (IWM)

Integrated management of weeds advocates adoption of more than one method, and has been considered as the most effective, eco-friendly and sustainable approach for weed management. Integration of at least two types of indirect or direct weed management strategies (i.e. preventive, cultural, mechanical, chemical and biological control measures) has been suggested to be effective in the management of weeds, without much damage to environment and biodiversity. Use of herbicides is an important component in IWM; however, doses should be low enough not to cause environmental degradation. IWM is considered as the most practical and effective approach on a long-term basis, since the combination of methods will take care of weeds in their totality and prevent build-up of a weed seed bank in the soil. It is considered an eco-friendly approach and can also address issues like weed shifts and development of herbicide-resistant weeds.

Allelopathy is a natural phenomenon in which plants produce certain secondary metabolites that affect the growth of neighbouring plants and can be explored as a major tool in the population control of invasive species. Allelopathy can play a significant role in the inhibition and suppression of growth of invasive alien species, and has been employed to control specific weeds in many parts of the world, with encouraging results. For example, in India, *Casia tora* is being used to suppress the population of *Parthenium*.

30.7.2.7 Utilization

Weeds can be used as food, feed, fodder, fibre, fuel, a source of nutrient for field crops, as a medicinal plant, as a source of important pharmaceuticals, and so on. The control of weeds by their utilization is also an effective method to keep them in check. Tessema (2012) advocated the exploitation of invasive species as a means of harnessing their economic potential for the welfare of society, and at the same time to keep their spread under control. This concept did not achieve momentum due to lack of scientific knowledge on the beneficial properties of weeds and their potential uses. However, in many parts of the world, including India, the rural population has been forced to exploit invasive species due to shortage of basic amenities. Recently, the need for diversification of food items has also drawn the attention of scientists to explore the potential of weed species having acceptable nutritional values. Competitiveness and hardiness against different biotic and abiotic factors are the two important inherent traits of invasive weeds. Due to these two characteristics, weeds can be a potential source of genetic material for trait-based crop improvement.

30.8 Success Stories of IAWS Management in India

Management of invasive weeds in India has been a big concern and always appeared to be a tough task. Despite this, there have been several successful cases on which future strategies can be built to tackle the problem of invasive weeds in India.

- Chemical management of *Ageratum conyzoides* with atrazine and alachlor at 1.5 kg/ha before flowering stage was found to be most effective in controlling this weed.
- Successful biological control was achieved in India when *Dactylopius ceylonicus* was introduced from Brazil in 1795. However, this was in the mistaken belief that it was the true carmine dye-producing insect, *D. coccus*. Later, two insects (*D. ceylonicus* and *D. opuntiae*) were found to be effective in controlling *O. vulgaris*, *O. stricta* and *O. elatior* to non-pest level, and these have been successfully employed as biological agents in India and Sri Lanka.
- Biological control of *Salvinia molesta* by the weevil *Cyrtobagous salviniae* was achieved in Kerala state. Successful establishment of weevils in several ponds, tanks, lakes and canals ensured that these waterbodies were clear of *Salvinia*.
- Water hyacinth has spread in water bodies throughout India and has been reckoned to be the most damaging aquatic weed. In 1982, three exotic insects (*Neochetina bruchi*, *N. eichhorniae* and *Orthogalumna terebrantis*) were introduced into India to control water hyacinth. Out of these, *N. bruchi* was found to be successful in the management of water hyacinth.
- The Mexican beetle *Zygogramma bicolorata* was found effective in controlling *Parthenium hysterophorus* population and is employed throughout the country.
- Competitive displacement of *Parthenium* with species such as *Cassia tora*, *Cassia occidentalis* and *Tephrosia purpurea* has shown strong allelopathic potential in several parts of the world, including India (Knox *et al.*, 2010; Shabbir and Javaid, 2010; Knox *et al.*, 2011).
- *Teleonemia scrupulosa*, a Mexican bug, has been shown to control *Lantana camara*; however, complete success is yet to be achieved.
- Safe and eco-friendly utilization of invasive weed species such as *Parthenium*, *Lantana* and water hyacinth is being encouraged at individual and community level and producing appreciable results, by using them as green manure, compost, in paper making, particle boards, baskets and so on.

30.9 Conclusion

Invasive alien weeds pose several threats which are huge in dimension and intensity, and so warrant immediate attention to devise effective strategies for management. Such threats are continuously on the rise and expected to be more severe in the near future, due to liberalized global trade practices, climate changes and other factors associated with modern agricultural systems. The consequences of invasion of new areas by alien plant species seem to be very serious, not only through loss of productivity but also in terms of adverse effects on biodiversity and ecosystem services, economic development, animal health and human welfare. Reduction in yields of production systems (crops, forests and fisheries); decrease in water availability; land degradation; blocking of transport routes; and the ability to host a number of exotic diseases are direct threats posed by invasive alien weed species. Several indirect impacts include social instability, constraints on sustainable development and economic growth, poverty alleviation and food security.

A developing country like India is more vulnerable to the invasion of alien weeds. Specific legislation to regulate the introduction of invasive species into the country is still far from its peak potential. Stringent measures need to be enforced in order to tackle the problem of intentional or unintentional entry of exotic invasive weeds. Multi-layer preparedness is required in terms of early detection facilities, establishment of a repository of invasive weed species, precise weed risk analysis, and a contingent action plan for management. Quarantine measures must also be adopted, not only at international borders but also at regional level in order to prevent the spread of any accidental introduction. Like other developed countries, a near-zero-tolerance policy must be adopted to prevent further invasions by weed species.

References

Alpert, P., Bone, E. and Holzapfel, C. (2000) Invasiveness, invasibility and the role of environmental stress in preventing the spread of non-native plants. *Perspectives in Plant Ecology and Evolution Systems* 3, 52–66.

Bakker, J. and Wilson, S.D. (2001) Using ecological restoration to constrain biological invasion. *Journal of Applied Ecology* 41(6), 1058–1064.

Bani, G. (2002) Status and management of *Chromolaena odorata* in Congo. In: Zachariades, C., Muniappan, R. and Strathie, L.W. (eds) *Proceedings of the Fifth International Workshop on Biological Control and Management of Chromolaena odorata, Durban, South Africa, October 2000, 34 Pietermaritzburg*. ARC-Plant Protection Research Institute, Pretoria, South Africa, pp. 34–39.

Belnap, J. and Phillips, S.L. (2001) Soil biota in an ungrazed grassland: response to annual grass (*Bromus tectorum*) invasion. *Ecological Applications* 11, 1261–1275.

Bhagwat, S.A., Breman, E., Thekaekara, T., Thornton, T.F. and Willis, K.J. (2012) A battle lost? Report on two centuries of invasion and management of *Lantana camara* L. in Australia, India and South Africa. *PLoS One* 7, e32407.

Bhatt, Y.D., Rawat, Y.S. and Singh, S.P. (1994) Changes in ecosystem functioning after replacement of forest by Lantana shrubland in Kumaun Himalaya. *Journal of Vegetable Science* 5, 67–70.

Booth, B.D., Murphy, S.P. and Swanton, C.J. (2003) *Weed Ecology in Natural and Agricultural Systems*. CAB International, Wallingford, Oxfordshire, UK.

Braithwaite, R.W., Lonsdale, W.M. and Estbergs, J.A. (1989) Alien vegetation and native biota in tropical Australia: the impact of *Mimosa pigra*. *Biological Conservation* 48, 189–210.

Brewer, J.S. and Cralle, S.P. (2003) Phosphorus addition reduces invasion of a longleaf pine savanna (Southeastern USA) by a non-indigenous grass (*Imperata cylindrica*). *Plant Ecology* 167, 237–245.

Brooks, M.L., D'Antonio, C.M., Richardson, D.M., Grace, J.B., Keeleu, J.E., DiTomaso, J.M., Hobbs, R.J., Pellant, M. and Pyke, D. (2004) Effects of invasive alien plants on fire regimes. *BioScience* 54, 677–688.

Burkart, A. (1976) A monograph of the genus *Prosopis* (Leguminosae subfam. Mimosoideae) (Parts 1 and 2). Catalogue of the recognized species of *Prosopis*. *Journal of the Arnold Arboretum* 57, 219–249; 450–525.

Burns, J.H. (2004) A comparison of invasive and non-invasive dayflowers (Commelinaceae) across experimental nutrient and water gradients. *Diversity and Distribution* 10, 387–397.

Burns, J.H. (2006) Relatedness and environment affect traits associated with invasive and noninvasive introduced Commelinaceae. *Ecological Applications* 16, 1367–1376.

Callaway, R.M. and Aschehoug, E.T. (2000) Invasive plants versus their new and old neighbors: a mechanism for exotic invasion. *Science* 290, 521–523.

Carlton, J.T. (2003) Community assemblage and historical biogeography in the North Atlantic Ocean: the potential role of human-mediated dispersal vectors. *Hydrobiologia* 503, 1–8.

Carlton, J.T. and Ruiz, G.M. (2005) Vector science and integrated vector management in bioinvasion ecology: conceptual frameworks. In: Mooney, H.A., Mack, R.N., McNeely, J.A., Neville, L.E., Schei, P.J. and Waage, J.K. (eds) *Invasive Alien Species: A New Synthesis*. CAB Direct, Wallingford, Oxfordshire, UK, pp. 36–58.

Castillo, J.M., Fernandez-Baco, L., Castellanos, E.M., Luque, C.J., Figueroa, M.E. and Davy, A.J. (2000) Lower limits of *Spartina densiflora* and *S. maritima* in a Mediterranean salt marsh determined by different ecophysiological tolerances. *Journal of Ecology* 88, 801–812.

Catford, J.A., Jansson, R. and Nilsson, C. (2009) Reducing redundancy in invasion ecology by integrating hypotheses into a single theoretical framework. *Diversity and Distribution* 15, 22–40.

Chakraborty, S., Tiedemann, A.V. and Teng, P.S. (2000) Climate change: potential impact on plant diseases. *Environmental Pollution* 108, 317–326.

Chatterjee, S., Chakraborty, M. and Saini, D.C. (2009) Preliminary investigation on invasive alien species of Uttar Pradesh. In: *Proceedings: National Conference on Invasive Alien Species: A Threat to Native Biodiversity*. Uttar Pradesh State Biodiversity Board, Uttar Pradesh, India, pp. 21–27.

Clavero, M. and García-Berthou, E. (2005) Invasive species are a leading cause of animal extinctions. *Trends in Ecology and Evolution* 20, 110.

Cline, J.F., Uresk, D.W. and Rickard, W.H. (1977) Comparison of water used by a sagebrush–bunchgrass community and a cheatgrass community. *Journal of Range Management* 30, 199–201.

CNAP (2000) *Creating an Integrated Weed Management Plan: A Handbook for Owners and Land Mangers of*

Lands with Natural Values. Colorado Natural Areas Program, Colorado Department of Agriculture. Available at: www.blm.gov/style/medialib/blm/wy/programs/invasiveplants/docs.Par.42434.File.dat/IWMhandbook.pdf (accessed 31 May 2016).

Cronk, Q.C.B. and Fuller, J.L. (1995) *Plant Invaders*. Chapman and Hall, New York, USA.

D'Antonio, C.M. (2000) Fire, plant invasions, and global changes. In: Mooney, H.A. and Hobbs, R.J. (eds) *Invasive Species in a Changing World*. Island Press, Washington, DC, USA, pp. 65–93.

D'Antonio, C.M. and Vitousek, P.M. (1992) Biological invasions by exotic grasses, the grass/fire cycle and global change. *Annual Reviews in Ecological Systems* 23, 63–87.

D'Antonio, C.M., Dudley, T.L. and Mack, R.M. (2001) Disturbance and biological invasions: direct effects and feedbacks. In: Walker, L.R. (ed.) *Ecosystems of Disturbed Ground. Ecosystems of the World*, Vol. 16. Elsevier Science, New York, USA, pp. 429–468.

Dascanio, L.M., Barrera, M.D. and Frangi, J.L. (1994) Biomass structure and dry matter dynamics of subtropical alluvial and exotic Ligustrum forests at the Rio de la Plata, Argentina. *Vegetation* 115, 61–76.

David, G.L., Dennis, W.F., Anne, I.F., Brianna, M. and Jay, O. (2005) The role of tree-fall gaps in the invasion of exotic plants in forests: the case of Wineberry, *Rubus phoenicolasius*, in Maryland. In: Gottschalk Kurt, W. (ed.) *Proceedings, 16th U.S. Department of Agriculture Interagency Research Forum on Gypsy Moth and Other Invasive Species*, Annapolis, Maryland, 18–21 January 2005. Gen. Tech. Rep NE-337. Northeastern Research Station, Forest Service, US Department of Agriculture, Newtown Square, Pennsylvania, USA.

Davis, M.A., Grime, P. and Thompson, K. (2000) Fluctuating resources in plant communities: a general theory of invasibility. *Journal of Ecology* 88, 528–534.

Dawson, K., Veblen, K.E. and Young, T.P. (2007) Experimental evidence for an alkali ecotype of *Lolium multiflorum*, an exotic invasive annual grass in the central valley, CA, USA. *Biological Invasions* 9, 327–334.

Dayal, V. (2007) Social diversity and ecological complexity: how an invasive tree could affect diverse agents in the land of the tiger. *Environment and Development Economics* 12(4), 553–571.

Derscheid, L.A., Wallace, K.E. and Nash, R.L. (1961) Russian knapweed control with cultivation, cropping and chemicals. *Weeds* 8, 268–278.

Dhileepan, K. and Strathie, L. (2009) *Parthenium hysterophorus* L. (Asteraceae). In: Muniappan, R., Reddy, G.V.P. and Raman, A. (eds) *Biological Control of Tropical Weeds Using Arthropods*. Cambridge University Press, Cambridge, UK, pp. 274–318.

Dickman, C.R. (2006) Species interactions: direct effects. In: Attiwell, P. and Wilson, B. (eds) *Ecology: An Australian Perspective*. Oxford University Press, South Melbourne, Australia, pp. 285–303.

Ding, J., Reardon, R., Wu, Y., Zheng, H. and Fu, W. (2007) Biological control of invasive plants through collaboration between China and the United States of America: a perspective. *Biological Invasions* 8, 1439–1450.

DiTomaso, J. (2000) Invasive weeds in Rangelands: species impacts and management. *Weed Science* 48, 255–265.

Dogra, K.S, Kohli, R.K. and Sood, S.K. (2009) An assessment and impact of three invasive species in the Shivalik hills of Himachal Pradesh, India. *International Journal of Biodiversity and Conservation* 1(1), 004–010.

Dukes, J.S. (2000) Will the increasing atmospheric CO_2 concentration affect the success of invasive species? In: Mooney, H.A. and Hobbs, R.J. (eds) *Invasive Species in a Changing World*. Island Press, Washington, DC, USA.

Dukes, J.S. and Mooney, H.A. (2004) Disruption of ecosystem processes in Western North America by invasive species. *Revista Chilena de Historia Natural* 77, 411–437.

Dyer, A.R. and Rice, K.J. (1999) Effects of competition on resource availability and growth of a native bunchgrass in two California grasslands. *Ecology* 80, 2697–2710.

Ehrenfeld, J.G. (2003) Effects of exotic plant invasions on soil nutrient cycling processes. *Ecosystems* 6, 503–523.

Ehrenfeld, J.G., Kourtev, P. and Huang, W. (2001) Changes in soil functions following invasions of exotic understory plants in deciduous forests. *Ecological Applications* 11, 1287–1300.

Enserink, M. (1999) Biological invaders sweep in. *Science* 285, 1834–1836.

Ervin, J. (2003) Rapid assessment of protected area management effectiveness in four countries. *BioScience* 53, 833–841.

Evans, H.C. (1997) *Parthenium hysterophorus*: a review of its weed status and the possibilities for biological control. *Biocontrol News Information* 18, 89–98.

Evans, R.D., Rimer, R., Sperry, L. and Belnap, J. (2001) Exotic plant invasion alters nitrogen dynamics in arid grassland. *Ecological Applications* 11, 1301–1310.

FAO (2008a) *Climate Change and Food Security: A Framework Document*. Food and Agriculture Organization, Rome, Italy.

FAO (2008b) *Climate Change Adaptation and Mitigation in the Food and Agriculture Sector: Technical Background Document (HLC/08/BAK/1)*. Food and Agriculture Organization, Rome, Italy.

Fisher, M.J., Rao, I.M., Ayarza, M.A., Lascano, C.E., Sainz, J.I., Thomas, R.J. and Vera, R.R. (1995) Scientific correspondence, reply. *Nature* 376, 473.

Fogarty, G. and Facelli, J.M. (1999) Growth and competition of *Cytisus scoparius*, an invasive shrub, and Australian native shrubs. *Plant Ecology* 144(1), 27–35.

Frew, M., Solomon, K. and Mashilla, D. (1996) Prevalence and distribution of *Parthenium hysterophorus* L. in Eastern Ethiopia. *Arem* 1, 19–26.

Funk, J.L. and Vitousek, P.M. (2007) Resource-use efficiency and plant invasion in low-resource systems. *Nature* 446, 1079–1081.
Gay, C., Estrada, F., Conde, C., Eakin, H. and Villers, I. (2006) Potential impacts of climate change on agriculture: a case of study of coffee production in Veracruz, Mexico. *Climate Change* 79, 259–288.
GISP (2004) *Africa Invaded: The Growing Danger of Invasive Alien Species*. Global Invasive Species Programme, Cape Town, South Africa. Available at: www.gisp.org/downloadpubs/gisp%20africa%202.pdf (accessed 31 May 2016).
Gordon, D.R. (1998) Effects of invasive, non-indigenous plant species on ecosystem processes: lessons from Florida. *Ecological Applications* 8, 975–989.
Grant, D.W., Peters, D.P.C., Beck, G.K. and Fraleigh, H.D. (2003) Influence of an exotic species, *Acroptilon repens* (L.) DC. on seedling emergence and growth of native grasses. *Plant Ecology* 166, 157–166.
Grice, A.C. (2004) Weeds and the monitoring of biodiversity in Australian rangelands. *Austral Ecology* 29, 51–58.
Grice, A.C., Radford, I.J. and Abbott, B.N. (2000) Regional and landscape-scale patterns of shrub invasion in tropical savannas. *Biological Invasions* 2, 187–205.
Grotkopp, E., Rejmánek, M. and Rost, T.L. (2002) Toward a causal explanation of plant invasiveness: seedling growth and lifehistory strategies of 29 pine (*Pinus*) species. *American Nature* 159, 396–419.
Gurevitch, J., Fox, G.A., Wardle, G.M., Inderjit and Taub, D. (2011) Emergent insights from the synthesis of conceptual frameworks for biological invasion. *Ecological Letters* 14, 407–418.
Hajra, P.K. and Mudgal, V. (1997) *Plant Diversity Hotspots in India: An Overview*. Botanical Survey of India, Calcutta, India.
Harris, M.R. and Facelli, J.M. (2003) Competition and resource availability in an annual plant community dominated by an invasive species, *Carrichtera annua* (L. Aschers.), in South Australia. *Plant Ecology* 167, 19–29.
Hellmann, J.J., Byers, J.E., Bierwagen, B.G. and Dukes, J.S. (2008) Five potential consequences of climate change for invasive species. *Conservation Biology* 22(3), 534–543.
Holm, L., Doll, J., Holm, E., Pancho, J. and Herberger, J. (1997) *World Weeds. Natural Histories and Distribution*. John Wiley & Sons, New York, USA.
Holmes, P.M. and Cowling, R.M. (1997) The effects of invasion by *Acacia saligna* on the guild structure and regeneration capabilities of South African fynbos. *Journal of Applied Ecology* 34, 317–332.
Holway, D.A., Lach, L., Tsutsui, N.D. and Case, T.J. (2002) The causes and consequences of ant invasions. *Annual Reviews on Ecological Systems* 33, 181–233.
Hooker, J.D. (1904) *A Sketch of the Flora of British India*. Eyre and Spottiswoode, London, UK.
Hughes, J.D. (2003) Europe as consumer of exotic biodiversity: Greek and Roman times. *Landscape Research* 28, 21–31.
Hulme, P.E. (2003) Biological invasions: winning the science battles but losing the conservation war? *Oryx* 37, 178–193.
Hulme, P.E. (2006) Beyond control: wider implications for the management of biological invasions. *Journal of Applied Ecology* 43, 835–847.
Hulme, P.E. (2008) Phenotypic plasticity and plant invasions: is it all Jack? *Functional Ecology* 22, 3–7.
IRIN (2010) *Swaziland: Aliens are Tough Adversaries*. UN Integrated Regional Information Networks. Available at: allafrica.com/stories/201006081095.html (accessed 22 April 2016).
Jacobs, J.S., Carpinelli, M.F. and Sheley, R.L. (1998) Revegetating weed-infested Rangeland: what we've learned. *Rangelands* 20(6), 10–15.
Jahodová, S., Trybush, S., Pysek, P., Wade, M. and Karp, A. (2007) Invasive species of Heracleum in Europe: an insight into genetic relationships and invasion history. *Diversity and Distribution* 13, 99–114.
Jeschke, J.M., Aparicir, L.G. and Haider, S. (2012) Support for major hypotheses in invasion biology is uneven and declining. *Neobiota* 14, 1–20.
Jongejans, E., Skarpaas, O., Tipping, P.W. and Shea, K. (2007) Establishment and spread of founding populations of an invasive thistle: the role of competition and seed limitation. *Biological Invasions* 9, 317–325.
Julien, M.H. and Griffiths, M.W. (1998) *Biological Control of Weeds: A World Catalogue of Agents and Their Target Weeds*, 4th edn. CAB International, Wallingford, UK.
Kaur, R., Gonzáles, W.L., Llambi, L.D., Soriano, P.J., Callaway, R.M., Rout, M.E., Gallaher, T.J. and Inderjit, T.Z. (2012) Community impacts of *Prosopis juliflora* invasion: biogeographic and congeneric comparisons. *PLoS One* 7(9), e44966.
Keane, R.M. and Crawley, M.J. (2002) Exotic plant invasions and the enemy release hypothesis. *Trends in Ecology and Evolution* 17, 164–170.
Keighery, G.J. and Keighery, B.J. (2006) Current status and potential spread of *Juncus acutus* in Western Australia. In: Longman, V. (ed.) *Managing Sharp Rush (Juncus acutus)*. Department of Environment and Conservation, Wollaston College Conference Centre, Mt Claremont, Perth, Australia.
Khuroo, A.A., Reshi, Z.A., Malik, A.H., Weber, E., Rashid, I. and Dar, G.H. (2012) Alien flora of India: taxonomic composition, invasion status and biogeographic affiliations. *Biological Invasions* 14, 99–113.
Knox, J., Jaggi, D. and Paul, M.S. (2010) Allelopathic effect of selected weeds on biochemical activity of *Parthenium hysterophorus*. *Current Research Journal of Biological Science* 2(4), 238–240.
Knox, J., Jaggi, D. and Paul, M.S. (2011) Population dynamics of *Parthenium hysterophorus* and its biological

suppression through *Cassia occidentalis*. *Turkish Journal of Botany* 35, 111–119.

Kolb, A., Alpert, P., Enters, D. and Holzapfel, C. (2002) Patterns of invasion within a grassland community. *Journal of Ecology* 90, 871–881.

Kour R., Kaur B., Bhatia S. and Sharma K.K. (2014) Documentation of aquatic invasive alien flora of Jammu region, Jammu & Kashmir. *International Journal of Interdisciplinary and Multidisciplinary Studies* 1(7), 90–96.

Kourtev, P.S., Huang, W.Z. and Ehrenfeld, J.G. (1999) Differences in earthworm densities and nitrogen dynamics in soils under exotic and native plant species. *Biological Invasions* 1, 237–245.

Kwiatkowska, A.J., Spalik, K., Michalak, E., Palinska, A. and Panufnik, D. (1997) Influence of the size and density of *Carpinus betulus* on the spatial distribution and rate of deletion of forest-floor species in thermophilous oak forest. *Plant Ecology* 129, 1–10.

Lamont, B.B. and Markey, A. (1995) Biogeography of fire-killed and resprouting banksias in southwestern Australia. *Australian Journal of Botany* 43, 283–303.

Lavergne, C., Rameau, J. and Figier, J. (1999) The invasive woody weed *Ligustrum robustum* subsp. walkeri threatens native forest on La Re'union. *Biological Invasions* 1, 377–392.

Leishman, M.R., Haslehurst, T., Ares, A. and Baruch, Z. (2007) Leaf trait relationships of native and invasive plants: community- and global-scale comparisons. *New Phytology* 176, 635–643.

Levine, J.M., Vila, M., D'Antonio, C.M., Dukes, J.S., Grigulis, K. and Lavorel, S. (2003) Mechanisms underlying the impacts of exotic plant invasions. *Proceedings of Research Society of London B* 270, 775–781.

Lippincott, C.L. (2000) Effects of *Imperata cylindrica* (L.) Beauv. (Cogongrass) invasion on fire regime in Florida sandhill (USA). *Natural Areas Journal* 20, 140–149.

Lonsdale, M. (1999) Global patterns of plant invasions and the concept of invasibility. *Ecology* 80, 1522–1536.

Low, T. and Booth, C. (2007) *The Weedy Truth about Biofuels*. Invasive Species Council, Melbourne, Australia.

Mack, M.C. and D'Antonio, C.M. (1998) Impacts of biological invasions on disturbance regimes. *Tree* 13, 195–198.

Mack, M.C. and D'Antonio, C.M. (2003) Exotic grasses alter controls over soil nitrogen dynamics in a Hawaiian woodland. *Ecological Applications* 13, 154–166.

Mack, R.N. and Lonsdale, W.M. (2001) Humans as global plant dispersers: getting more than we bargained for. *BioScience* 51, 95–102.

Mack, M.C., D'Antonio, C.M. and Ley, R. (2001) Alteration of ecosystem nitrogen dynamics by exotic plants: a case study of C4 grasses in Hawaii. *Ecological Applications* 11, 1323–1335.

Martin, P.H. (1999) Norway maple (*Acer platanoides*) invasion of a natural forest stand: understory consequence and regeneration pattern. *Biological Invasions* 1, 215–222.

McFadyen, R.E. (1992) Biological control against parthenium weed in Australia. *Crop Protection* 24, 400–407.

Milberg, P. and Lamont, B.B. (1995) Fire enhances weed invasion of roadside vegetation in southwestern Australia. *Biological Conservation* 73, 45–49.

Monaco, T.J., Weller, S.C. and Ashton, F.M. (2002) *Weed Science: Principles and Practices*, 4th edn. Wiley, New York, USA. 671 pp.

Monaco, T.A., Johnson, D.A. and Creech, J.E. (2005) Morphological and physiological responses of the invasive weed *Isatis tinctoria* to contrasting light, soil-nitrogen and water. *Weed Research* 45, 460–466.

Mueller-Dombois, D. and Whiteaker, L.D. (1990) Plants associated with *Myrica faya* and two other pioneer trees on a recent volcanic surface in Hawaii Volcanoes National Park. *Phytocoenologia* 19, 29–41.

National Focal Point for APFISN (2005) *India, Stocktaking of National Forest Invasive Species Activities (India Country Report 101005)*, Ministry of Environment and Forests, Delhi, India.

Navie, S.C., McFadyen, R.C., Panetta, F.D. and Adkins, S.W. (1996) The biology of Australian weed *Parthenium hysterophorus* L. *Plant Protection Q* 11, 76–88.

Nayar, M.P. (1977) Changing patterns of the Indian Flora. *Bulletin of Botanical Survey of India* 19, 145–155.

Nelson, G.C., Rosegrant, M.W., Koo, J., Robertson, R., Sulser, T., Zhu, T., Ringler, C., Msangi, S., Palazzo, A., Batka, M., Magalhaes, M., Valmonte-Santos, R., Ewing, M. and Lee, D. (2009) *Climate Change: Impact on Agriculture and Costs of Adaptation*. International Food Policy, Research Institute, Washington, DC, USA. DOI: 10.2499/0896295354.

Niklas, K.J. (2008) Functional adaptation and phenotypic plasticity at the cellular and whole plant level. *Journal of Bioscience* 33, 1–8.

NLWRA (2008) *The Distribution of Some Significant Invasive Plants in Australia* 2007. National Land and Water Resources Audit, Canberra, Australia.

Nuzzo, V.A. (1999) Invasion pattern of the herb garlic mustard (*Alliaria petiolata*) in high quality forests. *Biological Invasions* 1, 169–179.

Oerke, E.C., Dehne, D.W., Schonbeck, F. and Weber, A. (1994) *Crop Production and Crop Protection: Estimated Losses in Major Food and Cash Crops*. Elsevier, Amsterdam, the Netherlands.

Pandey, H.N. and Dubby, S.K. (1989) Growth and population of an exotic weed *Parthenium hysterophorus* Linn. *Plant Science* 99, 51–58.

Panetta, F.D. and Scanlan, J.D. (1995) Human involvement in the spread of noxious weeds. *Plant Protection Q* 10, 69–74.

Parmesan, C. (2006) Ecological and evolutionary responses to recent climate change. *Annual Reviews in Ecology Evolution Systems* 37, 637–669.

Parsons, W. and Cuthbertson, E. (2001) *Noxious Weeds of Australia*, 2nd edn. Inkata Press, Melbourne, Australia.

Pasiecznik, N.M., Felker, P., Harris, P.J.C., Harsh, L.N., Cruz, G., Tiwari, J.C., Cadoret, K. and Maldonado, L.J. (2001) *The Prosopis juliflora–Prosopis pallida Complex: A Monograph*. HDRA, Coventry, UK.

Pimentel, D., Lach, L., Zuniga, R. and Morrison, D. (2000) Environmental and economic costs of nonindigenous species in the United States. *BioScience* 50, 53–65.

Pimentel, D., McNair, S., Janecka, J., Wightman, J., Simmonds, C., O'Connell, C., Wong, E., Russel, L., Zern, J., Aquino, T. and Tsomondo, T. (2001) Economic and environmental threats of alien plant, animal, and microbe invasions. *Agriculture, Ecosystems & Environment* 84, 1–20.

Raghubanshi, A.S., Rai, L.C., Gaur, J.P. and Singh, J.S. (2005) Invasive alien species and biodiversity in India. *Current Science* 88(4), 539–540.

Randerson, J. (2003) Fungus in your tea sir? *New Science* 2401, 10.

Rangi, D. (2009) *Invasive Species and Poverty: The Missing Link. Environment Matters*. The World Bank, Washington, DC, USA, pp. 12–13.

Reddy, C.S. (2008) *Catalogue of Invasive Alien Flora of India*. Reddy, Catalogue of invasive alien flora of India. Available at: www.lifesciencesite.com/lsj/life0502/16_life0502_84_89_Catalogue.pdf (accessed 31 May 2016).

Redman, D.E. (1995) Distribution and habitat types for Nepal Microstegium [*Microstegium vimineum* (Trin.) Camus] in Maryland and the District of Columbia. *Castanea* 60, 270–275.

Reichard, S.H. and Hamilton, C.W. (1997) Predicting invasions of woody plants introduced into North America. *Conservation Biology* 11, 193–203.

Rejmánek, M. (1989) Invasibility of plant communities. In: Drake, J., Mooney, H.A., di Castri, F., Groves, R.H., Kruger, F.J., Rejmánek, M. and Williamson, M. (eds) *Biological Invasions: A Global Perspective*. Wiley, Chichester, UK, pp. 369–388.

Renz, M.J. and DiTomaso, J.M. (1998) The effectiveness of mowing and herbicides to control perennial pepperweed in Rangeland and roadside habitats. *Proceedings of Western Society of Weed Science* 129.

Richardson, D.M. (1998) Forestry trees as invasive aliens. *Conservation Biology* 12, 18–26.

Richardson, D.M. and Pysek, P. (2006) Plant invasions: merging the concepts of species invasiveness and community invasibility. *Progress in Physical Geology* 30, 409–431.

Richardson, D.M., Bond, W.J., Dean, W.R.J., Higgins, S.I., Midgley, G.F., Milton, S.J., Powrie, L.W., Rutherford, M.C., Samways, M.J. and Schulze, R.E. (2000) Invasive alien species and global change: a South African perspective. In: Mooney, H.A. and Hobbs, R.J. (eds) *Invasive Species in a Changing World*. Island Press, Washington, DC, USA.

Robbins, P. (2001) Tracking invasive land covers in India, or why our landscapes have never been modern. *Annual Association of American Geography* 91, 637–659.

Rojas, I., Becerra, P., Galvez, N., Laker, J., Bonacic, C. and Hester, A. (2011) Relationship between fragmentation, degradation and native and exotic species richness in an Andean temperate forest of Chile. *Gayana Botanica* 68, 163–175.

Rossiter, N., Setterfield, S., Douglas, M., Hutley, L. and Cook, G. (2006) The impact of exotic grass invasions on nitrogen cycling: a mini review. In: Preston, C., Watts, J.H. and Crossman, N.D. (eds) *15th Australian Weeds Conference Proceedings: Managing Weeds in a Changing Climate*, Adelaide, Australia, pp. 815–819.

Roundy, B.A. (1996) Revegetation of Rangeland for wildlife. In: Krausman, P.R. (ed.) *Rangeland Wildlife*. Society for Range Management, Denver, Colorado, USA, pp. 355–368.

Roux, L.J.J., Wieczorek, A.M., Ramadan, M.M. and Tran, C.T. (2006) Resolving the native provenance of invasive fireweed (*Senecio madagascariensis* Poir.) in the Hawaiian Islands as inferred from phylogenetic analysis. *Diversity and Distributions* 12, 694–702.

Sakai, A.K., Allendorf, F.W., Holt, J.S., Lodge, D.M., Molofsky, J., With, K.A., Baughman, S., Cabin, R.J., Cohen, J.E., Ellstrand, N.C., McCauley, D.E., O'Neil, P., Parker, I.M., Thompson, J.N. and Weller, S.G. (2001) The population biology of invasive species. *Annual Review of Ecology and Systematics* 32, 305–332.

Sankaran, K.V. (2002) Black wattle problem emerges in Indian forests. *Biocontrol News & Information* 23(1). Available at: cabweb.org/Journals/BNI/Bni23-1/Gennews.htm (accessed 31 May 2016).

SCBD (2006) *Global Biodiversity Outlook 2*. Secretariat of the Convention on Biological Diversity, Montreal, Canada. Available at: www.cbd.int/doc/gbo/gbo2/cbd-gbo2-en.pdf (accessed 31 May 2016).

Schardt, J.D. (1997) Maintenance control. In: Simberloff, D., Schmitz, D.C. and Brown, T.C. (eds) *Strangers in Paradise. Impact and Management of Non-indigenous Species in Florida*. Island Press, Washington, DC, USA, pp. 229–243.

Schussman, H., Geiger, E., Mau-Crimmins, T. and Ward, J. (2006) Spread and current potential distribution of an alien grass, *Eragrostis lehmanniana* Nees, in the southwestern USA: comparing historical data and ecological niche models. *Diversity and Distributions* 12(5), 582–592.

Scott, N.A., Saggar, S. and McIntosh, P. (2001) Biogeochemical impact of *Hieracium* invasion in New Zealand's grazed tussock grasslands: sustainability implications. *Ecological Applications* 11, 1311–1322.

SER (2004) *SER International Primer on Ecological Restoration*. Society for Ecological Restoration, Science & Policy Working Group. Available at: www.ser.org/resources/resources-detail-view/ser-international-primer-on-ecological-restoration (accessed 31 May 2016).

Shaanker, U.R., Ganeshaiah, K.N., Krishnan, S., Ramya, R., Meera, C., Aravind, N.A., Kumar, A., Rao, S., Vanaraj, G., Ramachandra, J., Gathier, R., Ghazoul, J., Poole, N. and Chinnapa Reddy, B.V. (2009) Livelihood gains and ecological costs of NTFP dependence: assessing the roles of dependence, ecological knowledge and market structure in three contrasting human and ecological setting in south India. *Environmental Conservation* 31, 242–253.

Shabbir, A. and Javaid, A. (2010) Phytosociological survey and allelopathic effects of *Parthenium* weed in comparison to other weeds in Pakistan. *Indian Journal of Agricultural Research* 44(2), 119–124.

Shadel, W.P. and Molofsky, J. (2002) Habitat and population effects on the germination and early survival of the invasive weed, *Lythrum salicaria* L. (purple loosestrife). *Biological Invasions* 4, 413–423.

Sharma, R. and Dakshini, K.M.M. (1996) Ecological implications of seed characteristics of the native *Prosopis cineraria* and the alien *P. juliflora*. *Vegetation* 124, 101–105.

Sharma, R. and Dakshini, K.M.M. (1998) Integration of plant and soil characteristics and the ecological success of two *Prosopis* species. *Plant Ecology* 139, 63–69.

Sheley, R.L., Svejcar, T.J. and Maxwell, B.D. (1996) A theoretical framework for developing successional weed management strategies on Rangeland. *Weed Technology* 10, 766–773.

Shimono, Y. and Konuma, A. (2008) Effects of human-mediated processes on weed species composition in internationally traded grain commodities. *Weed Research* 58, 10–18.

Silvertown, J. (2008) The evolutionary maintenance of sexual reproduction: evidence from the ecological distribution of asexual reproduction in clonal plants. *International Journal of Plant Science* 169, 157–168.

Singh, K.P., Shukla, A.N. and Singh, J.S. (2010) State level inventory of invasive alien plants, their source regions and use potential. *Current Science* 99(1), 107–114.

Standish, R.J., Williams, P.A., Robertson, A.W., Scott, N.A. and Hedderley, D.I. (2004) Invasion by a perennial herb increases decomposition rate and alters nutrient availability in warm temperate lowland forest remnants. *Biological Invasions* 6, 71–81.

Stohlgren, T.J., Bull, K.A. and Otsuki, Y. (1998) Comparison of rangeland vegetation sampling techniques in the central grasslands. *Journal of Range Management* 51, 164–172.

Stokes, V.L., Banks, P.B., Pech, R.P. and Spratt, D.M. (2009) Competition in an invaded rodent community reveals black rats as a threat to native bush rats in littoral rainforest of south-eastern Australia. *Journal of Applied Ecology* 46, 1239–1247.

Sundaram, B. and Hiremath, A.J. (2012) *Lantana camara* invasion in a heterogeneous landscape: patterns of spread and correlation with changes in native vegetation. *Biological Invasions* 14, 1127–1141.

Sutherst, R.W. (2000) Climate change and invasive species: A conceptual framework. In: Mooney, H.A. and Hobbs, R.J. (eds) *Invasive Species in a Changing World*. Island Press, Washington, DC, USA.

Tessema, Y.A. (2012) Ecological and economic dimensions of the paradoxical invasive species – *Prosopis juliflora* and policy challenges in Ethiopia. *Journal of Economics and Sustainable Development* 3(8), 62–70.

Thakur, M.L., Ahmad, M. and Thakur, R.K. (1992) Lantana weed (*Lantana camara* var. *aculeata* Linn.) and its possible management through natural insect pests in India. *Indian Forester* 118, 466–488.

Tiwari, J.W.K. (1999) Exotic weed *Prosopis juliflora* in Gujarat and Rajasthan, India: boon or bane? *Tigerpaper* 26, 21–25.

Tu, M., Hurd, C. and Randall, J.M. (2001) *Weed Control Methods Handbook: Tools and Techniques for Use in Natural Areas*. The Nature Conservancy. Available at: www.invasive.org/gist/handbook.html (accessed 31 May 2016).

Tunison, J.T., D'Antonio, C.M. and Loh, R. (2001) Fire, grass invasions and revegetation of burned areas in Hawaii Volcanoes National Park. In: Galley, K.E. and Wilson, T.P. (eds) *Proceedings of the Invasive Species Workshop: The Role of* Fire *in the Controls and Spread of Invasive Species*. Tall Timbers Research Station Publication No. 11. Allen Press, Lawrence, Kansas, USA, pp. 122–131.

UNEP (2004) Invasive aliens threaten biodiversity and increase vulnerability in Africa. *Call to Action* 1(1). United Nations Environment Programme, Nairobi, Kenya.

Usher, M.B., Kruger, F.J., Macdonald, I.A.W., Loope, L.L. and Brockie, R.E. (1988) The ecology of biological invasions into nature reserves: an introduction. *Biological Conservation* 44, 1–8.

Van der Putten, W.H., Klironomos, J.N. and Wardle, D.A. (2007) Microbial ecology of biological invasions. *ISME Journal* 1, 28–37.

Van Wilgen, B., Richardson, D. and Higgins, S. (2001) Integrated control of invasive alien plants in terrestrial ecosystems. *Land Use Water Resource Research* 1(5), 1–6.

Vermeij, G. (1996) An agenda for invasion biology. *Biological Conservation* 78, 3–9.

Vié, J.C., Hilton-Taylor, C. and Stuart, S.N. (2009) *Wildlife in a Changing World – An Analysis of the 2008 IUCN Red List of Threatened Species Gland*. International Union for Conservation of Nature, Gland, Switzerland.

Vitousek, P.M., D'Antonio, C.M., Loope, L.L., Rejmánek, M. and Westbrooks, R. (1997) Introduced species: a significant component of human-caused global change. *New Zealand Journal of Ecology* 21, 1–16.

Waage, J.K. and Greathead, D.J. (1988) Biological control: challenges and opportunities. *Philosophical Transactions of the Royal Society B* 318, 111–128.

Walker, L.R. and Vitousek, P.M. (1991) An invader alters germination and growth of a native dominant tree in Hawaii. *Ecology* 72, 1449–1455.

Walther, G.R., Roques, A., Hulme, P.E., Sykes, M.T., Pysûk, P., Kühn, I., Zobel, M., Bacher, S., Botta-Dukát, Z., Bugmann, H., Czúcz, B., Dauber, J., Hickler, T., Jaro,V., Kenis, M., Klotz, S., Minchin, D., Moora, M., Nentwig, W., Ott, J., Panov, V.E., Reineking, B., Robinet, C., Semenchenko, V., Solarz, W., Thuiller, W., Vilà, M., Vohland, K. and Settele, J. (2009) Alien species in a warmer world: risks and opportunities. *Trends in Ecology and Evolution* 24(12), 686–693.

Waters and Rivers Commission (2000) Wetlands and weeds. *Water Notes* WN1 (July). Waters and Rivers Commission, Government of Western Australia, Australia.

Watson, A.K. (1991) The classical approach with plant pathogens. In: TeBeest, D.O. (ed.) *Microbial Control of Weeds*. Chapman and Hall, New York, USA, pp. 3–23.

White, T.A., Campbell, B.D. and Kemp, P.D. (1997) Invasion of temperate grassland by a subtropical annual grass across an experimental matrix of water stress and disturbance. *Journal of Vegetation Science* 8, 847–854.

Williams, D.G. and Baruch, Z. (2000) African grass invasion in the Americas: ecosystem consequences and the role of ecophysiology. *Biological Invasions* 2, 123–140.

Wilson, F. and Huffaker, C.B. (1976) The philosophy, scope and importance of biological control. In: Huffaker, C.B. and Messenger, P.S. (eds) *Theory and Practice of Biological Control*. Academic Press, New York, USA, pp. 3–15.

Woods, K.D. (1993) Effects on invasion by *Lonicera tatarica* L. on herbs and tree seedlings in four New England forests. *The American Midland Naturalist Journal* 130, 62–74.

Wyckoff, P.H. and Webb, S.L. (1996) Understory influence of the invasive Norway maple (*Acer platanoides*). *Bulletin of the Torrey Botanical Club* 123, 197–205.

Yelenik, S.G., Stock, W.D. and Richardson, D.M. (2007) Functional group identity does not predict invader impacts: differential effects of nitrogen-fixing exotic plants on ecosystem function. *Biological Invasions* 9, 117–125.

Yurkonis, K.A., Meiners, S.J. and Wachholder, B.E. (2005) Invasion impacts diversity through altered community dynamics. *Journal of Ecology* 93, 1053–1061.

Zavaleta, E. (2000) The economic value of controlling an invasive shrub. *Ambio* 29, 462–467.

Zedler, J.B. and Kercher, S. (2004) Causes and consequences of invasive plants in wetlands: opportunities, opportunists, and outcomes. *Critical Reviews of Plant Sciences* 23, 431–452.

Zimdahl, R.L. (1999) *Fundamentals of Weed* Science, 2nd edn. Academic Press, New York, USA.

Ziska, L.H. (2005) Climate change impacts on weeds. *Climate Change and Agriculture: Promoting Practical and Profitable Responses*. Available at: www.uvm.edu/vtvegandberry/ClimateChange/ClimateChangeImpacts On Weeds. pdf (accessed 31 May 2016).

Index

Note: Page numbers in **bold** type refer to **figures**. Page numbers in *italic* type refer to *tables*